# THE EUROPEAN GARDEN FLORA

*A manual for the identification of plants cultivated
in Europe, both out-of-doors and under glass*

## VOLUME I

## Pteridophyta, Gymnospermae,
## Angiospermae – Monocotyledons (Part I)

*edited by*

S.M. Walters, A. Brady, C.D. Brickell,

J. Cullen, P.S. Green, J. Lewis, V.A. Matthews,

D.A. Webb, P.F. Yeo and J.C.M. Alexander

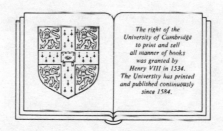

*The right of the
University of Cambridge
to print and sell
all manner of books
was granted by
Henry VIII in 1534.
The University has printed
and published continuously
since 1584.*

## CAMBRIDGE UNIVERSITY PRESS

*Cambridge*

*New York Port Chester*

*Melbourne Sydney*

Published by the Press Syndicate of the University of Cambridge
The Pitt Building, Trumpington Street, Cambridge CB2 1FP
40 West 20th Street, New York, NY 10011, USA
10 Stamford Road, Oakleigh, Melbourne 3166, Australia

First published 1986
Reprinted 1988, 1990

Printed in Great Britain by the Athenaeum Press Ltd, Newcastle upon Tyne.

*British Library cataloguing in publication data*

The European garden flora: a manual for the identification of
plants cultivated in Europe, both out of doors and under glass
Vol. 1: Pteridophyta, Gymnospermae,
Angiospermae – Monocotyledons (Part 1)

1. Plants, Cultivated – Europe – Dictionaries
I. Walters, S.M
635'.094   SB406.93.E85

*Library of Congress cataloguing in publication data*
(Revised for part 1)
Main entry under title:

The European garden flora.

Includes indexes.
Contents: v. 1. Pteridophyta, gymnospermae, angiospermae-
monocotyledons (pt. I) – v. 2. Monocotyledons (part II).
1. Plants, Ornamental – Europe – Identification – Collected works.
2. Fruit – Europe – identification – Collected works.
I. Walters, S. M. (Stuart Max)
SB406.93.E85E97 1984   635.9   83–7655

ISBN 0 521 24859 0 hardback

# CONTENTS

# MAP AND FIGURES

Map 1. Mean minimum January isotherms for Europe (hardiness codes)   xv

## FIGURES

# ORGANISATION AND ADVISERS

# EDITORS AND CONTRIBUTORS TO VOLUME I

The various sections of this volume were edited at the following institutions:

*Royal Botanic Garden, Edinburgh*: Pteridophyta, Gymnospermae (part), Liliaceae, Agavaceae, Amaryllidaceae, Hypoxidaceae, Velloziaceae, Taccaceae, Iridaceae.

*National Botanic Gardens, Glasnevin, Dublin*: Butomaceae, Hydrocharitaceae, Aponogetonaceae, Potamogetonaceae, Haemodoraccac, Tecophilaeaceae, Pontederiaceae.
*University Botanic Garden, Cambridge*: Alismataceae, Dioscoreaceae.
*Natural History Museum, London*: Gymnospermae (part).

## Contributors

J.C.M. Alexander (University Botanic Garden, Cambridge/RBG, Edinburgh)
P.G. Barnes (RHS, Wisley)
F.M. Bennell (RBG, Edinburgh)
J.J. Bos (Botanische Tuinen en Belmonte Arboretum, Wageningen, Netherlands)
A. Brady (National Botanic Gardens, Glasnevin, Dublin)
C.D. Brickell (RHS, Wisley)
E.J. Campbell (Edinburgh)
C.J. Couper (Department of Botany, University of Edinburgh)
J. Cullen (RBG, Edinburgh)
S.J.M. Droop (RBG, Kew)
C. Grey-Wilson (RBG, Kew)
E.H. Hamlet (RBG, Edinburgh)
C.J. King (University Botanic Garden, Cambridge)
S.G. Knees (RHS, Wisley)
J.M. Lees (Edinburgh)
A. Leslie (RHS, Wisley)

J. Lewis (Natural History Museum, London)
J. Lovett (Department of Forestry, University of Oxford)
B. Mathew (RBG, Kew)
V.A. Matthews (RBG, Edinburgh)
A. Mitchell (Forestry Commission, Alice Holt)
E.C. Nelson (National Botanic Gardens, Glasnevin, Dublin)
C.N. Page (RBG, Edinburgh)
E.M. Rix (London)
K.D. Rushforth (London)
M.J.P. Scannell (National Botanic Gardens, Glasnevin, Dublin)
W.T. Stearn (Natural History Museum, London)
D.C. Stuart (Belhaven, Dunbar)
T.J. Varley (Department of Forestry, University of Oxford)
D.A. Webb (Trinity College, University of Dublin)
H.J. Welch (Devizes)
T.C. Whitmore (Department of Forestry, University of Oxford)
P.F. Yeo (University Botanic Garden, Cambridge)

# ACKNOWLEDGEMENTS

During the writing of this volume, The European Garden Flora has received substantial support from the following:

(*a*) *The institutions to which members of the Editorial Committee belong*: staff time, support and services.

(*b*) *The Stanley Smith Horticultural Trust* (Director, Sir George Taylor, FRS): financial support which provided for the employment of the post-doctoral Research Associate from 1978 to 1982.

(*c*) *The Wolfson Industrial Fellowship Scheme*: financial support for the continuing employment of the Research Associate from 1982 to 1985.

(*d*) *The Cory Fund of Cambridge University Botanic Garden*: financial support for travel by the Research Associate and other uses.

(*e*) *The Council of the Royal Horticultural Society*: financial support from 1981.

The Editorial Committee gratefully acknowledges all this generous support.

Particular thanks are due to Dr D. Willis (New University of Ulster, Coleraine) and Mrs E. Molesworth-Allan (Cadíz) for specialist taxonomic advice. We are grateful to J.C.M. Alexander, Frances Bennell, Frances Hibberd, M. Hickey, Victoria A. Matthews, C.N. Page and Sally Rae for the preparation of the illustrations; and to Moira Watson, who prepared the whole of the typescript for this volume on a Rank-Xerox 640 word-processor.

# INTRODUCTION

Amenity horticulture (gardening, landscaping, etc.) touches human life at many points. It is a major leisure activity for a very large number of people, and is a very important means of improving the environment. The industry that has grown up to support this activity (the nursery trade, landscape architecture and management, public parks, etc.) is a large one, employing a considerable number of people. It is clearly important that the basic material of all this activity, i.e. plants, should be readily identifiable, so that both suppliers and users can have confidence that the material they buy and sell is what it purports to be.

The problems of identifying plants in cultivation are many and various, and derive from several sources, which may be summarised as follows:

(a) Plants in cultivation have originated in all parts of the world, many of them from areas whose wild flora is not well known. Many have been introduced, lost and then re-introduced under different names.

(b) Plants in gardens are growing under conditions to which they are not necessarily well adapted, and may therefore show morphological and physiological differences from the original wild stocks.

(c) All plants that become established in cultivation have gone through a process of selection, some of it conscious (selection of the 'best' variants, etc.), some of it unconscious (by methods of cultivation and particularly, propagation), so that, again, the populations of a species in cultivation may differ significantly from the wild populations.

(d) Many garden plants have been 'improved' by hybridisation (deliberate or accidental), and so, again, differ from the original stocks.

(e) Finally, and perhaps most importantly, the scientific study of plant classification (taxonomy) has concentrated mainly on wild plants, largely ignoring material in gardens.

Nevertheless, the classification of garden plants has a long and distinguished history. Many of the Herbals of pre-Linnaean times (i.e. before 1753) consist partly or largely of descriptions of plants in gardens, and this tradition continued, and perhaps reached its peak in the late eighteenth and early nineteenth centuries – the period following the publication of Linnaeus's major works, when exploration of the world was at its height. This is the period that saw the founding of *Curtis's Botanical Magazine* (1787) and the publication of J.C. Loudon's *Encyclopaedia of plants* (1829 and many subsequent editions).

The further development of plant taxonomy, from about the middle of the nineteenth century to the present, has seen an increasing divergence between garden and scientific taxonomy, leading on the one hand to such works as the *Royal Horticultural Society's dictionary of gardening* (1951 and reprinted, itself based on G. Nicholson's *Illustrated dictionary of gardening*, 1884–1888) and the very numerous popular, usually illustrated works on garden flowers available today, and, on the other hand, to the Floras, Revisions and Monographs of scientific taxonomy.

Despite this divergence, a number of plant taxonomists realised the importance of the classification and identification of cultivated plants, and produced works of considerable scientific value. Foremost among these stands L.H. Bailey, editor of *The standard cyclopedia of horticulture* (1900, with several subsequent reprints and editions), author of *Manual of cultivated plants* (1924, edn 2, 1949), and founder of the journals *Gentes Herbarum* and *Baileya*. Other important workers in this field are T. Rümpler (*Vilmorin's Blumengärtnerei*, 1879), L. Dippel (*Handbuch der Laubholzkunde*, 1889–93), A. Voss and A. Siebert (*Vilmorin's Blumengärtnerei*, edn 3, 1894–6), C.K. Schneider (*Illustriertes Handbuch der Laubholzkunde*, 1904–12), A. Rehder (*Manual of cultivated trees and shrubs*, 1927, edn 2, 1947), J.W.C. Kirk (*A British garden flora*, 1927), F. Encke (*Parey's Blumengärtnerei*, 1958), B.K. Boom (*Flora Cultuurgewassen*, 1959 and proceeding) and V.A. Avrorin & M.V. Baranova (*Decorativn'ie Travyanist'ie Rasteniya Dlya Otkritogo Grunta SSSR*, 1977).

The present Flora, which, of necessity, is based on original taxonomic studies by many workers, attempts to provide a scientifically accurate and up-to-date means for the identification of plants cultivated for amenity in Europe (i.e. it does not include crops, whether horticultural or agricultural, or garden weeds), and to provide what are currently thought to be their correct names, together with sufficient synonymy to make sense of catalogues and other horticultural works. The needs of the informed amateur gardener have been borne in mind at all stages of the work, and it is hoped that the Flora will meet his needs just as much as it meets the needs of the professional plant taxonomist. The details of the format and use of the Flora are explained in section 2 below (pp. xii–xiv).

In writing the work, the Editorial Committee has been fully aware of the difficulties involved. Some of these have been outlined above; others derive from that fact that herbarium material of cultivated plants is scanty and usually poorly annotated, so that material of many species is not available for checking the use of names, or for comparative purposes. Because of these factors, attention has been drawn to numerous problems which cannot be solved but can only be adverted to. The solution of such problems requires much more taxonomic work.

The form in which contributions appear is the responsibility of the Editorial Committee. The vocabulary and the technicalities of plant description are therefore not necessarily those endorsed by the contributors.

## 1. SELECTION OF SPECIES

The problem of determining which species are in cultivation is complex and difficult, and has no complete and final answer. Many species, for instance, are grown in botanic gardens but not elsewhere; others, particularly orchids, succulents and some alpines, are to be found in the collections of specialists but are not available generally. Yet others have been in cultivation in the past but are now lost, or perhaps linger in a few collections, unrecorded and unpropagated. Further problems arise from the fact that the identification of plants in collections is not always as good as it might be, and some less well-known species probably appear in published lists under the names of other, well-known species (and vice versa).

The Flora attempts to cover all those species that are likely to be found in general collections (i.e. excluding botanic gardens and specialist collections) in Europe, whether they are grown out-of-doors or under glass. In order to produce a basic working list of such species, a compilation of all European nursery catalogues available to us was made in 1978 by Margaret McDonald, a vacation student working at the Royal Botanic Garden, Edinburgh. Since then, numerous additions have been made by Hazel Hamlet. This list (known as the 'Commercial List'), which includes well over 12 000 specific names, forms the basis of the species included here. In addition to the 'Commercial List', several works on the flora of gardens have been consulted, and the species covered by them have been carefully considered for inclusion. These works are: Wehrhahn, H.R., *Die Gartenstauden* (1929–31); *The Royal Horticultural Society's dictionary of gardening*, edn 2 (1956, supplement revised 1969); Encke, F. (ed.), *Parey's Blumengärtnerei* (1956); Boom, B.K., *Flora der Cultuurgewassen Van Neder-*land (1959 and proceeding); Bean, W.J., *Trees and shrubs hardy in the British Isles* (edn 8, 1970–1981); Krüssmann, G., *Handbuch der Laubgehölze* (edn 2, 1976–1978); Encke, F., Buchheim, G. & Seybold, S. (eds.), *Zander's Handwörterbuch der Pflanzennamen* (edn 13, 1984). Most of the names included in these works are covered by the present Flora, though some have been rejected as referring to plants no longer in general cultivation.

As well as the works cited above, several works relating to plants in cultivation in North America have been consulted: Rehder, A., *Manual of cultivated trees and shrubs* (edn 2, 1947) and *Bibliography of cultivated trees and shrubs* (1949); Bailey, L.H., *Manual of cultivated plants* (edn 2, 1949); *Hortus Third* (edited by the staff of the L.H. Bailey Hortorium, Cornell University).

The contributors have also drawn on their own experience, as well as that of the family editors, European advisers and other experts, in deciding which species should be included.

Because some species are not very widely grown, but, by the criteria mentioned above, have had to be included, two levels of treatment have been used. Most species have a full entry, being keyed, numbered and described as set out under Section 2c, below (p. xiii). Less commonly cultivated species are not keyed or numbered individually, but are described briefly under the full-entry species to which they key out in the formal key (this, of course, will not necessarily be the one to which the additional species is thought to be most closely related). The name of any full-entry species which has additional species attached to it is distinguished in the formal keys by an asterisk (*); and the names of additional species are preceded by asterisks in the main descriptive text.

## 2. USE OF THE FLORA

a. *The taxonomic system followed in the Flora.* Plants are described in this work in a taxonomic order, so that similar genera and species occur close to each other, rendering comparison of descriptions more easy than in a work where the entries are alphabetical. The families (and higher groups) follow the Engler & Prantl system as expressed in H. Melchior's edition (edn 12, 1964) of *Syllabus der Pflanzenfamilien*. The exceptions are minor, apart from the placing of the monocotyledons before the dicotyledons; this has been done purely for convenience, and has no other implications. The assignment of genera to families also usually follows the *Syllabus*.

The order of the species within each genus has been a matter for the individual author's discretion. In general, however, some established revision of the genus has been followed, or, if no such revision exists, the author's own views on similarity and relationships have governed the order used.

b. *Nomenclature.* The arguments for using Latin names for plants in popular as well as scientific works are often stated and widely accepted, particularly for Floras such as this, which cover an area in which several languages are spoken. Latin names have therefore been used at every taxonomic level. A concise outline of the taxonomic hierarchy and how it is used can be found in C. Jeffrey's *An introduction to plant taxonomy* (1968, edn 2, 1982). Because of the difficulties of providing vernacular names in all the necessary languages (not to say dialects), they have not been included. A supplementary volume including them may be possible after the systematic part of the Flora has been completed. Meanwhile, S. Priszter's *Trees and shrubs of Europe, a dictionary in eight languages* (1983) is a useful source for vernacular names of woody plants.

Many horticultural reference works omit the authority which should follow every Latin plant name. Knowledge of this authority prevents confusion between specific names that may have been used more than once within the same genus, and makes it possible to find the original description of the species (using *Index Kewensis*, which lists the original references for all Latin plant names published since 1753). Authorities are therefore given for all names at or below the genus level. These are unabbreviated to avoid the obscure contractions which mystify the lay reader, and, on occasions, the professional botanist. In most cases we have not thought it necessary to include the initials or qualifying words and letters which often accompany author names, e.g. A. Richard, Reichenbach filius, fil. or f. (the exceptions involve a few, very common surnames).

In scientific taxonomic literature, the authority for a plant name sometimes consists of two names joined together by *ex* or *in*. Such formulae have not been used here; the authority has been shortened in accordance with *The international code of botanical nomenclature* (ed. E.G. Voss, 1983), e.g. *Capparis lasiantha* R. Brown ex de Candolle becomes *Capparis lasiantha* de Candolle; *Viburnum ternatum* Rehder in Sargent becomes *Viburnum ternatum* Rehder. The abbreviations *hort.* and *auct.*, which sometimes stand in place of the authority after Latin names, have not been used in this work as they are often obscure or misleading. The situations described by them can be clearly and unambiguously covered by the terms *invalid*, *misapplied* or *Anon*. *Invalid* implies that the name in question has never been validly published in accordance with the Code of Nomenclature, and therefore cannot be accepted. *Misapplied* refers to names which have been applied to the wrong species in gardens or in literature. *Anon*. is used with validly published names for which there is no apparent author.

Gardeners and horticulturists complain bitterly when long-used and well-loved names are replaced by unfamiliar ones. These changes are unavoidable if *The international code of botanical nomenclature* is adhered to. Taxonomic research will doubtless continue to unearth earlier names and will also continue to realign or split up existing groups, as relationships are further investigated. However, the previously accepted names are not lost; in this work they appear as synonyms, given in brackets after the currently accepted name; they are also included in the index. Dates of publication are not given either for accepted names or for synonyms.

c. *Descriptions and terminology*. Families, genera and species included in the Flora are generally provided with full-length descriptions. Shorter, diagnostic descriptions are, however, used for genera or species which differ from others already fully described in only a few characters, e.g.:

**3. P. vulgaris** Linnaeus. Like *P. officinalis* but leaves lanceolate and corolla red . . .

This implies that the description of *P. vulgaris* is generally similar to that of *P. officinalis* except in the characters mentioned; it should not be assumed that plants of the two species will necessarily look very like each other. Additional species (see p. xii), subspecies, varieties and cultivars (see p. xiv) are described very briefly and diagnostically.

Unqualified measurements always refer to length (though 'long' is sometimes added in cases where confusion might arise); simi-

larly, two measurements separated by a multiplication sign indicate length and breadth respectively.

The terminology used has been simplified as far as is consistent with accuracy. The technical terms which, inevitably, have had to be used are explained in the glossary (p. 393). Technical terms restricted to particular families or genera are defined in the observations following the family or genus description, and are also referred to in the glossary.

d. *Informal keys*. For most genera containing 5 to 20 species (and for most families containing 5 to 20 genera) an informal key is given; this will not necessarily enable the user to identify precisely every species included, but will provide a guide to the occurrence of the more easily recognised characters. A selection of these characters is given, each of which is followed by the entry-numbers of those species which show that character. In some cases, where only a few species of a genus show a particular character, the alternative states are not specified, e.g.:

*Leaves*. Terete: **18,19**.

This means that only species **18** and **19** in the particular genus have terete leaves; the other species may have leaves of various forms, but they are not terete. No distinction is made between full-entry and additional species. The occurrence of an entry-number merely means that one or more species found under that entry-number will have the particular character, but does not imply that they will all do so.

e. *Formal keys*. For every family containing more than one genus, and for every genus containing more than one species, a dichotomous key is provided. This form of key, in which a series of decisions must be made between pairs of contrasting character-states, should lead the user step by step to an entry-number followed by the name of a full-entry species. If that name is followed by an asterisk it will be necessary to read the descriptions of the additional species under that entry-number, as well as that of the full-entry species, in order to decide which one fits the plant in question. A similar key to all the families of monocotyledons included in this work (those in volume II as well as in the present volume) is also provided (p. 108).

f. *Horticultural information*. Notes on the cultural requirements and methods of propagation are usually included in the observation to each genus; more rarely, such information is given in the observations under the family description. These are necessarily brief and very generalised, and merely provide guidance. Reference to general works on gardening is necessary for more detailed information.

g. *Citation of literature*. References to taxonomic books, articles and registration lists are cited for each family and genus, as appropriate. No abbreviations are used in these citations (though very long titles have been shortened). The citation of a particular book or article does not necessarily imply that it has been used in the preparation of the account of the particular genus or family in this work.

h. *Citation of illustrations*. References to good illustrations are given for each species (or subspecies or variety); the names under which they were originally published (which may be different from those used here) are not normally given. They may be coloured or black and white, and may be drawings, paintings or photographs. Usually, up to four illustrations per species have been given, and an attempt has been made to choose pictures from widely available, modern works. Where no illustrations are cited, they either do not exist, as far as we know, or those that do are considered to be of doubtful accuracy.

In searching for illustrations, use was made of *Index Londoninensis* (1929–31, supplement 1941) and R.T. Isaacson's *Flowering plant index of illustration and information* (1979). Readers are referred to these works if they wish to find further pictures.

Several pages of figures of diagnostic plant parts are included with various groups in the Flora, and should be particularly helpful when plants are being identified by means of the keys. Some of these are original, others have either been redrawn from various sources, or, in the case of the pteridophytes, are 'Xerox' copies of pressed fronds or whole plants.

i. *Geographical distribution*. The wild distribution, as far as it can be ascertained, is given in italics at the end of the description of each species (or subspecies or variety). The choice and spelling of place names in general follows *The Times Atlas*, Comprehensive Edition (1980), except:

(1) Well-established English forms of names have been used in preference to unfamiliar vernacular names, e.g. Crete instead of Kriti, Naples instead of Napoli, Borneo instead of Kalimantan;

(2) New names or spellings will be adopted as soon as they appear in readily available works of reference.

j. *Hardiness* (see map, p. xv). For every species a hardiness code is given. This gives a tentative indication of the lowest temperatures that the particular species can withstand:

G2 – needs a heated glasshouse even in south Europe.

G1 – needs a cool glasshouse even in south Europe.

H5 – hardy in favourable areas; withstands 0 to −5 °C minimum.

H4 – hardy in mild areas; withstands −5 to −10 °C minimum.

H3 – hardy in cool areas; withstands −10 to −15 °C minimum.

H2 – hardy almost everywhere; withstands −15 to −20 °C minimum.

H1 – hardy everywhere; withstands −20 °C and below.

The map of mean January minima (p. xv) shows the isotherms corresponding to these codes. It must be understood that H4 includes H5, H3 includes H4 and H5, H2 includes H3–5 and H1 includes H2–5.

k. *Flowering time*. The terms spring, summer, autumn and winter have been used as a guide to flowering times in cultivation in Europe. It is not possible to be more specific when dealing with an area extending from northern Scandinavia to the Mediterranean. In cases where plants do not flower in cultivation, or flower rarely, or whose time of flowering is not recorded, no flowering time is given.

l. *Subspecies, varieties and cultivars*. Subspecies and varieties are described, where appropriate. This is done in various ways, depending on the number of such groups; all these ways are, however, self-explanatory.

No attempt has been made to describe the range of cultivars of a species, either partially or comprehensively. The former is scarcely worth doing, the latter virtually impossible. Reference to individual, commonly grown cultivars is, however, made in various ways:

(1) If a registration list of cultivars exists, it is cited in the 'Literature' paragraph (see section 2g) following the description of the genus.

(2) If a particular cultivar is very widely grown, it may be referred to, either in the description of the species to which it belongs (or which it most resembles), or in the observations to that species.

(3) If, in a particular species, cultivars are numerous and fall into reasonably distinct groups based on variation in some striking character, then these groups may be referred to, together with an example of each, in the observations to the species.

m. *Hybrids*. Many hybrids between species (interspecific hybrids) and some between genera (intergeneric hybrids) are in cultivation, and some of them are widely grown. Commonly cultivated interspecific hybrids are, where possible, included as though they were species. Their names, however, include the ' × ' sign indicating their hybrid origin; the names of their parents (when known or presumed) are given at the beginning of the paragraph of observations following the description. Other hybrids which are less frequently grown are mentioned in the observations to the individual parent species that they most resemble. In some genera (e.g. *Odontoglossum*), where the number of hybrids is very large, only a small selection of those most commonly grown is mentioned.

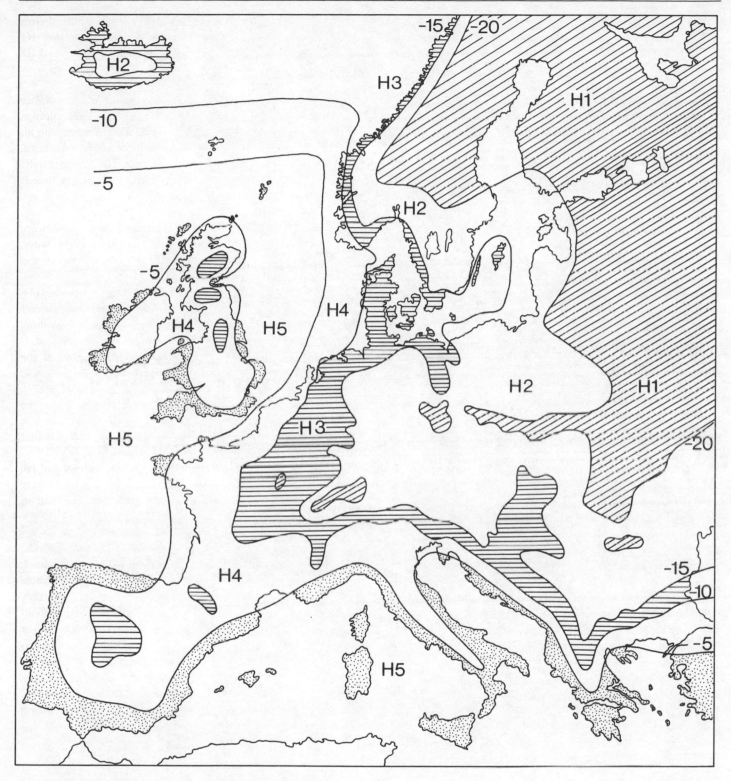

**Map 1.** Mean minimum January isotherms for Europe (hardiness codes). (After Krüssmann, *Handbuch der Laubgehölze*, 1960 and *Mitteilungen der Deutsche Dendrologische Gesellschaft* **75**, 1983.) Corrected from European Garden Flora volume II (1984).

# PTERIDOPHYTA

Plants with a dominant, vascular, sporophyte generation producing spores which grow into a separate, free-living, non-vascular prothallial gametophyte generation. New sporophytes arise from the prothalli, normally as a result of fertilisation of the archegonia by motile antherozoids which originate in antheridia.

The Pteridophyta cover a wide range of vascular plants which are essentially distinguished from the Gymnospermae (p. 68) and the Angiospermae (p. 108) in reproducing by spores rather than by seeds. The spores are borne in microscopic organs known as sporangia, which may be found grouped together on the leaf surfaces, or aggregated into cone-like structures; more rarely they are found in specialised bodies known as sporocarps. The spores are shed when ripe, and develop into small or microscopic, free-living plants (prothalli) which bear the sexual organs (antheridia and archegonia). Fertilisation of the archegonia leads to the production of a new spore-bearing plant. In some genera the spores are of 2 sizes, the smaller spores giving rise to male prothalli, the larger to female. The prothallus is rarely seen in the wild, and those of most genera look rather similar. It is, however, an important stage in the natural life-cycle of these plants, and of appropriate significance to the horticulturist rearing plants from spores, from which the majority of species are easily propagated in great numbers. Spores of exotic species which are not necessarily available as plants in the nursery trade are often available from cultivated and overseas sources. A spore-exchange scheme is operated between members of the British Pteridological Society (c/o British Museum (Natural History), Cromwell Road, London SW7 5BD), and annual lists are produced of species and cultivars available.

The classification and identification of the Pteridophyta are based entirely on the spore-bearing generation (the sporophyte); the keys and descriptions given below do not include the prothalli, which are rarely seen in gardens.

The whole group is considered here to consist of several classes, 4 of which are cultivated.

## KEY TO CLASSES

1a. Stems dichotomously branched            2
  b. Stems not dichotomously branched        3
2a. Stems with few or no leaves       **I. Psilopsida** (p. 3)
  b. Stems with numerous leaves       **II. Lycopsida** (p. 3)
3a. Plants with main leafy parts which unroll in a crozier-like manner       **IV. Filicopsida** (p. 6)
  b. Plants with main leafy parts which do not unroll in a crozier-like manner       4
4a. Plant growth frond-like or moss-like; spores of 2 sizes, borne in separate sporangia       **II. Lycopsida** (p. 3)
  b. Plants not as above; spores all of the same size       5
5a. Plants with leaves in whorls, reduced and forming sheaths at the nodes; branches in regular whorls

                    **III. Sphenopsida** (p. 5)
  b. Plants with a single, simple, entire leaf and a fertile spike, all unbranched       **IV. Filicopsida** (p. 6)

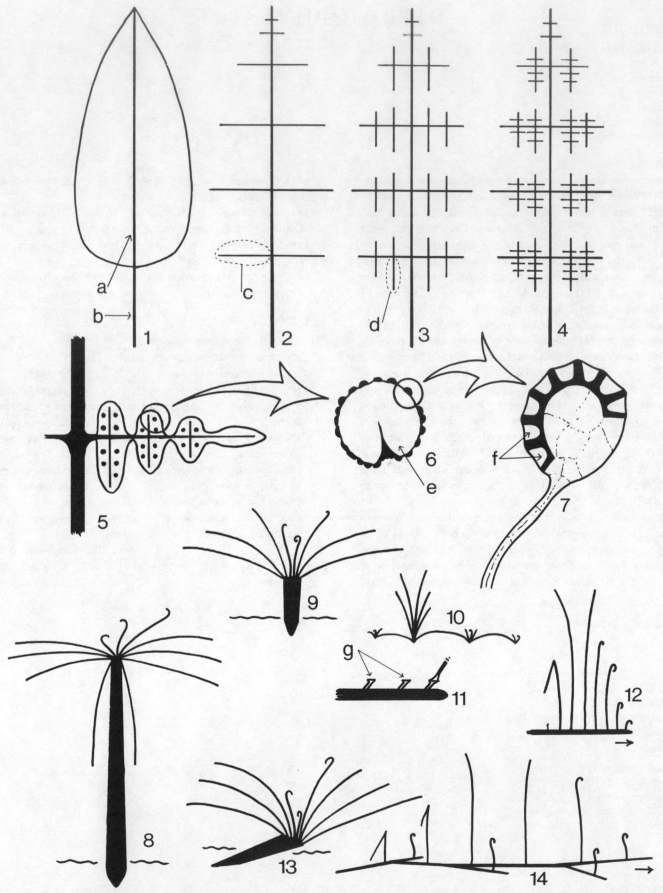

**Figure 1.** Diagram to illustrate the terminology used in the Ferns. 1, Entire frond. 2, Frond 1 × pinnate. 3, Frond 2 × pinnate. 4, Frond 3 × pinnate. 5, Underside of pinna showing sori. 6, Sorus covered by indusium. 7, Sporangium. 8, Rhizome erect, trunk-forming. 9, Rhizome erect. 10, Rhizome with stolons. 11, Rhizome with phyllopodia. 12, Rhizome short-creeping. 13, Rhizome ascending. 14, Rhizome long-creeping. (a, rachis; b, stipe; c, pinna; d, pinnule; e, indusium; f, ring of thickened cells (annulus); g, phyllopodia).

## I. PSILOPSIDA

Stems green, dichotomously branched, bearing very small leaves (microphylls) which are few and distant, some bearing grooved, spherical sporangia in their axils. Spores all of the same size.

Contains the single family, **Psilotaceae**, which contains a single genus.

### 1. PSILOTUM Swartz
*C.N. Page & F.M. Bennell*
Small to medium-sized plants. Rhizomes short, creeping, freely branching. Stems dichotomously forked, angular, green, tough, rigid, wiry and slender, erect or pendent, without obvious leaves but with sparse, inconspicuous scales at intervals. Sporangia usually frequent along the ultimate branches, fused into spherical or trilobed synangia (compound sporangia) which split when mature, each stalkless and solitary in the axil of a scale-leaf.

A genus of perhaps 3 species, widespread in moist tropical and subtropical regions. Easily cultivated in pots in glasshouses.

**1. P. nudum** (Linnaeus) Grisebach. Figure 2(1), p. 4.
Stems 20–60 cm, erect, spreading or hanging, triangular in section, branching profusely in old specimens, usually yellowish green. Synangia yellowish when mature, 2–3 mm in diameter. *Old World tropics and subtropics, extending to S Spain (1 locality).* G1–2.

This species is said to have been in cultivation for over 400 years in Japan, where several clones have been named.

## II. LYCOPSIDA

Stems dichotomously branched, bearing numerous, small, reduced leaves (microphylls). Sporangia usually borne in the axils of microphylls which are frequently grouped in erect, terminal cones. Spores all of the same size, or of 2 sizes.

This group comprises 3 orders, of which 2 are cultivated.

1a. Leaves with a small outgrowth (ligule) on the upper surface; spores of 2 different sizes  **IIb. Selaginellales**
 b. Leaves without a ligule; spores all the same size  **IIa. Lycopodiales**

## IIa. LYCOPODIALES

Leaves without ligules on their upper surfaces. Spores all of the same size. Contains the single family, **Lycopodiaceae**, which contains a single genus (divided by some authors into 5 or more separate genera).

### 1. LYCOPODIUM Linnaeus
*C.N. Page & F.M. Bennell*
Epiphytes (all the cultivated species) or free-living plants. Rhizome long or short, thick or thin, creeping, occasionally branched, fleshy, bearing stems in clusters. Stems rigid, slender, leafy, erect or pendent, dichotomously forked, rooting at the base. Leaves numerous, narrow, linear, each with a single central vein, arranged spirally all round the stem, spreading obliquely, evergreen. Sporangia spherical, solitary in the axils of leaves near the ends of the stems, sometimes in distinct cones. Spores yellow, all of the same size.

A genus of at least 100 species, cosmopolitan. The epiphytic species which are cultivated make distinct and handsome hanging-basket subjects, growing easily in a coarse compost in a sheltered, slightly shaded position. Propagation is by the splitting of established plants.

1a. Leaves 1–2 mm, more or less spreading; fertile leaves little different from sterile  **1. squarrosum**
 b. Leaves *c.* 2 cm, set obliquely forwards; fertile leaves smaller than sterile  **2. phlegmaria**

**1. L. squarrosum** Forster (*Urostachys squarrosus* (Forster) Hert). Figure 2(2), p. 4.
Stems to 60 cm, much branched. Leaves 1–2 mm, narrowly linear-lanceolate, acuminate, very crowded, widely spreading, stiff, pale yellow-green, the whole shoot somewhat prickly to the touch. Sporangia borne at the tips of the stems in the axils of leaves similar to the sterile leaves or very slightly smaller. *Old World tropics.* G2.

**2. L. phlegmaria** Linnaeus (*Urostachys phlegmaria* (Linnaeus) Hert). Figure 2(3), p. 4.
Stems to 90 cm, occasionally branched. Leaves to 2 cm, ovate-triangular, acute, all pointed obliquely forwards along the stem, crowded, leathery, shining bright green. Sporangia in distinctive, catkin-like, hanging cones 1–1.5 cm in diameter, with very small, closely overlapping leaves. *Old World tropics.* G2.

## IIb. SELAGINELLALES

Branching frequent, not dichotomous, forming moss-like or frond-like sprays. Leaves each with a ligule on the upper surface. Spores of 2 sizes, borne in separate sporangia.

Contains the single family, **Selaginellaceae**, which contains a single genus.

### 1. SELAGINELLA Beauvois
*C.N. Page & F.M. Bennell*
Usually small plants. Main stems long, creeping, subterranean or aerial and scrambling, occasionally to frequently branched, thin, rigid, wiry, producing root-like rhizophores at the base or along the length of the stem. Leaves very numerous, small (microphylls), usually triangular, overlapping, each with a single main vein and a ligule on the upper surface, spreading all round the shoot or in regular, flattened ranks of 2 sizes. Sporangia more or less spherical, solitary in the axils of leaves near the tips of the stems, usually in distinct, small cones. Spores of 2 sizes, in separate sporangia.

A genus of over 600 species, cosmopolitan but mostly from the wet tropics. The frond-like branching of many species superficially resembles a fern frond. Several species which eventually form extensive mats are grown, and make good basket or bottle-garden subjects, and are often grown under staging as ground-cover. They require a freely drained, moist compost, screened light and high humidity. Propagation is by cuttings or stem fragments.

1a. All stems trailing  2
 b. Stems of the branches erect or raised  4
2a. Branches with indefinite growth, themselves simply and sparsely branched, resembling the main stems  3
 b. Branches with definite growth, themselves compoundly branched, much more so than the main stem  **3. uncinata**
3a. Recent growth of branches with leaves crowded at the tip  **1. kraussiana***
 b. Recent growth of branches with leaves well spaced at the tip  **2. denticulata**
4a. Branches in tight, rosette-like clusters  5
 b. Branches distantly spread along the creeping main stem  6

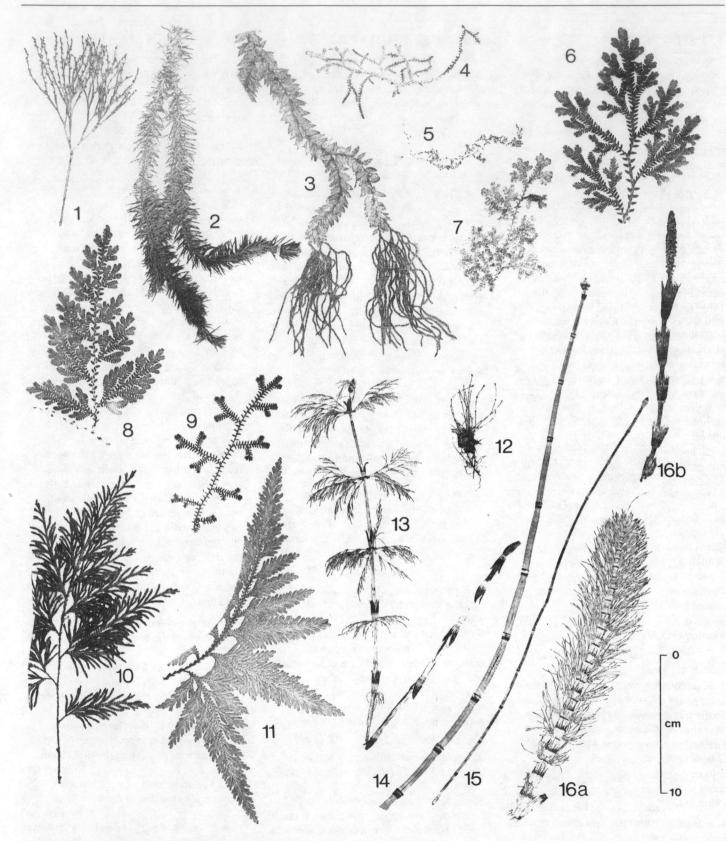

**Figure 2.** Silhouttes of fern allies. 1, *Psilotum nudum*.
2, *Lycopodium squarrosum*. 3, *L. phlegmaria*. 4, *Selaginella
denticulata*. 5, *S. lepidophylla*. 6, *S. martensii*. 7, *S. uncinata*.
8, *S. willdenowii*. 9, *S. kraussiana*. 10, *S. pulcherrima*.
11, *S. erythropus*. 12, *Equisetum scirpoides*. 13, *E. sylvaticum*.
14, *E. hyemale*. 15, *E. variegatum*. 16, *E. telmateia* (a, vegetative
shoot; b, fertile shoot).

5a. Stems of branches glossy green
beneath **5. martensii**
b. Stems of branches grey-green or
almost cream beneath **4. lepidophylla**
6a. Compound side branch systems with
a blue sheen **6. willdenowii**
b. Compound side branch systems
without a blue sheen 7
7a. Tips of branch systems irregularly
arranged, bluntly pointed or rounded
**7. pulcherrrima**
b. Tips of side branch systems all very
regularly arranged and acutely
pointed **8. erythropus**

**1. S. kraussiana** (Kunze) Brown. Figure
2(9), p. 4.
Stems trailing , more or less prostrate.
Branches simple, growing indefinitely,
brittle, leaves and branches closely spaced
and crowded behind the growing apex,
becoming more widely spaced with shoot
elongation; width of shoot (including
leaves) 8–9 mm. Lateral leaves *c.*
4 × 1–2 mm, vivid green. *Tropical and
southern Africa, Azores.* H5–G1.

Several cultivars exist, many with more
compact, moss-like growth than described
above.

**\*S. apoda** (Linnaeus) Fernald. Similar,
but much smaller, and always of a moss-
like growth. *Eastern N America, from Canada
to Texas.* H5–G1.

**2. S. denticulata** (Linnaeus) Spring. Figure
2(4), p. 4.
Stems trailing, more or less prostrate.
Branches simple, growing indefinitely, *c.*
7 mm in diameter (including the leaves),
the leaves and branches immediately well-
spaced behind the growing apex. Lateral
leaves *c.* 3 × 1.5 mm, yellowish green.
*Mediterranean area, Atlantic Islands.* G1.

A variegated form is occasionally seen in
cultivation.

**3. S. uncinata** (Desvaux) Spring. Figure
2(7), p. 4.
Stems trailing, more or less prostrate.
Branches of indefinite growth, 3–6 cm,
forming compound, bluntly rounded
sprays. Leaves pale green to straw-
coloured, slightly translucent, the whole
shoot with a metallic, blue iridescence. *S
China.* G2.

**4. S. lepidophylla** (Hooker & Greville)
Spring. Figure 2(5), p. 4.
Stems all ascending, short, wiry, freely
branched into bluntly rounded, frond-like,
compound branch systems which are
2–5 cm, emerald-green above, becoming
red-brown with age, grey-green or almost

cream beneath, all held in a distinct, rather
flat rosette. On drying the fronds roll
inwards to form a tight ball, opening out
again when watered. *Tropical and subtropical
America.* G2.

Only occasionally grown, but the dried,
dead plants are seen occasionally as
botanical curiosities, under the names
"Resurrection Plant" or "Rose of Jericho".

**5. S. martensii** Spring. Figure 2(6), p. 4.
Stems ascending, short, fleshy, freely
branched into bluntly rounded, ovate-
triangular, frond-like sprays, with all their
parts closely overlapping, the tips drooping,
the branches supported by strong, stilt-like
rhizophores near the base. Leaves bright
green, glossy above and beneath.
*C America.* G2.

Variegated forms occur from time to time
in cultivation.

**6. S. willdenowii** (Desvaux) Baker. Figure
2(8), p. 4.
Main stems of indefinite growth, creeping,
scrambling or climbing. Lateral branches of
definite growth, forming compact, flattened,
ascending or arching, broadly triangular
sprays with well-spaced, bluntly pointed
side branches and overlapping minor
branches. Leaves green, becoming pinkish
yellow or plum-coloured with age, and
with a blue sheen. Rhizophores stilt-like.
*Himalaya to S China and Indonesia.* H5–G1.

**7. S. pulcherrima** Liebmann & Fournier.
Figure 2(10), p. 4.
Main stem short, creeping, of indefinite
growth but not scrambling or climbing.
Lateral branches of definite growth,
forming compact, flattened, ascending to
spreading, broadly triangular-ovate, frond-
like branch systems with well-spaced,
irregularly arranged, bluntly pointed or
rounded side branches which are bright
green, with straw-yellow stems. *Mexico.*
G2.

**8. S. erythropus** (Martius) Spring. Figure
2(11), p. 4.
Main stem short, of indefinite growth but
not climbing or scrambling. Lateral
branches of definite growth, forming
compact, markedly flattened, ascending to
spreading, broadly triangular, frond-like
branch systems with closely spaced, very
regularly arranged, tapering, acutely
pointed, crowded side branches, pale green
or reddish throughout. *West Indies,
S America south to Chile.* G2.

## III. SPHENOPSIDA

Stems grooved. Leaves reduced
(microphylls), numerous, borne in regular
whorls at the nodes, those of each whorl
united into a sheath at the base. Sporangia
borne on the inner surfaces of peltate
sporangiophores which are aggregated into
terminal cones. Spores all of the same size,
each with 4 appendages (elaters).

Contains the single family, Equisetaceae,
with a single genus.

**1. EQUISETUM** Linnaeus
*C.N. Page & F.M. Bennell*
Terrestrial plants. Stems with very
conspicuous divisions into nodes and
internodes. Rhizomes thin, creeping, often
deeply subterranean, branching frequently,
giving rise to erect shoots singly or in
clusters. Shoots all of the same kind, or of 2
kinds (fertile and sterile), hollow, evergreen
or deciduous, simple or with regular whorls
of branches, mostly firm or harsh and
rough to the touch. Spores green, short-
lived, with elaters, borne in sporangia on
the inner faces of peltate sporangiophores
which form a cone.

A genus of about 25 species,
cosmopolitan except for Australia and New
Zealand, the majority in the temperate
northern hemisphere. They will grow in
almost any soil, especially if moist. Base-
rich heavy clays are particularly suitable.
Most thrive best with shelter and light
shade and readily naturalise; some are
noxious weeds. Propagation is by rhizome
fragments.

1a. Stems bearing whorls of branches 2
b. Stems without whorls of branches 3
2a. Internodes ivory-white, cone-bearing
shoots colourless; branches simple
**4. telmateia**
b. Internodes green, cone-bearing shoots
green; branches themselves branched
**5. sylvaticum**
3a. Stems 30–100 cm, at least 4 mm in
diameter **1. hyemale**
b. Stems 5–30 cm, not more than 3 mm
in diameter 4
4a. Stems mostly 2–3 mm in diameter,
somewhat rough **2. variegatum**
b. Stems mostly *c.* 1 mm in diameter,
smooth **3. scirpoides**

**1. E. hyemale** Linnaeus. Figure 2(14).
p. 4.
Stems 30–100 cm × 4–8 mm, unbranched,
evergreen, erect, rush-like, stiff, very rough
to the touch, dark blue-green. Leaf sheaths
usually lacking apical teeth. Internodes

somewhat inflated. Cone-bearing shoots similar to the vegetative. *N temperate Eurasia.* H2.

**2. E. variegatum** Weber & Mohr. Figure 2(15), p. 4.
Stems 10–30 cm × 2–3 mm, evergreen, unbranched, decumbent to erect, moderately rough, dark blue-green. Leaf sheaths with 4–10 more or less triangular teeth which have broad, white margins. Internodes not inflated. Cone-bearing shoots similar to the vegetative. *Circumboreal.* H1.

**3. E. scirpoides** Michaux. Figure 2(12), p. 4.
Stems 5–15 cm × *c.* 1 mm, evergreen, unbranched, prostrate to ascending with erect tips, smooth, dark blue-green. Leaf-sheaths with usually 3 teeth which are triangular and with narrow, white margins. Cone-bearing shoots similar to the vegetative. *Circumboreal.* H1.

**4. E. telmateia** Ehrhart. Figure 2(16), p. 4.
Stems 50–200 cm × 8–20 mm or more, winter-deciduous, freely branched throughout all but the lowermost and uppermost nodes, more or less smooth, succulent, ivory-white. Branches simple, ascending to spreading, very numerous. Leaf-sheaths with 20–40 long, more or less straight, slender, brown teeth. Cones borne on separate, short-lived, white, succulent, unbranched shoots in spring. *Atlantic islands, W Europe, Mediterranean area, Western N America.* H3.

**5. E. sylvaticum** Linnaeus. Figure 2(13), p. 4.
Stems 10–80 cm × 1–4 mm, winter-deciduous, green, freely branched throughout all but the extreme uppermost and lowermost nodes. Leaf-sheaths with 10–18 long, brown teeth, mostly partly united in 2s and 3s into a few, broad, papery lobes. Branches branched, spreading and drooping, 10–18 per whorl. Cone-bearing shoots similar to the vegetative, or with only rudimentary branches. *Temperate N Hemisphere.* H1.

# IV. FILICOPSIDA

Rhizomes present from which leaves arise directly. Leaves large (megaphylls), usually referred to as fronds, borne on a stalk (stipe), the blade frequently pinnately or palmately divided, often several times so, rarely simple or grass-like, usually unrolling in a crozier-like manner. Sporangia usually borne on the undersurfaces of the fronds, more rarely on fronds of specialised form (generally with reduced blades), forming groups (sori), each sorus often protected by an outgrowth (indusium) of the frond or its margin; very rarely the sori borne in specialised structures (sporocarps) which are borne in the frond axils. Spores usually all of the same size, very rarely of 2 different sizes.

This group, the true ferns, contains about 12 000 species in at least 400 genera; they are disposed in numerous families, but there is no general agreement as to what the families should be or should contain. As the families are of doubtful assistance in terms of identification, they are omitted here, and the genera are given in the sequence proposed by Crabbe, Jermy & Mickel (A new generic sequence for the Pteridophyte herbarium, Fern Gazette 11: 141–62, 1975).

The terminology of ferns is complex, and often rather different from that widely used for the flowering plants. In the descriptions which follow, some terms are used which require explanation:
*Size of plant*: in the generic descriptions, plants are described as *small* (fronds mostly no larger than 30 cm), *medium-sized* (fronds 30–150 cm) or *large* (fronds 150 cm or more).
*Rhizomes*: these may be erect, ascending or creeping. A few form trunks. Creeping rhizomes are described as *short-creeping* when contracted and bearing fronds or groups of fronds at close intervals of 2–5 cm; as *long-creeping* when extensive and bearing fronds or groups of fronds at intervals of more than 5 cm.
*Phyllopodia*: these are extensions or branches of the rhizome, which, in some species, rise above the ground and bear a single frond.
*Sporangia*: these are the spore-bearing structures found in the sori. Each has a ring of thickened cells, visible under a hand-lens or low-power microscope. The number of these cells is significant in *Polypodium* (p. 33).
*Fronds*: these are often very divided; the terminology of their various segments is elucidated in Figure 1, p. 2. Because of the difficulties of describing such complex objects, a silhouette of a frond or part-frond of almost every species included here has been provided; these form Figures 2–28.
Literature: The literature on ferns is very extensive; the following works, which deal mainly with ferns in cultivation, are particularly useful for identification and cultural advice. Copeland, E.B., *Genera Filicum: the genera of ferns* (1947); Macself, A.J., *Ferns for garden and glasshouse* (1952); Bruty, R., Cultivation of ferns under glass, *British Fern Gazette* 9: 23–5 (1959); Foster, F.G., *The gardener's fern book* (1964); Kaye, R., *Hardy ferns* (1968); Swindells, P., *Ferns for garden and greenhouse* (1971); Hoshizaki, B.J., *Fern grower's manual* (1976); Maatsch, R., *Das Buch der Freilandfarne* (1980); Page, C.N., *The ferns of Britain and Ireland* (1982).

A considerable number of ferns can be identified by individual, striking characteristics: a synopsis of these is given below, using the generic and specific numbers which follow in the main text. For example, 8(2) means that species 2 of genus 8 (*Anemia phyllitidis*) has the character in question; when a whole genus shows the character, a single number is given, e.g. 43, which means that all the species of genus 43 (*Cyathea*) show the character. The dichotomous key to the genera, which then follows, is divided into 3 subkeys, with a key to these subkeys at the beginning. This key is designed to allow for the numerous exceptional cases (particularly in deciding to which subkey any particular specimen should be referred).

*Frond fragrant*: 8(2), 68(1), 78(1,2,9).
*Plant of tree-like habit*: 6,43,44,45,67, 75(4), 88(8–10), 90.
*Plant with a long-climbing or twining growth*: 7,32(4), 81,82,83(1,2).
*Plant scrambling*: 32(4), 58,83(5), 86,92.
*Plant producing long runners, like those of the strawberry*: 87.
*Plant producing bulbils or plantlets on the upper surface or at the tip*: 9,16(3), 58,60(3,6,7,8,15), 67(1), 72(2,3), 75(4), 79,91.
*Plant amphibious, or rooted aquatic*: 9,32(3), 55,58,65,79,92,93,94,95.
*Plant free-floating aquatic*: 96,97.
*Frond extremely thin, only 1 cell thick, appearing membranous and translucent*: 6,24,25.
*Frond thick and fleshy*: 1,30,31,34(1), 36,37,60(5), 80(1), 94,95.
*Plant of moss-like habit*: 97.
*Frond kidney-shaped*: 21(16), 25.
*Frond surface with a bloomed, mealy texture*: 40,47(2), 68(1), 78(2).
*Frond conspicuously white, silver or gold beneath*: 13(3,4), 18,43(2).
*Undersurface of frond densely covered with scales*: 14,63(2,3), 78(1), 80(2).

*Frond either when emerging or expanded some other colour than green:* 4,21(2,5), 23(5,8), 78(11), 88(1–3,10), 89(3), 91,92(1), 93(1,4).

*Rhizome stout, creeping on the soil surface, densely covered with golden scales:* 26,27,28,39,40,41,47,83,85.

## KEY TO GROUPS

1a. Fronds completely undivided or only slightly lobed     **Group A**
  b. Fronds divided or at least regularly and conspicuously lobed     2
2a. Divisions of frond coarse (i.e. including all small plants which have 1 × pinnate or extensively pinnately lobed fronds to large plants which have fronds up to 3 × pinnate but with large, coarse, ultimate segments 1 cm or more in width)     **Group B**
  b. Divisions of frond fine (i.e. all plants with leaves more than 1 × pinnate with the ultimate segments less than 1 cm wide)     **Group C** (p. 9)

### Group A
1a. Fronds slender, cylindric, of thread-like or grass-like appearance   **95. Pilularia**
  b. Fronds with a flattened blade     2
2a. Blade of frond more or less linear     3
  b. Blade of frond relatively broader, of some other shape     5
3a. Blade of frond *c.* 5 mm broad     **22. Vittaria**
  b. Blade of frond at least 10 mm broad   4
4a. Frond leathery, jointed to the rhizome     **38. Dictymia**
  b. Frond fleshy, not jointed to the rhizome     **1. Ophioglossum**
5a. Blade of frond lanceolate     6
  b. Blade of frond of some other shape   16
6a. Fronds arising singly from different points along a long- or short-creeping rhizome     7
  b. Fronds all arising as a group from a single rhizome crown     10
7a. Rhizome rising and spreading well away from the ground surface at its growing apex     **86. Oleandra**
  b. Rhizome closely rooted to the ground surface throughout its length     8
8a. Frond with a conspicuous stipe     **30. Pyrrosia**
  b. Frond without a conspicuous stipe   9
9a. Sori mostly 2–3 mm in diameter     **34. Pleopeltis**
  b. Sori all less than 1.5 mm in diameter     **32. Microsorium**
10a. Tips of fronds rooting     11
  b. Tips of fronds not rooting     12

11a. Blade of frond broadest at base, gradually tapering    **61. Camptosorus**
  b. Blade of frond broadest above the base, tapering above and below     **60. Asplenium**
12a. Tip of frond drawn out into a long, rat-tail-like process    **37. Belvisia**
  b. Tip of frond not as above     13
13a. Sori confined to a few specialised, very narrow fronds    **88. Blechnum**
  b. Sori on fronds similar to those without     14
14a. Sporangia forming a felty mass all over the undersurface of some fronds     **80. Elaphoglossum**
  b. Sporangia along the veins only, forming linear sori     15
15a. Frond with a distinct stipe     **62. Phyllitis**
  b. Frond without a distinct stipe     **60. Asplenium**
16a. Blade of frond oval or rounded    17
  b. Blade of frond of some other shape     24
17a. Most fronds more than 3 cm, mostly 6–20 cm     18
  b. Most fronds less than 3 cm     20
18a. Blade covered in stiff, dark, bristle-like hairs    **80. Elaphoglossum**
  b. Blade without obvious hairs     19
19a. Blades with the main lateral veins widely spaced (mostly up to 5 mm apart)    **33. Selliguea**
  b. Blades with the main lateral veins closely spaced (mostly *c.* 1 mm apart)    **81. Humata**
20a. Plant floating on the surface of water, with 2 ranks of boat-shaped leaves    **96. Salvinia**
  b. Plant usually rooted in soil or on bark, not as above     21
21a. Blade thick and fleshy     22
  b. Blade not thick and fleshy     23
22a. Underside of blade covered in soft, rust-coloured scales    **30. Pyrrosia**
  b. Underside of blade without scales, or scales not obvious    **31. Drymoglossum**
23a. Most blades less than 1 cm, rounded in outline    **36. Lemmaphyllum**
  b. Most blades more than 1 cm, and at least twice as long as broad    **35. Microgramma**
24a. Blade of fertile frond broad and forked    **29. Platycerium**
  b. Blade of fertile frond not as above   25
25a. Blade of frond palmate     26
  b. Blade of frond not palmate     28

26a. Sporangia borne all around the frond margin    **16. Doryopteris**
  b. Sporangia borne along the veins, away from the margins     27
27a. Sporangia forming a net-like pattern on the undersurface of the frond    **19. Hemionitis**
  b. Sporangia forming separate, linear sori, not as above    **60. Asplenium**
28a. Blade of frond kidney-shaped    29
  b. Blade of frond arrowhead-shaped     30
29a. Blade extremely thin, membranous, translucent    **25. Cardiomanes**
  b. Blade of frond neither thin, membranous nor translucent    **21. Adiantum**
30a. Sporangia arranged all around the leaf margins    **16. Doryopteris**
  b. Sporangia along the veins, away from the leaf margins     31
31a. Sporangia forming a net-like pattern on the undersurface of the frond    **19. Hemionitis**
  b. Sporangia forming separate, long, linear sori     32
32a. Frond rooting at the tip    **61. Camptosorus**
  b. Frond not rooting at the tip    **62. Phyllitis**

### Group B
1a. Frond not pinnately constructed     2
  b. Frond basically pinnately constructed     6
2a. Fronds with leafy areas composed of 2 or 4 radiating leaflets     3
  b. Fronds with leafy areas basically palmate, not as above     4
3a. Each blade consisting of 2 leaflets    **94. Regnellidium**
  b. Each blade consisting of 4 leaflets    **93. Marsilea**
4a. Sporangia arranged all around the leaf margin    **16. Doryopteris**
  b. Sporangia along the veins, away from the margins     5
5a. Sporangia forming a net-like pattern on the undersurface of the frond    **19. Hemionitis**
  b. Sporangia forming long, linear, distinct sori    **60. Asplenium**
6a. Frond not divided down to the midrib in any part     7
  b. Frond divided to the midrib in at least some parts     18
7a. Fronds arising singly from different points along the rhizome     8
  b. Fronds all arising as a group from a single rhizome crown     12

8a. Sori always covered with cup-shaped
    indusia    **81. Humata**
  b. Sori rounded, without indusia    9
9a. Sori *c.* 1 mm in diameter
    **32. Microsorium**
  b. Sori more than 2 mm in diameter   10
10a. Sori partly sunken into the lower
    surface of the frond    **39. Phymatodes**
  b. Sori not sunken into the frond surface
    11
11a. Blade green, without bluish bloom
    **41. Polypodium**
  b. Blade with a distinct, bluish bloom
    **42. Phlebodium**
12a. Fronds all less than 50 cm    13
  b. Fronds all 50–200 cm    15
13a. Fronds 15–50 cm    **88. Blechnum**
  b. Fronds mostly less than 15 cm    14
14a. Fronds thick and leathery, somewhat
    fleshy    **63. Ceterach**
  b. Fronds not as above, rather
    membranous    **73. Quercifilix**
15a. Rhizomes with sparse, blackish scales
    **72. Tectaria**
  b. Rhizomes densely covered with a
    mass of felt-like, orange-brown scales
    16
16a. Bases of fronds surrounded by
    separate, lobed, oak-leaf-like 'nest
    leaves'    **26. Drynaria**
  b. Bases of the fronds themselves lobed,
    but not surrounded by separate nest
    leaves    17
17a. Sori uncovered, on the undersides of
    normal fronds    **28. Aglaomorpha**
  b. Sori never present on the undersides
    of normal fronds    **27. Merinthosorus**
18a. Plants free-floating on the surface of
    water    **96. Salvinia**
  b. Plants normally rooted in soil or bark
    19
19a. Plants with long climbing and
    twining fronds    **7. Lygodium**
  b. Fronds not climbing and twining   20
20a. Fronds arising singly from different
    points along a long-creeping rhizome
    21
  b. Fronds all arising in a group from a
    single rhizome crown or from a very
    short-creeping rhizome    32
21a. Rhizomes mostly well above the soil
    surface, scrambling and leaning on
    other plants    **92. Stenochlaena**
  b. Rhizomes beneath the soil surface or
    firmly attched to it    22
22a. Rhizomes beneath the soil surface
    23
  b. Rhizomes on top of the soil surface
    27
23a. Blade of frond not triangular    24
  b. Blade of frond triangular    25

24a. Sori covered by membranous
    marginal flaps    **15. Pellaea**
  b. Sori without indusia of any kind
    **55. Thelypteris**
25a. Sporangia borne on separate, tassel-
    like fronds, never on the backs of leafy
    fronds    **65. Onoclea**
  b. Sporangia borne on the backs of leafy
    fronds    26
26a. First (lowest) pair of pinnae very
    large, giving a 3-lobed appearance to
    frond    **68. Gymnocarpium**
  b. First pair of pinnae not longer than
    second pair    **56. Phegopteris**
27a. Frond 50–200 cm, soft and
    pendulous    **42. Goniophlebium**
  b. Frond less than 50 cm, rigid and not
    pendulous    28
28a. Fertile and sterile fronds differing, the
    fertile with much narrower segments
    than the sterile    29
  b. Fertile and sterile fronds similar in
    structure    30
29a. Rhizomes sparsely scaly; mature
    fronds rooting at their tips
    **79. Bolbitis**
  b. Rhizomes densely scaly; mature
    fronds not rooting    **82. Scyphularia**
30a. Sori covered by kidney-shaped indusia
    **85. Humata**
  b. Sori without indusia of any kind   31
31a. Fronds very irregularly lobed
    (especially the smaller)
    **32. Microsorium**
  b. Fronds very regularly lobed
    **41. Polypodium**
32a. Frond bearing small plantlets along its
    midrib    33
  b. Frond without small plantlets along
    its midrib    34
33a. Plantlets large, usually less than 10 to
    each frond    **58. Ampelopteris**
  b. Plantlets small, very numerous on
    each frond    **91. Woodwardia**
34a. Fertile and sterile fronds not alike
    throughout their whole length    35
  b. Fertile and sterile fronds alike
    throughout their whole length, rarely
    differing slightly at the extreme tips
    39
35a. Frond herring-bone-like, the main
    divisions not further divided    36
  b. Frond with the main divisions at least
    partially divided    37
36a. Margins of frond with regular, saw-
    like teeth    **89. Doodia**
  b. Margins of frond without any teeth
    **88. Blechnum**
37a. Fertile fronds with narrow, leaf-like
    segments with sporangia along their
    edges    **23. Pteris**

  b. Fertile frond entirely tassel-like,
    without any leafy part    38
38a. Frond with very short stipe, or stipe
    absent    **64. Matteuccia**
  b. Frond with a very long stipe
    **4. Osmunda**
39a. Fertile frond differing from the sterile
    only by the addition of a tassel-like tip
    40
  b. Fertile frond differing from the sterile
    throughout the whole length of the
    blade    43
40a. Frond without a stipe    41
  b. Frond with a stipe    42
41a. Main segments of fertile tassel linear
    and undivided    **27. Merinthosorus**
  b. Main segments of fertile tassel
    themselves segmented into small lobes
    **28. Aglaomorpha**
42a. Fertile tassel on a stalk which arises
    from the base of the leafy part of the
    frond    **8. Anemia**
  b. Fertile tassel an integral part of the
    frond    **4. Osmunda**
43a. Sporangia arranged along or very
    close to the margins on the underside
    of the frond    44
  b. Sporangia arranged in various
    patterns but always well in from the
    margins on the underside of the frond
    51
44a. Edge of blade folded over to protect
    the sori    45
  b. Edge of blade not folded over to
    protect the sori    46
45a. Indusium (folded over leaf margin)
    divided into many, small, separate
    sections    **21. Adiantum**
  b. Indusium more or less continuous
    **23. Pteris**
46a. Blade margin with very prominent,
    long, saw-like, spine-tipped teeth
    **75. Polystichum**
  b. Blade margin straight, wavy or
    very finely toothed but not as above
    47
47a. Frond more than once pinnate    48
  b. Frond once pinnate only    49
48a. Veins in ultimate segments
    transparent when viewed against the
    light    **3. Marattia**
  b. Veins in ultimate segments not
    transparent    **2. Angiopteris**
49a. Blades ovate    **59. Christella**
  b. Blades more or less linear    50
50a. Pinnae triangular-crescent-shaped
    **53. Lindsaea**
  b. Pinnae linear, only slightly tapering
    **87. Nephrolepis**
51a. Sporangia arranged in linear sori   52

b. Sporangia arranged in sori which do not form continuous lines on the frond 58

52a. Sori forming a pair of parallel lines lengthwise along the centre of each leafy segment 53

b. Sori forming single, radiating lines over each leafy segment 54

53a. Fronds once pinnate **88. Blechnum**

b. Fronds twice pinnatisect **90. Sadleria**

54a. Fronds covered with conspicuous fine hairs on all surfaces 55

b. Fronds without conspicuous hairs 56

55a. Pinnae rounded; sori without indusia **20. Gymnopteris**

b. Pinnae tapering; sori with indusia **70. Lunathyrium**

56a. Fronds leathery, upper surfaces glossy **60. Asplenium**

b. Fronds soft, upper surfaces not glossy 57

57a. Fronds less than 15 cm **60. Asplenium**

b. Fronds mostly 50–200 cm **67. Diplazium**

58a. Sori confined to patches near the midrib in the central part of the frond **5. Todea**

b. Sori forming a continuous area to the frond tip 59

59a. Fronds soft **56. Phegopteris**

b. Fronds harsh and leathery 60

60a. Sori rounded, covered by kidney-shaped or umbrella-shaped indusia 61

b. Sori squared, covered by squared flaps hinged at side 62

61a. Margins of blades with saw-like teeth **75. Polystichum**

b. Margins of blades without teeth **76. Cyrtomium**

62a. Fronds large, mostly more than 50 cm, divided more than once, often rooting at their tips **91. Woodwardia**

b. Fronds mostly less than 50 cm, divided once, never rooting at their tips **89. Doodia**

**Group C**

1a. Plant aquatic 2

b. Plant not aquatic 3

2a. Plant mostly less than 5 cm, moss-like, floating on the water surface **97. Azolla**

b. Plant 15–20 cm with distinct fronds, growing as a rooted aquatic **9. Ceratopteris**

3a. Plant with climbing fronds to 5 m or more **7. Lygodium**

b. Fronds not climbing, not as large 4

4a. Blades many times divided into narrow, linear segments, the whole blade arranged in a fan-like manner **10. Actiniopteris**

b. Blades not as above 5

5a. Fronds palmate 6

b. Fronds not palmate 8

6a. Fronds arising singly from different points along the creeping rhizome **81. Humata**

b. All fronds arising as a group from a single rhizome crown 7

7a. Blade mealy white beneath **13. Cheilanthes**

b. Blade green beneath **16. Doryopteris**

8a. Division of frond irregular, leafy areas formed from numerous fan-shaped units 9

b. Division of frond based on regular, simple or multiple herring-bone-like arrangement, and leafy areas of other shapes 14

9a. Fronds 15–150 cm or more 10

b. Fronds mostly less than 15 cm 11

10a. Fan-shaped segments more than 4 mm broad **21. Adiantum**

b. Fan-shaped segments up to 2 mm broad **54. Sphenomeris**

11a. Stipe dark brown or black **13. Cheilanthes**

b. Stipe green 12

12a. Fertile fronds with much narrower segments than the sterile **12. Cryptogramma**

b. Fertile fronds similar to the sterile 13

13a. Plant annual **17. Anogramma**

b. Plant perennial **60. Asplenium**

14a. Fronds arising singly from different points along a long-creeping rhizome 15

b. Fronds arising in a close group, usually from a single rhizome crown 31

15a. Rhizomes on top of the soil surface 16

b. Rhizomes well beneath the soil surface 22

16a. Fronds mostly less than 10 cm 17

b. Fronds mostly more than 10 cm 18

17a. Fronds thin, membranous **24. Hymenophyllum**

b. Fronds thick, of leathery texture **81. Humata**

18a. Sori linear **11. Onychium**

b. Sori not linear, of some other shape 19

19a. Sori in groups near the edges of each segment and partially covered by a cup-shaped flap arising from one side only 20

b. Sori in groups set in from the edge of each segment and partially covered by kidney- or umbrella-shaped flaps, each with a central stalk 21

20a. Blade of markedly soft texture **84. Leucostegia**

b. Blades not of markedly soft texture **83. Davallia**

21a. Mature fronds 30–80 cm **85. Rumohra**

b. Mature fronds 80–200 cm **71. Arachniodes**

22a. Fronds dying down to ground level in winter 23

b. Fronds remaining green and standing through the winter 24

23a. Fronds less than 20 cm **69. Cystopteris**

b. Fronds 20–100 cm or more **48. Dennstaedtia**

24a. Fronds less than 50 cm 25

b. Fronds 50–200 cm 27

25a. Ultimate segments of frond with rounded tips **21. Adiantum**

b. Ultimate segments of frond with pointed tips 26

26a. Fronds shining, glossy green above **51. Paesia**

b. Fronds dull green above **71. Lastreopsis**

27a. Stipe harsh and rough to the touch 28

b. Stipe very smooth to the touch 29

28a. Sori partly covered by a cup-shaped flap **49. Microlepia**

b. Sori partly covered by reflexed marginal teeth, or not covered at all **50. Hypolepis**

29a. Sori rounded **46. Culcita**

b. Sori linear 30

30a. Sori around the margins of each leafy segment **52. Histiopteris**

b. Sori in lines along the veins of each segment **67. Diplazium**

31a. Fronds more than 1.5 m 32

b. Fronds less than 1.5 m 38

32a. Fronds borne in a shuttlecock-like crown from the tops of massive, upright, trunk-like rhizomes (tree ferns) 33

b. Fronds borne irregularly from rhizomes which are not as above 35

33a. Bases of fronds covered with shaggy brown scales **43. Cyathea**

b. Bases of fronds covered in masses of woolly brown hairs 34

34a. Fronds all of 1 kind        **44. Dicksonia**

b. Fronds of 2 kinds        **45. Thyrsopteris**

35a. Bases of fronds covered in a mass of woolly brown hairs        **47. Cibotium**

b. Bases of fronds covered in shaggy brown scales                    36

36a. Sori linear and continuous along the margins of the leafy segments

**23. Pteris**

b. Sori discrete                          37

37a. Fronds all ascending; sori small, rounded, superficial, naked

**57. Macrothelypteris**

b. Fronds arching, sori large and squared, forming a row of box-like structures impressed into the frond

**91. Woodwardia**

38a. Blades very thin, membranous      39

b. Blades thicker, not as above        40

39a. Fronds long-oblong or linear, very finely dissected        **6. Leptopteris**

b. Fronds of other shapes, less finely dissected        **21. Adiantum**

40a. Undersides of fronds totally covered by dense masses of hairs or scales

**14. Notholaena**

b. Undersides of fronds not as above

41

41a. Undersides of fronds white- or yellow-mealy                          42

b. Undersides of fronds green        43

42a. Sori confined to the margins of the leafy segments        **13. Cheilanthes**

b. Sori set in from the margins of the leafy segments        **18. Pityrogramma**

43a. Plant with plantlets arising from existing fronds        **60. Asplenium**

b. Fronds not giving rise to new plantlets

44

44a. Sori grouped on separate, fertile, tassel-like parts of the frond

**8. Anemia**

b. Sori on the backs of normal, leafy parts of the frond                45

45a. Sori restricted to the margins of the leafy segments                  46

b. Sori borne away from the margins of the leafy segments                50

46a. Sori more or less continuous along the margins of the leafy segments

47

b. Sori discontinuous, broken into separate units along the margins of the leafy segments        48

47a. Stipe shining black        **15. Pellaea**

b. Stipe not shining black        **23. Pteris**

48a. Stipe shining black        **21. Adiantum**

b. Stipe not shining black        49

49a. Fronds 60–85 cm, 3–4 times pinnate

**54. Sphenomeris**

b. Fronds 50–150 cm, 2 times pinnate

**74. Didymochlaena**

50a. Indusial flap very short-lived, usually not visible at the time of spore maturity        **69. Cystopteris**

b. Indusial flap persistent        51

51a. Sori long, linear, along the veins

**60. Asplenium**

b. Sori not linear        52

52a. Sori J-shaped        **66. Athyrium**

b. Sori rounded        53

53a. Indusial flap kidney-shaped

**78. Dryopteris**

b. Indusial flap umbrella-shaped

**75. Polystichum**

## 1. OPHIOGLOSSUM Linnaeus

*C.N. Page & F.M. Bennell*

Small- to medium-sized terrestrial or epiphytic plants arising from tuberous corms or creeping rhizomes. Fronds simple, entire, more or less fleshy, with a network of veins visible against the light, evergreen or deciduous, ovate-elliptic or linear. Sori on a separate, cone-like, fertile spike originating from near the base of the frond, with 2 rows of spherical sporangia, each opening by a transverse slit.

A genus of about 30 species, almost world-wide, the epiphytic species mainly tropical. Plants are long-lived and slow-growing, and are cultivated as botanical curiosities. Terrestrial species require planting out in a damp, shaded place, preferably in basic soil. Epiphytes need rich compost in a hanging basket. Propagation is by division. All are slow to establish and are difficult and unpredictable in cultivation. Their unpredictable growth probably relates to their partly saprophytic nutrition, requiring the presence of appropriate species of soil fungi and conditions which encourage the formation of mycorrhizae.

1a. Fronds more or less ovate-lanceolate, erect-ascending, c. 10–15 cm

**1. vulgatum**

b. Fronds linear, pendulous, up to c. 90 cm        **2. pendulum**

**1. O. vulgatum** Linnaeus. Figure 3(2), p. 11.

Terrestrial. Rootstock a tuberous, spherical corm, situated several centimetres below the ground surface. Fronds c. 10–15 cm, rarely more, flexible, erect or ascending,

ovate to lanceolate. Fertile spike to 20–30 cm, eventually overtopping the leaf. *N Temperate Regions.* H3.

**2. O. pendulum** Linnaeus. Figure 3(1), p. 11.

Epiphytic. Rootstock a creeping rhizome. Fronds pendulous, to 90 × 2 cm, linear and strap-like, brittle. Fertile spike much shorter than the leafy blade, c. 15 cm. *Old World Tropics, from Madagascar to Polynesia.* G2.

## 2. ANGIOPTERIS Hoffmann

*C.N. Page & F.M. Bennell*

Very large terrestrial plants. Rhizomes very short, thick, ascending. Bases of the fronds surrounded by fleshy, stipule-like outgrowths. Fronds all of one kind, 2 × pinnate into numerous, regularly arranged, oblong-linear, leaf-like segments with conspicuous, forking, parallel venation. Fertile fronds with sori near the margin in double rows and elongate along the veins of each segment, each sorus with an indusium.

A genus of a single species (subdivided by some authors). It requires very warm, permanently moist soil conditions, still, humid air, and is cold-sensitive. This is the largest of living ferns, eventually requiring considerable space. Propagation is by spores or by rooting of the 'stipules'.

**1. A. evecta** (Forster) Hoffmann. Figure 3(4), p. 11.

Fronds with thick, fleshy, ascending stipes to 10 cm or more in diameter; blades spreading, commonly 2–3 m in cultivation, but up to 10 m is possible, evergreen, arising in large clusters, coarse, leathery, dark green. *Madagascar to Polynesia and Japan.* G2.

## 3. MARATTIA Swartz

*C.N. Page & F.M. Bennell*

Large to very large plants. Rhizome massive, stocky, spherical. Bases of the fronds surrounded by stipule-like outgrowths. Fronds all of one kind, 2 or 3 × pinnate into numerous, regularly arranged, oblong-linear, tapering, leaf-like segments with forking, parallel venation. Fertile fronds with sori c. 2 mm long, elongate, oblong.

A pantropical genus of about 60 somewhat ill-defined species. They are very long-lived and slow-growing, grown in moist shade, well protected from the cold or drying winds. Propagation is by separation of offsets from the rhizome.

**Figure** 3. Silhouettes of fern-fronds. 1, *Ophioglossum pendulum*. 2, *O. vulgatum*. 3, *Marattia fraxinea*. 4, *Angiopteris evecta*. 5, *Osmunda cinnamomea*. 6, *O. claytoniana*. 7, *O. regalis* (a, pinna from base of frond to show typical frond width; b, upper part of frond with fertile segments). 8, *Leptopteris superba* (a, base of frond; b, middle of frond). 9, *L. hymenophylloides*. 10, *Todea barbara*.

**1. M. fraxinea** J.E. Smith. Figure 3(3), p. 11.

Fronds spreading, commonly to 5 × 2 m in cultivation. Stipes to 4 cm thick. Pinnae leathery, dark glossy green. *Old World Tropics.* G2.

Other fairly similar species appear from time to time in cultivation but are mostly found only in the large glasshouses of botanical collections.

### 4. OSMUNDA Linnaeus
*C.N. Page & F.M. Bennell*

Large terrestrial plants. Rhizome thin, fibrous and woody, branching often to build up massive clumps, ascending to erect, bearing fronds in dense crowns. Fronds 1 or 2 × pinnate, often pinkish when expanding, deciduous, of 2 kinds, some entirely sterile, others partly fertile. Blades papery or leathery, hairless when mature but the young fronds usually with a soft brownish down which is shed on expanding. Sporangia large, spherical, each opening by a long slit, borne all over the backs of much-reduced pinnae but not in definite sori. Indusium absent.

A genus of 10–15 variable and widely distributed species, occurring in all parts of the world. The plants are slow-growing and long-lived. They thrive in permanently moist, peaty situations, especially by ponds or moving water when planted out to give a free root-run. They are tolerant of full light. Propagation is by spores which are short-lived and should be sown within about 3 weeks of collection.

  1a. Plants with some fronds entirely
       sterile, others fertile throughout
                                **1. cinnamomea**
   b. Plants with some fronds entirely
       sterile, others fertile only in parts
                                              2
  2a. Fertile areas present only in the
       middles of the fronds    **2. claytoniana**
   b. Fertile areas present only at the tips of
       the fronds                **3. regalis***

**1. O. cinnamomea** Linnaeus. Figure 3(5), p. 11.

Fronds 60–150 cm, ascending or erect. Blade ovate-lanceolate, papery, the sterile 1 × pinnate, the pinnae deeply pinnatisect, the fertile 2 × pinnate. Some fronds entirely sterile, others with all the pinnae reduced and fertile. *Eastern N America.* H2.

**2. O. claytoniana** Linnaeus. Figure 3(6), p. 11.

Fronds 40–80 cm, ascending to arching or almost erect. Blade ovate-lanceolate,

1 × pinnate with the pinnae deeply pinnatisect, papery. Plants with some fronds entirely sterile, others with the pinnae in the middle of the frond reduced and fertile, so that the outline of the frond is interrupted. *Eastern N America.* H2.

**3. O. regalis** Linnaeus. Figure 3(7), p. 11.

Fronds 1–2 m or more, ascending or erect. Blade broadly triangular-ovate, 2 × pinnate, leathery. Some fronds are entirely sterile, others have the pinnae at the tips reduced and fertile, forming a tassel-like tip. *Widespread in temperate and subtropical regions, more local in the tropics.* H4.

A plant in cultivation as *O. regalis* var. *gracilis* Anon. has much smaller fronds which are pinkish when young. This plant does not produce fertile fronds regularly, so it is impossible to be certain which species it belongs to. It is reputed to originate from N America, but this is also uncertain.

  **\*O. japonica** Thunberg. Similar but with fewer, smaller, broader, more spreading fronds with larger pinnules, rarely fertile. *Japan.* H4.

### 5. TODEA Willdenow
*C.N. Page & F.M. Bennell*

Medium to large terrestrial plants. Rhizome thick, stout, fibrous, very woody, ascending to almost erect, bearing fronds in clusters, eventually forming a dense crown. Fronds all of one kind, regularly 2 × pinnate, evergreen, thick, slightly leathery. Sporangia large, pale brown, spherical, not in definite sori, each opening by a longitudinal slit, borne all over the backs of normal or slightly reduced pinna-segments. Indusium absent.

A genus of 2 species from South Africa, New Guinea, Australia and New Zealand. Cultural conditions as for *Osmunda*, but the plants are less hardy.

**1. T. barbara** (Linnaeus) Moore. Figure 3(10), p. 11

Fronds 120–200 cm or more, ascending or erect, borne on strong, square stipes which are from one-quarter to one-tenth of the frond in length. Blade oblong-lanceolate, 2 × pinnately divided into very numerous, regularly arranged, oblong-lanceolate pinnules, those bearing sporangia situated in the middle of the frond or towards its base, all bright, glossy, mid to dark green. *South Africa,*

*E Australia (including Tasmania), New Zealand.* H5–G1.

### 6. LEPTOPTERIS Presl
*C.N. Page & F.M. Bennell*

Medium to large terrestrial plants. Rhizome thick, seldom branching, erect or trunk-forming, bearing fronds in regular, spreading rosettes. Fronds all of one kind, at least 2–3 × pinnate, evergreen, lanceolate-ovate, very thin but firm, delicate, membranous, mid to deep green. Sporangia large, pale brown, spherical, not in definite sori, covering much of the backs of the fertile segments, each sporangium opening by a longitudinal slit. Indusia absent.

A genus of 6 rather similar species from New Zealand, Australia (Queensland), New Guinea and Samoa. Delicate ferns of great beauty with fronds like ostrich plumes, combining the membranous texture of filmy ferns with a large, finely dissected frond and miniature tree-fern habit. They are extremely exacting, demanding highly specific growth conditions: low light, a very freely drained but constantly moist, acidic medium, shelter and constantly high humidity, such as is likely to be found only in closed culture. They are propagated by spores which are short-lived and should be sown within 3 weeks of collection.

  1a. Stipe about one-quarter of the length
       of the blade       **1. hymenophylloides**
   b. Stipe at most one-tenth of the length
       of the blade                  **2. superba**

**1. L. hymenophylloides** (Richard) Presl. Figure 3(9), p. 11.

Rhizome to 1 m or more, forming a trunk. Fronds to 100 cm, blade ovate-lanceolate, 2–3 × pinnate, all the ultimate divisions fairly widely spread in one plane. Stipe about one-quarter of the length of the blade. *New Zealand.* G1.

**2. L. superba** (Colenso) Presl. Figure 3(8), p. 11.

Rhizome to 30 cm, forming a short trunk. Fronds to 70–80 cm, the stipes very short. Blade ovate with ultimate segments closely spaced and in various planes, giving the whole frond a slightly crisped and fluffy texture and the appearance of an ostrich feather. *New Zealand.* G1.

### 7. LYGODIUM Swartz
*C.N. Page & F.M. Bennell*

Rhizome long- or short-creeping, below ground, thin, branching occasionally in a forking manner, bearing fronds in clusters. Fronds variously palmately or pinnately divided, mostly of 2 kinds, each frond normally of indefinite growth and becoming several metres in length, thin, soft, often papery but firm. Rachis thin, wiry, extensively climbing and tightly twining, giving off alternate frond-like branches (pinnae) to the right and left, each ending in a dormant bud and giving rise to a pair of further frond-like leaflets (pinnules). Fertile leaflets normally produced only high on the fronds, their segments often narrower than the sterile and their edges fringed with short, narrow, finger-like lobes, each lobe bearing 2 rows of large sporangia, each sporangium covered by a small indusium.

A genus of about 40 species from the tropics, extending south to South Africa and New Zealand and north to Japan and the eastern USA. They are long-lived and fast-growing, twining rapidly up strings and forming hanging curtains of foliage. Their climbing form of growth is distinctive among ferns. They are readily propagated from fresh spores, and grow best in bright, slightly filtered sunlight, with high humidity and a moist peaty soil. New growth is produced annually and the old fronds should be removed.

1a. Paired non-fertile pinnules dichotomously or palmately subdivided throughout  2
  b. Paired non-fertile pinnules mostly pinnately or pinnatisectly divided throughout  3
2a. Paired non-fertile pinnules usually palmately divided  1. palmatum
  b. Paired non-fertile pinnules dichotomously divided or occasionally partly pinnately divided  2. scandens*
3a. Paired non-fertile pinnules mostly 2 × pinnately divided  5. japonicum
  b. Paired non-fertile pinnules mostly 1 × pinnately divided  4
4a. All ultimate segments more or less rounded, not much longer than broad  4. microphyllum
  b. All ultimate segments linear, several times longer than broad  3. volubile

1. L. palmatum (Bernhardi) Swartz. Figure 4(4), p. 14.
Paired non-fertile pinnules each more or less palmately divided throughout into 3–7 lobes which are confluent at their bases. *Eastern USA from Florida to N Carolina.* H5–G1.

2. L. scandens (Linnaeus) Swartz. Figure 4(2), p. 14.
Paired non-fertile pinnules each

dichotomously divided or occasionally pinnately, then dichotomously divided into 2–4 (occasionally more) long, narrow, tapering segments. *Tropical E Asia.* G2.

  *L. flexuosum (Linnaeus) Swartz. Very similar, but larger in all its parts. *E Asia, Australia.* G2.

3. L. volubile Swartz. Figure 4(5), p. 14.
Paired non-fertile pinnules each pinnately divided into *c.* 6–12 stalked, linear segments which are several times longer than broad. *Tropical America, from Cuba to Brazil.* G2.

4. L. microphyllum (Cavanilles) R. Brown. Figure 4(3), p. 14.
Paired non-fertile pinnules each pinnately divided into *c.* 6–14 small, stalked segments, each more or less rounded, rarely more than 2 × longer than broad. *Tropical and subtropical Africa, Asia and Australia.* G1–2.

5. L. japonicum (Thunberg) Swartz. Figure 4(1), p. 14.
Paired non-fertile pinnules each 2–3 × pinnately divided into numerous fine, rounded or somewhat elongate, tapering segments. *India, China, Japan, Korea, southwards to Australia.* H5.

  The most finely divided species, and the most common in cultivation.

## 8. ANEMIA Swartz
*C.N. Page & F.M. Bennell*
Small to medium terrestrial plants. Rhizomes hairy, short-creeping, bearing fronds in clusters. Fronds with ascending stipes and spreading or ascending, evergreen blades of two kinds: the vegetative pinnately compound, the fertile similar to the vegetative in their upper parts but with the basal pinna-pair standing erect, each long-stalked and without blade, which is replaced by dense clusters of naked sporangia forming tall, fertile tassels.

A genus of about 90 species, all from the New World. They require bright, open conditions and are propagated by spores.

1a. Leafy part of frond 1 × pinnate  2. phyllitidis*
  b. Leafy part of frond 2 × pinnate  1. adiantifolia

1. A. adiantifolia (Linnaeus) Swartz. Figure 4(6), p. 14.
Fronds 50–70 cm. Blade triangular, broadest at the base; fertile portion of the frond usually shorter or slightly longer than the sterile portion. Blade leathery, mid-green, 1 × pinnate. *USA (Florida), West*

*Indies, central and western S America.* H5–G1.

2. A. phyllitidis (Linnaeus) Swartz. Figure 4(7), p. 14.
Fronds 30–60 cm. Blade oblong-lanceolate, broadest in the lower third, fertile portion usually longer than the sterile portion. Blade leathery, mid dark green, 2 × pinnate. Fronds odourless. *West Indies, C America, Brazil.* H5–G1.

  *A. underwoodiana Maxon. Very similar, but the bruised foliage has a pungent smell. *W Indies.* G1.

## 9. CERATOPTERIS Brongniart
*C.N. Page & F.M. Bennell*
Small to medium aquatic plants. Fronds of 2 kinds, pinnately compound, soft, the fertile with longer and narrower pinnules than the sterile, and often bearing plantlets in the pinna axils. Sporangia occupying the entire lower surface of the pinna but protected beneath a thick, continuous reflexed margin.

A genus of about 10 species, of which 1 is cultivated as an aquarium plant or an aquatic marginal in landscaped hothouse collections.

1. C. thalictroides (Linnaeus) Brongniart. Figure 4(11), p. 14.
Rhizomes short, becoming erect. Fronds 15–30 cm, ascending, blades 2 × pinnate, triangular, often with numerous plantlets. *New World Tropics.* G2.

  One of the very few 'annual' ferns, and also one of the few aquatic species. Plants can grow as free-floating aquatics or take root in shallow mud. The axillary buds persist over the winter after decay of the parent frond and can establish new plants in the following season.

## 10. ACTINIOPTERIS Link
*C.N. Page & F.M. Bennell*
Small terrestrial plants. Rhizome very short-creeping. Fronds all of one kind, the blades dichotomously divided into numerous radiating linear segments. Fertile fronds with one uninterrupted, long, narrow sorus per segment, each sorus protected by a continuous, reflexed margin.

A genus of about 3 species which require warm, dry conditions and are propagated readily by spores.

1. A. semiflabellata Pichi-Sermolli. Figure 4(8), p. 14.
Rhizome scaly, bearing fronds in densely grouped clusters. Fronds to 30 cm with ascending stipes and spreading blades,

**Figure 4.** Silhouettes of fern-fronds. 1, *Lygodium japonicum*
(a, vegetative frond; b, fertile frond). 2, *L. scandens*.
3, *L. microphyllum* (a, vegetative frond; b, fertile frond).
4, *L. palmatum* (a, vegetative pinnae; b, fertile pinnae).
5, *L. volubile*. 6, *Anemia adiantifolia*. 7, *A. phyllitidis*.
8, *Actiniopteris semiflabellata*. 9, *Cryptogramma crispa* (a, vegetative
frond; b, fertile frond). 10, *Onychium japonicum*. 11, *Ceratopteris
thalictroides*.

evergreen, leathery, dark green, fan-shaped, tips of segments each ending in a single hard point. *S & E Africa, Arabia to India.* G1–2.

**\*A. australis** Linnaeus (*A. radiata* (Swartz) Link). Similar, but with segments ending in several teeth. *S & E Africa, Arabia to India.* G1–2.

## 11. ONYCHIUM Kaulfuss
*C.N. Page & F.M. Bennell*
Small to medium terrestrial plants. Rhizome short-creeping, thin, compact, branching occasionally, ascending, covered with scales. Fronds all of one kind, 3 × or more pinnately divided into numerous, fine, ultimate segments, evergreen. Sori linear, continuous along the margins of the segments. A marginal or submarginal flap forms the indusium, those from the opposite margins of the slender segments often meeting.

A genus of about 7 species from E Asia and America. They are fast-growing and are cultivated for their fine, delicate, airy foliage. They require light shade and constantly moist but not wet compost. They can be divided or propagated by spores.

**1. O. japonicum** (Thunberg) Kunze. Figure 4(10), p. 14.
Fronds 20–60 × 10–15 cm, laxly arching, triangular, finely divided into very numerous, small, narrow, pointed segments. Stipe as long as or shorter than the blade. *Japan, China.* H5–G1.

## 12. CRYPTOGRAMMA R. Brown
*C.N. Page & F.M. Bennell*
Small terrestrial plants. Rhizomes small, creeping, congested, bearing fronds in crowded tufts. Fronds of 2 kinds, deciduous, hairless, finely dissected, the fertile taller and with longer and narrower pinnules than the sterile. Sori along the margins, protected by a continuous, reflexed margin.

A small genus of about 5 species from North temperate regions. They require light, well-drained, well-lit, lime-free, humid, open conditions and can eventually form large clumps, especially on screes. Propagation is by division of the clumps or from spores.

**1. C. crispa** (Linnaeus) R. Brown. Figure 4(9), p. 14.
Fronds 3 × pinnate, bright green, blade triangular with stipe up to twice as long as the blade. *Europe & SW Asia.* H1.

## 13. CHEILANTHES Swartz
*C.N. Page & F.M. Bennell*
Small terrestrial plants. Rhizomes short-creeping, dark, scaly. Fronds scaly in many species, mostly all of one kind or with the fertile differing only slightly from the sterile. Blades several times pinnately divided. Fertile fronds with confluent marginal sori, which are protected when young by the reflexed margins.

A genus of about 180 species occurring throughout the tropical and warm temperate zones, mainly in dry places. They grow in dry sunny conditions and are exacting and often short-lived in cultivation, resenting disturbance and over-watering. Most species grow rapidly from spores. Literature: Fuchs, H.P., The Genus Cheilanthes Swartz and its European species, *British Fern Gazette* **9**: 38–48 (1961).

1a. Undersides of fronds green      2
  b. Undersides of fronds white      3
2a. Ultimate segments of fronds short and rounded      **1. fragrans**
  b. Ultimate segments of fronds elongate and oblong      **2. pulchella**
3a. Fronds narrowly triangular, pinnately divided      **3. farinosa**
  b. Fronds broadly triangular, pedately divided      **4. argentea**

**1. C. fragrans** (Linnaeus) Swartz. Figure 5(10), p. 17.
Fronds arising in sparse clumps, mostly 6–12 cm, pinnately divided into many, small, rounded segments, green beneath, the surface with scattered scales. *Mediterranean area.* H5–G1.

**2. C. pulchella** Bory de St Vincent. Figure 5(11), p. 17.
Fronds mostly 10–15 cm, pinnately divided into a few long, elongate-oblong segments of leathery texture, green beneath. *Canary Islands.* H5–G1.

**3. C. farinosa** (Forsskal) Kaulfuss. Figure 5(12), p. 17.
Fronds to 70 cm, pinnately divided, the blade white beneath, narrowly triangular, almost twice as long as broad, stipe approximately equalling the blade in length. *E Africa, SW Asia.* H4.

**4. C. argentea** (Gmelin) Kunze. Figure 5(9), p. 17.
Fronds 15–25 cm, pedately divided, the blade white, creamy white or yellowish beneath, triangular in outline, as broad as long, stipe 3–4 × longer than blade. *E Asia.* H4.

## 14. NOTHOLAENA R. Brown
*C.N. Page & F.M. Bennell*
Small to medium terrestrial plants. Rhizomes short-creeping, scaly, bearing fronds in close, irregular clusters. Fronds all of one kind, rather few per plant, 1–2 × pinnate, mostly held more or less upright, more or less evergreen, ovate, harsh and leathery, densely covered over at least the lower surface with whitish or golden red scales or scales and hairs, dark green above. Sori small, linear, near the margins, scarcely visible among the scales, protected by an inrolled margin. Stipe one quarter to half the length of the blade.

A genus of an uncertain number of species, found in areas of both the Old and New Worlds which have a 'Mediterranean' type of climate. They are long-lived and slow-growing, cultivated because of their scaly appearance and their habit of inrolling tightly during dry periods. They require strong sun, a very freely draining, rocky or sandy compost and very free air movement, and are highly intolerant of over-watering. Propagation is by spores.

1a. Upper surface of the blade hairless      **1. marantae**
  b. Upper surface of the blade with woolly hairs      **2. vellea**

**1. N. marantae** (Linnaeus) Desvaux. Figure 5(7), p. 17.
Fronds 10–35 × 2–5 cm, strongly erect. Blade linear-lanceolate, 1 × pinnate, the pinnae pinnatisect, hairless above, densely covered beneath with red-brown or copper-coloured, closely overlapping scales. Stipe one-quarter to half the length of the blade. *C & S Europe, Atlantic islands, Ethiopia, Himalaya.* H4.

**2. N. vellea** (Aiton) Desvaux (*Cheilanthes vellea* (Aiton) Fuchs). Figure 5(8), p. 17.
Fronds 8–25 × 2–4 cm, ascending to erect. Blade linear-lanceolate, 1 × pinnate, the pinnae pinnatisect or scalloped, covered on both surfaces with whitish hairs forming a woolly covering which becomes rusty brown with age. Stipe sometimes one-third of the length of the blade, usually less. *Atlantic Islands, Mediterranean Europe, SW Asia to Afghanistan.* H4.

## 15. PELLAEA Link
*C.N. Page & F.M. Bennell*
Small to medium terrestrial plants. Rhizomes short-creeping, thin, covered in dark scales, bearing fronds individually or in close clusters. Fronds mostly 1–2 × pinnate, all of one kind, evergreen, linear-lanceolate in 1 × pinnate fronds, triangular in more divided fronds, more or less leathery, mostly hairless but occasionally with a few dark hairs or scales. Blade either rounded or ovate-

triangular or tapering. Leaflets jointed to the rachis which is always hard, dark and shining. Sori linear along the margins, protected by a continuously reflexed linear margin along the larger segments, often absent from their tips and bases.

A genus of about 80 species, mostly from dry places, extensively distributed in the southern hemisphere, especially South Africa and S America, but also in Australia and New Zealand, and in the northern hemisphere in Canada.

They are moderately long-lived and reasonably fast-growing, and can be tolerant of moderately dry air, but they require constantly slightly moist but very well-drained soils; they are highly tolerant of fluctuating soil moisture and waterlogging. Most require a lime-rich compost, and thrive in strong light. Propagation is by spores.

1a. Frond 1 × pinnate                                   2
 b. Fronds more than 1 × pinnate            4
2a. Fronds broadly ovate with less than
    10 pinna pairs                         **3. paradoxa**
 b. Fronds narrowly linear or linear-
    lanceolate with c. 10–40 pinna pairs
                                                            3
3a. Pinnae mostly rounded
                                         **1. rotundifolia**
 b. Pinnae somewhat elongate   **2. falcata**
4a. Most pinnae with only a single pair of
    basal pinnules                    **4. ternifolia**
 b. Most pinnae fully pinnate into several
    pinnules                                            5
5a. Blades broadly triangular, much less
    than 2 × longer than broad    **6. viridis**
 b. Blades narrowly triangular, at least
    2 × longer than broad
                                       **5. atropurpurea**

**1. P. rotundifolia** (Forster) Hooker.
Figure 5(1), p. 17.
Rhizome very short-creeping. Fronds to 40 cm, tufted, borne rather flaccidly in an arching-spreading manner with short, downy stipes. Blade 1 × pinnate, linear or very narrowly ovate-lanceolate, with 10–50 pairs of more or less rounded or very slightly elongate, stalkless or shortly stalked pinnae which are dark green, glossy and c. 1–2 cm long. *New Zealand*. H5–G1.

**2. P. falcata** (R. Brown) Fée. Figure 5(4), p. 17.
Rhizome moderately long-creeping. Fronds to c. 30 cm, more or less tufted, borne rather erectly on short, stiff, rigid, wiry stipes. Blade rather narrowly linear or linear-lanceolate, 1 × pinnate. Pinnae 10–40 pairs, 2–6 × 1–1.5 cm, rounded to

elongate, tapering, somewhat sickle-shaped, shortly stalked, dark green. *India to Australia and New Zealand*. H5–G1.

Plants are fairly slow-growing in cultivation. Pinnae are shed readily from the frond rachis if the plant is allowed to dry out.

**3. P. paradoxa** (R. Brown) Hooker. Figure 5(2), p. 17.
Rhizome medium to fairly long-creeping, often deeply subterranean. Fronds to c. 30 cm, occasionally more, rather few per plant, on stipes about equal in length to the blades, borne individually in a stiffly erect-ascending manner. Blade broadly ovate, 1 × pinnate. Pinnae 10 or fewer pairs, alternate, broadly based, tapering, 4–10 cm, deep green. *E Australia*. H5–G2.

Plants resent disturbance and are very slow to establish and to grow in cultivation.

**4. P. ternifolia** (Cavanilles) Link. Figure 5(3), p. 17.
Rhizome very short-creeping, branching. Fronds 15–45 cm, rather few per plant, stiffly erect with the stipes about as long as the blades. Blade narrowly oblong-lanceolate to linear-lanceolate, pinnate to pinnatisect at least in the lower part. Pinnae, except the uppermost, usually subdivided into a single (rarely more) basal pinna pair, each 2–3 cm, equalling the terminal pinna segment, giving each a 3-lobed appearance. In luxuriant plants a second, similar pinna pair is sometimes present. *C & S America to Argentina and N Chile*. H3.

**5. P. atropurpurea** (Linnaeus) Link. Figure 5(5), p. 17.
Rhizome very short, shallowly creeping and scaly. Fronds numerous, 20–45 cm, tufted, erect to erect-spreading or ascending, the stipes about as long as the blades, covered with rough scales. Blade rather narrowly ovate-triangular to oblong or oblong-lanceolate, usually at least 2 × longer than broad, 2 × pinnate. Pinnules ovate to elliptic-lanceolate, often rounded and auricled at the base, with an obtusely tapering tip, up to 1.5 × 1 cm, almost hairless to downy, leathery, dull or greyish green, often glaucous-reddish when young. Fertile fronds usually with slightly narrower pinnules than the sterile. *America from Canada to Mexico*. H4.

Requires relatively dry soil conditions.

**6. P. viridis** (Forsskal) Prantl. Figure 5(6), p. 17.
Rhizome short, shallowly creeping and scaly. Fronds numerous, to 65 cm or more,

tufted, suberect or arching. Stipes varying from quite short to as long as the blades. Blade broadly lanceolate to triangular, often nearly as broad as long, 2 × pinnate or occasionally 3 × pinnate. Pinnules to 15 cm, very variable in shape, lanceolate to triangular, oblong, obtuse or acutely tapering, sometimes deeply cut, fairly soft to leathery, bright green or sometimes glaucous, often vivid green when young. *Africa, especially the south, Mascarene Islands*. G1.

**16. DORYOPTERIS** J. Smith
*C.N. Page & F.M. Bennell*
Small- to medium-sized terrestrial plants. Rhizome short-creeping, moderately thin, congested, occasionally branching, clad with firm, narrow, black scales and bearing fronds in clusters. Fronds simple to deeply palmate or pinnatisect into tapering, spreading-ascending segments which are all of one kind, or the fertile more erect and on a longer stipe with a much reduced blade area, evergreen, hairless, leathery and waxy, mid to deep green, slightly glossy. Stipes wiry, polished, black, scaly near the base. Sori continuous, elongate along the frond margins and covered by the continuous reflexed edge of the frond which forms a firm indusium, bulging so that on fertile fronds a thick band of sporangia outlines the indented margin of the underside of the leaf.

A genus of about 35 species from the tropics. They are long-lived and rather slow-growing with long-persistent fronds, grown for their bold maple- or geranium-like foliage. Cultivation as for *Pellaea* (p. 15). The genus is unrelated to *Dryopteris* (p. 54), for which its name is frequently assumed to be a mis-spelling; species consequently sometimes appear under *Dryopteris* in catalogues and in gardens. All species of *Doryopteris* are variable in the degree of division of the fronds, which varies widely even in a single plant. In most species immature fronds are heart-shaped or arrowhead-shaped; identification is only possible with mature fertile fronds.

1a. Mature fronds simple, arrowhead-
    shaped                            **1. sagittifolia**
 b. Mature fronds palmatifid to
    palmatisect                                         2
2a. Fronds coarsely divided in a simple
    star-like manner, the first, lowermost
    pinnule with simple, tapering,
    downwardly pointing lobes, not
    further subdivided                                 3

**Figure 5.** Silhouettes of fern-fronds. 1, *Pellaea rotundifolia.*
2, *P. paradoxa.* 3, *P. ternifolia.* 4, *P. falcata.* 5, *P. atropurpurea.*
6, *P. viridis.* 7, *Notholaena marantae.* 8, *N. vellea.* 9, *Cheilanthes
argentea.* 10, *C. fragrans.* 11, *C. pulchella.* 12, *C. farinosa.*
13, *Doryopteris sagittifolia.* 14, *D. ludens.* 15, *D. palmata.*
16, *D. concolor.* 17, *D. pedata.*

b. Fronds more finely divided in a rather geranium-leaved manner, the first, lowermost pinnule itself usually further divided                                    4

3a. Mature fronds with proliferous buds at the base of the blade        **3. palmata**

b. Mature fronds without proliferous buds                                  **2. ludens**

4a. Lowermost, downwardly pointing pinnules of the first pinna pair deflexed and spreading, not overlapping each other        **4. concolor**

b. Lowermost downwardly pointing pinnules of the first pinna pair slightly deflexed and often overlapping each other                                        **5. pedata**

**1. D. sagittifolia** (Raddi) J. Smith. Figure 5(13), p. 17.
Fertile fronds with a simple blade, usually arrowhead-shaped, tapering to an acute apex, somewhat leathery, bright mid green. *Tropical S America* G2.

**2. D. ludens** (Wallich) J. Smith. Figure 5(14), p. 17.
Fronds scarcely of 2 kinds, to 15 cm with blade to 2 m and nearly as broad, deeply but simply palmately lobed into a small number (usually 3) of long, linear, acute, tapering segments of very thick leathery texture, dark green. *India to N Australia (Queensland).* G2.

More leathery than *D. palmata*, which it otherwise resembles in division of the frond, but not normally proliferous.

**3. D. palmata** J. Smith. Figure 5(15), p. 17.
Fronds somewhat of 2 kinds, the fertile deeply and compoundly palmatisect into 10–15 or more acutely tapering ultimate segments. Blade thick, leathery, dark green, with small proliferous buds arising at its base. *West Indies, C & W tropical S America to Peru & Brazil.* G2.

**4. D. concolor** (von Langsdorff & Fischer) Kuhn. Figure 5(17), p. 17.
Fronds somewhat of 2 kinds, to 12 cm, the fertile deeply and conspicuously palmatisect to pinnatisect into 10–20 or more acute or irregularly rounded ultimate segments, without proliferous buds. The lowermost downwardly pointing pinnules of the first pinna pair spreading, but usually overlapping one another, thin but leathery, bright mid green. *Australia (Queensland).* G2.

**5. D. pedata** (Linnaeus) Fée. Figure 5(16), p. 17.
Similar to *D. concolor* but fronds even more finely divided, the lowermost downwardly pointing pinnules of the first pinna pair

strongly developed and usually overlapping one another. *West Indies, tropical America south to Argentina.* G1.

**17. ANOGRAMMA** Link
*C.N. Page & F.M. Bennell*
Small terrestrial plants. Rhizome rudimentary. Fronds all of one kind, 2–3 × pinnate into small, fan-shaped segments. Fertile fronds with sori scattered along the forking veins, without indusia.

A genus of about 7 species, widespread in temperate and subtropical zones. All are of rapid growth and short-lived.

**1. A. leptophylla** (Linnaeus) Link. Figure 6(1), p. 19.
Fronds with ascending stipes and blades to 15 cm, ephemeral or annual, arising in closely grouped clusters. *Widespread in temperate zones.* H3.

One of the few annual ferns, growing rapidly in spring and dying by early summer. Propagated readily by spores.

**18. PITYROGRAMMA** Link
*C.N. Page & F.M. Bennell*
Small- to medium-sized terrestrial plants. Rhizome ascending, scaly with narrow brown scales, bearing fronds in clusters. Fronds all of one kind, 1–2 × pinnately divided, evergreen, ovate to triangular, sometimes somewhat leathery, bright green above, mealy and farinose beneath, silver-white or golden yellow. Sori on the ultimate divisions of the frond, linear, borne along the veins as thin rows, without indusia. Stipes scaly at the base, dark and polished above, usually about as long as the blades or longer.

A genus of about 40 species, mostly from tropical America, a few from Africa and Madagascar. Some species have been widely spread and naturalised through the tropics. They are fairly short-lived and fast-growing, cultivated for their brilliantly coloured fronds. They grow best in bright light or only slightly filtered, direct sunlight, in fairly dry, freely moving air and in a light, porous, sandy compost kept just damp but never wet. The colourful leaf undersurfaces dull greatly if the plants are kept in humid conditions. They grow easily from spores, often appearing as rogues in pots of other plants grown nearby.

1a. Fronds broadly triangular; basal pinna pair much larger than those above                            **1. triangularis**

b. Fronds narrowly triangular or ovate-triangular, basal pinna pair similar in size to those above                          2

2a. Undersides of fronds golden yellow                                **2. chrysophylla**

b. Undersides of fronds silver-white                                **3. calomelanos**

**1. P. triangularis** (Kaulfuss) Maxon. Figure 6(4), p. 19.
Fronds 5–15 cm, broadly triangular, 3 × pinnate with the basal pinna pair very much larger than those above. Undersurfaces of the fronds covered with a bright, creamy white powder. Stipes 1–3 × the length of the blades. *USA (California) to Mexico.* H5–G1.

**2. P. chrysophylla** (Swartz) Link. Figure 6(3), p. 19.
Fronds 20–60 cm, ovate to ovate-triangular, 2 × pinnate (or more finely divided), the undersurfaces covered with a bright golden yellow powder. Stipes up to 2 × longer than blades. *S America but widely naturalised (and weedy) throughout much of the Old World subtropics.* H5–G1.

**3. P. calomelanos** (Linnaeus) Link. Figure 6(2), p. 19.
Fronds 30–90 cm, ovate or ovate-triangular, 2 or more × pinnately divided, undersurface covered with brilliant silver-white powder. Stipes about as long as blades. *Tropical America, naturalised throughout the Old World tropics, Australia and the Pacific.* G1.

Var. *keyderi* Anon., occasionally seen in gardens, has more coarsely and irregularly lobed fronds. *P. argentea* (Willdenow) Domin, a name used widely in gardens in the USA, is very similar and possibly identical.

**19. HEMIONITIS** Linnaeus
*C.N. Page & F.M. Bennell*
Small- to medium-sized terrestrial plants. Rhizome short-creeping to ascending, thin, occasionally branching, clothed with linear, pale brown scales blending with similarly coloured hairs, bearing fronds in irregular crowded clusters. Fronds 1 × pinnate, or rounded, heart-shaped or arrowhead-shaped or palmately lobed, thick, flexible, slightly leathery, softly downy, mid to dark green, evergreen, all of one kind or the blades of the fertile fronds slightly reduced. Stripes dark red to black, stiff, usually as long as or much longer than the blades. Sporangia scattered as thin lines along the veins over the lower surface of each fertile segment. Indusia absent.

A genus of about 6 species from tropical Asia and America. They are short-lived and usually fast-growing, cultivated for their soft downy textures and unusual frond

**Figure 6.** Silhouettes of fern-fronds. 1, *Anogramma leptophylla*.
2, *Pityrogramma calomelanos*. 3, *P. chrysophylla*. 4, *P. triangularis*.
5, *Hemionitis palmata*. 6, *H. arifolia*. 7, *Gymnopteris rufa*.
8, *Adiantum caudatum*. 9, *A. macrophyllum*. 10, *A. polyphyllum*.
11, *A. trapeziforme*. 12, *A. peruvianum*.

shapes, variable even in a single plant. Cultivation as for *Gymnopteris*. Plants grow readily from spores and *H. palmata* can also be propagated by bulbils.

1a. Mature fronds palmate or palmatifid
                                          **1. palmata**
  b. Mature fronds arrowhead-shaped
                                          **2. arifolia**

**1. H. palmata** Linnaeus. Figure 6(5), p. 19.
Fronds mostly 10–15 cm, occasionally more. Blades mostly simple, but immature fronds variously and irregularly lobed, the mature fronds palmatifid into 3–5 broad, tapering, acute lobes resembling a 5-pointed star, each lobe variously cut, toothed or waved. *Tropical America, West Indies.* G2.

**2. H. arifolia** (Burman) Moore. Figure 6(6), p. 19.
Fronds mostly 10–15 cm, occasionally to 30 cm. Blades simple, immature fronds entire or variously lobed, mature fronds becoming strongly arrowhead-shaped, the apex and lower lobes all tapering to acute points. *E Asia (India to the Philippine Islands).* G2.

**20. GYMNOPTERIS** Bernhardi
*C.N. Page & F.M. Bennell*
Small to medium terrestrial plants. Rhizomes short-creeping to erect, thin, occasionally branching, clothed with pale brown scales blending with similarly coloured hairs, bearing fronds in irregular, crowded clusters. Fronds 1 × pinnate, all of one kind or with the blades of the fertile very slightly reduced, evergreen, thick, flexible, slightly leathery, softly downy, mid to dark green. Stipe pale to dark, stiff, as long as the blade or longer, scaly and hairy at the base, elsewhere with soft hairs. Sporangia scattered as thin lines along the veins over the whole surface of each fertile segment. Indusia lacking.

A genus of about 5 species from tropical America, India and China. They are short-lived and fast-growing, cultivated for their attractive, soft, very hairy fronds. They thrive in filtered but bright sunlight in slightly moist but very open, well-drained sandy-gravelly soil with some peat and are highly intolerant of draughts, desiccation and wet, stagnant conditions. They are propagated readily from spores.

**1. G. rufa** (Linnaeus) Bernhardi. Figure 6(7), p. 19.
Fronds mostly 10–20 cm. Blade pinnatisect

into 3–8 distant, horizontally arranged, ovate-oblong, rounded or pointed, shortly stalked, entire pinnae, on a stipe of about the same length as the blade, the whole covered in soft, red down. *Tropical America.* G2.

**21. ADIANTUM** Linnaeus
*C.N. Page & F.M. Bennell*
Medium to large terrestrial plants. Rhizomes mostly short-creeping or ascending, a few long-creeping. Fronds all of one kind, evergreen or deciduous, simple to 3–4 × pinnate into numerous finely cut, fan- or parallelogram-shaped segments, each entire to deeply lobed, firm, delicate, with radiating, forking veins. Fertile segments with reflexed flap-like indusia.

A genus of over 200 species mostly from the tropics and subtropics, a few extending into temperate zones. They are cultivated mainly for their billowing foliage and require warm, humid, shaded conditions; they grow readily from spores. Large plants can be divided.
Literature: Hoshizaki, B.J., The genus Adiantum in cultivation, *Baileya* 17: 97–191 (1970).

1a. Fronds entire           **16. reniforme***
  b. Fronds variously divided                2
2a. Mature fronds distinctly pedately divided and usually broader than long
                                              3
  b. Mature fronds essentially pinnately divided and usually longer than broad
                                              4
3a. Stipe and frond finely hairy
                                    **5. hispidulum**
  b. Stipe and frond not hairy
                                      **6. pedatum**
4a. Mature fronds 1 × pinnate              5
  b. Mature fronds 2 or more × pinnate     7
5a. Fronds rooting at their tips
                                    **1. caudatum***
  b. Fronds not rooting at their tips     6
6a. Ultimate segments small and delicate, less than 1.5 cm    **7. diaphanum**
  b. Ultimate segments robust, large, mostly 5–10 cm    **2. macrophyllum**
7a. Ultimate segments mostly distinctly fan-shaped                   12
  b. Ultimate segments mostly parallelogram-shaped                  8
8a. Fronds with 3 or fewer divisions
                                    **7. diaphanum**
  b. Fronds with more than 3 divisions   9
9a. Ultimate segments more than 3 cm
                                   **4. trapeziforme***
  b. Ultimate segments less than 3 cm   10
10a. Ultimate segments of mature fronds 2–3 cm          **3. polyphyllum**

  b. Ultimate segments of mature fronds mostly less than 2 cm      11
11a. Blade with bluish-bloomed undersurface       **8. cunninghamii***
  b. Blade undersurface not bluish
                                       **9. formosum**
12a. Ultimate segments 1 cm or less      13
  b. Ultimate segments more than 1 cm 16
13a. Mature fronds mostly 35–75 cm
                                     **10. concinnum**
  b. Mature fronds usually less than 35 cm                          14
14a. Ultimate segments with 1–2 indusia, rarely more       **11. venustum**
  b. Ultimate segments usually with 3–5 indusia                     15
15a. Mature rhizome short-creeping, 5–15 cm, sparsely branched
                                     **12. raddianum**
  b. Mature rhizome long-creeping, 10–100 cm or more, much branched
                                    **13. aethiopicum**
16a. Fronds usually over 30 cm, segments jointed to their stalks and falling readily in old, dry fronds
                                      **15. tenerum**
  b. Fronds usually less than 30 cm, segments continuous into their stalks and not falling readily
                                **14. capillus-veneris**

**1. A. caudatum** Linnaeus. Figure 6(8), p. 19.
Rhizome short-creeping. Fronds to 30 × 30 cm, arching to semi-prostrate in flattened, loose rosettes, rooting at the end of the long, leafless tips, evergreen, narrowly lanceolate, pinnae with fan-shaped, sometimes much cut ultimate segments, all pale yellowish green, finely hairy on both surfaces. *Old World Tropics.* G2.

***A. incisum** Forsskal. Similar, but with more deeply cut pinnules. *C Africa, SW Asia.* G1.

**2. A. macrophyllum** Swartz. Figure 6(9), p. 19.
Rhizome short-creeping, sparingly branched, bearing fronds in irregular clusters. Fronds to 30 × 20 cm, evergreen, ovate-oblong to ovate, pinnae with 3–8 pairs of large, entire, ultimate segments which taper towards their outer ends, thin but firm, pink when young, pale green when mature. Sori usually with long indusia. *C & S America.* G2.

**3. A. polyphyllum** Willdenow. Figure 6(10), p. 19.
Rhizomes creeping, occasionally branched, bearing fronds in well-spaced groups. Fronds 40–200 cm, with ascending stipes

and ascending to spreading, evergreen, broadly ovate blades which are irregularly forked into 6–8 or more pinnae, ultimate segments almost rectangular, 1–3 cm, with lobed upper margins, hairless, pale green. *Northern S America*. G2.

**4. A. trapeziforme** Linnaeus. Figure 6(11), p. 19.
Rhizomes short, branched, bearing fronds in small groups. Fronds to 1 m or more, to 50 cm wide, ascending to arching, evergreen, triangular, to 4 × pinnate but irregularly divided into large (3–4 cm), trapeziform ultimate segments with upper and lower margins partially cut, all pale green. *Tropical America*. G2.

*A. cultratum* Hooker. Differs mainly in its stouter rhizomes. *Tropical America*. G2.

*A. peruvianum* Klotzsch. Figure 6(12), p. 19. Similar, but the outer margins of the segments less cut, mostly slightly toothed. *Tropical S America*. G2.

**5. A. hispidulum** Swartz. Figure 7(1), p. 22.
Rhizomes short-creeping, much branched, bearing fronds in irregular groups. Fronds to 40 cm, with erect stipes and ascending to spreading, rather stiff blades which are evergreen, pedately divided, each segment being pinnately divided into almost rectangular ultimate segments which are 1.5 cm or less; the whole blade with fine but distinct short hairs on both surfaces, deep pink when young, yellow-green or blue-green when mature. *E Africa, Islands of Indian Ocean, SE Asia to Polynesia and New Zealand*. H5–G1.

A complex of several closely related taxa which differ in size, details of pinna shape, colour and hairiness. Most are easily grown.

**6. A. pedatum** Linnaeus. Figure 7(2), p. 22.
Rhizomes short-creeping, branched, bearing fronds in close groups. Fronds to 30 cm or more, ascending or arching, as wide as or wider than long, deciduous, pedate, first forking into 2 segments, then to 4–8 or more narrow, pinnately divided, almost rectangular ultimate segments which are sometimes deeply cleft along the upper margin, hairless, somewhat flaccid, dull pale blue-green. *N America, NE Asia*. H3.

Var. *aleuticum* Ruprecht differs mainly in its shorter rhizome and greater hardiness. 'Imbricatum' has smaller, more compact fronds with overlapping segments.

**7. A. diaphanum** Blume. Figure 7(3), p. 22.
Rhizomes very short, thin, roots with numerous small, ovoid tubers, bearing fronds in small groups. Fronds to 25 × 10 cm, ascending to arching, evergreen, basically 1–3 × divided, segments pinnately arranged, rectangular, all thin and delicate, pale to dark green. Sori in dense patches on outer segment margins. *Indonesia, Australasia, Pacific Islands*. G1.

**8. A. cunninghamii** Hooker. Figure 7(4), p. 22.
Rhizomes slender, thin, creeping, often much branched, shallowly subterranean, bearing fronds in irregular groups. Fronds to 90 × 25 cm, evergreen, irregularly ovate with moderately small, regularly arranged, diamond-shaped ultimate segments. All soft, thick, slightly fleshy, salmon-pink when young, with green upper, and bluish-bloomed lower surfaces when mature. *Tropical and subtropical Australia*. G1.

*A. sylvaticum* Tindale. Slightly smaller and more tolerant of dry conditions. *E Australia*. G1.

**9. A. formosum** R. Brown. Figure 7(5), p. 22.
Rhizomes slender, occasionally branched, very long-creeping, extremely deep (50–100 cm or more), bearing fronds at widely-spaced intervals. Fronds to 1 m × 60–80 cm, with erect, rather rough stipes and slightly spreading blades which are evergreen, broadly triangular, 4 × pinnate with numerously cleft, small, rhombic or parallelogram-shaped segments which are 8–10 mm wide, all soft but firm, mid green. *Tropical and subtropical Australia, northern New Zealand*. G1.

Plants of this species form extensive colonies in rich, deep soil in the wild. In cultivation they are slow-growing and the mature-sized frond is only produced several years after establishment, when given a deep soil and a free root run.

*A. affine* Willdenow and *A. fulvum* Raoul, both from New Zealand, the second also from Polynesia, are often confused with this species, but both are smaller with fronds at most 50 cm.

**10. A. concinnum** Willdenow. Figure 7(8), p. 22.
Rhizomes thin, short-creeping, often much branched, bearing fronds in irregular clusters. Fronds to 75 cm with short stipes and arching blades, evergreen, narrowly triangular, 2 × pinnate with fine, fan-

shaped ultimate segments, soft, thin, pale green. *C & northern S America*. G2.

A small number of cultivars exists, showing variation in segment size and frond habit. The large and vigorous 'Elatum' is frequently grown.

**11. A. venustum** Don (*A. microphyllum* Roxburgh). Figure 7(9), p. 22.
Rhizome slender, long-creeping, bearing fronds sparsely. Fronds to 15–30 cm with short stipes (usually one-third or less of the total length) and arching blades which are deciduous, triangular-ovate, 3–4 × pinnate and finely cut into uniformly fan-shaped ultimate segments each with rarely more than 2 indusia, all dull mid green, slightly bluish-bloomed. *Himalaya*. H5.

**12. A. raddianum** Presl (*A. cuneatum* von Langsdorff & Fischer; *A. gracillimum* Moore). Figure 7(11), p. 22.
Rhizome very short-creeping, much branched, bearing fronds in crowded groups. Fronds to 30 × 20–30 cm, with ascending stipes and spreading to arching blades which are evergreen, triangular to long-triangular, 2–4 × pinnate, moderately finely dissected into irregularly fan-shaped ultimate segments which are lobed or cleft with toothed outer margins, all thin and yellow-green. *American and African tropics*. H5–G2.

A variable species. Many cultivars are grown, including 'Decorum', 'Elegans', 'Fragrantissimum' and the large-segmented 'Pacific Maid', crowded 'Pacottii' and very finely dissected 'Micropinnulum' and 'Gracillimum'.

**13. A. aethiopicum** Linnaeus. Figure 7(6), p. 22.
Rhizome thin, wiry, long-creeping, much branched, bearing fronds in scattered groups. Fronds to 35 cm with erect stipes and steeply ascending, evergreen, irregularly triangular blades which are at least 2 × pinnate with the ultimate segments minutely toothed, thin but firm, pale yellow to bluish green. *Old World tropics and subtropics*. G1.

**14. A. capillus-veneris** Linnaeus. Figure 7(7), p. 22.
Rhizome thin, short-creeping, bearing fronds in scattered groups. Fronds to 50 cm, often smaller, with erect stipes and arching, evergreen, ovate to narrowly triangular blades with moderately finely cut, fan-shaped ultimate segments which do not overlap, thin, bright green. *Subtropical and warm temperate zones*. H4.

Several cultivars exist, including

**Figure 7.** Silhouettes of fern-fronds. 1, *Adiantum hispidulum*.
2, *A. pedatum*. 3, *A. diaphanum*. 4, *A. cunninghamii*.
5, *A. formosum*. 6, *A. aethiopicum*. 7, *A. capillus-veneris*.
8, *A. concinnum*. 9, *A. venustum*. 10, *A. tenerum*.
11, *A. raddianum*. 12, *A. reniforme*.

congested and imbricate forms such as 'Maivissii' and 'Magnificum'. Plants in cultivation as *A. chilense* probably belong to this species.

### 15. A. tenerum Swartz. Figure 7(10), p. 22.

Rhizome short-creeping, bearing fronds in dense tufts. Fronds 60–75 × 30–45 cm, with irregular, ascending to arching blades which are evergreen, broadly and irregularly triangular in outline, 3–4 × pinnate, with variable segments 1–3 cm wide, mostly asymmetric, deltoid or rhomboid, jointed to their stalks, falling readily on old and dry fronds, soft, mid green. *C America, West Indies, northern S America.* H5–G1.

Many greenhouse cultivars exist and include several with large ultimate segments, some of which are deeply cut, such as 'Farleyense', 'Lady Marham' and 'Scutum'.

### 16. A. reniforme Linnaeus. Figure 7(12), p. 22.

Rhizome thick, short-creeping, much branched, bearing fronds in dense clusters. Fronds to 10–20 × 3–5 cm, with erect or ascending stipes and spreading to ascending blades which are evergreen, simple and kidney-shaped with indusia set all around the outer margin, stiff, leathery, dark green. *Canary Islands, Madeira.* G1.

Madeiran plants are usually larger and have a more scalloped margin than those from the Canary Islands. This is a remarkably xerophytic species, succeeding particularly well under dry, bright, indoor conditions.

*\*A. asarifolium* Willdenow. Similar, but with larger fronds, to 50 cm, which are more robust. *Mascarene Islands.*

## 22. VITTARIA J.E. Smith
*C.N. Page & F.M. Bennell*

Small, medium or large epiphytes. Rhizome short-creeping, thin, bearing fronds in clusters. Fronds all of one kind, simple, linear, elongate, tapering, ribbon-like, firm, leathery, yellow-green to deep green. Sori linear in a continuous almost marginal line along either long edge of the frond, immersed into it and lacking an indusium.

A genus of about 80 species from tropical and subtropical regions. The plants are long-lived, slow-growing, cultivated for their very distinctive, pendulous, ribbon-like appearance. They are good subjects for hanging baskets, in a peaty, freely drained medium in a situation of medium to strong

light. They are propagated by spores or by division.

### 1. V. elongata Swartz. Figure 8(1), p. 25.

Fronds usually 30–60 cm, occasionally very much longer in the wild, 8–16 mm broad, hanging. Blades linear, of more or less uniform width throughout, leathery, dark green. *Malaysia to Australia.* G2.

Many other species have, from time to time, been introduced from many parts of the tropics, often accidentally with epiphytic orchids. Most are similar to *V. elongata.* **V. scolopendrina** (Bory) Thwaite is one of these with somewhat wider, fleshier fronds.

## 23. PTERIS Linnaeus
*C.N. Page & F.M. Bennell*

Rhizome short-creeping or ascending to erect, moderately thick, sparsely covered with scales, bearing fronds in irregular coarse tufts. Stipes ascending erect, grooved along the upper surface, more or less as long as the blades. Fronds all of one kind, or fertile and sterile, linear-lanceolate to triangular, soft to leathery, usually hairless and variable in colour, 1–3 × pinnate. Basal pair of pinnae with their lower, innermost pinnules reflexed and extensively developed to resemble a pinna. Sori continuous, linear along the margins of the frond, slightly bulging at maturity, covered by the continuous reflexed edge of the frond, which forms an indusium.

About 280 species, pantropical and extending north to Japan and USA and south to South Africa and New Zealand. They are short- to moderately long-lived, handsome ferns of rapid growth with bold and varying frond form and variegated foliage in some species. They enjoy a light, moist but well-drained soil which is only moderately acidic, fairly high humidity and bright but filtered sunlight. All are rather frost-tender and are propagated readily by spores which store well.

Literature: Walker, T.G., Species of Pteris commonly in cultivation, *British Fern Gazette* **10**: 143–51 (1970).

1a. Fronds 1 × pinnate or with only a few (1–5) linear, strap-like divisions to the pinnae      2
   b. Fronds more than 1 × pinnate on at least the lower part of the frond      6
2a. Pinnae on mature fronds more than 10 pairs      3
   b. Pinnae on mature fronds fewer than 10 pairs      4
3a. All pinnae simple, undivided
                 **1. vittata**

   b. Some pinnae, especially the lower, with a few divisions    **3. umbrosa**
4a. Pinnae decurrent to the rachis in at least the upper half of the frond
                 **4. multifida**
   b. Pinnae (except sometimes the terminal) not decurrent to the rachis   5
5a. Fertile pinnae 5 mm or more wide
                 **2. cretica**
   b. Fertile pinnae up to 5 mm wide
                 **5. ensiformis**
6a. Fronds with 3 approximately equal divisions    **9. tripartita**
   b. Fronds not divided as above    7
7a. Most pinnae markedly asymmetric, entire along the upper margin but pinnatisect along the lower
                 **6. semipinnata\***
   b. Most pinnae approximately symmetric, equally pinnatisect along upper and lower margins    8
8a. Lowest pair of pinnae with a single (first) larger, deflexed pinna on the lower side    9
   b. Lowest pair of pinnae with several larger deflexed pinnae on the lower side    10
9a. Pinnae uniformly green above
                 **7. quadriaurita\***
   b. Pinnae each with a broad white stripe along its length    **8. argyrea**
10a. Pinnule margins entire or only irregularly toothed, venation conspicuous, netted    **10. comans**
   b. Pinnule margins regularly and finely toothed (at least in non-fertile parts), venation inconspicuous, not netted
                 11
11a. Pinnules mostly 3–5 mm broad
                 **11. serrulata**
   b. Pinnules mostly less than 2 mm broad    **12. tremula**

### 1. P. vittata Linnaeus. Figure 8(2), p. 25.

Fronds to 60 cm or more, on very short stipes, ascending to erect, arching at the tips, ovate-lanceolate, 1 × pinnately divided throughout into numerous, regular, closely spaced, narrowly tapering pinnae; pinnae with auricled bases, becoming gradually shorter towards the base of the frond, more or less leathery, dark green, glossy. *Old World tropics and subtropics.* H5–G1.

### 2. P. cretica Linnaeus. Figure 8(3), p. 25.

Fronds to 30–60 cm, ascending to erect. Blade triangular to broadly ovate, 1 × pinnate. Pinnae papery, deep green, not decurrent to the rachis, 1–5 pairs, spreading, narrowly ovate-lanceolate, 5 mm or more broad, margins toothed; the lower pinnae also each bear a single, deflexed segment. Fertile fronds with narrower segments throughout. *Tropics and subtropics*. H5–G1.

One of the most widely sold ferns. 'Albo-lineata', which is as common as the type in cultivation, has a broad whitish streak down the centre of each pinna. Numerous forked, crested, congested and other monstrous cultivars occur.

**3. P. umbrosa** R. Brown. Figure 8(4), p. 25.
Fronds to 1 m or more, spreading, arching, obovate-lanceolate, 1–2 × pinnatisect, the lower pinnae longer than the upper, and divided into 3–5 linear, spreading segments from the lower margin, the bases all decurrent down the rachis. Blade thin, slightly leathery, wavy, hairless and shining, dark green. *E Australia*. G2.

**4. P. multifida** Poiret. Figure 8(5), p. 25.
Fronds to 30–60 cm or more, ascending to arching. Blade irregularly ovate, narrowing towards the base, 2 × pinnate into *c*. 5–8 narrow, linear, tapering pinnae, each bearing a small number of segments of similar form to the pinnae themselves, sometimes finely toothed. Bases of the pinnules and segments in the upper half of the frond variously and irregularly decurrent to the rachis into an irregular, somewhat interrupted wing, leathery, dark green. Fertile frond with generally narrower segments. *Japan, China*. H5–G1.

A common fern in the trade. Various crested, congested and pendulous-tipped cultivars occur. Distinguished from *P. ensiformis* particularly by the winged upper part of the rachis.

**5. P. ensiformis** Burman. Figure 8(8), p. 25.
Fronds to 35 cm, erect, of compact habit. Blade ovate, 1 × pinnatisect to 3 × pinnate, the pinnae lobed, usually more so on their lower margins than their upper, into fairly narrow segments, some stalked. Rachis not winged, lamina leathery, dark green often marked grey-white near the midrib. Fertile fronds with narrower segments. *Himalaya to Australia (Queensland) & Polynesia*. G1–2.

A number of variegated forms has been selected, in which the central silver-white markings contrast strongly with the dark green, e.g. 'Arguta'.

**6. P. semipinnata** Linnaeus. Figure 8(9), p. 25.
Fronds 20–40 cm, sharply ascending to erect. Blade ovate-lanceolate, 2 × pinnate or pinnatisect, the pinnae asymmetrically divided, each mostly entire

along the upper margin but deeply pinnatisect along the lower into 4 or more sickle-shaped, tapering, acute pinnules. Blade more or less leathery, deep green. *SE Asia, Japan*. G1.

**\*P. dispar** Kunze. Similar, but generally a much smaller plant. *Tropical Asia*.

**7. P. quadriaurita** Retzius. Figure 8(10), p. 25.
Fronds 1–1.5 m, ascending to spreading. Blade broadly ovate, mostly 2 × pinnate or pinnatisect, most pinnae with a well-developed, reflexed lower pinnule pair. Blade more or less leathery, deep green. *India, Sri Lanka*. G2.

This name is applied to a group of rather similar species; few of those in cultivation are, however, *P. quadriaurita* Retzius in the strictest sense (see the paper by Walker, cited at the beginning of the genus, particularly p. 150).

**\*P. faurei** Hieronymus. Similar in appearance, but of smaller, more delicate structure. *China*.

**8. P. argyrea** Moore. Figure 8(11), p. 25.
Fronds 1–1.5 m, ascending to spreading. Blade broadly ovate, mostly 2 × pinnate or pinnatisect, but with a single, well-developed, reflexed lower pinnule pair. Blade soft, mid green with a conspicuous lengthwise silver-white stripe along each pinna. *Origin uncertain, possibly Sri Lanka or Java*. G2.

Differs from *P. quadriaurita* in its generally softer texture and its white markings.

**9. P. tripartita** Swartz & Schrader. Figure 8(6), p. 25.
Fronds 1–3 m, ascending to spreading. Blade broadly triangular, borne on a long stipe, tripartite, the lateral pinnae as long as the remaining part of the frond and similar to it, each 3 × pinnatisect, the pinnae each with a large, secondary, bipinnatisect branch on the lower side reflexed towards the base, deeply and regularly lobed nearly to the axes. Blade thin to firm, hairless, yellow-green. *Old World tropics from Africa to Australia and Polynesia*. G2.

**10. P. comans** Forster. Figure 8(12), p. 25.
Fronds to 2 m or more, arching. Blades ovate, 2–3 × pinnate or pinnatisect, pinnules sickle-shaped, broad, 2–4 × 5 mm or more, tapering acutely to a rounded apex, the margins more or less entire or irregularly toothed, with conspicuous net veins. Blade slightly leathery, firm, hairless, dark green. *Australasia, Polynesia*. G2.

**11. P. serrulata** Forsskal (*P. dentata* Forsskal). Figure 8(7), p. 25.
Fronds to 1.5 m, arching-ascending. blade triangular-ovate, soft, hairless, pale green, 2–3 × pinnate or pinnatisect, pinnules oblong-tapering, mostly 2–3.5 cm × 3–5 mm, the margins very finely toothed. *Canary Islands, SW Europe, N Africa, Arabia*. G1.

**12. P. tremula** R. Brown. Figure 8(13), p. 25.
Fronds to 2 m, ascending. Blade more or less triangular, soft, hairless, yellow-green, 2–3 × pinnate or pinnatisect, pinnules narrow, oblong, mostly 1.5–3 cm × 2 mm, the margins very finely toothed. Fertile fronds with slightly narrower segments. *Australasia*. G2.

A highly frost-tender species.

**24. HYMENOPHYLLUM** J. Smith
*C.N. Page & F.M. Bennell*
Small plants growing epiphytically or on rocks. Rhizome long-creeping, very thin, thread-like, freely branching. Fronds all of one kind, extremely delicate, very thin, translucent, pale green, evergreen; when present in large numbers forming a moss-like carpet. Sori on the upper margins of the segments, flask-shaped, 2-valved.

A genus of at least 25 species (but thought to be as many as 300 by some authorities), found all over the world. They are long-lived and slow-growing, with exacting cutural requirements, demanding an acidic, very freely drained and permanently moist medium, low light, reasonably constant temperatures and high humidity such as can be achieved in a glazed case. A certain amount of free circulation of very humid air may also be necessary. Propagation is by division of established clumps.

Several filmy ferns (species of *Hymenophyllum* and the related *Trichomanes* Linnaeus) are sometimes found in botanic gardens and specialist collections. Three of these species (*H. tunbridgense* (Linnaeus) Smith, *H. wilsonii* W.J. Hooker and *T. speciosum* Willdenow) occur wild in Europe but are very rare; they should on no account be removed from the wild for transplantation into gardens. Further details of these plants are given in Page, C.N., The ferns of Britain and Ireland, 231–8, 304–7 (1982).

**1. H. demissum** (Forster) Swartz & Schrader (*Mecodium demissum* (Forster) Copeland). Figure 9(1), p. 27.

**Figure 8.** Silhouettes of fern-fronds. 1, *Vittaria elongata*. 2, *Pteris vittata*. 3, *P. cretica*. 4, *P. umbrosa*. 5, *P. multifida*. 6, *P. tripartita*. 7, *P. serrulata*. 8, *P. ensiformis*. 9, *P. semipinnata*. 10, *P. quadriaurita*. 11, *P. argyrea*. 12, *P. comans*. 13, *P. tremula*.

Fronds seldom more than 4 cm. Blade 3 × pinnate, very finely divided into narrow, tapering, widely spaced ultimate segments, giving the frond a lace-like appearance. Stipe one-quarter to one-half the length of the blade. *New Zealand.* H5–G1.

## 25. CARDIOMANES Presl
*C.N. Page & F.M. Bennell*
Small terrestrial plants or epiphytes. Fronds all of one kind, each simple, kidney-shaped, thin and membranous, on an elongate stipe. Sori in cylindric marginal receptacles surrounding projecting bristles.

A genus of a single species only occasionally seen in cultivation.

**1. C. reniforme** (Forster) Presl. Figure 9(2), p. 27.
Rhizome thin, creeping. Fronds erect, with entire, membranous, kidney-shaped blade. *New Zealand.* H5–G1.

A filmy fern of unusual appearance, requiring shade and high humidity.

## 26. DRYNARIA (Bory de St Vincent) J. Smith
*C.N. Page & F.M. Bennell*
Large plants growing epiphytically or on rocks. Rhizome short-creeping, thick, fleshy, branching occasionally to form compact, clumped masses, densely covered with golden brown scales and bearing fronds in close groups. Fronds 1 × pinnatisect or pinnate, of two kinds: 'nest leaves' forming a closely grouped mass around the rhizome, 10–40 cm, irregularly pinnatisect, permanent, clasping, stalkless, harsh and scarious; foliage leaves to 1.5 m or more, pinnatisect or pinnate, seasonally shed, sterile or fertile, green, sometimes becoming papery with age. Sori median, discrete, slightly or scarcely sunken, lacking an indusium.

A genus of about 20 species from Africa to the Pacific. They are long-lived and slow-growing. A very distinct genus, grown for its curious 'nests' of basal, lobed, papery fronds resembling the leaves of oak (*Quercus*). Propagation is by spores or division of the rhizome.

1a. Foliage leaves distinctly pinnate
                            **1. rigidula**
  b. Foliage leaves pinnatisect
                           **2. quercifolia**

**1. D. rigidula** (Swartz) Beddome. Figure 9(3), p. 27.
Rhizome *c.* 2–3 cm thick. Nest leaves narrow, to 35 × 8 cm, slightly lobed at the base. Foliage leaves to 1.5 m, divided into stalked, linear-lanceolate, completely separate pinnae. Sori rather few, in a single row on either side of each pinna. *India, S China, Malaysia to Polynesia & Australia (Queensland).* G1.

**2. D. quercifolia** (Linnaeus) J. Smith. Figure 9(4), p. 27.
Rhizome *c.* 4–5 cm thick. Nest leaves broad, to 40 × 30 cm, strongly lobed near the base. Foliage leaves to 1 m or more, pinnatisect. Sori very numerous along each pinna, arranged in rows along each side of small, secondary veins. *Indonesia (Sumatra) to Australasia & Polynesia.* G2.

## 27. MERINTHOSORUS Copeland
*C.N. Page & F.M. Bennell*
Large plants growing epiphytically or on rocks. Rhizome short-creeping, 2–3 cm thick, fleshy, occasionally branching, densely scale-clad, bearing fronds individually but in close groups, the basal parts of the fronds clasping. Fronds all similar, to 100 × 30 cm or more, pinnately lobed (pinnate in the upper parts of the fertile fronds), evergreen, linear-lanceolate, narrowed gradually towards the base, thinly leathery but becoming brittle, the lobes becoming smaller towards the base of the frond and with a pair of basal lobes clasping the stipe–rhizome junction. Sori borne on the lobes of the apical region of the fronds, which are there reduced to slender pinnae *c.* 20–30 cm × 3 mm, covered entirely with sporangia beneath.

A genus of a single species from Malaya to the Solomon Islands. The metre-long fronds, like single, enormous, lobed leaves give the plants an impressive appearance. Cutivation as for *Polypodium* (p. 33). Propagation by spores or by division of the rhizome.

**1. M. drynarioides** (Hooker) Copeland. Figure 9(5), p. 27.
*Malaya to the Solomon Islands.* G2.

*Merinthosorus* differs from the species of *Aglaomorpha* in the reduced, tassel-like fertile portion of the frond; juvenile specimens are difficult to separate with certainty.

## 28. AGLAOMORPHA Schott
*C.N. Page & F.M. Bennell*
Large epiphytes. Rhizome creeping, thick, fleshy, densely scaly. Fronds all of 1 kind, or of 2 kinds, the fertile with pinnae reduced towards the apex, the lower part of each usually dilated and scarious, humus-collecting, with conspicuous venation, the upper part deeply pinnatifid with broadly lanceolate, entire lobes. Sori borne on reduced pinnae at the apex of the fertile fronds, variously shaped, lacking indusia.

About 10 species in tropical Asia, chiefly in Malaysia. They need warm, humid conditions and their large ultimate size demands considerable space. Propagation is by spores or by division.

1a. Sporangium-bearing tips of fertile fronds with lobes of similar shape to those of non-sporangium-bearing parts         **1. heraclea\***
  b. Sporangium-bearing tips of fertile fronds with lobes narrowed and different in shape from the sterile parts         **2. meyeniana**

**1. A. heraclea** (Kunze) Copeland (*Polypodium heracleum* Kunze). Figure 9(8), p. 27. Illustration: Holttum, Flora of Malaysia, Ferns, 186 (1954).
Rhizome *c.* 2 cm in diameter, bearing fronds in persistent, spreading groups. Fronds to 2.5 m × 80 cm, ascending, all of one kind, lobed to within 1–2 cm of the midrib or less deeply towards the base; the largest lobes in the middle of the frond; margins waved but not toothed. Blade leathery to papery, mid green becoming pale brown at the base with age. Sori discrete in 2 fairly regular rows along some of the upper segments. *Malaysia.* G1.

**\*A. coronans** (Wallich) Copeland (*Polypodium coronans* Wallich). Figure 9(7), p. 27. Very similar vegetatively, but with a single row of larger sori. *S Himalaya to N Malaysia.* G1.

**2. A. meyeniana** Schott (*Polypodium meyenianum* (Schott) Hooker). Figure 9(6), p. 27.
Fronds to 50–60 cm, ascending, of two kinds, the vegetative similar to those of *A. heraclea*, the fertile with the segments in the upper third of the frond narrowed almost to the midribs into connected, bead-like segments, each segment occupied by a sorus. *Philippine Islands to Taiwan.* G1.

## 29. PLATYCERIUM Desvaux
*C.N. Page & F.M. Bennell*
Large, bizarre epiphytes. Rhizome short-creeping, thick, frequently branching, scale-clad, almost entirely concealed by the massive, clasping fronds and interwoven root masses, bearing fronds in dense masses. Fronds of two kinds: permanent 'nest leaves' which are erect and with broad blades, their upper regions often lobed, building up the bulk of the plant;

**Figure 9.** Silhouettes of fern-fronds. 1, *Hymenophyllum demissum*.
2, *Cardiomanes reniforme*. 3, *Drynaria rigida*. 4, *D. quercifolia*.
5, *Merinthosorus drynarioides*. (a, base of frond; b, tip of fertile
frond). 6, *Aglaomorpha meyeniana*. 7, *A. coronans*. 8, *A. heraclea*.

and impermanent fertile leaves which arch outwards, their tips hanging, dichotomously branched from the base into rather narrow branches. The bases of the fertile leaves are jointed to the rhizome, forming abscission zones by which they are shed. The sporangia cover large areas in a velvet-like mass (soral patches). All parts of the fronds when young covered with a thin, felty mass of stellate hairs. The fronds are leathery to fleshy and have conspicuous reticulate venation.

A genus of about 12 or more species from the tropics. They are extremely long-lived and slow-growing. Their large size, bizarre appearance and reliability in cultivation, thriving even with apparent neglect, make them particularly popular subjects, especially in warm climates. They require peaty humus to give a very freely draining medium, strong light and absence of cold, stagnant conditions. They die in heavy shade. Propagation is by spores or division of established plants.

  1a. Fertile fronds slender, dichotomously
      branched from the base, more or less
      linear, strap-like, the ultimate
      segments bearing small soral patches
      at their tips                              2
   b. Fertile fronds broad and massive at
      the base, bearing large soral patches
      on the webbed areas near the frond
      base, between the lobes                    4
  2a. Fronds silvery grey          **1. veitchii**
   b. Fronds green                              3
  3a. Ultimate segments of fertile fronds *c.*
      3 cm broad                   **2. bifurcatum**
   b. Ultimate segments of fertile fronds *c.*
      6 cm broad                      **3. hillii**
  4a. First forkings of fertile frond equal
                                   **4. superbum**
   b. First forkings of fertile frond very
      unequal                      **5. coronarium**

**1. P. veitchii** (Underwood) Christensen.
Fronds silvery-grey, covered with a very dense mass of 5-rayed stellate hairs. Nest leaves deeply lobed. Fertile leaves *c.* 50 cm, slender and very narrow at the base, a few times dichotomously forked near the apex into segments which are *c.* 4 cm broad, with soral patches near their tips. *Australia (C & N Queensland).* H5–G1.

Highly intolerant of shade and wet, stagnant conditions.

**2. P. bifurcatum** (Cavanilles) Christensen.
Fronds green. Nest leaves deeply lobed. Fertile leaves to 90 cm, narrowly wedge-shaped at the base, the upper half forking dichotomously 2 or 3 times into strap-shaped ultimate segments which are *c.* 3 cm broad and bear soral patches near their tips. *Australia (Queensland, New South Wales).* G1.

**3. P. hillii** Moore.
Fronds green. Nest leaves shallowly lobed. Fertile leaves *c.* 60 cm, broadly wedge-shaped at the base, the upper half irregularly dichotomously or even palmately lobed into ultimate segments which are *c.* 6 cm broad, bearing soral patches near their tips. *Australia (N Queensland).* G2.

**4. P. superbum** de Joncheere & Hennipman (*P. grande* (Fée) Presl).
Fronds green. Nest leaves with a deeply lobed and spreading upper margin. Fertile leaves to 2 m or more with a single large soral patch near the base, the remainder of the frond many times dichotomously forked into narrow, often twisted, hanging segments. *Australia (Queensland, northern New South Wales).* G1–2.

**5. P. coronarium** (Koenig) Desfontaines.
Fronds green. Nest leaves tall and deeply lobed. Fertile leaves to 4 m or more, hanging, many times forked with the first few forkings very unequal, soral patches entirely covering the lower surface of a single, specialised, semicircular or kidney-shaped lobe borne near the base of the frond. *Thailand, Malaya.* G2.

**30. PYRROSIA** Mirbel
*C.N. Page & F.M. Bennell*
Small- to medium-sized creeping epiphytes. Rhizome long-creeping, thin, branching occasionally, scale-clad, bearing fronds individually. Fronds simple, entire, all similar, or of two kinds, with stipes *c.* one-quarter of the length of the blade, evergreen, ascending to erect, linear-lanceolate, leathery, sometimes fleshy, clothed with persistent star-shaped hairs which readily rub off when touched. Fertile fronds, when distinct, narrower than the sterile. Sori small, round, scattered between the main veins or towards the margins, occasionally elongate and confluent, lacking indusia.

A genus of about 100 species ranging from Africa to Polynesia, Japan and New Zealand. They are long-lived and grow slowly and are fairly tolerant of strong light and dry air. Cultivation as for *Polypodium* (p. 33); they can be grown epiphytically on tree trunks or mossy logs. Propagation by spores or division of the rhizomes.

  1a. Mature fronds less than 8 cm (usually
      much less)                                 2
   b. Mature fronds more than 8 cm (often
      much more)                                 3
  2a. Undersurface of frond with a felty
      covering of fine hairs       **2. rupestris**
   b. Undersurface of frond almost without
      hairs                        **4. nummularia**
  3a. Sori few, mostly confluent around
      frond tips                   **5. confluens**
   b. Sori many, small, scattered              4
  4a. Fronds with stipes half the length of
      the blade or more              **1. lingua**
   b. Fronds almost stalkless        **3. serpens**

**1. P. lingua** (Thunberg) Farwell.
Figure 10(1), p. 30.
Fronds to 30 × 5 cm or more, all similar, tough, leathery, deep green with a buff-brown or whitish felt-like covering of hairs on the undersurface. Sori *c.* 2 mm in diameter, scattered in large numbers over much of the area of the frond undersurface. *S Japan.* H5–G1.

**2. P. rupestris** (R. Brown) Ching.
Figure 10(2), p. 30.
Fronds markedly of two types, the sterile spoon-shaped, 1–4 cm, the fertile linear-lanceolate, to 8 cm. Blade somewhat fleshy, dark green, thinly covered with buff or reddish hairs above, but with a denser, felt-like covering beneath. Sori mostly *c.* 1 mm in diameter, arranged in 1–4 rows on each side of the midrib near the tips of the fertile fronds. *E Australia.* H5–G1.

**3. P. serpens** (Forster) Ching. Figure 10(3), p. 30.
Fronds 40–60 × 1–2 cm, all similar, blades somewhat fleshy and with thickened margins, mid green with buff-brown, felt-like covering on the undersurface. Sori 0.5–1.5 mm in diameter, scattered abundantly in irregular rows over the upper half of the frond. *Australia (Queensland).* G2.

**4. P. nummularia** (Swartz) Ching.
Figure 10(4), p. 30.
Fronds of markedly different types, the sterile almost circular or broadly ovate, 1.5–3 cm, the fertile linear, narrowed basally to the stipe, 3–7 cm, blade fleshy, mid green, almost without hairs. Sori eventually covering much of the lower surface. *Himalaya to the Philippine Islands.* G2.

**5. P. confluens** (R. Brown) Ching.
Figure 10(5), p. 30.
Fronds all more or less similar, 10–25 × 1–2 cm, the fertile slightly longer than the sterile, blades fleshy, bright green with few hairs. Sori *c.* 1–5 mm in diameter

(rarely larger), restricted to the tips of the fronds where they form a continuous, confluent, single horseshoe-shaped row. *Australasia*. G1.

## 31. DRYMOGLOSSUM Presl
*C.N. Page & F.M. Bennell*

Small epiphytes. Rhizome long-creeping, thin, wiry, freely rooting, scale-clad, bearing fronds individually. Fronds of two kinds, jointed to the rhizome, all entire, the sterile ovate to almost circular, the fertile linear-elliptic, thick, fleshy, glossy green, with minute, scattered, stellate hairs. Sori in a continuous linear strip, often occupying much of the lower surface, lacking indusia.

A genus of about 6 closely related species from Madagascar, India, New Guinea and the Solomon Islands. The one cultivated species is long-lived and fast-growing, epiphytic on tree trunks in the wild; the plants form small creepers with small fronds and long rhizomes which can eventually spread to cover a whole tree trunk. Cultivation as for *Polypodium* (p. 33). Propagation is by spores or rhizome fragments.

**1. D. piloselloides** (Linnaeus) Presl
(*Microgramma piloselloides* (Linnaeus) Copeland). Figure 10(11), p. 30.
*India to New Guinea*. G2.

## 32. MICROSORIUM Link
*C.N. Page & F.M. Bennell*

Small, medium or large epiphytes or amphibious herbs. Rhizome short or long-creeping, thin, scale-clad, branching occasionally, bearing fronds individually. Fronds all of one kind, mostly simple and entire, sometimes lobed, evergreen, often rather thin, membranous or papery, main and lateral anastomosing veins often well defined. Sori median, rounded, discrete, superficial, small, sometimes very numerous, lacking indusia.

A genus of about 60 species found in tropical Asia, Australasia and the Pacific. Many of the species of *Phymatodes* (p. 31) have, from time to time, been placed in this genus. The cultivated species are long-lived and slow-growing, requiring tropical conditions and a coarse soil mixture. Species Nos. **1** & **2** may be grown in baskets; species No. **3** requires special treatment. Propagation of all is by spores or rhizome fragments.

1a. Mature fronds always simple, mostly
 more than 40 cm      **2**
 b. Mature fronds sometimes lobed, all
 much less than 40 cm      **3**

2a. Base of frond tapering gradually to
 the stipe     **1. punctatum**
 b. Base of frond abruptly narrowed to
 the stipe     **2. musifolium**
3a. All fronds 3-lobed     **3. pteropus**
 b. Mature fronds pinnatifid with several
 lobes     **4. scandens**

**1. M. punctatum** (Linnaeus) Copeland
(*Polypodium punctatum* Linnaeus; *P. irioides* Poiret; *Phymatodes punctatum* (Linnaeus) Presl). Figure 10(6), p. 30.
Rhizome medium-creeping, fleshy, to 1 cm wide, covered with dull brown scales. Fronds to 1.2 m, erect to semi-erect with short stipes; blade simple, entire, broadest towards the middle and tapering to a very slender wing at the base, apex rounded, leathery, light to yellowish green. Sori numerous, less than 1.5 mm in diameter, irregularly scattered through the upper part of the frond. *Tropical Australasia, Pacific Islands*. G2.

**2. M. musifolium** (Blume) Ching
(*Polypodium musifolium* Blume; *Pleopeltis musifolia* (Blume) Moore). Figure 10(7), p. 30.
Rhizome stout, very short-creeping, bearing fronds close together, covered with dull brown scales and with a large mass of brown roots. Fronds to 1.2 m, stalkless, erect to ascending; blade simple, entire, ovate, widening abruptly at the base, widest above the middle, with a tapering apex; thin and translucent, papery but firm with a conspicuously raised pattern of net veins easily visible between the parallel, spreading, main veins, often wavy and wrinkled, light green. Sori mostly less than 1 mm in diameter, very abundant on well-grown plants, but cultivated specimens are seldom fertile. *Malaysia to New Guinea*. G2.

**3. M. pteropus** (Blume) Copeland
(*Polypodium pteropus* Blume; *Pleopeltis pteropus* (Blume) Moore). Figure 10(8), p. 30.
Rhizome fleshy, creeping, the growing parts scaly, green when scales lost after *c.* 2 years. Fronds to 40 cm, mostly erect; blades of variable division, from simple to deeply 3-lobed; simple fronds obovate-lanceolate with a narrowly tapering apex; 3-lobed fronds with a long, tapering, acuminate central lobe and similar, but smaller and broader, lateral lobes, thin and rather membranous with conspicuous raised veins. *India, S China, Malaysia*. G2.

An amphibious species of stream banks, which becomes more or less completely submerged during the wet season in the wild. Being tolerant of such immersion, at least for periods in cultivation it is sometimes grown as a submerged aquatic in tropical aquaria.

**4. M. scandens** (Forster) Tindale
(*Polypodium scandens* Forster; *Phymatodes scandens* (Forster) Presl). Figure 10(9), p. 30.
Rhizome to 6 mm in diameter, long-creeping, seldom branching, often creeping or climbing considerable distances, fleshy, covered with papery brown scales. Fronds extremely variable in size and form, to 40 cm, often much less, arching to ascending; blade varying from entire to pinnatifid with a variable number of lobes, membranous. Sori rounded, 1–2 mm in diameter, scarcely sunken. *E Australia*. H5–G1.

In the wild this species climbs for considerable distances up tree trunks in an ivy-like manner.

## 33. SELLIGUEA Bory de St Vincent
*C.N. Page & F.M. Bennell*

Small epiphytes. Rhizome thin, long-creeping. Fronds of 2 kinds, the vegetative simple, entire, evergreen, jointed to the rhizome with stipe of the same length as or shorter than the blade; blade thinly leathery, shining, with main and lateral veins raised slightly on both surfaces, oblique and distinct as dark lines running almost to the margins. Fertile fronds similar to the vegetative but narrower with elongate-linear sori in single lines between adjacent main veins, sometimes interrupted, superficial, lacking indusia.

A genus of about 5 species extending from Malaysia to Polynesia. They are long-lived and moderately fast-growing, cultivated for their unusual fronds, the sterile very similar in size and texture to leaves of Beech (*Fagus sylvatica*). Cultivation as for *Polypodium* (p. 33). Propagation is by spores or division of the rhizome.

**1. S. plantaginea** Brackenridge.
Figure 10(10), p. 30.
Rhizome *c.* 2 mm in diameter, the upper parts densely scaly, bearing fronds at intervals of *c.* 5–10 mm. Fertile fronds to 12 cm, sterile fronds to 6 cm. Blade mostly ovate to elliptic, shortly pointed at the apex but very variable in shape and size, the fertile often broadest at the base and acuminate at the apex. *SE Asia, Polynesia*. G2.

Several very closely related species occur in collections from time to time, usually introduced with epiphytic orchids.

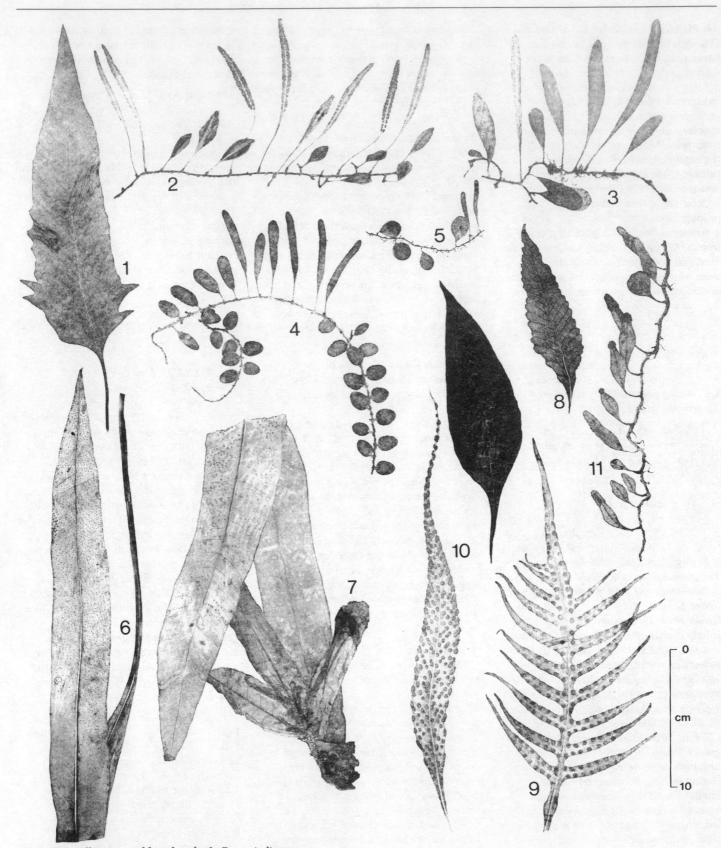

**Figure 10.** Silhouettes of fern-fronds. 1, *Pyrrosia lingua*.
2, *P. rupestris*. 3, *P. serpens*. 4, *P. nummularia*. 5, *P. confluens*.
6, *Microsorium punctatum*. 7, *M. musifolium*. 8, *M. pteropus*.
9, *M. scandens*. 10, *Selliguea plantaginea*. 11, *Drymoglossum
piloselloides*.

## 34. PLEOPELTIS Humboldt & Bonpland
*C.N. Page & F.M. Bennell*
Large epiphytes. Rhizome short-creeping, thick, woody, branching occasionally, densely hairy, bearing fronds in irregular clusters or individually. Fronds slightly ascending to erect, all similar, usually simple, entire, jointed to the rhizome; blade narrow, oblanceolate, tapering gradually to the extreme base, firm or leathery, bearing peltate scales and with the veins anastomosing freely and irregularly. Sori 2–3 mm in diameter, round, median, lacking indusia.

A genus of about 40 species from the tropics. The cultivated species are long-lived and fast-growing, grown for their large, showy leaves, often in baskets. Cultivation as for *Polypodium* (p. 33).

1a. Blade with the main veins standing out prominently on the lower surface **1. crassifolium**
b. Blade with the main veins only slightly prominent **2. phyllitidis**

**1. P. crassifolium** (Linnaeus) Moore (*Polypodium crassifolium* Linnaeus; *Pessopteris crassifolia* (Linnaeus) Underwood & Mason). Figure 11(3), p. 32.
Rhizome *c.* 1 cm thick. Fronds to 1 m, blade stiff, strap-shaped, very leathery, slightly fleshy and brittle when mature, stipe and main veins standing out prominently on the lower surface, the whole deep green and glossy. *West Indies, Mexico to Peru & Brazil*. G2.

**2. P. phyllitidis** (Linnaeus) Alston (*Polypodium phyllitidis* Linnaeus; *Campyloneuron phyllitidis* (Linnaeus) Presl). Figure 11(4), p. 32.
Rhizome *c.* 1 cm thick, short-creeping, densely clothed with brownish scales. Fronds to *c.* 1 m, blade stiff, strap-shaped, thin, leathery, brittle when mature, veins only slightly prominent on the lower surface, light green, glossy. *Tropical America*. G2.

## 35. MICROGRAMMA Presl
*C.N. Page & F.M. Bennell*
Small epiphytes. Rhizome long-creeping, thin, branching occasionally, scale-clad, bearing fronds individually and somewhat remotely, with short, basal phyllopodia. Fronds simple, of 2 kinds, the fertile longer and narrower than the sterile, which are ovate, all firm to leathery and pale green. Sori round, superficial or slightly impressed, discrete, rather few per frond, lacking indusia.

A genus of about 20 species from tropical America. The one cultivated species is long-lived and fast-growing, requiring the same conditions as *Polypodium* (p. 33). Propagation by spores or division of the rhizome.

**1. M. vacciniifolia** (von Langsdorff & Fischer) Copeland. Figure 11(1), p. 32.
Sterile fronds usually less than 2 cm, the fertile to 3 cm. *Brazil*. G2.

## 36. LEMMAPHYLLUM Presl
*C.N. Page & F.M. Bennell*
Small epiphytes. Rhizome long-creeping, very slender, branching occasionally, scale-clad and bearing fronds individually. Fronds jointed to the rhizome, of 2 kinds, both simple and entire, the sterile obovate, ovate or elliptic, the fertile linear or linear-oblanceolate, the tips drawn out to slender, tapering apices. Sori rounded in mostly continuous but not confluent, lengthwise, median rows on either side of the rachis.

A genus of 4 rather similar species from the Himalaya, Thailand, Indonesia and Japan. The 1 cultivated species is long-lived and fast-growing, of interest on account of the unusual leaves. Cultivation as for *Polypodium* (p. 33). Propagation by spores or division of the rhizome.

**1. L. microphyllum** Presl. Figure 11(2), p. 32.
Rhizome slender, creeping readily over rocky substrates. Sterile fronds 2–5 cm, fertile mostly 5–10 cm on a short stipe which is *c.* one-quarter of the length of the blade. Blades yellow-green, slightly shining, somewhat fleshy. *E Asia*. H5–G1.

## 37. BELVISIA Mirbel
*C.N. Page & F.M. Bennell*
Small plants, epiphytic or growing on rocks. Rhizome thin, short-creeping. Fronds fleshy, of two kinds, the blades entire, shortly stalked, the fertile similar to the sterile near the base, but bearing an additional, narrow, apical tail, the tightly inrolled margin of which protects the sporangia within.

A genus of 25 species, all very similar in appearance. The cultivated species require warm, humid conditions, and establishes and grows very slowly. Propagation is by spores.

**1. B. mucronata** (Fée) Copeland. Figure 11(5), p. 32.
Rhizome short-creeping, scaly, bearing fronds in small groups. Fronds with ascending stipes and arching or ascending blades, to 25 × 1.5 cm, narrowly linear, the narrow tail of the fertile fronds 10 cm or more; blades of all thick, somewhat fleshy, pale green. *Sri Lanka to Australia & Polynesia*. G2.

The most widespread of the species; others may occasionally be seen in specialist collections where they often are introduced among epiphytic orchids.

## 38. DICTYMIA J. Smith
*C.N. Page & F.M. Bennell*
Medium-sized epiphytes. Rhizome short-creeping, moderately thick, freely branching, densely clad when young with dark, spreading scales, bearing fronds individually but in closely grouped clusters. Fronds simple, entire, evergreen, jointed to the rhizome, narrowly strap-shaped or narrowly lanceolate, stiff, leathery, hairless all of 1 kind. Sori as a single median row on either side of the rachis, large, rounded or obliquely oblong, orange, slightly immersed, lacking indusia.

About 4 species from Australia, New Guinea, New Caledonia and Fiji. The 1 species in cultivation has large, distinctive, orange sori and makes a good basket subject, eventually building up considerable clumps. Cultivation as for *Polypodium* (p. 33). Propagation is by spores or division of the rhizome.

**1. D. brownii** (Wikström) Copeland. Figure 11(7), p. 32.
Fronds erect, more or less dark green, the midrib prominent. *Australia (Queensland)*. G2.

## 39. PHYMATODES Presl
*C.N. Page & F.M. Bennell*
Small- to medium-sized plants growing epiphytically or on rocks. Rhizomes medium-creeping, thick or thin, occasionally branching, shining or densely scale-clad, bearing fronds individually. Fronds all of one kind but varying in dissection even on a single plant, often deeply pinnatisect but occasionally simple and entire or irregularly lobed, evergreen, firm, membranous or leathery. Sori usually large, often deep orange-brown, in regular single rows along either side of the rachis or pinna mid-vein, usually well sunk in pits which appear as small, raised protuberances on the upper side, lacking indusia.

A genus of about 10 species from the Old World tropics. Long-lived and moderately fast-growing, they are cultivated for their

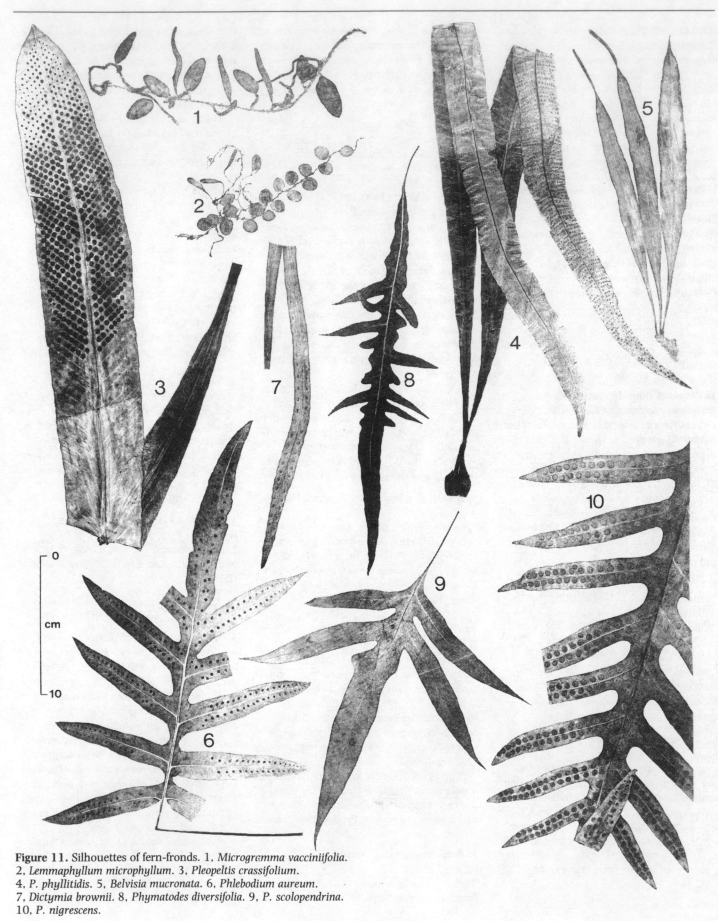

**Figure 11.** Silhouettes of fern-fronds. 1, *Microgræmma vacciniifolia*.
2, *Lemmaphyllum microphyllum*. 3, *Pleopeltis crassifolium*.
4, *P. phyllitidis*. 5, *Belvisia mucronata*. 6, *Phlebodium aureum*.
7, *Dictymia brownii*. 8, *Phymatodes diversifolia*. 9, *P. scolopendrina*.
10, *P. nigrescens*.

foliar diversity. Cultivation as for *Polypodium*. Propagation is by spores (slow) or by rhizome fragments.

1a. Fronds on mature plants varying from simple to lobed **3. diversifolia**
 b. Fronds on mature plants all lobed **2**
2a. Mature fronds less than 40 cm, with 2–4 pairs of lobes **1. scolopendrina**
 b. Mature fronds 40–180 cm, with *c.* 10 pairs of lobes **2. nigrescens**

**1. P. scolopendrina** (Bory de St Vincent) Christensen. Figure 11(9), p. 32.
Rhizome to *c.* 7 mm in diameter, long-creeping, fleshy, green, covered with scattered scales, bearing fronds *c.* 5 cm apart. Fronds to *c.* 40 cm, blade rarely simple, mostly pinnatifid with 2–4 (rarely more, pairs of strap-shaped lobes of even width, separated by sinuses of similar width, veins not visible, thin, leathery, rather flat, light green, glossy with dark rachis and pinna midribs. Sori 3–4 mm in diameter in 2 irregular rows on either side of the midribs of each lobe, with some also on the wing on either side of the midrib of the whole frond. *Tropical Asia.* G2.

A good basket plant, tolerant of fairly dry conditions.

**2. P. nigrescens** (Blume) J. Smith. Figure 11(10), p. 32.
Like *P. scolopendrina* but blades to 1.8 m, pinnatifid with *c.* about 10 pairs of long-acuminate lobes with the main veins raised as conspicuous dark lines forming a network, segment margins slightly wavy and crisped. *Malaya, Polynesia.* G2.

**3. P. diversifolia** (Willdenow) Pichi-Sermolli (*Polypodium diversifolium* Willdenow; *Microsorium diversifolium* (Willdenow) Copeland). Figure 11(8), p. 32.
Rhizome to 8 mm in diameter, long-creeping, freely branching, usually forming mats on rocks, fleshy, covered with adpressed, pale brown scales. Fronds variable in size and form, to 60 cm but usually much less, arching to descending. Blade varying from entire to pinnatifid, with a variable number of lobes, leathery. Sori round, 2–5 mm, sunk in depressions. *E Australia, New Zealand.* G1.

**40. PHLEBODIUM** (R. Brown) J.E. Smith
*C.N. Page & F.M. Bennell*
Large plants, growing epiphytically or on rocks. Rhizomes medium-creeping, thick, occasionally branched, densely clad with golden scales, bearing fronds individually. Fronds all similar, pinnatisect, stipe and blade of about equal lengths, blade to 1 m or more, ovate, usually hairless, often glaucous, firm, thin, leathery. Sori median, superficial, scarcely indented, rounded, conspicuously orange-brown, lacking indusia.

A genus of a few species confined to the American tropics. One species is grown for its golden-scaled rhizome and large, glaucous fronds. It requires strong light and a well-drained compost. Propagation is by spores or rhizome fragments.

**1. P. aureum** (Linnaeus) J. Smith (*Polypodium aureum* Linnaeus).
Figure 11(6), p. 32.
Fronds mostly 90–150 cm, ascending or arching, blade green to very glaucous. *Tropical America.* G1.

**41. POLYPODIUM** Linnaeus
*C.N. Page & F.M. Bennell*
Medium-sized plants growing on soil, epiphytically, or on rock. Rhizomes mostly long-creeping, thick, branching occasionally, densely scale-clad when young, sometimes losing scales with age, bearing fronds individually, with a distinct joint between the rhizome and stipe forming the point of abscission of old fronds. Fronds all of one kind, mostly 1 × pinnatisect or pinnate with strap-shaped, tapering or rounded segments, evergreen, mostly hairless, rarely downy. Sori median, discrete, rounded, superficial, often bright orange in colour when mature, entirely lacking an indusium but sometimes with branched, hair-like paraphyses set amongst the sporangia.

A genus of about 75 species mostly confined to N temperate regions. A great many tropical species were formerly included in this genus; they are treated in this work under the following generic names: *Aglaomorpha* (p. 26), *Drymoglossum* (p. 29), *Goniophlebium* (p. 35), *Microsorium* (p. 29), *Phlebodium* (above), *Phymatodes* (p. 31) and *Pleopeltis* (p. 31).

Species of *Polypodium* are long-lived, mainly slow but strong and reliable growers, and many are somewhat drought-tolerant. They require very freely draining compost and can easily be killed by over-feeding. Propagation is by spores or rhizome fragments.
Literature: Shivas, M.G., Contributions to the cytology and taxonomy of species of Polypodium in Europe and America, II, Taxonomy, *Journal of the Linnaean Society. Botany* 58: 27–38 (1961); Lloyd, R. & Lange, F.A., The Polypodium vulgare complex in North America, *British Fern Gazette* 9: 168–77 (1964); Roberts, R.H., Polypodium macaronesicum and Polypodium australe, a morphological comparison, *Fern Gazette* 12: 69–74 (1980).

1a. Blade narrowly linear or slightly tapering, with most pinnae of similar length **5. vulgare**
 b. Blade ovate-triangular, lower pinnae longer than upper **2**
2a. Blade thick and rigidly leathery **6. scouleri**
 b. Blade thin and flexible to slightly leathery **3**
3a. Sori without paraphyses **4. interjectum**
 b. Sori with paraphyses **4**
4a. Thickened cells of annulus of sporangium mostly 4–8, a few with up to 18 **1. australe**
 b. Thickened cells of annulus of sporangium mostly 9–20 **5**
5a. Frond slightly glossy above **3. macaronesicum**
 b. Frond highly glossy above **2. azoricum**

**1. P. australe** Fée (*P. vulgare* var. *cambricum* Linnaeus). Figure 12(1), p. 34.
Fronds mostly 10–30 cm. Blade very broadly triangular-ovate, thin, fairly flexible, bright green, somewhat glossy above. Sori with paraphyses. Thickened cells of the annulus of the sporangium mostly 4–8, occasionally to 18. *S Europe, Mediterranean region.* H4.

The shape of the frond varies with age and luxuriance; smaller fronds are more triangular, larger ones more ovate. Several cultivars occur, mainly with variously serrated pinna margins.

**2. P. azoricum** (Vasconcellos) Fernandez. Figure 12(2), p. 34.
Fronds mostly 10–35 cm. Blade very wide, broadly triangular to broadly ovate, thin but fairly rigid, bright yellow-green, highly glossy above. Paraphyses present. Thickened cells of the annulus of the sporangium 9–20. *Azores.* H5–G1.

Related to *P. australe* but distinguished by its bright green, glossier fronds.

**3. P. macaronesicum** Bobrov. Figure 12(3), p. 34.
Fronds mostly 10–50 cm. Blades broadly ovate, thin, highly flexible, green, scarcely glossy above. Paraphyses present. Thickened cells of the annulus of the sporangium mostly 9–20. *Extreme SW Europe, Madeira, Canary Islands.* H5–G1.

**Figure 12.** Silhouettes of fern-fronds. 1, *Polypodium australe*.
2, *P. azoricum*. 3, *P. macaronesicum*. 4, *P. interjectum*.
5, *P. vulgare*. 6, *P. scouleri*. 7, *Goniophlebium subauriculatum*.
8, *G. verrucosum*.

**4. P. interjectum** Shivas. Figure 12(4), p. 34.
Fronds mostly 20–40 cm. Blade ovate to lanceolate, moderately thick, slightly leathery, mid to bluish green, scarcely glossy. Paraphyses absent. Thickened cells of the annulus of the sporangium 8–10. *W Europe*. H4.

**5. P. vulgare** Linnaeus. Figure 12(5), p. 34.
Fronds mostly 10–25 cm. Blade usually narrowly linear or slightly tapering, moderately thick and leathery, mid to bluish green, scarcely glossy. Paraphyses absent. Thickened cells of the annulus of the sporangium 10–14. *Eurasia*. H3.

**6. P. scouleri** Hooker. Figure 12(6), p. 34.
Fronds usually 15–40 cm. Blade triangular-ovate, distinctly thick and very rigidly leathery with a cartilaginous border, dull, bluish green. Paraphyses absent. Thickened cells of the annulus of the sporangium 12–20. *Pacific coast of N America*. H3.

**42. GONIOPHLEBIUM** (Blume) Presl
*C.N. Page & F.M. Bennell*
Medium to large epiphytes. Rhizome medium- to long-creeping, thick, fleshy, branching occasionally, scale-clad when young, the scales shed with age, bearing fronds individually, often remotely from one another. Fronds 1 × pinnate or pinnatisect, all of 1 kind, evergreen; arched or hanging, linear-lanceolate, soft to firm, stipe one-tenth to one-third of the length of the blade, jointed to short phyllopodia. Blade with pinnae nearly stalkless or borne on short stalks, the whole linear-lanceolate, slightly sickle-shaped. Sori numerous, large, round, superficial or slightly impressed in a single row on either side of the pinna midrib, lacking indusia.

A genus of about 20 species formerly included in *Polypodium*. Cultivation as for that genus (p. 33). Handsome epiphytes, their long arching or hanging fronds rendering them ideal subjects for hanging baskets.

1a. Bases of the pinnae jointed
   **1. subauriculatum**
  b. Bases of the pinnae tapering
   **2. verrucosum**

**1. G. subauriculatum** (Blume) Presl (*Polypodium subauriculatum* Blume; *P. percussum* misapplied). Figure 12(7), p. 34.
Rhizome becoming hairless and chalky-white with age. Fronds to 1 m. Blade broad with basal auricles; pinnae closely spaced along the rachis, bases jointed. Sori shallowly impressed, sunken, producing only low papillae on the upper surface. *NE India and SW China to Australia (NE Queensland)*. G2.

**2. G. verrucosum** (Wallich) J. Smith (*Polypodium verrucosum* Wallich). Figure 12(8), p. 34.
Rhizome becoming hairless with age, but mostly green. Fronds to 1 m or more. Blade narrowed at the base and without auricles; pinnae widely spaced along the rachis. Sori fairly deeply impressed, producing distinct papillae on the upper surface. *Indonesia (Sumatra) to Australia (NE Queensland)*. G2.

**43. CYATHEA** J.E. Smith
*C.N. Page & F.M. Bennell*
Large, tree-like, terrestrial plants. Rhizomes erect, sometimes very tall and trunk-forming, bearing fronds in large, spreading crowns from the apex. Rhizome apex and stipe bases usually densely scale-clad. Fronds all of one kind, very large and several times pinnately divided to produce a very open, lace-like effect, ovate. Sori rounded, with or without indusia, these, when present, very variable, usually simple and scale-like.

A genus of several hundred species occurring in the tropics and subtropics of both the Old and New Worlds. The genus has been variously split into segregate genera, including *Alsophila* and *Sphaeropteris*. For convenience, all are treated here under *Cyathea*. They are long-lived plants, eventually forming attractive trees and are popular subjects in cultivation whenever space and conditions allow. They require frost-free, mostly warm, permanently moist, shaded conditions. Given such an environment, many are strong-growing and vigorous.

1a. Old stipe bases persistent along the trunk 2
  b. Stipe bases not persistent along the trunk, but breaking away to leave clean-cut leaf scars 5
2a. Fronds 2 × pinnate, pinnules simple with toothed margins **1. rebeccae**
  b. Fronds 3 × or more pinnate or pinnatisect, the pinnules always pinnate or pinnatisect 3
3a. Fronds bright silver-grey beneath **2. dealbata**
  b. Fronds pale green beneath 4
4a. Stipe bases covered with low, rounded or pointed tubercles **3. australis***
  b. Stipe bases covered with long, sharp, woody spines **4. leichardtiana**
5a. Stipe bases black **5. medullaris**
  b. Stipe bases brown **6. cooperi**

**1. C. rebeccae** (Mueller) Domin (*Alsophila rebeccae* Mueller). Figure 13(1), p. 36.
Rhizome forming a slender trunk, c. 10 cm in diameter and up to 7 m in height, clothed throughout with persistent stipe bases. Fronds to 3 m, 2 × pinnate, dark green, shining above, pinnules with long tips and toothed margins. Sori without indusia. *Australia (NE Queensland)*. G2.

An attractive species, unusual because of its coarsely divided, glossy fronds, reminiscent of those of some species of *Diplazium* (p. 48).

**2. C. dealbata** (Forster) Swartz (*Alsophila tricolor* (Colenso) Tryon). Figure 13(2), p. 36.
Rhizome forming a trunk to 20 cm in diameter, extending to 10 m in height, clothed throughout with persistent, glaucous-bloomed stipe bases and shining, dark brown scales. Fronds to almost 4 m, 3 × pinnate, bright green above but distinctly white or bright silver-grey beneath. Sori large, covered by cup-shaped indusia. *New Zealand*. H5–G1.

An attractive species, grown for the distinctive colour of the frond under-surface.

**3. C. australis** (R. Brown) Domin (*Alsophila australis* R. Brown). Figure 13(3), p. 36.
Rhizome forming a trunk to 1.5 m in diameter at the base, tapering gradually upwards, to 12 m in height, clothed throughout in persistent old stipe bases, buttressed at the base by masses of wiry, adventitious roots. Fronds to 4.5 m, 3 × pinnate, the stipe bases densely covered with rounded tubercles. Sori without indusia. *E Australia*. H5–G1.

***C. cunninghamii** Hooker (*Alsophila cunninghamii* (Hooker) Tryon). Similar, but with more slender, taller trunks (to 15 cm in diameter and up to 20 m tall), smaller, thinner-textured fronds (to 3 m) with stipe bases covered in more pointed tubercles, and sori with cup-shaped indusia. *E Australia, New Zealand*. H5–G1.

**4. C. leichardtiana** (Mueller) Copeland (*Sphaeropteris australis*) (Presl) Tryon). Figure 13(4), p. 36.
Rhizome forming a slender trunk, 10 cm in diameter and up to 7 m tall, clothed throughout with persistent, black stipe bases bearing long, sharp, woody spines and narrow, whitish scales. Fronds to 3 m,

**Figure 13.** Silhouettes of fern-fronds. 1, *Cyathea rebeccae*.
2, *C. dealbata*. 3, *C. australis*. 4, *C. leichardtiana*. 5, *C. medullaris*.
6, *C. cooperi*. 7, *Dicksonia squarrosa*. 8, *D. fibrosa*. 9, *D. antarctica*.

3 × pinnate. Sori without indusia. *E Australia*. H5–G1.

**5. C. medullaris** (Forster) Swartz (*Sphaeropteris medullaris* (Forster) Bernhardi). Figure 13(5), p. 36.
Rhizome forming a trunk 15–20 cm in diameter and up to 15 m or more in height, buttressed at the base by masses of adventitious roots, with stipe bases shed to leave distinct, hexagonal leaf scars, giving the trunk a clean-cut appearance. Fronds to 6 m, 3 × pinnate, the stipe and rachis black with a glaucous bloom, the stipe bases with numerous pointed tubercles. Sori large, circular, each covered by a circular indusium. *New Zealand*. H5–G1.

**6. C. cooperi** (Mueller) Domin (*Sphaeropteris cooperi* (Mueller) Tryon). Figure 13(6), p. 36.
Rhizome forming a trunk to 15 cm in diameter and 12 m or more in height, with stipe bases shed to leave distinct, oval leaf scars, giving the trunk a clean-cut appearance. Fronds to 6 m, 3 × pinnate, the stipe bases brown with numerous pointed tubercles and long whitish scales. Sori circular, covered by circular indusia. *Tropical and subtropical Australia*. G1.

This species, which under suitable conditions is vigorous and fast-growing, is cultivated for its conspicuous white scales.

**44. DICKSONIA** L'Héritier
*C.N. Page & F.M. Bennell*
Medium to large or tree-like terrestrial plants. Rhizome thick or thin, usually unbranched, covered with matted growths of old stipe bases, bristly hairs and interwoven roots, which build up the bulk of the girth, and with dense masses of rust-brown hairs amongst the crown, erect, bearing fronds in regular, spreading crowns. Fronds 3 × pinnate, all of 1 kind, evergreen, ovate, harsh, rather leathery and often slightly prickly, dark green. Sori marginal, on pinnules with deeply lobed margins, each protected by a 2-valved cup, formed on the inner side by an indusium and on the outer by a portion of the reflexed margin of the pinnule, both hard when mature.

About 25 or fewer species in Australasia, Polynesia and Mexico. They can be confused with *Cyathea*, but are distinguished by having hairs rather than scales. They are very long-lived and moderately slow-growing, cultivated for their spectacular, palm-like habit. They are mostly grown in moist, outdoor borders in warm, damp, frost-free climates. They are tolerant of strong light, but thrive best when lightly shaded, requiring rhizome (trunk) shelter and a well-drained but moist, humus-rich soil. They do well along streams and water should be sprayed directly on to the trunks in dry spells. The stalks of the old fronds persist for years beneath the crown of living fronds, and should not be removed, as they help to provide shade for the trunk. Young growth is easily damaged if creepers on the trunks spread into the crown. Propagation is by spores.

1a. Trunk 10 cm or less in diameter, even in mature plants; stipes mostly black with long, perpendicular hairs  
        **1. squarrosa**
  b. Trunk more than 10 cm in mature specimens; stipes mostly greenish brown with long, shaggy, non-perpendicular hairs at the extreme base only    2
2a. Trunk of mature specimens less than 7 m × 30 cm    **2. fibrosa**
  b. Trunk of mature specimens larger (up to 15 m × 60 cm)    **3. antarctica**

**1. D. squarrosa** (Forster) Swartz. Figure 13(7), p. 36.
Rhizome very slender, forming a stiff, black, woody trunk, at most 6 m × 10 cm, with numerous persistent stipe bases. Fronds to 120 × 50 cm, bright green. Stipe and rachis densely hairy, mostly black, the hairs standing out rigidly and perpendicularly. *New Zealand*. G1.

Established plants can spread by runners.

**2. D. fibrosa** Colenso. Figure 13(8), p. 36.
Rhizome moderately thick, brown, stiff, woody, to *c*. 7 m × 30 cm, usually entirely covered with matted roots. Fronds to 120 × 30 cm, dark green. Stipe and rachis sparsely hairy, greenish brown or green, with long, brown, shaggy hairs at the base which do not stand out perpendicularly. *New Zealand*. G1.

**3. D. antarctica** Labillardière. Figure 13(9), p. 36.
Rhizome stout, brownish, forming a tall, tough woody trunk to 15 m × 60 cm (or more), covered with numerous old stipe bases and densely matted roots. Fronds to 2 m × 60 cm, dark green. Stipe and rachis sparsely hairy, greenish brown or green with long, brown, softly shaggy and silky hairs, which do not stand out perpendicularly, at the base. *S & E Australia, including Tasmania*. H5–G1.

**45. THYRSOPTERIS** Kunze
*C.N. Page & F.M. Bennell*
Medium-sized terrestrial plants, forming small trees. Rhizome hairy. Fronds 2–3 × pinnate, of two kinds, the fertile with the upper pinnae sterile and the lower fertile and jointed to the axes. Base of stipe with long, soft, reddish hairs. Sori terminal on the segments, each enclosed within a spherical indusium which opens as a cup.

One species, endemic to Chile (Juan Fernandez). It is long-lived and moderately slow-growing, but only occasionally seen in cultivation. Growth requirements as for *Dicksonia*. Like very few other tree-ferns, this can be propagated by runners.

**1. T. elegans** Kunze. Figure 14(3), p. 38.
Fronds to 1.5 m × 65 cm, firm. *Chile (Juan Fernandez)*. G2.

**46. CULCITA** Presl
*C.N. Page & F.M. Bennell*
Medium to large terrestrial plants. Rhizomes long-creeping, hairy, subterranean, sometimes ascending to erect but not trunk-forming, bearing fronds individually. Fronds evergreen, all of 1 kind. Blades coarsely or finely pinnately divided, held horizontally on upright stipes which are almost as long as the blades. Sori marginal on the ultimate segments, each with a rounded, 2-valved indusium, the outer valve a continuation of the frond margin, the inner small and scale-like.

A genus of about 10 species, with a scattered distribution. Plants are easily cultivated in most types of well-drained soil.

1a. Rhizomes mostly ascending or erect; blade coarsely divided, triangular, as broad as long    **1. macrocarpa**
  b. Rhizomes mostly long-creeping; blade finely divided, ovate-triangular, about twice as long as broad    **2. dubia**

**1. C. macrocarpa** Presl (*Dicksonia culcita* L'Héritier). Figure 14(4), p. 38.
Rhizomes 1–3 cm or more in diameter, mostly ascending to erect, densely clothed with bright, reddish brown hairs, bearing fronds in lax tufts. Fronds 30–90 cm, blades triangular, nearly as broad as long, rather coarsely several times pinnate, thick, leathery, shining, green. Fertile fronds with very conspicuous sori 2–4 mm in diameter. *SW Europe, Atlantic Islands*. H5–G1.

**2. C. dubia** (R. Brown) Maxon. Figure 14(5), p. 38.
Rhizomes 1 cm or more in diameter, mostly long-creeping, densely clothed with buff-brown hairs, bearing fronds singly or in

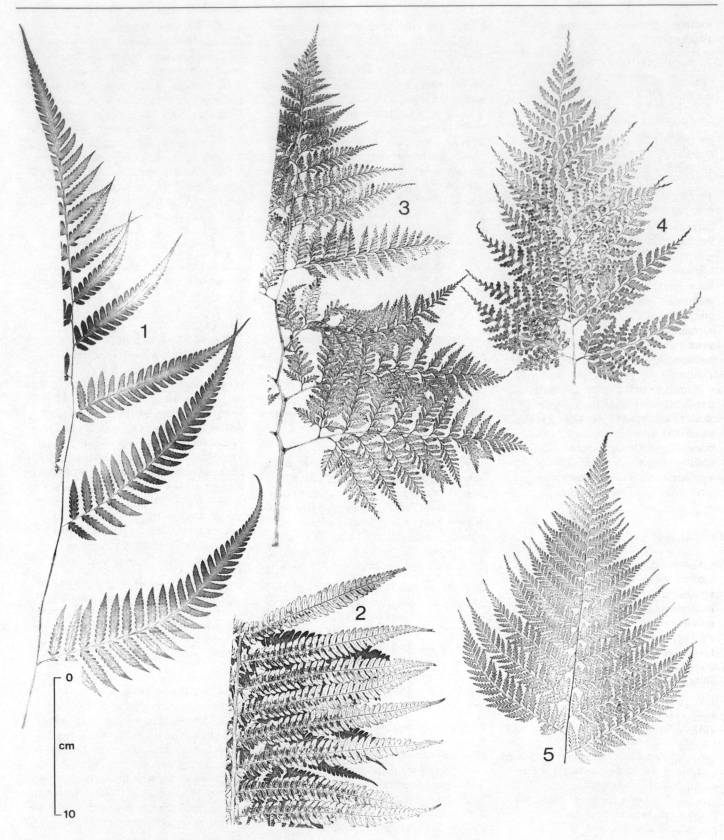

**Figure 14.** Silhouettes of fern-fronds. 1, *Cibotium barometz*.
2, *C. glaucum*. 3, *Thyrsopteris elegans*. 4, *Culcita macrocarpa*.
5, *C. dubia*.

scattered groups. Fronds to 1.5 m, the blades ovate-triangular, up to twice as long as broad, rather finely many times pinnate, thin, dull pale green. Fertile fronds with rather inconspicuous, small sori, mostly less than 1 mm in diameter. *E Australia.* H5–G1.

The finely divided fronds of this species are very like those of several species of *Dennstaedtia*, *Hypolepis* and *Macrolepia*, but are distinguishable by their sori.

## 47. CIBOTIUM Kaulfuss
*C.N. Page & F.M. Bennell*
Large to very large terrestrial plants forming tall and erect trunks or with prostrate rhizomes. Rhizome and frond bases covered with long, limp, soft, pale hairs. Fronds all of one kind, several times pinnately divided. Fertile fronds with marginal or almost marginal sori, each surrounded by a 2-valved indusium, the outer valve of which is a box-shaped continuation of the leaf margin bent at about right angles to the leaf surface.

A small genus of about 12 species, of tropical Pacific range. The cultivated species are long-lived and slow-growing, requiring a lime-free medium, moist tropical conditions and extensive space. In addition to their size, they are noted for the woolliness of their rhizomes, which gives them a shaggy appearance.

1a. Rhizome prostrate or decumbent
                  **1. barometz**
  b. Rhizomes trunk-forming  **2. glaucum\***

**1. C. barometz** (Linnaeus) J.E. Smith. Figure 14(1), p. 38.
Rhizomes large, thick, decumbent or prostrate, rhizome and stipe bases covered with a mass of shining brown hairs which are 1–1.5 cm long. Fronds 1.5–3 m, 3 × pinnatisect, the pinnules shortly stalked, pale glaucous green beneath. *China, Taiwan, SE Asia.* G2.

**2. C. glaucum** (J.E. Smith) W.J. Hooker. Figure 14(2), p. 38.
Rhizome large, thick, erect and trunk-forming, rhizome and stipe bases covered with a mass of long, shining, golden yellow hairs. Fronds 2–2.5 m, 3 × pinnatisect, pinnules stalkless, bright glaucous green beneath. *USA (Hawaii).* G2.

**\*C. chamissoi** Kaulfuss. Similar, but with pale fawn hairs, the undersurface of the frond green or dull glaucous green. *USA (Hawaii).*

## 48. DENNSTAEDTIA Bernhardi
*C.N. Page & F.M. Bennell*
Medium to large terrestrial plants. Rhizome long-creeping, thin but strong and woody, hairy, at the soil surface or shallowly subterranean, bearing fronds individually at well-spaced intervals. Fronds 3 × pinnate, ascending to spreading, all of one kind, triangular, soft. Stipe usually erect, hairless or finely downy, often as long as or longer than the blade. Sori marginal on the outer edges of the ultimate segments, but recessed into notches, with well-developed, slightly 2-valved, cup-shaped indusia.

A genus of about 70 species from the tropical zones, extending north to Japan and N America, south to Chile and Tasmania. *Hypolepis* and *Microlepia* (below) are very similar but are distinguished by their sori.

**1, D. punctiloba** (Michaux) Moore. Figure 15(1), p. 40.
Fronds 45–75 cm, triangular-lanceolate, 3 × pinnate. *N America.* H4.

## 49. MICROLEPIA Presl
*C.N. Page & F.M. Bennell*
Medium to large terrestrial plants. Rhizomes long-creeping, thin, hairy, bearing fronds individually but at short intervals. Fronds ascending to spreading, 2–3 × pinnate, triangular or broadly ovate, all of 1 kind, evergreen, soft, pale green, with obliquely cut ultimate segments which are acute-tipped. Stipes hairy, often as long as or longer than the blades, usually erect or semi-erect. Sori superficial, borne within the margin, protected by an often poorly developed, half cup-shaped, scale-like indusium, fixed by its rounded base, but sometimes lost at maturity.

A genus of about 45 pantropical species, ranging southwards to Madagascar and New Zealand and northwards to India and Jamaica. Cultivated for their finely divided fronds, they are grown as for *Hypolepis* (below).

1a. Veins on lower surface of segments
    distinctly raised      **1. strigosa**
  b. Veins on lower surface not distinctly
    raised           **2. speluncae**

**1. M. strigosa** (Thunberg) Presl. Figure 15(6), p. 40.
Fronds 70–90 cm, arching. Stipe rather roughly hairy towards the base. Veins raised on lower frond surface. *N India to Japan and Polynesia.* G2.

**2. M. speluncae** (Linnaeus) Moore. Figure 15(5), p. 40.
Fronds 80–150 cm, ascending to erect.

Stipe with short pale hairs. Veins on frond undersurface not distinctly raised. *SE Asia to Australia.* G2.

## 50. HYPOLEPIS Bernhardi
*C.N. Page & F.M. Bennell*
Medium to large terrestrial plants. Rhizome long-creeping at or near the surface of the soil, thin, hairy, especially at the growing tip, bearing fronds individually. Fronds 3 or more × pinnate, all of 1 kind, with finely dissected, broadly triangular blades. Stipes hairy, often as long as or longer than the blades and usually erect. Sori round, almost marginal, on the outer edge of the ultimate segments, each protected by a reflexed, tooth-like flap.

A pantropical genus of about 45 species, extending southwards to New Zealand and northwards to Japan. They are long-lived and fast-growing, requiring ordinary garden soil, preferably well-drained, in bright light. Once established, plants spread vigorously. They are propagated by spores or rhizome fragments.

1a. Stipe deep red, rough to the touch
                  **1. rugulosa**
  b. Stipe green or yellowish, more or less
    smooth to the touch         2
2a. Fronds to 1 m      **2. tenuifolia**
  b. Fronds 15–40 cm    **3. millefolium**

**1. H. rugulosa** (Labillardière) J. Smith. Figure 15(2), p. 40.
Rhizome with prominent reddish brown hairs. Fronds to 80 cm, more or less erect, 3 × pinnate. Stipe and rachis deep red and tuberculate, rough to the touch. *Temperate regions of the southern hemisphere.* H5–G1.

**2. H. tenuifolia** (Forster) Bernhardi & Schrader. Figure 15(4), p. 40.
Fronds to 1 m, 3 × pinnate, dark green. Stipes and rachis yellowish, covered with soft white hairs. *C Japan to Australia and New Zealand.* H5–G1.

**3. H. millefolium** Hooker. Figure 15(3), p. 40.
Fronds 15–40 cm, 3 × pinnate or more divided, presenting an extremely finely dissected appearance, uniformly mid green. *New Zealand.* H5–G1.

## 51. PAESIA St Hilaire
*C.N. Page & F.M. Bennell*
Small- to medium-sized terrestrial plants. Rhizome long-creeping, thin, frequently branching, hairy, bearing fronds individually. Fronds 3 × pinnate, very finely dissected, deciduous, triangular, all of 1 kind. Sori marginal, more or less

**Figure 15.** Silhouettes of fern-fronds. 1, *Dennstaedtia punctiloba*.
2, *Hypolepis rugulosa*. 3, *H. millefolium*. 4, *H. tenuifolia*.
5, *Microlepia speluncae*. 6, *M. strigosa*. 7, *Paesia scaberula*.
8, *Histiopteris incisa*.

continuous, each protected by a smaller inner and larger outer, indusium, the latter being formed by the reflexed margin of the frond.

A genus of about 12 species from SE Asia to New Zealand, Tahiti and tropical America. They are long-lived and fast-growing, thriving in deep, freely drained, moist, more or less sandy soils in light shade; propagated by spores and rhizome fragments. they can form extensive colonies when planted out, looking like miniature bracken (*Pteridium*).

**1. P. scaberula** (Richard) Kuhn. Figure 15(7), p. 40.
Fronds 25–60 cm. Pinnae and pinnules oblong, somewhat leathery and slightly firm, harsh to touch, bright yellow-green, glossy, more or less fragrant. Stipe and rachis wiry, shining, red-brown, typically zig-zag between alternate pinnae. *New Zealand*. H5–G1.

**52. HISTIOPTERIS** (Agardh) J. Smith
*C.N. Page & F.M. Bennell*
Medium to large terrestrial plants. Rhizome long-creeping, thin, branching occasionally, scale-clad. Fronds all of 1 kind, evergreen, 2–3 × pinnate with the pinnae always opposite. Sori along the margin, protected by a continuous, scarious, reflexed, false indusium formed by a fold of the margin.

A genus of 1–4 (perhaps more, and in need of revision) species from the tropics, southwards to South Africa, Tasmania and New Zealand. They are long-lived and fast-growing, eventually forming extensive colonies, requiring a moist atmosphere, well-drained, damp, sandy compost, moderately strong light and very free root-run. They grow rapidly from spores.

**1. H. incisa** (Thunberg) Smith. Figure 15(8), p. 40.
Blade triangular. Each pinna with a pair of reduced, stipule-like pinnules at the base, pinnules entire or deeply and regularly lobed into rounded segments, soft, firm, hairless and usually glaucous-bloomed. Stipe 2–3 × the length of the blade, erect, dark purple, shining, with a delicate glaucous bloom. *Tropics, S Hemisphere*. G1.

**53. LINDSAYA** Dryander
*C.N. Page & F.M. Bennell*
Small terrestrial plants. Rhizomes short-creeping, thin, bearing fronds at closely spaced intervals. Fronds erect or ascending, 1 × pinnate, with characteristically fan-shaped segments, the vegetative spreading to form a basal rosette, the fertile strongly erect. Blade thin, pale yellowish green. Sori almost marginal, linear, the indusium hinged on its inner side and opening towards the edge of the segment.

A genus of about 200 species (many fairly common and weedy) throughout the tropics. Many species have from time to time been introduced, but have proved extremely difficult to maintain. Only 1 species is usually grown; it seems to require strong light and a nutrient-poor, clay compost. All the species have a superficial similarity to *Adiantum* (p. 20), but are readily distinguished by the indusia opening in the opposite direction. The generic name is spelled '*Lindsaya*' or '*Lindsaea*'; both alternatives are widely used.

**1. L. linearis** Swartz. Figure 16(1), p. 42.
Stipe as long as blade or longer. Vegetative fronds 4–8 cm, more or less flexible; fertile fronds 10–20 cm, stiff. Blades of both very narrowly linear, 1 × pinnate, with numerous, regularly spaced, nearly opposite, triangular or more or less crescent-shaped segments. *Australasia*. H4.

**54. SPHENOMERIS** Maxon
*C.N. Page & F.M. Bennell*
Medium-sized terrestrial plants. Rhizome short-creeping to ascending. Fronds all of 1 kind, 3–4 × pinnate, yellowish green, more or less evergreen. Sori terminal, on the tips of the ultimate segments. The indusium is an outwardly opening flap situated close to the margin, fixed at the base and at the sides.

A genus of 20 rather indistinct species, occurring throughout the tropics and south to Polynesia and Madagascar, north to Japan and USA (Florida). They are fairly short-lived and fast-growing, cultivated for their finely dissected, fragile-looking fronds. They require bright, screened light, a freely ventilated atmosphere and a sandy, very freely draining medium which is kept moderately dry. They grow rapidly from spores, but are difficult to maintain and resent disturbance.

**1. S. chusana** Copeland. Figure 16(2), p. 42.
Stipe as long as or longer than the blade, hairless, shining, brown. Fronds 60–85 cm, arching, the tips spreading or descending, triangular-ovate to ovate-lanceolate, 3–4 × pinnate, pinnae oblique, finely dissected into a very large number of fairly widely spread, more or less tapered ultimate segments, all thin and hairless. *E & SE Asia to Madagascar & Polynesia*. G1.

**55. THELYPTERIS** Schmidel
*C.N. Page & F.M. Bennell*
Medium-sized terrestrial or subaquatic plants. Rhizome long-creeping, thin, branching frequently. Fronds 1 × pinnate with pinnatifid pinnae, deciduous, very soft, all of 1 kind or with the blades of the fertile fronds slightly reduced. Sori circular, median, scattered, black when mature, indusia very small, usually less than 1 mm in diameter, irregularly lobed, usually shed early.

Formerly a large genus, now much subdivided; the number of species currently remaining within it is not certain, but it is found all over the world. *Phegopteris*, *Oreopteris*, *Ampelopteris* and *Christella* are all closely related. Plants of *Thelypteris* are fairly short-lived and fast-growing; the 1 cultivated species requires semi-aquatic, fen conditions, and is propagated readily from spores.

**1. T. palustris** Schott. Figure 16(3), p. 42.
Rhizome 1–2 mm in diameter. Fronds borne at widely spaced intervals or in close groups, erect or arching. Blade lanceolate, 1 × pinnate, the pinnae pinnatisect, mostly 5–10 cm, linear, blunt, of more or less equal length throughout the greater part of the frond, the lowermost not much shorter than those above. Stipe mostly 25–75 cm, mostly greenish yellow. Fertile fronds often with longer stipes and narrower, shorter pinnae, produced mostly later in the season. *Most of Europe except Spain and the extreme north, Asia, N America*. H2.

**56. PHEGOPTERIS** (Presl) Fée
*C.N. Page & F.M. Bennell*
Small- to medium-sized terrestrial plants. Rhizome long- or short-creeping, subterranean or ascending, thin, branching frequently, bearing fronds individually or in clusters. Fronds 2 × pinnatisect, all of one kind, mostly deciduous. Stipe as long as or shorter than blade, rarely longer, with short, whitish hairs, and scattered pale brown, linear scales. Pinnae joined to the stipe across the whole width of their bases, the rachis winged between them, soft, downy, pale green. Sori small, scattered on the veins near the edges of the segments, rounded, black when mature. Indusia absent.

A genus of 3 species from temperate Eurasia. They are long-lived and moderately fast-growing, cultivated for their fresh, bright green, spring frond growth. Well-drained, humus-rich, moist soil and shelter are required, together with

**Figure 16.** Silhouettes of fern-fronds. 1, *Lindsaya linearis*.
2, *Sphenomeris chusana*. 3, *Thelypteris palustris*. 4, *Phegopteris connectilis*. 5, *P. decursive-pinnata*. 6, *Macrothelypteris torresiana*.
7, *Ampelopteris prolifera*. 8, *Christella dentata*.

moderate shade. They are easily propagated from spores.

1a. Rhizome long-creeping; blade ovate-triangular **1. connectilis**
 b. Rhizome short-creeping; blade linear-lanceolate **2. decursive-pinnata**

**1. P. connectilis** (Michaux) Watt (*Thelypteris phegopteris* (Linnaeus) Slosson). Figure 16(4), p. 42.
Rhizome slender, long-creeping, subterranean. Frond 20–40 cm, ovate-triangular. Stipe erect, blade more or less horizontal or descending. *Eurasia*. H1.

**2. P. decursive-pinnata** (van Hall) Fée. Figure 16(5), p. 42.
Rhizome short-creeping, more or less stocky, at the surface of the ground. Frond 20–60 cm. Blade ovate-triangular. Stipes and blades ascending to arching, in more or less the same plane. *Japan, Taiwan; ?China*. H1.

**57. MACROTHELYPTERIS** (Ito) Ching
*C.N. Page & F.M. Bennell*
Large terrestrial plants. Rhizomes short-creeping, thin, branching frequently, densely scale-clad, bearing fronds in close clusters. Fronds 3 × pinnate, all of 1 kind, evergreen. Stipe up to twice as long as the blade. Sori mostly marginal, small, rounded. Indusia absent.

A genus of 1 or more species from SE Asia, Polynesia, Australasia and tropical America. They are short-lived and fast-growing, reaching a large size very rapidly. Warm conditions, bright light and well-drained, rocky compost are required. Propagation by spores is very rapid.

**1. M. torresiana** (Gaudin) Ching. Figure 16(6), p. 42.
Fronds to 2 m, steeply ascending, broadly ovate-lanceolate, finely divided, thin, soft, pale green. Stipe whitish, to 1 m. *Mascarene Islands & SE Asia to Japan & Polynesia*. G2.

**58. AMPELOPTERIS** Kunze
*C.N. Page & F.M. Bennell*
Large, arching or sprawling terrestrial plants. Rhizomes creeping. Fronds all of 1 kind, 1 m or more, 1–2 × pinnate, of indefinite growth and producing smaller fronds or clusters of fronds from the axils of the pinnae; these grow into long secondary fronds while still attached. Pinnae firm, hairless, dark green. Fertile fronds scarce; sori elongate along the veins, lacking indusia.

A single species from the tropics and subtropics of the Old World. It requires moist, warm, sunny conditions, spreads extensively, and is propagated by division.

**1. A. prolifera** (Retzius) Copeland. Figure 16(7), p. 42.
*Old World tropics and subtropics*. G1.

**59. CHRISTELLA** Léveille
*C.N. Page & F.M. Bennell*
Medium-sized terrestrial plants. Fronds all of 1 kind. Blades 1 × pinnate, the pinnae pinnately lobed, long, linear-elliptic, each tapering to a narrow, pointed apex. Sori in small rounded groups. Indusia absent.

A small but widespread genus. The 1 cultivated species is noted for its rapid growth and softly textured fronds.

**1. C. dentata** (Forsskal) Brownsey & Jermy (*Cyclosorus dentatus* (Forsskal) Ching). Figure 16(8), p. 42.
Rhizome short-creeping, scaly. Fronds 50–100 × 9–19 cm, forming vigorous ascending or arching tufts, elliptic, widest near the middle, pinnae regularly decreasing in length towards the base of the frond, pale yellow-green, softly downy. *Tropics and subtropics of the Old World*. H5–G1.

A vigorously growing species, readily propagated from spores, often establishing spontaneously in pots of other ferns.

**60. ASPLENIUM** Linnaeus
*C.N. Page & F.M. Bennell*
Small to large plants, terrestrial, growing on rock, or epiphytic. Fronds mostly all of 1 kind, evergreen. Blades varying from almost simple to many times pinnately divided. Fertile fronds with narrow, linear, elongate, interrupted sori covered by a linear indusium which opens towards the midrib.

A genus of 650 or more species, of world-wide distribution. The cultivated species differ widely in their requirements, but almost all grow best in bright light and a freely draining compost in which the crowns are kept dry. Most species are readily propagated by spores; those bearing plantlets are easily propagated.

1a. Fronds simple, often variously lobed 2
 b. Fronds divided 4
2a. Fronds palmately or pedately lobed **2. hemionitis**
 b. Fronds not palmately or pedately lobed 3
3a. Fronds rooting at their tips **3. attenuatum**
 b. Fronds not rooting at their tips **1. nidus**

4a. Fronds 1 × pinnate 5
 b. Fronds 2–4 × pinnate 10
5a. Mature fronds more than 10 cm wide **4. lucidum**
 b. Mature fronds less than 10 cm wide 6
6a. Fronds rooting at their tips 7
 b. Fronds not rooting at their tips 8
7a. Fronds pinnately divided throughout **15. flabellifolium**
 b. Fronds pinnately divided near the base only, with a long, undivided terminal segment **3. attenuatum**
8a. Blade thick and fleshy **5. marinum**
 b. Blade thin, not fleshy 9
9a. Stipe and rachis black **9. trichomanes**
 b. Stipe and rachis green **10. viride**
10a. Fronds rooting at their tips **6. prolongatum**
 b. Fronds not rooting at their tips 11
11a. Mature fronds bearing small plantlets 12
 b. Mature fronds not bearing small plantlets 13
12a. Mature fronds 2–3 × pinnate, moderately finely divided **7. bulbiferum**
 b. Mature fronds at least 3 × pinnate, very finely divided **8. viviparum**
13a. Fronds triangular or ovate-triangular 14
 b. Fronds lanceolate or ovate-lanceolate 15
14a. Mature fronds mostly 15 cm or less; stipe green **11. ruta-muraria**
 b. Mature fronds usually 30 cm or more; stipe black **12. adiantum-nigrum**
15a. Mature fronds mostly less than 15 cm, of soft texture; stipe and rachis green, except at the extreme base **13. fontanum**
 b. Mature fronds mostly 15–60 cm, of leathery texture; stipe and rachis black throughout most of their lengths **14. aethiopicum***

**1. A. nidus** Linnaeus. Figure 17(1), p. 44.
Rhizomes short, broad, erect, bearing fronds in a shuttlecock-like rosette. Fronds with ascending, stout, almost black stipes. Blades to 150 × 20 cm, ovate-lanceolate, narrowing gradually towards the base, simple, entire, long-persistent, thin, slightly leathery, yellow-green, shining. Fertile fronds with long sori along every vein in the upper part of the frond. *Old World tropics*. G1.

Very frequent in cultivation, especially as a house-plant. Slow-growing and long-lived, reproducing by spores.

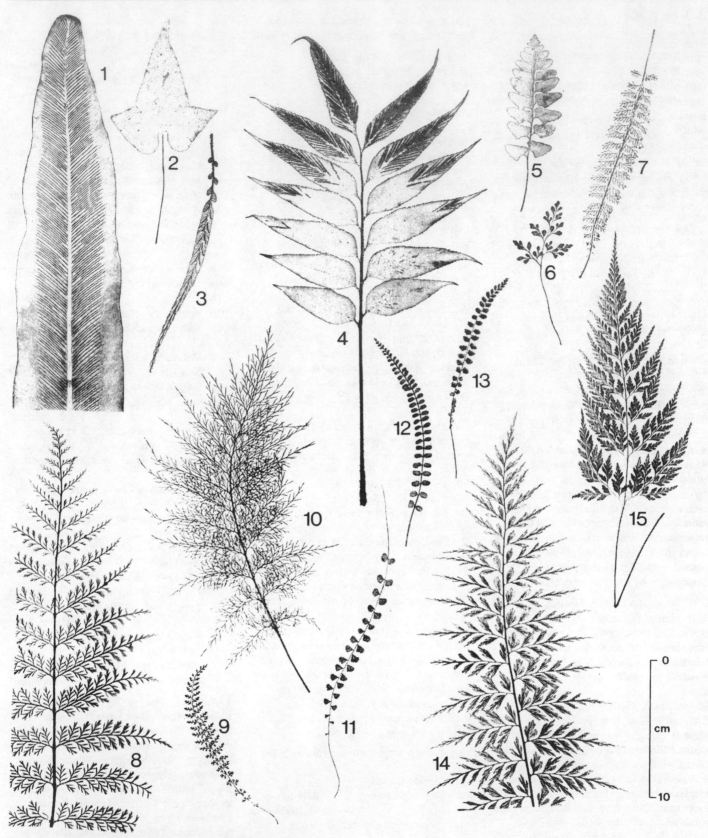

**Figure 17.** Silhouettes of fern-fronds. 1, *Asplenium nidus*.
2, *A. hemionitis*. 3, *A. attenuatum*. 4, *A. lucidum*. 5, *A. marinum*.
6, *A. ruta-muraria*. 7, *A. prolongatum*. 8, *A. bulbiferum*.
9, *A. fontanum*. 10, *A. viviparum*. 11, *A. flabellifolium*.
12, *A. trichomanes*. 13, *A. viride*. 14, *A. aethiopicum*.
15, *A. adiantum-nigrum*.

**2. A. hemionitis** Linnaeus. Figure 17(2), p. 44.

Rhizomes small, creeping or ascending, dark, bearing fronds in sparse groups. Fronds with ascending stipes which are dark purplish brown. Blades ascending or spreading, to 35 cm, simple, palmately lobed into 3-, 5- or 7-pointed segments of which the central is the longest, auricled at the junction with the stipe, slightly leathery, green, shining. Sori long, linear, often running the full length of a vein. *Atlantic islands, N Africa, Portugal.* H5–G1.

Very slow growing, slow to establish, requiring high humidity. Propagated readily by spores.

**3. A. attenuatum** R. Brown. Figure 17(3), p. 44.

Rhizomes small, inconspicuous, ascending, bearing fronds in closely grouped clusters. Fronds with short stipes and arching or spreading, leathery, dull dark green blades to 35 cm, simple or irregularly lobed in the basal third, with an acute, narrowly attenuate apex which bears plantlets at the tip. *Eastern Australia.* G1.

Plantlets are produced in abundance from the frond tip and the successive plants may remain linked for long periods.

**4. A. lucidum** Forster. Figure 17(4), p. 44. Rhizomes short-creeping or ascending, bearing fronds in small groups. Fronds with ascending or arching stipes which are brownish below, and spreading, arching or drooping blades which are 20–100 cm, broadly lanceolate, 1 × pinnate into round-based, pointed-tipped, finely toothed pinnae, soft but firm, dull green beneath, glossy green above, leathery, with a green rachis. *New Zealand.* H5–G1.

**5. A. marinum** Linnaeus. Figure 17(5), p. 44.

Rhizome short-creeping, dark, scaly, bearing fronds in closely grouped clusters. Fronds with short, arching, purple-brown stipes and spreading or ascending blades 15–40 cm, lanceolate or narrowly triangular, incompletely divided into lobed, oblong pinnae, thick, fleshy, stiff, shining glossy green above, paler green beneath. *Coasts of Europe, Atlantic islands.* G1.

A slow-growing and frost-sensitive species. Although it occurs naturally on rocks and walls exposed to sea spray, it survives without salt in cultivation.

**6. A. prolongatum** Hooker. Figure 17(7), p. 44.

Rhizomes short, almost erect, bearing fronds in rosette-like clusters. Fronds with arching or spreading stipes and spreading blades which are 10–35 cm, linear or narrowly lanceolate, 2 × pinnate, the pinnules 5–11 per pinna, distant, narrowly linear or a few forked with rounded apices, each with a single, linear sorus when fertile. Rachis of frond extended at the apex into a short, leafless section, the tip rooting and giving rise to new plants. *Himalaya to Taiwan & Japan.* H4.

**7. A. bulbiferum** Forster. Figure 17(8), p. 44.

Rhizomes short, ascending, bearing fronds in grouped clusters. Fronds with ascending stipes and ascending or arching, thin, soft, dark green blades which are 20–150 cm, ovate, 2–3 × pinnate, the pinnae producing plantlets towards their tips. *Australia, New Zealand.* H5–G1.

Plants produce bulbils very freely, sometimes over much of their upper surfaces. The bulbils readily produce new plants.

**8. A. viviparum** Linnaeus. Figure 17(10), p. 44.

Rhizomes short, stocky, creeping or ascending, bearing fronds in clusters. Fronds with ascending stipes and ascending or arching, somewhat delicate, green blades which are triangular-ovate, to 35 cm, 3 × pinnate into extremely narrow, linear, finely dissected segments, which each bear a single sorus when fertile. The pinnae bear large numbers of small bulbils on the upper surface, especially towards their tips. *Mascarene Islands.* G1.

A markedly lime-loving species, thriving best under still, humid conditions. Young plants have less finely divided fronds.

**9. A. trichomanes** Linnaeus. Figure 17(12), p. 44.

Rhizomes small, dark, creeping or ascending, bearing fronds in dense, often sinuous clusters. Fronds with spreading or ascending stipes which are blackish brown and shining, and spreading or ascending, arching blades 4–20 cm or more, very narrowly linear, 1 × pinnate into numerous, regularly arranged, rounded or squared, slightly stalked segments, leathery, slightly shining above, yellow to mid green. *N & S temperate regions.* H4.

Most specimens are strongly lime-loving and are often found in the mortar of walls.

**10. A. viride** Hudson. Figure 17(13), p. 44.

Rhizomes small, creeping or ascending, bearing fronds in clusters. Fronds with spreading or ascending stipes which are green except at the extreme base, and spreading, arching or descending blades 5–15 cm or more, very narrowly linear, 1 × pinnate into numerous, regularly arranged, rounded or fan-shaped, slightly stalked segments. *N temperate regions.* H1.

Another strongly lime-loving species.

**11. A. ruta-muraria** Linnaeus. Figure 17(6), p. 44.

Rhizome creeping, very small, bearing fronds in small groups. Fronds with ascending to erect green stipes and ascending to spreading, slightly leathery, dark green blades 3–15 cm, ovate-triangular, irregularly 2 × pinnate into a few, distant pinnae with a few (2–5) fan-shaped or rhombic segments. *N temperate regions.* H2.

Strongly lime-loving.

**12. A. adiantum-nigrum** Linnaeus. Figure 17(15), p. 44.

Rhizomes short, ascending, dark-coloured, bearing fronds in sparse groups. Fronds with strongly ascending stipes and spreading or ascending, leathery, dark green blades 10–50 cm, the stipe as long as or longer than the blade, blade triangular or ovate-triangular, with triangular-ovate pinnae subdivided into bluntly tapering pinnules. *Temperate regions of the Old World.* H3.

A very variable species throughout its range.

**13. A. fontanum** (Linnaeus) Bernhardi. Figure 17(9), p. 44.

Rhizomes very small, short-creeping, bearing fronds in clusters. Fronds with arching to spreading stipes which are green, becoming dark only at the extreme base, and spreading blades to 25 cm, lanceolate to narrowly elliptic, 2 × pinnate with short, oblong or slightly tapering pinnae, soft, delicate, light green. *S & C Europe.* H5–G1.

**14. A. aethiopicum** (Burman) Becherer. Figure 17(14), p. 44.

Rhizomes short-creeping, scaly, bearing fronds in closely grouped clusters. Fronds with short stipes and long, arching blades 30–60 cm, linear-lanceolate, 2–3 × pinnate, pinnae narrowly triangular and deeply divided into many, long, tapering segments, leathery, dark green, somewhat scaly, the segments folding together during drought. Stipe and rachis dark throughout most of the length of the frond. *Africa and adjacent islands.* H5–G1.

*****A. falcatum** Lamarck. Similar, differing in its more flaccid, hanging fronds with less divided pinnae. *SE Asia, Australasia.*

**15. A. flabellifolium** Cavanilles.
Figure 17(11), p. 44.
Rhizomes short-creeping, small, bearing fronds in sparse groups. Fronds with arching stipes and arching, spreading or prostrate blades 15–30 cm, narrowly lanceolate, tapering towards the tip and often terminating in a long, thin, undivided stolon-like section which bears plantlets at the tip, the lower half pinnately divided into 2, widely spaced, fan-shaped pinnae, soft, delicate, pale green. *Australia, New Zealand.* H5–G1.

### 61. CAMPTOSORUS Link
*C.N. Page & F.M. Bennell*
Small- to medium-sized plants. Fronds all of one kind, blade simple, lanceolate or linear, tapering to a long-attenuate apex, rooting at the tip, heart-shaped or arrowhead-shaped at the base. Sori linear, irregularly scattered.

A genus of 2 species, 1 of which is occasionally seen in cultivation, when it is usually short-lived.

**1. C. rhizophyllus** (Linnaeus) Link.
Figure 18(1), p. 47.
Rhizomes short, ascending to erect. Fronds 15–30 cm, spreading or arching, slightly leathery, each tip producing a single new plantlet where it touches the ground. *E USA.* H4.

Exacting in its growth requirements, needing lime and drought-free conditions; subject to slug damage.

### 62. PHYLLITIS Ludwig
*C.N. Page & F.M. Bennell*
Small- to medium-sized terrestrial plants. Rhizome ascending, sparsely scale-clad, bearing fronds in clusters. Fronds simple, all of 1 kind, evergreen, linear-lanceolate, auricled and sometimes somewhat lobed at the base. Stipe one-eighth to one-third of the length of the frond. Blade thinly leathery. Sori oblique, narrowly linear, along the veins, arranged as twinned pairs. Indusium linear, attached along the outer side of the sorus, those of each pair opening towards each other.

A genus of 3 or 4 species from temperate regions of the Old and New Worlds. They are long-lived and fast-growing, requiring shade, shelter, relatively moist, well-drained compost and lime. Propagation is by spores, or, in the case of a few cultivars, by leaf bases (Kaye, R., Leaf-base propagation of Phyllitis scolopendrium, *British Fern Gazette* **9:** 120–1, 1963).

1a. Fronds simple, linear
         **1. scolopendrium**
  b. Fronds with projecting lobes at the base     **2. sagittata**

**1. P. scolopendrium** (Linnaeus) Newman.
Figure 18(2), p. 47.
Fronds to 60 cm or more, stipe up to half the length of the frond, usually less. Blade linear-lanceolate, heart-shaped at the base, margins slightly wavy. *S, W & C Europe.* H3.

Many cultivars exist, mainly monstrous forms with contorted, forked or curled fronds.

**2. P. sagittata** (de Candolle) Guinea & Heywood (*P. hemionitis* Kuntze).
Figure 18(3), p. 47.
Fronds to 30 cm. Stipe as long as the blade or less. Blade oblong-lanceolate, deeply heart-shaped and auricled at the base, the auricles rounded or triangular and tapering, to 3–4 cm, giving the frond a 3-lobed appearance. *Mediterranean region.* H5–G1.

### 63. CETERACH Lamarck & de Candolle
*C.N. Page & F.M. Bennell*
Mostly small plants which grow on rocks. Fronds all of 1 kind, evergreen. Blades thick, slightly fleshy, pinnately lobed, most densely scale-clad beneath, the pinnae rolling inwards during dry spells. Sori linear along the veins, protected by a single indusial flap often hidden below the scales.

A genus of about 10 species from Eurasia and Africa. They require light, open, well-drained situations and are somewhat exacting subjects in cultivation, intolerant of stagnant conditions.

1a. Fronds sparsely scale-clad beneath
         **1. dalhousiae**
  b. Fronds densely scale-clad beneath   2
2a. Mature fronds less than 3 cm broad
         **2. officinarum**
  b. Mature fronds mostly 3–6 cm broad
         **3. aureum**

**1. C. dalhousiae** (Hooker) Christensen.
Figure 18(4), p. 47.
Rhizome short, erect. Fronds 8–12 cm, lanceolate, pinnatifid, spreading, sparsely scale-clad beneath. *Himalaya.* H5–G1.

**2. C. officinarum** de Candolle. Figure 18(5), p. 47.
Rhizome short, erect. Fronds 8–15 cm, mostly less than 3 cm broad, lanceolate, pinnatifid, spreading, densely scale-clad beneath. *Eurasia.* H3.

**3. C. aureum** (Cavanilles) Linnaeus.
Figure 18(6), p. 47.
Rhizome short, erect. Fronds 10–20 cm or more, mostly 3–6 cm broad, broadly

lanceolate, pinnatifid, ascending, densely scale-clad beneath. *Madeira, Canary Islands.* H5–G1.

### 64. MATTEUCIA Todaro
*C.N. Page & F.M. Bennell*
Medium to large terrestrial plants. Rhizome thick, scale-clad, more or less erect to erect, bearing fronds in regular, shuttlecock-like clusters. Fronds 2 × pinnate, of 2 kinds, the vegetative spreading, flexible, with very short stipes and broad, soft-textured, light green blades, the fertile erect, with stipe about equalling the blade, pinnae reduced, dark brown at maturity, stiff, leathery. Sori continuous in a single or double row on each contracted pinna, protected by a tighly inrolled leaf margin forming a spherical structure.

A genus of 2 or more species in Europe, E Asia and eastern N America. They are long-lived and moderately fast-growing, cultivated for their bold, ostrich-feather-like, large bright green fronds which expand early in spring, making them particularly showy plants for a shady corner where they will grow and spread in any good, moist, garden soil. Propagation is by division or spores sown immediately on collection.

**1. M. struthiopteris** (Linnaeus) Todaro.
Figure 18(7), p. 47.
Rhizome to 10 cm in diameter, erect, hairy. Vegetative fronds ovate- to obovate-lanceolate, arising first in spring and then dying rather suddenly in autumn. Fertile fronds narrowly lanceolate, arising in summer and persisting over winter, shedding spores the following spring. *Europe.* H2.

### 65. ONOCLEA Linnaeus
*C.N. Page & F.M. Bennell*
Medium-sized terrestrial and marginal aquatic plants. Rhizome long-creeping, subterranean, bearing fronds individually. Fronds deciduous, the stipe about equalling the blade; of 2 kinds, the vegetative erect to arching, 1 × deeply pinnatisect with broad blade and long pinnae, the lower lobes themselves often toothed, thin-textured, hairless, with conspicuously anastomosing venation and winged stipe, the whole frond dying in winter; the fertile erect, 1 × pinnate with blade greatly reduced and the contracted pinnae deeply divided into hard, tightly inrolled lobes, dark brown at maturity, stiff and leathery. Sori in compact groups on reduced pinnae, protected by the inrolled leaf margins,

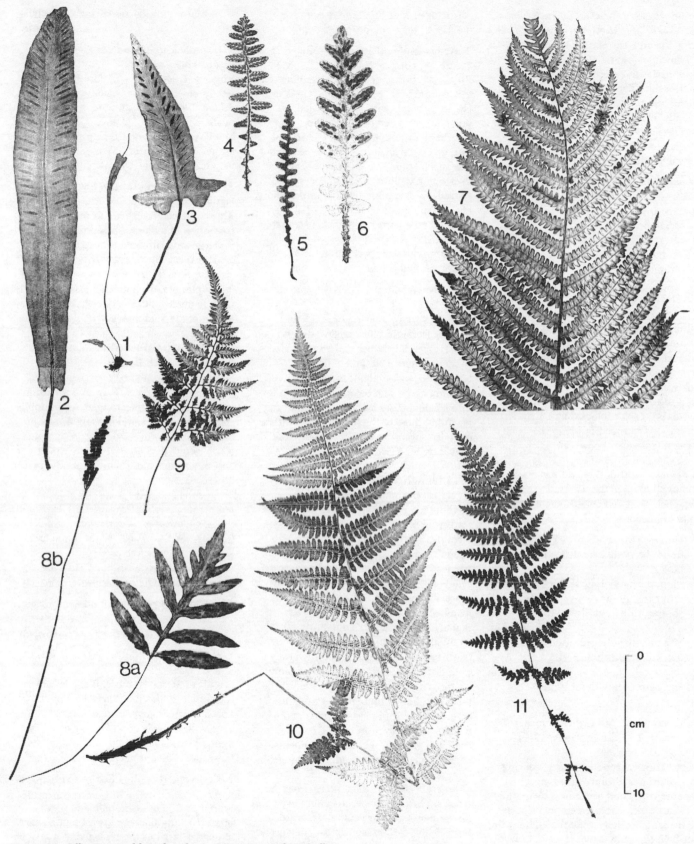

**Figure 18.** Silhouettes of fern-fronds. 1, *Camptosorus rhizophyllus*.
2. *Phyllitis scolopendrium*. 3, *P. sagittata*. 4, *Ceterach dalhousiae*.
5, *C. officinarum*. 6, *C. aureum*. 7, *Matteucia struthiopteris*.
8, *Onoclea sensibilis* (a, vegetative frond; b, fertile frond).
9, *Athyrium nipponicum*. 10, *A. filix-femina*. 11, *A. distentifolium*.

forming spherical structures. The fertile frond arises later than the vegetative and persists through the winter, shedding spores the following spring.

One species in eastern N America and E Asia. It is long-lived and fast-growing, thriving best in wet, boggy situations, especially by standing water; it has become naturalised in such habitats in Europe. Propagation is by division or by spores sown immediately on collection.

**1. O. sensibilis** Linnaeus. Figure 18(8), p. 47.
*Eastern N America, E Asia, naturalised in parts of Europe*. H2.

A few slightly monstrous forms, with more prominently lobed pinnae, are known in cultivation.

**66. ATHYRIUM** Roth
*C.N. Page & F.M. Bennell*
Medium to large terrestrial plants. Rhizomes short-creeping or ascending, branched, thick, scaly. Fronds all of 1 kind, winter-deciduous, pinnately divided into small, well-spaced segments. Fertile fronds with numerous, small, usually J-shaped, discrete sori, each protected by a membranous rounded or J-shaped indusium (ephemeral or absent in *A. distentifolium*).

A genus of 500-600 or more species, found in all parts of the world. They require moist, cool, sheltered conditions, and grow readily from spores.

1a. Longest pinnae near the middle of the frond, becoming smaller above and below                                          2
  b. Pinnae equally long throughout the lower half of the blade, becoming shorter only towards the top
                                          **1. nipponicum**
2a. Sori mostly J-shaped, each covered by a membranous indusium
                                          **2. filix-femina**
  b. Sori rounded, without indusia
                                          **3. distentifolium**

**1. A. nipponicum** (Mettenius) Hance (*A. goeringianum* Kunze). Figure 18(9), p. 47.
Rhizome short, ascending, hairy, bearing fronds in groups. Fronds with long, ascending stipes and ascending to arching blades 20–60 cm, soft, green or purplish, broadly ovate, equally wide throughout the lower half, tapering above, 2–3 × pinnate into small, stalked, oblong, pinnatisect pinnules, those of fertile fronds with several J-shaped or kidney-shaped sori with broad, brown indusia. *China, Japan, Taiwan*. H3.

'Pictum', with the frond delicately marked in pink and purple, is popular.

**2. A. filix-femina** (Linnaeus) Roth. Figure 18(10), p. 47.
Rhizomes short-creeping or ascending, freely branching, bearing fronds in dense clusters. Fronds with short, ascending, green or rarely red or purple stipes and ascending or arching blades to 150 cm, soft, green, linear-lanceolate, widest about the middle, tapering above and below, 2–3 × pinnate into small, stalkless, well-spaced, oblong, pinnatisect pinnules, those of the fertile fronds with mostly J-shaped sori and membranous indusia. *N temperate regions*. H2.

A very variable species in terms of dissection and colour.

**3. A. distentifolium** Tausch & Opitz. Figure 18(11), p. 47.
Rhizomes short-creeping or ascending, bearing fronds in clusters. Fronds with short, ascending stipes and arching blades 30 cm or more, soft, green, linear-lanceolate, widest about the middle, tapering above and below, 2–3 × pinnate into small, stalkless, oblong, pinnatisect pinnules, those of the fertile fronds with rounded sori, usually lacking indusia at maturity. *N temperate regions*. H1.

**67. DIPLAZIUM** Swartz & Schrader
*C.N. Page & F.M. Bennell*
Medium to large or almost tree-like terrestrial plants. Rhizome erect, trunk-forming, bearing fronds in clustered, irregular crowns. Fronds more or less hairless, 1–2 × pinnate into large, coarse segments, the degree of dissection often markedly increasing with the size of the frond. Fronds all of 1 kind, evergreen, usually more or less triangular or triangular-ovate, firm, more or less leathery, often glossy and slightly fleshy with distinct, linear, widely spread pinnules up to 2 cm wide and often auricled at the base. Sori linear, more or less straight, single to double in different parts of the frond, in radiating rows along either side of the main vein of the ultimate segments, each with a firm linear indusium hinged along one side.

A genus of about 350 species from the tropics. They are moderately long-lived and fast-growing, and are cultivated for their large, leafy, dark green fronds. They grow well in moist, well-drained, humus-rich compost in moderately strong indirect light. Propagation is by spores or bulbils.

1a. Mature fronds bearing numerous bulbils                          **1. proliferum**
  b. Mature fronds mostly lacking bulbils
                                          **2. esculentum**

**1. D. proliferum** (Lamarck) Kaulfuss. Figure 19(1), p. 49.
Rhizome erect, to 60 cm, forming a small tree in old specimens. Fronds to *c.* 80 × 30 cm. Blade 1–2 × pinnate, triangular-ovate, bright green, glossy, with numerous bulbils present along the upper side of the rachis. Stipe one-third to one-half of the length of the blade. *Old World tropics*. G2.

**2. D. esculentum** (Retzius) Swartz. Figure 19(2), p. 49.
Rhizome erect, to 60 cm, forming a dwarf tree when old. Plants spreading by underground runners. Fronds to 150 × 50 cm. Blade ovate-triangular when small to broadly triangular when large, 1–2 × pinnate, deep green, glossy. Bulbils usually absent. Stipe *c.* one-third of the length of the blade. *India to Polynesia*. G2.

**68. GYMNOCARPIUM** Newman
*C.N. Page & F.M. Bennell*
Small- to medium-sized terrestrial plants. Rhizomes long-creeping, thin, branching, wiry, shallowly subterranean, forming spreading colonies and bearing fronds individually. Fronds 2 or more × pinnate, all of 1 kind, deciduous, broadly triangular. Sori almost marginal, small, round, indusia absent.

A genus of at least 2 species from north temperate regions. They are long-lived and moderately fast-growing, requiring a very freely drained compost, slight to moderate shade and a very free root-run. *G. robertianum* requires lime. They are propagated rapidly by spores.

1a. Blade grey-green with a mealy surface
                                          **1. robertianum**
  b. Blade green, not mealy     **2. dryopteris**

**1. G. robertianum** Ching. Figure 19(3), p. 49.
Fronds to 30 cm or more. Blade broadly triangular, without a markedly tripartite appearance. Blade and stipe covered all over with a mealy, dust-like powder. The lightly bruised fronds have a balsam-like fragrance. *Circumboreal*. H2.

**2. G. dryopteris** Ching. Figure 19(4), p. 49.
Fronds to 20 cm or more. Blade very broadly triangular, the lowest pinna pair almost as large as the rest, giving it an almost tripartite appearance. Blade and stipe without mealy, dust-like powder.

**Figure 19.** Silhouettes of fern-fronds. 1, *Diplazium proliferum*.
2, *D. esculentum*. 3, *Gymnocarpium robertianum*. 4, *G. dryopteris*.
5, *Cystopteris dickeana*. 6, *C. fragilis*. 7, *Lunathyrium japonicum*.
8, *Lastreopsis decomposita*. 9, *L. hispida*.

Bruised fronds without any fragrance. *Circumboreal.* H2.

**69. CYSTOPTERIS** Bernhardi
*C.N. Page & F.M. Bennell*
Small terrestrial plants. Rhizomes short, much branched, creeping, bearing fronds in small groups. Fronds all of the same kind, winter-deciduous. Sori in small rounded groups, each with a delicate, membranous indusium.

A genus of 10 or more poorly distinguished species, cosmopolitan but mainly in temperate regions. They are fast-growing and have thin, brittle fronds which unfurl early in spring, thriving best in a cool, moist, permanently shaded situation in a lime-rich soil. Readily propagated from spores.
Literature: Green, F. & P., Cystopteris alpina and C. dickeana in the rock garden, *British Fern Gazette* 9: 85 (1962).

1a. Fronds spreading, pinnae mostly
    overlapping     **1. dickeana**
  b. Fronds usually ascending, pinnae
    mostly not overlapping   **2. fragilis***

**1. C. dickeana** Sim. Figure 19(5), p. 49.
Rhizome very small, much branched, eventually giving rise to a small cushion of many separate crowns. Fronds 6–12 cm, ovate, spreading, with mostly entire, broad, obtuse, overlapping, scalloped pinnae. Stipes short. *Europe.* H1.
Plants spread rapidly by vegetative means.

**2. C. fragilis** (Linnaeus) Bernhardi.
Figure 19(6), p. 49.
Rhizome 4–5 mm in diameter, short, branched. Fronds 5–45 cm, tufted, ascending or arching. Blade ovate or ovate-triangular, 2–3 × pinnate, pinnae narrow, acute, seldom overlapping. *N & S temperate regions.* H1.
Plants from different origins show much variation in frond form, especially in size and the degree of blade dissection.
  ***C. regia** Desvaux. Very similar and not always sharply distinguishable from *C. fragilis*, but with more finely dissected foliage. *Europe.*
Other, more extreme forms, including *C. alpina*, of European origin but uncertain status, occur occasionally.

**70. LUNATHYRIUM** Koidzumi
*C.N. Page & F.M. Bennell*
Small terrestrial plants. Rhizome short-creeping, bearing fronds individually or in irregular clusters. Fronds 1–2 x pinnatisect,

all of 1 kind. Sori linear, straight or slightly curved, single, in spreading rows along either side of the pinna midvein, each with an entire, simple indusium. Stipe short, rachis and lamina bearing long hairs.

Probably more than 30 species from E Asia to Australia and the Pacific. They are short-lived and fast-growing, succeeding well in damp, rocky places in fairly deep shade, in moist acidic compost. They are propagated readily from spores.

**1. L. japonicum** (Thunberg) Kurata (*Diplazium japonicum* (Thunberg) Makino; *Asplenium japonicum* Thunberg; *L. petersenii* (Kunze) Christ). Figure 19(7), p. 49.
Fronds to *c.* 30 cm. Blade lanceolate, deeply 2 × pinnatisect to almost 2 × pinnate, soft. *Japan, SE Asia, Australia, Polynesia.* H5–G1.

**71. LASTREOPSIS** Ching
*C.N. Page & F.M. Bennell*
Medium-sized terrestrial or rock-growing plants. Rhizome long- or short-creeping, thin, branching frequently, scale-clad, bearing fronds individually. Fronds 3 × or more pinnately divided, all of 1 kind, evergreen. Blades broadly triangular, spreading to horizontal; stipes mostly erect. Sori circular, median, on the ultimate divisions of the frond. Indusia kidney-shaped.

A genus of about 30 species from Asia, S America and Africa, with a concentration of species in Australia. They grow well in lightly shaded, open ground in moist well-drained soils. Propagation is by spores or by division.

1a. Stipes with long, stiff, bristle-like hairs
           **2. hispida**
  b. Stipes without such hairs
          **1. decomposita**

**1. L. decomposita** (R. Brown) Tindale.
Figure 19(8), p. 49.
Rhizome fairly short-creeping. Fronds to 90 cm, often fairly crowded. Blade 3 × pinnate, apices of ultimate segments acuminate, harsh, dull green to green. Stipe *c.* half the length of the blade, bearing a few brown scales towards the base. *E Australia (Queensland to New South Wales).* G1.

**2. L. hispida** (Swartz) Tindale. Figure 19(9), p. 49.
Rhizome long-creeping. Fronds to 100 cm. Blades 3 × or more very finely divided, the ultimate segments narrow and acute, thin but harsh, dark green. Stipes half or more of the length of the frond, densely clad with long, reddish brown, stiff, perpendicular, bristle-like hairs. *E Australia (including Tasmania), New Zealand.* H5–G1.

**72. TECTARIA** Cavanilles
*C.N. Page & F.M. Bennell*
Medium to large terrestrial plants. Rhizome short-creeping or ascending to erect, thick, fairly densely covered with large scales, bearing fronds in clusters. Fronds 1 × pinnately divided, coarse, all of 1 kind, evergreen, with pinnae typically large and entire. Blade broad, often triangular or linear, thickly papery, with clearly visible anastomosing veins, in some species with bulbils on the upper surface. Stipe often equal in length to the blade or shorter. Sori round, scattered over the back of the frond. Indusia round to kidney-shaped, often shed.

A genus of over 200 species, common throughout the wet tropics. They are moderately long-lived and fairly slow-growing, enjoying moderate light, high humidity and warmth in a permanently moist, peaty compost. Propagation is by spores or bulbils.

1a. Edges of blade extending downwards
    to form a broad wing to the upper
    part of the stipe   **1. decurrens**
  b. Stipe not winged       2
2a. Fronds varying from almost entire to
    deeply pinnate or pinnatisect into a
    few, acutely lobed segments  **2. incisa**
  b. Fronds pinnately divided at least in
    the lower part into several separated
    pinnae, ultimate segments with
    rounded lobes        3
3a. Pinnae mostly shallowly and
    irregularly lobed   **3. cicutaria**
  b. Pinnae deeply and regularly lobed
          **4. heracleifolia**

**1. T. decurrens** Copeland. Figure 20(1), p. 51.
Fronds to 60 cm, or more. Blade broadly triangular, pinnate or pinnatisect into a small number of acutely lobed segments, with the edges extending downwards to form a broad green wing to the upper part of the stipe. Sori deeply sunk. *Polynesia.* G2.

**2. T. incisa** Cavanilles. Figure 20(4), p. 51.
Fronds to 1 m or more. Blade broadly triangular, deeply pinnate or pinnatisect into a small number of acutely lobed, nearly entire, closely spaced segments, not decurrent on the stipe. Fronds with a small number of bulbils along the upper surface of the rachis (these sometimes absent). *Mexico & West Indies to Brazil.* G2.

**3. T. cicutaria** Copeland. Figure 20(2), p. 51.
Fronds to 80 cm. Blade triangular,

**Figure 20.** Silhouettes of fern-fronds. 1, *Tectaria decurrens*.
2, *T. cicutaria*. 3, *T. heracleifolia*. 4, *T. incisa*. 5, *Quercifilix zeylanica*.
6, *Polystichum lonchitis*. 7, *P. munitum*. 8, *P. aculeatum*.
9, *Didymochlaena truncatula*.

becoming ovate-oblong, mostly 2 × pinnate, the pinnae with deeply and regularly lobed segments. Fronds bearing abundant bulbils on both the rachis and the pinna midribs. *West Indies.* G2.

**4. T. heracleifolia** (Willdenow) Underwood. Figure 20(3), p. 51.
Fronds to 80 cm or more. Blade triangular-ovate or ovate-oblong, 1 × pinnate, with a large, almost entire terminal segment, the pinnae shallowly and irregularly lobed, without bulbils. *West Indies, C America south to Peru.* G2.

### 73. QUERCIFILIX Copeland
*C.N. Page & F.M. Bennell*
Small terrestrial plants. Rhizome short-creeping or ascending, thin, scale-clad. Stipes densely downy with jointed hairs. Fronds of 2 kinds, the vegetative with a short stipe and a broad blade which is *c.* 5 × 2.5 cm, more or less simple and deeply lobed like an oak leaf or with one pair of opposite pinnae divided off at the base and themselves lobed, more or less widened at the base, giving a trifoliolate appearance to large fronds; fertile with the stipe longer than in the sterile, blade much reduced, up to 3 cm × 2 mm, with lateral pinnae. Sori linear along the veins of the fertile fronds.

One species in the area of the Indian ocean. Probably short-lived and fast-growing. Its small size and unusual oak-leaf-like appearance make it attractive. It succeeds in very well-drained sandy soils in bright or lightly shaded conditions. Propagation is by spores.

**1. Q. zeylanica** (Houttuyn) Copeland. Figure 20(5), p. 51.
*Mauritius, Sri Lanka, S China, Taiwan, S Malaya, Borneo.* G2.

### 74. DIDYMOCHLAENA Desvaux
*C.N. Page & F.M. Bennell*
Large terrestrial plants. Rhizome thick, massive, unbranched, densely clad with dark scales, more or less erect, bearing fronds in clusters. Fronds 2 × pinnate, all of 1 kind, evergreen, thick, leathery and lightly waxy above, glossy, dark green. Stipe *c.* one-third of the length of the blade. Sori marginal, rounded. Indusium wide, kidney-shaped, opening towards the edge of the frond.

A genus of 2–4 pantropical species with many locally varying forms. They are long-lived and relatively fast-growing, having gracefully arching, glossy fronds on upright scaly rhizomes like a miniature tree-fern.

They need constantly warm, moist conditions with a freely drained, peaty compost.

**1. D. truncatula** (Swartz) J. Smith. Figure 20(9), p. 51.
Fronds 50–150 cm, 2 × pinnately divided into elongate, tapering pinnae with thick, close, fan-shaped or parallelogram-shaped pinnules jointed to the pinna midribs and shed from them with age. Blade bronze–red when expanding, becoming green with age. *Tropics.* G2.

### 75. POLYSTICHUM Roth
*C.N. Page & F.M. Bennell*
Small, medium or large terrestrial plants. Rhizome thick, woody, branching occasionally, densely scale-clad, ascending to erect or trunk-forming, woody, bearing fronds in more or less regular, shuttlecock-like clusters. Fronds 1–2 × pinnate, occasionally further divided, all of 1 kind, evergreen, typically ovate-lanceolate to linear with the ultimate divisions always toothed and ending in soft or sharp spines, of variable texture. Sori round, median on the ultimate segments mainly in the upper part of the frond, covered by rounded, eccentrically peltate indusia. Sporangia black at maturity. A few species bear bulbils on the upper surface of the rachis.

A genus of almost 200 species, occurring throughout the world, particularly rich in E Asia. They are long-lived and moderately fast-growing. They grow best in moist woodland, particularly where well-drained, and may benefit from the presence of lime. Most require shelter and shade. Propagation is by spores or bulbils. All the species are very variable, and those that have been in cultivation for a long period have given rise to many cultivars. Literature: Dyce, J.W., Variation in Polystichum in the British Isles, *British Fern Gazette* 9: 97–109 (1963).

1a. Fronds 1 × pinnate                                    2
  b. Fronds 2–3 × pinnate                          3
2a. Fronds very slender, pinnae markedly shortening towards the base
                                                **1. lonchitis**
  b. Fronds fairly broad, pinnae not markedly shortening towards the base
                                                **2. munitum***
3a. Fronds 3 × pinnate, at least at the base                        **7. tripteron**
  b. Fronds not more than 2 × pinnate    4
4a. Pinnae extending nearly to the base of the frond, gradually shortening
                                                **3. aculeatum**

  b. Pinnae not extending nearly to the base of the frond, not gradually shortening                                    5
5a. Stipe and rachis densely covered with dark brown to black scales which have paler margins; rhizome trunk-forming                        **4. vestitum***
  b. Stipe and rachis sparsely covered with uniformly pale brown scales; rhizome not trunk-forming                            6
6a. Stipe one-quarter to one-third of the length of the more or less ovate blade
                                                **5. setiferum**
  b. Stipe usually *c.* half the length of the blade, which is triangular
                                                **6. tsus-simense**

**1. P. lonchitis** (Linnaeus) Roth. Figure 20(6), p. 51.
Rhizome ascending. Fronds 15–30 × 3–5 cm, occasionally larger. Blade 1 × pinnate, pinnae often extending nearly to the base of the frond and gradually shortening, leathery with stiffly spiny margins, dark green, glossy. Stipe very short, up to one-quarter of the length of the frond. *N temperate regions.* H1.

**2. P. munitum** (Kaulfuss) Presl. Figure 20(7), p. 51.
Rhizome ascending. Fronds 40–120 × 3–15 cm, rarely larger. Blade 1 × pinnate, pinnae not or very slightly shortening towards the base, leathery with spiny margins, dark green, glossy. Stipe usually *c.* one-quarter of the length of the frond. *Western N America.* H2.

*\*P. acrostichoides* (Michaux) Schott. Very similar, but fronds mostly 30–60 cm. *Eastern N America.*

**3. P. aculeatum** (Linnaeus) Schott. Figure 20(8), p. 51.
Rhizome ascending. Fronds ovate-lanceolate, to 60 × 15 cm or more. Blade 2 × pinnate, pinnae often extending nearly to the base of the frond and gradually shortening, leathery, harsh, with fairly stiffly spiny margins, dark green, glossy. Stipe short, usually less than one-fifth of the length of the frond. *N temperate Eurasia.* H2.

**4. P. vestitum** (Forster) Presl. Figure 21(1), p. 53.
Rhizome erect, trunk-forming, to 1.5 m in old specimens. Fronds 30–100 × 10–30 cm, more or less ovate, 2 × pinnate, pinnae of approximately equal length throughout the lower half of the blade or only very slightly shortening, very harsh, leathery, prickly with spiny margins, dark green, dull. Stipe one-quarter or more of

**Figure 21.** Silhouettes of fern-fronds. 1, *Polystichum vestitum*.
2, *P. setiferum*. 3, *P. tsus-simense*. 4, *P. tripteron*. 5, *Cyrtomium*
*falcatum*. 6, *Arachniodes aristata*. 7, *Dryopteris fragrans*.
8, *D. submontana*. 9, *D. oreades*. 10, *D. affinis*. 11, *D. filix-mas*.

the length of the frond, it and the rachis densely covered with dark brown to black glossy scales which have paler margins. *New Zealand*. H4.

*P. proliferum (R. Brown) Presl. Very similar but rachis often with proliferous buds. *Temperate E Australia*.

5. P. setiferum (Forsskal) Woynar. Figure 21(2), p. 53.
Rhizome ascending. Fronds to 120 × 20 cm, more or less ovate, 2 × pinnate, pinnae of approximately equal length throughout the lower half of the blade or only very slightly shortening, very soft, flexible, with soft, finely spiny margins, pale yellow-green, dull. Stipe *c.* one-quarter to one-third of the length of the blade, with numerous, dense, chaffy pale brown, scales at the base, becoming more and more sparse along the rachis. *Atlantic islands, W Europe*. H3.

6. P. tsus-simense (Hooker) J.E. Smith (*P. mayebarae* Tagawa). Figure 21(3), p. 53.
Rhizome ascending. Frond more or less narrowly triangular, 40–60 × 8–15 cm. Blade 2 × pinnate, the longest pinnae at the base of the frond, rigid, stiff, rather leathery, with finely spiny margins, dark green. Stipe usually half the length of the blade, with chaffy pale brown scales at the base. *Warm-temperate Asia*. H4.

7. P. tripteron (Kunze) Presl. Figure 21(4), p. 53.
Rhizome ascending. Fronds to 60 cm. Blade usually 2 × pinnate, but 3 × pinnate at base, basically 3-lobed, the central lobe the longest, and with stalked pinnae, the paired lateral lobes formed from an enlarged pair of basal pinnae, *c.* one-third of the length of the central lobe and themselves pinnately divided, mid green, leathery. Stipe one-third to half of the length of the blade. *Japan and neighbouring islands*. H5–G1.

### 76. CYRTOMIUM Presl
*C.N. Page & F.M. Bennell*
Terrestrial plants. Rhizomes short, ascending to erect, densely scaly, bearing fronds in compact tufts. Fronds all of 1 kind, blades coarsely 1 × pinnate, ovate-lanceolate. Pinnae large, entire, linear to sickle-shaped, acuminate, leathery. Stipe one-quarter to half of the length of the blade. Sori discrete, scattered, rounded, each protected by a peltate indusium.

A widespread genus of about 20 species. They are relatively tough and tolerant of a wide range of conditions.

1. C. falcatum (Linnaeus) Presl. Figure 21(5), p. 53.

Rhizome short, ascending. Fronds to 1 m or more, pinnae large and broad with entire margins, glossy. *E Asia*. H5–G1.

*C. caryotideum (Wallich) Presl. Similar, but with drooping pinnae whose margins are conspicuously but minutely toothed. *E Asia*.

*C. fortunei J. Smith. Also similar but with smaller, narrower, less glossy fronds and smaller pinnae. *E Asia*.

### 77. ARACHNIODES Blume
*C.N. Page & F.M. Bennell*
Medium-sized terrestrial plants. Rhizomes long-creeping, becoming short-creeping or nearly erect in places, clothed with persistent, brown scales. Fronds all of 1 kind, evergreen, blades pinnately divided, the ultimate segments with toothed or bristly margins, fertile fronds with discrete, large, circular sori, each protected by a leathery, peltate indusium, the sori arranged in regular rows along each segment.

A genus of about 50 species widely spread throughout E Asia, Australasia and Polynesia. The fronds resemble those of *Rumohra* (p. 59) but differ in their smaller size and arise from thinner, darker rhizomes. They require warm, moist conditions and succeed when given a free root run. Propagation is by spores or by division of the rhizome.

1. A. aristata (Forster) Tindale. Figure 21(6), p. 53.
Rhizome subterranean, bearing fronds individually; rhizome and frond bases covered with dark, spreading scales. Fronds 30–100 cm with ascending stipes and ascending or spreading blades, the stipes *c.* half the total length. Blade broadly triangular, the ultimate segments shortly stalked, asymmetric, ending in long, sharp points. Lamina leathery, rigid, harsh, glossy, dark green, prickly to the touch. *Japan to Australia and Polynesia*. H4.

Plants in cultivation as *A. standishii* (Moore) Ohwi appear to be *A. aristata*.

### 78. DRYOPTERIS Adanson
*C.N. Page & F.M. Bennell*
Medium to large terrestrial plants. Rhizome short-creeping to ascending or almost erect, mostly woody, often stout, scale-clad, bearing fronds mostly in clusters, in many species forming regular, shuttlecock-like crowns. Fronds 2–3 × pinnate, all of 1 kind or the fertile with very slightly reduced blades, evergreen or deciduous, mostly firm; stipes scaly, sometimes densely so. Sori median on the ultimate segments, often

abundant, sometimes large, round to kidney-shaped. Indusium kidney-shaped.

About 150 species, mostly in the northern hemisphere. They are long-lived and fairly fast-growing. Most are large and vigorous, very hardy plants that make good subjects for naturalisation and establish well. They grow in almost any well-drained soil and thrive best with shelter, shade and moderate humidity. They propagate well from spores and a few form suitable clumps for division.

| | | |
|---|---|---|
| 1a. | Fronds 1–2 × pinnate or pinnatisect | 2 |
| b. | Fronds 2–3 × pinnate or pinnatisect | 7 |
| 2a. | Stipe and rachis densely scale-clad throughout | 3 |
| b. | Stipe and rachis densely scale-clad only at base, the scales becoming sparse along the rachis | 4 |
| 3a. | Blade more or less densely scale-clad beneath | **1. fragrans** |
| b. | Blade with few or no scales | **3. affinis*** |
| 4a. | Upper surface of frond mealy | **2. submontana** |
| b. | Upper surface of frond not mealy | 5 |
| 5a. | Edges of pinnules more or less flat | 6 |
| b. | Edges of pinnules turning upwards | **4. oreodes** |
| 6a. | Fronds ovate-lanceolate, 30–120 cm | **5. filix-mas** |
| b. | Fronds linear to oblong-lanceolate, 15–60 cm | **6. cristata** |
| 7a. | Edges of pinnae turning upwards | **9. aemula** |
| b. | Edges of pinnae not turning upwards | 8 |
| 8a. | Edges of pinnae mostly turning distinctly downwards | **10. dilatata** |
| b. | Edges of pinnae more or less flat | 9 |
| 9a. | Young fronds coloured coppery-pink | **11. erythrosora** |
| b. | Young fronds green | 10 |
| 10a | Young fronds in bud with whitish or very pale tan-coloured scales | **7. carthusiana** |
| b. | Young fronds in bud with distinctly ginger-coloured scales | **8. expansa** |

1. D. fragrans (Linnaeus) Schott. Figure 21(7), p. 53.
Rhizome spreading to ascending, much branched, covered with numerous curled and shrivelled old fronds. Fronds 10–25 × 3–4 cm, very numerous, strongly erect, stiff. Blade narrowly lanceolate or oblong, 2 × pinnate or pinnatisect, the pinnae more or less equal in length throughout most of the blade, short, oblong-lanceolate, often inrolling and densely covered with chaffy brown scales beneath, yellow-green above, glandular

and aromatic when bruised. Stipe up to one-quarter of the length of the frond. *High subarctic regions of both Old and New Worlds, south to N Urals and N Japan.* H1.

**2. D. submontana** Frazer-Jenkins (*D. villarii* (Bellardi) Woynar; *D. rigida* (Swartz) Gray). Figure 21(8), p. 53.
Rhizome spreading to ascending, sparsely branched. Fronds 15–40 × 5–15 cm, mostly erect, stiff, rather few in number. Blade oblong to triangular-lanceolate, 2 × pinnate or pinnatisect, the lower pinnae as long as or longer than those above, all oblong-lanceolate with pinnatisect pinnules, the upper surface covered with dense, whitish farinose glands giving the dull bluish green blade a mealy appearance. Blade aromatic when lightly bruised. Stipe one-third to half of the length of the frond. *C, S & W Europe.* H2.

**3. D. affinis** (Lowe) Fraser-Jenkins (*D. pseudomas* (Wollaston) Holub & Pouzar; *D. borreri* Newman). Figure 21(10), p. 53.
Rhizome stout, spreading to ascending, sparsely branched. Fronds 20–80 × 15–20 cm, 2 × pinnate or pinnatisect, ascending, more or less stiff, fairly numerous. Blade ovate-lanceolate, the tip gradually tapering, the widest pinnae at about the middle. Pinnae narrowly tapering with the tips of all the pinnules truncate, giving each a squared-off appearance, bright yellow-green, with a patch of dark colour at the junction of the pinna midrib and rachis. Stipe one-quarter or less of the length of the frond, rarely more. *Atlantic islands, Europe eastwards to Himalaya.* H3.

The size of the plants as well as details of scaliness and frond dissection vary considerably between plants from different areas, and several subspecies have been described.
*\*D. wallichiana* Hooker. Similar but the scales become very much darker, almost black, with age. *Himalaya.*

**4. D. oreades** Fomin (*D. abbreviata* Newman). Figure 21(9), p. 53.
Rhizome spreading to ascending, very much branched, eventually building up massive, crowded clumps. Fronds 30–50 × 10–15 cm or more, very numerous, strongly ascending, slightly stiff. Blade 2 × pinnate or pinnatisect, pinnules with rounded tips, their edges all turning upwards giving the frond a crisped appearance, widest pinnae at about the middle of the frond, oblong-lanceolate, mid-green. Stipe one-quarter to one-third of the length of the frond, its base densely clothed

with dull, tan-brown chaffy scales which thin rapidly upwards to become sparse along the rachis. *Europe, especially in the north.* H2.

**5. D. filix-mas** (Linnaeus) Schott. Figure 21(11), p. 53.
Rhizome ascending to erect, stocky, massive, rather seldom branched. Fronds 30–120 × 10–25 cm, fairly numerous, ascending to arching, flexible, often in regular shuttlecock-like clusters. Blade ovate-lanceolate, 2 × pinnate or pinnatisect, the widest pinnae at about the middle of the frond. Pinnae slender, often widely spaced, gradually tapering, pinnules with rounded tips, more or less flat, green. Stipe somewhat variable, usually one-fifth of the length of the frond or less, its base clothed with pale, tan-brown chaffy scales which thin rapidly upwards to become sparse along the rachis. *Europe, N America.* H2.

**6. D. cristata** (Linnaeus) Gray. Figure 22(1), p. 56.
Rhizome decumbent or creeping, slender, seldom branched. Fronds 15–60 × 5–20 cm, rather few, spreading to more or less erect, in irregular clusters, the fertile narrower and taller, more erect and with a reduced blade. Blade 1–2 × pinnate or pinnatisect, linear to oblong-lanceolate, narrow, with pinnae throughout the lower two-thirds of the frond of about equal length, oblong to triangular-ovate, often broad at the base in relation to length, pale yellow-green. Stipe one-third to half of the length of the frond. *C Europe to Siberia, N America.* H2.

**7. D. carthusiana** (Villars) Fuchs (*D. spinulosa* Watt). Figure 22(2), p. 56.
Rhizome creeping, slender, sparsely branched. Fronds 12–60 × 5–25 cm, rather few, mostly erect in irregular clusters of somewhat fragile appearance, the fertile usually taller. Blade almost 3 × finely pinnate, pinnae triangular-ovate, pinnules pinnate or deeply pinnatisect, more or less flat, the margins with small but conspicuous teeth which end in acuminate apices, light yellowish green. Young fronds in bud with shining, whitish or very pale tan-coloured scales. Stipe usually half the length of the frond or more, very slender, bearing a few, rather limp, pale tan-coloured scales. *Europe, especially the north.* H2.

**8. D. expansa** (Presl) Fraser-Jenkins & Jermy (*D. assimilis* Walker). Figure 22(3), p. 56.

Rhizome ascending, sparsely branched. Fronds 20–80 × 12–40 cm, mostly ascending to arching, rather few, in more or less regular clusters. Blades 3 × pinnate, triangular, the longest pinnae at the base, the lowermost with the first division on the basal side large and c. half the length of the pinna. Pinnules pinnate, more or less flat with wide spaces between the divisions, giving the whole frond a lace-like appearance, pale yellow-green. Young fronds in bud with distinctly ginger-coloured scales. Stipe c. one-third of the length of the frond, with scattered ginger-brown scales. *N Eurasia, N America.* H1.

**9. D. aemula** (Aiton) Kuntze. Figure 22(6), p. 56.
Rhizome ascending or more or less erect, occasionally branching. Fronds 15–60 × 10–40 cm, ascending to spreading, the blades more or less lax, arching to descending, 3 × pinnate, broadly triangular, the largest pinnae always the lowermost. Pinnae triangular, flexible, their tips descending, pinnules closely spaced, oblong, toothed, shortly stalked or stalkless, the edges of all the segments with prominently upturned margins giving the whole frond a crisped appearance. Surface of blade bright green with minute glands which give off a smell of new-mown hay on drying. Stipes c. half the length of the frond, relatively thick, purple-brown at the base and with a few, rather limp, reddish brown scales. *Atlantic islands, W Europe (from NW Spain to W Scotland).* H4.

**10. D. dilatata** (Hoffmann) Gray (*D. austriaca* (Jacquin) Woynar). Figure 22(4), p. 56.
Rhizome erect or ascending, seldom branched, massive and stocky. Fronds 10–150 × 5–40 cm, ascending to spreading, the blades more or less lax, arching, 3 × pinnate, triangular-ovate. Pinnae triangular-ovate, the lowermost basal pinnule large, sometimes half the length of the pinna, pinnules often closely spaced, toothed, the edges usually turning downwards, especially in exposed conditions, dark green. Young fronds green when expanding, densely clothed with 2-coloured scales. Stipe one-quarter to half of the length of the frond, densely clothed at the base with scales which vary from pale with a dark lengthwise blackish brown stripe to nearly entirely blackish brown, these rather sparse above the base. *Europe.* H2.

**Figure 22.** Silhouettes of fern-fronds. 1, *Dryopteris cristata*.
2, *D. carthusiana*. 3, *D. expansa*. 4, *D. dilatata*. 5, *D. erythrosora*.
6, *D. aemula*. 7, *Bolbitis heteroclita* (a, juvenile vegetative frond;
b, adult vegetative frond; c, fertile frond).

**11. D. erythrosora** (Eaton) Kuntze.
Figure 22(5), p. 56.
Rhizome decumbent. Fronds 25–60 × 8–15 cm or more, ascending to spreading, arching. Blades 2–3 × pinnate, somewhat narrowly triangular, pinnae triangular-ovate, more or less symmetric, widely spread along the frond, pinnules oblong-lanceolate, with rounded tips, pale green when mature. Young fronds coppery-pink when expanding. Sori small, indusia reddish. Stipe at least half the length of the frond, with scattered brown scales. *Japan, China.* H3.

## 79. BOLBITIS Schott
*C.N. Page & F.M. Bennell*
Small- to medium-sized terrestrial or aquatic plants. Rhizome surface-creeping. Fronds 1 × pinnate, of 2 kinds, the sterile with broad, leafy pinnae and stipe as long as the blade, the fertile with narrower pinnae and stipe twice as long as the blade or more. Sporangia covering the whole undersurface of each pinna in a felt-like mass.

A genus of 85 species occurring throughout the tropics. One species is found in cultivation as an aquarium plant.

**1. B. heteroclita** (Presl) Ching.
Figure 22(7), p. 56.
Rhizome *c.* 6–7 mm in diameter, stipes arising *c.* 1 cm apart. Sterile fronds 5–30 × 3–10 cm, of very varying form, simple on juvenile plants, 1 × pinnate when mature, pinnae elongate with acute apices, the terminal segment long, slender, often rooting at the tip. *Tropical SE Asia.* G2.

## 80. ELAPHOGLOSSUM Schott
*C.N. Page & F.M. Bennell*
Small- to medium-sized plants growing epiphytically or on rock. Rhizome short-creeping, thin, scale-clad. Fronds simple, entire, oval or oblong-lanceolate, all of 1 kind or the fertile slightly narrower, firm to hard and leathery, sometimes with a cartilaginous border, evergreen, pale to dark green. Fertile fronds rarely produced. Sporangia occupying the whole of the undersurface of the frond, forming a velvety mass. Indusia absent. Stipes short or long, merging into the blades and jointed near their bases.

A genus of well over 400 species in all tropical and subtropical parts of the world, especially tropical America. They are long-lived and fairly slow-growing, requiring a very freely drained compost and strong light. They reproduce by spores or by division of existing plants. They are usually accidentally introduced with tropical orchids.

1a. Fronds broadly oval, covered in numerous dark hairs **1. crinitum**
　b. Fronds varying but not broadly oval, without dark hairs **2. petiolatum**

**1. E. crinitum** (Linnaeus) Christ.
Figure 23(2), p. 58.
Fronds 20–30 × 3–5 cm. Blade broadly oval with distinct net-veins, thick and fleshy, pale yellow-green with conspicuous dark hairs fairly thickly scattered over all the surface. *West Indies, C America.* G2.

**2. E. petiolatum** (Swartz) Urban.
Figure 23(1), p. 58.
Fronds to *c.* 20–40 × 1–4 cm. Blade simple, entire, narrowly oblong-lanceolate, linear-lanceolate or ovate, with a pointed apex, thin but firm or leathery, dark green, without conspicuous dark hairs, although scattered, soft brown scales may be present. *SE Asia.* G2.

## 81. HUMATA Cavanilles
*C.N. Page & F.M. Bennell*
Small epiphytes. Rhizomes long-creeping or climbing, thin, freely branching, densely clad with narrow, pointed scales, bearing fronds individually. Fronds simple or 1–2 × pinnate, all of 1 kind or of 2 kinds, evergreen, often triangular but variable in shape, thick, leathery. Sori almost marginal, few per segment, usually near the apex. Indusia circular or broadly kidney-shaped, attached by the base and the lower part of the sides. Stipe slender, grooved, nearly as long as or longer than the blade, varying even in the same plant.

A genus of at least 50 species from Madagascar to the Himalaya, China and Japan, and to Australia and Polynesia. They are long-lived and slow-growing, and resemble miniature *Davallia* (p. 59). They require a freely drained, rocky substrate and strong light with a slightly moist compost which is allowed to become fairly dry between waterings. Propagation is by spores or rhizome fragments.

1a. Sterile fronds normally simple and entire **1. heterophylla**
　b. Sterile fronds pinnate 2
2a. Blades very narrowly triangular or almost linear, 1 × pinnate or pinnatisect only **2. pectinata**
　b. Blades broadly triangular, more than 1 × pinnate 3

3a. Fronds 2 × pinnate or pinnatisect **3. repens**
　b. Fronds mostly 3 × pinnate or pinnatisect **4. tyermannii**

**1. H. heterophylla** (J.E. Smith) Desvaux.
Figure 23(3), p. 58.
Fronds 6–12 × 2.5–3 cm, of two kinds, the sterile simple or occasionally slightly and irregularly lobed near the base with minutely scalloped edges, the fertile deeply pinnatisect with each sorus in a notch of the margin. *Indonesia (Sumatra) to Polynesia.* G2.

**2. H. pectinata** (J.E. Smith) Desvaux. Figure 23(5), p. 58.
Fronds to *c.* 20 × 7 cm. Fertile and sterile fronds not very different, narrowly triangular, deeply and regularly pinnatisect throughout, almost to the midrib. *Indonesia (Sumatra) to New Guinea & Australia (Queensland).* G2.

**3. H. repens** (Linnaeus) Diels. Figure 23(6), p. 58.
Fronds to *c.* 20 × 6 cm, but varying greatly in size. Fertile and sterile fronds not very different, narrowly to broadly triangular, very deeply pinnatisect, almost to the midrib throughout, the upper portion entire, the lower portion further deeply pinnatisect, especially along the lower edge. *Mascarene Islands to N India, Japan, Malaysia & Australia.* G2.

**4. H. tyermannii** Moore. Figure 23(4), p. 58.
Fronds to 23 × 10 cm. Fertile and sterile fronds not very different, finely and deeply 3 × pinnate or pinnatisect. *Himalaya, China.* G1.

## 82. SCYPHULARIA Fée
*C.N. Page & F.M. Bennell*
Small- to medium-sized plants growing epiphytically or on rocks. Rhizome long-creeping, occasionally branching, densely scale-clad with notably dark scales, thin, occasionally branching, bearing fronds individually. Fronds 1–2 × pinnate, all of 1 kind, evergreen, thin, soft, hairless. Stipe as long as or longer than blade. Sori marginal, each covered by an elongate, vase-shaped indusium.

A genus of about 8 species from SE Asia and Polynesia. They are fairly long-lived and slow-growing, making good basket subjects. They require a freely draining, rocky compost and strong light. Propagation is by spores or by division.

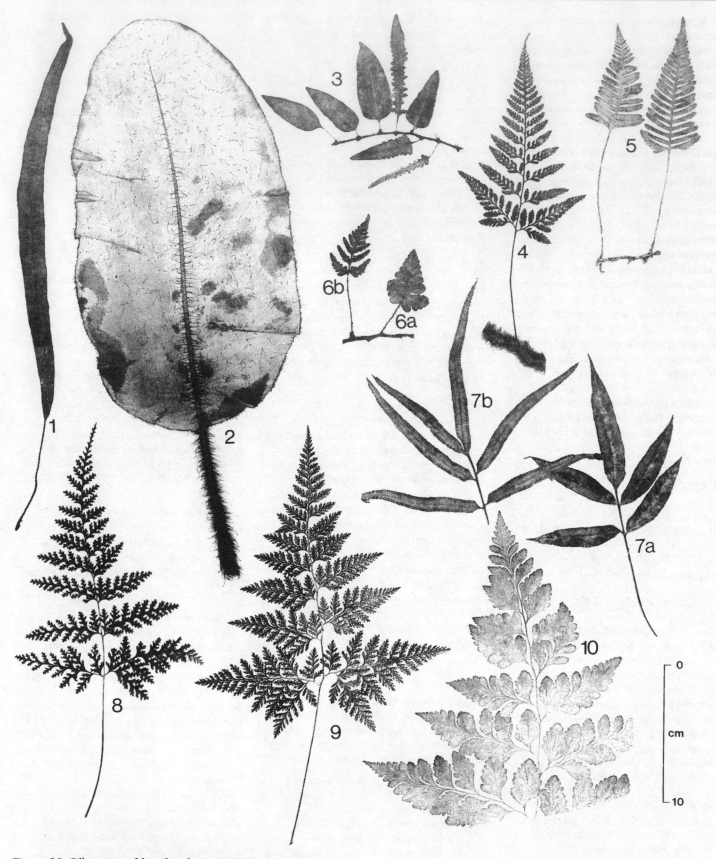

**Figure 23.** Silhouettes of fern-fronds. 1, *Elaphoglossum petiolatum*.
2, *E. crinitum*. 3, *Humata heterophylla*. 4, *H. tyermannii*.
5, *H. pectinata*. 6, *H. repens* (a, vegetative frond; b, fertile frond).
7, *Scyphularia pentaphylla* (a, vegetative frond; b, fertile frond).
8, *Davallia mariesii*. 9, *D. trichomanoides*. 10, *D. solida*.

**1. S. pentaphylla** Fée. Figure 23(7), p. 58.
Blade ovate-triangular, 1–2 × pinnate into a few, simple, entire pinnae, sometimes with a further single pinnule arising from the lower side of the first pinna and directed towards its base. Segments with irregular wavy or toothed margins. *Indonesia (Java) to New Guinea.* G2.

## 83. DAVALLIA J.E. Smith

*C.N. Page & F.M. Bennell*
Small to moderately large, mostly epiphytic plants. Rhizomes long-creeping, branching occasionally, mostly densely scale-clad, bearing fronds individually at regular intervals. Fronds all of 1 kind, semi-evergreen or deciduous, mostly broadly triangular to narrowly ovate, of soft, flexible texture. Blade typically rather finely dissected, with the ultimate segments decurrent. Stipes jointed to the rhizome. Sori terminal on veinlets or margins of ultimate segments, each covered by an elongate, vase-shaped indusium which opens near the margin.

A genus of about 40 species, mostly from the warm temperate or tropical Old World. They are long-lived and popular in cultivation, especially as subjects for hanging baskets. They are grown for their finely dissected foliage and handsome, scale-clad rhizomes, some of which resemble animals' paws. They require fairly well lit situations but not direct sunlight, and a well-drained compost which is allowed to become fairly dry between waterings. Propagation is by spores or division of the rhizomes. Most species shed their fronds completely once a year; this can happen rather suddenly, leaving only the scale-clad rhizomes, which can appear to be dead.

1a. Mature fronds less than 20 cm
        **1. mariesii**
  b. Mature fronds mostly more than
    20 cm        2
2a. Mature fronds 20–45 cm
        **2. trichomanoides**
  b. Mature fronds mostly more than
    45 cm        3
3a. Mature fronds 45–100 cm    4
  b. Mature fronds 100–150 cm
        **8. divaricata**
4a. Fronds not more than 2 × pinnate
        **3. solida**
  b. Fronds more than 2 × pinnate    5
5a. Rhizome tips always closely adpressed
    to surface of soil        6
  b. Rhizome tips often rising above the
    soil and protruding from the clumps
        **5. pyxidata**

6a. Fronds mostly 3 × pinnate
        **7. denticulata**
  b. Fronds mostly 4 × pinnate    7
7a. Blade usually longer than broad,
    pinnae usually not overlapping
        **4. fejeensis**
  b. Blade broader than long, pinnae
    overlapping    **6. canariensis**

**1. D. mariesii** Moore (*D. bullata* Wallich). Figure 23(8), p. 58.
Rhizome less than 5 mm thick, often very long-creeping and climbing. Fronds usually 15–20 cm, the stipe of about equal length to the blade. Blade about as broad as long, 3 × pinnate. *Japan, China.* H5–G1.

Probably the hardiest and smallest of the species.

**2. D. trichomanoides** Blume. Figure 23(9), p. 58.
Rhizome 3–4 mm thick, often very long-creeping and climbing. Fronds mostly 25–45 cm, stipe less than half the length of the blade. Blades mostly slightly longer than broad, 3 × pinnate with the ultimate segments deeply notched. *Japan to Sri Lanka & Malaysia.* H5–G1.

**3. D. solida** (Forster) Swartz. Figure 23(10), p. 58.
Rhizome *c*. 10 mm thick. Fronds usually 50–90 cm, stipe of about equal length to the blade. Blade about as broad as long, mostly 2 × pinnate with large (up to 40 × 15 mm), entire, toothed, mostly acutely pointed, stalked pinnules which have glossy upper surfaces. *SE Asia to Polynesia.* G1.

**4. D. fejeensis** Hooker. Figure 24(1), p. 60.
Similar to *D. solida* but blade much more finely dissected, 4 × pinnate into fine ultimate segments scarcely 1 mm broad. *Polynesia.* G1.

**5. D. pyxidata** Cavanilles. Figure 24(2), p. 60.
Rhizome *c*. 7 mm thick, ascending away from the soil and protruding from the clumps of leaves. Fronds to 1 m, the stipe one-third to half of the length of the blade. Blade nearly as broad as long, 3 × pinnate, the ultimate segments with scalloped margins. *E Australia.* H5–G1.

**6. D. canariensis** (Linnaeus) J. Smith. Figure 24(3), p. 60.
Rhizome 1–1.5 cm thick. Fronds mostly 50–80 cm or more, stipe as long as or slightly longer than blade. Blade about as broad as long, mostly 4 × pinnate. *Atlantic islands, SW Europe.* H5–G1.

**7. D. denticulata** (Burman) Mettenius. Figure 24(5), p. 60.
Rhizome 8–15 mm thick. Fronds commonly 60–100 cm or more, stipe usually one-quarter to one-third of the length of the blade. Blade about as broad as long, mostly 3 × pinnate, the lower pinnae sometimes further divided, the pinnules all ending in narrow apices. *Tropical Asia, Polynesia.* G2.

**8. D. divaricata** Blume. Figure 24(4), p. 60.
Rhizome 2 cm or more thick. Fronds commonly 1–1.2 m, stipe about as long as blade, 5–7 mm in diameter at the base. Blade about as wide as long, 3 or more × pinnate. *S China, Burma, Malaysia.* G2.

## 84. LEUCOSTEGIA Presl

*C.N. Page & F.M. Bennell*
Medium-sized plants growing on soil or on rock. Rhizome long-creeping, slender, less than 1 cm thick, fleshy, more or less scaly, surface-creeping. Fronds 3 or more × pinnate, finely divided into numerous, rhomboid, bluntly toothed segments. Sori in outwardly facing pockets set into the margins of the segments, with the upper surface raised. Indusia cup-shaped.

A genus of at least 2 species from India and tropical SE Asia and New Guinea. They are long-lived and moderately fast-growing; culture as for *Davallia* (above).

**1. L. immersa** (Wallich) Presl (*Davallia immersa* Wallich). Figure 24(6), p. 60.
Fronds 60–80 × *c*. 40 cm, arching. Blade broadly triangular, soft-textured, pale yellow-green. *Tropical SE Asia to New Guinea.* G2.

## 85. RUMOHRA Raddi

*C.N. Page & F.M. Bennell*
Medium-sized plants growing on soil or on rock or epiphytically. Rhizome long-creeping, thick, very densely clothed with golden-brown scales, branching infrequently. Fronds 2–3 or more × pinnate, all of 1 kind, triangular with a broad base. Stipe as long as the blade. Sori few, large, superficial, more or less median on most segments in the upper parts of the fertile fronds, covered by conspicuous, kidney-shaped indusia.

A genus of about 50 species of very scattered distribution. They are long-lived and fairly fast-growing, making attractive basket subjects. They thrive in strong, slightly screened light in very freely

**Figure 24.** Silhouettes of fern-fronds. 1, *Davallia fejeensis*. 2, *D. pyxi-data*. 3, *D. canariensis*. 4, *D. divaricata*. 5, *D. denticulata*. 6, *Leucoste-gia immersa*. 7, *Rumohra adiantiformis*.

draining, coarse compost, and are propagated by spores or division of the rhizomes. The plants look very similar to *Davallia* (particularly *D. solida*, p. 59), but are distinguished by their large, superficial, rounded, indusiate sori.

**1. R. adiantiformis** (Forster) Ching. Figure 24(7), p. 60.
Fronds 2–3 × pinnate, ultimate pinnules widely spread, narrowly rhomboid to oblong, coarsely toothed, thick, leathery, deep green, highly glossy above. *Southern hemisphere.* G2.

## 86. OLEANDRA Cavanilles
*C.N. Page & F.M. Bennell*
Medium-sized epiphytes. Rhizome long-creeping, usually sprawling or scrambling, thin but wiry and stiff, producing stilt-like roots and covered with distinctive, adpressed, regularly arranged, peltate scales with attenuate tips. Fronds borne individually at widely spaced intervals, jointed to the tops of the rhizome branches (phyllopodia) which project. Fronds simple, all of 1 kind, evergreen, usually lanceolate, with short stipes. Sori on the veins often near to the midrib in the upper parts of fertile fronds, each covered by a small, round to kidney-shaped indusium.

A genus of about 40 species, most in E Asia and Polynesia, a few in tropical America and Africa. They are long-lived and fast-growing, cultivated for their attractively scale-patterned rhizomes which are reminiscent of reptile skins. They thrive under moist, tropical conditions with good light, but the scrambling rhizomes can be difficult to contain. Propagation is by spores or rhizome fragments.

**1. O. articulata** (Swartz) Presl. Figure 25(1), p. 62.
Rhizome *c.* 5 mm in diameter. Fronds to 30 cm or more, of arching habit, bright to pale green, thin, papery but firm, blades often slightly wrinkled. *SE Asian tropics.* G2.

## 87. NEPHROLEPIS Schott
*C.N. Page & F.M. Bennell*
Medium to large terrestrial plants. Rhizome almost erect, condensed, commonly stoloniferous, bearing fronds in crowded clusters. Fronds 1 × pinnate (rarely 2 × pinnate), all of 1 kind. Blade long, narrow and arching, with stalkless pinnae often jointed to, and eventually shedding from, the rachis, hairy or becoming hairless, the whole frond often continuing to grow from the apex almost indefinitely. Sori almost marginal on the upper pinnae

of the frond with rounded to kidney-shaped indusia.

A genus of about 30 species from the tropics. Fast-growing and long-lived, these plants are cultivated because of their ease of growth and the existence of numerous monstrous forms. They succeed well in almost any well-drained medium, in good light. They are propagated by spores or by offsets from stolons. All the species are remarkably variable and hence somewhat ill-defined.

1a. Fronds, and especially the rachis, finely downy                    **4. hirsutula**
  b. Fronds hairless or becoming so with age                                      2
2a. Pinna tips mostly rounded
                                              **1. cordifolia**
  b. Pinna tips mostly pointed               3
3a. Blade thick, leathery, upper surface glossy                          **2. biserrata**
  b. Blade thin, soft, upper surface not markedly glossy                 **3. exaltata**

**1. N. cordifolia** (Linnaeus) Presl. Figure 25(2), p. 62.
Fronds mostly 50–80 × 3–4 cm, spreading to arching or hanging. Pinnae more or less oblong-linear, tapering slightly, usually without basal auricles, bluntly rounded and with fine marginal serrations in their outer parts, yellow-green, becoming hairless, fairly soft-textured. *Tropics & subtropics.* H5–G1.
Several cultivars exist, including congested, toothed and plumose forms.

**2. N. biserrata** (Swartz) Schott. Figure 25(5), p. 62.
Fronds to 2 m × 10–15 cm, arching to hanging, slender, tapering, fairly thick, becoming hairless, leathery, mid green, markedly shining above, with fairly widely spaced pinnae with toothed or twice-toothed margins, and with an auricle on the basal side. Stipe fairly flexible. *Tropical America.* G2.
A small number of cultivars exists, including variants with fronds which fork towards their tips.

**3. N. exaltata** (Linnaeus) Schott. Figure 25(3), p. 62.
Fronds to 150 × 12–15 cm, slender, ascending to arching. Pinnae pointed, overlapping occasionally, sickle-shaped with toothed or scalloped margins with an auricle on the basal side, thin, becoming hairless. Stipe rather stiff. *Tropical America & Africa, Polynesia.* G1.
A very large number of variants has been given cultivar status, including forms

with congested, bipinnate, tripinnate, curled, crisped, leathery, waxy or wavy fronds. Some are strong, fast growers; others are much slower, and their hardiness varies considerably.

**4. N. hirsutula** (Forster) Presl. Figure 25(4), p. 62.
Fronds to 1.8 m × 20–25 cm, spreading to arching. Pinnae widely spaced near the base, crowded near the tip, pointed, tapering, sickle-shaped, the basal side with an auricle and the margins irregularly scalloped, finely and persistently hairy, firm, pale green. Stipe densely and finely downy, especially when young. *Tropical SE Asia to the Pacific.* G1.

## 88. BLECHNUM Linnaeus
*C.N. Page & F.M. Bennell*
Medium to large, mostly terrestrial plants. Rhizomes usually short-creeping, occasionally long-creeping or becoming stocky and ascending to erect, a few trunk-forming. Blades coarsely pinnately or pinnatisectly divided, pinnae usually entire and linear in the sterile fronds, the fertile usually with the blade reduced and with paired, continuous, linear sori on either side of the main veins along the length of each pinna (or when entire-leaved, along the length of the frond). Indusium a continuous flap along each sorus, hinged on the outer side, opening towards the vein.

A cosmopolitan genus of 200 or more species. They are mostly long lived, some forming extensive colonies. They require warm or cool, permanently moist, sometimes wet conditions. Most are slow-growing, exacting and somewhat difficult subjects, requiring considerable root runs.

1a. Fronds mostly simple, entire
                                              **1. patersonii**
  b. Fronds pinnatifid or pinnate            2
2a. Fronds with most pinnae not stalked
                                                        3
  b. Fronds with pinnae stalked or at least not broadly joined to the rachis throughout much of the lower half of the frond                                 9
3a. Rhizome 2 cm or less in diameter, ascending or creeping                 4
  b. Rhizome 2–15 cm in diameter, erect, trunk-forming                       7
4a. Rhizome creeping     **2. cartilagineum**
  b. Rhizome ascending or more or less erect                                    5
5a. Mature fronds mostly 10–20 cm
                                            **5. penna-marina**
  b. Mature fronds mostly more than 20 cm                                      6

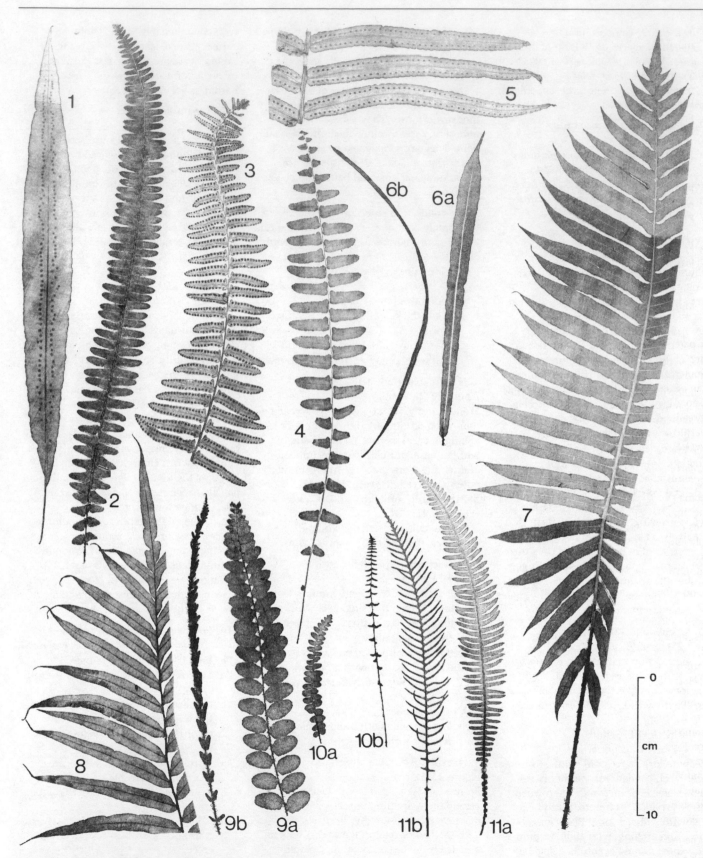

**Figure 25.** Silhouettes of fern-fronds. 1, *Oleandra articulata*.
2, *Nephrolepis cordifolia*. 3, *N. exaltata*. 4, *N. hirsutula*.
5, *N. biserrata*. 6, *Blechnum patersonii* (a, vegetative frond;
b, fertile frond). 7, *B. cartilagineum*. 8, *B. orientale*. 9, *B. fluviatile*
(a, vegetative frond; b, fertile frond). 10, *B. penna-marina*
(a, vegetative frond; b, fertile frond). 11, *B. spicant* (a, vegetative
frond; b, fertile frond).

6a. Vegetative fronds of mature plants mostly less than 5 cm wide
      **6. spicant**

b. Vegetative fronds of mature plants 5–10 cm wide    **7. nipponicum**

7a. Fronds of 2 kinds    8

b. Fronds of 1 kind    **10. brasiliense**

8a. Vegetative fronds 40–60 × 6–8 cm
      **8. nudum**

b. Vegetative fronds 50–200 × 8–10 cm
      **9. gibbum**

9a. Pinnae of vegetative fronds short and rounded    **4. fluviatile**

b. Pinnae of vegetative fronds linear and elongate    10

10a. Fronds of 1 kind    11

b. Fronds of 2 kinds    **12. tabulare**

11a. Mature fronds 20–40 cm
      **11. occidentale**

b. Mature fronds 40–120 cm or more
      **3. orientale**

**1. B. patersoni** (R. Brown) Mettenius. Figure 25(6), p. 62.
Rhizomes small, erect, tufted. Fronds to 40 cm or more, of 2 kinds, simple, entire or occasionally lobed, the vegetative arching or spreading, with broad dark green blades, the fertile more erect, very narrow. *New Zealand*. H5–G1.

**2. B. cartilagineum** Swartz. Figure 25(7), p. 62.
Rhizome short-creeping, occasionally branching, covered with stiff, black scales. Fronds erect, to 1.5 m, of 1 kind only, simply pinnate, broad, leathery. Stipe one-quarter to one-third of the length of the frond. *E Australia*. H5–G1.

**3. B. orientale** Linnaeus. Figure 25(8), p. 62.
Rhizome short, erect, densely covered with scales. Fronds tufted, 30–60 cm, all of 1 kind, pale green, soft. Stipe one-quarter to one-third of the length of the frond. *Tropical Asia, Australasia & the Pacific*. G2.

**4. B. fluviatile** (R. Brown) Lowe. Figure 25(9), p. 62.
Rhizome small, erect, densely covered with scales. Fronds to 50 cm, tufted, of 2 kinds, all narrowly linear, 1 × pinnate, the vegetative bright green with rounded pinnae, forming spreading rosettes, the fertile standing more erect. *SE Australia, New Zealand*. H4.

**5. B. penna-marina** (Poiret) Kuhn. Figure 25(10), p. 62.
Rhizome slender, creeping or ascending. Fronds mostly 10–20 × 1–1.5 cm, narrowly linear, of 2 kinds, all 1 × pinnate, dark green but often reddish when young; vegetative fronds forming low, compact rosettes, the fertile erect and with narrow pinnae. *SE Australia, New Zealand, southern S America, subantarctic islands*. H3.

Established plants from extensive, low-growing carpets. Specimens from different sources vary much in hardiness.

**6. B. spicant** (Linnaeus) Roth. Figure 25(11), p. 62.
Rhizome creeping to ascending. Fronds of 2 kinds. Vegetative fronds mostly 20–40 × 3–5 cm, narrowly linear, 1 × pinnate, dark green, spreading, forming rosettes and eventually large colonies. Fertile fronds slender, erect, with narrow pinnae throughout. *Europe, N Asia*. H2.

**7. B. nipponicum** (Kunze) Makino (*Struthiopteris nipponica* (Kunze) Nakai). Figure 26(1), p. 64.
Rhizome creeping to ascending. Fronds of 2 kinds. Vegetative fronds mostly 35 cm or more × 5–10 cm, broadly linear to linear-ovate, 1 × pinnate, bright green, arching, forming rosettes. Fertile fronds slender, erect, with narrow pinnae throughout. *N Japan*. H4.

**8. B. nudum** Mettenius (*B. discolor* (Forster) von Keyserling, in part). Figure 26(3), p. 64.
Rhizome erect, thick, eventually forming a dark, fibrous trunk to 1 m. Fronds of 2 kinds. Vegetative fronds 40–60 × 6–8 cm, narrowly to broadly linear, 1 × pinnate, bright green, ascending, forming rosettes. Fertile fronds slender, erect, the narrow pinnae in the lower part with broad, leaf-like bases. *E Australia, New Zealand*. H5–G1.

**9. B. gibbum** Labillardière. Figure 26(4), p. 64.
Rhizome erect, thick, eventually forming a dark fibrous trunk to 1 m or more. Fronds of 2 kinds. Vegetative fronds 50–200 × 8–10 cm, ovate, progressively narrowing towards the base, 1 × pinnate, bright green, ascending, forming rosettes. Fertile fronds erect with narrow pinnae throughout. *New Caledonia & New Hebrides*. G1.

**10. B. brasiliense** Desvaux. Figure 26(5), p. 64.
Rhizome erect, thick, eventually forming a dark fibrous trunk to 30 cm or more. Fronds all of 1 kind, 90–130 × 15–25 cm, tapering gradually towards the base, 1 × pinnate, bright pink when young, yellow-green when mature, ascending, forming compact, shuttlecock-like crowns. *Tropical S America*. G1.

**11. B. occidentale** Linnaeus. Figure 26(2), p. 64.
Rhizome thin, short-creeping. Fronds all of 1 kind, 20–35 × 5–8 cm, ovate-lanceolate, 1 × pinnate, coppery-pink when young, mid green when mature, arching and forming irregular clumps. *Tropical America, Pacific islands*. G1.

**12. B. tabulare** (Thunberg) Kuhn (*B. magellanicum* Desvaux; *B. borganum* Willdenow). Figure 26(6), p. 64.
Rhizome thin, long-creeping, branching freely, eventually forming large patches. Fronds of 2 kinds. Vegetative fronds 60–100 × 10–20 cm, broadly oblong, 1 × pinnate, leathery, deep green, glossy, arching and eventually forming extensive sprawling masses. Fertile fronds scattered amongst the vegetative, with narrow pinnae. *Southern S America, Falkland Islands*. H4.

### 89. DOODIA R. Brown

*C.N. Page & F.M. Bennell*

Small- to medium-sized terrestrial plants. Rhizomes long- or short-creeping, variable in thickness, branching frequently, harshly scale-clad, bearing fronds in irregular clusters. Fronds 1 × pinnate or pinnatisect, all of 1 kind or of 2 kinds, all evergreen, narrowly ovate-lanceolate, leathery and rough, deep green (often coppery-bronze when unrolling); veins forming 1 or more rows of meshes on either side of the midrib. Stipe short. Sori discrete, in 1–2 or more rows on either side of the midrib of each pinna and more or less impressed into it, forming a row of box-like structures, each with an almost square, flap-like indusium opening towards the midrib.

A genus of at least 10 species mainly from Australia and New Zealand, with a few from Sri Lanka, USA (Hawaii) and Polynesia. They can be grown in almost any kind of soil kept slightly moist, but freely draining. Tolerant of light and low humidity. Plants grow readily from spores.

1a. Fronds of 2 kinds    **1. caudata**

b. Fronds all of 1 kind    2

2a. Fronds pinnatisect    **3. aspera**

b. Fronds pinnate, at least near the base
      **2. media**

**1. D. caudata** (Cavanilles) R. Brown. Figure 26(7), p. 64.
Fronds 10–25 × 2–3 cm, of 2 kinds, a rosette of basal, spreading, vegetative

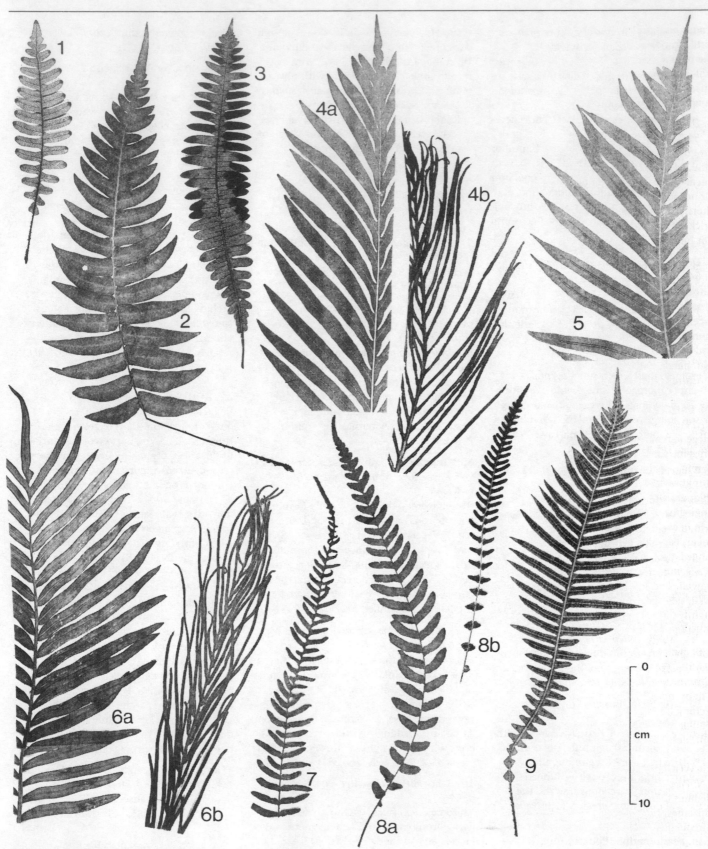

**Figure 26.** Silhouettes of fern-fronds. 1, *Blechnum nipponicum*.
2, *B. occidentale*. 3, *B. nudum*. 4, *B. gibbum* (a, vegetative frond;
b, fertile frond). 5, *B. brasiliense*. 6, *B. tabulare* (a, vegetative
frond; b, fertile frond). 7, *Doodia caudata*. 8, *D. media* (a, vegetative
frond; b, fertile frond). 9, *D. aspera*.

fronds and more erect, herringbone-like fertile fronds all 1 × pinnate or pinnatisect, the pinnae near the base very shortly stalked, moderately harsh and prickly. Stipes with a few sparse scales, finely downy. *E Australia*. H5–G1.

**2. D. media** R. Brown. Figure 26(8), p. 64.
Fronds mostly 20–30 × 3–4 cm, all of 1 kind, ascending or irregularly spreading, linear-lanceolate, 1 × pinnate or pinnatisect, the pinnae near the base very shortly stalked, moderately harsh and prickly. Stipe with sparse brownish scales, not harsh. *Australasia*. H5–G1.

**3. D. aspera** R. Brown. Figure 26(9), p. 64.
Rhizome creeping, ascending or erect. Fronds 30–40 × 10–15 cm, all of 1 kind, ascending or strongly erect, ovate-lanceolate to ovate, 1 × pinnatisect, harsh and prickly to touch. Stipes covered with stiff, harsh, blackish, prickly scales. *E Australia*. H5–G1.

## 90. SADLERIA Kaulfuss
*C.N. Page & F.M. Bennell*
Large to almost tree-like terrestrial plants. Rhizome thick, almost erect to erect, trunk-forming, scaly. Fronds 2 × pinnate or pinnatisect, all of 1 kind, evergreen, linear-oblong, thick, leathery. Sori linear in a single line along either side of the main vein of each ultimate segment, covered by a linear, flap-like, continuous indusium hinged along its outer side.

A genus of 6 or 7 species, all from Hawaii. The 1 cultivated species is long-lived, slow-growing and requires strong but slightly filtered light and a permanently moist but freely drained compost.

**1. S. cyatheoides** Kaulfuss. Figure 27(1), p. 66.
Rhizome eventually to 1.5 m. Fronds to 1 m or more. Blade linear-oblong, of arching habit, 2 × pinnate into neat, regular, closely spaced pinnae and pinnules. *USA (Hawaii)*. G2.

## 91. WOODWARDIA J.E. Smith
*C.N. Page & F.M. Bennell*
Medium to large terrestrial plants. Rhizomes short-creeping, thick, branching infrequently, scale-clad, ascending to erect, bearing fronds in clusters. Fronds all of 1 kind, evergreen, broadly ovate, spreading or laxly arching, coarsely 2 × pinnatisect, firm, leathery, with a conspicuous netted venation, deep green,

slightly glossy. Pinnae with entire or finely toothed margins and acute tips. Stipes stout. Sori discrete in a single row along either side of the midrib of each pinnule and impressed into it, forming a row of box-like structures, each with a thick, almost square, flap-like indusium opening towards the midrib. In some species bulbils are produced along the upper surface of the frond near its apex.

A genus of 10 species from the northern hemisphere. They require shade and well-drained, permanently moist soil, and grow well on sloping banks. Propagation is by spores or bulbils.

1a. Mature fronds with a single large plantlet developing from a bud below the apex **1. radicans***
  b. Mature fronds with many small plantlets developing from numerous buds scattered over the upper surface of the frond **2. orientalis**

**1. W. radicans** (Linnaeus) J.E. Smith. Figure 27(2), p. 66.
Fronds 60–200 cm, laxly arching, with usually a single scaly bud present below the tip of the frond developing into a single large plantlet which takes root where it touches the ground. Rarely a few other, similar, buds are present also. *Atlantic islands, SW Europe*. H5.
  ***W. unigemmata** (Makino) Nakai. Similar, always with a single bud and very widely spaced pinnae. *China, Japan, Taiwan*.

**2. W. orientalis** Swartz & Schrader. Figure 27(3), p. 64.
Fronds 30–120 cm, spreading to arching, with many small plantlets developing from numerous buds scattered all over the upper surface of the frond. *China, Japan, Taiwan*. G1.

## 92. STENOCHLAENA J.E. Smith
*C.N. Page & F.M. Bennell*
Large terrestrial or aquatic plants. Rhizome long-creeping, thin, freely branching, clad with a few sparse scales, indefinitely climbing or scrambling, bearing fronds individually. Fronds 1–2 × pinnate, distant, arching, linear-lanceolate, of 2 kinds, the fertile similar to the vegetative but with much reduced blade, all evergreen, with the pinnae jointed to the rachis, firm, hairless, leathery with cartilaginous margins, green to olive-green, rarely somewhat bronzed, shining. Sori narrow, linear, the linear surfaces of the fertile pinnae entirely covered by sporangia.

A genus of about 6 species from tropical America to Polynesia. They are long-lived, fairly fast-growing and require a peaty compost, strong light and an absence of cold draughts. Propagation by layering of the rhizome.
Literature: Joe, B., Ferns cultivated in California, *Lasca Leaves* 9: 61–7 (1959).

1a. Fronds 30–60 cm, the fertile 2 × pinnate **2. tenuifolia**
  b. Fronds 40–80 cm, the fertile 1 × pinnate **1. palustris**

**1. S. palustris** (Burman) Beddome. Figure 27(9), p. 66.
Fronds 40–80 cm, the vegetative and fertile 1 × pinnate. *Malaysia to Polynesia*. G2.

**2. S. tenuifolia** (Desvaux) Moore. Figure 27(10), p. 66.
Fronds 30–60 cm, the vegetative 1 × pinnate, the fertile 2 × pinnate. *Tropical Africa, Madagascar*. G2.

## 93. MARSILEA Linnaeus
*C.N. Page & F.M. Bennell*
Small aquatic or amphibious plants. Rhizome long-creeping, thin, branching occasionally, slightly hairy, bearing fronds individually. Fronds all of 1 kind, more or less evergreen, with small blades borne on very long stipes. Fronds coiled when young, 4-lobed, like a '4-leaved' clover (*Trifolium*) when expanded, the leaflets soft and often finely hairy, mostly bright green, some 2-coloured. Sori borne in specialised, bean-shaped, hard-shelled, blackish sporocarps at the bases of the stipes, each with a few megaspores or many microspores.

A genus of about 70 species, mainly from the tropics. They are long-lived and slow-growing, easily grown on damp sand or mud around ponds or in pots half-submerged in water. *M. mutica* is a submerged, floating-leaved aquatic. Propagation is by division.

1a. Upper surface of leaflets hairy 2
  b. Upper surface of leaflets hairless 3
2a. Hairs long and very dense; stipes to 30 cm **1. drummondii**
  b. Hairs short and moderately dense; stipes less than 15 cm **2. hirsuta**
3a. Undersurface of leaflets hairy **3. fimbriata**
  b. Undersurface of leaflets without obvious hairs 4
4a. Stipes to 90 cm; leaflets 2–4 cm **4. mutica**
  b. Stipes to 15 cm; leaflets 1.5–2 cm **5. quadrifolia**

**Figure 27.** Silhouettes of fern-fronds. 1, *Sadleria cyatheoides.*
2, *Woodwardia radicans.* 3, *W. orientalis.* 4, *Marsilea quadrifolia.*
5, *M. hirsuta.* 6, *M. mutica.* 7, *M. fimbriata.* 8, *M. drummondii.*
9, *Stenochlaena palustris.* 10, *S. tenuifolia* (a, vegetative frond;
b, fertile frond). 11, *Regnellidium diphyllum.* 12, *Pilularia
globulifera.* 13, *Salvinia auriculata.* 14, *S. natans.* 15, *Azolla
caroliniana.* 16, *A. filiculoides.*

**1. M. drummondii** Braun. Figure 27(8), p. 66.
Stipes to 30 cm. Leaflets broad, wedge-shaped, notched at the apex and sometimes with a waxy texture. Emerging leaflets *c.* 3 cm, with long fine whitish hairs densely over both surfaces, all bright green often becoming tinged with reddish brown. *Australia.* H5–G1.

**2. M. hirsuta** R. Brown. Figure 27(5), p. 66.
Stipes to 15 cm. Leaflets *c.* 2 cm, narrow to broad, wedge-shaped, rounded at apex. Emerging leaflets usually bright green with short whitish hairs moderately densely over both surfaces. *Australia.* H5–G1.

**3. M. fimbriata** Thunberg & Schumacher. Figure 27(7), p. 66.
Stipes 15–30 cm, very slender. Leaflets wedge-shaped, 1–2 cm, bright green, hairless above, finely hairy beneath with silky hairs protruding around their margins. *Tropical W Africa.* G2.

**4. M. mutica** Mettenius. Figure 27(6), p. 66.
Stipes to 90 cm. Leaflets 2–4 cm, broad, wedge-shaped, rounded at the apex, usually floating on the water, smooth, glossy, hairless, each of two shades of bright green separated by a transverse, purple-brown band. *Australia.* G1.

**5. M. quadrifolia** Linnaeus. Figure 27(4), p. 66.
Stipes 8–15 cm. Leaflets 1.5–2 cm, broad, wedge-shaped, apex rounded and minutely toothed, smooth, hairless. *Eurasia, N America.* H2.

## 94. REGNELLIDIUM Lindmann
*C.N. Page & F.M. Bennell*
Small aquatic or amphibious plants. Rhizome long-creeping, thick, branching occasionally, bearing fronds individually. Fronds with a single pair of opposite leaflets borne on each stalk-like stipe, evergreen to almost evergreen, rather fleshy, bright glossy green, each 2 cm or more, usually broader than long, notched on the outer margin. Stipe several times the length of the blade. Sori borne in sporocarps as in *Marsilea* (p. 65).

One species which is long-lived and slow-growing, succeeding well in soft mud in a pan half-submerged in water in fairly strong light. Propagation is by division.

**1. R. diphyllum** Lindmann. Figure 27(11), p. 66.
*S Brazil.* G2.

## 95. PILULARIA Linnaeus
*C.N. Page & F.M. Bennell*
Small aquatic or amphibious plants. Rhizome long-creeping, thin, occasionally branching, bearing fronds individually. Fronds all of 1 kind, bladeless, evergreen, linear, grass-like, coiled when young, brittle, bright green. Sori as in *Marsilea* (p. 65).

A genus of 6 species widely scattered in temperate zones. They are long-lived and slow-growing, thriving in the marginal mud of ponds but intolerant of strong competition. Propagation is by division.

**1. P. globulifera** Linnaeus. Figure 27(12), p. 66.
Fronds to 10 cm or more, soft, more or less fleshy, typically wavy, bright green. *W Europe.* H3.

## 96. SALVINIA Adanson
*C.N. Page & F.M. Bennell*
Small, free-floating aquatics. Rhizome thin, thread-like, branching occasionally, bearing 2 ranks of flattened, boat-shaped water-leaves (each a whole simple frond), each with a surface covering of stellate hairs, with a third, finely dissected leaf long-pendent in the water and substituting for roots, all bright green. Sori borne on the water-leaves in small, glandular sporocarps, seldom seen in cultivation.

A genus of about 10 mostly tropical species. Plants will grow vegetatively almost indefinitely, the growth breaking up to form new plants. They require regular, constantly warm water or very moist soil and strong or filtered light. They are weedy, and will often cover the surfaces of still water.

1a. Water-leaves mostly *c.* 2.5 × 2 cm, often nearly as wide as long **1. auriculata**
 b. Water-leaves up to 10 × 7 mm, usually considerably longer than broad **2. natans**

**1. S. auriculata** Aublet. Figure 27(13), p. 66.
Water-leaves mostly to *c.* 2.5 × 2 cm, often nearly as broad as long, often 40–50 per plant. *Originally from tropical America,* now widely introduced throughout the tropics. G2.

**2. S. natans** (Linnaeus) Allioni. Figure 27(14), p. 66.
Water-leaves mostly to 10 × 7 mm, usually conspicuously longer than broad, usually 20–30 per plant. *Old World tropics.* G2.

## 97. AZOLLA Lamarck
*C.N. Page & F.M. Bennell*
Very small, free-floating aquatic plants of moss-like appearance, laterally branching, flat. Rhizomes thin, spreading. Leaves reduced, simple, scale-like, overlapping, fleshy, arranged in regular, lengthwise rows.

About 6 species in temperate and subtropical zones. Sometimes grown as aquarium or pond plants; their appearance in ponds and ditches can be apparently spontaneous. Plants grow very vigorously vegetatively, fragmenting to new individuals, and can rapidly cover the surfaces of ponds with a green, moss-like carpet. They often die back in winter, as they require adequate light, warmth and still water. Plants contain the symbiotic blue-green alga *Anabaena*, and are readily propagated by fragmentation.

1a. Leaves, when viewed from above, with obtusely rounded tips and broad, membranous margins **2. filiculoides**
 b. Leaves, when viewed from above, with acute apices and narrow, membranous margins **1. caroliniana**

**1. A. caroliniana** Willdenow. Figure 27(15), p. 66.
Plants mostly 1–1.5 cm, of compact habit and fan-shaped growth, rounded in outline. Exposed part of the leaf very small, mostly 0.5–1 mm wide, often longer than broad, apex acute, margins membranous, narrow, olive green or brown. *Temperate America.* H5–G1.

**2. A. filiculoides** Lamarck. Figure 27(16), p. 66.
Plants 1–5 cm or more, of compact or loose habit, growth fan-shaped or irregularly elongate-elliptic in outline. Exposed part of leaf mostly 0.9–1.4 mm wide, about as long as broad, apex obtuse or rounded, margins broad, membranous, dull blue or yellow-green, slightly glaucous, turning slightly red in winter. *Tropical America.* H4.

# GYMNOSPERMAE

Trees or shrubs, often resinous or mucilaginous. Leaves mostly evergreen, frequently leathery, often linear, mostly with a midrib but usually without lateral veins. Plants monoecious or dioecious. Male inflorescence usually cone- or catkin-like, often fragile and ephemeral. Female inflorescence cone-or rarely berry-like, formed entirely from fertile scales or from fertile and sterile (bract) scales, sometimes very reduced or absent, usually robust and long-lived, taking 1 or more seasons to mature. Ovules naked, not enclosed in an ovary, usually borne on the face of a cone-scale, more rarely terminal on short branches. Seeds borne on the scales, often in pairs, sometimes more numerous. Cotyledons 2–4 or many and variable.

All gymnosperms are wind-pollinated plants with an extremely long fossil history, which includes many extinct families, genera and species. Their wood lacks vessels, and mucilages, gums or resins frequently exude from wounds to seal damaged areas. The leaves and parts of the cone may be arranged spirally or in opposite pairs. The form of the female cone is very variable: in some groups its scales are reduced or modified so that the cone-like nature of the organ is obliterated as it matures into a fleshy or berry-like structure.

Literature: Pilger, R., in Engler, A. & Prantl, K., *Die Naturlichen Pflanzenfamilien*, edn 2, **13**: 121–403 (1926); Pilger, R. & Melchior, H., *Syllabus der Pflanzenfamilien*, edn 12, **1**: 312–43 (1964); den Ouden, P. & Boom, B.K., *Manual of cultivated conifers* (1965); Dallimore, W. & Jackson, A.B., *Handbook of Coniferae and Ginkgoaceae*, edn 4 (1966); Krüssmann, G., *Handbuch der Nadelgehölze* (1972); Mitchell, A.F., *Conifers in the British Isles* (1972); Callen, G., *Les conifères cultivées en Europe* (1977); Welch, H.J., *Manual of dwarf conifers* (1979). Many of these books are well illustrated, and reference should be made to them for general illustrations not cited individually below.

1a. Plant palm-like or fern-like; trunk to 2 m, composed of leaf bases supporting a rosette of large compound leaves
    **I. Cycadales** (p. 70)

  b. Plant not as above     2

2a. Small shrubs of switch-like habit     **X. Gnetales** (p. 107)

  b. Trees or shrubs, not of switch-like habit     3

3a. Leaves fan-shaped     **IX. Ginkgoaceae** (p. 106)

  b. Leaves not fan-shaped     4

4a. Leaves stalked, linear-lanceolate, to $20 \times 3$ cm; seeds often inserted on a fleshy aril     **IV. Podocarpaceae** (p. 74)

  b. Leaves not stalked, scale-like, awl-shaped or needle-like, either less than 10 cm long or less than 1 cm broad     5

5a. Leaves relatively thick, closely spiralling the stem
    **II. Araucariaceae** (p. 72)

  b. Leaves scale-like (paired and small) awl-shaped or needle-like, or spreading in 2 lateral ranks     6

6a. Shrubs; leaves lanceolate, in 2 lateral ranks, tapering gradually to a sharp tip     **III. Cephalotaxaceae** (p. 73)

  b. Trees, or if shrubs with leaves in 2 lateral ranks, then leaves parallel-sided and rounded to an obtuse tip     7

7a. Leaves linear, deep green, stiff, in 2 ranks; seed solitary, more or less enclosed in a fleshy aril
    **VIII. Taxaceae** (p. 105)

  b. Leaves various, if linear and in 2 ranks, then pale green and flexuous; seeds in woody or fleshy cones     8

8a. Leaves awl-shaped (when juvenile) or scale-like, usually much less than 1 cm long; cones hard, or fleshy and berry-like, usually less than 2 cm
    **VI. Cupressaceae** (p. 79)

  b. Leaves linear, needle-like or awl-shaped, if scale-like then cones spherical, more than 2 cm long     9

9a. Leaves awl-shaped, linear or rarely needle-like; cone scales and bracts strongly fused; ovules 2–9 per scale; mature cone-scales usually swollen at their tips     **V. Taxodiaceae** (p. 77)

  b. Leaves linear or needle-like; bracts and cone scales distinct; ovules 2 per scale; mature cone-scales flat
    **VII. Pinaceae** (p. 87)

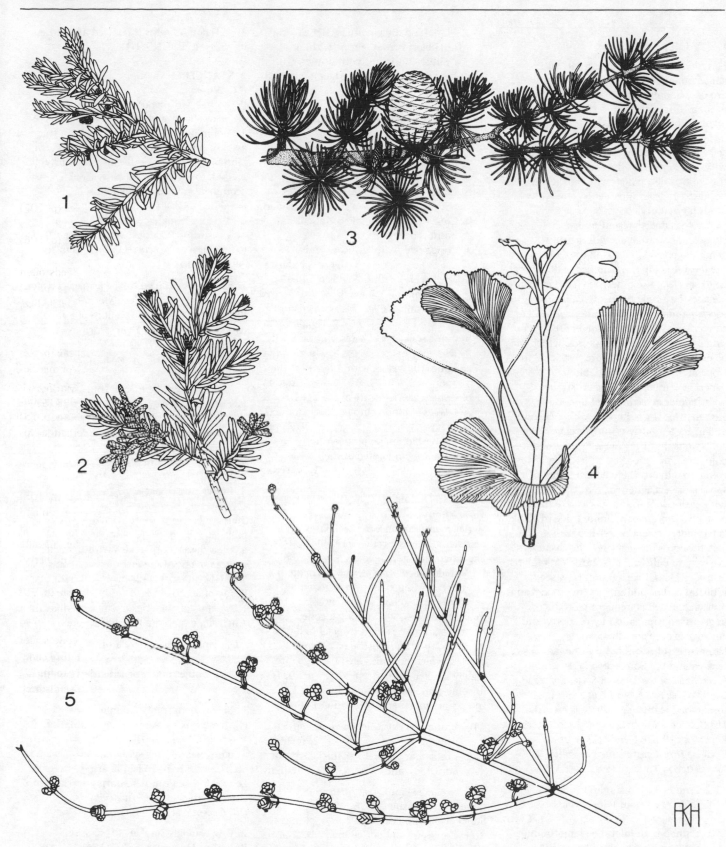

**Figure 28.** Gymnosperms, all × ⅔. 1, *Taxus baccata*: shoot with male inflorescences. 2, *Podocarpus andinus*: shoot with male inflorescences. 3, *Cedrus atlantica*: shoot with female cone. 4, *Ginkgo biloba*: vegetative shoot. 5, *Ephedra americana*: shoot with male inflorescences.

# I. CYCADALES

Long-lived, dioecious perennials with subterranean or aerial woody stems which are occasionally branched. Leaves (fronds) in a basal rosette or in a crown at the stem apex, pinnate or bipinnate. Cones axillary or terminal with spore-bearing scales spirally or vertically arranged. Male cones with pollen sacs on the lower surfaces of the cone scales; pollen wind-borne. Ovules stalkless, 1–several on each cone scale. Seeds large. Cotyledons 2.

The Cycadales present many classification problems, but are now generally considered to fall into 3 families, the largest being the **Zamiaceae** (8 genera, *c.* 80 species); the other two families each consist of a single genus: **Cycadaceae** (*Cycas*) and **Stangeriaceae** (*Stangeria*). For convenience they are treated here as a single entity. The living species, which may be extremely long-lived, are restricted to tropical and subtropical habitats in America, Africa, Asia and Australasia.

Cycadales are cultivated in botanic gardens (see IUCN Report **3**, cited below) and in larger gardens and specialist collections, but they are not generally cultivated, nor are they widely available commercially. In the wild a number of species are extremely rare or threatened with extinction; many species are therefore protected, and international trade in them is forbidden except under licence.

Although slow-growing, the Cycadales are easy to cultivate in subtropical regions, or under glass. Propagation is by seed; artificial pollination may be necessary and non-viable seed is frequently produced. Vegetative propagation by offshoots and cuttings is occasionally possible.
Literature: Johnson, L.A.S., The families of Cycads and the Zamiaceae of Australia, *Proceedings of the Linnean Society of New South Wales* **84**: 64–117 (1959); IUCN threatened plants committee secretariat, *The Botanic Gardens List of Cycads*, Interim Report **3** (1980); Hendricks, J. et al., Cycads, *Threatened Plants Committee Newsletter* **7**: 13–16 (1981).

1a. Pinnae with a single thickened midrib, without lateral veins (*Cycadaceae*)     **1. Cycas**
  b. Pinnae with lateral or longitudinal veins, with or without a midrib     **2**
2a. Plants fern-like; stem subterranean; pinnae with a definite midrib and numerous transverse, parallel, branched lateral veins, rolled in bud (*Stangeriaceae*)     **2. Stangeria**
  b. Plants usually palm-like; stems aerial or subterranean; pinnae without a definite midrib but with numerous longitudinal veins, overlapping in bud (*Zamiaceae*)     **3**
3a. Cone scales overlapping, spirally arranged; aerial stems clothed in persistent leaf bases     **4**
  b. Cone scales not overlapping, arranged in vertical rows; aerial stems sometimes naked     **7**
4a. Cones terminal; female cone scales with woolly ends     **5. Dioon**
  b. Cones axillary; female cone scales hairless or velvety     **5**
5a. Cone scales with truncate ends     **6. Encephalartos**
  b. Cone scales with the ends pointed or with terminal spines     **6**
6a. Cones stalkless; leaf bases with short hairs     **7. Lepidozamia**
  b. Cones stalked; leaf bases silky or woolly with long hairs     **8. Macrozamia**
7a. Cone scales each with 2 horns at the apex     **4. Ceratozamia**
  b. Cone scales truncate at the apex     **8**
8a. Leaves pinnate; pinnae clearly jointed at base     **9. Zamia**
  b. Leaves bipinnate; pinnae and pinnules not jointed at their bases     **3. Bowenia**

## 1. CYCAS Linnaeus
*E.C. Nelson*
Aerial stems unbranched. Leaf bases persistent. Leaves erect or spreading, crozier-like when young, pinnae each with a single, thick midrib. Female scales not forming a definite cone, but spirally arranged in a terminal mass, falling separately at maturity. Male scales forming a cone.

About 20 species from tropical habitats in Africa, Asia and Australasia. Some species, which are cultivated for sago, are widely planted in the tropics. The genus is in some confusion; fortunately, only 2 species are in general cultivation in Europe.

1a. Pinna margins thickened and curled inwards     **2. revoluta**
  b. Pinna margins neither thickened nor curled inwards     **1. circinalis**

**1. C. circinalis** Linnaeus. Illustration: Botanical Magazine, 2826, 2827 (1828). Stems to *c.* 7 m. Leaves to 2.5 m; pinnae to 30 cm, margins not curled inwards. *Asia*. H5–G1.

**2. C. revoluta** Thunberg. Illustration: Botanical Magazine, 2963, 2964 (1830). Stems to *c.* 4 m. Leaves to 1 m; pinnae 10–20 cm, margins thickened and curled inwards. *E Asia*. H5–G1.

## 2. STANGERIA Moore
*E.C. Nelson*
Perennials with woody, subterranean, occasionally branched, tuber-like stems to 10 cm in diameter. Leaves 1–4 from the apex of each stem or branch, 25–200 cm; pinnae 5–20 pairs, entire or irregularly incised, to 40 × 6 cm, each with a prominent midrib and dichotomously branched lateral veins. Male cones to 15 × 3 cm. Female cones to 18 × 8 cm.

A genus of 1 species, originally thought to be a fern and incorrectly placed in the genus *Lomaria*.

**1. S. eriopus** (Kunze) Nash (*Lomaria eriopus* Kunze; *Stangeria paradoxa* Moore). Illustration: Marloth, The flora of South Africa, t. 14 (1913); Dyer, Flora of southern Africa **1**: 2 (1966).
*South Africa (Cape Province to Natal)*. H5–G1.

A variety (var. *schizodon* (Bull) Marloth) has been proposed to include plants with incised leaves, but this character may be a cultural artefact. Plants from forest areas in the wild are very different from those growing in open grassland.

## 3. BOWENIA Hooker
*E.C. Nelson*
Stems subterranean. Leaves in a rosette, bipinnate; pinnae and pinnules not jointed at their bases. Cone scales with truncate apices, arranged in vertical rows; cones stalked or almost stalkless. Seeds with fleshy outer coat.

A genus of 2 species from eastern Australia which are fern-like in habit.

1a. Pinnules regularly toothed     **1. serrulata**
  b. Pinnules entire or irregularly toothed     **2. spectabilis**

**1. B. serrulata** (Bull) Chamberlain (*B. spectabilis* Hooker var. *serrata* Bailey; *B. spectabilis* var. *serrulata* Bull). Subterranean stems spherical, 20–25 cm in diameter with 5–20 leaf-bearing branches. Leaves 5–30. Pinnules sharply serrate with sharp teeth 1–3 mm. *Australia (SE Queensland)*. H5–G1.

**2. B. spectabilis** Hooker. Illustration: Botanical Magazine, 5398 (1863), 6008 (1872). Stems elongate, 2–10 cm in diameter, with 1–5 short, leaf-bearing branches. Leaves 1–7, erect, to 2 m. Pinnules 7–30 on each pinna, entire or with a few coarse teeth. *Australia (NE Queensland)*. H5.

## 4. CERATOZAMIA Brongniart
*E.C. Nelson*

Aerial stems unbranched. Leaves erect or spreading, gracefully recurved, with persistent bases. Cone scales arranged in vertical rows, not overlapping, each with 2 horns at the apex.

A genus of 2 species from Mexico, but only 1 is generally cultivated; it is easy to hand-pollinate, and may be grown in a glasshouse or outdoors in partial shade.

**1. C. mexicana** Brongniart. Illustration: Graf, Exotica, edn 3, 701, 704 (1963). Stems to 2 m. Leaves to 3 m; pinnae narrow, lanceolate, 10–40 × *c*. 8 cm. Male cones almost cylindric, to 15 cm. Female cone ellipsoid, to 15 cm. *Mexico*. H5–G1.

## 5. DIOON Lindley
*E.C. Nelson*

Stems aerial, covered with persistent leaf bases. Leaves woolly when young; pinnae with numerous veins running longitudinally, margins entire, apex pointed. Cones terminal, stalkless or shortly stalked. Cone-scales overlapping, spirally arranged; female scales with woolly, flattened ends.

About 4 species from C America, but only 1 is well-known in cultivation in Europe. It is difficult to pollinate, and female cones take about a year to mature. The seeds may be eaten if roasted or boiled; if not cooked they are poisonous.

**1. D. edule** Lindley. Illustration: Botanical Magazine, 6184 (1875); Graf, Exotica, edn 2, 493 (1959). Stems to 2 m. Leaves 1–1.5 m. Cones to 30 cm. *Mexico*. H5–G1.

## 6. ENCEPHALARTOS Lindley
*E.C. Nelson*

Stems subterranean or aerial, very rarely branched. Leaves erect or spreading, with persistent bases. Pinnae with longitudinal, parallel veins; margins entire, toothed or lobed. Cones axillary. Cone scales overlapping, spirally arranged. Female cone scales hairless or velvety, with truncate ends.

A genus of about 35 species widely distributed in Africa. There are about 26 species in South Africa, including several which are extremely rare, with wild populations composed of isolated individuals. They may be propagated from seeds or offshoots.

1a. Stems subterranean, or aerial part at most 50 cm     2
  b. Stems aerial and at least 50 cm     3
2a. Pinnae glaucous when young; median pinnae with 1 or 2 deep lobes on their lower margins    **2. horridus**
  b. Pinnae not glaucous; median pinnae with a few prickles on the margins and apex, not lobed    **4. villosus**
3a. Pinnae glaucous when young    **3. lehmannii**
  b. Pinnae not glaucous when young    **1. altensteinii**

**1. E. altensteinii** Lehmann. Illustration: Botanical Magazine, 7162, 7163 (1891). Aerial stems branched or unbranched, to 4 m (rarely to 7 m). Leaves 1–2 m, velvety when young, not glaucous. Pinnae inconspicuously veined, to 15 × 2.5 cm with 1–3 (rarely more) teeth on their margins. Cones 2–5 together, *c*. 50 cm, cylindric. *South Africa*. H5–G1.

**2. E. horridus** (Jacquin) Lehmann (*Zamia horrida* Jacquin). Illustration: Botanical Magazine, 5371 (1863); Graf, Exotica, edn 2, 493 (1959).
Stems usually subterranean, sometimes partly exposed for *c*. 30 cm (very rarely for more). Leaves glaucous, 50–100 cm, much recurved towards the apex. Pinnae to 10 × 2.5 cm, each with 1–3 sharply pointed lobes *c*. 4 cm on the lower margins. Cones solitary, to 40 cm. *South Africa (Cape Province)*. H5–G1.

**3. E. lehmannii** Lehmann. Illustration: Palmer & Pitman, Trees of southern Africa, 304–6 (1972).
Stems frequently branched from the base, to 1.5 m. Leaves glaucous when young but becoming dark green, 1–1.5 m, erect or slightly recurved. Pinnae to 18 × 2 cm, entire or very rarely with 1–2 small teeth on the lower margin. Cones blackish red, solitary, males to 35 cm, females to 50 cm. *South Africa*. H5–G1.

**4. E. villosus** Lemoine. Illustration: Botanical Magazine, 6654 (1882); Flowering plants of Africa 26: t. 1001, 1002 (1947); Dyer, Flora of southern Africa, 29 (1966).
Stems subterranean but sometimes exposed for *c*. 30 cm. Leaves almost hairless when young, becoming glossy green, 1–2.5 m, gracefully spreading. Pinnae 15–25 × 1.5–2.5 cm, usually with 1–3 forwardly pointing teeth on the margins, the apex sharply pointed and occasionally with 2 lateral spines. Cones solitary or in groups, borne on stalks *c*. 20 cm; males to 65 cm,

tapered towards the apex, females to 30 cm, ovoid. *South Africa*. H5–G1.

## 7. LEPIDOZAMIA Regel
*E.C. Nelson*

Aerial stems unbranched. Leaf bases persistent. Pinnae with numerous, parallel, longitudinal veins. Cones stalkless, axillary, usually solitary. Cone scales not overlapping, in vertical rows, with shortly velvety ends.

A genus of 2 species from eastern Australia, of which only 1 is in cultivation in Europe. The genus has sometimes been included in *Macrozamia*.

**1. L. peroffskyana** Regel (*Macrozamia peroffskyana* (Regel) Miquel; *M. denisonii* Moore & Mueller). Illustration: Graf, Exotica, edn 2, 494 (1959).
Stems 2–20 m. Leaves spreading, 2–3 m, bases with short hairs. Pinnae inserted on the upper midline of the leaf, numerous (160–200), entire, to 40 × 3 cm. *E Australia (Queensland to New South Wales)*. H5–G1.

## 8. MACROZAMIA Miquel
*E.C. Nelson*

Stems usually subterranean. Cones axillary, stalked. Cone scales overlapping, spirally arranged; female cone scales hairless.

A genus of 14 species from Australia; only 1 in general cultivation in Europe.

**1. M. communis** Johnson (*Zamia spiralis* misapplied; *Encephalartos spiralis* misapplied; *Macrozamia spiralis* misapplied). Illustration: Wrigley & Fagg, Australian native plants, 120 (1979).
Stem subterranean in the wild, but in cultivation the plant may form an aerial trunk 1–2 m. Leaves 50–100 in the crown, 70–200 cm. Pinnae 70–130, entire with sharply pointed apices, inserted near the edges of the leaf, 16–35 cm × 5–12 mm. Male cone cylindric, to 4 cm, female cylindric, to 45 cm. *SE Australia (New South Wales)*. H5–G1.

Plants offered as *M. spiralis* are usually this species; the true *M. spiralis* (Salisbury) Miquel has only 2–12 leaves in each crown.

## 9. ZAMIA Linnaeus
*E.C. Nelson*

Stems subterranean or partly exposed. Pinnae jointed at their bases. Cone-scales not overlapping, arranged in vertical rows, truncate.

A genus of perhaps 50 species from the Americas and Caribbean.

**1. Z. floridana** de Candolle (*Z. pumila* misapplied).
Leaves to 1 m. Pinnae *c.* 30, to 12 cm, margins curled inwards, occasionally toothed. Male cones *c.* 10 cm, in clusters of 4 or 5. Female cones to 15 cm, solitary. *USA (Florida).* H5–G1.

# II. ARAUCARIACEAE

Tall trees with columnar trunks; principal branchlet systems and often whole branches falling cleanly as the trees mature. Leaves long-persistent, leathery, simple, small, needle-like and sickle-shaped to large, flattened and broadly ovate or lanceolate. Male cones cylindric, catkin-like, axillary or terminal on short shoots, with 4–20 pollen sacs on the lower surface of each scale. Female cones usually large, more or less spherical, terminal on thick, short shoots, dismembering on the tree at maturity. Cone scales without distinct bracts and with a single reflexed ovule. Seed large.

A fairly small family of 2 genera and 32 species, found mainly in the tropics and south temperate regions extending north of the equator only in the Malay peninsula.

1a. Leaves opposite or almost so, ovate to lanceolate, usually shortly stalked, well separated on the shoot
**1. Agathis**
b. Leaves spirally arranged, triangular to sickle-shaped, stalkless or almost so, usually crowded and overlapping on the shoot **2. Araucaria**

## 1. AGATHIS Salisbury
*J. Lovett, C.N. Page & T.C. Whitmore*
Monoecious evergreen trees, often producing female cones several years before producing males. Trunk columnar, unbuttressed but often swollen near the base and sometimes with superficial roots. Bark smooth, flaking, exuding resin when damaged. Crown pyramidal with regularly whorled branches when young, becoming irregular at maturity. Branch-scars rapidly lost. Leaves shortly stalked, opposite or almost opposite, linear-lanceolate to broadly ovate, leathery, usually dark green, occasionally glaucous, with resin canals. Male cones solitary, stalkless, ovoid to cylindric, arising in the leaf axils. Female cones terminal, massive, more or less spherical, the scales scarcely spiny. Seed flattened-ovoid with a large, sometimes lobed wing.

A genus of 13 species of which 5 are endemic to New Caledonia; the remainder in Malaya and Sumatra eastwards to Fiji and from the Philippines to New Zealand, in tropical or subtropical rain forest. *A. australis* is grown in gardens; young and thus unidentifiable plants of several other species are grown as house plants. Propagation is by seed or by cuttings from erect shoots. Literature: Page, C.N., Leaf micromorphology in Agathis and its taxonomic implications, *Plant Systematics and Evolution* **135**: 71–9 (1980); Whitmore, T.C., A monograph of Agathis, *Plant Systematics and Evolution* **135**: 41–69 (1980); Whitmore, T.C. & Page, C.N., Evolutionary implications of the distribution and ecology of the tropical conifer Agathis, *New Phytologist* **84**: 407–16 (1980).

**1. A. australis** Salisbury. Illustration: Kirk, Forest Flora of New Zealand, f. 81 (1889); Dallimore & Jackson, Handbook of Coniferae, edn 4, 94 (1966); Krüssmann, Handbuch der Nadelgehölze, 55 (1972). Tree to 50 m. Leaves 1.5–6 × 1–1.5 cm, broadly lanceolate to linear-lanceolate, dark dull green, sometimes bronze or purplish on seedlings. Male cones cylindric, 1.5–5 cm × 7–10 mm. Female cones to 6 cm in diameter. *New Zealand (Auckland peninsula).* H5.

## 2. ARAUCARIA Jussieu
*J. Lovett, C.N. Page & T.C. Whitmore*
Evergreen, usually monoecious trees. Crown with a single main axis and with numerous whorled, horizontal or slightly ascending branches which fall or persist throughout the life of the tree. Trunk columnar, unbuttressed; bark dark grey-brown, thin and hard, sometimes flaking, wrinkled or lightly fissured, in rings or ring-like ridges in some species, exuding resin if damaged. Branch scars long-persistent. Leaves often closely clasping the branches or branchlets, leathery, spirally arranged, lanceolate, awl-shaped or triangular, sharply pointed, bright or dark green, changing considerably in form with age in some species. Male cones cylindric, solitary or in clusters. Female cones massive, spherical or ovoid, terminal, their scales usually with slender, curved spines. Seeds winged or unwinged.

A genus of about 19 species, of which 13 are endemic to New Caledonia, the rest in New Zealand, Australia, New Guinea and South America, in tropical, subtropical or warm temperate forests. Many of the

species are similar in their juvenile foliage. Propagation is by seed or cuttings of erect shoots.

The species are separated on details of their distinctive cones; however, as these are often not available or accessible only with difficulty, the species included here are distinguished on the basis of their general architecture.

1a. Crowns of mature trees pointed, tapering 2
b. Crowns of mature trees broadly domed, cylindric, irregular or flat-topped 3
2a. Crown extremely slender, the lower branches progressively falling and replaced by whorled secondary branches **1. columnaris**
b. Crown moderately broad, primary branches persisting usually throughout the life of the tree **2. heterophylla**
3a. Crown of mature trees domed 4
b. Crown cylindric, irregular or flat-topped 5
4a. Leaves dark green, scarcely flexible, closely spaced and overlapping, arranged all round all shoots **5. araucana**
b. Leaves often bright green, often fairly thin and flexible, widely spaced and spreading laterally on the weaker shoots **6. bidwillii**
5a. Crown fairly densely branched; leaves to 2 cm, awl-shaped on juvenile and weaker shoots, clasping on mature branches **3. cunninghamii**
b. Crown sparingly branched; leaves more than 2 cm, flattened and spreading on most shoots and branches **4. angustifolia**

**1. A. columnaris** (Forster) J.D. Hooker (*A. cookii* Endlicher). Illustration: de Laubenfels, Flore de la Nouvelle Caledonie et dépendances 4: 107 (1973).
Tree to 60 m. Crown pointed, tapering, becoming narrowly columnar with age. Branches short, horizontal, the primary branches soon shed and replaced, at least in the lower part of the tree by whorled secondary branches which are eventually similar in appearance to the primary. Branchlets in 2 ranks, horizontal or drooping, slender, whip-like, retained only along the outer halves of the branches. Leaves of 2 kinds, those on juvenile shoots 4–7 × 2–3 mm, awl-like, rigid, those on older shoots 5–7 × 5 mm, flattened, closely overlapping, ovate to triangular. Male cones 5–10 × 1.5–2.2 cm. Female cones

10–15 × 7–11 cm, spines *c.* 7 mm. *New Caledonia.* H5.

**2. A. heterophylla** (Salisbury) Franco (*Eutacta heterophylla* Salisbury; *A. excelsa* (Lambert) R. Brown; *Dombeya excelsa* Lambert). Illustration: Dallimore & Jackson, Handbook of Coniferae, edn 4, 115 (1966); Magrini, Le Conifère, 52–5 (1967); Hora, Oxford Encyclopaedia of trees of the world, 87 (1980).
Tree to 60 m. Mature crown pyramidal. Branches horizontal, eventually long, in whorls of 4–7, the primary branches usually persisting; branchlets in 2 ranks, horizontal or drooping, slender, whip-like, long-persistent. Leaves of 2 kinds, those on juvenile shoots or lateral branches 8–12 mm, soft, awl-shaped, incurved, those on older shoots 6 × 5–6 mm, flattened and somewhat rigid, overlapping, broadly ovate, with incurved, horny points; all bright green. Male cones 3.5–5 cm, borne in clusters. Female cones 7.5–10 × 9–11.5 cm, each scale terminating in a soft, flat, triangular spine of 1–1.2 cm. Seed 2.5–3 × 1.5 cm, exclusive of the well-developed wings. *New Zealand (Norfolk Island).* H5.

Several cultivars are recognised, differing in colour or foliage density. The compact 'Gracilis' is used as a pot plant.

**3. A. cunninghamii** Sweet. Illustration: Magrini, Le Conifère, 53 (1967); Krüssmann, Handbuch der Nadelgehölze, 59 (1972).
Tree to 45 m, usually dioecious. Mature crown flat topped, irregular or cylindric, fairly densely branched. Bark with conspicuous horizontal, circular hoops or bands. Branches long, in whorls of 4–7. Branchlets persistent only near tips of the branches, in dense tufts. Leaves of 2 kinds, those of juvenile and lateral branches 1.5–2 cm, spirally arranged, forwardly pointing, lanceolate to triangular, straight, sharply pointed, sometimes glaucous, those of older shoots shorter, more crowded, overlapping, incurved, shortly pointed. Male cones 5–7.5 cm. Female cones 10 × 7.5 cm, ovoid, symmetric. Seeds 6 mm, oblong, with a membranous wing on each side. *New Guinea, Australia (Queensland).* H5.
'Glauca' has silvery or glaucous leaves.

**4. A. angustifolia** (Bertoloni) Kuntze (*Colymbaea angustifolia* Bertoloni; *A. braziliana* Richard). Illustration: Magrini, Le Conifère, 52 (1967); Krüssmann, Handbuch der Nadelgehölze, 59 (1972).

Tree to 35 m with a sparingly branched, irregular or flat crown. Branches long, usually in whorls of 4–8; branchlets long and somewhat drooping. Leaves all of 1 kind, 3–6.5 cm × 5 mm, lanceolate, pointed, spreading, sometimes glaucous, stiff and leathery. Male cones 7–10.5 × 1.5–2 cm, usually on short shoots. Female cones 12.5 × 16.5 cm, narrowing from the middle upwards, the scales terminating in stiff, recurved appendages. Seeds to 5 × 2 mm, 8 mm thick, light brown. *Brazil.* H5.
'Elegans' has narrower leaves which are less stiff and pointed; 'Ridolfiana' is more robust in growth, with longer and broader leaves; 'Saviana' has narrower and more glaucous leaves.

**5. A. araucana** (Molina) Koch (*Pinus araucana* Molina; *A. imbricata* Pavon). Illustration: Den Ouden & Boom, Manual of cultivated conifers, 46–7 (1965); Magrini, Le Conifère, 51 (1967); Krüssmann, Handbuch der Nadelgehölze, 58 (1972).
Usually dioecious tree to 15 m; mature crown dome-shaped and often dense. Branches stout, spreading, in whorls of 4–8 (usually 5). Branchlets stout, usually in opposite pairs, horizontal or drooping, eventually retained only on the outermost parts of the branches. Leaves all of 1 kind, overlapping, ovate-lanceolate, 2.5–5 cm, spine-pointed, deep glossy green, margins thickened. Female cones usually axillary, solitary, erect, 7.5–13 × 5 cm, spherical or ovoid, taking 2–3 years to mature. Seeds oblong, slightly compressed, 2.5–4 cm, wingless. *Chile & Argentina (southern Andes).* H3.
The hardiest species, growing well in a moist soil in moist climates; it is possibly relatively short-lived. A few cultivars exist, but are infrequently grown.

**6. A. bidwillii** J.D. Hooker. Illustration: Den Ouden & Boom, Manual of cultivated conifers, 48, 49 (1965); Dallimore & Jackson, Handbook of Coniferae, edn 4, 115 (1966); Krüssmann, Handbuch der Nadelgehölze, 59 (1972).
Usually dioecious tree to 40 m. Crown densely branched, becoming broadly dome-shaped with age. Branches long, in whorls of 10–15, the lower ones eventually all falling cleanly beneath the dense crown, usually replaced on the upper part of the trunk by short, dense, secondary branches. Branchlets long, drooping. Leaves of 1 kind, 3–5 cm × 5–10 mm, overlapping, lanceolate-ovate, sharply pointed, thick, bright glossy green, widely spaced and

spreading on either side of vegetative shoots, shorter and more closely arranged on fertile shoots. Male cones to 15 × 1.5 cm, at the edge of the upper part of the crown. Female cones very massive, to 30 × 23 cm, spherical-ovoid, erect, borne only on the higher branches. Seeds 5–6.5 × 2.5 cm, pear-shaped, with rudimentary wings, maturing in the 3rd year. *Australia (SE Queensland).* H5.

## III. CEPHALOTAXACEAE

Small- to medium-sized evergreen trees or, in cultivation, irregular shrubs, with reddish, scaling or shredding bark. Branches opposite or whorled. Leaves spirally inserted but often twisted into 2 ranks on the lateral branchlets. Male cones stalked, with several scales, each scale with 3–8 pollen sacs; pollen without air-bladders. Female cones small, axillary, stalked; bracts cup-shaped, each surrounding a pair of ovules. Fertilisation delayed till the second year. Seed 1 to each scale, large, fleshy and drupe-like.

A family of 1 genus and 7 species, now restricted in the wild to Japan, China, Korea and northeastern India, but with fossil representatives in Europe and N America.

**1. CEPHALOTAXUS** Siebold & Zuccarini
*J. Lewis*
Evergreen trees or shrubs, 3–12 m, seldom producing secondary leaders in cultivation. Branchlets hairless, covered with leaf-bases. Buds numerous, with pointed scales. Leaves to 9 cm, spreading into 2 flat ranks, or 2 ranks with a V-shaped groove between them, sharply tipped. Male cones almost spherical; female of a few opposite scales. Seeds to 3 cm, green or brown.

Plants are very hardy, and grow well in a variety of soils except those which are very dry; exposed areas should be avoided.

1a. Leaves 2.5–6 cm, usually parallel-sided and straight; winter buds 2–5 mm **1. harringtonia**
  b. Leaves 5–9 cm, long-tapering, distinctly sickle-shaped; winter buds 1–2 mm **2. fortunei**

**1. C. harringtonia** (Forbes) Koch. Illustration: Harrison, Ornamental conifers, 34 (1975); Phillips, Trees in Britain, 100 (1978).
Small tree or more usually a shrub to 5 m. Leaves 2.5–6 cm, spreading in 2 ranks, more or less parallel-sided, apex sharply

acute, variable in colour, but with 2 broad, pale green bands beneath. Winter buds 2–5 mm, often grouped. Ripe seed 2.5–3 cm, broadly ovoid, often with a depressed apex, olive green. *Japan, China & Korea.* H1.

Var. **harringtonia** (*C. pedunculata* Siebold & Zuccarini) is a large shrub or bushy tree which is rare in cultivation and extinct in the wild (Korea & China). It has separated, bluish green leaves 3.5–6 cm, which spread either in 2 flat ranks or form a broad, outward-turned V.

Var. **drupacea** (Siebold & Zuccarini) Koidzumi (*C. drupacea* Siebold & Zuccarini). Illustration: Welch, Manual of dwarf conifers, t. 122 (1979). The variant from Japan and C China which is more commonly cultivated than var. *harringtonia*. It is a mound-like shrub, growing densely to *c.* 3 m, with the leaves yellowish green, 2–5 cm and forming an acute, inward-curved V, decreasing in size towards the branch tips.

'Fastigiata' can be grown into a strictly erect, spiral-leaved shrub if its laterals are pruned out. 'Prostrata', 'Nana' and 'Gnome' are dwarf variants.

**2. C. fortunei** Hooker. Illustration: Magrini, Le conifère, 74 (1967).
A slow-growing shrub in cultivation, rarely to 6 m. Leaves 5–9 cm, sickle-shaped, tapering and spreading into 2 flattened ranks, shiny yellow-green with 2 grey-white bands beneath; apex with a hard, sharp point. Winter buds 1–2 mm, usually solitary. Ripe seed *c.* 3 cm, ellipsoid, usually with a pointed apex. *C China.* H2.

'Prostrata' forms a widely spreading, dark green ground-cover; 'Grandis' is female, and has long leaves.

# IV. PODOCARPACEAE

Monoecious or dioecious evergreen trees or shrubs. Leaves spirally arranged or opposite, simple, ranging from minute and scale-like to large, linear or oblong; rarely replaced by flattened shoots (cladodes). Male cones catkin-like, each scale with 2 pollen sacs. Seeds borne in a cone in *Saxegothea*, but nut-like or drupe-like in the other genera, sometimes seated on and partially immersed in a fleshy aril or borne from a fleshy receptacle formed from swollen basal scales; seeds or arils often brightly coloured.

A family of at least 8 genera with about 180 species, mainly natives of the southern hemisphere.

The genera of the Podocarpaceae, and their boundaries, have recently been redefined, but for the small number of species included here, the established generic names have been retained for ease of reference, with the newer names cited as synonyms.

Most of the genera are occasionally in cultivation in Europe, but only 5 are widely available; many species remain to be introduced.

1a. Branchlets expanded to form cladodes; true leaves absent in adult plants
                    **3. Phyllocladus**
  b. Cladodes absent; adult plants with true leaves               2
2a. Leaves in adult plants generally more than 10 mm             3
  b. Leaves in adult plants generally much less than 10 mm       4
3a. Seeds numerous, in a cone formed of overlapping scales   **5. Saxegothea**
  b. Seeds arising singly or in small numbers only, nut-like or drupe-like, not in a cone    **4. Podocarpus**
4a. Prostrate shrub    **2. Microcachrys**
  b. Erect shrub or tree    **1. Dacrydium**

## 1. DACRYDIUM Forster
*E.C. Nelson*
Trees or shrubs with evergreen leaves, usually dioecious. Leaves of young plants with spreading tips, those of adult plants closely adpressed. Male cones catkin-like, in the axils of the upper leaves. Seeds arising at or near the tips of the branchlets, sometimes in a fleshy aril.

A genus of about 20 species distributed in countries around the Pacific Ocean, especially Australasia and Chile. Only 2 species are grown generally, though others may be found in botanic gardens. They all require relatively frost-free, moist, sheltered conditions and acid soil, and can eventually form elegant trees.
Literature: Quinn, C.J., Taxonomy of Dacrydium Sol. ex Lamb. emend. de Laub. (Podocarpaceae), *Australian Journal of Botany* **30**: 311–20 (1982); Edgar, E. & Conner, H.E., Nomina nova III, 1977–1982, *New Zealand Journal of Botany* **21**: 421–41 (1983).

1a. Leaves on adult plants *c.* 3 mm
                  **1. cupressinum**
  b. Leaves on adult plants *c.* 1 mm
                  **2. franklinii***

**1. D. cupressinum** Solander. Illustration: Salmon, The native trees of New Zealand, 74–7 (1980).

Tree to 50 m in the wild, rarely over 8 m in cultivation, pyramidal, with slender, hanging branchlets. Leaves of young plants to 7 × 1 mm, linear, sharply keeled; of adult plants to 3 mm, more rigid, closely adpressed, sometimes becoming bronzed in winter. Seed seated on a fleshy aril which is red when ripe. *New Zealand.* H5.

**2. D. franklinii** J.D. Hooker (*Lagarostrobus franklinii* (J.D. Hooker) Quinn). Illustration: Curtis & Stones, Endemic flora of Tasmania, t. 116 (1969).
Tree to 30 m with straight trunk in the wild, in cultivation forming a bushy shrub to *c.* 8 m. Branchlets slender, concealed by numerous minute leaves; in young plants leaves with pointed tips, in adults leaves scale-like, *c.* 1 mm, closely adpressed, with prominent, scattered white dots on the surface. Seeds seated in a non-fleshy aril, both seed and aril green when ripe. *Australia (Tasmania).* H5.

***D. bidwillii** J.D. Hooker (*Halocarpus bidwillii* (J.D. Hooker) Quinn). Illustration: Salmon, The native trees of New Zealand, 67–9 (1980). Ripe seed blackish, seated in a fleshy aril which is white when ripe. *New Zealand.* H5.

## 2. MICROCACHRYS J.D. Hooker
*E.C. Nelson*
Monoecious, low, evergreen shrub with prostrate, straggling branches to 1 m. Leaves scale-like, to 2 mm, opposite, adpressed and overlapping, keeled, producing tetragonal branches. Male cones to 3 mm. Seed seated in a large fleshy aril which is up to 8 mm long and bright red when ripe.

A genus of a single species native on exposed ridges and wet moors in Tasmania above 1200 m. It is not common in cultivation but is a useful, prostrate evergreen for rockeries. It is not very hardy, requiring moist, frost-free conditions and acid soil.

**1. M. tetragona** (J.D. Hooker) J.D. Hooker. Illustration: Curtis & Stones, Endemic flora of Tasmania, t. 61 (1969); Dallimore & Jackson, A manual of Coniferae, edn 4, 320 (1966).
*Australia (Tasmania).* H5.

## 3. PHYLLOCLADUS Richard & Richard
*E.C. Nelson*
Monoecious or dioecious evergreen trees or large shrubs. True leaves only present in seedlings or very young plants; in adult plants replaced by cladodes with thick, leathery, flattened segments. Male cones

stalked or stalkless, cylindric, clustered at the tips of short shoots. Seeds each borne on a fleshy aril, singly or in globular clusters, on the cladode margins or terminating short shoots.

A genus of 4 species, sometimes placed in a family of its own (Phyllocladaceae), distributed in Australasia and Malesia. The 3 Australasian species are grown in Europe; they grow best in moist, frost-free areas and prefer an acid soil. Literature: Hsuan Keng, The genus Phyllocladus (Phyllocladaceae), *Journal of the Arnold Arboretum* 59: 249–73 (1978).

1a. Cladodes mostly simple, sometimes lobed, very rarely truly pinnate
**1. aspleniifolius**
  b. Cladodes always pinnate    2
2a. Seeds in clusters of 10–20; cladodes 12–30 cm    **2. glaucus**
  b. Seeds in clusters of 2–3; cladodes less than 10 cm    **3. trichomanoides**

**1. P. aspleniifolius** (Labillardière) J.D. Hooker (*P. alpinus* J.D. Hooker).
Monoecious shrub or small tree, rarely over 5 m in cultivation. Cladodes variable in shape: lobes diamond- or fan-shaped, to 2 cm, deeply or shallowly lobed, margins toothed, base tapered; margins thickened in older plants. Male cone reddish, on a thick stalk. Arils crimson when ripe. *Australia (Tasmania), New Zealand.* H5.

Two varieties are recognised: var. **aspleniifolius** (Illustration: Curtis & Stones, Endemic flora of Tasmania, t. 60, 1969) from Tasmania has scale leaves 2–3 mm on the adult plants; var. **alpinus** (J.D. Hooker) Keng (Illustration: Salmon, The native trees of New Zealand, 84–5, 1980), from New Zealand, has scale leaves less than 1 mm.

**2. P. glaucus** Carrière. Illustration: Salmon, The native trees of New Zealand, 86–7 (1980).
Small, usually dioecious tree to 8 m. Cladodes pinnately compound, 12–30 cm with 5–10 segments which are diamond-or fan-shaped, bluish green above, glaucous beneath when young, 2–5 cm. Male cones in clusters of 10–20, catkin-like, 1–2 cm. Seeds usually densely clustered on the sides of the cladodes, 5–8 mm in diameter, the arils pinkish red when ripe. *New Zealand.* H5.

**3. P. trichomanoides** Don. Illustration: Salmon, The native trees of New Zealand, 88–9 (1980).
Usually dioecious tree to about 8 m in cultivation. Cladodes pinnately compound,

3–8 cm with 6–12 segments; segments diamond- or fan-shaped, 1.2–2.5 cm, margins toothed or deeply lobed. Male cones in clusters of 5–10, catkin-like, 8–10 mm. Seeds ovoid, 2–3 mm in diameter, sparsely clustered on the sides of normal cladodes or on cladodes reduced to stalk-like structures. *New Zealand.* H5.

**4. PODOCARPUS** Persoon
*E.C. Nelson*
Evergreen trees and shrubs, dioecious. Leaves simple, usually alternate. Male cones catkin-like, terminal or in axillary clusters or in spikes, each scale with 2 pollen sacs. Seeds drupe- or nut-like, usually terminal, solitary or occasionally in spikes, borne on a short, often scaly axis, a few scales often swollen to form a fleshy receptacle. Ovules 1 or 2, inverted.

A genus of about 100 species distributed widely in the southern hemisphere extending into tropical eastern Asia and Japan. Only about 12 species are widely cultivated in Europe, as specimen trees or shrubs in larger gardens. A few slow-growing species are suitable for rock gardens. They are generally frost-sensitive, and most grow well only in mild, frost-free areas. They thrive equally well on acid or alkaline soils, and are propagated easily by seed or by cuttings (taken in summer). In ideal conditions the larger species will form handsome, tall trees.
Literature: Bucholz, J.T. & Gray, N.E., A taxonomic revision of Podocarpus, *Journal of the Arnold Arboretum* 29: 46–76, 123–51 (1948), 32: 82–9 (1951), 37: 160–72 (1956), 39: 424–77 (1958); Hanan, A.M.S., Southern hemisphere conifers in Ireland, *Conifers in the British Isles*, 20 (1972).

*Habit.* Dwarf shrubs: **5,8**; shrubs 1–4 m: **1,2,5,8.**
*Bark.* Paper-like: **4**; flaky: **11**; stringy: **12.**
*Branchlets.* Hanging: **3,4,11**; spreading at wide angles: **11.**
*Leaves.* Sickle-shaped: **9,10,12.** With numerous veins: **7.** With blue-green bands beneath: **2,9.** With thickened margins: **8.** Ovate: **7.**
*Seeds.* Bluish green: **7**; green: **6,10**; whitish green: **2**; black: **3,10.**
*Receptacle.* Red: **1,3–5,8,12**; purple: **2.** Not fleshy: **2,7,11,12.**

1a. Leaves more than 3 cm    2
  b. Leaves less than 3 cm    6
2a. Leaves ovate to lanceolate, 1.5–3.5 cm broad, with numerous longitudinal veins    **7. nagi**

  b. Leaves lanceolate or sickle-shaped, less than 1 cm broad, with a midrib only    3
3a. Leaves more than 5 cm    4
  b. Leaves less than 5 cm    5
4a. Leaves lanceolate, 7–9 × c. 1 cm
**6. macrophyllus**
  b. Leaves sickle-shaped to lanceolate, less than 7 mm broad    **10. salignus**
5a. Leaves c. 4 cm, dark green above with pale blue-green bands beneath
**9. nubigenus**
  b. Leaves less than 4 cm, pale green, without bands beneath    **4. hallii**
6a. Leaves less than 5 × 1 mm, brownish green    **3. dacrydioides**
  b. Leaves larger, not brownish green    7
7a. Shrubs rarely over 3 m    8
  b. Trees, very rarely shrubs    10
8a. Leaves rigid with prominent, swollen midribs    **8. nivalis**
  b. Leaves not rigid, without swollen midribs    9
9a. Leaves with acute, sharply pointed apices    **i. acutifolius**
  b. Leaves with blunt apices
**5. lawrencei**
10a. Leaves 5–10 mm, in 2 distinct rows
**11. spicatus**
  b. Leaves generally more than 10 mm, spirally arranged    11
11a. Leaves c. 2 cm, pale green on both surfaces, grooved on the upper surface along the midrib
**12. totara**
  b. Leaves c. 3 cm, dark green on upper surface with pale blue-green bands beneath, midrib raised on upper surface    **2. andinus**

**1. P. acutifolius** Kirk. Illustration: Salmon, The native trees of New Zealand, 62–6 (1980).
Shrub or small tree to 9 m. Leaves lanceolate, 1.5–2.5 cm × 7–35 mm, pale green on both surfaces, apex acute, midrib indistinct. Seed ovoid; receptacle usually swollen, succulent, red. *New Zealand (South Island).* H5.

This species is seldom cultivated, but it is hardy in western and south-western Europe.

**2. P. andinus** Endlicher (*Prumnopitys elegans* Phillipi). Figure 28(2) p. 69.
Illustration: Dallimore & Jackson, Handbook of Coniferae, edn 4, t. 101 (1966).
Monoecious or rarely dioecious tree to 18 m, or a shrub. Leaves closely arranged on hairless branchlets, linear, straight or sickle-shaped, c. 3 cm × 3 mm, dark green above, with 2 glaucous, pale, blue-green

bands beneath, apex acute. Male cones ellipsoid, in terminal spikes. Seed whitish green, drupe-like; receptacle not swollen. *Andes of SC Chile & Argentina*. H4.

This species, sometimes placed in the separate genus *Prumnopitys* Phillipi, is similar in habit and appearance to *Taxus baccata* (p. 105) which has the leaves green beneath and the male cones spherical and solitary in the leaf-axils. It requires protection from cold winds, but otherwise is the hardiest species. It is also slow-growing and thrives in fertile soils including those rich in chalk.

**3. P. dacrydioides** Richard. Illustration: Salmon, The native trees of New Zealand, 51–4 (1980).
Tree to 50 m in the wild. Branchlets slender, hanging. In young plants, leaves more or less in 2 ranks, linear, decurrent, 3–7 × 0.5–1 mm; in adult plants, the leaves overlapping, adpressed, lanceolate, 1–2 mm, dull brownish green, sometimes acuminate. Seed black, nut-like; receptacle fleshy, swollen, orange-red. *New Zealand*. H5.

A distinct species with minute leaves, sometimes placed in a separate genus, *Dacrycarpus* (Endlicher) de Laubenfels. It forms a small elegant tree, but is suitable only for mild areas.

**4. P. hallii** Kirk. Illustration: Salmon, The native trees of New Zealand, 62–6 (1980).
Tree to 20 m; bark thin, paper-like. Branchlets hanging in young plants. Leaves linear to ovate, 2–5 cm × 3–5 mm, largest on juvenile plants, green on both surfaces, apex acute. Seed ovoid, nut-like; receptacle fleshy, swollen, usually red. *New Zealand*. H5.

Sometimes considered to be a variety of *P. totara*, this species differs in the form of the juvenile foliage, in the bark and in seed shape. It is found in habitats about 500 m higher in altitude than *P. totara* and is hardier. It grows well in southern and western Europe.

**5. P. lawrencei** J.D. Hooker (*P. alpinus* invalid). Illustration: Dallimore & Jackson, Handbook of Coniferae, edn 4, t. 100 (1966).
Prostrate or erect shrub to 2 m. Leaves linear, to 2 cm × 2–4 mm, green to dark green on both surfaces, apex acute. Seed nut-like; receptacle fleshy, red. *SE Australia*. H4.

A very slow-growing plant, attaining 2 m only after 50 years.

**6. P. macrophyllus** (Thunberg) Don (*P. chinensis* Sweet; *P. japonicus* Sieber). Illustration: Siebold, Flora Japonica, pl. 133, & 134 (1870); den Ouden & Boom, Manual of cultivated conifers, 364 (1965).
Tree to 20 m, occasionally a shrub. Leaves closely arranged on the branchlets, linear-lanceolate, straight, to 9 × *c.* 1 cm, bright green above, paler and slightly glaucous beneath except on the midrib, apex acute or obtuse. Seed green; receptacle fleshy, purple. *S Japan, S China*. H4.

Long cultivated in China and Japan. The species is represented in Europe by var. **maki** Endlicher, which is slow-growing and forms an erect-branched shrub or small tree rarely over 7 m. Four cultivars are listed, including 'Aureus' with leaves margined or striped with yellow, 'Argenteus' with leaves margined or striped with white, 'Hillier's Compact' which is very slow-growing, and 'Angustifolius' with longer and narrower leaves.

**7. P. nagi** (Thunberg) Makino (*P. nageia* R. Brown; *Nageia nagi* (Thunberg) Kuntze). Illustration: Siebold, Flora Japonica, pl. 135 (1870).
Tree to 25 m in the wild; bark smooth, brownish purple. Leaves more or less opposite, lanceolate to ovate, 3–8 × 1.2–3 cm, leathery, shining, dark green above, pale green or whitish green beneath with numerous, fine, longitudinal veins visible. Seed spherical, bluish green, slightly glaucous; receptacle not fleshy. *Japan, Taiwan*. H5.

Not common in cultivation except in the mildest parts of the British Isles and southern Europe. Two cultivars are available: 'Rotundifolia' with more rounded leaves, and 'Variegata' which has the leaves variegated with white.

**8. P. nivalis** J.D. Hooker. Illustration: Salmon, The native trees of New Zealand, 62–6 (1980).
Prostrate shrub with ascending branches, or a low shrub to *c.* 3 m. Leaves closely and spirally arranged, linear to oblong, 5–15 × 2–4 mm, rigid, margins and midrib thickened, apex acute or obtuse. Seed ovoid, nut-like; receptacle fleshy, red. *New Zealand*. H4.

Suitable for use as a ground-cover or rock garden plant. It is mostly slow-growing, and prostrate variants remain less than 30 cm tall. The cultivar 'Aureus' has bronze-green leaves.

**9. P. nubigenus** Lindley.
Tree, 15–25 m in the wild. Leaves linear-lanceolate, straight or sickle-shaped, 2–4 cm × 3–8 mm, rigid, green above and with 2 distinct blue-green bands beneath, apex mucronate. Seed spherical; receptacle fleshy. *S Chile*. H5.

Requires a high rainfall; in its native habitats rainfall exceeds 5000 mm per annum.

**10. P. salignus** Don. Illustration: Dallimore & Jackson, Handbook of Coniferae, edn 4, t. 106 (1966).
Pyramidal tree to 20 m, or shrubby. Branchlets green, hairless. Leaves more sparsely arranged on the branchlets than in *P. macrophyllus*, linear-lanceolate, often sickle-shaped, to 10 cm × 3–8 mm, dark green or bluish green above, paler green beneath, apex pointed. Seeds green, solitary or in pairs on a single stalk; receptacle fleshy. *Chile*. H4.

Forms an elegant tree suitable for gardens in mild, moist regions of western and south-western Europe.

**11. P. spicatus** Mirbel. Illustration: Salmon, The native trees of New Zealand, 55–8 (1980).
Tree to 25 m, bark flaking. Branches of young trees slender, flexuous, hanging and spreading at wide angles. Leaves of young plants 5–10 × 1–2 mm; in adult plants leaves 1–1.5 cm × 1–2 mm, in 2 opposite ranks at the tips of the branchlets, linear-oblong, dark or brownish green above, glaucous beneath, obtuse or sometimes apiculate. Seed black, drupe-like; receptacle not swollen. *New Zealand*. H5.

**12. P. totara** Lambert. Illustration: Salmon, The native trees of New Zealand, 62–6 (1980).
Tree to 30 m; bark thick, stringy, furrowed. Leaves linear-lanceolate in young plants, to 2 cm; in adult plants 1.5–3 cm × 3–4 mm, straight or sickle-shaped, dull green on both surfaces, sometimes tinted bronze, apex acute. Seed nut-like; receptacle usually fleshy, red. *New Zealand*. H5.

**5. SAXEGOTHEA** Lindley
*E.C. Nelson*
Monoecious evergreen trees to 15 m. Branchlets opposite or in whorls of 3–4. Leaves spirally arranged, spreading radially on leading shoots, in opposite ranks on lateral shoots, twisted, linear, 1–2 cm, shortly stalked, tapering to a sharply pointed apex; midrib prominent, lower surface with 2 bluish green bands. Ripe female cones to 8 mm in diameter, with 6–12 fleshy scales.

A genus of a single species, mainly grown in large gardens and arboreta, where it forms a fine evergreen like *Taxus baccata*. It has no particular cultural requirements, but is slow-growing and will tolerate lime.

**1. S. conspicua** Lindley. Illustration: Dallimore & Jackson, Handbook of Coniferae, edn 4, 571 (1966). *Chile*. H5.

## V. TAXODIACEAE

Medium-sized or large, monoecious, evergreen or deciduous trees, usually with shredding bark. Leaves usually linear or broadly awl-shaped, spirally arranged but sometimes appearing 2-ranked, or, more rarely, needle-like and in clusters. Male cones small, solitary or in clusters; pollen-sacs 2–9 per scale. Female cones with bracts and cone-scales free when young, later these closely fused; ovules 2–9 per scale; mature cones broadly ellipsoid, ovoid or almost spherical, the scales persistent, spirally arranged, leathery or woody. Seeds 2–9 on each scale, usually winged.

A family of 10 genera which contain, between them, only 14 species, mostly from eastern Asia but extending to Tasmania, and with 3 genera in N America. Of the Asiatic genera, 2 are rarely seen in cultivation in Europe, and then only in botanic gardens or specialist collections: **Glyptostrobus**, which is similar to *Taxodium*, but with 3-ranked leaves and stalked, pear-shaped cones, is somewhat tender (H4 or 5) but colours well in autumn; **Taiwania**, with *Cryptomeria*-like foliage when young, is hardier (H3), but has little to recommend it decoratively.

*Habit*. Deciduous: **1,2**; evergreen: **3–8**.
  Distinctly conical: **1,2,4–6**; irregular: **3,4,7,8**.
*Height (in cultivation)*. Well over 20 m: **1,2,5,6,8**; about 20 m: **1,4,7,8**; less than 20 m: **1,3,4,7**; dwarfs (cultivars): **6–8**.
*Leaves*. In 2 ranks: **1,2,6,7**; in spirals: **3,5,8**; in distinct whorls: **4**. Sheathing the branchlets: **3**. More than 2.5 cm and blunt: **4**; more than 2.5 cm and spine-tipped: **7**; less than 5 mm: **3,5**. Flat and narrowly elliptic: **1,2,6**; awl-shaped: **8**.
*Cones*. Distinctly ellipsoid: **1,6**; ovoid: **3,4,7,8**; irregularly spherical: **2**. More than 5 cm: **4,5**; *c*. 3 cm: **6,7**; 2–3 cm: **1–3,6,8**; less than 2 cm: **8**.

1a. Irregular bushy trees when young, with whorls of thick needles at the nodes; cones smooth, 5–10 cm
  **4. Sciadopitys**
  b. Regular or irregular conical trees with scale-like or linear leaves; cones sometimes rough, usually much less than 5 cm, rarely to 8 cm **2**
2a. Leaves in 2 spreading ranks along the branchlets **3**
  b. Leaves closely adpressed or in distinct spirals **6**
3a. Leaves usually more than 3 cm, stiff and strongly spine-tipped
  **7. Cunninghamia**
  b. Leaves to 3 cm, pliable and not spine-tipped **4**
4a. Outline of branchlets (with leaves) tapering from base to apex; leaves scarcely over 1 mm wide; cones almost spherical, lumpy
  **2. Taxodium**
  b. Outline of branchlets (with leaves) almost parallel sided or very narrowly elliptic; leaves more than 1 mm wide; cones usually broadly ellipsoid, regular **5**
5a. Leaves deciduous, parallel-sided, bright green; cones light brown, often with a terminal point
  **1. Metasequoia**
  b. Leaves evergreen, very narrowly elliptic, dull green; cones dark brown, rounded above **6. Sequoia**
6a. Leaves usually less than 1 cm, if more, then curved and rounded on the back **7**
  b. Leaves 1 cm or more, mostly stiff and angled on the back **8. Cryptomeria**
7a. Small irregular trees with incurved or scale-like leaves; mature cones ovoid, bristly, *c*. 2.5 cm **3. Athrotaxis**
  b. Tall, conical trees with short, divergent leaves; mature cones ellipsoid, smooth, to 8 cm
  **5. Sequoiadendron**

## 1. METASEQUOIA Miki
*J. Lewis*
Deciduous trees to 45 m, conical in cultivation, trunk deeply fluted below, branchlets ascending; bark reddish brown, peeling thinly in platelets. Branchlets hairless, some persistent and bearing shortly stalked, ovoid buds and few leaves, others *c*. 10 cm long, deciduous, bearing leaves twisted into 2 horizontal ranks. Leaves *c*. 1.5 cm (twice as long in seedlings), linear, slightly curved, colouring in autumn. Male cones *c*. 3 mm, ovoid, in drooping tassels; female cones *c*. 8 mm, terminal on short branchlets, cylindric. Mature female cones to 2.4 cm, made up of about 25 transversely elliptic, grooved scales, each usually bearing 5–8 ovules. Seed *c*. 5 mm, with 2 broad, pale wings.

A genus of a single species, first discovered as a fossil, later found living in C China. The plant is rapid in growth and hardy, tolerant of a variety of soils and of moist conditions, but requires good drainage and a sunny situation to grow well. Propagation is by (wild) seed and by cuttings.

**1. M. glyptostroboides** Hu & Cheng. Illustration: Magrini, Le conifère, 167 (1967). *C. China (NW Hubei)*. H1.

A number of cultivars is available, including 'National', which was selected in the USA.

## 2. TAXODIUM Richard
*J. Lewis*
Deciduous, irregularly branched trees to 40 m, with scaly, peeling bark; trunk sometimes buttressed below and with hollow protuberances extending above ground from the roots. Lateral shoots deciduous. Terminal buds rounded. Leaves spirally arranged, those on the lateral shoots pale green when young, colouring in autumn, twisted into 2 ranks, narrow, shortly pointed, 1–2 cm. Male cones in slender, purplish, drooping clusters. Female cones scattered, more or less terminal, of numerous fleshy scales, each scale bearing 2 erect ovules. Mature cones *c*. 2.5 cm, nearly spherical, lumpy, resinous; scales shield-shaped, falling separately. Seeds irregularly triangular, winged.

A genus of 2 very similar species from the lowlands of the eastern and southern USA, extending round the Gulf of Mexico to southern Mexico. The one cultivated species flourishes in moist, loamy soils and even in standing water, but requires warm summers to grow well. Propagation is by seed; young trees need protection from frost.

**1. T. distichum** (Linnaeus) Richard (*T. ascendens* Brongniart). Illustration: Bloom, Conifers for your garden, 125 (1972); den Ouden & Boom, Manual of cultivated conifers, 386–8 (1965). Tree to 40 m. Leaves linear-lanceolate, 1–1.5 cm, bright green. Mature cone ovoid or spherical, *c*. 2.5 cm in diameter. *S & SE USA (Delaware to Texas)*. H1.

Var. **nutans** (Aiton) Sweet (*T. ascendens* 'Nutans') has pendent branchlets. About 20 cultivars have been described, but only 'Pendens', with hanging branches, survives in the regular trade.

## 3. ATHROTAXIS D. Don
*J. Lewis*

Monoecious, evergreen trees, usually small, slow-growing in cultivation and usually poorly shaped in age. Bark shredding and flaking. Leaves in more or less overlapping spirals. Male cones solitary and terminal or nearly so. Female cones solitary and stalkless, with 15–25 scales, each bearing 3–6 inverted ovules. Cones maturing in 1 year, to 2.5 cm, almost spherical, the brown, woody, spine-tipped scales spreading to release the oblong, 2-winged seeds.

A genus of 3 species from Tasmania, found mainly in botanical collections. They grow in warm, light, loamy, moist, lime-reduced soils and require a sunny, sheltered site. Propagation is by seed, by cuttings, or by grafting (on *Cryptomeria* stocks).

1. **A. laxifolia** Hooker. Illustration: Mitchell, Conifers in the British Isles, t. 4 (1972); Phillips, Trees in Britain, 87 (1978). Tree to *c.* 20 m in cultivation. Leaves clasping the branchlets but with free, incurved tips. Cones *c.* 2 cm, yellow to orange-brown, spiky and open when mature. *Australia (Tasmania).* H4.

Rather slow to become established, but susceptible to training as a small tree, producing cones when young.

\***A. cupressoides** D. Don. Similar, but with small, adpressed leaves on terete branches. *Australia (Tasmania).* H4.

\***A. selaginoides** D. Don. Taller and less hardy, with longer, spreading leaves (like those of *Cryptomeria*). *Australia (Tasmania).* H5.

## 4. SCIADOPITYS Siebold & Zuccarini
*J. Lewis*

Evergreen, monoecious trees to 30 m, with short branches and branchlets. Leaves *c.* 10 cm, linear, needle-like, grooved, in whorls of 15–25 in the axils of rings of small scale leaves. Male cones *c.* 8 mm, clustered, terminal; female cones solitary on short stalks, made up of membranous bracts and small scales. Mature cones to 10 cm, ovoid, ripening in the 2nd year. Scales wedge-shaped, thickened, brown, furrowed, shortly hairy, ultimately with the margins downwardly curved. Seeds 5–9 per scale, narrowly winged.

A genus of a single species which is slow-growing even on fertile soils; it does best on a lime-free, peaty loam. Propagation is by seed.

1. **S. verticillata** (Thunberg) Siebold & Zuccarini. Illustration: Harrison, Ornamental conifers, t. 423–4 (1975). *Japan (C Honshu).* H1.

## 5. SEQUOIADENDRON Buchholz
*J. Lewis*

Very large, long-lived, monoecious, evergreen trees. Bark thick, fibrous, reddish brown. Leaves 3–6 mm, scale-like, spirally arranged, overlapping but spreading, on drooping branchlets. Male cones terminal, on short branchlets. Female cones terminal, with numerous scales bearing long spines when young, and 2 rows of erect ovules. Mature cones 5–8 cm, ellipsoid, ripening in the 2nd year. Scales woody, shield-like, diamond-shaped, with 3–9 broadly winged seeds.

The 1 species does best in sheltered habitats on fertile, well-drained soil under fairly heavy rainfall, preferring cool summers and mild winters. Propagation is by seed; cuttings are much less easy to root than those of *Sequoia*.
Literature: St Barbe Baker, R., *The Redwoods* (1943); Kaufmann, J.M., Giant sequoias, *National Geographic Magazine* **106**: 147–86 (1959).

1. **S. giganteum** (Lindley) Buchholz (*Sequoia gigantea* (Lindley) Decaisne; *Wellingtonia gigantea* Lindley). Illustration: Chandun, Ornamental conifers, f. 30, 31. *Western USA (mountains of California).* H2.

A number of colour variants have been selected and named, e.g. 'Argenteum', 'Flavescens' and 'Variegatum'. The unusual 'Pendulum' has branchlets hanging quite vertically from downwardly curved branches. 'Compactum' is a glaucous, tiered, conical tree to 20 m, and there is a dwarf variant, 'Pygmaeum', which forms a dense shrub to 4 m, but will grow vigorously upright if strong leading shoots are not pruned out.

## 6. SEQUOIA Endlicher
*J. Lewis*

Very tall, long-lived, monoecious, evergreen trees. Bark fibrous, spongy, reddish brown. Leaves on lateral branchlets 2-ranked, 1–2.5 cm, linear to lanceolate, shortly pointed, with 2 pale bands beneath, or (when spirally arranged on leading shoots), shorter and oblong with a horny point. Male cones *c.* 4 mm, terminal. Female cones with 15–20 shortly pointed, ovate bracts, each bearing 5–7 ovules. Mature cones *c.* 2.5 cm, ripening in the first year, ovoid, pendent, rounded at the apex, persistent, reddish brown. Scales peltate, woody. Seeds light brown, 2-winged.

The single species thrives in moist sites on deep, rich soil, especially in areas with cool summers, mild winters and high rainfall; it requires shelter from the wind to

develop its natural shape. Propagation is by seed, although fertility is low and the seedlings are very shade-sensitive. Cuttings of the cultivars usually root well.

1. **S. sempervirens** (Lambert) Endlicher (*Taxodium sempervirens* Lambert). Illustration: Mitchell, Conifers in the British Isles, t. 15 (1972). *Western N America (maritime highlands of N & C California, just extending into S Oregon).* H2.

Less given to colour variation than *Sequoiadendron*, though 'Glauca' and 'Variegata' have been selected and creamy young growth occurs in other variants. 'Pendula' (creeping) and 'Nana Pendula' (almost prostrate) have specialist uses and 'Prostrata', which has leaves twice the normal width, suits a high site on a rockery. 'Adpressa' is a popular variant; it is a slow-growing, dense bush with creamy white young growth, and can be trained by pruning. All the low-growing variants need to have any strong, erect leaders pruned out to maintain their character.

## 7. CUNNINGHAMIA Richard
*J. Lewis*

Monoecious, evergreen trees, conical, with whorled branches when young, to 40 m, bark shredding. Lateral branchlets deciduous. Leaves to 7 cm, spirally arranged but twisted into 2 ranks on lateral branches, surviving for 5 years or more, lanceolate, bright green, with 2 pale bands beneath, sharply tipped. Male cones shortly cylindric, clustered at the branch tips. Female cones ovoid, borne near the branch tips, enclosed in leaf-like scales. Mature cones with stiff, brown, long-spined scales. Seeds 3 per scale, winged.

A genus of 2 very similar species from China and Taiwan. Only the species originating from China is in cultivation; it thrives in moderately warm sites, on loamy soils, and is prone to frost damage when young. Propagation is by seed and by cuttings, especially of adventitious shoots, though these may be slow-growing at first.

1. **C. lanceolata** (Lambert) Hooker (*C. sinensis* Richard). Illustration: den Ouden & Boom, Manual of cultivated conifers, 130 (1965). Leaves 3–7 cm. Cones 3–4 cm. *C & SE China.* H4.

A variant with a silvery sheen has been selected as 'Glauca' and the dwarf 'Compacta', hardier than the wild variant of the species, is likely to gain in popularity when better known.

## 8. CRYPTOMERIA D. Don
*J. Lewis*

Evergreen, buttressed, monoecious trees to 50 m in the wild, to 30 m in cultivation. Bark reddish brown, shredding. Branchlets hairless, deciduous. Leaves to 2 cm, spirally arranged in 5 ranks, persisting for 4–5 years, awl-shaped, blunt, margins incurved. Male cones *c.* 7.5 mm, terminal in clusters of 20 or more, orange or red when ripe. Female cones on short branchlets, surrounded by small rosettes of leaves. Ovules erect. Mature cones solitary, spherical, *c.* 1.5 cm; scales 20–30, with 2–3 lobes and a recurved, triangular point on the back. Seeds triangular, very narrowly winged.

The single species thrives on good, moist soil in all but the coldest parts of Europe, but it is not suitable for exposed, dry sites. It tolerates heavy pruning, and is propagated by seed or cuttings. Literature: Tarazaki, W., The Distribution of wild Cryptomeria and its varieties, *Proceedings of the Japanese Association for the Advancement of Science* 4: 460 (1929); Kruse, H., Japans Sugi, *Deutsche Baumschule* 22: 66 (1970).

**1. C. japonica** (Linnaeus filius) D. Don. Illustration: Phillips, Trees in Britain, 109 (1978); Harrison, Ornamental conifers, 68 (1975).
*China, Japan.* H2.

Variable. Var. **sinensis** Siebold & Zuccarini, from S China, is a more tender tree with a looser, weaker habit, and cones with fewer seeds. Var. **radicans** Nakai, a denser, heavier variant, is widely cultivated in Japan ('Dai-sugi' and 'Ashio-sugi') and, less commonly, in Europe.

Over 50 different cultivars have been named but the distinctions between them are uncertain, and there is a confusing overlap between those with European and Japanese names.

## VI. CUPRESSACEAE

Tall or medium-sized trees, mostly with small and frequently scale-like leaves, usually in opposite pairs. Female cones small, to 2–3 cm when mature, with opposite, paired scales, finally woody or (in *Juniperus*) berry-like, resulting from the more or less complete fusion of the bracts and ovuliferous scales. Seeds usually 3–many per scale, erect.

A large, widespread family of about 20 genera which occur in both N & S temperate zones: a northern group which includes the almost continuous belt of *Juniperus* and a smaller, southern, discontinuous range in Africa and Australasia. **Tetraclinis articulata** (Vahl) Masters, which occurs wild in N Africa and S Spain is rarely cultivated at present, but it could prove a useful evergreen for dry areas; it forms a rather irregular, dark green tree.

*Habit.* Tree: **1,3,5–8**; small tree: **4,9**; shrub: **2,9,10**.
*Foliage.* In flattened sprays: **1,3–5,7,8**; on branchlets in different planes: **2,6,9,10**.
*Leaves.* Scale-like: **1,3–10**; oblong: **2**; needle-like: **9**. Spirally arranged: **6**; in whorls of three: **2**; in unequal pairs: **1,3,4,7,8**; regularly arranged: **2,5–10**.
*Mature cones.* Oblong or ovoid: **1,3,5,6**; spherical: **2,4,6–10**.
*Cone scales.* Erect: **1–5,9,10**; peltate: **6–8**. Three to five: **1,3,9,10**; six to eight: **3–10**; nine to twelve: **2,3,6–8**. Leathery: **1,3,4,9**; woody: **1,3,4,9,10**; fleshy: **9**.
*Seeds.* 1 per scale: **1,9,10**; 2 per scale: **1–3,5,7**; more than two per scale: **2–4,6,8**. Winged: **1,2,4–8**; unwinged: **3,9,10**.

1a. Foliage in flattened sprays; leaves scale-like in opposite, unequal pairs **2**
 b. Foliage on branchlets in different planes; leaves spirally arranged, in opposite, equal pairs, or in whorls of 3, either oblong and decurrent or scale-like, rarely needle-like **7**
2a. Lateral leaf-pairs strongly boat-shaped with large white areas beneath, spreading and with blunt, incurved tips; cones rough, spreading open **4. Thujopsis**
 b. Lateral leaf-pairs not boat-shaped, with white stripes beneath, the tips acute and spreading; cones oblong or ovoid, smooth, not spreading widely open **3**
3a. Cones oblong or ovoid; cone-scales erect, leathery, usually with smooth margins and nearly terminal bosses or spines **4**
 b. Cones spherical; cone-scales peltate, woody, with angular margins and short, small, central tips **6**
4a. Foliage usually aromatic when bruised; cone-scales 8–10; seeds unwinged **3. Thuja**
 b. Foliage not aromatic; cone-scales 4–6; seeds unequally winged **5**
5a. Foliage pale blue-green; lateral leaf-pairs blunt-tipped, much longer than the facial pairs **1. Austrocedrus**
 b. Foliage dark green; lateral leaf-pairs sharply tipped, almost as long as the facial pairs **5. Calocedrus**
6a. Mature cones 6–12 mm, smooth or sharply rough; cone-scales 4–6; seeds 2 per scale **7. Chamaecyparis**
 b. Mature cones 1–2 cm, bluntly tuberculate; cone-scales about 8; seeds *c.* 5 per scale **8. X Cupressocyparis**
7a. Leaves in whorls of 3, oblong, decurrent **2. Fitzroya**
 b. Leaves spirally arranged or in opposite, equal pairs, rarely needle-like **8**
8a. Cones distinctly woody and hard with adpressed, peltate scales, each bearing 6–10 seeds **6. Cupressus**
 b. Cones leathery, sometimes partly woody or fleshy and berry-like, enclosing 1–12 seeds altogether **9**
9a. Cones more than 3 mm, leathery (sometimes partially woody), fleshy and berry-like, remaining whole at maturity **9. Juniperus**
 b. Cones to 3 mm, leathery and somewhat woody, appearing berry-like but breaking up at maturity **10. Microbiota**

## 1. AUSTROCEDRUS Florin & Boutelje
*J. Lewis*

Evergreen, monoecious trees, approximately conical in shape with upcurved, grey-barked branches. Branchlets flattened with 4 series of very unequal leaves, the lateral pairs 3 × as long as the facial pairs, and grey-green with distinct, single, white bands beneath. Male cones profuse. Female cones to 2 cm, solitary, of 4 scales, of which 2 are fertile. Seeds 1 per scale, winged.

A genus of 1 species which grows slowly in cultivation but is tolerant as to soil and withstands some aridity. Propagation is said to be by cuttings.

**1. A. chilensis** (D. Don) Florin & Boutelje. Illustration: Harrison, Ornamental conifers, 27 (1975).
*Chile, Argentina.* H3.
Rarely fertile in cultivation.

## 2. FITZROYA Lindley
*J. Lewis*

Trees or shrubs with pendulous branches. Leaves in 3s, thick, obovate, with incurved tips, each distinctly marked with 2 white bands. Male cones to 2 mm, terminal, hanging. Female cones plentiful, to 5 mm, with spreading and erect scales which have raised tips. Mature cones *c.* 1 cm, spherical, brown, with the few scales widely spread. Seeds 2 or more per scale, 2-winged.

The single species is a loose shrub in cultivation, and has a dark bluish colour. It is propagated by cuttings or from seed collected in the wild.

**1. F. cupressoides** Johnston. Illustration: Phillips, Trees in Britain, 118 (1978). *Chile, Argentina.* H2.

### 3. THUJA Linnaeus
*J. Lewis*

Evergreen, monoecious trees or shrubs with thin, fissured, reddish bark which scales in irregular strips or patches. Branchlets slender, tough, the foliage in flattened sprays. Leaves triangular, scale-like, overlapping in 2 dissimilar paired ranks, each usually with a gland on the back, some persisting on the smaller branches. Male cones often reddish, very shortly cylindric, with 6–12 stamens. Female cones on short stalks near the limit of the previous year's growth. Mature cones erect, solitary, 8–16 mm, of 3–8 pairs of overlapping, flattish scales. Seeds maturing within a year, 2–3 (rarely 5) per scale, on 2–3 of the cone-scales, usually winged.

A genus of 6 species from Asia and N America; 3 of the 5 known in cultivation are very variable and have given rise to numerous cultivars. All are vegetatively similar to *Chamaecyparis*, but the genus is easily recognised by the cones. One species, *T. orientalis*, is more distinct than the others (it belongs to subgenus *Biota*, formerly recognised as the genus *Biota* Endlicher), but all are easily identifiable except for some permanently juvenile cultivars.

They are more or less hardy trees, flourishing in cool, moderately moist conditions; well-drained loam is ideal, though lighter soils are adequate if water is readily available. Propagation is by seed or by cuttings, less generally by grafting; seedlings may require winter protection.

1a. Foliage equally green on both surfaces, borne on the branchlets in vertical planes, almost scentless; seeds wingless **1. orientalis**
  b. Foliage marked with white, or paler green beneath, borne on horizontal branch systems, scented when crushed; seeds winged **2**
2a. Foliage strongly scented even without crushing; leaves 3–6 mm **4. plicata**
  b. Foliage scented when crushed; leaves to 3 mm **3**

3a. Foliage uniformly yellowish green beneath, smelling like fresh apples when crushed; fertile cone-scales unthickened, as long as the fragile open cone **5. occidentalis**
  b. Foliage marked with white beneath, with a sweet or fruity scent when crushed; fertile cone scales thickened, shorter than the tough, compact cone **4**
4a. Leaves flattened, with large, shining, white marks beneath, smelling like cooked fruit when crushed **2. koraiensis**
  b. Leaves half-round in section, with small, dull white marks beneath, with a sweet lemon scent when crushed **3. standishii**

**1. T. orientalis** Linnaeus (*Platycladus orientalis* (Linnaeus) Franco; *Biota orientalis* (Linnaeus) Endlicher; *T. chengii* Bordères-Rey & Gaussen). Figure 29(1), p. 81. Illustration: Harrison, Ornamental conifers, t. 476–7 (1975).
Small, slow-growing, conical or ovoid trees or shrubs mostly to 15 m. Foliage in flattened sprays, darkish green on both surfaces, almost scentless even when crushed. Leaves *c.* 2 mm, triangular, grooved on the back. Mature female cones *c.* 8 mm, ovoid, bluish brown, of 6 or 8 scales, the lowest few fertile; each scale with a near-terminal hooked process, ultimately thickly woody, the fertile bearing 2–3 ellipsoid, brown, wingless seeds. *N China.* H3.

Many cultivars exist, including dwarf and yellow-foliaged variants. Whiplash or thread-leaved forms have been called *T. flagelliformis* Knight.

**2. T. koraiensis** Nakai. Figure 29(2), p. 81. Illustration: den Ouden & Boom, Manual of cultivated conifers, 414, 415 (1965).
Low spreading shrub or rarely a tree to 9 m. Branches spreading and upturned at their ends. Foliage in flattened fronds, bright green above, bluish white beneath; leaves smelling of cooked fruit when crushed, to 3 mm, triangular, each with a distinct, round gland. Mature cones to 1 cm, ovoid, brown, of about 8 scales, the middle few fertile; scales ultimately woody with gnarled tips. Seeds winged on each side. *N & C Korea.* H1.

**3. T. standishii** (Gordon) Carrière (*Thujopsis standishii* Gordon; *Thuja japonica* Maximowicz). Figure 29(3), p. 81. Illustration: den Ouden & Boom, Manual of cultivated conifers, 442 (1965).

Broadly conical tree to 20 m, lower branches spreading and upturned. Branchlets forming irregularly flattened fronds, thread-like, yellowish to grey-green with small, dull white markings beneath; smelling sweetly of lemon when crushed. Leaves *c.* 2 mm, glands obscure; the facial leaves are somewhat swollen, giving a rounded upper surface to the branchlets. Male cones *c.* 1 mm, dark reddish. Mature cones 9–12 mm, narrowly ovoid, brown, of about 10 scales, the middle ones fertile; all scales partially woody with broad points, the fertile 3-seeded. Seeds winged more or less all round. *C & S Japan.* H1.

**4. T. plicata** D. Don (*T. lobii* Gordon; *T. menziesii* Endlicher; *T. gigantea* Nuttall). Figure 29(4), p. 81. Illustration: Bloom, Conifers for your garden, 138 (1972); Harrison, Ornamental conifers, t. 488 (1975); Mitchell & Wilkinson, Collins' hand guide to the trees of Britain and northern Europe, 20 (1978).
Conical tree to *c.* 50 m. Branches spreading on older trees, upturned, frequently self-layering. Foliage in flattened, open fronds, rich, glossy green above, paler beneath with distinct, white markings, giving off a strong, fruity scent when merely touched. Leaves 3–6 mm, with indistinct, round glands. Male cones 2 mm, dark red. Mature cones 1–1.5 cm, ovoid, reddish brown, of 8–10 scales, the middle pairs fertile. Scales ultimately thickened and woody with rounded apices. Seeds winged all round. *Western N America.* H1.

Numerous cultivars are available.

**5. T. occidentalis** Linnaeus (*T. obtusa* Moench). Figure 29(5), p. 81. Illustration: Phillips, Trees in Britain, Europe and North America, 204 (1978).
Irregular, conical tree to 20 m. Lower branches spreading. Foliage in rather congested fronds, darkish green above, uniformly yellow-green beneath, partially erect on young trees, drooping later. Leaves *c.* 2.5 mm, triangular, with distinct raised, oval or round glands, pungent when crushed. Male cones 1 mm, dark reddish. Mature cones *c.* 1 cm, of 8–10 scales, the middle ones fertile. The scales ultimately partially woody, slightly thickened and widely spreading. Seeds with narrow marginal wings. *Eastern N America.* H1.

Numerous cultivars, varying in size and form, are available.

### 4. THUJOPSIS Siebold & Zuccarini
*C.N. Page*

Tall, monoecious, evergreen trees of

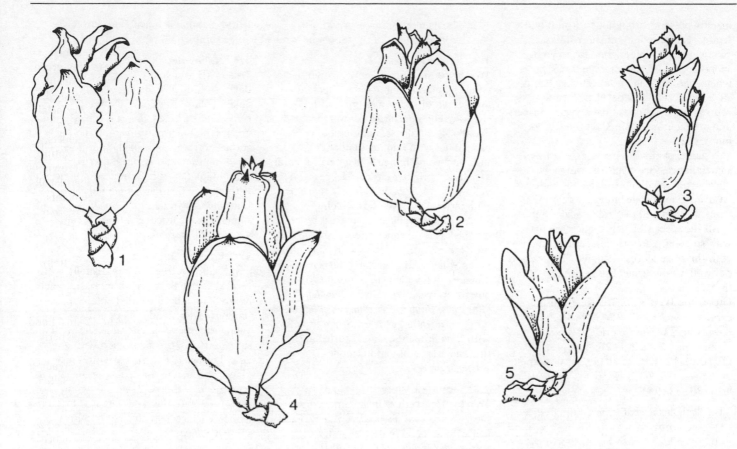

**Figure 29.** *Thuja* cones. 1, *T. orientalis*. 2, *T. koraiensis*. 3, *T. standishii*. 4, *T. plicata*. 5, *T. occidentalis*.

densely pyramidal-columnar habit, the lowermost branches often heavy and layering. Bark thin, yellowish brown, fissuring into vertical strips which are often spiralled, shedding in thin, stringy layers. Branches irregularly whorled, numerous, at first descending but sweeping upwards towards their tips, bearing numerous, flattened, fan-like branchlet systems held horizontally all around the shoot. Leaves scale-like, thick, leathery, usually 3–5 mm long, borne all round leading shoots and branch tips but in flattened, opposite pairs on branchlets, the laterals hatchet-shaped, the facials triangular, the upper surface of the shoot convex, deep, glossy green, the lower surface concave and marked with conspicuous, white, angular central depressions, the margins broad and green. Male cones 2–3 mm, numerous, blackish, inconspicuous. Female cones 8–12 mm, almost spherical to roundly cubic, blue-grey at first, turning grey-brown at maturity, the scales wedge-shaped, each bearing 3–5 small, winged seeds.

A genus of a single species endemic to Japan. It flourishes in moist, well-drained loam, and grows slowly at first, but eventually makes a large and long-lived tree. Propagation is by seeds and cuttings.

**1. T. dolabrata** Siebold & Zuccarini. Illustration: Clinton-Baker & Jackson, Illustrations of conifers 3: 83 (1935). *Japan*. H3.

### 5. CALOCEDRUS Kurz
*J. Lewis*
Tall, monoecious trees with flattened, dichotomous branchlets. Leaves in opposite pairs, strongly flattened, adpressed except for their sharp tips. Male cones solitary and terminal, longer than broad, with numerous, ovate bracts. Female cones pendulous with 6 erect, woody scales. Mature cones with spreading woody scales bearing sharp tips, the longest middle pair fertile. Seeds 2 to each scale, each with a large wing.

A genus of 3 species only, 1 of which is generally cultivated in Europe. It is very hardy and tolerates moist heavy soils in clean air; propagation by seeds is preferable.

**1. C. decurrens** (Torrey) Florin (*Libocedrus decurrens* Torrey). Illustration: Hora, Oxford Encyclopedia of trees, 95 (1981). *NW USA (Oregon to California)*. H1.

The narrow columnar variant, 'Columnaris' is commonest in cultivation; glaucous and variegated cultivars are known and a dense columnar variegated dwarf is commonly distributed as 'Intricata'.

### 6. CUPRESSUS Linnaeus
*A. Mitchell*
Evergreen trees of moderate size. Bark usually shallowly fissured, stringy, more rarely smooth and becoming scaly. Branches forming dense systems of fine, tough twigs encased in scale-leaves; branches usually radiating at all angles, more rarely the ultimate branches flattened into 1 plane. Foliage usually distinctively aromatic, leaves often with prominent resin-secreting glands. Male cones terminal on ultimate shoots, small, ovoid, yellow,

usually present throughout autumn and winter. Female cones on short stalks, yellow, green or blue-white when young, spherical, 1.5–4 cm across and woody when mature; scales relatively few, stout, each with a variably developed, raised central process. Cones take 2 years to ripen and are often maintained on the tree for many years.

A genus of about 15 species, mostly of somewhat restricted natural range, but widely cultivated and suitable for both large and small collections. All require a freely drained and neutral to acidic soil.

All the species have juvenile variants with spreading, free leaves, contrasting with the scale-leaves of adults. The genus is not well distinguished from *Chamaecyparis* (p. 83).
Literature: Wolf, C.B., The New World cypresses, *Aliso* **1**: 1–250 (1948); Little, E.D., Names of New World cypresses (Cupressus), *Phytologia* **20**: 429–45 (1970); Silba, J., Revised generic concepts of Cupressus L. (Cupressaceae), *Phytologia* **49**: 390–9 (1981).

1a.  Scale-leaves with spreading, free, acute tips                                    2
  b.  Scale-leaves with blunt, short tips   4
2a.  Foliage blue-green; branchlets conspicuously drooping
                                                **4. cashmeriana**
  b.  Foliage green; branchlets not conspicuously drooping            3
3a.  Sprays of foliage light green, flattened in 1 plane            **5. funebris**
  b.  Sprays of foliage deep green, not markedly flattened in 1 plane
                                                    **2. lusitanica**
4a.  Foliage spotted with white resin glands                                            5
  b.  Foliage without white resin glands
                                                                    6
5a.  Resin glands numerous and conspicuous; foliage blue-grey
                                                        **3. glabra**
  b.  Resin glands few; foliage deep or bright green       **6. goveniana**
6a.  Ultimate shoot systems remotely branched, forming long, fine, flattened sprays            **1. torulosa***
  b.  Ultimate shoot systems closely branched, forming short, thick, angular sprays                      7
7a.  Shoots branched perpendicularly
                                                    **6. goveniana**
  b.  Shoots branched at acute angles   8
8a.  Shoots roughened by the rounded leaf tips, sweetly aromatic when crushed
                                                **7. macrocarpa**

  b.  Shoots terete, scarcely resinously aromatic                    **8. sempervirens**

**1. C. torulosa** D. Don.
Tree to 20 m or rarely more. Crown ovoid, upswept, with hanging outer shoots. Bark dull brown or sometimes grey, with shallow ridges in spirals curling away with age. Foliage in dense, hanging bunches, each shoot with remote, slender, unbranched ultimate shoots; dark green at a distance, brighter close to, leaves rather yellowish green, each with pale margins and narrow, incurved tips which project slightly; fragrant like mown grass when crushed. Female cones often numerous all over the crown, *c.* 1 cm, spherical, whitish green when young, dark red-brown when mature, each scale with a minute, downwardly curved prickle. *Himalaya (Kumaon), Bhutan, SW China (Sichuan).* H5.

**\*C. corneyana** Carrière. Crown broader, with level branches, shoots curled at the tips and arising at various angles. *E Himalaya.* H5.

**2. C. lusitanica** Miller.
Tree to 30 m with slender, conic crown. Bark dark brown, fissured vertically. Each leaf with an acuminate, free outer half; leaves dark green, borne in rather sparse, slender, forwardly pointing sprays. Cones spherical, *c.* 1.5 cm across, bright green bloomed with bluish white during the first year, ripening shiny, dark purple-brown. *Mexico, Guatemala.* H5.

Young trees often have pale grey foliage and pink to purple shoots which are sinuous and are vigorous in growth (up to 1 m a year). 'Glauca' has terete shoots of a good blue-grey; it is somewhat hardier than the pure species, and so is more widespread. 'Glauca Pendula' has darker foliage hanging in dense, widely angled sprays from a broad crown of level branches.

**3. C. glabra** Sudworth (*C. arizonica* Greene var. *glabra* Sudworth).
Tree to 20 m with dense, ovoid, upswept crown. Bark dark purple and smooth at first, later blistering and flaking to reveal cream patches. Shoots orange-brown, bearing dense bunches of minor shoots at wide angles. Leaves blue-green, some with white resin glands, forming a terete shoot. Male cones profuse, yellow. Female cones solitary but clustered, remaining on the tree for many years, *c.* 1.5 cm across, green and brown at first, ripening dark brown. *USA (C Arizona).* H5.

'Pyramidalis' has thicker foliage with a

white spot on each leaf, borne in erect, outer sprays.

**4. C. cashmeriana** Carrière.
Tree to 10 m, usually not more than 6 m. Bark red-brown, stripping in long, thin flakes. Foliage in slender, hanging sprays. Leaves in opposite pairs, soft greyish blue-green with paler margins on the outer surface, blue-green broadly margined with white on the inner surface; tips free, acuminate. Cones *c.* 1 cm across, spherical, with about 10 scales. *Origin unknown.* H5. Unknown as an adult tree in the wild, and sometimes regarded as a fixed, juvenile variant of *C. torulosa.*

**5. C. funebris** Endlicher.
Small, upright tree to 12 m, usually with several stems. Bark grey-brown, shallowly fissured, on a deeply fluted bole. Sprays hanging, with slender, flattened foliage, yellowish green, the lateral leaves with fine, spreading tips, the facial each with a usually translucent (rarely white) resin gland; shoots pink-brown. Juvenile foliage blue-green with free leaves, sometimes found well into the crown. Cones 1–2 cm across, each scale with a short central spine. *C China.* H4.

**6. C. goveniana** Gordon.
Usually narrow trees in cultivation with erect, columnar crowns of great density. Bark pale grey-brown, shallowly stripping. Young shoots bright brown to purplish, bearing short, perpendicular systems of hard, dark green, terete minor shoots, producing long, slender but stiff sprays on the main shoots. Leaves obtuse, tips rounded and incurved, dark or bright green, usually with white resin glands. Male cones numerous, to 3 mm, cylindric, yellow. Female cones clustered, yellow-green when young. Mature cones 1.5–2 cm across, shiny green and brown, with a green ridge and a minute hook in the centre of each scale. *USA (California).* H5.

An extremely rare tree in the wild.

**7. C. macrocarpa** Hartweg.
Strongly branched tree to 35 m. Crown shape very variable. Bark pale brown, shallowly ridged, becoming grey with age. Shoots dull pinkish brown. Ultimate shoots more or less terete, slightly roughened by the incurving, convex tips of the leaves. Leaves at first bright green, later dark. Male cones sporadic, pale yellow. Female cones clustered, usually numerous on old trees, pale yellow when young. Mature cones to 3 cm across, more or less spherical, shining purple-brown; scales

7–8, each with a small central hook. *USA (California)*. H4.

In some areas, liable to attack by the fungus *Coryneum cardinale*, which kills the tree within a few years. 'Fastigiata' is narrowly columnar when young, later inverted conic with many large branches springing from the base. 'Lutea' is much bushier than the wild variant, and withstands sea winds; 'Donard Gold' is similar but brighter green and less bushy. 'Goldcrest' has juvenile foliage of a good yellow, and 'Golden Spire' is pale yellow and has a neat, slender, columnar crown.

**8. C. sempervirens** Linnaeus.
Tall, narrow or broadly crowned trees of dense habit. Bark in broad, shallow, often spiral ridges, tending to flake, light brown. Foliage dull, dark green, in nearly erect, dense sprays with the ultimate shoots in 1 plane, curved, terete. Male cones ovoid, greenish, to 3 mm. Female cones shiny green ripening dark brown and remaining on the tree for at least 1 year, weathering grey. *S Europe, Mediterranean area, SW Asia.* H4.

The wild variant of this species is scarcely, if ever, grown; it has a broad crown with spreading, level branches. The commonest, cultivated variant grows into a narrow, erect, columnar crown with an acute apex.

**7. CHAMAECYPARIS** Spach
*A. Mitchell*
Moderately large to very large trees, similar to *Cupressus*, but foliage arranged in a single plane, female cones small, scarcely woody when mature.

A genus of about 6 species, of somewhat doubtful distinctness from *Cupressus*; hybrids between the 2 genera are known.

1a. Sprays very short, dense, bunched
                      **6. thyoides**
  b. Sprays long, open, flat, not bunched
                                 2
2a. Leaves blunt, obtuse    **4. obtusa**
  b. Leaves acutely pointed        3
3a. Leaves with translucent spots seen when held against the light, tips incurved    **2. lawsoniana**
  b. Leaves without such spots, tips spreading               4
4a. Foliage thick, heavy, dark or medium green    **3. nootkatensis**
  b. Foliage fine, bright green or bronzed 5
5a. Foliage bright green, the sprays slightly upturned, marked with white beneath    **5. pisifera**
  b. Foliage pale, bronzed green, the sprays turned downwards at their tips, pale green beneath
                **1. formosensis**

**1. C. formosensis** Matsumura.
Tree to 16 m (or more?). Bark red-brown, shallowly ridged and peeling vertically. Crown broad-conic, open, with stout, low, level branches, their outer halves upswept. Foliage dull yellowish green, bronzed, lateral leaves with minutely mucronate, incurved tips; the crushed foliage smells like rotting seaweed. Cones less than 1 cm, green in summer. *Taiwan.* H5.

**2. C. lawsoniana** (Murray) Parlatore.
Tree to 40 m. Bark at first smooth, dark grey-green, then brown and fissured, finally dark purplish brown forming large plates with stringy margins. Crown narrowly conic, frequently forked or strongly branched low down and becoming broadly conic. Leading shoot arching over. Foliage thin in texture, in slender, dense sprays; shoots pinkish brown; leaves with acuminate, incurved tips, each with a central, translucent resin gland; upper side of spray dark or bluish green, under side dusted with white towards the base; the crushed foliage smells like parsley. Male cones usually profuse, ovoid, to 2 mm, scales black, broadly edged with white, opening to reveal dark red pollen sacs. Female cones with blue-black scales which are edged with white. Mature cones spherical, *c.* 7 mm, ripening purple-brown, each scale with a small, central, blunt spine; cones remaining on the tree for several months after the seeds have been shed. *USA (California & Oregon).* H4.

Very many variants have been selected and described. Most are susceptible to brown rot (*Phytophthora*) and some are also susceptible to honey fungus (*Armillaria*) and branch die-back (*Phomopsis*).

**3. C. nootkatensis** (D. Don) Spach.
Tree to 30 m with regular, conic crown and hanging shoots. Bark dark brown, stringy and with shallow plates. Shoots orange-brown bearing systems of minor shoots in a pinnate manner. Leaves of minor shoot systems in equal pairs, slender, acuminate, hard, dark green; the crushed leaves smell of turpentine. Male cones numerous, yellow. Female cones dark grey, later bluish, finally dark brown, taking 2 years to ripen, *c.* 1 cm across. Each cone-scale with a prominent beak. *Western N America (Alaska to N Oregon).* H4.

In 'Lutea' the young growth is yellowish,

and 'Pendula' has hanging sprays arising from upcurving branches.

**4. C. obtusa** (Siebold & Zuccarini) Endlicher.
Slow-growing trees to 25 m (or more?). Bark red-brown, fissured vertically into parallel ridges which strip off. Shoots orange-brown, bearing pinnate branch systems which are obovate in outline, the minor shoots at wide angles and curving backwards. Leaves in very unequal pairs, the facial showing only as broad, obtuse tips, the laterals longer, with blunt, incurved tips; all bright, shining green above, strongly patterned with white beneath. Crushed foliage sweetly aromatic. Male cones terminal on the shortest shoots, minute and inconspicuous. Female cones at first blue-grey, then bright green, finally brown, to 2 cm across when mature, scales few, minutely ridged. *C & S Japan.* H3.

Very variable; numerous cultivars have been selected, including 'Aurea' with dull golden foliage, 'Crippsii' with bright golden foliage, 'Filicoides' with foliage in dense, flat plates hanging from a gaunt crown, 'Lycopodioides' with congested, twisted bunches of foliage which show much blue-white colour, and 'Tetragona Aurea' with very bright golden foliage, ultimate shoots almost square in cross section near the tips.

**5. C. pisifera** (Siebold & Zuccarini) Endlicher.
Large tree with a broad, conic crown. Bark reddish brown, ridged vertically and shredding. Sprays thin, bright or dark green above, blue-white beneath; leaves acute and slightly spreading. The crushed foliage is sharply resinous. Male cones abundant, minute. Female cones abundant, to 5 mm across. *C & S Japan.* H4.

Many cultivars are known, including 'Aurea' with yellowish new growth, 'Filifera' with a thin crown and foliage in small bunches on thread-like shoots, 'Filifera Aurea' similar to 'Filifera' but foliage pale yellow, 'Plumosa' with free, spreading leaves in flattened, short sprays, 'Plumosa Aurea' with leaves similar to the above but leaves bright yellow when young, duller yellow with age and 'Squarrosa' with long, spreading, slender leaves which are dark blue-green above and white beneath.

**6. C. thyoides** (Linnaeus) Britton, Sterns & Poggenburg.
Small, short-lived tree to 16 m. Crown ovoid-conic. Bark dull red-brown or grey-brown, with thin, raised ridges in spirals which strip away. Shoots wiry, bearing

branches at all angles, lesser shoots generally branched pinnately, ultimate shoots about 1 mm in diameter. Foliage blue-green or yellowish green, dense, bunched, smelling of warm ginger when crushed. Cones to 5 mm across, bloomed blue-grey, each scale with a minute spine. *USA (Maine to Mississippi)*. H3.

## 8. × CUPRESSOCYPARIS Dallimore
*A. Mitchell*

Tall, evergreen trees produced by hybridisation between *Cupressus* and *Chamaecyparis*, generally similar to the latter but bark dark brown, shallowly ridged, and crown narrowly conic.

The genus contains 3 hybrids, all involving *Chamaecyparis nootkatensis*; only 1 is in general cultivation. Cultivation as for *Cupressus* (p. 81).

### 1. × C. leylandii (Jackson & Dallimore) Dallimore. *Chamaecyparis nootkatensis × Cupressus macrocarpa*. *Garden origin*. H3.

A very important landscape and garden tree, of rapid growth and considerable vigour. A few cultivars have been described, differing in foliage colour and form.

## 9. JUNIPERUS Linnaeus
*H.J. Welch*

Monoecious or dioecious evergreen trees with thin, shredding bark, or shrubs. Juvenile leaves needle-like, borne usually in whorls of 3 (rarely in opposite pairs), spreading, sharp at the apex, with 1 or 2 pale bands on the upper surface, the lower surface green, rounded or keeled, often with a gland or glands; these may persist on the mature plant or be replaced wholly or partly by adult leaves which are scale-like, minute, closely adpressed to the stem in opposite pairs, the outer surface rounded and bearing a gland. Male cones composed of numerous stamens attached to a central stalk. Female cones consisting of 3–8 scales which are opposite or in whorls of 3, some or all of them bearing 1–2 ovules on their upper surfaces. As the cone matures the scales become fleshy and coalescent, forming a fleshy, berry-like fruit which contains 1 or more seeds, taking 1–2 years to ripen.

A genus of about 60 species distributed throughout the northern hemisphere. Leaves on young plants are always of the juvenile form. With age the plant may develop scale-like leaves, depending on the species. Unfortunately, selection of cultivars has confused the picture in that, in some species which normally produce adult leaves, cultivars exist which have only juvenile leaves, or have adult leaves poorly developed. This makes the identification of garden material extremely difficult, a situation made worse by the fact that most plants are unisexual, so that in male plants the useful fruit characters are not available. The keys given here must be used with caution.

Junipers form a valuable group of garden plants; they will tolerate a wide range of conditions, including poor and alkaline soils. The various cultivars cover a full range of sizes, from trees to prostrate shrubs; a broad range of foliage colour is also available. All species can be propagated by seed, but selected clones must be propagated vegetatively: cuttings usually root readily, but in a few cases grafts must be used.

*Ultimate branchlets*. Erect to horizontal: 1,4,7,8,10–12,14–19; horizontal but with the tips upturned: 9; pendent: 2,3,5,6,13.
*Needle-like leaves*. Always present: 1–9; usually present, at least in isolated patches on some shoots: 10–14,16–19; absent: 15 – but see also *Cupressus* (p. 81). Usually in whorls of three: 1–14; usually in opposite pairs: 15–19. Borne at a joint in the stem: 1–5; decurrent and not borne at a joint in the stem: 6–19. With 2 whitish bands on the upper surface: 4,5,8,12–15; with 1 whitish band on the upper surface: 1–3,6,7,9–11,16–18. More than 1 cm: 1–5,14; less than 1 cm: 6–19.
*Scale-like leaves*. Entirely absent: 1–9,15 and juvenile plants of 10–19; present: 10–19. Margins minutely toothed: 10–12.
*Male and female cones*. Axillary: 1–5; terminal, sometimes on short shoots: 6–19.
*Fruit*. Ripening in the current year (or in the early part of the next year): 16,19; ripening later in the second year: 1–15,17,18. More than 2 cm in diameter: 4; 1–2 cm in diameter: 5,11–13; less than 1 cm in diameter: 1–3,5–19. Reddish brown when mature: 5,11–13; blue-black, purple or purplish brown when mature: 1–4,6–10, 14–19.
*Seeds (examine several fruits to ascertain number)*: 3, united into a large stone: 4; free, 1 only: 6–8,10,16,18,19; free, 2–3:

1–3,5,9,12,14,16–19; free, more than 3: 11–15,18.

1a. Leaves all needle-like, mostly borne in whorls of 3                                             2
 b. At least some of the leaves scale-like, and in pairs                                             10
2a. Leaves arising at a joint in the stem; male cones axillary                                       3
 b. Leaves decurrent, not arising at a joint in the stem; male cones terminal                        7
3a. Leaves with a single white band above                                                            4
 b. Leaves with 2 white bands separated by the midrib, above                                         6
4a. Leaves tapering from the base
                                         **1. communis**
 b. Leaves linear, parallel-sided for most of their length                                           5
5a. Tall tree with spreading branches with pendent tips; leaves 1.5–2 cm
                                           **2. rigida**
 b. Prostrate shrub; leaves 1.2–1.5 cm
                                         **3. conferta**
6a. Tree to 20 m; fruit 2–2.5 cm in diameter; leaves 1.5–2 cm
                                         **4. drupacea**
 b. Shrub or small tree to at most, 10 m; fruit 1–1.2 cm in diameter; leaves 1.2–1.8 cm                         **5. oxycedrus***
7a. Tall trees or plants tree-like                 8
 b. Shrubs                                          9
8a. Tips of the branchlets pendent; leaves 4–6 mm, not decurrent    **6. recurva**
 b. Branchlets entirely erect or ascending; leaves 3–4 mm, decurrent
                                     **7. morrisonicola**
9a. Bush upright or spreading; leaves 4–6 mm; shoots without angles, their tips pendent           **8. squamata**
 b. Bush prostrate or spreading; leaves 6–8 mm; shoots angled with a glaucous line along the ridges, each line starting from a white spot at the leaf base; tips of shoots not pendent
                                     **9. procumbens**
10a. Leaves with minutely toothed margins, visible only under a strong lens                          11
 b. Leaves with entire margins                      14
11a. Fruits black when fully ripe
                                   **10. wallichiana**
 b. Fruits red-brown or yellow-brown when ripe                                                       12
12a. Bark broken into a chequer pattern of scaly plates
                          **11. deppeana** var. **pachyphloea**
 b. Bark not as above                               13
13a. Upright tree with the branchlets frequently branched, appearing

'pinnate'; fruit yellow-brown when ripe **12. phoenicia**

b. Upright tree, branches not as above, pendent at their tips; fruit red-brown with a waxy bloom when ripe **13. flaccida**

14a. Upright small tree; fruits held erect or ascending, on erect or ascending stalks **15**

b. Spreading or prostrate shrubs; fruits held on curved or deflexed stalks **18**

15a. Scale-like leaves blunt and rounded; needle-like leaves in whorls of 3 in small, often sparse clusters **14. chinensis**

b. Scale-like leaves narrow and pointed; needle-like leaves usually in opposite pairs **16**

16a. Needle-like leaves with 1 whitish band on the upper surface; seeds more than 3 in each fruit **15. excelsa**

b. Needle-like leaves with 2 whitish bands above; seeds 1–3 in each fruit **17**

17a. Ultimate branchlets to 1 mm in diameter; fruit ripening in the first year **16. virginiana**

b. Ultimate branchlets more than 1 mm in diameter; fruit ripening in the second year **17. scopulorum**

18a. Leaves dark green, smelling unpleasantly when crushed **19. sabina**

b. Leaves blue-green, pleasantly aromatic when crushed **18. horizontalis***

**1. J. communis** Linnaeus. Illustration: Krüssmann, Die Nadelgehölze, 136 (1960); Mitchell, Conifers in the British Isles, 120 (1972).

Small upright tree (rarely to 5 m) or spreading to prostrate shrub. Young shoots pale brown, grooved, triangular in section. Leaves all needle-like, borne in whorls of 3, 1–1.2 cm, arising at a joint in the shoot and borne at a wide angle, tapering from a swollen base to a sharp point; upper surface concave and with a single whitish band, lower surface bluntly keeled. Male and female cones usually borne on separate plants, solitary in the leaf axils. Fruit 6–9 mm in diameter, almost stalkless, ripening in the second year to dark blue or black, bloomed, containing 2–3 triangular seeds. *Temperate northern Hemisphere*. H2.

Very variable, particularly in habit. Prostrate plants have been regarded as a separate species (*J. nana* Willdenow) or subspecies (subsp. *nana* (Willdenow) Syme),

but they occur sporadically throughout the species' geographical range, and are best regarded as varieties. Var. **depressa** Pursh (var. *canadensis* Loudon) is a widely spreading shrub with ascending branches, upturned leaves to 1.5 cm × 1.6 mm with a narrow whitish band; var. **hemisphaerica** Parlatore is bushy and low-growing with broad, thick leaves which are sharply upturned at the base and then straight; var. **hornibrookii** Hornibrook covers low-growing, bushy, prostrate or mat-forming plants with small, straight, loosely set leaves; var. **montana** Aiton (*J. nana* Willdenow; *J. communis* subsp. *nana* (Willdenow) Syme) is prostrate and often mat-forming, with thick, broad, short, boat-shaped leaves.

Many selections have been made from this variability and many cultivars have been named, including erect (from var. *communis*) as well as prostrate variants.

**2. J. rigida** Siebold & Zuccarini. Illustration: den Ouden & Boom, Manual of ornamental conifers, 186 (1965); Mitchell, Conifers in the British Isles, t. 10 (1972).

Tall shrub or eventually a tree to 12 m, dioecious, terminal branchlets pendent. Bark dull brown or grey, stripping in long shreds. Young shoots yellow-green, not noticeably angled. Leaves all needle-like, borne in whorls of 3 at a joint in the stem, 1.5–2 cm, widely spreading, linear, soft to the touch but sharply pointed; upper surface concave with a single narrow white band, lower surface glossy pale green. Cones numerous, axillary. Fruit to 1 cm in diameter, irregularly spherical, ripening in the second year to dark purple with a grey bloom. Seeds 2–3. *Japan*. H2.

**3. J. conferta** Parlatore (*J. maritima* Maximowicz; *J. litoralis* misapplied). Illustration: Bloom, Conifers for your garden, 78 (1972).

Dioecious shrub of dense, mat-forming habit, with crowded, pale glaucous green leaves which are 1.2–1.5 cm. Otherwise like *J. rigida* but fruit bloomy black when ripe. *Japan, USSR (Sakhalin)*. H3.

**4. J. drupacea** Labillardière. Illustration: Dallimore & Jackson, Handbook of Coniferae and Ginkgoaceae, edn 4, 254 (1966); Testu, Conifères de nos jardins, 123 (1970).

Dioecious conical tree to 20 m. Trunk orange-brown, bark peeling away and exposing the reddish young bark beneath. Young shoots green, prominently ridged, triangular in section. Leaves all needle-like, borne in whorls of 3 at joints in the stem,

1.5–2 cm × 2.5–3 mm, widespreading, lanceolate, hard; upper surface with a narrow green midrib separating 2 whitish bands, lower surface bright shining green. Cones axillary. Fruit 2–2.5 cm in diameter, ripening in the second year to brown-black or blue-black with a glaucous bloom. Seeds 3, united to form a hard stone. *Greece, Turkey, Syria*. H3.

This species has the largest leaves and fruits of any juniper; the fruits are edible. All the trees cultivated in the British Isles (at least), appear to be males.

**5. J. oxycedrus** Linnaeus (*J. rufescens* Link). Illustration: den Ouden & Boom, Manual of cultivated conifers, 182 (1965).

Dioecious shrub or small tree to 10 m, with open, loose habit. Young shoots slender, triangular in section, the ultimate branchlets pendent. Leaves all needle-like, arising in whorls of 3 at joints in the stem, 1.2–1.8 cm, linear, tapering to a short, prickle-like tip, widely spaced; upper surface with 2 white bands separated by a raised, green midrib, lower surface dark grey-green. Cones axillary, the males often 2–3 to a whorl. Fruit 6–20 mm in diameter, solitary, shortly stalked, irregularly spherical, ripening in the second year to a shining red-brown. Seeds 3. *Mediterranean area, Caucasus, Iran*. H5.

Subsp. **macrocarpa** (Sibthorp & Smith) Hall (*J. macrocarpa* Sibthorp & Smith) has fruits 1–2 cm in diameter, while subsp. *oxycedrus* has them 6–8 mm in diameter.

***J. formosana** Hayata. Leaves soft, fruit dark purple with a grey bloom. *Taiwan, China*. H3.

**6. J. recurva** Buchanan-Hamilton (*J. excelsa* Brandis not Bieberstein; *J. macropoda* Hooker not Boissier; *J. religiosa* Carrière). Dioecious, round-topped tree to 15 m. Bark grey-brown, peeling in long strips. Main branch systems upswept, the ultimate branchlets pendent. Leaves all needle-like, borne in whorls of 3 but not at joints in the stem, 3–6 mm, decurrent, pointing forwards and crowded, sharp; upper surface with a single whitish band, lower surface bright, dark green with pale margins; dead leaves persistent for several years. Cones terminal on short side shoots. Fruit 7–10 mm in diameter, ovoid, ripening in the second year to dark purple-brown. Seed solitary, pitted. *E Himalaya, SW China*. H2.

Var. **coxii** (Jackson) Melville (*J. coxii* Jackson). Growth more vigorous and habit more open and pendent, leaves 8–10 mm, more widely spaced, lighter green. *N Burma*.

In 'Castlewellan' the leaves are minute, giving the foliage a thread-like appearance, very attractive in maturity.

**7. J. morrisonicola** Hayata (*J. squamata* var. *morrisonicola* (Hayata) Li & Keng). Illustration: Li et al., Flora of Taiwan 1: 544 (1979).
Monoecious, erect, densely branched shrub or small tree. Young shoots erect, triangular in section. Leaves all needle-like, 3–4 mm, borne in whorls of 3 but not arising at a joint in the stem, sharp, spreading; upper surface concave with a glistening whitish band above (which gives the whole tree a distinctive, silvery appearance), glaucous beneath. Cones terminal on short branchlets. Fruit *c.* 6 mm in diameter, spherical to ovoid, smooth, ripening black. Seed 1. *Taiwan (Mt Morrison).* H2.

**8. J. squamata** Lambert (*J. recurva* var. *squamata* (Lambert) Buchanan-Hamilton). Illustration: Welch, Manual of dwarf conifers, 428 (1979).
Small trees to 5 m, or more usually upright or spreading shrubs, young branchlets thick. Leaves all needle-like, densely set in whorls of 3 but not arising from joints in the stem, decurrent, the free part 4–6 mm, straight or incurved; upper surface grey-green or bluish with 2 white bands, lower surface green, convex and grooved from the base to near the apex. Dead leaves persistent as brown scales. Cones terminal. Fruit 6–8 mm in diameter, ellipsoid, ripening purplish black. Seed 1. *Asia, from Iraq to Taiwan.* H3.

Var. **fargesii** Rehder & Wilson is a tree to 5 m, often with a divided trunk, the branches ascending and spreading, their tips pendent. It appears to form a link between this species and *J. recurva* and perhaps should belong to the latter.

Many cultivars, including trees, shrubs and prostrate shrubs are available and widely planted.

**9. J. procumbens** Miquel (*J. chinensis* var. *procumbens* (Miquel) Endlicher; *J. litoralis* misapplied). Illustration: Harrison, Ornamental conifers, 103 (1972).
Dioecious, procumbent shrub with long, stout, stiff main branches and with all the terminal growths slightly upturned. Young shoots ridged. Leaves all needle-like, 6–8 mm, borne in 3s but not arising at a joint in the stem, decurrent, pointing forwards, tapering to a point; upper surface concave with a single, broad whitish band, lower surface convex, green but with a

white spot at either side near the base, from which a white line runs down the shoot. Cones terminal. Fruit 8–9 mm in diameter, almost spherical or ovoid. Seeds 2 or more. *Japan.* H3.

It is possible that this species is not in cultivation in Europe, material so named being perhaps a shrubby, male clone of *J. chinensis* (below). Several rather similar, mat-forming clones are widely cultivated under the group name 'Nana' (the clone 'Bonin Islands' being one). Their dense growth and shorter leaves (5–6 mm) make them difficult to distinguish from prostrate clones of *J. squamata*, but the upturned tips of the young shoots, characteristic of *J. procumbens*, are generally conclusive.

**10. J. wallichiana** Brandis. Illustration: Mitchell, Conifers in the British Isles, t. 11 (1972).
Narrowly conical dioecious trees to 30 m in the wild (less in cultivation). Leaves of 2 kinds: needle-like leaves in whorls of 3 but not arising from a joint in the stem, sharply pointed, upper surface concave and glaucous with a single whitish band, scale-like leaves *c.* 2 mm, margins minutely toothed, adpressed and overlapping, the ultimate shoots quadrangular, *c.* 1 mm in diameter; the proportions of the 2 kinds of leaves vary from tree to tree, and leaves intermediate in character are found on the larger shoots; all, however, have a long, narrow furrow on the lower surface. Fruit *c.* 9 mm in diameter, ripening bluish in the second year. Seed 1, flattened. *Himalaya, SW China.* H4.

**11. J. deppeana** Steudel var. **pachyphloea** (Torrey) Martinez.
Dioecious, broadly pyramidal tree to 20 m. Bark red-brown, divided into a chequer pattern of square plates. Branchlets rounded in section. Leaves of 2 kinds: the needle-like leaves 3–6 mm, borne usually in whorls of 3, decurrent, spreading, ending in a hard tip, concave with a white line above, glaucous beneath, predominant on trees in cultivation; scale-like leaves 1–1.2 mm, bluish green, opposite, keeled, margins minutely toothed. Fruit 8–12 mm in diameter, almost spherical to ovoid, ripening in the second year, red-brown, bloomed. Seeds usually 4, angled, glossy, brown. *Mexico.* H4.

**12. J. phoenicia** Linnaeus (*J. bacciformis* Carrière; *J. tetragona* Moench).
Monoecious shrub or small tree to 10 m, branches erect, themselves much branched, appearing as if pinnately arranged, the

ultimate branches less than 1 mm in diameter, rounded in section. Leaves of 2 kinds: needle-like leaves decurrent, in whorls of 3, with 2 white bands on the upper surface and on the lower, found only in isolated patches on mature trees; scale-like leaves in opposite pairs or occasionally in whorls of 3, *c.* 1 mm, closely adpressed, dark green, margins minutely toothed. Fruit 6–12 mm in diameter, spherical or almost so, ripening to yellow-brown or red-brown in the second year. Seeds 3–4, brown. *Mediterranean area.* H4.

A very variable species.

**13. J. flaccida** Schlechtendahl.
Monoecious, round-topped tree to 12 m with widely spreading branches and pendent branchlets. Leaves of 2 kinds: needle-like leaves in whorls of 3, decurrent, the free part 1.5–2 mm, spreading, tapering to a horny point, the upper surface with 2 whitish bands, predominant on young trees, only in patches on mature trees; scale-like leaves 1.5–2 mm, ovate-lanceolate, the tips free and ending in a sharp point, outer surface rounded, margins minutely toothed. Male and female cones usually on different branches. Fruits almost spherical, 1–1.2 cm in diameter, ripening to red-brown and bloomed in the second year. Seeds 4–12. *Southern USA (Texas), NE Mexico.* H4?

Uncommon in cultivation and doubtfully hardy in Europe.

**14. J. chinensis** Linnaeus (*J. sphaerica* Lindley; *J. shepherdii* van Melle; *J. × media* van Melle; *J. davurica* misapplied). Illustration: Harrison, Ornamental conifers, 83 (1972).
Dioecious, conical tree to 20 m, or a widely spreading shrub. Trunk often divided or irregular. Bark dark brown, often peeling in strips. Leaves of 2 kinds: needle-like leaves 8–12 mm, usually in whorls of 3, rarely in opposite pairs, decurrent, spreading, apex pointed, upper surface with 2 greyish bands separated by a broad midrib, dark green and convex beneath, predominant in young trees and in small, isolated patches in mature trees; scale-like leaves *c.* 1.5 mm, narrowly diamond-shaped, closely adpressed, obtuse, with a glandular depression on the back. Cones on separate plants with either kind of foliage, the female inconspicuous, the males numerous. Fruit 6–8 mm in diameter, irregularly spherical, ripening to dark purple or blue in the second year. Seeds 2–5, glossy brown. *Himalaya, China, Japan.* H2.

A very variable species, mostly

represented in gardens by cultivars. Many of the shrubby cultivars (e.g. 'Blaauw', 'Globosa', 'Plumosa', 'Armstrong', 'Old Gold' and 'Pfitzeriana') are probably of hybrid origin (*J. chinensis* × *J. sabina*), but are traditionally placed under *J. chinensis*.

**15. J. excelsa** Bieberstein.
Monoecious, conical tree to 20 m, branching upright or spreading, branchlets less than 1 mm in diameter. Leaves of 2 kinds: needle-like leaves often absent on mature trees, 5–6 mm, borne in pairs, decurrent, spreading, with 2 pale bands above; scale-like leaves *c.* 1 mm, usually in pairs, occasionally in 3s on leading shoots, closely adpressed, each with a small groove on the back. Fruit 8–12 mm in diameter, spherical, ripening in the second year to dark purple-brown with a glaucous bloom. Seeds 4–6. *Balkan Peninsula & Cyprus to Afghanistan.* H5.

**16. J. virginiana** Linnaeus. Illustration: Harrison, Ornamental conifers, 100 (1972); Welch, Manual of dwarf conifers, 429 (1979).
Conical (ultimately spreading) trees to 30 m, branchlets less than 1 mm in diameter. Leaves of 2 kinds: needle-like leaves almost always present even in mature trees, 5–6 mm, in pairs, decurrent, spreading, sharply pointed, concave and with a white band above, convex and green beneath; scale-like leaves *c.* 1 mm, closely adpressed and overlapping or free at the tips, often with an inconspicuous glandular depression on the back. Male and female cones often borne on the same tree. Fruit *c.* 5 mm in diameter, ovoid or almost spherical, ripening in the first year to dark purple-brown, bloomed. Seeds 1–2. *Eastern & Central N America.* H2.

A very variable species from which numerous cultivars have been selected, mainly in the USA.

**17. J. scopulorum** Sargent (*J. dealbata* Endlicher; *J. virginiana* var. *scopulorum* (Sargent) Lemmon).
Similar to *J. virginiana* but with shorter, thicker, more crowded branchlets, leaves yellowish green or glaucous, fruit *c.* 6 mm in diameter, ripening in the second year to blue with a glaucous bloom. Seeds 2. *Western N America.* H2.

A very large number of cultivars has been selected, mainly in the USA.

**18. J. horizontalis** Moench (*J. prostrata* Persoon; *J. sabina* var. *procumbens* Pursh; *J. sabina* var. *prostrata* (Persoon) Loudon; *J. virginiana* var. *prostrata* (Persoon)

Torrey). Illustration: Harrison, Ornamental conifers, 94 (1972).
Dioecious procumbent or prostrate shrub, procumbent variants to 1 m, prostrate variants producing long, rooting shoots. Branchlets short and crowded. Leaves of 2 kinds: needle-like leaves 2–3 mm, in pairs, decurrent, slightly spreading; scale-like leaves seldom developing in young plants in cultivation, in 4 ranks, closely adpressed, ovate or oblong, sharply pointed, each with a glandular depression on the back, glaucous, often becoming blue or purple in winter. Fruit 7–9 mm in diameter, borne on a short stalk, ripening in the second year to glaucous blue. Seeds 1–4. *Eastern N America.* H2.

Many selections from the wild have received cultivar names.

**\*J. davurica** Pallas. An imperfectly known species. A widespreading, procumbent shrub with long, vigorous main branches bearing scale-like leaves; side branches bearing mainly needle-like leaves. *E Asia.* H2.

Three cultivars ('Expansa', 'Expansa Aureospicata' and 'Expansa Variegata') are traditionally referred to this species, but their foliage is very like that of *J. chinensis*, to which they may very well belong.

**\*J. sargentii** (Henry) Koidzumi (*J. chinensis* var. *sargentii* Henry). Leaves incurved to the stem, giving the branches a whipcord appearance. The bruised shoots have an unpleasant smell of camphor which is quite distinctive. *China, Japan.* H2.

**19. J. sabina** Linnaeus. Illustration: Welch, Manual of dwarf conifers, 426 (1979).
Monoecious or dioecious, spreading or procumbent shrubs, without a central stem. Ultimate branchlets *c.* 1 mm in diameter. Leaves of 2 kinds: needle-like leaves *c.* 4 mm, decurrent, in pairs, with 2 white bands separated by a midrib above, found on young plants and on occasional branchlets of mature plants; scale-like leaves in opposite pairs, 1–1.5 mm, ovate, blunt at the apex (sometimes sharply pointed on leading shoots), the lower surface convex, usually bearing a gland. Male and female cones usually on separate plants. Fruit 5–8 mm in diameter, ovoid or spherical, borne on a reflexed stalk, ripening in the first year (or very early in the spring of the second) to blue-black with a waxy bloom. Seeds 1–3, ovoid, furrowed. *S & C Europe to western Siberia.* H1.

Another variable species, represented in gardens mainly by cultivars. The unpleasant smell of the bruised foliage, and its bitter taste, are usually diagnostic.

**10. MICROBIOTA** Komarov
*J. Lewis*
Evergreen, dioecious shrubs to 1 m (in the wild), densely branched. Mature leaves in opposite pairs, scale-like, small, glandular and with acute tips. Male cones terminal, ovoid, not found on plants under 10 years old. Mature female cones berry-like, to 5 × 4 mm, composed of 4 spreading scales enclosing a single, dark brown, shiny seed.

A recently introduced genus of a single species, allied to *Juniperus*, which is likely to become popular for ground cover in shady areas where it does well and remains under 50 cm in height. It also tolerates full sun, is reliably hardy throughout Europe, colours well in winter and roots easily from cuttings.
Literature: van Hoey-Smith, J.R.P., Microbiota decussata, *International Dendrological Society Yearbook for 1972*: 51 (1973).

**1. M. decussata** Komarov. Illustration: International dendrological society yearbook for 1972: 51 (1973); Welch, Manual of dwarf conifers, t. 276 (1979). *USSR (SE Siberia).* H1.

## VII. PINACEAE

Resinous, evergreen or more rarely deciduous, monoecious trees, bark usually scaly. Buds scaly. Leaves with resin canals, linear or needle-like, borne singly and spirally arranged on long shoots, and, in some genera, arranged in apparent whorls on short, spur-like side shoots. Male cones ephemeral, catkin-like, with numerous, spirally arranged scales, each scale with 2 pollen sacs; pollen grains usually with 2 air bladders. Female cones usually thinly or thickly woody, with numerous, spirally arranged scales; bract scales and the fertile, ovuliferous scales distinct in most genera; ovules 2 per scale, maturing in the first or second season, occasionally long-persistent. Seeds usually winged.

The largest family of Conifers, and the most important economically, with numerous species grown for their softwood timber. The family includes 10 genera with over 200 species, and is widely spread throughout the northern hemisphere. Two genera, *Keteleeria* and *Cathaya* are not included here as they are not generally cultivated.

*Habit.* Deciduous: **5,6**; evergreen: **1–4,7,8**,

*Leaves*. In clusters on short shoots: **5–8**;
spirally arranged or in 2 ranks: **1–4**.
*Leaf apex*. Rounded: **1,3,5,6**; notched: **1–3**;
pointed: **1,2,4,7,8**.
*Mature female cones*. Always less than
10 cm: **3,4**. With deciduous scales:
**1,5,7**. Hanging: **2–4**.
*Bracts*. Projecting: **1,2,6**.

1a. Leaves needle-like, borne in clusters
or fascicles of 2, 3, 5 or many on
short, spur-like shoots                                 2
b. Leaves 2-surfaced or occasionally
needle-like, spirally arranged                          5
2a. Leaves united at the base into fascicles
of 2, 3 or 5, each fascicle surrounded
by a sheath at the base                    **8. Pinus**
b. Leaves on short shoots clustered, with
10–30 leaves in a cluster, the clusters
without sheaths at the base                             3
3a. Evergreen trees; cones 7–12 cm
long × more than 2 cm broad
                                          **7. Cedrus**
b. Deciduous trees; mature cones
1.5–6 cm long, or rarely longer and
then at most 2 cm broad                                 4
4a. Lateral shoots long and curved; bud
scales acuminate; male cones
clustered; female cones papery,
disintegrating on the tree
                                     **5. Pseudolarix**
b. Lateral shoots short and straight; bud
scales obtuse; male cones solitary;
female cones thinly woody, remaining
intact at maturity                         **6. Larix**
5a. Leaf scars circular in outline, and flat;
mature cones erect with deciduous
scales                                     **1. Abies**
b. Leaf scars raised from the stem on
ridges or peg-like extensions; mature
cones hanging, with persistent scales
                                                         6
6a. Trees to 55 m; buds fusiform, smooth,
reddish brown; female cones with
exposed, 3-lobed bracts
                                     **2. Pseudotsuga**
b. Trees to 45 m; buds ovoid to conical,
smooth or resinous; female cones with
the bracts hidden                                       7
7a. Foliage soft, delicate, on slender
shoots; buds 1–3 mm; cones to 3 cm,
usually with few scales                    **3. Tsuga**
b. Foliage harsh, spiny; buds 5–15 mm;
cones 4–15 cm, with many scales
                                          **4. Picea**

## 1. ABIES Miller
*K.D. Rushforth*
Evergreen trees, mostly 15–50 m. Leaves
linear, arranged spirally on the shoots but
usually twisted at their bases to give a

parted arrangement; leaf-scars rounded.
Female cones erect, usually on the upper
tiers of branches. Male cones hanging from
the undersides of the lower branches.
Mature female cones ripening in the
autumn of their first year, disintegrating to
scatter seeds and cone-scales, leaving the
central stalk attached to the shoot.

A genus of about 50 species, almost all of
which have been in cultivation, found in
mountainous regions of Eurasia, N Africa
and N America (as far south as
Guatemala). The characters used to
distinguish the species involve mainly the
cones and the foliage of non-coning shoots
from the lower crown. Leaves from the
upper crown and from coning shoots tend
to be shorter and more upstanding, and are
less reliable diagnostically.

All the species thrive on moist, well-
drained soils, while some will tolerate dry
or alkaline conditions. Some are susceptible
to damage by spring frosts. Most are
tolerant of shade and many grow faster
when young given light shading.
Literature: Liu, T.S., *A Monograph of the
genus Abies* (1971); Rushforth, K.D., Tree
genera, 5: The silver firs, Abies,
*Arboricultural Journal* 3(1): 37–46 (1976).

1a. Foliage beneath the shoot more or less
radial to it                                            2
b. Foliage beneath the shoot widely
parted, directed forwards along the
shoot, rising or parted and recurved
                                                        11
2a. Leaves with 2 white bands on the
lower surface only (occasionally with
a small whitish area on the
uppermost surface in the uppermost
fifth of the leaf)                                      3
b. Leaves with 2 white bands on the
lower surface and 1 or 2 (sometimes
irregular) whitish bands on the upper
surface                                                 7
3a. Leaf margins distinctly rolled down at
the sides                            **32. delavayi**
b. Leaf margins not or only slightly
rolled down at the sides                                4
4a. Shoots pale or whitish grey or pale
pinkish grey                                            5
b. Shoots red-brown, yellow-brown or
shining brown                                           6
5a. Leaves silvery glaucous beneath
                                      **25. koreana***
b. Leaves greenish white beneath
                                       **26. sibirica**
6a. Terminal buds conic or ovoid-conic,
moderately resinous; shoots shining
brown; leaves usually sharp
                                    **13. cephalonica**

b. Terminal buds spherical, thickly
encrusted with resin; shoots red-
brown or yellow-brown, usually matt;
leaves usually rounded and notched
                                      **31. forrestii***
7a. All or most leaves 2 cm or more       8
b. All leaves less than 2 cm             10
8a. Terminal bud conic or ovoid-conic,
moderately resinous
                                    **13. cephalonica**
b. Terminal bud spherical, thickly
resinous                                                9
9a. Leaves *c*. 1 mm broad      **26. sibirica**
b. Leaves 2–2.5 mm broad        **5. gamblei**
10a. Leaves of more or less the same
length all round the shoot
                                        **8. pinsapo**
b. Leaves above the shoot always shorter
than the others               **9. numidica**
11a. Terminal bud spindle-shaped,
1–2 cm, sharply pointed
                                      **1. bracteata**
b. Terminal bud not spindle-shaped or
pointed, less than 1 cm                                12
12a. Leaf with 2 whitish bands or 1 broad
whitish band nearly covering the
upper surface                                          13
b. Leaf without white bands above or
with only a scattered, incomplete
band                                                   17
13a. Leaf diamond-shaped in cross-
section                                                14
b. Leaf flat or oval in cross-section   15
14a. Leaf with a groove along the upper
surface                             **17. procera**
b. Leaf without a groove along the upper
surface                            **16. magnifica**
15a. Leaves green or yellow-green
                                        **5. gamblei**
b. Leaves grey or blue, at least for
2 years                                                16
16a. Leaves all less than 3 cm, densely
arranged, pressed down on the shoot
                                      **28. lasiocarpa**
b. Leaves all or most more than 3 cm,
loosely arranged               **19. concolor**
17a. Terminal buds encrusted with resin
                                                        18
b. Terminal buds either without resin or
with a small amount which does not
entirely cover the bud-scales           43
18a. White bands or patches present on
both leaf surfaces (sometimes merely
scattered rows on the upper surface)
                                                        19
b. White bands present only on the
lower leaf surface (occasionally a few
white spots above, near the tip)        23
19a. Shoot hairless to the naked eye     20
b. Shoot downy or hairy to the naked
eye                                                    22

20a. Bark flaking, red-brown or orange-
brown                  **36. squamata**
b. Bark not flaking, not red-brown or
orange-brown                      21
21a. Shoot olive-green in the first winter;
terminal buds blunt        **22. vejari**
b. Shoot dark brown in the first winter;
terminal buds pointed
                **12. bornmuelleriana**
22a. Leaves soft; shoot with whitish down
                        **26. sibirica**
b. Leaves firm; shoot with dark-coloured
down                    **29. balsamea***
23a. Foliage widely and clearly parted
above the shoot                    24
b. Foliage narrowly or not at all parted
above the shoot                    28
24a. Shoot olive-green in its first and
second years; terminal buds less than
3 mm                    **20. grandis**
b. Shoot brown or greyish white in its
1st and 2nd years; terminal buds
more than 3 mm                     25
25a. Shoots of previous summer hairless to
the naked eye                      26
b. Shoots (at least the side shoots)
obviously downy to the naked eye for
the first 2 years                  27
26a. Lateral leaves curving down below
the shoot; terminal buds spherical
                        **3. pindrow**
b. Lateral leaves widely parted but not
curving down below the shoot;
terminal buds ovoid or conic
                      **4. chensiensis**
27a. Hairs on shoots coarse, in linear tufts
in deep grooves, red-brown
                      **30. spectabilis**
b. Hairs fine, not in tufts in deep
grooves, blackish           **34. fabri***
28a. Terminal buds of strong shoots more
than 4 mm                          29
b. Terminal buds of strong shoots less
than 4 mm                          30
29a. Buds on strong shoots spherical, grey-
white                    **31. forrestii***
b. Buds on strong shoots conic or ovoid-
conic, yellow-brown        **33. chengii**
30a. Shoots deeply ridged and grooved,
hairless                  **7. homolepis**
b. Shoots not deeply ridged and grooved,
or, if so, downy in the grooves   31
31a. Terminal buds purple or purplish   32
b. Terminal buds brown or white        38
32a. Lateral leaves curving downwards at
the sides of the shoot     **21. religiosa**
b. Lateral leaves spreading flat or rising
at the sides of the shoot          33
33a. Shoots of the previous summer
(particularly strong ones) purple
                        **35. fargesii**

b. Shoots of the previous summer grey-,
green-, orange- or pink-brown or
grey-green                         34
34a. Bands on the leaves silvery white   35
b. Bands on the leaves grey-green or
greenish white                     37
35a. Leaf margins slightly rolled
downwards; midrib keeled beneath
                        **34. fabri***
b. Leaf margins and midrib flat beneath
                                   36
36a. Foliage above the shoot pointing more
or less forwards; leaves to 3 cm,
square-tipped              **23. veitchii**
b. Foliage above the shoot spreading at
right angles; leaves less than 2 cm,
round-tipped              **25. koreana***
37a. Hairs on the shoot long, reddish
brown, borne in grooves
                      **24. kawakamii**
b. Hairs on the shoot fine, dark-
coloured, not in distinctive lines in
grooves                 **27. nephrolepis***
38a. Shoots evenly downy for 1–2 years,
hairs red or grey-brown
                        **18. amabilis***
b. Shoots hairless or rapidly becoming
so, or the down uneven             39
39a. Leaves pointed at the tip          40
b. Leaves blunt and notched at the tip
                                   41
40a. Leaves blue-white beneath
                        **21. religiosa**
b. Leaves pale green beneath
                        **6. recurvata**
41a. Bands on the leaves greyish green or
greyish white           **27. nephrolepis***
b. Bands on the leaves silver or silvery
white                              42
42a. Terminal buds chocolate brown or
red; shoots pale yellow
                        **29. balsamea***
b. Terminal buds white or covered with
white resin; shoots pale grey- or
pinkish-brown             **25. koreana***
43a. Shoots of the previous summer hairless
or (under a lens) finely downy     44
b. Shoots of the previous summer
conspicuously downy to the naked
eye                                49
44a. Leaves less than 2 cm             45
b. Leaves mostly more than 2 cm       46
45a. Bands on leaves pale green
                        **6. recurvata**
b. Bands on leaves whitish green
                        **15. nebrodensis**
46a. Leaves yellow-green         **2. firma***
b. Leaves green or dark green         47
47a. Leaves loosely pressed down on the
upper surface of the shoot
                        **10. cilicica**

b. Leaves erect, parted or (rarely)
recurved on the upper surface of the
shoot                              48
48a. Bark brown, slightly flaky; cones
violet-blue while growing, to
8 × 3.5 cm                **6. recurvata**
b. Bark grey, smooth, becoming fissured
and scaly; cones green, 10–14 × c.
5 cm                    **4. chensiensis**
49a. Leaves on the upper surface of the
shoot pointing forwards            50
b. Leaves on the upper surface of the
shoot spreading                    51
50a. Leaves silvery white beneath; cones
with projecting bracts
                      **11. nordmanniana**
b. Leaves greenish white beneath; cones
with bracts not projecting
                        **10. cilicica**
51a. Leaves 2–4 mm wide, tip bifid or
rounded and deeply notched
                        **2. firma***
b. Leaves at most 2.5 mm wide, tip
rounded, notched or obscurely
pointed                            52
52a. Bands on leaves faint or greenish
white                   **27. nephrolepis***
b. Bands on leaves silvery white
                        **14. alba***

**1. A. bracteata** (D. Don) Nuttall (*A. venusta*
(Douglas) Koch). Illustration: Mitchell,
Conifers in the British Isles, 12 (1972);
Phillips, Trees in Great Britain, 60 (1978);
Rushforth, Mitchell Beazley pocket guide to
trees, 48 (1980).
Tree to 30 m with dark grey, smooth bark.
Shoot olive-brown at first, later red-brown,
hairless. Terminal buds ovoid-conic or
fusiform, sharply pointed, with many
overlapping scales, without resin, 1–2 cm.
Leaves spreading forwards above and
beneath the shoot, to 5 cm, glossy green
above and with 2 whitish green bands
beneath, tip sharply pointed, yellowish.
Mature cone 7–10 cm, ovoid, golden
brown, bracts distinctly projecting for *c.*
3 cm, reflexed, bearing blobs of resin. *USA
(California, Santa Lucia Mountains).* H3.
    Unique in its buds, spine-tipped leaves
and cones.

**2. A. firma** Siebold & Zuccarini. Illustration:
Mitchell, Conifers in the British Isles, 24
(1972); Phillips, Trees in Great Britain, 62
(1978).
Tree to 30 m, with pink-grey, finely flaky
bark which becomes grey and corky in old
trees. Shoots shiny pink-brown for
3–4 years, hairless or with dark hairs in
grooves. Terminal buds *c.* 8 mm, ovoid-
conic, shiny red-brown, faintly resinous,

the scales slightly free at their tips. Foliage on young trees slightly parted beneath and with a wide, V-shaped parting above, on older trees mostly spreading, without a V-shaped parting. Leaves on young trees *c.* 5 cm × 3 mm, the tips tapered to 2 sharp points, yellowish green above and with 2 pale, shiny green bands beneath; on older trees leaves 1.5–2 cm × 3–5 mm, the tips rounded and notched. Mature cone 5–12 × 5 cm, conic, yellowish green becoming brown, bracts projecting. *C & S Japan.* H2.

*****A. holophylla** Maximowicz. Illustration: Mitchell, Conifers in the British Isles, 28 (1972); Arboricultural Journal 3(1): t. 13 (1976). Bark flaking into papery scales; foliage upswept from beneath the shoot, upper leaves almost erect; leaves 2–4.5 cm, pointed; buds not resinous; cone 8–12 × 4 cm, cylindric, bracts not projecting. *NE China, Korea and adjacent USSR.* H1.

**3. A. pindrow** Royle. Illustration: Mitchell, Conifers in the British Isles, 38 (1972); Arboricultural Journal 3(1): t. 14 (1976). Tree to 30 m, with grey, shallowly fissured bark. Shoots shiny whitish grey for several years, hairless. Terminal buds to 8 mm, spherical, green-brown, thickly resinous. Foliage widely parted beneath the shoot, the lateral leaves curved down below the shoot, the upper leaves widely spreading or pointing forwards with a wide, V-shaped parting. Leaves 2–7 (rarely 9) cm, 2-pointed at the tips, glossy green above and with 2 whitish bands beneath. Mature cone 8–12 × 6 cm, violet-purple ripening brown, cylindric, bracts not projecting. *Afghanistan to W Nepal.* H2.

**4. A. chensiensis** van Tieghem. Illustration: Liu, Monograph of the genus Abies, 401 (1971).
Tree to 15 m (in cultivation; much taller in the wild), with grey, smooth or slightly fissured and scaly bark. Shoots ash-grey for at least 3 years, hairless to the naked eye, sometimes downy under a lens. Terminal buds *c.* 7 mm, ovoid, bluntly pointed, resinous, brown. Foliage widely parted beneath, narrowly parted above, angled slightly forwards. Leaves 2–3.5 cm × 2.5 mm (rarely less), tips acute or occasionally rounded and bifid, dark glossy green above, with 2 whitish green bands beneath separated by the broad, green midrib. Mature cone *c.* 10 cm, cylindric, green ripening brown, bracts not projecting. *N & W China (Shaanxi, NW Hubei, Yunnan).* H2.

**5. A. gamblei** Hickel (*A. pindrow* var. *brevifolia* Dallimore & Jackson). Illustration: Dallimore & Jackson, Handbook of Coniferae, edn 4, 75 (1966).
Tree to 15 m with grey bark. Shoots pale yellow-brown or pink-brown, hairless. Terminal buds rounded though lumpy, brown, resinous. Foliage radial around the shoot and directed slightly forwards, or parted beneath and slightly parted above. Leaves 2–3.5 cm × 2–2.5 mm, sickle-shaped, blunt, green or slightly yellow-green, with 2 greyish bands above, pale green with 2 whitish bands beneath. Mature cone 10–12 cm, cylindric, violet ripening brown, bracts not projecting. *NW Himalaya.* H3.

**6. A. recurvata** Masters. Illustration: Liu, Monograph of the genus Abies, 408 (1971); Mitchell, Conifers in the British Isles, 43 (1972).
Tree to 15 m with brown, slightly flaky bark. Shoot yellow-brown becoming brown and flaking slightly in the 4th year, hairless. Terminal buds *c.* 7 mm, rounded, pale brown, variously resinous. Foliage recurved on very vigorous shoots, more usually widely parted beneath with the lateral leaves at right angles to the shoot, the upper leaves erect or slightly recurved with a narrow, V-shaped parting between them. Leaves 1.5–2.5 cm × *c.* 2 mm, rounded and then pointed at the tips, dark glossy green above, shiny light green beneath with 2 faint greenish white bands. Mature cone *c.* 6 × 3 cm, ovoid, violet-blue ripening brown, the bracts with their tips just projecting and adpressed. *W China (NW Sichuan).* H2.

A very shade-tolerant tree. Var. **ernestii** (Rehder) Rushforth (*A. ernestii* Rehder; *A. beissneriana* Rehder & Wilson not Mottet; *A. chensiensis* var. *ernestiana* (Rehder) Liu) has whitish shoots with conical buds, leaves *c.* 4 cm × 2.5 mm, those on the upper part of the shoot arching upwards to form a U-shaped parting, each notched and with some white lines on the upper surface. *W China (NW Sichuan).*

**7. A. homolepis** Siebold & Zuccarini (*A. brachyphylla* Maximowicz). Illustration: Mitchell, Conifers in the British Isles, 28 (1972); Phillips, Trees in Great Britain, 64 (1978); Rushforth, Mitchell Beazley pocket guide to trees, 48 (1980).
Tree to 25 m with finely flaky or scaly, grey, grey-brown or pink-grey bark. Shoots white to yellow-grey, deeply grooved, hairless. Terminal buds *c.* 3 mm, conic, brown with clear (rarely clouded) resin.

Foliage parted or pointing forwards and downwards, lateral leaves at right angles to the shoot, the upper leaves erect or forming a narrow, U-shaped parting above. Leaves 1–3 cm × *c.* 2 mm, tips rounded, notched or acute, glossy green above and with 2 silvery green bands separated by the mid-green midrib beneath. Mature cone 7–12 × *c.* 3.5 cm, ovoid or cylindric-ovoid, brown, bracts not projecting. *Japan (Honshu, Shikoku).* H2.

**8. A. pinsapo** Boissier. Illustration: Rushforth, Mitchell Beazley pocket guide to trees, 42 (1980).
Tree to 30 m with grey (ultimately almost black) bark, fissuring into irregular plates. Shoots at first green-brown, later shiny pale brown, hairless. Terminal buds ovoid with resin covering the scales. Foliage almost at right angles all around the shoot. Leaves to 2 cm, rigid, tips obtuse, grey-green with 2 whitish bands beneath and 1 or 2 broad whitish bands above; leaf-base a sucker-like pad leaving a large, oval scar on the shoot. Mature cone 10–15 cm, cylindric-conic with a nipple-like tip, bracts not projecting. *S Spain.* H2.

Tolerant of dry, chalky sites. The natural populations vary in the blueness of the foliage; 'Glauca' is a commonly found cultivar, but it is no bluer than many other individuals.

**9. A. numidica** Carrière. Illustration: Mitchell, Conifers in the British Isles, 37 (1972); Phillips, Trees in Great Britain, 64 (1978).
Tree to 25 m with purplish or pinkish grey bark, at first smooth, later cracked into rounded scales. Shoots red-, orange- or pink-brown, hairless. Terminal buds ovoid-conic, not or only slightly resinous. Foliage almost at right angles all around the shoot, occasionally pointing slightly backwards, or with a V-shaped parting above. Leaves to 2 cm × 2 mm, those at the sides and beneath the shoot longer than those above, which form a flattish area; all rounded and blunt with 1 broad, incomplete whitish band above and 2 whitish bands beneath. Mature cone 10–15 cm, cylindric with a nipple-like tip, greenish white, bracts not projecting. *NE Algeria.* H2.

Tolerant of dry and alkaline sites. 'Glauca' and 'Glauca Pendula' are occasionally grown; they have grey foliage which is pendulous in the latter.

**10. A. cilicica** (Antoine & Kotschy) Carrière. Illustration: Mitchell, Conifers in the British Isles, 16 (1972).

Tree to 35 m with dark grey bark which is smooth except for wrinkles around the branch scars. Shoots pale brown or greenish brown, eventually orange-brown, downy. Terminal buds *c.* 4 mm, conic, not or only slightly resinous. Foliage rather open, the leaves beneath the shoot spreading, those at the sides pointing slightly forwards, those above pointing forwards and loosely pressed down on the shoot. Leaves 2–5 cm × *c.* 2 mm, with rounded and often notched tips, dark shiny green above, with 2 greenish white bands beneath. Mature cone 10–15 × *c.* 5 cm, cylindric, green becoming red-brown, bracts not projecting. *SE Turkey, Syria, Lebanon.* H2.

**11. A. nordmanniana** (Steven) Spach. Illustration: Mitchell, Conifers in the British Isles, 35 (1972); Phillips, Trees in Great Britain, 64 (1978); Rushforth, Mitchell Beazley pocket guide to trees, 41 (1980). Trees to 40 m with grey bark, smooth at first, ultimately fissuring into small squares. Shoot green-brown in the first winter, pink-brown in the second, finely downy. Terminal buds *c.* 5 mm, ovoid, acute, brown, without resin. Foliage parted beneath the shoot, the lateral leaves pointing forwards or curved down, the largest upper leaves pointing forward and pressed down on the shoot. Leaves 1.5–4 cm × *c.* 2 mm, tips rounded and notched, glossy green above and with 2 silvery bands separated by a green midrib beneath. Mature cone 10–15 × *c.* 5 cm, cylindric, green at first, later brown, resinous, the bracts projecting. *NE Turkey, USSR (Caucasus).* H2.

**12. A. bornmuelleriana** Mattfeld. Illustration: Mitchell, Conifers in the British Isles, 11 (1972); Liu, Monograph of the genus Abies, 445 (1971). Tree to 25 m with dark, finely roughened bark. Shoots dark brown in the first year, shiny pink-brown in the second, hairless. Terminal buds *c.* 8 mm, ovoid, pointed, resinous. Foliage parted or pointing forwards beneath the shoot, lateral leaves pointing forwards, upper leaves erect or narrowly parted. Leaves to 4 cm × 2.5 mm, tips rounded and notched, glossy green above with a few white spots or lines near the tip, and with 2 silver-glaucous bands separated by a bright green midrib beneath. Mature cones 10–15 × 4–6 cm, brown, bracts projecting and reflexed. *NW Turkey.* H2.

**13. A. cephalonica** Loudon. Illustration: Mitchell, Conifers in the British Isles, 14 (1972); Polunin & Everard, Trees and bushes of Europe, 23 (1976); Phillips, Trees in Great Britain, 61 (1978); Rushforth, Mitchell Beazley pocket guide to trees, 42 (1980). Tree to 40 m with a massive crown and grey or grey-brown bark which is fissured into small plates. Shoots shining dark brown in the first winter, later pink-brown or almost orange, often with greenish fissures. Terminal buds conic or ovoid-conic, brown, resinous. Foliage spreading radially from the shoot but with fewer leaves beneath, or parted beneath and with a narrow parting above. Leaves 1.5–3.5 cm × *c.* 2.5 mm, tips rounded to a sharp or blunt point, the margins not rolled downwards, glossy green and often with a few whitish lines above, and with 2 silvery bands separated by the green midrib beneath. Mature cone 10–15 × 4–5 cm, green-brown, bracts projecting and reflexed. *S Greece.* H2.

Tolerant of chalky sites but susceptible to damage from spring frosts. The wild tree is large, with a heavy crown; 'Meyer's Dwarf' is a slow-growing variant.

**14. A. alba** Miller (*A. pectinata* (Lambert) Lambert & de Candolle). Illustration: Polunin & Everard, Trees and bushes of Europe, 2 (1976); Phillips, Trees in Great Britain, 60 (1978); Rushforth, Mitchell Beazley pocket guide to trees, 41 (1980). Tree to 50 m with dull silvery grey bark which is cracked into plates. Shoots pink-brown or grey with dark hairs in grooves. Terminal buds ovoid, bluntly pointed, not or only slightly resinous. Foliage parted beneath, side leaves more or less at right angles to the shoot, upper leaves spreading forwards leaving a distinct parting. Leaves 1–2.5 cm × *c.* 2.5 mm, tips rounded and notched, glossy dark to yellow-green above, occasionally with a few white spots, beneath with 2 silver bands separated by the pale green midrib. Mature cone 10–15 × *c.* 5 cm, green ripening red-brown, bracts projecting, reflexed. *C & S Europe.* H2.

Tolerant of shade as a young tree, but susceptible to damage from spring frosts and to leaf-sap-sucking aphids. Various cultivars have been selected, including 'Fastigiata' with a narrow, erect crown, and 'Pendula' with hanging side branches.

***A. borisii-regis** Mattfeld. Illustration: Liu, Monograph of the genus Abies, 444 (1971); Mitchell, Conifers in the British

Isles, 11 (1972). Shoots pale with dense, dark brown hairs; foliage dense, leaves to 3 cm × 2 mm, tips rounded or acute. *Bulgaria, S Jugoslavia, S Greece.* H2.

**15. A. nebrodensis** (Lojacono-Pojero) Mattei. Illustration: Liu, Monograph of the genus Abies, 420 (1971); Mitchell, Conifers in the British Isles, 33 (1972). Tree 15–30 m with dark brown, fissured bark. Shoot yellowish grey or pale yellowish brown, hairless or with blackish hairs in faint grooves (visible only under a lens). Terminal bud ovoid-conic, acute, brown, not or slightly resinous. Foliage widely spreading and directed forwards beneath, upper leaves erect with a narrow, V-shaped parting. Leaves to 2 cm × 3 mm, tips rounded or pointed, usually fairly sharp, infrequently notched, dark glossy green above with a few white spots near the tip, beneath with 2 greenish white bands. Mature cone 8–12 cm, cylindric, greenish or yellow-brown, bracts projecting, reflexed. *Sicily.* H2.

A threatened species in the wild, reduced to small numbers by excessive felling.

**16. A. magnifica** Murray. Illustration: Bean, Trees and shrubs hardy in the British Isles, edn 8, t. 3 (1970); Mitchell, Conifers in the British Isles, 32 (1972); Phillips, Trees in Great Britain, 63 (1978); Rushforth, Mitchell Beazley pocket guide to trees, 45 (1980). Tree to 35 m with a narrow, columnar crown borne on a stout trunk with smooth, grey, resin-blistered bark which becomes thick and ultimately reddish. Shoots covered with dense, fine, red-brown hairs. Buds ovoid, lumpy, purple, covered with clear, whitish resin. Foliage spreading horizontally or rising on strong shoots, the upper leaves shorter than the others. Leaves 1.5–4.5 cm, length decreasing along the shoot, diamond-shaped in section, tips rounded, acute, all blue with 4 whitish bands. Mature cone 10–25 cm, ovoid or barrel-shaped, violet-purple or golden green, bracts not projecting. *W USA (Oregon to California).* H2.

Unusual for its narrow, columnar, conical crown with short, level branches; it quickly forms a stout trunk. Var. **shastensis** Lemmon has cones 10–15 cm with the bracts projecting and reflexed.

**17. A. procera** Rehder (*A. nobilis* Lindley). Illustration: Mitchell, Conifers in the British Isles, 41 (1972); Rushforth, Mitchell Beazley pocket guide to trees, 45 (1980). Tree to 40 m with smooth, grey, resin-blistered bark which is ultimately fissured. Shoots flattened at the nodes, covered in reddish brown hairs for several years. Terminal buds *c.* 3 mm, spherical, dark purplish brown with thick, whitish resin. Leaves all adpressed at their bases, those beneath the shoot spreading widely, the others rising and pointing forwards, 2–4 cm × *c.* 2 mm, diamond-shaped in section, tips rounded, blue-grey with 4 greyish white bands. Mature cone to 25 × 7 cm, cylindric, purplish brown, the bracts yellow-green and projecting and reflexed, almost covering the entire cone surface. *USA (Oregon to N California).* H2.

A fast-growing, shapely tree which produces its cones even when quite small. The foliage varies in the degree of blueness.

**18. A. amabilis** (Douglas) Forbes. Illustration: Mitchell, Conifers in the British Isles, 9 (1972); Phillips, Trees in Great Britain, 60 (1978); Rushforth, Mitchell Beazley pocket guide to trees, 44 (1980). Tree to 30 m with smooth, grey, resin-blistered bark which darkens with age. Shoots green-brown in the first year, later grey or purple-grey, covered with a dense, short, even, pale grey-brown down. Terminal buds *c.* 3 mm, spherical, coated with thick resin. Foliage parted and flat beneath the shoot, the upper leaves pointing forwards at an angle of *c.* 45°. Leaves to 4 cm × 2 mm, longest in the middle of the shoot, tips squarish, notched, grey-green or blue-green above, with 2 broad, grey-green or white bands separated by the narrow green midrib beneath. Mature cone to 15 × 5 cm, cylindric, domed, violet ripening brown, bracts not projecting *Western N America (British Columbia to California).* H2.

'Spreading Star' is a cultivar obtained from making a graft of a side-shoot, and is occasionally seen in collections of dwarf conifers.

\***A. mariesii** Masters. Illustration: Liu, Monograph of the genus Abies, 399 (1971); Mitchell, Conifers in the British Isles, 35 (1972). Shoot with red-brown down in the first year, the down later dark brown or black or grey; leaves 5–20 × *c.* 2.5 mm, glossy green; mature cone to 10 × 5 cm, ovoid. *C Japan.* H2.

**19. A. concolor** (Gordon) Hillebrand. Illustration: Mitchell, Conifers in the British Isles, 17 (1972); Phillips, Trees in Great Britain, 61 (1978); Rushforth, Mitchell Beazley pocket guide to trees, 43 (1980). Tree to 40 m with smooth, grey, resin-blistered bark which ultimately becomes furrowed and scaly. Shoot green-brown to grey-brown, hairless. Terminal buds spherical, brown, covered with clear or whitish resin. Foliage loose, all the leaves rising above the shoot without a parting. Leaves to 5.5 cm × 2 mm, flat or oval in section, the tips rounded, acute or slightly notched, blue-grey with 1 whitish band above and 2 beneath. Mature cone 7–12.5 cm, cylindric or ellipsoid, olive-green, violet-blue or yellowish ripening brown; bracts not projecting. *W USA, Mexico.* H2.

Tolerates dry sites. 'Violacea' and 'Candicans' are cultivars with brighter, silvery-blue leaves. Var. **lowiana** (Gordon) Lemmon reaches 50 m and has grey-green leaves with less obvious white bands above, all spreading flat or upstanding to produce a U-shaped parting. *W USA (California, Oregon).*

**20. A. grandis** (Douglas) Lindley. Illustration: Mitchell, Conifers in the British Isles, 27 (1972); Phillips, Trees in Great Britain, 62 (1978); Rushforth, Mitchell Beazley pocket guide to trees, 44 (1980). Tree to 60 m with smooth, brownish grey, resin-blistered bark which ultimately cracks into small squares. Shoot olive-green at first, later red-brown, green-brown or grey, finely downy at first but soon becoming hairless. Terminal buds *c.* 3 mm, conic, rounded, whitish purple, resinous. Foliage spreading nearly flat and slightly forwards above and beneath the shoot (spreading forwards at an angle of about 45° on strong shoots). Leaves to 5 cm, those above the shoot shorter than those beneath, tips rounded or square, notched, glossy green above with 2 whitish bands beneath. Mature cone 5–10 cm, cylindric but tapered at both ends, yellowish green ripening brown, bracts not projecting. *Western N America.* H2.

A fast-growing tree, moderately tolerant of shade.

**21. A. religiosa** (Humboldt, Bonpland & Kunth) Schlechtendahl & Chamisso. Illustration: Liu, Monograph of the genus Abies, 432 (1971). Tree 15–30 m with smooth, grey bark. Shoots greenish brown in the first winter, becoming brown or reddish brown, downy with hairs in faint grooves. Terminal buds *c.* 3 mm, broadly conic, blunt, purplish brown. Foliage widely parted beneath, lateral leaves at 90° to the shoot or angled slightly forwards, upper leaves narrowly parted, pointing slightly forwards. Leaves 2–4 cm × *c.* 2 mm, acute, green to grey-green above, with 2 bluish white bands beneath. Mature cone 10–11 × *c.* 5 cm, cylindric-ovoid, violet-blue ripening brown, bracts projecting, reflexed. *Mexico, N Guatemala.* H3.

**22. A. vejari** Martinez. Illustration: Liu, Monograph of the genus Abies, 434 (1971). Tree to 10 m (30 m in the wild), bark grey, smooth. Shoots blue-green in the first year, becoming red-brown by the third year, hairless. Terminal buds *c.* 5 mm, ovoid, blunt, resinous, whitish brown. Foliage widely parted beneath, lateral leaves pointing slightly forwards, upper leaves pointing forwards at *c.* 45° with a V-shaped parting between them, some leaves pressed down on the shoot with their lower surfaces uppermost. Leaves to 2.5 cm × 2 mm, pointed, dark glossy green with several scattered whitish rows of dots or an irregular whitish band above, and with 2 bluish white bands beneath. Mature cone 6–10 × *c.* 5 cm, oblong or ovoid, purple ripening brown, bracts projecting, reflexed. *NE Mexico.* H2.

**23. A. veitchii** Lindley. Illustration: Mitchell, Conifers in the British Isles, 48 (1972); Phillips, Trees in Great Britain, 65 (1978); Rushforth, Mitchell Beazley pocket guide to trees, 41 (1980). Tree to 20 m, bark grey or grey-brown, smooth or fluted, becoming plated in old trees. Shoots pale pink-brown or grey, with scattered, fine, black hairs in faint grooves. Terminal buds *c.* 3 mm, ovoid, reddish purple with thick, whitish resin. Foliage parted beneath, the lateral leaves directed forwards at *c.* 45° and slightly rising, the upper leaves erect or directed slightly forwards. Leaves to 2.5 cm, tips rounded and notched, margins and midrib flat, green to dark green and slightly bloomed above, with 2 silver bands beneath. Mature cone 5–8 cm, cylindric, violet-purple, bracts projecting for 2–3 mm. *Japan (Honshu).* H1.

More tolerant of atmospheric pollution than the other species.

**24. A. kawakamii** (Hayata) Ito. Illustration: Liu, Monograph of the genus Abies, 400 (1971). Tree to 15 m with a stout trunk, bark whitish or grey-brown with thick, corky scales. Shoot greyish brown, slightly grooved with lines of coarse, reddish brown hairs in the grooves. Terminal buds *c.* 3 mm, ovoid-conic, covered with clear or occasionally whitish resin. Foliage adpressed beneath, the lateral leaves pointing forwards, the upper leaves ascending, often slightly parted. Leaves to 2.5 cm, tips squarish and notched, dark green and often slightly bloomed above, with 2 whitish green bands beneath. Mature cone 6–8 cm, oblong or cylindric, violet-purple, bracts not projecting. *Taiwan*. H2.

**25. A. koreana** Wilson. Illustration: Phillips, Trees in Great Britain, 63 (1978); Rushforth, Mitchell Beazley pocket guide to trees, 46 (1980).
Tree to 10 m, bark smooth, grey. Shoot shining greyish pink, at first with fine, blackish hairs, later hairless. Terminal buds to 3 mm, ovoid-conic, purple or purple and white, thickly resinous. Foliage rather sparse, radial or pointing forwards beneath the shoot. Leaves 1.5–2 cm × *c.* 2.5 mm, tips rounded and notched, margins not rolled down, yellow-green above and with 2 broad, silvery-glaucous bands beneath. Mature cone *c.* 7 × 2 cm, cylindric, violet-purple or rarely green, finally brown, bracts projecting, reflexed. *Korea*. H2.

Often produces its ornamental cones when very small.

**26. A. sibirica** Ledebour. Illustration: Liu, Monograph of the genus Abies, 416 (1971); Mitchell, Conifers in the British Isles. 45 (1972).
Tree to 30 m, bark smooth, grey or greyish brown. Shoot smooth, grey or grey-brown, downy with pale hairs. Terminal buds spherical, brown or white, very resinous. Foliage arranged radially, or leaves pointing forwards at an angle of *c.* 45°, sometimes partly spreading above and beneath the shoot. Leaves 1.5–3.5 cm × *c.* 1 mm, the longest beneath the shoot, tips squarish, notched, margins not rolled downwards, glossy or yellow-green with a few white spots near the tip above, with 2 narrow, grey-green bands separated by the broad, green midrib beneath. Mature cone 5–7.5 cm, cylindric, violet ripening brown, bracts not projecting. *North Eurasia from 40° E to 135° E.* H1.

Very susceptible to damage by late spring frosts.

**27. A. nephrolepis** (Trautvetter) Maximowicz. Illustration: Mitchell, Conifers in the British Isles, 34 (1972).
Tree to 15 m with smooth, grey bark. Shoot pink-brown, smooth, covered with dark hairs. Terminal buds *c.* 4 mm, ovoid, bluntly pointed, purple-brown, resinous. Foliage widely parted beneath, the lateral leaves spreading or pointing forwards, the upper leaves narrowly parted or pressed down forwards. Leaves 1–2.5 cm × *c.* 1.5 mm, narrowly obovate, tips square and notched, grey or yellow-green above, with 2 grey-white bands beneath. Mature cone 4–7.5 × *c.* 3 cm, cylindric, violet or rarely green, ripening brown, bracts not projecting. *NE China, Korea, adjacent USSR.* H1.

*****A. sachalinensis** (Schmitt) Masters. Illustration: Liu, Monograph of the genus Abies, 412, 413 (1971). Shoot grey-green, darkening with age; foliage looser, leaves 1.5–2.5 cm, glossy green; cones 5–8 × *c.* 2 cm, bracts not projecting or projecting slightly. *N Japan (Hokkaido), USSR (Sakhalin).* H1.

**28. A. lasiocarpa** (Hooker) Nuttall. Illustration: den Ouden & Boom, Manual of cultivated conifers, 24 (1965); Mitchell, Conifers in the British Isles, 31 (1972); Phillips, Trees in Great Britain, 63 (1978). Spire-like tree to 15 m, with smooth, grey, resin-blistered bark which becomes hard and fissured into small scales. Shoot grey-brown with sparse hairs in grooves. Terminal buds ovoid, pale brown, covered with thick, whitish resin. Foliage parted beneath, lateral and upper leaves pointing forwards at *c.* 45°. Leaves to 3 cm × *c.* 1.5 mm, flat or oval in section, tips rounded and notched, grey green with 1 pale band above and 2 pale bands beneath. Mature cone to 10 cm, cylindric, dark violet-purple, bracts not projecting. *Western N America.* H1.

Var. **arizonica** Lemmon has a corky, cream-brown bark without resin blisters, shorter cones and more glaucous foliage.

**29. A. balsamea** (Linnaeus) Miller. Illustration: Liu, Monograph of the genus Abies, 440 (1971).
Tree to 15 m with dark grey, smooth, resin-blistered bark which becomes mealy in old trees. Shoots at first yellow-green, later grey, covered with short, dark hairs. Terminal buds ovoid, shiny red or chocolate brown, resinous. Foliage spreading widely beneath the shoot, less so above, all leaves pointing slightly forwards. Leaves 1.5–3 cm × *c.* 1.5 mm, tips rounded and notched, shining dark green above with rows of white dots, with 2 whitish grey bands beneath. Mature cone 5–10 × *c.* 2.5 cm, cylindric, purple later brown, bracts not projecting. *E Canada and E USA as far south as Virginia.* H1.

*****A. fraseri** (Pursh) Poiret. Illustration: Liu, Monograph of the genus Abies, 441 (1971). Buds purple or brown, lumpy; leaves 1–2 cm; mature cone 4–8 cm, oblong-ovoid or cylindric-conic, bracts dark purple, projecting and reflexed. *Eastern USA (Appalachian Mountains).* H2.

**30. A. spectabilis** (D. Don) Spach. Illustration: Mitchell, Conifers in the British Isles, 45 (1972).
Tree to 30 m with grey, smooth bark which becomes rough and fissured in old trees. Shoot red-brown or whitish brown, later darker, grooved, with coarse red-brown hairs in the grooves. Terminal buds grey-white or orange-brown with thick, usually clear resin. Foliage parted beneath, spreading slightly forwards at the sides, and pointing forwards with a clear, V-shaped parting above. Leaves 2–5 cm × 1.5–3 mm, tips rounded and deeply notched, green to dark grey-green or rarely blue-green above, with 2 silvery-grey bands beneath. Mature cone 8–15 cm, grey-blue to violet, bracts not projecting. *N Pakistan to E Nepal.* H2.

**31. A. forrestii** Rogers (*A. delavayi* var. *forrestii* (Rogers) Jackson). Illustration: Mitchell, Conifers in the British Isles, 21 (1972); Arboricultural Journal 3(1): 43 (1976); Rushforth, Mitchell Beazley pocket guide to trees, 47 (1980).
Tree to 20 m with smooth, grey bark which later splits into plates and with red-brown fissures at the base. Shoot red-brown, hairless or with coarse, red-brown hairs. Terminal buds 4 mm or more (to 1 cm on strong shoots), spherical, encrusted and often joined together with thick, grey resin. Foliage spreading forwards beneath the shoot, or occasionally radial or reflexed, upstanding above, forming a flat tier. Leaves to 3.5 cm, tips rounded, notched, green above and with 2 silver bands beneath; margins flat or slightly rolled down. Mature cone 6–10 × 4–5 cm, barrel-shaped or broadly cylindric, apex dimpled; bracts projecting 3–10 mm. *SW China (Yunnan, SE Xizang) & adjacent Burma.* H2.

An attractive, though short-lived tree, which will tolerate chalky soils.

Var. **smithii** Viguier & Gaussen (*A. delavayi* var. *georgei* misapplied) has shoots densely covered with red-brown hairs.

*****A. densa** Griffiths. Shoot yellow-brown, often shining, with coarse, red-brown hairs in grooves on the underside; buds reddish or greenish orange-brown; leaves 2–5 cm; cones 8–12 cm, cylindric-conic. *E Nepal, NE India, W China (SE Xizang).* H2.

**32. A. delavayi** Franchet. Illustration: Arboricultural Journal 3(1): 40 (1976).

Tree to 15 m with smooth, grey bark which eventually splits into plates. Shoots dark red or red-brown, hairless or occasionally with black or dark red hairs. Terminal buds ovoid-conic, orange-green or red-brown, resinous. Foliage more or less radial, rarely parted on shaded shoots. Leaves longest beneath the shoot, to 3.5 cm on vigorous shoots, curved into a shallow S-shape, tips squared, notched, dark glossy green above, with 2 silver bands beneath; margins strongly rolled downwards, midrib keeled beneath. Mature cone to 10 × 3.5 cm, cylindric-ovoid, violet-purple, ultimately brown, bracts projecting. *SW China, N Burma, NE India.* H4.

An attractive tree, growing best in the north west of Europe. A dwarf variant is occasionally grown.

**33. A. chengii** Rushforth (*A. fargesii* misapplied; *A. chensiensis* misapplied). Illustration: Mitchell, Conifers in the British Isles, 23 (1972) – as A. fargesii.
Tree to 15 m (or more ?) with smooth, grey bark which later becomes fissured. Shoot red-brown, becoming fissured, finely downy or hairless. Terminal buds conic-ovoid, yellow-brown, thickly resinous, lumpy. Foliage spreading or pointing forwards beneath the shoot, lateral leaves pointing forwards and rising above the shoot with a narrow parting in between them. Leaves 2–5 cm × c. 3 mm, apex bifid, the base yellow-green, dark glossy green above, with 2 whitish bands separated by the green midrib beneath. Mature cone to 7 × 3.5 cm, cylindric, violet-green, bracts not projecting or projecting slightly in the lower third of the cone. *SW China.* H4.

**34. A. fabri** (Masters) Craib (*Keteleeria fabri* Masters; *A. delavayi* misapplied; *A. delavayi* var. *fabri* (Masters) Hunt). Illustration: Mitchell, Conifers in the British Isles, 20 (1972); Rushforth, Nature library – Trees, 23 (1983).
Tree to 15 m with smooth, grey bark which eventually becomes brown and scaly. Shoot orange to pale brown, shining, side shoots with blackish hairs in faint grooves. Terminal buds to 3 mm, conic, purple or green-purple, resinous. Foliage widely parted beneath, spreading more or less at right angles at the sides, the upper leaves pointing forwards and with a V-shaped parting between them. Leaves to 3.5 cm, tips rounded and notched, green to dark glossy green above, with 2 silvery bands beneath; margins slightly rolled down, midrib keeled beneath. Mature cone

8–10 cm, violet-purple, bracts projecting, reflexed. *W China (W Sichuan).* H2.

***A. minensis** Bordères-Rey & Gaussen (*A. faxoniana* misapplied; *A. delavayi* var. *faxoniana* misapplied; *A. fargesii* var. *faxoniana* misapplied). Illustration: Liu, Monograph of the genus Abies, 406 (1971). Shoot pale golden brown becoming pinkish grey in the second year; foliage widely parted above; leaves with the margins flat and the midrib not keeled. *W China (W Sichuan).* H2.

**35. A. fargesii** Franchet (*A. sutchuenensis* (Franchet) Rehder & Wilson; *A. faxoniana* Rehder & Wilson). Illustration: Mitchell, Conifers in the British Isles, 47 (1972).
Tree to 15 m (to 60 m in the wild) with finely flaky, red-brown or pink-grey bark. Shoot purple or purplish brown, becoming red and shallowly fissured after 2 years, vigorous shoots hairless, weaker shoots with dense, red-brown hairs. Terminal buds ovoid-conic, purplish, with a thick coat of clear resin. Foliage widely parted beneath, lateral leaves spreading or pointing forwards, upper leaves pressed down on the shoot or narrowly parted. Leaves 1–2.5 cm × c. 2.5 mm, bases yellow-green, tips rounded and notched or acutely pointed (on vigorous shoots), dark glossy green above, with 2 silver bands beneath; margin pale, shining green. Mature cones 2.5–7 × c. 3 cm, ovoid or ovoid-cylindric, violet purple, bracts projecting. *China (Gansu, Hubei, Shaanxi, E Sichuan).* H2.

**36. A. squamata** Masters. Illustration: Mitchell, Conifers in the British Isles, 46 (1972); Arboricultural Journal 3(1): 45 (1976).
Tree to 15 m with red-brown or orange-brown bark which exfoliates in large, papery flakes. Shoot shining dark red or red-brown, hairless. Terminal buds c. 5 mm, spherical, coated with white resin. Foliage spreading beneath the shoot, mainly rising above it. Leaves 1–2.5 cm × c. 2 mm, decreasing in size along the shoot, tips rounded, notched or pointed, grey-green with 2 broad, whitish bands above and 2 whitish green bands beneath separated by the green midrib. Mature cone 5–6 cm, ovoid, violet-brown, resinous, bracts projecting, erect. *W China (W Sichuan).* H1.

**2. PSEUDOTSUGA** Carrière
*C.N. Page*
Tall, eventually massive, evergreen trees, of conic-pyramidal habit, with long, horizontally arranged, tapering branches

and drooping branchlets, some branches usually protruding irregularly from the crown. Leaves linear, narrow, flattened, spirally arranged or twisted at their bases to form fairly flattened, slightly forward-pointing ranks along horizontal shoots; leaves eventually shedding to leave slightly raised, oval scars. Short shoots absent. Buds narrow, tapering, acute. Male cones in groups, 2-ranked beneath young shoots. Female cones ovoid, hanging, solitary or in small groups, remaining intact at maturity, with blunt, thinly woody scales and long, projecting, trifid bracts. Seeds winged, 2 per scale. Cotyledons 6–12.

A genus of about 6 closely related species, 4 in eastern Asia and 2 in Pacific N America, distinguished from *Abies* (p. 88) by their pointed, *Fagus*-like buds and hanging cones which remain intact at maturity. They are suitable mainly for estates and large gardens, and thrive under moist conditions on well-drained, loamy soils. Only 1 species is widely cultivated in Europe as a fast-growing forest tree, commercially important for its timber. The remainder are infrequent due to their comparative tenderness and the infrequency of seed.

**1. P. menziesii** (Mirbel) Franco (*P. douglasii* Carrière; *P. taxifolia* (Poiret) Britton). Shoots usually downy. Cones 6–7 cm. *Western N America.* H2.
Several dwarf cultivars exist. Var. **glauca** (Mayr) Franco has bluish foliage.

***P. macrocarpa** (Torrey) Mayr. Cones 10–15 cm. *Western USA (mountains of southern California).* H5.

***P. japonica** (Shirasawa) Beissner. Shoots hairless; cones 3–4 cm. *Mountains of S Japan.* H5.

**3. TSUGA** Carrière
*J. Lewis*
Evergreen trees or shrubs with horizontal branches and drooping branchlets. Leaves short, linear, spread round the branchlets or in 2 ranks, persistent. Male cones axillary, clustered. Female cones terminal, with rounded scales enclosing the bracts. Mature cones usually small, hanging, persistent, releasing the seeds in their second year. Scales leathery, often ridged. Seeds 2 to each scale, resinous, winged.

A genus of about 10 species from north temperate regions, which form slow-growing, graceful trees. They flourish under moist conditions on well-drained, loamy soil. The species may be raised from seed, but the numerous cultivars require propagation by either cuttings or grafting. The Asiatic species are occasionally found in large estates, but are too rarely cultivated to be included here.

1a. Leaves set radially all around the
    shoot                          **1. mertensiana**
  b. Leaves in 2 lateral ranks          2
2a. Leaves tapering, borne in 2 ranks and
    with a single row of backwardly
    pointing leaves above the shoot
                                   **2. heterophylla**
  b. Leaves parallel-sided, borne in 2 ranks
    only                           **3. canadensis**

**1. T. mertensiana** Carrière. Illustration: Mitchell, Field guide to the trees of Britain and northern Europe, 145 (1974). Narrow conical tree to 20 m in cultivation. Branches horizontal, often congested below; branchlets pale brown, downy. Leaves spreading radially all around the branchlets, pointing forwards, to 2 cm, grey-blue, tips rounded. Male cones to 3 mm, several together, purplish. Female cones solitary, larger, green to dark violet. Mature cones to 5 cm, narrowly ellipsoid. Scales thin, downy, spreading, margins irregularly toothed. *Western N America (Alaska to California).* H1.

'Argentea' is a cultivar with even bluer leaves. The species requires cool conditions away from urban air pollution.

**2. T. heterophylla** Sargent. Illustration: Harrison, Ornamental conifers, 178 (1975). Trees to about 40 m in cultivation, spreading with age. Branches regular and ascending to 45°; branchlets pale brown, densely downy, always drooping at the tip. Leaves of mixed sizes, divided more or less into 2 ranks, those of the lower rank *c.* 1.6 cm, those of the upper shorter, parallel-sided, finely toothed, rounded at the tip. Male cones clustered, red. Female cones terminal on short shoots, purple. Mature cones ovoid, blunt, 2–3 cm, of few, entire scales, ripening pale brown. *Western N America (S Alaska to N California).* H1.

Grows strongly on moist, well-drained, chalk-free soils, requiring a large space; it may be clipped to form a dense hedge. T. × **jeffreyi** Henry is the hybrid with *T. mertensiana*; it is bright green, and much less commonly planted than its parents.

**3. T. canadensis** (Linnaeus) Carrière. Illustration: Dallimore & Jackson, Handbook of Coniferae, edn 4, 635 (1966). Broad-based tree to about 30 m, with a rounded crown. Branches irregular; branchlets grey-brown, very hairy. Leaves

in 2 ranks with a line of smaller ones reversed along the upper side of the branchlet, all tapering slightly to rounded tips. Male cones greenish yellow, female cones green. Mature cones to 2.5 cm, stalked; scales striped, entire. *Eastern N America.* H1.

Very variable, and including a large number of cultivars, some of which are slow-growing or dwarf. **T. caroliniana** Engelmann, which has narrower leaves and reddish male cones, is occasionally cultivated; it occurs in the eastern USA (N Carolina to Georgia).

### 4. PICEA Dietrich
*C.N. Page & K.D. Rushforth*
Fast-growing evergreen trees, 15–30 m or more, of narrowly conical and often dense, spire-like habit. Branches whorled. Bark usually thin and scaly. Leaves needle-like, diamond-shaped in section or flattened, usually stiff and spiny, originating all around the shoot but ranked and usually forwardly swept on horizontal branches (except where indicated), all leaves leaving peg-like bases attached to the shoot after falling. Male cones clustered. Female cones solitary, ellipsoid or elongate, hanging and intact at maturity. Cone scales numerous, bracts small, hidden, not projecting. Seeds winged, 2 per cone scale. Cotyledons 4–15.

A genus of about 40 species occurring throughout the temperate northern hemisphere, extending southwards to the mountains of north Burma and Mexico, some species covering extensive tracts of Canada and northern Asia. Many are cultivated, some as plantation forests of great commercial importance for sawn timber and paper pulp, others as estate and garden ornamentals of all sizes, yet others for the Christmas tree trade. Many tolerate poor, wet soils where they make better growth than almost any other kind of tree. Many are, however, very sensitive to urban pollution, and grow poorly in city gardens. Propagation is by seed.

1a. Leaves generally diamond-shaped in
    section, with 4 approximately equal
    sides (enabling the leaf to be rolled
    easily between finger and thumb    2
  b. Leaves flattened, the broad upper and
    lower surfaces ridged along their
    lengths (not easily rolled)       13
2a. Leaves of similar colour on both upper
    and lower surfaces                 3
  b. Leaves with lower surface clearly
    whiter than the upper surface     12

3a. Some leaves more than 3 cm, leaves
    spiralling radially all around the shoot
                                   **8. smithiana***
  b. All leaves less than 3 cm, leaves
    spreading widely beneath the shoot
                                        4
4a. Leaf apex sharp                     5
  b. Leaf apex blunt                    6
5a. Cones 5–8 cm, the scales flexible,
    spreading at their tips
                                   **2. pungens***
  b. Cones 10–15 cm, scales woody,
    adpressed at their tips
                                  **10. asperata***
6a. Bark grey-purple, exfoliating in thin
    flakes even on young trees
                                  **10. asperata***
  b. Bark of other colours, scaling only on
    older trees                         7
7a. Foliage blue or bluish green        8
  b. Foliage green or dark green        9
8a. Shoot of previous summer stout,
    yellow-brown, hairless; cones falling
    soon after maturity          **3. glauca***
  b. Shoot of previous summer slender,
    orange or pinkish brown, hairy; cones
    persisting on the tree for several years
                                   **4. mariana***
9a. Shoot shining greenish white in the
    first year, ashy grey in the second
    year                         **12. wilsonii***
  b. Shoot dull yellow-brown or pale
    orange-brown in the first year,
    deepening in colour in the second
    year                               10
10a. Leaves 6–8 mm, blunt     **5. orientalis**
  b. Leaves 1–1.5 cm, acute           11
11a. Buds not resinous, shining brown;
    leaves yellow-green; cones 13–20 cm
                                    **11. abies***
  b. Buds with white resin; leaves dark
    green; cones 4–7 cm
                               **7. maximowiczii**
12a. Bark grey, becoming furrowed into
    plates; cones 7–10 cm       **9. bicolor**
  b. Bark grey, finely flaky; cones less than
    7 cm                           **6. glehnii**
13a. Leaf arrangement on lateral branches
    radial                             14
  b. Leaf arrangement on lateral branches
    parted beneath                     15
14a. Shoot of previous summer hairless;
    leaves 2–2.5 cm           **13. spinulosa**
  b. Shoot of previous summer finely
    downy; leaves 2.5–3 cm
                                 **1. breweriana**
15a. Pale bands confined to the lower
    surface of the leaf                16
  b. Leaves with 2 very narrow pale bands
    on the upper surface of the leaf, as
    well as 2 on the lower surface    18

16a. Crown dense, forming a narrow spire; shoot of the previous summer downy
                                                  **19. omorika**
  b. Crown moderately dense or fairly open, broad; shoot of the previous summer hairless or nearly so          17
17a. Leaves blue-green above, bands bluish white; cone 2.5–3.5 cm
                                                  **17. jezoensis**
  b. Leaves green above, bands shining white; cone 5–13 cm  **14. brachytyla***
18a. Leaves mid to dark green above, tapering, acute; shoot hairless
                                                  **18. sitchensis**
  b. Leaves blue-green or greyish green above, not tapering, blunt; shoot downy                                  19
19a. Crown broadly conical, open; cones 10–15 cm          **15. likiangensis**
  b. Crown narrowly conical, fairly dense; cones less than 10 cm
                                                  **16. purpurea***

**1. P. breweriana** Watson. Illustration: Dallimore & Jackson, Handbook of Coniferae, edn 4, 349 (1966); Mitchell, Conifers in the British Isles, 158 (1972); Callen, Les conifères cultivées en Europe, 474–5 (1977); Rushforth, Mitchell Beazley pocket guide to trees, 57 (1980).
Conical trees, 10–15 m with pointed crowns and short, upcurving branches bearing heavy, dense, hanging branchlet systems. Bark grey, forming hard, curling plates. Leaves flattened, 2.5–3 cm, blunt, flexible, deep green, more or less shining above, arranged radially around the hanging shoots. Shoots of previous summer finely downy. Male cones large, globular, yellow. Female cones oblong, greenish red when young, red-brown when mature, 6–12 cm. Cone scales thinly woody, rounded, incurved, with some resin incrustation. *W USA (Siskiyou mountains of NW California and SW Oregon).* H3.
A very attractive, weeping species. Seedlings and young plants totally lack the characteristic weeping habit during their early years and may be incorrectly identified.

**2. P. pungens** Engelmann. Illustration: Mitchell, Conifers in the British Isles, 175 (1972); Callen, Les conifères cultivées en Europe, 515, 519, 521–3, 525 (1977).
Conical trees, 15–20 m. Bark grey, scaling. Leaves diamond-shaped in section, 1.5–2 cm, very stiff, stout, rigid, acute and harsh, green or blue-green, glaucous, projecting stiffly all around the shoot. Young shoots stout, hairless, ridged, yellow or orange-brown, glossy. Female cones pale

brown, cylindric, 5–8 cm. Cone scales thin, narrowed, spreading, wrinkled, finely toothed. *SW USA.* H2.
Plants vary widely in colour, even within a single seed-lot. The best blue forms have been selected as 'Glauca' and other cultivars of different colour and habit also exist.
***P. polita** (Siebold & Zuccarini) Carrière. Superficially similar (though probably not closely related) in its stiffly spiny leaves, but with a long-ovoid cone with rounded scales. *Japan.* H2.

**3. P. glauca** (Moench) Voss. Illustration: Dallimore & Jackson, Handbook of Coniferae, edn 4, 353 (1966); Mitchell, Conifers in the British Isles, 160 (1972); Callen, Les conifères cultivées en Europe, 479, 481–3 (1977); Edlin, The tree key, 227 (1978).
Conical trees 10–20 m. Bark pink-grey. Leaves diamond-shaped in section, 1–1.3 cm, blunt, stout, upswept, stiff and harsh, pale blue-green, glaucous. Shoot stout, yellow-brown, hairless, shining. Cones 3–6 cm, light brown, scales rounded, woody, shed soon after maturity. *N America (Alaska to Labrador).* H2.
A slow-growing species which is variable; var. **albertiana** (Brown) Sargent from Canada (Alberta) and USA (Montana) exists in cultivation as 'Conica', a narrower, taller tree with sparse foliage; other dwarf cultivars also exist.
***P. engelmannii** (Parry) Engelmann. Shoots more slender, finely downy, leaves grey-green, more flexible and forwardly pointing, cone scales papery. *N America.* H2.
Several cultivars exist, including 'Glauca' with bluer foliage and 'Fendleri' with longer leaves and hanging shoots.

**4. P. mariana** (Miller) Britton. Illustration: Dallimore & Jackson, Handbook of Coniferae, edn 4, 363 (1966); Mitchell, Conifers in the British Isles, 166 (1972); Callen, Les conifères cultivées en Europe, 495, 497–8 (1977).
Conical trees 10–20 m. Bark pink-grey, finely flaky. Leaves diamond-shaped in section, 1–1.5 cm, deep green, blunt, closely spaced, overlapping, pointing forwards above, widely parted beneath. Shoot slender, orange or purplish brown, hairy. Female cones often clustered, persisting for several years, deep purple-brown when mature, ovoid, 2.5–3 cm. Cone scales rounded smooth and glossy. *N America (Alaska to Newfoundland and south to British Columbia).* H1.
A slow-growing species, succeeding in

cold, wet soils. A number of cultivars, mostly dwarfs, exists.
***P. rubens** Sargent. Leaves more slender, acute, wiry, bright grass-green, curved inwards on the shoot, cones less persistent. *N America (Nova Scotia to N Carolina).* H1.

**5. P. orientalis** (Linnaeus) Link. Illustration: Dallimore & Jackson, Handbook of Coniferae, edn 4, 370 (1966); Mitchell, Conifers in the British Isles, 171 (1972); Callen, Les conifères cultivées en Europe, 513–515 (1977); Rushforth, Mitchell Beazley pocket guide to trees, 54 (1980).
Conical or columnar trees, 15–40 m (rarely more). Bark purple-brown. Crowns very dense, often with protruding, spike-like branch tips. Leaves diamond-shaped in section, 6–8 mm, flexible, blunt, the tips bevelled, deep shining green on all surfaces, dense on the shoot, pointing forwards above and parted beneath on horizontal branches. Young shoots yellow-brown or pale orange-brown, less brightly coloured in the second year, downy. Male cones deep red. Female cones purplish grey when young, pale ashen-brown when mature, narrowly ovoid, 6–10 cm. Cone scales rounded. *Turkey, SW USSR (Caucasus).* H2.
A number of cultivars, mostly dwarf forms, is available. 'Aurea' has the young foliage golden.

**6. P. glehnii** (Schmidt) Masters. Illustration: Callen, Les conifères cultivées en Europe, 485 (1977).
Conical trees 5–15 m, with dense crowns. Bark deep reddish brown, finely flaking and turning greyish. Leaves diamond-shaped in section, 1–1.5 cm, dark blue-green above, grey-green beneath, closely crowded and overlapping, pointing forwards above the shoot, parted beneath on horizontal branches, above with 1 or 2 pale bands, beneath with 2 broad white bands. Cones 4–5 cm. *N Japan, USSR (Sakhalin).* H1.

**7. P. maximowiczii** Masters. Illustration: Mitchell, Conifers in the British Isles, 167 (1972); Callen, Les conifères cultivées en Europe, 501 (1977).
Narrowly conical trees, 10–20 m, with dense crowns and projecting shoots. Bark orange-brown, finely flaking. Leaves diamond-shaped in section, 1–1.3 cm, glossy dark green on both surfaces, closely spaced and overlapping, pointing forwards above the shoot, parted beneath on horizontal shoots. Winter buds with white resin incrustation. Female cones pale red-brown when mature, 4–7 cm. Cone scales rounded, smooth. *C Japan.* H2.

**8. P. smithiana** (Wallich) Boissier (*P. morinda* Link). Illustration: Dallimore & Jackson, Handbook of Coniferae, edn 4, 381 (1966); Mitchell, Conifers in the British Isles, 180 (1972); Callen, Les conifères cultivées en Europe, 529–530 (1977); Rushforth, Mitchell Beazley pocket guide to trees, 57 (1980).
Conical-columnar trees 20–30 m or more with moderately dense crowns, horizontal branches and hanging branchlets. Bark purple-grey, forming plates. Leaves diamond-shaped in section, 3–4 cm, slender, acute, dark green, closely spaced, borne all round the shoot, spreading at the base, then pointing forwards. Young shoots cream-brown, more or less slender, shining, hairless. Buds long and spindle-shaped. Male cones large, terminal on hanging shoots. Female cones pale green when young, light brown when mature, 12–20 cm, obovoid-cylindric. Cone scales leathery, rounded, smooth, shining. *W Himalaya.* H4.

**\*P. schrenkiana** Fischer & Meyer (*P. tianshanica* Ruprecht). Habit narrower and more columnar, leaves 2.5–3.5 cm, greyer green and less regularly arranged. *USSR (Kazakhstan, Tien Shan mountains), China (W Xinjiang).* H2.

**9. P. bicolor** (Maximowicz) Mayr. Illustration: Mitchell, Conifers in the British Isles, 156 (1972); Callen, Les conifères cultivées en Europe, 469–70 (1977). Conical trees, 10–15 m (rarely more), with gaunt crowns and irregularly protruding branches. Bark grey, becoming fissured into plates. Leaves diamond-shaped in section, *c.* 1.5 cm, acute, somewhat stiff, more or less blue-green with 2 narrow, pale lines on the upper surface and 2 broad white bands on the lower, giving the leaf a bicolored appearance; leaves curved forwards above the shoot and parted beneath on horizontal branches. Young shoots stout, white to pale orange-brown, usually hairless. Male cones yellow tinged with purple. Female cones pale reddish purple when young, brown when mature, ovoid-cylindric, 7–10 cm. Cone scales thinly woody, more or less rounded. *C Japan.* H3.

**10. P. asperata** Masters. Illustration: Dallimore & Jackson, Handbook of Coniferae, edn 4, 343 (1966); Mitchell, Conifers in the British Isles, 154 (1977); Callen, Les conifères cultivées en Europe, 464–6 (1977).
Low, broadly conical trees, 5–20 m, with moderately dense crowns. Bark grey-purple, exfoliating in thin flakes even on young trees. Leaves diamond-shaped in section, 1.5–2 cm, thick, very stiff, acute, harsh and sharp in most specimens, occasionally blunt, spreading more or less all round the shoot, usually curved upwards to become dense above on horizontal shoots, green, grey-green or bluish green. Young shoots very stout, deeply grooved, orange-brown, hairless or slightly downy. Male cones yellow. Female cones green, tinged reddish when young, dull brown when mature, ovoid-cylindric, 10–15 cm, usually with abundant resin-incrustation. Cones scales thinly woody, more or less irregularly notched, waved. *NW China.* H1.

A variable species in which a number of named varieties (often regarded as species) exists.

**\*P. meyeri** Rehder & Wilson. Leaves blunter, always grey-green, bevelled, harsh and stiffly rising upwards from the shoots. *N China.* H1.

**11. P. abies** (Linnaeus) Karsten. Illustration: Dallimore & Jackson, Handbook of Coniferae, edn 4, 335 (1966); Mitchell, Conifers in the British Isles, 152 (1972); Callen, Les conifères cultivées en Europe, 441–3, 445–7, 449–51, 453–5, 457–9, 461–3 (1977).
Tall, conical-columnar trees, 10–40 m or more, with dense crowns. Bark red-brown. Leaves diamond-shaped in section, 1–1.5 cm, slender, acute, slightly flexible, somewhat shining, deep yellow-green on all surfaces, closely spaced and more or less overlapping, pointing forwards above the shoots, parted beneath on horizontal branches. Young shoots moderately stout, orange-brown, more or less hairless. Winter buds shining brown, non-resinous. Male cones yellow. Female cones green when young, pale brown when mature, ovoid-cylindric, 15–20 cm. Cone scales thinly woody, more or less rounded to irregularly notched, slightly waved. *Most of Europe, except the south.* H1.

Widely grown both for timber and the Christmas tree trade. Many cultivars exist, varying in habit, colour, size and crown density.

**\*P. obovata** Ledebour. A broadly conical or irregular, low tree, like *P. abies* but leaves very slender, more shiny, dispersed on the shoot, female cones 5–8 cm, with smooth, rounded scales. *N Europe (N Scandinavia), USSR.* H1.

**12. P. wilsonii** Masters. Illustration: Dallimore & Jackson, Handbook of Coniferae, edn 4, 383 (1966); Mitchell, Conifers in the British Isles, 183 (1972); Callen, Les conifères cultivées en Europe, 534 (1977).
Broadly conical trees, 10–15 m, with moderately dense crowns. Bark grey-brown, becoming flaky. Leaves diamond-shaped in section, sometimes slightly flattened, 1.5–2 cm, slender, flexible, acute, shining bright green on all surfaces, pointing forwards above the shoot and parted beneath on horizontal branches. Young shoots slender, shining greenish white in the first year, ash-grey later, ridged, hairless. Cones infrequent, the female ovoid-cylindric, 4–5 cm. Cone scales rounded or slightly toothed. *C & W China.* H2.

**\*P. morrisonicola** Hayata. Habit taller and narrower, leaves 1.4–1.5 cm, always very slender, pale green, pressed down on to the shoot. *Taiwan.* H4.

**13. P. spinulosa** (Griffith) Henry. Illustration: Mitchell, Conifers in the British Isles, 183 (1972); Callen, Les conifères cultivées en Europe, 535 (1977).
Conical trees 20 m or more with very open crowns and highly pendulous branchlets. Bark pinkish grey, becoming shallowly fissured. Leaves flattened, 2–2.5 cm, slender, straight, flexible, sharp, bright greyish green above, white beneath, rather widely spaced, arranged all around the shoot, pointing forwards. Young shoots white to pale brown, hairless, slender. Female cones shining brown at maturity, ovoid-cylindric, 5–8 cm. Cone scales thin, diamond-shaped, finely toothed. *E Himalaya.* H4.

**14. P. brachytyla** (Franchet) Pritzel. Illustration: Dallimore & Jackson, Handbook of Coniferae, edn 4, 347 (1966); Mitchell, Conifers in the British Isles, 156 (1972); Callen, Les conifères cultivées en Europe, 471, 474–5 (1977).
Broadly conical trees, 10–25 m, with open crowns and more or less drooping branchlets. Bark pale grey, cracked. Leaves flat, 1.2–1.6 cm, stiff but not harsh, broad at base, tapering, bright green above, shining white over the whole surface beneath, crowded, pointing forwards but arched downwards on either side of the shoot, widely parted beneath. Young shoots fairly stout, greenish white to very pale brown, hairless or sparsely hairy. Female cones green, tinged purple when young, dark brown at maturity, ovoid-cylindric, 6–13 cm, often with abundant resin incrustation. Cone scales thin, wavy. *NE India, W China.* H3.

\*P. farreri Page & Rushforth. Crown thinner and more open, branchlets long, slender and hanging, cones 6–9 cm with more rounded scales. *Burma.* H4.

**15. P. likiangensis** (Franchet) Pritzel. Illustration: Dallimore & Jackson, Handbook of Coniferae, edn 4, 361 (1966); Mitchell, Conifers in the British Isles, 164 (1972); Callen, Les conifères cultivées en Europe, 493–4 (1977).
Broadly conical trees 10–20 m with fairly open crowns. Bark grey, fissured. Leaves flattened, 1.5–1.7 cm, broad, blunt, blue-green or greyish green above, white beneath, pointing forwards and more or less overlapping above, widely parted beneath on horizontal shoots, their peg-like bases broad and curved. Young shoots greenish brown, sparsely downy. Male cones large, globular, bright dark red. Female cones bright purple-red when young, purple-brown when mature, ovoid-cylindric, 10–15 cm, usually with abundant resin incrustation. Cone scales thinly woody, rounded to irregularly notched, slightly waved. *Bhutan, W China.* H3.

A variable but particularly attractive ornamental tree with its bicolored leaves and spectacularly coloured, abundant cones.

**16. P. purpurea** Masters. Illustration: Mitchell, Conifers in the British Isles, 166 (1972); Callen, Les conifères cultivées en Europe, 441 (1977).
Narrowly conical trees 10–15 m with moderately dense crowns. Bark dull orange-brown, flaky. Leaves flattened, 1.2–1.5 cm, broad, blunt, with 2 narrow, white bands above and 2 broad, greyish white bands beneath, overlapping and pressed strongly against the shoot above, parted beneath on horizontal shoots. Young shoots pinkish grey, densely downy. Cones purple, 2–3.5 cm, with rounded, irregularly notched, flexible scales. *W China (Gansu, Sichuan).* H2.

\*P. balfouriana Rehder & Wilson has greyer foliage, grey bark and longer cones (5–9 cm), and in most respects it is intermediate between *P. purpurea* and *P. likiangensis.* *W China (Sichuan).* H2.

**17. P. jezoensis** (Siebold & Zuccarini) Carrière var. **hondoensis** (Mayr) Rehder. Illustration: Dallimore & Jackson, Handbook of Coniferae, edn 4, 357 (1966); Mitchell, Conifers in the British Isles, 162 (1972); Callen, Les conifères cultivées en Europe, 487, 489 (1977).
Columnar-conical trees, 20–30 m (rarely more), with fairly broad, moderately dense crowns. Leaves flat, *c.* 1.5 cm, blunt, deep green above and with 2 bright white bands beneath, closely spaced, overlapping, pointing forwards above and parted beneath on horizontal shoots. Young shoots stout, shining pale brown or yellow-brown, hairless. Female cones pale reddish brown when mature, cylindric, 2.5–3.5 cm. Cone scales narrowly oblong, thin, with waved and toothed margins. *Japan.* H3.

Var. *jezoensis* is not in general cultivation.

**18. P. sitchensis** (Bongard) Carrière. Illustration: Dallimore & Jackson, Handbook of Coniferae, edn 4, 378 (1966); Mitchell, Conifers in the British Isles, 177 (1972); Callen, Les conifères cultivées en Europe, 533 (1977); Edlin, The tree key, 227 (1978).
Conical-columnar trees 20–50 m with dense crowns. Leaves flattened, 2–2.5 cm, broad at base, tapering, acute, stiff, harsh, spine-pointed, deep green above, bluish white beneath, closely spaced, overlapping, pointing forwards, pressed flat or rising up from the shoot above, parted beneath on horizontal branches. Young shoots pale buff, hairless. Male cones yellow. Female cones green tinged with red when young, brown when mature, cylindric, 3.5–7 cm. Cone scales narrowly oblong, thin, hard, waved and with toothed margins. *Western N America from Alaska to N California.* H3.

Widely grown for timber and pulpwood.

**19. P. omorika** (Pančić) Purkyne. Illustration: Dallimore & Jackson, Handbook of Coniferae, edn 4, 368 (1966); Mitchell, Conifers in the British Isles, 169 (1972); Callen, Les conifères cultivées en Europe, 508–10 (1977); Edlin, The tree key, 227 (1978).
Narrow columnar trees 20–30 m, with sparse, spire-like crowns and short, downswept branches. Leaves flattened, 1–1.8 cm, blunt, boat-shaped, dark green above and with 2 broad white bands beneath, flexible, soft, pointing forwards and overlapping above, parted beneath on horizontal shoots. Young shoots slender, pale brown, downy. Female cones infrequent, at the top of the tree only, ovoid, dark purple-brown when mature, 5–6 cm, borne on long, curved stalks. Cone scales rounded with fine teeth. *Jugoslavia.* H1.

Apparently extremely sensitive to urban air pollution. Its very slender, spire-like crowns make it a useful species for group planting.

**5. PSEUDOLARIX** Gordon
*C.N. Page*
Deciduous trees, 10–20 m, of broad, pyramidal habit. Bark fissured. Leaves flattened, linear, flexible, spirally arranged and lying forwards on long primary shoots, clustered into whorls on blunt, spur-like, lateral secondary shoots. Male cones in umbellate clusters. Female cones 5–6 × 4–5 cm, with broad, pointed, triangular scales, dismembering at maturity. Seeds winged, 2 per scale. Cotyledons 5–7.

A genus of 1 species (occasionally divided into 2 by some authors) from China. Related to *Larix*, from which its broader leaves, clustered male cones and dismembering female cones distinguish it. Its scarcity in cultivation is due mainly to the lack of good seed from most cultivated specimens. It requires deep but well-drained soil, and is very slow-growing in northern Europe, where cones are rarely formed. Propagation is by seed.

**1. P. amabilis** (Nelson) Rehder (*P. fortunei* Mayr; *P. kaempferi* Gordon). *S & E China.* H3.

An attractive tree, deserving wider cultivation.

**6. LARIX** Miller
*C.N. Page*
Tall, deciduous trees of moderately narrow, conical habit. Bark more or less fissured. Leaves linear, needle-like, flexible, spirally arranged and lying forwards on long, primary shoots, clustered into brush-like tufts on short, spur-like, lateral, secondary shoots. Male cones borne singly. Female cones with blunt, thinly woody scales, intact at maturity, persisting thereafter; some species with projecting bracts. Seeds winged. Cotyledons 5–8.

A genus of about 12 species from the northern hemisphere. Several species are cultivated, many of them commercially for timber or as windbreaks. They can be grown on most types of soil; some will tolerate cold, wet, poorly drained sites. Propagation is by seed.
Literature: Ostenfeld, C.H. & Larsen, C.S., The species of the genus Larix and their geographical distribution, *Det Køniglich Danske Videnskabernas Selskab, Biologiske Medelelserie* 9(2): 1–106 (1930).

1a. Mature female cones with long, pointed bracts projecting between the cone scales    2

b. Mature female cones without such projecting bracts ................................... 3

2a. Female cones less than 4 cm ............ **6. occidentalis**

b. Female cones 4 cm or more ............ **7. griffithii***

3a. Female cone scales with margins rolled outwards ................................... 4

b. Female cone scales with the margins not rolled outwards ........................... 6

4a. Outward rolling of female cone scales slight but perceptible ............. **5. gmelini**

b. Outward rolling of female cone scales considerable ................................... 5

5a. Female cones, when open, as wide as long .................................... **3. kaempferi**

b. Female cones, when open, longer than wide ............................ **4. × eurolepis**

6a. Female cones to 1 cm broad, with 10–30 scales .................. **1. laricina**

b. Female cones 1–2.5 cm broad with 35–55 scales ................... **2. decidua**

**1. L. laricina** (du Roi) Koch (*L. alaskensis* Wright; *L. americana* Michaux; *L. microcarpa* Desfontaines). Illustration: Dallimore & Jackson, Handbook of Coniferae, edn 4, 309 (1966); Mitchell, Conifers in the British Isles, 140 (1972).

Slender, short-lived trees, 15–25 m in the wild, usually less in cultivation, with thin, narrowly conical crowns and short, level branch systems bearing very numerous, slender, twiggy shoots. Young shoots glaucous, more or less hairless. Leaves 1–3 cm, triangular in section, very slender, blunt. Female cones ovoid, 1.5–2 × up to 1 cm, with 10–30 scales, each cone erect and curving slightly towards the shoot, without projecting bracts. *N America (south to Minnesota and Illinois)*. H1.

This species is of little ornamental value, except in cold, wet, poorly drained sites, which it will tolerate.

**2. L. decidua** Miller (*L. europaea* de Candolle). Illustration: Dallimore & Jackson, Handbook of Coniferae, edn 4, 130 (1966); Mitchell, Conifers in the British Isles, 130 (1972).

Conical trees, 25–40 m, bark pale grey to reddish brown. Leaves 3–3.5 cm, triangular in section, bright green, similar above and beneath. Young shoots greenish white, pink-brown or pale yellow, hairless, not glaucous. Female cones often dark red when young, brown at maturity and of variable shape, usually ovoid, 1.5–3 × 1–2.5 cm, the scales with slightly waved outer margins which are often somewhat incurved, without projecting bracts; the cone has a compact appearance when viewed from above. *C & E Europe; widely planted elsewhere*. H2–3.

A variable species; var. **polonica** (Raciborsky) Ostenfeld & Larsen, which has shorter, broader cones with crowded, incurved scales, is sometimes seen. 'Corley' is a slow-growing bush of rounded habit derived from a witch's broom. *L. × pendula* (Solander) Salisbury is believed to be *L. decidua × laricina*; it originated in cultivation and has particularly long, hanging, branch systems. 'Repens', which has creeping branches, is probably a variant of this hybrid.

**3. L. kaempferi** (Lambert) Carrière (*L. leptolepis* (Siebold & Zuccarini) Endlicher). Illustration: Dallimore & Jackson, Handbook of Coniferae, edn 4, 311 (1966); Mitchell, Conifers in the British Isles, 138 (1972).

Conical trees 20–30 m or more. Bark dark red-brown to purple-brown. Leaves mostly 3.5–4 cm, flat, grey-green above, paler and keeled beneath, somewhat broader than those of *L. decidua*. Young shoots usually orange-brown or tinged with purple, somewhat glaucous. Female cones *c.* 3 × 3 cm when open, broadly ovoid, the scales with waved margins which are conspicuously rolled outwards, without projecting bracts; the cone has a distinctive rosette-like appearance when viewed from above. *Japan (mountains of C Honshu); widely planted elsewhere*. H2–3.

A variable species, depending on provenance; 'Minor' covers plants of bushy habit, presumably from montane situations; 'Nana' is a slow-growing congested bush; 'Varley' is a very compact bush, derived from a witch's broom, and 'Pendula' has particularly drooping branchlet systems.

**4. L. × eurolepis** Henry (*L. decidua × kaempferi*). Illustration: Mitchell, Conifers in the British Isles, 135 (1972).

Vigorous trees already exceeding 30 m in cultivation, intermediate between the parents in most respects. Leaves 3.5–4.5 cm, dark grey-green above, paler beneath. Young shoots variable in colour, slightly bloomed. Female cones longer than those of either parent, 3.5–4 cm, elongate-ovoid, combining the tall shape of those of *L. decidua* with the rolled out margins of *L. kaempferi*. *Originated in cultivation*. H2.

A fast-growing tree, soon outstripping both parents in size, and with better resistance to aphid and fungal attack.

**5. L. gmelini** (Ruprecht) Kuzeneva (*L. dahurica* Trautvetter). Illustration: Callen, Les conifères Cultivées en Europe. 235 (1976).

Trees to 25 m in the wild, usually less than 20 m in cultivation, with broadly conical crowns and level or descending branches with numerous, twiggy branchlet systems. Leaves mostly 1.5–2.5 cm, very slender, flat and bright shining green above, keeled and with conspicuous, greyish bands beneath. Young shoots pinkish brown, usually hairless. Female cones usually abundant, 2–2.5 × up to 1.8 cm, ovoid, stalked, shining pale brown, with thin, broad, curved scales which are rolled outwards slightly near the margin, without projecting bracts. *USSR (Siberia), NE China*. H1.

A variable species which, in Europe, flushes early, thus often sustaining damage from late frost. Var. **japonica** (Regel) Pilger has leaves 1.5–2 cm, downy young shoots, smaller cones and very dense branch systems; var. **olgensis** (Henry) Ostenfeld & Larsen has the young shoots densely covered in red-brown hairs, the leaves at most 1.5 cm, and the cones very small; var. **principis-ruprechtii** (Mayr) Pilger has more vigorous growth, leaves 3–4 cm, hairless young shoots and cones up to 4 × 2.5 cm.

**6. L. occidentalis** Nuttall. Illustration: Sudworth; Forest trees of the Pacific slope, 69 (1908); Mitchell, Conifers in the British Isles, 141 (1972).

Narrowly conical trees attaining 65 m in the wild, usually 15–20 m in cultivation. Bark dark grey-brown to yellow-brown, very thick and coarsely fissured in old specimens. Leaves to 3.5 cm, very slender, rounded and green above, paler and more or less keeled beneath. Young shoots whitish to pale orange-brown, hairless or only very slightly hairy. Female cones 2.5–4 × 2–3 cm, ovoid-cylindric, with conspicuously projecting, tapering bracts which may have down-curved tips. *Western N America*. H3–4.

The thick bark of this species confers considerable fire resistance.

**7. L. griffithii** Hooker (*L. griffithiana* (Lindley & Gordon) Carrière). Illustration: Mitchell, Conifers in the British Isles, 137 (1972).

Slender, conical trees to 20 m in cultivation, often with more or less downswept branches and long, hanging branch systems. Leaves 3–3.5 cm, flat and deep shining green above, deeply keeled beneath. Female cones 5–8 × 2.5–3 cm, ovoid-cylindric, violet-purple during growth, borne on short stalks, with conspicuously projecting, long-pointed, flexible bracts emerging from between the woody cone scales, the bract tips curved strongly downwards, at least in the lower part of the cone. *E Nepal, India (Sikkim), Bhutan.* H5.

A graceful tree requiring particularly mild, moist conditions for good growth.

**\*L. potaninii** Batalin has smaller cones (up to 5 × 2.5 cm) borne on shorter stalks. *China (Shaanxi, Sichuan, E Xizang, Yunnan).* H5.

## 7. CEDRUS Link
*J. Lewis & T.J. Varley*

Evergreen, often massive trees with greyish bark darkening and fissuring with age. Branchlets of 2 kinds: terminal shoots bearing spirally arranged leaves and short, spur-like shoots with condensed tufts of leaves. Winter buds with persistent, small, brown scales. Leaves needle-like, usually 3-angled, surviving for about 5 years. Male flowers in erect, catkin-like structures which contain numerous, densely crowded anthers. Female cones small, greenish, with 2 ovules to each scale; mature cones large, erect, very broadly ellipsoid, disintegrating on the tree.

A genus of 4 species from the western Himalaya, the Mediterranean area and Morocco. They are quite widely cultivated in parks and large gardens and are resistant to atmospheric pollution. The most suitable soil is a moist, well-drained loam, but they will grow on heavier soils which are not too wet. Propagation is by seed, which should be freshly ripe, and by grafting.
Literature: Maheshwari, P. & Biswas, C., *Cedrus* (1970).

1a. Leaves 3–5 cm; ends of branchlets hanging; young shoots densely and darkly downy                    **1. deodara**
 b. Leaves 7–30 mm; branchlets stiff, upright or spreading; young shoots hairless or sparsely and palely downy                                                        2

2a. Leaves 7–12 mm; branchlets very dense, and finely downy; cones to 12 × 5 cm, ellipsoid, tapered at the base, abruptly narrowed at the tip like a lemon                          **3. brevifolia**
 b. Leaves 1.5–3 cm; branchlets sparse, numerous, hairless or almost so; cones to 9 × 5 cm, broadly ellipsoid, rounded above to a flat or depressed apex                                                        3

3a. Bark red-brown; leaves 2–3 cm, slender, narrowing to a short translucent point which is green except at the extreme tip; crown

broad and slightly domed or low and flat, branches usually ascending; trunks often multiple               **2. libani**
 b. Bark greyish; leaves 1.5–2.5 cm, apices with translucent or almost colourless spines; crown broadly columnar, but usually tapering to a conical top, trunk usually single to above half the total height

**4. atlantica**

**1. C. deodara** (Roxburgh) G. Don (*Pinus deodara* Roxburgh). Illustration: Harrison, Ornamental conifers, 31 (1975); Phillips, Trees in Britain, 97 (1978).
A massive tree to 60 m, with a girth to 10 m, usually dioecious. Bark greyish. Leading shoots and tips of the branches hanging. Shoots grey-green with dense, black down. Leaves 3–5 cm, 15–20 per whorl, dark green or glaucous. Cones very broadly ellipsoid, 8–12 × 5–8 cm. *W Himalaya.* H3.

Some of the more commonly grown variants include: 'Albospica', in which the tips of the young shoots are white; 'Pygmy', an extremely slow-growing dwarf; 'Pendula', a widely spreading, low bush when correctly pruned; and 'Verticillata Glauca', a dense, small, bushy tree with horizontal branches.

**2. C. libani** Richard. Illustration: Phillips, Trees in Britain, 98 (1978).
A large tree to 30 m, with a girth to 8 m, trunks often divided. Branches erect or ascending, the numerous branchlets spreading to form wide, table-like surfaces. Bark red-brown. Shoots very pale brown with grey-brown down. Leaves 2.5–3 cm, 10–15 per whorl, narrowing to short, translucent points. Cones 7–10 × 4–6 cm, very broadly ellipsoid to ovoid, often tapering to the apex from about one-third of their total length. *Syria, Lebanon, SE Turkey.* H3.

Subsp. **libani**, described above, is from Syria and Lebanon; several cultivars of it are grown, e.g., 'Nana', a very slow-growing, dense, conical bush; 'Comte de Dijon', similar to 'Nana' but with fine leaves; and 'Sargentii', which has a spreading habit. Subsp. **stenocoma** (Schwarz) Davis occurs in SE Turkey, and has a more upright crown; it is intermediate between subsp. *libani* and *C. atlantica* in leaf and cone characters, and is rare in cultivation.

**3. C. brevifolia** (J.D. Hooker) Henry. Illustration: Phillips, Trees in Britain, 98 (1978).

Tree to 8 m (rarely to 15 m). Bark grey. Shoots pale greenish brown and finely downy. Leaves 7–12 mm, 20–30 per whorl, slender, green. Cones to 12 × 5 cm, tapered to the base, the apex abruptly pointed, like a lemon. *Cyprus.* H3.

**4. C. atlantica** (Endlicher) Carrière. Figure 28(3) p. 69. Illustration: Harrison, Ornamental conifers, 29 (1975); Phillips, Trees in Britain, 97 (1978).
Tree to 40 m, conical when young, but ultimately resembling *C. libani.* Bark grey. Young shoots and leaves similar to those of *C. libani* but shoots more densely downy and leaves 19–28 per whorl, each with a translucent or colourless spine at the apex. Cones like those of *C. libani*, to 8 × 5 cm. *Algeria & Morocco (Atlas mountains).* H2.

The most important cultivar is 'Glauca', with glaucous foliage, finely fissured, grey bark; it can be a large tree (to 32 m). Habit variants (e.g. 'Fastigiata' and 'Pendula') are also grown, and 'Aurea' has golden-yellow foliage.

## 8. PINUS Linnaeus
*C.N. Page & F.M. Bennell*

Mostly tall, evergreen trees, conical when young, usually with spreading, irregular or domed crowns when mature. Bark thick, rough and deeply furrowed. Branchlets woody, mostly flexible, resinous, variously fragrant when broken. Winter buds scaly, sometimes resinous. Leaves long, linear, needle-like, stiff or flexible, spirally borne on juvenile shoots but grouped into bundles of 2, 3 or 5 (occasionally variable in number) on adult shoots, each bundle united at the base by a fascicle sheath. Male cones small, catkin-like, yellow, usually in crowded clusters. Female cones large, maturing slowly, solitary or grouped, woody, intact and usually pendulous at maturity, without projecting bracts. Seeds usually winged, 2 per cone scale. Cotyledons 7–23.

A genus of about 100 species widely distributed throughout the northern hemisphere, of easy cultivation. Many species are cultivated as ornamentals, for windbreaks or for timber, and several are of commercial importance. They are propagated most readily by seed.
Literature: Shaw, G.R., *The Genus Pinus* (1914); Mirov, N.T., *The Genus Pinus* (1967); Critchfield, W.B. & Little, E.L., Geographic distribution of the pines of the world, *USDA Forest Service, Miscellaneous Publication* 991 (1969); Horsman, J., Pines in cultivation: a survey of the species, *The Plantsman* **2**: 225–56 (1981).

*Bark*. Multicoloured in patches: **8**.
*Leaves*. All or mostly in 5s: **1–7,10,20**; all
or mostly in 3s: **8,9,11,21–24**; all or
mostly in 2s: **12–19**. Less than 8 cm:
**1,3,7,8,10,14,24**. Foliage very drooping:
**2,4,6,11,20**. Some or many leaves
kinked: **2,6**.
*Cones*. Highly resinous: **2–6,10**. More than
15 cm: **4,6,24**. Remaining on shoots for
several years: **7,12,16,17,23**. Very light,
with more or less flexible, scarcely woody
scales: **2–8**.
*Seed*. Wingless or wing rudimentary:
**1–3,7,9,19**.

1a. Leaves mostly in pairs                          2
 b. Leaves in bundles of 3 or 5                     9
2a. Leaves 4–12 cm                                  3
 b. Leaves more than 12 cm                          7
3a. Cones nearly as wide as long                    4
 b. Cones much longer than wide                     5
4a. Leaves less than 8 cm; bark not
    reddish                         **13. tabuliformis**
 b. Leaves more than 8 cm; bark reddish
                                     **14. sylvestris***
5a. Cones mostly borne in 3s
                                     **18. halepensis**
 b. Cones mostly borne singly                       6
6a. Cones less than 5 cm, persisting on
    the tree for several years
                                     **17. contorta***
 b. Cones more than 5 cm, shed annually
                                       **15. nigra***
7a. Leaves more than 15 cm
                                     **12. pinaster**
 b. Leaves less than 15 cm                          8
8a. Seeds winged                     **16. muricata**
 b. Seeds unwinged                     **19. pinea**
9a. Leaves mostly in bundles of 3                  10
 b. Leaves mostly in bundles of 5                  16
10a. Leaves less than 15 cm                        11
 b. Leaves more than 15 cm                         13
11a. Seeds unwinged                **9. cembroides**
 b. Seeds winged                                   12
12a. Cones less than 5 cm, shed annually
                                     **8. bungeana**
 b. Cones more than 5 cm, persisting on
    the tree for several years
                                     **23. radiata***
13a. Leaves more than 1 mm broad; cones
     usually borne individually                    14
 b. Leaves less than 1 mm broad; cones
    usually borne in groups of 3
                                      **22. patula**
14a. Cones more than 20 cm
                                     **24. coulteri***
 b. Cones less than 20 cm                          15
15a. Cone scale tips without prickles
                                   **11. canariensis***
 b. Cone scale tips with prickles
                                      **21. jeffreyi***

16a. Leaves mostly less than 10 cm,
     spreading or forward-pointing                 17
 b. Leaves mostly more than 10 cm,
    drooping                                       21
17a. Cones more than 10 cm
                                     **5. monticola***
 b. Cones less than 10 cm                          18
18a. Each cone scale ending in a slender
     dark bristle                    **10. aristata***
 b. Each cone scale without a slender
    dark bristle                                   19
19a. Cones more than 6 cm            **3. flexilis***
 b. Cones less than 6 cm                           20
20a. Leaves curved and twisted; seed
     usually shed from cones remaining on
     the tree                        **7. parviflora***
 b. Leaves not curved and twisted; seed
    often retained in cones remaining on
    the tree                          **1. cembra***
21a. Cones less than 8 cm, with stiff,
     woody scales; buds more than 20 mm
                                    **20. montezumae***
 b. Cones more than 8 cm, with usually
    flexible scales; buds less than 20 mm
                                                   22
22a. Buds less than 6 mm; cone broadly
     cylindric-conical, straight; bark
     greenish                        **2. armandii**
 b. Buds more than 6 mm; cone
    fusiform, curved; bark grey to dark
    grey                                           23
23a. Cones 15–30 cm, scales throughout
     most of cone not markedly reflexed
                                    **6. wallichiana**
 b. Cones 20–35 cm or more, scales
    through much of the lower part of the
    cone markedly reflexed
                                    **4. ayacahuite***

**1. P. cembra** Linnaeus. Illustration:
Mitchell, A field guide to the trees of Britain
and northern Europe, 155 (1974);
Mitchell, Conifers in the British Isles, 199
(1972); Callen, Les conifères cultivées en
Europe, 559-61 (1976); Edlin, The tree
key, 219 (1978).
Dense-crowned, conical-columnar trees,
10–25 m, often with a long, clean trunk.
Bark red-brown, thin, fissured and shed in
large plates. Young shoots covered with
shaggy, orange-brown hairs. Leaves in 5s,
5–8 cm, rigid, crowded, spreading. Female
cones usually solitary, almost terminal,
bluntly ovoid, 5–6 × 3–5 cm, never
opening on the tree, scales woody with
thick, obtuse tips, brown tinged with
purple. Seeds with rudimentary wings.
*C Europe (Alps & Carpathians)*. H1.
An extremely hardy pine, particularly
useful as a windbreak species. It is variable,
and several cultivars are available, e.g.

'Aureovariegata' with golden-tinged leaves,
'Chlorocarpa' with yellow-green cones, and
'Compacta' which is a dense dwarf,
reaching 2 m, suitable as a rock-garden
plant.
***P. koraiensis** Siebold & Zuccarini. Cones
much longer (9–15 cm), highly resinous
and fragrant (often found chewed by
squirrels, rarely seen intact). *USSR
(Siberia), Korea, N Japan*. H1.
***P. pumila** (Pallas) Regel. Leaves
4–5 cm, cones 4–5 × 2–3 cm, and
with a shrubby habit (rarely exceeding
4 m). *USSR (Siberia, Sakhalin), C Japan*.
H1.

**2. P. armandii** Franchet. Illustration:
Mitchell, Conifers in the British Isles, 191
(1972); Mitchell, A field guide to the trees
of Britain and northern Europe, 159
(1974).
Broad, open-crowned trees, 10–20 m or
more. Bark thin, smooth, greenish,
cracking with age. Leaves in 5s, 10–14 cm,
slender, drooping, usually sharply kinked
near the base, bright, shining green. Shoots
hairless, bloomed, with numerous chaffy
sheaths when young, smooth and olive-
green with age. Buds *c*. 5 mm, ovoid.
Female cones solitary, almost terminal,
1–3 cm, stalked, broadly cylindric-conical
with rounded apex, 8–18 × 6–10 cm, the
scales thick, flexible, highly resinous and
sweetly fragrant. Seeds 8–11 mm,
unwinged, most liberated when the cone is
green. *W China, N Burma, Taiwan, S Japan*.
H3.
Plants vary in habit, needle length and
kinking, and cone size; var. **mastersiana**
Hayata, from Taiwan forms a more
narrowly conical tree, and var. **amamiana**
Koidzumi from Japan, is of lower growing
habit, and is easily killed by frost.

**3. P. flexilis** James. Illustration: Mitchell, A
field guide to the trees of Britain and
northern Europe, 156 (1974).
Slow-growing, conical trees, 10–14 m,
bark smooth or shallowly fissured, pinkish
grey. Leaves in 5s, 5–8 cm, curved, in tight
but distant bundles which are at first laid
forwards along the shoot, green, glossy.
Female cones solitary or in 2s and 3s,
almost terminal, 6–9 cm, ovoid-cylindric,
with a bluntly rounded apex, resinous,
orange-brown when mature, the scales
flexible. Seeds with rudimentary wings.
*S & C USA*. H2.
***P. albicaulis** Engelmann. Leaves shorter
(4–6 cm), darker green, more curved, in
less widely spaced bundles. *Western
N America, Mexico*. H4

**4. P. ayacahuite** Ehrenberg. Illustration: Mitchell, Conifers in the British Isles, 195 (1972); Mitchell, A field guide to the trees of Britain and northern Europe, 162 (1974); Callen, Les conifères cultivées en Europe, 551, 553 (1976).
Broadly conical, open-crowned trees, 15–30 m or more. Bark dark grey, flaking. Branches many, level. Leaves in 5s, 13–20 cm, slender, flexible, drooping, not kinked, shining pale green. Young shoots finely downy, green tinged with pink, with chaffy sheaths. Buds 6–7 mm, ovoid-conical. Female cones usually solitary, almost terminal, 2–4 cm, stalked, fusiform and curved, 20–35 × 8–15 cm, the scales thick, flexible, highly resinous, those in the lower part of the cone markedly reflexed. *Mexico, Guatemala*. H4.

P. × **holfordiana** Jackson (*P. ayacahuite × wallichiana*) differs mainly in its slightly more slender cones, 18–26 × 4–12 cm, the basal scales not reflexed, more fissured bark and a much more vigorous growth. H4.

*P. **lambertiana** Douglas. Leaves 8–11 cm, stiffer, cones 25–45 × 10–15 cm, basal cone scales not reflexed. *W USA*. H5.

**5. P. monticola** Don. Illustration: Mitchell, Conifers in the British Isles, 243 (1972); Mitchell, A field guide to the trees of northern Europe, 159 (1974); Callen, Les conifères cultivées en Europe, 588–9 (1976).
Narrowly conical to columnar trees, 15–25 m with dense crowns. Bark purple-grey, smooth to scaly. Leaves in 5s, 7–10 cm, slender, glaucous green, spreading or slightly drooping, without kinks. Female cones solitary or in small groups, shortly stalked, almost terminal, pendulous, usually frequent on trees of moderate size, narrowly conical-cylindric, often curved, 12–20 × 3–7 cm, scales pale yellow-brown, flexible, highly resinous, with spreading, pointed tips. *Western N America (British Columbia to California)*. H4.

*P. **strobus**. Linnaeus. Leaves more sparse, held forwards along the shoot, cones 8–15 × 2.5–4 cm, more tapering. *Eastern N America*. H2.

Formerly a widely planted species of rapid growth; suffers from pine blister rust. A number of cultivars were available, e.g. 'Compacta'. 'Densa' and 'Nana', which are small, dense bushes, and 'Pendula' with long, drooping branches.

*P. **peuce** Grisebach. With dense, dark blue-green leaves and cones 8–12 ×

3–4 cm; crown denser, and bark purple-grey. *SE Europe (mountains of Balkan Peninsula)*. H2.

**6. P. wallichiana** Jackson (*P. excelsa* Wallich). Illustration: Mitchell, Conifers in the British Isles, 252 (1972); Callen, Les conifères cultivées en Europe, 651 (1976).
Conical, open-crowned trees, 20–35 m or more. Bark thin, grey, becoming fissured with age. Leaves in 5s, 10–15 cm, slender, erect, spreading, lying forwards or drooping, often slightly kinked near the base, blue-green, the chaffy sheaths readily shed. Shoots hairless, sometimes bloomed, becoming grey with age. Buds cylindric, 6–12 mm. Female cones solitary or in small groups, almost terminal, 1–2 cm, stalked, fusiform, curved, 15–30 × 4–7 cm, the scales flexible, thinly woody, purple-brown, highly resinous. Seeds winged. *W Himalaya*. H3.

**7. P. parviflora** Siebold & Zuccarini (*P. pentaphylla* Mayr). Illustration: Mitchell, Conifers in the British Isles, 228 (1972); Callen, Les conifères cultivées en Europe, 607–9 (1976).
Low, broad-crowned, flat-topped trees, 8–15 m, with long, level branches. Bark dark grey, smooth to scaly. Leaves in 5s, 5–8 cm, widely spreading, slender, curved and twisted, blue- or grey-green with white-glaucous inner faces, chaffy basal sheaths shed early. Female cones usually clustered, erect, often numerous and dense, 4–6 × 2.5–3.5 cm, broadly ovoid. Scales broadly wedge-shaped, often twisted, leathery, flexible, resinous, pale orange-brown, spreading widely when mature. Seeds each with a readily detached, rudimentary wing. *Japan*. H3.

The widely spreading branches give the crown of this species a characteristic and picturesquely 'oriental' appearance in silhouette, when seen on a rocky bluff. Its small size makes it suitable for larger rock-gardens, and it is commonly used, as is *P. densiflora*, for bonsai purposes.

*P. **himekomatsu** Hsia. Taller, straighter growing, more tender. *S Japan*. H4.

**8. P. bungeana** Zuccarini. Illustration: Dallimore & Jackson, Handbook of Coniferae, edn 4, 410 (1966); Mitchell, Conifers in the British Isles, 199 (1972); Mitchell, A field guide to the trees of Britain and northern Europe, 169 (1974); Callen, Les conifères cultivées en Europe, 553–4 (1976).
Conical, thin-crowned trees, mostly 5–10 m, with ascending branches. Bark

smooth, white to pale grey mottled with grey-green, brown and purple as thin flakes scale away in irregular patches. Leaves in 3s, 5–8 cm, stiff, forward-pointing or spreading, dark shining yellow-green. Female cones infrequent, solitary or paired, almost terminal, shortly stalked, very broadly ovoid-spherical, 2–4 × 2.5–4.5 cm, brown, scales few, thick, woody, almost flexible. Seeds with small wing, readily shed. *C & N China (Sichuan to Hubei)*. H4.

An unusual and slow-growing tree, succeeding in warm sandy or limey soils in a sunny position. Its bark, scaling like that of *Platanus*, makes it worthy of much wider ornamental planting.

**9. P. cembroides** Zuccarini. Illustration: Dallimore & Jackson, Handbook of Coniferae, edn 4, 416 (1966); Mitchell, Conifers in the British Isles, 201 (1972); Callen, Les conifères cultivées en Europe, 561–2 (1976).
Small, densely crowned trees, 5–10 m. Leaves mostly in 3s (see below), 2.5–4.5 cm, rigid, sharply pointed, incurved, with bare lengths of shoot between annual growths, fascicle sheaths splitting, persistent and curling backwards round the bases of the needles. Female cones ovoid-spherical, 2.5–5 × 2–4.5 cm, yellowish brown, scales with thick, woody tips, spreading widely. Seeds wingless. *SW USA, Mexico*. H4.

A slow-growing tree; the number of needles in each fascicle varies, and this variation is sometimes given varietal rank: var. **parryana** (Engelmann) Voss has most of its leaves in 4s, var. **edulis** (Engelmann) Voss has the leaves mostly in 2s and var. **monophylla** (Torrey & Frement) Voss has the leaves borne singly.

**10. P. aristata** Engelmann. Illustration: Dallimore & Jackson, Handbook of Coniferae, edn 4, 403 (1966); Mitchell, Conifers in the British Isles, 190 (1972); Mitchell, A field guide to the trees of Britain and northern Europe, 164 (1974); Edlin, The tree key, 218 (1978).
Densely crowned, conical trees, 5–9 m. Bark smooth, grey-black, becoming ridged with age. Leaves in 5s, 2–5 cm, persisting, curved, in compact fascicles densely clustered on the shoot, dull dark green speckled with a covering of small, white, resin flakes. Female cones solitary, 4–9 cm, ovoid, resinous, the scales elongate, slender and flexible, and with dark bristles from their tips. Seeds winged. *USA (Colorado, New Mexico)*. H2.

**\*P. balfouriana** Murray. Leaves without resin flakes, cones narrow, ovoid-cylindric, with much reduced bristles. *W USA (California).* H3.

**\*P. longaeva** Bailey. Leaves like those of *P. balfouriana*, cones like those of *P. aristata*. *W USA (Utah, Nevada, California).* H2.

Some specimens of this species are thought to be the oldest known living plants.

**11. P. canariensis** Smith. Illustration: Notes from the Royal Botanic Garden Edinburgh 33: 318 (1974); Callen, Les conifères cultivées en Europe, 558 (1976). Broadly crowned trees, 10–30 m. Branches massive, spreading to ascending, branchlets drooping. Bark very thick, reddish, deeply fissured, exuding resin freely if damaged. Leaves in 3s, 15–30 cm, slender, pendulous, crowded, flexible, bright green, with glaucous juvenile leaves persisting for many years, particularly on shoots from the bole. Bud scale margins fringed. Female cones solitary, almost terminal and almost stalkless, broadly ovoid-conical but with a rounded apex, 9–20 × 5–8 cm, the scales numerous, hard and woody, shining, brown, with rounded bosses. Seeds prominently winged. *Canary Islands.* H5.

**\*P. roxburghii** Sargent. Bud scales less prominently fringed, cone scales with more prominent bosses. *Himalaya.* H5.

**\*P. yunnanensis** Franchet. Cones smaller (5–8 cm). *SW China.* H5.

**\*P. palustris** Miller. Leaves longer (to 45 cm), cone oblong-cylindric, 15–25 cm. *SE USA (Virginia to Florida and Texas).* H5.

**12. P. pinaster** Aiton (*P. maritima* Lambert). Illustration: Dallimore & Jackson, Handbook of Coniferae, edn 4, 467 (1966); Mitchell, Conifers in the British Isles, 232 (1972); Callen, Les conifères cultivées en Europe, 613, 615–7 (1976). Broadly domed trees, 10–20 m. Bark orange to purple-brown, thick and flaky when old, falling to leave dark red scars. Winter bud scales with rolled tips and fringed with long silver hairs. Leaves in 2s, 18–25 cm, stout, stiff, curved, shining, deep green, crowded. Female cones almost terminal, stalkless, usually in whorled clusters of 3–5 or more, persistent, ovoid-conical, 12–20 cm, remaining closed for many years, the scales orange-brown with pointed tips. Seeds winged. *Mediterranean area.* H4.

A fast-growing tree, particularly suited to very poor, dry, sandy soils in mild areas.

**13. P. tabuliformis** Carrière. Illustration: Callen, Les conifères cultivées en Europe, 644 (1976). Trees 8–10 m with domed crowns, branches mostly horizontal. Bark grey, fissured. Leaves in 2s, 8–12 cm, stout, rigid, spreading, grey-green. Female cones almost terminal, usually solitary or in pairs, broadly ovoid-conical, 3–5 × 2.5–5 cm, scales woody, grey-brown, each ending in a small, blunt spine. *China, Korea.* H2.

*P. yunnanensis*, mentioned under *P. canariensis*, is sometimes treated as a variety of *P. tabuliformis*; there is, however, little similarity between them.

**14. P. sylvestris** Linnaeus. Illustration: Mitchell, Conifers in the British Isles, 244 (1972); Mitchell, A field guide to the trees of Britain and northern Europe, 170 (1974); Callen, Les conifères cultivées en Europe, 634, 635 (1976); Edlin, The tree key, 210, 211 (1978). Trees with domed crowns, 20–35 m or more, often with low, heavy branches. Bark pinkish grey to orange-brown, finely cracked or fissured and scaling. Leaves in 2s, 4–7 cm, more or less flexible, spreading, often somewhat twisted, dull bluish grey-green. Female cones solitary or in pairs, mostly broadly ovoid-conical, 2.5–4 cm, scales dull greyish brown with protuding tips. Seeds winged. *Eurasia.* H1.

An extremely variable tree; numerous cultivars have been selected, e.g. 'Aurea' with yellow-green leaves, 'Fastigiata' with a columnar habit and 'Nana', 'Pumila' and 'Pygmaea', which are dwarf variants suitable for rock-gardens.

**\*P. densiflora** Siebold & Zuccarini. Leaves 9–10 cm, slender, bright green, cones in groups of 2 or 3, 3–5 cm. *Japan.* H2.

The most common pine used in bonsai by the Japanese.

**\*P. mugo** Turra. A small tree of densely bushy habit, 2–5 m. *Europe.* H1.

**\*P. resinosa** Aiton. Leaves 12–15 cm, slender, shining, dark green. *Eastern N America (Nova Scotia to Pennsylvania).* H1.

**15. P. nigra** Arnold. Illustration: Mitchell, Conifers in the British Isles, 225 (1972); Mitchell, A field guide to the trees of Britain and northern Europe, 171, 172 (1974). Callen, Les conifères cultivées en Europe, 596, 597, 599–601, 603–5, 607 (1976). Trees with irregularly domed crowns, 25–35 m, branches widely spreading or ascending, trunks clean. Bark dark brown to blackish, deeply fissured into scaly plates.

Leaves in 2s, 8–12 cm, straight or curved, stiff, densely set, deep blackish green. Shoots roughened after leaf-loss by persistent leaf bases. Winter buds ovoid-oblong, usually light brown, surrounded by brittle, papery, basal scales. Female cones usually solitary, almost stalkless, ovoid-conical, 5–8 × 2.5–5 cm, scales yellow-brown, with small, blunt points. *C & SE Europe, Turkey.* H2.

A vigorous and valuable large ornamental tree, suitable for large parks and windbreaks. Var. **caramanica** (Loudon) Rehder (var. *pattasiana* (Lambert) Ascherson & Graebner) has slightly longer, straighter leaves (12–15 cm), and a habit with the branches more obviously ascending. *USSR (Crimea), Turkey.* H2. Var. **cebennensis** (Grenier & Godron) Rehder has grey-green leaves which are 15–17 cm, and an almost smooth cone. *France, Spain (Pyrenees).* H3. Var. **maritima** (Aiton) Melville (var. *calabrica* (Loudon) Schneider) has a conical-columnar habit and short, level branches with less dense foliage; leaves flexible, 15–17 cm. *S Europe (Corsica, Italy, Sicily).* H4.

**\*P. leucodermis** Antoine. Leaves stiff, 7–9 cm, held pointing acutely forwards, buds with red-centred scales, bark smoother and greyer, habit more regularly conical. *SE Europe.* H2.

**\*P. thunbergii** Parlatore. Leaves 8–10 cm, slightly twisted, grey-green, branches slender, horizontal or upturned, buds white. *Japan.* H4.

**16. P. muricata** Don. Illustration: Mitchell, Conifers in the British Isles, 222 (1972); Mitchell, A field guide to the trees of Britain and northern Europe, 178 (1974) Trees with dense, broadly domed crowns, 10–25 m, with heavy branches. Bark dark grey, deeply fissured. Leaves in 2s, 12–15 cm, crowded, stiff, spreading widely, dull grey-green. Female cones solitary or in whorls of 2–5, 6–8 cm, ovoid-conical, asymmetric, obliquely set, remaining closed for long periods, scales pointed. Seeds winged, liberated only by forest fires. *USA (California).* H3.

**17. P. contorta** Douglas. Illustration: Mitchell, Conifers in the British Isles, 202 (1972); Mitchell, A field guide to the trees of Britain and northern Europe, 175 (1974); Edlin, The tree key, 213 (1978). Bushy, open-crowned trees, 5–20 m, with upswept branches. Bark dark reddish brown, fissured. Leaves in 2s, 4–5 cm, flattened, stiff, swept forwards on the shoot or speading, often somewhat twisted, pale to deep dark green, crowded. Female cones solitary, almost terminal, stalkless, ovoid-conical with a rounded apex, 2–5 cm, slightly prickly, not opening for several years. *Western N America.* H3.

Very variable, depending on provenance. Var. **latifolia** (Engelmann) Critchfield, from the more inland part of the species' range, has longer, looser foliage.

*__P. banksiana__ Lambert. Leaves 3–4 cm, more twisted, pointing upwards, growth habit irregular, producing sinuous branches and a gaunt crown. *NE Canada.* H1.

**18. P. halepensis** Miller. Illustration: Mitchell, Conifers in the British Isles, 211–2 (1972); Mitchell, A field guide to the trees of Britain and northern Europe, 172 (1974); Callen, Les conifères cultivées en Europe, 573–4 (1976).
Dense trees with domed crowns, to 20 m or more, with heavy, elevated branches and a clean trunk. Bark deep red-brown, fissured. Leaves in 2s, 6–9 cm, often curved and twisted, bright glossy green, with a conspicuous (*c.* 1 cm) basal sheath. Female cones mostly in 3s, thickly stalked, ovoid-conical, 2.5–11 cm, often not opening for several years, deflexed backwards along the shoot. Seeds winged. *Mediterranean area, SW Asia to Afghanistan.* H4.

Var. **brutia** (Tenore) Elwes & Henry, from the eastern Mediterranean, has leaves 10–15 cm, darker green, and the cones more spreading. H5.

**19. P. pinea** Linnaeus. Illustration: Mitchell, A field guide to the trees of Britain and northern Europe, 170 (1974); Callen, Les conifères cultivées en Europe, 621 (1976).
Very broadly dome-crowned trees, 10–15 m, the trunk usually first branching at a low level into stout, sharply ascending branches, then into numerous small branches to form a characteristically dense, umbrella-shaped crown. Bark thick, orange-brown, scaly. Leaves in 2s, 11–15 cm, stout, somewhat twisted, shining, dark green. Female cones almost terminal, solitary or in small groups, massive, ovoid-spherical, symmetric, 10–14 × 8–10 cm, their tightly packed scales shining brown, each with a protruding, rounded, cushion-like, woody tip. Seeds 1–1.3 cm, dark purple-brown, unwinged. *Mediterranean region.* H5.

The umbrella-like crown of this pine is characteristic of many Mediterranean coasts, and the large, edible seeds have some importance in commerce; 'Fragilis' is

said to have a thin-shelled seed, and is cultivated for that reason.

**20. P. montezumae** Lambert. Illustration: Dallimore & Jackson, Handbook of Coniferae, edn 4, 445 (1966); Mitchell, A field guide to the trees of Britain and northern Europe, 163 (1974); Callen, Les conifères cultivées en Europe, 585–8 (1976).
Broad, moderately dense, dome-crowned trees, 10–20 m or more, with stout, upcurved branches. Bark reddish brown, rough, deeply fissured. Leaves mostly in 5s (but varying from 3–8, especially in young trees), 25–30 cm, spreading or drooping, greyish blue-green. Buds cylindric, 2 cm or more. Female cones solitary or clustered, ovoid-conical to cylindric, 2.5–6.5 cm, reddish brown, woody. Seeds narrowly winged. *Mexico.* H4.

A bold and imposing tree, suitable only for very mild climates. A variable species in the wild; the following varieties have been cultivated: var. **hartwegii** (Lindley) Engelmann, with leaves 10–18 cm, green, more often in 3s and 4s than 5s, bark less fissured; var. **lindleyi** Loudon, with drooping, pale green leaves; var. **rudis** (Endlicher) Shaw with leaves 10–15 cm, yellowish green and crown thin and gaunt.

*__P. michoacana__ Martinez. Leaves 25–45 cm, bright green, drooping, cones more than 10 cm. *Mexico.* H5.

**21. P. jeffreyi** Murray. Illustration: Dallimore & Jackson, Handbook of Coniferae, edn 4, 435 (1966); Mitchell, A field guide to the trees of Britain and northern Europe, 167–8 (1974); Callen, Les conifères cultivées en Europe, 582 (1976).
Conical-columnar, fairly open-crowned trees 20–35 m or more, with short, horizontal branches and straight, clean trunks. Bark brown to brownish red, fissured. Shoots fragrant when broken, glaucous when young. buds non-resinous with free scale tips. Leaves mostly in 3s, 16–25 cm, stout, spreading, flexible, grey-green. Female cones solitary, ovoid-oblong, 12–20 × 8–12 cm, scale tips with small, recurved prickles. Seeds with large wings. *W USA (S Oregon, California).* H4.

*__P. ponderosa__ Douglas. Young shoots yellow-brown, leaves darker green, buds resinous with the scale tips adpressed, bark darker, often purple-grey. *Western N America (British Columbia to New Mexico).* H4.

**22. P. patula** Schlechtendahl & Chamisso. Illustration: Mitchell, A field guide to the trees of Britain and northern Europe, 167 (1974); Callen, Les conifères cultivées en Europe, 611 (1976).
Broad, low-crowned trees, 8–12 m, often forked into a few, long, ascending-spreading branches. Bark reddish brown, thin, scaling into flakes with rolled, papery margins. Leaves mostly in 3s, 18–20 cm, very slender, lying densely forwards then pendulous, not kinked, bright green with the basal sheath conspicuous, 2–2.5 cm. Female cones lateral, in clusters of 2–5 or more, persistent, ovoid-conical, curved, 8–12 cm. Seeds with large wings. *C & E Mexico.* H5.

**23. P. radiata** Don (*P. insignis* Douglas). Illustration: Mitchell, Conifers in the British Isles, 237 (1972); Mitchell, A field guide to the trees of Britain and northern Europe, 165 (1974); Callen, Les conifères cultivées en Europe, 627 (1976); Edlin, The tree key, 217 (1978).
Crown dense, irregularly domed, 25–40 m, with heavy branches. Bark purple-grey, thick and deeply fissured. Leaves in 3s, 10–14 cm, slender, densely set and spreading widely, bright shining deep green. Female cones in regular whorls of 4–7 or more, 8–10 × 7–9 cm, broadly ovoid, very asymmetric, remaining closed for a long period, red-brown, the outer, lower scales ending in rounded, dome-like knobs. Seeds winged. *W USA (California – Monterey Peninsula).* H5.

*__P. attenuata__ Lemmon. Bark often more conspicuously flaking, leaves 14–16 cm, grey-green, cones 11–14 × 5–6 cm, more pointed, yellow-brown, the lower scales with more pointed knobs. *W USA (S Oregon, California).* H3.

**24. P. coulteri** Don. Illustration: Mitchell, Conifers in the British Isles, 206 (1972); Mitchell, A field guide to the trees of Britain and northern Europe, 168 (1974); Callen, Les conifères cultivées en Europe, 566 (1976).
Large, open-crowned trees of somewhat gaunt appearance, 15–30 m, with thin, narrow, sinuous, seldom-branched branches to nearly ground level. Bark dark grey, deeply fissured. leaves in 3s, 15–30 cm, stout, rigid, densely set, widely spreading almost perpendicularly all round the apex of each branch, shed early to leave a long, leafless portion of branch, bluish grey-green. Buds large, ovoid-pointed, conspicuous, orange-red. Female cones

slightly asymmetrically ovoid or oblong-ovoid, 25–35 × 15–20 cm, massive, heavy, light brown, remaining closed on the tree, scales mostly ending in strong, woody claws. Seeds winged. *W USA (S California), Mexico.* H3.

*P. sabiniana* Douglas. Cones 15–25 × 10–15 cm. *W USA (California).* H4.

## VIII. TAXACEAE

Profusely branching, often dioecious, evergreeen shrubs or small trees. Leaves spirally inserted, often disposed in two ranks, linear or scale-like. Wood and leaves without resin canals. Stamens 6–14, each with 2–8 pollen sacs and peltate anthers. Female organs axillary, with a single erect ovule or seed, partly or wholly surrounded by a fleshy aril when ripe.

A small family of 5 genera and about 20 species; only 2 of the genera are common in cultivation.

1a. Midrib prominent on upper side of leaves; female cones solitary; ovule only partially covered by the aril  
**1. Taxus**

b. Midrib not showing on upper side of leaves; female cones in pairs; ovule completely covered by the aril  
**2. Torreya**

## 1. TAXUS Linnaeus
*T.J. Varley*

Evergreen, usually dioecious, small- to medium-sized trees and shrubs, with reddish-brown, thin, scaly bark. Winter buds small, usually with rounded scales. Leaves spreading, alternate, in two ranks or radial, linear, flat, sometimes sickle-shaped with two yellowish or greyish-green bands beneath. Male cones with 6–14 stamens and peltate anther scales. Female organs consisting of several overlapping scales, the uppermost bearing a single ovule. Seeds ovoid, slightly angular, surrounded by an aril which is open at the apex.

A genus of 9 very similar species, widely distributed in Europe, N America, Turkey and E Asia. Commonly grown as trees in parks, gardens and churchyards; very good hedge plants, often used for topiary work. The species are tolerant of most soils, and thrive in chalk, peat and light or heavy loams, as long as drainage is adequate; they also succeed in most situations, from heavy shade to full sun. Propagation is by seeds, whilst cultivars are increased by cuttings or grafting.

Literature: Bartkowiak, S. et al., The yew – Taxus baccata L., *Popular Scientific Monographs*, 3 (1978).

*Habit.* Medium-sized tree or shrub: **1–4,6,7**; low growing, prostrate shrub: **5**.  
*Bud scales.* Pointed: **2,4**; rounded: **1,3,5–7**.  
*Leaf colour.* Upper surface dark green or yellowish green: **1,2,4–7**; upper surface pale green: **3**.  
*Leaf arrangement.* 2-ranked, usually horizontal: **1,3–6**; upright and V-shaped: **2,7**.  
*Leaf size.* Always less than 20 × 2 mm: **4,5**.  
*Fruit.* Numerous and bunched: **2,6**; solitary: **1,3–5**. Up to 5 mm long: **3**; 5–9 mm long: **1,2**; 10 to 15 mm long: **4,6**.

1a. Leaves in 2 ranks, usually flat          2  
b. Leaves not ranked; parting irregularly V-shaped          6  
2a. Leaves slender and short, 7–15 × 0.5–2 mm          3  
b. Leaves broad and long, 2–3 cm × 2–3 mm          4  
3a. Leaves yellowish-green above; bud scales pointed          **4. brevifolia**  
b. Leaves dark green above, reddish in winter; bud scales rounded          **5. canadensis**  
4a. Foliage sparse; upper side of leaf pale green; lower surface, except midrib, densely covered with small papillae          **3. celebica***  
b. Foliage dense; upper side of leaf dark green to dark yellowish green; papillae absent from lower surface          5  
5a. Branchlets dull green, often reddish above; midrib very strongly raised on upper side of leaf; fruits 10 × 10 mm, numerous, bunched          **6. × media**  
b. Branchlets pale yellowish green; midrib only slightly raised on upper side of leaf; fruits 6–7 × 5 mm, solitary          **1. baccata***  
6a. Branchlets brown by 2nd year; leaves slender, to 2 mm wide; green margin on underside of leaf much broader than the pale band          **7. × hunnewelliana**  
b. Branchlets green in 2nd year; leaves broader, 2–3 mm wide; green margin on underside of leaf half as broad as the pale band          **2. cuspidata**

**1. T. baccata** Linnaeus. Figure 28(1) p. 69  
Illustration: Harrison, Ornamental conifers, 154 (1975), Phillips, Trees in Britain, 204 (1978).

A medium-sized, densely branched tree to 25 m with a girth up to a maximum of 10 m. Bark reddish-brown, thin and scaly. Branchlets spreading, alternate, with brownish scales at base. Winter buds green and obtuse. Leaves spirally inserted, radially arranged on erect shoots, 2-ranked on horizontal shoots; 2–3 cm × 2–2.5 mm, linear, tapering at the apex to a point, dark green and shining above with a prominent midrib, underside dull yellow-green. Stamens small, spherical, along the underside of the previous year's shoots; female organs solitary, green, from the leaf axils. Seeds solitary, slightly compressed, olive-brown; aril 6–7 × 5 mm, scarlet. *Europe and N Africa east to Iran.* H2.

There are over 100 cultivars of *T. baccata,* of which approximately 50 (including some which are dwarf) are commonly grown.

*T. floridana* Chapman. A shrub or small tree to 8 m, with numerous, stout, spreading branches. Leaves 2–2.5 cm × 1–2 mm, narrow, curved; midrib obscure, dark green. Seeds similar to *T. baccata. S USA (Florida).* H2.

**2. T. cuspidata** Siebold and Zuccarini. Illustration: Phillips, Trees in Britain, 204 (1978).

A broad bushy tree or shrub to 20 m with spreading branches. Bark reddish brown or greyish brown, scaly. Winter buds oblong and brown with concave, ovate scales. Leaves upright, parting V-shaped, 1.5–2.5 cm × 2–3 mm, linear, straight or sickle-shaped, apex abruptly pointed; dark green above, yellowish brown below. Seeds numerous, bunched; aril 7–8 × 6 mm, red. *Japan, Korea & China.* H1.

**3. T. celebica** (Warburg) Li (*T. chinensis* Rehder). Illustration: Phillips, Trees in Britain, 204 (1978).

A large shrub or tree to 15 m, wide and bushy in cultivation. Foliage sparse. Leaves 2-ranked, widely set, spreading, 2–3 cm × 3–4 mm, usually sickle-shaped, apex sharply pointed; shining, dark green above, undersurface greyish green with a dense covering of papillae, which are absent from midrib. Stamens sparse. Seeds sparse, 5 mm, ovoid, green, rarely ripening in cultivation. *China.* H1.

*T. wallichiana* Zuccarini. A tree or large shrub. Leaves 2–3 cm × 2 mm, linear, tapering gradually towards the apex, ending in a spiny point; undersurface including midrib, densely covered with papillae. *India (Himalaya).* H2.

**4. T. brevifolia** Nuttall. Illustration: Jepson, Trees of California, 126 (1909); Preston, North American trees, 118 (1976).
A shrub of 5–10 m, sometimes a tree to 20 m, bark dark red-purple, thin, scaly. Branches slender, spreading and slightly pendulous. Branchlets yellowish green. Bud scales pointed. Leaves distinctly 2-ranked, horizontal, 7–15 × 1.5 mm, abruptly pointed with a prominent raised midrib. Seeds 1–1.3 cm with a scarlet aril. *Western N America from California to British Columbia.* H1.

**5. T. canadensis** Marshall.
A monoecious, low-growing shrub to 2 m, often prostrate. Branches crowded and straggling. Branchlets short. Winter buds small, yellowish brown with overlapping, blunt scales. Leaves densely set in 2 ranks, 1–1.5 cm × 1–2 mm, midrib slightly prominent above, apex abruptly pointed. Seeds similar to those of *T. baccata* in size, but numerous and bunched with a lighter red aril. *Eastern N America from Virginia to Newfoundland.* H1.

**6. T. × media** Rehder (*T. baccata × T. cuspidata*). Illustration: Harrison, Ornamental conifers, 158 (1975).
A broad, pyramidal, spreading, branched bush. Branchlets dull green often reddish above. Bud scales obtuse with rounded apices. Leaves distinctly 2-ranked, often horizontal; stouter, broader and the upper side with a more strongly raised midrib than those of *T. baccata*. Seeds 10 × 10 mm, abundant; aril scarlet. A hybrid intermediate between the two parents. *Cultivated origin.* H1.

**7. T. × hunnewelliana** Rehder (*T. canadensis × T. cuspidata*).
Resembles *T. cuspidata*, but of more slender habit. Leaves narrower, partly pointing forwards, the green margin beneath more than half as broad as the pale band. *Cultivated origin.* H1.

**2. TORREYA** Arnott
*T.J. Varley*
Evergreen, usually dioecious trees or shrubs with furrowed bark. Branches opposite or whorled. Buds ovoid, acute; scales few, opposite, deciduous. Leaves spirally inserted, 2-ranked on lateral shoots, linear or linear-lanceolate, spiny-pointed; upper surface green, with an indistinct midrib, lower surface with a raised midrib and 2 glaucous bands sunk in longitudinal furrows. Stamens numerous, in whorls of four. Female organs in pairs. Seeds obovoid with a thick, woody shell entirely covered by a thin, fleshy aril, ripening the 2nd year.

Of the 6 species, 4 are grown as ornamental trees in gardens or parks, usually in mild climates. They succeed in most soils including chalk, loams and peats and tolerate shade. Propagation is either by cuttings or by imported seeds.

1a. Foliage not scented when bruised; shoots green in 2nd year   **3. grandis**
  b. Foliage scented when bruised; shoots reddish or brown in 2nd year   **2**
2a. Buds to 4 mm; leaves often slightly sickle-shaped with 2 white bands beneath, each as wide as the midrib; fruit 2–2.5 cm, obovoid   **2. nucifera***
  b. Buds to 9 mm; leaves linear and straight with 2 bands beneath, both much narrower than midrib; fruit 2.5–4 cm, elliptic or obovoid   **3**
3a. Shoots hairless; leaves 3–6 cm, dark shining green above, fruits green, streaked with purple   **1. californica**
  b. Shoots with occasional hairs; leaves 2–3 cm, dark yellowish green above; fruits purple-brown   **4. taxifolia**

**1. T. californica** Torrey. Illustration: Phillips, Trees in Britain, 209 (1978).
Small- to medium-sized tree, 15–20 m, conical. Bark pale red-brown or grey-brown. Branches straight, whorled, slightly deflexed. Branchlets green at first, reddish-brown by 2nd year. Winter buds to 9 mm, acute, scales overlapping, brown. Leaves widely spaced, irregularly pectinate, often curved upwards, 3–6 cm × 3–4 mm, rigid, linear, tapering to a spiny point; lustrous, dark green above, with 2 narrow shallow white bands beneath. The bruised foliage emits an oily, sage-like scent. Seeds 3–4 cm, ellipsoid, green, streaked with purple when mature. *USA (California).* H3.

**2. T. nucifera** (Linnaeus) Siebold & Zuccarini. Illustration: Harrison, Ornamental conifers, 174 (1975).
A large shrub or occasionally a small, slender, thinly foliaged tree in cultivation. Bark greyish brown. Branches crowded and whorled. Branchlets yellowish green in the first year, shining brown later. Winter buds 4 mm. Leaves spreading, regularly pectinate, downcurved, 2–3 cm × 2–3 mm, usually slightly sickle-shaped, apex sharply pointed; shining dark green above, with 2 white bands beneath each as wide as the midrib. Pungent smell when crushed. Seeds obovoid, 2–2.5 cm, green tinged with purple when mature. *Japan.* H1.

Two cultivars are known both of which are dwarfs. 'Prostrata' is an almost prostrate mat less than 1 m high and 'Spreadeagle', a more recent cultivar, is low-growing with long, spreading branches.

***T. jackii** Chun. A small tree to 10 m or more, or often a much branched shrub. Branches ascending. Branchlets spreading or pendulous, green in the 1st year, reddish brown by the 2nd year. Leaves to 9 × 5 mm, sickle-shaped, gradually tapering to a spine-pointed apex and with a smell like sandalwood when bruised. Seeds 2.5–3 cm, glaucous. *E China.* H3.

**3. T. grandis** Fortune.
A small tree or more often a large shrub. Bark greyish brown. Branchlets green. Leaves 2–2.5 cm × 3 mm, linear to linear-lanceolate, spiny-pointed; lustrous bright green above, with 2 white bands beneath both equalling midrib in breadth. Foliage without aroma. Seeds obovoid, 2.5–3 cm, brownish. *E China.* H3.

**4. T. taxifolia** Arnott. Illustration: Botanical Gazette **29**: 178 (1905); Coulter & Chamberlain, Morphology of Gymnosperms, 322 (1910).
A small tree to 15 m, conical. Bark brown tinged with orange. Branches spreading, slightly pendulous. Branchlets yellowish green at first, reddish brown by 2nd year. Leaves 2–3 cm × 2–3 mm, linear, tapering to a pointed apex; shining dark yellowish green above, lower surface pale green with 2 white shallow bands narrower than the midrib. Seeds obovoid, 2.5–3.5 cm, purple, the flesh with a foetid smell. *USA (Florida).* H2.

There is one colour variant, 'Argentea', which has white branchlets when young, irregularly dispersed over the plant.

# IX. GINKGOACEAE

Trees with 2 kinds of shoot: longer, extension shoots and short spurs. Leaves leathery, fan-shaped, often notched, with bifurcating, parallel veins. Male organs in axillary, catkin-like structures, each scale bearing 2–12 anther sacs. Ovules 2–10 on a single branched axis, developing into large, drupe-like seeds.

A family with a very long fossil history, of which only a single species survives today. This species has itself survived for 200 million years; its historic survival has probably depended on religious veneration since planted trees some 1000 years old are known in temple precincts in China. In cultivation it has few, if any, pests and diseases, and it withstands atmospheric pollution. Male plants are more suitable for ornamental plantings than the female, as the fruits of the latter have a disagreeable smell when fallen.

## 1. GINKGO Linnaeus
*J. Lewis*

Deciduous, dioecious trees, erect-growing when young, spreading irregularly later, with smooth, grey-brown bark until aged. Extension shoots frequently horizontal, bearing slow-growing spurs which produce clusters of leaves annually or themselves change to extension growth. Leaves long-stalked from small, brown, pointed buds; blade to 10 cm across, but sometimes larger in young trees, fan-shaped, notched above into 2 halves, yellow in autumn. Male catkins pendent from short shoots. Ovules paired on 2 or 3 long axes. One ovule of each pair developing into a seed. Seed to 2 cm, with a stony shell enclosed in yellow, evil-smelling flesh.

The only living species occurs (perhaps naturally) in a small region of southern China, but has long been cultivated in China and Japan for the edible kernels of its seeds. Once established, it grows well on a variety of soils. Propagation is by seed; the cultivars may be grafted on ordinary stocks, even out-of-doors. The preferable male seedlings may be detected by observing their earlier leaf-fall.

Literature: Li, H.L., Ginkgo, the maidenhair tree, *American Horticultural Magazine* **40**: 239–49 (1961).

**1. G. biloba** Linnaeus. Figure 28(4) p. 69. Illustration: Harrison, Ornamental conifers, 83 (1975).

*China (Zhejiang).* H1.

Columnar variants are known e.g. 'Fastigiata' and 'Fairmont', which are very erect, 'Mayfield' which is quite narrow and 'Tremonia', which forms a tall pillar with good autumn colour. Among the pyramidal variants, 'Laciniata' has large, divided leaves and 'Autumn Gold' colours bright yellow. 'Pendula' is broad with drooping branches, 'Aurea' has leaves which are yellow when young, and 'Variegata' has leaves which are longitudinally striped with yellow.

---

# X. GNETALES

Woody plants with unisexual flowers arranged in cone-like structures. Perianth segments 2, partly united; stamens 1 or several; ovule 1, bearing a tube developed around the micropyle and projecting above the perianth. Embryo with 2 cotyledons.

The Gnetales are an order of Gymnosperms comprising three very diverse families. The presence of vessels in the wood, a perianth in the flower and 2 cotyledons in the embryo, together with the appearance of the leaves in *Gnetum*, have in the past been taken as indications of an affinity with the Angiosperms; consequently the order has been the subject of intensive morphological and anatomical study. Any connection with the Angiosperms is now considered to be remote.

No members of the order are horticulturally important in Europe.

The family **Gnetaceae** consists only of the genus **Gnetum** Linnaeus, with about 40 species of tropical small trees or climbers. The leaves are opposite, flat and leathery and look like the leaves of a dicotyledonous angiosperm. The plants are usually monoecious and the flowers are arranged in whorls, each subtended by a ring of fused bracts. The ripe seed is surrounded by a hardened outer seed-coat and a fleshy perianth. Specimens are usually found in botanic garden collections (G2) and **G. gnemon** Linnaeus is cultivated in Malaysia for its edible seeds.

The family **Welwitschiaceae** comprises only **Welwitschia mirabilis** J.D. Hooker which grows in the Namib Desert, South West Africa. It has a short woody stem to 1 m in diameter, partly embedded in the soil. From this emerge a deep taproot and two strap-shaped leaves which grow from the base and rest on the ground. These are the only leaves which the plant produces and may endure for 100 years or more, their tips becoming split and eroded. The plants are dioecious and bear cones in dichasial cymes about 30 cm tall. The female perianth is flattened and becomes hardened and winged in fruit. The successful cultivation, usually in botanic gardens, of this remarkable plant has been described on a number of occasions (for example, by H. Teuscher in National Horticultural Magazine 30: 188–92, 1951).

## EPHEDRACEAE
*P.F. Yeo*

Shrubs, sometimes scrambling, or rarely small trees, often with underground rhizomes. Young shoots *Equisetum*-like, cylindrical, finely ribbed, green; older shoots woody. Leaves opposite, scale-like or sometimes with a subulate persistently green tip, their bases united to form a sheath around the stem. Plants usually dioecious. Flowers straw-coloured. Male flowers in dense spikes, each flower with 2–8 stamens, their filaments partly or completely fused into a column. Female flower-spikes solitary or clustered, consisting of a few pairs of bracts and 1 or 2 flowers. Ovule with the micropylar tube often twisted or bent and with an oblique opening. Perianth woody in fruit; inner bracts becoming fleshy, and white or red in colour, or drying out, and sometimes becoming winged.

The only genus, **Ephedra** Linnaeus, comprises about 40 species of the North temperate region, temperate South America and tropical Asia (Figure 28(5) p. 69). They are of little horticultural interest, though easily grown; one or two dwarf species (e.g. **E. gerardiana** Stapf) are occasionally grown on rock gardens. Identification usually requires all reproductive stages and even then is very difficult. Some species are a source of the drug ephedrine.

Literature: Stapf, O., Die Arten der Gattung Ephedra, *Denkschriften der Kaiserlichen Akademie der Wissenschaften, Wien, Mathematische-Naturwissenschaftliche Klasse* **56**: 1–112 (1899); Krüssmann, G., *Handbuch der Nadelgehölze*, 124–9 (1972).

# ANGIOSPERMAE

Plants herbaceous or woody. Seedling leaves (cotyledons) 1 or 2. Stamens and ovary borne in unisexual or bisexual flowers which generally have protective and/or pollinator-attracting envelopes (perianth, often composed of differentiated sepals and petals). Ovules borne inside closed ovaries; seeds enclosed, until ripe, inside a fruit.

The flowering plants, as generally understood, to which most garden plants belong. there are about 300 000 species arranged in about 11 000 genera in about 400 families; some 15 000–20 000 species are in general cultivation. The group is arranged in 2 large classes, the Monocotyledons (53 families in all, covered in this volume and in Volume II) and Dicotyledons (about 350 families in all, to be covered in Volume III and subsequent volumes).

## KEY TO CLASSES

1a. Cotyledon 1, terminal; leaves usually with parallel veins, sometimes these connected by cross-veinlets; leaves without stipules, opposite only in some aquatic plants; flowers usually with parts in 3s; mature root system wholly adventitious
**Monocotyledons**

b. Cotyledons usually 2, lateral; leaves usually net-veined, with or without stipules, alternate, opposite or whorled; flowers with parts in 2s, 4s or 5s or parts numerous; primary root (taproot) usually persistent, branched **Dicotyledons**

# MONOCOTYLEDONS
# (Part I)

Plants herbaceous or woody, frequently with bulbs, corms or rhizomes. Primary root usually quickly lost, the mature root system wholly adventitious, fibrous. Seedling leaf (cotyledon) 1. Leaves usually alternate or all basal, usually with several parallel veins, more rarely with a distinct midrib giving off parallel lateral veins, very rarely the veins forming a network. Parts of the flower usually in 3s or multiples of 3.

## KEY TO FAMILIES

*Families included in this volume are provided with page numbers; the others are to be found in volume II (1984).*

1a. Ovary superior or flowers completely without perianth (including all aquatics with totally submerged flowers)     2

b. Ovary inferior or partly so (if aquatic, flowers borne at or above water level)     27

2a. Trees, shrubs or prickly scramblers with large, pleated, usually palmately or pinnately divided leaves; flowers more or less stalkless in fleshy spikes or panicles with large basal bracts (spathes)     **XXI. Palmae**

b. Plant without the above combination of characters     3

3a. Small, usually floating aquatic plants not differentiated into stem and leaves     **XXIII. Lemnaceae**

b. Plants various, rarely floating aquatics, plant body differentiated into stem and leaves     4

4a. Perianth entirely scarious or reduced to bristles, hairs, narrow scales, or absent     5

b. Perianth well developed, though sometimes small, never entirely scarious     11

5a. Flowers in small, 2-sided or cylindric spikelets provided with overlapping bracts (spikelets sometimes 1-flowered)     6

b. Flowers arranged in heads, superposed spikes, racemes, panicles or cymes, never in spikelets as above     7

6a. Leaves alternate, in 2 ranks, on a stem which is usually hollow and with cylindric internodes; leaf-sheath usually with free margins, at least in the upper part; flowers arranged in 2-sided spikelets (sometimes 1-flowered) each usually subtended at the base by 2 sterile bracts (glumes); each flower usually enclosed by a lower lemma and an upper palea (sometimes absent); perianth of 2–3 concealed scales (lodicules), more rarely 6 or absent; styles generally 2, feathery     **XX. Gramineae**

b. Leaves usually spirally arranged on 3 sides of the cylindric or more usually 3-angled stems which usually have solid internodes; young leaf-sheath closed though sometimes splitting later; flowers arranged in 2-sided or cylindric spikelets often with a 2-keeled or 2-lobed glume at the base; each flower subtended only by a glume; perianth of several bristles, hairs or scales, or absent; style 1 with 2 or 3 papillose stigmas      **XXVII. Cyperaceae**

7a. Dioecious trees or shrubs with stiffly leathery, sharply toothed leaves, often supported by stilt-roots; fruit compound, often woody      **XXIV. Pandanaceae**

  b. Plant without the above combination of characters      8

8a. Inflorescence a simple, fleshy spike (spadix) of inconspicuous flowers subtended by or rarely joined to a large bract (spathe); leaves often net-veined or lobed (plant rarely a small, evergreen, floating aquatic)      **XXII. Araceae**

  b. Plant without the above combinations of characters      9

9a. Flowers bisexual; perianth-segments 6, scarious; ovary with 3–many ovules      **XVII. Juncaceae**

  b. Flowers unisexual; perianth-segments a few threads or scales; ovary with 1 ovule      10

10a. Flowers in 2 superposed, elongate, brownish or silvery spikes; ovary borne on a stalk with hair-like branches      **XXVI. Typhaceae**

  b. Flowers in spherical heads; ovary not stalked      **XXV. Sparganiaceae**

11a. Carpels free or slightly united at the base      12

  b. Carpels united for most of their length though the styles may be free, or carpel solitary      15

12a. Inflorescence a spike, sometimes bifid; perianth-segments 1–4      13

  b. Inflorescence not a spike; perianth-segments 6      14

13a. Stamens 6 or more; carpels 3–6; perianth-segments 1–3, petaloid      **IV. Aponogetonaceae** (p. 116)

  b. Stamens 4; carpels 4; perianth-segments 4, not petaloid      **V. Potamogetonaceae** (p. 117)

14a. Placentation marginal or parietal      **II. Butomaceae** (p. 113)

  b. Placentation basal      **I. Alismataceae** (p. 111)

15a. All perianth-segments similar      16

  b. Perianth-segments of the outer and inner whorls conspicuously different, the former usually sepal-like, the latter usually petal-like      25

16a. Inflorescence subtended by an entire, spathe-like sheath; plants aquatic      **XV. Pontederiaceae** (p. 331)

  b. Inflorescence not as above; plants terrestrial      17

17a. Perianth persistent, covered by branched hairs; stamens 3; sap usually orange      **VIII. Haemodoraceae** (p. 290)

  b. Plant without the above combination of characters      18

18a. Plant woody, or not woody and bearing rosettes of long-lived, fleshy or leathery leaves, at or near ground level      19

  b. Plant herbaceous, leaves usually not long-lived and in rosettes, if so, then deciduous and not fleshy      23

19a. Leaf-stalk bearing 2 tendrils; leaves net-veined      **VI. Liliaceae** (p. 117)

  b. Leaf-stalk without tendrils; leaves parallel-veined      20

20a. Leaves very small, scale-like or spiny, their function taken over by flattened stems (cladodes) on which the inflorescences are borne      **VI. Liliaceae** (p. 117)

  b. Plant with true leaves; cladodes absent      21

21a. Shrubs or woody climbers with scattered stem-leaves; flowers

solitary, usually large and hanging; placentation mostly parietal      **VI. Liliaceae** (p. 117)

  b. Plant without the above combination of characters      22

22a. Leaves leathery and more or less thin, if succulent, then with a spine-like or cylindric tip; flower usually green or whitish, bell- or cup-shaped or with a narrow tube and spreading lobes, often more than 1 to each bract      **VII. Agavaceae** (p. 271)

  b. Leaves succulent, usually without a spine-like or cylindric tip; flower usually red, yellow or orange, tubular, the lobes not spreading, always 1 to each bract      **VI. Liliaceae** (p. 117)

23a. Leaves very small, scale-like or spiny, their function taken over by flattened or needle-like stems (cladodes) on which the inflorescences are borne      **VI. Liliaceae** (p. 117)

  b. Plant with true leaves; cladodes absent      24

24a. Leaves evergreen, clearly stalked; flowers more than 1 to each bract, with a narrow tube as long as or longer than the spreading lobes      **VII. Agavaceae** (p. 271)

  b. Leaves deciduous, usually without distinct stalks; flowers of various shapes, rarely as above, always 1 to each bract      **VI. Liliaceae** (p. 117)

25a. Flowers solitary or in umbels; leaves broad, opposite or in a single whorl near the top of the stem      **VI. Liliaceae** (p. 117)

  b. Flowers in spikes, heads, cymes or panicles; leaves not as above      26

26a. Stamens 6 or 5–3 with 1–3 staminodes; anthers basifixed; leaves usually borne on the stems, often with closed sheaths, never grey with scales; bracts neither overlapping nor conspicuously coloured      **XIX. Commelinaceae**

  b. Stamens 6, staminodes 0; anthers dorsifixed; leaves mostly in basal rosettes, often rigid and spiny-margined, when on the stems usually grey with scales, bracts usually overlapping and conspicuously coloured      **XVIII. Bromeliaceae**

27a. Flowers radially symmetric or weakly bilaterally symmetric; stamens 6, 4, 3, or rarely many      28

  b. Flowers strongly bilaterally symmetric or asymmetric; stamens usually 5, 2 or 1 (very rarely 6)      40

28a. Unisexual climbers with heart-shaped or very divided leaves; rootstock tuberous or woody      **XIV. Dioscoreaceae** (p. 329)

  b. Plants without the above combination of characters      29

29a. Perianth persistent, variously hairy; sap usually orange      **VIII. Haemodoraceae** (p. 290)

  b. Plant without the above combination of characters      30

30a. Rooted or floating aquatics; stamens 2–12; ovules distributed all over the carpel walls (placentation diffuse-parietal)      **III. Hydrocharitaceae** (p. 114)

  b. Terrestrial or marsh plants, or epiphytes; stamens 3 or 6, rarely many; placentation axile or parietal (ovules restricted to a few rows on the carpel walls)      31

31a. Stamens 3, staminodes absent; leaves often sharply folded, their bases overlapping; style-branches often divided      **XVI. Iridaceae** (p. 332)

b. Stamens 6, or 3 plus 3 staminodes; leaves not usually as above; style-branches not divided                        32

32a. Placentation parietal; flowers in an umbel with the inner bracts long, thread-like and hanging

XIII. Taccaceae (p. 328)

b. Placentation usually axile; inflorescence and bracts not as above                                                                    33

33a. Perianth consisting of an outer, calyx-like whorl and an inner, corolla-like whorl; bracts usually overlapping and conspicuously coloured          XVIII. Bromeliaceae

b. Segments of the perianth not in 2 dissimilar whorls; bracts not as above                                                             34

34a. Ovary half-inferior                                              35

b. Ovary fully inferior                                               36

35a Anthers opening by pores        X. Tecophilaeaceae (p. 326)

b. Anthers opening by slits          VI. Liliaceae (p. 117)

36a. Leaves long-persistent, evergreen                           37

b. Leaves dying down annually                                    38

37a. Leaves fleshy or leathery, thick, rigid or flexible, spine-tipped, often with spines or teeth on the margins

VII. Agavaceae (p. 271)

b. Leaves neither fleshy, leathery nor spine-tipped, without spines or teeth on the margins

XII. Velloziaceae (p. 328)

38a. Flowers in a spike; leaves fleshy, often spotted with brown, the margins more or less rolled around each other in bud

VII. Agavaceae (p. 271)

b. Flowers in umbels or solitary; leaves not usually fleshy or spotted with brown but flat, pleated or with the margins folded outwards in bud                                          39

39a. Leaves mostly basal, densely hairy, pleated or with prominent veins               XI. Hypoxidaceae (p. 326)

b. Leaves various, not usually densely hairy, pleated or with prominent veins, basal or not   IX. Amaryllidaceae (p. 291)

40a. Fertile stamens 6; perianth-segments all similar, tube curved and unevenly swollen; stems below ground, fleshy

VII. Agavaceae (p. 271)

b. Fertile stamens 5, 2 or 1, very rarely 6; staminodes, which may be petal-like, often present; perianth-segments usually differing among themselves; fleshy underground stems rare

41

41a. Fertile stamens 2 or 1, united with the style to form a column; pollen usually borne in masses (pollinia); leaf-veins, when visible, all parallel to margins    XXXIII. Orchidaceae

b. Fertile stamens 5 or 1, rarely 6, not united to the style; pollen granular; leaves with a distinct midrib more or less parallel to the margins, the secondary veins parallel, running from the midrib to the margins                                       42

42a. Fertile stamens 5 or rarely 6                                  43

b. Fertile stamen 1, the remainder transformed into petal-like staminodes                                                          44

43a. Leaves and bracts spirally arranged; flowers unisexual

XXVIII. Musaceae

b. Leaves and bracts in 2 ranks; flowers bisexual

XXIX. Strelitziaceae

44a. Fertile stamen with normal structure, not petal-like

XXX. Zingiberaceae

b. Fertile stamen in part petal-like, and with only 1 pollen-bearing anther-lobe                                                  45

45a. Leaf-stalk with a swollen band (pulvinus) at the junction with the blade; ovary smooth, with 1–3 ovules

XXXII. Marantaceae

b. Leaf-stalk without a pulvinus at the junction with the blade; ovary usually warty, with numerous ovules

XXXI. Cannaceae

Many monocotyledonous plants which are woody, or form large or small rosettes of fleshy, leathery, long lived leaves, flower only irregularly, or at long intervals, or take some years to reach flowering size. These plants will therefore not be easily identifiable with the key above. The following informal key will provide some guidance for the identification of the families to which such plants belong.

*Habit*. Plants tree-like, with a distinct, woody trunk; **Liliaceae, Agavaceae, Palmae, Pandanaceae**; plants tree-like with a trunk which is not woody, but given rigidity by being made up of overlapping leaf-bases (sheaths): **Musaceae, Strelitziaceae**; plants shrubby, with woody stems but without a distinct, tall trunk: **Liliaceae, Agavaceae, Strelitziaceae**; plants forming large or small rosettes at substrate level, or borne on the ends of the branches of a short, erect or prostrate woody stem: **Liliaceae, Agavaceae, Bromeliaceae, Palmae** (young specimens). Plant with conspicuous stilt-roots: **Pandanaceae**.

*Leaf arrangement*. Leaves borne on 2 sides of the stem only: **Strelitziaceae**; leaves borne all round the stem, or in a crown at the stem apex, or in a basal rosette: **Liliaceae, Agavaceae, Bromeliaceae, Palmae, Musaceae**; leaves borne all round the stem in a crown at the apex, very regularly arranged, so that their spiral arrangement is clearly visible: **Pandanaceae**.

*Leaves*. Pleated, usually palmately or pinnately divided or lobed: **Palmae**; not pleated, entire, with a well-developed midrib running parallel with the margins, from which the parallel, lateral veins diverge: **Musaceae, Strelitziaceae**; not pleated, entire, usually leathery, fleshy and persistent, venation not visable, or not as above: **Liliaceae, Agavaceae, Pandanaceae**; neither pleated nor fleshy, but thin, leathery and persistent: **Agavaceae, Bromeliaceae**. With solid, spine-like or cylindric tips: **Agavaceae**. With spiny margins: **Agavaceae, Bromeliaceae, Pandanaceae**. Margins splitting off as threads: **Agavaceae**.

# I. ALISMATACEAE

Aquatic or marsh herbs, usually perennial. Leaves entire, alternate or basal. Flowers bisexual or unisexual, radially symmetric, borne in umbels, racemes or panicles. Sepals 3, free. Petals 3, free, often falling early. Stamens 3–many; anthers opening by slits. Ovary superior; carpels 3 to many, free or united at the base; ovules usually 1, with basal placentation. Fruit a group of achenes.

A small but cosmopolitan family of 13 genera and 90 species, some of which are cultivated in and around ornamental pools or in aquaria.
Literature: Buchenau, F., Alismataceae, *Das Pflanzenreich* **16**: 1–66 (1903).

*Leaves*. Sagittate: **1**.
*Flowers*. Some unisexual: **1**; all bisexual: **2,3,4,5**.
*Stamens*. Six: **3,4,5**; more than six: **1,2**.
*Carpels*. Spirally arranged: **1–3**; in a single, sometimes irregular whorl: **4,5**.
*Style*. Apical: **3,4**; lateral: **5**.
*Achenes*. Laterally compressed: **1,5**; ovoid: **3,4**; winged: **1**; ribbed: **2–4**.

1a. Leaves sagittate      **1. Sagittaria**
  b. Leaves not sagittate      **2**
2a. Some flowers unisexual; achenes winged      **1. Sagittaria**
  b. Flowers all bisexual; achenes ribbed or furrowed but not winged      **3**
3a. Stamens 9 or more      **2. Echinodorus**
  b. Stamens 6      **4**
4a. Achenes with 3 ribs on the back and a double inner rib, spirally arranged in a spherical head      **3. Baldellia**
  b. Achenes otherwise ribbed, in a single, sometimes irregular whorl      **5**
5a. Achenes 6–15, oblong-ovoid, in a hemispherical head; style apical      **4. Luronium**
  b. Achenes 11–28, laterally compressed, in a more or less flat head; style lateral      **5. Alisma**

## 1. SAGITTARIA Linnaeus
*C.J. King*
Rootstock often having stolons and tubers. Leaves aerial, floating or submerged. Some flowers unisexual, usually in whorls of 3, forming racemes or panicles with female or bisexual flowers at the base and male flowers above, or occasionally with the flowers all male or all female. Petals white. Stamens usually numerous. Carpels numerous, each with 1 ovule, spirally arranged on the receptacle, developing into

a head of laterally compressed, winged, beaked achenes.

A genus of about 20 species, mostly from America. Several species are in cultivation as waterside plants or in aquaria. They need a moist or wet loamy soil and are readily increased by division, seed or underground tubers when produced.
Literature: Bogin, C., Revision of the genus Sagittaria (Alismataceae), *Memoirs of the New York Botanical Garden* **9**: 179–233 (1955); Rataj, K., Revision of the genus Sagittaria, *Annotationes zoologicae et botanicae (Bratislava)* **76**: 1–31 (1972) & **78**: 1–61 (1972).

*Stolons*. Present: **2,3,5–7**.
*Leaf-blades*. Sagittate: **1–3**.
*Flowers*. 1.2 cm or less in diameter: **5**; 1.5–2.5 cm in diameter: **2, 6**; 3 cm or more in diameter: **1,3,4,7**.
*Petals*. With a purple blotch at base: **1,2**.
*Stamens*. eleven or fewer: **1,6**; twelve to nineteen: **1,4,5**; twenty or more: **2–5,7**.
*Filaments*. Hairy: **4,5**; dilated at base: **5,6**.

1a. Aerial leaves sagittate      2
  b. Aerial leaves not sagittate      4
2a. Stolons absent; petals usually with a purple blotch at base      **1. montevidensis**
  b. Stolons present; petals with or without a blotch      3
3a. Flowers to 2.5 cm in diameter; petals usually with a purple blotch at base      **2. sagittifolia**
  b. Flowers 3–4 cm in diameter; petals without a blotch      **3. latifolia**
4a. Stolons absent; filaments hairy      **4. lancifolia**
  b. Stolons present; filaments hairy or not      5
5a. Flowers to 1.2 cm in diameter; filaments hairy      **5. graminea**
  b. Flowers larger; filaments hairless      6
6a. Flowers 1.5–2 cm in diameter; stamens 7–10      **6. subulata**
  b. Flowers 3–4 cm in diameter; stamens 20–30      **7. macrophylla**

**1. S. montevidensis** Chamisso & Schlechtendal. Illustration: Botanical Magazine, 6755 (1884).
Rootstock without stolons. leaf-stalks to *c.* 20 cm, blades to 15 × 10 cm, broadly to narrowly sagittate, submerged leaves to 40 × 1 cm, linear. Stem 60–100 cm. Flowers *c.* 5 cm in diameter, in 5–9 whorls. Petals usually each with a dark purple blotch bordered with yellow at the base. Stamens 9–15; filaments rough with short,

stiff hairs. *S America*. G2. Summer.
Suitable for the tropical aquarium.

**2. S. sagittifolia** Linnaeus. Illustration: Hegi, Illustriertes Flora von Mittel-Europa **1**: t. 20 (1935); Ross-Craig, Drawings of British plants **31**: t. 10 (1973); Phillips, Wild flowers of Britain, 133(f) (1977); Grey-Wilson & Mathew, Bulbs, t. 1 (1981).
Rootstock with stolons and tubers which are 1–1.5 cm. Leaf-stalks to 45 cm, blades to 25 × 22 cm, broadly to narrowly sagittate, submerged leaves to 80 × 2 cm, linear. Stem to 1 m. Flowers to 2.5 cm in diameter, in 3–5 whorls. Petals usually with a purple blotch at the base. Stamens 20–25; filaments linear, hairless. *Europe, Asia*. H2. Summer.

The asiatic form with pure white petals is sometimes referred to as var. *leucopetala* Miquel; a double form of this and the typical variety bear the cultivar name 'Flore Pleno'.

**3. S. latifolia** Willdenow (*S. gracilis* Pursh; *S. hastata* Pursh; *S. simplex* Pursh). Illustration: Addisonia **2**: t. 54 (1917); Hess et al., Flora der Schweiz **1**: 212 (1967); Hay & Beckett, Reader's Digest encyclopaedia of garden plants and flowers, 636 (1978).
Rootstock with stolons and large tubers. leaf-stalks to 40 cm, blades 10–30 × 8–20 cm, broadly to narrowly sagittate, submerged juvenile leaves linear. Stem to 1.2 m. Flowers 3–4 cm in diameter, in 4–8 whorls. Stamens 20–40; filaments linear, hairless. *N America*. H2. Summer.

The most variable species in the genus with numerous names for the various wild variants. A double-flowered variant, 'Flora Pleno', is found in cultivation.

**4. S. lancifolia** Linnaeus (*S. falcata* Pursh). Illustration: Botanical Magazine, 1792 (1816); Addisonia **12**: t. 406 (1927).
Rootstock without stolons. leaf-stalks to 1 m, blades 15–45 cm, narrowly to broadly lanceolate, the midrib and 2 outer veins arising from the base, the other veins arising higher up the midrib. Stem to 2 m. Flowers 3–5 cm in diameter, in 5–12 whorls. Stamens 12–30; filaments linear, hairy. *SE USA, West Indies, C & northern S America*. H5. Summer.

**5. S. graminea** Michaux. Illustration: Das Pflanzenreich **16**: 55 (1903); Rickett, Wild flowers of the United States **2**: t. 16 (1967).
Rootstock with creeping stolons. leaves long-stalked, blades 5–20 × 1.5–6 cm, linear to narrowly elliptic, the veins all arising from the base, or entirely linear if

submerged. Stem to 50 cm or more. Flowers 1.2 cm in diameter, in 5–7 whorls, stamens 12–20; filaments dilated at the base, hairy. *E USA*. H3. Summer.

A variant with the submerged leaves terete instead of ribbon-like is known as var. **teres** (Watson) Bogin or is sometimes treated as a separate species, *S. teres* Watson.

**6. S. subulata** (Linnaeus) Buchenau (*S. filiformis* Smith; *S. natans* Michaux not Pallas). Illustration: Muenscher, Aquatic plants of the United States, 96 (1944). Rootstock with stolons. Leaves very variable, mostly linear, to *c.* 100 × 1.5 cm, submerged, rarely terminating in a floating blade, 5 × 2.5 cm. Stem to 30 cm. Flowers 1.5–2 cm in diameter, in 1–3 whorls. Stamens 7–10; filaments dilated at the base, hairless. *E USA*. H3. Summer.

**7. S. macrophylla** Zuccarini
Rootstock with stolons. Leaf-stalks to 50 cm, blades to 25 × 10 cm, lanceolate to ovate, sometimes truncate or hastate at base, submerged leaves to 40 × 1.5–4 cm, linear. Stem to 1 m. Flowers 3–4 cm in diameter, in 3–5 whorls. Stamens 20–30; filaments linear to spindle-shaped, hairless. *Mexico*. G1. Summer.

Suitable for tropical and subtropical aquaria.

**2. ECHINODORUS** Engelmann
*C.J. King*
Rootstock sometimes bearing stolons. leaves aerial, floating or submerged. Flowers bisexual, arranged in several simple whorls or the inflorescence compound with branches bearing whorls of flowers. Petals white. Stamens 6–30. Carpels several to many, spirally arranged on the receptacle, developing into a head of ribbed, beaked achenes.

A genus of about 47 species from the warmer regions of America. Several species are cultivated in heated aquaria. They require a loamy soil containing peat and sand, and are propagated by division or seed.

*Stolons*. Present: **1,2**.
*Leaf-blades*. Broadly ovate: **3**.
*Whorls of flowers*. one: **1,2**; two to eight: **3–5**; ten or more: **6**.
*Flowers*. Less than 1 cm in diameter: **1,2**; more than 2 cm in diameter: **5**.
*Flower-stalks*. to 1 cm: **2,4,6**; 1–2.5 cm: **1,5**; more than 2.5 cm: **3**.
*Stamens*. nine: **1,2**; ten to twelve: **3–5**; more than twelve: **3,5,6**.

*Achenes*. 1 mm: **1**; 1.5–3 mm: **2,3,5,6**; more than 3 mm: **3,4**.

1a. Stolons present                                    2
  b. Stolons absent                                     3
2a. Leaf-blades not more than 3 cm
                                    **1. tenellus**
  b. Leaf-blades 4–8 cm
                              **2. magdalenensis**
3a. Leaf-blades broadly ovate
                                  **3. cordifolius**
  b. Leaf-blades linear to lanceolate or elliptic                                            4
4a. All main veins arising from base of leaf-blade                                         5
  b. Some main veins arising higher up the midrib            **4. brevipedicellatus**
5a. Leaf-blades linear-lanceolate to elliptic, more than 3 times as long as broad                   **5. paniculatus**
  b. Leaf-blades elliptic to oblong, less than 3 times as long as broad
                                   **6. longistylis**

**1. E. tenellus** (Martius) Buchenau. Illustration: Muenscher, Aquatic plants of the United States, 83 (1944); de Wit, Aquarium plants, 151 (1964). Rootstock bearing stolons, mat-forming. Leaf-stalks to 5 cm, blades 2–3 × 1.2 cm, linear-lanceolate to oblong, submerged leaves to 10 cm × 2 mm, linear. Stem to 10 cm. Flowers to 1 cm in diameter, 3–6 in a single whorl. Flower-stalks 1.2–2.5 cm. Stamens 9. Achenes 1 mm, each with a short, hooked beak. *C & E USA, West Indies, C & S America*. G2. Summer.

**2. E. magdalenensis** Fassett. Illustration: de Wit, Aquarium plants, 144 (1964). Rootstock with many stolons. leaf-stalks 5–10 cm, sometimes longer, blades 4–8 × 1–4 cm, lanceolate to narrowly ovate. Stem to 30 cm. Flowers *c.* 5 mm in diameter, 4–7 in a single whorl. Flower-stalks 1 cm. Stamens 9. Achenes 1.5–2 mm, with a short beak. *Colombia*. G2. Summer.

**3. E. cordifolius** (Linnaeus) Grisebach (*E. radicans* (Nuttall) Engelmann). Illustration: Muenscher, Aquatic plants of the United States, 83 (1944); Rickett, Wild flowers of the United States 6: t. 15 (1973). Leaf-stalks 5–15 cm, blades 10–20 × 6–10 cm, broadly ovate, base cordate to truncate. Stem to 1 m or more. Flowers 1.2–1.8 cm in diameter, in 2–6 whorls, 5–15 flowers in each whorl. Flower-stalks to 6 cm. Stamens 12–20. Achenes 1.8–3.5 mm, with a short beak. *S, C & E USA, Mexico*. G1. Summer.

**4. E. brevipedicellatus** (Kuntze) Buchenau. Illustration: de Wit, Aquarium plants, 138 (1964). Leaf-stalks 5–15 cm, blades 20–30 × 2–4 cm, narrowly lanceolate, some main veins arising from the base, others arising higher up the midrib. Stem to 1 m. Flowers 1–2 cm in diameter, in 4–6 whorls, 3–8 flowers in each whorl. Flower-stalks 1 cm. Stamens 12. Achenes 3–4 mm. *Brazil*. G2. Summer.

**5. E. paniculatus** Micheli. Leaf-stalks 10–30 cm, winged, blades 25–45 × 3–8 cm, linear-lanceolate to elliptic, sometimes sickle-shaped, main veins all arising from the base. Stem to 2 m. Flowers 1.2–2.5 cm in diameter, in 5–8 whorls, 6–8 flowers in each whorl. Flower-stalks 1.5 cm. Stamens 10–24. Achenes 2 mm, each with a short beak. *Brazil & Venezuela to Paraguay*. G2. Summer.

**6. E. longistylis** Buchenau. Leaf-stalks to 50 cm, blades 15–25 × 10 cm, elliptic to oblong, main veins all arising from the base. Stem to 1 m. Flowers 1–2 cm in diameter, in 10–15 whorls, 6–9 flowers in each whorl. Flower-stalks 1 cm. Stamens numerous. Achenes 2.5 mm, each with a long beak. *Brazil*. G2. Summer.

**3. BALDELLIA** Parlatore
*C.J. King*
Rootstock sometimes bearing stolons. Leaves elliptic to lanceolate or linear-lanceolate. Flowers bisexual, in 1–3 whorls in umbels or racemes. Stamens 6. Carpels numerous, free, each with 1 ovule, spirally arranged in a spherical head. Achenes ovoid, longitudinally 5-ribbed (3 ribs on the back and an inner double rib), each with a short apical beak.

A genus of 2 species from Europe and N Africa; the more widespread species is cultivated in an unheated aquarium or in a bog garden.

**1. B. ranunculoides** (Linnaeus) Parlatore (*Alisma ranunculoides* Linnaeus; *Echinodorus ranunculoides* (Linnaeus) Engelmann). Illustration: Bonnier, Flore complète **10**: t. 570 (1929); Ross-Craig, Drawings of British plants **31**: t. 8 (1973); Grey-Wilson & Mathew, Bulbs, t. 1 (1981). Leaf-stalks to 25 cm, blades to 10 × 1.5 cm or sometimes larger, decurrent, submerged leaves linear. Stem to 30 cm. Flowers 1–1.5 cm in diameter. Petals white or pale pink with a yellow blotch at the base.

Achenes 2–3.5 mm. *Europe northwards to S Scandinavia, N Africa.* H2. Summer.

### 4. LURONIUM Rafinesque
*C.J. King*

Stems to 50 cm or more, floating or creeping and rooting at the nodes. Floating leaves long-stalked, blades to 4 × 1.5 cm, elliptic to ovate, submerged leaves linear. Flowers bisexual, 1.2–1.5 cm in diameter, long-stalked. Petals white with a yellow blotch at the base. Stamens 6. Carpels 6–15 in an irregular whorl, free, each with 1 ovule. Achenes oblong-ovoid, 2.5 mm, with 12–15 longitudinal ribs and a short, apical beak.

A genus of 1 species, formerly known as *Elisma* Buchenau, cultivated in unheated aquaria. It requires a loamy soil containing peat and sand, and is propagated by division.

**1. L. natans** (Linnaeus) Rafinesque (*Alisma natans* Linnaeus; *Elisma natans* (Linnaeus) Buchenau). Illustration: Hegi, Illustriertes Flora von Mittel-Europa **1**: t. 19 (1936); Ross-Craig, Drawings of British plants **31**: t. 9 (1973); Fitter, The Wild flowers of Britain and northern Europe, 259 (1974); Grey-Wilson & Mathew, Bulbs, t. 1 (1981).
*Europe.* H2. Summer.

### 5. ALISMA Linnaeus
*C.J. King*

Leaves aerial, floating or submerged. Flowers bisexual, in panicles or occasionally in racemes or umbels. Stamens 6. Carpels 11–28 in a single whorl, free, each with 1 ovule. Achenes laterally compressed, each with a short beak.

A genus of 9 species, almost cosmopolitan, but indigenous in the northern hemisphere. A few species are cultivated as waterside plants, and some are suitable for unheated aquaria. They will grow in any good wet soil, but prefer one that contains abundant organic matter. Propagation is by division of the rootstock or by seed.
Literature: Hendricks, A.J., A revision of the genus Alisma (Dill.) L., *American Midland Naturalist* **58**: 470–93 (1957); Bjorkqvist, I., Studies in Alisma L., *Opera Botanica* **17**: 1–128 (1967) & **19**: 1–138 (1968).

1a. Styles shorter than the ovaries, recurved          **1. gramineum**
 b. Styles as long as or longer than the ovaries, more or less erect          2
2a. Leaves lanceolate to elliptic, cuneate

at base; petals purplish pink
          **2. lanceolatum**
 b. Leaves ovate or elliptic-ovate to lanceolate, usually cordate or truncate at base; petals white or purplish white          **3. plantago-aquatica**

**1. A. gramineum** Lejeune (*A. plantago-aquatica* Linnaeus subsp. *graminifolium* (Steudel) Hegi). Illustration: Bonnier, Flore complète **10**: t. 571 (1929); Grey-Wilson & Mathew, Bulbs, t. 1 (1981).
Aerial leaves long-stalked, blades to 10 × 4.5 cm, elliptic or narrowly oblong, submerged leaves to 100 × 1.5 cm, linear. Stem to 1 m. Flowers about 6 mm in diameter. Petals white or purplish white. Anthers almost circular. Styles shorter than the ovaries, recurved. Achenes 2–2.75 mm. *Europe, Asia, N Africa.* H2. Summer.

A variant suitable for the aquarium, with entirely submerged leaves (forma **submersum** Gluck), is maintained by using sand as the base instead of soil.

**2. A. lanceolatum** Withering. Illustration: Bonnier, Flore complète **10**: t. 571 (1929); Ross-Craig, Drawings of British plants **31**: t. 7 (1973).
Leaves aerial, long-stalked, blades to 27 × 7.5 cm, lanceolate to elliptic, tapered at base. Stem 30–45 cm. Flowers 6–9 mm in diameter. Petals purplish pink. Anthers elliptic. Styles as long as or longer than the ovaries, more or less erect. Achenes 2–3 mm. *Europe, Asia, N Africa.* H2. Summer.

**3. A. plantago-aquatica** Linnaeus. Illustration: Bonnier, Flore complète **10**: t. 570 (1929); Ross-Craig, Drawings of British plants **31**: t. 6 (1973); Grey-Wilson & Mathew, Bulbs, t. 1 (1981).
Leaves usually aerial, long-stalked, blades to 30 × 12 cm, ovate or elliptic-ovate to lanceolate, mostly more or less cordate or truncate at base, but sometimes tapered. Stem 20–100 cm. Flowers to *c.* 1 cm in diameter. Petals white or purplish white. Anthers elliptic. Styles as long as or longer than the ovaries, more or less erect. Fruiting heads to 6 mm in diameter. Achenes usually 2–3 mm. *N temperate regions.* H2. Summer.

Two varieties are occasionally cultivated, var. **americanum** Schultes & Schultes (*A. triviale* Pursh) from N America and temperate E Asia, with white petals and fruiting heads 4–4.5 mm in diameter, and var. **parviflorum** (Pursh) Torrey (*A.*

*parviflorum* Pursh; *A. subcordatum* Rafinesque) from eastern N America, with flowers 3–3.5 mm in diameter and fruiting heads *c.* 3.5 mm in diameter.

## II. BUTOMACEAE

Annual or perennial, aquatic herbs, with rhizomes. Leaves basal, alternate, not submerged. Flowers bisexual, in axillary clusters or umbels which are long-stalked; more rarely flowers solitary. Sepals 3, sometimes petaloid. Petals 3, larger than the sepals. Stamens 6–many, anthers opening by longitudinal slits. Carpels 6–many, free or united at the base, with many ovules with marginal or parietal placentation. Fruit a whorl of follicles.

A family of 4 genera and 13 species with a scattered distribution in temperate and tropical regions of all continents. In this treatment, *Limnocharis* and *Hydrocleis* are included in the Butomaceae, although some authorities place them in a distinct family, *Limnocharitaceae*. Only 3 genera are in general cultivation in Europe, in ponds and aquaria.

1a. Leaves linear, without distinct blades; sepals petaloid, pink          **1. Butomus**
 b. Leaves with distinct blades; sepals green, not petaloid          2
2a. Leaves floating, stalks with transverse septa          **2. Hydrocleis**
 b. Leaves usually erect, above the water, stalks without transverse septa          **3. Limnocharis**

### 1. BUTOMUS Linnaeus
*A. Brady*

Perennial herbs with thick, creeping rhizomes. Leaves linear, triangular in cross-section in the lower part. Umbels many-flowered; flower-stalks 5–10 cm. Sepals petaloid, pink. Stamens 6–9.

A genus of a single species from temperate regions of Europe and Asia. It is suitable for cultivation in bog gardens, watercourses with still or slow-moving water, and at pool margins. Propagation is by division of the rootstock or by seeds.

**1. B. umbellatus** Linnaeus. Illustration: Smith & Sowerby, English Botany, t. 1441 (1805); Botanical Magazine, 9125 (1926); Keble Martin, The concise British flora in colour, 79 (1965); Cook, Water plants of the world, 166 (1974).
Leaves 50–100 cm × 3–10 mm, apex acuminate. Flowering stem to 1 m. Sepals and petals 1–1.5 cm; sepals greenish

outside, sepals inside and petals pinkish white with darker pink veins. *Eurasia*. H1. Summer.

## 2. HYDROCLEIS Richard
*E.C. Nelson*

Annual or perennial herbs, with stolons. Leaves basal or in groups along the stems; stalks with transverse septa. Flowers in axillary clusters or in terminal umbels or solitary. Petals yellow. Stamens 6–many; staminodes present or absent. Carpels 3 or 6, united at the base.

A genus of about 9 species from S America. The plants may be free-floating or sometimes rooted in shallow water. Only 1 species is grown in aquaria in Europe. It will thrive in a rich loam in about 20 cm of water; it is not frost-hardy, but may survive out of doors in mild winters. Propagation is by cuttings.

**1. H. nymphoides** (Willdenow) Buchenau. Illustration: Botanical Magazine, 3248 (1833); Cook, Water plants of the world, 298 (1974).
Perennial herb with stolons, sometimes rooting in shallow water. Leaves 5–7 cm in diameter, oval with cordate bases, thick, floating, deep green. Flowers yellow, bowl-shaped, to 5 cm in diameter. *Tropical S America*. H5–G1. Summer.

## 3. LIMNOCHARIS Humboldt & Bonpland
*E.C. Nelson*

Annual or perennial herbs without stolons. Leaves in rosettes; stalks triangular in cross-section, without transverse septa. Umbel with 2–15 flowers. Petals yellow. Stamens numerous, surrounded by staminodes. Carpels numerous, laterally flattened, closely adpressed and apparently forming a single ovary.

The 2 species are natives of the warmer regions of America, but are now naturalised elsewhere. They are tender, so should be grown indoors in aquaria or tubs, or in shallow pools. Propagation is by seed, offsets or suckers. Only 1 species is well known in European gardens.

**1. L. flava** (Linnaeus) Buchenau. Illustration: Cook, Water plants of the world, 299 (1974).
Perennial with erect stems. Leaves to 60 cm, blades *c.* 20 cm, ovate with cordate bases. Umbels with 6–12 flowers. Petals pale yellow bordered with white, to 2.5 cm in diameter. *West Indies, C & tropical S America*. G2. Summer.

# III. HYDROCHARITACEAE

Submerged or floating, annual or perennial herbs. Leaves alternate, opposite or whorled, sometimes with minute scales at the nodes. Flowers bisexual or unisexual, 1 or more together subtended by spathes formed from 1 or 2 united or free bracts. Sepals 3. Petals 3 or absent. Stamens 2–20, anthers usually opening by longitudinal slits. Ovary inferior, of 3–6 united carpels; ovules numerous on intrusive parietal placentas; styles 3–15. Fruit berry-like.

A family of 15 genera distributed throughout the world but with most of the species in tropical areas. Although most species grow in fresh water habitats, some are marine. Species of 9 genera are commonly cultivated in aquaria and water gardens.

*Habit*. Free-floating: **5,6,9**.
*Leaves*. All basal: **4,6–9**; arranged along the stem: **1–3,5**. Linear: **1–3,5,9**. Rigid: **8**. Stalked: **4,6,7**. Spirally arranged: **5**.
*Flowers*. Conspicuous: **1,4–8**. Petals white: **1,4,7,8**; petals pinkish: **5**.

1a. Leaves with distinct stalks, in basal rosettes  2
  b. Leaves arranged along the stem, or, if in basal rosettes, then stalkless  4
2a. Plants rooted in the substrate, without stolons; spathes tubular, ribbed or winged  **7. Ottelia**
  b. Plants floating, with stolons; spathes not ribbed or winged  3
3a. Leaves with spongy tissue beneath; petals absent, or, when present, at most $1\frac{1}{2}$ times as long as sepals  **6. Limnobium**
  b. Leaves without spongy tissue beneath; petals present and more than $1\frac{1}{2}$ times as long as the sepals  **4. Hydrocharis**
4a. Leaves in a basal rosette  5
  b. Leaves distributed along the stems  6
5a. Leaves rigid, margins with prominent, spiny teeth; plant aloe-like  **8. Stratiotes**
  b. Leaves not rigid but ribbon-like, margins without prominent, spiny teeth  **9. Vallisneria**
6a. All leaves spirally arranged  **5. Lagarosiphon**
  b. Leaves opposite or whorled, at least in the upper part of the stem  7
7a. Flowers conspicuous, held above the water surface, with nectaries  **1. Egeria**

  b. Flowers minute, submerged or borne at the water surface, without nectaries  8
8a. Leaves in whorls of 3–8, with fringed nodal scales  **3. Hydrilla**
  b. Leaves in whorls of 3–4, with inconspicuous, unfringed nodal scales  **2. Elodea**

## 1. EGERIA Planchon
*E.C. Nelson*

Dioecious or monoecious, aquatic, perennial herbs. Leaves submerged, whorled, stalkless, linear. Flowers conspicuous, with nectaries, held above the water surface. Male spathes 2–4-flowered; petals of male flowers much longer than sepals, stamens 9. Ovary stalkless, styles 3, free, bifid or occasionally 3-fid.

A genus of 2 species from warm temperate regions of S America, now naturalised in Europe and elsewhere. Both are cultivated in ponds and aquaria, and are sometimes listed in *Elodea*, but differ in their flowers: in *Egeria* these are conspicuous and held above the water surface. Propagation is by cuttings.

1a. Leaves to 4 cm; plants robust  **1. densa**
  b. Leaves at most 2.5 cm; plants very slender  **2. najas**

**1. E. densa** Planchon (*Elodea densa* (Planchon) Caspary; *Elodea canadensis* Michaux var. *gigantea* Bailey). Illustration: Cook, Water plants of the world, 257 (1974).
Stems stout, to 1 m. Leaves crowded in whorls of 3–5, to 4 cm × 5 mm, minutely toothed. Sepals 3–4 mm. Petals 8–11 mm, obovate, white. *S America*. H1. Summer.

A good aquarium plant, an excellent oxygenator.

**2. E. najas** Planchon (*Elodea najas* (Planchon) Caspary).
Stems very slender. Upper leaves in whorls of 4–6, linear, to 2.5 cm × 1.5 mm, sharply toothed. Petals *c.* 7 mm, white. *S America*. H5. Summer.

Not as commonly cultivated as *E. densa*.

## 2. ELODEA Michaux
*E.C. Nelson*

Dioecious or monoecious, aquatic perennial herbs, often with stolons. Leaves submerged, whorled or sometimes opposite, stalkless, usually linear, with 2 minute scales at each node. Flowers inconspicuous, without nectaries, borne and pollinated at the water surface. Spathes axillary, stalkless, with 1 or rarely 3 male flowers, or 1 female flower, or rarely 1 bisexual flower. Sepals 3. Petals 3. Stamens 3–9. Ovary stalkless, styles 3, free, bifid or occasionally entire.

A genus with poorly understood limits. Some plants commonly included in it belong to other genera (e.g. *Egeria*, *Hydrilla*). There are perhaps 15 species, all native to America, but now widely naturalised in other countries, where they are sometimes serious weeds of aquatic habitats. There are 3 species which are widely cultivated in ponds and aquaria; they are easily grown and propagated by cuttings or stolons.

1a. Leaves less than 2 mm broad; sepals of female flowers less than 2 mm
    **3. nuttallii**
  b. Leaves usually at least 2 mm broad; sepals more than 2 mm     **2**
2a. Leaves *c.* 10 × 2.5 mm, rounded at the apex     **1. canadensis**
  b. Leaves *c.* 2.5 cm × 2.5 mm, tapered to acute at the apex   **2. callitrichoides**

**1. E. canadensis** Michaux. Illustration: Cook, Water plants of the world, 257 (1974).

Leaves usually *c.* 10 × 2 mm, but the largest may be up to 17 × 4 mm, in whorls of 3, or the lower opposite, all linear to oblong, minutely toothed, rounded at the apex. Sepals of female flowers 2–2.7 mm. Petals white or pale purple. *N America*. H1. Summer.

A common aquarium plant, now widely naturalised in Europe, having been introduced into Ireland in 1834.

**2. E. callitrichoides** (Richard) Caspary (*E. ernstae* St John).

Leaves to 2.5 cm × 2.5 mm, in whorls of 3 or the lowest opposite, all gradually tapering to an acute apex, minutely toothed. Sepals of female flowers 3–3.5 mm. Petals white. *S America*. H4. Summer.

Like the other species, this is naturalised in Europe. It is not as hardy as *E. canadensis*. For the use of the name see Cook, Watsonia **15**: 117 (1984).

**3. E. nuttallii** (Planchon) St John.
Leaves to 1.5 cm, less than 2 mm broad, in whorls of 3 or occasionally 4, or the lowest opposite, all linear, tapering to an acute apex, minutely toothed. Sepals of female flowers 1–1.8 mm. Male flowers breaking free and floating to the water surface. *N America*. H1. Summer.

This plant is sometimes confused with *Hydrilla verticillata* and the name *E. nuttallii* has been misapplied to *Hydrilla*.

**3. HYDRILLA** Richard
*M.J.P. Scannell*
Dioecious, submerged, branched herbs with stolons, overwintering as turions. Leaves 1–4 cm × *c.* 2.5 mm, in whorls of 3–8 or the lowest opposite, lanceolate, each with a single midvein, stalkless, minutely toothed, with 2 brown, fringed scales, *c.* 0.5 mm in each axil. Flowers borne and pollinated at the water surface, solitary, emerging from a tubular spathe. Sepals 3. Petals 3. Male flowers breaking free and floating on the water surface at maturity; stamens 3. Female flowers stalkless, with an elongate perianth-tube (to 4.5 cm), sepals 1.5–3 × 0.3–0.5 mm, petals smaller, staminodes 3, styles not apparent, stigmas 3 (rarely to 5), ovary red. Fruit cylindric.

A misunderstood genus of a single species, widely distributed in isolated temperate and tropical habitats; sometimes cultivated in fresh-water aquaria or ponds. Propagation is by cuttings or by turions. Literature: Scannell, M.J.P. & Webb, D.A., The identity of the Renvyle Hydrilla, *Irish Naturalists' Journal* **18**: 327–30 (1976).

**1. H. verticillata** (Linnaeus) Royle (*Elodea nuttallii* misapplied). Illustration: Irish Naturalists' Journal **18**: t. 1 2 & f. 1 5 (1976).
*Europe (British Isles, Germany to USSR), E Asia, E Africa, Australia*. H2. Summer.

European plants are hardy and can be easily cultivated. The plant is a good oxygenator.

**4. HYDROCHARIS** Linnaeus
*E.C. Nelson*
Aquatic perennial herbs. Leaves stalked, all basal. Flowers unisexual, held above the water. Male spathes stalked, with 1–4 flowers. Female spathes stalkless, each 1-flowered. Petals much longer than sepals. Stamens 9–12, the outer usually reduced to staminodes. Styles 6, bifid, free.

A genus of 2 species, of which only 1 is widely cultivated in Europe. Propagation is by cuttings, stolons or seeds. The plants will grow floating in deep water, but will sometimes become rooted in mud in shallow water. They are suitable for still-water habitats, ponds and aquaria.

**1. H. morsus-ranae** Linnaeus. Illustration: Smith, English botany, 1444 (1805); Botanical Magazine, 3248 (1833); Keble

Martin, The concise British flora in colour, 79 (1965); Cook, Water plants of the world, 261 (1974).

Leaves circular to kidney-shaped, entire, cordate at base. Petals *c.* 1 cm, obovate, white, each with a yellow spot at the base. *Europe, W Asia, N Africa*. H1. Summer.

**5. LAGAROSIPHON** Harvey
*E.C. Nelson*
Dioecious, submerged plants. Leaves stalkless, each with 2 minute scales at the node. Flowers submerged until open, pollinated at the water surface. Male spathe with many flowers; stamens 3, staminodes 3. Female spathe usually with 1 flower; ovary stalkless, styles 3, free, bifid. Sepals 3, petals 3.

A genus of about 15 species, native in tropical and subtropical regions of Africa, Madagascar and India. Only 1 species is cultivated in Europe, where it has escaped from cultivation and become naturalised. It will grow easily in aquaria and ponds in milder regions, but thrives best when the water temperature is 12–18 °C throughout the year. Propagation is by cuttings.

**1. L. major** (Ridley) Moss (*L. muscoides* Harvey; *Elodea crispa* misapplied). Illustration: Cook, Water plants of the world, 259 (1974).
Stems to 60 cm, sometimes branched. Leaves dark green, linear, to 3 cm × 3 mm, densely arranged at the tips of the branches, reflexed, minutely toothed. Perianth-tube to 3.5 cm; sepals and petals tinted pink. *Southern Africa*. H4. Summer.

The plant resembles a large *Elodea* and is sometimes incorrectly called *E. crispa*. It is a good oxygenator, but requires plenty of light.

**6. LIMNOBIUM** Linnaeus
*E.C. Nelson*
Floating perennial herbs, with stolons. Leaves all basal, with distinct stalks; blade with a layer of spongy tissue beneath. Flowers unisexual, held above the water surface. Male spathes with 3–6 flowers; stamens 6–12, filaments united. Female spathes with 1 flower; ovary 6–9-celled, stigmas as many as cells, bifid. Sepals 3. Petals 3. Fruit many-seeded.

A genus of about 3 species, found in the warmer parts of America, but introduced and naturalised elsewhere. The plants are generally free-floating, but may occasionally root in shallow water. Propagation is by stolons. The genus is not common in cultivation, and only 1 species is seen in European gardens.

**1. L. laevigatum** (Willdenow) Heine (*L. stoloniferum* (Meyer) Grisebach). Illustration: Cook, Water plants of the world, 261 (1974).

Leaf blade to 3 cm, elliptic to ovate, wavy, pale green; stalk to 5 cm. Sepals oblong; petals only slightly longer than the sepals and narrower, linear. *West Indies, C & tropical S America*. G1. Summer.

This species needs plenty of light, but not direct sunlight. It grows best in shallow water over a peaty substrate.

**7. OTTELIA** Persoon
*E.C. Nelson*

Monoecious or dioecious aquatic herbs. Leaves submerged, basal, stalked. Flowers held above the water surface. Spathes stalked, tubular, ribbed and winged; male spathes with many flowers, the female (and occasional bisexual spathes) with 1 flower. Petals larger than sepals. Stamens 6–20, anthers opening laterally. Ovary stalkless; styles 3–15, bifid, free.

A genus of about 40 species from warmer regions of the world, mainly Africa and E Asia. Only 1 species is widely cultivated in Europe, and it is naturalised in N Italy. Propagation is by division or seed.

**1. O. alismoides** (Linnaeus) Persoon. Illustration: Botanical Magazine, 1201 (1809); Cook, Water plants of the world, 265 (1974).

Perennial with tuberous rootstock. Leaves varying greatly in size, generally submerged but sometimes partly emergent; juvenile leaves linear or with lanceolate blades; older leaves with broad, elliptic blades to 20 cm, thin, translucent pale green, margins wavy; stalk to 50 cm, triangular in cross-section. Petals 2–3 cm, obovate, white, each with a yellow basal spot. Stamens 6–9. Styles 6–9. *E Asia, NE Africa, Australia*. H5. Summer.

This species prefers warm water (22–28 °C) which is clear and not alkaline; it should be planted in a rich compost.

**8. STRATIOTES** Linnaeus
*E.C. Nelson*

Dioecious, submerged, aloe-like perennial herbs, with stolons. Leaves all basal, stalkless. Flowers held above the water. Male spathes with several flowers, female spathes with 1 flower. Petals larger than the sepals. Stamens 12, surrounded by numerous staminodes. Ovary stalkless, styles 6, bifid, free.

A genus of a single species which resembles an aloe in form. The plants are submerged, but rise to the surface at flowering time. It is easily cultivated in outdoor ponds and aquaria and is propagated very easily by means of the stolons. It can become a weed in suitable situations.

**1. S. aloides** Linnaeus. Illustration: Smith, English Botany, 379 (1791); Botanical Magazine, 1285 (1810); Keble Martin, A concise British flora in colour, 79 (1965); Cook, Water plants of the world, 266 (1974).

Leaves *c.* 50 × 2 cm, in a rosette, linear-lanceolate, brittle, tapered to the apex, with numerous marginal teeth. Petals white, 1.5–2 cm, broader than long. *Eurasia*. H1. Summer.

**9. VALLISNERIA** Linnaeus
*E.C. Nelson*

Dioecious, submerged plants, with stolons. Leaves all basal, alternate, stalkless. Flowers pollinated at the water surface. Male spathes almost stalkless or stalked, many-flowered; stamens 2–3. Female spathes on long stalks, 1-flowered; ovary linear, styles 3, bifid, free. Sepals 3. Petals 3.

A genus of about 10 species distributed throughout the world apart from the coldest regions. They grow in still or running water, and reproduce by means of stolons; some cause serious problems in irrigation canals. Only 3 species are cultivated in Europe and these are occasionally naturalised. Propagation is by stolons or offsets. In aquaria the plants will root in sand, and a layer of loam beneath the sand will improve their growth.

1a. Plants becoming dormant in winter, even in aquaria          **1. americana**
  b. Plants growing throughout the year 2
2a. Stalks of female spathes coiled after flowering          **3. spiralis**
  b. Stalks of female spathes straight and remaining so          **2. gigantea**

**1. V. americana** Michaux.
Leaves to 2 m × 2 cm, linear. Male spathes to 1.5 cm. Female spathes on stout stalks which curve or spiral in fruit. Fruit curved. *E USA*. H4. Summer.

**2. V. gigantea** Graebner.
Leaves to 1 m × 2 cm, linear, with 5–9 conspicuous veins and longitudinal brown or blackish stripes, indistinctly toothed. Female spathes on straight stalks. Fruits striped with brown or black. *S Asia to Australia*. H5–G1. Summer.

Although it will survive outdoors in the mildest parts of Europe, this is generally grown in indoor aquaria. It is adaptable, and can survive water temperatures of over 30 °C. It is a good oxygenator.

**3. V. spiralis** Linnaeus. Illustration: Botanical Magazine, 4894 (1856), 6912 (1886); Cook, Water plants of the world, 268 (1974).

Leaves linear, ribbon-like, to 1 cm broad, with minute teeth towards the apex. Male spathes on short stalks, longer than those of *V. americana*. Female spathes linear, their stalks becoming spirally coiled in fruit. Fruit straight. *S Europe, W Asia*. H5. Summer.

A common aquarium plant which grows continually and is a good oxygenator. A number of cultivars is available, including 'Torta', which has shorter and broader, spirally coiled leaves, and which may be grown outdoors in warm areas.

# IV. APONOGETONACEAE

Aquatic perennial herbs. Leaves alternate, basal. Flowers bisexual or unisexual in stalked spikes which may be branched. Perianth with 1–3, usually petaloid segments. Stamens 6–18, anthers opening by longitudinal slits. Ovary superior. Carpels 3–8, free; ovules 2–8 in each carpel. Fruit a whorl of follicles.

A family of a single genus from tropical and subtropical habitats in Asia and Africa.

**1. APONOGETON** Linnaeus
*E.C. Nelson*

A genus of about 20 species, formerly known as *Ouvirandra* Thouars. Only 2 species are well known in cultivation in Europe. Propagation is by division, offsets or seeds.

1a. Leaves entire          **1. distachyos**
  b. Leaves reduced to a lace-like structure          **2. madagascariensis**

**1. A. distachyos** Linnaeus. Illustration: Andrews' Botanical Repository, 290 (1803); Botanical Magazine, 1293 (1810). Leaves floating, to 25 × 7 cm, elliptic to oblong. Spike with 2 branches, each bearing *c.* 10 fragrant, white flowers. Perianth-segment 1, ovate, 1–2 cm. Stamens 6–18. *South Africa*. H4. Summer.

Widely cultivated in ponds out of doors in the British Isles and in south western Europe, where it is sometimes naturalised. It will grow in water to 60 cm deep. There are several cultivars listed, including 'Roseus', which has pink flowers.

**2. A. madagascariensis** (Mirbel) van Brugg (*A. fenestralis* (Poiret) Hooker). Illustration: Botanical Magazine, 4894 (1856).
Leaves submerged, to 20 × 12 cm, elliptic or oblong, reduced to a lace-like structure formed from the persistent network of veins. Spikes branched; flowers numerous, white; perianth-segments small, scale-like. Stamens 6. *Madagascar.* G2. Summer.

A botanical curiosity on account of its strange leaves. It is difficult to grow, as the water must be kept clean and free from sediment, must be maintained at *c.* 20 °C, and should be deep and continuously circulating. Shade is also required.

# V. POTAMOGETONACEAE

Submerged or floating, annual or perennial, aquatic herbs. Stems elongate, flexible, erect. Rhizomes floating or creeping, slender, elongate, often producing specialised winter buds (turions). Leaves opposite or alternate, rarely in whorls of 3, sheathing at their bases. Flowers bisexual, in axillary or terminal, bractless spikes, inconspicuous, radially symmetric. Perianth-segments 4. Stamens 4. Ovary superior. Carpels usually 4, rarely 1–3, free or shortly joined at the base, ovule 1 per carpel; style short. Fruit a small, green or brown achene or drupe.

A family of 2 genera, distributed throughout the world; the only genus of horticultural importance is *Potamogeton.*

### 1. POTAMOGETON Linnaeus
*A. Brady*
Rhizomes branching, sometimes producing turions. Leaves alternate, submerged or partly floating; submerged leaves thin, translucent, linear; floating leaves leathery, opaque. Stipules present, free or attached to the leaf base. Flowers wind- or water-pollinated. Fruit a drupe.

A genus of about 100 species distributed throughout the world; only 2 are widely cultivated in aquaria and ponds as oxygenating plants. They reproduce by rhizomes or turions.

1a. Margins of submerged leaves
    markedly wavy when mature
                              **1. crispus**
  b. Margins of submerged leaves flat
                              **2. natans**

**1. P. crispus** Linnaeus. Illustration: Keble Martin, The concise British Flora in colour, 90 (1965); Ross-Craig, Drawings of British plants **31**: t. 36 (1973); Cook, Water plants of the world, 495 (1974).

Branches to 100 cm. Submerged leaves stalkless, membranous, *c.* 4 cm × 8–15 mm, almost translucent, margins wavy when mature. Floating leaves stalked, pointed, leathery. Stipules small, membranous, evanescent. Spike 1–2 cm, loose, on a stalk *c.* 2.5 cm. Fruit 4–5 mm, with a long, curved beak which equals the rest of the fruit in length. Turions 1–5 cm with thick, horny, spiky-bordered leaves. *Most of the Old World to Australia and New Zealand.* H1. Summer.

**2. P. natans** Linnaeus. Illustration: Butcher, A new illustrated British flora, 597 (1961); Keble Martin, The concise British flora in colour, 89 (1965); Ross-Craig, Drawings of British plants **31**: t. 16 (1973).
Leafy stem more than 100 cm. Submerged leaves to 20 × 3 cm, stalked, lanceolate, obtuse at apex, margins flat; these leaves fall quickly. Floating leaves to 12.5 × 7 cm, ovate to elliptic or lanceolate, rounded to almost cordate at the base, opaque; stalk often longer than the blade. Stipules large, often fibrous, persistent. Spike 3–8 cm, dense, axillary, cylindric; stalk 5–12 cm. Fruit 3–5 mm, with a short, straight beak, green, ovoid, compressed. *Eurasia.* H1. Summer.

# VI. LILIACEAE

Perennial herbs or shrubs, sometimes climbing or scrambling, with rhizomes, fleshy or fibrous roots, bulbs or corms. Leaves very variable, borne on the stem or all basal, usually alternate, more rarely opposite or whorled, rarely absent, sometimes reduced to small scales when their function is taken over by needle-like, thread-like or flattened cladodes; veins usually parallel, more rarely net-like. Flowers usually with bracts, solitary or in racemes, spikes, panicles, clusters or umbels which are sometimes subtended by spathes. Perianth usually of 6 free or united segments, more rarely the segments 4 or more than 6. Stamens usually 6, more rarely 3 or 4 or more than 6, free or united, borne on the perianth or free from it, anthers basifixed or dorsifixed and often versatile, usually opening by slits. Ovary usually superior, usually 3-celled with 2–many ovules per cell with axile placentation, rarely 1-celled with numerous ovules with parietal placentation. Fruit a berry or capsule. Seeds 1–many.

A family of about 220 genera and 3500 species from all over the world. More than half of the genera are cultivated.

There are varying views as to how the family should be defined and classified. The classification followed here is that used by Melchior in his edition (edn 12) of *Syllabus der Pflanzenfamilien* (1964) except that *Phormium* is here placed in the Agavaceae (p. 271). The subfamilies *Alstroemerioideae* (genera Nos. **61** and **62**) and *Allioideae* (genera Nos. **86–96**) are often considered to form distinct families (Alstroemeriaceae, Alliaceae), or are included in the Amaryllidaceae. Some recent classifications pulverise the family into many smaller ones, the Liliaceae in this sense being restricted to a few genera closely related to *Lilium*; this approach is outlined in Dahlgren, R. & Clifford, R.T., *The Monocotyledons: a comparative survey* (1982).

### KEY TO GROUPS

1a. Stem woody, persistent through the
    winter                **Group 7** (p. 121)
  b. Stem not woody and persistent
    through the winter, though leaves
    sometimes evergreen                    2
2a. Ovaries below ground, hidden in the
    bulbs or corms        **Group 5** (p. 120)
  b. Ovaries exposed, above ground         3
3a. Ovary partly or fully inferior
                          **Group 6** (p. 121)
  b. Ovary completely superior             4
4a. Flowers in umbels which are
    subtended by spathes (rarely flower
    solitary and subtended by 2 united
    spathes)              **Group 4** (p. 120)
  b. Flowers in inflorescences of various
    kinds, or solitary, not subtended by
    spathes                                5
5a. Styles 3 (rarely 4), separate and
    distinct, or stigmas 3, separate, borne
    directly on the ovary
                                     **Group 1**
  b. Style 1, sometimes 3-branched at the
    apex, or style absent, stigmas united,
    borne directly on the ovary            6
6a. Perianth-segments free at the base
    or united for less than one-tenth of
    their total length
                          **Group 2** (p. 118)
  b. Perianth united at the base for more
    than one-tenth of its total length
                          **Group 3** (p. 119)

**Group 1**

1a. Plant a climber      **127. Smilax**
  b. Plant not a climber                   2

2a. Leaves 2, opposite or in 1 or 2 whorls on the flowering stem, never long and grass-like     3

b. Leaves several to many, all basal or borne on the flowering stem, if 2 and opposite then grass-like, at least 8 times longer than broad     6

3a. Flower solitary, terminal     4

b. Flowers in umbels     5

4a. Leaves 4 or more in a whorl; parts of the flower in 4s     **111. Paris**

b. Leaves in whorls of 3; parts of the flower in 3s     **112. Trillium**

5a. Leaves 2 at the base of each flowering stem; stamens 3     **109. Scoliopus**

b. Leaves in 2 whorls on each flowering stem; stamens 6     **110. Medeola**

6a. Some part of the plant obviously hairy or covered with scales     7

b. No part of the plant obviously hairy or covered with scales     10

7a. Leaves with a covering of scales which are more or less united, forming a skin which peels off in flakes or strips     **102. Astelia**

b. Leaves not as above     8

8a. Perianth-segments conspicuously clawed     **10. Melanthium**

b. Perianth-segments not clawed     9

9a. Leaves broadly ovate to elliptic with many conspicuous veins, narrowed towards the stalk-like, sheathing bases     **11. Veratrum**

b. Leaves narrow, without conspicuous veins, not narrowing towards the sheathing bases     **9. Zigadenus**

10a. Plant with a bulb or corm     11

b. Plant with a rhizome or fleshy roots     12

11a. Perianth-lobes linear, curled     **8. Stenanthium**

b. Perianth-lobes neither linear nor curled     **44. Wurmbea**

12a. Perianth-segments very unequal, the 2 or 3 lower ones small and thread-like     **7. Chionographis**

b. Perianth-segments all more or less equal     13

13a. Leaves narrowed towards the base, shortly stalked     14

b. Leaves not narrowed towards the base, not stalked     16

14a. Perianth-segments somewhat swollen or inflated towards the base; fruit a berry     **107. Disporum**

b. Perianth-segments not as above; fruit a capsule     15

15a. Flowers unisexual     **5. Chamaelirion**

b. Flowers bisexual     **4. Helonias**

16a. Stem-leaves numerous; flower-stalks very long     **3. Xerophyllum**

b. Stem-leaves few; flower-stalks very short or absent     **1. Tofieldia**

## Group 2

1a. Plants with bulbs or corms     2

b. Plants with rhizomes or fleshy or fibrous roots     23

2a. Leaves borne on the flowering stem     3

b. Leaves at the base of the scape     13

3a. Perianth-segments without nectaries at the base     4

b. Perianth-segments with nectaries at the base     5

4a. Flowers in panicles     **51. Gagea**

b. Flowers solitary or in racemes or umbels     **55. Tulipa**

5a. Style absent, stigmas borne directly on the ovary     **53. Calochortus**

b. Style present     6

6a. Anthers basifixed     7

b. Anthers dorsifixed, often versatile     9

7a. Perianth-segments reflexed from the base or from very near it     **54. Erythronium**

b. Perianth-segments not reflexed from the base, reflexed, if at all, at their tips only     8

8a. Perianth-segments in 2 distinct whorls; bulb-tunics usually absent, thin and white if present     **56. Fritillaria**

b. Perianth-segments not in 2 distinct whorls; bulb-tunics always brown     **52. Lloydia**

9a. Perianth-segments twisting together after flowering; flower-stalks jointed     **31. Chlorogalum**

b. Perianth-segments not twisting together after flowering; flower-stalks not jointed     10

10a. Perianth-whorls differing markedly in size or colour and markings; filaments swollen at their bases; nectary usually formed from a number of plates arranged in a fan-shape     **59. Nomocharis**

b. Perianth-whorls similar, or segments of the inner whorl slightly narrower than those of the outer; nectary various, not as above     11

11a. Bulb persistent after flowering, over-wintering; leaves various but neither long-stalked and heart-shaped nor linear and very rarely produced in autumn or winter     **57. Lilium**

b. Bulb dying after flowering, the plant persisting by offsets; leaves long-stalked and heart-shaped or linear and produced in autumn or winter 12

12a. Leaves long-stalked and heart-shaped     **60. Cardiocrinum**

b. Leaves not stalked, linear     **58. Notholirion**

13a. Mature plant leafless, consisting of a much-branched, climbing or scrambling stem     **43. Bowiea**

b. Mature plant with normal leaves, not as above     14

14a. Flowering stem ending in a tuft of bracts above the raceme     **69. Eucomis**

b. Flowering stem without a tuft of bracts above the raceme     15

15a. Perianth-segments reflexed from the base or from very near it     **54. Erythronium**

b. Perianth-segments not reflexed from the base, reflexed, if at all, at their tips only     16

16a. Inner perianth-segments erect, usually hooded or each with a swelling at the tip, outer segments spreading     **73. Albuca**

b. All perianth-segments similar, not as above     17

17a. Filaments flattened, usually broadest at the base, occasionally winged and toothed; flowers not blue     **68. Ornithogalum**

b. Filaments thread-like, or if flattened then lanceolate to elliptic, narrow at the base; flowers often blue     18

18a. Perianth withering and falling after flowering     **74. Urginea**

b. Perianth withering but not falling after flowering, persisting around or below the capsule     19

19a. Ovules 2 per cell; perianth-segments to 6 mm     **30. Schoenolirion**

b. Ovules several per cell; perianth-segments more than 6 mm     20

20a. Each perianth-segment 3-veined     **67. Camassia**

b. Each perianth-segment 1-veined     21

21a. Bracts 2 to each flower     **65. Hyacinthoides**

b. Bract 1 to each each flower, or bracts absent     22

22a. Perianth-segments spreading from the base; ovary not stalked     **63. Scilla**

b. Perianth-segments curved to form a cup at the base, spreading from the top of the cup; ovary stalked     **64. Ledebouria**

23a. Leaves borne on the flowering stem     24

b. Leaves at the base of the scape     39

24a. Fruit a berry     25

b. Fruit a capsule     30

25a. Anthers opening by pores
                              **16. Dianella**
   b. Anthers opening by slits          26
26a. Flowering stem leafy over most of its
      length below the inflorescence     27
   b. Flowering stem with only 2 or 3
      leaves along its length or with several
      leaves concentrated towards the base
                                          29
27a. Flowers solitary or in pairs, their
      stalks fused to the stem for some
      distance so that they appear to be
      borne well above their bract, the
      stalks with a downward bend at
      about the middle of the free part
                           **106. Streptopus**
   b. Flowers and flower-stalks not as
      above                               28
28a. Leafy part of the stem unbranched;
      perianth-segments neither swollen
      nor slightly inflated above the base
                            **104. Smilacina**
   b. Leafy part of the stem branched;
      perianth-segments often swollen or
      slighly inflated above the base
                             **107. Disporum**
29a. Flowering stem bearing 2 or 3 heart-
      shaped leaves; flowers with parts in 4s
                          **105. Maianthemum**
   b. Flowering stem with a few leaves
      which are not heart-shaped; flowers
      with parts in 3s      **98. Speirantha**
30a. Leaves with tendril-like tips; plants
      usually climbing or scrambling (if not
      climbing then style bent sharply to
      one side at its base)               31
   b. Leaves without tendril-like tips; plants
      not climbing or scrambling, style
      never bent sharply to one side at its
      base                                32
31a. Perianth bell-shaped; style not bent
      sharply to one side at its base
                             **49. Littonia**
   b. Perianth widely open, the segments
      spreading widely or reflexed; style
      bent sharply to one side at its base
                             **48. Gloriosa**
32a. Flowers solitary or in a head
      subtended by several bracts
                          **32. Aphyllanthes**
   b. Flowers not as above               33
33a. Leaves evergreen; seeds with linear or
      thread-like appendages at both ends
                           **6. Heloniopsis**
   b. Leaves not evergreen; seeds not as
      above                               34
34a. Flowers in long racemes             35
   b. Flowers in clusters, panicles or false
      umbels, or solitary                 36
35a. Leaves very numerous, not obviously
      in 2 ranks; stamens unequal,
      filaments hairless   **18. Asphodeline**

   b. Leaves few, obviously in 2 ranks;
      stamens equal, filaments woolly
                            **2. Narthecium**
36a. Filaments hairy; perianth variously
      coloured, but not pure yellow or
      spotted                             37
   b. Filaments hairless; perianth pure
      yellow, or spotted                  38
37a. Flower solitary      **14. Herpolirion**
   b. Flowers in panicles  **15. Stypandra**
38a. Perianth pure yellow, not spotted;
      flowers pendent; capsule splitting
      between the septa     **12. Uvularia**
   b. Perianth spotted; flowers ascending or
      erect; capsule splitting along the septa
                            **13. Tricyrtis**
39a. Leaves thick, succulent or leathery,
      evergreen, usually in dense rosettes
                                          40
   b. Leaves not as above                42
40a. Apex of perianth bilaterally
      symmetric, 2-lipped  **38. Haworthia**
   b. Apex of perianth radially symmetric,
      not 2-lipped                        41
41a. Perianth white or yellowish green;
      flowers borne all round the raceme-
      axis                   **39. Astroloba**
   b. Perianth orange-red; raceme 1-sided
                            **40. Poellnitzia**
42a. Fruit a berry                       43
   b. Fruit a capsule                    44
43a. Perianth-segments hairy outside
                            **103. Clintonia**
   b. Perianth-segments not hairy outside
                           **98. Speirantha**
44a. Stamens 3           **109. Scoliopus**
   b. Stamens 6                          45
45a. Anthers with tails at their bases
      which are papillose, hairy or warty,
      sometimes joined to the filaments for
      all or part of their length
                          **28. Arthropodium**
   b. Anthers without tails; filaments
      sometimes hairy                     46
46a. Outer perianth-segments narrower
      than the fringed, inner segments
                           **27. Thysanotus**
   b. All perianth-segments similar, none
      fringed                             47
47a. At least 3 of the filaments hairy    48
   b. None of the filaments hairy         49
48a. Perianth whitish inside, purplish
      outside              **23. Simethis**
   b. Perianth yellow or brownish inside
      and out               **22. Bulbine**
49a. Ovules 2 in each cell of the ovary
                                          50
   b. Ovules 4–8 in each cell of the ovary
                                          52
50a. Perianth yellow      **21. Bulbinella**
   b. Perianth white or purplish         51
51a. Perianth-segments curving upwards,

      forming a bell-shaped flower; seeds
      fleshy                **118. Liriope**
   b. Perianth-segments spreading, forming
      a star-shaped flower; seeds winged,
      not fleshy           **17. Asphodelus**
52a. Anthers dorsifixed; raceme 1-sided
                            **19. Paradisea**
   b. Anthers basifixed; raceme (or panicle)
      not 1-sided                         53
53a. Capsule conspicuously 3-angled;
      raceme or panicle often producing
      plantlets           **26. Chlorophytum**
   b. Capsule rounded; raceme not
      producing plantlets                 54
54a. Flowers 6–10 in a raceme; scape
      20–70 cm; perianth-segments white,
      each with 3 veins    **24. Anthericum**
   b. Flowers 50–800 in a raceme; scape to
      2.4 m, if less than 70 cm then each
      perianth-segment with 1 vein
                             **20. Eremurus**

### Group 3

1a. Flowering stem leafy or bearing
      cladodes                            2
   b. Flowering stem leafless, leaves all
      basal                               9
2a. True leaves small, scale-like, their
      function taken over by numerous
      cladodes which are borne singly or in
      clusters in the axils of the scale-leaves
                            **113. Asparagus**
   b. True leaves present, green; cladodes
      absent                              3
3a. Leaves with a covering of scales
      which are united to form a skin
      which peels off as flakes or strips;
      flowers in a panicle   **102. Astelia**
   b. Leaves not as above; flowers in spikes
      or racemes                          4
4a. Leaves distinctly narrowed to stalks at
      their bases; raceme more or less 1-
      sided                               5
   b. Leaves not narrowed to stalks; raceme
      not 1-sided                         6
5a. Perianth spherical to bell-shaped, the
      tube half to two-thirds of the total
      length; fruit a berry  **97. Convallaria**
   b. Perianth narrowly funnel-shaped, the
      tube less than half the total length;
      fruit a capsule        **33. Hosta**
6a. Flowers bilaterally symmetric;
      stamens deflexed     **18. Asphodeline**
   b. Flowers radially symmetric; stamens
      not deflexed                        7
7a. Leaves mostly towards the base of the
      stem, none with flowers in their axils
                             **120. Aletris**
   b. Leaves spread along the stem, the
      uppermost with flowers or flower-
      clusters in their axils             8

8a. Perianth spherical to urn-shaped, orange, the tube forming most of its length, the lobes small
**50. Sandersonia**

b. Perianth cylindric, white, pink, purple, greenish or yellowish, the tube making up not much more than half of the total length
**108. Polygonatum**

9a. Plant with a bulb    10

b. Plant with a rhizome or fleshy or fibrous roots    26

10a. Perianth-tube longer than the lobes    11

b. Perianth-tube shorter than to as long as the lobes    17

11a. Leaves 2, ovate, oblong or almost circular, spreading horizontally on the ground; scape absent or very short; perianth-lobes reflexed, almost as long as the tube    **84. Massonia**

b. Leaves various, not as above; perianth-lobes much shorter than the tube, not usually reflexed    12

12a. Bracts obvious, as long as or almost as long as the flower-stalks    13

b. Bracts minute or absent    14

13a. Leaves to 6 mm wide; flowers blue to violet-blue or white    **79. Brimeura**

b. Leaves much wider; flowers pink or pinkish purple, if whitish then spotted with red    **82. Veltheimia**

14a. Perianth constricted at the mouth of the tube    **80. Muscari**

b. Perianth not constricted at the mouth of the tube    15

15a. Capsule 3-angled    **77. Bellevalia**

b. Capsule rounded, not angled    16

16a. Perianth 4–9 mm    **78. Hyacinthella**

b. Perianth 1–3.5 cm    **76. Hyacinthus**

17a. Flowers very unequal within the corymbose raceme, the outer with bilaterally symmetric perianths, the inner with almost radially symmetric perianths    **85. Daubenya**

b. Flowers all similar within the raceme    18

18a. Outer perianth-lobes shorter than the inner    **81. Lachenalia**

b. All perianth-lobes of about the same length    19

19a. Raceme ending in a tuft of bracts above the flowers    **69. Eucomis**

b. Raceme not ending in a tuft of bracts    20

20a. Perianth disc-shaped, the lobes spreading from a short, obconical tube    21

b. Perianth cylindric or bell-shaped, the lobes spreading or not    22

21a. Filaments united into a corona with projecting lobes    **75. Puschkinia**

b. Filaments flat, not united    **66. Chionodoxa**

22a. Plant to 1 m; flowers 2.5 cm or more, white or greenish    **83. Galtonia**

b. Plant much smaller; flowers less than 2.5 cm, variously coloured    23

23a. Perianth to 6 mm, greenish white; leaves spotted    **71. Drimiopsis**

b. Perianth more than 1 cm, not greenish white; leaves not spotted    24

24a. Perianth blue or bluish purple    **77. Bellevalia**

b. Perianth white, greenish or brownish orange, or pink    25

25a. Perianth-tube bell-shaped; lobes spreading or reflexed    **70. Drimia**

b. Perianth-tube tubular to bell-shaped, the outer 3 lobes spreading, the inner 3 erect    **72. Dipcadi**

26a. Leaves evergreen, not stalked, often thick, fleshy or leathery    27

b. Leaves thin, usually deciduous, if evergreen then stalked    30

27a. Scape flattened; leaves thin    **25. Alectorurus**

b. Scape not flattened; leaves thick, fleshy or leathery    28

28a. Perianth-tube bowl-shaped, all the lobes shorter than it    **100. Rohdea**

b. Perianth-tube cylindric or swollen, sometimes somewhat 1-sidedly at the base, usually some, at least, of the lobes longer than it    29

29a. Perianth-tube cylindric or evenly swollen at the base    **42. Aloe**

b. Perianth-tube swollen 1-sidedly at the base    **41. Gasteria**

30a. Leaves distinctly stalked    31

b. Leaves not stalked    32

31a. Perianth funnel-shaped with 6 lobes; flowers in racemes borne above ground-level    **33. Hosta**

b. Perianth not funnel-shaped, with 6–8 lobes; flowers solitary, at ground-level or almost so    **101. Aspidistra**

32a. Flowers stalkless    33

b. Flowers stalked    34

33a. Fruit a berry; perianth-lobes longer than the tube    **99. Reineckea**

b. Fruit a capsule; perianth-tube longer than the lobes    **120. Aletris**

34a. Perianth cylindric or bell-shaped, the tube longer than the lobes    35

b. Perianth funnel-shaped, the lobes longer than the tube    36

35a. Ovary stalked; filaments attached near the middle of the perianth    **35. Blandfordia**

b. Ovary not stalked; filaments attached at the base of the perianth    **37. Kniphofia**

36a. Perianth blue    **29. Pasithea**

b. Perianth yellow or orange, sometimes reddish    **34. Hemerocallis**

## Group 4

1a. Plants with rhizomes or fleshy roots, bulbs absent    2

b. Plants with bulbs (which sometimes have a rhizome beneath them)    3

2a. Perianth with 6 outgrowths forming a corona    **87. Tulbaghia**

b. Perianth without outgrowths    **86. Agapanthus**

3a. Perianth with free segments or with a very short tube    4

b. Perianth with a distinct tube    7

4a. Plant smelling of onion or garlic    **88. Allium†**

b. Plant not smelling of onion or garlic    5

5a. Spathe 1    **90. Caloscordum**

b. Spathes 2 or more    6

6a. Spathes green, leaf-like; plant less than 20 cm    **51. Gagea**

b. Spathes papery; plant 20 cm or more    **89. Nothoscordum††**

7a. Flower solitary (rarely 2 together) subtended by 2 spathes which are united into a tube below    **92. Ipheion**

b. Flowers more than 2, in umbels subtended by 2 or 4 free spathes    8

8a. Stamens enclosed in the perianth-tube    **91. Leucocoryne**

b. Stamens projecting from the perianth-tube    9

9a. Spathes 4; flowers stalkless but appearing stalked because of the long, slender perianth-tube    **96. Milla**

b. Spathes 2; flowers stalked, not as above    10

10a. Leaves not keeled beneath; veins weak or not visible    **93. Brodiaea**

b. Leaves keeled beneath; veins distinct    11

11a. Anthers versatile; ovary shortly stalked    **94. Triteleia**

b. Anthers basifixed; ovary stalkless    **95. Dichelostemma**

## Group 5

1a. Style 1, branched above    2

b. Styles 3, free    3

† *Nectaroscordum* (see p. 392) also keys out here.

†† *Bessera, Bloomeria* and *Muilla* (see p. 392) also key out here.

2a. Perianth with a tube below
**36. Leucocrinum**
b. Perianth without a tube, the segments
clawed, with auricles at the junction
of the blade and claw which cohere,
making the perianth appear
superficially as though it has a tube
**46. Bulbocodium**
3a. Perianth with a tube below
**45. Colchicum**
b. Perianth-segments free to the base,
though the bases of the segments are
close and superficially appear to be
joined **47. Merendera**

### Group 6

1a. Flowers in panicles; styles 3, free
**9. Zigadenus**
b. Flowers in umbels, spikes or racemes
or solitary, style 1 2
2a. Stem-leaves twisted at the base
through 180°; flowers in umbels or
rarely solitary 3
b. Stem-leaves not twisted through
180°; flowers in racemes or spikes
4
3a. Perianth bilaterally symmetric;
flowers erect; plants not climbing or
scrambling **61. Alstroemeria**
b. Perianth radially symmetric; flowers
drooping; plant usually climbing or
scrambling **62. Bomarea**
4a. Leaves with conspicuous stalk and
blade; perianth united into a tube
below **119. Peliosanthes**
b. Leaves without distinct stalk and
blade; perianth-segments free to the
base **117. Ophiopogon**

### Group 7

1a. Leaves scale-like, their function taken
over by cladodes which are cylindric,
thread-like or flattened, borne singly
or in clusters in the axils of the scale-
leaves 2
b. Leaves normal, cladodes absent 5
2a. Stamens free; scale-leaves with
thickened, often spine-like or
projecting bases **113. Asparagus**
b. Stamens united into a tube; scale-
leaves not as above 3
3a. Plant climbing; flowers borne around
the edges of the cladodes
**115. Semele**
b. Plant not climbing; flowers not borne
around the edges of the cladodes 4
4a. Flowers in racemes of 5–8 at the tips
of the branches, separate from the
cladodes **114. Danaë**

b. Flowers in clusters on the surfaces of
the cladodes **116. Ruscus**
5a. Leaves with net veins, their stalks
bearing tendrils **127. Smilax**
b. Leaves with parallel veins, or veins
not visible, without tendrils 6
6a. Leaves thick, fleshy or leathery, often
with toothed or spiny margins;
flowers in racemes or panicles
**42. Aloe**
b. Leaves thin, sometimes leathery,
margins neither toothed nor spiny;
flowers solitary in the leaf-axils or in
clusters or cymes 7
7a. Ovary 3-celled; flower less than 4 cm
8
b. Ovary 1-celled; flowers more than
4 cm 10
8a. Inner perianth-segments densely
fringed with long hairs; filaments
united into a tube **122. Eustrephus**
b. Inner perianth-segments not fringed;
filaments free 9
9a. Anthers opening by pores; flowers in
cymes **123. Geitonoplesium**
b. Anthers opening by slits; flowers
solitary in the leaf-axils
**121. Luzuriaga**
10a. Plant a small, upright shrub
**124. Philesia**
b. Plant climbing or twining 11
11a. Perianth-segments all equal in length
**126. Lapageria**
b. Perianth-segments unequal, the outer
somewhat shorter than the inner
**125. × Philageria**

### 1. TOFIELDIA Hudson
*J.C.M. Alexander*

Perennial herbs with rhizomes. Leaves
narrowly sword-shaped, densely tufted in 2
opposite ranks. Flowering stems erect, with
or without leaves. Flowers small, in
terminal racemes or heads. Flower-stalks
short, bearing a bract at the base and
sometimes a bracteole just below the
flower. Perianth persistent; segments 6,
free, narrow, ascending or spreading.
Stamens free; filaments hairless; anthers
broadly ovate, basifixed. Ovary superior;
carpels 3, free above, gradually narrowed
into 3 short persistent styles. Fruit a
spherical to ellipsoid capsule; seeds very
small, numerous, ellipsoid to narrowly
oblong.

A North temperate genus of about 18
species, not regarded as ornamental but
sometimes grown in gardens devoted to
native or local floras. Only the 2 European
species are described here. They perform

best in acid, peaty soils in cool, moist
situations such as the banks of ponds or
streams. Propagation can be effected by
division of the rhizome or from seed.
*T. pusilla* is notoriously difficult to establish
and seed is best sown in soil taken from its
wild habitat; it must be moved without
disturbing the roots.

1a. Bract at base of flower-stalk unlobed;
bracteole just below flower 3-lobed
**1. calyculata**
b. Bract at base of flower-stalk 3-lobed;
bracteole absent **2. pusilla**

**1. T. calyculata** (Linnaeus) Wahlenberg
(*T. palustris* Hudson, in part; *T. alpina*
Smith, not Sternberg & Hoppe; *T. rubra*
Braun; *T. glacialis* Gaudin). Illustration:
Bonnier, Flore complète 10: pl. 573 (1929);
Journal of the Linnean Society, Botany, **53**:
197 (1947); Fitter et al., Wild flowers of
Britain and northern Europe, 261 (1974).
Leaves 1.5–15 cm, with 4–10 veins. Stem
5–35 cm, usually bearing 1–3 leaves.
Flowers yellowish, rarely reddish, up to 30
in a 2–6 cm raceme, or rarely in a
3 flowered head *c.* 5 mm long. Bract at
base of flower-stalked unlobed; bracteole
just below flower 3-lobed. Perianth-
segments 2–3.5 mm. Capsule 3.5 mm.
*Gotland (Sweden) & Estonia south to the
Pyrenees, Jugoslavia & the Ukraine.* H1.
Summer.

**2. T. pusilla** (Michaux) Persoon (*T. palustris*
Hudson in part; *T. borealis* (Wahlenberg)
Wahlenberg; *T. alpina* Sternberg & Hoppe
not Smith). Illustration: Journal of the
Linnean Society, Botany, **53**: 197 (1947);
Ross-Craig, Drawings of British Plants 29:
pl. 32 (1972); Fitter et al., Wild flowers of
Britain and northern Europe, 261 (1974).
Leaves 1–8 cm, with 3–7 veins. Stem
1.5–25 cm, leafless. Flowers white or
greenish, in a 3–20 mm head. Bract at
base of flower-stalk 3-lobed; bracteole
absent. Perianth-segments 1.5–2.5 mm.
Capsule 2.5–3 mm. *N Europe, Alps,
W Carpathians & C Urals.* H2. Summer.

### 2. NARTHECIUM Hudson
*J. Cullen*

Perennial herbs with rhizomes and fibrous
roots. Leaves basal and also borne on the
stem. Flowers in a usually dense raceme;
bracts about as long as the flower-stalks.
Perianth-segments 6, free, spreading in
flower, erect and persistent in fruit.
Stamens 6, versatile, opening to the outside
of the flower; filaments woolly. Ovary
superior, 3-celled, 3-sided, ovules
numerous in each cell. Style simple. Fruit a

capsule splitting between the septa. Seeds numerous, tailed.

A genus of about 8 species from north temperate regions, of which 1 is occasionally grown in places where the soil is boggy and very acidic.

**1. N. ossifragum** (Linnaeus) Hudson. Illustration: Keble-Martin, The concise British flora in colour, t. 86 (1965); Everard & Morley, Wild flowers of the world, t. 24 (1970).
Leaves 4–30 cm × 2–5 mm, rigid, often curved. Stems to 40 cm. Perianth-segments 6–8 mm, linear-lanceolate, yellow. Anthers orange. Stem, perianth and ovary becoming deep orange after flowering. *W Europe, from Scandinavia to N Spain and Portugal.* H3. Summer.

**3. XEROPHYLLUM** Michaux
*J.C.M. Alexander*
Perennial, deciduous herbs, with stout, woody, stem-like rhizomes. Leaves numerous, densely tufted, glaucous on backs, mostly basal, linear, slightly broader at base and finely tapered at tip; margins very hard and rough or finely toothed. Flowering stems stout, unbranched, erect, bearing smaller and shorter stem-leaves with membranous margins. Flowers white or yellowish white, numerous in long, dense terminal racemes. Perianth-segments 6, free, spreading, ovate, in 2 similar whorls, persistent. Stamens 6, free. Styles 3, thread-like, erect, becoming curled. Ovary superior. Fruit a 3-lobed, 3-celled capsule with 2–4 wedge shaped seeds per cell.

A genus of 2 or 3 species from N America, which perform best if given plenty of space in a damp, peaty soil. They should be left undisturbed for as long as possible and are easily propagated by seed or division of the rhizome.

1a. Basal leaves to 45 cm × 2.5 mm; flowers *c.* 9 mm across; stamens shorter than perianth-segments
     **1. asphodeloides**
 b. Basal leaves to 90 cm × 6 mm; flowers *c.* 15 mm across; stamens as long as or longer than perianth-segments
      **2. tenax**

**1. X. asphodeloides** (Linnaeus) Nuttall (*X. setifolium* Michaux). Illustration: Botanical Magazine, 748 (1804); Gardener's Chronicle **13**: 433 (1880); Rickett, Wild flowers of the United States **2**: pl. 39 (1967).
To 1.5 m. Basal leaves to 45 cm × 2.5 mm. Raceme to 30 cm; flowers yellowish white,

*c.* 9 mm across, fragrant. Perianth-segments *c.* 6 mm; stamens shorter than perianth-segments. *Eastern N America.* H4. Summer.

**2. X. tenax** (Pursh) Nuttall (*X. douglasii* Watson; *X. setifolium* misapplied). Illustration: Botanical Magazine, 8964 (1922); RHS Dictionary of gardening, 2294 (1956); Journal of the Royal Horticultural Society **91**: pl. 7 (1966); Rickett, Wild flowers of the United States **5**: pl. 41 (1971).
To 1.8 m. Basal leaves to 90 cm × 6 mm. Raceme to 60 cm; flowers white to cream-coloured, *c.* 1.5 cm across. Perianth-segments 8–9 mm; stamens as long as or longer than perianth-segments. *Western N America.* H4. Summer.

**4. HELONIAS** Linnaeus
*S.G. Knees*
Evergreen perennial producing leaves and inflorescences from a horizontally branched, tuberous rhizome. Leaves 15–45 × 3–5 cm, entire, shortly stalked, lanceolate, in basal rosettes. Racemes dense, spike-like, conical, 2–3 cm long, borne terminally. Scapes 4–6 cm × 8–10 mm, stout, hollow, with bracts. Flowers small (*c.* 5 mm), 25–30, bisexual, star-shaped, sweetly scented, subtended by bracts. Perianth of 6 persistent, pink segments. Stamens 6, with long filaments and projecting, blue-grey anthers. Ovary superior, 3-celled. Styles 3, divided to the base. Fruit a 3-celled, dehiscent capsule, 3–4 mm. Seeds many, with white appendages.

A genus of a single species, grown for its spring flowering, pink, orchid-like inflorescences, distinguished from the closely related *Heloniopsis* by its dense spikes of smaller flowers. It is best suited to a moisture-retentive, but humus-enriched soil, reflecting its native habitat of swamp lands. Showing a preference for shaded conditions, it succeeds well in the woodland or bog garden. Propagation is by seed or division.

**1. H. bullata** Linnaeus. Illustration: Botanical Magazine, 747 (1804); Loddiges' Botanical cabinet **10**: t. 961 (1824); Britton & Brown, Flora of the northern United States and Canada **1**: 402 (1896), edn 2, **1**: 488 (1913); Marshall Cavendish encyclopedia of gardening **26**: 877 (1968). *Eastern N America.* H4. Spring.

**5. CHAMAELIRION** Willdenow
*A. Leslie*
Dioecious herb with a bitter, tuberous rootstock. Male plants 30–70 cm, female to 1.2 m, both erect, unbranched, hairless. Lower leaves in a loose rosette, 5–20 × 1–4 cm, obovate to spathulate, stalked; stem-leaves smaller, the uppermost linear to lanceolate, stalkless. Inflorescence a dense, cylindric raceme, the male 4–12 cm, often drooping at the tip at first, the female more slender, to 30 cm, erect. Flower-stalks to 5 mm, without bracts. Flowers *c.* 7 mm wide; perianth-segments 3 mm, linear-spathulate, white. Stamens with almost spherical anthers. Female flowers usually with 6 staminodes. Ovary superior, oblong or obovoid, with 3 short styles; stigmas decurrent. Fruit a capsule. Seeds numerous, linear-oblong, broadly winged.

A genus of a single species, uncommon in cultivation. The flowers become yellowish when dried. It requires a rich, damp soil and is shade-tolerant.

**1. C. luteum** (Linnaeus) Gray. Illustration: Botanical Magazine, 1062 (1808); Gleason, Illustrated flora of the north-eastern United States and adjacent Canada **1**: 405 (1952). *Eastern N America.* H2. Summer.

**6. HELONIOPSIS** Gray
*P.G. Barnes*
Perennial evergreen herbs with short rhizomes. Leaves basal, scape arising from the centre of the rosette, erect, unbranched, with reduced, bract-like lanceolate leaves crowded below. Racemes few-flowered, often condensed and umbel-like. Flowers pendent. Perianth-segments 6, spreading, equal. Stamens 6, filaments subulate. Ovary superior, 3-celled. Style simple with a capitate stigma. Fruit a 3-lobed capsule; seeds numerous, linear, each with a thread-like appendage at each end.

Three or four species in Japan, Korea and Taiwan. They are easily cultivated in humus-rich soil in light shade and sometimes form colonies by means of viviparous leaf-tips.

**1. H. orientalis** (Thunberg) Tanaka (*H. japonica* Maximowicz). Illustration: Botanical Magazine. 6986 (1888).
Leaves oblanceolate, acute, sometimes viviparous at the apex, gradually narrowed to the base. Flower-stalks 1–1.5 cm, at first curved, elongating and becoming erect in fruit. Perianth-segments 1–1.5 cm, narrowly spathulate or oblong, obtuse, pink or violet, persistent in fruit and

becoming greenish. Filaments subulate, *c.*
1.5 cm, anthers bluish violet. Fruit
1.2–1.5 cm. *Japan, Korea, USSR (Sakhalin).*
H3. Spring.

Cultivated material is mostly var.
**orientalis**; var. **breviscapa** (Maximowicz)
Ohwi, differs only in the smaller, pale pink
or whitish flowers.

### 7. CHIONOGRAPHIS Maximowicz
*J.C.M. Alexander*
Hairless herbaceous perennials with short,
thick rhizomes. Basal leaves elliptic to ovate
or oblanceolate, often with long stalks;
stem-leaves smaller and stalkless. Flowers
many, in a terminal spike, white, bisexual,
bilaterally symmetric. Perianth-segments
3–6, unequal; the upper 3 or 4 petal-like,
linear to thread-like; the lower 2 or 3 very
small or absent. Stamens 6, filaments very
short or absent. Styles 3, free; ovary
superior, 3-celled. Fruit a capsule with 2
seeds per cell.

A genus of about 7 species from China,
Japan & S Korea grown for their delicate
brush-like flower-spikes. They do well in
moist, humus-rich loam in partial shade
such as found in woodland gardens. Some
winter protection may be needed. They are
easily divided in early spring or just after
flowering and can also be grown from seed.

**1. C. japonica** Maximowicz. Illustration:
Botanical Magazine, 6510 (1880); RHS
Dictionary of gardening, 463 (1951);
Hutchinson, The families of flowering
plants, 595 (1959).
Rhizome oblong. Basal leaves several in a
loose rosette, to 8 × 3 cm, narrowly ovate
to oblong, entire, wavy or irregularly
toothed, with short or long stalks; stem-
leaves to 4 cm, erect, linear to linear-
lanceolate. Flowering stem 15–45 cm,
erect. Flowers to 1.8 cm across in a spike
5–20 cm long. *Japan & Korea.* H3.
Spring–early summer.

### 8. STENANTHIUM (Gray) Kunth
*J.C.M. Alexander*
Perennial herbs with small bulbs. Basal
leaves 2–4, long and narrow, bright green,
arched above, hairless, acute. Flowering
stem slender, erect; stem-leaves few, bract-
like above. Flowers bisexual or unisexual,
in dense or loose, terminal racemes or
panicles. Perianth with a tube below, fused
to the base of the ovary, lobes 6, linear,
curled. Stamens 6, short, borne at base of
perianth-lobes. Styles 3, free. Ovary
superior. Fruit a 3-beaked capsule with
many narrow oblong seeds.

A genus of about 5 species from
N America and Sakhalin which generally
perform well in a dry, sunny, well-drained
position on slighly acid humus-rich loam.
*S. occidentale* (and its close relative
*S. sachalinensis* Schmidt) are rather different
from the other species and are sometimes
regarded as the separate genus *Stenanthella*
Rydberg. They prefer cooler moister
conditions with some lime in the soil.

1a. Flowers star-like, white to pale
greenish or purplish in a dense
panicle; perianth-lobes fused at base
only                         **1. gramineum**
 b. Flowers bell-shaped, brownish or
greenish purple in a loose raceme;
perianth-lobes *c.* half fused
                              **2. occidentale**

**1. S. gramineum** (Ker Gawler) Morong
(*S. angustifolium* Kunth). Illustration:
Botanical Magazine, 1599 (1813); Rickett,
Wild flowers of the United States 2: pl. 37
(1967).
Bulb oblong. Basal leaves 4, linear,
30–40 × up to 1.8 cm, keeled and
channelled, spreading. Flowering stem to
1.5 m, bearing a dense panicle to 60 cm
long, with ascending, spreading or
drooping branches; branches and upper
part of stem spike-like. Flowers many,
variable in size on a single plant,
1.2–1.8 cm across, white to greenish or
purplish, fragrant, star-like, lower ones
often male only. Perianth-lobes *c.* 6 mm,
linear-lanceolate, acuminate, shortly fused
at base. Stamens very short. Style *c.* half
filament length; stigmatic branches very
short, spreading. Fruit hanging. *SE United
States.* H3. Late summer.

Var. **robustum** (Watson) Fernald
(*S. robustum* Watson). Illustration: Rickett,
Wild flowers of the United States 2: pl. 37
(1967); Everett, Encyclopedia of
horticulture, 3226 (1982). Leaves to
2.5 cm wide. Flowering stem to 1.8 m
bearing a denser panicle with ascending or
spreading branches. Fruit often erect. *SE
United States.* H3. Summer.

**2. S. occidentale** Gray. Illustration: Rickett,
Wild flowers of the United States 5: pl. 41
(1971); Clark, Wild flowers of the Pacific
northwest, 62 (1976).
Bulb ovoid. Basal leaves 2–4, linear to very
narrowly oblanceolate, 15–30 × *c.* 1.8 cm,
slightly keeled, ascending. Flowering stem
to 60 cm, bearing a loose raceme to 20 cm
long, rarely with a few short basal
branches. Flowers few, bell-shaped,
1.2–1.8 cm long, brownish or greenish

purple, slightly fragrant, hanging.
Perianth-lobes linear, fused for *c.* half their
length. Stamens *c.* half segment length.
Stigmatic branches long and slender, erect.
*Western N America.* H2. Summer.

### 9. ZIGADENUS Michaux
*S.J.M. Droop*
Perennial herbs with rhizomes or bulbs.
Leaves usually narrow, linear, hairless.
Stems simple, leafy. Flowers in terminal
panicles or racemes, greenish to yellowish
white. Perianth-segments 6, free,
spreading, persistent, ovate to lanceolate,
each with 1 or 2 greenish glands near the
base. Ovary superior or partly inferior,
3-celled with several ovules in each cell.
Styles 3, persistent. Fruit a 3-lobed
capsule which splits between the septa to
the base.

A genus of about 18 species distributed
across N America and with 1 species in
N Asia. A few species are cultivated; they
are easily grown in most garden soils.
Propagation is by division of the rootstock
or by seed. The genus name is sometimes
wrongly spelled 'Zygadenus'. The plants are
poisonous to animals.

1a. At least the lower one-fifth to one-
third of the ovary inferior; gland at
the base of each perianth-segment
reversed heart-shaped or divided into
2 equal, round glands            2
 b. Ovary completely superior; gland at
the base of each perianth-segment
ovate, its outer margin drawn out
into long, green lines            4
2a. Each perianth-segment with 2 distinct
glands at the base; plant with a
rhizome                **3. glaberrimus**
 b. Each perianth-segment with a single
gland at the base; plant with a bulb
                                      3
3a. Inflorescence usually a simple raceme,
rarely branched below, but not
branched throughout; capsule twice
as long as the persistent perianth-
segments                    **1. elegans**
 b. Inflorescence usually a panicle,
branched throughout; capsule only
slightly longer than the persistent
perianth-segments            **2. glaucus**
4a. Perianth-segments more than 1 cm,
stamens about half as long; panicle
more or less corymbose, the lower
flower-stalks much longer than the
upper                      **4. fremontii**
 b. Perianth-segments mostly less
than 8 mm, stamens as long as them
or longer; panicle not corymbose    5

5a. Perianth-segments 3–6 mm, distinctly clawed at the base; leaves not sickle-shaped  **6. venenosus**

b. Perianth-segments 6–8 mm, not clawed; leaves sickle-shaped  **5. nuttallii**

**1. Z. elegans** Pursh. Illustration: Botanical Magazine, 1680 (1814) – as Helonias glaberrima; Hitchcock et al., Vascular plants of the Pacific northwest **1**: 816 (1969).

Plant with a bulb. Stem to *c*. 80 cm. Leaves mostly basal, to 1 cm broad, sharply pointed. Inflorescence usually a simple, loose raceme, sometimes branched at the base. Upper bracts with scarious margins. Perianth-segments 8–12 mm, each with a heart-shaped gland near the base. Ovary with its lower fifth to third inferior. Capsule conical, about twice as long as the persistent perianth-segments. *Western N America*. H4. Summer.

**2. Z. glaucus** Nuttall. Illustration: Rickett, Wild flowers of the United States **2**: t. 13 (1967).

Like *Z. elegans*, but leaves blunt, bracts fleshy throughout, inflorescence a panicle, the capsule only slightly exceeding the persistent perianth-segments. *Eastern N America*. H3. Summer.

The differences between this and *Z. elegans* are not always clear-cut.

**3. Z. glaberrimus** Michaux. Illustration: Botanical Magazine, 1703 (1815) – as Helonias bracteata; Rickett, Wild flowers of the United States **2**: t. 13 (1967).

Plant with a rhizome. Stems to 1.2 m, leafy. Inflorescence a loosely pyramidal panicle. Perianth-segments 1–1.5 cm, clawed at the base, each with 2 distinct glands at the base. Ovary with its lower fifth to third inferior. Capsule usually shorter than the persistent perianth-segments. *Southeast N America*. H4. Summer.

**4. Z. fremontii** (Torrey) Watson. Illustration: Botanical Magazine, n.s., 361 (1962); Rickett, Wild flowers of the United States **5**: t. 8 (1971).

Plant with a bulb. Basal leaves to 3 cm wide. Stem smooth, leafy. Inflorescence racemose or paniculate, often corymbose, the lower flower-stalks much longer than the upper. Perianth-segments 1–1.6 cm, each with a single gland near the base, the margins toothed towards the apex. Stamens about half as long as the perianth-segments. Ovary completely superior. Capsule 1.5–3 cm, oblong. *Western N America*. H4. Summer.

**5. Z. nuttallii** Gray. Illustration: Rickett, Wild flowers of the United States **6**: t. 8 (1973).

Plant with a bulb. Stems to 75 cm, leafy. Leaves mostly basal, *c*. 1 cm wide, keeled and sickle-shaped. Inflorescence usually a raceme, sometimes branched below. Perianth-segments 6–8 mm, not clawed, with a single gland at the base, the margins toothed towards the apex. Stamens about equal to the perianth-segments. Ovary completely superior. Capsule ovoid to ellipsoid, 3–4 times longer than the persistent perianth-segments. *Western N America*. H4. Summer.

**6. Z. venenosus** Watson. Illustration: Rickett, Wild flowers of the United States **5**: t. 8 (1971).

Plant with a bulb. Stem to 60 (rarely 70) cm. Leaves mostly basal, stem leaves reduced. Inflorescence usually a raceme, occasionally branched below. Perianth-segments to 3–6 mm, clawed, those of the outer whorl slightly shorter than those of the inner, each with a single gland at the base, margins toothed towards the apex. Stamens exceeding the perianth-segments. Ovary completely superior. Capsule 8–15 mm, cylindric. *Western N America*. H4. Summer.

**10. MELANTHIUM** Linnaeus
*J. Cullen*

Perennial herbs with rhizomes and fibrous roots. Leaves basal and on the stem. Inflorescence a panicle with its axis, branches and the backs of the flowers roughly hairy. Flowers bisexual and male in the same inflorescence. Perianth of 6, free, widely spreading segments, each with a conspicuous, narrow claw and an expanded blade which has 2 glands (appearing as dark spots) at the base. Stamens 6, borne on the claws of the perianth-segments. Ovary superior, cylindric, 3-celled with several ovules in each cell; styles 3, free, divergent. Fruit an inflated capsule, 3-lobed in section, 3-beaked at the apex. Seeds flat, broadly winged.

A genus of about 5 species from N America. One or two are occasionally grown as ornamentals. Cultivation as for *Veratrum*.

**1. M. virginicum** Linnaeus. Illustration: Botanical Magazine, 985 (1807); House, Wild flowers, t. 10 (1961); Rickett, Wild flowers of the United States **1**(1): t. 8 (1965).

Stem to 1.7 m. Leaves linear, folded but not obviously tapering towards the base, 1.5–3 cm broad. Perianth-segments at first cream, later green, the blade broadly oblong to ovate, flat, blunt, 2–3 times longer than the claw. *Eastern USA, from New York to Texas, inland to Indiana*. H1. Summer.

**\*M. hybridum** Walter (*M. latifolium* Desrousseaux). Illustration: Rickett, Wild flowers of the United States **1**(1): t. 8 (1965). More slender, leaves tapering to the base, perianth-segments with an almost circular blade which has crisped-undulate margins, and is about as long as the claw. *Eastern USA from Connecticut to Georgia*. H1. Summer.

**11. VERATRUM** Linnaeus
*S.J.M. Droop*

Robust perennial herbs with rhizomes. Stems simple, leafy, usually hairless below, downy or coarsely hairy above. Leaves alternate, broadly ovate to elliptic, many-veined, pleated, more or less erect and overlapping, narrowed into long, sheathing bases. Flowers numerous in a terminal panicle or raceme, usually bisexual, occasionally male only. Perianth-segments 6, free, widely spreading, elliptic or lanceolate, persistent in fruit, white, green, reddish brown or almost black. Stamens 6, free. Ovary superior, 3-celled, ovules numerous. Fruit a 3-celled capsule containing numerous seeds.

A genus of about 45 species widely distributed in the northern hemisphere. Only a few are grown; they do best in moist, rich soils and are propagated by division of the rhizome.

Literature: Loesener, O., Die Gattung Veratrum, *Feddes Repertorium* **24**: 61–72 (1927), **25**: 1–10 (1928).

1a. Flowers reddish brown to black; leaves hairless  **2. nigrum\***

b. Flowers white to green; leaves shortly downy beneath  **1. album\***

**1. V. album** Linnaeus (*Helonias viridis* misapplied). Illustration: Botanical Magazine, 1096 (1808); Polunin, Flowers of Europe, t. 162 (1969).

Plant to 2 m. Leaves hairless above, hairy beneath, the lower elliptic to broadly elliptic, the upper progressively narrower. Flowers in a much-branched panicle with erect to drooping branches; flower-stalks 2–3 mm. Perianth-segments to 1.8 cm, white to pale green inside, sparsely to densely hairy outside, margins wavy to irregularly toothed. Capsule hairless to sparsely downy. *North temperate regions*. H1. Summer.

Variable and sometimes the N American species split off as below:

**\*V. californicum** Durand. Illustration: Rickett, Wild flowers of the United States **5**: t. 9 (1971). Leaves oblong to ovate, loosely hairy beneath, the hairs mostly near the margins; lower panicle branches not drooping; perianth-segments white to pale green, very sparsely hairy to hairless outside, margins entire. *Northwest USA*. H5. Summer.

**\*V. viride** Aiton. Illustration: Rickett, Wild flowers of the United States **5**: t. 9 (1971). Leaves oblong to ovate, loosely hairy beneath, panicle with distinctly drooping lower branches, perianth-segments pale to bright green, hairy outside at least on the midrib and margins, the margins of at least some irregularly toothed. *N America*. H5. Summer.

**2. V. nigrum** Linnaeus. Illustration: Botanical Magazine, 963 (1806); Polunin, Flowers of Europe, t. 162 (1969). Plant to 1.3 m. Leaves hairless, the lower to 35 cm, broadly elliptic, the upper progressively smaller and narrower. Flowers in a raceme or sparsely branched panicle. Perianth-segments usually less than 7 mm, reddish brown to black, hairless or sparsely hairy outside. Capsule hairless. *Eurasia*. H1. Summer.

**\*V. maackii** Regel. Similar, but with smaller, narrower, lanceolate leaves. *Temperate Asia*. H1.

## 12. UVULARIA Linnaeus
*V. A. Matthews*

Rootstock a rhizome. Stem simple or branched, leafy. Leaves alternate, stalkless, sometimes perfoliate. Flowers drooping, yellow, narrowly bell-shaped, the perianth-segments free. Stamens much shorter than perianth-segments; anthers opening by slits. Ovary superior. Style with 3 branches. Fruit a 3-lobed or 3-winged capsule which splits between the septa.

A genus of 5 species native to eastern N America. The species with non-perfoliate leaves were formerly placed in the genera *Oakesia* Watson and *Oakesiella* Small. They are best grown in shade, in a well-drained soil. Propagation is by division.

1a. Leaves perfoliate                              2
  b. Leaves not perfoliate     **3. sessilifolia\***
2a. Leaves hairless beneath; perianth-segments 2–3.5 cm, pale yellow with tiny projecting glands on the inner surface               **2. perfoliata**
  b. Leaves downy beneath; perianth-

segments 2.5–5 cm, bright yellow and smooth on the inner surface
                                  **1. grandiflora**

**1. U. grandiflora** Smith. Illustration: Botanical Magazine, 1112 (1808); Everard & Morley, Flowers of the world, 167 (1970); The Garden **100**: 527 (1975); Klein, Gartenblumen 1: Frühlingsblumen, 8 (1979). Stem to 75 cm, hairless. Leaves perfoliate, 5–13 cm, oblong to narrowly ovate, downy beneath. Perianth-segments 2.5–5 cm, bright yellow, inner surface smooth. Stamens longer than the style. *Eastern N America*. H3. Spring–summer.

**2. U. perfoliata** Linnaeus. Illustration: Smith, Exotic Botany **1**: t. 49 (1804); Rickett, Wild flowers of the United States **1**: pl. 27 (1965); Peterson & McKenny, Field guide to wildflowers of north-eastern and north-central North America, 103 (1968). Stem 20–60 cm, hairless. Leaves perfoliate, 5–9 cm, oblong to narrowly ovate, hairless, paler beneath. Perianth-segments 2–3.5 cm, pale yellow, inner surface with tiny projecting glands. Stamens shorter than the style. *Eastern N America*. H3. Spring–early summer.

**3. U. sessilifolia** Linnaeus (*Oakesiella sessilifolia* (Linnaeus) Small). Illustration: Smith, Exotic Botany **1**: t. 52 (1804); Botanical Magazine, 1402 (1811); Rickett, Wild flowers of the United States **1**: pl. 27 (1965); Niering & Olmstead, Audubon Society field guide to North American wildflowers: Eastern region, pl. 318 (1979). Stem to 40 cm, hairless. Leaves 3.5–7.5 cm, narrowly oblong, hairless, paler beneath, margin rough. Perianth-segments 1.5–3 cm, greenish yellow, inner surface smooth. Ovary and capsule borne on a stalk. *Eastern N America*. H3. Spring–summer.

**\*U. caroliniana** (Gmelin) Wilbur (*U. pudica* (Walker) Fernald; *U. puberula* Michaux). Illustration: Loddiges' Botanical Cabinet **13**: t. 1260 (1827). Stem with lines of downy hairs, leaves bright green on both surfaces and ovary and capsule not borne on a stalk. *SE USA*. H4. Spring–summer.

## 13. TRICYRTIS Wallich
*P.F. Yeo*

Perennial herbs. Stems (in cultivated species) 20–110 cm, simple or few-branched, leafy. Leaves 5–20 cm, alternate, sometimes in 2 rows, ovate to lanceolate, stalkless, lightly pleated, sometimes spotted

with dark green. Flowers bisexual. Perianth-segments erect at base, spotted, 3–6 times as long as broad, the three outer each with a conspicuous, sometimes bilobed, spherical or ovoid nectarial pouch. Stamens curved outwards at the top. Ovary superior, 3-celled, narrowly oblong; style single, with 3 recurved bifid stigmatic branches. Perianth-segments (mainly on upper surfaces), stamen filaments and stigmatic branches usually with violet to reddish purple spots. Fruit a fusiform, furrowed capsule opening along the septa. Seeds flat.

A genus of 11–16 species from the mountains of Japan, China, Taiwan, Philippines and E Himalaya, often growing in woods. They are suitable for semi-shade or full sun and benefit from a moist and humus-rich soil. Autumn-flowering species need a warm position in northern countries to ensure flowering before the first frosts. The flowers are unusual in both form and colour.

Literature: Masamune, G., *Journal of Society of Tropical Agriculture (Taiwan)* **2**: 38–48 (1930); Mathew, B., *The Plantsman* **6**: 193–225 (1985).

*Stems*. Arching and drooping: **1,4**. Not more than 15 cm: **2**.
*Flowers*. Bell-shaped: **1**.
*Ground-colour of flowers*. Yellow: **1–5**; white or pinkish: **6–8**.
*Flowers*. On separate stalks: **1–4,7**; in cymes: **5–8**.

1a. Ground-colour of flowers yellow      2
  b. Ground-colour of flowers white or pinkish                            6
2a. Stems arching and drooping; perianth-segments not mucronate      3
  b. Stems erect; perianth-segments mucronate, conspicuously so in bud
                                        4
3a. Leaves cordate and embracing stem; flowers drooping, bell-shaped
                              **1. macrantha**
  b. Leaves perfoliate; flowers ascending, funnel-shaped      **4. perfoliata**
4a. Flowers 2 to several in branched cymes                **5. latifolia**
  b. Flowers 1 or few on their own stalks in the leaf-axils                 5
5a. Stems 5–15 cm; flower-stalk much shorter than flower      **2. nana**
  b. Stems 20–50 cm; flower stalks longer than or slightly shorter than flower
                                **3. flava**
6a. Perianth-segments not more than 2 cm, with the upper two-thirds spreading abruptly or recurved
                              **7. macropoda\***

b. Perianth-segments 2.5–3 cm,
   diverging gradually                              7
7a. Plant tufted, thickly hairy; leaves pale
   green, not glossy, usually evenly
   tapered from near the base      **8. hirta**
 b. Plant with stolons, moderately or
   sparsely hairy; leaves deep green,
   more or less glossy, ovate to elliptic
                                         **6. formosana**

**1. T. macrantha** Maximowicz. Illustration:
Makino, An illustrated flora of Japan, edn
2, 756 (1961); Kitamura et al., Coloured
illustrations of herbaceous plants of Japan
(Monocotyledoneae), t. 39, f. 247 (1970);
Garden **107**: 228 (1982).
Stems arching and drooping, with coarse
pale brown hairs (except in subsp.
**macranthopsis** (Masamune) Kitamura).
Leaves ovate-oblong to lanceolate, cordate,
embracing the stem. Flowers 3–4 cm long,
in upper axils, 2 or few on separate stalks,
bell-shaped, drooping; ground-colour
yellow; nectarial pouch ovoid, bent
outwards. *Japan*. H2. Early autumn.

**2. T. nana** Yatabe. Illustration: Makino, An
illustrated flora of Japan, edn 2, 755
(1961); Kitamura et al., Coloured
illustrations of herbaceous plants of Japan
(Monocotyledoneae), t. 39, f. 249 (1970).
Stems 5–15 cm, erect, shortly hairy. Leaves
few, lanceolate to ovate, embracing the
stem, heavily blotched. Flower-stalks much
shorter than flower and fruit. Flowers few,
solitary in leaf-axils, erect, funnel-shaped.
Perianth-segments about 2 cm, with apical
bristle to 2 mm; ground-colour yellow.
Anthers 2 mm. *Japan*. H2.

**3. T. flava** Maximowicz. Illustration:
Makino, An illustrated flora of Japan, edn
2, 755 (1961); Kitamura et al., Coloured
illustrations of herbaceous plants of Japan
(Monocotyledoneae), t. 39, f. 250, 251
(1970).
Stems 20–50 cm, erect, hairy. Leaves
embracing the stem, blotched, the upper
more or less broadly ovate. Flower-stalks
longer than flower and fruit or only slightly
shorter. Flowers 1 or few on separate stalks
in leaf-axils, erect, funnel-shaped. Perianth-
segments about 2.5 cm; apical bristle of outer
segments 3–4 mm; ground-colour yellow.
Anthers 2.5–3 mm. *Japan*. H2. Autumn.

Subsp. **ohsumiensis** (Masamune)
Kitamura (Illustration: Kitamura et al.,
Coloured illustrations of herbaceous plants
of Japan (Monocotyledoneae), t. 39, f. 251,
1970) has perianth-segments 2.5–3.5 cm,
with apical bristle not more than 2 mm
and anthers about 4 mm. *Japan*. H2.

**4. T. perfoliata** Masamune. Illustration:
Kitamura et al., Coloured illustrations of
herbaceous plants of Japan
(Monocotyledoneae), t. 39, f. 248 (1970);
The Plantsman **6**: 204 (1985).
Stems 50–70 cm, arching and drooping,
hairless. Leaves perfoliate. Flowers
ascending, solitary in leaf-axils, funnel-
shaped. Perianth-segments 2.5–3 cm, with
bright yellow ground-colour. *Japan*. H2.

**5. T. latifolia** Maximowicz (*T. bakeri*
Koidzumi; *T. macropoda* misapplied;
*T. puberula* Nakai & Kitagawa). Illustration:
Botanical Magazine, 6544 (1881);
Kitamura et al., Coloured illustrations of
herbaceous plants of Japan
(Monocotyledoneae), t. 40, f. 252 (1970);
The Plantsman **6**: 193 (1985).
Stems erect, hairy above. Leaves broadly
ovate, cordate, thin, embracing the stem.
Flowers 2–2.5 cm, erect, 2 or several in
branched cymes, funnel-shaped, with
bracts. Perianth-segments diverging from
the middle; ground-colour yellow. *Japan,
China*. H2. Summer.

**6. T. formosana** Baker (*T. stolonifera*
Matsumura). Illustration: Botanical
Magazine, 8560 (1914).
Plant forming a dense even stand of erect,
somewhat hairy stems and spreading by
stolons. Leaves more or less glossy, deep
green with darker green spots, especially
when young, the lower narrowed towards
base, continued downwards for about
1–1.5 cm as a stem-clasping sheath; upper
leaves ovate, cordate, embracing the stem.
Flowers 2.5–3 cm, erect, several in a
loosely branched terminal cyme, widely
funnel-shaped; perianth-segments with a
yellowish zone above the nectarial pouches,
otherwise more or less brownish purple
outside in bud and pinkish purple above the
yellow zone when open, nearly white inside
but densely purple-spotted. *Taiwan*. H3.
Early autumn.

 Some plants in cultivation in Britain
under the names *T. stolonifera* and
*T. formosana* have cordate-clasping, non-
sheathing leaf-bases throughout. They
have well-developed stolons but are
probably hybrids between *T. formosana* and
*T. hirta*. They vary in hairiness, leaf-colour
and intensity of spotting of the flower.

**7. T. macropoda** Miquel. Illustration:
Makino, An illustrated flora of Japan, edn
2, 754 (1961); Kitamura et al., Coloured
illustrations of herbaceous plants of Japan
(Monocotyledoneae), t. 40, f. 253 (1970).
Plants sometimes hairy. Stems erect. Leaves

rounded or slightly cordate at base and
forming a short open sheath. Flowers
1.5–2 cm long, erect, several in terminal
and axillary cymes which may be
glandular-hairy; perianth-segments with
the upper two-thirds reflexed; ground-
colour white. *Japan, China*. H2. Early
autumn.

 *****T. maculata** (D. Don) Macbride.
Illustration: Botanical Magazine, 4955
(1855). Hardly different from *T. macropoda*.
Cymes loose, repeatedly branched, with
flowers held well above the leaves. Upper
two-thirds of perianth-segments spreading.
*C Nepal to Burma*. H2. Summer.

 *****T. affinis** Makino (*T. macropoda* Miquel
subsp. *affinis* (Makino) Kitamura).
Illustration: Makino, An illustrated flora of
Japan, edn 2, 754 (1961); Kitamura et al.,
Coloured illustrations of herbaceous plants
of Japan (Monocotyledoneae), t. 40, f. 254
(1970). Plant hairy. Flowers like those of
*T. macropoda* but on separate stalks and
with upper two-thirds of perianth-segments
spreading, not reflexed. *Japan*. H2. Autumn.

**8. T. hirta** (Thunberg) W.J. Hooker.
Illustration: Botanical Magazine, 5355
(1863); Makino, An illustrated flora of
Japan, edn 2, 754 (1961); Kitamura et al.,
Coloured illustrations of herbaceous plants
of Japan (Monocotyledoneae), t. 40, f. 255
(1970).
Plant tufted, without stolons, thickly hairy.
Stems slightly divergent from the rootstock.
Leaves cordate, the base forming a short
open sheath, pale green, usually evenly
tapered from near base. Flowers 2.5–3 cm
long, erect, 1–5 in axils on separate stalks
usually shorter than flowers, or the upper
sometimes in cymes with short internodes,
funnel-shaped; perianth-segments
spreading or recurved only at the tips;
ground-colour white, spots violet or purple,
sometimes lacking. *Japan*. H2. Early
autumn.

 Sometimes cultivated and illustrated in
books as *T. macropoda*.

**14. HERPOLIRION** J.D. Hooker
*E.H. Hamlet*
Rootstock a creeping, wiry rhizome. Leaves
6–8, in 2 ranks, 2–6 cm × 1.5–4 mm,
bluish green, strongly keeled towards the
base. Flowering stem short, erect, covered
with the sheathing bases of the leaves,
bearing a solitary terminal flower.
Perianth-segments free, 1.1–1.5 cm × c.
5 mm, pale blue to white, or cream, linear
to narrowly obovate, spreading from near
the middle, the inner 3 slightly narrower.

Filaments finely downy, except at the apex; anthers almost basifixed, 2–3 mm long. Ovary superior, more or less spherical with many ovules per cell. Style 1, cylindric, narrowing to small stigma. Capsule with 1 or 2 seeds per cell. Seeds black, more or less flattened.

A genus of one species from SE Australia and New Zealand. In cultivation it prefers a moist soil and needs protection in winter. Propagation is by seed or division of the rhizome.

**1. H. novae-zelandiae** J.D. Hooker. Illustration: Salmon, Field guide to the alpine plants of New Zealand, 170 (1968); Cochrane et al., Flowers and plants of Victoria, 167 (1973); Moore & Unwin, The Oxford book of New Zealand plants, 184 (1978); Costin et al., Kosciusko alpine flora, pl. 137 (1979). *SE Australia, New Zealand.* H5. Summer.

**15. STYPANDRA** R. Brown
*E.H. Hamlet & V.A. Matthews*
Differs from *Dianella* mainly in the filaments being woolly at the top, and in having a capsule rather than a berry.

A genus of 6 species from temperate Australia; one is sometimes cultivated. A sunny position is best, as is a well-drained soil. Propagation is by seed or division of the rhizome.

**1. S. caespitosa** R. Brown. Illustration: Cochrane et al., Flowers and plants of Victoria, 30 (1968); Willis et al., Field guide to the flowers and plants of Victoria, 41 (1975); Galbraith, Collins field guide to the wild flowers of south-east Australia, pl. 2 f., 7 (1977).
Flowering stem 30–60 cm. Leaves 10–35 cm × 5–8 mm. Flowers blue or yellowish, occasionally white, erect on 2.5–5 cm stalks. Perianth-segments 8–15 mm. *SE Australia.* H4. Summer.

**16. DIANELLA** Lamarck
*E.H. Hamlet & V.A. Matthews*
Rhizomatous perennials. Leaves in 2 ranks crowded at the base of the stem, linear or lanceolate, keeled beneath. Flowers borne in a panicle. Perianth-segments free, spreading, with 3–8 veins. Filaments thickened; anthers basifixed, opening by terminal pores. Ovary superior; style 1. Fruit a berry.

A genus of about 30 species found in tropical Asia, Australia, New Zealand and Polynesia. They need a sheltered position or cool greenhouse conditions in the colder parts of Europe. They are propagated by division or from seed.

1a. Leaves with smooth margins  2
 b. Leaves with rough margins  3
2a. Anthers twice as long as filaments  **4. laevis**
 b. Anthers three or more times as long as filaments  **5. revoluta**
3a. Flowers greenish white or purplish white  **6. intermedia\***
 b. Flowers pure white or blue  4
4a. Anthers brownish yellow, shorter than or equal in length to the filaments  **3. tasmanica**
 b. Anthers yellow, longer than the filaments  5
5a. Flowering stem less than 60 cm; leaves 6–13 mm broad, rarely to 1.5 cm  **2. caerulea**
 b. Flowering stem more than 90 cm; leaves 1.3–3.3 cm broad  **1. ensifolia**

**1. D. ensifolia** Redouté (*D. nemorosa* Lamarck). Illustration: Botanical Magazine, 1404 (1811); Yatabe, Iconographia Florae Japonicae 1(3): t. 58 (1893); Koorders, Exkursionsflora von Java 1: 290 (1911). Flowering stem 90–180 cm, leafy in lower part. Leaves up to 30 × 1.3–3.3 cm, lanceolate, margins and keel rough. Inflorescence a spreading panicle. Perianth-segments white or blue, 7.5–8 mm. Anthers yellow, longer than the filaments. Berry yellow when unripe, turning to bluish purple. *Tropics of the Old World.* G1–H5. Summer.

**2. D. caerulea** Sims. Illustration: Botanical Magazine, 505 (1801); Addisonia 18: pl. 585 (1933).
Flowering stem 50–60 cm, sometimes branched. Leaves 10–60 cm × 6–13 mm, rarely to 1.5 cm, margins and keel rough. Inflorescence a loose panicle but sometimes clustered, 10–30 cm. Perianth-segments white or blue, *c.* 7.5 mm. Anthers pale yellow, 2–3 times as long as the filaments. Berry blue. *E & S Australia.* H5. Summer.

**3. D. tasmanica** Hooker. Illustration: Burbidge & Gray, Flora of the Australian Capital Territory, pl. 96 (1970); Hay et al., Dictionary of indoor plants, pl. 182 (1974) – fruit; Costin et al., Kosciusko alpine flora, pl. 134–6 (1979); Australian Plants 12: 156 (1983) – fruit.
Flowering stem to 150 cm. Leaves 30–120 × 1–3 cm, rigid, broadly lanceolate, keel rough, margins revolute and very sharply toothed. Panicle robust, much branched, loose, 30–60 cm. Perianth-segments pale blue. Anthers brownish yellow, shorter than, or equal in length to the filaments. Berry blue to violet-

blue, 1–2 cm. *SE Australia, Tasmania.* H4. Summer.

**4. D. laevis** Brown. Illustration: Black, Flora of South Australia, edn 3, 1: f. 307 (1978).
Flowering stem to 90 cm. Leaves 20–60 cm × 6–10 mm, margins and keel smooth. Panicle loose and spreading, 20–50 cm. Perianth-segments blue. Anthers yellow, about twice as long as the filaments. Berry blue. *Australia (New South Wales).* H4. Summer.

**5. D. revoluta** Brown. Illustration: Edwards's Botanical Register 13: t. 1120 (1828); Rotherham et al., Flowers and plants of New South Wales and South Queensland, pl. 137 (1975).
Flowering stem to 120 cm. Leaves 20–90 cm × 4–15 mm, linear, margins and keel smooth, margins revolute. Panicle loose, 20–60 cm. Perianth-segments blue, 7–10 mm. Anthers yellow, brown or black, three or more times as long as filaments. *S & E Australia, Tasmania.* H4. Summer.

**6. D. intermedia** Endlicher.
Leaves 30–150 × 1.2–1.8 cm, linear to lanceolate, margins and keel rough. Panicle 15–60 cm. Perianth-segments more than 5 mm, greenish white or purplish white. Anthers yellow. Berry blue, 8–15 mm. *Norfolk Island, Fiji.* G1. Summer.
**\*D. nigra** Colenso (*D. intermedia* Endlicher var. *norfolkensis* Brown). Illustration: Moore & Irwin, The Oxford book of New Zealand plants, 183 (1978). Differs from *D. intermedia* in having a more slender flowering stem and perianth-segments less than 5 mm. *New Zealand.* H4.

**17. ASPHODELUS** Linnaeus
*V.A. Matthews*
Annuals, or perennials with rhizomes. Leaves basal, linear, with a membranous sheathing base. Inflorescence a dense raceme or panicle. Bracts membranous. Flower-stalks jointed. Perianth-segments free or united at the extreme base, 1-veined, spreading. Stamens shorter than perianth-segments, free. Anthers dorsifixed. Ovary superior, 3-celled. Fruit a capsule. Seeds winged, 6.

A genus of about 12 species growing from the Mediterranean to the Himalaya. They will grow in a sunny position in any well-drained soil and are best propagated by division.

*Leaves.* More or less round, hollow: **1.**
*Flowering stem.* Absent or very short: **5**: Up to 1 m: **1–3**; more than 1 m: **3,4**.
  Hollow: **1**; solid: **2–4**.

*Bracts*. Pale green: **3**; brown: **2**.
*Perianth-segments*. 5–14 mm: **1,4**;
  1.5–2 cm: **2,3**; 3.5–4 cm: **5**.

1a. Flowering stem absent or very short;
    perianth-segments 3.5–4 cm
                                   **5. acaulis**
  b. Flowering stem present; perianth-
     segments 5–20 mm                    **2**
2a. Perianth-segments 5–14 mm            **3**
  b. Perianth-segments 1.5–2 cm          **4**
3a. Flowering stem 15–70 cm, hollow;
    leaves more or less round
                                   **1. fistulosus**
  b. Flowering stem 1–2 m, solid; leaves
     flat                          **4. aestivus**
4a. Inflorescence unbranched or with a
    few short branches; leaves without a
    keel beneath; capsule 1.6–2 cm, if less
    then bracts dark brown         **2. albus**
  b. Inflorescence much-branched with
     the side-branches almost as long as
     the terminal; leaves with a shallow
     keel beneath; capsule 8–14 mm;
     bracts whitish to pale green
                                    **3. ramosus**

**1. A. fistulosus** Linnaeus (*A. tenuifolius*
Cavanilles). Illustration: Reichenbach,
Icones florae Germanicae et Helveticae **10**:
pl. 513 (1848); Polunin & Huxley, Flowers
of the Mediterranean, f. 238 (1965).
Annual or short-lived perennial. Leaves
5–35 cm, more or less round, hollow.
Flowering stem 15–70 cm, hollow,
branched or unbranched. Bracts whitish.
Perianth-segments 5–12 mm, white or pale
pink with a pinkish or brownish midvein.
Capsule 5–7 mm. *Mediterranean area,
SW Europe, SW Asia, India*. H4. Summer.

**2. A. albus** Miller (*A. delphinensis* Grenier &
Godron). Illustration: Taylor, Wild flowers
of Spain and Portugal, 95 (1972); Grey-
Wilson & Blamey, Alpine flowers of Britain
and Europe, 269 (1979); Polunin, Flowers
of Greece and the Balkans, pl. 55 (1980).
Perennial. Leaves 15–60 cm, flat.
Flowering stem 30–100 cm, solid, usually
unbranched or with a few short branches.
Bracts membranous, whitish or dark
brown. Perianth-segments 1.5–2 cm, white
or pale pink with a darker midvein. Capsule
8–20 mm. *S & C Europe*. H4.
Spring–summer.

**3. A. ramosus** Linnaeus (*A. cerasiferus* Gay).
Illustration: Reichenbach, Icones florae
Germanicae et Helveticae **10**: pl. 514
(1848).
Perennial. Leaves 15–40 cm, flat, with a
shallow keel beneath. Flowering stem
40–150 cm, solid, usually much-branched

with the side branches almost as long as
the terminal. Bracts whitish to pale green.
Perianth-segments 1.5–2 cm, white with a
reddish brown midvein. Capsule 8–14 mm.
*S Europe, N Africa*. H4. Summer.

**4. A. aestivus** Brotero (*A. microcarpus*
Salzmann & Viviani). Illustration: Polunin,
Flowers of Europe, pl. 163 (1969); Megaw,
Wild flowers of Cyprus, pl. 30 (1973);
Nehmeh, Wild flowers of Lebanon, pl. 21,
22 (1978).
Perennial. Leaves 25–45 cm, flat.
Flowering stem 1–2 m, solid, branched
with short side branches. Bracts whitish.
Perianth-segments 1–1.4 cm, white with a
pink or brownish midvein. Capsule
5–7 mm. *Canary Islands, through S Europe
and N Africa to Turkey*. H4. Spring–early
summer.

**5. A. acaulis** Desfontaines. Illustration:
Botanical Magazine, 7004 (1888); Bulletin
of the Alpine Garden Society **34**: 172
(1966); RHS Lily Year Book **30**: f. 24
(1966).
Perennial. Leaves 15–30 cm, flat.
Flowering stem absent or very short. Bracts
whitish. Perianth-segments 3.5–4 cm,
white or pink with a greenish midvein.
*Algeria, Morocco*. H5. Spring.

**18. ASPHODELINE** Reichenbach
*V.A. Matthews*
Biennial or perennial herbs with clusters of
fleshy roots. Leaves numerous, linear, with
wide, membranous sheathing bases and
often rough margins. Inflorescence
branched or unbranched with the flowers
in racemes. Bracts membranous. Flower-
stalks jointed. Flowers more or less
bilaterally symmetric. Perianth united at
the base, each lobe with 3 central veins
(superficially appearing as 1 vein), the
outer 3 narrower than the inner. Stamens
unequal; stamens and style curved
downwards; anthers dorsifixed. Ovary
superior, 3-celled. Fruit a capsule with 6
unwinged, sharply angled seeds.
   A genus with 19 species occurring from
the Mediterranean area to the Caucasus.
The plants should be grown in a sunny
position. Propagation is by division.

1a. Flowers white            **3. taurica***
  b. Flowers yellow                      **2**
2a. Stem leafy up to the raceme; capsule
    1–1.5 cm                    **1. lutea**
  b. Leaves confined to the lower half of
     the stem; capsule 6–9 mm
                              **2. liburnica***

**1. A. lutea** (Linnaeus) Reichenbach
(*Asphodelus luteus* Linnaeus). Illustration:
Polunin, Flowers of Europe, pl. 162 (1969);
Huxley & Taylor, Flowers of Greece, f. 335
(1977).
Perennial. Stems to 1 m, leafy up to the
raceme. Leaves with smooth margins,
sometimes rough towards the apex.
Inflorescence unbranched. Bracts longer
than the fruit-stalks. Perianth-lobes
2–3 cm, yellow. Capsule 1–1.5 cm, ovoid
to spherical. *Mediterranean area*. H4.
Spring–summer.

**2. A. liburnica** (Scopoli) Reichenbach.
Illustration: Polunin, Flowers of Greece and
the Balkans, f. 1594 (1980).
Perennial. Stems to 1 m, leafy in the lower
half. Inflorescence usually unbranched.
Bracts shorter than the fruit-stalks.
Perianth-lobes 2.3–3 cm, yellow. Capsule
7–8 mm, spherical. *S Europe, Turkey*. H4.
Summer.
   *****A. brevicaulis** (Bertoloni) Baker. Stems
only to 50 cm. Perianth-lobes 1.5–2 cm,
pale yellow. *SW Asia*. H4. Spring–summer.

**3. A. taurica** (Pallas) Kunth. Illustration:
Gardener's Chronicle **21**: 175 (1897); Flora
SSSR **4**: pl. 3 f. 1 (1935).
Perennial. Stems to 1 m, leafy up to the
raceme. Bracts longer than the fruit-stalks.
Perianth-lobes 1.2–1.7 cm, white. Capsule
8–12 mm, ovoid to oblong. *E Mediterranean
area, USSR (Caucasus)*. H4. Summer.
   *****A. tenuior** (Fischer) Ledebour. Stem
15–40 cm, leafy in the lower half or all
leaves basal, leaf sheaths with ciliate
margins. Inflorescence branched or
unbranched. Perianth-lobes 1.7–2 cm,
whitish, sometimes tinged with pink.
Capsule 6–8 mm, spherical to obovoid.
*Turkey, Iran, USSR (Caucasus)*. H4. Summer.
   *****A. damascena** (Boissier) Baker
(*A. balansae* Baker; *A. isthmocarpa* Baker).
Biennial. Stem to 1.5 m, leafy to slightly
above the middle or all leaves basal, leaf
sheaths not ciliate. Inflorescence branched
or unbranched. Perianth-lobes 1.5–2.5 cm,
white to pale pink. Capsule pear-shaped to
cylindric. *SW Asia*. H4. Summer.

**19. PARADISEA** Mazzucato
*V.A. Matthews*
Similar to *Anthericum* but flowers always
borne in a raceme, perianth trumpet-shaped,
the segments with a claw, style and stamens
curved upwards and anthers dorsifixed.
   There are 2 species, native to the
mountains of S Europe. Cultivation as for
*Anthericum* (p. 131). Propagation is by
division of the rootstock, or by seed.

**1. P. liliastrum** (Linnaeus) Bertoloni (*Anthericum liliastrum* Linnaeus). Illustration: Robert, Alpine flowers, pl. 8 (1938); Uberto, Mountain flowers, 45 (1978); Grey-Wilson & Blamey, Alpine flowers of Britain and Europe, 265 (1979); Grey-Wilson & Mathew, Bulbs, pl. 12 (1981).

Stem 30–60 cm. Leaves 4–7, 12–25 cm. Raceme with 3–10 flowers, usually rather 1-sided. Perianth-segments 3–5 cm, white often tipped with green. Capsule 1.3–1.5 cm. *Mountains of S Europe.* H3. Summer.

'Major' has larger flowers and is a more robust plant.

## 20. EREMURUS Bieberstein
*V.A. Matthews*

Perennials with thick rhizomes, the rhizome neck bearing fibrous or membranous remains of old leaves. Leaves basal, forming a tuft or rosette, usually narrow and often keeled beneath. Flowers many, borne in a raceme on an unbranched stem. Bracts membranous. Perianth-segments 6, almost free, with 1, 3 or 5 dark central veins, the inner segments often broader than the outer. Anthers basifixed. Ovary superior; style 1. Fruit a 3-celled, usually spherical capsule. Seeds usually winged.

A genus of 40–50 species native to W and C Asia. They grow best in a rich well-drained soil in a sunny position. Propagation is by division or seed. Literature: Wendelbo, P., On the genus Eremurus in south-west Asia, *Acta Universitatis Bergensis. Series Mathematica Rerumque Naturalum* No. 5 (1964); Fedtschenko, O., Eremurus, Kritische Übersicht über die Gattung, *Cramer Plant Monograph Reprints* 3 (1968); Wendelbo, P. & Furse, P., Eremurus of south-west Asia, *RHS Lily Year Book* 1969: 56–69 (1968).

*Leaves.* With rough margin: **1,3–8**; with smooth margin: **1,2,4,7,8**.
*Scape.* Hairless: **1–3,5–8**; downy, at least at the base: **2–4,8**.
*Bracts.* Entirely hairless: **3,4**.
*Outer perianth-segments.* With 1 vein: **3–8**; with 3 or 5 veins: **1,2**. Bright yellow: **3**; yellowish green: **1,2,4**; white: **4–8**; pink or pinkish brown: **2,4–6,8**.
*Flower length:* 8–11 mm: **1–4,8**; 1.2–1.6 cm: **3,4,8**; 1.7–2.5 cm: **4–7**.
*Stamens.* Shorter than perianth: **6,8**; equal to perianth: **2,4,5,7**; longer than perianth: **1–4,7**.
*Capsule length.* 5–14 mm: **1–4,7,8**; 1.5–3.5 cm: **2,5–8**.

1a. Flowers bright yellow  **3. stenophyllus**
b. Flowers white, pink, brown, pale yellow or yellowish green   2
2a. Outer perianth-segments with 3 or 5 veins   3
b. Outer perianth-segments with 1 vein   4
3a. Perianth-segments pale yellow, often flushed with green   **1. spectabilis**
b. Perianth-segments whitish, yellowish green or pinkish brown   **2. turkestanicus\***
4a. Stamens more or less equal to or longer than perianth-segments   5
b. Stamens shorter than perianth-segments   8
5a. Perianth-segments 1–1.4 cm  **4. olgae\***
b. Perianth-segments 1.5–2.5 cm   6
6a. Scape hairless for entire length   7
b. Scape downy, at least at the base   **4. olgae\***
7a. Perianth-segments white, sometimes yellow at the base; flower-stalks up to 3 cm   **7. himalaicus\***
b. Perianth-segments pink, rarely white; flower-stalks 3–4 cm   **5. robustus**
8a. Scape 35–100 cm; perianth-segments 1–1.5 cm   **8. bucharicus\***
b. Scape 90–250 cm; perianth-segments 1.8–2.5 cm   **6. aitchisonii**

**1. E. spectabilis** Bieberstein. Illustration: Botanical Magazine, 4870 (1855); RHS Lily Year Book 1969: f. 16 (1968).
Leaves narrowly strap-shaped, to 4.5 cm broad, hairless, margins smooth or rough. Scape to 1.5 m, hairless. Raceme to 80 cm. Bracts linear, ciliate. Flower-stalks *c*. 1 cm, recurving in fruit. Perianth-segments 9–10 mm, pale yellow flushed with green, with 3 veins. Stamens *c*. 2 × as long as perianth; anthers brick red, filaments orange at the base. Capsule *c*. 1.2 cm, transversely ribbed. Seeds with very narrow wings. *SW Asia, USSR (Soviet Central Asia).* H4. Late spring–summer.

**2. E. turkestanicus** Regel. Illustration: Regel, Turkestan Flora, t. 2 (1876); Cramer Plant Monograph Reprints 3: t. II (1968).
Leaves broadly linear, hairless, margins smooth. Scape to 1 m, hairless. Raceme 40–60 cm. Bracts narrowly linear, hairy and ciliate. Flower-stalks *c*. 1 cm. Perianth-segments *c*. 1 cm, curling inwards with age, the outer greenish yellow with a whitish margin and a central brownish stripe and with 5 veins, the inner white with a yellowish green stripe and with 3 veins. Stamens equal in length to the perianth, projecting as the perianth-segments curl inwards. Capsule smooth. *USSR (Soviet Central Asia).* H4. Summer.

**\*E. comosus** Fedtschenko. Illustration: Cramer Plant Monograph Reprints 3: t. VII (1968). Differs in having downy leaves, broadly lanceolate to ovate bracts, flower-stalks 2–3 cm and pinkish brown perianth-segments. *USSR (Soviet Central Asia), Afghanistan, W Pakistan.* H4. Summer.

**3. E. stenophyllus** (Boissier & Buhse) Baker (*E. bungei* Baker; *E. aurantiacus* Baker). Illustration: Botanical Magazine, 7113 (1890); Morley & Everard, Wild flowers of the world, pl. 50 (1970); Huxley (ed), Garden perennials and water plants, f. 108 (1971); Gartenpraxis 1981 (8): 360.
Leaves linear, hairless or finely downy, margins rough. Scape 30–150 cm, hairless or finely downy. Raceme densely-flowered. Bracts narrowly linear, hairless or ciliate. Flower-stalks to 2 cm. Perianth-segments 9–12 mm, bright yellow, turning orange, then brownish with age, with 1 vein. Stamens longer than the perianth, with orange anthers. Capsule 5–8 mm, wrinkled. Seeds with broad wings. *USSR (Soviet Central Asia), Afghanistan, Iran, W Pakistan.* H4. Summer.

Plants may be divided into 2 subspecies: subsp. **stenophyllus** (*E. bungei* Baker) which has hairless leaves and scapes, and subsp. **aurantiacus** (Baker) Wendelbo (*E. aurantiacus* Baker) in which the scape and leaves are finely downy.

Hybrids between *E. stenophyllus* and *E. olgae* are offered by the trade under names E. × **isabellinus** Vilmorin, E. × **warei** Mottet, E. × **shelfordii** Hort., E. 'Shelford Hybrids' and E. 'Highdown Hybrids'. They have flowers which vary from white to yellow, orange, bronze or pink.

E. × **tubergenii** van Tubergen is an hybrid between *E. stenophyllus* and *E. himalaicus*, and has pale yellow flowers borne on a scape up to 240 cm in height.

**4. E. olgae** Regel. Illustration: Gartenflora 30: t. 1048 (1881); RHS Lily Year Book 1969: f. 15 (1968); Morley & Everard, Wild flowers of the world, pl. 50 (1970); Rix & Phillips, The bulb book, 148–9 (1981).
Leaves narrowly linear, to 1.5 cm broad, hairless, margin rough. Scape 60–160 cm, downy at the base. Bracts narrowly linear, hairless. Flower-stalks 1.5–4 cm. Perianth-segments 1.2–1.6 cm, pale pink, or sometimes white with a yellow base, with 1 vein. Stamens equal in length to the perianth. Capsule 1–1.2 cm. Seeds with narrow wings. *N Iran, N Afghanistan, USSR (Soviet Central Asia).* H4. Late summer–autumn.

**\*E. sogdianus** (Regel) Bentham & Hooker.

Illustration: Cramer Plant Monograph Reprints 3: t. V (1968). Differs from *E. olgae* in having hairy bracts, longer flower-stalks (4–7.5 cm), flowers with the outer segments yellowish green with 1 vein and the inner segments white with 3 green veins, the stamens longer than the perianth and the capsule only 6–7 mm. *Afghanistan, USSR (Soviet Central Asia)*. H4. Summer.

*E. kaufmannii Regel. Illustration: Journal of the Royal Horticultural Society 93: f. 58 (1968); RHS Lily Year Book 1969: f. 15 (1968). Leaves to 3.5 cm broad, downy, margins smooth. Scape downy. Bracts narrowly ovate-triangular, hairy. Flower-stalks 1–2 cm. Perianth-segments 1.5–2.2 cm, white with a greenish yellow base and 1 vein. Stamens more or less equal in length to perianth-segments. *Afghanistan, USSR (Soviet Central Asia)*. H4. Summer.

**5. E. robustus** (Regel) Regel. Illustration: Gartenflora 22: t. 769 (1873); Botanical Magazine, 6726 (1883); Huxley (ed), Garden perennials and water plants, f. 107 (1971); Petrová, Flowering bulbs, 171 (1975).
Leaves strap-shaped, up to 4 cm broad, hairless, margins rough. Scape to 3 m, hairless. Raceme to 1 m, bearing up to 800 flowers. Bracts linear, widening at the base, densely hairy. Flower-stalks 3–4 cm. Perianth-segments 1.8–2.1 cm, pink, occasionally white, with 1 vein. Stamens equal in length to the perianth. Capsule *c.* 2 cm. *Afghanistan, USSR (Soviet Central Asia)*. H4. Summer.

**6. E. aitchisonii** Baker (*E. elwesii* Micheli; *E. robustus* (Regel) Regel var. *elwesii* (Micheli) Leichtlin). Illustration: Journal of the Linnean Society, Botany 19: t. 27 (1882); Revue Horticole 69: 280 (1897); RHS Lily Year Book 1969: f. 15 (1968). Leaves narrowly lanceolate, to 8 cm broad, hairless, margins rough. Scape 90–250 cm, hairless. Raceme loosely-flowered, *c.* 30 cm. Bracts linear, widening at the base, ciliate. Flower-stalks 2–3.5 cm. Perianth-segments 1.8–2.5 cm, pale pink, occasionally white, with 1 vein. Stamens a little shorter than the perianth. Capsule 1.5–2 cm. *E Afghanistan, USSR (Soviet Central Asia)*. H4. Summer.

**7. E. himalaicus** Baker. Illustration: Botanical Magazine, 7076 (1889); Blatter, Beautiful flowers of Kashmir 2: pl. 60 (1928); Journal of the Royal Horticultural Society 73: f. 43 (1948); Gartenpraxis 1981 (8): 361.

Leaves strap-shaped, to 4 cm broad, hairless. Scape to 2.5 m, hairless. Raceme densely-flowered, to 90 cm. Bracts narrowly linear, ciliate. Flower-stalks to 3 cm. Perianth-segments 1.7–2 cm, white, with 1 vein. Stamens equal in length to the perianth. Capsule 1.2–1.4 cm, wrinkled. *Afghanistan, NW Himalaya*. H4. Late spring–summer.

*E. lactiflorus Fedtschenko. Illustration: Cramer Plant Monograph Reprints 3: t. XVI (1968); Rix & Phillips, The bulb book, 148 (1981). A smaller plant, rarely exceeding 1 m, with a loosely flowered raceme. The white perianth-segments are yellow at the base and the stamens are slightly longer than the perianth. The capsule is *c.* 3.5 cm, smooth. *USSR (Soviet Central Asia)*. H4. Summer.

**8. E. bucharicus** Regel. Illustration: Gartenflora 39: t. 1315 (1890), 44: t. 1420 (1895); Cramer Plant Monograph Reprints 3: t. XVIII (1968).
Leaves linear, up to 5 mm broad, hairless or somewhat downy, margins rough. Scape 80–100 cm, hairless, or downy at the base. Raceme loosely flowered. Bracts narrowly linear, widening at the base, sparsely ciliate. Flower-stalks 2–3 cm. Perianth-segments 1.3–1.5 cm, white or pale pink with 1 vein. Stamens shorter than the perianth-segments. Capsule 1–1.8 cm. Seeds with broad wings at the top. *Afghanistan, USSR (Soviet Central Asia)*. H4. Summer.

*E. anisopterus (Karelin & Kirilow) Regel (*E. korolkowii* Regel). Illustration: Regel, Turkestan flora, t. 1 (1876); Cramer Plant Monograph Reprints 3: t. XVII (1968). Differs mainly in having the perianth-segments *c.* 1 cm, concave, the scape always hairless and the leaves usually with smooth margins. *Iran, USSR (Soviet Central Asia)*. H4. Summer.

**21. BULBINELLA** Kunth
*J. Cullen*
Roots fleshy. Plants stemless, leaves in rosettes, somewhat fleshy. Scape well-developed, overtopping the leaves. Raceme with many flowers, terminal. Perianth of 6, free segments, persistent in fruit. Stamens 6, all the filaments hairless. Ovary superior, 3-celled, with 2 ovules in each cell. Fruit a capsule opening between the septa.

A genus of about 20 species from Africa and New Zealand. Only 2 species, both from New Zealand, are in general cultivation. One of these has functionally unisexual flowers, the males and females

being borne on separate plants. Cultivation as for *Bulbine*.

1a. Flowers bisexual; ovary and fruit borne on a short but distinct stalk
                                        **2. hookeri**
  b. Flowers functionally unisexual (i.e. the anthers small and without pollen, or the ovary small and without ovules), plants dioecious; ovary and capsule stalkless        **1. rossii**

**1. B. rossii** (J.D. Hooker) Cheeseman. Plant large, robust. Leaves oblong or oblong-lanceolate, to 30 × 6 cm, tapering, the tips recurved and obtuse to acute. Scape exceeding the leaves. Raceme cylindric with many closely packed flowers. Flowers yellow, functionally unisexual, 1–1.4 cm in diameter. Perianth-segments spreading in male flowers, more erect in the female flowers, remaining erect and becoming hardened as the capsule develops. Ovary and capsule stalkless. *New Zealand*. H5–G1.

**2. B. hookeri** (J.D. Hooker) Cheeseman. Illustration: Salmon, Field guide to the alpine plants of New Zealand, 107 (1968). Plant to 1 m (usually much less in cultivation). Leaves to 30 × 3 cm, erect. Scape exceeding the leaves. Raceme cylindric with many flowers, which are bisexual and yellow. Perianth spreading, to 1.4 cm in diameter, shrivelling and hanging below the capsule as it develops. Ovary and capsule borne on a short stalk. *New Zealand*. H5–G1.

**22. BULBINE** Wolf
*J. Cullen*
Herbaceous perennials with somewhat to very swollen, tuber- or bulb-like bases, usually stemless. Leaves in rosettes, fleshy, terete or flat, occasionally subterranean. Scape well-developed. Flowers few to many in terminal racemes. Perianth of 6, free, yellow or brownish, spreading segments. Stamens 6, free, the filaments of all, or of 3 only, with a patch of hairs. Ovary superior, 3-celled with several ovules in each cell. Fruit a capsule containing few seeds.

A genus of about 30 species, mostly from Africa, a few from Australia. The plants are easily cultivated, though may require protection in winter in colder areas. Propagation is by seed, though *B. caulescens* can be propagated by cuttings.

*Leaves*. Only 2, subterranean except for the apex: **4**; several, aerial, terete: **1**; several, aerial, flat: **2,3,5**.
*Stem*. Present: **1**; absent: **2–5**.

*Filaments*. All with hairy patches: **1–4**; 3 hairless: **5**.

1a. 3 filaments hairy, 3 hairless
           **5. semibarbata**
  b. All 6 filaments hairy       **2**
2a. Leaves 2, mostly subterranean, their
    apices at soil level, shallowly concave,
    translucent    **4. mesembrianthoides**
  b. Leaves several, most of their length
    above ground, the apex not as above
                   **3**
3a. Plant with a branched, woody stem
    above ground, its branches ending in
    rosettes of terete leaves
               **1. caulescens**
  b. Plant stemless, the flat leaves borne
    on a swollen, tuber- or bulb-like base
                   **4**
4a. Leaves 4–6 mm broad near the base,
    tapering gradually to a fine, acute
    apex            **2. alooides**
  b. Leaves to 2 mm broad, parallel-sided
    for most of their length, the apex
    rounded        **3. bulbosa**

**1. B. caulescens** Linnaeus (*B. fruticosa*
Anon.; *B. frutescens* Anon.). Illustration:
Eliovson, Wild flowers of southern Africa,
xviii, f. 18 (1980).
Plant with short (to 25 cm), often
branched, creeping, slightly woody stems.
Scapes to 25 cm. Leaves several in rosettes
at the ends of the branches, terete,
somewhat glaucous. Raceme loose.
Perianth brownish yellow, segments to
5 mm. Each filament with a patch of hairs
above the middle. *South Africa (Cape
Province)*. H5. Summer.

**2. B. alooides** (Linnaeus) Willdenow.
Illustration: Botanical Magazine, 1317
(1810).
Plant stemless, the rosettes of leaves borne
on a swollen, tuber-like base. Leaves
several, flat, lanceolate, 5–25 cm ×
4–6 mm, tapering to an acute apex. Scape
12–45 cm. Raceme loose, with many
flowers. Perianth yellow, segments to
10 mm. Each filament with a dense patch
of hairs from just above the middle to
immediately beneath the anther. *South
Africa*. H5. Spring.

**3. B. bulbosa** (R. Brown) Haworth.
Illustration: Botanical Magazine, 3017
(1830), 3129 (1832) – as B. semibarbata.
Plant stemless, the rosettes of leaves borne
on a swollen, tuber-like base. Leaves
several, flat, linear, to 35 cm × 2 mm, apex
rounded. Scape 20–70 cm. Raceme loose,
with many flowers. Perianth yellow,
segments 8–18 mm. Each filament with a

patch of hairs above the middle. *E Australia*.
H5–G1. Spring.

**4. B. mesembrianthoides** Haworth.
Illustration: Hooker's Icones Plantarum **28**:
t. 2528 (1897); Flowering plants of Africa
**10**: t. 377 (1930).
Plant stemless, with tuberous roots. Leaves
2, very fleshy, unequal, borne below ground
with only the shallowly concave,
translucent apex exposed. Scape to 10 cm.
Raceme with up to 8 flowers, loose.
Perianth bright yellow, segments to 7 mm.
Each filament with hairs over most of its
length. *South Africa (Cape Province)*. H5–G1.

A remarkable example of a 'window
plant', in which the chlorophyll-containing
parts are subterranean; light is transmitted
to them through the translucent, exposed
leaf apex.

**5. B. semibarbata** (R. Brown) Willdenow
(*B. annua* misapplied). Illustration:
Rotherham et al., Plants of western New
South Wales and southern Queensland,
130 (1975); Cunningham et al., Plants of
western New South Wales, 183 (1981).
Plant stemless, the base scarcely swollen.
Leaves linear or lanceolate, acute,
8–19 cm × 1–5 mm. Scape to 35 cm.
Raceme with 3–many flowers. Perianth
yellow, segments to 6 mm. Three of the
filaments bear patches of hairs just above
the middle, the other 3 are hairless.
*Australia*. H5–G1. Spring–summer.

### 23. SIMETHIS Kunth
*V.A. Matthews*
Roots fleshy, from a short rhizome.
Flowering stem 12–40 cm, hairless. Leaves
mostly basal, 15–60 cm × 2.5–7 mm, linear
with sheathing bases. Flowers borne in a
terminal panicle, on non-jointed stalks.
Perianth-segments free, 8–11 mm, white
above, purplish beneath. Filaments white,
hairy; anthers yellow, dorsifixed. Style
single, straight. Ovary superior, Ovules 2 per
cell. Fruit a capsule. Seeds 3–6, black, with
a whitish swelling at one end.

A genus containing one species mainly
occurring in the west Mediterranean area,
which is occasionally cultivated. It grows
best in peaty soil in a semi-shaded position.
Propagation is by seed.

**1. S. planifolia** (Linnaeus) Grenier (*S. bicolor*
(Desfontaines) Kunth). Illustration: Ross-
Craig, Drawings of British plants, **29**: pl. 17
(1972); Fitter et al., Wild flowers of Britain
and Northern Europe, 261 (1974); Rose,
The wild flower key, 409 (1981).
*W & S Europe, N Africa*. H3. Summer.

### 24. ANTHERICUM Linnaeus
*V.A. Matthews*
Perennial hairless herbs with rhizomes and
somewhat fleshy roots. Leaves basal, linear
with a membranous sheathing base.
Flowers in a loose raceme or panicle. Bracts
usually small and membranous. Flower-
stalks usually jointed. Perianth-segments
spreading, free or shortly joined at the
extreme base, each with 3 central veins.
Stamens equal, straight; filaments hairless;
anthers basifixed. Ovary superior, 3-celled,
with 4–8 ovules in each cell. Fruit a
capsule.

A genus of about 300 species, occurring
mainly in tropical and southern Africa and
Madagascar, but also present in Europe,
America and E Asia. Only 2 are in
cultivation. They are best grown in a well-
drained soil, in a sunny position.
Propagation is by division of the rootstock,
or by seed.

1a. Flowers usually borne in a raceme;
    perianth-segments 1.5–2 cm; style
    curved upwards; capsule 8–10 mm
    with a pointed apex   **1. liliago**
  b. Flowers borne in a panicle; perianth-
    segments 8–14 mm; style straight;
    capsule *c.* 5 mm with a blunt apex
               **2. ramosum**

**1. A. liliago** Linnaeus (*A. algeriense* invalid).
Illustration: Felsko, Portraits of wild
flowers, pl. 121 (1959); Grey-Wilson &
Blamey, Alpine flowers of Britain and
Europe, 265 (1979); Grey-Wilson &
Mathew, Bulbs, pl. 12 (1981).
Stem 20–70 cm. Leaves 12–40 cm. Raceme
with 6–10 flowers, rarely shortly branched
at the base. Perianth-segments 1.5–2 cm,
white, narrowly elliptic. Stamens
9–13 mm. Style curved upwards. Capsule
8–10 mm, egg-shaped with a pointed apex.
*Europe, Turkey*. H4. Late spring–summer.

**2. A. ramosum** Linnaeus. Illustration:
Huxley, Mountain flowers, f. 893 (1967);
Grey-Wilson & Blamey, Alpine flowers of
Britain and Europe, 265 (1979); Polunin,
Flowers of Greece and the Balkans, pl. 57
(1980).
Differs from *A. liliago* in having the flowers
in a panicle, shorter perianth-segments
(8–14 mm), a straight style and the
capsule only *c.* 5 mm with a blunt apex.
*Europe, Turkey*. H4. Summer.

### 25. ALECTORURUS Makino
*V.A. Matthews*
Rootstock a short rhizome. Leaves
evergreen, basal, in 2 ranks, narrowly

strap-shaped, curved, 10–50 cm. Flowering stem 15–40 cm, flattened with a slightly winged margin, bearing a many-flowered panicle. Flowers white tinged with lilac, to pale pink, of 2 forms which occur on different plants: 1 form has the stamens projecting, the other has the stamens equal in length to the perianth-segments. Flower-stalks jointed. Perianth joined at the base. Anthers dorsifixed. Style 1, with a slightly thickened stigma. Ovary superior. Ovules 2 per cell. Seeds 3-angled, brown with a tuft of white hairs at one end.

A genus of 1 species from Japan. In cultivation it is tolerant of shade, but not very free-flowering. Propagation is by seed, or by division of the rhizome in spring.

**1. A. yedoensis** (Franchet & Savatier) Makino. Illustration: Botanical Magazine, 8336 (1910); Bulletin of the Alpine Garden Society **8**: 5 (1940); Makino, Makino's new illustrated flora of Japan, 834 (1963); Kitamura et al., Coloured illustrations of herbaceous plants of Japan (Monocotyledoneae), pl. 35 f. 231 (1981). *Japan*. H3. Summer.

**26. CHLOROPHYTUM** Ker Gawler
*S.G. Knees*
Evergreen perennials with rhizomes and fibrous or fleshy roots. Leaves all basal, stalkless and linear or stalked with ovate or lanceolate blades. Scape bearing a raceme or panicle of bisexual, stalked flowers. Perianth star-shaped, of 6 free segments, each with 3–7 veins, white, greenish or yellowish. Stamens with basifixed anthers which open inwardly. Ovary spherical to ovoid, 3-celled with several ovules per cell; style thread-like, stigma capitate. Fruit a deeply 3-angled capsule containing black, round slightly flattened seeds.

A genus of 215 species native of tropical America, Africa, India and Australia but only 1 is widely cultivated in Europe. Propagation is by seed, offsets or division in early spring. In most areas greenhouse conditions are necessary; a rich sandy loam provides an ideal growing medium for the majority of species.
Literature: Dress, W.J., Chlorophytum in cultivation, *Baileya* **9**: 29–50 (1961).

**1. C. comosum** (Thunberg) Jacques. Illustration: Graf, Exotica, edn 3, 1068 (1963).
Leaves linear to linear-lanceolate, 20–45 cm × 5–20 mm. Inflorescence branched or simple, 30–80 cm. Flowers 1–3 per node on jointed stalks. Perianth-segments ovate, acute, to 8 mm, white.

Sometimes rosettes of leaves, often with roots, also arise from several of the inflorescence nodes. Capsule greenish, *c.* 9 mm, containing 3–5 seeds per cell. *South Africa*. G1. Flowering sporadically throughout the year.

In cultivation this species is most commonly represented by its variegated cultivars: 'Mandaianum' has a central creamish white band along the leaf while 'Variegatum' has white or creamish margins. The name *C. capense* (Linnaeus) Voss is often wrongly applied to the species.

**27. THYSANOTUS** R. Brown
*V.A. Matthews*
Rootstock fibrous, rhizomatous or tuberous. Leaves basal, linear, with membranous, sheathing wings at the base. Flowers in panicles or umbels. Perianth-segments free, the outer much narrower than the inner which (in cultivated species) are fringed. Stamens 3 or 6; anthers basifixed or almost so. Style 1, unlobed. Ovary superior. Capsule enclosed within the persistent perianth-segments. Seeds black, with an aril.

There are 47 species, the majority of which come from Australia; 2 species extend into SE Asia. Two species occur uncommonly in cultivation; other species have been introduced from time to time but always die out quickly. *Thysanotus* species are not frost-hardy and need a lot of sun to flower well. They will grow in any well-drained soil and water should be more or less withheld after flowering. Propagation is by seed.
Literature: Brittan, N.H., Revision of the genus Thysanotus R. Br. (Liliaceae), *Brunonia* **4**: 67–181 (1981).

**1. T. tuberosus** R. Brown. Illustration: Edwards's Botanical Register **8**: pl. 655 (1822); Galbraith, Field guide to the wild flowers of south-east Australia, pl. 1 f. 6 (1977); Australian Plants **12**(94): front cover (1983).
Roots tuberous. Leaves 5–15, 20–30 cm, linear, almost terete towards the tip, hairless. Flowering stem 20–60 cm bearing a panicle of 1–5-flowered umbels. Perianth-segments 7–20 mm, purple. Outer segments lanceolate, with 3 or 4 veins; inner segments ovate, fringed. Stamens 6, unequal to almost equal. Style straight or curved. *E Australia*. H5–G1. Summer.
**\*T. multiflorus** R. Brown (*T. proliferus* Lindley; *T. multiflorus* var. *prolifer* (Lindley) Bentham). Illustration: Edwards's Botanical Register **24**: pl. 8 (1838); Flore des Serres,

ser. 2, **8**: t. 1911 (1869–70); Australian Plants **12**(94): 77 (1983). Differs mainly in that the flowering stem usually bears a single, 6–20-flowered umbel of blue-violet flowers which have only 3 stamens which are twisted to one side. *W Australia*. H5–G1. Summer.

**28. ARTHROPODIUM** R. Brown
*J. Cullen*
Perennial herbs with rhizomes, usually forming tufts, roots fibrous but often also some swollen and tuber-like. Leaves in a basal rosette, evergreen or deciduous. Inflorescence a terminal panicle with few or many branches; flowers often 2 or 3 in the axil of each bract. Flower-stalks conspicuously jointed at or about the middle. Perianth of 6 free, similar, spreading or slightly reflexed segments, persisting in fruit. Stamens 6, the anthers bearing hairy 'tails' which are attached to the filament for all or most of their lengths. Ovary superior, 3-celled, ovules several. Style 1. Fruit a capsule containing several black seeds, opening between the septa.

A genus of about 12 species from Australia, New Zealand, New Guinea, New Caledonia and Madagascar, of which 4 are occasionally seen in gardens. They will grow in any well-drained garden soil, and can be propagated by seed or by division.

1a. Leaves evergreen, more than 3 cm wide; flowers 2 cm or more in diameter                **1. cirratum**
  b. Leaves deciduous, mostly less than 1 cm wide, occasionally to 3 cm wide; flowers to 1.5 cm in diameter                **2**
2a. Flowers deep purple, borne singly in the bract axils                **4. minus**
  b. Flowers white or mauve, at least the lower bracts with 2 or 3 flowers in their axils                **3**
3a. Perianth-segments to 5 mm, white; anther tails white                **2. candidum**
  b. Perianth-segments 6–10 mm, mauve or white, mauve in bud; anther tails yellow                **3. milleflorum**

**1. A. cirratum** (Forster) R. Brown. Illustration: Botanical Magazine, 2350 (1822); Baileya **1**: 43 (1953).
Roots thick. Leaves evergreen, 30–60 × 3–10 cm, fleshy, green above, whitish green or glaucous beneath. Scape robust, to 1 m, bearing a panicle with numerous, rather short lateral branches which are spreading or ascending. Flowers 2 or 3 together in the axils of conspicuous bracts. Perianth white, 2–4 cm in diameter. Anther tails free and curled at

their lower ends, purplish at the top, white in the middle, yellow at their free ends. *New Zealand*. G1.

Variable in size of all its parts. The name was spelled as above by Forster, but it is often found in the literature as 'cirrhatum'.

**2. A. candidum** Raoul (*A. reflexum* Colenso). Illustration: Salmon, Field guide to the alpine plants of New Zealand, 197 (1968).

Rhizome short, roots swollen. Leaves dying down in winter, $10-30$ cm $\times 3-10$ mm. Scape slender, bearing a few, erect branches; at least the lower bracts with 2 or 3 flowers in their axils. Flower-stalks $8-20$ mm. Perianth to 1 cm in diameter, white. Anther tails white, joined to the filament for the whole of their lengths. *New Zealand*. G1.

**3. A. milleflorum** (de Candolle) MacBryde (*A. paniculatum* R. Brown; *A. pendulum* invalid). Illustration: Botanical Magazine, 1421 (1812); Rotherham et al., Flowers and plants of New South Wales and southern Queensland, 35 (1975); Elliott & Jones, Encyclopaedia of Australian plants 2: 235 (1982).

Rhizome short, most roots fibrous. Leaves dying down in winter, to $30 \times 1-3$ cm. Scape to 1 m, bearing few to many erect or ascending branches. Flower-stalks $8-20$ mm, usually 2 or 3 together in the axil of each bract. Perianth mauve in bud, white or mauve when open, segments $6-10$ mm. Anther tails yellow, attached to the filament for all of their length or the lower part free and curled. *E Australia (Queensland to Tasmania and S Australia)* G1.

**4. A. minus** R. Brown. Illustration: Cunningham et al., Plants of western New South Wales, 181 (1981).

Rhizome short, some roots swollen and tuberous. Leaves dying down in winter, to $20$ cm $\times 2-4$ mm. Scape to 30 cm, bearing few, erect branches. Flowers fragrant, solitary in the bract axils; flower-stalks to 8 mm. Perianth deep purple, to 1 cm in diameter. Anther tails purple, attached to the filament for the whole of their lengths. *Australia*. G1.

**29. PASITHEA** D. Don
*V.A. Matthews*

Rootstock a short erect rhizome with swollen roots. Leaves mostly basal, in 2 ranks, linear, to 25 cm. Flowers borne in a terminal panicle to 60 cm. Perianth joined at the base, blue, each lobe with a darker central stripe; flowers *c.* 2.5 cm in diameter. Stamens 6, 3 shorter than the others; anthers versatile. Style 1, stigma 3-lobed. Ovary superior, with 4 ovules per cell. Capsule spherical.

This genus contains a single species from SW South America. Cultivation as for *Camassia* (p. 215). Propagation by seed.

**1. P. caerulea** (Ruiz & Pavon) D. Don. Illustration: Botanical Magazine, 7249 (1892); RHS Dictionary of gardening, 1489 (1951).
*Chile, Peru*. H5–G1. Summer.

**30. SCHOENOLIRION** Durand
*V.A. Matthews*

Bulbs with membranous or fibrous tunics. Leaves basal, linear. Flowers in a raceme which is sometimes branched at the base. Flower-stalks jointed. Perianth-segments free, becoming papery with age. Stamens 6, joined to base of perianth-segments; anthers versatile. Style 1, short. Ovary superior, with 2 ovules in each cell. Capsule ovoid, with black seeds.

A genus of 5 species found in SW and SE United States. Only one of species is occasionally cultivated. It grows best in a sunny position, in a fairly moist soil. Propagation is by seed.

**1. S. album** Durand. Illustration: Abrams, Illustrated flora of the Pacific states 1: 412 (1923); Rickett, Wild flowers of the United States 5: pl. 13 (1971); Niehaus & Ripper, Field guide to Pacific states wildflowers, 13 (1976).

Flowering stem $25-150$ cm. Leaves $15-60$ cm $\times 4-12$ mm. Bracts $2-3$ mm, papery. Flower-stalks *c.* 1 mm. Perianth-segments $3-6$ mm, white tinged with green or pink. Stamens $3-5$ mm. *W USA (N California, S Oregon)*. H4. Summer.

**31. CHLOROGALUM** (Lindley) Kunth
*V.A. Matthews*

Bulbs with a fibrous or membranous tunic. Leaves mostly basal, linear; stem-leaves much reduced. Flowers borne on jointed stalks, in a terminal panicle. Perianth-segments free, with 3 central veins, twisting together over the ovary after pollination. Stamens with versatile anthers. Ovary superior. Style 1, slightly 3-lobed at the apex. Capsule with 1 or 2 black seeds in each cell.

A genus of 5 species from west coastal USA, extending into N Mexico; 2 are sometimes cultivated. The flowers do not open until late afternoon which is somewhat disadvantageous to the gardener. Cultivation as for *Camassia* (p. 215). Propagation is by seed. Literature: Hoover, R.F., A monograph of the genus Chlorogalum, *Madroño* 5: 137–47 (1940).

1a. Perianth-segments $1-2.5$ cm; stamens *c.* two-thirds the length of the segments; leaves with wavy margins      **1. pomeridianum**
 b. Perianth-segments $8-12$ mm; stamens more or less equal in length to segments; leaves not wavy      **2. angustifolium**

**1. C. pomeridianum** (de Candolle) Kunth. Illustration: Edwards's Botanical Register 7: pl. 564 (1821); Rickett, Wild flowers of the United States 4: pl. 11 (1970); Niehaus & Ripper, Field guide to Pacific states wildflowers, 13 (1976); Spellenberg, Audubon Society field guide to North American wildflowers – western region, f. 79 (1979).

Leaves to $75 \times 2.5$ cm, margins wavy. Flowering stem $30-250$ cm. Flower-stalks $5-35$ mm. Perianth-segments $1-2.5$ cm, white with a purple or green central stripe. Stamens about two-thirds of the length of the segments. Style $1-1.5$ cm. *W USA (S Oregon to N California)*. H4. Summer.

**2. C. angustifolium** Kellogg. Illustration: Proceedings of the California Academy of Natural Sciences 2: 105 (1863); Rickett, Wild flowers of the United States 5: 48 (1971); Niehaus & Ripper, Field guide to Pacific states wildflowers, 13 (1976).

A smaller plant than *C. pomeridianum*. Leaves shorter and narrower with flat margins, flower-stalks *c.* 3 mm, perianth-segments $8-12$ mm, white with a greenish yellow central stripe, stamens more or less equal to segments, style $3-4$ mm. *W USA (N & C California)*. H4. Summer.

**32. APHYLLANTHES** Linnaeus
*D.A. Webb*

Perennial with a dense, fibrous rootstock. Leaves reduced to reddish brown sheaths, which surround the lower part of the numerous slender, wiry, glaucous stems $15-40$ cm high. Flowers terminal, solitary or in compact groups of 2 or 3, subtended by about 5 reddish brown, scarious bracts. Perianth-segments free, erect in the lower half, spreading above. Anthers dorsifixed, introrse. Ovary superior; style 1. Stigma 3-lobed. Fruit a 3-seeded capsule enclosed by the bracts.

A genus of a single species with very distinctive habit. It requires a sunny position in well-drained soil. In the less sunny areas, even if the winter is mild, it is best grown in a frame or a cool house.

**1. A. monspeliensis** Linnaeus. Illustration: Coste, Flore de la France **3**: 349 (1906); Bonnier, Flore complète **10**: t. 593 (1929); Reisigl & Danetsch, Flore mediterranéenne, 101 (1979).

Flowers 2 cm across, varying from fairly deep to very pale blue. *SW Europe and Morocco.* H5. Early summer.

**33. HOSTA** Trattinick
*J.C.M. Alexander*

Perennial, clump-forming herbs with short thick fleshy rhizomes bearing dense clusters of roots. Leaves mostly basal. Leaf-stalks fleshy, erect or ascending, channelled, often winged, sometimes purple-spotted. Leaf-blades elliptic or oblong to ovate or almost circular, acuminate, acute or obtuse; with tapered, rounded, truncate or cordate bases, often decurrent, membranous to leathery, flat or with marginal undulations or wrinkles, yellowish or mid-green to dark green (with many yellow and variegated varieties and cultivars), dull, shiny or glaucous, with 2–14 pairs of side veins. Leaf-surface usually raised between the veins and sometimes also puckered, with wrinkles at right-angles to the veins. Flowering stems fleshy, erect or ascending, to *c.* 1.75 m, almost always unbranched, occasionally with a few leaves or large leaf-like bracts below; upper bracts erect or spreading, often channelled, persistent or withering at flowering. Flowers few to many in a loose to dense head, bell- or trumpet-shaped, white or various shades of bluish purple; perianth fused below into a narrow tube, which broadens upwards; perianth-lobes 6, spreading or recurved. Stamens 6, not or slightly projecting, free from perianth-tube (except in No. 1), curved upward just below anthers; anthers versatile, white, yellow or purplish. Ovary superior, 3-celled; style projecting beyond stamens, curved upwards. Seeds many, membranous, winged.

A genus variously regarded as having 20–50 species, known as *Funkia* Sprengel until early this century, mostly native to Japan, with a few representatives in Korea and China. Cultivated for centuries in Japan, they were first imported into Europe early in the nineteenth century, since when a vast and confusing array of varieties and cultivars has arisen from sports or from spontaneous or deliberate hybridisation. Many of these cannot be assigned to any known wild species. Revisions of the asiatic species are of very limited use when attempting to put names on plants growing in Europe. The account that follows is not a revision, it merely attempts to account for and describe most of what is grown in Europe and keeps as separate entities several species that might best be combined.

Hostas do particularly well if grown in partial shade where the roots can get down to a steady supply of water beside a stream or pond, but will also perform well in an open herbaceous border. They flower from early summer onwards and are easily propagated by dividing the clump with a spade. This can be done by cutting out a segment without digging up the clump. Though many clones are sterile, others reproduce freely by seed; however, cross-pollination is rife and offspring will very often not be true to the female parent. Horticultural (and taxonomic) interests will best be served by not growing hostas from seed.

Literature: Grey, C.H., *Hardy bulbs* **3**: 292–305 (1938); Maekawa, F., The genus Hosta, *Journal of the Faculty of Science, Tokyo (Botany)* **5**: 317–425 (1940); Hylander, N., The genus Hosta in Swedish gardens, *Acta Horti Bergiani* **16**: 331–420 (1954); Hylander, N., The genus Hosta, *Journal of the Royal Horticultural Society* **85**: 356–66 (1960); Thomas, G.S., Notes on the genus Hosta, *Bulletin of the Hardy Plant Society* **2**: 92–101 (1960), photograph captions all transposed; Hensen, K.J.W., De in Nederland Gekweekte Hostas, *Mededelingen. Directeur Tuinbouw* **26**: 725–53 (1963); Brickell, C.D., Notes from Wisley, *Journal of the Royal Horticultural Society* **93**: 365–72 (1968); Hylander, N., *RHS Dictionary of Gardening, Supplement* 347–50 (1969); Fujita, N., The genus Hosta in Japan, *American Hosta Society Bulletin* **10**: 14–41 (1979), a very stilted translation of *Acta Phytotaxonomica et Geobotanica* **27**: 66–96 (1976); Grenfell, D., A survey of Hosta and its availability in commerce, *Plantsman* **3**: 20–44 (1981).

The key that follows is based almost entirely on vegetative characters because of the relative uniformity of the flowers, and will thus be useful even when plants are not flowering. Some of the characters used in the key and descriptions need clarification.

*Leaf Dimensions.* These have generally been taken from the largest leaves on mature plants in late summer. In species with clearly defined blade and stalk, measurements are not a problem; however, many have tapered, decurrent blades which merge with the winged stalks. In some of these, the sides of the base of the blade are curled forward across the top of the stalk; this point where the sides approach each other is taken to be the base of the blade (Figure 30(8), p. 135). For flat, tapering leaves, the position of the base has to be decided arbitrarily.

*Leaf surfaces.* Many species have a grey or bluish bloom on one or both leaf-surfaces; they are referred to as *glaucous.* The description of a leaf as dull implies that it is not glaucous.

*Number of pairs of side-veins.* This has been used extensively, and there can be uncertainty as to how many of the smaller side-veins should be included. In this account, only those veins cutting a line running forwards from the widest part of the leaf have been counted (Figure 30(9), p. 135).

1a. Leaves not variegated                               2
  b. Leaves variegated                                 52
2a. Leaves glaucous or bloomed below        3
  b. Leaves dull or shiny below, not glaucous or bloomed                            24
3a. Leaves yellow to bright greenish yellow at least in spring and early summer, not densely white-bloomed below                                           4
  b. Leaves mid-green to dark or slightly yellowish green, sometimes densely white-bloomed below                    6
4a. Leaves becoming pale to mid-green in summer; blades 19 × 12 cm or more, with 8 or 9 pairs of side-veins; leaf-stalks 30 cm or more
     **7. fortunei** var. **albopicta** forma **aurea**
  b. Leaves remaining yellow; blades 15 × 11 cm or less, with 7 pairs of side-veins or fewer; leaf-stalks less than 25 cm                                         5
5a. Leaves ovate, with rounded, truncate or cordate bases, slightly leathery, with 6 or 7 pairs of side-veins
                    **24. see 'Golden Nakaiana'**
  b. Leaves narrowly lanceolate to elliptic, tapered at base, membranous, with *c.* 3 pairs of side-veins
                         **16. see 'Wogon Gold'**
6a. Leaves with 7 pairs of side-veins or fewer                                              7
  b. Leaves with 8 pairs of side-veins or more                                             9
7a. Leaves lanceolate, glaucous above and below                    **11. × tardiana**

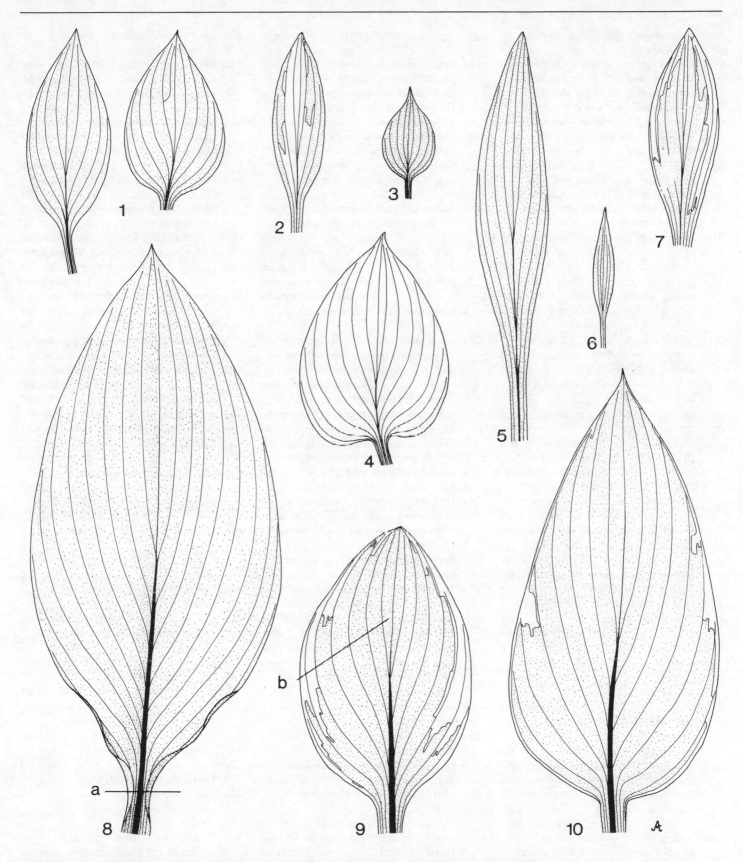

**Figure 30.** *Hosta* leaf-blades ( × ½). 1, *H. minor*. 2, *H. lancifolia* 'Kabitan'. 3, *H. gracillima*. 4, *H.* 'Golden Nakaiana'. 5, *H. longissima*. 6, *H. tortifrons*. 7, *H. helonioides* forma *albopicta*. 8, *H. rectifolia* (a, point taken as base of leaf-blade). 9, *H. decorata* (to show number of side-veins cut by a line (b) running forwards from the widest point of the leaf: in this case, 6). 10, *H.* 'North Hills'.

b. Leaves ovate, dull mid-green to dark green above                           8

8a. Leaves mid-green, acuminate; leaf-stalks spotted        **24. nakaiana**

b. Leaves dark green, very shortly acuminate; leaf-stalks not spotted
        **24. see 'Pastures New'**

9a. Leaves slightly shiny yellowish green above, densely white-bloomed below
        **8. hypoleuca**

b. Leaves dull or glaucous and mid-green to dark green above, if slightly yellowish then not densely white-bloomed below        10

10a. Leaves very glaucous above        11

b. Leaves not or slightly glaucous above        18

11a. Leaves not puckered, though often raised between the veins        12

b. Leaves puckered, with wrinkles at right-angles to the veins        14

12a. Leaf-blades to *c*. 26 × 17.5 cm, erect or steeply ascending, only slightly arched at the tips
        **2. see 'Krossa Regal'**

b. Leaf-blades to *c*. 15 × 12 cm, spreading, horizontal or strongly arched        13

13a. Leaf-stalks spotted; blades with 8 or 9 pairs of side-veins, not cupped
        **11. × tardiana 'Halcyon'**

b. Leaf-stalks not spotted; blades with 10–12 pairs of side-veins, cupped
        **5. see 'Buckshaw Blue'**

14a. Leaves ovate to oblong; length usually at least 10 cm more than width
        **6. see 'Snowden'**

b. Leaves broadly ovate; length usually 5 cm more than width or less        15

15a. Leaf-blades with a very thin, white 'pencil-line' along margin
        **7. fortunei** var. **hyacinthina**

b. Leaf-blades without such a line        16

16a. Leaf-blades 25 × 15 cm or more, not cupped, floppy, greyish green, with 12–14 pairs of side-veins
        **6. sieboldiana**

b. Leaf-blades 20 × 15 cm or less, cupped, stiff, rather bluish, with 10–12 pairs of side-veins        17

17a. Leaf-blades almost circular, strongly puckered, bluish green        **5. tokudama**

b. Leaf-blades broadly ovate, weakly puckered, very blue
        **5. see 'Buckshaw Blue'**

18a. Young leaves yellow or bright greenish yellow with a dark green margin and cross-veins, later entirely green, though margin remaining darker        **7. fortunei** var. **albopicta**

b. Leaves not variegated when young,

lacking a darker margin at any time        19

19a. Young leaves yellow or bright greenish yellow, later pale green
        **7. fortunei** var. **albopicta** forma **aurea**

b. Leaves mid-green to dark or slightly yellowish green        20

20a. Mature leaves slightly glaucous above, on and between the veins        21

b. Mature leaves not glaucous above or only glaucous in vein-depressions
        22

21a. Leaves not puckered, with a very thin, white 'pencil-line' along margin
        **7. fortunei** var. **hyacinthina**

b. Leaves strongly puckered, without such a line        **7. fortunei** var. **rugosa**

22a. Flowering-head just above leaf-mound
        **7. fortunei** var. **stenantha**

b. Flowering-head well above leaf-mound        23

23a. Leaf-blades lustrous dark green above, very broadly ovate to almost circular, with *c*. 8 pairs of side-veins
        **7. fortunei** var. **obscura**

b. Leaf-blades dull mid-green or slightly yellowish green, ovate to elliptic, with 10–13 pairs of side-veins        **4. elata**

24a. Leaves yellow        **16. see 'Wogon Gold'**

b. Leaves green or yellowish green        25

25a. Leaf-blades 7 cm long or less        26

b. Leaf-blades 8 cm long or more        28

26a. Leaf-blades more than 3 times longer than broad, gradually tapered at base, often slightly twisted        **12. tortifrons**

b. Leaf-blades *c*. 2 times longer than broad, rounded or shortly tapered at base, not twisted        27

27a. Leaves dark green; margin sometimes finely wrinkled; back of leaf-stalk spotted throughout        **14. gracillima**

b. Leaves mid-green; margin with 1 or 2 undulations; back of leaf-stalk spotted only in lower half        **23. venusta**

28a. Leaf-blades with 5 or fewer pairs of side-veins        29

b. Leaf-blades with 6 or more pairs of side-veins        35

29a. Leaf-blades less than 2 times longer than broad        30

b. Leaf-blades 2 or more times longer than broad        31

30a. Leaf-blades *c*. 16 × 11 cm; flowers up to 25 per head
        **20. decorata** var. **normalis**

b. Leaf-blades *c*. 8.5 × 5.5 cm; flowers up to 10 per head        **22. minor**

31a. Leaves shiny dark green above
        **16. lancifolia**

b. Leaves dull mid-green, yellowish or dark green above        32

32a. Leaves membranous        33

b. Leaves leathery        34

33a. Leaves erect or ascending
        **18. helonioides**

b. Leaves spreading or arched
        **15. sieboldii**

34a. Leaf-tips acuminate, stalks strongly spotted        **10. tardiflora**

b. Leaf-tips acute, stalks not spotted
        **17. longissima**

35a. Leaves pale to bright yellowish green
        36

b. Leaves mid-green to dark or slightly yellowish green        37

36a. Leaves elliptic; flowers very pale bluish purple        **1. see 'Honey Bells'**

b. Leaves ovate; flowers white
        **1. plantaginea**

37a. Leaves 13 cm wide or more        38

b. Leaves 12 cm wide or less        44

38a. Leaves dark green        39

b. Leaves mid-green or slightly yellowish green        41

39a. Leaves membranous
        **1. see 'Royal Standard'**

b. Leaves slightly leathery        40

40a. Leaf-blades broadly elliptic to almost circular, abruptly acuminate, lustrous dark green above, not rolled at base
        **21. ventricosa**

b. Leaf-blades elliptic to ovate, acuminate, slightly shiny dark green above, sides rolled forwards at base
        **2. nigrescens**

41a. Leaves with 10–13 pairs of side-veins
        **4. elata**

b. Leaves with 9 or fewer pairs of side-veins        42

42a. Leaves broadly ovate, rounded to cordate at base
        **1. see 'Royal Standard'**

b. Leaves lanceolate or narrowly ovate to elliptic, shortly tapered at base        43

43a. Flowering stems bearing large leaf-like bracts; leaves spreading
        **26. undulata** var. **erromena**

b. Flowering stems lacking large leaf-like bracts; leaves erect or steeply ascending        **19. see 'Tall Boy'**

44a. Leaves very shiny above
        **16. lancifolia**

b. Leaves dull or slightly shiny above        45

45a. Leaf-bases tapered        46

b. Leaf-bases rounded, truncate or cordate        48

46a. Leaves broadly ovate to broadly elliptic, *c*. 1.5 times longer than broad
        **20. decorata** var. **normalis**

b. Leaves lanceolate, narrowly ovate or narrowly elliptic, *c*. 2 times longer than broad        47

47a. Leaf-blades erect or steeply ascending
**19. rectifolia**
b. Leaf-blades spreading or arched
**26. undulata** var. **erromena**
48a. Leaves membranous 49
b. Leaves leathery 51
49a. Leaf-blades less than 6 cm wide
**22. minor**
b. Leaf-blades more than 8 cm wide 50
50a. Leaf-tips obtuse
**20. decorata** var. **normalis**
b. Leaf-tips acuminate **9. longipes**
51a. Leaves shiny below, rounded to
truncate and rolled forwards at base
**13. rupifraga**
b. Leaves dull below, cordate and not
rolled forwards at base **25. capitata**
52a. Paler colour on leaf margin 53
b. Paler colour in middle of leaf-blade, or
leaf mottled 63
53a. Leaves with 4 pairs of side-veins or
fewer 54
b. Leaves with 5 pairs of side-veins or
more 57
54a. Leaf-blade plus stalk less than 18 cm;
leaf-tips 45° or less
**15. see 'Ginko Craig' & 'Louisa'**
b. Leaf-blade plus stalk more than
20 cm; leaf-tips more than 45° 55
55a. Leaf-margins pale yellow, more than
8 mm wide in places; leaves erect or
ascending
**18. helonioides** forma **albopicta**
b. Leaf margins white to cream-
coloured, less than 6 mm wide; leaves
spreading or arched 56
56a. Leaf-stalks 20 cm or more, less than
1 cm wide **15. sieboldii**
b. Leaf-stalks 15 cm or less, more than
1 cm wide **15. see 'Emerald Isle'**
57a. Leaves glaucous below 58
b. Leaves dull or shiny below 59
58a. Leaves with 10 pairs of side-veins or
more
**6. sieboldiana 'Aureo Marginata'**
b. Leaves with *c.* 8 pairs of side-veins
**7. fortunei** var. **obscura**
forma **marginata**
59a. Leaves strongly curled and often
twisted at tip **3. crispula**
b. Leaves flat or with shallow
undulations 60
60a. Leaf-tips obtuse **20. decorata**
b. Leaf-tips acute or acuminate 61
61a. Leaves less than 10 cm wide
**26. undulata** var. **albo-marginata**
b. Leaves more than 10 cm wide 62
62a. Leaves lanceolate, dull above, dull to
slightly shiny below; pale margin less
than 1 cm wide
**7. fortunei 'North Hills'**

b. Leaves broadly ovate, lustrous above,
very shiny below; pale margin more
than 1 cm wide in places
**21. ventricosa 'Variegata'**
63a. Leaf-bases gradually tapered 64
b. Leaf-bases truncate, rounded or
cordate 65
64a. Leaves with *c.* 3 pairs of side-veins,
4 cm wide or less, yellow in middle
**16. lancifolia 'Kabitan'**
b. Leaves with 6 or more pairs of side-
veins, 6 cm wide or more, white in
middle **26. undulata**
65a. Leaves predominantly yellow; green
margin less than 1 cm wide
**7. see 'Gold Leaf' & 'Gold Standard'**
b. Green margin to 2 cm wide in places
or leaves green with paler mottling 66
66a. Leaves cupped, abruptly pointed,
glaucous above and below
**5. tokudama 'Variegata'**
b. Leaves not cupped, slightly
acuminate, dull above, shiny or
slightly glaucous below 67
67a. Leaves yellow or bright greenish
yellow with green margins when
young, later entirely green, though
margin remaining darker, slightly
glaucous below, puckered, rounded at
base **7. fortunei** var. **albopicta**
b. Leaves green, mottled with pale
yellowish green when young, later
almost entirely green, shiny below,
not puckered, truncate or slightly
cordate at base
**21. ventricosa 'Aureomaculata'**

**1. H. plantaginea** (Lamarck) Ascherson
(*Hemerocallis japonica* Redouté, not Thunberg;
*Hemerocallis alba* Andrews; *Funkia subcordata*
Sprengel; *Hosta* 'Thomas Hogg' misapplied).
Illustration: Botanical Magazine, 1433
(1812); Journal of the Faculty of Science,
Tokyo (Botany) **5**: 343 (1940); Acta Horti
Bergiani **16**: pl. 24 (1954).
Leaf-stalks 15–28 cm, deeply channelled,
winged, often with overlapping edges in
upper part. Leaf-blades 14–22 ×
9.5–17 cm, ovate, shortly acuminate at tip,
rounded to slightly cordate at base,
decurrent, membranous, flat or with
several small undulations, bright yellowish
green, slightly shiny above, very shiny
below, with 6–9 pairs of side-veins.
Flowering stems to 65 cm, mostly erect,
yellowish green; lower bracts leaf-like,
upper bracts 3–8.5 cm, greenish white.
Flowers *c.* 11.5 cm, white, fragrant;
anthers yellow; filaments fused to perianth-
tube at base. *China.* H2. Late
summer–autumn.

Plants with larger flowers and broader
perianth-segments have been called var.
**grandiflora** (Siebold & Zuccarini) Ascherson
& Graebner (*H. grandiflora* Siebold &
Zuccarini). Illustration: Flore des Serres, t.
158 & 159 (1846); Thomas, Plants for
ground cover, f. 34 & 35 (1970). They are
variously described as having leaves
broader than, or narrower than the species.
**'Honey Bells'** (*H. plantaginea* ×
?*H. lancifolia*) has longer pale green, elliptic,
scarcely acuminate leaves with 6 or 7 pairs
of side-veins and very pale bluish purple,
fragrant flowers. **'Royal Standard'**
(*H. plantaginea* × ?) has broadly ovate mid-
green to dark green leaves and fragrant,
pure white flowers. These cultivars flower a
little earlier than *H. plantaginea.*

**2. H. nigrescens** (Makino) Maekawa
(*H. sieboldiana* (Hooker) Engler var.
*nigrescens* Makino). Illustration: Journal of
Japanese Botany **13**: 902 (1937); Journal
of the Faculty of Science, Tokyo (Botany) **5**:
353 & 354 (1940).
Young plant with dense greyish bloom
which later vanishes. Leaf-stalks
20–50 cm, channelled, winged in upper
half, purple-spotted. Leaf-blades
20–29.4 × 13–21 cm, ovate to elliptic,
acuminate at tip, tapered to rounded or
cordate at base, sides rolled forward at base
forming a funnel, decurrent, slightly
leathery, flat or with a few shallow
undulations, dark green and slightly shiny
above, mid-green and very shiny below,
with 8–13 pairs of side-veins. Flowering
stems to 142 cm, arched, bearing 2–4
spreading sterile bracts below, and 30–40
flowers. Flowers 4.5–5 cm, white flushed
with violet. *Japan.* H2. Summer.

Forma **elatior** Maekawa is larger in all
its parts; flowering stems to 170 cm;
flowers *c.* 6.5 cm. Illustration: Journal of
the Faculty of Science, Tokyo (Botany) **5**:
323 (1940).
**'Krossa Regal'** (*H. nigrescens* forma *elatior*
misapplied) is a statuesque plant with erect
or steeply ascending, slightly arched leaves.
The whole plant is densely bluish-glaucous
when mature. Illustration: The Garden
**107**: 197 (1982).

**3. H. crispula** Maekawa (*H. latifolia*
Matsumura var. *albimarginata* Wehrhahn;
*H. undulata* var. *albomarginata* misapplied; *H.*
'Thomas Hogg' misapplied; *H. albomarginata*
misapplied). Illustration: Acta Horti Bergiani
**16**: pl. 13 (1954); Journal of the Royal
Horticultural Society **85**: f. 104 (1960) &
**93**: f. 187 (1968); Thomas, Plants for
ground cover, pl. XIIB & XIIA (1970).

Leaf-stalks to 45 cm, channelled, winged in upper half. Leaf blades 13–27 × 6–10.5 cm, lanceolate to narrowly elliptic, attenuate at tip, tapered, decurrent and often rolled at base, slightly leathery, strongly curled and often twisted at tip, mid-green to dark green with an irregular white margin to *c.* 2 cm wide, dull above, shiny below, with 7–9 pairs of side-veins. Flowering stems 50–85 cm bearing 30–40 flowers, without leaf-like bracts below; bracts 1.3–2 cm, pale, spreading. Flowers *c.* 4 cm, pale violet; anthers yellow, becoming purple. *Unknown in wild.* H2. Summer.

Numbers 4–6 are sometimes considered to be a single species, *H. sieboldiana.*

**4. H. elata** Hylander (*H. montana* invalid; *H. fortunei* (Baker) Bailey var. *gigantea* Bailey; *H. fortunei* (Baker) Bailey 'Robusta'). Illustration: Gentes Herbarum 2: 134 (1930); Journal of the Faculty of Science, Tokyo (Botany) 5: 361 (1940); Acta Horti Bergiani 16: pl. 6, 10–12 (1954); Journal of the Royal Horticultural Society 81: f. 130 (1956).
Leaf-stalks 15–47 cm, channelled, winged in upper half. Leaf-blades 18–32 × 10–24 cm, broadly ovate, rarely almost elliptic, acuminate at tip, rounded to cordate at base, decurrent, slightly leathery, with several shallow undulations, mid-green or slightly yellowish green, slightly glaucous, though becoming dull, especially above, with 10–13 pairs of side-veins. Flowering stems to *c.* 90 cm, rarely with a leaf-like bract below; bracts pale yellowish green. Flowers 4.5–6 cm, pale bluish violet; anthers pale yellow. *Japan.* H2. Early summer.

Sometimes considered part of *H. sieboldiana* though much less glaucous and somewhat yellower in the leaf, and with a much longer flowering stem. It has been suggested that some clones grown as *H. montana* are distinct enough to be maintained as a separate species. This seems unlikely as Hylander published *H. elata* because he discovered that the name *H. montana* was invalid; he was not combining *H. montana* with anything else.

**5. H. tokudama** Maekawa (*H. sieboldiana* Engler var. *fortunei* Voss not Ascherson; *H. sieboldiana* Engler var. *glauca* Makino & Matsumura; *H. glauca* misapplied). Illustration: Journal of the Faculty of Science, Tokyo (Botany) 5: 367 (1940); Journal of the Royal Horticultural Society 93: f. 184 (1968).

Leaf-stalks to 35 cm, shallowly channelled, slightly glaucous. Leaf-blades to *c.* 20 × 15 cm, broadly ovate to almost circular, cupped, abruptly acuminate, cordate at base, not decurrent, stiff and leathery, puckered, dark bluish green, densely glaucous below, less so above, with 10–12 pairs of side-veins. Flowering stem 30–45 cm, barely exceeding leaves, glaucous; bracts stiff, purplish. Flowers 4–4.5 cm, very pale purple; anthers yellowish. *Japan.* H2. Summer.

Sometimes considered part of *H. sieboldiana* though generally compacter with more cupped leaves; probably close to *H. sieboldiana* var. *elegans.*

**'Variegata'** (*H. tokudama* Maekawa forma *aureonebulosa* Maekawa) has breaks of pale or yellowish green in the middle of the leaf-blades. This name is often misapplied to another unnamed cultivar which has pale or yellowish green margins and resembles a small form of *H. sieboldiana* 'Aureo Marginata' (*H.* 'Frances Williams').

**'Buckshaw Blue'** (*H. tokudama* × ?) has narrower, intensely glaucous, more deeply cupped leaves.

**6. H. sieboldiana** Engler (*H. glauca* (Siebold) Stearn; *H. caerulea* misapplied; *H. sieboldii* misapplied). Illustration: Botanical Magazine, 3663 (1838); Gentes Herbarum 2: 131 & 132 (1930); Journal of the Faculty of Science, Tokyo (Botany) 5: 369 & 370 (1940); Acta Horti Bergiani 16: pl. 1, 2 & 10 (1954).
Leaf-stalks to 70 cm, deeply channelled, not winged, glaucous. Leaf-blades 25–36 × 14.5–32 cm, ovate (some very broadly), acuminate, rounded to cordate at base, shortly decurrent, leathery, puckered, margin with many shallow or a few deeper undulations, mid-green and glaucous above, paler, greyish green and more densely glaucous below, with 12–14 pairs of side-veins. Flowering stems to 50 cm, arched above, about equalling leaves; bracts 2.5–6.5 cm, pale yellowish green, lanceolate, persistent. Flowers dense, 5–5.7 cm, white to pale greyish purple; anthers pale yellow. *Japan.* H2. Early summer.

Var. **elegans** Hylander has rounder, more abruptly tipped, greyer leaves with deeper puckering, similar to *H. tokudama* though less cupped. Illustration: Acta Horti Bergiani 16: pl. 3 & 4 (1954); Journal of the Royal Horticultural Society 85: f. 103 (1960); Thomas, Plants for ground cover, pl. XIIA & XXXII & f. 32 (1970).

**'Aureo Marginata'** (*H.* 'Frances Williams', *H.* 'Gold Edge', not *H.* 'Gold Edger') has an irregular broad margin up to 4 cm wide, of cream, pale yellow or yellowish green to the leaves. Illustration: Thomas, Plants for ground cover, f. 33 (1970); The Garden 107: 197 (1982).
**'Snowden'** (*H. sieboldiana* Engler × *H. fortunei* (Baker) Bailey var. *albopicta* (Miquel) Hylander forma *aurea* (Wehrhahn) Hylander) has mid-green to dark green, ovate to oblong, puckered leaves, 28.5–35 × 17.6–20 cm, with 10 or 11 pairs of side-veins and rounded bases which are more decurrent than in *H. sieboldiana.*

**7. H. fortunei** (Baker) Bailey (*H. sieboldiana* Engler var. *fortunei* Ascherson & Graebner). Leaf-stalks channelled and winged, often glaucous. Leaf-blades lanceolate to broadly ovate, gradually or abruptly acuminate, rounded to cordate at base, decurrent, membranous to slighly leathery, flat or with a few shallow undulations, sometimes puckered, mid-green to dark green, glaucous below and sometimes above, often variegated, with 7–10, occasionally to 13, pairs of side-veins. Some clones are bright yellow to pale yellowish green in spring but become darker green later in the season. Flowering stem just exceeding or well clear of the leaves, usually with 1 or 2 leaf-like bracts below and many densely packed flowers; bracts purplish, often glaucous, boat-shaped. Flowers pale violet, rarely reddish, with narrow lobes; anthers bluish. *Origin uncertain.* H2. Summer.

This name has long been applied to a group of similar cultivars and varieties, most of which are probably the result of hybridisation in cultivation both in Japan and Europe, with *H. sieboldiana* being a likely parent. *H. fortunei* is therefore a 'species of convenience' (Brickell 1968) rather than a true botanical species. Among the most popular cultivars and varieties are:
Var. **albopicta** (Miquel) Hylander (*H. fortunei* 'Aureomaculata'; *H. lancifolia* (Thunberg) Engler var. *aureomaculata* Wehrhahn). Illustration: Acta Horti Bergiani 16: pl. 6 & 8 (1954); Thomas, Plants for ground cover, f. XIA (1970). Leaves in spring yellow to pale greenish yellow with irregular mid-green to dark green margins to 2 cm, and dark cross-veins, becoming wholly green later in the season, though margin remaining slightly darker. Leaf-blades 17–20 × 10–12 cm, ovate, slightly acuminate, rounded at base,

slightly leathery, with a few shallow undulations, dull above, slightly glaucous below, with 8 or 9 pairs of side-veins. Some recently developed seedlings or budsports such as 'Gold Leaf' and 'Gold Standard' which also have dark margins, keep their intense yellow colour to the end of the season.

Var. albopicta (Miquel) Hylander forma aurea (Wehrhahn) Hylander (*H. fortunei* 'Albopicta Aurea'; *H. fortunei* 'Aurea'; *H. lancifolia* (Thunberg) Engler var. *aurea* Wehrhahn). Illustration: Thomas, Plants for ground cover, f. XIB (1970). Similar to var. *albopicta* but lacking the darker green margin to the leaf and becoming pale green later in the season. Leaf-stalks to 45 cm; leaf-blades 19–26 × 12–18 cm, lanceolate to ovate or narrowly elliptic, acute, very slightly acuminate, rounded or tapered at base, membranous, slightly puckered, dull or slightly glaucous above, more strongly glaucous below.

Var. hyacinthina Hylander (*H. fortunei* var. *stenantha* misapplied; *H. glauca* misapplied). Illustration: Acta Horti Bergiani 16: pl. 5, 9 & 10 (1954); Thomas, Plants for ground cover, f. 32 (1970); Thomas, The art of planting, f. 50 (1984). Leaves not variegated, very glaucous below, less so above, with a very narrow 'pencil-line' of white along the margin, appearing bluish to greyish green *en masse*. Leaf-blades 18.5–22 × 14–19 cm, broadly ovate, shortly acuminate, rounded to cordate at base, slightly leathery, flat or with 1 or 2 shallow undulations, slightly puckered, with 9 or 10 pairs of side-veins.

'North Hills' has dark green leaves with a narrow white margin to c. 8 mm wide. Leaf-blades 20–25 × 10–14 cm, lanceolate to oblong, shortly acuminate, rounded to slightly cordate at base, dull to slightly shiny on both surfaces, with 7 or 8 pairs of side-veins (Figure 30(10), p. 135).

Var. obscura Hylander. Illustration: Acta Horti Bergiani 16: pl. 7 (1954). Leaves not variegated, dark green and lustrous above, paler and slightly glaucous below. Leaf-blades 20–23 × 14–17 cm, very broadly ovate to almost circular, acuminate or with a short abrupt tip, cordate at base, leathery, with c. 8 pairs of side-veins. Flowering stems much taller than leaf-mound.

Var. obscura Hylander forma marginata Hylander (*H. fortunei* 'Obscura-Marginata'; *H. fortunei* 'Aureo-marginata'). Illustration: Thomas, Plants for ground cover, f. 31 (1970); Thomas, The art of planting, f. 43 (1984). Differs from var. *obscura* in having leaves with irregular yellow margins,

becoming paler later in the season. A form with pure white margins is known as 'Marginato-Alba'.

Var. rugosa Hylander. Illustration: Acta Horti Bergiani 16: pl. 5 (1954). Leaves strongly puckered between the veins, slightly glaucous above, more densely glaucous below. Leaf-blades to 30 × 23 cm, ovate, shortly acuminate, cordate at base, with up to 10 shallow undulations, leathery, with 8–13 pairs of side-veins.

Var. stenantha Hylander (*H. fortunei* var. *hyacinthina* misapplied). Illustration: Acta Horti Bergiani 16: pl. 7 & 8 (1954). Similar to var. *rugosa* but leaves less puckered. Often confused with var. *hyacinthina* from which it differs in its less rigid leaves, lacking the marginal 'pencil-line', which are dull or glaucous only in vein-depressions. In full sun the leaves can become rather yellowish green. Leaf-blades to 35 × 25 cm, broadly ovate, acuminate, cordate at base, slightly leathery, with 6–8 shallow undulations and 9–11 pairs of side-veins. Flowering stems barely exceeding leaf-mound.

**8. H. hypoleuca** Murata. Illustration: Acta Phytotaxonomica et Geobotanica 19: 67 (1962).
Leaf-stalks to 35 cm, deeply channelled, glaucous, purple-spotted below. Leaf-blades 20–45 × 10–30 cm, broadly ovate, obtuse or very slightly acuminate, cordate at base, not or slightly decurrent, leathery, with up to 20 small marginal wrinkles, yellowish green and slightly shiny above, very densely white-bloomed below, with 9–12 pairs of side-veins. Flowering stems not exceeding leaves, 25–35 cm, arched or bent sideways, bearing 10–18 flowers; bracts 1.4–2 cm, ovate to oblong, very pale; flowers 3.5–4.5 cm, white; anthers c. 3 mm, yellow to purple. *Japan*. H2. Summer.

**9. H. longipes** (Franchet & Savatier) Bailey (*H. sieboldiana* var. *longipes* (Franchet & Savatier) Matsumura; *H. minor* misapplied). Illustration: Gentes Herbarum 2: 138 (1930); Journal of Japanese Botany 13: 901 (1937); Journal of the Faculty of Science, Tokyo (Botany) 5: 386 & 387 (1940).
Leaf-stalks to 16 cm, broadly channelled, winged in upper half, purple-spotted. Leaf-blades to 14 × 8–10 cm, ovate to elliptic, shortly acuminate, truncate or slightly cordate at base, decurrent, membranous, flat or with a few undulations, mid-green to dark green, dull above, shiny below, with 6–8 pairs of side-veins. Flowering stems to

c. 30 cm, usually much exceeding leaves, often branched, purple-spotted; bracts membranous, pale purple; flowers c. 4 cm, pale violet; anthers c. 2.5 cm, bluish purple. *Japan*. H2. Late summer–autumn.

**10. H. tardiflora** (Irving) Stearn (*Funkia tardiflora* Irving; *H. lancifolia* (Thunberg) Engler var. *tardiflora* (Irving) Bailey; *H. sparsa* Nakai). Illustration: Acta Horti Bergiani 16: pl. 20 (1954); Botanical Magazine, n.s., 204 (1953); Journal of the Royal Horticultural Society 93: f. 186 (1968).
Leaf-stalks 12–27 cm, very shallowly and broadly channelled, not winged, strongly purple-spotted. Leaf-blades 10–15 × 2.8–6.5 cm, lanceolate to narrowly elliptic, acuminate, tapered at base, not decurrent, firm and leathery, flat or with 1–3 very shallow undulations, dull mid-green to yellowish green above, dull to slightly shiny mid-green below, with 4 or 5 pairs of side-veins. Flowering stems to c. 35 cm, ascending at c. 45°, bearing up to 50 flowers; bracts 1–1.5 cm, ovate to oblong, pale purple; flowers 4–4.5 cm, pale bluish purple; anthers 3.5–4 mm, greyish purple. *Known only in cultivation*. H2. Autumn.

**11. H. × tardiana** Anon (*H. tardiflora* (Irving) Stearn × *H. sieboldiana* Engler var. *elegans* Hylander). Illustration: The Garden 107: 198 (1982).
Differs from *H. tardiflora* in its thicker, often broader, very glaucous leaves which have 6–9 pairs of side-veins and very tough stalks. Flowers pale violet. Differs from *H. sieboldiana* var. *elegans* in its much smaller, narrower, unpuckered leaves. *Garden origin*. H2. Late summer.

This cross has given rise to many named and unnamed clones. Among the bluest and most popular are 'Halcyon' and 'Harmony'.

**12. H. tortifrons** Maekawa (*H. lancifolia* (Thunberg) Engler var. *tortifrons* invalid) Figure 30(6), p. 135. Illustration: Journal of the Faculty of Science, Tokyo (Botany) 5: 323 (1940).
Leaf-stalks less than 5 cm, almost flat, slightly spotted. Leaf-blades to c. 7 × 1.3 cm, oblong to very narrowly elliptic, tapered at base and tip, slightly decurrent, leathery, often twisted through c. 90°, slightly shiny dark green above, shiny mid-green below, with 2 or 3 pairs of very indistinct side-veins. Flowers not seen, and not described by Maekawa. *Japan*. H2.

**13. H. rupifraga** Nakai (*H. longipes* (Franchet & Savatier) Bailey var. *latifolia* Maekawa). Illustration: Journal of the Faculty of Science, Tokyo (Botany) **5**: 390 (1940); Botanical Magazine, Tokyo **44**: 28 (1930).
Leaf-stalks to 12(23) cm, shallowly channelled, winged above, purple-spotted below. Leaf-blades 9–13(16) × 4–8.5(12) cm, ovate, acuminate, rounded to truncate and funnel-shaped at base, shortly decurrent, leathery, with a few marginal undulations, mid-green to dark green and dull above, slightly paler and shiny below, with 6–8 pairs of side-veins. Flowering stems to *c.* 50 cm, bearing many flowers; bracts pale purple; flowers 3.2–4.1 cm, very pale violet. *Japan.* H2. Late summer.

One clone in cultivation under this name has much larger leaves than those described in previous publications (including the original description). These sizes are given in brackets. The name is often wrongly spelled 'ruprifraga'.

**14. H. gracillima** Maekawa (*H. longipes* (Franchet & Savatier) Bailey var. *gracillima* (Maekawa) Fujita). Figure 30(3), p. 135. Illustration: Iconographia Plantarum Asiae Orientalis **1**(4): pl. 33 (1936); Journal of the Faculty of Science, Tokyo (Botany) **5**: 396 (1940).
Leaf-stalks 2–8 cm, very narrow, shallowly channelled, slightly winged above, spotted on back from bottom to top. Leaf-blades 2.5–6.2 × 1.1–3 cm, lanceolate to ovate, acute or slightly acuminate, rounded to slightly tapered at base, slightly decurrent, membranous to slightly leathery, usually with several small marginal wrinkles, rarely flat, dark green, dull above, slightly shiny below, with *c.* 3 pairs of side-veins. Flowering stems erect or spreading, to 25 cm, bearing *c.* 10 flowers; bracts 1–2 cm, narrowly triangular; flowers 3–3.5 cm, pale pinkish purple. *Japan.* H2. Late summer-early autumn.

This name is widely misapplied to several clones with a range of leaf-shapes and sizes, particularly to a probably unnamed clone with narrow, wrinkled, almost oblong leaf-blades to *c.* 18 cm, and leaf-stalks to 15 cm. This is probably due to the name 'gracillima' being loosely applied to many dwarf or slender hybrids and sports.

Numbers 15–19 are sometimes considered to be a single species, *H. sieboldii.*

**15. H. sieboldii** (Paxton) Ingram (*Hemerocallis sieboldii* Paxton; *Hosta lancifolia* (Thunberg) Engler var. *albomarginata* (Hooker) Stearn; *H. albomarginata* (Hooker) Ohwi; *H. sieboldiana* misapplied). Illustration: Botanical Magazine, 3657 (1838); Journal of the Royal Horticultural Society **85**: f. 103 (1960); Thomas, Plants for ground cover, f. 29 (1970).
Leaf-stalks to 26 cm, shallowly channelled, broadly winged, often spotted. Leaf-blades to 15 × 6 cm, lanceolate to elliptic, acute or obtuse, not acuminate, tapering at base, decurrent, membranous, flat or with a few broad undulations, mid-green (yellowish in full sun), dull above, shiny below, usually with a narrow cream-coloured or very pale yellow margin (many clones have plain green leaves), with *c.* 4 pairs of side-veins. Flowering stems erect, slender, to *c.* 30 cm, exceeding leaves, with a few small leaf-like bracts below; upper bracts green, concave; flowers 5–6 cm, bluish purple or white; anthers yellow. *Japan.* H2. Summer.

Plants with white flowers and plain leaves are known as **H. sieboldii 'Alba'** (*H. albomarginata* (Hooker) Ohwi var. *alba* (Robinson) Hylander). Illustration: Acta Horti Bergiani **16**: pl. 19 (1954). All combinations of white or purple flowers with plain or variegated leaves are found in cultivation. The variegated form has not been found in the wild. Several clones similar to *H. sieboldii*, though generally smaller, are becoming popular; among these are **'Louisa'**, **'Ginko Craig'** and **'Emerald Isle'**.

**16. H. lancifolia** (Thunberg) Engler (*H. japonica* (Thunberg) Ascherson & Graebner not Redouté; *Hemerocallis lancifolia* Thunberg; *Hosta lancifolia* (Thunberg) Engler var. *thunbergii* Stearn). Illustration: Acta Horti Bergiani **16**: pl. 17 (1954); Journal of the Royal Horticultural Society **85**: f. 102 (1960); Thomas, Plants for ground cover, f. 28 (1970).
Leaf-stalks to 31 cm, broadly channelled, winged above, purple-spotted. Leaf-blades 10–17 × 4.7–8.5 cm, lanceolate to elliptic, acute, slightly acuminate, shortly tapered at base, shortly decurrent, membranous to slightly leathery, flat or with a few undulations, shiny dark green above, very shiny and slightly paler below, with 5 or 6 pairs of side-veins. Flowering stems to 45 cm, erect, well exceeding leaf-mound, with 2 or 3 sterile bracts and 20–30 flowers; upper bracts 1.5–3 cm, shiny green; flowers *c.* 5.3 cm, strong bluish

purple; anthers bluish purple. *Japan.* H2. Late summer–autumn.

Clones with slenderer, more acuminate leaves have been called var. *gracilis* Anon.

**'Kabitan'** (*H. lancifolia* forma *kabitan* Maekawa). Illustration: Journal of the Faculty of Science, Tokyo (Botany) **5**: 401 (1940). This is smaller in all its parts and has leaves which are yellow to pale yellowish green with an irregular dark green margin. Leaf-stalks to 10 cm; leaf-blades 8–10.5 × 2.5–3.5 cm, elliptic, acute, tapered at base, membranous, with a few strong undulations, dull above, shiny below, with *c.* 3 pairs of side-veins. (Figure 30(2), p. 135).

**'Wogon Gold'** with entirely golden yellow, broader, elliptic, flimsy leaves and broadly winged leaf-stalks may be a 'Kabitan' hybrid.

**17. H. longissima** (Honda) Honda (*H. lancifolia* (Thunberg) Engler var. *longifolia* Honda; *H. lanceolata* invalid). Figure 30(5), p. 135. Illustration: Acta Horti Bergiani **16**: pl. 21 (1954).
Leaf-stalks to 22 cm, shallowly channelled, not spotted, narrowly winged. Leaf-blades to *c.* 20 × 4 cm, narrowly elliptic to oblong, acute, gradually tapered at base, narrowly decurrent, slightly leathery, flat, mid-green to dark green, dull above, slightly shiny below, with 3 or 4 pairs of side-veins. Flowering stems to 55 cm, well exceeding leaves, erect, slender, few-flowered; bracts shiny, green, concave; flowers *c.* 3.3 cm, pale reddish. *Japan.* H2. Summer.

Var. **brevifolia** Maekawa (*Funkia lancifolia* (Thunberg) Engler var. *angustifolia* Gaerdtner) is described as having narrower leaf-blades to *c.* 19 × 3 cm, and smaller flowers, but may not be distinct.

**18. H. helonioides** Maekawa. Illustration: Journal of Japanese Botany **13**: 896 (1937); Journal of the Faculty of Science, Tokyo (Botany) **5**: 402 (1940).
Leaf-stalks *c.* 16 cm, channelled, broadly winged. Leaf-blades to *c.* 12 × 5 cm, elliptic, acute to obtuse, tapered at base, strongly decurrent, mid-green, dull above, slightly shiny below, thin and membranous, flat or with a few undulations, with 3 or 4 pairs of side-veins. Flowering stems well exceeding leaves, purple-spotted below; lower bracts clasping stem. Flowers *c.* 5 cm, pale bluish purple; anthers *c.* 3.2 mm, blue. *Japan.* H2. Late summer.

Most commonly seen in cultivation as forma **albopicta** Maekawa with irregular pale yellow leaf-margins to 1.5 cm wide. (Figure 30(7), p. 135).

**19. H. rectifolia** Nakai (*H. longipes* misapplied). Figure 30(8), p. 135. Illustration: Journal of the Faculty of Science, Tokyo (Botany) **5**: 411 (1940); Acta Horti Bergiani **16**: pl. 21 (1954); Botanical Magazine, n.s., 138 (1951). Leaf-stalks to *c.* 50 cm, shallowly channelled, broadly winged, unspotted. Leaf-blades erect or steeply ascending, to *c.* 25 × 12 cm, narrowly ovate to elliptic, shortly acuminate, tapered at base, decurrent, slightly leathery, dull mid-green above and below, with 6–9 pairs of side-veins. Flowering stems to *c.* 60 cm, sometimes spotted, bearing 15–35 flowers; bracts green, concave. Flowers to *c.* 4.8 cm, violet; anthers *c.* 3 mm, yellow. *Japan & Sakhalin.* H2. Summer–early autumn.

Many of the clumps seen in British gardens are somewhat mixed, containing some plants with much broader cordate leaves. These do not conform with the original description of *H. rectifolia*. Some clones have fewer pairs of side-veins.

'**Tall Boy**' (*H. rectifolia* × ?) is larger in all its parts, with leaves to *c.* 15 cm wide and 8 or 9 pairs of side-veins; flowering stems to *c.* 80 cm. Illustration: Thomas, Perennial garden plants, pl. 140 (1976).

**20. H. decorata** Bailey (*H. decorata* var. *marginata* Stearn; *H.* 'Thomas Hogg' misapplied). Figure 30(9), p. 135. Illustration: Gardeners' Chronicle **90**: 88 (1931); Botanical Magazine, 9395 (1935). Leaf-stalks to 30 cm, broadly channelled, winged, spotted at base. Leaf-blades to 16 × 11 cm, broadly ovate to broadly elliptic, obtuse, not acuminate, rounded to tapered at base, decurrent, membranous to slightly leathery, flat or with a few shallow undulations, mid-green to dark green with an irregular white to cream-coloured margin to *c.* 1.5 cm, dull above, slightly shiny below, with 5–7 pairs of side-veins. Flowering stems to *c.* 60 cm, occasionally with 1 or 2 leaf-like bracts below, bearing up to 25 flowers; flowers 4–5 cm, dark lilac, paler at base; anthers *c.* 4 mm, yellow. *Known only in cultivation.* H2. Late summer.

Var. **normalis** Stearn has plain green leaves which are sometimes slightly acuminate. Illustration: Journal of the Faculty of Science, Tokyo (Botany) **5**: 418 (1940).

**21. H. ventricosa** Stearn (*Funkia ovata* Sprengel; *H. caerulea* (Trattinick) Andrews, not Jacquin). Illustration: Botanical Magazine, 894 (1806); Gentes Herbarum **2**: 128 (1930); Parey's Blumengärtnerei **2**: 241 (1958).

Leaf-stalks to 40 cm, broadly channelled, winged, spotted below. Leaf-blades to 25 × 18 cm, broadly ovate to elliptic or circular, abruptly acuminate, rounded to truncate or slightly cordate at base, decurrent, leathery, flat or with several shallow undulations, dull dark green with a metallic lustre above, very shiny mid-green to dark green below, with 7–9 pairs of side-veins, and conspicuous small transverse veins. Flowering stems to *c.* 90 cm, bearing 20–30 flowers; bracts 1.6–1.8 cm; flowers 5.5–5.7 cm, bell-shaped, bluish purple; anthers *c.* 5 mm, purple-spotted. *China.* H2. Summer.

'**Aureomaculata**' has leaves variegated with patches of yellow or yellowish green which become darker and less distinct later in the season.

'**Variegata**' ('Aureomarginata') has broad, irregular, cream-coloured to yellow margins. Illustration: Journal of the Royal Horticultural Society **93**; f. 185 (1968); Thomas, Plants for ground cover, f. 34 (1970).

**22. H. minor** (Baker) Nakai (*Funkia ovata* Sprengel var. *minor* Baker). Figure 30(1), p. 135. Illustration: Gentes Herbarum **2**: 140 (1930); Journal of the Faculty of Science, Tokyo (Botany) **5**: 419 (1940). Leaf-stalk to *c.* 8 cm, broadly channelled, winged. Leaf-blades *c.* 8.5 × 5.5 cm, broadly ovate, acute to obtuse, very slightly acuminate, rounded to truncate at base, decurrent, membranous, with very shallow undulations or none, dark green, slightly lustrous with 5 or 6 pairs of side-veins. Flowering stems to *c.* 60 cm, bearing up to 10 flowers; flowers 4.5–5 cm, bluish purple. *Japan & Korea.* H2.

Forma **alba** (Nakai) Maekawa has white flowers.

The name *H. minor* is much misused in gardens and reference works for diminutive varieties of several species. True *H. minor* is probably not commonly grown, and was originally considered a variety of *H. ventricosa*.

**23. H. venusta** Maekawa. Illustration: Journal of the Faculty of Science, Tokyo (Botany) **5**: 415 (1940); Botanical Magazine, n.s. 499 (1967). Leaf-stalks to *c.* 6.5 cm, slender, channelled, winged above, spotted with pinkish purple below. Leaf-blades to 6.5 × 2.7 cm; lanceolate to narrowly elliptic, acute, slightly acuminate, rounded or tapered at base, decurrent, membranous to slightly leathery, with 1 or 2 broad shallow undulations, mid-green, dull

above, shiny below, with 4–6 pairs of side-veins. Flowering stems 10–35 cm, erect, with 3–8 flowers; bracts to 1 cm, ovate to elliptic; flowers 2.5–3.2 cm, bluish purple; anthers *c.* 2.7 mm, greyish purple. *Quelpaert Island (Korea) & Japan.* H2. Summer.

**24. H. nakaiana** Maekawa. Illustration: Journal of Japanese Botany **11**: 688 (1935); Journal of the Faculty of Science, Tokyo (Botany) **5**: 419 (1940), flowers only. Leaf-stalks to *c.* 10(21) cm, deeply channelled, spotted below, winged. Leaf-blades 6–7.5(14) × 2.5–4.5(8) cm, broadly ovate, acuminate to attenuate, truncate at base, decurrent, slightly leathery, with several small marginal undulations, mid-green, dull above, slightly glaucous below, with 5–7 pairs of side-veins. Flowering stem to 45 cm, much exceeding leaves, erect or curved; bracts overlapping, elliptic to circular, 1.4–1.7 cm. Flowers in dense almost spherical heads, 4.6–5.2 cm, pale bluish purple; anthers white, 3–3.2 mm. *Japan & Korea.* H2. Summer.

At least one clone in cultivation under this name has leaves considerably larger than those described in previous publications. These measurements are given in brackets.

'**Golden Nakaiana**' (*H. nakaiana* × ?) has broader, more leathery, slightly cordate leaves which are bright greenish yellow throughout the season, and have a rough colourless margin. (Figure 30(4), p. 135).

'**Pastures New**' (*H. nakaiana* × ?) has small flat rather bluish slightly acuminate leathery leaves to *c.* 10 × 6 cm.

**25. H. capitata** (Koidzumi) Nakai (*H. coerulea* (Trattinick) Andrews var. *capitata* Koidzumi). Similar to *H. nakaiana* though generally larger with abruptly acuminate leaves which are cordate rather than truncate at the base and have 7–9 pairs of side-veins. *Japan.* H4. Early summer.

**26. H. undulata** (Otto & Dietrich) Bailey. Var. **undulata** (*H. japonica* (Thunberg) Ascherson & Graebner forma *undulata* (Otto & Dietrich) Voss 'Albo-Variegata'; *H. lancifolia* (Thunberg) Engler var. *undulata* Bailey 'Medio-Variegata'). Illustration: Gentes Herbarum **2**: 136 (1930); Journal of the Royal Horticultural Society **85**: f. 105 (1960); The Garden **107**: 197 (1982). Leaf-stalks to 25 cm, deeply channelled, winged, sometimes faintly spotted below, striped. Leaf-blades to 17 × 6.8 cm, lanceolate to elliptic, acute, slightly acuminate, tapered at base, decurrent, membranous to slightly leathery, with a few strong undulations, tip sometimes twisted, mid-green with a large irregular area of white or very pale yellow in the middle, dull above, very shiny below, with 6 or 7 pairs of side-veins. Flowering stems to 1 m, bearing large, leaf-like bracts below and up to 10 flowers above; upper bracts 2–2.6 mm, ovate to elliptic, variegated; flowers 5.5–6 cm, pale bluish purple; anthers purple. *Known only in cultivation.* H2. Summer.

Var. **univittata** (Miquel) Hylander. Illustration: Acta Horti Bergiani 16: pl. 15 (1954); Thomas, Plants for ground cover, f. 27 (1970). Very similar though generally more vigorous. The pale area is confined to the centre of the leaf and ends in a narrow stripe towards the leaf-tip. Plants often sport back to var. *undulata*.

Var. **erromena** (Stearn) Maekawa (*H. erromena* Stearn; *H. japonica* (Thunberg) Ascherson & Graebner var. *fortis* Bailey). Illustration: Journal of the Faculty of Science, Tokyo (Botany) 5: 422 & 423 (1940); Acta Horti Bergiani 16: pl. 14 & 16 (1954). Differs from var. *undulata* in its larger, plain mid-green leaves. Leaf-stalks to 50 cm. Leaf-blades to 22.5 × 13 cm, mid-green, with 8 or 9 pairs of side-veins. Flowering stems to 1.5 m.

Var. **albo-marginata** Maekawa (*H.* 'Thomas Hogg'). Illustration: Journal of the Faculty of Science, Tokyo (Botany) 5: 423 (1940); Thomas, Plants for ground cover, pl. XIB (1970). Leaves broader than in var. *undulata*, flat or with a few very shallow undulations, dark green with a narrow cream-coloured to pale yellow margin to *c.* 1 cm wide, dull above, shiny below. *H. undulata*, like *H. fortunei*, is a group of related, largely sterile clones, probably hybrid in origin, with *H. decorata* being a possible parent.

The name *H. japonica* is still sometimes encountered in gardens, catalogues and reference works. Its correct application was unclear from the time it was first published, and it was never unequivocally linked to a particular species.

## 34. HEMEROCALLIS Linnaeus

*E. Campbell & J. Cullen*
Herbaceous perennials with rhizomes, roots fibrous, often swollen and fleshy at their ends. Leaves in 2 ranks, deciduous or persisting well into winter, linear, tapered gradually to the apex, usually recurved, flat or folded at the base. Flowers in a raceme or panicle borne on a scape, bracts small or large. Perianth united into a tube at the base, more or less funnel-shaped or openly funnel-shaped, 6-lobed, yellow or orange, rarely reddish. Stamens 6, deflexed downwards, attached to the top of the perianth-tube. Ovary superior, 3-celled, with many ovules, style thread-like. Fruit a 3-angled or 3-winged capsule.

A genus of about 15 species from E Asia, cultivated in China and Japan for several centuries. Its classification is difficult, as many of the plants in cultivation (as well as plants naturalised from gardens in China and Japan) are selections or hybrids of unknown parentage. Hybridisation has continued and many cultivars are available (see the list cited below). Twelve species are included here, and many of these are probably involved in the parentage of both old and new cultivated hybrids.
Literature: Bailey, L.H., Hemerocallis: the day lilies, *Gentes Herbarum* 2: 143–56 (1930); Stout, A.B., *Daylilies* (1934); Stuntz, M.F. et al., *Hemerocallis Checklist, 1893–1957* (1957).

*Leaves.* Persisting well into winter: **9.** Erect or spreading, not recurved or bent over in the upper half: **7.**
*Scape.* Branched above: **3–10;** not branched above: **1,2.**
*Bracts.* Large and cup-shaped: **1,2;** small, not cup-shaped: **3–10.**
*Flowering time.* Spring: **2–4,7;** summer: **1,5,6,8–10.**
*Flowers.* Yellow: **3–6;** orange, brownish orange or reddish: **1,2,7–10.** More than 20 in each inflorescence: **5,8;** less than 20 in each inflorescence: **1–4,6,7,9,10.** Fragrant: **4–7;** not fragrant: **1–3,7–10.**

1a. Scape not forked or branched above, the flowers apparently stalkless and close together, their stalks hidden by the large, enfolding, cup-shaped bracts  2
  b. Scape forked and sometimes further branched above, or bearing only a single flower, the stalks clearly visible, not hidden by the bracts, which are small and not cup-shaped  3
2a. Flowers 8–10 cm; inner perianth-lobes 2–2.5 cm wide  **1. middendorffii**
  b. Flowers 5–7 cm; inner perianth-lobes less than 2 cm wide  **2. dumortieri**
3a. Flowers yellow  4
  b. Flowers orange, brownish orange or reddish  7
4a. Tube *c.* one-quarter of the total length of the perianth; spring-flowering  5
  b. Tube of perianth longer, about one-third to half of the total length; summer-flowering  6
5a. Leaves 30–45 cm × 5–9 mm; flowers not fragrant  **3. minor**
  b. Leaves 50–65 × 1–1.4 cm; flowers fragrant  **4. lilio-asphodelus**
6a. Flowers opening at night; leaves robust, 1–1.6 cm broad; raceme with 20–65 flowers  **5. citrina**
  b. Flowers opening in the morning; leaves 5–10 mm broad; raceme with 3–15 flowers  **6. thunbergii**
7a. Leaves rather stiff, erect or arching but not bent over in the upper half, to 30 cm; scape not or scarcely exceeding leaves  **7. forrestii***
  b. Leaves flaccid, arching and bent over in the upper half, 30 cm or more; scape clearly exceeding the leaves  8
8a. Scape with many long branches bearing 20–75 flowers, glaucous above  **8. multiflora**
  b. Scape forked, but without other long branches, bearing up to 20 flowers, not glaucous above  9
9a. Leaves persisting well into winter; flowers orange, sometimes flushed with red-purple but without bands or stripes of a darker colour  **9. aurantiaca**
  b. Leaves dying down in autumn; flowers brownish orange or rarely reddish, usually with bands or stripes of darker colour  **10. fulva**

**1. H. middendorffii** Trautvetter & Meyer. Illustration: Addisonia 14: t. 463 (1929). Roots mostly fibrous, some slightly swollen. Leaves recurved, soft, to 30 × 1–2.5 cm. Scape erect, exceeding the leaves. Flowers few, in a compressed raceme, the flower-stalks hidden by the cup-shaped, enfolding bracts. Flowers 8–10 cm, yellow, inner perianth-lobes 2–2.5 cm broad, tube to 2 cm. *Japan, E USSR, Korea, N China.* H2. Summer.

**2. H. dumortieri** Morren. Illustration: Addisonia 14: t. 462 (1929); Bloom, Perennials for your garden, 80 (1975). Roots mostly swollen. Leaves recurved, to 35 × 1.5 cm. Scapes shorter than or slightly longer than the leaves. Flowers few in a compressed raceme, the flower-stalks hidden by cup-shaped bracts. Flowers 5–7 cm, orange-yellow, the inner perianth-lobes less than 2 cm broad, the tube *c.* 1 cm. *Japan, E USSR, Korea.* H2. Spring.

**3. H. minor** Miller (*H. graminea* Andrews; *H. graminifolia* Schlechtendahl). Illustration: Botanical Magazine, 873 (1805); Addisonia **14**: t. 458 (1929).
Roots mostly fibrous, rarely a few swollen. Leaves recurved, 30–45 cm × 5–9 mm. Scapes exceeding the leaves, branched above at least once, erect, raceme with 2–5 flowers. Flowers 5–7 cm, tube about one-quarter of the total length, yellow, lobes narrow. *Japan, China, E USSR*. H2. Spring.

**4. H. lilio-asphodelus** Linnaeus (*H. flava* Linnaeus). Illustration: Botanical Magazine, 19 (1788); Addisonia **14**: t. 456 (1929).
Rhizome spreading. Roots mostly swollen. Leaves recurved, 50–65 × 1–1.4 cm. Scapes ascending, exceeding the leaves, branched above, raceme with 8–12 flowers. Flowers fragrant, 7–8 cm, yellow, tube about one-quarter of the total length. *E Asia*. H3. Spring.

'Major', with larger flowers, darker leaves and wavy perianth-lobes, reflexed at their tips, is widely grown.

**5. H. citrina** Baroni. Illustration: Addisonia **15**: t. 482 (1930).
Roots very fleshy. Leaves recurved, 70–80 × 1–1.6 cm. Scapes erect, exceeding the leaves, branched above, raceme with 20–65 flowers. Flowers fragrant, opening at night, 9–12 cm, lemon yellow, tube about one third of the total length, lobes narrow. *China*. H3. Summer.

'Baroni' is often referred to *H. citrina* but correctly covers a hybrid between it and *H. thunbergii*; it is like *H. citrina* but has broader perianth-lobes.

**6. H. thunbergii** Baker. Illustration: Addisonia **14**: t. 459 (1929); Kitamura et al., Herbaceous Plants of Japan 3: f. 241 (1978).
Roots slender, cylindric, sometimes slightly swollen. Leaves recurved, 30–60 cm × 5–8 (rarely to 10) mm. Scapes erect, branched above, exceeding the leaves, raceme with 3–15 flowers. Flowers fragrant, opening in the morning, yellow, 9–11 cm, tube about one-third of the total length, lobes narrow. *China, Korea*. H2. Summer.

There is a related species, *H. vespertina* Hara, in Japan; this is often included in *H. thunbergii* or *H. citrina*, but has broader peianth-lobes than either. It is uncertain whether or not this is in cultivation in Europe.

**7. H. forrestii** Diels.
Roots cylindric but markedly swollen at their ends. Leaves erect or ascending, not recurved, to 30 × 1.5 cm. Scapes erect, not

or scarcely exceeding the leaves, branched, raceme with 4–8 flowers. Flowers orange, *c.* 7 cm, tube less than one-quarter of the total length. *China (mainly the SW)*. H5. Spring.

\***H. plicata** Stapf. Very similar, but leaves narrower and raceme with usually 8 or more flowers. The leaves are reputedly folded at the base, but this requires confirmation. *SW China*. H5. Spring.

\***H. nana** Smith & Forrest. Also similar, but the whole plant much smaller, leaves spreading, raceme with usually 1 (rarely as many as 3) flowers, which are fragrant. *SW China*. H5. Spring.

**8. H. multiflora** Stout. Illustration: Addisonia **14**: t. 464 (1929).
Roots fleshy. Leaves recurved, 60 × 1 cm or more. Scapes erect, exceeding the leaves, glaucous above, richly branched with 20–75 widely spaced flowers. Flowers orange, 7–8 cm, tube about one-quarter of the total length. *China (Honan)*. H4. Summer.

**9. H. aurantiaca** Baker. Illustration: The Garden **23**: additional plate (1895); Addisonia **14**: t. 461 (1929); Kitamura et al., Herbaceous Plants of Japan 3: f. 244 (1978).
Rhizomes spreading, roots swollen. Leaves persisting well into winter, recurved, 60–80 × 1.5–2.5 cm, glaucous. Scape exceeding the leaves, forked but not extensively branched above, flowers up to 20, closely grouped, orange, sometimes flushed with red-purple but without bands or stripes of a darker colour, 10–13 cm, the tube much less than one-quarter of the total length. *China*. H3. Summer.

**10. H. fulva** Linnaeus. Illustration: Botanical Magazine, 64 (1798); Addisonia **15**: t. 483, 484 (1930).
Rhizomes spreading. Roots fleshy at their ends. Leaves recurved, 30–90 × 1–2.5 cm. Scapes exceeding the leaves, erect, forked and slightly further branched above, racemes with up to 20 flowers. Perianth brownish orange or rarely reddish, usually with zones or stripes of darker colour, 7–10 cm, the tube one-quarter to half the total length. *Origin uncertain, perhaps China or Japan, naturalised there and elsewhere*. H2. Summer.

This is the most widely cultivated and variable of the species. Var. **longituba** Miquel (*H. longituba* (Miquel) Maximowicz) has narrower leaves and the perianth-tubes about one-third of the total length. Many cultivars are grown, including 'Rosea' with

reddish purple flowers, and 'Kwanso' which has double flowers.

## 35. BLANDFORDIA J.E. Smith
*S.G. Knees*

Deciduous, hairless perennial herbs with fibrous roots and often with rhizomes. Leaves sheathing, linear, triangular in cross-section, often 2-ranked. Perianth tubular to bell-shaped with 6 equal lobes, varying in colour from red to orange and yellow. Scapes solitary, bearing 3–20 bisexual flowers with bracts, in terminal racemes. Stamens 6, included in perianth and attached at or below the middle of the tube. Ovary superior, 3-celled with many ovules in each cell; style 1. Fruit a stalked, 3-celled, ovoid, capsule. Seeds many.

A genus of 4 species, all Australasian in distribution. None is widely cultivated, and one, **B. punicea**, is reputedly difficult to grow. A sandy loam should provide an ideal growing medium. Although readily grown as pot plants, care should be taken to ensure that they do not dry out. This can be facilitated by planting in partial shade. Propagation is usually from seed, but large plants can be divided in early spring.

1a. Leaf-margins smooth **1. cunninghamii**
 b. Leaf-margins toothed 2
2a. Flowers 4–6 cm **2. grandiflora**
 b. Flowers usually less than 4 cm 3
3a. Leaves 8–12 mm wide; raceme with 15–25 flowers **4. punicea**
 b. Leaves 1–5 mm wide; raceme with 3–15 flowers **3. nobilis**

**1. B. cunninghamii** Lindley. Illustration: Botanical Magazine, 6209 (1876).
Leaves with smooth margins, 35–70 cm × 4–9 mm. Scape to 85 cm bearing 5–20 flowers in compact racemes. Flower-stalks 1.5–5 cm, equalling bracts. Perianth 4–5 cm, tubular, brownish red with yellow, obtuse lobes *c.* 3.8 mm long. Stamens attached below middle of perianth-tube. *Australia (New South Wales)*. H5. Spring.

**2. B. grandiflora** R. Brown (*B. flammea* Hooker). Illustration: Botanical Magazine, 5374 (1868); Elliot & Jones, Encyclopaedia of Australian plants, 325 (1982); RHS dictionary of gardening, edn 2, 287 (1956).
Leaves linear, bearing small rounded teeth, 60–80 cm × 1–4 mm. Scape to 60 cm, bearing 7–10 flowers. Corolla tubular to bell-shaped, 4–6 cm, varying in colour from red with yellow apical lobes to pure yellow. *Australia (Queensland and New South Wales)*. H5. Spring and early summer.

**3. B. nobilis** Smith. Illustration: Botanical
Magazine, 2003 (1818); Elliot & Jones,
Encyclopaedia of Australian plants, 326
(1982); Edwards's Botanical Register 4: t.
286 (1818).
Leaves linear, finely toothed on margins
40–80 cm × 1–5 mm, crowded at the base.
Scape 40–70 cm, bearing narrowly tubular
flowers 2–3 cm × 8–12 mm, brick red or
orange, often streaked with yellow, 3–15
per raceme. Fruits narrowly ovoid, 4–5 cm.
*Australia (New South Wales)*. H5. Spring
and early summer.

**4. B. punicea** (Labillardière) Sweet
(*B. marginata* Herbert). Illustration:
Edwards's Botanical Register **31**: t. 18
(1845); Elliot & Jones, Encyclopaedia of
Australian plants, 326 (1982).
Leaves finely toothed, 70–90 cm ×
8–12 mm, strap-like, leathery, coarsely
sheathed at base, with prominent midrib.
Scape to 80 cm, stout, bearing 3–4 bracts,
and a loose raceme of 15–25 flowers in
almost opposite pairs. Flower-stalks
2–5 cm. Perianth 3–4 cm, narrowly bell-
shaped, pinkish scarlet, with orange-yellow
tips. Stigmas slightly projecting, stamens
not projecting. Fruits 8–10 × 3.5–4.5 cm.
*Australia (Tasmania)*. H5. Summer.

## 36. LEUCOCRINUM A. Gray
*V.A. Matthews*
Rootstock a cluster of short, fleshy roots.
Leaves narrowly linear, forming a tuft,
10–20 cm × 2–6 mm, surrounded at the
base by membranous bracts. Stem absent to
very short, below ground level, flowers
borne on subterranean 5–30 mm stalks,
appearing at ground level, white, fragrant.
Perianth joined into a slender tube
5–12 cm, the free lobes 1.4–2 cm,
narrowly oblong. Stamens 6, borne near
the top of the perianth-tube; anthers
4–6 mm. Style 1, slender, stigma slightly 3-
lobed. Ovary subterranean with several
black, angled seeds in each cell.

A genus with one species native to
western USA. Plants grow best in a sunny
position but need a soil which never dries
out completely. Propagation is by seed.

**1. L. montanum** A. Gray. Illustration:
Bulletin of the Alpine Garden Society **39**:
125 (1971); Rickett, Wild flowers of the
United States **5**: pl. 18 (1971); Niehaus &
Ripper, Field guide to Pacific states
wildflowers, 9 (1976); Cronquist et al.
Intermountain Flora **6**: 483 (1977).
*Western USA*. H4. Spring–early summer.

## 37. KNIPHOFIA Moench
*S.J.M. Droop*
Plants with thick rhizomes, rarely with
aerial stems. Leaves usually basal, in
several ranks, linear, tapering gradually to
the apex, usually keeled beneath, margins
smooth or very finely toothed. Scape
simple, erect, with a few small, sterile
bracts beneath the raceme. Raceme usually
dense, flowers usually opening from the
bottom upwards, occasionally from the top
downwards, spreading or drooping, white
or red, the buds nearly always red; bracts
scarious, exceeding the flower-stalks.
Perianth with a cylindric or narrowly bell-
shaped, often curved tube to 4.5 cm and 6
small lobes. Stamens 6, often becoming
spirally twisted and withdrawn after
shedding pollen, filaments attached to the
perianth near the base. Ovary superior, 3-
celled with numerous ovules in each cell.
Style usually projecting. Capsule spherical
or ovoid, sometimes 3-angled.

A genus of about 70 species distributed
mainly in eastern and southern Africa,
with 1 species in Madagascar and 1 in
southern Arabia. About 11 species are
found in cultivation, along with many
hybrids and cultivars (mainly derived from
*K. uvaria*). They are relatively hardy in the
more sheltered parts of Europe, where they
can be grown out of doors in a light soil in
a sunny position, with plenty of water
during spring and summer. Some
protection against frost, usually by means
of leaf litter or straw, is helpful even in mild
areas. Propagation is by division of the
rhizomes or by seed.
Literature: Codd, L.E., The South African
species of Kniphofia (Liliaceae), *Bothalia* 9:
363–511 (1968); Marais, W., The tropical
species of Kniphofia (Liliaceae), *Kew Bulletin*
28: 465–83 (1973).

1a. Leaves 8–12 cm broad, not keeled
　　　　　　　　　　　　　**5. northiae**
　b. Leaves to 8 cm broad, keeled beneath
　　　　　　　　　　　　　　　　　2
2a. Plants with a distinct simple or
　　branched stem to 60 cm; leaves very
　　glaucous　　　　　**4. caulescens**
　b. Plants without a stem; leaves not or
　　slightly glaucous　　　　　　　3
3a. Inflorescence opening from the top
　　downwards　　　　　**10. pumila**
　b. Inflorescence opening from the
　　bottom upwards　　　　　　　4
4a. Open perianth white, spreading; buds
　　ascending　　　　　　**1. gracilis**
　b. Open perianth cream to red or
　　orange, or, if white, then perianth
　　pendulous and buds spreading　　5

5a. Bracts acute, obtuse, rounded　　6
　b. Bracts acuminate, sharply pointed or
　　drawn out into a tail　　　　　7
6a. Stamens not or scarcely projecting at
　　flowering, withdrawn later　**8. uvaria***
　b. Stamens projecting at flowering, often
　　more than 5 mm beyond perianth,
　　usually withdrawn later　**1. gracilis**
7a. Stamens projecting by more than
　　5 mm at flowering and not usually
　　withdrawn later　　　　　　8
　b. Stamens not projecting or projecting
　　by less than 5 mm at flowering,
　　usually withdrawn later　　　10
8a. Bract apices acute　　**11. foliosa**
　b. Bract apices acuminate　　　　9
9a. Leaf margins and keel conspicuously
　　finely toothed　　　**3. ensifolia**
　b. Leaf margins and keel smooth
　　　　　　　　　　　　**10. pumila**
10a. Perianth constricted above the ovary
　　　　　　　　　　　　　　　11
　b. Perianth not constricted above the
　　ovary　　　　　　　　　　12
11a. Bract apices long-acuminate;
　　inflorescence rather loose to loose
　　　　　　　**9. thomsonii** var. **snowdenii**
　b. Bract apices sharply rounded;
　　inflorescence dense　　**8. uvaria***
12a. Inflorescence loose (axis visible
　　between the individual flowers),
　　usually 8–25 cm long　　**2. rufa**
　b. Inflorescence denser (axis not visible
　　between the flowers), less than 8 cm
　　long　　　　　　　　　　13
13a. Buds redder than the open flowers
　　　　　　　　　　　　**6. galpinii**
　b. Buds and open flowers dull red to
　　orange-yellow　　**7. triangularis**

**1. K. gracilis** Baker (*K. modesta* in the sense
of Baker; *K. sparsa* N.E. Brown).
Illustration: Botanical Magazine, 7293
(1893); Bothalia 9: t. 7 (1968).
Leaves usually less than 2 cm wide,
channelled above, keeled beneath, margin
smooth or sparingly toothed towards the
apex. Scape 25–80 cm, usually
overtopping the recurved leaves. Raceme
oblong to narrowly cylindric, dense to
loose, 9–35 cm. Buds ascending, exceeded
by the bracts, flowers spreading or deflexed.
Bracts ovate to lanceolate, acute, obtuse or
rounded. Perianth almost cylindric to
narrowly funnel-shaped, to 2.2 cm,
constricted above the ovary. Stamens
not projecting or projecting for up to 5 mm
at flowering, usually withdrawn later.
*South Africa (E Cape Province, Natal)*.

Usually with yellow or orange buds and

creamy yellow flowers and ovate, obtuse or rounded bracts. The variant with pinkish buds, whitish flowers and lanceolate bracts was named *K. sparsa*, but this intergrades with typical *K. gracilis*.

**2. K. rufa** Baker (*K. natalensis* Baker var. *angustifolia* Baker). Illustration: Botanical Magazine, 7706 (1900); Bothalia **9**: t. 2a (1968).
Leaves soft, at first erect, later arching over, usually less than 1 cm broad, dull green to glaucous, keeled beneath; margins and keel remotely and finely toothed towards the apex. Scape slender, 40–65 cm. Raceme loose, cylindric, 8–25 cm; buds cream or dull yellow to orange-red, flowers pendulous, white to cream, yellow or dull red. Bracts lanceolate to linear-lanceolate, acute to acuminate. Perianth more or less cylindric, not constricted above the ovary, tube to 3.3 cm. Stamens not or slightly projecting at flowering, later withdrawn. *South Africa (Natal)*. H5. Summer.

**3. K. ensifolia** Baker (*K. pumila* Baker not (Aiton) Kunth). Illustration: Botanical Magazine, 7644 (1899); Flowering plants of South Africa **22**: t. 866 (1942); Bothalia **9**: t. 9 (1968).
Leaves glaucous, usually less than 4 cm broad, V-shaped in section; margins and keel usually conspicuously but finely toothed. Scape 60–80 cm. Raceme very dense, cylindric, 9–20 cm; buds spreading, greenish white or dull to bright red, flowers pendulous, greenish white to cream. Bracts linear-lanceolate, acute to acuminate. Perianth narrowly funnel-shaped. Stamens projecting more than 5 mm at flowering. *South Africa (Cape Province, Transvaal, Orange Free State)*. H5. Summer.

Only subsp. **ensifolia** (described above) is in cultivation.

**4. K. caulescens** Baker. Illustration: Botanical Magazine, 5946 (1872).
Plants with stems to 60 cm crowned with a large rosette of leaves. Leaves glaucous, fleshy, V-shaped in section, to 8 cm broad. Scape stout, 30–60 cm. Raceme oblong to cylindric, very dense; buds dull orange to red, flowers pale greenish yellow to creamy yellow. Bracts linear-oblong to linear, acute to acuminate. Perianth slightly constricted above the ovary. Stamens conspicuously projecting at flowering. *South Africa (Cape Province, Natal, Orange Free State)*. H5. Summer.

**5. K. northiae** Baker. Illustration: Botanical Magazine, 7412 (1895).
Plants usually with stems 20–170 cm, with a rosette of leaves at the stem apex. Leaves 8–12 cm broad, leathery, recurved, not keeled beneath, channelled above, margins conspicuously finely toothed. Scape 20–30 cm, up to 2.5 cm in diameter. Raceme cylindric to ovoid, very dense, 10–22 cm, buds and flowers more or less spreading, buds pinkish or orange-red, flowers whitish or yellow. Bracts oblong, acute to obtuse. Perianth more or less cylindric, to 3.2 cm, slightly constricted above the ovary. Stamens conspicuously projecting at flowering. *South Africa (Cape Province, Natal, Orange Free State), Lesotho*. H5. Summer.

**6. K. galpinii** Baker. Illustration: Flowering plants of South Africa **20**: t. 783 (1940); Bothalia **9**: t. 14 (1968).
Leaves narrow, grass-like, usually less than 6 mm broad, dull green, markedly veined, triangular in section, margins smooth or minutely toothed towards the apex. Scape 30–60 cm. Raceme almost spherical to ovoid, dense, 5–8 cm; buds and flowers pendulous, buds bright red to orange-red, flowers yellow to orange. Bracts linear-lanceolate, acuminate or tailed. Perianth to 3 cm, not constricted above the ovary. Stamens not or scarcely projecting at flowering, later withdrawn. *South Africa (Transvaal, Natal), Swaziland*. H5. Summer.

**7. K. triangularis** Kunth (*K. macowanii* Baker; *K. nelsonii* Masters). Illustration: Botanical Magazine, 6167 (1875); Flowering plants of Africa **30**: t. 1184 (1954–5); Bothalia **9**: t. 15 (1968).
Leaves soft, somewhat fibrous, to 8 mm broad, triangular in section, margin sparingly to conspicuously finely toothed. Scape 30–60 cm. Raceme more or less cylindric or ovoid, pyramidal at the apex, dense, 5–8 cm; buds and flowers pendulous, all dull red to orange-yellow. Bracts ovate-lanceolate to lanceolate, acute to long-acuminate. Perianth to 4 cm, not constricted above the ovary, expanded abruptly at the mouth. Stamens projecting for up to 3 mm at flowering, later withdrawn. *South Africa (Cape Province, Natal, Orange Free State), Lesotho*. H5. Summer.

Subsp. **triangularis** is described above; other subspecies are known in the wild but are not cultivated.

**8. K. uvaria** (Linnaeus) Hooker (*K. alooides* Moench; *K. burchellii* (Lindley) Kunth; *Tritoma uvaria* (Linnaeus) Ker Gawler). Illustration: Botanical Magazine, 758 (1804); Flowering plants of Africa **33**: t. 1289 (1959); Bothalia **9**: t. 24 (1968).
Plants to 1.2 m. Leaves stiffly erect to arched and spreading, dull green to glaucous, tough, to 2 cm wide, keeled beneath, margins and keel smooth or very sparingly finely toothed towards the apex. Scape to 1 m. Raceme oblong to almost spherical, dense, 4.5–11 cm; buds spreading, bright red to greenish, tinged with red, flowers becoming pendulous, orange-yellow to greenish yellow. Bracts ovate, rounded or obtuse. Perianth to 4 cm, slightly constricted above the ovary. Stamens not or slightly projecting at flowering, withdrawn later. *South Africa (Cape Province)*. H4. Summer–autumn.

Long cultivated, the parent of many hybrids; it has also given rise to numerous cultivars. All of these are very difficult to distinguish.

*****K. rooperi** (Moore) Lemoine. Illustration: Bothalia **9**: t. 23 & f. 72 (1968). Leaves clear green, usually more than 2 cm wide. *South Africa (Cape Province, Natal)*. H5. Summer–autumn.

**9. K. thomsonii** Baker var. **snowdenii** (Wright) Marais (*K. snowdenii* Wright). Illustration: Botanical Magazine, 8867 (1920).
Leaves soft, somewhat glaucous, less than 4 cm wide, keeled but not V-shaped in section, the upper surface almost flat, margins and keel smooth. Scape 60–300 cm. Inflorescence rather loose, to 40 cm. Flowers yellow, orange or red. Bracts ovate, long-acuminate, green, fleshy and keeled. Perianth to 3.5 cm, inflated above the ovary, much constricted above, curved, gradually widening towards the mouth. Stamens not or scarcely projecting at flowering, later withdrawn. *Kenya, Uganda*. G1. Summer.

**10. K. pumila** (Aiton) Kunth (*K. leichtlinii* Baker; *Tritoma pumila* (Aiton) Ker Gawler). Illustration: Botanical Magazine, 764 (1804), 6716 (1883).
Leaves to 2 cm broad, slightly glaucous, prominently keeled, margins smooth or minutely papillose, keel smooth. Scape 50–150 cm. Raceme 4–12 cm, very dense, oblong-cylindric, flowers opening from the top downwards or the bottom upwards, yellow to red. Bracts lanceolate to ovate-lanceolate, acuminate. Perianth to 3 cm, bell-shaped, slightly constricted above the ovary. Stamens much projecting at flowering, the filaments yellow to red. *Ethiopia, Zaire, Sudan & Uganda*. G2. Summer.

*K. leichtlinii* is the name given to the variant known only in cultivation, in which the flowers open from the top of the raceme downwards; otherwise it is essentially the same as *K. pumila*.

**11. K. foliosa** Hochstetter (*K. quartiniana* Richard). Illustration: Botanical Magazine, 6742 (1884).

Leaves to 7 cm broad, keeled, the upper surface almost flat, margins finely toothed, the keel smooth below, finely toothed towards the apex. Scape 30–120 cm. Raceme 20–40 cm, very dense, cylindric. Flowers yellow, orange or red. Bracts oblong or ovate-lanceolate, acute or rarely the upper ones rounded. Perianth to 3 cm, constricted above the ovary, narrowly funnel-shaped. Stamens conspicuously projecting at flowering. *Ethiopia*. G2. Summer.

**38. HAWORTHIA** Duval
*V.A. Matthews*

Plants stemless or with a short stem. Leaves numerous, in a rosette or densely clothing the stem and overlapping, succulent, variable in shape, smooth or with tubercles or ridges, margin often with teeth, often with a keel (sometimes 2 keels in broad leaves), sometimes with a recurved flattened end-area (retused), or with a translucent tip (window). Inflorescence a raceme or panicle. Flowers bilaterally symmetric, tubular, the perianth-segments free at the base, the outer sometimes adherent to the inner above, white or greenish, each segment with a green, brown or pink central stripe. Anthers dorsifixed. Ovary superior, 3-celled. Fruit a capsule, splitting between the septa.

A genus of *c.* 70 species native to South Africa. The plants are not frost-hardy and require protection. They grow best in a well-drained medium and should not be over-watered; protection from strong sun is advisable for most species. Propagation is best from offsets; because the species tend to cross, seed may produce hybrid plants. Literature: Bayer., M.B., *The New Haworthia Handbook*, (1982); Pilbeam, J., *Haworthia and Astroloba, a collector's guide*, (1983); Scott, C.L., *The genus Haworthia, a taxonomic revision* (1985) – not seen.

1a. Perianth triangular or rounded-triangular at the base; flowers borne in a raceme; style upcurved (Subgenus *Haworthia*)                            2
  b. Perianth hexangular or rounded-hexangular at the base; flowers borne in a simple or branched inflorescence; style straight                            17
2a. Leaves truncate                            3
  b. Leaves not truncate                            4
3a. Leaves arranged in 2 opposite rows, leaf-end rounded-oblong                            **1. truncata**
  b. Leaves arranged spirally, leaf-end more or less irregularly circular                            **2. maughanii**
4a. Leaves with a recurved flattened end-area (retused), or a convex upper surface above the middle                            5
  b. Leaves lacking a recurved, flattened end-area, or with a convex upper surface at the middle                            6
5a. Leaves usually smooth                            **3. retusa***
  b. Leaves rough, with tubercles or papillae, or if smooth then leaves not acuminate                            **5. mirabilis***
6a. Leaves lacking marginal teeth                            7
  b. Leaves with marginal teeth or bristles, sometimes rather sparse, sometimes only at the base                            8
7a. Leaves obovoid, sometimes attenuate at the tip, upper surface usually convex, terminal awn 1–5 mm                            **6. cymbiformis**
  b. Leaves long-triangular, upper surface flat to concave, if convex then with a terminal awn 10 mm or more, or leaves strongly incurved                            **4. turgida***
8a. Leaves bearing at least some tubercles or raised spots                            **9. angustifolia***
  b. Leaves lacking tubercles                            9
9a. Leaf margin bearing usually dense, long, often wispy bristles which often overlap and interlace between the leaves                            10
  b. Leaf margin bearing short or rather stout bristles or teeth, sometimes minute                            11
10a. Leaves green                            **11. arachnoidea***
  b. Leaves blue-green or grey-green                            **10. translucens***
11a. Leaves of 1 colour, or spotted, never with lines                            12
  b. Leaves with longitudinal lines or a net-like pattern, sometimes also with flecks                            13
12a. Upper surface of leaf smooth                            **13. variegata***
  b. Upper surface of leaf with scattered translucent teeth                            **12. herbacea***
13a. Leaves light to dark green, often turning red, orange, purple or brown in full sun                            14
  b. Leaves blue-green, grey-green, yellow-green or brownish                            16
14a. Leaves with longitudinal lines, and with obvious translucent flecks or clear areas towards the tip                            **6. cymbiformis**
  b. Leaves with a net-like pattern, obvious flecks absent                            15
15a. Leaves gradually tapered to a long tip                            **7. batesiana***
  b. Leaves almost spherical to shortly pointed                            **8. reticulata**
16a. Leaves spreading                            **4. turgida***
  b. Leaves erect or incurved                            **10. translucens***
17a. Perianth-tube curved, hexangular at the base, narrowing gradually to the junction with the flower-stalk; inflorescence simple or branched (Subgenus *Hexangulares*)                            18
  b. Perianth-tube straight, rounded-hexangular at the base, narrowing suddenly to the junction with the flower-stalk; inflorescence branched (Subgenus *Robustipedunculares*)                            33
18a. Plants with columnar leafy stems                            19
  b. Plants with leaves in flattish, stemless rosettes                            25
19a. Leaves in 3 vertical or sometimes spiral rows                            **14. viscosa***
  b. Leaves in more than 3 rows                            20
20a. Leaves with tubercles confined to the margin and keel                            **18. armstrongii**
  b. Leaves lacking tubercles, or with tubercles at least on the lower surface                            21
21a. Leaves 4–8 cm                            22
  b. Leaves 1.5–3 cm                            24
22a. Leaves with rounded tubercles, or occasionally tubercles absent                            23
  b. Leaves with flattened tubercles                            **16. reinwardtii**
23a. Leaves 3–6 cm, green                            **15. coarctata**
  b. Leaves 7–8 cm, brownish green or yellowish green                            **20. fasciata**
24a. Leaves grey-green or blue-green, with 5–7 darker ridges or stripes                            **17. glauca**
  b. Leaves green or grey-green, lacking ridges or stripes, but sometimes with a net-like pattern                            **16. reinwardtii***
25a. Leaf margin with teeth, or lower surface of leaf with green tubercles or ridges                            26
  b. Leaf margins untoothed, although often tuberculate; lower surface of leaf with white tubercles or ridges, or smooth                            30
26a. Leaves 10–25 cm, margin never toothed                            27
  b. Leaves 2–10 cm, if longer then margin toothed                            29
27a. Leaf with irregularly arranged tubercles                            **24. limifolia**
  b. Leaf with tubercles in patches, or in rows which may be indistinct                            28

28a. Leaves rough, bright or bluish green, 10–15 cm, the tubercles in distinct rows **21. glabrata**

b. Leaves almost smooth, shiny dark green, 15–25 cm, the tubercles in patches and often in indistinct longitudinal rows **23. longiana**

29a. Leaves pale or yellowish green, with tubercles in rows, margin lacking teeth **25. ubomboensis**

b. Leaves dark green, the tubercles scattered or merging together into ridges, margin with teeth
**24. limifolia***

30a. Upper surface of leaves covered in dense, tiny white tubercles, or with white ridges 31

b. Upper surface of leaves smooth or with sparse tubercles 32

31a. Upper surface of leaves tuberculate **22. radula**

b. Upper surface of leaves with ridges **24. limifolia**

32a. Leaves bright or dark green, upper surface with sparse tubercles
**19. attenuata**

b. Leaves light green, grey-green, yellow-green or brownish, rarely dark green, upper surface smooth
**20. fasciata***

33a. Leaves dark grey-green, blue-green or yellow-green **26. minima***

b. Leaves dark to bright green or brownish green **27. pumila***

Subgenus **Haworthia**. Inflorescence simple. Perianth-tube curved, triangular or rounded-triangular at the base. Style upcurved.

**1. H. truncata** Schoenland. Illustration: van Laren, Succulents, f. 16 (1935); Rauh, Schöne Kakteen und andere Sukkulenten, f. 246 (1978); Rowley, Illustrated encyclopedia of succulents, 165 (1978); Bayer, The new Haworthia handbook, f. 39 (1982).
Rosettes solitary or forming clusters, with 7–13 leaves. Leaves in 2 ranks, more or less erect, c. 2 cm, oblong, truncate at the top which is translucent at first, both surfaces with tubercles. *South Africa*. G1. Spring–summer.
von Poellnitz split this species into 3 forms, using the thickness of the rounded-oblong truncated leaf-ends: forma **truncata** has leaf-ends 6–8 mm thick, forma **tenuis** von Poellnitz has leaf-ends only 3–5 mm thick, and in forma **crassa** von Poellnitz, the leaf-ends are 9–11 mm thick, often with white radiating lines. These forms are not recognised by Bayer.

**2. H. maughanii** von Poellnitz. Illustration: Rauh, Schöne Kakteen und andere Sukkulenten, f. 247 (1978); Rowley, Illustrated encyclopedia of succulents, 165 (1978); Bayer, The new Haworthia handbook, f. 24 (1982); Pilbeam, Haworthia and Astroloba, 94–5 (1983). Rosette usually solitary. leaves spirally arranged, more or less erect, c. 2.5 cm, cylindric, truncate at the top giving an irregularly circular leaf-end 1–1.2 cm across. *South Africa*. G1. Spring–summer.
A hybrid between *H. maughanii* and *H. truncata* is available commercially.

**3. H. retusa** (Linnaeus) Duval (*H. fouchei* von Poellnitz; *H. geraldii* C.L. Scott, *H. longebracteata* G.G. Smith; *H. retusa* var. *multilineata* G.G. Smith). Illustration: Botanical Magazine, 455 (1799); van Laren, Succulents, f. 4 (1935); Bayer, The new Haworthia handbook, f. 33a (1982); British Cactus and Succulent Journal 1: 32, f. 3 (1983).
Rosette stemless, 7–10 cm in diameter, offsetting from the base, with 15–20 leaves. Leaves 3–8 cm, recurved, ovate-triangular, end-area more or less tapered-triangular and translucent, with 5–8 paler lines, leaf margin with sparse teeth, terminal bristle 3–6 mm. *South Africa*. G1. Spring.
Var. **acuminata** Bayer (Illustration: Bayer, the new Haworthia handbook, f. 33b, 1982) has leaves which are long-pointed with slightly roughened end-areas. It often appears in cultivation under the names *H. retusa* var. *densiflora* G.G. Smith, *H. magnifica* or *H. paradoxa*.
Var. **dekenahii** (G.G. Smith) Bayer (Illustration: Bayer, The new Haworthia handbook, f. 33c, 1982) has silver-spotted leaves.

***H. comptoniana** G.G. Smith. Illustration: Excelsa 4: 29 (1974); Bayer, The new Haworthia handbook, f. 9 (1982); Pilbeam, Haworthia and Astroloba, 56–7 (1983). Leaves c. 12, glossy with white flecks, and 5–7 white lines connected to make a net-like pattern. *South Africa*. G1. Spring.

***H. heidelbergensis** G.G. Smith. Illustration: Bayer, The new Haworthia handbook, f. 18 (1982); Pilbeam, Haworthia and Astroloba, 81 (1983). Rosette to 7 cm in diameter with c. 35 leaves. Leaves acuminate, end-areas with 3 or 4 lines. *South Africa*. G1. Spring.

**4. H. turgida** Haworth (*H. laetivirens* Haworth; *H. caespitosa* von Poellnitz; *H. turgida* var. *pallidifolia* G.G. Smith; *H. turgida* var. *suberecta* von Poellnitz). Illustration: Bayer, The new Haworthia

handbook, f. 40 (1982); Pilbeam, Haworthia and Astroloba, 132–3 (1983). Rosette stemless, 5–6.5 cm in diameter, offsetting from the base. Leaves 1–4 cm, usually grey-green, long-triangular with 3–7 longitudinal lines and sometimes with whitish flecks, margin and keel smooth or margin with teeth. *South Africa*. G1. Spring
Var. *erecta* invalid, of nurserymen's catalogues is probably var. *suberecta* which is not distinguished here.

***H. aristata** Haworth (*H. unicolor* von Poellnitz; *H. venteri* von Poellnitz). Illustration: National Cactus and Succulent Journal **35**: 11–12 (1980); Bayer, The new Haworthia handbook, f. 41a (1982); Pilbeam, Haworthia and Astroloba, 39 (1983). Rosette c. 4 cm in diameter, leaves green to brownish, 4–6 cm, margin and terminal bristle smooth. *South Africa*. G1. Spring.
Var. **helmiae** (von Poellnitz) Pilbeam (*H. unicolor* von Poellnitz var. *helmiae* (von Poellnitz) Bayer). Illustration: Bayer, The new Haworthia handbook, f. 41b (1982). Rosette more compact, and the leaf margin and terminal bristle with teeth.

***H. rycroftiana** Bayer. Illustration: Bayer, The new Haworthia handbook, f. 34 (1982). Rosette 5–8 cm in diameter with up to 50 leaves. Leaves pale green with an inconspicuous net-pattern, margin and keel smooth. *South Africa*. G1. Spring.
Often grown under the names *H. aristata* or *H. unicolor*.

***H. zantneriana** von Poellnitz. Illustration: Bayer, The new Haworthia handbook, f. 45 (1982); Pilbeam, Haworthia and Astroloba, 144 (1983). Leaves dull grey-brown, often turning lilac in full sun, with irregular translucent blotches and longitudinal stripes, margin and keel smooth, margin pale. *South Africa*. G1. Spring.

***H. habdomadis** var. **inconfluens**, and var. **morrisiae** may both key out here (see under 10. *H. translucens*). They differ in having bright green, yellow-green or pale brownish green leaves.

**5. H. mirabilis** Haworth (*H. mirabilis* forma *beukmannii* (von Poellnitz) Pilbeam; *H. nitidula* von Poellnitz; *H. triebneriana* von Poellnitz including var. *napierensis* Triebner & von Poellnitz, var. *rubrodentata* Triebner & von Poellnitz and var. *sublineata* von Poellnitz). Illustration: Botanical Magazine, 1354 (1811); Das Pflanzenreich **33**: 100 (1908); Bayer, The new Haworthia handbook, f. 25a (1982); Pilbeam,

Haworthia and Astroloba, pl. 5, 97–8 (1983).
Rosette stemless, 5–6 cm in diameter. leaves lanceolate, to 8 cm, recurved, end-area somewhat translucent, with a prominent central line extending almost to the tip and 2 shorter lines, lower surface with tubercles, margin with white teeth, terminal bristle to 1 cm. *South Africa*. G1. Spring.

Subsp. **badia** (von Poellnitz) Bayer. Illustration: Cactus and Succulent Journal (U.S.) **48**: 105 (1976); Bayer, The new Haworthia handbook, f. 25b (1982); Pilbeam, Haworthia and Astroloba, pl. 5, 98 (1983). Leaves strongly recurved, usually reddish brown, not tuberculate beneath, margin smooth or slightly rough.

Subsp. **mundula** (G.G. Smith) Bayer. Illustration: Bayer, the new Haworthia handbook, f. 25c (1982); Pilbeam, Haworthia and Astroloba, pl. 5, 99 (1983). Leaves strongly recurved, short, thick and blunter than the other subspecies, not tuberculate beneath, margin with minute greenish white teeth.

*H. emelyae von Poellnitz. Illustration: National Cactus and Succulent Journal **34**: 28–31 (1979); Court, Succulent flora of Southern Africa, 160 (1981); Bayer, The new Haworthia handbook, f. 14a (1982); Pilbeam, Haworthia and Astroloba, pl. 3, 69–70 (1983). Rosette *c*. 4 cm in diameter. Leaves 3–4 cm, recurved, end-areas usually marked with brown, lilac or pink in full sun, tuberculate, margin and keel with tiny sparse teeth. *South Africa*. G1. Spring.

Var. **multiflora** Bayer. Illustration: Bayer, The new Haworthia handbook, f. 14b (1982); Pilbeam, Haworthia and Astroloba, 71 (1983). Leaves to 4.5 cm, almost erect, narrow, sharply pointed, scarcely retused, upper surface with longitudinal lines towards the leaf-tip, margin and keel with obvious teeth.

*H. magnifica von Poellnitz. *South Africa*. G1. Spring.

Var. **magnifica**. Illustration: Bayer, The new Haworthia handbook, f. 22a (1982); Pilbeam, Haworthia and Astroloba, pl. 4, 89 (1983). Leaves *c*. 3.5 cm, dark green, stiff, triangular-elliptic, acuminate with a long terminal bristle, end-area triangular with 4 or 5 longitudinal lines, almost smooth or sometimes tuberculate, margin and keel with whitish teeth. Flowers with green lines and a green throat.

Var. **atrofusca** (G.G. Smith) Bayer. Illustration: Bayer, The new Haworthia handbook, f. 22b (1982); Pilbeam, Haworthia and Astroloba, 90 (1983).

Leaves *c*. 4 cm, dark blackish green in full sun, upper surface with small green tubercles, end-area rather swollen, with 3–5 indistinct longitudinal lines, margin and tip of keel with small, transparent teeth.

Var. **major** (G.G. Smith) Bayer. Illustration: Bayer, The new Haworthia handbook, f. 22c (1982); Pilbeam, Haworthia and Astroloba, pl. 4, 90 (1983). Rosettes larger than in var. *magnifica* and forming clusters reluctantly, leaves colouring to brown or pink in full sun, with many tubercles each with a small bristle.

Var. **maraisii** (von Poellnitz) Bayer (*H. maraisii* von Poellnitz; *H. schuldtiana* von Poellnitz). Illustration: Bayer, The new Haworthia handbook, f. 22d (1982); Pilbeam, Haworthia and Astroloba, pl. 4, 90 (1983); British Cactus and Succulent Journal **1**: 32, f. 7 (1983). Rosettes readily forming clusters. Leaves *c*. 3.5 cm with irregular tubercles, end-area with 3 longitudinal lines, margin with prominent teeth.

Var. **meiringii** Bayer. Illustration: Bayer, The new Haworthia handbook, f. 22e (1982); Pilbeam, Haworthia and Astroloba, pl. 5, 90 (1983). Leaves erect, green, end-area poorly defined, both surfaces with bristly tubercles, margin and keel with pronounced teeth.

Var. **notabilis** (von Poellnitz) Bayer. Illustration: Bayer, The new Haworthia handbook, f. 22f (1982); Pilbeam, Haworthia and Astroloba, 91 (1983). Leaves spreading-erect, green becoming brown-purple in full sun, end-area not markedly separate, both surfaces with bristly tubercles, margin and keel toothed. Flower usually with a yellow throat.

Var. **paradoxa** (von Poellnitz) Bayer (*H. paradoxa* von Poellnitz). Illustration: Bayer, The new Haworthia handbook, f. 22g (1982); Pilbeam, Haworthia and Astroloba, 91 (1983). Leaves *c*. 3 cm, dark green, similar to var. *magnifica* but with irregular lines of tubercles and pale longitudinal lines on the end-area, and pale tubercles and transparent flecks on the lower surface towards the tip. Terminal bristle *c*. 1 mm.

*H. parksiana von Poellnitz. Illustration: Bayer, The new Haworthia handbook , f. 28 (1982); Pilbeam, Haworthia and Astroloba, 103 (1983). Rosette 2–3 cm in diameter, leaves dark green to almost black, slightly to strongly recurved, covered with minute tubercles the same colour as the leaf. *South Africa*. G1. Spring.

*H. pygmaea von Poellnitz. Illustration: Excelsa **5**: 82 (1975); Bayer, The new Haworthia handbook, f. 31 (1982); Pilbeam, Haworthia and Astroloba, 106–7 (1983). Rosette to 3 cm in diameter. Leaves 2.5–3 cm, strongly recurved, more or less papillose, end-area rounded-triangular with 4 or 5 pale longitudinal lines, margin smooth. *South Africa*. G1. Spring.

Pilbeam recognises two forms: forma **major** Pilbeam in which the rosettes are up to 6 cm in diameter, and forma **crystallina** Pilbeam in which the leaves have heavily papillose end-areas.

*H. springbokvlakensis C.L. Scott. Illustration: Journal of South African Botany 36: 289 (1970); Excelsa 4: 29 (1974), 5: 82 (1975); Bayer, The new Haworthia handbook, f. 37 (1982); Pilbeam, Haworthia and Astroloba, 124 (1983). Rosette *c*. 7 cm in diameter. Leaves 3.5–4 cm, often greyish pink, very swollen, with minute tubercles the same colour as the leaf, end-area with long and short branched lines, margin with minute teeth. *South Africa*. G1. Spring.

6. **H. cymbiformis** (Haworth) Duval (*H. obtusa* Haworth; *H. lepida* G.G. Smith; *H. planifolia* Haworth var. *exulata* von Poellnitz; *H. planifolia* Haworth var. *setulifera* von Poellnitz). Illustration: Botanical Magazine, 802 (1805); van Laren, Succulents, f. 22 (1935); Bayer, The new Haworthia handbook, f. 11a (1982); Pilbeam, Haworthia and Astroloba, pl. 1, 61–5 (1983).
Rosette stemless, 3–10 cm in diameter. Leaves 2–4 cm, green, turning yellow or orange-red in full sun, obovoid, upper surface flat to somewhat convex, with longitudinal lines and often with translucent flecks towards the leaf-tip, lower surface convex usually with a prominent keel towards the tip, margin and keel smooth, terminal bristle 1–5 mm. *South Africa*. G1. Spring.

There is a variegated form in cultivation.

Var. **incurvula** (von Poellnitz) Bayer (*H. incurvula* von Poellnitz). Illustration: Marloth, Flora of South Africa 4: pl. 223 (1915); Bayer, The new Haworthia handbook, f. 11c (1982); Pilbeam, Haworthia and Astroloba, pl. 3, 66 (1983). Leaves narrow, incurved, with rounded, translucent teeth.

Var. **transiens** (von Poellnitz) Bayer (*H. planifolia* Haworth var. *translucens* Triebner & von Poellnitz. Illustration: Bayer, The new Haworthia handbook, f. 11d (1982); Pilbeam, Haworthia and Astroloba, 66

(1983). Leaves 1.5–2.5 cm, translucent for entire length with 8–10 longitudinal stripes, margin with minute teeth. The leaves turn brownish red in full sun.

Var. **umbraticola** (von Poellnitz) Bayer (*H. umbraticola* von Poellnitz). Illustration: Bayer, The new Haworthia handbook, f. 11e (1982); Pilbeam, Haworthia and Astroloba, 66 (1983). The ends of the leaves have large clear areas separated by dark green lines.

Forma **gracilidelineata** (von Poellnitz) Pilbeam. Illustration: Pilbeam, Haworthia and Astroloba, 63 (1983). Rosettes to 3 cm across, in dense clusters. Leaves incurved, almost completely translucent.

Forma **ramosa** (G.G. Smith) Bayer (*H. ramosa* G.G. Smith). Illustration: Bayer, The new Haworthia handbook, f. 11b (1982); Pilbeam, Haworthia and Astroloba, 65 (1983). Rosettes smaller than other forms or varieties, with elongated stems.

**7. H. batesiana** Uitewaal. Illustration: Bayer, The new Haworthia handbook, f. 5 (1982); Pilbeam, Haworthia and Astroloba, 45–6 (1983).
Rosettes stemless, 4–5 cm in diameter, readily clustering to form dense cushions. Leaves 2–2.5 cm, green, ovate-lanceolate tapering to a long wispy tip, indistinctly tessellated beneath the surface, margin and keel with minute teeth. *South Africa*. G1. Spring.

*II. **marumiana** Uitewaal. Illustration: Bayer, The new Haworthia handbook, f. 23 (1982), Pilbeam, Haworthia and Astroloba, 95 (1983). Leaves 1.2–2 cm, green, turning purple in full sun, terminal bristle *c.* 2 mm, both surfaces tessellated, smooth or with bristles which arise from translucent flecks. *South Africa*. G1. Spring.

**8. H. reticulata** Haworth (*H. haageana* von Poellnitz; *H. haageana* von Poellnitz var. *subreticulata* von Poellnitz; *H. reticulata* Haworth var. *acuminata* von Poellnitz; *H. reticulata* Haworth var. *subregularis* (Baker) Pilbeam; *H. subregularis* Baker; *H. hurlingii* von Poellnitz var. *ambigua* Triebner & von Poellnitz). Illustration: Botanical Magizine, 1314 (1810); National Cactus and Succulent Journal 27: 10–11 (1972); Bayer, The new Haworthia handbook, f. 32a (1982); Pilbeam, Haworthia and Astroloba, 114–15 (1983).
Rosettes stemless, 4–7 cm in diameter, eventually forming large clumps. Leaves 2–3.5 cm, green, turning orange or red in full sun, narrowly triangular-ovate, both surfaces with a net-like pattern, pointed,

with a terminal bristle to 1 cm, margin minutely toothed. *South Africa*. G1. Spring.

Var. **hurlingii** (von Poellnitz) Bayer (*H. hurlingii* von Poellnitz). Illustration: Flowering Plants of Africa 29: pl. 1149 (1953); Bayer, The new Haworthia handbook, f. 32b (1982); Pilbeam, Haworthia and Astroloba, pl. 7, 115 (1983). Rosette 2–3 cm across, leaves to 2 cm, relatively fatter and less pointed than in var. *reticulata*, margin and keel with tiny teeth towards the base.

**9. H. angustifolia** Haworth (*H. monticola* Fourcade; *H. angustifolia* var. *albanensis* (Schoenland) von Poellnitz; *H. angustifolia* var. *subfalcata* von Poellnitz). Illustration: Bayer, The new Haworthia handbook, f. 1a (1982); Pilbeam, Haworthia and Astroloba, 35 (1983).
Rosettes stemless, 2–6 cm in diameter, clustering readily. Leaves 2–10 cm, dull or dark green, lanceolate, tapered, tip with a whitish bristle, upper surface with a prominent central line and 5–7 indistinct longitudinal lines, rough from a covering of tiny tubercles, margin with tiny teeth. *South Africa*. G1. Spring.

*H. **emelyae** may key out here (see under 5. *H. mirabilis*). It differs in the margin and keel having obvious teeth.

*H. **marumiana** may key out here (see under 7. *H. batesiana*). It differs in the leaves being 2 cm or less, with a tessellated pattern.

**10. H. translucens** Haworth (*H. gracilis* von Poellnitz; *H. translucens* var. *deliculata* (Berger) von Poellnitz). Illustration: Botanical Magazine, 1417 (1811); Bayer, The new Haworthia handbook, f. 38a (1982); Pilbeam, Haworthia and Astroloba, 129 (1983).
Rosette stemless, 4–6 cm in diameter, offsetting from the base. Leaves *c.* 3 cm, erect to slightly incurved, grey-green, becoming lilac-grey in full sun, translucent with green net-like lines, narrowly lanceolate, tapering to a fine bristle, margin and keel with translucent teeth. *South Africa*. G1. Spring.

Subsp. **tenera** (von Poellnitz) Bayer (*H. tenera* von Poellnitz). Illustration: Bayer, The new Haworthia handbook, f. 38b (1982); Pilbeam, Haworthia and Astroloba, 131 (1983). Rosette 1–3 cm in diameter with strongly incurved leaves.

*II. **bolusii** Baker. Illustration: Desert Plant Life 7: 119 (1935); Bayer, The new Haworthia handbook, f. 7a (1982); Pilbeam, Haworthia and Astroloba, 48

(1983). Rosette stemless, *c.* 5 cm across, not readily offsetting. Leaves 2–4 cm, blue-green or grey-green, incurved, with large translucent areas at the tips, margin and keel with teeth 5–6 mm. *South Africa*. G1. Spring.

Var. **blackbeardiana** (von Poellnitz) Bayer (*H. blackbeardiana* von Poellnitz). Illustration: Court, Succulent flora of Southern Africa, 162 (1981); Bayer, The new Haworthia handbook, f. 7b (1982); Pilbeam, Haworthia and Astroloba, pl. 2, 49 (1983). Rosette to 15 cm in diameter.

*H. **cooperi** Baker (*H. vittata* Baker; *H. pilifera* Baker; *H. cooperi* forma *pilifera* (Baker) Pilbeam; *H. obtusa* Haworth var. *pilifera* (Baker) Uitewaal). Illustration: Cactus and Succulent Journal (U.S.) 53: 70–71 (1981); Bayer, The new Haworthia handbook, f. 10a (1982). Pilbeam, Haworthia and Astroloba, pl. 3, 57–9 (1983). Rosette to 9 cm across, leaves 2–2.5 cm, blue-green or grey-green with translucent tips, round-ended or with a terminal bristle to 6 mm, both surfaces convex with dark longitudinal lines, margin and keel with translucent teeth. *South Africa*. G1. Spring.

Var. **leightonii** (G.G. Smith) Bayer (*H. leightonii* G.G. Smith). Illustration: Bayer, The new Haworthia handbook, f. 10b (1982). The longitudinal lines towards the end of the leaf are reddish.
This variety often appears in nurserymen's lists as *leightoniae*.

*H. **decipiens** von Poellnitz. Illustration: Bayer, The new Haworthia handbook, f. 12 (1982); Pilbeam, Haworthia and Astroloba, pl. 3, 67–8 (1983). Rosette *c.* 7 cm or more in diameter. Leaves 2–4 cm, spreading with incurved tips, grey-green with darker tessellation beneath the surface, turning yellow or red in full sun, terminal bristle *c.* 1 cm, margin with curved white teeth 2–2.5 mm, keel smooth. *South Africa*. G1. Spring

*H. **habdomadis** von Poellnitz. Illustration: Bayer, The new Haworthia handbook, f. 17a (1982); Pilbeam, Haworthia and Astroloba, pl. 3, 78 (1983). Leaves 2.5–3.5 cm, pale greyish green to brownish green with darker tessellation, terminal bristle 3–4 mm, margin and keel with translucent teeth. Wide leaves may have 2 keels. *South Africa*. G1. Spring.

Var. **inconfluens** (von Poellnitz) Bayer (*H. altilinea* Haworth var. *limpida* (Haworth) von Poellnitz forma *inconfluens* von Poellnitz). Illustration: Court, Succulent flora of Southern Africa, 164 (1981); Bayer, The new Haworthia handbook, f.

17b (1982); Pilbeam, Haworthia and Astroloba, 79 (1983); British Cactus and Succulent Journal 1: 32, f. 6 (1983). Leaves pale brownish green, strongly incurved, margin and keel usually untoothed, terminal bristle *c*. 6 mm.

Var. **morrisiae** (von Poellnitz) Bayer. Illustration: Court, Succulent flora of Southern Africa, 164 (1981); Bayer The new Haworthia handbook, f. 17c (1982); Pilbeam, Haworthia and Astroloba, 80 (1983). Leaves 2–5 cm, bright or yellowish green with a brown tip, margin and keel untoothed, sometimes with tiny teeth on the margin of young leaves.

**11. H. arachnoidea** (Linnaeus) Duval (*H. setata* Haworth). Illustration: van Laren, Succulents, f. 3 (1935); Rauh, Schöne Kakteen und andere Sukkulenten, f. 244 (1978); Bayer, The new Haworthia handbook, f. 2 (1982); Pilbeam, Haworthia and Astroloba, 37 (1983).
Rosette stemless, 3–4.5 cm in diameter or more, eventually offsetting to form clumps. Leaves *c*. 1.5 cm, lanceolate, opaque green with darker longitudinal lines, terminal bristle 6–7 mm, margin and keel with whitish bristles about 1.5 mm apart. *South Africa*. G1. Spring.

**\*H. aranea** (Berger) Bayer (*H. bolusii* Baker var. *aranea* Berger). Illustration: Das Pflanzenreich 33: 114 (1908); Bayer, The new Haworthia handbook, f. 3 (1982); Pilbeam, Haworthia and Astroloba, 38 (1983). Rosette to 15 cm in diameter, flattish with a depressed centre. Leaves uniformly green, margin and keel with fine, dense bristles less than 1.5 mm apart. *South Africa*. G1. Spring.

**\*H. semiviva** (von Poellnitz) Bayer (*H. bolusii* Baker var. *semiviva* von Poellnitz). Illustration: Cactus and Succulent Journal (U.S.) 48: 256 (1976); Bayer, The new Haworthia handbook, f. 35 (1982); Pilbeam, Haworthia and Astroloba, pl. 8, 122 (1983). Rosette 7–8 cm across. Leaves green, incurved, the tip translucent and flattened, drying up in the resting period, margin and keel with long, translucent bristles. *South Africa*. G1. Spring.

**\*H. aristata** may key out here (see under 4. *H. turgida*). It differs in the margin having relatively sparse teeth and the terminal bristle being toothed.

**12. H. herbacea** (Miller) Stearn (*H. atrovirens* (de Candolle) Haworth; *H. submaculata* von Poellnitz; *H. luteorosea* Uitewaal). Illustration: van Laren, Succulents, f. 14 (1935); National Cactus and Succulent Journal 27: 51–52 (1972);

Bayer, The new Haworthia handbook, f. 19 (1982); Pilbeam, Haworthia and Astroloba, pl. 4, 81 (1983).
Rosette stemless, 2–8 cm in diameter, eventually offsetting to form small clumps. leaves 1–4 cm, pale to dark green, erect to incurved, broadly lanceolate, both surfaces convex, with translucent flecks especially towards the leaf-tip, often each fleck giving rise to a tooth, margin and keel with translucent teeth. *South Africa*. G1. Spring.

**\*H. maculata** (von Poellnitz) Bayer. Illustration: Bayer, The new Haworthia handbook, f. 21 (1982); Pilbeam, Haworthia and Astroloba, pl. 4, 89 (1983). Differs from *H. herbacea* in its fewer and more turgid leaves which turn violet in full sun, and are spreading to somewhat recurved. *South Africa*. G1. Spring.

**13. H. variegata** L. Bolus. Illustration: Bayer, The new Haworthia handbook, f. 42 (1982); Pilbeam, Haworthia and Astroloba, 134 (1983).
Rosette stemless, 3–4 cm in diameter, with 20–30 leaves. Leaves 5–6 cm, linear-lanceolate, long-tapering, dark green or brownish green with pale spots, margin and keel with sharp, translucent teeth. *South Africa*. G1. Spring.

**\*H. chloracantha** Haworth. Illustration: Das Pflanzenreich 33: 110 (1908); Bayer, The new Haworthia handbook, f. 8a (1982); Pilbeam, Haworthia and Astroloba, 51 (1983). Rosette *c*. 6 cm, leaves pale to yellowish green, 3.5–4 cm, margin and keel with small teeth the same colour as the leaf or darker. *South Africa*. G1. Spring.

Var. **denticulifera** (von Poellnitz) Bayer (*H. angustifolia* Haworth var. *liliputana* Uitewaal). Illustration: Bayer, The new Haworthia handbook, f 8b (1982); Pilbeam, Haworthia and Astroloba, 51 (1983). Rosette smaller, leaves dark green or purplish green.

Var. **subglauca** von Poellnitz. Illustration: Bayer, The new Haworthia handbook, f. 8c (1982); Pilbeam, Haworthia and Astroloba, 51 (1983). Rosette more robust, leaves blue-green.

Subgenus **Hexangulares** Bayer. Inflorescence simple or branched. Perianth-tube curved, hexangular at the base, narrowing gradually to the junction with the flower-stalk. Style straight.

**14. H. viscosa** (Linnaeus) Haworth (*H. asperiuscula* Haworth; *H. beanii* G.G. Smith; *H. viscosa* var. *concinna* (Haworth) Baker; *H. viscosa* var. *caespitosa* von Poellnitz). Illustration: van Laren, Succulents, f. 8, 12

(1935); Barkhuizen, Succulents of Southern Africa, f. 191 (1978); National Cactus and Succulent Journal 36: 99–100 (1981); Bayer, The new Haworthia handbook, f. 62 (1982).
Rosette columnar, 7–20 cm tall, offsetting at the base. Leaves in 3 straight or spiral rows, erect to projecting, to 4 cm, triangular-lanceolate, dark green, both surfaces covered in small tubercles. *South Africa*. G1. Summer.

Many variants of *H. viscosa* have been given names and nurserymen's catalogues often list several varieties (var. *pseudotortuosa* (Salm-Dyck) Baker with spirally arranged leaves, var. *torquata* (Haworth) Baker with long, twisted leaves and var. *subobtusa* von Poellnitz with shorter blunt leaves) but Bayer maintains that the variation in leaf size and arrangement is influenced by the environment, and so they are not recognised here.

**\*H. nigra** (Haworth) Baker (*H. nigra* var. *schmidtiana* (von Poellnitz) Uitewaal). Illustration: Cactus and succulent Journal (U.S.) 48: 105 (1976); Court, Succulent flora of Southern Africa, 170 (1981); Bayer, the new Haworthia handbook, f. 55 (1982); Pilbeam, Haworthia and Astroloba, 100–1 (1983). Rosette 6–12 cm tall. Leaves dark green, turning almost black in full sun, to 7 cm, triangular-ovate, both surfaces with large, dense tubercles. *South Africa*. G1. Summer.

As in *H. viscosa*, many varieties and forms have been named and some are offered commercially, but they are not generally considered to be worthy of recognition.

**15. H. coarctata** Haworth (*H. fulva* G.G. Smith; *H. musculina* G.G. Smith; *H. greenii* Baker; *H. chalwinii* Marloth & Berger; *H. reinwardtii* (Salm-Dyck) Haworth var. *conspicua* von Poellnitz: *H. reinwardtii* var. *fallax* von Poellnitz; *H. reinwardtii* var. *committeesensis* G.G. Smith). Illustration: Marloth, Flora of South Africa 4: pl. 22A (1915); Desert Plant Life 7: 108 (1935), 18: 68 (1946); Bayer, The new Haworthia handbook, f. 48a–b (1982); Pilbeam, Haworthia and Astroloba, pl. 2, 53–5 (1983).
Rosette columnar, 5–20 cm tall, offsetting from the base. Leaves 4–6 cm, triangular-lanceolate, tips incurved, shining green, turning red in full sun, with greenish white rounded tubercles in longitudinal or sometimes transverse rows, density of tubercles very variable. *South Africa*. G1. Summer.

Var. **tenuis** (G.G. Smith) Bayer (*H. reinwardtii* (Salm-Dyck) Haworth var. *tenuis* G.G. Smith; *H. reinwardtii* var. *riebeckensis* G.G. Smith). Illustration: Bayer, The new Haworthia handbook, f. 48c (1982). Pilbeam, Haworthia and Astroloba, pl. 2, 55 (1983). Rosettes to 45 cm, usually trailing and rooting where they touch the soil, leaves to *c.* 3 cm.

Subsp. **adelaidensis** (von Poellnitz) Bayer (*H. reinwardtii* (Salm-Dyck) Haworth var. *adelaidensis* von Poellnitz; *H. reinwardtii* var. *bellula* G.G. Smith). Illustration: Bayer, The new Haworthia handbook, f. 48d (1982); Pilbeam, Haworthia and Astroloba, pl. 2, 50 (1983). Rosette to 13 cm with narrow leaves.

**16. H. reinwardtii** (Salm-Dyck) Haworth (*H. reinwardtii* var. *major* Baker; *H. reinwardtii* var. *pulchra* von Poellnitz; *H. reinwardtii* var. *archibaldiae* von Poellnitz; *H. reinwardtii* var. *peddiensis* G.G. Smith). Illustration: van Laren, Succulents, f. 13 (1935); Flowering plants of Africa **29**: pl. 1148 (1953); Bayer, The new Haworthia handbook, f. 57a (1982); Pilbeam, Haworthia and Astroloba, 108–9 (1983). Very similar to *H. coarctata*, differing mainly in having flattened tubercles on the leaves, the leaves themselves being more densely arranged on the stem and less thick. *South Africa*. G1. Summer.

Forma **chalumnensis** (G.G. Smith) Bayer (*H. reinwardtii* var. *chalumnensis* G.G. Smith). Illustration: Court, Succulent flora of Southern Africa, 167 (1981); Bayer, The new Haworthia handbook, f. 57b (1982); Pilbeam, Haworthia and Astroloba, 109, 112 (1983). Rosette to 13 cm or more × 6–7 cm with the leaves tending to stand out from the stem. Leaves becoming red-purple in full sun, and with prominent transverse bands of almost confluent white tubercles on the back.

Forma **kaffirdriftensis** (G.G. Smith) Bayer. Illustration: Bayer, The new Haworthia handbook, f. 57c (1982); Pilbeam, Haworthia and Astroloba, 109 (1983). Rosette to 12 × 3.5 cm with incurved leaves. Leaves becoming dark green to red-purple in full sun and with longitudinal rows of clear white, very large tubercles on the back which tend to coalesce into vertical ridges.

Forma **olivacea** (G.G. Smith) Bayer. Illustration: Bayer, The new Haworthia handbook, f. 57d (1982); Pilbeam, Haworthia and Astroloba, 110 (1983). Rosettes *c.* 8 cm tall with rather narrow leaves. Leaves olive-green becoming orange-brown in full sun, with greenish white tubercles.

Forma **zebrina** (G.G. Smith) Bayer. Illustration: Bayer, The new Haworthia handbook, f. 57e (1982); Pilbeam, Haworthia and Astroloba, 110–11 (1983). The incurved leaves have transverse bands of coalescing white tubercles on the back.

Var. **brevicula** G.G. Smith. Illustration: Bayer, The new Haworthia handbook, f. 57f (1982); Pilbeam, Haworthia and Astroloba, 111 (1983). A smaller and more compact plant with rosettes to 8 × 2–3 cm. The backs of the dark green leaves bear bright white scattered tubercles.

The name *H. reinwardtii* var. *minor* Baker appears in catalogues, but the original description of this plant is inadequate and the variety cannot be assigned either to *H. reinwardtii* or *H. coarctata*.

\*H. venosa subsp. **granulata** may key out here (see under 24. *H. limifolia*). It differs in the leaf surface having a faint net-pattern and grey-green tubercles.

**17. H. glauca** Baker. Illustration: Baker, The new Haworthia handbook, f. 51a (1982); Pilbeam, Haworthia and Astroloba, 75 (1983).
Rosette columnar, 5–12 cm tall, *c.* 5 cm wide, offsetting from the base. Leaves 2–2.5 cm, grey-green turning reddish in full sun, narrowly ovate to lanceolate, somewhat incurved, with 5–7 darker longitudinal ridges on the back. *South Africa*. G1. Summer.

Var. **herrei** (von Poellnitz) Bayer (*H. herrei* von Poellnitz; *H. eilyae* von Poellnitz; *H. jacobseniana* von Poellnitz; *H. jonesiae* von Poellnitz). Illustration: Bayer, the new Haworthia handbook, f. 51b (1982); Pilbeam, Haworthia and Astroloba, 76–7 (1983). Rosettes to 20 cm or more, usually trailing when they exceed 10 cm. Leaves blue-green, lacking tubercles or with irregular rows of whitish tubercles on the back.

**18. H. armstrongii** von Poellnitz (*H. glauca* Baker var. *herrei* (von Poellnitz) Bayer forma *armstrongii* (von Poellnitz) Bayer). Illustration: Pilbeam, Haworthia and Astroloba, 40 (1983).
Rosettes columnar, to 15 cm tall, 3–4 cm wide, offsetting from the base and lower part of the stem. Leaves 3–4 cm, dark green, grey-green or brownish green, triangular-lanceolate, with whitish tubercles confined to the margin and keel. *South Africa*. G1. Summer.

**19. H. attenuata** Haworth (*H. fasciata* (Willdenow) Haworth var. *caespitosa* Berger; *H. attenuata* var. *caespitosa* (Berger) Farden; *H. attenuata* var. *argyrostigma* (Baker) Berger; *H. attenuata* var. *deltoidea* Farden; *H. attentuata* var. *linearis* Farden; *H. attenuata* var. *minissima* Farden; *H. attenuata* var. *odonoghueana* Farden; *H. attenuata* var. *uitewaaliana* Farden). Illustration: Botanical Magazine, 1345 (1811); Rauh, Schöne Kakteen und andere Sukkulenten, f. 243 (1978); Bayer, The new Haworthia handbook, f. 46a (1982). Rosette stemless or almost so. Leaves 3–6.5 cm, dark green, erect to spreading-recurved, lanceolate, lower surface with sparse to dense white tubercles, usually unevenly distributed, upper surface with fewer tubercles. *South Africa*. G1. Summer.

Forma **britteniana** (von Poellnitz) Bayer. Illustration: Bayer, The new Haworthia handbook, f. 46b (1982). Leaves with many large, white separate tubercles.

Forma **clariperla** (Haworth) Bayer. Illustration: Bayer, The new Haworthia handbook, f. 46c (1982). Leaves with large white tubercles merging together in horizontal bands on the lower surface.

Pilbeam's concept of the forms of *H. attenuata* does not agree with that of Bayer, and is not followed here.

There is a plant in cultivation under the name H. 'Kuentzii' which seems to be a hybrid of *H. attenuata* (Illustration: Pilbeam, Haworthia and Astroloba, 83, 1983). The rosettes eventually tend to become shortly columnar and the leaves are shorter than those of *H. attenuata*, with tubercles which are separate or partly merged together.

**20. H. fasciata** (Willdenow) Haworth (*H. fasciata* forma *sparsa* von Poellnitz; *H. fasciata* var. *subconfluens* von Poellnitz; *H. fasciata* forma *vanstaadensis* von Poellnitz). Illustration: van Laren, Succulents, f. 2 (1935); Rowley, Illustrated encyclopedia of succulents, 165 (1978); Bayer, The new Haworthia handbook, f. 49a (1982); Pilbeam, Haworthia and Astroloba, 71–2 (1983).
Rosette stemless, 5–7 cm in diameter. Leaves 3–5.5 cm, incurved, grey-green, lanceolate, upper surface smooth, lower surface with white tubercles which merge together into horizontal bands. *South Africa*. G1. Summer.

Forma **browniana** (von Poellnitz) Bayer (*H. browniana* von Poellnitz). Illustration: Bayer, The new Haworthia handbook, f. 49b (1982); Pilbeam, Haworthia and Astroloba, 71–2 (1983). Rosette with a tendency to become columnar, to 12 cm, leaves brownish green or yellowish green, longer than in forma *fasciata*.

Var. *staadensis* sometimes appears in catalogues and is probably an error for forma *vanstaadensis*, a synonym of *H. fasciata*.

*****H. starkiana** von Poellnitz. Illustration: Barkhuizen, Succulents of Southern Africa, f. 189 (1978); Court, Succulent flora of Southern Africa, 169 (1981); Bayer, The new Haworthia handbook, f. 60a (1982); Pilbeam, Haworthia and Astroloba, 125 (1983). Rosette stemless or more or less so. Leaves 4–5 cm, spreading, smooth, yellowish green, margin slightly thickened. *South Africa*. G1. Summer.

Var. **lateganiae** (von Poellnitz) Bayer. Illustration: Bayer, The new Haworthia handbook, f. 60b (1982); Pilbeam, Haworthia and Astroloba, 125 (1983). Leaves dull green, longer (to 10 cm) and narrower.

**21. H. glabrata** (Salm-Dyck) Baker. Illustration: Bayer, The new Haworthia handbook, f. 50 (1982); Pilbeam, Haworthia and Astroloba, 74 (1983). Rosette stemless, to 18 cm across. Leaves 10–15 cm, often spreading-recurving, lanceolate, bright green or bluish green with rows of many tiny tubercles of the same colour. *Not known in the wild*. G1. Summer.

**22. H. radula** (Jacquin) Haworth. Illustration: Das Pflanzenreich 33: 91 (1908); National Cactus and Succulent Journal 23: 92 (1968); Bayer, The new Haworthia handbook, f. 56 (1982); Pilbeam, Haworthia and Astroloba, 107 (1983).
Rosette stemless, offsetting readily. Leaves 6–8 cm, pointing in all directions, green, lanceolate, densely covered with tiny, white, evenly distributed tubercles. *South Africa*. G1. Summer.

**23. H. longiana** von Poellnitz. Illustration: Excelsa 4: 29 (1974); Court, Succulent flora of Southern Africa, 168 (1981); Bayer, the new Haworthia handbook, f. 54 (1982); Pilbeam, Haworthia and Astroloba, 88 (1983).
Rosette stemless, offsetting from the base. Leaves 15–25 cm, rigid, often dying back from the tips, lanceolate, shiny dark green with small tubercles of the same colour, in patches and often forming indistinct longitudinal rows. *South Africa*. G1. Summer.

**24. H. limifolia** Marloth (*H. limifolia* var. *stolonifera* Resende). Illustration: National Cactus and Succulent Journal 22: 80 (1967); Bayer, The new Haworthia handbook, f. 53a (1982); Pilbeam, Haworthia and Astroloba, 84 (1983). Rosette stemless, 8–10 cm in diameter, with stolons which produce offsets up to 10 cm or more from the parent rosette. Leaves to 7.5 cm, borne in a tight spiral, spreading, dark green, broadly triangular-ovate, both surfaces with 15–20 transverse ridges the same colour as the leaf. *South Africa*. G1. Summer.

Var. **gigantea** Bayer. Illustration: Journal of South African Botany 28: pl. ix (1962); Bayer, The new Haworthia handbook, f. 53b (1982); Pilbeam, Haworthia and Astroloba, pl. 4, 86 (1983). Rosettes to 23 cm across, leaves 7–11 cm with small pale green tubercles, usually arranged irregularly.

Var. **striata** Pilbeam. Illustration: Journal of South African Botany 28: pl. x (1962); Pilbeam, Haworthia and Astroloba, pl. 1, 85 (1983). Similar to var. *limifolia* but the transverse ridges on the leaves are white.

*****H. venosa** (Lamarck) Haworth. Illustration: Botanical Magazine, 1353 (1811); Bayer, The new Haworthia handbook, f. 61a (1982); Pilbeam, Haworthia and Astroloba, 135 (1983). Rosettes not producing stolons. Leaves 2–12 cm, recurved, upper surface with 5 or 6 longitudinal lines, usually with net-like connexions lower surface rough with irregular tubercles, margins with small white teeth. *South Africa*. G1. Summer.

Subsp. **granulata** (Marloth) Bayer (*H. granulata* Marloth). Illustration: Cactus and Succulent Journal (U.S.) 50: 74, 76 (1978); Bayer, The new Haworthia handbook, f. 61b (1982); Pilbeam, Haworthia and Astroloba, 136 (1983). Rosette columnar, to 9 × 4 cm. Leaves 2–3 cm, lower surface with many green tubercles more or less in transverse rows.

Subsp. **tessellata** (Haworth) Bayer (*H. tessellata* Haworth; *H. recurva* (Haworth) Haworth). Illustration: Marloth, Flora of South Africa 4: pl. 22c (1915); van Laren, Succulents, f. 6 (1935); Bayer, The new Haworthia handbook, f. 61c (1982); Pilbeam, Haworthia and Astroloba, 136–8 (1983). Leaves 3–5 cm, broadly triangular-ovate, strongly recurved, upper surface with a strong, net-like pattern, lower surface with tiny tubercles, margin with recurved white teeth.

**25. H. ubomboensis** Verdoorn (*H. limifolia* Marloth var. *ubomboensis* (Verdoorn) G.G. Smith). Illustration: Flowering plants of South Africa 21: pl. 818 (1941); Bayer, The new Haworthia handbook, f. 53c (1982); British Cactus and Succulent Journal 1: 32, f. 4 (1983); Pilbeam, Haworthia and Astroloba, pl. 8, 133 (1983).
Rosette stemless, with offsets produced at the ends of stolons. Leaves 3–6.5 cm, pale or yellowish green, becoming lilac-green in full sun, broadly lanceolate, margin hard and horny, lower surface with sparse green tubercles in rows. *South Africa*. G1. Summer.

Subgenus **Robustipedunculares** Bayer. Inflorescence branched. perianth-tube straight, rounded-hexagonal, narrowing suddenly to the junction with the flower-stalk. Style straight.

**26. H. minima** (Aiton) Haworth. Illustration: Excelsa 4: 29 (1974); Cactus and Succulent Journal (U.S.) 53: 227 (1981); Bayer, The new Haworthia handbook, f. 66 (1982); Pilbeam, Haworthia and Astroloba, pl. 5, 96 (1983). Rosette stemless. Leaves blue-green or grey-green, triangular-ovate, with dense, prominent white tubercles on both sides. *South Africa*. G1. Summer.

*****H. marginata** (Lamarck) Stearn (*H. albicans* Haworth; *H. uitewaaliana* von Poellnitz). Illustration: Das Pflanzenreich 33: 95 (1908); Barkhuizen, Succulents of Southern Africa, f. 187 (1978); Bayer, The new Haworthia handbook, f. 65 (1982); Pilbeam, Haworthia and Astroloba, pl. 5, 92–3 (1983). Leaves yellow-green, smooth or with occasional tubercles beneath. *South Africa*. G1. Summer.

**27. H. pumila** (Linnaeus) Duval (*H. margaritifera* (Linnaeus) Haworth; *H. papillosa* (Salm-Dyck) Haworth; *H. semiglabrata* Haworth). Illustration: Rauh, Schöne Kakteen und andere Sukkulenten, f. 245 (1978); Bayer, The new Haworthia handbook, f. 68 (1982); Pilbeam, Haworthia and Astroloba, pl. 6, 106 (1983); Cactus and Succulent Journal (U.S.) 55: 164 (1983).
Rosette stemless, to 15 cm across and 30 cm tall. Leaves 7–8 cm, triangular-ovate to broadly lanceolate, dark brownish-green, both surfaces with large white tubercles in indistinct rows. *South Africa*. G1. Summer.

*****H. kingiana** von Poellnitz. Illustration: Excelsa 4: 29 (1974); Barkhuizen, Succulents of Southern Africa, f. 185

(1978); Bayer, The new Haworthia handbook, f. 64 (1982); Pilbeam, Haworthia and Astroloba, 83 (1983). Rosette to 7 cm across and 8 cm tall. Leaves bright green, with rather indistinct, flattened tubercles, occasionally lacking tubercles. *South Africa*. G1. Summer.

### Names of doubtful application

All these names are found either in catalogues, or are widely used in the labelling of plants:

**H. antilinea** Haworth. The original description cannot be applied to any known species and in view of its mottled history, the name is rejected as being confused.

**H. asperula** Haworth. This name has been applied at various times to *H. pygmaea*, *H. magnifica*, *H. magnifica* var. *maraisii*, *H. pubescens* Bayer, *H. emelyae* and *H. mutica* Haworth; it is now such a source of confusion that it should not be used.

**H. cassytha** Baker. Plants bearing this name are offered for sale and appear to be near to *H. viscosa*. However, Baker himself was unsure that the plant which he originally described from material at Kew, was in fact a *Haworthia*. The name therefore cannot be validly applied to any known species.

**H. columnaris** Baker. Baker considered this species to be related to *H. affinis* Baker and *H. bilineata* Baker; both these names have now been discarded due to confusion. later workers related *H. columnaris* to *H. cooperi*, but its identity is still a mystery, so the name cannot be used.

**H. cuspidata** Haworth. The name was applied originally by Haworth to a plant related to *H. mucronata* Haworth. It is now often used erroneously for a hybrid of garden origin between *H. cymbiformis* and *H. retusa*. Bayer recommends that the name should be discarded.

**H. ferox** von Poellnitz. Bayer considers that the plant originally described under this name is probably *Aloe humilis* (Linnaeus) Miller. Plants sold by nurserymen as *H. ferox* usually turn out to be *H. herbacea*, *H. magnifica* var. *paradoxa*, robust forms of *H. turgida*, or *H. aristata* var. *helmiae*

**H. integra** von Poellnitz. This species was allied originally to *H. reticulata* and *H. cymbiformis*. Plants cultivated under this name differ from these two species in their longer, narrower leaves with longitudinal lines and a terminal white bristle; they cannot be assigned to any named species.

**H. janseana** Uitewaal. The origin of this plant is unknown; it is thought that it may be a garden hybrid involving *H. turgida*.

Plants sold or grown under this name often turn out to be *H. angustifolia*.

**H. mucronata** Haworth. It is not known where this plant was originally collected and there are no original specimens. So much confusion surrounds the name that Bayer has suggested that it should not be used.

**H. otzenii** G.G. Smith. Although the place where this plant was first collected is known, there are no original specimens, and plants subsequently collected there do not fit Smith's description. Nevertheless, Bayer regards it as a synonym of *H. mutica* Haworth. Plants in cultivation under this name appear to be hybrids of *H. mirabilis*.

**H. subfasciata** (Salm-Dyck) Baker. This name is sometimes applied erroneously to plants in cultivation, but is not known to which plant the name should apply.

The following names apply to garden hybrids:

**H. × icosiphylla** Baker (possibly *H. tortuosa* (see below) × *H. glabrata*); **H. kewensis** von Poellnitz; **H. rigida** (Lamarck) Haworth (possibly *H. tortuosa* (see below) is involved); **H. rugosa** (Salm-Dyck) Baker (possibly *H. attenuata* × *H. radula*); **H. ryderiana** von Poellnitz; **H. sampaiana** Resende; **H. subattenuata** (Salm-Dyck) Baker (possibly *H. pumila* × *H. attenuata*, most commonly seen in its variegated form); **H. subulata** (Salm-Dyck) Baker; **H. tisleyi** Baker (possibly *H. tortuosa* (see below) or *H. glabrata* is involved); **H. tortuosa** Haworth (very similar to and probably a hybrid of *H. viscosa*; it is commonly cultivated, as are its varieties, and there is also a variegated form).

## 39. ASTROLOBA Uitewaal

*V.A. Matthews*

Similar to *Haworthia* but with radially symmetric, usually white or yellow-green flowers, and spirally arranged thick ovoid-triangular, pointed leaves crowded onto elongated stems.

A genus of about 10 species native to South Africa. Cultivation as for *Haworthia* (p. 146).

Literature: Pilbeam, J., *Haworthia and Astroloba – a collector's guide* (1983).

1a. Leaves bearing tubercles on the lower surface      **3. aspera**
  b. Leaves lacking tubercles, except sometimes on the keel or margin

2a. Leaves spreading more or less horizontally     **1. foliolosa***
  b. Leaves borne at an angle of *c.* 45°       **2. pentagona***

**1. A. foliolosa** (Haworth) Uitewaal (*Apicra foliolosa* (Haworth) Willdenow). Illustration: Botanical Magazine, 1352 (1811); Jacobsen, Lexicon of Succulent plants, pl. 22 (1974); Pilbeam, Haworthia and Astroloba, 151 (1983).
Stems to 20 cm or more, eventually sprawling, 4–5 cm wide, producing offsets from the base. leaves in 5 ranks, dark green, smooth, spreading, upper surface flat to convex, lower surface with a keel, the margin with minute tubercles. Flowers yellow-green. *South Africa (Cape Province)*. G1. Summer.

*A. deltoidea* (J.D. Hooker) Uitewaal (Illustration: Botanical Magazine, 6071, 1873; Jacobsen, Lexicon of succulent plants, pl. 22, 1983) is a commonly cultivated species which is so similar to *A. foliolosa* that it cannot be separated satisfactorily. Recent unpublished work has shown that it should probably be regarded as synonymous.

***A. congesta** (Salm-Dyck) Uitewaal (*Apicra congesta* (Salm-Dyck) Baker). Illustration: Pilbeam, Haworthia and Astroloba, 150 (1983). Stems *c.* 7 cm wide, leaves tending to become concave above. *South Africa (Cape Province)*. G1. Summer.

Recent unpublished work suggests that it may be best to consider *A. congesta* as a subspecies of *A. foliolosa*.

***A. herrei** Uitewaal. Illustration: Jacobsen, Lexicon of succulent plants, pl. 22 (1974); Pilbeam, Haworthia and Astroloba, 152 (1983). Stems branching from the base and along the stem. Leaves grey-green, with darker longitudinal lines below the surface. *South Africa (Cape Province)*. G1. Summer.

**2. A. pentagona** (Aiton) Uitewaal. Illustration: Jacobsen, Lexicon of succulent plants, pl. 22 (1974); Pilbeam, Haworthia and Astroloba, 153 (1983).
Stems to 20 cm or more × 3–4 cm, erect, offsetting from the base. leaves in 5 ranks, green, becoming reddish in full sun, borne at an angle of *c.* 45°, both surfaces smooth, upper surface flat or convex, lower surface with 1 or 2 keels, keels and margins minutely tubercled. Flowers greenish. *South Africa (Cape Province)*. G1. Summer.

***A. dodsoniana** Uitewaal. Illustration; Pilbeam, Haworthia and Astroloba, 151 (1983). Stems to 25 cm or more × 4–6 cm, with blue-green or grey-green leaves, appearing whitish when grown in full sun, upper surface somewhat concave. Flowers white, striped with green. *South Africa (Cape Province)*. G1. Summer.

**\*A. spiralis** (Linnaeus) Uitewaal (*Apicra spiralis* (Linnaeus) Baker). Illustration: Botanical Magazine, 1455 (1812); Pilbeam, Haworthia and Astroloba, 153–4 (1983). Stems *c*. 3 cm wide, eventually sprawling, with grey-green or blue-green leaves which have a finely roughened keel and margin. *South Africa (Cape Province)*. G1. Summer.

**3. A. aspera** (Haworth) Uitewaal (*Apicra aspera* (Haworth) Willdenow). Illustration: Pilbeam, Haworthia and Astroloba, 148 (1983).

Stems to 20 × 2–3 cm, sprawling, producing offsets from the base and along the stem. Leaves in 3 or 4 ranks, green, turning reddish in full sun, upper surface smooth, flat to slightly convex, lower surface keeled, with irregular lines of small green tubercles. Flowers pinkish white. *South Africa (Cape Province)*. G1. Summer.

Var. **major** Haworth. Illustration: Pilbeam, Haworthia and Astroloba, 149 (1983). Stems more upright, 5–6 cm wide.

The hybrid genus × **Astroworthia** Rowley contains hybrids between *Astroloba aspera* and *Haworthia pumila* and occurs naturally in the wild in South Africa.

× **Astroworthia bicarinata** (Haworth) Rowley (*Astroloba bicarinata* (Haworth) Uitewaal) (Illustration: Jacobsen, Lexicon of succulent plants, pl. 22, 1974; Pilbeam, Haworthia and Astroloba, 149, 1983) has stems to *c*. 20 cm and leaves which are obliquely curved towards the tip, with 1 or 2 keels beneath and green tubercles. The keel and margin are minutely toothed.

The plant which is generally seen in cultivation under the name *Astroloba skinneri* (Berger) Uitewaal, has now been proved to belong to × *Astroworthia* and has been given the name × **Astroworthia bicarinata** nothomorph **skinneri** (Berger) Rowley (Illustration: Jacobsen, Lexicon of succulent plants, pl. 22 f. 5, 1974). Pilbeam suggests that its status should be raised to × *Astroworthia skinneri* but has not published this name correctly. The stems grow 10–15 cm tall and are 7–8 cm wide at the base, with leaves which usually have 2 paler green keels beneath and have paler green tubercles and margin. The name *Astroloba bulbulata* (Jacquin) Uitewaal appears to be a synonym of this hybrid plant.

**40. POELLNITZIA** Uitewaal

*V.A. Matthews*

Differs from *Astroloba* in having larger, orange-red flowers which tend to be borne on only 1 side of the raceme. Stems to 30 × 7 cm, with ovoid-triangular leaves in 5 spiral ranks. Flowers to 2.5 cm. Perianth-segments incurved at the tips; margins of the outer segments fused to the midribs of the inner ones.

A genus of 1 species from South Africa. Cultivation as for *Haworthia* (p. 146).

**1. P. rubriflora** (L. Bolus) Uitewaal (*Apicra rubriflora* L. Bolus; *Aloe rubriflora* (L. Bolus) Rowley). Illustration: Pilbeam, Haworthia and Astroloba, 147, 154 (1983). *South Africa (Cape Province)*. G1. Summer.

**41. GASTERIA** Duval

*V.A. Matthews*

Succulent perennials, usually stemless. Leaves in rosettes, arranged spirally or in 2 ranks, fleshy, flat or somewhat keeled, with spots or tubercles, margin entire. Flowers borne in loose racemes or panicles, red or pinkish, often green at the apex. Perianth with a tubular, curved, inflated tube, formed by the joining of the outer lobes. Stamens dorsifixed. Ovary superior, 3-celled; style 1. Fruit a 3-celled capsule opening between the septa and containing flat, winged seeds.

About 50 species, native to South Africa. They will grow in any well-drained soil in a cool greenhouse, and need semi-shady conditions. Propagation is by seed or leaf cuttings; the plants form clumps of rosettes which can be separated to produce more plants.

| | | |
|---|---|---|
| 1a. | Leaves 3–5.5 cm | 2 |
| b. | Leaves 6 cm or more | 3 |
| 2a. | Leaves linear, 1–1.5 cm wide at the base | **14. liliputana** |
| b. | Leaves tongue-shaped to triangular-ovate, 2–3.5 cm wide at the base | **12. armstrongii\*** |
| 3a. | Mature rosettes with leaves in 2 ranks or almost so, occasionally the ranks arranged spirally | 4 |
| b. | Mature rosettes with all leaves arranged spirally, or in several rows | 18 |
| 4a. | Leaves with green spots | 5 |
| b. | Leaves with white or greenish white spots or tubercles | 6 |
| 5a. | Stem 1–1.5 cm; leaves 1.8–2 cm wide at the base | **11. caespitosa** |
| b. | Stem 15–25 cm; leaves 3–4 cm wide at the base | **10. marmorata** |
| 6a. | Leaves 25–35 cm | 7 |
| b. | Leaves to 25 cm | 8 |
| 7a. | Leaves 1.3–2.5 cm wide at the base | **9. pulchra\*** |
| b. | Leaves 4–5 cm wide at the base | **7. picta** |
| 8a. | Stem 10–30 cm | 9 |
| b. | Stem absent or very short | 11 |
| 9a. | Leaves *c*. 34, 4.5–5 cm wide at the base | **8. maculata** |
| b. | Leaves 12–20, 2–3 cm wide at the base | 10 |
| 10a. | Leaves with white spots arranged in transverse bands | **9. pulchra\*** |
| b. | Leaves with scattered white spots | **6. planifolia\*** |
| 11a. | Leaves 1.3–2 cm wide at the base | **1. verrucosa\*** |
| b. | Leaves 2–5 cm wide at the base | 12 |
| 12a. | Leaves with scattered spots or tubercles | 13 |
| b. | Leaves with spots or tubercles arranged in transverse rows or more or less so | 15 |
| 13a. | Leaf margin with tubercles | 14 |
| b. | Leaf margin lacking tubercles | **5. nigricans\*** |
| 14a. | Leaves 4–10 | **1. verrucosa\*** |
| b. | Leaves 12–16 | **13. excavata\*** |
| 15a. | Flowers to 2.2 cm | **5. nigricans\*** |
| b. | Flowers 2.4 cm or more | 16 |
| 16a. | Leaves 4–6, 18–20 cm; flowers *c*. 3.5 cm | **4. neliana** |
| b. | Leaves 8–12, 20–25 cm; flowers *c*. 2.5 cm | 17 |
| 17a. | Leaves with a straight margin | **2. angulata** |
| b. | Leaves somewhat wavy along the margin | **3. disticha** |
| 8a. | Leaves 6–22 cm | 19 |
| b. | Leaves 25–35 cm | 28 |
| 19a. | Leaves 1–1.5 cm wide at the base | **14. liliputana** |
| b. | Leaves 2–6.5 cm wide at the base | 20 |
| 20a. | Leaves with white tubercles on the upper surface | 21 |
| b. | Leaves with white or greenish white spots or blotches on the upper surface | 22 |
| 21a. | Leaves 2.5–3 cm wide at the base | **17. subcarinata** |
| b. | Leaves 3.5–5 cm wide at the base | **16. carinata** |
| 22a. | Leaves 25 or more | **8. maculata** |
| b. | Leaves 8–20 | 23 |
| 23a. | Leaves convex on the upper surface | **5. nigricans\*** |
| b. | Leaves concave on the upper surface | 24 |

24a. Leaves 8–10 cm **15. humilis**
 b. Leaves 10–22 cm 25
25a. Leaves 10–15 cm 26
 b. Leaves 15–22 cm 27
26a. Young rosettes with leaves in 2 ranks
 **20. schweickerdtiana***
 b. Young rosettes with leaves in a spiral
 **19. obtusa**
27a. Leaves 8–12 **20. schweickerdtiana***
 b. Leaves 12–15 **18. nitida**
28a. Leaves curved, acute
 **21. acinacifolia**
 b. Leaves straight, blunt
 **22. candicans**

**1. G. verrucosa** (Miller) Duval. Illustration: Step, Favourite flowers of garden and greenhouse **4**: pl. 270 (1897); Das Pflanzenreich **33**: f. 42, 43 (1908); Lamb & Lamb, Illustrated reference on cacti and succulents **1**: 228 (1955); Jacobsen, Lexicon of succulent plants, pl. 87, f. 4 (1974).
Rosettes with a very short stem. Leaves 6–10, in 2 ranks, 10–22 × 2–3.5 cm, lanceolate, shortly acuminate at the apex, grooved above, convex beneath, dull green with scattered white tubercles; margin thickened, tuberculate. Flowering stems to 60 cm; flowers 1.8–2.5 cm. *South Africa.* G1.
*G. prolifera Lemaire. Illustration: Riha & Subik, Illustrated encyclopedia of cacti and other succulents, f. 369 (1981). Differs in the rosettes being stemless, and having leaves 20–30 × 1.3–1.5 cm, spotted with white. *South Africa.* G1.
*G. pillansii Kensit. The rosettes are stemless and the obtuse, mucronate leaves are 4–4.5 cm wide at the base, spotted with white. *South Africa.* G1.
*G. × repens Haworth. Leaves 6–8 cm with small white tubercles in transverse rows. A hybrid between G. *verrucosa* and G. *carinata.*

**2. G. angulata** (Willdenow) Haworth.
Rosettes stemless. Leaves 8–10, in 2 ranks, 20–25 × c. 5 cm, grooved above, convex beneath, green with small white spots in transverse rows. Flowering stems to 90 cm; flowers c. 2.5 cm. *South Africa.* G1. Summer.

**3. G. disticha** (Linnaeus) Haworth (*G. lingua* (Thunberg) Berger). Illustration: Das Pflanzenreich **33**: f. 44 (1908); Jacobsen, Lexicon of succulent plants, pl. 86 f. 6 (1974).
Rosettes stemless. leaves 10–12, in 2 ranks, 20–25 × 3–5 cm, tongue-shaped, green with white spots in transverse rows;

margin wavy. Flowering stems to 90 cm; flowers c. 2.5 cm. *South Africa.* G1. Spring.

**4. G. neliana** von Poellnitz. Illustration: Lamb & Lamb, Illustrated reference on cacti and succulents **2**: 497 (1959).
Rosettes stemless. leaves 4–6, in 2 ranks, 18–20 × 4–4.5 cm, tongue-shaped, pointed at the tip, both surfaces slightly convex, emerald green with whitish green blotches in more or less transverse rows; margin minutely toothed. Flowers c. 3.5 cm. *South Africa.* G1.

**5. G. nigricans** (Haworth) Duval.
Illustration: Das Pflanzenreich **33**: f. 45 (1908).
Rosettes with a very short stem. Leaves 10–20, in 2 ranks or somewhat spiralled, 6–20 × 4–5 cm, tongue-shaped, pointed at the tip, both surfaces convex, dark green with white or greenish white, often rather indistinct spots, in irregular transverse rows; margin horny. Flowering stems to 90 cm; flowers 1.8–2 cm. *South Africa.* G1.
*G. brevifolia Haworth. Illustration: Hay et al., Dictionary of indoor plants in colour, f. 250 (1974). Rosettes stemless, leaves with white spots which merge into transverse bands, margin minutely toothed. *South Africa.* G1. Summer.
*G. triebneriana von Poellnitz. Illustration. The Cactus Journal **7**: f. 27 (1939). Leaves fewer, with whitish spots which may be scattered or coalesced into indistinct transverse bands. Older leaves much curved inwards, the upper side with a distinct central rib. Flowers 2–2.2 cm. *South Africa.* G1.

**6. G. planifolia** (Baker) Baker. Illustration: Das Pflanzenreich **33**: f. 47 (1908).
Rosettes columnar with a stem 15–25 cm. Leaves 12–20, in 2 ranks, lanceolate, 15–20 × 2–2.5 cm, glossy, dark green with scattered, confluent white spots. Flowering stems to 180 cm; flowers c. 1.8 cm. *South Africa.* G1.
*G. bicolor Haworth. Illustration: Das Pflanzenreich **33**: f. 46 (1908). Rosettes only 10–15 cm, leaves dull light green with scattered white spots especially at the base. *South Africa.* G1.

**7. G. picta** Haworth.
Rosettes becoming short-stemmed with age. leaves 12–20, in 2 spiral ranks, 25–35 × 4–5 cm, triangular-lanceolate, slightly convex above, glossy blackish green, with confluent white spots in transverse rows. Flowering stems to 90 cm; flowers c. 1.8 cm. *South Africa.* G1. Summer.

**8. G. maculata** (Thunberg) Haworth.
Illustration: Rowley, Illustrated encyclopedia of succulents, 163 (1978); Riha & Subik, Illustrated encyclopedia of cacti and other succulents, f. 367 (1981).
Rosettes columnar with a stem 20–30 cm. Leaves c. 34, in 2 ranks or arranged spirally, 15–20 × 4.5–5 cm, tongue-shaped, with 2 keels beneath, glossy dark green with white spots which are in rows towards the tip. Flowering stems to 120 cm; flowers 1.8–2 cm. *South Africa.* G1. Summer.

**9. G. pulchra** (Aiton) Haworth. Illustration: Addisonia **23**: pl. 752 (1955); Rauh, Die Grossartige Welt der Sukkulenten, t. 68 (1967).
Rosettes columnar with a stem 15–30 cm. Leaves 15–20, more or less in 2 ranks, 20–30 × c. 2.5 cm, linear, often with 2 keels beneath, shiny grey-green with confluent white spots in transverse rows; margins horny, white. Flowering stems to 90 cm; flowers c. 1.8 cm. *South Africa.* G1. Spring–summer.
*G. prolifera which differs in its stemless rosettes, and leaves only 1.3–1.5 cm wide at the base with scattered spots, may key out here (see under 1. G. verrucosa).

**10. G. marmorata** Baker. Illustration: Jacobsen, Lexicon of succulent plants, pl. 87 f. 2 (1974).
Rosettes columnar with a stem 15–25 cm. Leaves 20–30, in 2 ranks, 13–15 × 3–4 cm, lanceolate, blunt, dark green mottled with darker green. Flowering stems to 75 cm; flowers c. 1.8 cm. *South Africa.* G1.

**11. G. caespitosa** von Poellnitz.
Rosettes with a short stem 1–1.5 cm. Leaves in 2 ranks, 10–14 × 1.8–2 cm, lanceolate, green spotted with light green above, the spots coalescing into indefinite transverse bands beneath. Flowers 1.9–2 cm. *South Africa.* G1.

**12. G. armstrongii** Schönland. Illustration: Lamb & Lamb, Illustrated reference on cacti and succulents **3**: 798 (1963); Court, Succulent flora of Southern Africa, 175 (1981); Riha & Subik, Illustrated encyclopedia of cacti and other succulents, f. 368 (1981).
Rosettes stemless, or with a short stem to 4 cm. Leaves few, in 2 ranks when young, spirally arranged in older rosettes, 3–5.5 × 2.8–3.5 cm, tongue-shaped to triangular-ovate, apex pointed, green with white tubercles in tranverse rows. Flowers c. 2 cm. *South Africa.* G1.

*G. batesiana Rowley. Illustration: National Cactus and Succulent Society Journal 10: 32 (1955); Rauh, Die Grossartige Welt der Sukkulenten, t. 67 (1967); Jacobsen, Lexicon of succulent plants, pl. 81 f. 4 (1974); Barkhuizen, Succulents of Southern Africa, f. 179 (1978). Rosettes stemless, with spirally arranged leaves. Leaves blackish green with large green and small white tubercles in irregular transverse rows especially in the lower half. Flowers *c*. 3.5 cm. *South Africa*. G1.

*G. stayneri von Poellnitz. Illustration: The Cactus Journal 7: 18 (1938). Leaves dark green with many dark green tubercles when young, becoming spotted with greenish white when mature. *South Africa*. G1.

**13. G. excavata** (Willdenow) Haworth. Illustration: Das Pflanzenreich 33: f. 48 (1908).
Rosettes stemless. leaves 12–16, in 2 spiral ranks, 10–15 × 2.5–3.5 cm, lanceolate, pale green with small whitish spots; margin tuberculate. Flowering stems to 60 cm; flowers *c*. 2.5 cm. *South Africa*. G1. Summer.

*G. pseudonigricans (Salm-Dyck) Haworth (*G. subnigricans* Haworth). Leaves in 2 ranks, 15–20 × 3.5–4 cm, glossy green with indistinct white spots. Flowering stems 90–100 cm; flowers 1.8–2 cm. *South Africa*. G1.

**14. G. liliputana** von Poellnitz. Illustration: Flowering plants of South Africa 9: pl. 360 (1929); Jacobsen, Lexicon of succulent plants, pl. 87 f. 3 (1974); Barkhuizen, Succulents of Southern Africa, f. 180 (1978).
Rosettes with short stems. Leaves arranged in a spiral, 3.5–6.5 × 1–1.5 cm, lanceolate, dark shiny green with white spots which often coalesce into transverse rows; margin with tiny teeth. Flowering stems to 10 cm; flowers *c*. 1.5 cm. *South Africa*. G1.

**15. G. humilis** von Poellnitz.
Rosettes stemless. leaves 8–12, arranged in a spiral, 8–10 × 2–4 cm, long-triangular, concave above, glossy dark green with whitish spots which may be scattered or coalescent into transverse rows; margin tuberculate. Flowering stems to 60 cm; flowers *c*. 2.3 cm. *South Africa*. G1.

**16. G. carinata** (Miller) Duval. Illustration: Botanical Magazine, 1331 (1811); Das Pflanzenreich 33: f. 50 (1908); Flowering plants of South Africa 8: pl. 291 (1928).
Rosettes stemless. leaves 4–8, arranged in a spiral, 12–15 × 3.5–5 cm, triangular-lanceolate narrowed to a spine at the tip, concave above, dull green with white tubercles which may be scattered or coalescent into transverse rows. Flowering stems to 90 cm; flowers 2.5–4 cm. *South Africa*. G1. Summer.

**17. G. subcarinata** (Salm-Dyck) Haworth. Rosette with a short stem. leaves 10–15, in several rows. 10–15 × 2.5–3 cm, triangular-lanceolate, concave above, green with flat, white tubercles; margin minutely toothed. *South Africa*. G1.

Possible a hybrid between *G. carinata* and *G. disticha*.

**18. G. nitida** (Salm-Dyck) Haworth. Illustration: Marloth, Flora of South Africa 4: pl. 21 (1915).
Rosette with a short stem. Leaves 12–15, arranged in a spiral, 15–22 × *c*. 6.5 cm, broadly triangular-lanceolate, glossy green with whitish spots which coalesce into indistinct transverse rows on the lower surface. Flowering stems to 90 cm; flowers 2.2–2.5 cm. *South Africa*. G1. Summer.

**19. G. obtusa** (Salm-Dyck) Haworth. Leaves 11–13, arranged in a spiral, 10–15 × *c*. 3.5 cm, triangular, glossy green with indistinct spots in transverse rows; margin minutely toothed. Flowers *c*. 2.5 cm. *South Africa*. G1.

**20. G. schweickerdtiana** von Poellnitz. Rosettes stemless or with a very short stem. Leaves 8–12, in 2 ranks in young plants, later arranged in a spiral, 14–18 × 3.5–4 cm, concave above, convex beneath, light glossy green with white spots which are usually scattered on the upper surface, coalescent into transverse rows beneath. *South Africa*. G1.

*G. trigona Haworth. Leaves spirally arranged in many rows, 3–3.5 cm wide, with white spots in transverse bands. *South Africa*. G1.

**21. G. acinacifolia** (Jacquin) Haworth. Illustration: Das Pflanzenreich 33: f. 52 (A–F), (1908); Rauh, Die Grossartige Welt der Sukkulenten, t. 67 (1967); Olmos, Los Cactus y las otras plantas succulentas, 91 (1977); Court, Succulent Flora of Southern Africa, 175 (1981).
Rosettes stemless. leaves arranged in many rows, to 35 × 5 cm, lanceolate, concave above, dark green with small greenish white spots in transverse rows, lower part with tubercles. Flowering stems to 120 cm; flowers to 5 cm. *South Africa*. G1. Summer.

**22. G. candicans** Haworth.
Rosettes stemless. Leaves 12–20, arranged in many rows, 25–30 × *c*. 7 cm, broadly triangular-lanceolate, concave above, glossy green with tiny spots in indistinct transverse rows; margin white. Flowers 4–5 cm. *South Africa*. G1. Summer.

**Names of doubtful application**

The names *G. nova* and *G. variegata* have been found in nurserymen's catalogues; it is not known to which species these names apply.

**42. ALOE** Linnaeus
*J. Cullen*
Plants herbaceous or, more frequently, woody, succulent, with or without distinct stems, sometimes with trunks and forming small trees. Leaves usually in distinct rosettes either at ground level or at the ends of the branches, more rarely dispersed along the branches, fleshy, often tough and leathery, and usually with soft, firm or sharp spines, at least along the margins. Inflorescence a raceme or panicle, one to several from each rosette; flowers stalked, each subtended by a bract, the inflorescence-stalk often bearing sterile bracts. Perianth of 6 lobes in 2 whorls, those of the outer whorl free or united for part of their length, those of the inner whorl variably joined to those of the outer, yellow, orange or red, their tips often greenish. Stamens 6, the anthers usually projecting from the perianth one after another. Style and stigma usually projecting from the perianth. Ovary superior, 3-celled, with many ovules in each cell. Fruit a papery or woody capsule with many seeds.

A genus of about 200 species, mostly from southern Africa, but also occurring in Madagascar, the Canary Islands, east Africa and the southern part of the Arabian peninsula; several species are naturalised in southern Europe. Its classification is difficult, made so by the occurrence of natural hybrids, and the fact that material in cultivation may often differ morphologically from the wild stock. Reynolds (Aloes of South Africa, edn 4, 104, 1974) goes so far as to say: 'In the wild state, and in their various geographical situations, plants vary considerably in size, length and width of leaf, leaf markings, racemes, length of pedicels and flowers, while plants grown in gardens and greenhouses under conditions very different from their natural habitats, frequently modify out of almost all recognition'. In spite of the last part of this sentence, an attempt is made here to provide means of identification for all the species likely to be found in general cultivation; the existence of hybrids and the difficulties mentioned by Reynolds must, however, be borne in mind when attempts at identification are made. The elaborate hierarchy of Sections and Series proposed by Reynolds (references below) is not used here, as its value for identification seems limited. Various members of the Agavaceae (p. 271–90) may be incorrectly keyed out to this genus; all of these have leaves with hard, spiny tips, a feature uncommon in the cultivated Aloes. Some hybrids have been raised between species of *Gasteria* (p. 154) and *Aloe*; These are known as × *Gastrolea* Walther.

In the descriptions, the leaves must be considered to be acute unless the opposite is stated; the measurements of bracts and flower-stalks refer to those of the lowest open flowers in the raceme.

Aloes are easily grown in well-drained, rich soils; watering should be reduced in the resting season and resumed when new growth appears; most species can tolerate or benefit from full exposure to sunlight. Propagation is mainly by seed.
Literature: Berger, A., *Das Pflanzenreich* 33: 159–330 (1908); Reynolds, G.W., *The Aloes of tropical Africa and Madagascar* (1966); Reynolds, G.W., *The Aloes of South Africa*, edn 4 (1974).

1a. Plants without distinct stems or trunks, or stems not exceeding 50 cm                                                2
  b. Plants with obvious and distinct stems or trunks which exceed 50 cm           22
2a. Outer perianth-lobes free for more than half of their length                           3
  b. Outer perianth-lobes united for more than half of their length                        14
3a. Perianth 3.5 cm or more                    4
  b. Perianth less than 3.4 cm                 9
4a. Leaves 30–60 × 10–15 cm                    5
  b. Leaves to 20 × 4 cm                        6
5a. Racemes inclined almost to the horizontal, the flowers borne on the upper side; flower-stalks *c.* 8 mm
                                    **29. ortholopha**
  b. Racemes erect or spreading, the flowers borne all round the axis; flower-stalks 3–4 cm          **12. glauca**
6a. Leaves broadly triangular, *c.* 3 × longer than the basal breadth; plant producing numerous offsets from the base                       **11. brevifolia**
  b. Leaves lanceolate, usually much more than 3 × longer than the basal breadth;

plant not producing offsets                7
7a. Spines on the margins of mature leaves black, *c.* 1 cm
                                    **10. melanacantha**
  b. Spines on the margins of mature leaves white, 2–3 mm or absent          8
8a. Leaves at most 2 cm broad, tuberculate or prickly on the upper and/or lower surfaces; flower-stalks 2.5–3.5 cm                **8. humilis**
  b. Leaves 2–4 cm broad, neither tuberculate nor prickly; flower-stalks up to 1.8 cm              **9. krapohliana**
9a. Leaves 30–50 × 5–10 cm; flower-stalks less than 5 mm or more than 2.5 cm                                  10
  b. Leaves 7–13 cm × 8–20 mm; flower-stalks more than 5 mm and less than 2.5 cm                                  12
10a. Flower-stalks 2.5–3 cm, leaves with H-shaped spots          **27. microstigma**
   b. Flower-stalks to 5 mm; leaves unspotted or with obscure lines      11
11a. Inflorescense usually branched; flowers loose, their perianths clearly exceeding the bracts          **17. vera**
   b. Inflorescense simple; flowers very dense and touching, bracts exceeding the perianths             **5. broomii**
12a. Stem present, 10–20 cm; perianth orange at the base, shading to yellow at the mouth, the tips of the lobes green                       **3. bakeri**
   b. Stem absent; perianth orange-pink
                                                13
13a. Leaves to 1 cm broad, both surfaces rough and with round or lens-shaped pale green spots; perianth to 1.5 cm
                                    **1. bellatula**
   b. Leaves 1.5–2 cm broad, both surfaces smooth and with pale, H-shaped spots; perianth *c.* 2.5 cm      **2. rauhii**
14a. Leaves copiously spotted on at least the upper surface, the spots sometimes forming the bases of spines        15
   b. Leaves unspotted or very rarely sparsely spotted, without spines on the surface                        19
15a. Leaves with white spots forming the tuberculate bases of spines; perianth *c.* 5.5 cm, upcurved in the upper third; style projecting *c.* 2.5 cm
                                    **4. longistyla**
   b. Leaves with spots which do not form the bases of spines; perianth top 4.5 cm, not upcurved in the upper third; style not projecting, or projecting at most for 5 mm        16
16a. Leaves 100–150 in each rosette, each leaf tapering to a long, dry 'awn'
                                    **6. aristata**

  b. Leaves to 25 in each rosette, acute or acuminate, but not as above        17
17a. Racemes crowded at the ends of the inflorescence branches; flower-stalks 3.5–4.5 cm; leaves 8–12 cm broad
                                    **13. saponaria**
   b. Racemes elongate along the inflorescence branches; flower-stalks 6–15 mm; leaves 6–7 cm broad      18
18a. Perianth conspicuously constricted above the ovary, forming a rounded bulge at the base; flower stalks 6–7 mm                      **14. zebrina**
   b. Perianth not as above, very slightly and gradually narrowed above the ovary; flower-stalks 1–1.5 cm
                                    **15. grandidentata**
19a. Leaf margin entire, untoothed, forming a pink or red zone 2–3 mm wide                              **16. striata**
   b. Leaf margin toothed (sometimes distantly so), not as above        20
20a. Leaves 2–3 cm broad; inflorescence simple, unbranched          **7. virens**
   b. Leaves 7.5–12 cm broad; inflorescence branched            21
21a. Perianth bright red, ultimately fading yellow; leaves to 35 × 7.5 cm
                                    **18. perryi**
   b. Perianth orange or yellow; leaves 50–60 × 8–12 cm          **19. camperi**
22a. Leaves borne strictly in 2 ranks, their margins entire and their apices rounded                       **38. plicatilis**
   b. Leaves in more than 2 ranks, their margins usually toothed and their apices usually acute            23
23a. Plant tree-like with a richly, apparently dichotomously branched crown                       **37. dichotoma***
   b. Plant not as above            24
24a. Outer perianth-lobes united for half or more of their length            25
   b. Outer perianth-lobes free for half or more of their length            29
25a. Leaves 40–65 × 10–20 cm, forming a distinct, tight rosette, the stem scarcely visible between the leaves
                                                26
   b. Leaves 10–15 × 1–2.5 cm, forming an open rosette dispersed along the stem, which is clearly visible      28
26a. Leaf margin entire, untoothed, forming a pink or red zone 2–3 mm wide                              **16. striata**
   b. Leaf margin distinctly toothed, not as above                        27
27a. Flowers distinctly curved; leaves dull green to almost glaucous
                                    **31. africana**
   b. Flowers straight; leaves reddish or

purple near the base, shading to pink near the apex, the whole with a violet bloom     **30. rubroviolacea**

28a. Leaf-base conspicuously auricled and ciliate; flowers *c.* 3 cm   **21. ciliaris**
  b. Leaf-base not auricled and ciliate; flowers 1.1–1.4 cm   **20. tenuior**

29a. Flower-stalks 3–4.5 cm   30
  b. Flower-stalks 1–6 mm   34

30a. Leaves 15–25 cm, marginal teeth golden yellow, brown or black; bracts 8–10 mm   31
  b. Leaves 30 cm or more, marginal teeth white or reddish brown; bracts 1.5–3 cm   32

31a. Leaves not, or very sparingly, white-spotted   **24. perfoliata**
  b. Leaves conspicuously and densely white-spotted, especially on the lower surface   **25. distans**

32a. Marginal teeth of leaves hard and sharp; leaves 10–15 cm broad; bracts *c.* 3 cm   **12. glauca**
  b. Marginal teeth of leaves firm but not hard and sharp, leaves 5–10 cm broad; bracts to 2 cm   33

33a. Leaves widely spreading, deflexed towards their apices; perianth orange-red   **28. arborescens**
  b. Leaves spreading at the extreme base, then curving upwards; perianth red to pink   **26. succotrina**

34a. Leaves to 25 × 2.5 cm, with conspicuously striped sheaths   **22. striatula**
  b. Leaves 50–160 × 7–25 cm, sheaths not conspicuously striped   35

35a. Racemes held horizontally or obliquely, the flowers borne on 1 side of the axis   **36. marlothii**
  b. Racemes erect or ascending, the flowers borne all round the axis   36

36a. Rosette loose, open, the leaves spread along the stem, which is clearly visible between them   **23. divaricata**
  b. Rosette tight, dense, the stem scarcely visible between the leaves   37

37a. Raceme 15–25 cm; flower-stalks 1–2 mm   38
  b. Raceme 50–80 cm; flower-stalks 4–6 mm   39

38a. Marginal teeth of the leaves 5–6 mm; perianth reddish orange   **35. excelsa**
  b. Marginal teeth of the leaves *c.* 2 mm; perianth pale yellow   **34. thraskii**

39a. Leaves spreading, flat or only slightly channelled; inner perianth-lobes with brown tips   **32. ferox**
  b. Leaves much recurved, deeply channelled; inner perianth-lobes without brown tips   **33. candelabrum**

**1. A. bellatula** Reynolds. Illustration: Reynolds, Aloes of tropical Africa and Madagascar, t. 89 & f. 407–9 (1966). Plants small, stemless, suckering and forming dense tufts. Leaves *c.* 16 in a basal rosette, 10–13 cm × 9–10 mm, linear but tapering, both surfaces roughened with papilla-like points, dark green, copiously spotted on both surfaces with pale, lens-shaped spots, margins with soft, triangular teeth. Raceme to 60 cm, rarely branched; flower-stalks *c.* 1.2 cm, bracts 4–6 mm. Perianth *c.* 1.3 cm, cylindric to bell-shaped, pale orange-red, outer lobes free for little more than half their length. Anthers and style scarcely projecting. *Madagascar.* H5–G1.

**2. A. rauhii** Reynolds. Illustration: Reynolds, Aloes of tropical Africa and Madagascar, f. 422–3 (1966); Cactus & Succulent Journal of America **51**: 49 (1979).
Stem absent or very short. Rosettes *c.* 10 cm in diameter, with about 20 leaves. Leaves 7–10 × 1.5–2 cm, triangular-lanceolate, greyish brown or brownish with numerous H-shaped spots on both surfaces, margins with soft, white teeth. Raceme to 30 cm, rarely branched; flower-stalks *c.* 1 cm, bracts 4–5 mm. Perianth *c.* 2.5 cm, constricted above the ovary then cylindric, orange-pink, outer lobes free to the base. Anthers and style projecting *c.* 1 mm. *Madagascar.* H5–G1.

**3. A. bakeri** Scott-Elliott. Illustration: Reynolds, Aloes of tropical Africa and Madagascar, t. 91 & f. 424–6 (1966). Stem 10–20 cm, branched, producing offsets at the base. Rosettes with about 12 leaves. Leaves *c.* 7 cm × 8 mm, lanceolate, green with a reddish tinge, usually unspotted, rarely with a few pale green spots, margins with firm, white teeth. Racemes to 30 cm, simple; flower-stalks 1–1.2 cm, bracts *c.* 3 mm. Perianth *c.* 2.3 cm, cylindric, orange shading to yellowish at the mouth, outer lobes free to the base. Anthers and style projecting *c.* 1 mm. *Madagascar.* H5–G1.

**4. A. longistyla** Baker. Illustration: Jeppe, South African Aloes, 35 (1969); Reynolds, Aloes of South Africa, edn 4, t. 6 & f. 142–3 (1974).
Plant stemless. Rosettes with 20–30 leaves. Leaves arching-erect, incurved, 12–15 × *c.*

3 cm, glaucous, both surfaces with soft or firm spines which arise from white, tubercular bases, margins with similar spines. Inflorescence to 20 cm, simple; flower-stalks 6–8 mm, bracts 2.5–3 cm. Perianth *c.* 5.5 cm, cylindric, upcurved in the upper third, pink to red, outer lobes united for more than half their length. Anthers and style projecting 2–2.5 cm. *South Africa (Cape Province).* H5–G1.

**5. A. broomii** Schönland. Illustration: Jeppe, South African Aloes, 54 (1969); Reynolds, Aloes of South Africa, edn 4, f. 147–9 (1974); Flowering plants of South Africa **16**: t. 605 (1936).
Plant stemless or stem very short. Rosette to 1 m in diameter, with numerous leaves. Leaves about 30 × 10 cm, broadly ovate-lanceolate, terminating in a sharp spine, unspotted but with obscure lines, margins with pale, triangular teeth. Inflorescence a very dense, simple raceme to 1.5 m; flower-stalks 1–2 mm, bracts exceeding the perianths. Perianth 2–2.5 cm, cylindric, slightly curved, pale yellow, outer lobes free for more than half their length. Anthers and style projecting 1.2–1.5 cm. *South Africa (Cape Province).* H5–G1.

Easily recognised by its tight, dense, cylindric racemes, the flowers with short stalks and long bracts.

**6. A. aristata** Haworth. Illustration: Jeppe, South African Aloes, 15 (1969); Reynolds, Aloes of South Africa, edn 4, f. 157–9 (1974).
Plants small, stemless, forming compact groups of rosettes. Leaves 100–150 in each rosette, 8–10 × 1–1.5 cm, incurved, narrowly lanceolate, tapering to a long, dry 'awn', green with white spots on both surfaces, margins with soft, white teeth. Inflorescence with 2–6 branches, to 50 cm; flower-stalks to 3.5 cm, bracts 1–1.5 cm. Perianth *c.* 4 cm, cylindric, curved, swollen around the ovary, red, the outer lobes united for most of their length. Anthers and style projecting 1–2 mm. *South Africa.* H5–G1. Summer.

**7. A. virens** Haworth. Illustration: Botanical Magazine, 1355 (1811).
Plant stemless, rosettes producing offshoots and forming groups. Leaves numerous in each rosette, *c.* 20 × 2–3 cm, narrowly lanceolate, acuminate, marked with lines and faint spots, the margins with firm, white teeth. Inflorescence simple, to 60 cm, most of its axis without bracts; flower-stalks *c.* 3 cm, bracts *c.* 1.5 cm. Perianth *c.* 4 cm, cylindric, broadening upwards, red, the

outer lobes united for more than half their length. Stamens and style scarcely projecting. *Origin uncertain.* H5–G1. Spring–summer.

This species has been in cultivation for almost 200 years, but it is not definitely known in the wild. Reynolds (Aloes of South Africa, edn 4, 173, 1974) suggests that it may be a hybrid, with *A. humilis* as one of the parents.

**8. A. humilis** (Linnaeus) Miller. Illustration: Jeppe, South African Aloes, 14 (1969); Reynolds, Aloes of South Africa, edn 4, f. 166–8 (1974).
Plants stemless, forming dense groups. Leaves 20–30 in each rosette, *c.* 10 × 1.2–1.8 cm, erect or incurved, ovate-lanceolate, glaucous, both surfaces usually tuberculate and with some soft, white spines, the margins with soft, white teeth. Inflorescence simple, to 35 cm; flower-stalks 2.5–3.5 cm, bracts 2.5–3.5 cm. Perianth cylindric but slightly swollen in the middle, 3.5–4.2 cm, almost straight, orange-red or orange, the outer lobes free for most of their length. Anthers and style scarcely projecting. *South Africa (Cape Province).* H5–G1. Spring.

A variable species; var. **echinata** (Willdenow) Baker, which has leaves with soft prickles (rather than tubercles) on the upper surface, is widely grown.

**9. A. krapohliana** Marloth. Illustration: Jeppe, South African Aloes, 34 (1969); Reynolds, Aloes of South Africa, edn 4, t. 8 & f. 170–1 (1974).
Plants stemless, or stem to 20 cm, solitary. Leaves 20–30 in each rosette, incurved, to 20 × 2 4 cm, glaucous, smooth, the lower surface sometimes with differently coloured bands, the margins with soft, white teeth. Inflorescence usually simple, to 40 cm; flower-stalks to 1.8 cm, bracts somewhat shorter. Perianth cylindric, orange-red, *c.* 3.5 cm, the outer lobes free to the base. Anthers and style projecting 2–3 mm. *South Africa.* H5–G1.

**10. A. melanacantha** Berger. Illustration: Flowering plants of South Africa 11: t. 433 (1931); Jeppe, South African Aloes, 25 (1969); Reynolds, Aloes of South Africa, edn 4, t. 9 & f. 172–3 (1974).
Plants with short stems, ultimately to 50 cm. Rosettes to 30 cm in diameter. Leaves numerous, incurved, triangular-lanceolate, to 20 × 4 cm, dull green, the upper surface smooth, the lower with a few black prickles, the margins with hard, sharp, black spines. Inflorescence simple, to

1 m; flower-stalks *c.* 1.5 cm, bracts to 2.5 cm. Perianth *c.* 4.5 cm, cylindric, orange-red becoming yellow, the outer lobes free almost to the base. Anthers and style projecting *c.* 5 mm. *South Africa, South West Africa.* H5–G1.

Easily recognised by the sharp, black spines on the leaf margins.

**11. A. brevifolia** Miller. Illustration: Jeppe, South African Aloes, 13 (1969); Reynolds, Aloes of South Africa, edn 4, f. 177–9 (1974).
Plants stemless with many offsets from the base, forming large groups. Rosettes 7–8 cm in diameter, with 30–40 leaves. Leaves broadly triangular-lanceolate, *c.* 6 × 2 cm, glaucous, the upper surface without spots or spines, the lower with a few soft spines, the margins with firm, white teeth. Inflorescence simple, to 40 cm; flower-stalks *c.* 1.5 cm; bracts *c.* 1.5 cm. Perianth *c.* 3.8 cm, cylindric, orange-yellow, the outer lobes free to the base. Anthers and style projecting 4–5 mm. *South Africa (Cape Province).* H5–G1.

**12. A. glauca** Miller. Illustration: Jeppe, South African Aloes, 54 (1969); Reynolds, Aloes of South Africa, edn 4, f. 197–200 (1974).
Stems variable in height, to 60 cm or rarely more. Rosette open, with many, spreading leaves. Leaves to 40 × 15 cm, broadly lanceolate, glaucous, lined on both surfaces, margins with dark brown, hard, sharp teeth. Inflorescence simple, to 1 m; flower-stalks to 4 cm, bracts 2.5–3 cm. Perianth *c.* 3.5 cm, cylindric, pink-orange fading yellow, outer lobes free to the base. Anthers and style projecting 2–3 mm. *South Africa.* H5–G1.

**13. A. saponaria** (Aiton) Haworth. Illustration: Lamb, The illustrated reference to cacti and other succulents 2: 443 (1959); Jeppe, South African Aloes, 67 (1969); Reynolds, Aloes of South Africa, edn 4, t. 12 & f. 230–3 (1974).
Plant stemless or with a stem to 50 cm. Rosettes open, with 12–20 spreading leaves. Leaves 25–30 × 8–12 cm, narrowly or broadly lanceolate, the upper surface pale to dark green with numerous, dull, white, oblong spots arranged in irregular, transverse bars, lower surface paler and not spotted, margins with hard, sharp, triangular teeth. Inflorescence branched, 40–60 cm; flower-stalks 3.5–4.5 cm, bracts 1.7–2.2 cm; flowers congested at the ends of the branches. Perianth conspicuously constricted above the ovary,

cylindric above, 3.5–4.5 cm, yellow to orange, outer lobes united for more than half their length. Anthers and style projecting 1–5 mm. *South Africa.* H5–G1. Summer.

**14. A. zebrina** Baker. Illustration: Jeppe, South African Aloes, 87 (1969); Reynolds, Aloes of South Africa, edn 4, f. 302–3 (1974).
Plants stemless or with very short stems, suckering freely and forming groups. Rosettes with 15–25 leaves. Leaves 15–30 × 6–7 cm, linear-lanceolate, tapering from about the middle, dark green with a glaucous bloom and with large, whitish spots arranged in irregular transverse bands above, margins with hard, sharp, brownish teeth. Inflorescence to 1.5 m, branched; flower-stalks 6–7 mm, bracts 1.2–1.5 cm. Perianth 3–3.5 cm, conspicuously and abruptly narrowed above the ovary, then cylindric and curved downwards, dull red, the outer lobes united for most of their length. Anthers and style not or scarcely projecting. *Southern Africa, from Angola to Mozambique southwards.* H5–G1.

A very variable species.

**15. A. grandidentata** Salm-Dyck. Illustration: Jeppe, South African Aloes, 63 (1969); Reynolds, Aloes of South Africa, edn 4, f. 305–7 (1974).
Like *A. zebrina* but leaves 10–15 cm, the apex dried and twisted, flower-stalks 1–1.5 cm, perianth 2.8–3 cm, cylindric to club-shaped, not conspicuously and abruptly constricted above the ovary. *South Africa.* H5–G1. Summer.

**16. A. striata** Haworth. Illustration: Lamb, The illustrated reference to cacti and other succulents 1: 167 (1955); Jeppe, South African Aloes, 63 (1969); Reynolds, Aloes of South Africa, edn 4, t. 20 & f. 316–17 (1974).
Plants prostrate, usually stemless, more rarely with a stem to 1 m. Leaves 12–20 in a dense rosette, lanceolate, spreading or incurved, to 50 × 20 cm, glaucous, sometimes tinged with red, unspotted, the margins entire, forming a pink or red zone 2–3 mm wide. Inflorescence branched, to 1 m; flower-stalks 1.5–2.5 cm, bracts *c.* 5 mm. Perianth 2.5–3 cm, constricted abruptly above the ovary, then cylindric and curving downwards, orange-red, outer lobes united for most of their length. Anthers and style projecting 1–2 mm. *South Africa.* H5–G1. Spring.

**17. A. vera** (Linnaeus) Burmann (*A. perfoliata* var. *vera* Linnaeus; *A. barbadensis* Miller). Illustration: Reynolds, Aloes of tropical Africa, f. 150–1 (1966).
Plants stemless or with very short stems, suckering freely. Leaves *c.* 16 in each rosette, 40–50 × 6–7 cm, lanceolate, erect-spreading and compact, grey-green or reddish, striped, unspotted, margins with firm, pale teeth. Inflorescence usually branched, to 1 m; flower-stalks *c.* 5 mm, bracts *c.* 10 mm. Perianth 2.8–3 cm, cylindric or slightly swollen below, yellow, outer lobes free for more than half their length. Anthers and style projecting 3–5 mm. *Canary Islands, widely cultivated and escaped elsewhere.* H5–G1. Winter–spring.

This species, which has a long history in cultivation, was introduced to Britain from Barbados; hence the inappropriate species name given by Miller.

**18. A. perryi** Baker. Illustration: Botanical Magazine, 6596 (1881).
Stem to 30 cm. Leaves 12–20 in a rosette, *c.* 35 × 7.5 cm, lanceolate, spreading, glaucous and reddish-tinged, margins with small, pale brown teeth. Inflorescence to 60 cm, branched; flower-stalks *c.* 8 mm, bracts 4–6 mm. Perianth 2–2.5 cm, cylindric, bright red turning yellow, outer lobes united for more than half their length. Anthers and style slightly projecting. *Socotra.* G1–2.

**19. A. camperi** Schweinfurth (*A. eru* Berger). Illustration: Lamb, The illustrated reference to cacti and other succulents 2: 436 (1959); Reynolds, Aloes of tropical Africa, t. 45 & f. 216–17 (1966).
Stems to 50 cm. Leaves 12–16 in each rosette, 50–60 × 8–12 cm, lanceolate, spreading and arching downwards, channelled, dark green, unspotted or occasionally with a few pale spots, margins reddish with hard, sharp, brownish red teeth. Inflorescence to 1 m, branched; flower-stalks 1.2–1.8 cm, bracts *c.* 2 mm. Perianth 2–2.2 cm, cylindric, orange-yellow, outer lobes united for more than half their length. Anthers and style projecting 3–4 mm. *Ethiopia.* G1–2. Spring.

Material in cultivation as *A. abyssinica* Lamarck (a name of doubtful application) probably belongs to this species or a hybrid of it.

**20. A. tenuior** Haworth. Illustration: Jeppe, South African Aloes, 111 (1969); Reynolds, Aloes of South Africa, edn 4, f. 374–5 (1974).
Stem slender but fleshy, 1–3 m. Leaves in a very loose rosette, the stem clearly visible between them, 10–15 × 1–1.5 cm, linear-lanceolate, glaucous unspotted, not auriculate at the base, margins with very small, white teeth. Inflorescence usually simple; flower-stalks 3–5 mm, bracts 4–6 mm. Perianth 1–1.4 cm, tubular or slightly bell-shaped, yellow or red, outer lobes united for more than half of their length. Anthers and style projecting 4–6 mm. *South Africa.* H5–G1. Winter.

A variable species; Reynolds (pp. 347–52) divides it into 3 varieties but it is uncertain which of these is in cultivation.

**21. A. ciliaris** Haworth. Illustration: Jeppe, South African Aloes, 110 (1969); Reynolds, Aloes of South Africa, edn 4, t. 29 & f. 380 (1974).
Plants with somewhat fleshy and scrambling stems to 5 m. Leaves dispersed along the stem, 10–15 × 1.5–2.5 cm, linear-lanceolate, tapering, the base broadly auricled, the auricles ciliate; margins with firm, white teeth. Inflorescence simple; flower-stalks *c.* 5 mm, bracts *c.* 3 mm. Perianth *c.* 3 cm, cylindric, orange-red, the outer lobes united for most of their length. Anthers and style projecting 2–4 mm. *South Africa (Cape Province).* H5–G1. Winter.

**22. A. striatula** Haworth. Illustration: Jeppe, South African Aloes, 114 (1969); Reynolds, Aloes of South Africa, edn 4, t. 31 & f. 392–7 (1974).
Stem to 1.75 m. Leaves dispersed along the stem, *c.* 25 × 2.5 cm, linear-lanceolate, tapering, green, margins with firm, white teeth, the base not auriculate but the sheath prominently striped. Inflorescence simple; flower-stalks 3–5 mm, bracts 1.5–2.5 mm. Perianth 4–4.5 cm, cylindric, slightly curved, yellowish or reddish-orange, the outer lobes free almost to the base. Anthers and style projecting 4–7 mm. *South Africa.* H5–G1.

**23. A. divaricata** Berger. Illustration: Reynolds, Aloes of tropical Africa, t. 105 & f. 530–2 (1966).
Stems 2–3 m. Leaves dispersed along the stems for the upper 50–100 cm, to 65 × 7 cm, broadly lanceolate, dull grey-green tinged with red, the base not auriculate, the sheath not striped; margins with sharp teeth. Inflorescence branched, to 1 m; flower-stalks *c.* 6 mm, bracts *c.* 4 mm. Perianth 2.8–3 cm, constricted sharply above the ovary, then cylindric, curved, red, the outer lobes free to the base (they may cohere in the lower half but are not united). Anthers and style projecting 2–4 mm. *Madagascar.* G2.

**24. A. perfoliata** Linnaeus (*A. mitriformis* Miller). Illustration: Botanical Magazine, 1270 (1810); Jeppe, South African Aloes, 20 (1969); Reynolds, Aloes of South Africa, edn 4, f. 406–7 (1974); Il Giardino Fiorito 47: 624 (1981).
Stems 1–2 m, mostly sprawling on the ground. leaves many in a rosette, ovate-lanceolate, spreading or incurved, to 20 × 15 cm, glaucous, margins with hard, sharp, golden yellow, brown or black teeth. Inflorescence branched, to 60 cm, flower-stalks 4–4.5 cm, bracts *c.* 10 mm. Perianth 4–4.5 cm, cylindric, dull orange-red, the outer lobes free to the base (they may cohere to some extent in the lower half, but are not united). Anthers and style projecting at most 3 mm. *South Africa (Cape Province).* H5–G1. Summer.

**25. A. distans** Haworth. Illustration: Lamb. The illustrated reference to cacti and other succulents 1: 165 (1955); Jeppe, South African Aloes, 21 (1969); Reynolds, Aloes of South Africa, edn 4, f. 414–15 (1974).
Stems to 3 m, sprawling on the ground. Leaves in elongate rosettes, to 15 × 7 cm, broadly lanceolate or lanceolate, glaucous, with whitish spots towards the base, margins with golden yellow teeth. Inflorescence branched, to 60 cm; flower-stalks 3–4 cm, bracts *c.* 8 mm. Perianth *c.* 4 cm, cylindric, slightly curved, orange-red, the outer lobes free to the base. Anthers and style little projecting. *South Africa (Cape Province).* H5–G1. Summer.

*A. stans* Berger is a name of doubtful application; the species was placed close to *A. distans* by Berger, and specimens in cultivation under the former name may well be the latter or a hybrid of it.

**26. A. succotrina** Allioni. Illustration: Botanical Magazine, 472 (1800); Jeppe, South African Aloes, 50 (1969); Reynolds, Aloes of South Africa, edn 4, f. 433–4 (1974).
Stems usually to 2 m. Leaves numerous in a dense rosette, curved upwards or erect, to 50 × 7–10 cm, broadly lanceolate, dull to greyish green, sometimes with a few whitish spots, the margins dull with firm, white teeth. Inflorescence usually simple, to 1 m; flower-stalks *c.* 3 cm, bracts to 2 cm. Perianth *c.* 4 cm, cylindric, red or pink, glossy, the outer lobes free to the base. Anthers and style projecting 3–5 mm. *South Africa (Cape Province).* H5–G1. Winter.

This species was once thought to have originated in Socotra (hence the species name); it is not, however, found there, being restricted to the Cape peninsula in South Africa.

**27. A. microstigma** Salm-Dyck. Illustration: Jeppe, South African Aloes, 33 (1969); Reynolds, Aloes of South Africa, edn 4, f. 436–7 (1974).
Stems to 50 cm. Leaves numerous in a dense rosette, spreading and arching upwards, to 30 × 6.5 cm, broadly lanceolate, green, usually with H-shaped, whitish spots, the margins with hard, sharp teeth. Inflorescence simple, to 80 cm; flower-stalks 2.5–3 cm, bracts 1.2–1.5 cm. Perianth 2.5–3 cm, cylindric, orange fading to yellow, the outer lobes free to the base. Anthers and style projecting 1–3 mm. *South Africa, South West Africa.* H5–G1.

**28. A. arborescens** Miller. Illustration: Botanical Magazine, 8663 (1916); Jeppe, South African Aloes, 48 (1969); Reynolds, Aloes of South Africa, edn 4, f. 450–2 (1974); Il Giardino Fiorito 47: 624 (1981).
Stems 2–3 m. Leaves numerous in each rosette, widely spreading and curved downwards towards their apices, 50–60 × 5–7 cm, lanceolate, dull green, margins with firm, pale teeth. Inflorescence usually simple; flower-stalks 3.5–4 cm, bracts 1.5–2 cm. Perianth *c.* 4 cm, cylindric, slightly constricted above the ovary, orange-red, outer lobes free to the base. Anthers and style projecting *c.* 5 mm. *South Africa north to Malawi.* H5–G1. Winter–spring.

A. × **principis** (Haworth) Stearn is the hybrid between this species and *A. ferox* (no. 32). It is intermediate between its parents, flowering somewhat later than *A. arborescens.*

**29. A. ortholopha** Christian & Milne-Redhead. Illustration: Flowering plants of South Africa 23: t. 882 (1943); Reynolds, Aloes of tropical Africa, t. 51 & f. 237 (1966).
Stems solitary, less than 50 cm. Leaves 30 or more in a rosette, spreading and curving upwards, to 50 × 12–14 cm, broadly lanceolate, dull grey-green tinged with pink, margins with sharp teeth. Inflorescence branched, its stalk arching so that the branches are almost horizontal, the flowers borne on the upper sides of the axes. Flower-stalks *c.* 8 mm, bracts 1–1.5 cm. Perianth *c.* 4 cm, cylindric, somewhat swollen in the middle, orange-red to deep red, the outer lobes free for most

of their length. Anthers and style projecting 1–1.5 cm. *Zimbabwe.* H5–G1.

**30. A. rubroviolacea** Schweinfurth. Illustration: Botanical Magazine, 7882 (1903).
Stems to 1 m. Leaves numerous in each rosette, dense, spreading and recurved, to 60 × 10–11 cm, broadly lanceolate, reddish or purplish at the base, shading to pink towards the apex, margins with small, hooked teeth. Inflorescence to 1 m or more, branched with erect branches; flower-stalks 3–4 mm, bracts 2–2.5 cm. Perianth 3–4 cm, cylindric, bright red, the outer lobes united for half their length. Anthers and style projecting 1–1.5 cm. *Yemen.* G1.

**31. A. africana** Miller. Illustration: Flowering plants of South Africa 9: t. 333 (1925); Jeppe, South African Aloes, 41 (1969); Reynolds, Aloes of South Africa, edn 4, f. 503–4 (1974).
Stems to 2 m. Leaves *c.* 30 in a rosette, widely spreading, to 65 × 12 cm, broadly lanceolate, dull green to glaucous, with a few spines on both surfaces, margins with sharp, reddish brown teeth. Inflorescence to 80 cm, branched; flower-stalks 5–6 mm, bracts 1–1.2 cm. Perianth to 5.5 cm, cylindric, conspicuously curved, yellow to orange-yellow, the outer lobes united for more than half their length. Anthers and style projecting 1.5–2 cm. *South Africa (Cape Province).* H5–G1. Spring.

**32. A. ferox** Miller. Illustration: Botanical Magazine, 1975 (1818); Lamb, The illustrated reference to cacti and other succulents 1: 166 (1955); Jeppe, South African Aloes, 40 (1969); Reynolds, Aloes of South Africa, edn 4, f. 509–13 (1974).
Stems to 3 or rarely 5 m. Leaves 50–60 in a dense rosette, spreading, to 100 × 15 cm, lanceolate, flat or slightly channelled, dull green with a reddish tinge, the surface usually spiny, the margins with stout, reddish brown spines. Inflorescence branched, the individual racemes 50–80 cm; flower-stalks 4–6 mm, bracts 8–10 mm. Perianth *c.* 3.3 cm, cylindric, straight, orange-red, the outer lobes free for most of their length, inner lobes with brown tips. Anthers and style projecting 2–2.5 cm. *South Africa.* H5–G1.

A variable species, varying particularly in the size and spininess of the leaves.

**33. A. candelabrum** Berger. Illustration: Flowering plants of South Africa 24: t. 945 (1944); Jeppe, South African Aloes, 39 (1969); Reynolds, Aloes of South Africa, edn 4, f. 520–2 (1974).

Like *A. ferox*, but leaves deeply channelled and recurved, perianth pink to orange-red, the inner lobes without brown tips. *South Africa (Natal).* H5–G1.

**34. A. thraskii** Baker. Illustration: Jeppe, South African Aloes, 42 (1966); Reynolds, Aloes of South Africa, edn 4, t. 54 & f. 531 (1974).
Stems to 2 m. Leaves numerous in dense rosettes, recurved and channelled, to 1.6 m × 22 cm, lanceolate, dull green or glaucous, sometimes with a few spines on the surface, marginal spines reddish, *c.* 2 mm. Inflorescence branched, individual racemes to 25 cm; flower-stalks 1–2 mm, bracts *c.* 9 mm. Perianth 2.5–3 cm, cylindric, pale yellow, the outer lobes free for most of their length. Anthers and style projecting 1.5–2 cm. *South Africa (Natal).* H5–G1.

**35. A. excelsa** Berger. Illustration: Flowering plants of South Africa 2: t. 62 (1922); Reynolds, Aloes of tropical Africa, t. 72 & f. 310–11 (1966).
Stem to 4 m. Leaves *c.* 30 in a rosette, spreading or reflexed, to 80 × 15 cm, broadly lanceolate, channelled, dull green, usually with some spines on the lower surface, marginal spines sharp, 5–6 mm, reddish brown. Inflorescence to 1 m, branched; flower-stalks *c.* 1 mm, bracts 4–6 mm. Perianth *c.* 3 cm, cylindric, reddish orange, the outer lobes free for most of their length. Anthers and style projecting 8–10 mm. *Malawi, Zimbabwe.* G1–2.

**36. A. marlothii** Berger. Illustration: Flowering plants of South Africa 6: t. 171 (1925); Jeppe, South African Aloes, 37 (1969); Reynolds, Aloes of South Africa, edn 4, t. 56 & f. 535–9 (1974); Il Giardino Fiorito 47: 624 (1981).
Stems 2–6 m. Leaves 40–50 in a rosette, spreading, 1–1.5 m × 20–25 cm, broadly lanceolate, grey-green or glaucous, the surfaces usually spiny, the margins with stout, sharp, reddish brown spines. Inflorescence branched, the branches held almost horizontally, the flowers borne on their upper sides; flower-stalks *c.* 5 mm, bracts 8–9 mm. Perianth cylindric, 3–3.5 cm, orange to yellowish orange, the outer lobes free for most of their length. Anthers and style projecting 1.5–2 cm. *Botswana.* H5–G1.

**37. A. dichotoma** Masson. Illustration: Jeppe, South African Aloes, 57 (1969); Reynolds, Aloes of South Africa, edn 4, f. 551–5 (1974).

Plant tree-like, with a single trunk and a richly and apparently dichotomously branched crown, total height 6–9 m. Leaves in rosettes at the ends of the branches, spreading, 25–35 × 4–6 cm, linear-lanceolate, glaucous, margins with brownish yellow teeth. Racemes *c.* 15 cm; flower-stalks 5–10 mm, bracts 5–7 mm. Perianth *c.* 3.3 cm, swollen near the base then gradually tapered to the mouth, bright yellow, the outer lobes free for most of their length. Anthers and style projecting 1.2–1.5 cm. *South Africa, South West Africa.* H5–G1.

*A. bainesii* Dyer. Crown less branched, leaves 60–90 cm, dull green, perianth pink, 3.3–3.7 cm. *South Africa.* H5–G1.

**38. A. plicatilis** (Linnaeus) Miller. Illustration: Botanical Magazine, 457 (1799); Lamb, The illustrated reference to cacti and other succulents **2**: 441 (1959); Jeppe, South African Aloes, 67 (1969); Reynolds, Aloes of South Africa, edn 4, f. 570–1 (1974).
Stem to 5 m. Leaves borne in dense rosettes, on 2 sides of the stem and forming a flattened 'fan', *c.* 30 × 4 cm, linear or oblong-linear, glaucous, flat, margins entire, the apex very rounded. Inflorescence simple, to 50 cm; flower-stalks *c.* 1 cm, bracts *c.* 8 mm. Perianth 4.5–5.5 cm, cylindric, orange-red, the outer lobes united for more than half their length. Anthers and style projecting 1–5 mm. *South Africa (Cape Province).* H5–G1.

Very distinct with its fans of glaucous, tongue-shaped, entire, blunt leaves.

**43. BOWIEA** J.D. Hooker
*J. Cullen*
Plants with green bulbs, composed of fleshy scales, which are mostly exposed above the soil. Leaves absent, apart from 1 or 2 small, quickly deciduous scale-leaves produced on the bulb. Stem scrambling, richly branched to several orders of branching, green, photosynthetic. Flowers borne singly at the apices of the short, upper branches. Perianth-segments 6, free, green. Stamens 6, shorter than the perianth-segments. Ovary superior, 3-celled, style simple, 3-lobed at the apex. Fruit a capsule.

A genus of 2 species from southern Africa, 1 of them occasionally grown as a novelty (and sometimes also as a crop for the sake of the medicinally active chemicals it contains). It is easily grown in light, well-drained soil, but requires protection from frost throughout Europe. Propagation is by seed or by offsets from the bulb.

**1. B. volubilis** J.D. Hooker. Illustration: Botanical Magazine, 5619 (1867); Flowering plants of South Africa **21**: t. 815 (1941).
Bulb to 15 cm in diameter. Stem to 2 m or more. Perianth to 1.3 cm in diameter. *South Africa, South West Africa.* G1. Summer.

**44. WURMBEA** Thunberg
*J. Cullen*
Plants with corms deep in the soil. Stems solitary from each corm. Leaves usually 3, borne on the stem, loosely sheathing, the blades channelled. Flowers several, crowded or distant, in an erect spike. Perianth of 6 lobes united into a tube below, each lobe with a nectary (appearing as a fold or ridge) at its base. Stamens 6, borne on the perianth-lobes, the base of the filament just below the nectary. Ovary superior, 3-celled and 3-lobed, with several ovules in each cell. Styles 3, free, rigid, spreading. Fruit a 3-horned capsule.

A genus of several species from South Africa and Australia, whose classification is poorly understood. Such few plants as are cultivated are generally found under the names *W. capensis* Thunberg, *W. spicata* (Burman) Durand & Schinz and *W. campanulata* Willdenow; all these names may well apply to the same species, but it is not certain that the plants in cultivation belong to it. For further information see Nordenstam, B., Botaniska Notiser **117**: 173–82 (1964) and Notes from the Royal Botanic Garden, Edinburgh **36**: 211–33 (1978).

**45. COLCHICUM** Linnaeus
*C.D. Brickell*
Perennial stemless herbs with corms which occasionally have long, horizontal, rhizome-like outgrowths; tunics membranous, papery or leathery, frequently extended into a tubular, persistent false-stem or neck. Developing leaves and flowers enclosed within a membranous sheath. Leaves basal, partially developed at flowering or developing after flowering (occasionally developing as the flowers fade). Flowers solitary or in clusters, each subtended by a small bract and very shortly stalked, the stalks elongating as the capsules ripen. Perianth bell-shaped, funnel-shaped or star-shaped, purple, pink or white, sometimes tessellated; with a tube at the base and 6 lobes in 2 equal or almost equal whorls, bases of the lobes occasionally with auricles, the throat of the perianth sometimes ridged on either side of

the filament-bases to form a short 'filament-channel', the ridges hairless or downy. Stamens borne near the bases of the perianth-lobes, in 1 or 2 series; filaments slender, sometimes thickened at the base; anthers dorsifixed and versatile, or basifixed and rigid. Styles 3, free, stigmas point-like or unilaterally decurrent along the style; ovary subterranean. Capsule 3-celled, splitting along the septa. Seeds numerous, spherical or almost so.

A genus of about 45 species extending from E Europe and N Africa to W Asia, through Afghanistan to N India and W China (Xizang). The large-flowered species whose leaves develop after flowering thrive in any sunny, well-drained site, with the exception of *C. variegatum* which, like the majority of those with leaves partially developed at flowering, is better grown in a bulb-frame or alpine house.
Literature: Stefanoff, B., Monographie der Gattung Colchicum, *Sbornik Bulgariskata Akademiya na Naukita* **22**: 1–100 (1926); Bowles, E.A., *A handbook of Crocus and Colchicum for gardeners,* edn 2 (1952); Feinbrun, N., The genus Colchicum of Palestine and neighbouring countries, *Palestine Journal of Botany, Jerusalem series* **6, 2**: 71–95 (1953); Feinbrun, N., Chromosome numbers and evolution in the genus Colchicum, *Evolution* **12**: 173–88 (1958); Burtt, B.L., Meikle, R.D. & Furse, J.P.W., Colchicum and Merendera, *RHS Lily Year Book* **31**: 90–103 (1968).

1a. Anthers basifixed, rigid 2
 b. Anthers dorsifixed, versatile 3
2a. Flowers white with pale to deep red-purple stripes **2. kesselringii**
 b. Flowers pale to deep yellow, sometimes purple-brown at base **1. luteum**
3a. Leaves partly developed at flowering or appearing immediately after flowering; flowering in autumn, winter or spring 4
 b. Leaves undeveloped at flowering and appearing from December to April; flowering from August to November 14
4a. Corms with more or less horizontal, underground stolons 5
 b. Corms erect, without underground stolons 6
5a. Leaves 2 or 3, narrowly linear to linear-lanceolate, 7–15 cm × 2–15 mm at maturity; flowers star-shaped to narrowly funnel-shaped; anthers black or purplish black **3. psaridis**

b. Leaves 3, narrowly lanceolate to very narrowly lanceolate, 20–32 × 2.4–4.5 cm at maturity; flowers bell-shaped or funnel-shaped; anthers yellow **4. baytopiorum**

6a. Corm-tunics membranous, frequently short-lived; leaves 3 (rarely 4) 7

b. Corm-tunics leathery to papery, persistent; leaves 2–12 8

7a. Leaves narrowly to very narrowly lanceolate, recurving, 20–32 × 2.4–4.5 cm at maturity; anthers yellow; flowering from October to November **4. baytopiorum**

b. Leaves linear-lanceolate, erect-spreading, 11–15 cm × 5–11 mm at maturity; anthers purplish black or purplish green; flowering from February to June **5. triphyllum**

8a. Corm-tunics strongly leathery, with distinct longitudinal corrugations; filaments hairy at least near the base **6. burttii**

b. Corm-tunics papery, somewhat leathery or leathery, lacking distinct longitudinal corrugations; filaments hairless 9

9a. Flowering from September to December 10

b. Flowering from December to June 12

10a. Leaves 2 (rarely 3), 2–18 mm wide at maturity; corm-tunics leathery **7. cupanii**

b. Leaves 3–12, 1–5 mm wide at maturity; corm-tunics papery to somewhat leathery 11

11a. Anthers yellow; perianth-lobes bright purplish pink, 1.5–3 cm × 2–9 mm **8. stevenii**

b. Anthers purplish black to dark brown or greyish brown; perianth-lobes white to pale rosy lilac, 1–2 cm × 1.5–4 mm **9. pusillum**

12a. Leaves 3–6, narrowly linear, 1–7 mm wide at maturity; flowers star-shaped or narrowly funnel-shaped **10. falcifolium***

b. Leaves 2 or 3, strap-shaped to very narrowly linear-lanceolate, 1–3.5 cm wide at maturity; flowers ovoid-bell-shaped to funnel-shaped 13

13a. Leaves hairless, strap-shaped to very narrowly linear-lanceolate, usually 2–3.5 cm wide at maturity **11. szovitsii**

b. Leaves ciliate, the upper surface partially or entirely covered with short hairs, narrowly linear-lanceolate, 1–2 cm wide at maturity **12. hungaricum**

14a. Flowers lightly to strongly tessellated 15

b. Flowers not tessellated 21

15a. Stigmas point-like or decurrent for not more than 0.5 mm; flowers usually 3–10, rarely fewer or more, lightly tessellated **21. cilicicum**

b. Stigmas decurrent for at least 1.5 mm; flowers up to 8, lightly to strongly tessellated 16

16a. Perianth-lobes spreading; pollen grey-green; leaves strongly pleated longitudinally, ovate to elliptic-ovate, 24–42 × 11–15.5 cm **26. macrophyllum**

b. Perianth-lobes spreading or forming a bell- or funnel-shape; pollen yellow; leaves never strongly pleated longitudinally, strap-shaped to narrowly lanceolate, never more than 35 × 7 cm 17

17a. Leaves 3–7, spreading, margins usually wavy; perianth-lobes hairless, spreading or forming a bell- or funnel-shape 18

b. Leaves 4–11, usually 5–9, more or less erect, margins not wavy, but sometimes twisted towards the apex; flowers broadly bell- to funnel-shaped; perianth-lobes downy along the ridges of the filament-channels 19

18a. Flowers lightly tessellated, funnel shaped to bell-shaped; anthers yellow **28. lingulatum***

b. Flowers strongly tessellated, perianth-lobes spreading; anthers purplish black or purplish brown **25. variegatum***

19a. Flowers strongly tessellated, broadly bell-shaped; anthers purplish black or purplish brown **27. bivonae**

b. Flowers lightly tessellated, bell-shaped to funnel-shaped; anthers yellow 20

20a. Flowers narrowly bell-shaped, purplish pink to white; leaves 3–5, linear-lanceolate to broadly lanceolate, 14–35 × 1–7 cm, hairless, green **22. autumnale***

b. Flowers bell- to funnel-shaped, deep red-purple, occasionally paler; leaves 5–9, strap-shaped to narrowly lanceolate, 9–19 × 1–3.5 cm, ciliate (sometimes obscurely so), somewhat glaucous **29. turcicum**

21a. Stigmatic surface decurrent for at least 1.5–2 mm 22

b. Stigmatic surface point-like or shortly decurrent, usually to not more than 1 mm (if to 1.5 mm, then anthers purple or purplish brown) 28

22a. Anthers (undehisced) purple or purplish brown **24. bornmuelleri**

b. Anthers yellow 23

23a. Leaves 3–9, strap-shaped to oblong, occasionally narrowly lanceolate, usually not more than 7.5–19 × 1–3.5 cm 24

b. Leaves 3–5, elliptic, oblong-elliptic or oblong-lanceolate (occasionally the innermost leaf strap-shaped), 10–35 × 3–9.5 cm 25

24a. Leaves 3–7, spreading, margins hairless, usually wavy; flowers 1–3 (rarely 4), pale to deep purplish pink, perianth-lobes hairless **28. lingulatum***

b. Leaves 5–9, more or less erect, often twisted near the apex, ciliate (sometimes obscurely so), not wavy; flowers 3–8 (rarely 1), deep red-purple, occasionally paler, perianth-lobes downy along the ridges of the filament-channels **29. turcicum**

25a. Perianth-lobes narrowly oblong-elliptic, hairless; leaves elliptic; corm-tunic with a very long, persistent, strongly fibrous neck to 25 cm, obscuring the membranous sheath **19. balansae**

b. Perianth-lobes oblanceolate, oblong elliptic to elliptic, downy along the ridges of the filament-channels; neck of corm to 14 cm, not strongly fibrous; membranous sheath usually projecting well beyond the neck 26

26a. Flowers funnel-shaped; leaves 10–12 (rarely to 16) × 3–5 cm **20. kotschyi**

b. Flowers bell-shaped, sometimes narrowly; leaves 18–35 × 2–9.5 cm (rarely less) 27

27a. Anthers 5–8 mm; flowers narrowly bell-shaped; leaves linear-lanceolate to broadly lanceolate **22. autumnale***

b. Anthers 1–1.2 cm; flowers bell-shaped; leaves narrowly elliptic to oblong-lanceolate **23. speciosum**

28a. Corms with more or less horizontal, underground stolons **13. boissieri**

b. Corms erect, without underground stolons 29

29a. Perianth-lobes lilac-purple to deep rosy purple, usually 5–7.5 × 1.1–2.5 cm; leaves very narrowly elliptic to narrowly elliptic-lanceolate 30

b. Perianth-lobes white to purplish pink, usually 1.5–4.5 cm × 2–12 mm; leaves linear, linear-oblanceolate, very narrowly linear-lanceolate or strap-shaped 31

30a. Anthers yellow; stigmatic surface point-like or shortly decurrent to

0.5 mm; flowers usually 3–25
                               **21. cilicium***
b. Anthers purple or purple-brown;
   stigmatic surface 0.5–1.5 mm;
   flowers usually 1–3, rarely to 6
                               **24. bornmuelleri**
31a. Leaves 4–10, usually 6–8, linear to
   narrowly linear, 7–10 cm × 1–4 mm;
   flowers funnel- or bell-shaped
                               **17. parlatoris**
b. Leaves usually 2–6, strap-shaped to
   very narrowly linear-lanceolate; if
   linear, then leaves only 2 or 3, more
   than 10 cm and flower funnel-shaped,
   becoming star-shaped             32
32a. Flowers narrowly bell-shaped to
   funnel-shaped; anthers 2–3 mm;
   leaves 2–5, strap-shaped to linear-
   lanceolate, usually 8–15 cm
                               **14. alpinum***
b. Flowers funnel-shaped to star-shaped;
   anthers 3–8 mm; leaves 2–8, strap-
   shaped or linear to very narrowly
   linear-oblanceolate, usually
   12–30 cm                        33
33a. Corm-tunics membranous, reddish
   brown, neck often only slightly
   developed; leaves 2 or 3, linear to
   linear-oblanceolate; flowers 1–2,
   rarely to 5          **15. micranthum**
b. Corm-tunics leathery or almost so,
   dark blackish brown, neck persistent;
   leaves 3–8, strap-shaped to very
   narrowly lanceolate; flowers 1–12
                                    34
34a. Flowers 1–3, rarely to 6; perianth-
   lobes narrowly oblanceolate to linear-
   elliptic, usually 1.5–2.5 cm ×
   2–6 mm; filaments 5–8 mm; anthers
   3–4 mm              **16. umbrosum**
b. Flowers 2–6 or more; perianth-lobes
   narrowly oblong-lanceolate,
   2.8–4.5 cm × 4–11 mm; filaments
   1.5–2 cm; anthers 6–8 mm
                               **18. troodii**

**1. C. luteum** Baker. Illustration: Rix &
Phillips, The bulb book, 33 (1981).
Corm 1.5–3 × 1–2.5 cm, oblong-ovoid,
tunic dark chestnut brown, papery,
produced into a short neck. Leaves 2–5,
linear-lanceolate, more or less erect and
1–3 cm at flowering, 10–30 cm ×
6–20 mm at maturity, hairless. Flowers
1–4, funnel-shaped to narrowly bell-
shaped. Perianth-lobes pale to deep yellow,
narrowly oblong-lanceolate, 1.5–2.5 cm ×
2–6 mm, tube sometimes purple-brown.
Filaments 3–4 mm, hairless; anthers
yellow, 6–13 mm, basifixed. Pollen yellow.
Styles straight, stigmas point-like. Capsule

ovoid, 2–3 cm. *USSR (Soviet Central Asia),
Afghanistan, N India, SW China (Xizang).*
H3. Spring–summer.

**2. C. kesselringii** Regel (*C. regelii* Stefanoff;
*C. crociflorum* Regel). Illustration: Rix &
Phillips, The bulb book 32 (1981).
Corm 1–3 × 1–2 cm, oblong-ovoid, tunics
dark brown, leathery, produced into a short
neck. Leaves 2–7, linear-lanceolate, more
or less erect and 1–2 cm at flowering,
7–10 cm × 3–10 mm at maturity, hairless
or margins slightly rough. Flowers 1–4,
narrowly bell-shaped to funnel-shaped.
Perianth-lobes with pale to deep red-purple
central stripes, narrowly linear-lanceolate
to narrowly elliptic, 1.5–3 cm × 2–7 mm;
tube red-purple. Filaments 3–4 mm,
hairless; anthers yellow, 8–10 mm,
basifixed; pollen yellow. Styles straight,
stigmas point-like. *USSR (Soviet Central
Asia) & N Afghanistan.* H3. Early
spring–summer.

**3. C. psaridis** Halacsy. Illustration: Grey-
Wilson & Mathew, Bulbs, pl. 10 (1981).
Corm more or less horizontal with runners,
2–5 cm × 4–12 mm, irregular in shape,
sometimes swollen; tunics pale brown,
papery or membranous. Leaves 2 or 3,
usually partly developed at flowering, erect
or almost so, narrowly linear to narrowly
linear-lanceolate, extending to 7–9 cm at
flowering, 7–15 cm × 2–15 mm at
maturity, margin hairless or partly ciliate.
Flowers 1–3, star-shaped to narrowly
funnel-shaped. Perianth-lobes white to
pinkish purple, very narrowly elliptic or
narrowly elliptic-oblong, 1.1–2.7 cm ×
2–6 mm. Filaments 4–8 mm, hairless;
anthers black or purplish black, 2–3 mm,
dorsifixed; pollen yellow. Styles straight,
stigmas point-like. Capsule oblong-ovoid,
1.5 cm. *S Greece & W Turkey.* H4. Autumn-
winter.

**4. C. baytopiorum** Brickell. Illustration:
Mathew & Baytop, The bulbous plants of
Turkey, f. 68 (1984).
Corm 2.5–3.5 × 1.5–2.5 cm, narrowly
ovoid to almost spherical, sometimes more
or less horizontal with runners,
4–6 × 1.3–2 cm, irregular in shape; tunics
reddish brown, membranous to papery,
frequently short-lived. Leaves 3, more or
less erect at first, then recurving, narrowly
to very narrowly lanceolate, extending to
1–8 cm at flowering, 20–32 × 2.5–4.5 cm
at maturity, hairless. Flowers 1–3, rarely to
5, bell-shaped or funnel-shaped. Perianth-
lobes bright purplish pink, 2.2–4.2 cm ×
5–11 mm, elliptic or oblong-elliptic to

oblanceolate. Filaments 1–1.4 cm,
hairless; anthers yellow, 5–7 mm,
dorsifixed; pollen lemon yellow. Styles
straight, stigmas point-like. Capsule
narrowly ellipsoid, 2–2.5 cm. *W Turkey.*
H4. Autumn.

**5. C. triphyllum** Kunze (*C. bulbocodioides*
Bieberstein not Brotero; *C. biebersteinii*
Rouy; *C. catacuzenium* Stefanoff;
*C. ancyrense* Burtt; *C. montanum*
misapplied). Illustration: Botanical
Magazine, 9652 (1943); Grey-Wilson &
Mathew, Bulbs, pl. 10 (1981); Mathew &
Baytop, The bulbous plants of Turkey, f. 66
(1984).
Corm 1.5–2.5 × 1–1.5 cm, oblong-ovoid,
tunic chestnut brown, membranous, short-
lived, lacking an extended neck. Leaves 3,
linear-lanceolate, erect-spreading,
extending to 2–9 cm at flowering,
11–15 cm × 5–11 mm at maturity,
margins rough (sometimes obscurely so) or
smooth. Flowers 1–6, bell-shaped to
funnel-shaped. Perianth-lobes purplish pink
or white flushed with purplish pink,
narrowly elliptic to oblanceolate,
1.5–3 cm × 5–12 mm, occasionally with
thread-like outgrowths or auricles at the
base. Filaments 7–9 mm, hairless; anthers
2.5–3.5 mm, purplish black or purplish
green, dorsifixed; pollen yellow. Styles
straight, stigmas point-like. Capsule ovoid
to oblong-ovoid, 2–3 cm. *NW Africa, Spain,
Greece, Turkey, USSR (S Russia).* H3.
Early–late spring.

**6. C burttii** Meikle. Illustration: Botanical
Magazine, n.s., 735 (1977); Mathew &
Baytop, The bulbous plants of Turkey, f. 63
(1984).
Corm 3–5 × 1.5–2 cm, narrowly ovoid to
almost spherical, tunics blackish brown,
very leathery, with prominent longitudinal
corrugations, apex fringed with rigid fibres
extending for 1–5 cm. Leaves 2–4,
narrowly linear to linear-lanceolate, erect-
spreading and extending for 1–4 cm at
flowering, recurved and
10–15 cm × 8–10 mm at maturity, shortly
hairy at least on the margins. Flowers 1–4,
funnel-shaped, becoming star-shaped.
Perianth-lobes white or pale purplish pink,
narrowly oblanceolate,
1.5–4 cm × 3–8 mm. Filaments 8–10 mm,
thinly hairy at least near the base; anthers
dark purplish black or black, 2–3 mm,
dorsifixed; pollen yellow. Styles straight,
stigmas point-like. Capsule (immature)
narrowly ellipsoid, *c.* 1 cm, roughly hairy
at the apex. *W Turkey.* H4. Early spring.

**7. C. cupanii** Gussone (*C. glossophyllum* Heldreich). Illustration: Mathew, Dwarf bulbs, pl. 21 (1973); Grey-Wilson & Mathew, Bulbs, pl. 11 (1981); Rix & Phillips, The bulb book, 184, 185 (1981). Corm 1–2 × 1–1.5 cm, ovoid, tunics dark brown, leathery, with a short neck. Leaves 2 (very rarely 3), to 10 cm at flowering, 7–15 cm × 2–18 mm at maturity, linear to linear-lanceolate, hairless or rarely ciliate at the base. Flowers 1–12, funnel-shaped to star-shaped. Perianth-lobes pale to deep purplish pink, narrowly elliptic, 1.8–2.5 cm × 3–5 mm. Filaments 6–10 mm, hairless; anthers purplish black, 2–3 mm, dorsifixed; pollen yellow. Styles straight, stigma point-like. Capsule oblong-ovoid, 1–1.3 cm. *N Africa, France, Italy, Greece, Crete.* H3. Autumn–early winter.

**8. C. stevenii** Kunth. Illustration: Botanical Magazine, 8025 (1905).
Corm 1.6–3 × 1–1.5 cm, ovoid to almost spherical; tunics dull, dark brown, papery to almost leathery, produced into a neck 1–6 cm long. Leaves 4–12, more or less erect, extending for 1–12 cm at flowering, recurved and often wavy, 8–18 cm × 1–5 mm at maturity, hairless or ciliate. Flowers 1–10, funnel-shaped. Perianth-lobes bright purplish pink, oblong-elliptic to oblanceolate, 1.5–3 cm × 2–9 mm. Filaments 8–12 mm, hairless; anthers yellow. Styles straight, stigmas point-like. Capsule oblong-ovoid, 1–1.5 cm. *Cyprus, Turkey, W Syria.* H4. Autumn–early winter.

Similar plants from Greece and the islands have been described as *C. peloponnesiacum* Rechinger & Davis and *C. andrium* Rechinger & Davis. Their status is uncertain.

**9. C. pusillum** Sieber (*C. montanum* Linnaeus var. *pusillum* (Sieber) Fiori). Illustration: Rix & Phillips, The bulb book, 177 (1981).
Corm 1.5–2.5 cm × 9–17 mm, ovoid to almost spherical; tunics dull, dark brown, papery, the neck 1–6 cm. Leaves 3–8, narrowly linear, more or less erect and extending for 1–4 cm at flowering, recurved and 8–11 cm × 1–5 mm at maturity, hairless or sometimes ciliate. Flowers 1–6, funnel-shaped, opening star-shaped. Perianth-lobes pale rosy lilac to white, narrowly oblanceolate to narrowly oblong-oblanceolate, 1–2 × 1.5–4 cm. Filaments 5–8 mm, hairless; anthers purplish black, brownish black or grey-brown, 1.5–4.5 mm, dorsifixed; pollen yellow. Styles straight, stigma point-like.

Capsule ovoid, 1–1.3 cm. *Greece, Crete, Cyprus.* H4. Autumn.

**10. C. falcifolium** Stapf (*C. varians* Freyn & Bornmüller; *C. serpentinum* Mishchenko; *C. tauri* Stefanoff; *C. hirsutum* Stefanoff; *C. szovitsii* misapplied; *C. szovitsii* Fischer & Meyer var. *freynii* Stefanoff). Illustration: Rix & Phillips, The bulb book, 32 (1981) – as C. fasciculare; Mathew & Baytop, The bulbous plants of Turkey, f. 64 (1984). Corm 2–4 × 1.5–2 cm, ovoid; tunics dark reddish brown, papery to almost leathery, usually persistent, neck absent or vestigial. Leaves 3–6, narrowly linear, channelled, 1–8 cm at flowering, 9–20 cm × 1–7 mm at maturity, hairless, roughly hairy or with margins and upper surface sparsely to densely stiffly hairy. Flowers 1–8, star-shaped or narrowly funnel-shaped. Perianth-lobes white to purplish pink, very narrowly to narrowly elliptic or narrowly oblanceolate, 1.3–2.5 cm × 2–6 mm. Filaments 8–12 mm, hairless; anthers black, greenish black or blackish brown, 2.5–3.5 mm, dorsifixed; pollen yellow. Styles straight, stigmas point-like. Capsule narrowly ovoid to almost spherical, 1–2 cm. *S Russia, Turkey, Iran, Iraq, W Syria.* H4. Late winter–early spring.

\*C. **fasciculare** (Linnaeus) R. Brown from Syria, and C. **crocifolium** Boissier from Syria, Iran and Iraq are occasionally cultivated. Both have numerous, star-shaped flowers which are white and pink. The leaves of *C. fasciculare* are hairless and broader than those of *C. crocifolium* which may be harshly downy or hairless.

**11. C. szovitsii** Fischer & Meyer (*C. szovitsii* var. *nivale* Boissier & Huet; *C. bifolium* Freyn & Sintenis; *C. hydrophilum* Siehe; *C. szovitsii* var. *bifolium* (Freyn & Sintenis) Bordzilowski; *C. armenum* Fedtschenko; *C. nivale* Stefanoff). Illustration: Botanical Magazine, 8040 (1905); Rix & Phillips, The bulb book, 32 (1981); Mathew & Baytop, The bulbous plants of Turkey, f. 65 (1984). Corm 1.5–4 × 1–3 cm, ovoid; tunic blackish brown, papery to almost leathery, neck absent or vestigial. Leaves 2 or 3, strap-shaped or very narrowly linear-lanceolate, more or less erect, extending for 2–12 cm at flowering, 12–25 × 1.2–3.5 cm at maturity, hairless. Flowers 1–7, ovoid-bell-shaped or sometimes bell-shaped or narrowly so. Perianth-lobes deep to pale purplish pink or white strongly suffused with purplish pink, oblanceolate to very narrowly elliptic, 2.1–3.5 cm × 4–10 mm, occasionally with basal auricles. Filaments 7–11 mm, hairless; anthers

purplish black or greenish black, 2–5 mm, dorsifixed; pollen yellow. Styles straight, stigmas point-like. Capsule ellipsoid to spherical, 3.5–4 cm. *Turkey, Iran, USSR (Caucasus).* H3. Early spring–summer.

Plants under the name C. **brachyphyllum** Boissier, from Syria and Lebanon are occasionally cultivated; they resemble *C. szovitsii* but have 4–6 narrowly ovate to lanceolate leaves and up to 15 flowers.

**12. C. hungaricum** Janka (*C. doerfleri* Halacsy). Illustration: Grey-Wilson & Mathew, Bulbs, pl. 11 (1981); Rix & Phillips, The bulb book, 32 (1981). Corm 2–3 × 1–2 cm, oblong-ovoid; tunics dark brown, papery to almost leathery, with a short neck. Leaves 2, very rarely 3, narrowly linear-lanceolate, more or less erect, 3–10 cm at flowering, up to 20 × 1–2 cm at maturity, ciliate; upper surface sometimes partially or entirely covered with short hairs. Flowers 1–8, bell-shaped to funnel-shaped. Perianth-lobes purplish pink to white, elliptic-lanceolate or narrowly elliptic, 2–3 cm × 6–7 mm. Filaments 7–10 mm, hairless; anthers purplish black, 2–3.5 mm, dorsifixed; pollen yellow or orange. Styles straight or slightly curved at the apex, stigmas point-like. Capsule almost spherical, c. 1 cm. *Hungary, Jugoslavia, Albania, Bulgaria, Greece.* H3. Late winter–early spring.

White variants are referred to as forma **albiflorum** Maly.

**13. C. boissieri** Orphanides (*C. procurrens* Baker). Illustration: Grey-Wilson & Mathew, Bulbs, pl. 10 (1981); Rix & Phillips, The bulb book, 175 (1981); Mathew & Baytop, The bulbous plants of Turkey, f. 70, 77 (1984) – the latter as C. siehanum.
Corm with horizontal runners, 2.5–3 cm × 3–13 mm, irregular, often with tooth-like projections, occasionally irregularly ovoid or oblong-ovoid and more or less vertical; tunics pale reddish brown, membranous. Leaves 2 or 3, developing after flowering, erect or almost so, linear, 11–22 cm × 2–8 mm, margins hairless or partly ciliate. Flowers 1 or 2, bell-shaped to narrowly funnel-shaped. Perianth-lobes bright rosy lilac, very narrowly elliptic to narrowly elliptic-obovate, 2.5–5 cm × 5–15 mm. Filaments 1–2 cm, hairless; anthers yellow, 5–10 mm, dorsifixed; pollen yellow. Style straight or slightly curved at the apex, stigmas point-like. Capsule ellipsoid, c. 2 cm. *S Greece, W Turkey.* H3. Autumn.

**14. C. alpinum** de Candolle. Illustration: Grey-Wilson & Mathew, Bulbs, pl. 9 (1981).
Corm 1–2.5 × 1–1.5 cm, almost spherical to ovoid; tunics dark reddish brown, membranous or somewhat leathery, neck 1–2 cm. Leaves 2 or 3, developing after flowering, strap-shaped to linear-lanceolate, 8–15 cm × 2–14 mm, hairless. Flowers 1 or 2, narrowly bell-shaped to funnel-shaped. Perianth-lobes purplish pink, occasionally white, narrowly oblong-elliptic, 1.7–3 cm × 4–10 mm. Filaments 2–9 mm, hairless; anthers yellow, 2–3 mm, dorsifixed; pollen yellow. Styles straight, stigmas point-like. Capsule oblong-ellipsoid, 1.5–2 cm. *France, Switzerland, Italy, Corsica, Sardinia*. H2. Late summer–autumn.

*\*C. arenarium* Waldstein & Kitaibel. Leaves 3–5, somewhat longer and broader, flowers slightly larger. *Eastern C Europe*. H2. Summer–autumn.

*\*C. corsicum* Baker. Illustration: Rix & Phillips. The bulb book, 175 (1981). Leaves 3 or 4, shorter and broader, stigmas shortly decurrent. *Corsica*. H2. Summer–autumn.

**15. C. micranthum** Boissier. Illustration: Grey-Wilson & Mathew, Bulbs, pl. 10 (1981); Mathew & Baytop, The bulbous plants of Turkey, f. 75 (1984).
Corm 2.5–3 × 1.3–2.7 cm, almost spherical to ovoid; tunics reddish brown, membranous, neck 1–4 cm but often only slightly developed. Leaves 2 or 3, developing after flowering, almost erect, linear to very narrowly linear-oblanceolate, 10–20 cm × 3–10 mm, hairless. Flowers 1 or 2, narrowly bell-shaped or funnel-shaped. Perianth-lobes white to pale purplish pink, narrowly oblanceolate, oblanceolate or narrowly linear-elliptic, 1.8–4 cm × 3–11 mm. Filaments 6–10 mm, hairless; anthers yellow, 3–5 mm, dorsifixed; pollen yellow. Styles straight, shortly curved and slightly swollen at the apex, stigmas decurrent for 1 mm. Capsule narrowly ellipsoid to ovoid, 1.8–2.5 cm. *NW Turkey*. H4. Autumn.

**16. C. umbrosum** Steven (*C. arenarium* Waldstein & Kitaibel var. *umbrosum* Ker Gawler). Illustration: Darnell, Hardy and half-hardy plants 1: 139 (1930); Flora SSSR 4: t.2 f. 8 (1935); Mathew & Baytop, The bulbous plants of Turkey, f. 80 (1984).
Corm 1.5–3 × 1.2–2.5 cm, ovoid to almost spherical; tunics dark blackish brown, usually leathery, usually with a persistent neck to 7 cm. Leaves 3–5, developing after flowering, more or less erect, strap-shaped

to very narrowly lanceolate, 8–17 × 1–2.7 cm, hairless. Flowers 1–6, funnel-shaped at first, star-shaped when fully open. Perianth-lobes white to purplish pink, narrowly oblanceolate to linear-elliptic, 1.5–3 cm × 2–6 mm. Filaments 5–8 mm, hairless; anthers yellow, 3–4 mm, dorsifixed; pollen yellow. Styles straight or slightly curved and swollen at the apex, stigmas decurrent for 0.5–0.75 mm. Capsule oblong-ellipsoid, 2–3 cm. *Romania, USSR (Crimea), N Turkey*. H3. Late summer–autumn.

**17. C. parlatoris** Orphanides. Illustration: Rix & Phillips, The bulb book, 184 (1981).
Corm 2–4 × 1.5–2 cm, ovoid; tunics dark brown, leathery, with a neck 3–4 cm. Leaves 4–10, usually developing after flowering, linear to narrowly linear, 7–10 cm × 1–4 mm, hairless. Flowers 1 or 2, narrowly bell-shaped. Perianth-lobes purplish pink, narrowly elliptic, 8–50 × 4–12 mm. Filaments 1–1.2 cm, hairless; anthers yellow, 4–6 mm, dorsifixed; pollen yellow. Styles straight, stigmas point-like. Capsule ovoid-oblong, c. 1 cm. *S Greece*. H4. Late summer–autumn.

**18. C. troodii** Kotschy (*C. decaisnei* Boissier). Illustration: Botanical Magazine, 6901 (1886); Megaw, Wild flowers of Cyprus, pl. 37 (1973).
Corm 3–6 × 2.5–4 cm, ovoid; tunics dark blackish brown, papery to almost leathery, apex produced into a persistent neck to 9 cm. Leaves 3–8, developing after flowering, erect-spreading, strap-shaped, 10–30 × 1.2–4.5 cm, hairless, ciliate or thinly hairy. Flowers 2–12, narrowly funnel-shaped to star-shaped. Perianth-lobes white to pale purplish pink, narrowly oblong-lanceolate, 2.8–4.5 cm × 4–11 mm. Filaments 1.5–2 cm, hairless; anthers yellow, 6–8 mm, dorsifixed, with membranous longitudinal margins; pollen yellow. Styles erect or slightly curved at the apex, stigmas point-like or obliquely decurrent for 0.5 mm. Capsule ellipsoid, 1.5–3 cm. *Cyprus, Turkey, W Syria; ? Israel, Lebanon*. H4. Autumn–winter.

**19. C. balansae** Planchon (*C. candidum* Boissier). Illustration: Mathew & Baytop, The bulbous plants of Turkey, f. 67 (1984).
Corm 4–5 × 2–3 cm, ovoid; tunics dull, dark blackish brown, membranous to almost leathery, apex produced into a very long, thick, persistent, strongly fibrous neck to 25 cm. Leaves 4 or 5, developing after flowering, erect-spreading, elliptic to oblong-elliptic (the outer) to strap-shaped

(the inner), 18–24 × 4–7.5 cm. Flowers 3–11, funnel-shaped. Perianth-lobes white to purplish pink, narrowly oblong-elliptic, 4.5–7.5 cm × 4–13 mm. Filaments 1–1.9 cm, hairless; anthers yellow, 1–1.7 cm, dorsifixed, with membranous longitudinal margins; pollen yellow. Styles curved at the apex, stigmas decurrent for 1.5–3 mm. Capsule oblong-ovoid, 3–3.5 cm. *Turkey, Greece (Rhodes)*. H4. Late summer–autumn.

**20. C. kotschyi** Boissier (*C. candidum* Boissier var. *hirtiflorum* Boissier; *C. imperatoris-frederici* Siehe; *? C. persicum* Baker). Illustration: Botanical Magazine, n.s., 520 (1967); Wendelbo, Tulips and irises of Iran, 13, t. 2 (1977); Mathew & Baytop, The bulbous plants of Turkey, f. 73 (1984).
Corm ovoid, 3–5 × 2–3 cm; tunics dark brown, membranous to almost leathery, apex produced into a persistent neck 3–14 cm. Leaves 3–5, developing after flowering, erect-spreading, elliptic-lanceolate to oblong-elliptic, 10–16 × 3–5 cm, hairless, margins sightly wavy. Flowers 3–12, funnel-shaped. Perianth-lobes white to purplish pink, narrowly oblanceolate to oblong-elliptic, 2.3–5.5 cm × 4–12 mm, downy along the bases of the ridges of the filament-channels. Filaments 7–12 mm, hairless; anthers yellow, 6–12 mm, dorsifixed, with membranous longitudinal margins; pollen yellow. Style curved at the apex, stigmas decurrent for 2–4 mm. Capsule oblong-ovoid, 3–4.5 cm. *Turkey, Iran, Iraq*. H3. Late summer–autumn.

**21. C. cilicicum** (Boissier) Dammer (*C. byzantinum* Ker Gawler var. *cilicicum* Boissier; *C. balansae* Planchon var. *macrophyllum* Siehe; *C. decaisnei* misapplied; *C. byzantinum* misapplied). Illustration: Botanical Magazine, 9135 (1928); Bowles, Crocus and Colchicum, opposite 156 (1952); Rix & Phillips, The bulb book, 175 (1981); Mathew & Baytop, The bulbous plants of Turkey, f. 72 (1984).
Corm 4–6 × 3–4.5 cm, almost spherical to ovoid; tunics dull, dark brown, papery to almost leathery, apex produced into a persistent neck 5–17 cm. Leaves 4 or 5, developed after flowering, more or less erect, very narrowly elliptic to narrowly elliptic-lanceolate, 30–40 × 4–11.5 cm, hairless. Flowers 3–25, funnel-shaped to bell-shaped. Perianth-lobes pale lilac-purple to deep rosy purple, occasionally lightly tessellated, oblanceolate to elliptic, 4–7.5 × 1.2–2.5 cm, downy along the

ridges of the filament-channels. Filaments 2.5–3.5 cm; anthers bright yellow, 6–10 mm, dorsifixed; pollen bright yellow. Styles straight or slightly curved at the apex, often with a dull crimson tip, very frequently projecting from the perianth, stigmas point-like or decurrent for not more than 0.5 mm. Capsule ellipsoid to obovoid, 3–4 cm. *Turkey, Syria, Lebanon*. H2. Autumn.

A deeper-coloured clone is offered as 'Purpureum'.

**\*C. byzantinum** Ker Gawler. Illustration: Botanical Magazine, 1122 (1808). Probably a hybrid derived from *C. cilicicum*. The foliage does not appear until the spring, unlike that of *C. cilicicum* in which the leaves begin growth soon after flowering; the leaves are also broader and more strongly ribbed than those of *C. cilicicum*.

**22. C. autumnale** Linnaeus. Illustration: Grey-Wilson & Mathew, Bulbs, pl. 9 (1981); Rix & Phillips, The bulb book 178 (1981). Corm 2.5–6 × 2–4 cm, ovoid to almost spherical; tunics dark brown, membranous to leathery, neck 2–4 cm. Leaves 3–5, occasionally more, developing after flowering, linear-lanceolate to broadly lanceolate, erect or erect-spreading, 14–35 × 1–7 cm, hairless. Flowers 1–6, narrowly bell-shaped to bell-shaped. Perianth-lobes purplish pink, occasionally tessellated, to white, narrowly elliptic to oblong-elliptic, 4–6 × 1–1.5 cm. Filaments 1–1.6 cm, hairless; anthers yellow, 5–8 mm, dorsifixed; pollen yellow. Styles curved at the apex, stigmas decurrent for 3–5 mm. Capsule oblong-ovoid, 2–6 cm. *S, W & C Europe to USSR (Belorussiya & Ukraine)*. H1. Late summer–autumn.

Very variable; several closely related species are cultivated:

**\*C. tenorii** Parlatore. Flowers slightly tessellated, anthers yellow, crooks of the stigmas purplish. *Italy*.

**\*C. lusitanicum** Brotero. Flowers lightly tessellated, anthers pale or deep purplish black. *SW Europe, N Africa*.

**\*C. longiflorum** Castagne (*C. neapolitanum* Tenore). Leaves shorter, linear-lanceolate; flowers slightly smaller. *S Europe*.

**\*C. parnassicum** Boissier. Corm-tunics membranous, leaves arched. *Greece*.

Plants cultivated as *C. laetum*, with numerous star-shaped, pale purplish pink flowers are akin to *C. byzantinum* (above). True *C. laetum* Steven is similar to *C. autumnale* but has 1–3 smaller flowers and shortly decurrent stigmas.

Cultivars of *C. autumnale* available include the double white 'Alboplenum' and the double lilac-pink 'Pleniflorum'.

**23. C. speciosum** Steven. Illustration: Botanical Magazine, 6078 (1874); Wendelbo, Tulips and irises of Iran, 11, t. 1 (1977); Mathew & Baytop, The bulbous plants of Turkey, f. 78 (1984). Corm 5–8 × 2.5–4 cm, oblong-ovoid; tunics dull mid-brown, papery to almost leathery, apex produced into a persistent neck to 12 cm. Leaves 3–5, developing after flowering, more or less erect, narrowly elliptic to oblong-lanceolate, 18–25 × 5.5–9.5 cm, hairless. Flowers 1–3, narrowly bell-shaped to bell-shaped. Perianth-tube green or white flushed with purple; lobes pale to deep rosy purple, sometimes with a white throat, occasionally white throughout, oblanceolate to oblong-oblanceolate or elliptic, 4.5–8 × 1–2.7 cm, downy along the ridges of the filament-channels. Filaments 1–1.8 cm, hairless; anthers orange-yellow, 1–1.2 cm, dorsifixed; pollen deep yellow. Style curved but not or only very slightly swollen at the apex; stigmas decurrent for 2–4 mm. Capsule ellipsoid, 4–5 cm. *N Turkey, Iran, USSR (Caucasus)*. H2. Autumn.

Variants of *C. speciosum* with green perianth tubes and conspicuously white-throated flowers and yellow anthers are frequently, but incorrectly grown as *C. bornmuelleri*, which has purple-brown anthers. White variants of *C. speciosum* occur scattered through populations of purple-flowered plants in the wild, and are available in cultivation. The plant grown under the name *C. giganteum* Arnott (*C. illyricum superbum* invalid) is very closely related to *C. speciosum* but has broadly funnel-shaped, not bell-shaped flowers.

**24. C. bornmuelleri** Freyn. Illustration: Gardeners' Chronicle 52: 157 (1912). Corm 2.5–4.5 × 2.5–4 cm, ovoid to almost spherical; tunics dull mid-brown, papery to membranous, apex produced into a neck 1–8 cm. Leaves 3 or 4, developing after flowering, more or less erect, very narrowly elliptic, 17–25 × 2.6–4.5 cm, hairless. Flowers 1–6, bell-shaped, sometimes narrowly so. Perianth-lobes rosy purple (throat usually white), oblanceolate to narrowly elliptic, 4.5–7 × 1.1–2.4 cm, downy along the ridges of the filament-channels. Filaments 1.3–2.8 cm, hairless; anthers purple or purple-brown, 8–12 mm, dorsifixed; pollen yellow. Styles slightly

curved, stigmatic surface 0.5–1.5 mm, confined to the swollen apex of the style. Capsule narrowly ovoid or narrowly ellipsoid, 4–5 cm. *Turkey*. Late summer–autumn.

**25. C. variegatum** Linnaeus (*C. parkinsonii* Hooker; *C. variegatum* var. *desii* Pampanini; *C. agrippinum* misapplied). Illustration: Grey-Wilson & Mathew, Bulbs, pl. 9 (1981); Mathew & Baytop, The bulbous plants of Turkey, f. 81 (1984). Corm 2–5 × 2–3.5 cm, ovoid to almost spherical; tunics dark brown, more or less leathery, apex produced into a persistent neck 5–15 cm. Leaves 3 or 4, developing after flowering, linear-lanceolate or strap-shaped, 9–15 cm × 7–25 mm, hairless, margins cartilaginous, wavy. Flowers 1 or 2. Perianth-lobes spreading, deep red- or violet-purple, occasionally paler or white at the base, strongly tessellated, lanceolate to elliptic or oblanceolate, 2–2.7 cm × 5–25 mm, frequently slightly twisted near the obtuse or acute apex, hairless. Filaments 1.5–4 cm, hairless; anthers purplish black or purplish brown, 4–9 mm, dorsifixed; pollen yellow. Styles straight, occasionally slightly curved and swollen at the apex, stigmas decurrent for 1.5–2 mm. Capsule oblong-ovoid, 2 cm. *Greece, SW Turkey*. H4. Autumn.

**\*C. agrippinum** Baker is probably a hybrid between *C. variegatum* and *C. autumnale*. It resembles *C. variegatum* closely, but is distinguished by its upright flowers and the narrower-based, less tessellated and wavy perianth-lobes. It increases quickly and is a very easy garden plant.

**26. C. macrophyllum** Burtt (*C. latifolium* misapplied; ? *C. latifolium* Sibthorp & Smith var. *longistylum* Pampanini). Illustration: Goulimis, Wild flowers of Greece, 143 (1968); Rix & Phillips, The bulb book, 177 (1981); Grey-Wilson & Mathew, Bulbs, pl. 9 (1981); Mathew & Baytop, The bulbous plants of Turkey, f. 76 (1984). Corm 5–7 × 4–6 cm, ovoid to almost spherical; tunics dull, dark brown, leathery, apex produced into a persistent, somewhat fibrous neck to 24 cm. Leaves 3 or 4, developing after flowering, erect-spreading, ovate to elliptic-ovate, 24–42 × 11–15.5 cm, almost acute to bluntly acuminate, strongly pleated, hairless. Flowers 1–5. Perianth-lobes spreading, lilac-purple to rosy purple, tessellated, often paler or white in the throat, elliptic or oblong-elliptic, 4.5–7 × 1.5–3 cm, hairless. Filaments

2–2.5 cm, hairless; anthers purple, 8–12 mm, dorsifixed; pollen grey-green, drying paler. Styles shortly curved at the apex, stigmas decurrent for 1.5–2.5 mm. Capsule ovoid, 4–5 cm. *Greece (Crete, Rhodes), SW Turkey*. H3. Autumn.

**27. C. bivonae** Gussone (*C. latifolium* Sibthorp & Smith; *C. visianii* Parlatore; *C. bowlesianum* Burtt; *C. sibthorpii* misapplied). Illustration: Goulimis, Wild flowers of Greece, 141 (1968); Polunin, Flowers of Greece and the Balkans, pl. 55 (1980); Rix & Phillips, The bulb book, 180 (1981).
Corms 2.5–6 × 2.5–4 cm, ovoid to almost spherical; tunics mid-brown, papery to almost leathery, apex produced into a short neck 1–2 cm. Leaves 4–11, developing after flowering, more or less erect, strap-shaped or linear-lanceolate, 12–30 × 1–4.5 cm, hairless. Flowers 1–6, bell-shaped, sometimes broadly so. Perianth-lobes rosy purple, strongly tessellated, sometimes white at the base, narrowly to broadly elliptic, occasionally obovate-elliptic, 4–8.5 cm × 8–35 mm, downy along the ridges of the filament-channels. Filaments 1–2.5 cm, hairless; anthers purplish black or purplish brown, 7–12 × 2–3 mm, dorsifixed; pollen bright yellow. Styles curved and slightly swollen at the apex, stigmas decurrent for 2–4 mm. Capsule oblong-ellipsoid, 3.5–4 cm. *S Europe from Corsica and Sardinia to W Turkey*. H3. Late summer–autumn.

A parent of many garden hybrids with *C. speciosum*, *C. bornmuelleri* and *C. autumnale*.

**28. C. lingulatum** Boissier (*C. sibthorpii* misapplied). Illustration: Grey-Wilson & Mathew, Bulbs, pl. 10 (1981); Mathew & Baytop, The bulbous plants of Turkey, f. 74 (1984).
Corm 3–6 × 2–3.5 cm, ovoid to almost spherical; tunics blackish brown, more or less leathery, apex produced into a persistent neck to 8 cm. Leaves 3–7, developing after flowering, spreading, oblong to strap-shaped, hairless, margins cartilaginous and usually wavy. Flowers funnel-shaped to bell-shaped. Perianth-lobes pale to deep purplish pink, sometimes lightly tessellated, narrowly oblanceolate to narrowly oblong-elliptic, 2.5–5 cm × 2–12 mm, hairless. Filaments 6–14 mm, hairless; anthers yellow, 4–8 mm, dorsifixed; pollen yellow. Styles curved at the apex, stigmas decurrent for 2–3 mm. Capsule oblong-ovoid, 1.5–2.5 cm. *Greece, NW Turkey*. H4. Late summer–autumn.

**\*C. chalcedonicum** Aznavour. Foliage similar, flowers deep rosy purple, tessellated. *NW Turkey*.

There are records of *C. lingulatum* from Greece with slightly tessellated flowers, so the two species may not be distinct.

**29. C. turcicum** Janka. Illustration: Polunin, Flowers of Greece and the Balkans, pl. 55 (1980); Grey-Wilson & Mathew, Bulbs, pl. 10 (1981).
Corm 3–5 × 2.5–4 cm, almost spherical to ovoid; tunics dark blackish brown, leathery, with an extended, persistent neck to 11 cm. Leaves 5–9, developing after flowering, more or less erect and often twisted at the apex, strap-shaped to narrowly lanceolate, 9–19 × 1–3.5 cm, somewhat glaucous, margins thinly cartilaginous, ciliate (sometimes obscurely so). Flowers 1–8, bell-shaped to funnel-shaped. Perianth-lobes red-purple, occasionally paler, sometimes lightly tessellated, elliptic to narrowly obovate, 3–6 cm × 3.5–13 mm, downy along the ridges of the filament-channels. Filaments 1–2.5 cm, hairless; anthers 5–8 mm, yellow (?sometimes purplish), dorsifixed; pollen yellow. Styles straight, curved at the apex, stigmas decurrent for 3–4 mm. Capsule ovoid to oblong-ovoid, 2–3 cm. *Balkans, NW Turkey*. H3. Late summer–autumn.

The closely related *C. atropurpureum* Stearn, with small, dark magenta-red flowers is occasionally cultivated.

**46. BULBOCODIUM** Linnaeus
*J.C.M. Alexander*
Perennials with corms. Leaves linear to linear-lanceolate, obtuse. Perianth of 6 free segments, each divided into a blade and a long, narrow claw; claws held together by teeth or auricles at base of blade, thus forming a tube. Stamens 6, with slender filaments borne at base of blade. Ovary superior, but below ground, 3-celled, containing many ovules. Style 3-fid above, undivided below. Seeds spherical.

A genus of 2 species from S & E Europe, sometimes considered part of *Colchicum* but differing in the undivided lower style, and distinctly clawed perianth-segments. Cultivation conditions are similar to those for *Colchicum* (see p. 162).

**1. B. vernum** Linnaeus (*Colchicum bulbocodium* Ker Gawler; *C. vernum* (Linnaeus) Ker Gawler). Illustration: Botanical Magazine, 153 (1791); Polunin, Flowers of Europe, t. 163 (1969); Polunin & Smythies, Flowers of SW Europe, 389

(1973); Rix & Phillips, The bulb book 111 (1981).
Corm oblong, to 3 cm across; scales almost black. Leaves 3 or 4, appearing with the flowers but reaching full size later, to 15 × 1.5 cm, with sheathing base and spreading blade. Flowers solitary (rarely 2 or 3), borne close to the ground, 4–8.5 cm, pinkish purple, rarely white. Blade linear-lanceolate, acute, toothed at base. Capsule 1.5–2 cm. *Pyrenees, SW & SC Alps*. H3. Winter–spring.

**47. MERENDERA** Ramond
*C.D. Brickell*
Perennial, stemless herbs. Corms oblong-ovoid, occasionally narrow and more or less horizontal, enclosed by a membranous or leathery tunic which is usually extended into a short neck. Developing leaves and inflorescences enclosed within a membranous, cylindric sheath. Leaves basal, partially developed at flowering or developing after flowering. Flowers purple to pink or white, solitary or in clusters. Perianth-segments free, each with a long, narrow claw and a broader, linear or narrowly elliptic to narrowly obovate blade, with or without 2 auricles at the base. Stamens borne close to the base of the blade. Anthers versatile or basifixed, pollen yellow. Styles 3, free to the base; stigmas point-like. Ovary subterranean. Capsule opening along the septa with 3 abruptly pointed flaps, maturing at or just above ground-level by the elongation of the flower-stalk. Seeds numerous, spherical or almost so.

A genus of about 10 species extending from S Europe, N Africa and W Asia to Afghanistan, with outliers in Israel, Jordan and Ethiopia. It is very closely related to *Colchicum* (and included in it by some authors), but differs in the free perianth-segments. *M. montana* may be grown successfully in the open in a well-drained, sunny site. The remaining species are best cultivated as bulb-frame or alpine house plants in sharply drained, but fairly rich compost.
Literature: Stefanoff, B., Monographie der Gattung Colchicum, *Sbornik Bulgariskata Akademiya na Naukita* 22: 1–100 (1926).

1a. Anthers basifixed                                              2
  b. Anthers versatile                                             4
2a. Leaves partially developed at
     flowering; anthers greenish yellow;
     flowering in spring           **5. robusta\***
  b. Leaves developing after flowering;
     anthers yellow; flowering in autumn
                                                                   3

3a. Leaves 3 or 4, 4–10 mm wide;
anthers usually 8–12 mm

            **1. montana**

  b. Leaves 5–15, 1–3 mm wide; anthers
usually 5–8 mm     **2. filifolia**

4a. Plant with narrow, more or less
horizontal corms    **3. sobolifera**

  b. Plant with erect corms       5

5a. Leaves 3, lanceolate to narrowly
lanceolate, to 17 × 3 cm at maturity;
corm-tunics brown, membranous;
anthers yellow      **4. kurdica**

  b. Leaves 2–6, linear to linear-
lanceolate, to 18 cm × 3–8 (rarely to
18) mm; corm-tunics black or
blackish brown, leathery; anthers
black, purple or brown       6

6a. Leaves 2–4, hairless; flowers 1–3;
blade of perianth-segment
oblanceolate to narrowly
oblanceolate, 2–3 cm × 4–9 mm,
auricles usually present    **6. trigyna**

  b. Leaves 2–6, ciliate or with margins
roughened towards the base; flowers
2–6; blade of perianth-segment linear
to narrowly elliptic, 1.5–2.7 cm ×
2–5 mm, auricles usually absent

            **7. attica***

**1. M. montana** Lange (*M. bulbocodium*
Ramond; *M. pyrenaica* (Pourret) Fournet).
Illustration: Rix & Phillips, The bulb book,
176f (1981); Grey-Wilson & Mathew,
Bulbs, pl. 11 (1981).
Corm oblong-ovoid, 2–3 × 1.5–3 cm,
tunics black or dark brown, leathery, neck
1–2 cm. Leaves 3 or 4, developing after
flowering or as the flowers fade, narrowly
linear, to 22 cm × 4–10 mm at maturity.
Flowers 1 or 2, purple to red-purple. Blade
of perianth-segment narrowly elliptic to
narrowly oblong-oblanceolate,
2–6.5 cm × 3–11 mm. Filaments 4–6 mm;
anthers yellow, 5.5–12 mm, basifixed.
Capsule oblong-ellipsoid, 1–2.5 cm. *Iberian
peninsula, C Pyrenees.* H2. Late
summer–autumn.

**2. M. filifolia** Cambessedes. Illustration:
Mathew, Dwarf bulbs, f. 38 (1973).
Corm oblong-ovoid, 1.5–2 × 1–2 cm,
tunics black or dark brown, leathery, neck
1–5 cm. Leaves 5–15, developing after
flowering or as the flowers fade, thread-like,
to 15 cm × 1–3 mm at maturity. Flowers
solitary, purple to red-purple. Blade of
perianth-segment narrowly elliptic to
narrowly oblong-elliptic, 2–4 cm ×
2–8 mm. Filaments 4–7 mm; anthers
yellow, 5–10 mm, basifixed. Capsule
oblong-ellipsoid, 8–12 mm. *SW Europe,
N Africa.* H5. Autumn.

**3. M. sobolifera** Meyer (*Colchicum
soboliferum* (Meyer) Stefanoff). Illustration:
Botanical Magazine, 9576 (1939); Polunin,
Flowers of Greece and the Balkans, 497, f.
5 (1980).
Corms narrow, horizontal, 2–8 cm long,
tunics brown, membranous. Leaves 3,
linear to strap-shaped, 2–10 cm at
flowering, 10–21 cm × 3–13 mm at
maturity. Flowers 1 or 2, white, pink or
pale purple. Blade of perianth-segment
linear to narrowly elliptic,
2–4 cm × 2.5–8 mm, frequently with basal,
usually linear auricles. Filaments to 7 mm;
anthers blackish violet, 1.5–3.5 cm,
versatile. Capsule oblong-ovoid, 1.5–2 cm.
*E Europe to Iran & USSR (Turkestan).* H3.
Spring–early summer.

**4. M. kurdica** Bornmüller (*Colchicum
kurdicum* (Bornmüller) Stefanoff).
Corm oblong-ovoid, 2–4 × 1.5–2 cm,
tunics brown, membranous, neck 1–3 cm.
Leaves 3, lanceolate to narrowly lanceolate,
4–12 cm at flowering, to 17 × 3 cm at
maturity. Flowers 1 or 2, pale to deep
purple. Blade of perianth-segment
oblanceolate to narrowly oblanceolate,
3–3.5 cm × 5–10 mm, with auricles at the
base. Filaments 8–9 mm; anthers yellow,
3–4 mm, versatile. Capsule ellipsoid,
2–3 cm. *SE Turkey, N Iraq.* H5. Spring.

**5. M. robusta** Bunge (*M. persica* Boissier &
Kotschy; *M. aitchisonii* J.D. Hooker).
Illustration: Botanical Magazine, 6012
(1873).
Corm oblong-ovoid, 3–6 × 2–3 cm, tunics
dark brown, leathery, neck 1–3 cm. Leaves
3–6, partially developed at flowering, linear
to linear-lanceolate, to 5 cm at flowering,
to 25 × 2.5 cm at maturity. Flowers 1–4,
deep pink to white. Blade of perianth-
segment oblong-elliptic to linear-elliptic,
1.8–4 cm × 2–9 mm. Filaments 8–10 mm;
anthers greenish yellow, 8–12 mm,
basifixed. Capsule oblong-ovoid, 1–3 cm.
*Iran, Afghanistan, USSR (Turkmeniya) &
N India.* H4. Spring.

  ***M. hissarica** Regel. Illustration: Rix &
Phillips, The bulb book, 32 (1981). Differs
in having basifixed anthers, solitary white
flowers and 2 leaves per corm. *USSR (Soviet
Central Asia), Afghanistan.* H4. Spring.

**6. M. trigyna** (Adam) Stapf (*Bulbocodium
trigynum* Adam not Janka; *M. caucasica*
Bieberstein; *Colchicum caucasicum*
(Bieberstein) Sprengel; *M. raddeana* Regel;
*M. eichleri* Boissier; *M. manissadjianii*
Aznavour; *M. navis-noae* Markgraf).
Illustration: Botanical Magazine, 3690

(1839); Rix & Phillips, The bulb book, 32
(1981); Mathew & Baytop, The bulbous
plants of Turkey, f. 99 (1984).
Corm oblong-ovoid, 1.5–3.5 × 1–2 cm,
tunics black or blackish brown, leathery,
neck 2–3 cm. Leaves 2–4, linear to linear-
lanceolate, 1–4 cm at flowering, to
17 cm × 3–18 mm at maturity, acute,
green, hairless. Flowers 1–3, white to
purplish pink. Blade of perianth-segment
oblanceolate to narrowly oblanceolate,
2–3 cm × 4–9 mm, base usually with
auricles. Filaments 1.5–4 mm, anthers
black, purplish black, greenish black or
olive brown, 2–4 mm, versatile. Capsule
ellipsoid to ovoid, 1.5–3 cm. *Turkey, Iran,
USSR (Caucasus).* H2. Spring–early
summer.

**7. M. attica** (Tommasini) Boissier &
Spruner (*Colchicum atticum* Tommasini;
*M. rhodopea* Velenovsky; *M. brandigiana*
Markgraf). Illustration: Stojanoff &
Stefanoff, Flora Bulgarica, 220, f. 243
(1925); Grey-Wilson & Mathew, Bulbs, pl.
11 (1981).
Corm oblong-ovoid, 1.5–3 × 1–2 cm,
tunics black or blackish brown, leathery,
neck 1–2 cm. Leaves 2–6, linear to linear-
lanceolate, 3–5 cm at flowering, to
18 cm × 3–8 mm at maturity. Flowers 2–6,
white to purplish pink. Blade of perianth-
segment linear to narrowly elliptic,
1.5–2.7 cm × 2–5 mm, base not usually
with auricles. Filaments to 8 mm, anthers
black, deep purplish black or greenish
black, 2–4 mm, versatile. Capsule ellipsoid
to ovoid, 1.5–2.5 cm. *Balkan peninsula,
Turkey.* H2. Late winter–spring.

  ***M. androcymbioides** Valdes. Resembles
*M. attica* but anthers yellow, basifixed,
leaves slightly broader. *SW Spain.* H5.

## 48. GLORIOSA Linnaeus
*V.A. Matthews*

Rootstock a tuberous rhizome. Stems often
climbing, to 2.5 m. Leaves alternate,
opposite or whorled, stalkless, often with a
tendril at the tip. Flowers borne in the axils
of the upper leaves, pendulous. Perianth-
segments free, reflexed, yellow to red or
purplish or variously bicolored, margins
wavy or not. Stamens spreading, shorter
than the perianth. Ovary superior, 3-celled.
Style bent at a right angle at the base, with
3 branches. Fruit a capsule. All parts of the
plant are poisonous.

  A genus of 1 species, although in the
past about 30 species have been
recognised. Recent work by D.V. Field
maintains that the various forms cannot be

distinguished in the wild, and they are best regarded as forms of G. superba. He recommends that they should be given cultivar names for horticultural purposes but so far this has not been done. The main variation is in the height of the stem, possession of leaf tendrils, colour and width of perianth-segments and their degree of crisping or waving, and the width of the leaves. Flower colour is further complicated because it may change as the flower ages.

Gloriosa is generally grown in a cool greenhouse, although, in warm areas, plants may be put out-of-doors during the summer. Any well-drained soil is suitable. Propagation is by offsets, division of the rhizome, or by seed.
Literature: Field, D.V., The genus Gloriosa, Lilies and other Liliaceae 1973: 93–5 (1972).

**1. G. superba** Linnaeus (G. simplex Linnaeus; G. virescens Sims; G. carsonii Baker; G. abyssinica Richard; G. minor Rendle; G. lutea Anon; G. rothschildiana O'Brien; G. verschuurii Hoog). Illustration: Journal of the Royal Horticultural Society **94**: pl. 1 (1969); Perry, Flowers of the world, 176 (1972); Hay et al., Dictionary of indoor plants in colour, pl. 258 (1974); Herklots, Flowering tropical climbers, 127 (1976).
Tropical Africa, tropical Asia. G1. Summer–autumn.

**49. LITTONIA** Hooker
*D.A. Webb*
Perennial herbaceous climbers with tubers. Leaves stalkless, often whorled below, usually opposite or alternate above, with the apex drawn out into a tendril. Flowers nodding, single in the leaf-axils, or sometimes with the stalk arising beside or just below the leaf. Perianth bell-shaped, the lobes not opening widely but free almost to the base. Anthers basifixed, opening outwards. Ovary superior; style simple, straight, continuing the axis of the flower. Stigma of 3 linear lobes. Fruit a many-seeded capsule.

A genus of about 8 species, ranging from South Africa to Senegal and Arabia, best grown in a peaty loam and dried off during the dormant season. A conservatory plant in all but the hottest regions of Europe.

**1. L. modesta** Hooker. Illustration: Botanical Magazine, 4723 (1853); Jeppe, Natal wild flowers, pl. 3 (1975).
Tuber 3 cm across. Stems slender, unbranched, to 1.2 m. Lower leaves ovate-lanceolate, tapering from a broad base to

the tendril; upper leaves usually narrower. Flower-stalks short. Perianth orange-yellow, the lobes lanceolate, acuminate. South Africa (Transvaal, Natal, Orange Free State). H5–G1. Early summer.

Var. **keitii** Leichtlin, with a branched stem and more numerous flowers, makes a more decorative plant than var. *modesta*.

**50. SANDERSONIA** Hooker
*V.A. Matthews*
Rootstock a tuberous rhizome. Stem up to 75 cm, bearing scattered stalkless leaves, up to 10 cm and usually with a tendril at the tip. Flowers solitary in the axils of the upper leaves, pendulous on 2–3 cm stalks. Perianth 2–2.5 cm, orange, spherical-urn-shaped, the perianth-lobes fused except at the tips which curl outwards, the base of the perianth with 6 short nectar-bearing spurs. Stamens c. one-third as long as perianth, borne at the base of the lobes; anthers opening by slits. Ovary superior. Style with 3 branches. Fruit a capsule.

A genus with 1 species, native to South Africa. Cultivation as for Gloriosa (see p. 169).

**1. S. aurantiaca** Hooker. Illustration: Botanical Magazine, 4716 (1853); Flore des Serres, ser. 1, **9**: t. 862 (1853–4); Jeppe, Natal wild flowers, pl. 3 (1975); Mathew, The larger bulbs, 127 (1978).
SE South Africa. G2–H5. Summer.

**51. GAGEA** Salisbury
*E.M. Rix*
Perennial herbs with small bulbs which are sometimes closely entwined with persistent, thickened roots. Basal leaves 1 or 2, arising from the bulb, linear or linear-lanceolate, hollow, solid or flat. Flowers usually yellow, rarely white, either in an umbel subtended by an opposite pair of leaf-like spathes on an otherwise leafless stem, or in a few-flowered panicle with at least 1 leaf on the stem below the point of branching. Perianth-segments free, equal, without nectaries. Anthers basifixed. Ovary superior, style 1. Seeds flat or pear-shaped.

A genus of about 50 species, mainly found in C Asia. Most species have small, yellow, star-shaped flowers and are not considered worthy of cultivation. The Mediterranean species are grown in shallow, heavy soil, wet in spring, dry in the summer; the alpine and woodland species need cool, humus-rich soil.
Literature: Uphof, J.C.T., A review of the genus Gagea Salisb., *Plant Life* **14**: 124–32 (1958), **15**: 151–61 (1959), **16**: 163–76 (1960).

*Basal leaves.* Flat: 1,2,6; thread-like: 5; hollow: 3; solid, sometimes channelled: 2,4,5.
*Inflorescence.* An umbel: 1–4; a panicle: 5,6.
*Perianth:* Yellow: 1–5; white: 6.

1a. Flowers white       **6. graeca**
  b. Flowers yellow       2
2a. Inflorescence a panicle, with leaves on the stem below the first branch
           **5. peduncularis***
  b. Inflorescence an umbel with no stem-leaves below the point of branching
           3
3a. Basal leaves hollow    **3. fistulosa**
  b. Basal leaves flat or solid, sometimes channelled       4
4a. Basal leaves linear-lanceolate, flat, 7–15 mm wide       **1. lutea**
  b. Basal leaves linear, flat or channelled, 1–4 mm wide       5
5a. Basal leaf 1 per bulb    **2. pratensis**
  b. Basal leaves 2 per bulb    **4. villosa**

**1. G. lutea** (Linnaeus) Ker Gawler (G. silvatica (Persoon) Loudon. Illustration: Reichenbach, Icones Florae Germanicae et Helveticae **10**: 477, f. 1045 (1848); Ross-Craig, Drawings of British plants **29**: pl. 28 (1972).
New bulb formed within the old. Basal leaf 1 per bulb, linear-lanceolate, flat, 7–15 mm wide. Inflorescence an umbel of 1–7 flowers. Spathes ciliate. Flower-stalks hairy or not. Perianth-segments 1.5–1.8 cm, greenish yellow. Seeds pear-shaped. *Europe, from N England and Russia southwards to Spain, Greece and the Caucasus.* H1. Spring.

**2. G. pratensis** (Persoon) Dumortier (G. stenopetala Reichenbach). Illustration: Reichenbach, Icones Florae Germanicae et Helveticae **10**: 474, f. 1038 (1848).
New bulb formed below old, outside the tunic, sometimes with a second new bulb present. Basal leaf 1 per bulb, linear, flat or channelled, 2–4 mm wide. Inflorescence an umbel of 1–4 flowers. Spathes lanceolate, sometimes ciliate. Flower-stalks 1–2 cm, hairless. Perianth-segments 9–16 mm, to 2.2 cm in fruit, yellow inside, green outside. Seeds pear-shaped. *Europe, from Denmark and Finland southwards to Spain, Turkey and the Crimea.* H1. Spring.

**3. G. fistulosa** Ker Gawler (G. liotardii (Sternberg) Schultes & Schultes; G. anisanthos Koch). Illustration: Grey-Wilson & Mathew, Bulbs, pl. 16 (1981).
New bulb formed beside old. Basal leaves 1, or 2 and unequal, per bulb, hollow,

6–20 × 3–5 mm when flattened. Inflorescence an umbel of 1–6 flowers, or replaced by bulbils. Spathes unequal, lanceolate. Flower-stalks 1.5–5 cm, usually hairy. Perianth-segments yellow, 1–1.8 cm, curling and elongating after flowering. Seeds pear-shaped. *S Europe, Greece, Turkey & Iran*. H1. Spring.

**4. G. villosa** (Reichenbach) Duby (*G. arvensis* (Persoon) Dumortier). Illustration: Reichenbach, Icones Florae Germanicae et Helveticae **10**: 479, f. 1049 (1848).

New bulb formed beside old. Basal leaves 2 per bulb, linear, solid, D-shaped in section, 1–2.5 mm wide. Inflorescence more or less umbellate with 1–15 flowers. Spathes without bulbils in their axils. Flower-stalks usually hairy. Perianth-segments yellow, 7–10 mm, to 1.8 cm in fruit. Seeds pear-shaped. *Europe, from the Netherlands and Finland southwards to Spain, N Africa and Iran*. H1. Spring.

This species is a common weed of cultivation in the Mediterranean area; bunches of very small bulbils are often formed at ground-level, but not at the bases of the flower-stalks, as in the rather similar *G. granatelli* Parlatore.

**5. G. peduncularis** (Presl & Presl) Pascher. Illustration: Rix & Phillips, The bulb book, 44, 45 (1981).

New bulb formed beside old. Basal leaves 2 per bulb, linear, solid, 1–2 mm wide, exceeding the inflorescence. Inflorescence a panicle with 1–7 flowers. Stem-leaf 1, lanceolate, acuminate, 2–5 mm wide. Flower-stalks to 6.5 cm, usually 3–4 cm, densely woolly above, elongating during flowering. Perianth-segments yellow, 8–16 mm, to 1.8 cm in fruit. Seeds pear-shaped. *S Europe, N Africa & Turkey*. H2. Spring.

One of the most showy species very common in the eastern Mediterranean area.

**\*G. bohemica** (Zauschner) Schultes & Schultes (*G. saxatilis* (Mertens & Koch) Schultes & Schultes). Illustration: Reichenbach, Icones Florae Germanicae et Helveticae **10**: 480, f. 1052 (1848). Basal leaves thread-like, flower-stalks not elongating during flowering. *From C England and Germany south to Portugal & Israel*. H2. Spring.

**6. G. graeca** (Linnaeus) Terraciano (*Lloydia graeca* (Linnaeus) Kunth). Illustration: Polunin & Huxley, Flowers of the Mediterranean, pl. 239 (1965).

New bulb formed within the old; bulbs

usually crowded together, surrounded with thickened roots. Basal leaves 2 per bulb, linear, flat, 1–2 mm wide, hairless. Inflorescence a panicle, usually with 1–5 flowers. Stem-leaves usually 3, alternate. Flowers pendent in bud, never opening flat. Perianth-segments 7–16 mm, oblanceolate, obtuse, white, with 3 purplish veins. Seeds flat. *Greece, Turkey*. H3. Spring.

## 52. LLOYDIA Reichenbach
*V.A. Matthews*

Perennial hairless herbs possessing bulbs with a brown, fibrous membranous tunic. Leaves basal and on the stem, linear. Flowers 1 or 2 at the top of the stem. Perianth-segments free, each usually with a gland at the base. Stamens borne at the base of the perianth-segments; anthers basifixed. Ovary superior, 3-celled. Fruit a capsule with 3 ribs.

About 12 species native to the temperate northern hemisphere. They grow best in a well-drained peaty soil in a shady position. Propagation is by seed.

**1. L. serotina** (Linnaeus) Reichenbach. Illustration: Polunin, Flowers of Europe, pl. 164 (1969); Morley & Everard, Wild flowers of the world, pl. 7 (1970); Fitter et al., Wild flowers of Britain and northern Europe, 261 (1974); Grey-Wilson & Mathew, Bulbs, pl. 16 (1981).

Stem 5–15 cm. Leaves 7–20 cm, basal usually 2, usually exceeding stem, stem-leaves 2–5. Flowers 1–1.5 cm, erect, white with reddish or purplish veins. *Temperate northern hemisphere*. H2. Late spring–summer.

**\*L. longiscapa** Hooker. Illustration: Hooker's Icones Plantarum **9**: pl. 834 (1852). Flowers larger, usually pendulous, each perianth-segment with a brownish purple or orange blotch at the base. *Himalaya, W China*. H2. Summer.

## 53. CALOCHORTUS Pursh
*S.J.M. Droop*

Perennial herbs with bulbs. Bulbs ovoid, with membranous or fibrous coats. Stems leafy or sometimes scape-like, often branched, frequently bearing bulbils. Leaves usually linear, the basal one large and conspicuous at flowering, or withering before flowering. Inflorescence cymose or more or less umbellate. Flowers conspicuous, often marked with colour contrasts. Perianth of 2 distinct whorls. Sepals usually lanceolate, hairless. Petals obovate to oblanceolate, usually more or less hairy on the inner surface and with a depression or gland near the base. Stamens

6, anthers oblong to linear, attached to the bases of the perianth-segments, the filaments broadened at the base. Ovary superior, 3-celled with numerous ovules in each cell, triangular or 3-winged in section. Fruit spherical to linear, 3-angled or 3-winged in section, erect or drooping, containing many seeds.

A genus of about 57 species from western N & C America, mainly from California. Probably about half the species have been in cultivation from time to time, but only a few are commonly grown. Cultivation is quite difficult in most of Europe, as the plants are easily killed by excessive moisture. They may be grown in a cool house or frame, in deep rich soil in a sunny position. Propagation is by seed, bulblets or bulbils.

Literature: Ownbey, M., A monograph of the genus Calochortus, *Annals of the Missouri Botanical Garden* **27**: 371–560 (1940).

*Bulb-coats*. Membranous: 1–19; thickly fibrous: 20,21.
*Stem*. Usually bearing bulbils: 9,12–16; sometimes bearing bulbils: 11,17–19, 21; never bearing bulbils: 1–8,10,20.
*Basal leaf*. Large and conspicuous at flowering: 1–10, 21; withered before flowering: 11–20.
*Stem leaves*. Tightly curled under at the tip: 16.
*Flowers*. Spherical to narrowly bell-shaped: 1–4; bell-shaped: 5–21. Drooping: 1–4,21; erect or spreading: 2,5–21.
*Sepals*. Usually exceeding petals: 3,16; usually about equalling the petals: 5,20,21; usually shorter than petals: 1,2,4,6–15,17–19. Tightly curled under at the tip: 12–15. Sparsely hairy on the lower surface: 21.
*Petals*. Fringed: 1–8,20,21; not fringed: 9–21. Hairs on the upper surface club-shaped: 5,17,18. Hairs on upper surface branched and glandular: 19.
*Gland* (see figures 31 & 32, p. 173 & 175). Traversed by several membranes: 1,2; densely covered with simple or branched hair-like processes: 11–19; naked or very nearly so: 3–11,20,21; naked but covered from below by a single, fringed membrane: 5–10; not associated with any membrane: 3,4,11–15,19–21; completely surrounded by a ring of hairs: 19–21; completely surrounded by a membrane: 19,20.
*Anthers*. Usually longer than filaments: 7,8,16,17,19; usually shorter than filaments: 1,3,4,6,9,11,15,21; about

equalling filaments: 2,5,10, 12–14,18,20.

*Fruit.* Drooping: 1–9; erect: 10–21. Spherical to oblong, 3-winged: 1–10; linear to linear-oblong, 3-angled: 11–21.

1a. Fruit more or less spherical to oblong, 3-winged, usually less than 3 times longer than broad and drooping; inflorescence more or less umbellate  2
b. Fruit linear to linear-oblong, 3-angled, more than 3 times as long as broad, erect; inflorescence cymose  11
2a. Flowers spherical to narrowly bell-shaped, usually drooping  3
b. Flowers openly bell-shaped, erect or spreading  6
3a. Flowers white to deep purplish pink; gland with several associated membranes  4
b. Flowers yellow; gland not associated with membranes  5
4a. Flowers white or tinged with pink, always drooping; lowest gland membrane less than two-thirds of the width of the petal at the level of the gland  **1. albus**
b. Flowers deep purplish pink, sometimes erect or spreading; lowest gland membrane as wide as the petal at gland level  **2. amoenus**
5a. Petals 1.6–2 cm, inner face hairless or with a few thick hairs near the gland, margin densely fringed, the fringe 1–1.5 mm deep  **3. amabilis**
b. Petals 2.5–3.3 cm, inner face hairy to the tip, margin less densely fringed, fringe 1–2 mm deep  **4. pulchellus**
6a. Flowers yellow; hairs on surface of petal club-shaped  **5. monophyllus**
b. Flowers white to greenish, bluish or pale purple; hairs on surface of petal, if any, not club-shaped  7
7a. Fruit erect; petals hairless except for a single gland membrane and occasionally a few hairs near it  **10. nudus**
b. Fruit nodding; petals with hairs associated with the gland and also near it  8
8a. Stem barely reaching above the surface of the ground, with many bulbils  **9. uniflorus**
b. Stem more than 3 cm, without bulbils  9
9a. Stem usually branched, if unbranched then with a single stem leaf; anthers shorter than filaments  **6. tolmiei**
b. Stem unbranched and without leaves; anthers slightly longer than filaments  10

10a. Anthers oblong, acute; petals 8–12 mm  **7. coeruleus**
b. Anthers lanceolate, tapering to a fine point; petals 1.2–2.5 cm  **8. elegans**
11a. Gland densely covered with short, thick, simple or branched processes, or, if naked, then not surrounded by a ring of hairs or a membrane; bulb-coats membranous  12
b. Gland naked or nearly so, but surrounded by a ring of hairs which are joined at the base or not; bulb-coats thickly fibrous-netted  21
12a. Gland not surrounded by a fringed membrane, or, if the outermost hairs are more or less joined at the base to form a membrane, then hairs on the petals gland-tipped  13
b. Gland obviously surrounded by a fringed membrane; hairs on petals sometimes club-shaped but never gland-tipped  18
13a. Petal hairs gland-tipped; anthers longer than filaments  **19. gunnisonii**
b. Petal hairs not gland-tipped; anthers equalling or shorter than filaments  14
14a. Gland broader than long, not isodiametric  15
b. Gland more or less isodiametric, circular or diamond-shaped  17
15a. Gland more or less oblong, slightly arched above; petals yellow or yellow with reddish brown markings  **15. luteus**
b. Gland distinctly arched above, more or less linear; petals with a reddish brown marking surrounded by yellow, never uniformly yellow  16
16a. Gland like an inverted V, without a notch at the apex; petals with a reddish brown mark surrounded by a bright yellow zone  **13. superbus**
b. Gland more or less doubly crescent-shaped; petals with a pale yellow zone  **14. vestae**
17a. Gland circular, less than one gland diameter from base of petal, sometimes naked; inflorescence distinctly cymose  **11. splendens**
b. Gland more or less circular or diamond-shaped, usually more than 2 gland diameters from the base of the petal, never naked; inflorescence umbellate  **12. venustus**
18a. Gland sagittate at base; petals acute to acuminate, purple, each with a

median longitudinal green stripe; leaves tightly curled under at the tip  **16. macrocarpus**
b. Combination of characters not as above  19
19a. Flowers vermilion to orange  **17. kennedyi**
b. Flowers yellow  20
20a. Stem 10–50 cm; petals without zig-zag marking above the gland; anthers usually longer than filament  **17. kennedyi**
b. Stem 50–100 cm; petals with zig-zag marking above the gland; anthers about as long as filaments  **18. clavatus**
21a. Flowers drooping, usually yellow, often purplish; basal leaf conspicuous at flowering; hairs on petals purple; anthers oblong, much shorter than filaments  **21. barbatus**
b. Flowers erect, orange-yellow; basal leaf usually withering before flowering; hairs on petals yellow; anthers linear, about as long as filaments  **20. weedii**

Section **Calochortus**. Bulb-coats membranous. Stem usually without bulbils. Basal leaf conspicuous at flowering, tapering at both ends. Flowers spherical to openly bell-shaped. Ovary 3-winged, contracted to a short style and persistent, 3-fid stigma. Fruit spherical or oblong, 3-winged, drooping (erect in *C. nudus*).

**1. C. albus** Bentham. Figure 31(4), p. 173. Illustration: Rickett, Wild flowers of the United States **5**: t. 5 (1971).
Stem stout, often branched, 20–80 cm. Basal leaf 30–70 cm, present at flowering. Flowers white to pink, spherical to spherical-bell-shaped, drooping. Sepals 1–1.5 cm, ovate to lanceolate, acuminate. Petals 1.5–3 cm, elliptic-obovate, acute to obtuse, fringed, sparsely covered above the gland with long, slender hairs. Gland depressed, traversed by *c.* 5 deeply fringed membranes, extending for one-third to two-thirds of the width of the petal at gland level. Anthers oblong, obtuse or acute, usually shorter than filaments. *USA (California).* H4. Spring–summer.

**2. C. amoenus** Greene (*C. albus* Bentham var. *amoenus* (Greene) Purdy & Bailey). Illustration: Rickett, Wild flowers of the United States **5**: t. 5 (1971).
Like *C. albus* but stems slender, to 50 cm; basal leaf 20–50 cm; flowers dark purplish pink, narrowly bell-shaped to spherical-bell-shaped, erect or drooping; sepals lanceolate;

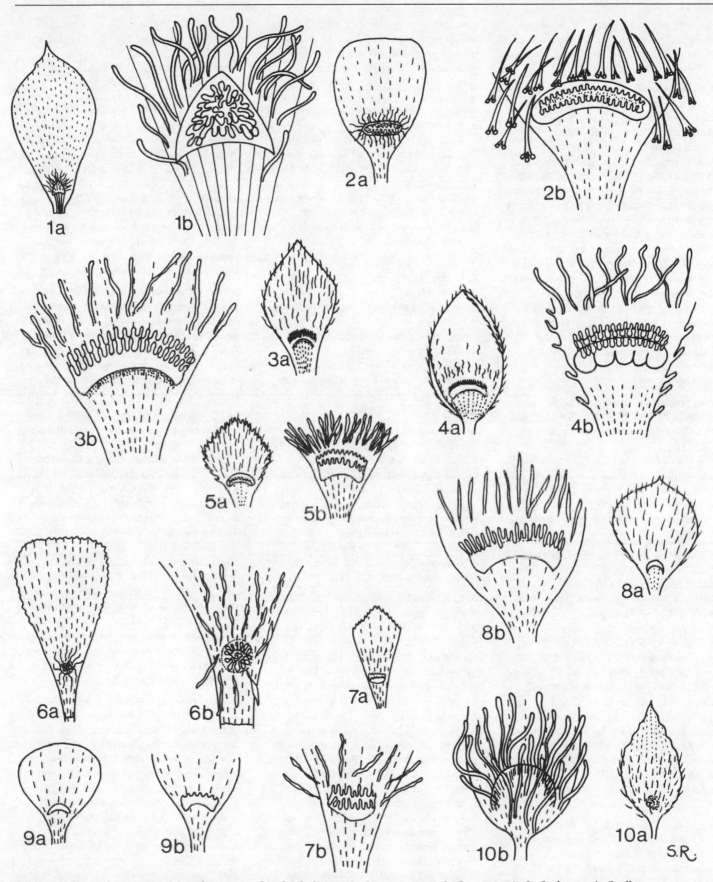

**Figure 31.** Perianth-segments (a) and nectaries (b) of *Calochortus*. 1, *C. macrocarpus*. 2, *C. gunnisonii*. 3, *C. elegans*. 4, *C. albus*. 5, *C. coeruleus*. 6, *C. kennedyi*. 7, *C. uniflorus*. 8, *C. tolmiei*. 9, *C. nudus*. 10, *C. barbatus*.

gland broad, arched, slightly depressed, extending right across the base of the petal; anthers obtuse, about as long as filaments. *USA (California)*. H5. Late spring.

**3. C. amabilis** Purdy (*C. pulchellus* Bentham var. *amabilis* (Purdy) Jepson; *C. pulchellus* misapplied). Figure 32(17), p. 175. Illustration: Rickett, Wild flowers of the United States **5**: t. 4 (1971).
Stem erect, 10–30 cm, often branched. Basal leaf 20–50 cm, exceeding the stem. Flowers deep yellow, spherical or spherical-bell-shaped, drooping. Sepals more or less spreading, usually exceeding the petals, ovate to lanceolate, acute to acuminate. Petals 1.6–2 cm, diamond-shaped, clawed, apex sharply rounded, densely fringed to 1–1.5 mm, the inner face hairless or with a few thick hairs near the gland. Gland deeply depressed, arched and bounded above by a transverse line of hairs 2–3 mm long, pointing into the centre of the depression. Anthers oblong, obtuse, usually shorter than the filaments. *USA (California)*. H5. Late spring.

**4. C. pulchellus** Bentham (*Cyclobothra pullchella* Anon.). Figure 32(14), p. 175. Illustration: Gardeners' Chronicle, 765 (1883).
Like *C. amabilis* but sepals not usually exceeding petals, petals 2.5–3.3 cm, less deep yellow and hairy within almost to the tip; margin more deeply but less densely fringed. *USA (California)*. H5. Late spring.

**5. C. monophyllus** (Lindley) Lemaire (*C. benthamii* Baker). Figure 32(15), p. 175. Illustration: Rickett, Wild flowers of the United States **5**: t. 4 (1971).
Stem erect, 8–20 cm, usually branched. Basal leaf 10–30 cm, greatly exceeding the stem. Flowers yellow, openly bell-shaped, usually erect. Sepals 1.6–2 cm, elliptic, acuminate, about as long as petals. Petals obovate, densely covered above the gland with club-shaped hairs, and sparsely fringed with similar hairs. Gland naked, spreading across the entire base of the petal, bounded below by a deeply fringed membrane and above by a few short hairs, hairs and membrane densely and shortly papillose. Anthers lanceolate, abruptly pointed, about as long as filaments. *USA (California)*. H5. Late spring.

**6. C. tolmiei** Hooker & Arnott (*C. elegans* J.D. Hooker not Pursh; *C. maweanus* Baker; *C. coeruleus* (Kellogg) Watson var. *maweanus* (Baker) Jepson). Figure 31(8), p. 173 & 32(21), p. 175. Illustration: The Garden **87**: 474 (1923).

Stem erect, 5–30 cm, usually branched, with 1 stem leaf. Basal leaf 10–40 cm, usually exceeding the stem. Inflorescence more or less umbellate. Flowers 1–10, openly bell-shaped, erect or spreading, white or cream tinged with pink or purple. Sepals 8–20 mm, oblong-lanceolate, acute to acuminate. Petals 1.2–2.5 cm, obovate, hairy above the gland, sparsely fringed. Gland transverse, arched above, naked, bounded below by an irregularly toothed to fringed membrane, and above by a band of short, thick processes, these and the fringes of the membrane papillose. Anthers lanceolate, acute to abruptly pointed, shorter than filaments. *USA (Oregon & California)*. H5. Late spring–early summer.

**7. C. coeruleus** (Kellogg) Watson. Figures 31(5) & 32(16), p. 173 & 175. Illustration: Rickett, Wild flowers of the United States **5**: t. 5 (1971).
Like *C. tolmiei* but stem unbranched, scape-like, 3–15 cm; flowers bluish, sepals 8–10 mm, petals 8–12 mm, broadly obovate; anthers oblong, acute, usually slightly longer than filaments. *USA (California)*. H5. Summer.

**8. C. elegans** Pursh. Figure 31(3), p. 173. Illustration: Rickett, Wild flowers of the United States **5**: t. 5 (1971).
Like *C. tolmiei* but stem unbranched and scape-like, flowers greenish white, often with a purple crescent on each segment; gland membrane deeply fringed; anthers finely pointed, usually longer than filaments. *USA (C Idaho and adjacent Washington & Oregon)*. H5. Late spring–early summer.

**9. C. uniflorus** Hooker & Arnott (*C. lilacinus* Kellogg). Figure 31(7), p. 173. Illustration: Rickett, Wild flowers of the United States **5**: t. 5 (1971).
Stem short, simple, barely reaching above ground surface, with many bulbils. Inflorescence more or less umbellate with 1–5 flowers. Basal leaf 10–40 cm, much exceeding the inflorescence. Flowers pale purple, openly bell-shaped, erect on very long stalks. Sepals 1.2–1.6 cm, elliptic-lanceolate, acuminate. Petals 1.5–2.8 cm, rounded and irregularly toothed above, with a few hairs near the gland. Gland naked, acute below, truncate above, bounded below by a broad, ascending, fringed membrane and above by a band of slender processes. Anthers oblong, obtuse or acute, shorter than the filaments. *USA (Oregon & California)*. H5. Late spring.

**10. C. nudus** Watson. Figure 31(9), p. 173. Illustration: Rickett, Wild flowers of the United States **5**: t. 6 (1971).
Stem erect, scape-like, 5–15 cm. Basal leaf 10–20 cm, usually shorter than inflorescence. Flowers 1–6, openly bell-shaped, erect, white to pale purple. Sepals 1–1.6 cm, elliptic-lanceolate, acuminate. Petals 1.4–2 cm, rounded or acute and irregularly finely toothed, not fringed, hairless except occasionally for very few hairs near the gland. Gland naked, transverse, shallowly arched above, bounded below by an ascending, fringed membrane, the fringe minutely papillose. Anthers linear-oblong, obtuse to acute, about as long as filaments, pale blue at first. *USA (California)*. H5. Summer.

Section **Mariposa** Wood. Bulb-coats membranous. Stems often with bulbils in the axils of the stem leaves. Leaves linear, tapering, the basal one withering before flowering. Flowers bell-shaped, erect. Ovary linear, tapering to a persistent, 3-fid stigma. Fruit linear-oblong to linear, 3-angled, erect.

**11. C. splendens** Bentham. Figure 32(18), p. 175. Illustration: Rickett, Wild flowers of the United States **5**: t. 6 (1971).
Stem erect, branched, rarely with bulbils, 20–60 cm. Inflorescence cymose. Flowers 1–4, erect, bell-shaped, pale purple often with a darker purple spot on each segment. Sepals 2–2.5 cm, ovate-lanceolate, acuminate. Petals 2.5–5 cm, obovate, rounded and irregularly finely toothed above, sparsely covered with silky hairs from above the gland to about the middle. Gland small, circular, very near the base of the petal, naked or more usually covered with thick, branched processes. Anthers linear-oblong, obtuse to pointed, blue to purple, shorter than filaments. *USA (California), Mexico (Baja California)*. H5. Late spring–early summer.

**12. C. venustus** Bentham. Figure 32(12), p. 175. Illustration: Rickett, Wild flowers of the United States **5**: t. 6 (1971).
Stem erect, usually branched, with bulbils, 20–60 cm. Inflorescence more or less umbellate, flowers 1–3, bell-shaped, erect, white to yellow, purple or dark red, each petal with a conspicuous, median, dark red mark, often with a second, paler mark above the first. Sepals 2–3.5 cm, linear-lanceolate, acuminate, curled under at the tip. Petals 2–4.5 cm, obovate, rounded above, sparsely hairy near the gland with long, slender hairs. Gland more or less

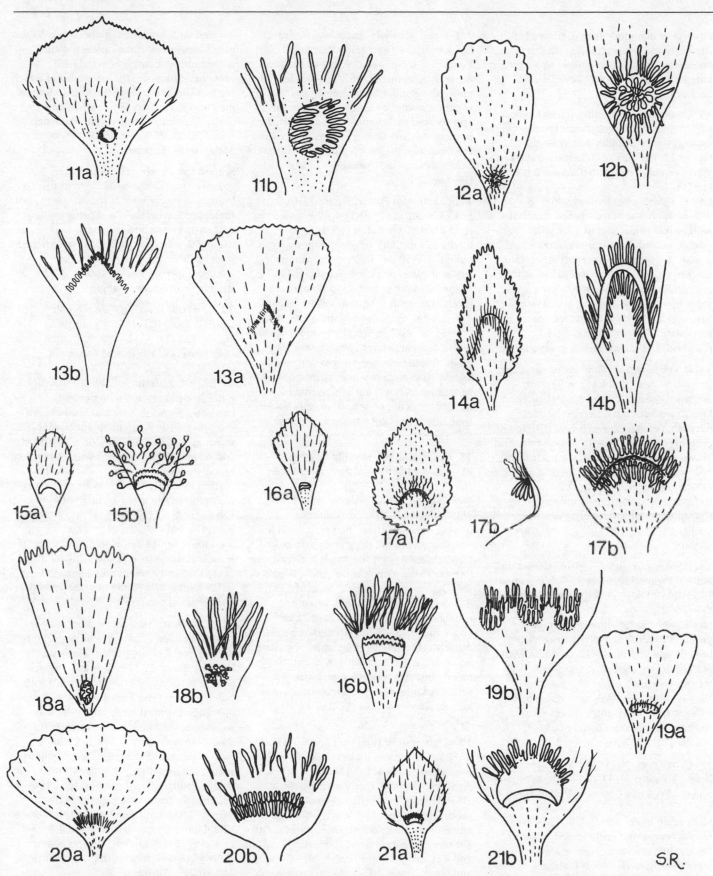

**Figure 32.** Perianth-segments (a) and nectaries (b) of *Calochortus* (continued). 11, *C. weedii*. 12, *C. venustus*. 13, *C. superbus*. 14, *C. pulchellus*. 15, *C. monophyllus*. 16, *C. coeruleus*. 17, *C. amabilis* (nectary in side and face view). 18, *C. splendens*. 19, *C. vestae*. 20, *C. luteus*. 21, *C. tolmiei*.

circular or diamond-shaped, covered with simple, hair-like appendages. Anthers linear-oblong, acute to obtuse, about as long as filaments. *USA (California)*. H5. Summer.

**13. C. superbus** Howell (*C. luteus* Lindley var. *citrinus* Watson; *C. luteus* Lindley var. *oculatus* Purdy & Bailey not Watson). Figure 32(13), p. 175. Illustration: Rickett, Wild flowers of the United States **5**: t. 6 (1971).
Like *C. venustus* but flowers white to yellowish or pale purple, petals streaked with purple below and usually with a median reddish brown mark surrounded by a zone of bright yellow; sepals lanceolate, tapering, usually not curled under at the tip; petals rounded to acute, with a few short hairs near the gland; gland more or less inverted V-shaped; anthers linear-lanceolate to linear-oblong. *USA (California)*. H5. Late spring–early summer.

**14. C. vestae** Purdy (*C. venustus* var. *vestae* (Purdy) Wilson; *C. luteus* Lindley var. *oculatus* Watson). Figure 32(19), p. 175.
Like *C. venustus* but flowers white to purplish, petals streaked below with red or purplish and usually with a median reddish brown mark surrounded by a zone of pale yellow; sepals lanceolate, tapering, usually not curled under at the tip; gland transverse, more or less doubly crescent-shaped, covered with short, thick processes. *USA (California)*. H5. Late spring–early summer.

**15. C. luteus** Lindley (*C. citrinus* Anon. not Baker). Figure 32(20), p. 175. Illustration: Rickett, Wild flowers of the United States **5**: t. 5 (1971).
Like *C. venustus* but flowers 1–4, dark yellow, the petals usually with reddish brown radiating lines and often also with a reddish brown median mark; sepals sometimes curled under at tip. Gland transverse, more or less crescent-shaped, covered with short, thick processes. Anthers shorter than filaments. *USA (California)*. H5. Summer.

**16. C. macrocarpus** Douglas (*C. cyaneus* Nelson). Figure 31(1), p. 173. Illustration: Rickett, Wild flowers of the United States **6**: t. 4 (1973).
Stem erect, often with bulbils, 20–50 cm. Leaves becoming tightly curled under at the tip. Inflorescence more or less umbellate, flowers 1–3, bell-shaped, erect, purple, each petal with a median longitudinal green stripe and sometimes a dark purple band above the gland. Sepals

3.5–6 cm, narrowly lanceolate, tapering, hairless, usually exceeding the petals. Petals 3.5–6 cm, oblanceolate or broader, acute to acuminate with a few slender hairs above the gland. Gland more or less sagittate, bound by a fringed membrane and covered with slender, usually branched process. Anthers linear, obtuse, longer than the filaments. *Western N America (British Columbia south to Nevada and California)*. H4. Summer.

**17. C. kennedyi** Porter. Figure 31(6), p. 173. Illustration: Rickett, Wild flowers of the United States **4**: t. 4 (1970).
Stem erect, usually simple, rarely with bulbils, 10–50 cm. Inflorescence more or less umbellate, flowers 1–6, erect, bell-shaped, usually vermilion to orange, more rarely (var. **munzii** Jepson) yellow. Sepals 1.5–2.5 cm, ovate-lanceolate, acute. Petals 2.5–5 cm, obovate, usually rounded above, with a few club-shaped hairs near the gland. Gland circular, covered with simple or branched processes and surrounded by a fringed membrane. Anthers narrowly lanceolate, acute or obtuse, usually longer than filaments. *USA (Arizona, California)*. H5. Summer.

**18. C. clavatus** Watson. Illustration: Rickett, Wild flowers of the United States **5**: t. 4 (1971).
Stem zig-zag, simple or branched, rarely with bulbils, 50–100 cm. Inflorescence more or less umbellate, flowers 1–6, erect, bell-shaped yellow, sometimes with red-brown markings on the sepals and a zig-zag line above the gland on the petals. Sepals 2.5–3.5 cm, ovate-lanceolate, acute. Petals 3.5–5 cm, broadly obovate, usually rounded above, with club-shaped hairs above the gland. Gland circular, depressed, densely covered with short, much-branched processes, and surrounded by a fringed membrane. Anthers oblong, obtuse or acute, reddish or purplish brown, about as long as filaments. *USA (California)*. H5. Summer.

**19. C. gunnisonii** Watson. Figure 31(2), p. 173. Illustration: Rickett, Wild flowers of the United States **4**: t. 3 (1970).
Stem erect, simple, rarely with bulbils, 20–60 cm. Inflorescence more or less umbellate, flowers 1–3, erect, bell-shaped, white to purple, greenish within, often with a transverse purple band above the gland and a purple spot on the claw of each petal and sepal. Sepals 2.5–3 cm, lanceolate, acute. Petals 3–4 cm, obovate, obtuse and rounded above, near the gland densely

covered with branched, gland-tipped hairs; gland transverse, more or less oblong, arched above, densely covered with branched processes, the outermost of which are sometimes joined at the base to form a membrane. Anthers lanceolate, acute to abruptly pointed, longer than filaments. *USA (Dakota, Montana, south to New Mexico)*. H5. Summer.

Section **Cyclobothra** (D. Don) Baker. Bulb-coats thickly fibrous-netted. Stems usually branched, rarely with bulbils in the axils of the upper leaves; basal leaf conspicuous, withering before flowering or not. Flowers usually 2, narrowly to broadly bell-shaped, erect or drooping. Ovary linear, not winged, tapering to a persistent, 3-fid stigma. Fruit linear, 3-angled, erect.

**20. C. weedii** Wood (*C. luteus* Lindley var. *weedii* (Wood) Baker; *C. citrinus* Baker). Figure 32 (11), p. 175. Illustration: Rickett, Wild flowers of the United States **4**: t. 2 (1970).
Stem erect, without bulbils, 30–60 cm. Basal leaf usually withering before flowering. Flowers 2, broadly bell-shaped, erect. Sepals 2–3 cm, ovate-lanceolate, tapering. Petals 2.5–3 cm, orange-yellow, flecked and often margined with red-brown, broadly obovate, rounded to acute, the margin minutely toothed to fringed, hairy over most of the inner surface with long, yellow hairs. Gland circular, nearly naked, bound by a dense ring of hairs which are sometimes united below to form a more or less continuous membrane. Anthers linear, acute to shortly acuminate, about as long as filaments. *USA (California), Mexico (Baja California)*. H5. Summer.

**21. C. barbatus** (Humboldt, Bonpland & Kunth) Painter (*C. flavus* Schultes). Figure 31(10), p. 173. Illustration: Mathew, Dwarf bulbs, t. 17 (1973) – as C. luteus.
Stem erect, usually branched, rarely with bulbils, 30–60 cm. Flowers usually 2, openly bell-shaped, drooping, yellow, or sometimes the petals and frequently the sepals purplish. Sepals 1–2 cm, elliptic-lanceolate, acute or obtuse, sparsely hairy within, often with a gland like those on the petals. Petals 1.5–2 cm, obovate, obtuse or acuminate, entire or fringed laterally, densely hairy with tapering, purple hairs. Gland more or less circular, naked, bound by a ring of small hairs. Anthers oblong, shortly pointed, shorter than the filaments. *Mexico*. H5. Autumn.

## 54. ERYTHRONIUM Linnaeus
*V.A. Matthews & J.M. Lees*

Perennial herbs with membranous-coated corms. Leaves usually 2, basal or almost so, often mottled. Flowers nodding, solitary or several in a raceme, rarely in an umbel whose stem is below ground-level. Perianth-segments free, usually recurved, one or both whorls usually with 2 or 4 inflated appendages near the base. Stamens shorter than the perianth-segments; anthers basifixed. Stigma entire to 3-lobed. Ovary superior. Fruit a 3-celled capsule.

A genus of about 25 species from temperate N America with 1 species in Eurasia. The plants grow best in semi-shade in a soil which has plenty of peat or leaf-mould and does not dry out in summer. They are especially useful for planting beneath trees. Literature: Applegate, E.I., The genus Erythronium: a taxonomic and distributional study of the western north American species, *Madroño* 3: 58–113 (1935); Parks, C.R. & Hardin, J.W., Yellow Erythroniums of the eastern United States, *Brittonia* 15: 245–59 (1963).

*Leaves.* Mottled: 1,4–12; not mottled: 2,3,11. *Perianth-segments.* White, cream or pale yellow: 1–5,8–11; golden yellow: 2,3,12; pink to purple: 1,6–8,11. Inner segments with basal appendages: 1–3, 5–12; inner segments lacking basal appendages: 4,7,11.
*Anthers.* White: 3–5,8–11; yellow: 1–3, 7,8,11,12; dark red: 3; blue, purple or brown: 1,6,12.
*Stigma.* Entire or almost so: 2,4–7; lobed: 1,3,7–12.

1a. Leaves mottled   2
  b. Leaves not mottled   13
2a. Flowers golden yellow
      **12. americanum**
  b. Flowers white, cream, pale yellow, pink, lavender or purple   3
3a. Anthers blue, purple or brown   4
  b. Anthers white or yellow   5
4a. Stigma deeply lobed   **1. dens-canis**
  b. Stigma entire or almost so
      **6. hendersonii**
5a. Flowers pink   6
  b. Flowers white, cream or pale yellow   8
6a. Flowers pale pink, tips of perianth-segments twisted   **8. oregonum**
  b. Flowers bright or rose pink, tips of perianth-segments not twisted   7
7a. Perianth-segments with a band of brown or purplish markings towards the base   **1. dens-canis**
  b. Perianth-segments lacking a brown or purplish band   **7. revolutum***

8a. Stigma entire or almost so   9
  b. Stigma 3-lobed   10
9a. Inner perianth-segments with basal appendages   **5. citrinum**
  b. Inner perianth-segments lacking basal appendages   **4. howellii**
10a. Anthers white, filaments not broadened at the base   11
  b. Anthers yellow, or if white then filaments broadened at the base   12
11a. Leaves 10–15 cm; flower with a greenish yellow centre which is surrounded by a band of yellow or brown; stigma with short, erect or spreading lobes   **9. californicum**
  b. Leaves 3.5–10 cm; flower with a greenish yellow centre which is not surrounded by a band of yellow or brown; stigma with long, thread-like, recurved lobes   **10. multiscapoideum**
12a. Filaments broadened at the base; perianth-segments with basal appendages   **8. oregonum**
  b. Filaments not broadened at the base; perianth-segments lacking basal appendages, or if appendages present then stigma lobes erect to spreading and perianth-segments not twisted at the tips   **11. albidum***
13a. Inner perianth-segments lacking basal appendages   **11. albidum***
  b. Inner perianth-segments with basal appendages   14
14a. Stigma entire or almost so   **2. tuolumnense***
  b. Stigma deeply 3-lobed   **3. grandiflorum***

**1. E. dens-canis** Linnaeus. Illustration: Synge, Collins guide to bulbs, pl. 11, f. 9 (1961); Lippert, Fotoatlas der Alpenblumen, f. 12 (1981); Grey-Wilson & Mathew, Bulbs, pl. 4 (1981); Rix & Phillips, The bulb book, 100, 101 (1981).
Leaves 10–15 cm, oblong, mottled. Flower-stem 15–30 cm, bearing a solitary flower. Perianth-segments *c.* 2.5 cm, rose-pink to purple, sometimes white, with a pale yellow or brownish base inside, above which is a band of brown or purplish markings. Inner segments with basal appendages. Anthers blue to purple. Stigma deeply lobed. *Eurasia.* H2. Spring.

There are several named cultivars based on differences in flower colour, e.g. 'Album' and 'Snowflake' with white flowers and 'Purpureum' with purple-violet flowers.

Var. **japonicum** (Decaisne) Baker (*E. japonicum* Decaisne). Illustration: Kitamura et al., Coloured illustrations of herbaceous plants of Japan (Monocotyledoneae) 3: pl. 31 (1978); Rix &

Phillips, The bulb book, 100 (1981). Perianth-segments 5–6 cm, violet-purple with a dark, often 3-lobed mark at the base inside. *Japan, Korea.*

Var. **sibiricum** Fischer & Meyer. Perianth-segments 2.5–5.5 cm; anthers yellow. *USSR (Siberia), Mongolia.*

**2. E. tuolumnense** Applegate. Illustration: Synge, Collins guide to bulbs, pl. 11, f. 7 (1961); Journal of the Royal Horticultural Society 89: f. 106 (1964); Rickett, Wild flowers of the United States 5: pl. 11 (1971); Rix & Phillips, The bulb book, 100 (1981).
Leaves 20–30 cm, lanceolate to broadly oblanceolate, yellow-green, not mottled. Flower-stem to 30 cm, bearing 1–several flowers. Perianth-segments *c.* 3 cm, golden yellow with a greenish base and 1 or 2 pairs of basal appendages. Anthers yellow. Stigma entire or almost so. *USA (C California).* H3. Spring.

The often-grown cultivars 'Pagoda' and 'Kondo' are hybrids between *E. tuolumnense* and a species with mottled leaves.

***E. klamathense** Applegate. Illustration: Rickett, Wild flowers of the United States 5: pl. 12 (1971). Flowers usually solitary, sometimes a few. Perianth-segments 2–2.5 cm, white or cream, yellow in the lower third. *USA (SW Oregon, NW California).* H3. Spring.

**3. E. grandiflorum** Pursh (*E. grandiflorum* subsp. *candidum* Piper, subsp. *chrysandrum* Applegate and var. *pallidum* St. John; *E. giganteum* misapplied). Illustration: Synge, Collins guide to bulbs, pl. 11, f. 4 (1961); Rickett, Wild flowers of the United States 5: pl. 11 (1971); Clark, Wild flowers of the Pacific northwest, 50 (1976); Rix & Phillips, The bulb book, 101 (1981).
Leaves 10–20 cm, oblong-elliptic, not mottled. Flower-stem 30–60 cm, bearing 1–several flowers. Perianth-segments 2–3.5 cm, golden yellow with a paler base and well-developed basal appendages. Anthers dark red, or sometimes yellow or white. Stigma with long, recurved lobes. *Western N America from British Columbia S to Oregon and Utah.* H3. Spring–summer.

Several cultivars are commercially available, such as 'Album' which has white flowers with yellow centres, and 'Robustum' which is a variant with taller stems and larger leaves.

***E. montanum** Watson. Illustration: Hitchcock et al., Vascular plants of the Pacific northwest 1: 789 (1969); Clark, Wild flowers of the Pacific northwest, 35

(1976); Rix & Phillips, The bulb book, 99 (1981). Leaf-blade abruptly narrowed at the base. Flowers white with a yellow or orange centre. Anthers always yellow. *Western N America (British Columbia, Oregon, Washington).* H3. Spring–summer.

**4. E. howellii** Watson. Illustration: Abrams, Illustrated flora of the Pacific states **1**: 430 (1923); Synge, Collins guide to bulbs, pl. 11, f. 10 (1961).
Leaves to 15 cm, lanceolate to lanceolate-oblong, mottled. Flower-stem to 20 cm, bearing 1–4 flowers. Perianth-segments *c.* 2.5 cm, white or pale yellow, with an orange base or yellow bars inside, lacking basal appendages. Anthers white. Stigma almost entire. *W USA (S Oregon, NW California).* H3. Spring.

**5. E. citrinum** Watson. Illustration: Synge, Collins guide to bulbs, pl. 11, f. 8 (1961); Neihaus & Ripper, Field guide to Pacific states wildflowers, 11 (1976).
Leaves 10–15 cm, lanceolate to oblong, mottled, margin crisped. Flower-stem to 30 cm, bearing 1–3 flowers. Perianth-segments 2.5–4 cm, creamy white to pale yellow with a greenish yellow base inside. Inner segments with basal appendages. Anthers white. Stigma entire or slightly lobed. *W USA (SW Oregon, NW California).* H3. Spring.

**6. E. hendersonii** Watson. Illustration: Botanical Magazine, 7017 (1888); Synge, Collins guide to bulbs, pl. 11, f. 6 (1961); Rickett, Wild flowers of the United States **5**: pl. 12 (1971); Rix & Phillips, The bulb book, 99 (1981).
Leaves 10–20 cm, lanceolate-oblong, mottled, margin crisped. Flower-stem to 30 cm, bearing 1–10 flowers. Perianth-segments 2.5–4 cm, lavender, with a purple base inside, above which is an indistinct yellowish band. Inner segments with basal appendages. Filaments purple, anthers brownish or bluish. Stigma entire or almost so. *W USA (SW Oregon, NW California).* H3. Spring–summer.

**7. E. revolutum** Smith (*E. revolutum* var. *smithii* (Hooker) Orcutt and var. *johnsonii* (Bolander) Purdy). Illustration: Synge, Collins guide to bulbs, pl. 11, f. 3 (1961); Rickett, Wild flowers of the United States **5**: pl. 12 (1971); Clark, Wild flowers of the Pacific northwest, 46 (1976); Rix & Phillips, The bulb book, 96–7 (1981).
Leaves 15–20 cm, lanceolate-ovate, mottled, margin crisped. Flower-stem 15–40 cm, bearing 1–several flowers. Perianth-segments 3.5–4.5 cm, rose-pink

with 1 or 2 yellow bands at the base inside, and with basal appendages. Filaments broadened at the base, anthers yellow. Stigma with recurved lobes. *Western N America from British Columbia to N California.* H3. Spring–summer.

The cultivar 'Pink Beauty' is thought to be either a robust variant of *E. revolutum* or a hybrid.

**\*E. propullans** Gray. Illustration: American Naturalist **5**: 299 (1871); Britton & Brown, Illustrated flora of the northern United States **1**: 421 (1896). Leaves 5–10 cm, flowers solitary, perianth-segments 1–1.4 cm, not or scarcely recurved, bright pink with a yellow base and no basal appendages. Stigma entire. *USA (Minnesota).* H3.

**8. E. oregonum** Applegate (*E. giganteum* misapplied). Illustration: Botanical Magazine, 5714 (1868); Synge, Collins guide to bulbs, pl. 11, f. 1 (1961); Rickett, Wild flowers of the United States **5**: pl. 11 (1971); Rix & Phillips, The bulb book, 98–9 (1981).
Leaves 12–15 cm, lanceolate to oblong-lanceolate, mottled. Flower-stem 15–30 cm, bearing 1–6 flowers. Perianth-segments 3.5–5 cm, twisted at the tips, white to pink, tinged with brown or reddish at the base outside, and with a yellow base inside, above which are zig-zag marks of orange-brown. Basal appendages present. Filaments broadened at the base, anthers yellow, or white (subsp. **leucandrum** (Applegate) Applegate). Stigma with long, recurved lobes. *Western N America from British Columbia to Oregon.* H3. Spring.

'White Beauty' which is frequently offered by nurserymen may be a variant of of *E. oregonum* or a hybrid.

**9. E. californicum** Purdy. Illustration: Abrams, Illustrated flora of the Pacific states **1**: 428 (1923); Rickett, Wild flowers of the United States **5**: pl. 11 (1971); Rix & Phillips, The bulb book, 98–9 (1981).
Leaves 10–15 cm, narrowly oblong to ovate-oblong, strongly mottled. Flower-stalk 10–35 cm, bearing 1–3 flowers. Perianth-segments 2.5–3.5 cm, white to cream with a greenish yellow base inside, above which is a band of yellow, orange or brown; basal appendages present. Anthers white. Stigma with short, erect or spreading lobes. *W USA (N California).* H3. Spring.

**10. E. multiscapoideum** (Kellogg) Nelson & Kellogg (*E. purdyi* Anon.; *E. hartwegii* Watson). Illustration: Botanical Magazine,

7583 (1898); Niehaus & Ripper, Field guide to Pacific states wildflowers, 11 (1976).
Leaves 3.5–10 cm, oblanceolate, mottled. Flower-stems to 20 cm, apparently bearing solitary flowers (the flowers are borne in an umbel whose stem is below ground-level). Perianth-segments 2.5–4 cm, white with a pale greenish yellow base inside, the inner segments with small basal appendages. Anthers white. Stigma with long, recurved lobes. *W USA (N California).* H3. Spring.

**11. E. albidum** Nuttall (*E. mesochoreum* Knerr). Illustration: Addisonia **18**: pl. 593 (1934); Rickett, Wild flowers of the United States **2**: pl. 10 (1967); Dean et al., Wildflowers of Alabama, 23 (1983).
Leaves to 15 cm, elliptic, green or mottled. Flower-stalk to 30 cm, bearing a solitary flower. Perianth-segments 2.5–5 cm, white with a pale bluish or pinkish tinge outside and a yellow base inside; basal appendages absent. Anthers yellow. Stigma with spreading lobes. *Eastern C USA and adjacent S Canada.* H3. Spring–summer.

**\*E. helenae** Applegate. Illustration: Synge, Collins guide to bulbs, pl. 11, f. 5 (1961); Botanical Magazine, n.s., 834 (1981); Rix & Phillips, The bulb book, 98 (1981). Leaves always strongly mottled. Flowers 1–4, white with a yellow centre, segments with basal appendages. *W USA (N California).* H3. Spring.

**\*E. purpurascens** Watson. Illustration: Niehaus & Ripper, Field guide to Pacific states wildflowers, 11 (1976). Leaves narrowly lanceolate to lanceolate, bright yellow-green. Flowers to 8, perianth-segments 1–1.5 cm, white or cream with a yellow base and turning pink or purplish with age. Anthers white. *W USA (California).* H3. Spring.

**12. E. americanum** Ker Gawler. Illustration: Botanical Magazine, 1113 (1808); Peterson & McKenny, Field guide to wildflowers, 103 (1968); The Garden **102**: 18 (1977); Rix & Phillips, The bulb book, 97 (1981).
Leaves to 15 cm, elliptic, mottled. Flower-stalks to 30 cm, bearing a solitary flower. Perianth-segments to 5 cm, golden yellow, often spotted with brownish purple at the base inside, the inner ones with small basal appendages. Anthers yellow or brown. Stigma-lobes short. *Eastern N America.* H3. Spring–summer.

## 55. TULIPA Linnaeus

*V.A. Matthews & C. Grey-Wilson*

Bulbous perennials. Bulb-tunics of various textures, hairless inside or lined with hairs. Leaves few, rather fleshy, alternate, decreasing in size up the stem. Flowers usually solitary, rarely to 12. Perianth-segments free. Nectaries absent. Filaments broadened towards the base, hairy or not. Anthers basifixed, dehiscing inwards. Ovary superior. Style very short or absent. Stigma 3-lobed. Fruit a spherical or ellipsoid capsule containing numerous, flat seeds.

A genus of *c.* 100 species from the north temperate Old World, especially Soviet Central Asia. Most species require a sunny position where the bulbs can be baked during the summer. In cooler, wetter areas it is advisable to lift the bulbs after the leaves have died down, store them and replant in the autumn.

Literature: Hall, A.D., *The Genus Tulipa* (1940); Classified list and international register of tulip names (revised 1971); Botschantzeva, Z.P. (translated and edited by H.Q. Varekamp), *Tulipa* (1982); Stork, A.L., *Tulipes sauvages et cultivées*, Série documentaire 13 des Conservatoire et Jardin botaniques de Genève (1984).

1a. Main colour of perianth-segments pink, mauve or purple    **Group A**
 b. Main colour of perianth-segments not pink, mauve or purple    2
2a. Main colour of perianth-segments white or cream    **Group B**
 b. Main colour of perianth-segments not white or cream    3
3a. Main colour of perianth-segments yellow or greenish    **Group C**
 b. Main colour of perianth-segments orange, brownish red, red or crimson-red    **Group D**

**Group A** (Perianth-segments pink, mauve or purple)

1a. Filaments hairless    2
 b. Filaments hairy    3
2a. Anthers twisted    **25. kaufmanniana***
 b. Anthers straight    **13. gesneriana*** (including **platystigma**)
3a. Stem 15–45 cm; bud nodding; leaves green    **4. saxatilis**
 b. Stem 5–20 cm; bud erect; leaves somewhat glaucous    **5. humilis**

**Group B** (Perianth-segments white or cream)

1a. Anthers twisted    **25. kaufmanniana***
 b. Anthers straight    2
2a. Perianth-segments with a basal blotch    3
 b. Perianth-segments lacking a basal blotch    9
3a. Filaments hairy    4
 b. Filaments hairless    7
4a. Leaves glaucous    5
 b. Leaves green    **8. tarda***
5a. Stem hairy    **7. turkestanica**
 b. Stem hairless    6
6a. Perianth-segments with a yellow basal blotch    **6. biflora***
 b. Perianth-segments with a bluish basal blotch    **5. humilis**
7a. Perianth-segments with a deep red, purple, reddish purple or purplish brown basal blotch    8
 b. Perianth-segments with a yellow basal blotch    **9. clusiana**
8a. Outer perianth-segments 1.5–3.2 cm; 2 or 3 bracts present below flowers    **32. edulis**
 b. Outer perianth-segments 3–6.5 cm; flowers lacking bracts    **9. clusiana**
9a. Filaments hairy, yellow    **1. sylvestris**
 b. Filaments hairless, blue-black    **marjolettii** (see **13. gesneriana**)

**Group C** (Perianth-segments yellow or greenish)

1a. Filaments hairy    2
 b. Filaments hairless    4
2a. Visible stem less than 10 cm    **2. urumiensis***
 b. Visible stem 10 cm or more    3
3a. Bud nodding    **1. sylvestris**
 b. Bud erect    **3. orphanidea***
4a. Perianth-segments greenish    **viridiflora** (see **13. gesneriana**)
 b. Perianth-segments yellow, sometimes striped or marked with green or another colour    5
5a. Perianth-segments pure yellow (excluding basal blotch if present)    6
 b. Perianth-segments yellow, flushed or striped with another colour    11
6a. Filaments yellow, at least at the base    7
 b. Filaments purple or black    9
7a. Anthers olive-green    **12. urumoffii**
 b. Anthers yellow or blackish    8
8a. Stem to 30 cm    **10. linifolia***
 b. Stem 30–60 cm    **13. gesneriana*** (including **retroflexa**)
9a. Leaves streaked or mottled with reddish or purplish brown    **23. greigii**
 b. Leaves not so marked    10
10a. Stem to 25 cm; leaf margin wavy    **29. armena***
 b. Stem 30–60 cm; leaf margin not wavy    **13. gesneriana***
11a. Perianth-segments with a red margin    **grengiolensis** (see **13. gesneriana**)

 b. Perianth-segments variously marked, but not with a red margin    12
12a. Perianth-segments with a basal blotch or V-shaped mark    13
 b. Perianth-segments lacking a basal blotch or V-shaped mark    16
13a. Leaves streaked and mottled with reddish or purplish brown    **23. greigii**
 b. Leaves not so marked    14
14a. Perianth-segments with a black, purple or olive-green basal blotch    **13. gesneriana***
 b. Perianth-segments with a yellow or orange basal blotch, or red V-shaped mark    15
15a. Stem 3–10 cm, hairless    **9. clusiana***
 b. Stem 10–35 cm, downy    **25. kaufmanniana***
16a. Perianth-segments 7.5–13 cm, long-pointed, twisted    **acuminata** (see **13. gesneriana**)
 b. Perianth-segments 1.5–7.5 cm, neither long-pointed nor twisted    17
17a. Bud nodding    **27. kolpakowskiana***
 b. Bud erect    18
18a. Inside of bulb-tunics more or less hairless, or with a few short hairs towards the apex    **13. gesneriana*** (including **galatica**)
 b. Inside of bulb-tunics with short hairs at least at the base and apex, sometimes with a tuft of hairs protruding from the neck    19
19a. Leaf margin wavy    **29. armena***
 b. Leaf margin not wavy    **9. clusiana***

**Group D** (Perianth-segments orange, brownish red, red or crimson-red)

1a. Perianth-segments with a basal blotch or V-shaped mark    2
 b. Perianth-segments lacking a basal blotch or V-shaped mark    58
2a. Basal blotch or mark yellow or red    3
 b. Basal blotch black, dark blue, dark purple, brownish or olive-green, bordered or not with a contrasting colour    10
3a. Stem smooth, hairless    4
 b. Stem rough, bristly or downy at least in upper part    5
4a. Leaf margin wavy    **29. armena***
 b. Leaf margin not wavy    **13. gesneriana** (including **fulgens, mauritiana**)
5a. Anthers twisted    **25. kaufmanniana***
 b. Anthers straight    6
6a. Outer perianth-segments to 5.8 cm    7
 b. Outer perianth-segments 6–10.5 cm    8
7a. Stem 7.5–25 cm; perianth-segments 1–3 cm wide    **29. armena***

b. Stem 20–55 cm; perianth-segments *c.* 4 cm wide                    **22. carinata**

8a. Leaf margin wavy towards the apex                    **22. carinata**

b. Leaf margin not wavy                    9

9a. Filaments brown or yellow                    **30. vvedenskyi***

b. Filaments purple                    **elegans** (see **13. gesneriana**)

10a. Basal blotch with a contrasting border                    11

b. Basal blotch lacking a border                    35

11a. Leaves streaked and mottled with reddish or purplish brown                    12

b. Leaves not so marked                    13

12a. Anthers black                    **23. greigii**

b. Anthers violet or yellow                    **24. micheliana**

13a. Border of basal blotch white                    **10. linifolia***

b. Border of basal blotch not white                    14

14a. Leaf margin wavy, at least towards the apex                    15

b. Leaf margin not wavy                    25

15a. Filaments yellow, reddish or brownish                    16

b. Filaments black or dark blue, sometimes yellow or white at the apex                    17

16a. Bud erect; outer perianth-segments 2.5–5.8 cm; anthers 5.5–9 mm                    **29. armena***

b. Bud nodding; outer perianth-segments 5–8 cm; anthers *c.* 12 mm                    **28. ostrowskiana**

17a. Leaves downy, margin ciliate or not                    18

b. Leaves hairless, margin ciliate or not                    22

18a. Filaments white at the apex; pollen reddish purple                    **18. lanata**

b. Filaments not white at the apex; pollen yellow, brown, pink, purple or green                    19

19a. Inside of bulb-tunics hairy at the base, sparsely hairy elsewhere                    **14. undulatifolia**

b. Inside of bulb-tunics woolly or adpressed-hairy                    20

20a. Leaf margin wavy only towards the apex                    **22. carinata**

b. Leaf margin wavy from the base to the apex                    21

21a. Stem to 25 cm; anthers yellow or blackish                    **29. armena***

b. Stem 30–60 cm; anthers purple                    **20. tubergeniana**

22a. Anthers purple                    **10. linifolia***

b. Anthers black, blue, yellow or cream                    23

23a. Filaments blue, sometimes yellow at the apex                    **16. agenensis***

b. Filaments black or yellow                    24

24a. Inside of bulb-tunics lined with woolly matted hairs                    **16. agenensis***

b. Inside of bulb-tunics lined with short or adpressed hairs                    **29. armena***

25a. Filaments yellow, at least at the base                    26

b. Filaments black, dark blue, purplish, reddish or olive-green, sometimes yellow at the apex                    29

26a. Stem 7.5–15 cm; anthers yellow                    **29. armena***

b. Stem 15–60 cm; anthers usually dark purple or olive-green                    27

27a. Leaves hairless although cilate near the apex; inside of bulb-tunics sparsely hairy or hairless                    28

b. Leaves downy on upper surface; inside of bulb-tunics densely hairy                    **21. fosteriana**

28a. Stem 15–30 cm; anthers greenish; leaves glaucous                    **12. urumoffii**

b. Stem 30–60 cm; anthers yellow or purple; leaves green                    **13. gesneriana***

29a. Perianth-segments orange with a brownish green basal blotch                    **15. praecox**

b. Perianth-segments orange, red or crimson; basal blotch black, dark blue, dark purple or dark olive-green                    30

30a. Filaments purplish or greenish, sometimes yellow at the apex                    31

b. Filaments black, dark blue or reddish                    32

31a. Filaments hairless                    **13. gesneriana***

b. Filaments hairy                    **3. orphanidea***

32a. Stem slightly downy                    **21. fosteriana**

b. Stem hairless                    33

33a. Outside of perianth-segments yellowish or paler than the inside                    34

b. Outside of perianth-segments neither yellowish nor paler                    **17. hoogiana***

34a. Leaves green; bulb-tunics lined with densely matted woolly hairs                    **16. agenensis***

b. Leaves glaucous; bulb-tunics lined with long hairs, or with short hairs at the base and top, hairs never matted or woolly                    **29. armena***

35a. Anthers twisted                    **25. kaufmanniana***

b. Anthers straight                    36

36a. Filaments yellow, sometimes black or green at the apex, or red with a black base                    37

b. Filaments black, purple, greenish or brown                    41

37a. Stem hairless                    38

b. Stem downy or bristly                    39

38a. Filaments red with a black base                    **11. montana**

b. Filaments yellow, sometimes black at the apex                    **29. armena***

39a. Stem downy                    **19. ingens**

b. Stem bristly, or at least with rough hairs above                    40

40a. Perianth-segments 2–6 cm                    **29. armena***

b. Perianth-segments more than 6 cm                    **30. vvedenskyi***

41a. Outer perianth-segments 8 cm or more                    42

b. Outer perianth-segments to 8 cm                    46

42a. Leaves streaked with reddish or purplish brown                    **24. micheliana**

b. Leaves not streaked                    43

43a. Stem downy, at least above; leaf margin wavy at least towards the apex                    44

b. Stem bristly; leaf margin not wavy                    **30. vvedenskyi***

44a. Inner perianth-segments more or less rounded at the apex; basal blotch of outer perianth-segments *c.* 5.5 × 3 cm                    **20. tubergeniana**

b. Inner perianth-segments markedly pointed at the apex; basal blotch of outer perianth-segments 8–45 × 6–10 mm                    45

45a. Outer perianth-segments 4–9 cm; inside of bulb-tunics woolly                    **22. carinata**

b. Outer perianth-segments 8–15 cm; inside of bulb-tunics sparsely hairy                    **19. ingens**

46a. Filaments hairy                    **3. orphanidea***

b. Filaments hairless                    47

47a. Stem to 25 cm                    48

b. Stem 30–60 cm                    54

48a. Leaf margin wavy or crisped, at least towards the apex                    49

b. Leaf margin neither wavy nor crisped                    53

49a. Leaves downy on upper surface, sometimes sparsely so                    50

b. Leaves hairless on upper surface                    **29. armena***

50a. Leaves streaked with reddish or purplish brown                    **24. micheliana**

b. Leaves not streaked                    51

51a. Anthers to 9 mm                    **29. armena***

b. Anthers 10 mm or more                    52

52a. Inside of bulb-tunics woolly                    **22. carinata**

b. Inside of bulb-tunics sparsely hairy, more densely hairy at the base                    **14. undulatifolia**

53a. Perianth-segments 2–6 cm                    **29. armena***

b. Perianth-segments more than 6 cm
    **30. vvedenskyi***

54a. Leaf margin wavy or crisped, at least towards the apex    **55**

b. Leaf margin neither wavy nor crisped    **57**

55a. Leaves streaked with reddish or purplish brown    **24. micheliana**

b. Leaves not streaked    **56**

56a. Inside of bulb-tunics woolly    **22. carinata**

b. Inside of bulb-tunics sparsely hairy, more densely hairy at the base    **14. undulatifolia**

57a. Outer perianth-segments to 10.5 cm, with a hairy midrib on the back; filaments yellow or brown    **30. vvedenskyi***

b. Outer perianth-segments 4–7.5 cm, the midrib not hairy; filaments purple or black    **13. gesneriana*** (including **didieri**)

58a. Perianth-segments red, brownish red or orange on both sides    **59**

b. Perianth-segments red or orange, marked or flushed with another colour, or outer segments yellowish buff on the outside    **65**

59a. Stem hairless    **60**

b. Stem downy    **62**

60a. Leaf margin at least slightly wavy; inside of bulb-tunics lined with adpressed hairs or densely woolly    **29. armena***

b. Leaf margin not wavy; inside of bulb-tunics hairless or with a few silky hairs    **61**

61a. Stem 15–30 cm; filaments yellow at the base, olive-green at the apex    **12. urumoffii**

b. Stem 30–60 cm; filaments purple, white or wholly yellow    **13. gesneriana*** (including **acuminata**)

62a. Anthers twisted    **25. kaufmanniana***

b. Anthers straight    **63**

63a. Flowers 1–5; filaments red, shading to yellow at the base    **26. praestans**

b. Flowers solitary, rarely 2; filaments yellow, purple, grey or black    **64**

64a. Leaf margins wavy    **29. armena***

b. Leaf margins not wavy    **13. gesneriana***

65a. Perianth-segments 7.5–13 cm, long-pointed, twisted    **acuminata** (see **13. gesneriana**)

b. Perianth-segments to 7.5 cm, neither long-pointed nor twisted    **66**

66a. Outside of outer perianth-segments differing in colour from inner ones; filaments red    **31. sprengeri**

b. Outside of outer perianth-segments not differing in colour from inner ones; filaments black, purple or yellow    **67**

67a. Leaf margins wavy    **29. armena***

b. Leaf margins not wavy    **13. gesneriana***

**1. T. sylvestris** Linnaeus (*T. australis* Link; *T. celsiana* de Candolle). Illustration: Hall, The genus Tulipa, pl. 3 (1940); Grey-Wilson & Mathew, Bulbs, pl. 5 (1981); Rix & Phillips, The bulb book, 122 (1981). Inside of bulb-tunics with short hairs at the base and top. Stem 10–45 cm, hairless. Leaves 2–4, hairless. Bud nodding. Flowers usually solitary, sometimes 2, often fragrant; perianth-segments yellow, occasionally cream; outer segments 2–6.5 cm × 4–25 mm, usually tinged with green or red on the back; inner segments 2–7 cm × 6–25 mm. Filaments hairy, yellow; anthers and pollen yellow. *Origin unknown: naturalised from Europe and N Africa to the Middle East and Siberia.* H3. Spring.

In the wild, *T. sylvestris*, *T. australis* and *T. celsiana* cannot be distinguished satisfactorily from one another, although plants named *T. sylvestris* tend to be lowland plants of more northerly latitudes and plants named *T. australis* and *T. celsiana* tend to occur further south and at higher altitudes. Plants in cultivation represent only parts of the range of variation found in the wild; as offered commercially they show the following characteristics:

*T. sylvestris* has a stout stem, at least 2.5 mm across and the outer perianth-segments are 3.5–6.5 cm, often tinged with green on the back. The filaments are 9–14 mm and the anthers 4–9 mm.

*T. australis* (*T. sylvestris* ssp. *australis* (Link) Pampanini) has a more slender stem not more than 2 mm across and the outer perianth-segments are 2–3.5 cm, tinged with red on the back. The filaments are 5–8 mm and the anthers 2.5–4 mm.

*T. celsiana* tends to have a shorter stem (to 15 cm) but the flowers are similar to those of *T. australis*. The leaves differ in that they tend to twist and lie flat on the ground, rather than being erect-spreading. *T. celsiana* is usually late-flowering, often 2–3 weeks after *T. sylvestris* and *T. australis*.

**2. T. urumiensis** Stapf. Illustration: Synge, Collins guide to bulbs, pl. 29, f. 18 (1971); Bulletin of the Alpine Garden Society **39**: 265 (1971); Journal of the Royal Horticultural Society **100**: 504 (1975); Rix & Phillips, The bulb book, 124 (1981). Inside of bulb-tunics sparsely hairy towards the base. Stem 10–20 cm, mostly below ground, hairless. Leaves 2–4, green or slightly glaucous, usually in a flat rosette, hairless. Bud nodding. Flowers 1 or 2; perianth-segments to 4 × 1 cm, yellow, flushed with lilac or reddish brown on the back. Filaments hairy; anthers and pollen yellow. *NW Iran.* H4. Spring.

***T. dasystemon** (Regel) Regel. Illustration: Rix & Phillips, The bulb book, 117 (1981). Differs in the solitary flower having perianth-segments 1.5–3 cm × 4–7 mm, yellow with a central band of brownish or greenish. *USSR (Soviet Central Asia – Pamir Alai, Tien Shan).* H4. Spring.

**3. T. orphanidea** Heldreich (*T. hageri* Heldreich; *T. whittallii* Hall). Illustration: Hay & Synge, Dictionary of garden plants in colour, f. 907 (1969); Synge, Collins guide to bulbs, pl. 29, f. 2, 6 (1971); Grey-Wilson & Mathew, Bulbs, pl. 5 (1981); Rix & Phillips, The bulb book, 124 (1981). Inside of bulb-tunics with some hairs at the base, denser hairs at the top. Stem 10–35 cm, hairless or downy. Leaves 2–7, green, hairless, margin often reddish purple. Bud erect. Flowers 1–4, perianth-segments 3–6 cm, pale to bright orange or brownish red, rarely yellow, flushed with red, the outer segments buff and usually tinged with green and sometimes purple, basal blotch blackish, sometimes with a yellow margin. Filaments hairy, purplish or greenish, anthers 5–12 mm, dark green or brown; pollen brownish, olive-green or yellow. *Bulgaria, Greece, Crete, W Turkey.* H4. Spring.

In addition to *T. orphanidea*, *T. hageri* and *T. whittallii* are listed in commercial catalogues. The plants grown in gardens can usually be roughly separated, but in the wild the three species cannot be distinguished. The clones grown in gardens show the following features: **T. orphanidea** has dull orange-brown flowers tinged with green and purple outside; **T. orphanidea** 'Flava' has yellow flowers flushed with red; **T. hageri** has dull red flowers, tinged with green outside; **T. hageri** var. **nitens** Anon. has orange-scarlet flowers and rather glaucous leaves; **T. hageri** 'Splendens' has flowers which are brownish red inside and crimson-scarlet outside; **T. whittallii** is taller than the other species (usually 30–35 cm) and the flowers are bright bronzy orange.

*T. orphanidea* is closely related to *T. sylvestris* and may, in fact, be a hybrid which has *T. sylvestris* as one of the parents. *T. orphanidea* 'Flava' is almost indistinguishable from *T. sylvestris*.

*T. kurdica* Wendelbo. Illustration: Rix & Phillips, The bulb book, 118 (1981). Differs in usually having a shorter stem (6–15 cm), and the solitary flower with bright brick-red to orange-red perianth-segments with a greenish black basal blotch. Anthers 1.3–1.4 cm. *NE Iraq*. H4.

*T. primulina* may key out here (see under 6. *T. biflora*). Differs in having grey-green leaves and 1 or 2 flowers which often droop slightly. Perianth-segments whitish to creamy yellow, outer segments flushed with pink or green. Anthers yellow.

**4. T. saxatilis** Sprengel (*T. bakeri* Hall). Illustration: Journal of the Royal Horticultural Society **85**: pl. 118 (1960); Grey-Wilson & Mathew, Bulbs, pl. 5 (1981); Rix & Phillips, The bulb book, 126 (1981); Baytop & Mathew, The bulbous plants of Turkey, pl. 120 (1984). Inside of bulb-tunics densely hairy at the base and the top. Stem 15–45 cm, hairless. Leaves 2–4, shiny green, hairless. Bud nodding. Flowers 1–4, fragrant; perianth-segments pink to lilac-purple, basal blotch yellow bordered with white; outer segments 3.8–5 × 1–2 cm; inner segments 4–5.5 × 1.5–3 cm. Filaments hairy, yellow; anthers yellow, purple or brown; pollen yellow. *Crete, W Turkey*. H4. Spring.

The clones introduced into cultivation differ from one another in that '*T. saxatilis*' has clear lilac-pink flowers and the flowers of '*T. bakeri*' are a deeper, more purplish pink colour.

**5. T. humilis** Herbert (*T. violacea* Boissier & Buhse; *T. violacea* var. *pallida* Bornmüller; *T. pulchella* (Regel) Baker; *T. pulchella* var. *albocaerulea-occulata* van Tubergen; *T. aucheriana* Baker). Illustration: Hall, The genus Tulipa, pl. 6–9 (1940); Hay & Synge, Dictionary of garden plants in colour, f. 906 (1969); Synge, Collins guide to bulbs, pl. 29, f. 17, 19, 21 (1971); Rix & Phillips, The bulb book, 118, 125, 126 (1981). Inside of bulb-tunics with some hairs near the base and top. Stem 5–20 cm, hairless. Leaves 2–5, somewhat glaucous, hairless, usually in a basal rosette. Bud erect. Flowers usually solitary, occasionally to 3; perianth-segments pale pink to purplish pink or magenta, rarely whitish flushed with pink, basal blotch yellow, olive-green or blue-black often bordered with white or yellow; outer segments 2.3–4.7 cm × 5–18 mm, sometimes tinged with greyish green on the back, inner segments 2.5–5 cm × 9–20 mm. Filaments hairy,

yellow or sometimes dark bluish purple; anthers yellow, brown, purple or blackish; pollen yellow, bluish or greenish. *SE Turkey, N & W Iran, N Iraq, USSR (Azerbaidjan)*. H4. Late winter–spring.

*T. humilis*, *T. violacea*, *T. pulchella* and *T. aucheriana* are commonly offered by nurserymen as separate species. As grown in gardens, they represent only part of the variation displayed in the wild where they cannot be separated. Commercially available plants usually have the following characteristics: **T. humilis** has pinkish magenta flowers with a yellow centre and yellow filaments; **T. violacea** has violet-pink flowers with a yellow or blue-black centre and deep purplish filaments; **T. pulchella** has purplish flowers with a yellow blue-black centre and yellow filaments; **T. aucheriana** has pink flowers with a yellow centre and yellow filaments and tends to flower later than the other 'species'. Plants sold as *T. pulchella* var. *albocaerulea-occulata* or *T. violacea* var. *pallida* have whitish flowers with a dark blue centre and usually flower rather early.

**6. T. biflora** Pallas (*T. polychroma* Stapf). Illustration: Synge, Collins guide to bulbs, pl. 29, f. 16 (1971); Lodewijk, The book of tulips, 107 (1979); Grey-Wilson & Mathew, Bulbs, pl. 5 (1981); Rix & Phillips, The bulb book, 119 (1981). Inside of bulb-tunics hairy, sometimes densely so. Stem 5–10 cm, sometimes more, hairless. Leaves 2, grey-green, hairless, sometimes with a ciliate margin. Bud erect. Flowers 1–5, scented; perianth-segments white, flushed with greenish grey or greenish pink on the back and yellow at the base; outer segments 1.5–3.5 cm; inner segments 2–3 cm. Filaments hairy, yellow; anthers yellow, often dark purple or blackish at the tip; pollen yellow. *E Turkey, Iran, Afghanistan, SW USSR*. H4. Late winter–spring.

*T. primulina* Baker. Illustration: Botanical Magazine, 6786 (1884); Hall, the genus Tulipa, pl. 1 (1940); Synge, Collins guide to bulbs, pl. 29, f. 8 (1971); Rix & Phillips, The bulb book, 124 (1981). Differs in having 1 or 2 flowers which are often slightly drooping; perianth-segments whitish to creamy yellow, deeper yellow at the base inside, outer segments *c*. 2 cm broad, flushed with pink or greenish on the back. *Algeria*. H4. Spring.

**7. T. turkestanica** Regel. Illustration: Lodewijk, The book of tulips, 107 (1979); Rix & Phillips, The bulb book, 117, 118 (1981); Botschantzeva, Tulips, pl. 39 (1982).

Similar to *T. biflora* but with a hairy stem, 10–30 cm, and 2–4 leaves. There may be up to 12 slightly smaller flowers on a stem, which usually smell somewhat unpleasant, and have a yellow or orange centre. The anthers are yellow with a purplish tip, or completely purple or brown. *USSR (Soviet Central Asia), NW China*. H4. Late winter–spring.

Very close to *T. biflora*; further study may show that these two species are the same.

**8. T. tarda** Stapf (*T. dasystemon* misapplied). Illustration: Hall, The genus Tulipa, pl. 13 (1940); Hay & Synge, Dictionary of garden plants in colour, f. 904 (1969); Reader's Digest encyclopaedia of garden plants and flowers, 721 (1971); Rix & Phillips, The bulb book, 126 (1981). Inside of bulb-tunics hairless or with sparse hairs at the base and the top. Stem to 15 cm, hairless. Leaves 3–7, shiny green, hairless, margin often ciliate. Buds erect. Flowers 4–6; perianth-segments 3–4 × 1.2–2 cm, white, tinged with greenish and sometimes red on the back, and yellow in the lower half inside. Filaments hairy, yellow; anthers and pollen yellow. *USSR (Soviet Central Asia – Tien Shan)*. H4. Spring.

*T. cretica* Boissier & Heldreich. Illustration: Hall, The genus Tulipa, pl. 9 (1940); Grey-Wilson & Mathew, Bulbs, pl. 5 (1981); Rix & Phillips, The bulb book, 122 (1981). Differs in having only 2 or 3 leaves and 1–3 flowers. Perianth-segments 1.5–3 cm × 4–11 mm. *Crete*. H4. Spring.

**9. T. clusiana** de Candolle (*T. aitchisonii* Hall). Illustration: Hall, The genus Tulipa, pl. 14, 17 (1940); Hay & Synge, Dictionary of garden plants in colour, f. 892 (1969); Synge, Collins guide to bulbs, pl. 29, f. 1 (1971); Wendelbo, Tulips and irises of Iran, 35 (1977). Bulb-tunics with hairs protruding from the apex. Stem 3.5–30 cm, hairless. Leaves 2–5, glaucous, *c*. 1 cm broad, margin sometimes wavy. Bud erect. Flowers 1 or 2; perianth-segments white, the outer deep pink on the back except at the edge, basal blotch purple or crimson; outer segments 3–6.5 cm; inner segments 2–5 cm. Filaments hairless, purple. Anthers and pollen purple. *NW Pakistan, N India to Iran; naturalised in S Europe and Turkey*. H4. Spring.

Var. **stellata** (Hooker) Regel (*T. stellata* Hooker). Illustration: Hall, The genus Tulipa, pl. 16 (1940); Synge, Collins guide to bulbs, pl. 29, f. 3 (1971). The basal blotch of the perianth-segments and the stamens are yellow.

Var. **chrysantha** (Hall) Sealy (*T. stellata* Hooker var. *chrysantha* Hall; *T. aitchisonii* Hall subsp. *cashmeriana* Hall). Illustration: Hall, The genus Tulipa, pl. 15, 16 (1940); Botanical Magazine, n.s., 13 (1948); Synge, Collins guide to bulbs, pl. 29, f. 11, 14 (1971); Reader's Digest encyclopaedia of garden plants and flowers, 720 (1971). Flowers 1–3. Perianth-segments yellow, except usually for red or brownish purple coloration on the back of the outer segments; stamens yellow.

There is some controversy about the relationship of var. *stellata* and var. *chrysantha* to *T. clusiana*. Further work may show that it would be better to raise these varieties to subspecific or even specific level.

*\*T. altaica* which has solitary yellow flowers tinged with red or green on the outside and leaves *c.* 3 cm broad, may key out here (see under 29. *T. armena*).

**10. T. linifolia** Regel (*T. batalinii* Regel; *T. maximowiczii* Regel). Illustration: Dykes, Notes on tulip species, pl. 19 (1930); Hall, The genus Tulipa, pl. 18 (1940); Synge, Collins guide to bulbs, pl. 29 f. 9, 12 (1971); Rix & Phillips, The bulb book, 123 (1981).
Inside of bulb-tunics hairy towards the top. Stem 10–30 cm, hairless. Leaves 3–9, usually up to 10 mm wide, greyish green, hairless, margin often wavy and often ciliate, usually red. Bud erect. Flower solitary; perianth-segments 2–6 × 1.2–3.5 cm, red with a blackish purple basal blotch often bordered with white, cream or yellow, or pale yellow with a deep yellow or brownish basal blotch. Filaments hairless, black or yellow; anthers and pollen dark purple or yellow. *N Iran, Afghanistan, USSR (Soviet Central Asia – Uzbekistan).* H4. Spring.

A variable species which is cultivated as clones which can be distinguished to some extent. As grown in gardens, **T. linifolia** has bright red flowers with a blackish purple centre bordered with cream or yellow, and black filaments with dark purple anthers containing purple pollen; **T. batalinii** has flowers of pale yellow with a deep yellow or brownish centre and yellow stamens and pollen; **T. maximowiczii** has red flowers but the black centre has a white border, the black filaments are white at the apex and the purplish or yellow anthers contain purple pollen.

Hybrids have been produced between *T. linifolia* and *T. batalinii* which have apricot or bronze flowers, e.g. 'Bronze Charm', 'Bright Gem' and 'Apricot Jewel'.

*\*T. julia* Koch. Illustration: Rix & Phillips, The bulb book, 123 (1981); Botschantzeva, Tulips, pl. 4 (1982); Baytop & Mathew, The bulbous plants of Turkey, pl. 112 (1984). Bulb-tunics lined with dense matted hairs, leaves 1–3 cm wide, and perianth-segments red to orange-red, pinkish or orange on the back with a yellow-bordered black basal blotch. *USSR (Transcaucasia), NW Iran, E Turkey.* H4. Spring.

*\*T. albertii* which differs in having red, orange or yellow perianth-segments, always with a yellow-bordered black basal blotch, and segments 4–6 cm wide may key out here (see under 29. *T. armena*).

**11. T. montana** Lindley (*T. wilsoniana* Hoog). Illustration: Dykes, Notes on tulip species, pl. 17, 18 (1930); Hall, The genus Tulipa, pl. 19 (1940); Botanical Magazine, n.s., 433 (1963); Botschantzeva, Tulips, pl. 20 (1982).
Inside of bulb-tunics hairy towards the top and with hairs protruding from the apex. Stem 20–30 cm, hairless. Leaves 3–6, glaucous, hairless, margin of basal leaves wavy, often red. Bud erect. Flower solitary; perianth-segments 4–5.5 × 2–2.5 cm, red with a greenish black basal blotch. Filaments hairless, red, black towards the base; anthers and pollen yellow. *N Iran, USSR (Soviet Central Asia – Turkmeniya).* H4. Spring.

**12. T. urumoffii** Hayek (*T. rhodopea* (Velenovsky) Velenovsky). Illustration: Hall, The genus Tulipa, pl. 22 (1940); Grey-Wilson & Mathew, Bulbs, pl. 6 (1981).
Inside of bulb-tunics lined with sparse silky hairs. Stem 15–30 cm, hairless. Leaves 3–5, glaucous, hairless, margin ciliate towards the apex. Bud erect. Flowers usually solitary, sometimes to 3; perianth-segments 3.5–6 × 1.7–2.5 cm, yellow or red to brownish red, basal blotch absent or black with a yellow border. Filaments hairless, yellow shading to greenish at the apex; anthers greenish; pollen pale brownish or greenish. *S Bulgaria.* H4. Spring.

**13. T. gesneriana** Linnaeus (*T. billetiana* Jordan; *T. bonarotiana* Reboul; *T. connivens* Levier; *T. etrusca* Levier; *T. passeriniana* Levier; *T. pubescens* Willdenow; *T. scabriscapa* Fox-Strangways; *T. sommieri* Levier; *T. spathulata* Bertoloni; *T. strangulata* Reboul; *T. variopicta* Reboul). Illustration: Revue Horticole 1890: 475 (1890); The Garden 44: 534 (1893); Grey-Wilson & Mathew, Bulbs, pl. 6 (1981).

Inside of bulb-tunics without hairs or with a few hairs towards the top. Stem 30–60 cm, hairless or finely downy. Leaves 2–7, green, hairless, margin ciliate towards the apex. Bud erect. Flower solitary; perianth-segments 4–8.2 cm, inner broader than outer, red, orange, yellow or purplish, sometimes 'broken' (see below), with or without a yellow or blackish basal blotch which may have a yellow border. Filaments hairless, yellow or purple; anthers purple or yellow. *Origin unknown, possibly SW & SC Asia; naturalised in SW Europe.* H4. Spring.

A 'broken' flower is one in which the normal self-colour is variegated with another colour. The 'breaking' is caused by a virus. 'Broken' flowers were very common and new 'breaks' much sought-after during the period of tulipomania in Holland in the seventeenth century and in Britain from *c.* 1835 to the end of the nineteenth century.

*T. gesneriana* is an extremely complex species from which most of the garden cultivars have been derived. From the fifteenth century onwards there was a great influx of tulips from western and central Asia and some of the selections from these eventually became naturalised in Europe: they were known as the Neo-tulipae. Several of these selections were given species names and still appear in nurserymen's catalogues, but cannot really be separated satisfactorily from the mass of material included under *T. gesneriana*. Some of these, as sold commercially and grown in gardens, are described below.

**T. acuminata** Hornemann (*T. cornuta* Delile). Illustration: Revue Horticole 1890: 476 (1890); Everett, Illustrated encyclopedia of horticulture 10: 3426 (1982); Mathew, P.J. Redouté: Lilies and related flowers, 103 (1982). Stem 30–45 cm. Perianth-segments 7.5–13 cm, long-pointed and tapering to a twisted point, pale red or yellowish, usually tinged with green or red. Filaments yellow or white; anthers reddish brown.

**T. didieri** Jordan. Stem 30–40 cm. Perianth-segments red, tinged with crimson on the back, basal blotch dark purple, margin irregular, creamy white or yellow, segment tips reflexed. Anthers black, occasionally yellow.

**T. elegans** Baker. Illustration: The Garden 32: 514 (1887). Stem to 45 cm, downy. Perianth-segments red with a yellow basal blotch. Filaments and anthers purple.

**T. fulgens** Baker differs from *T. elegans* in its hairless stem and yellow anthers.

**T. galatica** Freyn. Illustration: Hall, The genus Tulipa, pl. 21 (1940). Stem *c.* 15 cm. Perianth-segments pale yellow tinged with olive-green on the back and with an olive-green basal blotch. Filaments yellow.

**T. grengiolensis** Thommen. Stem 25–40 cm. Perianth-segments pale yellow, margined with red.

**T. marjolettii** Perrier & Songeon. Stem 40–50 cm. Perianth-segments creamy-white, edged with deep pink and tinged with purple on the back. Filaments blue-black, anthers yellow.

**T. mauritiana** Jordan (*T. mauriana* Jordan & Fournier). Stem 30–40 cm. Perianth-segments scarlet, yellow at the base. Filaments yellow, anthers purple.

**T. platystigma** Jordan. Stem to 55 cm. Perianth-segments rose-pink, blue at the base and suffused with orange towards the margins.

**T. retroflexa** Baker. Illustration: The Garden **32**: 514 (1887), **85**: 241 (1921). Perianth-segments 7.5–10 cm, bright yellow, long-pointed, reflexing from the middle as the flower opens. Filaments yellow.

**T. viridiflora** Anon. Illustration: The Garden **32**: 514 (1887). Stem to 60 cm. Perianth-segments pale green, yellowish or whitish at the edge, acuminate.

*__T. cypria__ Turrill. Illustration: Botanical Magazine, 9363 (1934); Megaw, Wild flowers of Cyprus, pl. 33 (1973). Differs in the bulb-tunics being densely woolly inside and stem only to 35 cm. Perianth-segments dark red with a blue-black basal blotch usually bordered with yellow, outside green at the base. *Cyprus.* H4. Spring.

*__T. lehmanniana__ Bunge. Illustration: Botschantzeva, Tulips, pl. 21, 22 (1982). Differs in the bulb having a long neck and the tunics being densely woolly inside and the stem only to 25 cm. Leaf margin wavy, not ciliate. Bud nodding. Perianth-segments yellow, orange or red, flushed with orange-red or reddish brown, basal blotch black, olive-green or violet. *USSR (Soviet Central Asia), Afghanistan, NE Iran.* H4. Spring.

*__T. stapfi__ Turrill. Illustration: Journal of the Royal Horticultural Society **82**: pl. 45 (1957), **88**: pl. 198 (1963). Differs in the bulb-tunics being densely woolly inside and the stem up to 30 cm. Leaves glaucous. Perianth-segments red with a yellow-bordered purplish basal blotch. *W Iran, N Iraq.* H4. Spring.

*__T. tschimganica__ may key out here (see under 25. *T. kaufmanniana*). Differs in having a stem to 25 cm and glaucous

leaves. Perianth-segments yellow with red V-shaped marks towards the base inside, the outer with red midribs on the back, or segments entirely red, or red with a yellow base. Filaments yellow with a brownish tip.

*__T. albertii__ which has stems to 20 cm and red, orange or yellow flowers with a black, yellow-bordered basal blotch, may key out here (see under 29. *T. armena*).

**14. T. undulatifolia** Boissier (*T. baeotica* Boissier & Heldreich; *T. eichleri* Regel). Illustration: Journal of the Royal Horticultural Society **88**: pl. 158 (1963); Hay & Synge, Dictionary of garden plants in colour, f. 893 (1969); Grey-Wilson & Mathew, Bulbs, pl. 5 (1981); Rix & Phillips, The bulb book, 126 (1981). Inside of bulb-tunics hairy at the base, sparsely hairy above. Stem 15–50 cm, usually minutely downy. Leaves 3 or 4, glaucous, usually minutely downy on the upper surface, margin wavy or crisped, ciliate. Bud erect. Flower solitary; perianth-segments orange-red to red, paler red, bluish red or buff on the back, downy, basal blotch dark olive-green or black, often with a yellow border; outer segments 3–7.5 × 1.5–2.5 cm, inner segments slightly longer. Filaments hairless, black; anthers dark purple; pollen dark green or dirty yellow. *Jugoslavia, Greece, W Turkey, Iran, USSR (W Soviet Central Asia).* H4. Spring.

**15. T. praecox** Tenore. Illustration: Dykes, Notes on tulip species, pl. 29 (1930); Hall, The genus Tulipa, pl. 25, 27 (1940); Grey-Wilson & Mathew, Bulbs, pl. 6 (1981); Baytop & Mathew, The bulbous plants of Turkey, pl. 114 (1984). Inside of bulb-tunics densely woolly. Stem 20–65 cm, hairless or slightly downy. Leaves 3–5, glaucous, hairless. Bud erect. Flowers solitary; perianth-segments orange tinged with green on the back, basal blotch brownish green with a yellow border; outer segments 4–10 × 2–5 cm; inner segments 3.5–7.5 × 1.5–3.5 cm, with a yellowish central stripe. Filaments hairless, black or dark green; anthers black or dark green; pollen dark green or yellow. *Origin unknown: naturalised in S Europe & W Turkey.* H4. Spring.

**16. T. agenensis** de Candolle (*T. oculis-solis* St. Amans; *T. lortetii* Jordan). Illustration: Hall, The genus Tulipa, pl. 23 (1940); Megaw, Wild flowers of Cyprus, pl. 32 (1973); Grey-Wilson & Mathew, Bulbs, pl. 6 (1981); Mathew, P.J. Redouté: Lilies and related flowers, 107 (1982).

Inside of bulb-tunics lined with dense matted woolly hairs. Stem to 20 cm, hairless. Leaves 3–5, green, margins flat or somewhat wavy, often ciliate. Bud erect. Flower solitary; perianth-segments dull red inside, lighter red or yellowish on the back, with a black basal blotch bordered with yellow; outer segments 4.2–8.5 × 1.8–3 cm; inner segments 3.7–7 × 1.5–2.3 cm. Filaments 8–12 mm, hairless, black or dark blue; anthers black; pollen yellow. *W & S Turkey, NW Iran; naturalised in S France & Italy.* H4. Spring.

*__T. sharonensis__ Dinsmore. Differs in the perianth-segments having a very large deep olive-green yellow-bordered basal blotch covering almost the basal half of the segment; outer segments *c.* 3 × 1 cm. *Israel.* H4. Spring.

*__T. systola__ Stapf. Illustration: Journal of the Royal Horticultural Society **88**: pl. 198 (1963), **100**: 257 (1975); Rix & Phillips, The bulb book, 125 (1981). Differs in having the red perianth-segments slightly greyish on the back. Filaments 6–7 mm, dark blue with a yellow apex. *Iran.* H4. Spring.

**17. T. hoogiana** Fedtschenko. Illustration: Gardeners' Chronicle **48**: 53 (1910); Hall, The genus Tulipa, pl. 29 (1940); Rix & Phillips, The bulb book, 125 (1981). Inside of bulb-tunics lined with thick matted wool. Stem to 45 cm, hairless. Leaves 3–5, greyish green, hairless, margin long-ciliate. Flower solitary, at first strongly scented; perianth-segments 7.5–10 cm, orange-red, basal blotch black or dark olive-green with a yellow border. Filaments black or reddish; anthers black, brown, pinkish or yellow; pollen brown, pinkish or yellow. *USSR (Soviet Central Asia – mountains of Turkmeniya).* H4. Spring.

*__T. aleppensis__ Regel. Illustration: Hall, The genus Tulipa, pl. 26, 27 (1940). Differs in having stems 10–25 cm, and perianth-segments 5–5.5 cm, red-crimson. *Syria.* H4. Spring.

*__T. kuschkensis__ Sealy. Illustration: Botanical Magazine, 9370 (1934); Gartenflora **84**: 295 (1935); Hall, The genus Tulipa, pl. 28 (1940); Botschantzeva, Tulips, pl. 2 (1982). Differs in having scentless flowers and outer perianth-segments whose margins roll back. *USSR (Soviet Central Asia – Turkmeniya), NW Afghanistan.* H4.

**18. T. lanata** Regel. Illustration: Botanical Magazine, 9151 (1926); Hall, The genus Tulipa, pl. 29 (1940); Botschantzeva, Tulips, pl. 5 (1982).

Inside of bulb-tunics thickly lined with straight hairs. Stem 15–60 cm, downy. Leaves 4 or 5, glaucous, downy, margin wavy, red, ciliate. Bud erect. Flower solitary; perianth-segments 6–12 × 3–6 cm, red, somewhat silvery on the back and downy on the upper half, basal blotch olive-green with a yellow border. Filaments hairless, black, white at the tip; anthers and pollen reddish purple. *NE Iran, USSR (Soviet Central Asia – Turkmeniya), Afghanistan*. H4. Spring.

Very closely related to the three following species; further research may show that they all belong to a single species.

**19. T. ingens** Hoog. Illustration: Gardeners' Chronicle **32**: 14 (1902); Dykes, Notes on tulip species, pl. 51 (1930); Hall, The genus Tulipa, pl. 30 (1940); Botschantzeva, Tulips, pl. 6 (1982).
Inside of bulb-tunics sparsely hairy. Stem to 25 cm, downy, often reddish or purplish in upper part. Leaves 3–5, glaucous, downy, margin wavy and ciliate. Flower solitary; perianth-segments 8–15 cm, red with a black basal blotch. Filaments hairless, blackish or dirty yellow; anthers and pollen dark purple. *USSR (Soviet Central Asia – Pamir Alai)*. H4. Spring.
Very close to *T. lanata*.

**20. T. tubergeniana** Hoog. Illustration: Gardeners' Chronicle **35**: 358 (1904); Flora and Sylva **3**: 94 (1905); Hall, The genus Tulipa, pl. 30 (1940); Botschantzeva, Tulips, pl. 1 (1982).
Inside of bulb-tunics thickly hairy. Stem 30–60 cm, downy. Leaves 3–5, glaucous, downy, margin wavy and ciliate. Flower solitary; perianth-segments to 10 × 3.5 cm, red, basal blotch olive-green or blackish with or without a narrow yellow border. Filaments hairless, black; anthers dark purple; pollen brownish purple or dirty yellow. *USSR (Soviet Central Asia)*. H4. Spring.
Very close to *T. lanata*.

**21. T. fosteriana** Irving. Illustration: Hall, The genus Tulipa, pl. 32 (1940); Botschantzeva, Tulips, pl. 7 (1982).
Inside of bulb-tunics densely hairy. Stem 20–45 cm, slightly downy. Leaves 3–6, glaucous, downy on upper surface, margin ciliate. Bud erect. Flower solitary, slightly fragrant; perianth-segments 4.5–8 cm, occasionally to 12.5 cm, bright red, basal blotch purplish black with a yellow border. Filaments hairless, black or yellow; anthers purplish black; pollen purplish brown or yellow. *USSR (Soviet Central Asia)*. H4. Spring.

This species is closely related to *T. lanata*. A number of named cultivars is available as well as hybrids with *T. greigii*.

**22. T. carinata** Vvedensky. Illustration: Botschantzeva, Tulips, pl. 8 (1982).
Inside of bulb-tunics sparsely to densely woolly. Stem 20–55 cm, downy in upper part. Leaves 3 or 4, greyish green, downy, keeled below, margin wavy towards the apex, ciliate. Flower solitary; perianth-segments red, 4–9 × c. 4 cm, with the basal blotch yellow, black, or black bordered with yellow. Filaments hairless, blackish, c. 1 cm; anthers violet or yellow, c. 1.8 cm; pollen yellow, pinkish or buff. *USSR (Soviet Central Asia – Pamir Alai)*. H4. Spring.

**23. T. greigii** Regel. Illustration: Botanical Magazine, 6177 (1875); Journal of the Royal Horticultural Society **97**(4): front cover (1972); Witham Fogg, Bulbs, 42 (1980); Rix & Phillips, The bulb book, 124 (1981).
Inside of bulb-tunics sparsely, or sometimes densely, hairy towards the top. Stem 20–45 cm, downy. Leaves 3–5, streaked and mottled with reddish or purplish brown, margin sometimes wavy, upper leaves often downy with ciliate margins. Bud erect. Flower solitary; perianth-segments red or yellow, basal blotch blackish with a yellow border; outer segments 3–10 cm, inner segments 4–11 cm. Filaments hairless, black; anthers black; pollen yellow. *USSR (Soviet Central Asia – Tadjikistan)*. H4. Spring.

Numerous named cultivars of *T. greigii* are available, as well as hybrids between *T. greigii* and *T. fosteriana*.

**24. T. micheliana** Hoog. Illustration: Gardeners' Chronicle **31**: 353 (1902); Revue Horticole 1903: 206 (1903); Journal of the Royal Horticultural Society **90**: pl. 233 (1965); Botschantzeva, Tulips, pl. 18 (1982).
Inside of bulb-tunics lined with adpressed hairs which are denser at the base and the top. Stem 25–35 cm, sometimes to 60 cm, downy. Leaves 3–5, glaucous, streaked with reddish or purplish brown, more or less downy, margin ciliate and usually wavy. Bud erect. Flower solitary; perianth-segments 5.5–10 cm, bright red to crimson, the long basal blotch purplish black with a white or yellow border which may be very narrow or absent. Filaments hairless, black; anthers violet or yellow; pollen violet, yellow or brown. *NE Iran, USSR (Soviet Central Asia – Turkmeniya)*. H4. Spring.

**25. T. kaufmanniana** Regel. Illustration: Hall, The genus Tulipa, pl. 34 (1940); Hay & Synge, Dictionary of garden plants in colour, f. 897 (1969); Journal of the Royal Horticultural Society **100**: 504 (1975); Rix & Phillips, The bulb book, 119, 121 (1981).
Inside of bulb-tunics lined with adpressed hairs, denser at the base and top. Stem 10–35 cm, slightly downy. Leaves 3–5, often in a basal rosette, greyish green, hairless, margin minutely ciliate. Bud erect. Flowers 1–5, often fragrant; perianth-segments 3–10 × 1–5 cm, cream or yellow flushed with pink or greyish green on the back, or pink, orange or red, often with a basal blotch of a different colour. Filaments hairless, yellow; anthers twisted, yellow; pollen yellow. *USSR (Soviet Central Asia – Tadjikistan, Kirgizia)*. H4. Late winter–spring.

A large range of cultivars is available, as well as hybrids with *T. fosteriana* and *T. greigii*.

***T. dubia** Vvedensky. Illustration: Rix & Phillips, The bulb book, 121 (1981). Differs in the leaves being more or less downy, with a wavy margin. Flower solitary; perianth-segments 2–4 × c. 1.5 cm, yellow, streaked on the back with bluish pink, basal blotch orange. Anthers not twisted. *USSR (Soviet Central Asia – Tien Shan)*. H4.

***T. tschimganica** Botschantzeva. Illustration: Rix & Phillips, The bulb book, 120, 121 (1981). Differs in having downy leaves and solitary flowers. Perianth-segments yellow with red V-shaped marks towards the base inside, the outer with red midribs on the back, or segments entirely red, or red with a yellow base. Anthers not twisted. *USSR (Soviet Central Asia – Tien Shan)*. H4. Spring.

***T. subpraestans** Vvedensky. Illustration: Botschantzeva, Tulips, pl. 11 (1982). Differs in the bulb-tunics being hairless or almost so inside, and flowers usually solitary. Perianth-segments red, lacking a basal blotch. Filaments red; anthers yellow or black, twisted. *USSR (Soviet Central Asia – Pamir Alai)*. H4. Spring.

**26. T. praestans** Hoog. Illustration: Hall, The genus Tulipa, pl. 35 (1940); Hay & Synge, Dictionary of garden plants in colour, f. 901 (1969); Rix & Phillips, The bulb book, 118, 125 (1981); Botschantzeva, Tulips, pl. 10 (1982).
Inside of bulb-tunics sparsely hairy at the top. Stem 10–45 cm, minutely downy. Leaves 3–6, greyish green, downy, margin ciliate. Flowers 1–5; perianth-segments 5.5–6.5 × c. 3 cm, scarlet-orange. Filaments hairless, red shading to yellow at

the base; anthers yellow or purplish; pollen yellow, greyish brown or reddish. *USSR (Soviet Central Asia – Pamir Alai)*. H4. Spring.

The following cultivars are more frequently offered in the trade than is the species itself:

'Tubergen's Variety'. Illustration: Botanical Magazine, 7920 (1903). Has 2–5 flowers, often yellow at the base, and increases very freely.

'Fuselier'. Illustration: Lodewijk, The book of tulips, 107 (1979); Rix & Phillips, The bulb book, 122 (1981). Produces 3–5 flowers and blooms a little earlier than the species itself.

**27. T. kolpakowskiana** Regel. Illustration: Hay & Synge, Dictionary of garden plants in colour, f. 899 (1969); Synge, Collins guide to bulbs, pl. 29, f. 13 (1971); Journal of the Royal Horticultural Society **100**: 505 (1975); Rix & Phillips, The bulb book, 118, 121 (1981).
Inside of bulb-tunics lined with adpressed hairs. Stem 10–30 cm, hairless. Leaves 2–4, greyish green, hairless, margin wavy and ciliate. Bud nodding. Flowers usually solitary, sometimes 2–4; perianth-segments 3.7–5.5 × 1.5–2 cm, yellow, marked with crimson, orange or olive-green on the back. Filaments hairless, yellow; anthers yellow. *USSR (Soviet Central Asia), Afghanistan*. H4. Spring.

*\*T. heterophylla* (Regel) Baker (*Eduardoregelia heterophylla* (Regel) Popov). Illustration: Rix & Phillips, The bulb book, 117 (1981). Differs in having the bulb-tunics hairless inside and the stem 5–15 cm. Flower solitary; perianth-segments 1.5–3 cm, yellow, marked with dull violet or green on the back. *USSR (Soviet Central Asia – Tien Shan)*. H4. Spring.

*\*T. iliensis* Regel. Illustration: The Garden **17**: 311 (1880). Differs in having a downy stem and 1–5 flowers. Perianth-segments 1.5–3.5 cm × 4–10 mm, yellow, the outer segments tinged with red or olive-green on the back. *USSR (Soviet Central Asia)*. H4. Spring.

*\*T. tetraphylla* Regel. Illustration: Gartenflora **28**: t. 964 (1879). Differs in having 1–4 flowers; outer perianth-segments 2–4 cm, yellow tinged with violet and green on the back. *USSR (Soviet Central Asia – Tien Shan)*. H4.

**28. T. ostrowskiana** Regel. Illustration: Gartenflora **33**: t. 1144 (1884); The Garden **45**: 486 (1894); Botschantzeva, Tulips, pl. 27 (1982).
Inside of bulb-tunics densely hairy

especially at the base and top. Stem 20–35 cm, hairless. Leaves 2–4, slightly glaucous, hairless, margin wavy and sometimes sparsely ciliate. Bud nodding. Flower solitary; perianth-segments 5–8 × 1.8–3 cm, red, basal blotch dark olive-green with an irregular yellow border; outer segments with a slight purplish sheen on the back. Filaments yellow, brown or reddish; anthers yellow or black; pollen yellow, black, brown or reddish. *USSR (Soviet Central Asia – Tien Shan)*. H4. Spring.

Closely related to *T. kolpakowskiana*; further research may show that these two species should be united.

**29. T. armena** Boissier (*T. suaveolens* Roth). Illustration: Hall, The genus Tulipa, pl. 20 (1940); Rix & Phillips, The bulb book, 124 (1981).
Inside of bulb-tunics lined with long hairs, or with short hairs at the base and the top. Stem 7.5–25 cm, hairless or with rough hairs at least in the upper part. Leaves 3–6, glaucous, hairless, or downy on the upper surface, usually recurved, margin wavy and often ciliate. Bud erect. Flower solitary, often fragrant; perianth-segments red to crimson, often lighter red or yellowish on the back, with a black or dark blue, or sometimes yellow or greenish basal blotch which may be bordered with yellow. Sometimes the segments are yellow, or partly yellow and partly red, with or without the basal blotch. Outer segments 2.3–5.8 cm × 9–27 mm, inner segments 2–4.5 cm × 7–20 mm. Filaments hairless, black or yellow; anthers yellow or blackish. *Turkey, NW Iran, Iraq, USSR (Transcaucasia)*. H4. Spring.

There is much confusion about *T. schrenkii* Regel, and where it stands in relation to other species. In the wild it is extremely variable in flower colour, white, yellow, pink and red flowers being known. In gardens it is usually represented by a variant in which the perianth-segments are red edged with deep yellow or orange. Recent publications (Hortus Third, 1976 and Stork, 1984) as well as Hall (1940) equate *T. schrenkii* and *T. suaveolens*, and Baytop & Mathew in The bulbous plants of Turkey, 103, 1984 have reduced *T. suaveolens* to a synonym of the variable *T. armena*. It would therefore seem reasonable to assume that *T. schrenkii* should be sunk into *T. armena*.

*\*T. albertii* Regel. Illustration: Revue Horticole **53**: 430 (1881); Botanical Magazine, 6761 (1884); Botschantzeva, Tulips, pl. 15 (1982). Differs in having red,

orange or yellow perianth-segments, always with a yellow-bordered black basal blotch, and segments 4–6 cm wide. *USSR (Soviet Central Asia)*. H4. Spring.

*\*T. altaica* Sprengel. Illustration: Hall, The genus Tulipa, pl. 37 (1940). Differs in having yellow perianth-segments without a basal blotch and filaments always yellow. *USSR (Soviet Central Asia to W Siberia)*. H4. Spring.

*\*T. borszczowii* Regel. Illustration: Gartenflora **33**: t. 1175 (1884); Botschantzeva, Tulips – frontispiece (1982). Differs in having orange-red perianth-segments, or the inner segments deep yellow, or the outer segments red with a yellow margin. The filaments are grey or dark purple. *USSR (Soviet Central Asia – Kazakstan), ?SW Iran*. H4. Spring.

*\*T. butkovii* Botschantzeva. Illustration: Botschantzeva, Tulips, pl. 12 (1982). Differs on having the bulb-tunics lined with short hairs and a downy stem. The perianth-segments are brick-red with a dark red basal blotch. *USSR (Soviet Central Asia – Tien Shan)*. H4. Spring.

*\*T. ferganica* Vvedensky. Illustration: Rix & Phillips, The bulb book, 121 (1981). Differs in having downy leaves and 1 or 2 flowers. Perianth-segments yellow, flushed with bluish pink or pinkish brown on the back. Filaments deep yellow. *USSR (Soviet Central Asia – Pamir Alai, Tien Shan)*. H4. Spring.

Closely related to and possibly the same as *T. altaica*.

*\*T. goulimyi* Sealy & Turrill. Illustration: Goulandris, Wild flowers of Greece, 159 (1968). Differs in the bulb-tunics being densely woolly inside and the bright orange to brownish red perianth-segments lacking a basal blotch. Filaments hairy. *S Greece*. H4. Spring.

*\*T. julia* may key out here (see under 10. *T. linifolia*). Differs in the red or orange-red perianth-segments being pinkish or orange on the back.

*\*T. lehmanniana* may key out here (see under 13. *T. gesneriana*). Differs in the bud nodding and the leaf margin being hairless. Perianth-segments yellow, orange or red, flushed with orange-red or reddish brown, basal blotch black, olive-green or violet. The bulb has a long neck.

**30. T. vvedenskyi** Botschantzeva. Illustration: Journal of the Royal Horticultural Society **91**: pl. 162 (1966); Rix & Phillips, The bulb book, 124 (1981); Botschantzeva, Tulips, pl. 13, 14 (1982). Inside of bulb-tunics sparsely lined with hairs, more densely so at the top. Stem

15–45 cm, bristly, glaucous, sometimes tinged with purple. Leaves 4 or 5, glaucous, often downy, margin slightly ciliate. Flower solitary; perianth-segments red with a black or yellow basal blotch; outer segments to 10.5 × 6.2 cm; inner segments to 1.5 × 5.2 cm. Filaments hairless, brown or yellow; anthers violet or yellow; pollen brownish, violet or yellow. *USSR (Soviet Central Asia – Tien Shan).* H4. Spring.

*T. cypria Turrill may key out here (see under 13. *T. gesneriana*). Differs in the bulb-tunics being densely woolly inside, the stem hairless and the blue-black blotch at the base of the dark red perianth-segments being bordered with yellow. The filaments are purple.

*T. tschimganica may key out here (see under 25. *T. kaufmanniana*). Differs in having a downy stem and yellow perianth-segments with red V-shaped marks towards the base inside, the outer with red midribs on the back, or segments entirely red, or red with a yellow base.

**31. T. sprengeri** Baker. Illustration: Gartenflora 44: t. 1411 (1895); Synge, Collins guide to bulbs, pl. xxiii (1971); Rix & Phillips, The bulb book, 134 (1981); Baytop & Mathew, The bulbous plants of Turkey, pl. 108 (1984).
Inside of bulb-tunics hairless or more or less so. Stem 30–45 cm, hairless. Leaves 5 or 6, green, hairless. Bud erect. Flower solitary; perianth-segments orange-red to red, outer segments 4.5–6.5 × 1.5–2 cm, yellowish buff on the back; inner segments slightly larger. Filaments hairless, red, 3 long and 3 short; anthers and pollen yellow. *Turkey.* H4. Late spring–early summer.

**32. T. edulis** (Miquel) Baker (*Amana edulis* (Miquel) Honda). Illustration: Hall, The genus Tulipa, pl. 40 (1940); Botanical Magazine, n.s., 293 (1956); Mathew, Dwarf bulbs, pl. 82 & f. 57 (1973); Rix & Phillips, The bulb book, 117 (1981).
Inside of bulb-tunics densely woolly. Stem 5–15 cm, hairless, bearing 2 or 3 linear bracts below the flowers. Leaves *c.* 6, 15–25 cm × 5–10 mm. Flowers 1 or 2; perianth-segments 1.5–3.2 cm × 4–6 mm, white with reddish brown or purplish veins on the back, basal blotch deep purple with a yellow border. Filaments hairless. *S Japan, NE China, Korea.* H5–G1. Late winter–spring.

Var. **latifolia** Makino (*Amana latifolia* (Makino) Honda; *Tulipa latifolia* (Makino) Makino) differs in the leaves being shorter and generally broader, 10–15 cm × 7–15 mm.

## 56. FRITILLARIA Linnaeus
*E.M. Rix*
Spring ephemerals, dormant in summer. Bulbs of 2 or more thick, closely wrapped scales, usually spherical or spindle-shaped, rarely with a thin, white, papery tunic, or bulbs of thick, separate scales, often surrounded at the base with numerous, white, rice-like bulbils. Basal leaves (on non-flowering shoots) 1, lanceolate or rarely linear. Stem-leaves usually alternate, more rarely opposite or whorled. Bracts similar to the leaves but smaller, alternate or whorled. Flowers usually nodding, bell-shaped to tubular, rarely saucer-shaped or conical. Perianth-segments free, in 2 whorls, those of the inner whorl wider than those of the outer, each with a usually conspicuous nectary. Stamens more or less equal to the style; anthers more or less basifixed, rarely versatile. Style 1, entire or 3-fid. Capsule erect (nodding in *F. japonica*), flat-topped, sometimes 6-winged, the style falling. Seeds flat (spherical in *F. japonica*).

A genus of about 100 species found throughout the temperate regions of the northern hemisphere (excluding eastern North America), with most species in Mediterranean area, the mountains of SW Asia and in California. Most species succeed in a well-drained soil kept dry in summer, but some species will tolerate damp in summer and a few (e.g. *F. camschatcensis* and *F. cirrhosa*) require it
Literature: Beck, C., *Fritillaries* (1953); Furse, P., Fritillaries in Britain, *RHS Lily Year Book, 1960*, 92–100 (1959); Macfarlane, R.F., Fritillaria in California, *Lilies, 1975*, 53–66 (1975); Rix, E.M., Fritillaries in Iran, *Iranian Journal of Botany* 1: 75–95 (1977); Turrill, W.B. & Sealy, J.R., Fritillaries, *Hooker's Icones Plantarum* 39: parts 1 & 2 (1980).

1a. Style papillose, sometimes minutely so, or only in part                     2
  b. Style smooth, except for the stigmatic surface                           15
2a. Style undivided or 3-lobed with lobes less than 1 mm                        3
  b. Style 3-fid with branches more than 1 mm                                 8
3a. Flowers pink; nectaries visible as bulges on the outside of the perianth; bracts paired            **6. stenanthera**
  b. Flowers not pink; nectaries not visible as bulges on the outside of the perianth; bracts solitary or in a group of 3                                 4
4a. Bracts in a group of 3; flowers green, unmarked                **34. alfredae**

  b. Bract solitary; flowers not green and unmarked                             5
5a. Flowers pure yellow, unmarked
                                                **31. carica\***
  b. Flowers purplish, greyish or striped outside                               6
6a. Leaves linear                 **28. elwesii\***
  b. Leaves lanceolate                           7
7a. Leaves glaucous; style 1–2 mm in diameter                        **32. pinardii\***
  b. Leaves shining green; style 3–4 mm in diameter                  **33. uva-vulpis**
8a. Flowers in an umbel surmounted by a tuft of leaves; nectary circular, white, at the base of each perianth-segment
                                                          9
  b. Flowers usually solitary or in a raceme; nectary not white and at the base of each segment                    10
9a. Flowers orange or red, sometimes deep yellow in cultivated plants; nectary *c.* 5 mm in diameter
                                                **1. imperialis\***
  b. Flowers greenish yellow; nectary *c.* 2 mm in diameter           **2. raddeana**
10a. Leaves linear                             11
  b. Leaves lanceolate                          13
11a. Flowers narrowly bell-shaped
                                                **28. elwesii\***
  b. Flowers broadly bell-shaped             12
12a. Flowers purple or pink, tessellated with purple, or pure white
                                                  **7. meleagris**
  b. Flowers greenish, heavily tessellated with purplish brown or black
                                                **27. montana\***
13a. Flowers greenish, tessellated with purplish or brown         **21. hermonis**
  b. Flowers purplish at base, yellowish at apex, rarely all yellow               14
14a. Bracts paired; flowers usually in a long raceme               **26. reuteri**
  b. Bracts not paired; flowers usually solitary                  **23. davisii\***
15a. Style undivided, or lobed at apex for less than 1 mm                       16
  b. Style 3-fid, the branches more than 1 mm                                22
16a. Flowers commonly more than 4 in a raceme                               17
  b. Flowers solitary or rarely more than 4 in an umbel                         20
17a. Bracts in pairs; flowers whitish green
                                                  **5. bucharica**
  b. Bracts solitary or absent; flowers not whitish green                        18
18a. Flowers pink; style lobed at the apex
                                                **42. pluriflora\***
  b. Flowers black to brown, green or yellowish; style not lobed                19

19a. Flowers not flared at the mouth;
nectary triangular or rectangular
**3. persica**
  b. Flowers flared at the mouth; nectary
linear                **4. sewerzowii**
20a. Flowers purplish brown inside and
out                    **32. pinardii***
  b. Flowers yellow to green inside    21
21a. Flowers yellow outside; bulb
surrounded at the base with more
than 10 bulbils          **41. pudica**
  b. Flower green or purplish outside;
bulbils, when present, fewer than 10
**35. bithynica***
22a. Leaves mostly in whorls of 3 or more
23
  b. Leaves mostly alternate or opposite,
only the bracts sometimes in a whorl
of 3                      28
23a. Bulb surrounded at the base with
more than 10 bulbils      24
  b. Bulb not surrounded at the base with
more than 10 bulbils      26
24a. Flowers black; nectaries
inconspicuous  **38. camschatcensis**
  b. Flowers not black; nectaries
conspicuous              25
25a. Flowers orange or red  **46. recurva***
  b. Flowers brownish and green
**45. affinis***
26a. Flowers whitish, lightly tessellated, or
pinkish          **11. thunbergii***
  b. Flowers brown and green, tessellated,
sometimes pale yellowish green    27
27a. Leaves linear in 3–5 whorls of 2 or 3;
bracts coiled, tendril-like  **13. cirrhosa**
  b. Leaves lanceolate in 6 or 7 whorls of
4 or 5; bracts not tendril-like
**12. roylei**
28a. Bracts in a group of 3        29
  b. Bracts solitary or alternate    36
29a. Flowers tessellated          30
  b. Flowers not tessellated        35
30a. Leaves linear, less than 10 mm wide
31
  b. Leaves lanceolate, more than 10 mm
wide                      33
31a. Plant very small, with about 5 leaves;
flowers whitish; nectary linear
**37. japonica**
  b. Plant tall with 7–10 leaves; flowers
green or brown; nectary ovate    32
32a. Flowers mainly green; nectary less
than 10 mm long    **15. involucrata**
  b. Flowers mainly brown; nectary more
than 10 mm long    **17. messanensis**
33a. Flowers tessellated, purple on a white
ground              **8. latifolia***
  b. Flowers greenish or yellowish    34
34a. Flowers greenish yellow, lightly

tessellated with red-purple; nectary
2 mm long          **10. pallidiflora**
  b. Flowers green, tessellated, sometimes
heavily, with brown to purple;
nectary 4–5 mm long      **22. graeca**
35a. Flowers broadly bell-shaped, green
marked with red-brown; capsule
winged              **16. pontica**
  b. Flowers narrowly bell-shaped, red-
brown or sometimes purplish outside;
capsule not winged    **32. pinardii***
36a. Nectary at the base of the perianth-
segment; flower narrowly bell-shaped
or conical, rarely saucer-shaped or
flat                      37
  b. Nectary at the angle of the bell, above
the base of the perianth-segment;
flowers broadly bell-shaped      42
37a. Flowers pale pink, sometimes saucer-
shaped              **36. alburyana**
  b. Flowers not pink, narrowly bell-
shaped or conical        38
38a. Flowers pure yellow      **30. conica**
  b. Flowers not yellow        39
39a. Flowers brownish; leaves linear
**28. elwesii***
  b. Flowers blackish or whitish; leaves
lanceolate                40
40a. Flowers white or pale green
**40. liliacea***
  b. Flowers blackish          41
41a. Leaves glaucous, scattered up the
stem                **29. tuntasia***
  b. Leaves shining green, predominantly
basal                **39. biflora**
42a. Nectary linear to linear-lanceolate or
oblong, more than 4 times as long as
wide                      43
  b. Nectary ovate to lanceolate, less than
4 times as long as wide        49
43a. Leaves linear to linear-lanceolate
**18. lusitanica**
  b. Leaves lanceolate to ovate      44
44a. Flowers purple at the base, yellow at
the apex            **25. michailovskyi**
  b. Flowers green, tessellated with brown
or black, rarely yellowish      45
45a. Nectary linear, extending from the
corner of the bell to near the apex of
the segment          **24. crassifolia**
  b. Nectary linear-lanceolate or oblong,
less than half as long as the apical
portion of the segment        46
46a. Leaves usually glaucous; perianth-
segments striped        **22. graeca**
  b. Leaves usually shining green;
perianth-segments not striped    47
47a. Flowers yellow, lightly tessellated
**9. aurea***
  b. Flowers greenish or brownish purple
48

48a. Flowers greenish, mottled and veined;
style c. 1.5 cm        **43. purdyi**
  b. Flowers purplish brown, heavily
tessellated; style 7–9 mm
**23. davisii***
49a. Flowers green, not tessellated but
sometimes shaded with brown    50
  b. Flowers tessellated, not usually green
52
50a. Leaves lanceolate to ovate, clasping
the stem            **20. gussichae***
  b. Leaves linear to linear-lanceolate, not
clasping the stem        51
51a. Perianth-segments strongly recurved
at the apex, the inner much wider
than the outer        **19. acmopetala**
  b. Perianth-segments not recurved at
the apex, the inner slightly wider
than the outer        **17. messanensis**
52a. Flowers yellow, usually lightly
tessellated              53
  b. Flowers purplish or blackish, heavily
tessellated              55
53a. Flowers pale greenish yellow, usually
more than 2 per stem; capsule winged
**10. pallidiflora**
  b. Flowers pure yellow, sometimes
greenish outside, rarely more than 1
per stem; capsule not winged    54
54a. Style-branches 2–5 mm, not recurved
**9. aurea***
  b. Style-branches 6–9 mm, very slender,
recurved                **44. glauca**
55a. Flowers very broadly bell-shaped,
with angular shoulders to the bell,
ground colour white, tessellated with
purple              **8. latifolia***
  b. Flowers broadly bell-shaped, with
rounded shoulders to the bell; ground
colour greenish yellow, tessellated
with brown or black    **14. pyrenaica**

**1. F. imperialis** Linnaeus. Illustration:
Journal of the Royal Horticultural Society
**88**: f. 65 (1963); Wendelbo, Tulips and
irises of Iran, f. 23 (1977).
Bulb a compressed sphere to 10 cm in
diameter with a pungent, foxy smell. Stem
50–150 cm. Leaves lanceolate, in 3 or 4
whorls of 4–8. Bracts 10–20 in a close
group above the flowers. Flowers 3–5 in an
umbel, broadly bell-shaped, red or orange
(to yellow in cultivated clones). Perianth-
segments 4–5.5 cm, each with a large,
circular nectary c. 5 mm in diameter at its
base. Style 3–4.5 cm, papillose towards the
apex, 3-fid, the branches 1–4 mm. Capsule
winged. *S Turkey eastwards to NW India
(Kashmir)*. H1. Spring.
  Several cultivars are grown, differing in
size and flower colour.

**\*F. eduardii** Regel. Illustration: Rix & Phillips, The bulb book, 89 (1981). Differs in its more conical flowers, smaller nectaries and scentless bulb. *USSR (Soviet Tadzhikistan)*. H1. Spring.

**2. F. raddeana** Regel (*F. askabadensis* Micheli). Illustration: Journal of the Royal Horticultural Society **90**: f. 210, 215 (1965); Rix & Phillips, The bulb book, 79 (1981). Like *F. imperialis* but with 6–20 smaller, pale yellow, conical flowers. Perianth-segments 3–3.7 cm, with nectaries *c.* 2 mm in diameter. *NW Iran, USSR (SW Soviet Turkmenia)*. H2. Spring.

    **F. imperialis** var. **chitralensis** Anon. has 1–4 flowers of deeper yellow, but is otherwise similar to *F. raddeana*.

**3. F. persica** Linnaeus (*F. libanotica* (Boissier) Baker; *F. eggeri* Bornmüller). Illustration: Wendelbo, Tulips and irises of Iran, f. 24 (1977); Rix & Phillips, The bulb book, 88 (1981) – F. 'Adiyaman'. Bulb spindle-shaped, to 6 cm high, of closely fitting scales. Stem 20–150 cm. Leaves lanceolate, all alternate. Bracts 1 at the base of each flower-stalk, or absent. Flowers 7–20 or more in a raceme, narrowly bell-shaped, greyish, purplish, blackish or greenish. Perianth-segments 1.5–2 cm, all alike, each with a small, triangular to rectangular nectary *c.* 2 mm long above the base. Style 6–8 mm, slender, smooth. Capsule narrowly winged. *S Turkey to W Iran and south to Jordan and Israel*. H2. Spring.

    'Adiyaman' is the finest and most commonly grown variant.

**4. F. sewerzowii** Regel (*Korolkowia sewerzovii* (Regel) Regel). Illustration: Rix & Phillips, The bulb book, 85 (1981). Bulb a compressed sphere to 6 cm in diameter, of 1 or 2 closely fitting scales. Stem 15–20 cm. Leaves broadly lanceolate, alternate or the lowest opposite. Bracts to 7 × 2 cm, 1 at the base of each flower-stalk. Flowers 4–12 (rarely 1) in a long raceme, narrowly bell-shaped, widely flared at the mouth, greenish or purplish outside, yellowish to brick-red inside. Perianth-segments 2.5–3.5 cm, all alike, each with a grooved, linear nectary 1–1.5 cm long. Anthers versatile, fixed *c.* 2 mm above the base. Style 1.5–2 cm, slender, undivided, smooth. Capsule winged. *USSR (Soviet Central Asia), NW China*. H1. Spring.

**5. F. bucharica** Regel. Illustration: Rix & Phillips, The bulb book, 86, 87 (1981). Bulb spherical, to 4 cm in diameter, of 2 large scales. Stem 10–35 cm, papillose.

Leaves lanceolate to ovate, the lower in 2 pairs, the rest alternate. Bracts 2 at the base of each flower-stalk. Flowers 1–10, usually 4–7 in a raceme, cup-shaped, horizontal or pendulous, whitish with green veins. Perianth-segments 1.4–2 cm, lanceolate, each with a deeply indented nectary about 3 mm long. Style 3–4 mm, slender, smooth. Capsule winged. *USSR (Soviet Central Asia), N Afghanistan*. H1. Spring.

**6. F. stenanthera** (Regel) Regel. Illustration: Rix & Phillips, The bulb book, 55 (1981). Like *F. bucharica* but usually shorter, with the lowest leaves ovate, opposite, the rest much narrower. Flowers pinkish, conical, widely flared at the mouth, the outer perianth-segments larger than the inner, the nectaries visible as bulges on the outside of the perianth. Style *c.* 5 mm, slender, papillose below. Capsule with 6 horn-like wings on the corners. *USSR (Soviet Central Asia)*. H1. Spring.

**7. F. meleagris** Linnaeus. Illustration: Rix & Phillips, The bulb book, 77 (1981); Hooker's Icones Plantarum **39**: t. 3801 (1980). Bulb spherical, to 2.5 cm in diameter, of 2 large scales. Stem 12–30 cm. Leaves 4–6, linear, alternate. Flower 1, very broadly bell-shaped, purple or pinkish, tessellated all over, to pure white. Perianth-segments 3–4.5 cm. the inner slightly wider than the outer, all acute, each with a linear (7–10 mm long), green nectary at the angle of the bell. Style 1.3–1.6 cm, 3-fid, the branches 2–5 mm, papillose. Capsule not winged. *Europe, from S England and Sweden (naturalised) to N Yugoslavia, Romania and W Russia*. H1. Spring.

    Various cultivars, some with greenish flowers, have been named, but are now seldom offered. The white-flowered plants appear to be most common in England, rare in southern Europe.

**8. F. latifolia** Willdenow (*F. nobilis* Baker). Illustration: Mathew & Baytop, The bulbous plants of Turkey, t. 89 (1984). Bulb spherical, to 2.5 cm in diameter. Stem 4–35 cm. Leaves 5–9, usually 6 or 7, ovate to lanceolate, alternate, shining green. Flower 1, very broadly bell-shaped, dark reddish purple, tessellated. Perianth-segments 3.5–5 cm, obtuse or rounded, incurved at the apex, each with a narrowly ovate nectary, 3–4 mm long. Style 1–1.5 cm, smooth, 3-fid; branches 2–5 mm. Capsule not winged. *NE Turkey, USSR (Caucasus); ?NW Iran*. H1. Spring.

**\*F. tubiformis** Grenier & Godron (*F. delphinensis* Grenier & Godron). Illustration: Rix & Phillips, The bulb book, 86 (1981). Like *F. latifolia*, but with narrower, glaucous leaves, flowers greyish outside and style 1.2–1.3 cm, branches *c.* 1 mm. *France, Italy (SW Alps)*. H3. Spring.

**9. F. aurea** Schott (*F. bornmulleri* Haussknecht). Illustration: Rix & Phillips, The bulb book, 81 (1981). Like *F. latifolia* but with usually smaller, yellow flowers. Perianth-segments 2–4.8 cm, with diamond-shaped nectaries *c.* 2 mm across. *Turkey*. H1. Spring.

**\*F. collina** Adams (*F. lutea* Bieberstein). Illustration: Rix & Phillips, The bulb book, 85 (1981). Leaves narrower, flowers with acute perianth-segments, the inner fringed on the margin, nectaries deeply indented. *USSR (S Russia, Georgia)*. H1. Spring.

**\*F. tubiformis** Grenier & Godron subsp. **moggridgei** (Boissier & Reuter) Rix. Flowers greenish yellow, perianth-segments rounded with linear-lanceolate nectaries 4–5 mm long, style 1.3–1.7 cm. *France, Italy (SW Alps)*. H3. Spring.

    These are not merely semi-albinos of *F. latifolia* and *F. tubiformis* (see also under No. **8**); they occur in separate populations and differ in minor characters. Pure white albinos of *F. tubiformis* are recorded, though very rare.

**10. F. pallidiflora** Schrenk. Illustration: Journal of the Royal Horticultural Society **85**: f. 127 (1960). Bulb spherical to spindle-shaped, to 5 cm in diameter. Stem 10–80 cm. Leaves glaucous, broadly lanceolate, opposite or alternate. Flowers 1–6 (rarely to 12), very broadly bell-shaped, pale greenish yellow, faintly tessellated with reddish brown. Perianth-segments 2.5–4.5 cm, rounded, each with an ovate, deeply indented nectary *c.* 2 mm long, at the angle of the bell. Style 1.4–1.7 cm, smooth, 3-fid, the branches 2–4 mm. Capsule with 6 wide wings. *USSR (E Siberia), NW China*. H2. Spring.

**11. F. thunbergii** Miquel (*F. verticillata* var. *thunbergii* (Miquel) Baker). Illustration: Rix & Phillips, The bulb book, 87, 91 (1981). Bulb spindle-shaped, to 4 cm in diameter. Stem 30–80 cm. Leaves many, linear, the lowest opposite, the rest in whorls or alternate, the bracts coiled and tendril-like. Flowers 2–6 (rarely 1), broadly bell- or cup-shaped, whitish, faintly tessellated or veined with green. Perianth-segments 2.3–3.5 cm, obtuse, each with a linear-lanceolate nectary 3–4 mm long *c.* 2 mm above the base. Style *c.* 10 mm, smooth, 3-fid, the branches 2 mm. Capsule winged. *?C China; widely cultivated in China and Japan.* H1. Spring.

**\*F. verticillata** Willdenow. Illustration: Botanical Magazine, 3083 (1831) – as *F. leucantha*; Hooker's Icones Plantarum **39**: t. 3827 (1980). With fewer, larger flowers which are usually unmarked, and fewer leaves. Nectaries deeply indented. *NW China, USSR (E Siberia).* H1. Spring.

**\*F. walujewii** Regel. Illustration: Rix & Phillips, The bulb book, 87 (1981). Flowers reddish inside, reddish or white outside, nectary larger, lanceolate. *USSR (Soviet Tadzhikistan), NW China.* H1. Spring.

**12. F. roylei** J.D. Hooker. Illustration: Hooker's Icones Plantarum, t. 860 (1852). Bulb to 3 cm in diameter, spindle-shaped. Stem 20–60 cm. Leaves many in 6 or 7 whorls of 4 or 5, lanceolate. Flowers 1–4, broadly bell-shaped, yellowish green, spotted and streaked with dull purple. Perianth-segments 3.5–5 cm, rounded, each with a large, ovate nectary at the angle of the bell. Style *c.* 1.2 cm, 3-fid, the branches *c.* 5 mm. Capsule narrowly winged. *N India (Kashmir, Punjab).* H1. Late spring.

**13. F. cirrhosa** D. Don. Illustration: Botanical Magazine, n.s., 255 (1955). Like *F. roylei* but with fewer, usually linear leaves in 3–5 whorls of 2 or 3, the upper 3 bracts tendril-like. Flowers purplish, greenish or whitish, variously marked. Perianth-segments acute, nectary at the angle of the bell, smaller than in *F. roylei*. Capsule winged. *Himalaya, China.* H1. Late spring.

**14. F. pyrenaica** Linnaeus (*F. nigra* Miller). Illustration: Hooker's Icones Plantarum **39**: t. 3817 (1980); Rix & Phillips, The bulb book, 91, 93 (1981). Bulbs to 3 cm in diameter, spherical. Stem 15–30 cm. Leaves 7–10, lanceolate to linear-lanceolate, alternate. Flowers solitary (rarely 2 together), broadly bell-shaped, outside blackish purple to brownish, heavily tessellated; inside yellowish green, the basal half tessellated with brown (sometimes pure yellow – var. **lutescens** Baker). Perianth-segments 2.5–3.5 cm, the inner obtuse, recurved at the apex, each with a triangular to ovate-lanceolate nectary at the angle of the bell. Style 8–9 mm, smooth, 3-fid, the branches

2–4 mm. Capsule not winged. *SC France, NW Spain.* H1. Spring.

**15. F. involucrata** Allioni. Illustration: Hooker's Icones Plantarum **39**: t. 3807 (1980); Rix & Phillips, The bulb book, 92 (1981). Bulbs to 3 cm in diameter, spherical. Stem 15–25 cm. Leaves 7–10, linear-lanceolate to linear, opposite or almost so, the bracts in a whorl of 3. Flower solitary, broadly bell-shaped, green, variously tessellated with purplish brown, or not tessellated. Perianth-segments 2.5–4 cm, all obtuse and not recurved at the apex, each with an ovate, blackish nectary at the angle of the bell. Style 1.2–1.5 cm, smooth, 3-fid, the branches 5–7 mm. Capsule not winged. *SE France, NW Italy.* H1. Spring.

**16. F. pontica** Wahlenberg. Illustration: Botanical Magazine, 8865 (1920); Rix & Phillips, The bulb book, 92 (1981). Bulb to 3 cm in diameter, spherical, often with an antler-like projection when dormant. Stem 15–45 cm. Leaves about 8, lanceolate to linear-lanceolate, opposite or almost so, the bracts in a whorl of 3, close to the flower. Flowers solitary or rarely 2 together, broadly bell-shaped, green, marbled with reddish brown or unmarbled, not tessellated. Perianth-segments 2.4–4.5 cm, the inner obtuse, somewhat recurved at the apex, each with a round, blackish nectary at the angle of the bell. Style 1.2–1.5 cm, smooth, 3-fid, the branches 5–7 mm. Capsule with 6 narrow wings on the angles. *S Albania, S Yugoslavia, N Greece, NW Turkey.* H1. Spring.

**17. F. messanensis** Rafinesque (*F. oranensis* Pomel). Illustration: Rix & Phillips, The bulb book, 91 (1981). Bulb to 4 cm in diameter, spherical. Stem 15–45 cm. Leaves 7–10, linear, the lowest usually opposite, the bracts in a whorl of 3. Flowers solitary or rarely up to 3 together, broadly bell-shaped, green, tessellated with reddish brown. Perianth-segments 2.2–4.2 cm, the inner pointed and recurved at the apex, each with a large, ovate-lanceolate, green nectary near the base. Style *c.* 10 mm, smooth, 3-fid, the branches 5–6 mm. Capsule not winged. *N Africa, Sicily, S Italy, Crete, Greece.* H1. Spring.

Subsp. **gracilis** (Ebel) Rix (*F. gracilis* (Ebel) Ascherson & Graebner). Illustration: Rix & Phillips, The bulb book, 90 (1981). Bracts usually alternate, perianth untessellated, the segments not recurved. *Yugoslavia, Albania.* H1. Spring.

Var. **atlantica** Maire (*F. messanensis* forma *oranensis* misapplied) has very glaucous, shorter leaves. *Morocco (Atlas mountains).* H1. Spring.

**18. F. lusitanica** Wikström (*F. hispanica* Boissier & Reuter). Illustration: Rix & Phillips, The bulb book, 73 (1981). Bulb to 2.5 cm in diameter. Stem 10–50 cm. Leaves usually 6–9, alternate, linear to narrowly lanceolate, the bracts alternate. Flowers 1–3, broadly bell-shaped, green to brown outside, yellowish inside, marked or tessellated with brown. Perianth-segments 2–4 cm, the inner usually acute, sometimes recurved at the apex, each with a linear to linear-lanceolate nectary which is 1–1.2 cm long, 3–4 mm above the base. Style 8–11 mm, smooth, 3-fid, the branches 3–4 mm. Capsule not winged. *C & S Spain, Portugal.* H2. Spring.

The fritillaries of the Iberian peninsula (except for the north where *F. pyrenaica* is found) do not seem to be satisfactorily classifiable. They range from the coastal variants from Portugal (illustration cited above) to high alpine plants from the Sierra Nevada, which have shorter stems and much longer leaves and flowers, but all have alternate, narrow leaves and long, narrow nectaries.

**19. F. acmopetala** Boissier. Illustration: Botanical Magazine, 9148 (1928); Rix & Phillips, The bulb book, 92 (1981). Bulb to 3 cm in diameter, often with bulblets. Stem 14–45 cm. Leaves 7–11, alternate, linear. Flowers solitary or rarely 3 together, broadly bell-shaped, green, marked but not tessellated with reddish brown. Perianth-segments 2.5–4 cm, the inner pointed, recurved at the apex, each with a large ovate to ovate-lanceolate, green or blackish nectary *c.* 5 mm above the base. Style 8–12 mm, smooth, 3-fid, the branches 3–5 mm. Capsule not winged. *S Turkey, Syria, Lebanon, Cyprus.* H2. Spring.

**20. F. gussichae** (Degen & Doerfler) Rix (*F. graeca* Boissier & Spruner var. *gussichae* Degen & Doerfler). Illustration: Hooker's Icones Plantarum **39**: t. 3822 (1980); Rix & Phillips, The bulb book, 92 (1981). Bulb to 25 cm in diameter. Stem 20–30 cm. Leaves 5–8, very glaucous, alternate, ovate to lanceolate, stem-clasping. Flowers 1–3 in a long raceme, broadly bell-shaped, green, marked but not tessellated with reddish brown. Perianth-segments *c.* 3 cm, obtuse, not recurved at the apex, each with an ovate, green nectary *c.* 6 mm above the base. Style *c.* 9 mm, smooth, 3-fid, the branches 4 mm. Capsule narrowly winged. *Yugoslavia, N Greece, Bulgaria.* H2. Spring.

**\*F. olivieri** Baker. Illustration: Journal of the Royal Horticultural Society **88**: f. 70 (1980); Hooker's Icones Plantarum **39**: t. 3818 (1980). Leaves usually green, narrowly lanceolate, capsule unwinged. *Iran.* H2. Spring.

**21. F. hermonis** Fenzl.
Bulb to 2 cm in diameter, often with bulblets or stolons. Stem 8–35 cm. Leaves 5 or 6 (rarely to 9), alternate, lanceolate or oblong. Flowers 1 or 2, broadly bell-shaped, green, faintly tessellated with brown or purple. Perianth-segments 2.5–3.5 cm, the inner obtuse, each with an ovate, green or blackish nectary at the angle of the bell. Style 8–12 mm, papillose, 3-fid for 3–4 mm. Capsule not winged. *S Turkey to Lebanon.* H1. Spring.

Subsp. **amana** Rix (*F. crassifolia* misapplied). Illustration: Kew Bulletin **29**: 649 (1974). This is the commonly cultivated variant, described above. Subsp. *hermonis* which is found only on Mt. Hermon, is not in cultivation.

**22. F. graeca** Boissier & Spruner.
Illustration, Rix & Phillips, The bulb book, 83, 90 (1981).
Bulbs to 2.5 cm in diameter, spherical or spindle-shaped. Stem 6–20 cm. Leaves 5–12, usually 7–10, glaucous, alternate or the lowest opposite, bracts alternate. Flowers 1–3, broadly bell-shaped, green, tessellated, often heavily, with black or brown, usually with a clear, green, central stripe to each segment. Perianth-segments 1.8–2.4 cm, the inner obtuse, each with a lanceolate nectary at the angle of the bell. Style 7–10 mm, smooth, 3-fid, the branches 3–6 mm. Capsule not winged. *S Greece.* H1. Spring.

Subsp. **thessala** (Boissier) Rix. (Illustration: Hooker's Icones Plantarum **39**: t. 3806, 1980 – as F. pontica var. ionica; Rix & Phillips, The bulb book, 90, 1981) is generally larger, not glaucous and has the lower leaves opposite, the intermediate leaves opposite or alternate, the bracts in a whorl of 3; the perianth-segments are 2.8–3.8 cm, green, usually lightly tessellated. *S Albania, Yugoslavia, NW Greece.*

Subsp. *thessala* is clearly distinguished from *F. pontica* (p. 190) by its unwinged capsule, tessellated flowers and longer flower-stalks.

**23. F. davisii** Turrill. Illustration: Hooker's Icones Plantarum, t. 3427 (1943); Rix & Phillips, The bulb book, 83 (1981). Like *F. graeca* but the leaves shining green, the lowest wider, spreading. Flowers without a clear, green stripe on each segment. Style 7–9 mm, 3-fid, branches 3–7 mm. *S Greece.* H3. Spring.

**\*F. rhodocanakis** Orphanides. Flowers dark purple, not tessellated, each segment with a yellow apex, style often papillose. *S Greece (Idhra).* H3. Spring.

Intermediates between this and *F. davisii* are found on the Greek mainland close to Idhra.

**24. F. crassifolia** Boissier & Reuter.
Like *F. graeca* but leaves always alternate, nectary linear, 8–12 × 1–2 mm, forming a raised ridge from the angle of the bell to near the apex of the segment, flowers often yellowish or greenish with smaller brownish tessellations and less distinct green stripes. *SE Turkey.* H1. Spring.

Subsp. **crassifolia** is rarely cultivated. Subsp. **kurdica** (Boissier & Noë) Rix (*F. karadaghensis* Turrill) is described above (Illustration: Rix & Phillips, The bulb book, 82, 1981)

**25. F. michailovskyi** Fomin. Illustration: RHS Lily Year Book **33**: f. 25 (1969); Rix & Phillips, The bulb book, 82 (1981); Gartenpraxis **11**: t. 4 (1982).
Bulb to 2.5 cm in diameter. Stem 6–24 cm. Leaves 5–9, alternate or the lowest opposite or almost so, lanceolate. Flowers 1–4, more or less in an umbel, broadly bell-shaped, purplish brown or greenish outside, the apical one-quarter or one-third bright yellow. Perianth-segments 2–3.2 cm, the inner obtuse, each with a linear nectary from the angle of the bell to near the apex of the segment. Style 7–9 mm, smooth, 3-fid, the branches 2–3 mm. Capsule not winged, the dried perianth remaining attached to it. *NE Turkey.* H1. Spring.

Hybridises very easily with *F. crassifolia* subsp. *kurdica*; the hybrids have the flowers shading to greenish yellow at the apex and not with a clearly defined zone of yellow.

**26. F. reuteri** Boissier. Illustration: Journal of the Royal Horticultural Society **94**: f. 61–2 (1969); Botanical Magazine, n.s., 658 (1973).
Like *F. michailovskyi* in flower colour, but usually taller (15–25 cm), with up to 8 cup-shaped flowers with perianth-segments 1.5–2.5 cm; style minutely papillose. *C Iran.* H1. Spring.

**27. F. montana** Hoppe (*F. racemosa* Ker Gawler; *F. orientalis* misapplied; *F. tenella* misapplied; *F. nigra* misapplied). Illustration: Hooker's Icones Plantarum **39**: t. 3803 (1980); Rix & Phillips, The bulb book, 86 (1981).
Bulb to 2.5 cm in diameter. Stem 16–40 cm. Leaves 8–20, linear, the lowest usually opposite, the rest alternate or in whorls of 3, bracts not coiled at the tips. Flowers 1–3, broadly bell-shaped, green, heavily tessellated with blackish purple or brown. Perianth-segments 1.8–3.2 cm, not recurved at the apex, each with a linear nectary 1–1.5 cm long, borne *c.* 5 mm above the base. Style 8–10 mm, papillose, 3-fid, the branches 2–7 mm. Capsule not winged. *S France, Italy, Yugoslavia, N Greece.* H1. Spring.

**\*F. ruthenica** Wikström. Illustration: Hooker's Icones Plantarum **39**: t. 3829 (1980). Bracts coiled; capsule winged. *USSR (European Russia, W Siberia).* H1. Spring.

**\*F. orientalis** Adam (*F. tenella* Bieberstein). Very like *F. montana* in leaf arrangement, but similar to *F. meleagris* (p. 189) in flower. *USSR (Caucasus).*

**28. F. elwesii** Boissier. Illustration: Botanical Magazine, 6321 (1877) – as F. acmopetala; Hooker's Icones Plantarum **39**: t. 3820 (1980).
Bulb to 3 cm in diameter. Stem 15–55 cm. Leaves 4–8, alternate, linear. Flowers 1–4, narrowly bell-shaped, brownish purple with clear, green stripes. Perianth-segments 2–3 cm, the inner obtuse or abruptly pointed, each with a small (2–3 mm), lanceolate nectary at the base. Style 7–11 mm, very papillose, stout, 3-fid, the branches 1–3.5 mm. Capsule not winged. *S Turkey.* H2. Spring.

**\*F. latakiensis** Rix. Style slender, smooth or minutely papillose. *S Turkey, Syria.*

Both this and *F. elwesii* are variable, and probably originated by hybridisation between *F. acmopetala* (p. 190) and *F. assyriaca.*

**\*F. assyriaca** Baker. Illustration: Rix & Phillips, The bulb book, 78 (1981). Plant much shorter at flowering, stem elongating in fruit, flowers 1.2–2.5 cm, greyish outside and not clearly striped, the style always undivided. *C Turkey.* H2.

**29. F. tuntasia** Heldreich. Illustration: Rix & Phillips, The bulb book, 86 (1981).
Bulb to 2.5 cm in diameter. Stem 10–35 cm. Leaves 8–25, alternate, narrowly lanceolate. Flowers 1–6, conical to bell-shaped, blackish purple and

glaucous outside. Perianth-segments 1.5–2.5 cm, each with a linear nectary c. 4 mm long at the base. Style c. 10 mm, smooth, 3-fid, the branches 1–1.5 mm. Capsule not winged. *Greece (Kithnos, Serifos)*. H2. Spring.

*****F. obliqua** Ker Gawler. Plant shorter, with fewer leaves, flowers slightly larger (perianth-segments 2–3 cm). *S Greece (Attica)*. H2. Spring.

**30. F. conica** Boissier. Illustration: Rix & Phillips, The bulb book, 80 (1981).
Bulb to 2 cm in diameter. Stem 7–25 cm. Leaves 5–7, shining green, the lowest broadly lanceolate, more or less opposite, the rest alternate. Flowers 1 or 2, conical or bell-shaped, yellow, not tessellated. Perianth-segments 1.2–2 cm, each with a greenish nectary c. 1 mm at the base. Style 7–9 mm, smooth, 3-fid, the branches 2–4 mm. Capsule not winged. *S Greece*. H2. Spring.

**31. F. carica** Rix (*F. sibthorpiana* misapplied). Illustration: Botanical Magazine, n.s., 129 (1950); Rix & Phillips, The bulb book, 80 (1981).
Bulb to 3 cm in diameter. Stem 5–15 cm. Leaves 6–8, lanceolate, all alternate, glaucous. Flowers 1–3, narrowly bell-shaped, yellow. Perianth-segments 1.3–2 cm, obtuse, each with a small (2–4 mm), blackish, narrowly ovate nectary at the base. Style 6–10 mm, stout, papillose, entire or 3-lobed at the apex, the lobes to 1 mm. Capsule not winged. *SW Turkey*. H2. Spring.

*****F. sibthorpiana** (Smith) Baker. Leaves 2 or 3, ovate-lanceolate. *Greece (Symi), SW Turkey*. H2. Spring.

*****F. forbesii** Baker. Leaves 5–10, linear or linear-lanceolate. *SW Turkey*. H2. Spring.

**32. F. pinardii** Boissier (*F. syriaca* Hayek & Siehe). Illustration: Botanical Magazine, n.s., 227 (1954); Rix & Phillips, The bulb book, 80, 81 (1981).
Bulb to 3 cm in diameter, often with bulblets. Stem 6–20 cm. Leaves 3–8, glaucous, lanceolate, alternate. Flowers 1 or 2 (rarely to 4 together), narrowly bell-shaped, greyish or purplish with a glaucous bloom outside, orange, yellow or greenish inside, not tessellated. Perianth-segments 1.5–2.5 cm, the inner obtuse, each with a small (3–5 mm), linear-lanceolate nectary at the base. Style 7–10 mm, slender, less than 3 mm in diameter, papillose, undivided or 3-fid at the apex, the branches 1–2 mm. Capsule not winged. *Turkey*. H1. Spring.

*****F. armena** Boissier. Flowers dark purplish brown inside and outside. *NE Turkey*.

*****F. caucasica** Adams. Flowers dark purplish brown inside and outside, style 9–17 mm, smooth, undivided. *NE Turkey, adjacent Caucasus*.

*****F. minuta** Boissier & Noë (*F. carduchorum* Rix). Flowers brownish orange, style smooth, deeply 3-fid. *SE Turkey*.

*****F. drenovskii** Degen & Stojanoff. Illustration: Rix & Phillips, The bulb book, 92 (1981). Flowers purple-brown or greenish inside, sometimes greyish outside, leaves scattered, narrowly lanceolate. *NE Greece, SW Bulgaria*.

**33. F. uva-vulpis** Rix (*F. assyriaca* misapplied). Illustration: Iranian Journal of Botany **1**: t. 5 (1977); Hooker's Icones Plantarum **39**: t. 3831 (1980); Rix & Phillips, The bulb book, 80 (1981).
Bulb to 3 cm in diameter, usually with a few large bulblets. Stem 10–35 cm. Leaves 3–5, shining green, alternate, lanceolate. Flowers solitary (rarely 2 together), narrowly bell-shaped, rounded, outside purplish grey, glaucous, inside yellowish. Perianth-segments 2–2.8 cm, the inner obtuse, each with a large, ovate nectary at the base. Style 5–7 mm, papillose, stout (3–4 mm in diameter), undivided. Capsule not winged. *SE Turkey, N Iraq, W Iran*. H1. Spring.

**34. F. alfredae** Post.
Bulb to 3 cm in diameter. Stem 10–36 cm. Leaves 9–11, the lowest oblanceolate to ovate, often opposite; bracts usually in a whorl of 3. Flowers 1–3, narrowly bell-shaped, green, glaucous outside, yellowish green inside, unmarked. Perianth-segments 1.2–3 cm, the inner spathulate, each with an ovate, green nectary, c. 5 × 2 mm at the base. Style 5–10 mm, stout (to 3 mm in diameter), papillose, undivided or 3-lobed. Capsule broadly 6-winged. *S Turkey*. H2. Spring.

Subsp. **glaucoviridis** (Turrill) Rix (Illustration: Botanical Magazine, 9462, 1936; Hooker's Icones Plantarum **39**: t. 3830, 1980) is described above; subsp. *alfredae* from the Lebanon is a smaller, more slender plant, and is not cultivated.

**35. F. bithynica** Baker (*F. dasyphylla* Baker; *F. schliemannii* Sintenis; *F. citrina* Baker; *F. pineticola* Schwarz). Illustration: Botanical Magazine, 6321 (1877).
Bulb to 2 cm in diameter. Stem 7–20 cm. Leaves 5–12, usually 6–8, the lowest

oblanceolate to ovate, usually opposite; bracts usually in a whorl of 3. Flowers 1 or 2 (rarely to 4), narrowly bell-shaped, outside glaucous or yellowish green, rarely with purple markings, inside greenish yellow. Perianth-segments 1.7–2.7 cm, the inner obovate, cuneate, each with a lanceolate nectary, c. 3 × 1 mm at the base. Style 7–10 mm, slender, smooth, undivided. Capsule usually winged. *W Turkey, Greek Islands (Samos, Khios)*. H2. Spring.

*****F. ehrhartii** Boissier & Orphanides. Illustration: Rix & Phillips, The bulb book, 84 (1981). Flowers purple outside, apex yellow; capsule always unwinged. *Greece (Aegean area)*. H2. Spring.

**36. F. alburyana** Rix. Illustration: Rix & Phillips, The bulb book, 83 (1981); Gartenpraxis, January 1982, cover.
Bulb to 2 cm in diameter, often with bulblets. Stem 4–10 cm. Leaves 3 or 4, alternate, lanceolate. Flowers 1 or 2, upright, horizontal or drooping, cup-shaped to almost flat, pale pink, tessellated. Perianth-segments 2–3 cm, each with an elliptic, green nectary, c. 1 mm at the base. Anthers semi-versatile. Style 9–15 mm, slender, smooth, 3-fid, the branches 1–2 mm. Capsule not winged. *NE Turkey*. H1. Spring.

**37. F. japonica** Miquel.
Bulb to 1 cm in diameter. Leaves 5–7, the lower opposite, linear-lanceolate; bracts in a whorl of 3. Flowers solitary (rarely 2 together), drooping, broadly bell-shaped, whitish, lightly marked and tessellated with brown and purple. Perianth-segments 2–2.5 cm (fringed on the margin in var. **koidzumiana** Nakai), each with a linear nectary, 5–8 mm long, from the angle of the bell to near the apex. Style 8 mm, 3-fid, the branches to 5 mm. Capsule drooping, not winged. *Japan*. H2. Spring.

**38. F. camschatcensis** (Linnaeus) Ker Gawler. Illustration: Hooker's Icones Plantarum **39**: t. 3844 (1980); Rix & Phillips, The bulb book, 159 (1981).
Bulb to 2.5 cm in diameter, of numerous closely-packed scales, with many small bulblets around the base. Stem 15–75 cm. Leaves many, in 1–3 whorls of 5–7 below, alternate above, all lanceolate. Flowers 1–8, broadly bell-shaped or cup-shaped, purple-brown to black, rarely green or yellowish. Perianth-segments 2–3 cm, acute, each with c. 12 raised lines inside and a rather inconspicuous, narrowly oblong nectary. Anthers versatile. Style

8–10 mm, smooth, deeply 3-fid, the branches 6–8 mm, recurved. Capsule angled but not winged. *Japan, E Siberia, N America from Alaska to Washington.* H1. Late spring.

**39. F. biflora** Lindley. Illustration: Lilies, 1975, 10 (1975); Hooker's Icones Plantarum **39**: t. 3843 (1980).
Bulb to 2 cm in diameter, of about 3, thick, loose scales. Stem 15–40 cm. Leaves shining green, mostly basal, alternate or opposite, ovate-lanceolate. Flowers 1–6 (rarely to 12), bell-shaped, blackish brown to purple, suffused with green. Perianth-segments 2–3.5 cm, each with raised, longitudinal lines inside, and an inconspicuous linear nectary. Anthers versatile. Style *c.* 9 mm, deeply 3-fid, the branches *c.* 5 mm, recurved. Capsule 6-angled, not winged. *USA (California).* H3. Spring.

**40. F. liliacea** Lindley. Illustration: Botanical Magazine, 9541 (1939); Hooker's Icones Plantarum **39**: t. 3842 (1980); Rix & Phillips, The bulb book, 94 (1981).
Bulbs to 2.5 cm in diameter, of 4–6 fleshy, ovate, loose scales. Stem 15–35 cm. Leaves 2–20, mostly basal, shining green, the lower opposite, oblong-lanceolate or obovate; bracts alternate. Flowers 2–4 (rarely to 8), openly conical to bell-shaped, white with a greenish or yellow throat. Perianth-segments 1.2–2.5 cm, each with a small, oblong, greenish yellow, sometimes purple-spotted nectary. Style *c.* 8 mm, smooth, 3-fid to more than halfway, the branches 4–5 mm, recurved. Capsule angled but not winged. *USA (California).* H4. Spring.

**\*F. agrestis** Greene. Illustration: Rix & Phillips, The bulb book, 94 (1981). Leaves often glaucous, flowers dull yellow-green, usually spotted with purple-brown inside, nectaries conspicuous. *USA (California).* H3. Spring.

**\*F. grayana** Reichenbach & Baker (*F. roderickii* Knight). Illustration: Rix & Phillips, The bulb book, 93 (1981). Leaves narrower, oblanceolate; flowers greenish to reddish brown, with a cream or green patch at the apex outside; perianth-segments with dark brown raised lines and each with an inconspicuous, linear nectary almost as long as the segment. *USA (California).* H3. Spring.

**41. F. pudica** (Pursh) Sprengel. Illustration: Botanical Magazine, 9617 (1942); Hooker's Icones Plantarum **39**: t. 3840 (1980); Rix & Phillips, The bulb book, 94 (1981).

Bulb a thick disc of 2–4 scales, surrounded at the base by numerous bulblets. Stem 7–30 cm. Leaves 2–7, linear to narrowly lanceolate, the lowest usually opposite, the rest alternate. Flowers 1 or 2 (rarely to 6), narrowly bell-shaped, yellow to orange-yellow. Perianth-segments 1–2.5 cm, each with a small dark nectary at the base. Style 8 mm, slender, smooth, undivided. Capsule not winged. *Western N America.* H1. Spring.

**42. F. pluriflora** Torrey. Illustration: Botanical Magazine, 7631 (1898); Rix & Phillips, The bulb book, 95 (1981).
Bulb of up to 8 ovate, fleshy scales. Stem 18–40 cm. Leaves 6–13, mostly basal, oblong-lanceolate. Flowers 1–4 (rarely to 12), drooping or facing outwards, pink to pinkish purple, unmarked. Perianth-segments 2–3.5 cm, each with a linear nectary running for two-thirds of the length of the segment. Style 1–1.5 cm, entire or 3-lobed at the apex. Capsule 3-ridged, not winged. *USA (California).* H3. Spring.

**\*F. striata** Eastwood. Illustration: Rix & Phillips, The bulb book, 95 (1981). Flowers sweetly scented, pale, perianth-segments recurved at their tips, each with a circular nectary at the base. *W USA.*

**43. F. purdyi** Eastwood. Illustration: Rix & Phillips, The bulb book, 94 (1981).
Bulb of 3 or 4 ovoid, fleshy scales 8–15 mm long. Stem 10–40 cm. Leaves mainly basal, forming a rosette, elliptic to oblong-lanceolate, margins usually wavy. Flowers 1–4 (rarely to 14), bell-shaped, drooping, pale greenish or whitish heavily mottled and veined with purplish brown. Perianth-segments *c.* 1.5 cm, each with an oblong nectary, 2 × 10 mm. Style *c.* 1.5 cm, 3-fid, the branches to 1 cm, spreading and recurved. Capsule 6-angled, not winged. *USA (California).* H2. Spring.

**44. F. glauca** Greene. Illustration: Hooker's Icones Plantarum **39**: t. 3840 (1980) – poor specimen.
Bulb 5–10 mm in diameter, of 2 or 3 large scales and a few smaller ones at the base. Stem 12–18 cm. Leaves 2–4, glaucous, alternate, mainly basal, recurved. Flowers solitary (rarely up to 4 together), broadly bell-shaped, drooping, yellow mottled with brown or rarely all brown. Perianth-segments 1.5–2 cm, each with a lanceolate or elliptic nectary 5–6 mm long. Style to 1.2 cm, 3-fid, the branches 6–9 mm, recurved. Capsule 6-angled, not winged. *USA (California, Oregon).* H2. Spring.

**45. F. affinis** (Schultes) Sealy (*F. lanceolata* invalid). Illustration: Hooker's Icones Plantarum **39**: t. 3847 (1980); Rix & Phillips, The bulb book, 94 (1981).
Bulb to 2 cm in diameter with numerous bulblets around its base. Stem 15–120 cm. Leaves in 1–3 whorls of 3–5, linear-lanceolate to ovate; bracts 1 per flower. Flowers 1–4 (rarely to 12) in a raceme, broadly bell-shaped, drooping, greenish yellow to purple, tessellated. Perianth-segments 2–4 cm, each with a prominent, triangular to ovate-lanceolate nectary 1–2 cm long. Style 3-fid, the branches recurved. Capsule 6-winged. *Western N America.* H2. Spring.

**\*F. atropurpurea** Nuttall. Illustration: Rix & Phillips, The bulb book, 95 (1981). Leaves linear, flowers smaller, more open, perianth-segments 1–2 cm, each with an indistinct, circular nectary near the base. *Western N America.* H1. Spring.

**46. F. recurva** Bentham. Illustration: Botanical Magazine, 9353 (1934); Rix, Growing bulbs, pl. 4 (1983).
Bulb 2–3 cm in diameter, a fleshy disc studded with bulblets around the base. Stem 20–90 cm. Leaves in 3–5 whorls of 3–5, linear to linear-lanceolate, often glaucous. Flowers 3–6 (rarely to 12), narrowly bell-shaped, borne in a raceme, orange-scarlet to red, tessellated with yellow. Perianth-segments 2–3.5 cm, recurved at the tip, each with a lanceolate nectary 5–6 mm long near the base. Style to 1.3 cm, 3-fid, the branches to 3 mm, not recurved. Capsule 6-winged. *USA (S Oregon, California).* H2. Spring.

**\*F. gentneri** Gilkey. Illustration: Rix & Phillips, The bulb book, 94 (1981). Perianth-segments 3.5–4 cm, not recurved at their tips, style more deeply divided with recurved branches. *USA (S Oregon).* H2. Spring.

**57. LILIUM** Linnaeus
*V.A. Matthews*
Rootstock a bulb composed of fleshy overlapping scales, sometimes stoloniferous. Stem erect, unbranched, leafy. Leaves scattered or in whorls, usually linear or lanceolate, sometimes with bulbils in the axils. Flowers in a terminal raceme or umbel, or sometimes solitary, funnel-shaped, cup-shaped, bowl-shaped or bell-shaped, with the perianth-segments usually spreading or reflexed to a greater or lesser extent, or of turk's-cap type (see Figure 33, p. 194). Perianth-segments free, each with a nectar-bearing furrow or gland at the

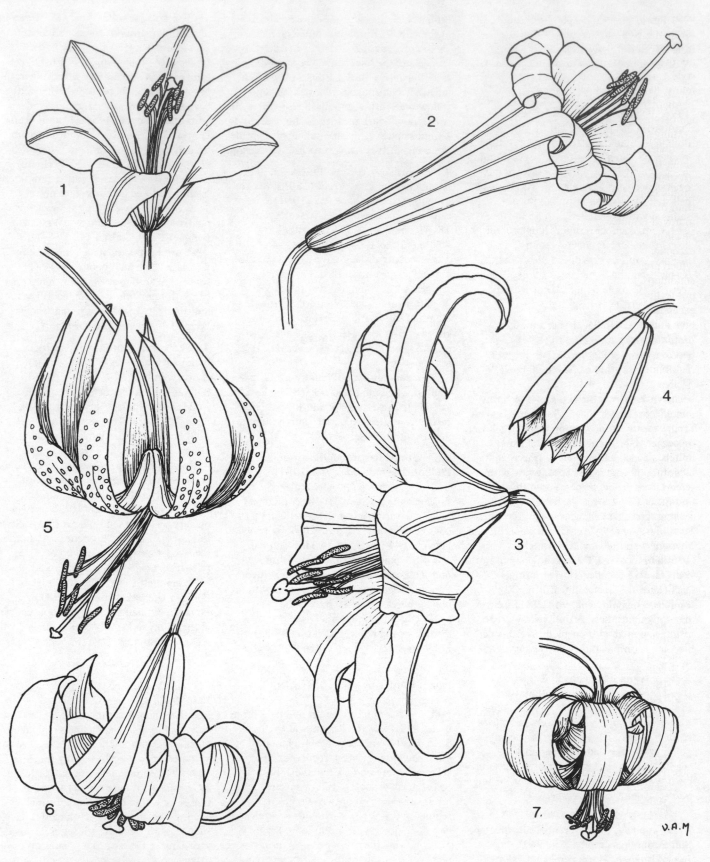

**Figure 33.** Types of *Lilium* flowers. 1, cup-shaped. 2, narrowly funnel-shaped. 3, bowl-shaped. 4, bell-shaped. 5–7, turk's-cap types.

base, papillose or sometimes hairy at the apex. Stamens borne at the base of the perianth-segments, usually free from one another; anthers versatile. Ovary superior; style 1. Fruit a capsule with 3 cells. Seeds many, flat, 2 rows in each cell.

About 100 species native to the temperate northern hemisphere.

Many species produce annual roots from the underground part of the stem which is above the bulb; these species are referred to as 'stem-rooting' in the descriptions, and should be planted more deeply (at *c.* 20 cm) than the others. In general, lilies grow best in a well-drained soil in semi-shade or sun, never in a very hot, dry position. There are exceptions e.g. *L. catesbaei* and *L. grayi* which prefer damp, but not stagnant conditions. Most species will grow in acid or alkaline soils but again, there are exceptions such as *L. henryi*, which requires lime and *L. speciosum*, *L. superbum* and *L. auratum* which will not tolerate it. It is best to consult the literature for the requirements of particular species.

Lilies are especially susceptible to aphid-borne virus diseases. So far there is no cure, and affected plants should be discarded. Propagation is by seed, bulb scales, or in those species which produce them, stem bulbils.

Literature: Elwes, H.J., *A Monograph of the genus Lilium* (1877–1880); Grove, A. & Cotton, A.D., *Supplement to Elwes' Monograph of the genus Lilium*, parts 1–7 (1933–1940); Turrill, W.R., *Supplement to Elwes' Monograph of the genus Lilium*, parts 8 & 9 (1960–1962); Woodcock, H.B.D. & Stearn, W.T., *Lilies of the world* (1950); Synge, P.M., *Lilies* (1980); A.C. Leslie, *The International Lily Register (1982)*, and supplements.

1a. Basic colour of perianth-segments yellowish green 2
b. Basic colour of perianth-segments not yellowish green 6
2a. Leaves with minutely ciliate margin 3
b. Leaves with hairless margin 4
3a. Perianth-segments 4–6.5 cm, with dark purple spots and lines inside, and purplish warts; flowers of turk's-cap type **23. pyrenaicum***
b. Perianth-segments 6–15 cm, unmarked or variously blotched or suffused with purplish red inside, and lacking warts; flowers funnel-shaped **14. primulinum**
4a. Perianth-segments 2.5–5 cm; flowers of turk's-cap type **41. kelleyanum***
b. Perianth-segments 6–15 cm; flowers funnel-shaped 5

5a. Leaves with 5 or 7 veins; filaments and stigma green; perianth-segments thin with deep longitudinal ribs and transverse corrugations **13. nepalense**
b. Leaves with 1 or 3 veins; filaments and stigma purple or with purple spots; perianth-segments thick, lacking longitudinal ribs and corrugations **14. primulinum**
6a. Flowers of turk's-cap type 7
b. Flowers not of turk's-cap type 56
7a. Basic colour of perianth-segments white, greenish white, pink, crimson or purple 8
b. Basic colour of perianth-segments cream, yellow, orange or red 15
8a. At least some leaves in whorls 9
b. All leaves scattered 10
9a. Perianth-segments 3–5.5 cm **25. martagon***
b. Perianth-segments 6–10 cm, if less then upper leaves scattered **45. michauxii**
10a. Perianth-segments with wavy margins; pollen brown or dark red **37. speciosum**
b. Perianth-segments without wavy margins; pollen orange, yellow or lilac 11
11a. Perianth-segments white or greenish white, sometimes flushed with purple, or turning red or purple with age 12
b. Perianth-segments pink, lilac, purple or crimson-maroon 14
12a. Anthers yellow or orange-red; nectary-furrows usually green, warty **31. duchartrei***
b. Anthers purplish or grey; nectary-furrows purple or brown, not warty 13
13a. Leaves up to 6 mm broad with 1 or 3 veins which are minutely hairy beneath; perianth-segments white with purple spots **32. taliense**
b. Leaves 2–3.5 cm broad with 5 or 7 veins which are hairless beneath; perianth-segments greenish white, often reddish inside towards the base **13. nepalense**
14a. Nectary-furrows hairless **34. wardii**
b. Nectary-furrows bordered with warts or hairs **33. lankongense***
15a. At least some leaves in whorls 16
b. All leaves scattered 28
16a. Basic colour of perianth-segments yellow 17
b. Basic colour of perianth-segments orange-yellow, orange or red 20
17a. Perianth-segments 2.5–5 cm 18

b. Perianth-segments 6.5–10 cm 19
18a. Leaves oblanceolate; anthers orange **40. columbianum***
b. Leaves lanceolate to broadly elliptic; anthers brown **41. kelleyanum***
19a. Leaves with veins hairless beneath; anthers orange **40. columbianum***
b. Leaves with veins downy beneath; anthers purplish or reddish brown **39. humboldtii***
20a. Perianth-segments with some spots encircled with yellow **42. pardalinum**
b. Perianth-segments with spots not encircled with yellow 21
21a. Stem usually downy; leaves with veins downy beneath **39. humboldtii***
b. Stem hairless; leaves with veins hairless or minutely roughened beneath 22
22a. Perianth-segments 2.5–5.5 cm 23
b. Perianth-segments 6–10 cm 25
23a. Anthers 2–2.2 cm **28. hansonii**
b. Anthers 5–13 mm 24
24a. Anthers orange **40. columbianum***
b. Anthers purplish, brown or reddish brown, if orange then stem roots present **41. kelleyanum***
25a. Anthers 1.6–2 cm, red **44. superbum**
b. Anthers 6–11 mm, orange, purplish brown or red 26
26a. Leaves lanceolate to broadly elliptic **43. michiganense***
b. Leaves oblanceolate to obovate 27
27a. Stem roots present; flowers usually 1 or 2, occasionally up to 5; anthers purplish brown **45. michauxii**
b. Stem roots absent; flowers usually 6–10, occasionally up to 40; anthers orange **40. columbianum***
28a. Basic colour of perianth-segments cream, yellow or orange 29
b. Basic colour of perianth-segments red or orange-red 45
29a. Perianth-segments up to 5.5 cm 30
b. Perianth-segments 6 cm or more 34
30a. Anthers red **23. pyrenaicum***
b. Anthers brown, reddish brown or maroon 31
31a. Margin of leaves hairless **41. kelleyanum***
b. Margin of leaves ciliate or minutely rough 32
32a. Leaves with veins hairless beneath **23. pyrenaicum***
b. Leaves with veins downy beneath 33
33a. Stem hairless; leaves with 3–15 veins **23. pyrenaicum***
b. Stem downy; leaves usually with 3 veins **38. amabile**

34a. Leaves with veins hairless beneath
35
  b. Leaves with veins minutely rough or
downy beneath                                     38
35a. Perianth-segments recurved in the
upper half to two-thirds
**14. primulinum**
  b. Perianth-segments strongly recurved
36
36a. Anthers 4–10 mm    **23. pyrenaicum\***
  b. Anthers 1.2–1.7 cm                         37
37a. Stem hairless; leaves with many
veins, 2–3 cm broad           **36. henryi**
  b. Stem usually cobwebby-hairy; leaves
with 5 or 7 veins, 1–2 cm broad
**30. lancifolium**
38a. Basic colour of perianth-segments
yellow-orange or orange             39
  b. Basic colour of perianth-segments
cream or yellow                           41
39a. Anthers 1.2–1.7 cm, orange
**36. henryi**
  b. Anthers 4–10 mm, red                   40
40a. Leaves linear; flowers 6–12; perianth-
segments with reddish spots
**22. × testaceum**
  b. Leaves lanceolate; flowers usually 1–6;
perianth-segments with brownish
purple spots        **23. pyrenaicum\***
41a. Pollen red or reddish brown       42
  b. Pollen yellow, orange or brown     43
42a. Perianth-segments 3–7 cm; anthers
4–10 mm           **23. pyrenaicum\***
  b. Perianth-segments 6.5–12 cm;
anthers 1–2.5 cm    **29. leichtlinii\***
43a. Perianth-segments bearing warts
towards the base           **36. henryi**
  b. Perianth-segments without warts   44
44a. Leaves with 1 or 3 veins
**14. primulinum**
  b. Leaves with 9, 11 or 13 veins
**26. monadelphum\***
45a. Bulbils present in the leaf axils
**30. lancifolium**
  b. Bulbils absent                           46
46a. Upper leaves adpressed to stem, lower
leaves spreading   **21. chalcedonicum**
  b. Upper and lower leaves not changing
in aspect                               ' 47
47a. Stem minutely rough or downy,
sometimes only in the lower part, or
hairy at the leaf axils                  48
  b. Stem hairless                            51
48a. Perianth-segments 3–3.5 cm
**27. pumilum**
  b. Perianth-segments 5–8.5 cm        49
49a. Flowers with an unpleasant scent;
nectary-furrows warty    **38. amabile**
  b. Flowers unscented; nectary-furrows
hairy                                      50
50a. Flowers 1–6; stem with sparse hairs

at the leaf axils, otherwise hairless
**29. leichtlinii\***
  b. Flowers 5–40; stem downy, especially
in the lower part           **35. davidii**
51a. Perianth-segments bright red     52
  b. Perianth-segments orange-red     55
52a. Leaves 1–3 mm broad               54
  b. Leaves 4–17 mm broad              53
53a. Leaves downy on the veins beneath
**23. pyrenaicum\***
  b. Leaves hairless on the veins beneath
**24. pomponium**
54a. Perianth-segments 3–3.5 cm
**27. pumilum**
  b. Perianth-segments 4–7 cm
**24. pomponium**
55a. Perianth-segments 3–7 cm; anthers
4–10 mm           **23. pyrenaicum\***
  b. Perianth-segments 6.5–8.5 cm;
anthers 1–1.7 cm    **29. leichtlinii\***
56a. Basic colour of perianth-segments
white, greenish white, cream, pink,
purple or purplish brown            57
  b. Basic colour of perianth-segments
yellow, orange, red or crimson     78
57a. At least some leaves in whorls   58
  b. All leaves scattered                   59
58a. Flowers nodding        **51. mackliniae\***
  b. Flowers erect or borne horizontally
**47. rubescens\***
59a. Flowers open bowl-shaped, 20–30 cm
in diameter; perianth-segments
basically white with a yellow or red
central band, spotted or not
**15. auratum**
  b. Flowers funnel-shaped, trumpet-
shaped or bell-shaped, if bowl-shaped
then less than 20 cm in diameter   60
60a. Perianth-segments up to 8 cm     61
  b. Perianth-segments 9–30 cm        64
61a. Perianth-segments white, sometimes
flushed with green outside
**1. candidum\***
  b. Perianth-segments pink to purple or
brownish purple, if white then flushed
with pinkish purple                     62
62a. Perianth-segments 6–8 cm
**12. rubellum**
  b. Perianth-segments 1–5.5 cm       63
63a. Nectary-furrows fringed   **52. nanum\***
  b. Nectary-furrows hairless
**51. mackliniae\***
64a. Perianth-segments pure white    65
  b. Perianth-segments cream, pink or
purplish, if white then flushed with
green, yellow, brownish purple or
pinkish purple at least towards the
base                                       66
65a. Filaments smooth       **8. longiflorum**
  b. Filaments papillose towards the base
**3. formosanum**

66a. Pollen yellow                          67
  b. Pollen orange, reddish brown or
brown                                   72
67a. Perianth-tube yellow inside, at least at
the base                                 68
  b. Perianth-tube not yellow inside   69
68a. Leaves with 1 vein which is minutely
rough beneath; perianth-segments
12–15 cm, pinkish purple outside
**2. regale**
  b. Leaves with 3 or 5 veins which are
hairless beneath; perianth-segments
15–30 cm, white or creamy outside,
tinged with green or yellow
**7. wallichianum**
69a. Perianth-tube greenish at the base
inside           **7. wallichianum**
  b. Perianth-tube not greenish inside   70
70a. Nectary-furrows hairless
**8. longiflorum**
  b. Nectary-furrows bordered with hairs
and/or warts                           71
71a. Filaments smooth     **9. philippinense\***
  b. Filaments papillose towards the base
**3. formosanum**
72a. Nectary-furrows hairless            73
  b. Nectary-furrows bordered with hairs
and/or warts                           76
73a. Bulbils present in the leaf-axils
**10. sulphureum\***
  b. Bulbils absent                           74
74a. Perianth-segments pink inside and
outside                  **11. japonicum**
  b. Perianth-segments flushed with
purplish red or brownish purple
outside, white inside                  75
75a. Leaves of flowering plants with 1
vein; pollen reddish brown to brown
**6. leucanthum**
  b. Leaves of flowering plants with 3 or 5
veins; pollen orange     **8. longiflorum**
76a. Bulbils present in the leaf axils
**5. sargentiae\***
  b. Bulbils absent                           77
77a. Leaves 5–10 mm broad, margin
recurved               **3. formosanum**
  b. Leaves 1–2.5 cm broad, margin not
recurved                    **4. brownii**
78a. Flowers bell-shaped, funnel-shaped or
trumpet-shaped, usually drooping or
borne horizontally                     79
  b. Flowers cup-shaped or star-shaped,
erect                                    83
79a. Nectary-furrows fringed   **52. nanum\***
  b. Nectary-furrows hairless or with short
hairs                                    80
80a. Veins of leaves minutely rough
beneath                **50. canadense\***
  b. Veins of leaves hairless beneath   81
81a. Anthers 3–4 mm        **46. parvum\***
  b. Anthers 5–9 mm                       82

82a. Leaves in whorls; perianth-segments
3–4.5 cm; pollen deep yellow
**48. bolanderi**

b. Leaves usually scattered, sometimes 1
or more whorls present towards base
of stem; perianth-segments 7–10 cm;
pollen brownish red **49. parryi**

83a. At least some leaves in whorls
**19. philadelphicum\***

b. Leaves scattered 84

84a. Perianth-segments 3–4 cm; flowers
star-shaped **20. concolor**

b. Perianth-segments 5–12 cm; flowers
cup-shaped 85

85a. Stem hairless to sparsely hairy or
rough **18. maculatum\***

b. Stem woolly or cobwebby-hairy at
least towards the top 86

86a. Perianth-segments with yellow
towards the base, or completely
yellow with black spots
**17. dauricum\***

b. Perianth-segments lacking yellow
**16. bulbiferum**

**1. L. candidum** Linnaeus. Illustration:
Elwes, Monograph of the genus Lilium, pl.
9 (1877); Hay & Synge, Dictionary of
garden plants in colour, 101 (1969); Perry,
Flowers of the world, 167 (1972); Grey-
Wilson & Mathew, Bulbs, pl. 3 (1981).
Stem roots absent. Stem 90–200 cm, dark
reddish purple, hairless. Basal leaves up to
22 × 5 cm with 3 or 5 veins, hairless. Stem
leaves scattered, up to 7.5 × 1 cm,
lanceolate, hairless. Flower-stalks erect or
spreading. Flowers 5–20, broadly funnel-
shaped, fragrant. Perianth-segments
5–8 × 1–4 cm, white, yellow at the base
inside, obovate, recurved slightly at the
tips. Anthers 5–18 mm; pollen yellow. W
& S Greece, S Yugoslavia, SW Turkey,
Lebanon, Syria, Israel. H2. Summer.

L. candidum differs from almost all other
species of Lilium in producing basal leaves
in the autumn which over-winter. Fruits
are not produced in cultivation.

Var. **salonikae** Stoker differs in being
usually fertile, having a green stem and
more widely spreading perianth-segments.
W Greece. H2. Summer.

Var. **plenum** Weston has double flowers
and is now extremely rare.

**\*L. bakerianum** Collett & Hemsley.
Illustration: Grove & Cotton, Supplement to
Elwes' Monograph of the genus Lilium, pl.
7 (1935); Woodcock & Stearn, Lilies of the
world, f. 26 (1950). Stem roots present.
Stem usually shortly hairy or minutely
rough below. Leaves linear to lanceolate
with 1 or 3 veins, minutely rough on the

upper surface, margin and veins beneath.
Flowers 1–6, bell-shaped, nodding.
Perianth-segments white, flushed with
green outside and with reddish brown spots
inside. Nepal, N Burma, W China. H3.
Summer.

**2. L. regale** Wilson. Illustration: Woodcock
& Stearn, Lilies of the world, f. 99 (1950);
Journal of the Royal Horticultural Society
**95**: pl. 264 (1970); Perry, Flowers of the
world, 168 (1972); Synge, Lilies, pl. xv
(1980).
Stem roots present. Stem 50–200 cm, grey-
green, spotted and flushed with reddish
purple, with tiny rough hairs, leafless at the
base. Leaves scattered, 5–13 cm ×
4–6 mm, linear with 1 vein, minutely
rough on margin and vein beneath.
Flower-stalks spreading or somewhat
ascending, 1.8–13 cm. Flowers sometimes
solitary, usually up to 25 or more, funnel-
shaped, fragrant. Perianth-segments
12–15 cm, outer c. 2 cm wide, inner up to
3.7 cm wide, pinkish purple outside, white
inside, recurved towards the tips; perianth-
tube flushed with yellow inside. Anthers
and pollen yellow. W China (W Szechuan).
H2. Summer.

**L. × imperiale** Wilson (Illustration:
Reader's Digest encyclopaedia of garden
plants and flowers, 402, 1971) is a hybrid
between L. regale and L. sargentiae. Stems
up to 120 cm, grey-green. Leaves hairless,
with 1, 3 or 5 veins, the upper leaves with
bulbils in their axils. Flowers similar to
L. regale but the tips of the perianth-
segments are broader and more recurved.
Anthers orange-brown.

**3. L. formosanum** Wallace. Illustration:
Botanical magazine, 9205 (1930);
Woodcock & Stearn, Lilies of the world, f.
47 (1950); Leeburn, Garden lilies, pl. 20
(1963); Synge, Lilies, pl. 9 (1980).
Stem roots present. Stem 30–150 cm,
purplish brown especially towards the base,
usually somewhat rough-hairy. Leaves
scattered, often crowded near the base of
the stem, 7.5–20 cm × 5–10 mm, linear to
narrowly oblong with 3, 5 or 7 veins,
hairless, margin recurved. Flower-stalks
erect or ascending, 5–15 cm. Flowers
usually 1 or 2, sometimes up to 10,
narrowly funnel-shaped, fragrant, borne
horizontally. Perianth-segments 12–20 cm,
outer 2.5–3 cm wide, inner up to 5 cm
wide, flushed with reddish purple outside,
rarely white, white inside, recurved at tips.
Nectary-furrows green, bordered with hairs
and warts. Filaments papillose towards the
base; anthers c. 9 mm, yellow to purplish,

pollen brown or yellow. Taiwan. H5.
Summer–autumn.

Var. **pricei** Stoker. Illustration: Woodcock
and Stearn, Lilies of the world, f. 46
(1950). An earlier-flowering, shorter plant
(stems usually under 60 cm) with solitary
flowers which are more deeply flushed with
reddish purple. It is generally hardier than
var. formosanum. Taiwan. H3. Summer.

**4. L. brownii** Miellez. Illustration: RHS Lily
Year Book **27**: f. 5 (1963); Leeburn, Garden
lilies, pl. 17 (1963); Synge, Lilies, pl. vii
(1980).
Stem roots present. Stem up to 1 m, flushed
with dark purplish brown, hairless. Leaves
scattered, 17–30 × 1–2.5 cm, lanceolate,
with 5 or 7 veins, hairless. Flowers 1–4,
narrowly funnel-shaped, slightly fragrant,
borne horizontally or slightly drooping.
Perianth-segments 12–15 × 2.5–6.5 cm,
pinkish purple outside, sometimes tinged
with green, white inside, recurved at the
tips. Nectary-furrows bordered with hairs
and warts. Anthers 1–1.8 cm, brown with
reddish brown pollen. China, Upper Burma.
H4. Summer.

Var. **australe** (Stapf) Stearn is taller (to
3 m), with narrower leaves, and flowers
which are white or slightly flushed with
green outside. SE China, Hong Kong.

Var. **viridulum** Baker (L. brownii var.
colchesteri (van Houtte) Elwes) usually has
a green stem and the flowers tinged outside
with yellowish green and sometimes with
pale pinkish purple, and yellow inside the
tube fading to white with age. The flowers
are strongly scented. China.

**5. L. sargentiae** Wilson. Illustration:
Woodcock & Stearn, Lilies of the world, f.
104 (1950); Synge, Lilies, pl. xvii (1980).
Stem roots present. Stems up to 1.5 m,
grey-green, spotted and tinged with pinkish
purple, with tiny sparse rough hairs,
leafless at the base. Leaves scattered,
10–20 cm × 5–20 mm, narrowly oblong
with 3, 5 or 7 veins, hairless, bearing
bulbils in their axils. Flower-stalks
spreading or somewhat ascending,
3–10 cm. Flowers 2–5, funnel-shaped,
fragrant, borne horizontally or slightly
drooping. Perianth-segments 12–15 cm,
outer 1.5–2 cm wide, inner 3.5–5 cm
wide, pinkish or brownish purple or
sometimes greenish outside, white inside,
recurved at the tips; perianth-tube yellow
inside. Nectary-furrows with sparse warts
on the margins. Filaments somewhat
warty-pubescent in lower one-third;
anthers purplish with brown pollen.
W China (W Szechuan). H4. Summer.

**\*L. alexandrae** which differs in having the flowers borne horizontally to almost erect, and has white perianth-segments flushed with green towards the base and sometimes tinged pinkish, may key out here (see entry under 10. *L. sulphureum*).

**6. L. leucanthum** (Baker) Baker var. **centifolium** (Stapf) Stearn (*L. centifolium* Stapf). Illustration: Botanical Magazine, 8960 (1923); Woodcock & Stearn, Lilies of the world, f. 72 (1950); RHS Lily Year Book **30**: f. 2 (1966); Synge, Lilies, pl. xii (1980).
Stem roots present. Stem 2–3 m, glaucous, densely leafy. Leaves scattered, 15–20 cm, linear to lanceolate with 1 vein (in non-flowering plants the leaves are shorter and have 3 veins), hairless. Flower-stalks horizontal to ascending, 6–12 cm. Flowers up to 18, funnel-shaped, fragrant, spreading or slightly drooping. Perianth-segments strongly recurved at the tips, outer 15–18 × c. 3 cm, inner 14–17 × 4.5–6 cm, flushed with purplish red outside and greenish towards the base, white inside; perianth-tube yellow inside. Nectary-furrows green, hairless. Filaments warty and downy towards the base; anthers 1.8–2 cm, reddish brown with reddish brown or dark brown pollen. *W China (W Kansu)*. H4. Summer.

**7. L. wallichianum** Schultes & Schultes. Illustration: Botanical Magazine, 4561 (1851); Woodcock & Stearn, Lilies of the world, f. 117 (1950); Synge, Lilies, pl. 16 (1980).
Bulb stoloniferous. Stem roots present. Stem up to 2 m, green or tinged with purple, hairless. Leaves scattered, up to 25 × 1.2 cm, linear or lanceolate with 3 or 5 veins, hairless. Flowers 1–4, funnel-shaped, fragrant, borne horizontally. Perianth-segments 15–30 cm, white or creamy with a greenish or yellowish tinge outside towards the base, and yellow or greenish at the base inside, recurved at the tips. Stamens shorter than style; pollen yellow. *Himalaya*. H5. Summer–autumn.
Var. **neilgherrense** (Wight) Hara (*L. neilgherrense* Wight). Illustration: Elwes, Monograph of the genus Lilium, pl. 6 (1877); Woodcock & Stearn, Lilies of the world, f. 82 (1950). Usually a shorter plant (*c.* 90 cm) with broadly lanceolate leaves up to 12 × 1.5–3 cm. *S India*. H5. Summer–autumn.

**8. L. longiflorum** Thunberg. Illustration: Reader's encyclopaedia of garden plants and flowers, 406 (1971).

Stem roots present. Stem 30–100 cm, green, hairless. Leaves scattered, 10–20 cm × 5–15 mm, lanceolate to oblong-lanceolate with 3 or 5 veins. Flower-stalks spreading, up to 12 cm. Flowers 1–6, funnel-shaped, fragrant, borne horizontally. Perianth-segments 13–20 × c. 3.5 cm, white, slightly recurved at the tips. Nectary-furrows green, hairless. Anthers c. 9 mm, pollen yellow. *S Japan, Taiwan*. H4. Summer.
Var. **eximium** (Courtois) Baker. Flowers with a narrower tube and perianth-segments more strongly recurved.
Var. **takeshima** Duchartre. Stems tall, brownish purple. Flowers flushed with brownish purple outside. Pollen orange.

**9. L. philippinense** Baker. Illustration: Botanical Magazine, 6250 (1876); Elwes, Monograph of the genus Lilium, pl. 3 (1877).
Bulb stoloniferous. Stem roots present. Stem usually 30–45 cm, occasionally to 100 cm, green sometimes marked with purple towards the base, hairless. Leaves scattered, up to 15 cm × 5 mm, linear with 3 veins, hairless. Flowers usually 1 or 2, sometimes up to 6, trumpet-shaped, fragrant, borne horizontally. Perianth-segments 18–25 × 4–5 cm, white with a slight green flush and sometimes with purple streaking outside towards the base, spreading at the tips. Nectary-furrows with downy margins. Anthers and pollen yellow. *Philippine Islands*. H4. Summer.

**\*L. nobilissimum** (Makino) Makino. Illustration: Woodcock & Stearn, Lilies of the world, f. 87 (1950); RHS Lily Year Book, **30**: f. 27 (1966). Stem to 170 cm. Leaves with short stalks and bearing pale green bulbils in their axils. Flowers erect. Perianth-segments up to 15 cm. Anthers brown. *S Japan*. H5. Summer.

**10. L. sulphureum** Baker (*L. myriophyllum* Franchet). Illustration: Botanical Magazine, 8102 (1906); Grove & Cotton, Supplement to Elwes' Monograph of the genus Lilium, pl. 21 (1939); Woodcock & Stearn, Lilies of the world, f. 82 (1950).
Stem roots present. Stem 1.5–2 m, green, marked with purple, minutely rough especially below, slightly ribbed. Leaves scattered, to 20 × 1.2–2 cm, shorter in lower part of stem, narrowly lanceolate with 3, 5 or 7 veins, hairless, bearing brown bulbils in their axils. Flower-stalks spreading or drooping, up to 20 cm. Flowers up to 15, trumpet-shaped, strongly fragrant, spreading or drooping. Perianth-segments 14–25 cm, outer up to 4 cm

wide, inner up to 5.7 cm wide, cream to greenish white, flushed with pink outside, inside deep yellow in the tube, recurved at the tips. Nectary-furrows hairless. Anthers 1.8–2 cm, purplish brown with orange or brownish pollen. *W China (Yunnan), N Burma*. H4. Late summer.

**\*L. alexandrae** (Wallace) Coutts. Illustration: Grove & Cotton, Supplement to Elwes' Monograph of the genus Lilium, pl. 15 (1936); RHS Lily Year Book **30**: f. 26 (1966); Synge, Lilies, pl. 4 (1980). Stem only to 1 m. Leaves lanceolate to narrowly ovate, shortly stalked, up to 4 cm wide. Flowers 1–5, bowl-shaped, horizontal or almost erect. Perianth-segments up to 17 cm, outer c. 3 cm wide, inner up to 4.5 cm wide, white flushed with green at the base inside and outside, sometimes tinged with pink. Pollen reddish brown. *Japan*. H5. Summer.

**11. L. japonicum** Houttuyn. Illustration: Woodcock & Stearn, Lilies of the world, f. 66 (1950); Hay & Synge, Dictionary of garden plants in colour, 102 (1969); Synge, Lilies, pl. x (1980).
Stem roots present. Stem 45–100 cm, hairless. Leaves scattered, 5–20 cm × 5–25 mm, narrowly lanceolate with 3 or 5 veins, hairless except for the minutely rough margin. Flowers 1–5, broadly funnel-shaped, fragrant, borne horizontally. Perianth-segments 9–17 × 2–3.5 cm, pink, oblanceolate to oblong. Nectary-furrows greenish, hairless. Anthers brown with orange-brown or reddish pollen. *Japan*. H4. Summer.
Var. **platyfolium** Anon. has wider leaves and is rather more vigorous.

**12. L. rubellum** Baker. Illustration: Woodcock & Stearn, Lilies of the world, f. 102 (1950); Hay & Synge, Dictionary of garden plants in colour, 103 (1969); Reader's Digest encyclopaedia of garden plants and flowers, 408 (1971); The Garden **100**: f. 51 (1975).
Stem roots present. Stem 30–80 cm, green, marked with brownish purple, hairless. Leaves scattered, c. 10 × 4–5 cm, each with a short stalk, narrowly elliptic to narrowly ovate with 5 or 7 veins, hairless on veins beneath, margin minutely rough. Flowers up to 9, usually less, broadly funnel-shaped, fragrant, borne horizontally. Perianth-segments 6–8 × 1.7–2.5 cm, pink, unspotted or with maroon dots at the base, oblanceolate to oblong. Anthers yellow. *Japan*. H4. Summer.

**13. L. nepalense** Don. Illustration: Woodcock & Stearn, Lilies of the world, f. 84 (1950); Everard & Morley, Wild flowers of the world, 107 (1970); The Garden **103**: 311 (1978); Synge, Lilies, pl. xiv (1980). Bulb stoloniferous. Stem roots present. Stem 70–100 cm, pale green, hairless. Leaves scattered, 9–15 × 2–3.5 cm, lanceolate to oblong-lanceolate with 5 or 7 veins, hairless. Flower-stalks spreading to ascending up to 10 cm. Flowers 1–3, funnel-shaped, sometimes unpleasantly scented, more or less drooping. Perianth-segments up to 15 × 4 cm, greenish white to pale greenish yellow, reddish inside towards the base, deeply ribbed and with transverse corrugations outside, spreading in the upper two-thirds and recurved at the tips. Nectary-furrows brown or purple, hairless. Anthers purplish with yellow or yellow-brown pollen. *Bhutan, Nepal, N India (Kumaon).* H4. Summer.

Var. **concolor** Cotton. Illustration: Woodcock & Stearn, Lilies of the world, f. 85 (1950). Flowers entirely greenish yellow, and lacking the reddish throat. *Bhutan, ?Assam.*

**14. L. primulinum** Baker. Bulb stoloniferous or not. Stem roots present. Stem to 2.4 m, brown, or green spotted with brown, hairless or minutely rough below. Leaves scattered, to 15 × 4 cm, lanceolate with 1 or 3 veins, margin hairless or minutely ciliate. Flower stalks ascending or horizontal, decurved below flower, up to 15 cm. Flowers usually 2–8, occasionally to 18, funnel-shaped, usually fragrant, drooping. Perianth-segments 6–15 × 1.3–4.5 cm, yellow or greenish yellow, unmarked or variously blotched or suffused with deep purplish red inside, oblong-lanceolate, varying in the degree to which they are recurved. Nectary-furrows green, hairless. Pollen orange-brown. *W China (Yunnan), N Burma, Thailand.* H4. Summer–autumn.

Var. **primulinum** Illustration: Botanical Magazine, 7227 (1892). Flowers entirely yellow or greenish yellow with the perianth-segments recurved for about half their length. *N Burma.*

Var. **burmanicum** (Smith) Stearn. Illustration: Grove & Cotton, Supplement to Elwes' Monograph of the genus Lilium, pl. 19 (1937); Woodcock & Stearn, Lilies of the world, f. 95 (1950). Stem 1.2–2.4 m, leaves to 4 cm wide, flowers purplish red in the throat with perianth-segments 6.5–15 cm, recurved in the upper third. *N Burma, Thailand.*

Var. **ochraceum** (Franchet) Stearn. Illustration: RHS Lily Year Book **30**: f. 12 (1966). Stem to 1.2 m, leaves to 2 cm wide, flowers purplish red in the throat with perianth-segments *c.* 6 cm, recurved in the upper two-thirds. *W China (Yunnan).*

**15. L. auratum** Lindley. Illustration: Elwes, Monograph of the genus Lilium, pl. 15 (1878); Perry, Flowers of the world, 169 (1972); Witham Fogg, Bulbs, 58 (1980); Rix & Phillips, The bulb book, 165 (1981). Stem roots present. Stem 60–150 cm, occasionally to 225 cm, purplish green, hairless. Leaves scattered, 10–22 × 1.5–4 cm, narrowly or broadly lanceolate with a short stalk and 5 or 7 veins, hairless. Flowers usually up to 10, sometimes to 30, open bowl-shaped, 20–30 cm in diameter, fragrant, borne horizontally or slightly drooping. Perianth-segments 12–18 × 4–5 cm, white with a yellow or crimson central stripe, and usually yellow or crimson spots, slightly recurved at the tips and with scattered warts towards the base. Pollen brownish red. *Japan.* H4. Summer–autumn.

The following variants are in cultivation.

Var. **platyphyllum** Baker. Illustration: de Graaff & Hyams, Lilies, pl. 36 (1967); Synge, Lilies, pl. vi (1980). The leaves are especially broad and the large flowers have fewer spots than var. *auratum* and a central yellow band to each segment.

Var. **virginale** Duchartre. Illustration: de Graaff & Hyams, Lilies, pl. 37 (1967). Flowers white with yellow spots and bands.

Several cultivars have been selected and are commonly offered by nurserymen.

*L. auratum* crossed with *L. speciosum* has produced L. × **parkmannii** Moore, the name given to a range of hybrids in which the perianth-segments are usually crimson or deep pink with a paler or white margin.

**16. L. bulbiferum** Linnaeus (*L. bulbiferum* var. *chaixii* (Maw) Stoker). Stem roots present. Stem 40–150 cm, green, sometimes with purple spots and woolly above, ribbed. Leaves scattered, 5–15 × *c.* 1.5 cm, narrowly to broadly lanceolate with 3, 5 or 7 veins, margin ciliate, otherwise hairless, with or without bulbils in the axils. Flowers usually up to 5, sometimes up to 50, shallowly cup-shaped, erect. Perianth-segments 6–8.5 × 2–3 cm, bright orange, marked at the base with deeper orange and often orange-red towards the tips, warty and with maroon or blackish spots inside, narrowly ovate, narrowed to a claw at the base. Nectary-furrows greenish, hairy. Anthers 6–8 mm,

brown with orange pollen. *C Europe south to N Spain, S Italy and N Yugoslavia.* H2. Summer.

Var. **bulbiferum**. Illustration: Elwes, Monograph of the genus Lilium, pl. 23 (1877); Grey-Wilson & Blamey, Alpine flowers of Britain and northern Europe, 271 (1979); Grey-Wilson & Mathew, Bulbs, pl. 3 (1981). Flowers redder and bulbils usually present.

Var. **croceum** (Chaix) Persoon. Illustration: Robert & Schroeter, Alpine flowers, pl. 6 (1938); Huxley, Mountain flowers, 17 (1967); Grey-Wilson & Blamey, Alpine flowers of Britain and northern Europe, 271 (1979); Synge, Lilies, pl. i (1980). This variety has orange flowers and usually lacks bulbils.

**17. L. dauricum** Ker Gawler. (*L. pensylvanicum* Ker Gawler). Illustration: Botanical magazine, 872 (1805); Elwes, Monograph of the genus Lilium, pl. 21 (1877); Woodcock & Stearn, Lilies of the world, f. 40 (1950). Bulb stoloniferous. Stem roots present. Stem to 75 cm, green spotted with reddish brown, ribbed and covered with white cobwebby hairs towards the top. Leaves scattered, 5–15 cm × 5–35 mm, linear to lanceolate with 3 or 5 veins, margin often cobwebby-hairy. Flowers to 6, open cup-shaped, erect. Perianth-segments 5–10 × 1.5–3.5 cm, red to orange-red, yellow towards the base, with warts and usually with brownish red spots, oblanceolate, slightly recurved at the tips, narrowed to a claw at the base. Nectary-furrows hairy. Anther 4–10 mm, pollen red. *NE Asia.* H1. Summer.

*****L. wilsonii** Leichtlin. Illustration: Woodcock & Stearn, Lilies of the world, f. 120 (1950). Stem to 1 m, leaves hairless, usually with 5 or 7 veins. Perianth-segments 7.5–10 × 2.5–4 cm, reddish orange or apricot, with a yellow stripe from the centre of the segment to the base, densely dark-spotted. *Japan.* H1. Summer.

Very similar to and possibly only a variety of *L. dauricum*. 'Luteum' has yellow flowers spotted with black.

**18. L. maculatum** Thunberg (*L. elegans* Thunberg; *L. thunbergianum* Schultes & Schultes). Stem roots present. Stem to 60 cm, ribbed, hairless to sparsely hairy or minutely rough. Leaves scattered, 4–10 × 1–2 cm, lanceolate to elliptic with 3, 5 or 7 veins, hairless. Flowers 1 or several, cup-shaped, erect. Perianth-segments 8–10 cm, yellow, orange or red with sparse or dense spots, clawed at the base. Nectary-furrows densely hairy. Anthers reddish. *Japan*. H2. Summer.

Thought by some botanists to be a hybrid between *L. dauricum* and *L. concolor*. A large number of variants is cultivated.

**L. × hollandicum** Woodcock & Stearn. Illustration: Reader's Digest encyclopaedia of garden plants and flowers, 399 (1971). This is the name given to hybrids between *L. maculatum* and *L. bulbiferum*. They usually grow 70 cm or more in height and generally have leaves with 3 veins. The flowers are red, orange or yellow. Many cultivars have been raised and selected.

***L. catesbaei** Walter. Illustration: Botanical Magazine, 259 (1794); Elwes, Monograph of the genus Lilium, pl. 25 (1877). Stem roots absent. Stem hairless. Flowers usually solitary, occasionally 2. Perianth-segments 5–12 × 1–2.5 cm, deep yellow, scarlet towards the recurved tip, blotched with maroon near the base, claw half the length of the segment, greenish outside. Pollen orange-red. *SE USA*. H5. Late summer.

**19. L. philadelphicum** Linnaeus. Illustration: Elwes, Monograph of the genus Lilium, pl. 17 (1877); Woodcock & Stearn, Lilies of the world, f. 91 (1950). Stem roots present. Stem to 125 cm, pale green, hairless, usually leafless in the lower one-third. Leaves in whorls of 4–9, usually with some scattered between the whorls, 5–10 × 1.2–1.5 cm, oblanceolate with 3 or 5 veins, hairless, sometimes minutely rough on the margin. Flowers usually 1 or 2, occasionally to 5, cup-shaped, erect. Perianth-segments 4–7.5 × c. 2 cm, broadly oblanceolate, narrowed into a claw at the base, deep yellow to orange-crimson, usually yellow towards the base with purple or maroon spots. Nectary-furrows hairless. Anthers 6–9 mm; pollen red. *Eastern N America*. H4. Summer.

Var. **andinum** (Nuttall) Ker Gawler. Leaves mostly scattered and narrower (up to 5 mm wide). The distribution is more westerly.

***L. tsingtauense** Gilg. Illustration: Grove & Cotton, Supplement to Elwes' Monograph of the genus Lilium, pl. 16 (1936); Woodcock & Stearn, Lilies of the world, f. 114 (1950); Synge, Lilies, pl. 15 (1980); Rix & Phillips, The bulb book, 162 (1981). Leaves in whorls of 5–14, the upper usually scattered, 4–12.5 × 1.5–3 cm, hairless. Flowers 1–5, often slightly bilaterally symmetric. Perianth-segments up to 5 × 2 cm, orange to orange-red with maroon spots, slightly warty. Anthers and pollen orange. *NE China, EC & S Korea*. H3. Summer.

**20. L. concolor** Salisbury. Stem roots present. Stem 30–90 cm, green, variably flushed with purple, slightly hairy. Leaves scattered, 7–9 cm × 5–15 mm, linear to narrowly lanceolate with 3, 5 or 7 veins, slightly ciliate on margins and veins beneath. Flowers up to 10, star-shaped, fragrant, erect. Perianth-segments 3–4 cm × 7–10 mm, orange-red, unspotted, oblanceolate, slightly recurved at the tips. Anthers 7–9 mm, orange-red. *C China*. H4. Summer.

Var. **pulchellum** (Fischer) Regel. Illustration: Botanical Magazine, 6005 (1872); Elwes, Monograph of the genus Lilium, pl. 18 (1879); Woodcock & Stearn, Lilies of the world, f. 39 (1950). Stems green, hairless. Flower buds hairy and flowers spotted. *NE Asia*. H3. Summer.

**21. L. chalcedonicum** Linnaeus (*L. heldreichii* Freyn). Illustration: Elwes, Monograph of the genus Lilium, pl. 43 (1877); RHS Lily Year Book **31**: pl. 2 (1967); Synge, Lilies, pl. ii (1980); Grey-Wilson & Mathew, Bulbs, pl. 4 (1981). Stem roots present. Stem 45–150 cm, hairless, occasionally rough-hairy below. Leaves scattered, lower 5–12 cm × 5–20 mm, lanceolate to obovate with 3 or 5 veins, margin ciliate and veins hairless or minutely rough beneath; there is an abrupt change between the lower spreading leaves and the smaller upper adpressed leaves. Flowers up to 12, turk's-cap type, faintly scented, nodding. Perianth-segments 5–7 cm, red or orange-red, unspotted, with warts inside towards the base. Nectary-furrows hairless. Anthers and pollen reddish. *Greece, S Albania*. H3. Summer.

Var. **maculatum** Stoker. Illustration: Woodcock & Stearn, Lilies of the world, f. 35 (1950). The flowers are spotted with deep purple or maroon.

**22. L. × testaceum** Lindley. Illustration: RHS Lily Year Book **22**: f. 7 (1958); Harrison, Know your lilies, 48 (1970); Rix & Phillips, The bulb book, 165 (1981). Stem 100–150 cm, purplish and with a grey-green bloom, hairless. Leaves scattered, 5–10 cm, linear with ciliate margin and hairy on the veins beneath, often twisted. Flowers 6–12, turk's-cap type, fragrant, nodding. Perianth-segments c. 8 cm, tawny yellow to pale orange, usually with sparse reddish spots inside. Anthers red with orange pollen. *Garden origin*. H3. Summer.

A hybrid between *L. chalcedonicum* and *L. candidum*.

**23. L. pyrenaicum** Gouan. Stem roots often present. Stem 15–135 cm, green, sometimes with sparse purple spots, hairless. Leaves scattered, 3–15 cm × 3–20 mm, linear to lanceolate or narrowly elliptic with 3–15 veins, margin very minutely cilate, veins hairless to densely downy beneath. Flowers to 12, turk's-cap type, nodding, faintly to strongly and unpleasantly scented. Perianth-segments 3–7 cm × 4–13 mm, yellow, orange or red with dark purplish or reddish brown spots and lines inside, especially towards the base, and usually with purplish warts. Filaments greenish or yellowish, smooth or papillose. Anthers 4–10 mm, with orange, red or brown pollen. *N Spain & Pyrenees, SE Alps, Balkans, NE Turkey, USSR (Georgia)*. H3. Summer.

1a. Leaf-veins hairless beneath     2
  b. Leaf-veins with at least some hairs beneath     3
2a. Flowers spotted   subsp. **pyrenaicum**
  b. Flowers unspotted   subsp. **carniolicum**
3a. Perianth-segments with warts towards the base inside; filaments smooth   subsp. **carniolicum**
  b. Perianth-segments lacking warts; filaments papillose   subsp. **ponticum**

Subsp. **pyrenaicum**. Illustration: Woodcock & Stearn, Lilies of the world f. 98 (1950); Synge, Lilies, pl. 3 (1980); Grey-Wilson & Mathew, Bulbs, pl. 3 (1981); Rix & Phillips, The bulb book, 167 (1981). Stem 30–135 cm, green with sparse purple spots. Leaves 7–15 cm × 3–20 mm with 3–15 veins which are hairless beneath. Flowers to 12. Perianth-segments 4–6.5 cm, yellow to greenish yellow with dark purple spots and lines inside and purplish warts. Filaments smooth. *N Spain, SW France*.

Forma **rubrum** Stoker (*L. pyrenaicum* 'Rubrum') has orange-red flowers.

'Aureum' with deep yellow flowers is sometimes offered commercially.

Subsp. **carniolicum** (W. Koch) Matthews. Stem to 120 cm, green. Leaves 3–11 cm × 4–17 mm, with 3–9 veins which are hairless or downy beneath. Flowers usually to 6, sometimes to 12. Perianth-segments 3–7 cm, yellow, orange or red, often spotted with brownish purple, and with purplish or rarely yellow warts. Filaments smooth. *SE Europe*.

Three varieties of subsp. *carniolicum* are in cultivation.

Var. **carniolicum** (*L. carniolicum* W. Koch). Illustration: Elwes, Monograph of the genus Lilium, pl. 45 (1877); The Garden **103**: 209 (1978); Grey-Wilson & Mathew, Bulbs, pl. 4 (1981). Stem to 120 cm. Leaf-veins densely downy beneath. Flowers red or orange, spotted. *SE Alps (Austria & Italy), N Yugoslavia.*

Var. **albanicum** (Grisebach) Matthews (*L. albanicum* Grisebach; *L. carniolicum* W. Koch subsp. *albanicum* (Grisebach) Hayek). Illustration: Botanical Magazine, n.s., 839 (1982). Stem to 40 cm. Leaf-veins hairless beneath. Flowers yellow, unspotted. *SW Yugoslavia, Albania, NW Greece.*

Var. **jankae** (Kerner) Matthews (*L. jankae* Kerner; *L. carniolicum* W. Koch subsp. *jankae* (Kerner) Ascherson & Graebner). Stem to 80 cm. Leaf-veins downy beneath. Flowers yellow, spotted or unspotted. *NW Italy, NW Yugoslavia, W Bulgaria, C Romania.*

Subsp. **ponticum** (C. Koch) Matthews. Stem 15–90 cm, green. Leaves 3–8 cm × 8–20 mm, with 7–15 veins which are downy beneath. Flowers to 5, occasionally to 12. Perianth-segments 3.5–5.5 cm, deep yellow or deep orange with dense red-brown or purplish stripes and spots towards the base inside, but lacking warts. Filaments papillose. *NE Turkey, USSR (Georgia).*

Var. **ponticum** (*L. ponticum* C. Koch; *L. georgicum* Mandenova; *L. carniolicum* W. Koch subsp. *ponticum* (C. Koch) Davis & Henderson). Illustration: RHS Lily Year Book **27**: f. 26, 27 (1963), **33**: pl. 29e (1969). Stem usually stout. Perianth-segments deep yellow, 7–10 mm wide. *NE Turkey.*

Var. **artvinense** (Mischenko) Matthews (*L. artvinense* Mishchenko; *L. ponticum* C. Koch var. *artvinense* (Mishchenko) Davis & Henderson). Illustration: RHS Lily Year Book **33**: pl. 29f (1969). Stem more slender than in var. *ponticum* and with fewer, less crowded leaves. Perianth-segments deep orange, 5–6 mm wide. *NE Turkey, USSR (Georgia).*

**\*L. ledebourii** (Baker) Boissier. Illustration: RHS Lily Year Book **32**: pl. 5 (1968); Wendelbo, Tulips and irises of Iran and their relatives, 32, 33 (1977). Flowers up to 5, fragrant, perianth-segments 6–7.5 cm, creamy-white, becoming greenish yellow, and dotted with purple or red towards the base. Stamens with the filaments joined together at base; anthers

and pollen orange-red. *USSR (Azerbaidjan), NW Iran.* H3. Summer.

**\*L. ciliatum** Davis. Illustration: Woodcock & Stearn, Lilies of the world, f. 94 (1950) – as L. ponticum; RHS Lily Year Book **33**: pl. 29 a & c (1969); Synge, Lilies, pl. 2 (1980). Stem 60–160 cm, sometimes with cobwebby hairs in the upper part. Leaves long-ciliate on the margin, veins hairless beneath. Flowers up to 20. Perianth-segments 4–5.5 cm, cream or pale yellow, marked with purplish brown towards the base, apex hairy. *NE Turkey.* H3. Summer.

**\*L. callosum** Siebold & Zuccarini. Illustration: Elwes, Monograph of the genus Lilium., pl. 41 (1877); Woodcock & Stearn, Lilies of the world, f. 30 (1950). Stem to 90 cm. Leaves 8–13 cm × 3–8 mm, linear with 3 or 5 veins, hairless. Flowers up to 10, scentless. Perianth-segments 3–4 cm, orange-red with faint black dots towards the base. *USSR (Amur region), N & C China, Korea, Japan, Taiwan.* H3. Late summer.

**24. L. pomponium** Linnaeus. Illustration: Botanical Magazine, 971 (1807); Woodcock & Stearn, Lilies of the world, f. 93 (1950); Grey-Wilson & Mathew, Bulbs, pl. 4 (1981); Rix & Phillips, The bulb book, 161 (1981).

Stem roots present. Stem 20–100 cm, green with purple spots below, hairless. Leaves scattered, 9–15 cm × 2–5 mm, linear with 1 or 3 veins, margin ciliate, otherwise hairless. Flowers usually 1–6, sometimes to 10, turk's-cap type, unpleasantly scented, nodding. Perianth-segments 4–7 cm, bright red with purplish or black spots and lines and purplish warts inside. Anthers 4–7 mm, sometimes to 1.4 cm, red. *Maritime Alps of France and Italy.* H3. Summer.

**25. L. martagon** Linnaeus. Stem roots present. Stem 90–200 cm, purplish or green, hairy or hairless. Leaves in whorls of 6–10, up to 16 × 6.5 cm, oblanceolate with 7 or 9 veins, hairless or downy on the veins beneath. Flowers to 50, turk's-cap type, scented, nodding. Perianth-segments 3–4.5 cm × 6–10 mm, white, pink or dark purplish red to maroon, often spotted. Nectary-furrows with warty margins. Anthers orange-yellow or reddish purple. *Europe, Asia east to Mongolia.* H1. Summer.

1a. Flowers white; stem green
     var. **album**
  b. Flowers not white; stem purplish     2
2a. Stem hairless or more or less so     3

  b. Stem hairy     4
3a. Flowers maroon
     var. **sanguineo-purpureum**
  b. Flowers dull purplish pink
     var. **martagon**
4a. Flowers maroon, unspotted
     var. **cattaniae**
  b. Flowers purplish pink to wine red,
     spotted     5
5a. Buds hairy; leaves hairless on veins
     beneath     var. **pilosiusculum**
  b. Buds not hairy; leaves downy on
     veins beneath     var. **hirsutum**

Var. **album** Weston. Illustration: de Graaff & Hyams, Lilies, pl. 1 (1967); RHS Lily Year Book **33**: pl. 8 (1969). Stem green and hairless, flowers white, unspotted.

Var. **cattaniae** Visiani (var. *dalmaticum* Elwes). Illustration: Elwes, Monograph of the genus Lilium, pl. 33 (1877); Synge, Lilies, pl. iii (1980); Grey-Wilson & Mathew, Bulbs, pl. 3 (1981). Stem purplish, hairy. Buds hairy, opening to dark maroon unspotted flowers.

Var. **hirsutum** Weston. Stem purplish, hairy. Leaves downy beneath. Flowers purplish pink, spotted.

Var. **martagon**. Illustration: Robert & Schroeter, Alpine flowers, pl. 5 (1938); Polunin, Flowers of Europe, pl. 167 (1969); Aichele, Wild flowers, 259 (1975); Grey-Wilson & Mathew, Bulbs, pl. 3 (1981). Stem purplish, stem and leaves hairless or almost so, flowers dull purplish pink, spotted.

Var. **pilosiusculum** Freyn. Stem purplish, cobwebby-hairy. Leaves narrow with hairs only on the margin. Bracts and buds hairy. Flowers wine red, with sparse spots.

Var. **sanguineo-purpureum** Beck. Stem purplish, hairless. Flowers dark maroon with heavy spotting.

The following varieties have been in cultivation and may still persist in some gardens: var. **albiflorum** Vukotinovic with white flowers which are spotted with pink; var. **caucasicum** Mishchenko which grows to 2.5 m and has lilac-pink flowers.

There are several hybrids and hybrid groups derived from *L. martagon* crossed with *L. hansonii*. *L.* × **marhan** Baker (Illustration: Reader's Digest encyclopaedia of garden plants and flowers, 399, 1971) was raised from *L. martagon* var. *album* and *L. hansonii* and has flowers with thick perianth-segments which are dull orange with reddish brown spots especially towards the tips. *L.* × **dalhansonii** Powell (Illustration: Garden **44**: t. 927, 1893) resulted from a cross between *L. martagon* var. *cattaniae* and *L. hansonii*. The abundant flowers are maroon, flushed with orange at the middle of the segments and heavily spotted. The 'Backhouse Hybrids' have flowers varying from cream, yellow or pink to maroon, all with dark spots; a number of clones have been named.

*L. polyphyllum** which differs in having less recurved perianth-segments, greenish white outside, inside creamy-white, with green or purplish streaks, and yellow anthers with orange pollen, may key out here (see entry under 31. *L. duchartrei*).

*L. kelloggii** Purdy. Illustration: Woodcock & Stearn, Lilies of the world, f. 67 (1950); Synge, Lilies, pl. 20 (1980); Rix & Phillips, The bulb book, 160 (1981). Stem roots absent. Stem glaucous. Leaves in whorls of 12–18, 6–10 × 1–2 cm, with 1 or 3 veins, hairless. Flowers to 20, very fragrant. Perianth-segments 3–5.5 cm, mauvish, pink or white, usually with a yellow band below the middle, spotted with purple in lower part. Anthers reddish to orange with orange pollen. *USA (S Oregon, NW California)*. H4. Summer.

**26. L. monadelphum** Bieberstein (*L. szovitsianum* Fischer & Avé-Lallement). Illustration: RHS Lily Year Book **27**: pl. 4 (1963); de Graaff & Hyams, Lilies, pl. 2 (1967); Synge, Lilies, pl. iv, v (1980); Rix & Phillips, The bulb book, 166, 167 (1981). At least a few stem roots present. Stem 50–200 cm, hairless. Leaves scattered, 5–14 × 1–3 cm, lanceolate to oblanceolate or ovate with 9, 11 or 13 veins, margins ciliate, downy on veins beneath. Flowers usually 1–5, occasionally to 30, turk's-cap type, fragrant, nodding. Perianth-segments 6–10 × 1–2.2 cm, pale to mid-yellow, usually spotted with purple or maroon inside, flushed with purplish brown at the base outside, varying in the degree to which they are recurved. Stamens with the filaments joined at the base or free; anthers 8–13 mm, brown or maroon, pollen orange or yellow. *USSR (Caucasus), NE Turkey*. H3. Summer.

Var. **armenum** (Mishchenko) Davis & Henderson. Illustration: RHS Lily Year Book **33**: pl. 29b (1969). Inner perianth-segments acute, 1–1.6 cm wide. *USSR (Armenia), NE Turkey*.

There is a lily from the Caucasus which is occasionally grown under the names *L. kosa* or *L. kossii*. Both names are invalid because they have not been properly published. The lily was originally said to be closely related to *L. monadelphum*, and said to differ in having pink-tinged flowers, a

feature which was observed in wild plants. However, plants grown in gardens have produced only pale yellow flowers, and do not appear to differ from *L. monadelphum*.

*L. kesselringianum** Mishchenko. Illustration: Woodcock & Stearn, Lilies of the world, f. 68 (1950); RHS Lily Year Book **33**: pl. 29d (1969). Flowers usually 1–3. Perianth-segments cream to pale yellow, spotted with purple or reddish brown towards the base, recurved in the upper two-thirds but not reflexed. Stamens with free filaments. *USSR (Georgia), NE Turkey*. H3. Summer.

**27. L. pumilum** de Candolle (*L. tenuifolium* Schrank). Illustration: Elwes, Monograph of the genus Lilium, pl. 42 (1877); Woodcock & Stearn, Lilies of the world, f. 97 (1950); Reader's Digest encyclopaedia of garden plants and flowers, 407 (1971). Stem roots present. Stem 15–45 cm, occasionally to 90 cm, green, hairless or minutely rough, especially below. Leaves scattered, 3–10 cm × 1–3 mm, linear with 1 vein, hairless or slightly rough on margin and veins beneath. Flowers to 30, turk's-cap type, slightly fragrant, nodding, buds sometimes white-hairy. Perianth-segments 3–3.5 cm, scarlet, unspotted or with sparse black dots at the base. Nectary-furrows with a papillose border. Anthers 4–8 mm; pollen scarlet. *USSR (Siberia), Mongolia, N China, N Korea*. H2. Summer.

**28. L. hansonii** Moore. Illustration: Elwes, Monograph of the genus Lilium, pl. 34 (1877); Woodcock & Stearn, Lilies of the world, f. 49 (1950). Stem roots present. Stem 1–1.5 m, green. Leaves in whorls of 12–20, up to 18 × 4 cm, oblanceolate to elliptic with 3 or 5 veins. Flowers 3–12, turk's-cap type, fragrant, nodding. Perianth-segments very thick, 3–4 × c. 1.5 cm, deep orange-yellow spotted with purplish brown towards the base, recurved from about the middle. Nectary-furrows hairless. Anthers 2–2.2 cm, purplish with yellow pollen. *E USSR, Korea, Japan*. H3. Summer.

**29. L. leichtlinii** Hooker. Illustration: Botanical Magazine, 5673 (1867); Elwes, Monograph of the genus Lilium, pl. 39 (1877); Woodcock & Stearn, Lilies of the world, f. 71 (1950); Jefferson-Brown, Modern lilies, 48 (1965). Bulb stoloniferous. Stem roots present. Stem up to 120 cm, green, hairless or with sparse white hairs especially at the leaf axils. Leaves scattered, up to 15 × 1.25 cm, linear to narrowly lanceolate with 1 or 3

veins, margin minutely rough and midrib slightly minutely rough beneath. Flowers 1–6, turk's-cap type, scentless, nodding; buds usually white-hairy. Perianth-segments 6.5–8.5 × 1.3–2.7 cm, yellow, spotted with purplish maroon. Nectary-furrows hairy. Anthers 1–1.7 cm, reddish brown; pollen reddish brown. *Japan*. H3. Summer.

Var. **maximowiczii** (Regel) Baker (*L. maximowiczii* Regel; *L. leichtlinii* var. *tigrinum* (Regel) Nicholson). Illustration: Elwes, Monograph of the genus Lilium, pl. 40 (1879). Stem to 2.5 m. Flowers reddish orange with purplish brown spots. *Japan, C Korea, E USSR*.

*L. rhodopaeum** Delipavlov. Illustration: RHS Lily Year Book **32**: pl. 6 (1968); Grey-Wilson & Mathew, Bulbs, pl. 4 (1981). Stem 80–100 cm. Leaves with ciliate margin, downy on the veins beneath. Perianth-segments 8–12 cm, yellow, unspotted, less recurved than in *L. leichtlinii*. *Greece, Bulgaria*. H4. Summer.

**30. L. lancifolium** Thunberg (*L. tigrinum* Ker Gawler). Illustration: Rix & Phillips, The bulb book, 164 (1981); Mathew, P.J. Redouté, Lilies and related flowers, 45 (1982). Stem roots present. Stem 60–150 cm, dark purple, usually with white cobwebby hairs. Leaves scattered, 12–20 × 1–2 cm, narrowly lanceolate with 5 or 7 veins, margins minutely rough, veins hairless beneath; axils with bulbils. Flowers to 40, turk's-cap type, scentless, nodding. Perianth-segments 7–10 × 1–2.5 cm, pinkish orange to orange-red with deep purple spots and purplish warts inside. Nectary-furrows bordered by warty hairy ridges. Anthers 1.2–1.4 cm, orange-red to purplish with reddish brown or purplish brown pollen. *Japan, Korea, E China*. H4. Summer–early autumn.

Var. **flaviflorum** Makino (*L. tigrinum* var. *flaviflorum* (Makino) Stearn). Flowers yellow.

Var. **fortunei** (Standish) Matthews (*L. tigrinum* var. *fortunei* Standish). Illustration: Synge, Lilies, pl. xix (1980). A larger plant with densely woolly stem and many orange-red flowers.

Var. **splendens** (van Houtte) Matthews (*L. tigrinum* var. *splendens* van Houtte). A vigorous plant with larger flowers of a brighter orange-red.

'Flore-pleno' is a selection with double flowers.

*L. lancifolium* sold by nurseries often turns out to be *L. speciosum*.

**31. L. duchartrei** Franchet (*L. farreri* Turrill; *L. duchartrei* var. *farreri* (Turrill) Grove). Illustration: Botanical Magazine, 8847 (1920); Woodcock & Stearn, Lilies of the world, f. 43 (1950); Synge, Lilies, pl. 8 (1980); Rix & Phillips, The bulb book, 162 (1981).

Bulb stoloniferous. Stem roots present. Stem 60–100 cm, green or flushed with purplish brown, slightly ribbed, hairless or slightly papillose; leaf axils often hairy. Leaves scattered, 5–10 × 1–1.5 cm, lanceolate with 3 or 5 nerves, margin minutely rough and veins sometimes minutely rough beneath. Flowers to 12, turk's-cap type, fragrant, nodding. Perianth-segments up to 5.5 × 3 cm, white, often slightly flushed with purple outside and green or purplish at the base, streaked and spotted with purple inside. Nectary-furrows green with warty margins. Anthers 1–1.3 cm, yellow or orange, with orange pollen. *W China*. H4. Summer.

**\*L. ledebourii** which differs in its narrower perianth-segments and orange-red anthers may key out here (see entry under 23. *L. pyrenaicum*).

**\*L. polyphyllum** Don. Illustration: Elwes, Monograph of the genus Lilium, pl. 48 (1877); Woodcock & Stearn, Lilies of the world, f. 109 (1950); Journal of the Royal Horticultural Society 93: pl. 126 (1968); Rix & Phillips, The bulb book, 158 (1981). Stem green, hairless. Leaves scattered or with the lower in a whorl, 5–12 cm × 3–15 mm, margin minutely rough. Flowers usually 1–6, sometimes to 30, very fragrant. Perianth-segments less recurved than in *L. martagon*, greenish white outside, inside creamy white, with green or purplish streaks. Anthers 8–9 mm. *E Afghanistan, W Himalaya*. H4. Summer.

**32. L. taliense** Franchet. Illustration: Grove & Cotton, Supplement to Elwes' Monograph of the genus Lilium, pl. 24 (1939); Woodcock & Stearn, Lilies of the world, f. 111 (1950); RHS Lily Year Book 27: f. 8 (1963); Everard & Morley, Wild flowers of the world, 108 (1970).

Bulb shortly stoloniferous. Stem roots present. Stem to 140 cm, deep purple or green marked with purple, hairless or minutely rough towards the top. Leaves scattered, 7–13 cm × 4–6 mm, narrowly lanceolate with 1 or 3 veins, margin minutely rough, veins minutely rough beneath. Flowers to 12, turk's-cap type, fragrant, nodding. Perianth-segments 4–6 cm × 7–20 mm, white, spotted with purple except at the tips, strongly recurved. Nectary-furrows purple, hairless. Anthers 7–11 mm, mauve or greyish with yellow pollen. *W China*. H4. Summer.

**33. L. lankongense** Franchet. Illustration: Woodcock & Stearn, Lilies of the world, f. 69 (1950); de Graaff & Hyams, Lilies, pl. 5 (1967); RHS Lily Year Book 34: front cover (1970); Synge, Lilies, pl. xi (1980).

Bulb stoloniferous. Stem roots present. Stem to 120 cm, green, minutely rough, slightly ribbed. Leaves scattered, 3–10 cm × 4–8 mm, lanceolate with 3, 5 or 7 veins, margin minutely rough, veins hairless beneath. Flowers to 15, turk's-cap type, fragrant, nodding. Perianth-segments 4–6.5 × 1–2.5 cm, pink sometimes flushed with mauve, spotted with reddish purple especially on the incurved margins and with a green median band, strongly recurved. Nectary-furrows green, bordered with white warts. Anthers 8–12 mm, purplish with orange-brown pollen. *W China (SE Xizang)*. H4. Summer.

**\*L. papilliferum** Franchet. Illustration: de Graaff & Hyams, Lilies, pl. 4 (1967); Synge, Lilies, pl. 11 (1980). Stem minutely rough except near the base. Flowers 1–3, sometimes more. Perianth-segments 4–5 × c. 1 cm, deep crimson-maroon, becoming paler towards the base and bearing warts. Anthers brown. *W China (Yunnan)*. H4. Late spring–late summer.

**\*L. cernuum** Komarov. Illustration: Woodcock & Stearn, Lilies of the world, f. 34 (1950); de Graaff & Hyams, Lilies, pl. 3 (1967); Synge, Lilies, pl. 6 (1980). Stem to 60 cm, green or marked with purplish brown, base and upper part more or less leafless. Leaves scattered, densely crowded in middle part of stem, 10–18 cm × 1–5 mm, linear with 1 or 3 veins, hairless. Perianth-segments 3.5–5 cm × 7–12 mm, lilac, pinkish purple or sometimes white, spotted with purple. Pollen lilac. *Korea north to E USSR*. H3. Summer.

**34. L. wardii** Stern. Illustration: Woodcock & Stearn, Lilies of the world, f. 118 (1950); Synge, Lilies, pl. xx (1980).

Bulb stoloniferous. Stem roots present. Stem to 1.5 m, dark green tinged with purple, hairless or minutely rough. Leaves scattered, 3–8 cm × 6–20 mm, narrowly lanceolate to elliptic with 3 veins, margin hairless or minutely rough, veins usually hairless beneath. Flowers up to 35, turk's-cap type, fragrant, nodding. Perianth-segments 5–6.5 cm × 7–15 mm, pink or pinkish purple spotted inside with purple and with a purple median line in the lower half. Nectary-furrows hairless. Anthers 10–11 mm, yellow or purple with orange pollen. *W China (SE Xizang)*. H4. Summer.

**35. L. davidii** Elwes. Illustration: RHS Lily Year Book 18: f. 17 (1954); Synge, Lilies, pl. 7 (1980).

Bulb stoloniferous or not. Stem roots present. Stem 1–1.4 m, green, usually marked with brownish purple, downy especially in lower part, usually with cobwebby hairs at the leaf axils. Leaves scattered, 6–10 cm × 2–4 mm, linear with 1 vein which is minutely rough beneath, margin usually inrolled and minutely rough. Flowers 5–20, turk's-cap type, unscented, nodding; buds usually hairy. Perianth-segments 5–8 cm × 8–25 mm, red or orange-red, densely spotted with deep purple raised spots except at the apex, warty in the lower part, strongly recurved. Nectary-furrows bordered with white hairs. Anthers 9–12 mm, purplish orange with orange or red pollen. *W China (Szechwan, Yunnan)*. H4. Summer.

Var. **macranthum** Raffill. Stems taller (to 2 m) and flowers larger, borne 2 or 3 together on a stalk.

Var. **unicolor** (Hoog) Cotton. Stems shorter (to 1 m) and leaves longer and very crowded. Flowers paler, unspotted or with reddish or mauve spots.

Var. **willmottiae** (Wilson) Raffill. Illustration: Woodcock & Stearn, Lilies of the world, f. 41 (1950); Synge, Lilies, pl. viii (1980). Bulb stoloniferous. Stem to 2 m, arching under the weight of the flowers. Leaves to 6 mm wide. Flowers to 40 per stem.

**36. L. henryi** Baker. Illustration: Botanical Magazine, 7177 (1891); Woodcock & Stearn, Lilies of the world, f. 60 (1950); Hay & Synge, Dictionary of garden plants in colour, 101 (1969); Synge, Lilies, pl. ix (1980).

Stem roots present. Stem 1–3 m, green with reddish purple markings, hairless. Leaves scattered, 8–15 × 2–3 cm, lanceolate with many veins, margin hairless, hairless or sparsely hairy on the veins beneath, sometimes with bulbils in the axils. Flowers usually 4–20, occasionally up to 70, 1 or 2 on each flower-stalk, turk's-cap type, nodding. Perianth-segments 6–8 × 1–2 cm, orange with dark spots towards the base and long warts, shortly clawed. Nectary-furrows green. Anthers 1.2–1.7 cm, orange. *C China*. H3. Summer.

Var. **citrinum** Anon. Flowers pale yellow lightly spotted with brown or reddish brown.

*L. henryi* crossed with *L. sargentiae* gave rise to L. × **aurelianense** Debras. The original cross is similar to *L. henryi*, but with the flowers borne almost horizontally, pale yellow to yellowish orange, unspotted and trumpet-shaped with the segments recurved at the tips. Later back-crosses show more variation. Illustration: Synge, Collins guide to bulbs, pl. 19 (1961); Reader's Digest encyclopaedia of garden plants and flowers, 402 (1971).

**37. L. speciosum** Thunberg. Illustration: Elwes, Monograph of the genus Lilium, pl. 13 (1878); Everard & Morley, Wild flowers of the world, 107 (1970); Perry, Flowers of the world, 170 (1972).
Stem roots present. Stem 120–170 cm, green or flushed with purple, hairless. Leaves scattered, shortly stalked, up to 20 × 6 cm, broadly lanceolate to oblong-lanceolate with 7 or 9 veins, hairless. Flowers usually to 12, occasionally to 40, turk's-cap type, fragrant, nodding. Perianth-segments to 10 × 4.5 cm, white or pale pink tinged with deep pink towards the base and spotted with pink or crimson, warty in the lower part, strongly recurved, margins slightly wavy. Nectary-furrows green, hairless. Anthers 1.3–2.2 cm with brown or dark red pollen. *Japan, China, Taiwan.* H4. Late summer.

Var. **gloriosoides** Baker. The spots and warts are almost scarlet and the warts are confined to an area towards the base but are absent from the basal one-quarter of the segments.

Several cultivars which vary mainly in flower colour are commonly grown.

**38. L. amabile** Palibin. Illustration: Grove & Cotton, Supplement to Elwes' Monograph of the genus Lilium, pl. 6 (1935); Woodcock & Stearn, Lilies of the world, f. 20 (1950).
Stem roots present. Stem 30–90 cm, downy, leafless in the lower part. Leaves scattered, 4–9 cm × 5–10 mm, lanceolate, usually with 3 veins, hairy. Flowers to 10, turk's-cap type, scent rather unpleasant, nodding. Perianth-segments 5–5.5 cm, red with black spots, strongly recurved. Nectary-furrows warty. Anthers 1.1–1.3 cm, brown with reddish pollen. *Korea.* H4. Summer.

Var. **luteum** Anon. Flowers yellow.

**39. L. humboldtii** Duchartre. Illustration: Elwes, Monograph of the genus Lilium, pl. 32 (1877); RHS Lily Year Book **33**: pl. 19 (1969).
Stem roots absent. Stem up to 225 cm, green often tinged with purple, often slightly downy. Leaves in whorls of 10–20, 9–12.5 × 1.5–3 cm, oblanceolate with 5 veins, margin wavy and downy. Flowers usually 10–15, occasionally to 80, turk's-cap type, nodding. Perianth-segments 6.5–10 × 1.5–2.5 cm, yellow to orange, spotted with maroon or purple towards the centre, strongly recurved. Anthers 1.2–1.3 cm, purplish with orange pollen. *USA (C California).* H4. Summer.

Var. **ocellatum** (Kellogg) Elwes (var. *magnificum* Purdy). Illustration: Woodcock & Stearn, Lilies of the world, f. 62 (1950); Synge, Lilies, pl. xxii (1980). Stem roots present. Perianth-segments have red tips and the spots are encircled with crimson.

*L. humboldtii* has been crossed with *L. pardalinum* to produce L. × **pardaboldtii** Woodcock & Coutts. The stem reaches 150 cm and the leaves are in whorls; the turk's-cap type flowers are reddish orange, spotted with dark crimson.
The 'Bellingham Hybrids' were also developed from this cross.

\***L. iridollae** Henry. Illustration: Bartonia 24: pl. 1–5 (1946); Woodcock & Stearn, Lilies of the world, f. 64, 65 (1950); RHS Lily Year Book **17**: f. 4 (1953). Stem up to 175 cm. Leaves usually in whorls, downy on margin and veins beneath. Flowers to 8. Perianth-segments yellow, spotted with brown. Anthers reddish brown. *SE USA (S Alabama, N Florida).* H4. Late summer.

**40. L. columbianum** Baker. Illustration: Elwes, Monograph of the genus Lilium, pl. 31 (1877); Woodcock & Stearn, lilies of the world, f. 38 (1950); Synge, Lilies, pl. 18 (1980); Rix & Phillips, The bulb book, 160 (1981).
Stem roots absent. Stem to 2.5 m, hairless. Lower leaves in whorls of 5–13, upper scattered, 5–14 × 1.2–4 cm, oblanceolate with 3 or 5 veins, hairless. Flowers usually 6-10, occasionally to 40, turk's-cap type, nodding. Perianth-segments 3.5–6.5 cm × 8–12 mm, yellow to orange-red with purplish red spots towards the base, strongly recurved. Anthers 6–11 mm, orange, with yellow to brownish pollen. *N America (Pacific coast from S Canada to N California).* H3. Summer.

\***L. wigginsii** Beane & Vollmer. Stem roots present. Stem to 90 cm. Leaves scattered or in 2–4 whorls half-way up the stem, to 22 cm, narrowly lanceolate. Perianth-segments deep yellow, spotted with purple. Pollen yellow. *W USA (S Oregon, N California).* H4. Summer.

**41. L. kelleyanum** Lemmon (*L. nevadense* Eastwood; *L. nevadense* var. *shastense* Eastwood). Illustration: Woodcock & Stearn, Lilies of the world, f. 86 (1950); RHS Lily Year Book **24**: f. 10 (1960).
Bulb stoloniferous. Stem roots absent. Stem to 1 m, sometimes to 2 m. Leaves in 2 or 3 whorls of 7–16, scattered above and below, or all leaves scattered, 5–15 × 1.5–3.5 cm, lanceolate to broadly elliptic, hairless. Flowers up to 25, turk's-cap type, fragrant, nodding. Perianth-segments 2.5–5 × *c.* 1 cm, greenish yellow, yellow or orange, spotted with purplish red or brown towards the base, tips sometimes reddish. Anthers 5–7 mm, brown. *W USA (California, S Oregon).* H4. Summer.

\***L. occidentale** Purdy. Illustration: Grove & Cotton, Supplement to Elwes' Monograph of the genus Lilium, pl. 25 (1940); Woodcock & Stearn, Lilies of the world, f. 88 (1950); RHS Lily Year Book **31**: pl. 9 & f. 19 (1967); Journal of the Royal Horticultural Society **92**: pl. 245, 247 (1967). Stem to 2 m or more. Leaves in whorls, scattered above. Flowers usually 1–5, sometimes up to 20, unscented. Perianth-segments 3.5–6.25 cm, crimson, green or yellow at the base, or orange-red with an orange base and spotted with purplish brown. Anthers 7–8 mm, purple to reddish brown. *W USA (coastal S Oregon & N California).* H4. Summer.

\***L. medeoloides** Gray. Illustration: Elwes, Monograph of the genus Lilium, pl. 35 (1879) – as L. avenaceum; Woodcock & Stearn, Lilies of the world, f. 80 (1950). Stem roots present. Stem to 75 cm, hollow, hairless or minutely rough below. Leaves in 1 or 2 whorls about half-way up the stem, with scattered leaves above. Flowers scentless. Perianth-segments apricot or orange-red, unspotted or with dark spots. Anthers purplish with orange-red pollen. *NE USSR, S Korea, China, Japan.* H3. Summer.

\***L. distichum** Kamibayashi. Illustration: Woodcock & Stearn, lilies of the world, f. 42 (1950). Stem roots present. Stem hollow, ribbed. Leaves in 1 or 2 whorls with upper leaves scattered. Flowers 3–8. Perianth-segments pale orange, usually with darker spots. Anthers 8–10 mm. *E USSR, NE China, N & C Korea.* H3. Summer.

\***L. pitkinense** Beane & Vollmer. Illustration: Turrill, Supplement to Elwes' Monograph of the genus Lilium 9: pl 3

(1962); RHS Lily Year Book **27**: pl. 6 (1963). Leaves linear. Flowers to 8. Perianth-segments orange-red, yellow towards the base with tiny maroon spots. Anthers brownish purple with reddish brown pollen. *W USA (California)*. H4. Summer.

**42. L. pardalinum** Kellogg. Illustration: Elwes, Monograph of the genus Lilium, pl. 28, 29 (1877–9); Woodcock & Stearn, Lilies of the world, f. 108 (1950); Hay & Synge, Dictionary of garden plants in colour, 102 (1969), Witham Fogg, Bulbs, 48 (1980).
Bulb stoloniferous. Stem roots absent. Stem 2–3 m, green, hairless. Leaves in whorls of 8–19, with a few scattered above and below, to 20 × 5.5 cm, narrowly elliptic to oblanceolate with 3 veins, hairless. Flowers to 10, turk's-cap type, often fragrant, nodding. Perianth-segments 5–9 × 1.2–1.8 cm, dark red in upper half, orange to orange-red in lower half and heavily spotted with maroon, some spots encircled with yellow, strongly recurved. Anthers reddish or brown with orange pollen. *W USA (S Oregon, N California)*. H4. Summer.

Var. **angustifolium** Kellogg (*L. roezlii* Regel). Leaves narrow.

Var. **giganteum** Woodcock & Coutts. Illustration: Synge, Lilies, pl. xxiii (1980). Tall-growing with up to 30 crimson and yellow, densely spotted flowers.

*L. pardalinum* has been crossed with *L. parryi* to produce L. × **burbankii** Anon. The leaves are whorled and the strongly reflexed perianth-segments are yellow with brown spots and often flushed red at the tips.

**43. L. michiganense** Farwell. Illustration: Woodcock & Stearn, Lilies of the world, f. 81 (1950); RHS Lily Year Book **19**: f. 13 (1955); Turrill, Supplement to Elwes' Monograph of the genus Lilium 8: 15 (1960).
Bulb stoloniferous. Stem 60–150 cm, green, hairless. Leaves in whorls of 4–8, 9–12 cm, lanceolate to narrowly elliptic with 3 veins, usually minutely rough on margin and veins beneath. Flowers usually 3–6, sometimes to 25, turk's-cap type, nodding. Perianth-segments 6.5–8 cm × 7–18 mm, orange-red with a green zone at the base, spotted with deep red or purple, strongly recurved. Anthers 6–9 mm, orange with orange pollen. *Eastern N America*. H3. Summer.

**\*L. vollmeri** Eastwood. Illustration: Woodcock & Stearn, Lilies of the world, f.

115 (1950). Leaves scattered or with a few in whorls, hairless. Perianth-segments deep orange, tinged with red towards the tip at the margin, spotted with blackish maroon. *W USA (S Oregon, N California)*. H4. Summer.

**L. occidentale** which differs in having leaves with usually 7 veins, perianth-segments 3.5–6.25 cm, and purple anthers with orange-red pollen, may key out here (see under 41. *L. kelleyanum*).

**44. L. superbum** Linnaeus. Illustration: Elwes, Monograph of the genus Lilium, pl. 26 (1877); Woodcock & Stearn, Lilies of the world, f. 110 (1950); Reader's Digest encyclopaedia of garden plants and flowers, 408 (1971); Synge, Lilies, pl. xxv (1980).
Bulb stoloniferous. Stem roots present. Stem 1.5–3 m, green, usually flushed with purple, hairless. Leaves in whorls of 4–20, sometimes the upper ones scattered, 3.5–11 cm × 8–28 mm, lanceolate to elliptic with 3, 5 or 7 veins, sometimes minutely rough on margin and veins beneath. Flowers to 40, turk's-cap type, nodding. Perianth-segments 6–10 × 1–2 cm, orange flushed with red, especially at the tips, and spotted towards the green base with maroon. Nectary-furrows green, hairy. Anthers 1.6–2 cm, red with orange-brown pollen. *E USA*. H3. Summer.

**45. L. michauxii** Poiret. Illustration: Edwards's Botanical Register 7: t. 580 (1821) – as L. carolinianum; Elwes, Monograph of the genus Lilium, pl. 25 (1877); Woodcock & Stearn, Lilies of the world, f. 79 (1950).
Bulb stoloniferous. Stem roots present. Stem to 120 cm, green or slightly spotted, hairless. Leaves in whorls of 3–7, upper ones usually scattered, up to 12 × 4 cm, oblanceolate to obovate, glaucous, hairless. Flowers usually 1 or 2, sometimes up to 5, turk's-cap type, scented, nodding. Perianth-segments up to 10 cm, orange-red to pale crimson, yellowish towards the base, spotted with purple or black. Anthers purplish brown with orange-red pollen. *SE USA*. H4. Summer–early autumn.

**L. occidentale** which differs in having up to 20 flowers, with deep crimson perianth-segments 3.5–6.25 cm, may key out here (see under 41. *L. kelleyanum*).

**46. L. parvum** Kellogg. Illustration: Botanical Magazine, 6146 (1875) – as L. canadense var. parvum; Elwes, Monograph of the genus Lilium, pl. 30 (1877); Rix & Phillips, The bulb book, 160 (1981).

Bulb stoloniferous. Stem roots absent. Stem to 2 m. Leaves usually in whorls of 7–10, occasionally scattered, 5–12.5 cm × 5–30 mm, linear or lanceolate with 1 or 3 veins, hairless, margin occasionally minutely rough. Flowers to 30, open bell-shaped, borne horizontally or almost erect. Perianth-segments 3–4 cm, red or crimson, usually heavily spotted with dark red or purple, slightly recurved at the tips. *W USA (SE Oregon to C California)*. H4. Summer.

Forma **crocatum** Stearn. Illustration: RHS Lily Year Book **31**: pl. 9 (1967). Flowers yellow or orange.

**\*L. maritimum** Kellogg. Illustration: Elwes, Monograph of the genus Lilium, pl. 12 (1879); Woodcock & Stearn, Lilies of the world, f. 75 (1950); Synge, Lilies, pl. 21 (1980). Stem 30–70 cm, occasionally to 200 cm. Leaves usually scattered, sometimes with 1 whorl near the middle of the stem. Flowers up to 12, nodding. Perianth-segments deep red or reddish orange, spotted inside with maroon, recurved at the tips. *W USA (coastal N California)*. H4. Summer.

**47. L. rubescens** Watson. Illustration: Woodcock & Stearn, Lilies of the world, f. 101 (1950); RHS Lily Year Book **19**: f. 19 (1955); Synge, Lilies, pl. 22 (1980).
Bulb sometimes stoloniferous. Stem roots absent. Stem usually to 180 cm, sometimes to 300 cm, green suffused with purple, hairless. Leaves in whorls of 5–12, with some leaves scattered in lower part of stem, 5–12 × 1.5–3.5 cm, oblanceolate with 3 or 5 veins, hairless. flower-stalks more or less erect, 3–6 cm. Flowers usually to 30, occasionally 100 or more, trumpet-shaped, fragrant, more or less erect. Perianth-segments 3.5–6.5 × 7–10 mm, at first white or whitish, changing to pinkish purple or purple, unspotted or sometimes with deep purple dots towards the base, spreading or recurved at the tips. Anthers and pollen yellow. *W USA (S Oregon, NW California)*. H4. Summer.

**\*L. washingtonianum** Kellogg. Illustration: Elwes, Monograph of the genus Lilium, pl. 10 (1878); Woodcock & Stearn, Lilies of the world, f. 119 (1950); RHS Lily Year Book **33**: pl. 20 (1969). Stem to 2.5 m. Leaves with wavy margins. Flowers broadly funnel-shaped, borne horizontally on almost erect stalks. Perianth-segments 8–10 × 1–2.5 cm, white or lilac with purplish dots towards the base, recurved towards the tips. Anthers pinkish. *W USA (S Oregon, NW California)*. H4. Summer.

Var. **purpurascens** Stearn (var. *purpureum* Purdy). Illustration: RHS Lily Year Book **24**: pl. 3 (1960) & **31**: pl. 9 (1967); Synge, Lilies, pl. 23 (1980). Flowers more open, almost bowl-shaped and changing colour as they age from white to pink, finally becoming purplish.

**48. L. bolanderi** Watson. Illustration: Woodcock & Stearn, Lilies of the world, f. 28 (1950); RHS Lily Year Book **22**: f. 4 (1958); Synge, Lilies, pl. 17 (1980); Rix & Phillips, The bulb book, 160 (1981). Stem roots absent. Stem to 140 cm, usually less, grey-green, sometimes marked with brown. Leaves in whorls of up to 12, 3–6 × 1–2.5 cm, oblanceolate with 1 or 3 veins, grey-green. Flower-stalks spreading or ascending, decurved below flower, up to 6.5 cm. Flowers to 18, funnel-shaped, scentless, usually somewhat drooping. Perianth-segments 3–4.5 × *c*. 1 cm, slightly recurved at the tips, dark crimson with darker red or purple spots, greenish at the base outside, the throat and tube yellowish spotted with purple or brown. Anthers purple with deep yellow pollen. *W USA (S Oregon, N California).* H4. Summer.

**49. L. parryi** Watson. Illustration: Botanical Magazine, 6650 (1882); Woodcock & Stearn, Lilies of the world, f. 89 (1950); Synge, Lilies, pl. xxiv (1980); Rix & Phillips, The bulb book, 162 (1981). Bulb somewhat stoloniferous. Stem roots absent. Stem to 2 m, hairless. Leaves mostly scattered, sometimes with 1 or more whorls towards base of stem, 8–15 cm × 5–22.5 mm, linear-lanceolate to narrowly ovate with 3 veins, margins minutely rough. Flowers usually *c*. 12, occasionally up to 50, narrowly trumpet-shaped, fragrant, drooping or almost horizontal. Perianth-segments 7–10 cm × 8–12 mm, yellow, sometimes with deep red or brown dots near the base, narrowly oblanceolate with narrow claws, recurved at the tips. Anthers 7–9 mm, orange-brown with brownish red pollen. *SW USA (S California, S Arizona).* H4. Summer.

**50. L. canadense** Linnaeus (*L. canadense* var. *flavum* Pursh). Illustration: Elwes, Monograph of the genus Lilium, pl. 27 (1878); Hay & Synge, Dictionary of garden plants in colour, 101 (1969); Synge, Lilies, pl. xxi (1950). Bulb stoloniferous. At least some stem roots present. Stem to 1.5 m, green. Leaves mostly in whorls of 6–10 but some scattered, up to 15 × 2 cm, lanceolate to oblanceolate with 5 or 7 veins, minutely rough on margin and veins beneath. Flowers usually *c*. 10, sometimes up to 20, bell-shaped, nodding. Perianth-segments 5–7.5 × 1.2–2.5 cm, yellow, usually with maroon or purplish brown spots towards the base, tips recurved. Anthers 6–7 mm with yellow to reddish brown pollen. *Eastern N America.* H3. Summer.

Var. **coccineum** Pursh (var. *rubrum* Moore). Flowers red with a yellow throat.

Var. **editorum** Fernald. Flowers red with narrow perianth-segments (8–13 mm broad). Leaves broader than in var. *canadense*.

\*L. **grayi** Watson. Illustration: Botanical Magazine, 7234 (1892); Woodcock & Stearn, Lilies of the world, f. 48 (1950); RHS Lily Year Book **23**: pl. 5 (1959); Synge, Lilies, pl. 19 (1980). Stem roots absent. All leaves in whorls, 5–12 × 1.5–3 cm. Perianth-segments 4–4.5 × *c*. 1 cm, deep red outside, inside paler and often yellow at the base, spotted with purple, tips not recurved. *E USA.* H3. Summer.

**51. L. mackliniae** Sealy. Illustration: Woodcock & Stearn, Lilies of the world, f. 74 (1950); Reader's Digest encyclopaedia of garden plants and flowers, 406 (1971); Synge, Lilies, pl. xiii (1980); Rix & Phillips, The bulb book, 165 (1981). Bulb not stoloniferous. Stem roots present. Stem to 40 cm, green or tinged with brownish purple. Leaves scattered or in whorls near the stem apex, 3–6 cm × 4–10 mm, narrowly lanceolate to narrowly elliptic with usually 3 veins, hairy at the base. Flowers 1–6, broadly bell-shaped, more or less nodding. Perianth-segments 4.5–5 × 1.6–1.8 cm, purplish pink. Nectary-furrows green, hairless. Anthers purple with yellow, orange or brown pollen. *NE India (Manipur).* H4. Late spring–summer.

\*L. **oxypetalum** which differs in having fringed nectary-furrows, may key out here (see under 52. *L. nanum*).

\*L. **sherriffiae** Stearn. Illustration: Journal of the Royal Horticultural Society **75**: f. 108 (1950); Botanical Magazine, n.s., 141 (1951); Synge, Lilies, pl. 13 (1980). Leaves scattered, linear to narrowly lanceolate. Flowers usually solitary, narrowly bell-shaped. Perianth-segments deep brownish purple outside, chequered with deep yellow and green inside towards the tips. Nectary-furrows with warty-crested margins. *Bhutan, Nepal.* H4. Summer.

\*L. **henrici** Franchet (*Nomocharis henrici* (Franchet) Wilson). Illustration: Woodcock & Stearn, Lilies of the world, f. 59 (1950); Turrill, Supplement to Elwes' Monograph of the genus Lilium 8: 3 (1960); RHS Lily Year Book **27**: f. 51 (1963); Rix & Phillips, The bulb book, 162 (1981). Stem to 90 cm. Leaves 10–15 cm, rough on the margin. Perianth-segments whitish flushed with pinkish outside especially at the base; inner segments dark purple at the base with many longitudinal veins, slightly 3-lobed at the apex. Anthers yellow. *W China (Yunnan).* H4. summer.

**52. L. nanum** Klotzsch (*Nomocharis nana* (Klotzsch) Wilson). Illustration: Turrill, Supplement to Elwes' Monograph of the genus Lilium 9: 9 (1962). Stem 6–45 cm, green, hairless. Leaves scattered, 4.5–12 cm × 2–5 mm, narrowly linear to linear with 3, 5 or 7 veins, margin minutely rough, hairless beneath. Flowers solitary, bell-shaped, fragrant, nodding. Perianth-segments 1–3.8 cm × 3–16 mm, pale purple to pink often with tiny reddish purple or brown dots, hairy or papillose towards the base inside. Nectary almost circular on the outer segments, more elongated on the inner. Anthers yellowish brown. *Himalaya, W China (SE Xizang, Yunnan).* H4. Summer.

Var. **flavidum** (Rendle) Sealy. Illustration: Botanical Magazine, n.s., 218 (1953). Flowers pale yellow.

\*L. **oxypetalum** (Royle) Baker (*Nomocharis oxypetala* (Royle) Wilson). Illustration: Synge, Lilies, pl. 10 (1980). Stem to 25 cm. Leaves scattered, below the flower often in a whorl, to 7 × 1.2 cm. Flowers broadly bell-shaped. Perianth-segments 3.5–5.5 cm, yellow with or without purplish spots towards the base; inner segments with fringed glands at the base. *NW Himalaya.* H4. Summer.

Var. **insigne** Sealy. Illustration: Botanical Magazine, n.s., 274 (1956). Flowers purple.

**58. NOTHOLIRION** Boissier
*V.A. Matthews*
Differs from *Lilium* in flowering once and then dying (the plant perennates by offsets) and having bulbs with few scales covered by a ribbed tunic; long basal leaves produced in autumn and winter; style with 3 narrow branches.

A genus of 6 species occurring from Afghanistan to W China. Most of the species like a cool position in peaty but not waterlogged soil; *N. thomsonianum* seems to require a sunnier place than the other species. They can also be grown in pots in an unheated greenhouse, especially in those districts where the leaves are subject to frost-damage.

Literature: Woodcock, H.B.D. & Stearn, W.T., *Lilies of the world*, 373–80 (1950).

1a. Perianth-segments with green tips; raceme with 10–40 flowers    2
  b. Perianth-segments without green tips; raceme with up to 6 flowers
       **3. macrophyllum**
2a. Perianth-segments 4–6 mm wide
       **1. thomsonianum**
  b. Perianth-segments 8–18 mm wide
       **2. bulbuliferum***

**1. N. thomsonianum** (Royle) Stapf.
Illustration: Botanical Magazine, 4725 (1853) – as Lilium roseum; Grove & Cotton, Supplement to Elwes; Monograph of the genus Lilium, t. 28 (1940); Woodcock & Stearn, Lilies of the world, f. 121 (1950); Rix & Phillips, The bulb book, 139 (1981).
Stem to 1 m. Raceme with 10–30 flowers. Flowers funnel-shaped, fragrant, borne horizontally. Perianth-segments 5–6.5 cm × 4–6 mm, pale pinkish lilac tipped with green, recurved at the tips. *Afghanistan, W Himalaya.* H4. Spring–summer.

**2. N. bulbuliferum** (Lingelsheim) Stearn (*N. hyacinthinum* (Wilson) Stapf).
Illustration: Grove & Cotton, Supplement to Elwes' Monograph of the genus Lilium, t. 29 (1940); Woodcock & Stearn, Lilies of the world, f. 124 (1950); RHS Lily Year Book **31**: pl. 7 (1967).
Stem 60–100 cm. Raceme with 10–30 flowers. Flowers funnel-shaped at the base, spreading wide towards the mouth, slightly fragrant, borne horizontally. Perianth-segments 2–4 cm × 8–15 mm, pale lilac tipped with green. *Nepal to W China.* H4. Summer.

**\*N. campanulatum** Cotton & Stearn.
Illustration: Grove & Cotton, Supplement to Elwes' Monograph of the genus Lilium, t. 30 (1940); Woodcock & Stearn, Lilies of the world, f. 123 (1950). Flowers 4–5 cm, drooping, scentless, deep crimson tipped with green. *N Burma, W China (Xizang, Yunnan).* H4. Summer.

**3. N. macrophyllum** (Don) Boissier.
Illustration: Woodcock & Stearn, Lilies of the world, f. 125 (1950); Journal of the Royal Horticultural Society **94**: f. 101 (1969); Everard & Morley, Wild flowers of the world, pl. 108 (1970); Rix & Phillips, The bulb book, 163 (1981).
Stem 20–40 cm. Raceme bearing up to 6

flowers. Flowers broadly funnel-shaped, *c.* 5 cm across at the mouth, horizontal or slightly drooping. Perianth-segments 3–5 cm, pale lavender-pink spotted with purple inside and without green tips. *Nepal, Bhutan, N India, (Sikkim), W China (Xizang).* H4. Summer.

## 59. NOMOCHARIS Franchet
*V.A. Matthews*

Similar to *Lilium* but with the flowers opening flat or sometimes shallowly cup-shaped, and with the inner perianth-segments entire to fringed and bearing basal nectaries which usually have ridges of tissue arranged in a fan shape. Filaments swollen, with a needle-like appendage (aciculus) at the top.

A genus of 7 species from W China (including Xizang), Burma and N India where they are found in the alpine zone of high mountains. They grow best in a semi-shady position in which the soil never completely dries out. They resent root disturbance and attempts to move them are rarely successful. Propagation is by seed or bulb scales.

Literature: Sealy, J.R., A revision of the genus Nomocharis Franchet, *Botanical Journal of the Linnean Society* **87**: 285–323 (1983).

1a. Leaves in whorls; filaments cylindric or broader at the top    2
  b. Leaves scattered; filaments subulate, often flattened below or on one side    3
2a. Style 7–10 mm; perianth-segments with spots from base to apex, or if spotted only near the base then anthers 3–4 mm    **1. pardanthina***
  b. Style 12–15 mm; perianth-segments spotted only near the base; anthers 4–6 mm    **2. farreri**
3a. Style 8–13 mm; flowers opening flat    **3. aperta***
  b. Style 4–6 mm; flowers opening shallowly cup-shaped    **4. saluenensis**

**1. N. pardanthina** Franchet (*N. mairei* Léveillé; *N. leucantha* Balfour). Illustration: RHS Lily Year Book **10**: f. 23, 24, 26 (1946); Synge, Lilies, pl. xxix, xxx (1980); Botanical Journal of the Linnean Society **87**: 293 (1983).
Stem 25–90 cm. Leaves in whorls. Flowers 1–8, occasionally to 24, erect, horizontal or nodding, opening flat, 5–8.5 cm in diameter. Outer perianth-segments 2.5–4.5 × 1.5–2.5 cm, entire, white to pale pink with purple or maroon-crimson blotches, the base dark purple or dark

maroon-crimson. Inner perianth-segments 2.6–4.5 × 2.4–3.6 cm, the upper half with fringed margins, white to pale pink with purple or maroon-crimson blotches which are deeper and more numerous than the blotches of the outer segments, the base dark purple. Nectaries with erect ridges of tissue in a fan shape on each side. Filaments 6–8 mm, cylindric or broader at the top, purplish, aciculus 1.5–3 mm; anthers 3–4 mm, crimson-maroon; pollen yellow or orange. Style 7–10 mm; style and ovary sometimes absent. *W China (Yunnan, Szechuan).* H3. Summer.

Forma **punctulata** Sealy (Illustration: Gardeners' Chronicle ser. 3, **59**: f. 138, 1916; RHS Lily Year Book **10**: f. 25, 1946) has perianth-segments with fine spots only towards the base, and the margins of the inner segments with tiny irregular teeth rather than fringed.

**\*N. meleagrina** Franchet. Flowers nodding, 7–9.5 cm in diameter, white or white flushed with pink on the back, uniformly blotched with purple. Outer perianth-segments 4–5.5 cm, dark purple at the base. Inner perianth-segments 4–6 cm, margin entire or wavy to irregularly toothed or slightly fringed, dark purple at the base. Anthers 4–7 mm. Style and ovary sometimes absent. *W China (Yunnan, SE Xizang).* H3. Summer.

**2. N. farreri** (Evans) Harrow (*N. pardanthina* Franchet var. *farreri* Evans). Illustration: Botanical Magazine, 9557 (1939); Synge, Lilies, pl. xxviii (1980).
Stem 35–100 cm. Leaves in whorls. Flowers to 20, nodding at first, eventually horizontal or erect, opening flat, 5.5–9.5 cm in diameter. Outer perianth-segments 2.5–5.5 × 1–2.5 cm, entire, white to pink, blotched or spotted with purple or dark crimson for up to 1 cm at the base and with a larger basal blotch of the same colour. Inner perianth-segments 2.5–5.5 × 1.4–3.5 cm, entire or with small irregular teeth, white to pink, the base with a purplish crimson or reddish brown blotch, above which is a band of yellow-green, and then some reddish purple spots for up to 1 cm from the base. Nectaries with erect ridges of tissue in a fan shape on each side. Filaments 6–11 mm, cylindric or broader at the top, deep pink or purplish, aciculus 2–2.5 mm; anthers 4–6 mm; pollen yellow. Style 12–15 mm. *NE Burma.* H3. Summer.

There is a tendency for hybrids to occur when certain species of *Nomocharis* are grown together, and for them to combine the characters of the parents in varying degrees. Hybrids between *N. farreri* and *N. saluenensis* have been named N. × **notabilis** Sealy (Illustration: RHS Lily Year Book **32**: pl. 3, 1968). Hybrids between *N. farreri* and *N. pardanthina* are known as N. × **finlayorum** Synge (Illustration: RHS Lily Year Book **32**: pl. 4, 1968).

**3. N. aperta** (Franchet) Wilson. Illustration: Botanical Magazine, 9533 (1938); RHS Lily Year Book **10**: f. 16 (1946) & **29**: f. 37 (1965); Synge, Lilies, pl. xxvii (1980). Stem 30–80 cm. Leaves scattered. Flowers 1–6, horizontal or nodding, opening flat, 5–10 cm in diameter. Outer perianth-segments 2.6–5 × 1–2.3 cm, entire, pale to deep pink, spotted with crimson in the lower part or all over, sometimes unblotched, the base with a dark purplish or crimson blotch, above which is a greenish or creamy area. Inner perianth-segments 2.5–4.1 × 1.3–3.2 cm, entire, similar in colour to the outer segments, but never unblotched. Nectaries with a fan-shaped swelling on each side, or more usually with a fleshy hump on one or both sides. Filaments 10–15 mm, subulate, plano-convex in section, purplish except for the yellow apex, aciculus 2–2.5 mm; anthers 3–6 mm; pollen yellow. Style 8–13 mm; style and ovary sometimes absent. *W China (Yunnan, Szechuan, SE Xizang), N Burma.* H3. Summer.

**\*N. synaptica** Sealy. Illustration: Kew Bulletin **1950**: 275 (1950). Flowers white, flushed with purplish pink and spotted all over with dark purple, the base of each perianth-segment dark purple with a narrow yellow border. Nectaries with erect ridges of tissue in a fan shape on each side. *India (Assam).* H3. Summer.

**4. N. saluenensis** Balfour. Illustration: New Flora and Silva **1**: f. 22 (1929); Botanical Magazine, 9296 (1933); RHS Lily Year Book **10**: f. 19 (1946); Botanical Journal of the Linnean Society **87**: 313 (1983). Stem 20–85 cm. Leaves scattered. Flowers 2–5, horizontal or nodding, opening shallowly cup-shaped, 5–6.5 cm in diameter. Outer perianth-segments 3.3–4.3 × 1.3–2.3 cm, entire, white to pink, whitish or yellowish in lower part with a dark purple basal blotch, spotted with crimson near the base and sometimes more faintly so over the lower half. Inner perianth-segments 3.2–4.3 × 1.6–2.5 cm, entire, similar in colour to the outer segments. Nectaries with a fleshy hump on one or both sides. Filaments 8–11 mm, slightly flattened in lower part, purplish,

aciculus 1.5–3 mm, anthers 7.5–8 mm, dark or blue-green; pollen yellow or greenish. Style 4–6 mm; ovary and style sometimes absent. *W China (Yunnan, SE Xizang), NE Burma.* H3. Summer.

**60. CARDIOCRINUM** (Endlicher) Lindley
*V.A. Matthews*
Differs from *Lilium* in the monocarpic bulbs which die after flowering (the plant perennates by offsets), the large, long-stalked, heart-shaped leaves and the capsule which splits into 3 valves which are fringed with teeth.

A genus of 3 species from the Himalaya, China and Japan, formerly included in *Lilium*. The plants grow best in semi-shade, in a rich soil with plenty of leaf-mould. The bulbs should not be planted too deeply.

1a. Stem 1.5–4 m, leafy for whole length; flowers 15–20 cm          **1. giganteum**
  b. Stem 1.2–2 m, leafless in lower part; flowers 15 cm or less          2
2a. Leaf-blade longer than broad; raceme with 1–5 flowers          **2. cathayanum**
  b. Leaf-blade as broad as long; raceme with 4–10 or more flowers
          **3. cordatum**

**1. C. giganteum** (Wallich) Makino (*Lilium giganteum* Wallich).
Stem 1.5–4 m. Basal leaves in a rosette, blade up to 45 × 40 cm, stem leaves decreasing in size up the stem, all dark shiny green above, paler beneath. Raceme with up to 25 flowers. Flowers 15–20 cm, regularly funnel-shaped, fragrant, pure white or with a green tinge, striped with reddish purple inside. Capsule 5–6.5 cm.

Var. **giganteum.** Illustration: Woodcock & Stearn, Lilies of the world, frontispiece & f. 54 (1950); Synge, Collins guide to bulbs, pl. III (1961); Reader's Digest encyclopaedia of garden plants and flowers, 116 (1971); Rix & Phillips, The bulb book, 163 (1981). Stem 1.5–4 m, green, flowers white, only greenish outside when young, and the lowest flowers in the raceme opening first. *Himalaya, NW Burma, China (SE Xizang).* H3. Summer.

Var. **yunnanense** (Elwes) Stearn. Illustration: Synge, Lilies, pl. 26 (1980). Stem 1.5–2 m, deep purple, flowers often tinged with green and the flowers at the top of the raceme opening first. *W & C China.* H3. Summer.

**2. C. cathayanum** (Wilson) Stearn (*Lilium cathayanum* Wilson). Illustration: Journal of the Royal Horticultural Society **70**: f. 27, 28 (1945).

Stem to 1.5 m, green, often dark purple below. Leaves absent in lower part of stem, blade up to 20 × 13 cm, dark shiny green. Raceme with 1–5 flowers. Flowers 10–13 cm, irregularly funnel-shaped, white or greenish white, the lower segments marked with reddish brown inside and often with reddish spots at the tips of the perianth-segments. Capsule *c.* 5.5 cm. *E & C China.* H3. Summer.

**3. C. cordatum** (Thunberg) Makino (*Lilium cordifolium* Thunberg; *L. cordatum* (Thunberg) Koidzumi). Illustration: Botanical Magazine, 6337 (1878); Woodcock & Stearn, Lilies of the world, 371 (1950).
Stem 1.2–2 m, green. Basal leaves in a rosette, blade up to 30 × 30 cm, marked with reddish brown when young; stem leaves absent in lower part of stem. Raceme usually with 4–10 flowers, occasionally with up to 24. Flowers to 15 cm, irregularly funnel-shaped, fragrant, creamy white, the lower segments with a yellow blotch and reddish marks and spots. Capsule *c.* 5.5 cm. *Japan, USSR (Sakhalin Islands).* H3. Summer.

**61. ALSTROEMERIA** Linnaeus
*V.A. Matthews*
Rootstock fibrous, with clusters of tubers or creeping rhizomes. Leaves alternate or scattered up the stem, larger on the sterile shoots, with short stalks which are twisted through 180°, many-veined. Flowers usually bilaterally symmetric, borne usually in a terminal, simple or compound umbel, rarely solitary. Perianth-segments free, clawed at the base, almost equal or the outer 3 broader and shorter than the inner. Stamens often unequal, usually curved downwards, then upwards at the tips; anthers basifixed. Style slender, curved downwards, with a 3-fid stigma. Ovary inferior, 3-celled with numerous ovules in each cell. Fruit a capsule containing numerous seeds.

A genus of about 50 species from S America. They will grow in any well-drained fertile soil, in semi-shade or sun. The more tender species should be given winter protection or the roots may be lifted and stored during the winter; alternatively they can be grown under glass. Propagation is by seed or division of the rootstock.

Literature: Garaventa, A., El genero Alstroemeria en Chile, *Anales del Museo de Historia Natural de Valparaiso* **4**: 63–108 (1971).

*Leaf-margin.* Hairy: 3,6; hairless: 1,2,4–7.
*Umbel.* Simple: 4,6; compound: 1–3,5–7.
*Flowers solitary:* 1,6.
*Basic colour of flowers.* Whitish or white:
  2,6; yellow: 1,4,5; orange: 1–4; red:
  2–6; pink: 2,6; lilac to purple: 2,7.
*Spots or stripes of perianth-segments.* On all
  segments: 4; on inner segments only:
  1–3,5–7. Yellow: 2,6; brown: 5,6;
  purplish brown or reddish brown: 4;
  purple: 2–4,7; reddish purple: 3, 6; red:
  1,2.

1a. All perianth-segments spotted
  **4. psittacina\***
  b. Only inner perianth-segments spotted
  2
2a. Flowers whitish or white, often
  flushed and splashed with mauve or
  pink, or mauve, purple, pink or
  crimson 3
  b Flowers yellow, orange, or red 5
3a. Flowers solitary or with simple
  umbels **6. pelegrina\***
  b. Flowers in compound umbels with
  3–15 branched rays 4
4a. Leaves narrowly linear to linear-
  lanceolate or narrowly obovate;
  flowers white to lilac, pink or crimson,
  if purple then upper inner perianth-
  segments yellow **2. ligtu\***
  b. Leaves ovate-oblong; flowers violet;
  inner upper perianth-segments
  lacking any yellow, although
  sometimes orange at the base
  **7. violacea**
5a. Leaf margins hairy **3. haemantha\***
  b. Leaf margins hairless 6
6a. Inner perianth-segments with brown
  spots **5. brasiliensis**
  b. Inner perianth-segments spotted or
  streaked with white, yellow, red or
  purple 7
7a. Flowers yellow to orange **1. aurea\***
  b. Flowers reddish **2. ligtu\***

**1. A. aurea** Graham (*A. aurantiaca* D. Don;
*A. aurea* Meyen). Illustration: Edwards's
Botanical Register 22: t. 1843 (1836); Hay
& Synge, Dictionary of garden plants in
colour, 120 (1969).
Stems to 1 m. Leaves 7–10 cm, lanceolate,
hairless. Umbel with 3–7 rays, each ray
bearing 1–3 flowers. Perianth-segments
4–5 cm, orange, the outer segments tipped
with green, the upper inner segments
spotted and streaked with red, the lower
inner segment spotted or unspotted.
Stamens shorter than perianth-segments.
*Chile.* H3. Summer.
  Several cultivars are available with paler,
deeper or larger flowers.

**\*A. pygmaea** Herbert differs in having its
stem below the ground and a solitary
flower growing from a tuft of grey-green
leaves which are *c.* 2.5 cm. The perianth-
segments are *c.* 5 cm, yellow with the inner
ones spotted with red. *Argentina.* H5.
Summer.

**2. A. ligtu** Linnaeus. Illustration: Loddiges'
Botanical Cabinet 1: t. 17 (1818);
Edwards's Botanical Register 25: pl. 13
(1839); Mathew, P.J. Redouté, Lilies and
related flowers, 121 (1982).
Stems 45–60 cm. Leaves 5–8 cm,
narrowly lanceolate, hairless. Umbel with
3–8 rays, each ray bearing 2 or 3 flowers.
Perianth-segments 4–5 cm, varying from
whitish to lilac, pink or reddish, the upper
inner segments spotted and streaked with
white or yellow and red, or purple. Stamens
shorter than perianth-segments. *Chile,
Argentina.* H4. Summer.
  The commonly grown 'Ligtu hybrids'
were bred from *A. ligtu* var. *angustifolia*
Anon. and *A. haemantha.* They show great
variation in height and habit and have
pink, orange or yellow flowers.

**\*A. revoluta** Ruiz & Pavon differs in
having very narrowly linear leaves and
purple or crimson flowers only 8–20 mm
with the upper inner perianth-segments
yellow, spotted with purple. *Chile.* H5–G1.
Summer.

**\*'A. chilensis'** may key out here (see
p. 210).

**3. A. haemantha** Ruiz & Pavon.
Illustration: Sweet's Ornamental Flower
Garden 1: t. 14 (1854).
Stems 60–90 cm. Leaves 7–15 cm,
narrowly lanceolate, with hairs on the
margin. Umbel with 3–15 rays, each ray
bearing 1–4 flowers. Perianth-segments
4–5 cm, orange to deep orange-red, outer
segments tipped with green, upper inner
segments yellowish with purple stripes,
lower inner segment similar in colour to
outer segments but striped with dark red.
Stamens shorter than perianth-segments.
*Chile.* H3. Summer.

**\*'A. chilensis'** may key out here (see
p. 210).

**4. A. psittacina** Lehmann. Illustration:
Edwards's Botanical Register 18: t. 1540
(1832); Sweet's Ornamental Flower Garden
3: t. 183 (1854).
Stems up to 90 cm, spotted with pale
purple. Leaves to 7.5 cm, lanceolate,
hairless. Umbels with 4–6 unbranched
rays. Perianth-segments 4–4.5 cm, deep
red, tipped with green and spotted with

reddish brown or purplish brown. Stamens
more or less equal in length to perianth-
segments. *N Brazil.* H4. Summer.
  Plants sold under the name *A. pulchella*
often turn out to be *A. psittacina* or *A. ligtu.*
*A. pulchella* Linnaeus filius seems to be
similar to *A. psittacina* but the name is
surrounded by confusion and work needs
to be done on this problem.

**\*A. versicolor** Ruiz & Pavon differs in
having shorter stems (20–60 cm) and
umbels with 2–4 rays. The perianth-
segments are yellow or orange, spotted
with purple. *Chile.* H5–G1.
Summer–autumn.

**5. A. brasiliensis** Sprengel.
Stems to 120 cm. Leaves 5–10 cm,
lanceolate, hairless. Umbel with *c.* 5 rays,
each ray bearing 1–3 flowers. Perianth-
segments *c.* 4 cm, reddish yellow, inner
segments with brown spots. Stamens
shorter than perianth-segments. *Brazil.* H4.
Summer–early autumn.

**6. A. pelegrina** Linnaeus. Illustration:
Reader's Digest encyclopaedia of garden
plants and flowers, 30 (1971).
Stems 30–60 cm. Leaves 5–8 cm,
lanceolate, hairless. Flowers solitary or in
umbels with 2 or 3 unbranched rays.
Perianth-segments *c.* 5 cm, whitish flushed
with mauve or pink, and with a darker
central patch, the upper inner segments
yellow at the base and spotted with brown
or reddish purple. Stamens shorter than
perianth-segments. *Peru.* H5–G1.
Summer–autumn.
  'Alba' (var. *alba* Anon.). Illustration:
Paxton's Magazine of Botany 1: 199
(1841). The flowers are white, flushed with
green and the upper inner segments are
spotted with green and often have a yellow
blotch.

**\*A. hookeri** Loddiges. Illustration:
Loddiges' Botanical Cabinet 13: t. 1272
(1827); Maxwell, Flowers: A garden note
book, pl. 11 (1923). The flowers are pink,
with the upper inner segments blotched
with yellow and spotted with reddish
purple. *Chile.* H4. Summer.

**7. A. violacea** Philippi. Illustration:
Botanical Magazine, n.s., 42 (1948).
Stems 50–100 cm. Leaves 6–9 cm, ovate-
oblong, hairless. Umbel with 3–6 rays,
each ray bearing 3–5 flowers. Perianth-
segments 3.5–5.5 cm, violet, the upper
inner white below, spotted with purple,
sometimes orange at the base. Stamens
shorter than perianth-segments. *N Chile.*
H4. Summer.

*A. violacea* has been crossed with *A. aurea* to produce 'Walter Fleming' which has deep yellow flowers flushed with pink and spotted with maroon.

**Name of doubtful application**

There is a plant in cultivation which is known as '*A. chilensis*', which grows to *c.* 75 cm and has narrowly obovate leaves with very short marginal hairs. The umbel has 5 or 6 rays, each ray usually bearing 2 flowers. The perianth-segments vary from pale pink to bright pink or red, and the upper inner segments are yellow with reddish purple stripes. The plant is said to come from Chile, but great confusion surrounds the name 'chilensis' which cannot be assigned with certainty to any species.

**62. BOMAREA** Mirbel
*V.A. Matthews*
Herbs, usually climbing, growing from a rhizome with slender roots which bear tubers at their tips. Stem leafy. Leaves with short stalks which are twisted through 180°, many-veined. Flowers in terminal, simple or compound umbels. Perianth narrowly bell-shaped with free segments in 2 whorls which are usually rather dissimilar. Stamens borne at the bases of the segments and equal in length to them; anthers basifixed. Style slender, with 3 branches. Ovary superior, 3-celled. Capsule 3-celled, with 6 ribs. Seeds orange or red, numerous.

About 120 New World species occurring mainly from Mexico to the Andes. The stems normally die back to ground level in the winter. They require a well-drained soil and almost no water in the resting season. They grow best with complete frost-protection, but in northern countries can be planted out-of-doors during the summer in a sheltered position. Propagation is by seed or division of the rhizome.

*Perianth-segments.* Equal or almost so: **1,4–7**; unequal: **2,3**. Outer segments 1–1.5 cm: **7**; 2.5–3.5 cm: **1–4,6**; 5–6.5 cm: **3,5,6**. Outer segments crimson, pink or purplish: **3–5,7**; scarlet, orange or yellow: **1–3,6**. Inner segments spotted: **1–6**; inner segments unspotted: **2,7**.
*Umbel.* Simple: **1–3**; compound: **4–7**.

1a. Umbel simple                                          2
 b. Umbel compound                                       4
2a. Perianth-segments more or less equal
    in length                             **1. multiflora***
 b. Inner perianth-segments longer than
    outer                                                 3

3a. Inner perianth-segments 2.5–3.5 cm
                                            **2. caldasii**
 b. Inner perianth-segments 3.5–7.5 cm
                                          **3. patacocensis***
4a. Flowers 1–1.5 cm, inner perianth-
    segments not spotted              **7. salsilla**
 b. Flowers more than 1.5 cm; inner
    perianth-segments spotted                             5
5a. Outer perianth-segments orange-red
                                          **6. shuttleworthii***
 b. Outer perianth-segments crimson-
    purple or pink                                        6
6a. Flowers 2.5–3 cm                        **4. edulis**
 b. Flowers 5.5–6.5 cm                      **5. cardieri**

**1. B. multiflora** (Linnaeus filius) Mirbel. Illustration: The Garden **41**: 449 (1892); Menninger, Flowering vines of the world, f. 21 (1970); Harrison, Climbers and trailers, f. 23 (1973).
Stem downy. Leaves oblong, 7–10 cm. Flowers 20–40 in a simple umbel. Perianth-segments more or less equal, *c.* 2.5 cm. Outer segments tinged with red; inner segments reddish yellow with brown spots. *Colombia, Venezuela.* H5–G1.

***B. oligantha** Baker. Umbel with only 6–8 flowers which are up to 3.3 cm. *Ecuador.* H5–G1.

***B. acutifolia** which has umbels with 13–20 flowers which are 2.5–4 cm, may key out here (see entry under *6. B. shuttleworthii*).

**2. B. caldasii** (Humboldt, Bonpland & Kunth) Ascherson & Graebner (*B. caldasiana* Herbert; *B. kalbreyeri* misapplied, not Baker). Illustration: Gartenflora **32**: 21 (1883); Illustration Horticole **32**: pl. 550 (1885); Botanical Magazine, n.s., 465 (1965); Harrison, Climbers and trailers, f. 21 (1973).
Stem to 4 m. Leaves oblong, to 15 × 2.5 cm, almost hairless or downy on the veins beneath. Flowers 5–40 or more, in a simple umbel. Perianth-segments unequal. Outer segments 2–2.5 cm, yellow to red or reddish brown; inner segments 2.5–3.5 cm, yellow to orange, unspotted or spotted with green, brown or red. *Andes of Colombia and Ecuador.* H5–G1.

**3. B. patacocensis** Herbert (*B. conferta* Bentham).
Stem to 5 m, downy. Leaves oblong-lanceolate, to 15 cm, downy beneath. Flowers 20–60 in a simple umbel. Perianth-segments unequal. Outer segments *c.* 3.3 cm, orange or crimson; inner segments to 6.5 cm, yellow or crimson with brown or violet spots, orange at the tip. *Andes of Colombia and Ecuador.* H5–G1.

Plants grown under this name often turn out to be *B. racemosa*.

***B. racemosa** Killip. Stem densely red-hairy, outer segments 5–6.3 cm, scarlet; inner segments 6.3–7.5 cm, scarlet, spotted with brown, yellow at the base. *Colombia.* H5–G1.

***B. frondea** Masters. Illustration: Gardeners' Chronicle **17**: 669 (1882); the Garden **41**: 444 (1892); Botanical Magazine, 7247 (1892). Outer segments yellow, inner segments yellow with dark red spots. *Colombia.* H5–G1.

**4. B. edulis** (Tussac) Herbert. Illustration: Addisonia **2**: pl. 65 (1917); Herklots, Flowering tropical climbers, f. 28 (1976). Leaves lanceolate, to 12.5 × 2.5 cm, hairless or downy beneath. Flowers borne in a compound umbel. Perianth-segments more or less equal, 2.5–3 cm. Outer segments pink; inner segments green or yellowish, spotted with purple or dark pink. *West Indies, C America south to Brazil & Peru.* G1.

**5. B. cardieri** Masters. Illustration: Revue Horticole **49**: 187 (1877); The Garden **55**: 130 (1899); Botanical Magazine, 9601 (1940); Herklots, Flowering tropical climbers, f. 31 (1976).
Stem to 4 m, hairless. Leaves oblong-lanceolate, acuminate, to 18 × 7 cm. Flowers up to 40 in a compound umbel. Perianth-segments equal, 5.5–6.5 cm. Outer segments pink, finely speckled inside with deep purple, and with a short, green horn-like projection on the back just below the tip; inner segments pink or sometimes green, spotted with deep purple. *Colombia.* H5–G1.

**6. B. shuttleworthii** Masters. Illustration: Gardeners' Chronicle **17**: 85 (1882). Leaves ovate-lanceolate, to 15 × 5 cm, hairless beneath. Flowers borne in a compound umbel. Perianth-segments equal, to 5 cm. Outer segments orange-red with dark spots towards the tip; inner segments yellow at the base merging into green at the tip, with dark spots towards the tip and a red midrib. *Colombia.* H5–G1.

***B. acutifolia** (Link & Otto) Herbert. Illustration: Botanical Magazine, 3871 (1841), 6444 (1879); Herklots, Flowering tropical climbers, f. 26 (1976). Leaves downy beneath, flowers 2.5–4 cm, outer segments not spotted. *Mexico.* G1.

**7. B. salsilla** (Linnaeus) Herbert. Illustration: Revue Horticole series 4, **5**: 61 (1856); The Garden **29**: 260 (1886). Stem to 2 m, hairless. Leaves lanceolate to oblong. Flowers 6–18, borne in a compound umbel. Perianth-segments equal, 1–1.5 cm, crimson or purplish red, tinged with green at the tips. *Chile*. H5–G1. Summer.

## 63. SCILLA Linnaeus

*S.G. Knees & P.G. Barnes*

Perennial herbs with ovoid to spherical bulbs composed of numerous free scales which are progressively renewed annually. Roots annual or perennial. Leaves few to several, all basal, linear to elliptic, sometimes channelled, usually hairless, appearing before, with or after the flowers. Scapes hairless, erect, usually 1–4 per bulb. Flowers few to many, in terminal racemes. Bracts absent or 1 subtending each flower. Perianth-segments 6, free, although occasionally close at the base giving the appearance of a short tube, blue, purple, pink or white, often each with a darker midrib. Filaments 6, free, equal, inserted at the base of the perianth; anthers dorsifixed. Ovary superior, almost spherical to ovoid, 3-celled, with 1–12 ovules per cell; style 1, straight, stigma small, truncate. Fruit a capsule, seeds spherical or oblong, sometimes angled, pale brown to black, each occasionally bearing an appendage.

A genus of up to 90 species occurring in Europe, Asia and southern Africa, valued for its ornamental, usually blue flowers and suited to a variety of sites in the garden, particularly screes, rock-gardens or raised beds. Some can be naturalised in grass or the wild garden and a few are best suited to lightly shaded positions. They are occasionally grown as specimen plants in the alpine house, whilst the more tender species require cool greenhouse conditions. Propagation is by offsets or by seed which is freely produced.

Literature: Anderson, E.B. & Synge, P.M., Hardy Scillas, *RHS Lily Year Book* **24**: 27–30 (1961); Anderson, E.B. & Meikle, R.D., A Lily Group discussion of Scillas and Chionodoxas, *RHS Lily Year Book* **25**: 116–33 (1962); Speta, F., Die frühjahrsblühenden Scilla-Arten des östlichen Mittelmeerraumes, *Naturkundliches Jahrbuch der Stadt Linz* **25**: 19–198 (1980).

*Perianth length*. More than 1 cm: **1**,**5–10**, **16**; less than 1 cm: **2–4**,**11–15**,**17**.

*Number of flowers*. More than 25: **2–5**,**11**; fewer than 25: **1**,**6–10**,**12–17**.

*Flowering season*. Spring: **3**,**6–9**,**12**,**14–17**; late spring and summer: **1**,**2**,**4**,**5**,**10**,**11**; autumn: **13**.

1a. Lower bracts more than 4 mm long 2
b. Lower bracts less than 4 mm long or absent 10
2a. Inflorescence with more than 20 flowers 3
b. Inflorescence with fewer than 20 flowers 6
3a. Leaves 9–15; lower bracts 3–8 cm **16. peruviana***
b. Leaves 3–8; bracts less than 2 cm 4
4a. Summer-flowering; bracts *c*. 1 cm 5
b. Spring-flowering; lower bracts 2–7 mm **5. persica**
5a. Flowers more than 40 in a long raceme; bracts longer than the flower-stalks **11. natalensis**
b. Flowers 15–35 in a conical raceme; bracts and flower-stalks similar in length **14. litardierei**
6a. Bracts equalling or longer than flower-stalks 7
b. Bracts much shorter than flower-stalks 8
7a. Bulb 3–5 cm, scales loose; leaves 1–3 cm wide; perianth-segments 9–12 mm **15. lilio-hyacinthus**
b. Bulb 1–2.5 cm, scales tight; leaves 2–5 mm wide; perianth-segments 5–8 mm **12. verna***
8a. Leaves usually 1–1.3 cm wide; perianth-segments less than 1 cm, light violet-blue **13. monophyllos**
b. Leaves 3–20 mm wide; perianth-segments more than 1 cm 9
9a. Perianth-segments 2–3 mm wide, pale blue, about as long as the flower-stalks **8. hohenackeri**
b. Perianth-segments 4–8 mm wide, whitish blue, half as long as the flower-stalks **9. mischtschenkoana**
10a. Perianth-segments less than 1 cm 11
b. Perianth-segments 9–25 mm 14
11a. Summer- or autumn-flowering; perianth-segments 3–5 mm 12
b. Spring-flowering; perianth-segments 5–10 mm 13
12a. Flowers more than 40; leaves 4–7 mm wide **7. scilloides**
b. Flowers 5–25; leaves 1–2 mm wide **17. autumnalis**
13a. Leaves usually 2, 3–15 mm wide; flowers 1–10, perianth-segments shorter than the lower flower-stalks **1. bifolia**
b. Leaves 5–7, 1.2–2.5 cm wide; flowers 7–20, perianth-segments similar in length to the lower flower-stalks **2. messeniaca**

14a. Flowers 1–4, pendent; perianth-segments usually more than 1.2 cm, exceeding the lower flower-stalks 15
b. Flowers 2–8, erect or horizontal; perianth-segments 9–12 mm, equalling or shorter than the flower-stalks 16
15a. Perianth-segments strongly recurved, pale blue **6. rosenii**
b. Perianth-segments straight, bright deep blue (occasionally pale blue or white) **3. siberica**
16a. Flowers mauve-blue; bracts absent **4. amoena**
b. Flowers whitish or pale blue; bracts 1–5 mm 17
17a. Perianth-segments 2.5–3.5 mm wide; flower-stalks 3–10 mm **10. puschkinioides**
b. Perianth-segments 4–8 mm wide; flower-stalks 1–2.5 cm **9. mischtschenkoana**

**1. S. bifolia** Linnaeus. Illustration: Botanical Magazine, 746 (1804); Polunin & Huxley, Flowers of the Mediterranean, pl. 240 (1965); Grey-Wilson & Mathew, Bulbs, pl. 17 (1981); Rix & Phillips, The bulb book, 40 (1981).

Bulb 5–25 mm in diameter, pinkish beneath the brown tunics. Leaves usually 2, occasionally 3–5, 5–20 cm × 3–15 mm, broadly linear, appearing with the flowers. Scape solitary, 5–30 cm, terete, partially sheathed at the base by the leaves. Raceme slightly 1-sided, with 1–10 flowers, the lower stalks 1–4 cm, the upper less than 1 cm. Bracts absent, or, if present, ovate-lanceolate, *c*. 1 mm. Perianth-segments ovate to elliptic, 5–10 × 1–3 mm, slightly hooded at the apex, blue to purplish blue. Seeds almost spherical, *c*. 2 mm, brownish, each with an irregular, white appendage. *C & S Europe & Turkey*. H4. Spring.

White- and pink-flowered variants are grown under the cultivar names 'Alba' and 'Rosea' (Illustration: Botanical Magazine, 746, 1804; Synge, Collins guide to bulbs, pl. 24, 1961). A smaller variant, with fewer, violet flowers has been called **S. nivalis** Boissier.

**2. S. messeniaca** Boissier. Illustration: Botanical Magazine, 8035 (1905); Rix & Phillips, The bulb book, 41 (1981).

Bulb 2–3 cm in diameter, ovoid, tunic pale brown. Leaves 5–7, 15–25 × 1.2–2.5 cm, broadly linear, appearing with the flowers and sheathing the scapes at the base. Scapes 1 or 2, 5–15 cm, angled. Flowers 7–20 in a dense, conical raceme 4–12 cm long; bracts *c*. 1 mm, linear, occasionally

bifid; lower flower-stalks 4–8 mm, ascending or spreading. Perianth-segments 6–8 mm, lilac-blue, somewhat spreading, linear, apex obtuse. Stamens 4–6 mm, with pale blue filaments, anthers dark violet-blue.
Ovary almost spherical, ovules 2 per cell. Capsule almost spherical, broadly 3-angled, c. 7 mm. *S Greece.* H4. Spring.

**3. S. siberica** Haworth. Illustration: Botanical Magazine, 1025 (1807); Synge, Collins guide to bulbs, pl. 24 (1961); Hay & Synge, Dictionary of garden plants, pl. 881 (1969); Rix & Phillips, The bulb book, 44 (1981).
Bulb 1.5–2 cm in diameter, ovoid, tunic dark purplish brown. Leaves 2–4, 10–15 cm × 5–15 mm, broadly linear, slightly hooded at the tips, shorter than the scape at flowering. Scapes 1–4, 10–20 cm, purplish above, finely ribbed. Flowers 1–4, occasionally more, broadly bowl-shaped, pendent, in loose racemes; bracts 1–2 mm, white; flower-stalks 8–12 mm, purple. Perianth-segments 1.2–1.6 cm × 4–6 mm, elliptic-oblong, obtuse, bright blue with a darker central stripe. Stamens 5–8 mm, filaments lanceolate, white below, blue above, anthers grey-blue. Ovary ovoid, pale green, ovules 2–12 per cell, style 4–5 mm, white. Capsule almost spherical, 8–10 mm in diameter, seeds spherical, pale brown, each with a long appendage. *S USSR; naturalised in C Europe.* H2. Spring.

Somewhat variable in colour: 'Spring Beauty' ('Atrocaerulea') is a clone with particularly bright, deep blue flowers (Illustration: Addisonia 16: pl. 516, 1931; Rix & Phillips, The bulb book, 40, 43, 1981). The white-flowered cultivar 'Alba' is less vigorous.

**4. S. amoena** Linnaeus. Illustration: Botanical Magazine, 341 (1796); Loddiges' Botanical Cabinet 11: t. 1015 (1825); Reichenbach, Icones Florae Germanicae et Helveticae 4: t. 464 f. 1014 (1847); Bonnier, Flore complète 10: pl. 578 (1929).
Bulb 1.5–2 cm in diameter, ovoid, whitish beneath a dark, purplish brown tunic. Leaves 4–7, 15–22 × 1–2 cm, linear, often reddish near the base, usually appearing before the flowers. Scapes 1–5, 15–20 cm, angled. Inflorescence without bracts, with 3–6 flowers; flower-stalks 6–15 mm, the lower ones longer than the perianth; flowers erect. Perianth-segments 9–12 mm, mauve-blue with deeper blue midribs. Filaments lanceolate, anthers blue. Capsule spherical, containing 6–8 seeds per cell,

each seed with a scarcely developed appendage. *Origin uncertain but naturalised in SE Europe.* H4. Spring–summer.

**5. S. persica** Haussknecht. Illustration: Botanical Magazine, n.s., 486 (1966); Mathew, Dwarf bulbs, f. 53 (1973); Rix & Phillips, The bulb book, 73 (1981).
Bulb 2.5–3 cm in diameter, broadly ovoid to almost spherical, outer tunic dark brown, inner violet-purple. Leaves 3–7, 30–45 × 1–1.5 cm, slightly keeled, linear with acute tips and rolled bases forming a tube, appearing with the flowers. Scapes 1–3, 20–40 cm. Flowers 20–80 in a compact, conical raceme 5–10 cm long, elongating to 20–25 cm in fruit; bracts 2–7 mm, fringed or 2-lobed above, flower-stalks 1.5–4 cm, thickening in fruit. Perianth-segments 7–8 mm, lanceolate to elliptic, slightly concave, spreading, bright blue. Ovary almost spherical, 3 mm in diameter, ovules 2 or 3 per cell. Seeds 4 × 2.5 mm, dark brown. *W Iran, N Iraq.* H4. Spring.

**6. S. rosenii** C. Koch. Illustration: Flora SSSR 4: pl. 9 (1935); Bulletin of the Alpine Garden Society 30: 126 (1962).
Bulb 1–2.5 cm in diameter, ovoid, tunic dark brownish violet. Leaves 2–3, 10–15 cm × 6–15 mm, linear, apex hooded, appearing with the flowers. Scapes 1–4, terete, 10–25 cm. Flowers usually 1 or 2, occasionally more, pendent, in short racemes; bracts 2–3 mm, 2-lobed; flower-stalks 1–1.5 cm, arching. Perianth-segments 1.5–2.5 cm × 4–8 mm, oblong, obtuse, becoming recurved, pale blue with a deeper median line on the outside, whitish near the base within. Stamens 8–12 mm, filaments slightly widened towards the base, white. Ovary obovoid, c. 5 mm, style longer than it. Seeds to 3 × 2 mm, pale brown, each with a white appendage. *USSR (Caucasus).* H3. Spring.

**7. S. scilloides** (Lindley) Druce (*S. chinensis* Bentham; *S. sinensis* (Loureiro) Merrill; *Barnardia scilloides* Lindley). Illustration: Edwards's Botanical Register 12: t. 1029 (1826); Botanical Magazine, 3788 (1840); Bulletin of the Alpine Garden Society 20: 180 (1952); Flora of Taiwan 5: t. 1285 (1978).
Bulb 1.5–2 cm in diameter, ovoid, tunic blackish. Leaves 2–7, 15–25 cm × 4–7 mm, linear, flaccid, appearing with the flowers, apex obtuse and slightly hooded. Scape 20–40 cm, slightly angled. Flowers 40–80 in a dense, oblong raceme 7–12 cm long;

bracts 1–2 mm, linear; flower-stalks 4–8 mm, ascending. Perianth-segments 3–4 mm, narrowly oblong, acute, spreading, mauve-pink. Filaments widened at their bases, finely hairy. Ovary ovoid, finely hairy on the angles, ovules few. Capsule c. 5 mm, obovoid-spherical. Seeds black. *China, Korea, Taiwan, Japan & Ryukyu islands.* H3–H5, depending on origin. Summer.

**8. S. hohenackeri** Fischer & Meyer. Illustration: Botanical Magazine, n.s., 453 (1964).
Bulb 1.5–2 cm in diameter, ovoid, tunic papery, greyish brown. Leaves 3–5, 10–25 cm × 3–10 mm, linear, V-shaped in cross-section, appearing before the flowers. Scape 5–20 cm, bearing a loose raceme of 4–12 flowers; flower-stalks 1–1.5 cm, spreading or recurved, equalling the perianths, but elongating to c. 2.5 cm in fruit. Bracts conspicuous, 5–6 mm, spurred below, fringed and 2-lobed above. Perianth-segments 1.2–1.5 cm × 2–3 mm, oblanceolate, pale blue, recurved at maturity. Filaments pale blue, anthers greenish blue. Capsule 3-angled, pale brown, containing 2–4 glossy, black seeds per cell. *S Iran, USSR (Azerbaidjan).* H3. Spring.

**9. S. mischtschenkoana** Grossheim (*S. tubergeniana* Stearn). Illustration: Botanical Magazine, n.s., 106 (1950); Marshall Cavendish encyclopedia of gardening, 2003, pl. 1 (1967); Hay & Synge, Dictionary of garden plants, f. 882 (1969); Rix & Phillips, the bulb book, 40 (1981).
Bulb 1.5–3 cm in diameter, ovoid or almost spherical, tunic papery, grey-brown. Leaves 3–5, 4–10 cm × 4–20 mm, linear to oblanceolate, appearing with the flowers. Scapes 1–3, 5–10 cm, terete. Flowers 2–6 in a loose raceme 6–12 cm long. Bracts oblong, 1–5 mm; flower-stalks to 2.5 cm, ascending. Perianth-segments 1–1.2 cm × 4–8 mm, spreading, oblong-elliptic, obtuse, whitish blue with deeper blue mid-veins outside. Stamens 6–11 mm, slightly widened at the base, white, anthers blue. Ovary almost spherical, with usually 6 ovules per cell; style longer than the ovary. Capsule c. 9 mm in diameter, almost spherical. Seeds spherical, brown or black, each with a conspicuous, pale appendage. *Iran, USSR (Caucasus).* H4. Spring.

**10. S. puschkinioides** Regel. Illustration: Gartenflora 30: t. 1050 (1881); Rix & Phillips, The bulb book, 43 (1981).

Bulb 1–2 cm in diameter, ovoid, tunic greyish. Leaves 2–5, 10–15 cm × 3–6 mm, narrowly oblanceolate or linear, obtuse, appearing with the flowers. Scape 10–15 cm, terete. Flowers 2–8 in a short raceme, stalks 3–10 mm, longer than the bracts. Perianth-segments 1–1.2 cm × 2.5–3.5 mm, oblong, acute, whitish or pale blue, each with a darker central line. Stamens 8–12 mm, filaments subulate, white, anthers blue. Ovary ovoid. Capsule flattened-spherical. *USSR (Soviet Central Asia)*. H2. Spring.

**11. S. natalensis** Planchon. Illustration: Flore des Serres 10: t. 1043 (1855); Botanical Magazine, 5379 (1863); Eliovson, South African flowers for the garden, edn 2, 50 (1957); Fabian & Germuishuizen, Transvaal wild flowers, pl. 11a (1982).
Bulb 7–10 cm in diameter, ovoid or almost spherical, tunic brown. Leaves 4–8, lanceolate, acuminate, to 20 cm at flowering, later extending to 30–60 × 7–10 cm. Scape 30–45 cm, terete. Flowers 50–100 in a raceme 15–30 cm long. Bracts *c.* 1 cm, linear; flower-stalks 2–4 cm, ascending, blue. Perianth-segments 6–10 × 3–4 mm, elliptic-oblong, obtuse, spreading, light violet-blue or occasionally pink or white. Filaments widened towards the base, white. Ovary ovoid, usually white, ovules 10–12 per cell; style equalling the ovary. *E South Africa, Lesotho*. H5–G1. Summer.

**12. S. verna** Hudson. Illustration: Sowerby, English botany 1: t. 23 (1790); Ross-Craig, Drawings of British plants 29: pl. 23 (1972); Rix & Phillips, The bulb book, 47 (1981).
Bulb 1–2 cm in diameter, ovoid. Leaves 2–7, 3–20 cm × 2–5 mm, linear, slightly channelled, obtuse, appearing before the flowers. Scape 5–20 cm, terete. Flowers 2–12 in a short, corymbose raceme; bracts 5–15 mm, linear. Lower flower-stalks 5–12 mm, ascending, elongating slightly in fruit. Perianth-segments 5–8 × 2–3 mm, narrowly oblong-ovate, light violet-blue. Stamens 2–4 mm, filaments lanceolate, white, anthers purplish blue. Ovary oblong, blue. Capsule almost spherical, seeds oblong, black, each with a small appendage. *W Europe*. H3. Spring.

*****S. ramburei** Boissier. Illustration: Rix & Phillips, The bulb book, 73 (1981). Differs in its more robust habit, bulb 1.8–2.5 cm in diameter, broader leaves, 6–30 flowers and lower flower-stalks 1.5–5 cm. *Portugal, S Spain*. H3. Spring.

Sometimes regarded as a subspecies of *S. verna*.

**13. S. monophyllos** Link. Illustration: Botanical Magazine, 3023 (1830); Polunin & Smythies, Flowers of SW Europe, pl. 59 f. 1636 (1973); Rix & Phillips, The bulb book, 73 (1981).
Bulb 1–2 cm in diameter, ovoid, tunic light brown. Leaf solitary, 10–25 × 1–3 cm, broadly lanceolate to elliptic, sheathing the scape at the base, appearing with or slightly before the flowers. Scape 5–15 cm. Flowers 3–15 in a loose raceme 5–10 cm long. Bracts 4–7 mm, linear-lanceolate, acuminate; lower flower-stalks 1–1.5 cm, ascending. Perianth-segments 6–9 mm, elliptic, pale violet-blue. Stamens 3–5 mm, filaments linear, pale violet, anthers blue. Ovary almost spherical, ovules few per cell. *Spain, Portugal, NW Africa*. H4. Spring.

**14. S. litardierei** Breistroffer (*S. pratensis* Waldstein & Kitaibel; *S. amethystina* Visiani). Illustration: Reichenbach, Icones Florae Germanicae et Helveticae 4: t. 463 f. 1011 (1848); RHS Lily Year Book 24: f. 6 (1961); Polunin, Flowers of Greece and the Balkans, pl. 58 (1980); Grey-Wilson & Mathew, Bulbs, pl. 17 (1981).
Bulb 1.5 cm in diameter, ovoid, tunic brown. Leaves 3–6, 25–30 cm × 4–8 mm, linear, appearing with the flowers. Scapes usually solitary, occasionally 2 or 3 together, 10–25 cm. Flowers 15–35 in narrow, conical racemes 5–15 cm long. Bracts *c.* 1 cm, ovate; flower-stalks 8–12 mm, pale bluish violet. Perianth-segments 4–6 mm, ovate, pinkish or bluish violet. Stamens 3–5 mm, filaments pale blue, anthers deep violet-blue. Ovary with 3 or 4 ovules per cell. Capsule *c.* 4 mm in diameter. Seeds brown or black, to 3 × 1–2 mm. *Yugoslavia*. H4. Summer.

**15. S. lilio-hyacinthus** Linnaeus. Illustration: The Garden 73: 265 (1909); Bonnier, Flore complète 10: pl. 578 (1929); Polunin & Smythies, Flowers of SW Europe, 133, f. 16 (1973); Rix & Phillips, The bulb book, 68 (1981).
Bulb 3–5 cm in diameter, ovoid, with loose, overlapping, yellow scales, without a tunic. Leaves 6–10, 15–30 × 1–3 cm, glossy, oblanceolate, acute, appearing with the flowers. Scapes 2 or 3, 5–10 cm. Flowers 5–15 in a dense conical raceme 3–8 cm long. Bracts 1–2.5 cm, ovate, papery, whitish; flower-stalks 8–12 mm, ascending, bluish violet. Perianth-segments 9–12 mm, ovate to elliptic, bright violet-blue, rarely white. Stamens 7–9 mm, filaments pale,

1–2 mm wide, lanceolate, anthers deep blue. Ovary almost spherical, with 6–10 ovules per cell. Capsule slightly 3-angled, seeds obovoid, blackish. *SW France, Spain*. H4. Summer.

**16. S. peruviana** Linnaeus. Illustration: Reichenbach, Icones Florae Germanicae et Helveticae 4: t. 465, f. 1017 (1848); Polunin & Huxley, Flowers of the Mediterranean, 243 (1965); Polunin, Flowers of Europe, pl. 168 (1969); Grey-Wilson & Mathew, Bulbs, pl. 18 (1981).
Bulb 6–8 cm in diameter, ovoid, tunic brown with woolly outer scales. Leaves 9–15, occasionally fewer, 40–60 × 1–4 cm, linear to lanceolate in a dense rosette, sometimes ciliate. Scapes solitary, rarely 2 together, 15–25 cm, stout. Flowers 40–100 in a broadly conical raceme 5–20 cm long. Lower bracts 3–8 cm, subulate, papery; lower flower-stalks 3–5 cm, greenish, elongating in fruit to 6–10 cm. Perianth-segments 8–15 mm, elliptic, deep violet-blue or white. Stamens lanceolate-elliptic, pale blue, anthers yellow. Ovary almost spherical. Capsule ovoid, acuminate, with 3 or 4 seeds per cell. *SW Europe, NW Africa*. H4. Summer.

*****S. hughii** Gussone is larger in all its parts, with leaves 4–6 cm wide and deep violet flowers. *Sicily*. H5. Summer.

**17. S. autumnalis** Linnaeus. Illustration: Reichenbach, Icones Florae Germanicae et Helveticae 4: t. 463 f. 1012 (1848); Polunin, Flowers of Europe, pl. 168 (1969); Grey-Wilson & Mathew, Bulbs, pl. 17 (1981).
Bulb 1.5–3 cm in diameter, ovoid, tunic brown or pinkish, occasionally with a few membranous scale-leaves. Leaves 5–12, 2–18 cm × 1–2 mm, linear, usually appearing after the flowers. Scapes 1–3, 5–30 cm. Raceme with 5–25 flowers, conical, without bracts, elongating in fruit; flower-stalks 3–10 mm, ascending. Perianth-segments spreading, lilac to pink, 3–5 mm. Capsules *c.* 4 mm, ovoid-spherical, obscurely 3-angled. Seeds 2 per cell, *c.* 2 × 1 mm, black, finely roughened. *S, W & C Europe, NW Africa to S Russia, Iran & Iraq*. H4. Summer–autumn.

**64. LEDEBOURIA** Roth
*P.G. Barnes*
Perennial herbs with bulbs. Leaves often spotted or striped. Inflorescence a slender, usually flexuous raceme. Flowers small (*c.* 6 mm in the species described here), greenish or purplish, the perianth-segments recurved. Filaments free to base. Ovary

superior, broadly conical, abruptly flared at the base and stalked, with 2 basal ovules per cell. Seed very rarely produced.

A genus of 16 species in southern Africa and 1 (not cultivated) in India. Formerly a subgenus of *Scilla*, distinguished mainly by the stalked, conical ovary and usually spotted or striped leaves. Easily cultivated in a cool greenhouse for the attractive foliage. Propagation is by offsets.
Literature: Jessop, J.P., Studies in the bulbous Liliaceae: 1, *Journal of South African Botany* **36**(4): 233–66 (1970).

1a. Bulb ovoid, more or less above soil-level, narrowed into a neck; leaves narrowed to the stalk-like base, silvery above    **2. socialis**
  b. Bulb spherical, subterranean, without an elongated neck; leaves not silvery above    **2**
2a. Leaves narrowed to the base, green or spotted above    **3. revoluta**
  b. Leaves scarcely narrowed to the base, green or striped with dark brown    **1. cooperi**

**1. L. cooperi** (J.D. Hooker) Jessop (*Scilla adlamii* Baker). Illustration: Botanical Magazine, 5580 (1866); Bulletin of the Alpine garden Society **42**: 341 (1974).
Bulb spherical. Leaves 1–5, narrowly oblong-lanceolate, scarcely narrowed to the base, 4–20 × 1–2 cm, both surfaces green or with brown longitudinal stripes. Perianth-segments pale purple or with a green keel. *South Africa*. H5–G1. Spring–summer.

Highly variable in the wild, but mainly represented in cultivation by a variant with stolons and heavily brown-striped leaves (*Scilla adlamii* Baker).

**2. L. socialis** (Baker) Jessop (*Scilla violacea* Hutchinson). Illustration: Baileya **18**: 52 (1971).
Bulb ovoid, 2–4 cm, produced into a neck 1-2 cm long, often purple. Leaves lanceolate, 4–15 × 1–3 cm, narrowed to a channelled stalk-like base 1–2 cm long. Upper surface silvery with copious dark green blotches, lower surface green or purple. Perianth-segments pale purple with green keels. *South Africa*. H5–G1. Spring–summer.

Variable in leaf colour. The commonly cultivated variant has the leaf underside and the bulb deep purple, but plants lacking purple pigment in leaves and bulb are also grown.

**3. L. revoluta** (Linnaeus filius) Jessop (*Scilla lancifolia* (Jacquin) Baker). Illustration:

Saunders Refugium Botanicum **3**: t. 182 (1870); Loddiges'; Botanical Cabinet, t. 1041 (1825).
Bulb spherical, 2–4 cm in diameter. Leaves 4–8, lanceolate to narrowly ovate, 4–15 × 1–3 cm; upper surface green with or without darker spots. Perianth-segments greenish. *South Africa*. H5–G1. Spring–summer.

## 65. HYACINTHOIDES Medikus
*A.C. Leslie*
Perennial herbs. Bulbs with tubular, coalescent scales, completely renewed each year. Flowers in a raceme, each subtended by 2 bluish, linear-lanceolate bracts. Perianth-segments free to the base, erect or spreading, usually blue. Filaments free, attached to the perianth-segments; anthers versatile. Ovary superior, 3-celled. Style simple, stigma capitate. Fruit a capsule. Seeds spherical, black.

A genus of 3–5 species in W Europe and N Africa. They are easily grown, and *H. non-scripta*, *H. hispanica* and hybrids between them will naturalise readily in deciduous woodland. Propagation is by offsets or seed.
Literature: Adolfi, K., Zure Unterscheidung von Hyacinthoides non-scripta (L.) Chouard und Hyacinthoides hispanica (Mill.) Rothm., *Gottinger Floristische Rundbriefe* **11**(2): 33–4 (1977); Quere-Boterenbrood, A.J., Over het vorkomen van Scilla non-scripta (L.) Hoffmannsegg & Link, S. hispanica Miller en hur hybrids in Nederland, *Gorteria* **12**(5): 91–104 (1984).

1a. Perianth-segments less than 1 cm    **3. italica**
  b. Perianth-segments more than 1 cm    **2**
2a. Raceme 1-sided, flowers pendent; anthers cream    **1. non-scripta**
  b. Racemes not 1-sided, flowers erect; anthers blue    **2. hispanica**

**1. H. non-scripta** (Linnaeus) Rothmaler (*Endymion non-scriptus* (Linnaeus) Garcke; *Scilla nutans* Smith; *S. non-scripta* Hoffmannsegg & Link). Illustration: Coste, Flore de la France **3**: 323 (1906); Keble-Martin, The concise British Flora in colour, pl. 85 (1965); Ross-Craig, Drawings of British plants **29**: pl. 4 (1972).
Leaves linear to linear-lanceolate, 20–45 cm × 7–15 mm. Flowering stem 20–50 cm, Racemes loose, 1-sided, drooping at the tip. Flowers erect in bud, finally pendent, scented. Perianth-segments 1.8–2 cm, oblong-lanceolate, erect below (so that the flower appears tubular),

reflexed at the tips, violet blue or rarely pink or white. Outer filaments inserted at about the middle of the perianth, the inner inserted near the base. Anthers cream. *W Europe*. H2. Spring.

**2. H. hispanica** (Miller) Rothmaler (*Endymion hispanicus* (Miller) Chouard; *E. patulus* Dumortier; *Scilla campanulata* Aiton). Illustration: Coste, Flore de la France **3**: 322 (1906); Bonnier, Flore complète **10**: pl. 588 (1929); Polunin & Smythies, Flowers of Southwest Europe, 69 (1973).
Like *H. non-scripta* but leaves often broader; racemes erect, not 1-sided; at least the upper flowers remaining erect or spreading, unscented; perianth-segments spreading (so that the flower is openly bell-shaped), not reflexed at their tips; filaments all inserted below the middle of the perianth; anthers blue. *SW Europe, N Africa*. H2. Spring.

Hybrids between *H. non-scripta* and *H. hispanica* are frequent in gardens and a complete range of intermediates can be found. Plants offered as *H. hispanica* are often hybrids.

**3. H. italica** (Linnaeus) Rothmaler (*Scilla italica* Linnaeus). Illustration: Botanical Magazine, 663 (1803); Coste, Flore de la France **3**: 321 (1906); Bonnier, Flore complète **10**: pl. 578 (1929).
Leaves linear to linear-lanceolate, 10–25 cm × 3–12 mm. Flowering stem 10–40 cm. Racemes dense, conical, erect, not 1-sided. Perianth-segments 5–7 mm, spreading widely, bluish violet (rarely white). All filaments inserted at base of perianth. Anthers blue. *SE France, NW Italy, Spain, Portugal*. H2. Spring.

## 66. CHIONODOXA Boissier
*E.M. Rix*
Perennial herbs with small bulbs, tunics brown. Leaves usually 2. Scapes with 1–15 flowers in a loose raceme. Flowers blue or pinkish, often with a white central zone. Perianth with a short, obconical tube at the base, the lobes spreading. Stamens borne at the apex of the perianth-tube; filaments of unequal length, flattened. Anthers dorsifixed. Ovary superior, style 1. Fruit an almost spherical capsule.

A genus of 6 species from western Turkey, Crete and Cyprus, grown for their very early, rich blue flowers. The genus is very close to *Scilla* (especially *S. bifolia*) and is sometimes united with it. Hybrids between the two are frequent; those between *C. forbesii* (and probably other species) and *Scilla bifolia* are grown as

× **Chionoscilla allenii** Nicholson, and are intermediate between their parents, with very short perianth-tubes, a small pale central zone to the flower and pale blue filaments.

Chionodoxas grow in well-drained but rich soil, and a position under deciduous shrubs is ideal.
Literature: Meikle, R.D., Chionodoxa luciliae, a taxonomic note, *Journal of the Royal Horticultural Society* **95**: 21–24 (1970); Speta, F., Über Chionodoxa Boiss., ihre Gliederung und Zugehörigkeit zu Scilla L., *Naturkundliches Jahrbuch der Stadt Linz* **21**: 9–79 (1976).

*Flowers.* 1 or 2, rarely to 4 on each scape: **3–5**; 4 or more on each scape: **1,2**. Bright blue with pale centre: **2,3**; bright blue without pale centre: **1,4**; not bright blue: **5** (and cultivars of other species).
*Perianth-lobes.* More than 1.2 cm: **2–4**; less than 1.2 cm: **1,5**.

1a. Flowers bright blue, without a pale central zone                                  2
  b. Flowers bright blue, lavender-blue or pink, with a pale central zone or entirely pale                              3
2a. Scape with 2–4 flowers      **4. lochiae**
  b. Scape with 4–12 flowers
                                  **1. sardensis**
3a. Scape with 4–12 flowers     **2. forbesii**
  b. Scape usually with 1–3 flowers     4
4a. Perianth-lobes 1.2 cm or more
                                   **3. luciliae**
  b. Perianth-lobes less than 1.2 cm
                                   **5. nana***

**1. C. sardensis** Whittall. Illustration: Botanical Magazine, n.s., 50 (1949); Rix & Phillips, The bulb book, 44 (1981).
Leaves erect or spreading. Scape usually 1 per bulb, to 40 cm, with 4–12 flowers. Flowers slightly pendent. Perianth bright blue, without a white central zone; tube 3–5 mm, almost spherical; lobes 8–10 × 2–4 mm. Filaments 2 mm and 3 mm, white. *W Turkey*. H2. Spring.

**2. C. forbesii** Baker (*C. luciliae* misapplied; *C. tmolusi* Whittall; *C. siehei* Stapf). Illustration: Botanical Magazine, 6433 (1879) – as C. luciliae & 9068 (1925).
Leaves erect or spreading, 7–28 cm. Scape usually 1 per bulb, to 30 cm, with 4–12 flowers. Flowers slightly pendent. Perianth rich blue with a white central zone, or pink with a white centre (in 'Pink Giant'); tube 3–5 mm, almost spherical; lobes 1–1.5 cm × 4–5 mm. Filaments 2 mm and 3 mm, white. *W Turkey*. H2. Spring.

**3. C. luciliae** Boissier (*C. gigantea* Whittall). Illustration: Journal of the Royal Horticultural Society **95**: f. 19, 20 (1970); Rix & Phillips, The bulb book, 40, 44 (1981).
Leaves often recurved, 7–20 cm. Scape usually 1 per bulb, to 14 cm, with 1 or 2 flowers. Flowers erect. Perianth lavender-blue with a white central zone; tube 2.5–4 mm; lobes 1.2–2 cm × 3–8 mm. Filaments 2.5 mm and 3 mm, white. *W Turkey*. H2. Spring.

**4. C. lochiae** Meikle. Illustration: Botanical Magazine, n.s., 281 (1956).
Leaves erect, 7–18 cm. Scape usually 1 per bulb, to 18 cm, usually with 2–4 flowers. Flowers horizontal or pendent. Perianth bright blue without a white central zone; tube 5–7 mm; lobes 1.2–1.3 cm × 4–6 mm. Filaments 3 mm and 4 mm, white. *Cyprus*. H2. Spring.

**5. C. nana** (Schultes & Schultes) Boissier (*C. cretica* Boissier). Illustration: Rix & Phillips, The bulb book, 42e (1981).
Leaves spreading, 8–18 cm. Scape usually 1 per bulb, to 20 cm, usually with 1–3 flowers. Flowers erect. Perianth blue with a white central zone; tube 3–5 mm, narrowly conical; lobes 6–11 × c. 3 mm. Filaments c. 3 mm, the outer longer and narrower, white. *Crete*. H2. Spring.

*\*C. albescens* (Speta) Rix. Flowers c. 10 mm, pale pinkish or lavender. *Crete*. H2. Spring.

# 67. CAMASSIA Lindley
*C.J. King*
Perennial herbs. Bulbs usually large, ovoid or spherical, with brown or black tunics. Leaves basal, linear, keeled, sheathing at the base, hairless, with entire margins. Scape terete, bearing a dense or loose raceme of flowers. Flower-stalks in axils of lanceolate-acuminate bracts. Perianth-segments spreading, persistent, blue, purple or white. Stamens 6, attached at the base of the segments; filaments thread-like; anthers versatile. Ovary 3-celled; style thread-like; stigma 3-lobed. Fruit a 3-lobed capsule, opening by 3 slits, with several black seeds in each cell.

A genus of 5 species from N America, and possibly 1 from S America. The bulbs do not increase much by division, but seed is freely produced and plants will reach flowering size in 3 or 4 years. They grow best in a rather moist border of good, well-drained loam, in full or at least half sun, and the bulbs should be left undisturbed. While Camassias are generally quite hardy,

they should be protected from severe frost in colder areas.
Literature: Gould, F.W., A systematic treatment of the genus Camassia Lindl., *American Midland Naturalist* **28**: 712–42 (1942).

*Bulbs.* Clustered, and with an unpleasant smell: **5**.
*Leaves.* More than 7: **5**. Glaucous above: **1,5**.
*Flowers.* With one perianth-segment deflexed: **1,5**; regular: **1–4**.
*Perianth-segments.* White: **1,2,4**. Twisting together as they wither: **1–3**; remaining separate: **1,4,5**.

1a. Perianth-segments twisting together around and above the capsule after flowering                                   2
  b. Perianth-segments withering separately at base of capsule after flowering                                   4
2a. Flowers bilaterally symmetric, one perianth-segment deflexed; flower-stalks erect in fruit     **1. quamash**
  b. Flowers radially symmetric; flower-stalks not erect in fruit     3
3a. Capsule narrowly ovoid or oblong, dull green     **2. leichtlinii**
  b. Capsule broadly ovoid to nearly spherical, glossy green     **3. howellii**
4a. Capsule ovoid or oblong     5
  b. Capsule nearly spherical
                                 **4. scilloides**
5a. Bulbs clustered, with an unpleasant smell; leaves more than 7, 2–4 cm broad     **5. cusickii**
  b. Bulbs rarely clustered, without an unpleasant smell; leaves fewer than 7, 5–20 mm broad     **1. quamash**

**1. C. quamash** (Pursh) Greene (*C. esculenta* Lindley not (Ker-Gawler) Colville; *C. teapeae* St. John). Illustration: Edwards's Botanical Register **18**: t. 1486 (1832); Flore des Serres **3**: t. 275 (1847); The Garden **20**: t. 302 (1881); Rickett, Wild flowers of the United States **5**(1): t. 12 (1971), **6**(1): t. 11 (1973).
Bulb broadly ovoid to spherical, 2–5 × 1–3 cm, with usually black tunics. Leaves 5 or 6, 20–50 cm × 5–20 mm, green above and beneath or glaucous above. Scape 20–80 cm, deep green. Raceme dense or loose; 5–30 cm. Bracts narrow, acuminate, shorter or longer than the flower-stalks. Flower-stalks 5–30 mm, spreading horizontally or curving upwards and becoming erect in fruit. Flowers either regular, or irregular with 1 perianth-segment deflexed. Perianth-segments

linear-oblong, pale blue to deep violet-blue or white, 1–3.5 cm, either remaining separate as they wither, or twisting together over the young capsule. Stamens two-thirds as long as the segments; filaments pale blue or white; anthers yellow or blue. Ovary oblong-ovoid, bright green; style as long as or longer than the stamens, pale purple or white. *Western N America east to Montana and Utah.* H3. Late spring–early summer.

A very variable species, with up to 8 subspecies or varieties recognised by a combination of geographical locality and flower and foliage characters. The white form (Illustration: Botanical Magazine, 2774, 1827) has been called 'Flore Albo'.

**2. C. leichtlinii** (Baker) Watson.
Bulb broadly ovoid to spherical, 1.5–4 cm in diameter, with brown tunics. Leaves 4–6, 20–60 cm × 5–25 mm, bright green. Scape 20–130 cm, light green. Raceme loose, 10–30 cm. Bracts linear-lanceolate, *c.* 20 mm. Flower-stalks 1.5–5 cm, directed slightly upwards. Flowers regular. Perianth-segments oblong-lanceolate, creamy white or blue to violet, 2–5 cm, twisting together as they wither. Stamens half to three-quarters as long as the segments; filaments white or bluish; anthers yellow. Ovary narrowly ovoid to oblong, dull green: style rather shorter to slightly longer than the stamens, white. *Western N America.* H3. Late spring–early summer.

Subsp. **leichtlinii** (Illustration: Botanical Magazine, 6287, 1877; The Garden 46, t. 983, 1894) has creamy white flowers and is found wild only in Oregon.

Subsp. **suksdorfii** (Greenman) Gould (Illustration: Rickett, Wild flowers of the United States 5(1): t. 12, 1971) has blue to violet flowers and extends from British Columbia to California.

Several of the colour forms have received cultivar names, and are perhaps the most desirable garden plants in the genus.

**3. C. howellii** Watson. Illustration: Abrams, Illustrated flora of the Pacific States 1: f. 1019 (1940).
Bulb ovoid, *c.* 1.5 cm in diameter. Leaves 3–5, *c.* 30 cm × 5–10 mm. Scape 30–100 cm. Raceme 20–25 cm. bracts linear, less than 1.2 cm. Flower-stalks 1.5–2.5 cm, spreading horizontally. Flowers regular. Perianth-segments oblong, pale blue or purple, 1–2 cm, twisting together as they wither. Stamens slightly shorter than the segments; filaments pale purple; anthers yellow. Ovary broadly ovoid

to nearly spherical, glossy green; style about as long as stamens, pale purple. *W United States (SW Oregon).* H3. Late spring.

**4. C. scilloides** (Rafinesque) Cory (*C. esculenta* (Ker-Gawler) Colville not Lindley; *C. fraseri* Torrey; *C. hyacinthina* (Rafinesque) Palmer & Steyermark). Illustration: Botanical Magazine, 1574 (1813); Addisonia 7: t. 239 (1922); Rickett, Wild flowers of the United States 6(1): t. 11 (1973).
Bulb ovoid, 4 × 3 cm, with an almost black outer tunic. Leaves 20–60 cm × 5–20 mm, bright green. Scape 20–80 cm. Raceme fairly loose, 8–10 cm. Bracts linear, 1–2 cm. Flower-stalks 5–30 mm, directed slightly upwards. Flowers regular. Perianth-segments narrowly oblong, blue, blue-violet or white, 5–15 mm, remaining separate as they wither. Stamens slightly shorter than the segments; filaments coloured as the perianth-segments; anthers yellow. Ovary nearly spherical, green; style as long as the segments, white or bluish. *C & E North America.* H3. Late spring–early summer.

**5. C. cusickii** Watson. Illustration: Garden and Forest 1: f. 32 (1888); Botanical Magazine, n.s., 319 (1958).
Bulbs clustered, ovoid to almost spherical, up to 9 × 5 cm, with an unpleasant smell. Leaves 8–20, 40–80 × 2–4 cm, wavy, glaucous above, bright green beneath. Scape 60–80 cm, green. Raceme up to 40 cm. Bracts narrowly triangular-acuminate, the lowermost *c.* 8.8 cm, decreasing in size upwards to 1.5 cm, green and white at first, becoming membranous. Flower-stalks 1.1–2.3 cm, green, spreading at first, curving inwards towards the scape after flowering, and finally erect. Flowers irregular, with one perianth-segment deflexed. Perianth-segments linear-oblong, pale blue, 2.5–2.7 cm × 4–5 mm, remaining separate as they wither. Stamens shorter than perianth; filaments 1.5–1.6 cm, white; anthers yellow. Ovary oblong-ovoid, light green; style 1.5 cm, straight at first, becoming curved, white. *W United States (NE Oregon).* H3. Late spring–early summer.

**68. ORNITHOGALUM**
*J. Cullen*
Plants with bulbs which are usually subterranean with whitish or brownish papery tunics, more rarely partly exposed and green and fleshy. Leaves in a rosette, linear to lanceolate or obovate, margins

ciliate or not, sometimes marked with a whitish line above. Raceme corymbose, pyramidal, almost spherical or cylindric, with 2–many flowers. Bracts usually conspicuous. Perianth-segments 6, equal, usually white, more rarely orange, yellow or red, each usually marked with a green stripe on the outside, usually widely spreading, more rarely erect. Stamens 6, filaments flattened, often broadened to the base, sometimes winged or abruptly widened towards the base. Ovary superior, cylindric to spherical, yellow, green or purplish black, 3-celled. Ovules numerous. Style terminal, long or short. Fruit a capsule with many seeds.

A genus of about 80 species with 2 main centres of distribution: southern Africa and the Mediterranean area. Species from both of these centres are grown. The South African species are rather tender and mostly require glasshouse protection (at least in northern Europe); in mild areas they can be stood outside in the summer. All can be grown in pots or in the open ground in well-drained soil on a sunny site. Several species reproduce by both seed and bulblets, and can become invasive.
Literature: Publications on the Mediterranean species are numerous and confusing: helpful accounts can be found in *Flora Europaea* 5: 35–40 (1980) and in Davis, P.H. (ed), *Flora of Turkey* 8: 227–45 (1984). The South African species are covered in Obermeyer, A.A., Ornithogalum: a revision of the South African species, *Bothalia* 12: 323–76 (1978).

*Bulb.* Upper part exposed, with succulent, green tunics: 6. Surrounded by numerous bulblets: 1,6,10,13.
*Leaves.* With a white line above: 10,13,14. Margins ciliate: 3,4.
*Raceme.* Corymbose: 11–14; cylindric: 1,3,5–10; triangular to almost spherical: 2,3,4,6. With 2–20 flowers: 1,10–14; with 20–300 flowers: 1–10.
*Perianth-segments.* Without a green stripe outside: 1–5. Orange, red or yellow: 3,9.
*Ovary.* Black, greenish black or purplish: 1,2.
*Filaments.* Some or all of them winged, the wings projecting as teeth beside the anther: 10.

1a. Ovary black, greenish black or purplish                                2
 b. Ovary yellowish or green                                             3
2a. Perianth-segments 1.5–3.2 cm, erect in fruit                   **1. arabicum**
 b. Perianth-segments 1–1.5 cm, reflexed in fruit                **2. saundersiae**

3a. Filaments of at least 3 of the stamens winged for their whole length, the wings ending in teeth at either side of the anther; flowers drooping; perianth-segments not widely spreading **10. nutans**

b. Filaments various but not as above; flowers horizontal to erect; perianth-segments widely spreading **4**

4a. Inflorescence a cylindric, triangular or almost spherical raceme with 20–300 flowers **5**

b. Inflorescence a corymbose raceme with 2–20 flowers **11**

5a. Perianth-segments without a green stripe outside, sometimes dark-coloured at the base inside **6**

b. Perianth-segments with a conspicuous green stripe outside, not dark-coloured at the base inside **8**

6a. Style very short or almost absent; perianth-segments yellow, red or orange, rarely white. **3. dubium**

b. Style as long as the ovary or longer; perianth-segments white or creamy white **7**

7a. Filaments of the inner stamens with conspicuous, expanded bases which curve around the ovary; raceme pyramidal to almost spherical; scape 20–50 cm **4. thyrsoides**

b. Filaments of the inner stamens thread-like or with small broadened bases; raceme cylindric; scape 40–100 cm **5. conicum**

8a. Bulb with its upper part exposed, with succulent, green tunics, surrounded by numerous bulblets; bracts much exceeding the flowers **6. longibracteatum**

b. Bulb well below ground, with brownish or whitish tunics, without bulblets; bracts shorter than to a little longer than the flowers **9**

9a. Perianth-segments pale yellow **9. pyrenaicum**

b. Perianth-segments white **10**

10a. Ovary 2–2.5 mm; style usually shorter than the ovary, thickened and conical at its base **7. pyramidale**

b. Ovary 2.5–5 mm; style longer than the ovary, thread-like throughout its length **8. narbonense**

11a. Leaves broadest at ground level, tapering from there to an acute apex **11. montanum***

b. Leaves broadening above ground level, or equally broad for most of their length, tapering abruptly to a rounded apex **12**

12a. Leaves usually 2–4, conspicuously broadened above, without a white line on the upper surface **12. oligophyllum**

b. Leaves usually more than 4, parallel-sided for most of their length, with a white line on the upper surface **13**

13a. Fruit-stalks robust, 5–9 cm, rigid and horizontal; bulb surrounded by bulblets **13. umbellatum**

b. Fruit-stalks slender, 2–3.5 cm, weak, curved or ascending; bulb without bulblets **14. orthophyllum**

**1. O. arabicum** Linnaeus. Illustration: Botanical Magazine, 3179 (1832) – as O. corymbosum; Grey-Wilson & Mathew, Bulbs, t. 15 (1981).
Bulb below ground, surrounded by numerous bulblets. Scapes 30–80 cm. Leaves 5–8, more or less erect, dark green. Raceme cylindric with 6–25 flowers. Perianth-segments 1.5–3.2 cm, white or creamy white. Filaments broadening below. Ovary black or purplish black. Perianth-segments withering but not reflexing in fruit. *Mediterranean region.* H5–G1. Spring.

A species of uncertain origin, widely introduced in southern Europe from former extensive cultivation as a cut-flower crop.

**2. O. saundersiae** Baker. Illustration: Jeppe, Natal wild flowers, t. 6 (1975).
Bulb below ground. Scapes 30–100 cm. Leaves 6–8, erect, dark green and glossy above. Raceme triangular, with many flowers. Perianth-segments white or creamy white, 1–1.5 cm. Stamens with filaments conspicuously broadened below. Ovary black or greenish black. Perianth-segments persistent and reflexed in fruit. *South Africa (Transvaal, Natal, Swaziland, Zululand).* G1. Winter–spring.

**3. O. dubium** Houttuyn (*O. miniatum* Jacquin; *O. aureum* Curtis). Illustration: Botanical Magazine, 190 (1792); Marloth, The Flora of South Africa 4: t. 28 (1915); Bothalia 12: f. 8 (1978).
Bulb below ground. Leaves 3–8, yellowish green, margins ciliate. Raceme cylindric to almost spherical with 20 or more flowers. Perianth-segments 1–2 cm, orange, red, yellow or rarely white, often greenish or brownish at the base within. Stamens with filaments broadened below. Ovary yellowish green. Style much shorter than the ovary. *South Africa (Cape Province).* G1. Winter–spring.

**4. O. thyrsoides** Jacquin. Illustration: Marloth, the Flora of South Africa 4: t. 28 (1915); Hay & Synge, Dictionary of garden plants in colour, t. 870 (1969); Mason, Western Cape sandveld flowers, 35 (1972).

Bulb below ground. Scapes 20–50 cm. Leaves 6–12, ascending, with ciliate margins. Raceme triangular to almost spherical, with many flowers. Perianth-segments 1–2 cm, translucent white. Filaments of the inner stamens with broadly and abruptly expanded bases which curve around the ovary. Ovary yellowish green. Style as long as the ovary. *South Africa (Cape Province).* G1. Winter.

**5. O. conicum** Jacquin (*O. lacteum* Jacquin). Illustration: Botanical Magazine, 1134 (1808).
Very similar to *O. thyrsoides*, but scapes 40–100 cm, leaves without ciliate margins, filaments only slightly expanded at their bases. *South Africa (Cape Province).* G1. Winter.

**6. O. longibracteatum** Jacquin (*O. caudatum* Aiton). Illustration: Botanical Magazine, 805 (1805).
Bulb with the upper part exposed and with green, succulent tunics, surrounded by many bulblets. Scapes 1–1.5 m. Leaves 8–12, flaccid and succulent. Raceme triangular to cylindric with up to 300 flowers. Bracts conspicuous, much longer than the flowers. Perianth-segments to 9 mm, white with a green stripe outside. Filaments broadened towards the base. Ovary yellowish green. *South Africa (Cape Province, Natal).* G1. Flowering intermittent.

**7. O. pyramidale** Linnaeus.
Bulb below ground. Scapes 30–120 cm. Leaves 4–7, glossy green. Raceme cylindric with many flowers. Bracts shorter than the flower-stalks. Perianth-segments 1.1–1.5 cm, translucent white with a green stripe outside. Ovary 2–2.5 mm, green. Style shorter than the ovary, thickened and conical at the base. *C Europe, Yugoslavia, Romania.* H4. Spring.

**8. O. narbonense** Linnaeus. Illustration: Botanical Magazine, 2510 (1824); Huxley & Taylor, Wild flowers of Greece and the Aegean, t. 364 (1977); Grey-Wilson & Mathew, Bulbs, t. 15 (1981).
Very similar to *O. pyramidale* but ovary 2.5–5 mm, style longer than the ovary, thread-like throughout. *Mediterranean area, USSR (Caucasus), NW Iran.* H4. Spring.

**9. O. pyrenaicum** Linnaeus (*O. flavescens* Lamarck). Illustration: Grey-Wilson & Mathew, Bulbs, t. 15 (1981).
Very similar to *O. pyramidale* but generally smaller and more delicate, perianth-segments pale yellow with a narrow green stripe outside, 9–13 mm, ovary 2.4–3 mm,

style longer than ovary. *Europe, Turkey, USSR (Caucasus)*. H2. Spring.

**10. O. nutans** Linnaeus. Illustration: Botanical Magazine, 269 (1794); Polunin & Huxley, Flowers of the Mediterranean, t. 241 (1965); Hay & Synge, Dictionary of garden plants in colour, t. 869 (1969); Huxley & Taylor, Wild flowers of Greece and the Aegean, t. 361 (1977).
Bulb below ground, with numerous bulblets. Scapes to 60 cm. Leaves several, pale green, each with a white line on the upper surface. Raceme cylindric with 10–20 drooping flowers. Perianth-segments 2–3 cm, translucent white with a broad green stripe outside, not widely spreading. Filaments of at least the 3 inner stamens winged, the wings terminating in teeth at either side of the anther. Style longer than the green ovary. *Europe (naturalised in many areas), SW Asia*. H3. Spring.

**11. O. montanum** Cyrillo. Illustration: Polunin & Huxley, Flowers of the Mediterranean, t. 237 (1965); Grey-Wilson & Mathew, Bulbs, t. 15 (1981).
Bulb below ground. Scapes to 20 cm. Leaves several, arching outwards, broadest at the ground level and tapering evenly to an acute apex. Raceme corymbose with 7–14 flowers. Perianth-segments 1.1–1.6 cm, white with a green stripe outside. Fruit-stalks erect-spreading, 4–7 cm. *S Europe from Italy eastwards, Turkey, Lebanon, Israel*. H4. Spring.

**\*O. lanceolatum** Labillardière. Very similar, but leaves broader and less acute, raceme stalkless, borne directly on the leaf rosette. *Turkey, Syria, Lebanon*. H4. Winter–spring.

**12. O. oligophyllum** Clarke (*O. balansae* Boissier). Illustration: Grey-Wilson & Mathew, Bulbs, t. 15 (1981).
Bulb below ground. Scapes to 15 cm. Leaves usually 2–4, obovate, widening upwards then abruptly rounded to the apex. Raceme with 2–5 flowers, corymbose. Perianth-segments 1–1.6 cm, white with a green stripe outside. Fruit-stalks erect-spreading, 1–3 cm. *Balkans, Turkey, USSR (Georgia)*. H4. Spring.

**13. O. umbellatum** Linnaeus. Illustration: Hay & Synge, Dictionary of garden plants in colour, t. 871 (1969); Grey-Wilson & Mathew, Bulbs, t. 15 (1981).
Bulb below ground, with numerous bulblets. Scapes 10–30 cm. Leaves several, linear, parallel-sided, tapering abruptly to the apex, with a white line on the upper surface. Raceme corymbose with 6–20 flowers. Perianth-segments 1.5–2.2 cm, white with a green stripe outside. Fruit-stalks horizontal, rigid, the lower 5–9 cm. *Most of Europe, N Africa, Turkey, Syria, Lebanon, Israel, Cyprus*. H2. Spring.

Can become an invasive weed, difficult to remove from rock gardens or screes.

**14. O. orthophyllum** Tenore (*O. tenuifolium* Gussone). Illustration: Matthews, Lilies of the field, opposite p. 25 (1968).
Similar to *O. umbellatum*, but bulb without bulblets, fruit-stalks spreading and curving upwards, the lower 2–3.5 cm. *S & C Europe to N Iran*. H3. Spring.

**69. EUCOMIS** L'Héritier
*C. J. King*
Perennial herbs. Bulbs usually large, ovoid, with shiny tunics. Leaves many, basal, strap-shaped to obovate, glossy, appearing with the flowers. Scape terete, bearing a simple, dense raceme of flowers, topped by a tuft of green bracts. Perianth fused at base, persistent; lobes 6–17 mm, spreading. Stamens 6, attached near base of lobes; filaments broadened and fused below into a shallow cup; anthers versatile. Ovary rounded to obovoid, 3-angled, with many ovules. Fruit a triangular capsule, inflated and membranous, or dry, hard and compact, opening by 3 slits. Seeds rounded or ovoid, black or brown.

About 10 species, ranging from south tropical Africa to South Africa. Most species are easily raised from seed and will flower in 2–5 years, or they can be increased by offsets. The plants require a sunny position in a rich, well-drained soil. They may be grown out-of-doors as border plants in favourable areas, but elsewhere they need the protection of a cool greenhouse. The single tropical species, *E. zambesiaca*, requires warm glasshouse conditions.
Literature: Reyneke, W.F., Three subspecies of Eucomis autumnalis, *Bothalia* 13: 140–2 (1980) (part of an unpublished MSc thesis, *'n Monografiese studie van die genus Eucomis L'Herit. in Suid-Afrika* (1972), submitted to the University of Pretoria).

1a. Perianth-lobes green with distinct purple margin (except cultivar 'Alba') **1. bicolor**
b. Perianth-lobes entirely green **2**
2a. Scape and leaves spotted or striped with purple; ovary purple **2. comosa**
b. Scape and leaves green; ovary green **3**
3a. Scape not more than 30 cm; raceme not more than 15 cm **3. autumnalis**
b. Scape 40 cm or more; raceme 20 cm or more **4. pallidiflora\***

**1. E. bicolor** Baker. Illustration: Botanical Magazine, 6816 (1885); Everard & Morley, Wild flowers of the world, t. 85 (1970).
Bulb about 5 cm across. Leaves oblong, 30–50 × 8–10 cm, with wavy margins. Scape cylindric, to 60 cm, sometimes spotted with dark purple. Raceme dense, many-flowered, 8–10 cm. Perianth pale green; lobes 1.2–1.5 cm, with a distinct purple margin. Ovary green. Bracts 30–40, ovate or oblong, acute, 5–8 cm, with crisped, sometimes purple, margins. *South Africa (Griqualand East, Natal)*. G1. Late summer.

In cultivar 'Alba' the perianth is uniformly green.

**2. E. comosa** (Houttuyn) Wehrhahn (*E. punctata* L'Héritier). Illustration: Botanical Magazine, 913 (1806); Flore des Serres **22**: t. 2307 (1877); Batten & Bokelmann, Wild flowers of the eastern Cape Province, t. 9(2) (1966).
Bulb about 7 cm across. Leaves oblong-lanceolate, 30–77 × 3–8 cm, with numerous small dark purple spots near the base of the under-surface, and a slightly wavy margin. Scape cylindric, 45–70 cm, spotted with dark purple. Raceme dense, many-flowered, to 30 cm. Perianth pale green, lobes *c*. 1.2 cm. Ovary purple. Bracts 8–20, lanceolate, 2.5–8 cm, sometimes with purple margins. *South Africa (Eastern Cape Province, Natal)*. G1. Late summer.

A variety where the spotting on the leaves and scape merges to form stripes is called var. **striata** (Houttuyn) Wehrhahn.

**3. E. autumnalis** (Miller) Chittenden.
Bulb 5–10 cm across. Leaves usually linear, sometimes lanceolate or ovate, with wavy margins. Scape cylindric or club-shaped, green. Raceme more or less dense, many-flowered, 5–15 cm. Perianth at first white or yellowish green, later green; lobes 6–17 mm. Ovary green. Bracts 10–45, oblong, ovate, or lanceolate, 2.5–8 cm, with crisped margins. G1. Late summer.

Subsp. **autumnalis** (*E. undulata* Aiton). Illustration: Botanical Magazine, 1083 (1808); Flowering plants of South Africa **6**: t. 220 (1926); Batten & Bokelmann, Wild flowers of the eastern Cape Province, t. 9(1) (1966); Graf, Tropica, 603 (1978). Leaves linear to ovate, 15–55 × 6–13 cm. Scape cylindric, sometimes slightly club-shaped, 6–30 cm. Raceme to 15 cm. Perianth-lobes 1–1.3 cm. bracts 10–45, oblong, 3–7 cm. *South Africa (Eastern Cape Province, Transvaal), Rhodesia and Malawi*.

Subsp. **clavata** (Baker) Reyneke (*E. clavata* Baker; *E. robusta* Baker; *E. albomarginata* Barnes). Leaves lanceolate to ovate, sometimes linear, 15–60 × 6–13 cm. Scape usually club-shaped, 7–13 cm, raceme 8–15 cm. Perianth-lobes 1.2–1.7 cm. Bracts 15–30, ovate, 5–8 cm. *South Africa (Transvaal, Orange Free State, Natal), Swaziland and Botswana.*

Subsp. **amaryllidifolia** (Baker) Reyneke (*E. amaryllidifolia* Baker). Leaves linear, 13–50 × 1.5–4 cm. Scape club-shaped, 6–23 cm. Raceme 5–8 cm. Perianth-lobes 6–8 mm. bracts 13–20, lanceolate, 2.5–8 cm. *South Africa (Eastern Cape Province, Orange Free State).*

**4. E. pallidiflora** Baker (*E. punctata* var. *concolor* Baker). Illustration: Botanical Magazine, n.s., 561 (1970).
Bulb about 10 cm across. Leaves broadly sword-shaped, 50–70 × 10–12.5 cm, with minutely crisped margins. Scape stout, cylindric, 45–75 cm, green. Raceme fairly dense, many-flowered, 24–43 cm. Perianth greenish white or green; lobes 1.2–1.6 cm. Ovary green. Bracts about 20, elliptic or lanceolate, acute, 3–4 cm, with minutely crisped margins. *South Africa (Natal, ? Transvaal).* G1. Summer.

**\*E. pole-evansii** N.E. Brown. Illustration: Flowering plants of South Africa 24: t. 953 (1944); Graf, Tropica, 603 (1978). Like *E. pallidiflora* in floral and other characters, differing principally in its larger leaves (75–120 × 10–17.5 cm), taller scape (90–180 cm), and larger bracts (5–10 cm). It may prove to be simply a large form of *E. pallidiflora* and therefore not worthy of separate specific rank. At present the range of variation within the 2 species is not known.

**70. DRIMIA** Willdenow
*C.J. King*
Perennial herbs. Bulbs with loose or tightly packed scales. Leaves 1–many, terete or flattened, usually appearing after the flowers. Inflorescence-stalk erect, terete, bearing a loose or dense raceme with usually many flowers. Flower-stalks each subtended by a membranous persistent bract. Flowers horizontal or pendent. Perianth white, greenish or greenish brown, usually with a short tube and reflexed lobes. Stamens 6, attached at the base of the perianth-tube. Ovary ovoid or ellipsoid, 3-celled. Fruit a capsule, opening by 3 slits. Seeds angled or flattened, black.

A genus with about 15 species in southern Africa, or, if enlarged to include

*Urginea*, comprising about 120 species in Europe, Asia and Africa. Only one South African species, *D. haworthioides*, is normally found in cultivation in Europe, and this requires greenhouse protection. It will grow well in sandy loam and peat, but must be kept dry or almost dry during the summer resting period.
Literature: Jessop, J.P., The taxonomy of Drimia and certain allied genera, *Journal of South African Botany* **43**: 265–319 (1977).

**1. D. haworthioides** Baker. Illustration: Marloth, Flora of South Africa 4: t. 24 (1915); Jacobsen, Handbook of Succulent Plants 1: f. 370 & 371 (1960).
Bulb 2–3 cm high, with loose, stalked scales giving it a diameter of 5–6 cm. Leaves up to 10, oblong-lanceolate, usually ciliate, 6–8 × 1–1.5 cm. Inflorescence-stalk 20–40 cm bearing 15–30 flowers. Bracts 1–4 mm. Flower-stalks 3–8 mm. Perianth greenish or greenish brown, 1–1.5 cm, fused for 2–4 mm into a basal tube, lobes reflexed. Stamens projecting. Capsule ellipsoid, 7–10 mm. *South Africa (Eastern Cape Province).* G1. Winter.

This species is occasionally cultivated, usually because of its peculiar bulb which sits on the soil surface. When the leaf-blades wither, the leaf-stalks lengthen and become considerably enlarged at their tips, forming a group of purplish grey storage organs which may persist for several years. The plant may be increased readily by means of these swollen bulb-scales.

**71. DRIMIOPSIS** Lindley
*J. Cullen*
Herbaceous perennials. Bulb more or less spherical, pale, green where exposed. Leaves in a loose rosette, stalked or not, pale green, spotted with darker or brownish green above. Scape exceeding the leaves, bearing a many-flowered raceme; bracts absent. Perianth united into a tube below, the lobes little spreading, hooded at their apices. Stamens 6, filaments triangular, attached to the top of the perianth-tube. Ovary superior, 3-celled, ovules 2 in each cell. Fruit a few-seeded capsule.

A genus of about 7 species from tropical and southern Africa, of which 1 or 2 are occasionally grown. They require rather warm, moist conditions, but are not particularly ornamental. Propagation is by seed.

**1. D. maculata** Lindley & Paxton. Illustration: Saunders' Refugium Botanicum, t. 191 (1870); Flowering plants of South Africa 24: t. 957 (1944).

Leaves with a distinct stalk, which is channelled and as long as or longer than the blade; blade ovate or ovate-oblong, to 10 × 5 cm. Scape to 40 cm. Flowers at first white, later greenish. Perianth to 4 mm. *South Africa.* G1–2. Spring.

**\*D. kirkii** Baker. Illustration: Botanical Magazine, 6276 (1877). Plant larger, leaves oblong, tapering slightly to the base but not stalked, perianth *c.* 6 mm, persistently white. *Tanzania (Zanzibar).* G2.

**72. DIPCADI** Medikus
*S.G. Knees*
Hairless perennial herbs with bulbs. Leaves linear, often sheathing at the base. Flowers bisexual in loose racemes. Perianth tubular to bell-shaped, united in the lower third, the inner 3 lobes held erect, the outer 3 spreading. Stamens 6, included in perianth. Ovary superior; style 1. Fruit a 3-celled capsule.

A genus of 55 species, from South and tropical Africa, SW Europe and the East Indies. Only the European species is cultivated to any extent. Although easily grown in any light sandy soil, protection from winter wet may be necessary in the northern European countries. Propagation is by seed or offsets, which should be taken in spring.

**1. D. serotinum** (Linnaeus) Medikus. Illustration: Botanical Magazine, 859 (1805); Grey-Wilson & Mathew, Bulbs, t. 20 (1981); Polunin & Smythies, Flowers of Southwest Europe, t. 59 (1973).
Herbaceous plant with scape 5–45 cm. Leaves deciduous, few, all basal, usually shorter than the scape, 5–20 × 1–2 cm, broadly V-shaped in cross-section. Racemes 1-sided, of 4–20 stalked flowers, subtended by scarious bracts. Perianth 1–1.6 cm, dull brownish orange. *SW Europe, from S Italy & France to Spain.* H5. Spring.

**\*D. fulvum** (Cavanilles) Webb & Berthelot. Illustration: Botanical Magazine, 1185 (1809). Similar but with brownish pink perianth-lobes. *S Spain, Morocco & Canary Islands.* H5. Autumn.

**73. ALBUCA** Linnaeus
*E.H. Hamlet & V.A. Matthews*
Bulbous perennials. Leaves basal, linear or lanceolate, usually hairless. Flowers in loose, terminal racemes. Perianth-segments free, often with a central band of a contrasting colour, the outer spreading, the inner erect, hooded at the tips. Outer whorl of stamens sometimes sterile. Ovary superior; style 1. Fruit a capsule; seeds black, flattened.

A genus of *c.* 30 species, from tropical and South Africa. The plants are not frost-hardy and need cool-house protection in the colder parts of Europe. Propagation is by seed.

*Bulb.* With fibres at the top: 9.
*Leaves.* Downy: 4. To 15 cm: 2,4,9; 15–60 cm: 3,5–10; 90–120 cm: 1.
*Perianth-segments.* Outer white: 1–3; green: 4,10; yellow: 5–9. All striped: 1,3,5–9; inner unstriped: 2,4; outer unstriped: 4,10; all unstriped: 4.
*Stamens.* All fertile: 1,2,5,6,10; outer sterile: 3,4,7–9.

1a. Basic colour of perianth-segments white                                      2
 b. Basic colour of perianth-segments yellow to green                       4
2a. Leaves 45–120 cm; perianth-segments 1.5–4 cm, all with a central green or red-brown band      3
 b. Leaves 7.5–15 cm; perianth-segments *c.* 1.2 cm, the inner ones lacking a central band              **2. humilis**
3a. Perianth-segments 3–4 cm; outer stamens fertile            **1. nelsonii**
 b. Perianth-segments 1.5–2.5 cm; outer stamens sterile         **3. altissima**
4a. Basic colour of outer perianth-segments green                       5
 b. Basic colour of outer perianth-segments yellow                       6
5a. Leaves 4 or 5, 30–45 cm; inner perianth-segments white with a green stripe; outer stamens fertile; style cylindric               **10. wakefieldii**
 b. Leaves 10–20, 10–15 cm; inner perianth-segments green with yellow tips; outer stamens sterile; style 3-angled                 **4. namaquensis**
6a. Leaves 6–9; outer stamens fertile      7
 b. Leaves 3–6; outer stamens sterile      8
7a. Perianth-segments 1.5–2 cm, with a central green band which remains green; flowers more or less pendulous, fragrant              **6. fragrans**
 b. Perianth-segments 2–3 cm, with a central green band which turns red-brown with age; flowers erect, not fragrant                **5. aurea**
8a. Flowering stem 15–50 cm; leaves to 20 cm, if more then bulb with a tuft of fibres at the top              9
 b. Flowering stem 40–100 cm; leaves 20–60 cm                      **8. major**
9a. Leaves lanceolate; central band of perianth-segments green
                                                 **7. canadensis**
 b. Leaves linear; central band of perianth-segments green, turning red-brown with age             **9. cooperi**

**1. A. nelsonii** N.E. Brown. Illustration: Botanical Magazine, 6649 (1882). Leaves 4–6, 90–120 cm, lanceolate. Flowering stem 60–150 cm, bearing a raceme *c.* 30 cm long. Perianth-segments 3–4 cm, white with a green or red-brown central band. All stamens fertile. Style 3-angled. *South Africa.* H5. Early summer.

**2. A. humilis** Baker.
Leaves 1–3, 7.5–15 cm, narrowly linear. Flowering stem to 10 cm, bearing 1–3 flowers. Perianth-segments *c.* 1.2 cm, white, outer segments with a green central band, inner segments unstriped and often tipped with yellow. All stamens fertile. Style 3-angled. *South Africa.* H5.

**3. A. altissima** Dryander. Illustration: le Roux & Schelpe, Namaqualand and Clanwilliam: South African wild flower guide **1**: 31 (1981).
Leaves 5 or 6, 45–60 cm, lanceolate, with recurved margins. Flowering stem 50–100 cm, bearing a raceme 30–45 cm long, with *c.* 50 flowers. Perianth-segments 1.5–2.5 cm, white, with a broad, green central band. Outer stamens sterile. Style 3-angled. *South Africa.* H5.

**4. A. namaquensis** Baker. Illustration: Flowering plants of Africa **32**: pl. 1255 (1957).
Leaves 10–20, 10–15 cm, almost cylindrical and very narrow, minutely downy. Flowering stem to 30 cm, bearing a raceme 10–20 cm long, with 6–10 flowers. Perianth-segments 1.5–2 cm, green, the inner with yellow tips. Outer stamens sterile. Style 3-angled. *South Africa.* H5.

**5. A. aurea** Jacquin.
Leaves 6–9, 45–60 cm, linear-lanceolate. Flowering stem 30–60 cm, bearing a raceme to 30 cm long. Flowers erect. Perianth-segments 2–3 cm, yellow, with a green central band which turns red-brown with age. All stamens fertile. Style 3-angled. *South Africa.* H5. Early spring.

**6. A. fragrans** Jacquin. Illustration: Kidd, Wild flowers of the Cape peninsula, pl. 74 (1973).
Simlar to *A. aurea* but flowers more or less pendulous, very fragrant with perianth-segments 1.5–2 cm, yellow, with a green central band which remains green. *South Africa.* H5.

**7. A. canadensis** (Linnaeus) Leighton (*A. minor* Linnaeus). Illustration: Levyns, Guide to the flora of the Cape peninsula, f. 25 (1966); Mason, Western Cape sandveld flowers, pl. 10 (1972); Kidd, Wild flowers of the Cape peninsula, pl. 53 (1973).
Leaves 3–6, *c.* 15 cm, lanceolate. Flowering stem 20–40 cm, bearing a raceme with up to 7 flowers. Perianth-segments 1.5–2 cm, yellow to greenish yellow with a broad, green central band. Outer stamens sterile. Style 3-angled. *South Africa.* H5. Late winter–spring.

**8. A. major** Linnaeus.
Similar to *A. canadensis* but flowering stem 40–100 cm, leaves 20–60 cm, raceme with 6–12 flowers, and perianth-segments with a green or brownish central band. *South Africa.* H5. Late winter–spring.
    Possibly not distinct from *A. canadensis.*

**9. A. cooperi** Baker. Illustration: Kidd, Wild flowers of the Cape peninsula, pl. 66 (1973); le Roux & Schelpe, Namaqualand and Clanwilliam: South African wild flower guide **1**: 31 (1981).
Bulb with a tuft of fibres at the top. Leaves 2–4, 15–30 cm, linear. Flowering stem 15–50 cm, bearing a raceme with 4–8 flowers. Perianth-segments 1.5–2.5 cm, yellow with a broad, green central band which turns red-brown with age. Outer stamens sterile. Style 3-angled. *South Africa.* H5. Spring.

**10. A. wakefieldii** Baker. Illustration: Botanical Magazine, 6429 (1879).
Leaves 4 or 5, 30–45 cm, linear. Flowering stem longer than the leaves, bearing a raceme 15–23 cm long, with 10–12 flowers. Perianth-segments *c.* 2.5 cm, outer segments green, inner segments white with a green central stripe. All stamens fertile. Style cylindric. *E Africa.* H5.

**74. URGINEA** Steinheil
*P.G. Barnes*
Perennial herbs. Bulbs with numerous free scales. Leaves narrow, deciduous. Flowers stalked, in long, erect racemes, the perianth falling after flowering, the stalks with small, often persistent bracts. Perianth-segments 6, free, spreading. Filaments thread-like, sometimes dilated at the base, inserted at the base of the perianth-segments. Style equalling or exceeding the stamens. Fruit a 3-angled capsule, seeds many, flattened and winged.

    A genus of about 100 species, mostly African, 3 occurring in the Mediterranean region.

**1. U. maritima** (Linnaeus) Baker.
Illustration: Botanical Magazine, 918 (1806); Rix & Phillips, The bulb book, 173 (1981).

Bulb 5–15 cm diameter. Leaves narrowly lanceolate, 30–100 × 3–10 cm, somewhat glaucous and fleshy. Scape 50–150 cm, flowers very numerous, appearing after the leaves, in a dense raceme up to 40 cm long. Flower-stalks up to 3 cm long, spreading, exceeding the normally quickly deciduous, subulate bracts. Perianth-segments oblong, 6–8 mm, white with green or purple mid-veins. *Mediterranean coasts and Portugal*. H4. Summer.

Easily cultivated in a well-drained soil, but needing a sunny position and not free-flowering in northern Europe.

## 75. PUSCHKINIA Adams
*E.M. Rix*
Perennial herbs with small bulbs, tunics brown. Leaves usually 2. Scapes 5–20 cm, leafless, with 4–10 flowers in a loose raceme. Flowers almost stalkless or the lowest with stalks to 10 mm, pale blue with darker stripes, rarely white or greenish. Perianth 7–10 mm, with a short tube at the base, lobes erect or slightly spreading, not widely open; throat with a 6-lobed corona, its lobes alternating and projecting between the anthers. Anthers almost stalkless, borne on the corona, dorsifixed. Ovary superior, style 1. Fruit an almost spherical capsule.

A genus with a single, variable species, grown for its early, ice-blue flowers. It does well in any well-drained soil and benefits from being dry in summer.

**1. P. scilloides** Adams (*P. libanotica* Zuccarini; *P. hyacinthoides* Baker).
Illustration: Botanical Magazine, 2244 (1821); Rix & Phillips, The bulb book, 42, 43 (1981).
*USSR (Caucasus), Turkey, N Iran, N Iraq & Lebanon*. H1. Spring.

## 76. HYACINTHUS Linnaeus
*P.F. Yeo*
Perennial herbs with bulbs. Leaves present in spring, 2 or more on each flowering bulb, with smooth or rough margins. Flowers in a raceme arising between the leaves. Bracts small, 2-lobed. Flower-stalks much shorter than perianth. Flowers bisexual. Perianth with a narrow tube for about half to two-thirds of its length, its lobes divergent, spreading or recurved. Stamens attached to perianth at or below middle of tube, not exceeding perianth-lobes. Ovary superior, 3-celled, with 2–several ovules in each cell; style much shorter than perianth-tube. Capsule nearly spherical. Seeds black, wrinkled.

Three species from W and C Asia, all cultivated. Only *H. orientalis*, is commonly grown and is one of the most important of the ornamental spring bulbous plants of Europe. The numerous cultivars are far more showy than the wild plant, and exhibit an extended range of colours. For bedding purposes or indoor display, new bulbs from the producers have to be used and are planted in autumn. For domestic decoration, cold-treated bulbs are supplied which flower earlier. For informal garden use, bulbs continue to give a display and to increase in number from year to year, but a rather rich soil is needed to sustain plants with well-formed racemes.
Literature: Bentzer, B. & von Bothmer, R., Cytology and morphology of the genus Hyacinthus L. s. str. (Liliaceae), *Botaniska Notiser* **127**: 297–301 (1974); Persson, K. & Wendelbo, P., The artificial hybrid Hyacinthus orientalis × transcaspicus (Liliaceae), *Botaniska Notiser* **132**: 207–9 (1979).

1a. Filaments shorter than the anthers
    **1. orientalis**
  b. Filaments longer than the anthers  2
2a. Anthers held outside throat of flower; perianth-lobes shorter than tube
    **2. transcaspicus**
  b. Anthers just reaching throat of flower; perianth-lobes longer than tube  **3. litwinowii**

**1. H. orientalis** Linnaeus.
Plant to about 30 cm. Leaves 4–6, 15–35 cm × 5–40 mm, erect at first, channelled, bright green. Scape thick, hollow, collapsing in fruit. Raceme loose to dense, with 2–40 very fragrant flowers. Perianth 2–3.5 cm, pale to deep violet-blue, pale to deep pink, white or yellowish; tube constricted above ovary, as long as or longer than the lobes; lobes spreading, recurved at tips. Anthers longer than filaments, their tips well below throat of flower. Capsule about 1–1.5 cm, conical-spherical, fleshy. Seeds with an appendage. *C & S Turkey, NW Syria, Lebanon*. H1. Early spring.

Subsp. **orientalis**. Illustration: Reichenbach, Icones Florae Germanicae et Helveticae **10**: t. 450, f. 1005 (1848); Huxley & Taylor, Flowers of Greece and the Aegean, t. 365 (1977); Everett, Encyclopedia of horticulture **5**: 1736 (1981). Leaves 4–5 mm (rarely to 11 mm) wide, linear. Flowers usually 2–12; perianth 2–3 cm, pale violet-blue at base, nearly white above; lobes half to four-fifths as long as tube. *SC Turkey, NW Syria, Lebanon; naturalised in S Europe*.

Plants naturalised in Europe may have broader leaves and up to about 18 flowers.
Susp. **chionophilus** Wendelbo.
Illustration: Rix & Phillips, The bulb book, 46 (1981); Mathew & Baytop, The bulbous plants of Turkey, t. 96 (1984). Leaves 1.2–1.5 cm (rarely to 3.6 cm) wide, linear-lanceolate, perianth-lobes about as long as tube, otherwise like subsp. *orientalis*. *C Turkey*.

The numerous cultivars of *H. orientalis* (Illustration: Botanical Magazine, 937 (1806); Synge, Collins guide to bulbs, t. 16 (1961); Rix & Phillips, The bulb book, 196 (1981) have broad leaves (about 2–4 cm) and 20–40 more crowded flowers in the raceme, and they show all the colours listed in the description of the species above. The perianth-lobes are usually shorter than the tube, as in subsp. *orientalis*.

**2. H. transcaspicus** Litwinow. (*H. kopetdaghi* Czerniakowska). Illustration: Botaniska Notiser **127**: 298 (1974); Wendelbo, Tulips and irises of Iran, 45 (1977).
Leaves 2 or 3, 7–16 cm × 4–15 mm, linear to lanceolate, channelled, green. Flowers 4–10, 1.3–1.5 cm, dark or medium violet-blue; tube cylindric; lobes shorter than tube, abruptly divergent or spreading. Filaments longer than anthers; anthers not at all enclosed by perianth, blue-black. Seeds without an appendage. *USSR (Turkmenistan: Kopet Dagh), NE Iran (E Elburz)*. H1. Early spring.

**3. H. litwinowii** Czerniakowska.
Illustration: Komarov, Flora SSSR 4: 306 (1935); Botaniska Notiser **127**: 298 (1974); Rix & Phillips, The bulb book, 46 (1981).
Plant 10–25 cm. Leaves 3 or 4, 15–17 × 1.5–5 cm, ovate-lanceolate to ovate, glaucous above, sometimes becoming spreading and wavy. Flowers 3–13, unscented. Perianth 1.8–2.5 cm, tube slender, constricted above ovary, dark greenish blue; lobes longer than tube, their basal parts continuing the tube upwards, their upper parts narrow, flared and recurved above the level of the anthers, very pale blue with deeper blue central band outside. Filaments longer than anthers. Seeds without an appendage. *USSR, E Iran (Kopet Dagh)*. H1. Early spring.

## 77. BELLEVALIA Lapeyrouse
*P.F. Yeo*
Perennial herbs with bulbs. Leaves 2–several on each flowering bulb, with margin cartilaginous and sometimes ciliate,

appearing in autumn or spring. Flowers in a raceme arising between the leaves. Bracts usually minute, 2-lobed. Flowers bisexual, radially or slightly bilaterally symmetric, nodding or horizontal. Perianth with a tube for one-quarter of its length or usually more, violet-blue to brownish, forming a 6-lobed bell which is not constricted at the mouth though (rarely) the lobes close the mouth. When withered the perianth is sometimes carried on the fruit as a cap. Stamens attached to perianth at bases of lobes, rarely lower, not projecting beyond them; filaments shorter or longer than anthers, flattened, broadly to narrowly triangular, shortly united with each other at base. Ovary 3-celled, with 2–6 ovules in each cell; style 1. Fruit a capsule, obtuse at apex, the cells in cross-section forming prominent acutely to acuminately angled lobes. Seeds rounded or occasionally pear-shaped, smooth, dull, black, but sometimes with a bluish bloom.

About 50 species in the Mediterranean and Black Sea regions, extending eastwards to NE Afghanistan and Tadjikistan (USSR). They are small- to medium-sized spring-flowering bulbs of varying appearance, some resembling *Muscari* or *Hyacinthella*; most of them are or have been cultivated for research purposes, but only a few have attractively coloured flowers. The limits of the genus have long been uncertain but the number and appearance of the chromosomes are decisive in the placement of borderline species. *Bellevalia* species may be grown in pots or planted in the open.

Literature: Feinbrun, N., A monographic study on the genus Bellevalia Lapeyr., 1: Caryological part, *Palestine Journal of Botany, Jerusalem Series*, **1**: 42–54 (1938), II: Taxonomical-geographical parts, *Ibid.* **1**: 131–42 (1938), 336–409 (1940); Freitag, H. & Wendelbo, P., The genus Bellevalia in Afghanistan, *Israel Journal of Botany* **19**: 220–4 (1970); Wendelbo, P., Notes on Hyacinthus and Bellevalia (Liliaceae) in Turkey and Iran, *Notes from the Royal Botanic Garden, Edinburgh* **38**: 423–34 (1980); Bothmer, R. von & Wendelbo, P., Cytological and morphological variation in Bellevalia, *Nordic Journal of Botany* **1**: 4–11 (1981).

*Flower-stalks*. Almost none at flowering time: 9:
*Raceme*. Broadly conical, loose, flower-stalks when with young fruits 7 cm or more: 1,2; cylindric or ovoid, loose, flower-stalks never more than 5 cm: 2,3,4, 5,6,8; narrowly conical, dense: 7,8,9.
*Leaf-margin*. Ciliate: 1,2,4,7,9; rough: 1,2,4,7; hairless and smooth: 1,3,6,8.
*Flower-colour*. White: 3; pale to dark blue, not changing: 7,8,9; bright violet-blue, becoming brownish: 6; white to dull violet in bud, becoming brownish or purplish green: 1,2,4,5.

1a. Flowering and fruiting parts of raceme loose with the flower-stalks as long as or longer than the flowers and easily visible                                           2
  b. Flowering and usually fruiting parts of raceme dense with the flowers crowded, concealing the flower-stalks which are about as long as flowers or shorter                                           7
2a. Flower-stalks more or less straight, diverging from main axis of raceme above the horizontal and remaining the same from late bud stage to fruit; perianth white or greenish white
                                           **3. romana**
  b. Flower-stalks with a different posture, or curved, or changing during the life of the flower; perianth at least partly pigmented from the time of opening onwards                                           3
3a. Flower-stalks greatly elongating during and after flowering, becoming 4–15 cm in fruit and making the raceme conical                                           4
  b. Flower-stalks elongating only slightly during flowering and not afterwards, not exceeding 2 cm (except in **4. B. warburgii**); raceme becoming cylindric                                           5
4a. Longest flower-stalks 7–14 cm in fruit; leaves not glaucous above; flowers 9–13 mm                                   **1. ciliata***
  b. Longest flower-stalks 6 cm or less in fruit; leaves glaucous above; flowers 7–10 mm                                   **2. saviczii**
5a. Flower-stalks in fruit 2.5–5 cm; flowers 1–1.3 cm                                   **4. warburgii**
  b. Flower-stalks in fruit not more than 1.5 cm; flowers 6–10 mm                                           6
6a. Flower-buds white or nearly so, changing to greenish brown when open; leaves usually prostrate and wavy, with rough or shortly ciliate margins                                   **5. flexuosa**
  b. Flower-buds bright violet-blue, changing to greenish brown when open; leaves channelled, with smooth margins                                   **6. dubia**
7a. Flowers bright or pale blue or deep purplish blue with widely spreading lobes and exposed anthers
                                   **9. hyacinthoides**
  b. Flowers blue-black; lobes not widely spreading; anthers not exposed          8
8a. Flowers erect in bud, nodding when open, not yellowish inside, bell-shaped
                                   **7. atroviolacea**
  b. Flowers nodding in bud and when open, yellowish inside and on edges of lobes, irregularly urceolate
                                   **8. pycnantha**

**1. B. ciliata** (Cyrillus) Nees (*Hyacinthus ciliatus* Cyrillus). Illustration: Edwards's Botanical Register **5**: t. 394 (1819); Coste, Flore de la France **3**: 325 (1905); Polunin & Huxley, Flowers of the Mediterranean, t. 242 (1965).

Plant 20–50 cm. Leaves usually 4–6, to 30 × 3 cm, lanceolate, glossy, densely ciliate on margins. Racemes very loose, in flower 10–12 cm, broadly conical, in fruit normally to 15 × 20 cm. Flower-stalks erect in bud stage, spreading and becoming recurved during flowering, ultimately becoming straighter and horizontal. Flowers 9–11 mm, narrowly bell-shaped, with tube brown or lilac and the lobes greenish or white. Capsule 1.5–2 cm × 8 mm, oblong with shortly tapered base and notched apex. *Aegean region, NW Africa*. H2. Spring.

**\*B. longipes** Post. Illustration: Rix & Phillips, The bulb book, 75 (1981). Differs from *B. ciliata* mainly in having 3 or 4 leaves, rough or smooth but hairless leaf-margins and even longer fruit-stalks, the raceme in fruit being up to 30 cm wide. Flower-stalks recurved, horizontal or ascending immediately after flowering. Flowers 9–13 mm. *N Syria, S Turkey, NW and W Iran, USSR (Transcaucasia)*. H1. Spring.

**2. B. saviczii** Woronow.
Plant to 40 cm. Leaves 3–6, 15–25 cm, glaucous above, with shortly ciliate or rough margins, the outermost 5–20 mm wide. Racemes loose, in flower cylindric-oblong to ovate, in fruit broadly conical, then to 15 × 12 cm. Flower-stalks widely spreading during flowering, recurved afterwards. Flowers 7–10 mm, white in bud, bell-shaped when open and then greyish brown to pale greyish green. Capsule 1–1.2 cm, obovate-oblong, slightly notched at apex. *E Iran, Afghanistan and adjacent USSR*. H1. Spring.

**3. B. romana** (Linnaeus) Reichenbach (*Hyacinthus romanus* Linnaeus). Illustration: Botanical Magazine, 939 (1806); Reichenbach, Icones Florae Germanicae et Helveticae **10**: t. 458, f. 1002, 1003

(1848); Grey-Wilson & Mathew, Bulbs, t. 19 (1981).

Plant to 30 cm. Leaves usually 3–5, emerging in autumn, reaching to top of racemes or beyond, channelled, with smooth margins, outermost to 1.5 cm wide. Racemes rather loose, in flower oblong or ovoid, to about 4 cm wide, in fruit cylindric, not widening. Flower-stalks 7–20 mm, diverging above the horizontal from late bud stage onwards, nearly straight. Flowers 8–10 mm, widely bell-shaped or broadly obconical, held in line with the stalk, white, becoming tinged with green, violet or brown, with conspicuous blue-black anthers. Capsule 1–1.5 cm, ellipsoid. *C & S Mediterranean area, SW France*. H2. Spring.

**4. B. warburgii** Feinbrun. Illustration: Palestine Journal of Botany, Jerusalem Series 1: t. 16 (1940).

Plant to 60 cm. Leaves usually 3–6, not longer than scape, with shortly ciliate or rough margins, outermost to 3.5 cm wide. Racemes loose, soon becoming cylindric and longer than scape, 7–11 cm wide. Flower-stalks usually 1–1.5 cm in flower and to 5 cm in fruit, ascending in bud, becoming recurved in flower and young fruit, finally nearly straight. Flowers 1–1.3 cm, narrowly bell-shaped, white with green veins in bud, tube becoming purplish brown when open. Capsule about 1 cm, ovate in profile. *Israel*. H3. Spring.

**5. B. flexuosa** Boissier. Illustration: Palestine Journal of Botany, Jerusalem Series 1: 371 & t. 17 (1940); Feinbrun & Zohary, Flora of the Land of Israel, Iconography, t. 12 (1940).

Plant to 50 cm. Leaves usually 4 or 5, usually longer than the scape, with shortly ciliate or rough margins, prostrate, wavy and up to 1.5 cm wide or, in shade, sometimes erect, not wavy and up to 3 cm wide. Racemes rather loose, soon becoming cylindric, about as long as scape, to 5 cm wide in fruit. Flower-stalks 8–20 mm, not lengthening after flowering, spreading or recurved in flower, ascending or horizontal in fruit. Flowers usually 7–8 mm, rarely to 10 mm, widely bell-shaped or broadly obconical, white in bud with green veins, sometimes tinged with lilac, tube becoming purplish brown when open. Capsule 8–10 mm, circular or ovate in profile. *Syria, Israel, Egypt*. H3. Early spring.

**6. B. dubia** (Gussone) Reichenbach (*Hyacinthus dubius* Gussone). Illustration: Huxley & Taylor, Flowers of Greece and the Aegean, t. 369 (1977); Grey-Wilson & Mathew, Bulbs, t. 19 (1981); Rix & Phillips, The bulb book, 75 (1981).

Plant to 40 cm. Leaves 2–5, longer than the scape, channelled, prostrate, with smooth margins, outermost to 1.3 cm wide. Racemes rather loose, narrowly cylindric, about 3 cm wide, not widening in fruit. Flower-stalks erect in young bud, in flower as long as perianth, soon spreading and then lightly recurved, more recurved after flowering but approximately horizontal in fruit and then about equalling the fruit in length. Flowers 6–8 mm, ellipsoid to weakly bell-shaped, bright violet-blue with green tips in bud, when open with tube gradually turning brown and lobes yellowish. Capsule 6–10 mm, obovate in profile. *Sicily to Balkan Peninsula and W Turkey*. H2. Spring.

*B. dubia* superficially resembles some species of *Muscari* allied to *M. comosum* (p. 225), but is easily distinguished by the unconstricted mouth of the perianth. In addition, the palest part of flowers which have turned brown is the apex, whereas in *Muscari* it is the base.

**7. B. atroviolacea** Regel. Illustration: Rix & Phillips, The bulb book, 47 (1981).

Plant 8–30 cm. Leaves 3–6, as long as or longer than scape, slightly channelled and slightly wavy, glaucous, with margins shortly ciliate or rough, the outermost to 2 cm wide. Racemes 3–6 cm in flower, densely conical, longer and looser in fruit. Flower-stalks at flowering time much shorter than perianth, mostly concealed, becoming 1–2 cm long in fruit. Flowers 8–9 mm, erect in bud, overlapping and sharply nodding when open, bell-shaped, uniformly blackish violet in all stages. *USSR (Turkestan & Tadjikistan), NE Afghanistan*. H1. Spring.

**8. B. pycnantha** (C. Koch) Losina-Losinskaya (*Muscari paradoxum* misapplied; *M. pycnantha* C. Koch). Illustration: Botanical Magazine, 6763 (1903); RHS Lily Year Book **30**: 113 (1967); Wendelbo, Tulips and irises of Iran, 47 (1977); Rix & Phillips, The bulb book, 72 (1981); Mathew & Baytop, The bulbous plants of Turkey, t. 61 (1984).

Plant to 40 cm. Leaves usually 3, much longer than scape, channelled, deep green, glaucous on upper surface, with smooth margins, outermost to about 2.5 cm wide. Racemes densely conical at first, with numerous overlapping flowers, later becoming ellipsoid or cylindric and loose. Flower-stalks about as long as flowers, concealed by the flowers in young racemes, to 1.2 cm in fruit. Flowers about 6 mm, nodding from late bud stage onwards, irregularly urceolate, angled in cross-section, with somewhat folded lobes, blue-black in all stages, greenish yellow on edges and inner surfaces of lobes. Capsule 8 mm, circular in profile with notched apex *USSR (Transcaucasia), E Turkey, N & W Iran, N Iraq*. H1. Spring.

**9. B. hyacinthoides** (Bertoloni) Persson & Wendelbo (*Hyacinthus spicatus* Sibthorp & Smith not Moench; *Strangweja spicata* (Boissier) Boissier). Illustration: Botaniska Notiser **132**: 66 (1979); Grey-Wilson & Mathew, Bulbs, t. 19 (1981); Rix & Phillips, The bulb book, 63 (1981).

Plant to 15 cm. Leaves 4–8, emerging in autumn, much longer than the raceme, prostrate, slightly channelled, glaucous above, with shortly ciliate margins, to 5.5 mm wide. Racemes dense, ellipsoid to cylindric, with few (less than 20) crowded flowers. Flower-stalks not developed. Flowers 7–12 mm, horizontal or turned upwards, bell-shaped, with tube one-third to half as long as lobes, very pale blue with deeper blue veins to deep purplish blue. Filaments very short, with a prominent tooth on either side at base; anthers exposed. Capsule about 7 mm, wider than long in profile, notched. Seeds pear-shaped. *Greece*. H2. Spring.

## 78. HYACINTHELLA Schur
*P.F. Yeo*

Small perennial herbs with bulbs, leafless in summer and winter. Bulb-scales coated with powdery crystals. Leaves 2 or sometimes 3 on each (undivided) flowering bulb, the second (or third if there are 3) narrower than the first. Flowers in a raceme on a stalk arising between the leaves and overtopping them. Bracts minute. Flower-stalks not longer than flowers. Flowers not more than 1.2 cm, bisexual, radially symmetric or nearly so, more or less bell-shaped, blue to whitish or pinkish. Perianth-tube longer than the lobes, persistent in fruit. Stamens (in our species) attached just below the perianth-lobes and with filaments shorter than to slightly longer than the anthers. Ovary superior, 3-celled with 2–4 ovules in each cell. Style simple. Fruit a rounded capsule to 5 mm wide with the cells rounded on the back. Seeds black; seed-coat wrinkled.

Sixteen species of E and SE Europe and W Asia, which have mostly been included in *Hyacinthus* or *Bellevalia* in the past. They

have all been brought into cultivation for study at Göteborg, Sweden. They are spring-flowering bulbs, usually grown in pots under glass or in frames on account of their small size and inconspicuous flowers; their ornamental value is slight.
Literature: Feinbrun, N., Revision of the Genus Hyacinthella Schur, *Bulletin of the Research Council of Israel, Section D: Botany* **10D**: 324–47 (1961); Persson, K. & Wendelbo, P., Taxonomy and cytology of the genus Hyacinthella (Liliaceae-Scilloideae) with special reference to the species in SW Asia, Part 1, *Candollea* **36**: 513–41 (1981), Part 11, *Candollea* **37**: 157–75 (1982).

1a. Flowers 7–9 mm; flower-stalks less than 1 mm      **4. nervosa**
  b. Flowers 4–6 mm; flower-stalks 1 mm or more                              2
2a. Leaves with distinct short or usually long hairs on margins and often also with hairs on surface      **2. lineata\***
  b. Leaves without distinct hairs but often rough on margins                3
3a. Leaf-margins not rough      **1. glabrescens**
  b. Leaf-margins rough                                                      4
4a. Leaves to 1.2 cm wide, erect, nearly flat, with a hooded tip      **3. leucophaea**
  b. Leaves to 6 mm wide, arching outwards, gutter-shaped      **5. pallens**

**1. H. glabrescens** (Boissier) Persson & Wendelbo (*H. lineata* (Steudel) Chouard var. *glabrescens* (Boissier) Feinbrun). Illustration: Bulletin of the Research Council of Israel, Botany, **10D**: t. 3, f. 9 (1961); Candollea **36**: 521 (1981).
Plant to *c.* 15 cm. Leaves hairless, with smooth margins, spreading. First leaf 8–20 mm wide, narrowly oblong-lanceolate to narrowly oblong-ovate to elliptic; second leaf half as wide as first or more. Inflorescence-stalk scarcely elongating in fruit. Flower-stalks 2–7 mm, scarcely elongating in fruit. Flowers 15–30, 4.5–6 mm, cylindric or slightly bell-shaped, medium to deep blue-violet, with lobes half as long as tube. *S Central Turkey*. H2. Spring.

**2. H. lineata** (Steudel) Chouard. Illustration: Bulletin of the Research Council of Israel, Botany, **10D**: 341 and t. 2, f. 5 & 6 (1961); Candollea **36**: 527 (1981).
Plant to 14 cm in flower. leaves with usually long hairs on margins and sometimes some on the surface near base, spreading, often twisted and often purplish at base. First leaf usually 5–15 mm wide,

narrowly oblong-lanceolate to elliptic-ovate; second leaf more than half as wide as first. Inflorescence-stalk lengthening to 20–29 cm in fruit. Flower-stalks 2–6 mm, increasing to 4–12 mm in fruit. Flowers 6–25, usually 4.5–5.5 mm, weakly bell-shaped, deep blue to violet-blue, with lobes half as long as tube. *W Turkey*. H2. Spring.

**\*H. hispida** (Gay) Chouard. Illustration: Grey, Hardy bulbs 3: 311 (1938); Candollea 36: 519, f. 1A–C (1981). Leaves copiously stiff-hairy on margins and surfaces, erect or spreading, the first to 1.2 cm wide, the second less than half as wide as first. Flowers 5–6 mm in a very loose raceme, dark blue-violet, cylindric or weakly bell-shaped. *SC Turkey*. H2. Spring.

**3. H. leucophaea** (C. Koch) Schur. Illustration: Polunin, Flowers of Greece and the Balkans, t. 59, f. 1643d (1980); Grey-Wilson & Mathew, Bulbs, t. 18 (1981) – as H. dalmatica; Rix & Phillips, The bulb book, 46 (1981).
Plant to 15 cm in flower. Leaves hairless but with rough margins, erect, nearly flat. First leaf to *c.* 1.2 cm wide, linear-lanceolate; second leaf more than half as wide as the first. Inflorescence-stalk lengthening to 25 cm in fruit. Flower-stalks 1–3 mm. Flowers up to 25, 4–5 mm, broadly cylindric or bell-shaped, pale or very pale blue, often green when old, with lobes one-third as long as tube. *E Europe*. H2. Spring.
  The figure cited above from Rix & Phillips is stated in the text to be *H. atchleyi* (Jackson & Turrill) Feinbrun, but it does not show the dark blue flowers of that species.

**4. H. nervosa** (Bertoloni) Chouard. Illustration: Grey, Hardy bulbs 3: 313 (1938); Candollea 36: 535 (1981).
Plant to *c.* 17 cm in flower. Leaves hairless but with rough margins, erect. First leaf to 2 cm wide; second leaf half as wide as the first. Flower-stalk less than 1 mm. Flowers 10–25, 7–9 mm, urn-shaped or narrowly bell-shaped, pale blue or violet-blue, with lobes about half as long as tube. *SC Turkey to Jordan and eastwards to NE Syria & Iraq*. H3. Spring.

**5. H. pallens** Schur (*H. dalmatica* (Baker) Chouard). Illustration: Bulletin of the Research Council of Israel, Botany **10D**: 341 f. 1B & t. 1 f. 3 (1961); Polunin, Flowers of Greece and the Balkans, t. 61, f. 1643e (1980); Grey-Wilson & Mathew, Bulbs, t. 18 (1981) - as H. leucophaea; Rix & Phillips, The bulb book, 135 (1981).
Plant to 10 cm. Leaves hairless but with

rough margins, arching and spreading outwards, channelled. First leaf to 6 mm wide; second leaf half as wide as first. Flower-stalks *c.* 2 mm. Flowers usually 6–20, 4–5 mm, rather narrowly bell-shaped, medium blue, with lobes *c.* half to two-thirds as long as tube. *Yugoslavia*. G2. Early spring.

## 79. BRIMEURA Salisbury
*P.F. Yeo*
Small hairless perennial herbs with bulbs, leafless in summer and winter (?). Leaves several on each flowering bulb. Flowers in a raceme arising between the leaves. Bracts about as long as flower-stalks or longer. Flowers bisexual, radially symmetric. Perianth with a long tube and 6 shorter lobes. Stamens concealed or exposed. Ovary superior, 3-celled with 2–4 ovules in each cell. Style simple. Fruit a capsule, 5 mm long (*B. amethystina*). Seeds black, shining, finely wrinkled (*B. amethystina*).

  Two highly dissimilar species, each with an interrupted distribution in S Europe, both cultivated. *B. fastigiata* (Viviani) Chouard, of slight horticultural value, is deceptively like a species of *Scilla*; *B. amethystina* is a delightful late spring dwarf bulb suitable for the rock garden, open border and positions under shrubs.

**1. B. amethystina** (Linnaeus) Chouard (*Hyacinthus amethystinus* Linnaeus). Illustration: Edwards's Botanical Register **5**: 398 (1819); Polunin & Smythies, Flowers of Southwest Europe, t. 59, f. 1643 (1973); Grey-Wilson & Mathew, Bulbs, t. 19 (1981).
Stem 10–25 cm. Leaves becoming slightly longer than stem, to 6 mm wide, channelled, bright green. Flower-stalks half to one and a quarter times as long as flower, longer in fruit. Flowers 5–15, *c.* 10 mm, horizontal or nodding and turned to one side, narrowly bell-shaped, bright blue or sometimes violet-blue or pure white. Perianth-tube cylindric or slightly constricted below throat; perianth-lobes one-third to half as long as tube. Stamens attached about half-way up the tube, small, not reaching the throat. *Pyrenees and NE Spain, NW Yugoslavia*. H3. Late spring–early summer.

## 80. MUSCARI Miller
*D.C. Stuart*
Herbaceous perennials with bulbs, with or without offsets. Leaves 1–8, basal. Scape present, flowers in racemes with minute bracts. Perianth united for most of its

length into a cylindric, tubular, bell-shaped or urceolate tube, the 6 lobes small, blue, brownish, yellow, whitish or white, not changing as the flower dies, often of a different colour from the tube. Stamens 6, anthers not projecting from perianth-tube. The raceme often has sterile flowers near the apex, which may differ in colour or tone from the fertile, sometimes forming a tuft (comus). Ovary superior, 3-celled, with 2 ovules in each cell. Perianth carried up on the developing capsule and falling from there. Capsule strongly angled, with 2 seeds per cell. Seeds black, often shiny, often minutely wrinkled.

A genus of about 30 species from the Mediterranean area and SW Asia; only 12 of the species are in general cultivation. The genus is variable, and is sometimes split into 4 small genera: *Muscari* in the strict sense, *Leopoldia* Parlatore, *Botryanthus* Kunth and *Pseudomuscari* Garbari & Greuter. However, as each group contains at least 1 species which seems intermediate to those in the other groups, it seems sensible to retain the widest possible circumscription of *Muscari* here. Chromosome morphology and number, floral, capsule and seed characters clearly separate *Muscari* in this wide sense from *Bellevalia* (p. 221) and *Hyacinthella* (p. 223). The commonly cultivated *M. comosum* and *M. armeniacum* both have wide geographical ranges, are very variable (with many synonyms, most of which are not cited here) and have given rise to monstrous forms which are frequently cultivated.

Literature: Schschian, A., Sistematika i geographiy kavkazskich vidov roda Muscari Miller, *Trudy Tbilisskii Botanicheskogo Instituta* 5 (10): 203–5 (1946); Stuart, D.C., Muscari and allied genera, *RHS Lily Year Book* 29: 125–38 (1966); Speta, F., Uber die Abgrenzung und Gliederung der Gattung Muscari, und uber ihre Beziehungen zu anderen Vertreten der Hyacinthaceae, *Botanische Jahrbucher* 103: 247–291 (1982).

*Leaves*. Single: 11; two, broadened above: 7,8.
*Fertile flowers*. White: 6–8; yellow: 2; greenish: 1,3; buff or light brown: 4; blackish blue: 9–11; mid to light blue, sometimes purplish: 5–8.
*Inflorescence*. Branched or monstrous: cultivars of 4 & 6.
*Perianth-lobes of fertile flowers*. Of the same colour as the tube: 5,10,11; of a different colour from the tube: 1–4,6–9.

1a. Raceme simple, most flowers fertile,
   sterile flowers at the apex                          2
 b. Inflorescence branched, most flowers
   reduced to coloured threads                         12
2a. Perianth bell-shaped, not constricted
   towards the apex, blue or rarely white
                                          **5. azureum**
 b. Perianth constricted towards the
   apex, blue, white, purplish, yellow,
   buff, brown or greenish                              3
3a. Mature fertile flowers blue, rarely
   purplish or white                                    4
 b. Mature fertile flowers ivory, yellow,
   greenish, buff or brown                              9
4a. Perianth-lobes of the same colour as
   the tube                                             5
 b. Perianth-lobes white or pale blue, the
   tube a different colour                              6
5a. Leaf single, 1–3 cm broad, linear-
   oblanceolate                      **11. latifolium**
 b. Leaves 2 or more, 5–15 mm broad,
   linear                           **10. commutatum**
6a. Leaves spathulate (sometimes
   narrowly so)                                         7
 b. Leaves linear to linear-oblanceolate
                                                        8
7a. Raceme becoming loose in fruit;
   leaves not sickle-shaped, often ribbed
                                       **7. botryoides**
 b. Raceme remaining dense in fruit;
   leaves sickle-shaped, not ribbed
                                          **8. aucheri**
8a. Perianth of fertile flowers blackish
   blue                                **9. neglectum**
 b. Perianth of fertile flowers sky-blue,
   rarely purplish or white
                                       **6. armeniacum**
9a. Perianth yellow             **2. macrocarpum**
 b. Perianth buff, greenish or ivory        10
10a. Perianth buff, lobes cream or
   yellowish                            **4. comosum**
 b. Perianth greenish to ivory, lobes
   blackish purple or brown                            11
11a. Flowers narrowly urceolate, with
   prominent shoulders towards the
   apex, forming a brownish corona;
   sterile flowers absent or minute and
   violet                               **1. muscarimi**
 b. Flowers oblong-urceolate, without a
   corona; sterile flowers conspicuous,
   pinkish violet                    **3. massayanum**
12a. Inflorescence blue
                      **6. armeniacum** 'Blue Spike'
 b. Inflorescence violet
                      **4. comosum** 'Plumosum'

**1. M. muscarimi** Medikus (*M. moschatum* Willdenow). Illustration: Botanical Magazine, 734 (1804); Redouté, Liliacées 3: t. 132 (1805).
Bulbs 2–4 cm across, roots thick. Leaves 3–6, 10–20 cm × 5–15 mm, pale greyish green. Scape erect to prostrate, shorter than the leaves. Fertile flowers 8–14 mm, narrowly urceolate, at first purple, becoming pale greenish to ivory, strongly contracted towards the apex, with expanded shoulders which form a distinct, brownish corona. The flowers smell strongly of musk. Sterile flowers rarely present, if so, small and purplish. *SW Turkey*. H5. Spring.

**2. M. macrocarpum** Sweet (*M. luteum* Todaro). Illustration: Botanical Magazine, 1565 (1813); Griffith, Collins guide to alpines, t. 32 (1964); RHS Lily Year Book 29: f. 29 (1966).
Bulbs as in *M. muscarimi*. Leaves to 30 cm. Fertile flowers 8–12 mm long, oblong-urceolate, at first purplish, later bright yellow, shoulders expanded to form a deep brown or rarely yellow, corona, smelling of bananas. Sterile flowers mostly absent, rarely few (when purplish). *Greek islands & W Turkey*. H5. Early spring.

**3. M. massayanum** Grunert. Illustration: The Garden 107: 335 (1982).
Bulbs 2–6 cm across, without offsets, tunics ivory. Leaves 2–4, linear, sickle-shaped or sinuous, to 25 × 1–2.5 cm, thick, glaucous, apex hooded. Scape stout, to 22 cm, mauvish. Raceme cylindric, dense. Fertile flowers at first pink to violet, becoming light greenish or yellowish brown, narrowly urceolate, to 1.1 cm long. Perianth-lobes dark brown to blackish. Stalks of the sterile flowers ascending to spreading, forming a dense, fleshy, pink or violet tuft. *E Turkey*. H5. Spring.

**4. M. comosum** (Linnaeus) Miller (*M. pinardii* Boissier). Illustration: Polunin, Flowers of Europe, t. 169 (1969); Rix & Phillips, The bulb book, 75 (1981).
Bulbs to 3.5 cm across, without offsets, tunics pink. Leaves 3–7, erect-spreading, generally shorter than the scape. Fertile flowers oblong-urceolate, pale brown, the shoulders rounded and darker brown, lobes cream or yellowish. Stalks of sterile flowers fleshy, ascending, bright violet, to 4 mm. Sterile flowers almost spherical or obovoid, rarely tubular, bright violet, many, forming a conspicuous tuft. *S & C Europe, N Africa, SW Asia*. H5. Spring.

In the wild the tuft of sterile flowers is very variable in development. In horticultural variants it is the main reason for growing the plant. In *M. comosum* 'Plumosum' (illustration: Puttock, Bulbs and corms, 96, 1958) the branched inflorescence consists only of sterile flowers and their stalks, and is extremely showy.

**5. M. azureum** Fenzl (*Bellevalia azurea* (Fenzl) Boissier; *Hyacinthella azurea* (Fenzl) Chouard). Illustration: Rix & Phillips, The bulb book, 70 (1981); Mathew, Dwarf bulbs, 73 (1973) – M. azureum 'Album'. Leaves 2 or 3, erect or spreading, narrowly oblanceolate, 6–20 cm × 3–15 mm, often rather hooded at the apex, the upper surface pale and glaucous. Raceme dense, ovoid, 1.5–3 cm. Fertile flowers bell-shaped, not constricted, 4–5 mm, pale to bright blue, each lobe with a darker stripe, or rarely all white. Lobes of the same colour as the tube. Anthers blue. Sterile flowers few, smaller and paler than the fertile. *E Turkey*. H5. Spring.

An attractive species, widely grown. Pale blue variants are very similar to *M. coeleste* Fomin, a species from the Caucasus and E Turkey which is not yet in cultivation.

**6. M. armeniacum** Baker (*M. szovitsianum* Baker; *M. cyaneo-violaceum* Turrill). Illustration: Botanical Magazine, 6855 (1886); Rix & Phillips, The bulb book, 68, 71 (1981). Bulbs often with bulbils. Leaves 3–7 (usually 3–5), linear to linear-oblanceolate, upper surface sometimes pale. Inflorescence dense, flowers overlapping or contiguous. Fertile flowers obovoid to oblong-urceolate, bright sky-blue, sometimes with a weak or strong purplish cast, rarely all white, 3.5–5.5 × 2.3–3.5 mm. Lobes always paler than the tube or white. Sterile flowers few, smaller and paler than the fertile, rarely of the same colour. *SE Europe to the Caucasus*. H3. Spring.

An attractive, variable and prolific species, found in quantity both in gardens and in the wild. A rare, purple variant from Varna (Bulgaria) was at one time given specific rank as *M. cyaneo-violaceum*. The monstrous variant, now known as 'Blue Spike' is an ancient garden plant.

**7. M. botryoides** (Linnaeus) Baker. Illustration: The Garden 26: 136 (1884); RHS Lily Year Book 29: f. 27 (1966); Polunin, Flowers of Europe, t. 170 (1969); Rix & Phillips, The bulb book, 68 (1981). Leaves 2–4, erect, spathulate, rarely narrowly so, 5–25 cm × 5–13 mm, apex abruptly contracted, hooded or acuminate, upper surface often prominently ribbed. Raceme dense at first, generally becoming loose and cylindric. Fertile flowers almost spherical, 2.5–5 mm long, sky-blue, rarely white; lobes white. *C & SE Europe*. H2. Spring–early summer.

**8. M. aucheri** (Boissier) Baker (*Botryanthus aucheri* Boissier; *M. tubergenianum* Turrill). Illustration: Gartenflora 77: 56 (1928); Die Gartenstauden 1: 165 (1929). Leaves 2 (rarely 3 or 4), erect-spreading, sickle-shaped, often symmetric, spathulate (sometimes narrowly so), 5–20 cm × 2–15 mm, upper surface pale green, glaucous, apex hooded. Raceme dense, ovoid or cylindric. Fertile flowers almost spherical or obovoid, 3–5 × 2.3–5 mm, bright sky-blue, rarely white, lobes paler blue or white. Sterile flowers larger or smaller than the fertile, paler, as many as them or fewer. *Turkey*. H3. Spring.

*M. aucheri* is most often cultivated in the attractive variant formerly known as *M. tubergenianum*, which has a prominent bunch of sterile flowers (which are paler than the fertile) in the upper part of the raceme.

**9. M. neglectum** Gussone (*M. racemosum* Lamarck & de Candolle; *M. atlanticum* Boissier & Reuter). Illustration: The Garden 26: 136 (1884); Polunin, Flowers of Europe, t. 170 (1969); Rix & Phillips, The bulb book, 71 (1981). Bulbs 1–2.5 cm across, sometimes with offsets. Leaves 3–6, erect-spreading to prostrate, channelled to almost terete, bright green, sometimes reddish beneath, 6–40 cm × 2–8 mm. Raceme dense, becoming loose in fruit. Fertile flowers ovoid to oblong-urceolate, strongly contracted towards the apex, 1.5–3.5 × 3.5–7.5 mm, very dark to blackish blue, lobes white, recurved. Sterile flowers smaller and paler, rarely white. *Europe, N Africa, SW Asia*. H4. Early spring.

Quite an attractive garden plant, of variable size, the perianth-lobes making a strong contrast with the tube. Most cultivated variants increase readily by means of bulblets.

**10. M. commutatum** Gussone (*Botryanthus commutatus* (Gussone) Kunth). Illustration: Sweet, British Flower Garden, ser. 2, 4: t. 369 (1837); Polunin, Flowers of Europe, t. 169 (1969); Rix & Phillips, The bulb book, 71 (1981). Leaves 2–5, erect-spreading, linear to linear-lanceolate. raceme dense, ovoid, flowers often overlapping. Fertile flowers obovate-urceolate, strongly contracted and with prominent shoulders, 3.5–7 mm long, deep violet-black, the lobes of the same colour, recurved. *Italy, Yugoslavia, Greece*. H5. Early spring.

**11. M. latifolium** Kirk. Illustration: Botanical Magazine, 7843 (1902); Mathew, Dwarf bulbs, 168 (1973). Leaf single (rarely 2 together), erect, broadly oblanceolate, 7–30 × 1–3 cm, apex acuminate and often hooded. Raceme dense at first, becoming loose. Fertile flowers oblong-urceolate, strongly contracted towards the apex and with well-developed shoulders, 5-6 mm long, deep violet-black, lobes of the same colour, recurved. *S & W Turkey*. H5. Spring.

A distinctive and handsome species.

**81. LACHENALIA** Murray

*E. Campbell*

Herbaceous perennials with bulbs. Leaves basal, usually 2 but occasionally up to 10, linear, usually with clasping bases, sometimes spotted or pustulate. Flowers in a raceme or spike, borne on a scape. Perianth united into a tube at the base, bell-shaped, persistent, the 3 outer lobes usually with apical swellings and usually shorter than the 3 inner lobes, variously coloured. Stamens 6, attached to the top of the perianth-tube; anthers versatile, opening towards the centre of the flower. Ovary superior, 3-celled, with many ovules. Style simple, stigma capitate. Fruit a 3-celled capsule opening between the septa.

A genus of about 50 species endemic to South Africa. Identification of the cultivated species is difficult as many of the plants grown are selections or hybrids. Fifteen species are included here; they are easily grown as pot-plants.

Literature: Ingram, J., Notes on the cultivated Liliaceae 4: Lachenalia, *Baileya* 14: 123–36 (1966).

*Leaves*. More than two: 1,2; one or two: 3–15. Spotted or patched with different colours: 2,7–9,11,12; not spotted or patched: 1,3–6,8–10,13–15. Pustulate: 3–6; not pustulate: 1,2,7–15. With thickened margins: 5,8–10; without thickened margins: 1–4,6,7,11–15.

*Scape*. Spotted or patched with different colours: 1,3,5,7–9,11,12,15; not spotted or patched: 2–4,6,8,10,13,14.

*Flowering time*. January–May: 1–11,14,15; August–September: 12,13.

*Flowers*. White or yellowish: 1,2,4–6,8,13,14; blue or bluish: 3, 7–10; red and/or yellow: 6,11,12,15. Fragrant: 3,5,6,8,9; not fragrant: 1,2,4,7,10–15.

*Outer perianth-lobes*. More than three-quarters of the length of the inner lobes: 1–5,7–9,12,14,15; up to three-quarters of the length of the inner lobes: 1,6,10,11,13. With apical swellings: 1,4–8,11,12,14,15; without apical swellings: 2,3,9,10,13.

*Stamens.* Projecting from the perianth: 6,13; not projecting: 1–5,7–12,14,15.

*Style.* Projecting from the perianth: 2,3,6,8,12,13,15; not projecting: 1,4,5,7,9–11,14.

1a. Leaves more than 2                                  2
 b. Leaves 1 or 2                                        3
2a. Scape shorter than to as long as leaves; leaves 6–10; outer perianth-lobes with apical swellings
                                     **1. contaminata**
 b. Scape longer than leaves; leaves 4 or 5; outer perianth-lobes without apical swellings          **2. orthopetala**
3a. Leaves pustulate                                    4
 b. Leaves not pustulate                                7
4a. Flowers purplish blue
                               **3. purpureo-caerulea**
 b. Flowers white or yellowish                          5
5a. Leaves narrow, less than 1.5 cm broad          **4. liliiflora**
 b. Leaves wider, more than 2 cm broad
                                                        6
6a. Leaves with thickened margins; neither stamens nor style projecting from the perianth          **5. pallida**
 b. Leaves without thickened margins; stamens and style projecting
                                        **6. pustulata**
7a. Leaf 1, less than 1 cm wide, with distinct, transverse, dark purple bands on the clasping base   **7. unifolia**
 b. Leaves usually 2, more than 1 cm wide, without distinct, transverse bands on the clasping base    8
8a. Leaves with thickened margin                       9
 b. Leaves without thickened margin     11
9a. Outer perianth-lobes with apical swellings; style projecting from the perianth          **8. glaucina**
 b. Outer perianth-lobes without apical swellings; style not projecting     10
10a. Flowers closely packed, the uppermost fertile          **9. orchioides**
 b. Flowers loosely packed, the uppermost sterile and abortive     **10. mutabilis**
11a. Leaves with purple patches on the upper and/or lower surface     12
 b. Leaves without purple patches on either surface     13
12a. Outer perianth-lobes much shorter than the inner; flowering in winter and spring          **11. aloides**
 b. Outer perianth-lobes almost as long as the inner; flowering in autumn
                                         **12. rubida**

13a. Flowers less than 1.5 cm; stamens longer than perianth     **13. unicolor**
 b. Flowers more than 2 cm; stamens shorter than to as long as the perianth          14
14a. Leaves becoming reflexed, with margins finely wavy; outer perianth-lobes with red apical swellings
                                         **14. reflexa**
 b. Leaves not becoming reflexed, margin not wavy; outer perianth-lobes with green apical swellings     **15. bulbifera**

**1. L. contaminata** Aiton (*L. angustifolia* Jacquin). Illustration: Botanical Magazine, 1401 (1811).
Leaves 6–10, to 20 cm × 3 mm, channelled, nearly erect, without spots. Scapes shorter than or just equalling leaves, spotted. Flower-stalks to 2 mm. Flowers many, to 5 mm, broadly bell-shaped. Perianth-lobes all spreading, white, the inner tipped with greenish brown, the outer often flushed with red, bearing reddish apical swellings, and about three-quarters of the length of the inner. Stamens as long as the inner perianth-lobes. Style not projecting. *South Africa (Cape Province, Natal).* H5–G1. Winter–spring.

**2. L. orthopetala** Jacquin. Illustration: Baileya 14: f. 43 (1966).
Leaves 4–5, 10–15 cm × *c.* 6 mm, channelled, spotted. Scape to 25 cm, exceeding the leaves, flexuous, without spots. Flowers numerous, to 1.3 cm, cylindric, in a spike-like raceme; flower-stalks *c.* 2 mm. Perianth white or tinged with red, the outer lobes without apical swellings and almost as long as the inner. Stamens as long as perianth. Style projecting. *South Africa.* H5–G1. Winter–spring.

**3. L. purpureo-caerulea** Jacquin. Illustration: Botanical Magazine, 745 (1804).
Leaves 2, 15–20 × *c.* 1 cm, pustulate, without spots. Scape to 23 cm, exceeding the leaves, patched with purple. Flowers 30–40 in the raceme, bell-shaped, fragrant, to 1 cm. Perianth-lobes all spreading, purplish blue, the outer without apical swellings and about four-fifths of the length of the inner. Stamens as long as perianth. Style projecting. *South Africa.* H5–G1. Spring.

**4. L. liliiflora** Jacquin.
Leaves 2, to 23 × 1 cm, pustulate, without spots. Scape to 30 cm, exceeding the leaves, without spots. Flowers 12–20 in a spike-like raceme, to 2 cm, cylindric; flower-

stalks *c.* 4 mm. Perianth-lobes all spreading, white, the outer with green apical swellings, more than three-quarters of the length of the inner. Stamens shorter than the perianth. Style not projecting. *South Africa.* H5–G1. Winter–spring.

**5. L. pallida** Aiton (*L. lucida* Sims; *L. racemosa* Sims). Illustration: Saunders Refugium Botanicum, t. 170 (1870).
Leaves 2, 15–23 × 2–3 cm, pustulate but without purple spots and with thickened margins. Scape to 25 cm, exceeding the leaves, minutely spotted. Flowers to 1.8 cm, 12–30 in a dense raceme, bell-shaped, fragrant; stalks very short. Perianth white, the outer lobes with green apical swellings and about four-fifths of the length of the inner. Stamens shorter than the perianth. Style not projecting. *South Africa.* H5–G1. Spring.

**6. L. pustulata** Jacquin. Illustration: Botanical Magazine, 817 (1805).
Leaves 2, to 28 × 2.5 cm, pustulate, without purple spots, margin not thickened. Scape to 26 cm, exceeding the leaves, without spots. Flowers many in a dense spike-like raceme, *c.* 8 mm, bell-shaped, fragrant; stalks 2–5 mm. Perianth white or yellowish with spreading inner lobes, the outer lobes with pale green apical swellings and three-quarters of the length of the inner. Stamens and style projecting. *South Africa.* H5–G1. Winter–spring.

**7. L. unifolia** Jacquin. Illustration: Botanical Magazine, 766 (1804).
Leaf 1, 15–30 cm × 5–10 mm, patched with purple and with distinct, transverse purple banding on the clasping base. Scape to 32 cm, exceeding the leaves, spotted. Flowers few to several in a loose raceme, to 1.5 cm, bell-shaped; stalks to 1 cm. Perianth blue near the base, white often tinged with pink towards the apex, the inner lobes spreading, the outer with green apical swellings and five-sixths of the length of the inner. Stamens as long as perianth. Style not projecting. *South Africa.* H5–G1. Spring.

**8. L. glaucina** Jacquin (*L. sessiliflora* Andrews). Illustration: Botanical Magazine, 3552 (1837).
Leaves 2, to 27 × 2.5 cm, with thickened margins, sometimes spotted with purple. Scape to 40 cm, exceeding the leaves, sometimes spotted. Flowers to 2 cm, many in a moderately dense spike, bell-shaped, fragrant; stalks very short or absent. Perianth iridescent, in various colours, usually the inner lobes pale purple at the

base, white at the apex, slightly spreading, with wavy margins, the outer blue at the base, pink at the tip, with darker apical swellings, more than three-quarters of the length of the inner. Stamens as long as the perianth. Style projecting. *South Africa (Cape Province)*. H5–G1. Spring.

**9. L. orchioides** Aiton (*L. pulchella* Kunth). Illustration: Botanical Magazine, 1269 (1810).

Leaves 2, to 28 × 2 cm, dark green, sometimes spotted with purple, with thickened margins. Scape to 35 cm, exceeding the leaves, sometimes spotted. Flowers to 1 cm, fragrant, 10–50 in a dense spike. Perianth iridescent, in various colours, usually the inner lobes cream tinged with pale purple, spreading and with wavy margins, the outer blue or blue-green, without apical swellings, four-fifths of the length of the inner. Stamens shorter than perianth. Style not projecting. *South Africa (Cape Province)*. H5–G1. Winter–spring.

**10. L. mutabilis** Loddiges. Illustration: Loddiges' Botanical Cabinet **9**: t. 1076 (1825); Botanical Magazine, 9433 (1936).

Leaves 2, to 20 × 2 cm, bright green, unspotted, with thickened margins. Scape to 30 cm, exceeding the leaves, unspotted. Flowers to 1.2 cm, numerous in a loose spike, tubular, the uppermost flowers abortive and sterile, borne on long stalks, the fertile flowers more or less stalkless. Perianth with the inner lobes spreading, yellow-green at the base shading to reddish brown at the tips, the outer lobes blue to grey-green at the base, without apical swellings and three-quarters of the length of the inner. Stamens shorter than perianth. Style not projecting. *South Africa (Natal)*. H5–G1. Winter–spring.

**11. L. aloides** (Linnaeus) Ascherson & Graebner (*Phormium aloides* Linnaeus; *L. tricolor* Jacquin). Illustration: Botanical Magazine, 82 (1789), 1704 (1815), 5992 (1872).

Leaves 2, to 20 × 2 cm, spotted with purple. Scape to 29 cm, usually exceeding the leaves, spotted, arching or hanging. Flowers to 2.5 cm, 10–20 in a loose raceme, tubular, hanging; stalks to 5 mm. Perianth with the inner lobes yellow at the base shading to green with purple tips, all overlapping, the outer lobes bright red at the base, shading to yellow at the tips and with green apical swellings, half to two-thirds of the length of the inner. Stamens as long as perianth. Style not projecting. *South*

*Africa (Cape Province, Natal)*. H5–G1. Winter–spring.

This species is very variable in flower-colour, and many selections and hybrids of it have been given specific names, e.g. *L. aurea* Lindley, *L. luteola* Jacquin, *L. nelsonii* Anon. and *L. quadricolor* Jacquin.

**12. L. rubida** Jacquin (*L. punctata* Jacquin). Illustration: Botanical Magazine, 993 (1807).

Leaves 2, to 14 × 2.5 cm, spotted or patched with purple on one or both sides. Scape to 26 cm, exceeding the leaves, spotted. Flowers to 3 cm, few to several in a loose raceme, tubular, hanging; stalks 2–4 mm. Perianth with inner lobes deep red at base shading to pale yellow at the tips, all overlapping, the outer lobes deep red with yellow-green apical swellings, seven-eighths of the length of the inner. Stamens as long as the perianth. Style projecting. *South Africa (Cape Province)*. H5–G1. Autumn.

**13. L. unicolor** Jacquin. Illustration: Botanical Magazine, 1373 (1811).

Leaves 2, to 15 × 1.5 cm, without spots. Scape about equalling the leaves, without spots. Flowers 30–60, to 8 mm, held horizontally in a loose raceme, shortly cylindric; stalks 2–3 mm. Perianth white or tinged with pink, the inner lobes spreading, the outer without apical swellings, two-thirds of the length of the inner. *South Africa*. H5–G1. Autumn.

**14. L. reflexa** Thunberg (*L. pusilla* Kunth). Illustration: Wiener Illustrierte Garten Zeitung **17**: t. 418 (1892).

Leaves 2, to 15 × 3 cm, bright green, spreading just above the ground, then reflexed, with wavy margins, unspotted. Scape not exceeding the leaves, without spots. Flowers to 2.8 cm, few to many in a loose spike, bell-shaped, erect; stalks to 2 mm. Perianth white or yellowish green with a swollen tube, the inner lobes overlapping, each with a red median stripe, the outer lobes with red apical swellings, five-sixths of the length of the inner. Stamens as long as perianth. Style not projecting. *South Africa*. H5–G1. Winter–spring.

**15. L. bulbifera** (Cyrillo) Ascherson & Graebner (*Phormium bulbiferum* Cyrillo; *L. pendula* Aiton). Illustration: Loddiges' Botanical Cabinet **3**: t. 267 (1818); Flowering Plants of South Africa **4**: t. 158 (1924).

Leaves 2, to 30 × 4 cm, without spots. Scape shorter than leaves, faintly spotted.

Flowers 3–4 cm, few to many in a loose raceme, tubular, hanging; stalks to 5 mm. Perianth with the inner lobes pale yellow with a purple and green apex and a red median stripe, all overlapping, the outer purplish red with green apical swellings, eight-ninths of the length of the inner. Stamens as long as the perianth. Style just slightly projecting. *South Africa*. H5–G1. Winter–spring.

**82. VELTHEIMIA** Gleditsch
*C.J. King*

Perennial herbs with ovoid bulbs. Leaves several, in a basal rosette, lanceolate, oblong or strap-shaped, rather thick, with wavy margins. Flowers in a dense raceme on a stout scape, nodding, bisexual. Flower-stalks short, subtended by bracts. Perianth tubular with tooth-like lobes, persistent. Stamens 6, attached at about the middle of the perianth-tube. Fruit a 3-celled, winged capsule. Seeds 2 in each cell, pear-shaped, black.

Two species from South Africa with a long history of nomenclatural confusion. Both have been cultivated in Europe for about 2 centuries, either in cool greenhouses or, in suitably mild areas, out-of-doors. They prefer a sunny position, a light, loamy soil, and good drainage. Literature: Obermeyer, A.A., *Flowering Plants of Africa* **34**: t. 1356 (1961); Marais, W., *Journal of the Royal Horticultural Society* **97**: 483–4 (1972).

1a. Leaves glossy, dark green, strap-shaped, to 35 × 10 cm; perianth 3–4 cm **1. bracteata**
 b. Leaves glaucous, lanceolate, to 30 × 4 cm; perianth 2–3 cm **2. capensis**

**1. V. bracteata** Baker (*V. capensis* misapplied; *V. viridifolia* Jacquin). Illustration: Botanical Magazine, 501 (1799) & n.s., 215 (1953); Journal of the Royal Horticultural Society **97**: f. 239 (1972).

Bulb broadly ovoid, about 7 × 6 cm, with fleshy scales. Leaves 8–12, to 35 × 10 cm, strap-shaped, oblong, shining green on both surfaces. Scape to 45 cm, dark purple, finely spotted with yellow, fleshy. Raceme 7.5–12.5 cm. Flowers numerous, up to 60, pendent; perianth 3–4 cm, pinkish purple with some pale yellow speckling. Fruit narrowly obovate in profile, conspicuously 3-winged. *South Africa (Eastern Cape Province)*. H5. Winter–early spring.

This is the more robust of the 2 species and the easier to cultivate. It stays in growth most of the year and therefore does not require a dormant period. A paler coloured variant with yellowish flowers tinted with red has been named 'Rosalba' (Illustration: Journal of the Royal Horticultural Society 97: f. 240, 1972).

**2. V. capensis** (Linnaeus) de Candolle (*Aletris capensis* Linnaeus; *V. glauca* (Aiton) Jacquin; *V. roodeae* Phillips; *V. viridifolia* Jacquin). Illustration: Botanical Magazine, 1091 (1808), 3456 (1835) – as V. glauca; Flowering plants of South Africa 4: t. 126 (1924) – as V. roodeae; Flowering plants of Africa 34: t. 1356 (1961).
Bulb ovoid, with a flattened base, long neck, and thin scales, up to $13 \times 6$ cm. Leaves about 10, to $30 \times 4$ cm, lanceolate, glaucous green, shortly mucronate at the apex. Scape to 30 cm or more, glaucous, mottled with purple. Raceme 5–15 cm. Perianth 2–3 cm, varying from whitish spotted with red, to pink with a greenish tip or wine-red. Fruit a membranous, inflated, 3-furrowed capsule, rounded in profile. *South Africa (SW Cape Province)*. G1. Winter–early spring.

A more delicate-looking plant than *V. bracteata*, less easy to grow and propagate, and requiring a dormant period during the summer when the bulbs should be kept dry. The darker colour-form of this species has been named 'Rubescens'. A more compact plant, growing up to about 25 cm high with a sturdy stem and short, crisped margined, greyish leaves, has been called *V. deasii* Barnes. But this is now considered to be merely a smaller form of *V. capensis* and, although horticulturally distinct, should be treated only as a variety or cultivar (Illustration: Botanical Magazine, 8931 (1938) as *V. deasii*).

## 83. GALTONIA Decaisne
*C.J. King*
Perennial herbs: the bulbs with membranous tunics. Leaves few, basal, flaccid, flat. Scape terete, bearing a loose raceme of white or green flowers. Flower-stalks more or less spreading at flowering time, curving upwards in fruit, subtended by large bracts. Flowers bell-shaped, horizontal or slightly nodding. Perianth persistent, the tube slightly to much shorter than the obovate or oblong lobes. Stamens 6, attached just below the mouth or at about the middle of the perianth-tube, shorter than the lobes. Ovary oblong, 3-angled. Fruit a 3-celled capsule. Seeds numerous, angled, brownish or black.

Three species of medium to tall bulbous plants from the eastern region of South Africa. One of these, *G. candicans*, is grown much more frequently than the others, being considered the most attractive and also the hardiest. All species require a sunny position in a light, rich, well-drained soil, and once planted should not be frequently disturbed. Some protection during the winter may be necessary in colder areas. They are well adapted for growing in clumps in borders, but may also be potted up for the conservatory or greenhouse.

1a. Leaves to 5 cm broad, narrowing gradually to the apex; flowers usually white or with some green at the base　2
　b. Leaves to 10 cm broad, the lower narrowing abruptly to form a broad apex with a distinct point; flowers pale green　**1. viridiflora**
2a. Flowers *c.* 3 cm, perianth-lobes distinctly longer than the tube　**2. candicans**
　b. Flowers less than 3 cm, perianth-lobes about as long as the tube　**3. princeps**

**1. G. viridiflora** Verdoorn. Illustration: Flowering plants of Africa 30: t. 1188 (1955).
Leaves about 7, forming a short column at the base, pale yellowish green, in parts thinly glaucous, up to about $60 \times 10$ cm, at least the lower narrowing abruptly at the apex, mucronate. Scape to 1 m, bearing a 15–30-flowered raceme. Bracts greenish, becoming papery, to 3 cm, lanceolate, acuminate. Flower-stalks pale green, 2–5 cm. Perianth pale green, about 2–5 cm, the lobes longer than the tube, with whitish margins. Stamens all attached about the middle of the perianth-tube, or the 3 inner slightly higher. Capsule erect, 3-lobed, ovate in profile, $2.5 \times 1.4$ cm. Seeds brownish, weakly 3-angled. *South Africa (Orange Free State, Natal), Lesotho*. H4. Late summer.

**2. G. candicans** (Baker) Decaisne (*Hyacinthus candicans* Baker). Illustration: Flore des Serres, t. 2172–3 (1875); Rix & Phillips, The bulb book, 169 (1981).
Leaves 4–6, broadly lanceolate, slightly greyish green, fleshy, $50–100 \times 5$ cm. Scape stout, to 120 cm in cultivation, bearing up to 30 flowers. Bracts ovate-lanceolate, acute, $3–4 \times 1$ cm. Flower-stalks rigid, green, 2.5–6 cm. Perianth white, faintly tinged with green at the base of the tube, *c.* 3 cm, the lobes about twice as long as the tube. Stamens attached just below the mouth of the tube. Capsule erect, about 3 cm. Seeds black. *South Africa (Orange Free State, Natal), Lesotho*. H4. Late summer.

**3. G. princeps** (Baker) Decaisne (*Hyacinthus princeps* Baker). Illustration: Botanical Magazine, 8533 (1914).
Leaves 4–6, narrowly lanceolate, sheathing at the base, $40 \times 4$ cm. Scape up to 90 cm. Flowers fewer than in *G. candicans*. Bracts ovate-lanceolate, acute, papery, about $3 \times 1$ cm. Flower-stalks green, about 3 cm, lengthening to 8 cm in fruit. Perianth to 3 cm, tube green, lobes greenish white, only slightly longer than the tube. Stamens attached about the middle of the tube. Seeds black. *South Africa (Natal, Transkei)*. H4. Late summer.

## 84. MASSONIA Houttuyn
*C.J. King*
Perennial herbs; the bulbs with whitish scales surrounded by brown membranous scales. Leaves 2, ovate, oblong or almost circular, spreading horizontally on the ground. Scape absent or very short. Inflorescence usually surrounded by large bracts. Flowers stalked, usually pleasantly scented. Perianth-lobes fused in the lower part, usually spreading or reflexed above. Stamens attached at the mouth of the perianth-tube. Ovary superior. Style usually slightly longer than the stamens. Fruit a winged or deeply lobed capsule, opening by slits. Seeds numerous, black.

A genus now reduced to about 8 species, native to South Africa. The plants may be grown outside in essentially frost-free, mediterranean-type climates, and are suitable in those regions for a sunny, well-drained position in the rock-garden. They can also be easily grown in sunny greenhouses, where the soil should be kept dry during the dormant period but moderately moist while they are in leaf. Propagation is normally by seed, plants reaching flowering size in about 3 years. Literature: Jessop, J.P., Studies in the bulbous Liliaceae in South Africa: 6, The taxonomy of Massonia and allied genera, *Journal of South African Botany* 42: 401–37 (1976).

1a. Anthers more than 2 mm; leaves usually hairless, or hairy only on the margins　**1. depressa**
　b. Anthers less than 2 mm or, if more, than the leaves hairy or papillose　2
2a. Leaves usually hairless, rarely papillose; perianth-lobes about half as long as the tube　**2. jasminiflora**

b. Leaves usually hairy or papillose; perianth-lobes about as long as the tube                            3

3a. Leaves papillose on upper surface

   **3. pustulata**

b. Leaves usually hairy, at least on the margins, but not papillose

   **4. echinata**

**1. M. depressa** Houttuyn (*M. brachypus* Baker; *M. latifolia* Linnaeus filius; *M. sanguinea* Jacquin). Illustration: Botanical Magazine, 848 (1805); Flowering plants of South Africa **2**: t. 46 (1922); Everett, Encyclopedia of horticulture **7**: 2154 (1981).
Bulb usually ovoid, 2–3.5 cm. Leaves circular to oblong, acute, 7–15 × 4–10 cm, hairless or with hairs only on the margins. Inflorescence capitate, with *c.* 20–30 flowers. Perianth green, yellow, white, cream, pink, red or brown, occasionally flecked with purple; tube 1–1.5 cm; lobes 8–10 mm. Filaments 1–1.6 cm, often cream, yellow or green, less often purplish; anthers 2.5–4 mm, yellow or purple. *South Africa (Cape Province).* H5. Late autumn–early winter.

**2. M. jasminiflora** Baker (*M. bowkeri* Baker). Illustration: Botanical Magazine, 7465 (1896); Flowering plants of South Africa **10**: t. 367 (1930).
Bulb ovoid to ellipsoid, 1–2 cm. Leaves ovate to broadly oblong, acute, 3–6 × 1.5–5 cm, hairless or occasionally with hairs only on the margins, rarely papillose. Inflorescence capitate, generally with fewer than 15 flowers. Perianth white or pink; tube 8–20 mm; lobes 4–8 mm. Filaments 2–4.5 mm, white or pink; anthers 1–1.5 mm, green, blue, dark purple or black. *South Africa (Cape Province, Orange Free State), Lesotho.* H5. Late autumn–early winter.

**3. M. pustulata** Jacquin. Illustration: Botanical Magazine, 642 (1803); Grey, Hardy bulbs **3**: 428 (1938); Flowering plants of South Africa **23**: t. 915 (1943); The Garden **109**: 418 (1984).
Bulb spherical or ovoid, 1–2.5 cm. Leaves ovate to oblong, acute, 3–10 × 2–7 cm, papillose on the upper surface, especially towards the apex, and with the margins often minutely hairy or minutely toothed. Inflorescence capitate, with 15–25 flowers. Perianth pink, white, yellow or greenish; tube 6–11 mm; lobes 4.5–10 mm. Filaments 4–12 mm, white; anthers 1–1.75 mm, yellow or reddish. *South Africa*

*(Cape Province).* H5. Late autumn–early winter.

Closely allied to *M. echinata*, but distinguished by the relatively uniformly papillose upper surface of the leaves.

**4. M. echinata** Linnaeus filius (*M. amygdalina* Baker; *M. bolusiae* Barker; *M. longifolia* Jacquin var. *candida* Ker Gawler; *M. muricata* Ker Gawler; *M. scabra* Thunberg). Illustration: Botanical Magazine, 559 (1802); Edwards's Botanical Register **9**: t. 694 (1823); Flowering plants of South Africa **11**: t. 429 (1931).
Bulb usually ovoid, 1–2 cm. Leaves ovate to oblong, acute or obtuse, 2–8 × 1–4 cm, usually hairy, at least on the margins. Inflorescence capitate, with 5–20 flowers. Perianth yellow, white or pink; tube 5–7 mm; lobes 4–8 mm. Filaments 4–8 mm, white; anthers 0.5–1.25 mm, yellow or purple. *South Africa (Cape Province).* H5. Late autumn–early winter.

**85. DAUBENYA** Lindley
*P.F. Yeo*
Hairless, perennial, summer-dormant herbs with bulbs. Leaves 2, 5–15 × 3.5–7.5 cm, ovate or oblong, strongly parallel-veined. Inflorescence-stalk *c.* 4 cm, concealed in neck of bulb. Inflorescence a head-like corymb of *c.* 10 flowers, held close to the leaf-bases. Bracts to 3 × 2.2 cm, acute, overlapping, the inner smaller than the outer. Flowers erect, strongly unequal within the inflorescence. Outer flowers to 6.5 cm, with tube to 2.5 cm and *c.* 4 mm in diameter, and bilaterally symmetric limb to 4 cm, consisting of a large lower lip with 3 blunt, hooded lobes and a minute, 3-lobed upper lip; the lower lips encircle the corymb. Innermost flowers nearly radially symmetric with tubes similar to those of outer and lobes 2–6 mm. Transitional flowers also occur. Stamens inserted at different levels near the bases of the perianth-lobes; filaments free from each other; anthers dorsifixed. Style unbranched, about as long as the perianth-tube, with no stigmatic enlargement. Ovary superior, 3-celled with numerous ovules. Seeds spherical, 2 mm in diameter, black.
One species in the winter-rainfall area of South Africa, closely related to *Massonia*. It should be grown in a bulb-frame or greenhouse and dried off in summer. Literature: Hall, H., Daubenya Lindley, *Journal of the Botanical Society of South Africa* **56**: 13–16 and cover and frontispiece (1970); Jessop, J.P., Studies in

the bulbous Liliaceae in South Africa, 6, *Journal of South African Botany* **42**: 401–37 (1976).

**1. D. aurea** Lindley (*D. fulva* Lindley). Illustration: Edwards's Botanical Register **21**: t. 1813 (1835), **25**: t. 53 (1839); Botanical Magazine, n.s., 700 (1975).
Outer flowers usually red or yellow but sometimes orange, inner yellowish. *South Africa (SW Cape Province).* G1. Spring.

**86. AGAPANTHUS** L'Héritier
*C.J. King*
Perennial herbs with fleshy roots arising from a tuberous rootstock. Leaves 10–70 cm × 8–60 mm, usually basal, linear to strap-shaped with entire margins. Flowers few to many in an umbel on a usually stout scape 30–180 cm high. Perianth tubular to spreading, united towards the base, ranging from dark violet or deep blue to white. Stamens 6, inserted on the perianth-tube. Ovary of 3 cells, superior, developing into a many-seeded capsule. Seeds flat and winged, black.

All (except *A. africanus*) prefer a sunny position in a light, rich, well-drained soil, copiously watered in summer but kept fairly dry during the winter. The hardier forms may survive the winter in the open with a slight covering for the crowns, but, except in the mildest areas, the tenderer forms are best placed outside in tubs for the summer and brought into a light but frost-free place in autumn. Propagation is by offsets, by divison, or by seed.

Leighton recognised 10 species and 15 subspecies in her monograph of *Agapanthus* (see below). She found that plants hybridise quite freely when grown near each other and that natural variation was so great that it would have been possible for the genus to be broken down into innumerable species or subspecies.

Subsequent research by other investigators tends to support this view. Indeed, Palmer (Journal of the Royal Horticultural Society **92**, 1967) was of the opinion that, as cross-pollination was so rife amongst cultivated plants, practically all garden forms of *Agapanthus* should be regarded as hybrids, and McNeil (Journal of the Royal Horticultural Society **97**, 1972) concluded that the whole *Agapanthus* population should be considered as one unstable mega-species.

In spite of the difficulties, a modification of Leighton's key and descriptions may enable a number of cultivated plants to be identified with some precision. However, the large numbers of individual clones that have been distinguished by cultivar names are often of complex parentage, and it may not be possible to assign these to a particular species or subspecies. Some of these cultivars were formerly included in the group of hardier plants raised or selected from existing garden forms by Palmer and known collectively as 'Headbourne Hybrids'.

Literature: Leighton, F.M., The Genus Agapanthus L'Héritier, *Journal of South African Botany*, Supplementary Volume No. IV (1965).

*Leaves*. Arising from an obvious stem: **5**.
   More than 4.5 cm broad: **1,5,6**.
*Scape*. More than 100 cm: **1,5**.
*Flower-stalks*. More than 7 cm: **6**.

*Perianth* (see figure 34, below). More than 5 cm: **2,6**. Tubular: **1**; semi-open: **2,3**. Thick in texture: **4**.
*Stamens*. Projecting: **3,4,6**.

1a.  Perianth tubular (see figure 34)
                                    **1. inapertus**
  b.  Perianth open or semi-open (see
       figure 34)                                      2
2a.  Perianth always semi-open, more
       than 3.5 cm                        **2. nutans**
  b.  Perianth open, or if semi-open, less
       than 3.5 cm                                     3
3a.  Perianth semi-open, up to 3.5 cm
                                 **3. campanulatus**
  b.  Perianth open                                    4
4a.  Perianth thick and waxy in texture;
       leaves not usually more than
       35 × 2 cm                       **4. africanus**
  b.  Perianth thin in texture, leaves often
       longer and broader                              5

5a.  Perianth to 3.5 cm; leaves not usually
       more than 40 × 2.5 cm
                               **3. campanulatus**
  b.  Perianth usually more than 3.5 cm;
       leaves often longer and broader        6
6a.  Perianth not more than 5 cm; flower-
       stalks not more than 7 cm; stamens
       and style shorter than perianth
                                   **5. caulescens**
  b.  Perianth to 7 cm; flower-stalks to
       12 cm; stamens and style usually as
       long as perianth              **6. praecox**

**1. A. inapertus** Beauverd. Figure 34(2)
Illustration: Botanical Magazine, 9621 (1942); Letty, Wild flowers of the Transvaal, t. 13 (1962); Leighton, The genus Agapanthus, 42 & t. 12 (1965); Flowering plants of Africa **37**: t. 1479 & 1480 (1965–6).
Leaves to 70 × 6 cm. Scape to 1.8 m; flowers many. Flower-stalks 2–5.5 cm,

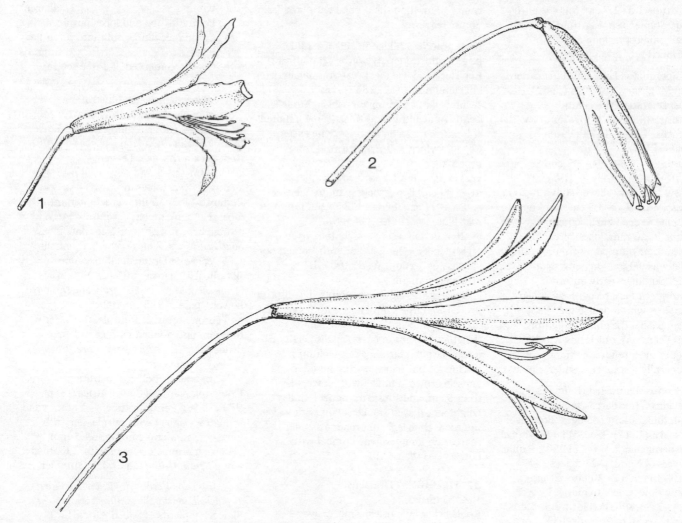

**Figure 34.** Flowers of *Agapanthus*. 1, Open flower of *A. praecox* subsp. *minimus*. 2, Tubular flower of *A. inapertus* subsp. *inapertus*. 3, semi-open flower of *A. nutans*.

erect at first, later nodding. Perianth 2.5–5 cm, tubular, bright blue to dark violet, sometimes white; tube at least as long as the lobes; lobes slightly spreading. Stamens about as long as perianth. Style usually slightly projecting when mature. *South Africa.* H4. Autumn.

Of the 5 subspecies, the 2 most frequently grown in European gardens are subsp. **inapertus**, with leaves *c.* 2.5 cm broad, and flowers deep blue or violet-blue, occasionally white, and subsp. **pendulus** (Bolus) Leighton (*A. pendulus* Bolus) which has leaves 5 cm or more broad, and very deep blue or dark violet flowers.

**2. A. nutans** Leighton. Figure 34(3), p. 231. Illustration: Leighton, The genus Agapanthus, 38 (1965).
Leaves 20–50 × 1–4 cm, often glaucous. Scape to 90 cm; flowers few to many. Flower-stalks spreading or nodding. Perianth 3.5–6 cm, nodding, semi-open, pale blue; tube 1.3–2.7 cm; lobes slightly spreading. Stamens shorter than perianth. Style finally almost as long as perianth. *South Africa.* H4. Late summer.

**3. A. campanulatus** Leighton. Illustration: Botanical Magazine, n.s., 380 (1962) – as A. patens; Leighton, The genus Agapanthus, 30, 32 (1965); Flowering plants of Africa **37**: t. 1478 (1965–6) – subsp. patens.
Leaves usually 15–40 × 1–2.5 cm. Scape usually 40–70 cm, rarely to 100 cm; flowers few to many. Flower-stalks 2–7 cm, spreading. Perianth 2–3.5 cm, semi-open or open, pale to deep blue, sometimes white; tube 5–10 mm; lobes spreading. Stamens shorter than perianth, projecting. Style finally as long as stamens. *South Africa.* H4. Summer–early autumn.

Two subspecies are recognised: subsp. **campanulatus** with a bell-shaped perianth, and subsp. **patens** (Leighton) Leighton (*A. patens* Leighton) which has widely spreading or even reflexed perianth-lobes, and is generally a smaller, slenderer plant.

**4. A. africanus** (Linnaeus) Hoffmansegg (*A. umbellatus* L'Héritier; *A. minor* Loddiges). Illustration: Loddiges' Botanical Cabinet **1**: 42 (1817); Kidd, Wild flowers of the Cape peninsula, t. 3 f. 2 (1950); Hulme, Wild flowers of Natal, t. 21 f. 6 (1954).
Leaves 10–35 cm × 8–20 mm. Scape 30–60 cm; flowers few to many. Flower-stalks 1.5–5 cm, stiff. Perianth 2.5–5 cm, thick and waxy in texture, deep blue or blue-violet; tube 9–14 mm; lobes spreading. Stamens shorter than perianth.

Style finally as long as stamens. *South Africa.* H5. Autumn.

The most difficult species to cultivate successfully since it requires watering in winter and to be kept moderately dry in summer.

**5. A. caulescens** Sprenger. Illustration: Gartenflora **50**: 281 (1901).
Leaves arising from an obvious stem, the basal sheathing leaves often only 5–15 cm, the upper 25–60 × 1.5–5 cm, green and glossy, acute or obtuse at apex. Stem 60–130 cm; umbel dense; flowers many. Flower-stalks 3–7 cm, spreading and somewhat drooping. Perianth 3–5 cm, bright to deep blue; tube 1–1.9 cm; lobes widely spreading, sometimes recurved. Stamens and style shorter than perianth. *South Africa.* H4. Late summer–early autumn.

Of the 3 subspecies, the one usually cultivated is subsp. **caulescens**, a more robust plant than the others, with broader, obtuse leaves.

**6. A. praecox** Willdenow. Figure 34(1), p. 231. Illustration: Edwards's Botanical Register, t. 699 (1823) – as A. umbellatus var. minimus & t. 7 (1843) – as A. umbellatus var. maximus; Leighton, The genus Agapanthus, 20 & 22 (1965); Batten & Bokelmann, Wild flowers of the eastern Cape Province, t. 9 f. 3 (1966) – subsp. praecox.
Leaves 20–70 × 1.5–5.5 cm. Scape 40–100 cm; flowers few to many. Flower-stalks 4–12 cm. Perianth 3–7 cm, bright to pale blue, sometimes white; tube 7–26 mm; lobes spreading. Stamens and style usually as long as perianth, sometimes shorter, projecting. *South Africa.* H5. Summer–autumn.

Three subspecies are recognised: subsp. **praecox**, a large stout plant with flowers at least 5 cm in length, subsp. **orientalis** (Leighton) Leighton (*A. orientalis* Leighton), a smaller plant forming dense clumps, having a stout scape bearing an often densely crowded umbel with flowers less than 5 cm, and subsp. **minimus** (Lindley) Leighton, an even smaller plant with leaves not more than 2.5 cm broad, a slender scape, and an uncrowded umbel with flowers less than 5 cm.

**87. TULBAGHIA** Linnaeus
*V.A. Matthews*
Rootstock a bulb or rhizome. Leaves 4–8, strap-shaped to linear, basal. Flowering stem solitary, erect, bearing an umbel of 6–40 flowers which is subtended by 2

bracts. Perianth united into a tube for about half its length; mouth of tube with a fleshy corona which is cylindric or composed of 3 free scales. Anthers stalkless on the perianth-tube, in 2 whorls 1 above the other. Style short. Stigma capitate. Capsule 3-celled, containing triangular black seeds.

A genus of 21 species, 5 of which are cultivated. *Tulbaghia* is native to southern temperate and tropical Africa. With the exception of *T. violacea*, the bulbs require protection from frost. They will grow in any soil and prefer a sunny position if planted out-of-doors.
Literature: Burbidge, R.B., A revision of the genus Tulbaghia (Liliaceae), *Notes from the Royal Botanic Garden, Edinburgh* 36(1): 77–103 (1978); Burbidge, R.B., The genus Tulbaghia, *Lilies 1978/9 and other Liliaceae* 78–84 (1978/9).

1a. Perianth-tube and lobes green or white     **1. alliacea\***
  b. Perianth-tube and lobes purple    **2**
2a. Corona a cylinder split into 3 toothed lobes     **2. fragrans**
  b. Corona composed of 3 free scales     **3. violacea**

**1. T. alliacea** Linnaeus. Illustration: Flowering plants of South Africa **17**: pl. 653 (1937); Rice & Compton, Wild flowers of the Cape of Good Hope, pl. 244 (1950).
Rootstock a rhizome. Leaves 15–25 cm × 3–5 mm, linear. Flowering stem 15–30 cm, bearing 6–10 flowers. Perianth-tube *c.* 6 mm, green; perianth-lobes 2–4 mm, green; corona 2–3 mm, orange-brown with 3 or 6 shallow lobes. *Southern Africa, Zimbabwe.* H5. Summer.

**\*T. capensis** Linnaeus. Illustration: Botanical Magazine, 806 (1805) – as T. alliacea; Mason, Western Cape sandveld flowers, pl. 8 (1972). Perianth olive-green and corona brownish purple, the latter up to 5 mm and 3-lobed with each lobe divided into 2. *South Africa (Cape Province).* H5–G1. Summer.

**\*T. natalensis** Baker. Illustration: Flowering plants of South Africa **25**: pl. 979 (1945). Perianth-tube 3–4 mm, white, sometimes tinged with purple, perianth-lobes 5–7 mm and corona 3–4.5 mm, yellowish orange or green with 3 toothed lobes. *South Africa (Natal).* H5. Summer.

**2. T. fragrans** Verdoorn (*T. daviesii* Grey; *T. pulchella* misapplied). Illustration: Flowering plants of South Africa **11**: pl. 438 (1931); Letty, Wild flowers of the Transvaal, pl. 11 (1962).
Rootstock a bulb. Leaves

30–60 × 1.5–2.5 cm, strap-shaped.
Flowering stem to 60 cm, bearing 20–40
flowers. Perianth-tube 8–9 mm, light
purple; perianth-lobes 6–8 mm, light
purple, sometimes with ragged and inrolled
margins; corona a 3–4 mm cylinder split
into 3 toothed lobes, purple tinged with
pink. *South Africa (Transvaal)*. H5–G1. Late
spring–summer.

*T. fragrans* differs from the other species
in not or scarcely smelling of onion.

**3. T. violacea** Harvey. Illustration:
Botanical Magazine, 3555 (1837); Perry,
Flowers of the world, 19 (1972).
Rootstock a rhizome. Leaves
15–30 cm × 4–7 mm, linear. Flowering
stem 30–65 cm, bearing 8–30 flowers.
Perianth-tube 7–13 mm, purple; perianth-
lobes 5–12 mm, purple; corona
1.5–3.5 mm, composed of 3 free fleshy
scales, purple tinged with reddish, or white.
*South Africa (Cape Province, Transvaal)*. H4.
Early summer–late autumn.

'Silver Lace' is a large-flowered form.

**88. ALLIUM** Linnaeus
*W.T. Stearn assisted by E. Campbell*
Perennial herbs, mostly smelling of onion,
usually with well-formed bulbs, arising in
some species from a short rhizome. Leaves
linear to elliptic, basal or sheathing the
stem, flat to channelled, terete or semi-
terete, solid or hollow. Flowers usually
many (unless replaced by bulbils) in an
umbel on a solid or hollow stem and
enclosed at first within a spathe consisting
of 1 or more bracts. Perianth star-shaped,
cup-shaped or bell-shaped, with 6 usually
free segments. Stamens 6; filaments free or
united at base, simple or variously toothed;
anthers opening by longitudinal slits. Ovary
3-celled, each cell with 1–8 ovules, usually
2; style slender; stigma usually simple but
3-lobed in a few species. Fruit a 3-celled
capsule splitting between the septa. Seeds
angled or rounded, blackish, in a few
species with a spongy appendage.

A genus of about 690 species in Europe,
Asia and America, most abundant in
central Asia, but extending into tropical
Africa, Sri Lanka and Mexico. Some species
have wide ranges, notably
*A. schoenoprasum*; some are very local.
Nearly all are easy to cultivate, given good
drainage and freedom from competition.
Hybrids are almost unknown, the only
well-authenticated cross being
*A. cepa × fistulosum*. The diversity of bulb
structure, particularly the nature of the
outer covering (tunic), provides important

characters for the distinction of sections
and species, but these have been kept to a
minimum in the following key owing to the
natural reluctance of gardeners to dig up
their plants when in flower. Since species
are continually being introduced into
cultivation, which may not be included in
the following account, reference should be
made to *Flora Europaea* **5** (1980) for
European species; to *Flora Iranica* **76**
(1971) and *Flora SSSR* **4**: 112–280 (1935)
translated and annotated in *Herbertia* **11**:
65–218 (1946) for central Asiatic species.
Some species are of culinary importance,
i.e. *A. cepa* (onion), *A. sativum* (garlic),
*A. porrum* (leek), *A. fistulosum* (Japanese
leek), *A. schoenoprasum* (chives),
*A. tuberosum* (Chinese chives) and
*A. chinense* (rakkyo or ch'iaot'ou).
Literature: Regel, F., Alliorum adhuc
cognitorum monographia, *Acta Horti
Petropolitani* **3**: 1–266 (1875); Stearn,
W.T., Notes on the genus Allium in the Old
World, *Herbertia* **11**: 11–34 (1946);
Vvedensky, A.I., The genus Allium in the
USSR, *Herbertia* **11**: 65–218 (1946);
Moore, H.E., The cultivated Alliums, *Baileya*
**2**: 103–23 (1954), **3**: 137–49, 156–67
(1955); Ownbey, M. & Aase, H.C., The
Allium canadense alliance, *Research studies
of the State College of Washington
Monographic supplement* **106** (1955);
Stearn, W.T., Allium and Milula in the
central and eastern Himalaya, *Bulletin of
the British Museum (Natural History),
Botany* **2**: 1159–91 (1960); Bothmer, R.
von, Biosystematic studies in the Allium
ampeloprasum complex, *Opera Botanica* **34**:
1–104 (1974); Wilde-Duyfjes, B.E.E. de,
Revision of the genus Allium L. (Liliaceae)
in Africa, *Belmontia* **7**: 1–237 (1977);
Stearn, W.T., European species of Allium
and allied genera of Alliaceae: a synonymic
enumeration, *Annales Musei Goulandris* **4**:
83–198 (1978); Stearn, W.T., The genus
Allium in the Balkan peninsula, *Botanische
Jahrbücher* **102**: 201–13 (1981); Pastor, J.
& Valdés, B., *Revision del Genero Allium en la
Peninsula Iberica e Islas Baleares* (1983).

1a. Leaves completely withered or
   absent at flowering time          2
 b. Leaves present although sometimes
   mostly withered at flowering time
                                      3
2a. Stem completely covered with
   adpressed dry sheaths
                        **49. callimischon**
 b. Stem naked          **23. tricoccum**
3a. Leaves clearly contracted into a
   stalk at base                      4

 b. Leaves scarcely or not stalked, in
   most species linear or thread-like
                                      6
4a. Leaves sheathing the lower one-
   third to half of the stem; stamens
   longer than the pale yellow or
   greenish white perianth-segments
                        **22. victorialis**
 b. Leaves all basal; stamens shorter
   than the white perianth-segments
                                      5
5a. Bulbs densely covered with netted
   fibres; leaves mostly withered at
   flowering time; stem terete
                        **23. tricoccum**
 b. Bulbs without netted fibres, only a
   few parallel fibres at base; leaves
   completely green at flowering time;
   stem angled          **39. ursinum**
6a. Plant rarely producing flowering
   stems but forming clumps of hollow
   leaves, semi-circular in section,
   scarcely 1 cm wide but to 40 cm
   long                 **25. cepa 'Perutile'**
 b. Plant producing flowering stems
   with flowers or bulbils or both    7
7a. Umbel with bulbils only: flowers
   absent or abortive                 8
 b. Umbel with flowers, sometimes also
   with bulbils                       13
8a. Leaves basal, solitary, not sheathing
   the 3-angled stem
                        **42. paradoxum**
 b. Leaves sheathing lower part of the
   terete stem                        9
9a. Leaves and stem inflated, 5–20 mm
   broad; bulbils large, 2–4 cm       10
 b. Leaves and stem slender, not
   inflated; the leaves if flat and keeled
   to 3 cm broad, if hollow to 4 mm
   broad, bulbils less than 2 cm      11
10a. Leaves circular in section
                        **26. fistulosum 'Viviparum'**
 b. Leaves semi-circular in section
                        **25. cepa 'Viviparum'**
11a. Leaves flat and keeled, 5–30 mm
   broad                **59. sativum**
 b. Leaves hollow, terete, to 4 mm
   broad                              12
12a. Spathes 2, persistent, long
                        **54. oleraceum**
 b. Spathe 1, soon falling  **63. vineale**
13a. Umbel almost stalkless in a clump or
   rosette of spreading, linear, ciliate
   leaves               **38. chamaemoly**
 b. Umbel on an evident stem          14
14a. Leaves hairy along the margin,
   sometimes also beneath             15
 b. Leaves hairless                   20
15a. Leaves 1–4 cm broad; umbel
   10–15 cm or more across            16

b.  Leaves to 2 cm broad; umbel
    4–8 cm across                18
16a. Stem 15–40 cm; perianth-segments
    1.2–2 cm, becoming rigid after
    flowering            **69. christophii**
b.  Stem 50–120 cm; perianth-
    segments 8–10 mm, reflexing and
    twisted (not rigid) after flowering
                              17
17a. Stem distinctly ribbed; bulbs
    blackish            **72. stipitatum**
b.  Stem lightly ribbed; bulbs grey
                    **73. hirtifolium**
18a. Stamens about as long as or longer
    than perianth-segments
                    **33. subvillosum**
b.  Stamens one-third to two-thirds as
    long as perianth-segments     19
19a. Flower-stalks 3–5 times as long as
    perianth-segments
                    **32. subhirsutum**
b.  Flower-stalks half to 3 times the
    length of perianth-segments;
    anthers yellowish    **34. trifoliatum**
20a. Stem and leaves inflated, hollow;
    stem 2–4 cm in diameter      21
b.  Stem not inflated, usually slender
    and solid                    23
21a. Leaves semi-circular in section,
    grooved above; perianth-segments
    3–4.5 mm; stamens 4–6 mm
                        **25. cepa**
b.  Leaves circular in section; perianth-
    segments 6–9 mm; stamens
    8–12 mm                      22
22a. Flower-stalks 2–3 times as long as
    the perianth; bulb cylindric
                    **26. fistulosum**
b.  Flower-stalks slightly shorter than
    the perianth to 1½ times as long;
    bulb ovoid          **27. altaicum**
23a. Filaments of outer and inner
    stamens markedly unlike, the outer
    simple or 3-toothed, the inner
    always with an expanded basal part
    bearing 3–5 teeth, the central tooth
    anther-bearing, the lateral teeth
    filamentous and longer      24
b.  Filaments of all stamens simple or
    sometimes with short teeth at base
    very much shorter than the anther-
    bearing filament             29
24a. Leaves hollow, 1–4 mm broad    25
b.  Leaves flat, keeled, 5–40 mm broad
                              27
25a. Spathe 1, soon falling    **63. vineale**
b.  Spathes 2, persistent         26
26a. Perianth-segments 3.5–5 mm;
    stamens projecting
                **62. sphaerocephalon**

b.  Perianth-segments *c*. 10 mm;
    stamens included     **64. heldreichii**
27a. Flowers mostly abortive and bulbils
    present; basal part of inner stamens
    shorter than or equalling the
    central anther-bearing tooth
                        **59. sativum**
b.  Flowers normal; basal part of inner
    stamens 2–3 times as long as the
    central anther-bearing tooth    28
28a. Bulblets (and bulbils if present)
    blackish; flower-stalks not more
    than 2 cm; stamens shorter than
    perianth            **61. scorodoprasum**
b.  Bulblets yellowish, not blackish;
    some flower-stalks more than 2 cm
    long; stamens equalling or slightly
    longer than perianth with anthers
    projecting          **60. ampeloprasum**
29a. Leaves basal or almost so, sheathing
    not more than the lower one-
    quarter of the above-ground stem
    and parting from it at the same level
    or not much apart            30
b.  Leaves sheathing more than the
    lower one-quarter of the above-
    ground stem and parting from it at
    different heights            96
30a. Perianth bright yellow        31
b.  Perianth white, greenish, yellowish,
    pink, brownish, blue or purple    32
31a. Leaves 1.5–3.5 cm broad; ovary not
    prominently angled or crested
                        **35. moly**
b.  Leaves 3–7 mm broad; ovary
    prominently angled, 6-crested
                **36. scorzonerifolium**
32a. Stigma 3-lobed                33
b.  Stigma entire                  36
33a. Perianth white with longitudinal
    green stripes; stem triangular in
    section, flaccid after flowering    34
b.  Perianth rose to purple-red; stem
    terete, rigid after flowering    67
34a. Leaf solitary          **42. paradoxum**
b.  Leaves 2–5                     35
35a. Umbel 1-sided; flowers always
    pendent and bell-shaped; perianth-
    segments 1-veined    **40. triquetrum**
b.  Umbel loose; flowers spreading and
    star-shaped, later pendent and bell-
    shaped with perianth-segments
    converging around the capsule;
    segments 3-veined
                    **41. pendulinum**
36a. Leaves 1–8 cm broad           37
b.  Leaves 0.5–10 mm broad         57
37a. Flower-stalks markedly unequal, the
    longest at least 3 times as long as
    the shortest; umbel loose, *c*.
    20–30 cm across              38

b.  Flower-stalks almost equal; umbel
    dense or loose, less than 20 cm
    across                       39
38a. Stem 30–60 cm; perianth purplish
                    **78. schubertii**
b.  Stem 10–25 cm; perianth pale
    brown              **79. protensum**
39a. Perianth bell-shaped, the segments
    convergent and completely hiding
    the ovary; ovary with 6 well-
    marked apical swellings
                    **15. macranthum**
b.  Perianth cup-shaped to star-shaped,
    the ovary visible and without well-
    marked apical swellings       40
40a. Stamens distinctly longer than the
    perianth-segments            41
b.  Stamens shorter than or equalling
    the perianth-segments        45
41a. Leaves ascending or erect; stem
    angled; umbel pendulous before
    opening                **2. nutans**
b.  Leaves spreading outwards; stem
    not angled; umbel erect before
    opening                      42
42a. Stem 10–30 cm (rarely more),
    shorter than the leaves      43
b.  Stem 30–200 cm, much longer
    than the leaves              44
43a. Leaves 3–6, linear    **80. caspium**
b.  Leaves 1–2, elliptic or narrowly
    ovate                  **81. mirum**
44a. Stem feebly ribbed; perianth-
    segments obtuse      **75. giganteum**
b.  Stem strongly ribbed; perianth-
    segments almost acute
                      **76. macleanii**
45a. Perianth blue        **13. sikkimense**
b.  Perianth not blue              46
46a. Stem angled; leaves keeled    47
b.  Stem terete; leaves not keeled    48
47a. Perianth-segments reddish purple,
    spreading then reflexing
                      **14. wallichii**
b.  Perianth-segments white,
    convergent after flowering
                    **31. neapolitanum**
48a. Leaves 1–3, elliptic, broadly elliptic
    or lanceolate; stems 10–40 cm,
    often shorter than the leaves    49
b.  Leaves 2–7, broadly linear; stem
    40–100 cm, longer than the leaves
                              51
49a. Perianth-segments obovate
                        **81. mirum**
b.  Perianth-segments narrowly oblong
    to lanceolate                50
50a. Perianth-segments lanceolate,
    acute; stamens almost equalling the
    perianth-segments, the filaments

thread-like except for broadened bases          **77. karataviense**

b. Perianth-segments narrowly oblanceolate, obtuse; stamens half as long as the perianth-segments; filaments narrowly triangular almost to apex          **68. akaka**

51a. Ovary on a short stalk visible after removal of perianth and stamens          52

b. Ovary stalkless          54

52a. Bulb-tunic elongated into a collar around leaves at base; perianth-segments 4–5 mm          **70. suworowii**

b. Bulb-tunic not elongated; perianth-segments 6–10 mm          53

53a. Stem feebly ribbed          **74. aflatunense**

b. Stem strongly ribbed          **71. rosenbachianum**

54a. Perianth-segments and filaments of stamens dark purple          **67. atropurpureum**

b. Perianth-segments and filaments white, pink or greenish          55

55a. Leaves sheathing the stem only to soil level; perianth-segments becoming reflexed or spreading; ovules 4–8 in each cell          56

b. Leaves sheathing the stem a little above soil level; perianth-segments erect and touching after flowering; ovules 2 in each cell          77

56a. Perianth-segments 1.5–3 mm broad, narrowly elliptic; filaments subulate above the base          **65. nigrum**

b. Perianth-segments 1–1.5 mm broad, broadly linear; filaments narrowly triangular          **66. cyrilli**

57a. Leaves thread-like, 0.5–2 mm broad          58

b. Leaves linear, flat, channelled or terete, 1–10 mm broad          65

58a. Stamens longer than perianth          59

b. Stamens shorter than perianth          63

59a. Perianth-segments white or yellowish          60

b. Perianth-segments blue or purplish          61

60a. Bulb-tunics ultimately breaking into fibres; flower-stalks smooth          **4. ericetorum**

b. Bulb-tunics membranous, never fibrous; flower-stalks rough          **5. albidum**

61a. Perianth-segments blue, obtuse; filaments of all stamens subulate, without teeth          **11. cyaneum**

b. Perianth-segments purplish mauve or rose; filaments of outer stamens subulate, of inner stamens with a short tooth on each side          62

62a. Bulb-tunic with netted fibres; perianth-segments acute, purplish; filaments of inner stamens broadened in lower half          **16. przewalskianum**

b. Bulb-tunic membranous; perianth-segments obtuse, rose; filaments of inner stamens not distinctly broadened          **47. virgunculae**

63a. Spathes 2; tips of perianth-segments erect          **43. moschatum**

b. Spathe 1; tips of perianth-segments recurving          64

64a. All flowers stiffly erect or ascending; perianth pale rose          **9. mairei**

b. Some flowers nodding; perianth deep rose          **10. amabile**

65a. Outer and inner perianth-segments markedly dissimilar, the outer entire and recurved at the tips, the inner erect with minutely toothed upper margin          **90. acuminatum**

b. Outer and inner perianth-segments not markedly dissimilar, the inner not minutely toothed          66

66a. Leaves cylindric, hollow          **24. schoenoprasum**

b. Leaves flat or channelled or, if hollow, then semi-circular, flattened or 3–5 angled          67

67a. Perianth blue          **13. sikkimense**

b. Perianth not blue          68

68a. Perianth 9–18 mm, reddish purple; stigma slightly or distinctly 3-lobed          69

b. Perianth 3–13 mm, usually white, lilac or pale purple; stigma entire          72

69a. Leaves 3–5; bulbs oblong, clustered on a short rhizome; perianth cylindric-bell-shaped; filaments of stamens subulate          70

b. Leaves 2; bulbs ovoid, without rhizome; perianth broadly bell-shaped or almost star-shaped; filaments of stamens narrowly or broadly triangular          71

70a. Bulb with persistent elongated fibrous tunic; umbel at first pendent, later erect          **20. narcissiflorum**

b. Bulb without fibrous tunic or with only a few fibres at base; umbel always pendent          **21. insubricum**

71a. Leaves 2–8 mm broad, longer than flower stem; stamens converging and covering ovary, inner stamens broadly triangular, outer narrowly triangular          **37. oreophilum**

b. Leaves 1–3 mm broad, equalling or shorter than flower stem; all

stamens very narrowly triangular, not converging          113

72a. Stamens distinctly longer than perianth-segments          73

b. Stamens shorter than or only slightly longer than perianth-segments          76

73a. Umbel nodding at flowering time, the stem curved just below the umbel; ovary with 6 crests          **82. cernuum**

b. Umbel erect at flowering time; ovary rounded above, not crested          74

74a. Leaves hollow, 3–5 angled; perianth light violet; filaments 7.5–8 mm          **46. chinense**

b. Leaves solid, flat; perianth white or yellowish; filaments 3–6 mm          75

75a. Bulb-tunics ultimately breaking into fibres; flower-stalks smooth          **4. ericetorum**

b. Bulb-tunics membranous, never fibrous; flower-stalks rough          **5. albidum**

76a. Perianth white, sometimes with a green, brown or reddish stripe down the middle of each perianth-segment          77

b. Perianth light purple, lilac, rose or reddish brown          81

77a. Perianth star-shaped; segments 4–7 mm, with faint green or brownish mid line externally, reflexing or withering away from the capsule          **18. tuberosum**

b. Perianth open, cup-shaped or bell shaped; segments 6–14 mm, with or without a reddish mid-line externally, erect after flowering          78

78a. Bulbs with membranous or crustaceous (not fibrous) outer tunic, ovoid or almost spherical; leaves solid          79

b. Bulbs with netted, fibrous outer tunic; if leaves solid, then ovary with apical projections (crests); if leaves hollow, then ovary without projections          114

79a. Spathes 2; perianth with purplish keel          **30. massaessylum**

b. Spathe 1 but often becoming 3 or 4 lobed; perianth without purplish keel          80

80a. Bulb-tunic minutely and profusely pitted          **28. roseum**

b. Bulb-tunic not pitted          **29. zebdanense**

81a. Flower-stalks rough; perianth narrowly bell-shaped, red-purple, 6–8 mm; stamens united at base

into a tube *c.* 3 mm long around the
ovary            **19. cyathophorum**
   b. Flower-stalks smooth; perianth
cup-shaped or star-shaped; stamens
free or only shortly united at base
       82
82a. Ovary rounded at the top without
crests but sometimes with 6 small
rounded knobs or 3 flat ridges    83
   b. Ovary with 3 or 6 prominent crests
       90
83a. Perianth star-shaped; segments
purplish brown, contracted and not
overlapping at base; bulb solitary
but putting out slender rhizomes
each terminated by a bulblet
      **96. glandulosum**
   b. Perianth cup-shaped or star-shaped;
segments not brown, slightly
overlapping at base          84
84a. Bulbs cylindric or narrowly conical,
clustered on a short persistent
rhizome; stem 2-edged above    85
   b. Bulbs narrowly ovoid to almost
spherical, without a persistent
rhizome; stem terete or only slightly
angled above                86
85a. Leaves smooth and rounded
beneath; stamens distinctly
projecting          **1. senescens**
   b. Leaves sharply keeled beneath;
stamens only slightly projecting
      **3. angulosum**
86a. Bulb-tunic crustaceous, minutely
pitted; umbel often with bulbils;
perianth-segments 7–12 mm
      **28. roseum**
   b. Bulb-tunic fibrous, leathery or
membranous, not pitted; umbel
rarely or never with bulbils;
perianth-segments 4–9 mm (rarely
10–15 mm as in *A. unifolium*)    87
87a. Leaves usually 1 or 2          111
   b. Leaves 3 or more                88
88a. Bulb-tunics leathery; filaments of
inner stamens 2–3 times as broad
as outer ones          **44. rubellum**
   b. Bulb-tunics fibrous or membranous;
filaments of inner and outer
stamens equally broad          89
89a. Flower-stalks often less than twice
as long as the perianth, nearly
equal; ovary with 6 rounded knobs
      **84. geyeri**
   b. Flower-stalks 1–3 times as long as
the perianth; ovary without knobs
but sometimes crested          90
90a. Stamens equalling or slightly longer
than the perianth-segments
      **83. stellatum**

   b. Stamens two-thirds to three-
quarters as long as the perianth-
segments                91
91a. Leaves much longer than the
somewhat flattened stem which is
5–10 cm                92
   b. Leaves equalling or shorter than the
terete stem which is 20–30 cm    93
92a. Perianth-segments 1–1.6 cm, long-
acuminate          **88. falcifolium**
   b. Perianth-segments *c.* 8 mm, acute
      **89. crenulatum**
93a. Perianth-segments 6–8 mm;
filaments thread-like, almost equally
slender from the apex to the slightly
broadened base
      **91. campanulatum**
   b. Perianth-segments 1–1.4 cm;
filaments gradually broadened
downwards                94
94a. Outer perianth-segments broadly
elliptic, obtuse and mucronate,
inner entire          **92. dichlamydeum**
   b. Outer perianth-segments lanceolate
or narrowly ovate, acuminate, inner
often finely toothed above        95
95a. Bulb-tunic with network of 4- or
6-sided meshes; flower-stalks
1–2 cm          **90. acuminatum**
   b. Bulb-tunic with close horizontal
zigzag patterning; flower-stalks
2–3 cm          **93. peninsulare**
96a. Spathes persistent, 2, 1 or both
much longer than the umbel    97
   b. Spathe often deciduous; if persistent,
then 1; if 2, then both shorter than
or just equalling the umbel    104
97a. Stamens long-projecting        98
   b. Stamens included or with anthers
just protruding                100
98a. Perianth purple (rarely white);
ovary narrowly obovoid, much
longer than broad      **56. carinatum**
   b. Perianth yellow or yellowish; ovary
almost spherical, about as long as
broad                99
99a. Flower-stalks unequal, 3–5 times as
long as the perianth, 3–25 cm
      **57. flavum**
   b. Flower-stalks almost equal, 1½–2
times as long as the perianth, to
8 mm          **58. phrygium**
100a. Umbel loose, with unequal flower-
stalks, the longest 1.5–7 cm    101
   b. Umbel compact, the longest flower-
stalks usually less than 2 cm    102
101a. Umbel with bulbils, few-flowered
      **54. oleraceum**
   b. Umbel without bulbils, many-
flowered          **52. paniculatum**

102a. Perianth yellow    **6. scabriscapum**
   b. Perianth white or rose        103
103a. Leaves sheathing the lower seven-
eigths of stem; anthers whitish
      **55. tardans**
   b. Leaves sheathing the lower one-
third to half of stem; anthers yellow
      **53. pallens**
104a. Leaf-sheaths covering the stem
almost up to the umbel; perianth-
segments truncate
      **49. callimischon**
   b. Leaves sheathing the lower one-
third to two-thirds of the stem;
perianth-segments acute or obtuse
      105
105a. Umbel compressed with erect or
ascending flower-stalks; leaves
thread-like, to 0.4 mm broad
      **48. cupani**
   b. Umbel almost spherical or
hemispherical or with pendulous
flowers; leaves linear          106
106a. Leaves with sparse spreading hairs
      **45. chrysonemum**
   b. Leaves hairless                107
107a. Perianth 1–1.7 cm, blue
      **12. beesianum**
   b. Perianth 2.5–6 mm, not blue    108
108a. Perianth greenish yellow
   **6. scabriscapum & 8. obliquum**
   b. Perianth blue or purplish pink    109
109a. Perianth purplish pink
      **7. hymenorhizum**
   b. Perianth light blue              110
110a. Leaves 3-angled in section;
filaments of inner stamens simple or
with 2 short teeth near base
      **50. caeruleum**
   b. Leaves semi-circular in section;
filaments of inner stamens widened
for lower two-thirds and with 2
short teeth          **51. caesium**
111a. Bulb-tunic membranous with
horizontal undulating markings;
perianth-segments 1–1.5 cm
      **95. unifolium**
   b. Bulb-tunic with netted fibres;
perianth-segments 4–8 mm    112
112a. Ovary and capsule with short apical
projections (crests)      **86. textile**
   b. Ovary and capsule without apical
projections          **87. drummondii**
113a. Inner perianth-segments with
margin wavy or finely toothed in
upper part          **94. crispum**
   b. Inner perianth-segments flat at
margin          **93. peninsulare**
114a. Leaves hollow; perianth-segments
white with broad outer reddish

stripe; ovary without apical projections (crests)

**17. ramosum**

b. Leaves solid; perianth-segments without outer reddish stripe; ovary with apical projections      115

115a. Leaves several (up to 10); perianth-segments 1 cm      **85. plummerae**

b. Leaves 2; perianth-segments 6–8 mm      **86. textile**

**1. A. senescens** Linnaeus. Illustration: Botanical Magazine, 1150 (1808); Coste, Flore de la France 3: 341 (1905).
Bulbs oblong to narrowly conical, *c.* 1 cm thick, clustered, on a short rhizome; outer tunics membranous. Leaves 4–9, 4–30 × 1 cm, basal, almost flat, not keeled beneath. Stem 7–60 cm, 2-angled above. Spathes 2 or 3 lobed, 5–8 mm. Umbel 2–5 cm, hemispherical; flower-stalks 8–20 mm. Perianth cup-shaped, lilac; segments 3.5–8 × 2–2.5 mm, the outer shorter. Stamens projecting, 4–8 mm, the inner with broader base. Capsule 4 mm. *Europe & USSR (Siberia)*. H1. Summer–autumn.

This species comprises two main geographical races: (a) subsp. **senescens**, a robust plant, with leaves 5–10 mm broad and stems 3–60 cm, of northern Asia; (b) subsp. **montanum** (Fries) Holub (*A. montanum* Schmidt not Schrank; *A. lusitanicum* Lamarck; *A. fallax* Schultes & Schultes; *A. senescens* var. *calcareum* Wallroth) with leaves 1–6 mm broad and stems 7–45 cm, from western USSR (Ukraine) to northern Portugal. There are numerous intermediate plants in gardens, some with very glaucous leaves (*A. glaucum* Schrader) and twisted leaves (*A. spirale* Willdenow).

**2. A. nutans** Linnaeus. Illustration: Mathew, P.J. Redouté: Lilies and related flowers 51 (1981).
Bulbs narrowly conical, to 2 cm thick, on a short rhizome; outer tunics membranous; leaves 6–8, to 30 × 1.5 cm, basal, not keeled beneath. Stem 30–60 cm, 2-angled above. Spathe 2-lobed. Umbel *c.* 6 cm, almost spherical, nodding before flowering. Perianth cup-shaped, rose or lilac; segments 4–6 mm, the outer shorter. Stamens projecting, the inner nearly 3 times as broad as the outer and with a short tooth each side. *USSR (Siberia)*. H1. Summer.

**3. A. angulosum** Linnaeus (*A. acutangulum* Schlechtendal). Illustration: Redouté, Liliacées 5: 281 (1809); Reichenbach, Icones Florae Germanicae 10: 500

(1848); Coste, Flore de la France 3: 341 (1905).
Bulbs cylindric or narrowly conical, 5 mm thick, clustered, on a short rhizome; outer tunics membranous. Leaves 4–6, 10–25 × 1.5–6 mm, basal, channelled, sharply keeled beneath. Stem 20–45 cm, 2-angled above. Spathe 2–5-lobed, *c.* 1 cm. Umbel 2.5–4.5 cm, hemispherical, many-flowered; flower-stalks 1–3 cm, almost equal. Perianth cup-shaped, pale purple; segments 4–6 × 1.5–2.5 mm, the outer shorter. Stamens just included *c.* 6 mm, anthers becoming dark purple. Capsule 3.5 mm. *Europe to Siberia*. H1. Summer–autumn.

**4. A. ericetorum** Thore (*A. ochroleucum* Waldstein & Kitaibel). Illustration: Coste, Flore de la France 3: 336 (1905).
Bulbs oblong to very narrowly conical, to 1.5 cm thick, clustered on a short rhizome; outer tunics breaking into parallel fibres. Leaves 3 or 4, to 25 cm × 3 mm, basal. Stem 10–40 cm, terete. Spathe 2-lobed, to 1 cm. Umbel 1–2.5 cm, hemispherical, many-flowered; flower-stalks 4–10 mm, almost equal, smooth. Perianth cup-shaped, white or yellowish; segments 3.5–4 × 1.5–2 mm. Stamens long-projecting, 4.5–6.5 mm; anthers brownish. Capsule 3.5–4 mm. *Yugoslavia and Carpathians to SW France & N Portugal*. H4. Summer.

**5. A. albidum** Bieberstein (*A. flavescens* Besser; *A. ammophilum* Heuffel). Illustration: Reichenbach, Icones Florae Germanicae 10: 499 (1848).
Bulbs oblong to very narrowly conical, to 1 cm thick, clustered on a short rhizome; outer tunics membranous. Leaves 5–9, to 14 cm × 2.5 mm, basal. Stem 10–30 cm, slightly ribbed. Spathe 2-lobed, to 1 cm. Umbel 1.5–2.5 cm, hemispherical or compressed, many-flowered; flower-stalks 8–15 mm, almost equal, rough. Perianth cup-shaped or star-shaped, white or yellowish; segments 3.5 × 1–1.5 mm. Stamens projecting, 3–4 mm, anthers yellow. Capsule 3 mm. *Bulgaria, Romania, European USSR*. H5. Summer.

**6. A. scabriscapum** Boissier & Kotschy. Bulbs very narrowly conical to 1 cm thick, clustered on a short rhizome; outer tunics with netted fibres. Leaves 4–6, shorter than the stem, 2–4 mm broad, sheathing the lower one-quarter to one-third of stem. Stem 15–50 cm, rough or smooth. Spathe 2-lobed, to 1.3 cm. Umbel hemispherical or almost spherical, many-flowered; flower-

stalks 1.5–2.5 cm, almost equal. Perianth cup-shaped, yellow; segments 4–6 mm. Stamens slightly longer than perianth, the inner filaments twice as broad at base as the outer ones. *Turkey, Caucasus, C Asia, Iran, Iraq*. H4. Summer.

**7. A. hymenorhizum** Ledebour. Illustration: Ledebour, Icones 4: 359 (1833).
Bulbs very narrowly conical, to 2 cm thick, clustered on a short rhizome; outer tunics breaking into leathery glossy strips. Leaves 4–6, to 30 cm × 6 mm, sheathing lower half of stem. Stem 30–90 cm. Spathe 2-lobed, not longer than flower-stalks. Umbel hemispherical to almost spherical, 2–3.5 cm, many-flowered, dense; flower-stalks 1–1.5 cm, almost equal. Perianth bell-shaped, purplish pink; segments 4–6 × 1.5–2 mm. Stamens long-projecting, *c.* 8 mm; anthers yellow. *USSR (Soviet Central Asia, W Siberia), Iran, Afghanistan*. H3. Summer.

**8. A. obliquum** Linnaeus. Illustration: Botanical Magazine, 1408 (1811); Redouté, Liliacées 7: 364 (1812).
Bulbs oblong, to 2 cm thick, usually solitary, on a short rhizome; tunics membranous. Leaves 4–10, to 35 × 2 cm, sheathing lower half of stem. Stem 60–100 cm. Spathe 1, not longer than flower-stalks. Umbel hemispherical to almost spherical, 3.5–4 cm, many flowered, dense; flower-stalks 1–2 cm, almost equal. Perianth cup-shaped or almost spherical, pale yellowish green; segments 4–5 × 2.5 mm. Stamens projecting, *c.* 8–9 mm; anthers yellow. *Romania, Siberia, C Asia*. H3. Summer.

**9. A. mairei** Léveillé (*A. yunnanense* Diels). Illustration: Rix & Phillips, The bulb book, 146f (1981).
Bulb very slender; outer tunics with netted fibres. Leaves 13–25 cm × 1–10 mm, thread-like, sheathing lower half of stem. Stem 10–40 cm, 2-angled. Spathes 2. Umbel with flowers stiffly erect; flower-stalks 1.5–3 cm, unequal, papillose. Perianth bell-shaped, light to bright pink or white with red spots; segments 8–12 × 1.5–3.2 mm, recurving above, the outer longer. Stamens included, 5–6 mm, united at base. *SW China (Xizang, Yunnan)*. H5. Autumn.

**10. A. amabile** Stapf. Illustration: Botanical Magazine, 9257 (1931).
Bulb very slender, on a short rhizome, outer tunics with netted fibres. Leaves 2–4, 10–18 cm × 1 mm, thread-like, sheathing lower half of stem, erect. Stem 10–20 cm,

finely striped. Spathe 1, to 2 cm. Umbel
2–5-flowered, with some flowers pendent;
flower-stalks 1.2–1.6 cm. Perianth bell-
shaped, deep rose to magenta-crimson;
segments 1.2–1.6 cm × 1.5–2 mm, keeled,
recurving above. Stamens included,
5–6.5 mm, united at base. Capsule 4 mm.
*SW China (Yunnan).* H5. Summer–autumn.
    Doubtfully distinct as a species from
*A. mairei.*

**11. A. cyaneum** Regel (*A. purdomii* W.W.
Smith). Illustration: Gartenflora, t. 1317
(1890); Botanical Magazine, 9483 (1937);
Baileya **3**: 161 (1955); Rix & Phillips, The
bulb book, 147h (1981).
Bulbs cylindric, clustered, on a short
rhizome, outer tunics fibrous but
inconspicuously netted. Leaves 1–3,
15 cm × 1–4 mm, sheathing stem near
base, semi-cylindric. Stem 10–45 cm.
Spathe 1, 5–8 mm. Umbel 5–18-flowered,
pendent; flower-stalks 4–10 mm. Perianth
violet-blue to purplish; segments with
darker blue or green mid-veins,
4–6.5 × 2–2.5 mm. Stamens projecting,
5–9 mm, united at base, the inner broader
with a small tooth on either side, violet or
blue. *W China.* H5. Summer.

**12. A. beesianum** W.W. Smith. Illustration:
Botanical Magazine, 9331 (1933); Hay &
Synge, Dictionary of garden plants, 83
(1969).
Bulbs narrowly cylindric, 4–8 mm,
clustered, outer tunics with netted fibres.
Leaves 2–4, 15–20 cm × 4–10 mm,
sheathing the lower half to one-third of
stem, erect. Stem 25–50 cm. Spathe 1, to
1.2 cm. Umbel 6–12-flowered,
hemispherical, more or less pendent;
flower-stalks 7–10 mm. Perianth bell-
shaped to tubular, bright blue to purplish-
blue, segments 1.1–1.7 cm × 2.5–5 mm.
Stamens included, 7–11 mm, blue. Capsule
3-angled. *W China.* H4. Summer.

**13. A. sikkimense** Baker (*A. kansuense*
Regel; *A. tibeticum* Rendle). Illustration:
Gartenflora, t. 1317 (1890); Botanical
Magazine, 7290 (1893), 8858 (1920).
Bulbs cylindric, 3–5 mm, clustered on a
short rhizome, outer tunics with netted
fibres. Leaves 2–5, 7–30 cm × 2–5 mm,
sheathing base of stem, flat. Stem
10–40 cm. Spathe 1, *c.* 2 cm. Umbel few to
many-flowered, pendent; flower-stalks
2–6 mm, unequal. Perianth bell-shaped,
blue-purple, segments 6–10 × 3–4 mm,
acuminate, the outer shorter. Stamens
included, the inner shortly toothed at base.
*Himalaya to W China.* H4. Summer.

**14. A. wallichii** Kunth.
Bulb hardly developed, cylindric to ovoid;
outer tunics membranous. Leaves
60–90 cm × 8–20 mm, all sheathing the
stem to the same level in the lower part,
flat, keeled. Stem 30–75 cm, 3-angled.
Spathes 2, 2.5–3.8 cm. Umbel 5–7.5 cm,
many-flowered, loose; flower-stalks
2.5–3.8 cm. Perianth star-shaped, magenta
to purple, papery; segments 7–11 mm,
reflexed. Stamens included, 4–7 mm.
Capsule 6 mm. *Nepal to W China.* H4.

**15. A. macranthum** Baker (*A. oviflorum*
Regel). Illustration: Gartenflora **23**: t. 1134
(1883); Botanical Magazine, 6789 (1884).
Bulb oblong to oblong-cylindric, 3–7 mm,
outer tunics membranous. Leaves 3–many,
15–45 cm × 3 mm, channelled. Stems
many, 20–30 cm, 3-angled. Spathe 2 cm.
Umbel 7.5–10 cm, 3–12 (rarely to
50)-flowered, loose; flower-stalks 2.5–5 cm.
Perianth bell-shaped, dark purple, segments
8–12 × 4 mm. Stamens included or just
equalling perianth. Capsule 4 mm,
6-crested. *Himalaya to SW China.* H4.
Summer.

**16. A. przewalskianum** Regel
(*A. jacquemontii* Regel not Kunth).
Bulbs cylindric, 6–10 mm, clustered; outer
tunics with netted fibres, reddish. Leaves 3
or 4, to 35 cm × 1 mm, almost basal. Stem
15–30 cm, terete. Spathe to 1 cm. Umbel
hemispherical 1.5–2.5 cm, many-flowered;
flower-stalks 5–10 mm. Perianth star-
shaped; segments mauve to lilac, spreading
or reflexed, 4–5 mm. Stamens projecting,
*c.* 7 mm, the inner broadened below the
middle with a small tooth each side.
*Himalaya to W China.* H4.

**17. A. ramosum** Linnaeus (*A. odorum*
Linnaeus; *A. tataricum* Linnaeus filius).
Illustration: Redouté, Liliacées **2**: 98
(1804); Botanical Magazine, 1142
(1808); Rix & Phillips, The bulb book,
145d (1981).
Bulb cylindric, 1 cm, clustered, on a short
rhizome; outer tunics with netted fibres.
Leaves 4–9, to 35 cm, sheathing stem for
one-third of its length, semi-terete, hollow.
Stem 24–50 cm, terete to slightly angled.
Spathes 1 or 2. Umbel usually compressed,
3–5 cm, few- to many-flowered; flower-
stalks 1–3 cm. Perianth bell-shaped;
segments white with dark red mid-veins,
8–10 mm, lanceolate-oblong. Stamens
included, half the length of perianth-
segments. Capsule 5–6 mm, broadest below
middle. *C. Asia.* H3. Summer.

**18. A. tuberosum** Sprengel (*A. uliginosum*
G. Don; *A. tuberosum* Roxburgh).
Illustration: Herbertia **11**: 231, 234, 235
(1946); Botanical Magazine, n.s., 386
(1962); Rix & Phillips, The bulb book, 144a
(1981).
Bulbs cylindric, *c.* 1 cm broad, clustered on
a short rhizome; outer tunics with netted
fibres. Leaves 4–9, to 35 cm × 8 mm,
sheathing the stem for about one-eighth of
its length, solid, slightly keeled. Stem
25–50 cm, slightly angled. Spathe 1,
usually becoming 2-lobed, shorter than the
flower-stalks. Umbel hemispherical or
compressed, many-flowered, 3–5 cm;
flower-stalks 1–3 cm. Perianth star-shaped;
segments white with a faint green or brown
mid-l;ine on the back, 4–7 mm. Stamens
included, four-fifths of the length of
perianth-segments. Capsule 4–5 mm,
broadest above middle. *SE Asia (native or
cultivated from Nepal to Japan).* H3. Late
summer–autumn.
    Although grown in Europe usually for its
white scented flowers and known as
'Chinese Chives' or 'Kiu tsai', this is an
important culinary herb in China and
Japan.

**19. A. cyathophorum** Bureau & Franchet.
Illustration: Baileya **2**: 119 (1954);
Botanical Magazine, n.s., 252 (1955); Rix
& Phillips, The bulb book, 147 (1981).
Bulbs cylindric, 3–5 mm, clustered; outer
tunics breaking down into parallel fibres.
Leaves 3–6, 18–24.5 cm × 1–5 mm,
almost basal. Stem 19–41.5 cm, 3-angled.
Spathe 1, entire or splitting into 2 or 3
lobes, 1 cm. Umbel 6–30-flowered, loose,
without bulbils; flower-stalks 8–20 mm,
unequal. Perianth bell-shaped, light reddish
purple; segments 6–8 × 1.5–2 mm, the
outer slightly narrower. Stamens included,
5 mm, filaments fused into a tube for half
their length. Capsule 4 mm. *NW China.* H2.
Summer.
    The plants in cultivation belong to var.
**farreri** (Stearn) Stearn (*A. farreri* Stearn).

**20. A. narcissiflorum** Villars
(*A. grandiflorum* Lamarck; *A. pedemontanum*
Willdenow). Illustration: Reichenbach,
Icones Florae Germanicae **10**: 504 (1848);
Coste, Flore de la France **3**: 340 (1905);
Baileya **2**: 119 (1954).
Bulbs oblong, 5 mm, clustered, on a short
rhizome; outer tunic consisting of layers of
parallel fibres. Leaves 3–5,
9–28 cm × 2–6 mm, sheathing stem all to
about the same level, flat. Stem 15–35 cm,
2-edged. Spathe 1, often 2 or 3-lobed.
Umbel 5–8-flowered, erect in flower;

flower-stalks 8–18 mm. Perianth bell-shaped, purple, segments 1–1.5 cm, the outer narrower. Stamens included, filaments 5 mm. Capsule 5 mm. *SW Alps (France, Italy), N Portugal*. H4.

A. insubricum is often cultivated as A. narcissiflorum.

**21. A. insubricum** Boissier & Reuter (*A. narcissiflorum* var. *insubricum* (Boissier & Reuter) Fiori). Illustration: Botanical Magazine, 6182 (1875); Rix & Phillips, The bulb book, 147 (1981).
Bulbs oblong, 5 mm, clustered on a short rhizome, outer tunics membranous, sometimes with a few parallel fibres at the base. Leaves 3 or 4, 12–20 cm × 2–5 mm, all sheathing the stem to the same level in the lower part, flat. Stem 14–30 cm, 2-edged. Spathe 1, to 1.8 cm. Umbel 3–5-flowered, pendent; flower-stalks 8–20 mm. Perianth purple, segments *c*. 18 mm, the outer narrower. Stamens included, 6 mm. *N Italy*. H5.

**22. A. victorialis** Linnaeus. Illustration: Botanical Magazine, 1222 (1809); Redouté, Liliacées 5: 265 (1809); Reichenbach, Icones Florae Germanicae 10: 508 (1848); Coste, Flore de la France 3: 335 (1905).
Bulbs cylindric, 1–2 cm thick, clustered on a short rhizome; outer tunics densely fibrous. Leaves 2 or 3, 8–25 × 1.8–9 cm, sheathing the stem for up to half its length, narrowly lanceolate to broadly elliptic, narrowed at base into a short stalk. Stem 30–60 cm, 2-edged below. Spathes 1 or 2, shorter than flower-stalks, persistent. Umbel 3–5 cm, spherical or hemispherical, many-flowered; flower-stalks 1–3 cm, unequal. Perianth star-shaped, pale or greenish yellow; segments 4–5 × 2.25 mm, the outer narrower, all ultimately deflexed. Stamens simple, projecting. Ovary with distinct nectary pits. Capsule *c*. 4 mm. *S Europe, N Asia*. H3. Summer.

**23. A. tricoccum** Aiton. Illustration: Baileya 2: 111 (1954); Rickett, Wild flowers of the United States 1: t. 31 (1966).
Bulbs ovoid, clustered on a short rhizome, outer tunics with netted fibres. Leaves 2 or 3, 10–30 × 2.5–5 cm, basal, narrowed into long stalks, deciduous, withering before the flowers appear. Stem 10–40 cm, terete. Spathes 2. Umbel many-flowered, loose, hemispherical, without bulbils; flower-stalks 1–2 cm. Perianth white, star-shaped; segments 4–6 mm, obtuse. *E & C North America from Quebec to Virginia*. H2. Summer.

**24. A. schoenoprasum** Linnaeus (*A. sibiricum* Linnaeus; *A. purpurascens* Losa). Illustration: Reichenbach, Icones Florae Germanicae 10: 496 (1848); Sowerby & Smith, English Botany, Supplement 4: 2934 (1849); Pratt, Grasses, sedges and ferns of Great Britain, edn 3, 5: 225 (1873); Rix & Phillips, The bulb book, 143 (1981).
Bulbs very narrowly conical, clustered, borne on a short rhizome, outer tunics membranous. Leaves 1 or 2 (rarely more), up to 35 cm × 1–6 mm, sheathing up to one-third of stem, cylindric, hollow. Stem 5–50 cm, hollow. Spathe with 2 or 3 lobes, to 1.5 cm. Umbel 1.5–5 cm, dense, 8–30-flowered, without bulbils; flower-stalks 2–20 mm, more or less unequal. Perianth lilac or pale purple, occasionally white, bell-shaped; segments 7–15 × 2.5–4 mm, lanceolate. Stamens included, 5–7.2 mm. Capsule 4 mm. *Northern hemisphere*. H1.

A species with many local variants, some dwarf, some tall, long cultivated as a culinary herb (chives).

**25. A. cepa** Linnaeus. Illustration: Botanical Magazine, 1469 (1812); Reichenbach, Icones Florae Germanicae 10: 495 (1848); Belmontia 7: 85 (1977).
Bulb often flattened-spherical, to 10 cm thick, outer tunics membranous. Leaves up to 10, to 40 × 2 cm basal or sheathing the lower part of stem, semi-circular in section. Stem to 100 cm, hollow, inflated in lower half, thick. Spathes often 3, shorter than umbel. Umbel 4–9 cm, dense, many-flowered, with bulbils, or with only bulbils ('Viviparum'); flower-stalks to 4 cm. Perianth star-shaped, white, segments 3–4.5 × 2–2.5 mm, white with green mid-veins, slightly unequal. Stamens projecting, *c*. 4–6 mm, the inner shortly toothed. Capsule 5 mm. *Not known in the wild*. H1. Summer.

Widely cultivated for its edible bulbs (onions); apparently derived from *A. oschaninii* B. Fedtschenko (*A. cepa* var. *sylvestre* Regel) of C Asia. 'Perutile' which is sometimes cultivated, rarely produces flowering stems.

**26. A. fistulosum** Linnaeus. Illustration: Botanical Magazine, 1230 (1809); Reichenbach, Icones Florae Germanicae 10: 495 (1848); Belmontia 7: 88 (1977).
Bulb cylindric, borne on a short rhizome; outer tunics membranous. Leaves 2–6, 6–30 cm × 5–15 mm, sheathing the lower one-quarter to one-third of stem, hollow, circular in section. Stem 12–70 cm, hollow, inflated in the middle. Spathes 1 or

2, almost equalling umbel. Umbel 1.5–5 cm, dense, many-flowered or with only bulbils ('Viviparum'); flower-stalks 3–20 mm, unequal. Perianth conically bell-shaped, yellowish white, segments unequal, outer 6–7 × 2 mm, inner 7–9 × 3 mm. Stamens long-projecting, 2–3 times as long as the perianth; filaments 8–12 mm. Capsule 4 mm. *Not known in the wild*. H1. Summer.

Widely cultivated, particularly in the Far East, where it originated, but unknown in a wild state.

**27. A. altaicum** Pallas.
Very like *A. fistulosum* but with more developed, narrowly ovoid bulbs to 4 cm thick, the membranous tunics reddish brown, the almost equal flower-stalks slightly shorter than or at most 1½ times as long as the yellowish, bell-shaped perianth. *USSR (Siberia), Mongolia*. H1. Summer.

**28. A. roseum** Linnaeus (*A. amoenum* G. Don; *A. incarnatum* Hornemann). Illustration: Botanical Magazine, 978 (1806); Redouté, Liliacées 4: 213 (1808); Reichenbach, Icones Florae Germanicae 10: 486, 504 (1848); Belmontia 7: 173–4 (1977); Rix & Phillips, The bulb book, 139 (1981).
Bulb ovoid to spherical, 1.5 cm, with numerous bulblets, outer tunics crustaceous, with numerous minute pits or perforations. Leaves 2–7, 12–35 (rarely to 60) cm × 1–15 mm, sheathing the lower one-fifth of the stem. Stem 10–65 cm, solid. Spathe 1, deeply 3 or 4-lobed, 1.2–2.5 cm. Umbel 2–8 cm, 5–many-flowered, without bulbils (var. **roseum**), or with bulbils (var. **bulbiferum** de Candolle); flower-stalks 7–45 mm. Perianth bell-shaped to cup-shaped, pink or white, segments 7–12 × 3–5 mm. Stamens included, filaments 7 mm. Capsule 4 mm. *S Europe, N Africa, Turkey*. H4. Spring–summer.

**29. A. zebdanense** Boissier & Noë. Illustration: Rix & Phillips, The bulb book, 139 (1981).
Bulbs ovoid, 1–1.5 cm; outer tunics membranous, not pitted. Leaves 2, slightly shorter than stem, 3–6 mm broad, flat to channelled. Stem 25–40 cm. Spathe 1, 1–1.5 cm. Umbel 3–10-flowered; flower-stalks 8–13 mm, equal. Perianth bell-shaped, white; segments 9–13 mm, very obtuse. Stamens included, 4–7 mm. Capsule 4 mm. *Lebanon*. H5. Spring.

**30. A. massaessylum** Battandier & Trabut. Illustration: Belmontia 7: 166 (1977); Pastor & Valdés, Revision del Genero Allium, 135 (1983).

Bulbs ovoid or almost spherical, 1–2 cm; outer tunics papery, with sinuate patterning. Leaves 1–3, hairless, to 30 × 1 cm, basal, flat to channelled. Stem 15–40 cm. Spathes 2, persistent, 1–1.5 cm. Umbel compressed, 2–6 cm, many-flowered; flower-stalks 1–2.5 cm. Perianth cup-shaped; segments white, with purple mid-vein, 1–1.4 cm, somewhat acute. Stamens included, 5–6.5 mm. Capsule 3–5 mm. *Algeria, Morocco, Portugal & Spain*. H5. Spring.

**31. A. neapolitanum** Cyrillo (*A. cowanii* Lindley; *A. album* Santi; *A. lacteum* Smith). Illustration: Edwards's Botanical Register 9: t. 758 (1823); Botanical Magazine, 3531 (1836); Reichenbach, Icones Florae Germanicae 10: 507 (1848); Belmontia 7: 161 (1977); Rix & Phillips, The bulb book, 139, 143 (1981).
Bulbs almost spherical, 1–2 cm, outer tunics membranous, or crustaceous, not pitted. Leaves usually 2, 8–35 cm × 5–20 mm, sheathing lower one-fifth to one-quarter of stem, keeled beneath, hairless. Stem 20–50 cm, slightly angled, solid. Spathe 1, 1.2–2.5 cm. Umbel 5–11 cm, many-flowered, without bulbils; flower-stalks 1.5–3.5 cm. Perianth cup-shaped or star-shaped, white, segments 7–12 × 4–6 mm. Stamens included, 6 mm. Capsule 5 mm, enclosed within shiny perianth-segments. *S Europe, N Africa*. H4. Spring.

**32. A. subhirsutum** Linnaeus (*A. hirsutum* Lamarck; *A. ciliatum* Cyrillo). Illustration: Botanical Magazine, 774 (1804); Loddiges' Botanical Cabinet 10: 943 (1824); Reichenbach, Icones Florae Germanicae 10: 502 (1848); Belmontia 7: 127–8 (1977); Rix & Phillips, The bulb book, 143 (1981).
Bulbs almost spherical, to 1.5 cm, outer tunics membranous, not pitted. Leaves 2 or 3 (rarely to 7), 8–50 cm × 1–20 mm, flat, hairy at margin, almost basal. Stem 7–30 cm, solid. Spathe 1, 8–25 mm. Umbel 2–8 cm, with few to many flowers, loose; flower-stalks to 4 cm, unequal, 3–5 times as long as perianth. Perianth star-shaped, white, segments 7–9 mm, unequal. Stamens included, 4–6 mm. Capsule 3 mm. *Mediterranean area*. H5. Spring.

**33. A. subvillosum** Schultes & Schultes (*A. subhirsutum* var. *subvillosum* (Schultes & Schultes) Ball). Illustration: Redouté, Liliacées 6: t. 305 (1812); Belmontia 7: 127 (1977); Pastor & Valdés, Revision del genero Allium, 119 (1983).
Bulbs almost spherical, to 2.5 cm; outer tunics membranous, not pitted. Leaves 2–5, 6–40 cm × 2–20 mm, ciliate at

margin, almost basal. Stems 10–60 cm, terete. Spathe 1, becoming 3–4-lobed, persistent. Umbel 2.5–5 cm, hemispherical or compressed, many-flowered, flower-stalks to 2 cm. Perianth cup-shaped; segments white, 5–9 mm, acute or obtuse. Stamens slightly shorter to slightly longer than the perianth-segments; filaments 6–11 mm; anthers yellow. Capsule 3–4 mm. *W Mediterranean area, Portugal, Canary Islands*. H5. Spring.

**34. A. trifoliatum** Cyrillo (*A. subhirsutum* subsp. *trifoliatum* (Cyrillo) Arcangeli). Illustration: Bicknell, Flowering Plants of the Riviera, t. 78 (1885).
Like *A. subhirsutum* but with flower-stalks 1.5–3 times as long as perianth-segments, which are tinged with pink or pink-veined and often redden on ageing, and anthers yellowish. *Mediterranean area*. H5.

**35. A. moly** Linnaeus. Illustration: Botanical Magazine, 499 (1800); Redouté, Liliacées 2: 97 (1805); Reichenbach, Icones Florae Germanicae 10: 501 (1848); Rix & Phillips, The bulb book, 143 (1981).
Bulb almost spherical, to 2.5 cm, outer tunics parchment-like. Leaves 1–3, 20–30 × 1.5–3.5 cm, almost basal, keeled beneath. Stem 12–35 cm. Spathes 2, less than 3.5 cm. Umbel 4–7 cm, many-flowered, rarely with bulbils (var. **bulbilliferum** Rouy); flower-stalks 1.5–3.5 cm, unequal. Perianth star-shaped, segments bright yellow, with greenish keels outside, 9–12 × 4–5 mm. Stamens included; filaments 5–6 mm. Capsule covered by the persistent perianth-segments. *E Spain & SW France*. H3. Summer.

**36. A. scorzonerifolium** de Candolle. Illustration: Redouté, Liliacées 2: t. 99 (1804); Pastor & Valdés, Revision del genero Allium, 133 (1983).
Bulbs almost spherical, 1.5 cm; outer tunics parchment-like. Leaves 1–3, 18–40 cm × 3–7 mm, almost basal, keeled beneath. Stem 14–30 cm, angled. Spathe 1. Umbel 4–5 cm, few-flowered with bulbils (var. **scorzonerifolium**) or without bulbils and up to 18 flowers (var. **xericiense** (Perez Lara) Fernandes); flower-stalks 15–30 cm. Perianth star-shaped; segments yellow, 7–12 × 2–5 mm. Stamens included; filaments 4–7 mm. *Portugal & Spain*. H4. Summer.

**37. A. oreophilum** Meyer (*A. ostrowskianum* Regel). Illustration: Gartenflora, t. 775 (1873), 1089 (1882); Botanical Magazine, 7756 (1901); Rix & Phillips, The bulb book, 143 (1981).

Bulb ovoid-spherical, 1–2 cm; outer tunics papery, grey. Leaves 2, longer than umbel, 2–8 mm broad, linear. Stem 5–20 cm. Spathe 8–15 mm. Umbel few-flowered, loose; flower-stalks 7–15 mm, almost equal. Perianth broadly bell-shaped, segments rose purple with darker mid-vein, 8–11 mm, ovate. Stamens included; filaments 3–5.5 mm, united into a ring for half their length. Stigma 3-lobed. Capsule 4 mm. *USSR (Caucasus, C Asia)*. H4. Spring–summer.

**38. A. chamaemoly** Linnaeus. Illustration: Botanical Magazine, 1203 (1809); Redouté, Liliacées 6: 325 (1812); Reichenbach, Icones Florae Germanicae 10: 501 (1848); Belmontia 7: 155 (1977).
Bulbs ovoid, 5–10 mm, outer tunics parchment-like, pitted. Leaves 2–5 (rarely to 11), 3–27 cm × 3–10 mm, basal, usually spreading, flat, with margins ciliate, sometimes hairy on upper surface. Stem very short, solid. Spathe 1, 2–4-lobed. Umbel 1.5–2.5 cm, 2–20-flowered, almost stalkless on the rosette of leaves; flower-stalks 5–10 mm. Perianth star-shaped, segments white with green or purplish mid-veins. Stamens included, 5.5 mm. Capsule 4 mm. *S Europe, N Africa*. H5. Winter–spring.

**39. A. ursinum** Linnaeus. Illustration: Reichenbach, Icones Florae Germanicae 10: 507 (1848); Hogg & Johnson, Wild flowers of Great Britain 3: 190 (1866); Ross-Craig, Drawings of British plants 29: pl. 20 (1972); Rix & Phillips, The bulb book, 139 (1981).
Bulb more or less cylindric; outer tunics papery with a few parallel fibres at the base. Leaves 2 or 3, 6–20 × 1.5–8 cm, basal, narrowed at base into a stalk. Stem 10–50 cm, 2 or 3-angled. Spathes 2, shorter than flower-stalks, persistent. Umbel 2.5–6 cm, 6–20-flowered; flower-stalks 1–4.5 cm, rough in subsp. **ursinum**, smooth in subsp. **ucrainicum** Kleopow & Oxner. Perianth star-shaped, segments white, 7–12 × 2–2.5 mm, lanceolate, acute. Stamens 5 mm. Capsule 3–4 mm. *Europe, USSR (Caucasus)*. H1. Spring.

**40. A. triquetrum** Linnaeus. Illustration: Botanical Magazine, 869 (1805); Redouté, Liliacées 5: 319 (1812); Reichenbach, Icones Florae Germanicae 10: 503 (1848); Belmontia 7: 115–16 (1977); Rix & Phillips, The bulb book, 139 (1981).
Bulb spherical to ovate, 1.2–2 cm; outer tunics membranous. Leaves 2–5, up to

50 × 1.7 cm, keeled, basal. Stems 1–4, 10–35 cm, 3-angled. Spathes 2, to 4 cm. Umbel 4–7 cm, 3–15-flowered, never with bulbils, usually 1-sided, loose; flower-stalks to 2.5 cm, 3-angled. Perianth bell-shaped, drooping; segments white with distinct green mid-veins, 1–1.9 cm × 2–5 mm, the outer broader. Stamens 6–7 mm, united at base. Stigma 3-lobed. Capsule 6–7 mm; seeds with an appendage. *W Mediterranean area, naturalised in Britain.* H4. Spring.

**41. A. pendulinum** Tenore (*A. triquetrum* var. *pendulinum* (Tenore) Regel). Illustration: Loddiges' Botanical Cabinet **11**: t. 1087 (1825); Reichenbach, Icones Florae Germanicae **10**: t. 503 (1848); Coste, Flore de la France **3**: 340 (1905).
Bulbs almost spherical, *c.* 1 cm; outer tunics membranous. Leaves 2, to 25 cm × 7 mm, keeled. Stem 6–25 cm, 3-angled. Spathes 2. Umbel not 1-sided, loose; flower-stalks to 4 cm, ascending, later drooping. Perianth star-shaped; segments white with distinct green mid-vein, 3–5 × 1–1.5 mm. Stamens 4–5 mm. Stigma 3-lobed. Capsule 4–6 mm; seeds with an appendage. *C Mediterranean area.* H4. Spring.

**42. A. paradoxum** (Bieberstein) G. Don. Illustration: Flora Iranica **76**: 18 (1971); Rix & Phillips, The bulb book, 71 (1981). Bulbs almost spherical, 1 cm; outer tunics membranous. Leaf 1, to 30 × 2.5 cm, keeled. Stem 15–30 cm, 3-angled. Spathes 2, 1–2.5 cm. Umbel with bulbils and usually 1-flowered, or without flowers (var. **paradoxum**) or without bulbils and up to 10 flowers (var. **normale** Stearn); flower-stalks 2–4.5 cm. Perianth bell-shaped; segments white with a faint green mid-vein, 8–12 × 6 mm. Stamens included, united at base. Stigma 3-lobed. Capsule 5 mm; seeds with a white appendage. *USSR (Caucasus), Iran.* H4. Spring.

This is usually represented in gardens by the weedy, bulbil-bearing var. *paradoxum* introduced early in the nineteenth century and sometimes naturalised; the floriferous var. *normale* was introduced in the mid-twentieth century.

**43. A. moschatum** Linnaeus. Illustration: Reichenbach, Icones Florae Germanicae **10**: t. 498 (1848); Pastor & Valdés, Revision del genero Allium, 97 (1983).
Bulbs narrowly ovoid, clustered, 1–1.5 cm; outer tunics breaking into fibres. Leaves 3–6, to 14 cm × 0.5 mm, thread-like, sheathing the lower one-quarter to one-third of the stem. Stem 10–35 cm. Spathes

2, short. Umbel compressed with 3–12 flowers, loose; flower-stalks 1–1.5 cm, almost equal. Perianth bell-shaped; segments 5–7 × 1.5–2 mm, whitish or pink with darker mid-vein, acute. Stamens included. *S Europe from E Spain to the Balkan Peninsula.* H4. Summer.

**44. A. rubellum** Bieberstein.
Bulbs ovoid, 7.5–15 mm, outer tunics leathery; bulblets yellow to brown. Leaves 3–5, to 30 cm × 1–5 mm, sheathing lower one-quarter of stem, hollow. Stem 10–60 cm. Spathe shorter than umbel. Umbel 2.5–4 cm, hemispherical, many-flowered; flower-stalks 5–15 mm, almost equal. Perianth bell-shaped, segments pink with darker mid-veins, 5–6 × 1.5–2 mm. Stamens included, 3–4 mm, inner filaments 1½–2 times wider at base than the outer. Capsule 3 mm. *USSR (SE Russia, Caucasus), Afghanistan, W Himalaya.* H4.

**45. A. chrysonemum** Stearn. Illustration: Annales Musei Goulandris **4**: 150 (1978); Pastor & Valdés, Revision del genero Allium, 103 (1983).
Bulbs ovoid, *c.* 1.5 cm; outer tunics membranous. Leaves 3 or 4, to 15 cm × 2 mm, sheathing lower one-third of the stem, hairy. Stem 30–60 cm. Spathes 2. Umbel *c.* 4 cm, loose, many-flowered; flower-stalks 1–1.5 cm. Perianth cup-shaped; segments 4.5–5 × 2–2.4 mm, pale greenish yellow. Stamens slightly projecting; filaments yellow. Capsule 4 mm. *Spain.* H4. Summer.

**46. A. chinense** G. Don (*A. bakeri* Regel). Illustration: Economic Botany **14**: 70, 78 (1960).
Bulbs narrowly ovoid, 1–1.5 cm, clustered; outer tunics membranous. Leaves usually 3, to 30 cm × 1–3 mm, almost basal, hollow, 3–5-angled. Stem 28–30 cm. Spathes 2. Umbel hemispherical, loose, with up to 18 flowers; flower-stalks 1–1.5 cm. Perianth cup-shaped; segments light violet, 5 × 3.5–4 mm, obtuse. Stamens projecting; inner 3 filaments with broadened bases and each with 2 short teeth. *China, cultivated in Japan as economic crop (rakkyo).* H3. Autumn.

**47. A. virgunculae** Maekawa & Kitamura.
Bulbs 5–7 mm, clustered; outer tunics membranous. Leaves 3–5, 10–20 cm × 1 mm, thread-like, with numerous minute white dots. Stem 8–22 cm. Spathe 1. Umbel loose, 2–12-flowered; flower-stalks 1.2–1.5 cm. Perianth star-shaped; segments *c.* 5 × 3–4 mm, rose, obtuse. Stamens projecting,

*c.* 6 mm, the inner 3 each with a tooth each side at base. Capsule *c.* 4 mm. *Japan (Kyushu).* H3.

**48. A. cupani** Rafinesque. Illustration: Fiori & Paoletti, Iconographia Florae Italicae, 80 (1898); Belmontia **7**: 19 (1977).
Bulbs narrowly ovoid, 7–15 mm; outer tunics with netted fibres. Leaves 3–5, to 10 cm × 0.4 mm, thread-like, sheathing lower one-third to two-thirds of stem. Spathe 1, tubular at base. Umbel compressed, few-flowered; flower-stalks to 4 cm, very unequal. Perianth bell-shaped, segments 5.5–9 × 1.5–2 mm, whitish or pink with darker mid-vein. Stamens included, equal. Capsule 4 mm. *Mediterranean area.* H4. Summer.

**49. A. callimischon** Link. Illustration: Rix & Phillips, The bulb book, 177 (1981).
Bulbs narrowly ovoid, 1 cm, outer tunics leathery, breaking down into parallel fibres. Leaves 3–5, the lowest to 30 cm × 1 mm, sheathing stem almost up to umbel. Stem 9–38 cm. Spathe 1, 2–4 cm. Umbel 8–25-flowered, clustered; flower stalks 7–25 mm, very unequal. Perianth cup-shaped, segments white with brown or reddish mid-veins, often becoming pink, 5–7 × 1.5 mm. *Greece, Crete & W Turkey.* H5.

This consists of two geographical variants: subsp. **callimischon**, with segments with a distinct mid-vein but unspotted, and subsp. **haemostictum** Stearn, with segments with dark red spots in upper part.

**50. A. caeruleum** Pallas (*A. azureum* Ledebour) Illustration: Edwards's Botanical Register **26**: t. 51 (1840); Rix & Phillips, The bulb book, 137 (1981).
Bulbs almost spherical, 1–2 cm, outer tunics membranous. Leaves 2–4, 7 cm × 2–4 mm, sheathing lower one-quarter to one-third of stem, 3-angled. Stem 20–80 cm. Spathes 2, to 1.5 cm. Umbel 3–4 cm, almost spherical to hemispherical, many-flowered; flower-stalks 1–2 cm, almost equal. Perianth cup-shaped, segments blue with darker mid-vein, 3.5–4.5 × 1.2–1.4 mm. Stamens equalling perianth, or slightly projecting, united at base; anthers blue. Capsule 3 mm. *N to C Asia.* H3. Summer.

**51. A. caesium** Schrenk.
Bulbs ovoid, outer tunics almost leathery. Leaves 2–4, 1–3 mm broad, sheathing up to half the stem, semi-cylindric, hollow. Stem 25–65 cm (rarely less). Spathes 2, to 1.5 cm. Umbel more than 1.5 cm, many-flowered, with some bulbils, or rarely all

bulbils; flower-stalks 1.2–2 cm. Perianth-segments violet-blue with a brownish-green mid-vein, or more rarely white, 4–5 × 2 mm, the outer slightly narrower. Stamens included, united at base. Capsule 3–4 mm. *USSR (Siberia to C Asia)*. H3. Summer.

**52. A. paniculatum** Linnaeus (*A. longispathum* Delaroche). Illustration: Redouté, Liliacées **6**: 316 (1812); Reichenbach, Icones Florae Germanicae **10**: t. 487 (1848).
Bulbs ovoid, 1–2.5 cm; outer tunics membranous. Leaves 3–5, to 25 cm × 1.5 mm, sheathing lower one-third to half of stem, prominently ribbed beneath. Stem 30–70 cm. Spathes 2, unequal, each contracted into a long slender appendage very much longer than the flower-stalks. Umbel 3.5–7 cm, loose, without bulbils; flower-stalks very unequal, 1–7 cm. Perianth bell-shaped, white, pink, yellowish or greenish brown; segments 4.5–7 × 1.5–2.5 mm, obtuse. Stamens included or slightly projecting. Ovary ellipsoid, twice as long as broad. Capsule *c.* 5 mm. *Europe*. H1. Summer.

A very variable, widespread species divided into several subspecies, including subsp. **paniculatum** with white or pink flowers, and subsp. **fuscum** (Waldstein & Kitaibel) Arcangeli, with greenish, brownish or dirty white flowers.

**53. A. pallens** Linnaeus (*A. coppoleri* Tineo; *A. paniculatum* var. *pallens* (Linnaeus) Grenier & Godron). Illustration: Pastor & Valdés, Revision del genero Allium, 91 (1983).
Bulbs 1–1.5 cm; outer tunics membranous. Leaves 3 or 4, to 30 cm × 5 mm, thread-like to linear, sheathing lower one-third to half of the stem. Stem 10–30 cm, rarely more. Spathes 2, unequal, each contracted into a long slender appendage. Umbel 1.5–3 cm, compact, few- or many-flowered; flower-stalks 5–20 mm, almost equal. Perianth cup-shaped, white or rose-tinged; segments 3.5–5 × 1.5–2 mm, truncate or obtuse. Stamens included or with projecting anthers. Ovary ellipsoid, longer than broad. *S Europe*. H4. Summer.

**54. A. oleraceum** Linnaeus. Illustration: Sowerby & Smith, English Botany **7**: 488 (1798); Reichenbach, Icones Florae Germanicae **10**: 487 (1848); Butcher, New illustrated British flora **2**: 652 (1961).
Bulbs 1–2 cm; outer tunics membranous. Leaves 2–4, to 25 cm × 6 mm, thread-like to linear, sheathing the lower half or more

of the stem. Stem 25–100 cm. Spathes 2, unequal, each contracted into a slender appendage, the longest to 20 cm. Umbel with bulbils, loose; flower-stalks 1.5–6 cm, slightly flattened. Perianth cup-shaped, whitish but variably tinged with brown, green or pink; segments 5–7 × 2–3 mm, obtuse. Stamens included. Ovary narrowly obovoid or cylindric, truncate. *Most of Europe*. H1. Summer.

**55. A. tardans** Greuter & Zahariadi.
Bulbs 1–1.5 cm; outer tunics membranous, breaking into strips. Leaves 3 or 4, to 16 cm × 0.5 mm, thread-like, sheathing the lower five-sixths of stem. Stem 10–30 cm. Spathes 2, unequal, each contracted into a long slender appendage. Umbel 1.2–5 cm, compressed, flower-stalks 9–15 mm. Perianth cup-shaped, segments pale pink with green mid-veins, 5–6.5 × 1.8–2 mm, almost truncate. Stamens included or with anthers projecting. *Crete*. H5. Autumn.

**56. A. carinatum** Linnaeus. Illustration: Sowerby & Smith, English Botany **24**: 1658 (1807); Redouté, Liliacées **10**: 368 (1813); Reichenbach, Icones Florae Germanicae **10**: 482 (1848); Bonnier, Flore complète **10**: 584 (1929).
Bulbs ovoid to oblong, *c.* 1 cm; outer tunics membranous, sometimes breaking into parallel fibres. Leaves 2–4, to 20 cm × 1–2.5 mm, sheathing stem for half to three-fifths of its length. Stem 30–60 cm. Spathes 2, unequal, each contracted into a slender appendage to 12 cm. Umbel *c.* 30-flowered, usually with bulbils; flower-stalks 1–2.5 cm, unequal, the outer curving downwards. Perianth cup-shaped, purple, segments 4–6 × 1.5–2 mm. Stamens long-projecting, 6.5–9 mm. Ovary oblong, much longer than broad. Capsule 5 mm, rarely produced. *W, C & S Europe, S Russia, Turkey*. H3. Summer.

Subsp. **carinatum** has numerous greenish bulbils in the umbel and few flowers and can become a weed. Subsp. **pulchellum** Bonnier & Layens (*A. pulchellum* G. Don) has no bulbils but numerous flowers on purplish stalks and is an attractive garden plant.

**57. A. flavum** Linnaeus (*A. tauricum* Reichenbach). Illustration: Redouté, Liliacées **7**: 119 (1805); Botanical Magazine, 1330 (1810); Reichenbach, Icones Florae Germanicae **10**: 485 (1848); Rix & Phillips, The bulb book, 147 (1981).
Bulbs ovoid, 1–1.5 cm, outer tunics membranous. Leaves 2 or 3, to 20 cm × 2 mm, sheathing lower one-third

to half of stem. Stem 8–50 cm. Spathes 2, unequal, each contracted into a long slender beak to 11 cm. Umbel 9–60-flowered, hemispherical, never with bulbils; flower-stalks 3–25 mm, unequal, erect in fruit, the outer usually curving downwards. Perianth bell-shaped, lemon-yellow, segments 4.5–5 × 2 mm. Stamens projecting, 7–8 mm, yellow. Ovary spherical, as long as wide. Capsule 4 mm. *C Europe to W Asia*. H3. Summer.

**58. A. phrygium** Boissier.
Bulbs *c.* 1 cm; outer tunics membranous. Leaves sheathing half to two-thirds of stem, thread-like. Stem 23–30 cm. Spathes 2, unequal, each contracted into a long slender appendage. Umbel 2–2.5 cm, compact, spherical; flower-stalks 6–8 mm. Perianth cup-shaped, straw-coloured; segments 4 mm, acute. Stamens projecting. *Turkey*. H5.

**59. A. sativum** Linnaeus. Illustration: Reichenbach, Icones Florae Germanicae **10**: 488 (1848); Bentley & Trimen, Medicinal Plants **4**: 280 (1875); Belmontia **7**: 61 (1977).
Bulbs depressed ovoid, 2–7 mm thick, composed of 5–18 (rarely to 60) bulblets, outer tunics membranous to papery. Leaves 6–12, to 60 × 3 cm, sheathing lower half of stem, linear, flat or keeled. Stem 25–100 cm or more, solid. Spathes 4–20 cm, long-beaked. Umbel 2.5–5 cm, usually with few flowers (aborting in bud) and many bulbils; flower-stalks 1–2 cm, unequal. Perianth greenish-white, pink or rarely white or purple, segments 3–5 mm, the outer narrower. Stamens included or equalling perianth, the outer filaments simple or 3-toothed, the inner with the basal blade broad, one-third as long as central tooth, lateral teeth 2–4, much longer than central tooth. Capsules abortive, without seeds. *Not known in the wild*. H4.

Widely cultivated as a crop (garlic).

**60. A. ampeloprasum** Linnaeus. Illustration: Botanical Magazine, 1385 (1811); Redouté, Liliacées **7**: 385 (1813); Reichenbach, Icones Florae Germanicae **10**: 489 (1848); Belmontia **7**: 67–8 (1877).
Bulbs 2–6 cm; outer tunics membranous, bulblets usually numerous, yellowish. Leaves 4–10, to 50 × 4 cm, sheathing lower one-third to half of stem, flat with rough margin. Stem 45–180 cm. Spathe 1, 4–11.5 cm, deciduous. Umbel 5–9 cm, dense, with up to 500 flowers, with or without bulbils; flower-stalks 4–5 cm, unequal. Perianth cup-shaped or bell-

shaped, pink or dark red, with large papillae especially on keels; segments 4–5.5 × 1.3–2.4 mm. Stamens slightly to distinctly projecting, the outer filaments usually simple, 4–6 mm, the inner with basal blade as wide as perianth, and 2 times as long as the central tooth. Capsule 4 mm. *S Europe, USSR (Caucasus), Iran, Turkey, N Africa.* H2. Summer.

Var. **babingtonii** (Borrer) Syme, found in Ireland, SW England and Channel Islands, has numerous bulbils and few flowers and is apparently a relic of former cultivation.

*A. porrum* Linnaeus, the cultivated leek is derived from the above species.

**61. A. scorodoprasum** Linnaeus. Illustration: Reichenbach, Icones Florae Germanicae 10: 490 (1848); Pratt, Grasses, sedges and ferns of Great Britain, edn 3, 5: 224 (1873); Bonnier, Flore complete 10: 2615 (1929); Rix & Phillips, The bulb book, 145 (1981).
Bulbs 1–2 cm, outer tunics membranous, sometimes breaking into parallel fibres; bulblets reddish black. Leaves 2–5, to 27 × 2 cm, linear, sheathing up to half of stem. Stem 25–90 cm. Spathe 1.5 cm, shortly beaked, deciduous. Umbel 1.5 cm, 0–many-flowered, usually with bulbils; flower-stalks to 2 cm, unequal. Perianth ovoid, lilac to dark purple, segments 4–7 × 1.5–3.5 mm. Stamens included, the outer filaments simple, 2.5–4.5 mm, the inner with basal blade more than 3 times length of central tooth, lateral teeth 2–3 times as long as central tooth. *C & E Europe, USSR (Caucasus), Turkey, N Iran.* H3. Spring–summer.

Subsp. **scorodoprasum**, with purplish bulbils and few flowers, of northern and central Europe, is apparently a relic of former cultivation. It is evidently a bulbilliferous variant of subsp. **rotundum** (Linnaeus) Stearn, from southern Europe, which has an almost spherical or hemispherical umbel without bulbils.

**62. A. sphaerocephalon** Linnaeus (*A. descendens* Linnaeus). Illustration: Redouté, Liliacées 7: 391 (1813); Botanical Magazine, 1764 (1815); Reichenbach, Icones Florae Germanicae 10: 492 (1848); Belmontia 7: 48 (1977); Rix & Phillips, The bulb book, 141 (1981).
Bulbs spherical to ovoid, 1–3.5 cm thick, outer tunics membranous, sometimes breaking into parallel fibres. Leaves 2–6, 7–35 cm × 4–5 mm, sheathing stem for up to half, hollow, semi-cylindric. Stem 5–90 cm, hollow, terete. Spathes 2, to 2 cm, persistent. Umbel 1.6 cm, many-

flowered, sometimes with bulbils; flower-stalks 2–30 mm, unequal. Perianth cylindric to narrowly ovoid, pink to dark reddish-brown, segments 3.5–5.5 × 1.2–2.5 mm, keeled, the outer broader. Stamens 4–7 mm, the outer, simple (rarely 3-toothed), the inner 3-toothed with basal blade 1–2 times as long as central tooth, which is 1.5–3 mm, the lateral teeth shorter or longer than central tooth. Capsule 4 mm. *Europe, N Africa, W Asia.* H1. Summer.

Subsp. **sphaerocephalon** has red flowers, but subsp. **trachypus** (Boissier) Richter and subsp. **arvense** (Gussone) Arcangeli, have white, green-striped flowers.

**63. A. vineale** Linnaeus. Illustration: Sowerby & Smith, English Botany 28: t. 1974 (1809); Reichenbach, Icones Florae Germanicae 10: t. 490 (1848); Ross-Craig, Drawings of British plants 29: t. 19 (1972).
Bulbs ovoid, 1–2 cm; outer tunics breaking into parallel fibres. Leaves 2–4, 15–60 cm × 1.5–4 mm, hollow, sheathing lower one-third to two-thirds of stem. Stem 30–120 cm. Umbel with numerous greenish bulbils and few or no flowers or 2–5 cm with numerous flowers and few or no bulbils; flower-stalks 5–30 mm. Spathe 1, deciduous. Perianth bell-shaped, greenish white, pink or dark red (var. **purpureum** H.P.G. Koch); segments 2–4.5 × 1.2–1.5 mm, somewhat acute or obtuse. Stamens slightly to distinctly projecting, the outer 3.5–4 mm, the inner 3-toothed, with basal blade slightly or much longer than central tooth, the lateral teeth much longer than central tooth. Capsule 3–3.5 mm. *Europe, N Africa, W Asia.* H1. Summer.

A pernicious weed which, in botanic gardens, may usurp the place of other species and be labelled accordingly.

**64. A. heldreichii** Boissier. Illustration: Strid, Wild flowers of Mount Olympus, t. 47 (1980).
Bulbs broadly ovoid; outer tunics membranous. Leaves 2–4, 5–30 cm × 0.5–3 mm, hollow, sheathing lower one-quarter to one-third of stem. Stem 20–60 cm. Spathes 2, 1–1.5 cm, persistent. Umbel 2.5–4.5 cm, spherical or hemispherical, many-flowered. Perianth bell-shaped, pink; segments about 10 × 2.5–3 mm, acute. Stamens included, the outer simple, *c.* 5.5 mm, the inner 3-toothed, with basal blade 4 times as long as central tooth, the lateral teeth 2–3 times as long as central tooth. Capsule 4–5 mm. *Northern Greece.* H5. Summer.

**65. A. nigrum** Linnaeus (*A. multibulbosum* Jacquin). Illustration: Redouté, Liliacées 2: 102 (1805); Botanical Magazine, 1148 (1809); Reichenbach, Icones Florae Germanicae 10: 506 (1848); Belmontia 7: 195 (1977); Rix & Phillips, The bulb book, 143 (1981).
Bulbs ovoid, 2.5–5 cm, outer tunics membranous. Leaves 3–6, to 50 × 8 cm, basal. Stem 40–90 cm. Spathe 1, becoming 2–4-lobed, to 3 cm. Umbel 5–10 cm, compressed or hemispherical, many-flowered, rarely with bulbils; flower-stalks 2.5–4.5 cm. Perianth star-shaped, segments white or pale lilac with greenish mid-veins, 6–9 × 1.5–3.5 cm, later deflexed. Stamens included, 5 mm, united at base into ring 1–1.5 mm broad; anthers yellow. Capsule 6–8 mm. *S Europe, N Africa, W Asia.* H4. Spring.

**66. A. cyrilli** Tenore.
Bulb broadly ovoid, 1–2 cm, outer tunics membranous. Leaves 3–5, 20–30 × 1–3.5 cm, basal. Stem 50–60 cm. Spathes 2 or 3, to 2 cm. Umbel 4–7 cm, many-flowered, compressed or hemispherical; flower-stalks 1.5–2.5 cm, almost equal. Perianth cup-shaped, segments white with broad green mid-veins, 6–7 × 1–1.5 mm, later deflexed. Stamens just included, united at base into a ring 1.5–2 mm broad. *Italy to Turkey.* H4.

**67. A. atropurpureum** Waldstein & Kitaibel. Illustration: Reichenbach, Icones Florae Germanicae 10: 505 (1848); Rix & Phillips, The bulb book, 143 (1981).
Bulbs almost spherical, 1.5–3 cm; outer tunics membranous. Leaves 3–7, 15–35 × 1–4 cm, basal. Stem 40–100 cm. Spathes 2, to 12 cm. Umbel 3–7 cm, compressed; flower-stalks 2–4 cm. Perianth star-shaped, dark purple, segments 7–9 × 1 mm, later deflexed. Stamens 4–5 mm, united at base; anthers purple. Capsule 6–8 mm. *S Europe.* H4. Summer.

**68. A. akaka** Gmelin (*A. latifolium* Jaubert & Spach).
Bulbs almost spherical, 1.5–3 cm; outer tunics membranous. Leaves 1–3, to 20 × 6 cm, elliptic-oblong, basal. Stem 5–15 cm. Spathe 3 or 4-lobed. Umbel hemispherical, many-flowered, almost stalkless in the leaves; flower-stalks 8–30 mm. Perianth almost star-shaped; segments lilac, very narrowly elliptic, almost linear, rigid after flowering. Stamens erect, half as long as segments. Capsule 4–5 mm. *Turkey, USSR (Caucasus), Iran.* H5. Summer.

**69. A. christophii** Trautvetter
(*A. albopilosum* C.H. Wright). Illustration: Botanical Magazine, 798 (1805); Baileya 3: 143 (1955); Rix & Phillips, The bulb book, 145 (1981).
Bulbs spherical, 2–3 cm, outer tunics papery. Leaves 2–7, 15–40 × 1–3.8 cm, erect, somewhat glaucous with stiff spreading hairs beneath and on margins. Stem 15–40 cm, buried in ground, ribbed. Spathe 2.5 cm. Umbel many-flowered, loose; flower-stalks to 11 cm, almost equal. Perianth star-shaped, purple-violet, segments 1–1.8 cm, erect, rigid after flowering due to their thickened mid-veins. Stamens included, united at base; filaments dark purple. Capsule 5–8 mm. *Iran to Turkey*. H5. Summer.

**70. A. suworowii** Regel. Illustration: Gartenflora 30: t. 1062 (1881); Botanical Magazine, 6994 (1888).
Bulbs almost spherical, 2–3 cm; outer tunics leathery, elongated above. Leaves 2–6, to 3 cm broad, basal, rough at margin. Stem 30–100 cm, indistinctly ribbed. Spathe 3- or 4-lobed. Umbel hemispherical, many-flowered; flower-stalks 1–4 cm, almost equal. Perianth star-shaped, rose-violet; segments 4–5.5 mm, almost linear, reflexed and twisted after flowering. Stamens slightly shorter to slightly larger than segments. Ovary almost stalkless, smooth. *C Asia, Afghanistan*. H4.

**71. A. rosenbachianum** Regel. Illustration: Flora Iranica **76**: 27 (1971).
Bulbs almost spherical, 1.5–2.5 cm, outer tunics papery. Leaves 2–4, much shorter than stem, 1–5 mm broad, basal. Stem 30–100 cm, ribbed. Spathe shorter than umbel. Umbel many-flowered, loose; flower-stalks 2–6 cm, unequal. Perianth star-shaped, segments dark violet, rarely white, with darker mid-veins, 7–10 mm, later reflexed and twisted. Stamens 7–10 mm, united at base; anthers violet. Ovary on a short stalk, papillose. Capsule 5 mm. *SW Asia*. H4. Spring.

A. jesdianum Boissier, from Iraq and Iran is sometimes cultivated as the above but it differs in having larger bulbs with netted fibres, shorter stamens with a wider ring, and a larger capsule.

**72. A. stipitatum** Regel. Illustration: Gartenflora 30: t. 1062 (1881).
Bulbs almost spherical, 3–6 cm, outer tunics papery, blackish. Leaves 4–6, 2–4 cm broad, hairy beneath. Stem 60–150 cm, distinctly ribbed. Spathes 2, *c.* 3 cm. Umbel hemispherical to almost spherical, many-flowered; flower-stalks 2.5–5 cm, almost equal. Perianth star-shaped, lilac-purple, rarely white, segments 8–9 mm, tapered for most of their length, later becoming reflexed and twisted. Stamens equalling perianth-segments, united at base. Ovary on a short stalk, papillose. Capsule 5 mm. *C Asia*. H4. Summer.

**73. A. hirtifolium** Boissier.
Bulbs almost spherical, 2.5–4 cm; outer tunics papery, breaking into strips and fibres. Leaves 4 or 5, to 5 cm broad, basal, hairy towards the base. Stem 80–120 cm, slightly ribbed. Spathe 2-lobed. Umbel spherical, many-flowered; flower-stalks 3.5–5 cm. Perianth star-shaped, purple, rarely white; segments 8–9 mm, almost linear, reflexed and twisted after flowering. Stamens almost equalling segments, united at base. Ovary on a short stalk, papillose. *Iran*. H5. Summer.

**74. A. aflatunense** B. Fedtschenko.
Illustration: Hay & Synge, Dictionary of garden plants, 658 (1969); Reader's Digest encyclopaedia of garden plants, 28 (1972). Bulb ovoid, 2–6 cm, outer tunics papery. Leaves 6–8, much shorter than stem, 2–10 cm broad, basal, strap-shaped. Stem 80–150 cm, feebly ribbed. Spathes 4. Umbel many-flowered, spherical; flower-stalks 1.4–3.2 cm, almost equal. Perianth star-shaped; segments light violet with darker mid-veins, 7–8 mm, later reflexed and twisted. Stamens slightly projecting, not united at base. Ovary with a short stalk, papillose. Capsule 5 mm. *C Asia*. H4. Spring–summer.

**75. A. giganteum** Regel. Illustration: Gartenflora, t. 1113 (1883); Botanical Magazine, 6828 (1885); Flora Iranica **76**: 26 (1971); Rix & Phillips, The bulb book, 153 (1981).
Bulbs ovoid, 4–6 cm, outer tunics leathery, splitting into parallel fibres. Leaves 30–100 × 5–10 cm, basal. Stem 80–200 cm, slightly ribbed. Spathe half as long as umbel. Umbel spherical, many-flowered, dense; flower-stalks 6 cm, almost equal. Perianth star-shaped, purple-violet, rarely white, segments 5–7 mm, elliptic, unchanged after flowering. Stamens projecting, up to 1½ times as long as perianth, united at base, the inner filaments broader. Ovary stalkless. Capsule 4 mm. *C. Asia*. H4. Spring.

**76. A. macleanii** Baker (*A. elatum* Regel). Illustration: Botanical Magazine, 6707 (1883); Gartenflora, t. 1251 (1887); Rix & Phillips, The bulb book, 153 (1981).

Bulb ovoid-spherical, 2–6 cm, outer tunics papery. Leaves 2–4, shorter than stem, 2–8 cm broad, basal. Stem 60–100 cm, prominently ribbed. Spathe one-third shorter than umbel. Umbel spherical, many-flowered, dense; flower-stalks 1.8–5.6 cm. Perianth star-shaped, bright violet, segments 6–8 mm, linear-lanceolate, unchanged after flowering. Stamens slightly projecting, united at base. Ovary stalkless. Capsule 5 mm. *SW Asia*. H4. Spring–summer.

**77. A. karataviense** Regel. Illustration: Botanical Magazine, 6451 (1879), 7294 (1893); Rix & Phillips, The bulb book, 141, 153 (1981).
Bulb depressed-spherical, *c.* 5 cm, outer tunics membranous. Leaves 2 or 3, 15–23 × 3–15 cm, basal, oblong to elliptic. Stem 10–25 cm, sometimes buried in ground to half its length. Spathes 2, less than 3.5 cm. Umbel 7.5–10 cm, many-flowered, spherical, dense; flower-stalks 1.5–4 cm, equal. Perianth star-shaped, segments light purple-pink with purple mid-veins, 5–8 mm, later reflexed and twisted. Stamens just projecting, with triangular bases. Capsule large, reversed heart-shaped. *C Asia*. H4. Spring.

**78. A. schubertii** Zuccarini. Illustration: Botanical Magazine, 7587 (1898); Belmontia 7: 191 (1977); Rix & Phillips, The bulb book, 141 (1981).
Bulbs spherical to ovoid, 3–4 cm, outer tunics leathery, sometimes breaking down into parallel fibres. Leaves 4–8, 20–45 × 6 cm. Stem 30–60 cm, hollow, not distinctly tapering towards base. Spathes 2 or 3, 2–3 cm. Umbel 2–4 cm, many-flowered; flower-stalks 2–20 cm, very unequal. Perianth star-shaped, segments white to pink or violet, with purplish mid-veins, 6–10 × 2–2.5 mm, the outer longer, all becoming more or less rigid after flowering. Stamens included, 5.5–7.5 mm. Capsule 6–8 mm. *E Mediterranean area to C Asia*. H4. Spring.

**79. A. protensum** Wendelbo. Illustration: Flora Iranica **76**: 27 (1971); Rix & Phillips, The bulb book, 141 (1981).
Bulb spherical, 2–3 cm, outer tunics leathery, splitting into fibres. Leaves 2–5, longer than stem, 1–5 cm broad, flat. Stem 10–30 cm, tapering towards base. Spathes 3, *c.* 2 cm. Umbel many-flowered, loose; flower-stalks 6–16 cm, very unequal, the longer ones with sterile flowers. Perianth broadly bell-shaped, segments pale brown with darkish mid-veins, 7–8 mm, erect

after flowering. Stamens included, united at base. Ovary of fertile flowers on a short stalk, papillose. Capsule 4 mm. *Afghanistan, Iran, C Asia.* H4.

**80. A. caspium** (Pallas) Bieberstein. Illustration: Botanical Magazine, 4598 (1851).
Bulbs 2–4.5 cm, almost spherical; outer tunics membranous. Leaves 3–6, to 25 × 3.5 cm, shorter than the stem, basal. Spathes 2. Umbel 5–20 cm, many-flowered, loose; flower-stalks 3–15 cm, elongating in fruit, unequal. Perianth broadly bell-shaped, lilac; segments green, 5–11 × 2–3 mm. Stamens projecting, $1\frac{1}{2}$–2 times as long as segments, united at base. Ovary on a short stalk. *SE Russia, Caucasus, C Asia.* H4. Summer.

**81. A. mirum** Wendelbo. Illustration: Flora Iranica **76**: t. 28 (1971).
Bulbs almost spherical, to 4 cm; outer tunics papery. Leaves 2, to 30 × 8 cm, elliptic or narrowly ovate, glaucous. Stem 8–40 cm. Umbel 5–9 cm, spherical, many-flowered; flower-stalks 1.5–3 cm. Perianth broadly bell-shaped, segments pale brownish purple to almost white with dark veins, 8–12 mm, narrowly obovate to elliptic, remaining erect and unchanged after flowering. Stamens about the length of the segments, united at base. *Afghanistan.* H4.

**82. A. cernuum** Roth. Illustration: Botanical Magazine, 1324 (1810); Redouté, Liliacées **6**: 345 (1812); Cronquist et al., Intermountain Flora **6**: 513 (1977); Rix & Phillips, The bulb book, 145 (1981).
Bulbs narrowly ovoid, 1–2 cm, clustered, often on a short rhizome; outer tunics membranous. Leaves 4–6, 10–20 cm × 5–7 mm (rarely more), flat, basal. Stem 30–70 cm, terete or angled, abruptly curved at tip. Spathes 2, to 1 cm. Umbel 30–40-flowered, nodding; flower-stalks 8–18 mm, in fruit abruptly bent upwards at attachment point. Perianth cup-shaped, white to pink; segments 4–6 mm. Stamens long-projecting. Ovary and capsule with 6 conspicuous crests. *America from Canada to Mexico.* H2. Summer.

A very deep rose variant has been distinguished as **A. allegheniense** Small and an almost white variant as **A. oxyphilum** Wherry.

**83. A. stellatum** Ker Gawler. Illustration: Botanical Magazine, 1576 (1813); Gleason, New Britton & Brown Illustrated Flora **1**: 415 (1952).
Bulbs ovoid; outer tunics membranous.

Leaves 3–6, 1–2 mm broad, basal. Stem 30–70 cm. Spathes 2. Umbel many-flowered, hemispherical, erect; flower-stalks 1–2 cm. Perianth cup-shaped, pink; segments 4–7 mm, ovate. Stamens equalling or longer than segments. Ovary and capsule with 6 conspicuous crests. *N America.* H2. Summer.

**84. A. geyeri** S. Watson Illustration: Cronquist et al., Intermountain Flora **6**: 511 (1977).
Bulbs ovoid or elongate, 1.5–2.5 cm, usually clustered, outer tunics with netted fibres. Leaves 3 or more, 10–20 cm × 2–4 mm. Stem 15–50 cm, terete to angled. Spathes 2 or 3, 1.5 cm. Umbel 10–25-flowered, sometimes with bulbils; flower-stalks 3–4 mm, becoming rigid and stiffly spreading in fruit. Perianth pink to white, segments 4–10 mm, often obscurely toothed. Stamens included. Capsule inconspicuously crested, enclosed within the persistent perianth-segments. *USA (Texas, New Mexico, Washington, Oregon, Nevada, Arizona).* H3. Spring.

**85. A. plummerae** S. Watson.
Bulb narrowly ovoid, 2 cm, borne on a short rhizome; outer tunic becoming netted-fibrous. Leaves numerous (up to 10), to 27 cm × 6 mm, channelled, linear, solid, almost basal. Stem 30–50 cm. Umbel several-flowered, erect. Perianth white or pink, star-shaped; segments 10 mm, lanceolate, acute. Ovary with apical projection. *USA (Arizona), N Mexico.* H4. Summer.

**86. A. textile** Nelson & Macbride. Illustration: Hitchcock et al., Vascular Plants of the Pacific Northwest **1**: 259 (1969); Cronquist et al., Intermountain Flora **6**: 513 (1977).
Bulbs ovoid, 1–1.5 cm, outer tunics netted-fibrous. Leaves 1–3, usually 2, up to 30 cm × 5 mm, almost basal. Stem 5–30 cm. Spathes 2. Umbel loose, many-flowered, hemispherical; flower-stalks to 2 cm, unequal. Perianth bell-shaped, usually white, sometimes pink; segments 5–7 mm, acute or acuminte. Stamens included. Ovary inconspicuously crested. *USA (central States).* H2. Summer.

**87. A. drummondii** Regel (*A. nuttallii* S. Watson).
Bulbs ovoid, 1.5 cm, outer tunics with netted fibres. Leaves 10–15 cm × 1–4 mm, basal, flat or channelled. Stem 10–20 cm (rarely more). Spathes 2 or 3, 1–1.5 cm. Umbel several-flowered, erect. Perianth pink to white, segments 4–7 mm, ovate to

ovate-lanceolate, becoming rigid in fruit. Stamens included. Capsule not crested. *N America (Texas to new Mexico and N Mexico to Nebraska).* H4. Spring.

**88. A. falcifolium** Hooker & Arnott. Illustration: Jepson, Flora of California **6**: 273 (1922); Abrams, Illustrated flora of the Pacific States, 385 (1923).
Bulb ovoid to spherical, 1.5 cm; outer tunics membranous. Leaves 1 or 2, to 25 cm × 9 mm, recurving, sickle-shaped. Stem 5–12 cm, flattened. Spathes 2, 1.2–1.9 cm. Umbel many-flowered, compact; flower-stalks 6–16 mm. Perianth bell-shaped, deep rose; segments 8–15 mm, often glandular. Stamens included, 4–9 mm. Ovary and capsule 3-crested. *W USA (California, Oregon).* H4. Summer.

**89. A. crenulatum** Weigand. Illustration: Abrams, Illustrated flora of the Pacific States, 389 (1923).
Bulbs oblique, narrowly ovoid; outer tunics membranous. Leaves 1 or 2, 3–10 cm × 3–6 mm, basal, recurved with margins toothed. Stem 5–8 cm, 2-angled, with edges toothed. Spathes 2, 8–10 mm. Umbel hemispherical, few-flowered; flower-stalks 8–15 mm. Perianth pink, segments 6–12 mm lanceolate, acute. Stamens included, 4 mm. Ovary and capsule 6-crested. *Western N America (British Columbia, Washington, Oregon).* H4. Summer.

**90. A. acuminatum** Hooker (*A. murrayanum* Regel). Illustration: Gartenflora, t. 770 (1873); Botanical Magazine, n.s., 213 (1953); Cronquist et al., Intermountain Flora **6**: 517 (1977).
Bulbs ovoid to almost spherical, 1–1.5 cm; outer tunics membranous, netted, the meshes almost square or hexagonal. Leaves 2–4, shorter than stem, c. 1–3 mm broad, channelled. Stem 10–30 cm. Spathes 2, 1–3 cm. Umbel 12–30-flowered, loose; flower-stalks 1.2–2.4 cm, subequal. Perianth deep rose to white; segments 8–15 mm, acuminate, the outer with recurved tips and entire, the inner shorter, narrower and with slightly toothed tips, becoming rigid and keeled in fruit. Stamens included. Ovary and capsule inconspicuously 3-crested. *Western N America (British Columbia to N California, east to Montana, Colorado & Arizona).* H2. Spring.

**91. A. campanulatum** S. Watson (*A. bidwelliae* Watson). Illustration: Abrams, Illustrated flora of the Pacific States, 391 (1923); Cronquist et al., Vascular Plants of the Pacific Northwest **6**: 746 (1969).

Bulb ovoid, 1.5–2 cm; outer tunics membranous, faintly netted with oblong, sinuous meshes. Leaves 2–3, equalling or shorter than stem, often withered at flowering time, 3–5 mm broad. Stem 10–30 cm. Spathes 2. Umbel hemispherical, many-flowered; flower-stalks 1.5–3.5 cm. Perianth cup-shaped or almost star-shaped, pale rose or white; segments 6–8 mm, acute or acuminate, becoming rigid and keeled in fruit. Stamens included. Ovary and capsule conspicuously 6-crested. *USA (California and Nevada).* H4. Summer.

**92. A. dichlamydeum** Greene. Illustration: Abrams, Illustrated flora of the Pacific States, 394 (1923).
Bulbs ovoid, *c.* 1.5 cm; outer tunics membranous, with close horizontally zigzag impressed surface. Leaves 1–3, equalling stem, 1–2 mm broad. Stem 10–30 cm. Spathes 2. Umbel hemispherical; flower-stalks 1–1.5 mm. Perianth openly bell-shaped, deep rose-purple; segments 9–11 mm, oblong, the outer somewhat spreading, the inner erect. Stamens included. Ovary and capsule minutely crested. *USA (California).* H4.

**93. A. peninsulare** Lemmon. Illustration: Abrams, Illustrated flora of the Pacific States, 395 (1923).
Bulbs ovoid, about 1.5 cm; outer tunics membranous, with close horizontally zigzag impressed surface. Leaves 2–4, about equalling stem, 1–6 mm broad. Stem 20–40 cm. Spathes 2. Umbel hemispherical, loose; flower-stalks 1.5–3 cm. Perianth bell-shaped, deep rose-purple, segments 1–1.3 cm, acuminate, the outer entire, longer and broader than the entire and crisped inner segments. Stamens included. Ovary and capsule minutely crested. *USA (California).* H4. Summer.

**94. A. crispum** Greene (*A. peninsulare* var. *crispum* (Greene) Jepson). Illustration: Abrams, Illustrated flora of the Pacific States, 395 (1923).
Bulbs ovoid, about 1 cm; outer tunics membranous, with close horizontally zigzag impressed surface. Leaves 2, mostly shorter than stem, basal, about 1.5 mm broad. Stem 10–30 cm. Spathes 2. Umbel hemispherical, loose, many-flowered; flower-stalks 1–3 cm. Perianth bell-shaped; segments 9–12 mm, acute, the outer entire, the inner narrower with distinctly crisped margin. Stamens included. Ovary and capsule slightly crested. *USA (California).* H4. Summer.

**95. A. unifolium** Kellogg. Illustration: Botanical Magazine, 6320 (1877); Rix & Phillips, The bulb book, 147 (1981).
Bulbs obscure, with 2 bulblets, on a short rhizome, outer tunics cellular-netted. Leaf 1, shorter than stem, 7 mm broad, sickle-shaped, flat to slightly channelled. Stem 30 cm or more, terete. Spathes 2. Umbel 5 cm, 15–20-flowered; flower-stalks 2.5–3.8 cm, thickened at tips. Perianth widely bell-shaped, pink to lavender pink, segments 1–1.4 cm. Stamens included, 7–9 mm. *USA (Oregon, California).* H4. Spring–summer.

**96. A. glandulosum** Link & Otto. Illustration: Edward's Botanical Register 12: t. 1034 (1826).
Bulbs *c.* 1.5 cm, almost spherical, ending a long slender rhizome; outer tunics membranous. Leaves 2–5, to 30 cm × 6 mm, equalling or longer than stem, 1–7 mm broad, almost basal, strongly veined. Stem 20–55 cm, 2-edged. Spathe 1, becoming 2-lobed. Umbel 4–9 cm, compressed, loose; flower-stalks unequal, 1.5–4 cm. Perianth star-shaped, brownish purple, segments 6–9 mm, lanceolate, acute. Stamens outspread, half the length of segments, united at base. Ovary without crests. *Mexico, USA (Texas & Arizona).* H4. Autumn.

**89. NOTHOSCORDUM** Kunth
*V.A. Matthews*
Perennial herbs with bulbs. Leaves linear, basal. Scape erect, bearing an umbel of 4–15 flowers which is subtended by 2 papery bracts (spathe valves). Flower-stalks unequal. Perianth shortly united at the base. Ovary superior, 3-celled. Capsule 3-celled, containing angular black seeds.

A genus of 20 species, similar to *Allium* but with a terminal style and no smell of onion. Native to N and S America. Cultivation and propagation as for *Allium* (p. 233).

**1. N. gracile** (Aiton) Stearn (*N. fragrans* (Ventenat) Kunth); *N. inodorum* misapplied. Illustration: Rix & Phillips, The bulb book, 143 (1981).
Bulb ovoid with membranous outer tunics. Leaves 6–8, 20–40 cm, linear. Scape 30–75 cm bearing 8–15 fragrant flowers. Perianth-lobes 9–15 mm, white or lilac with brownish or pink streaks and a pink or mauve central vein. Stamens shorter than perianth-lobes. Capsule 6–10 mm, obovoid. *Subtropical S America, Mexico.* H5. Spring–summer.

**N. bivalve** (Linnaeus) Britton. Illustration: Rickett, Wild flowers of the United States 2: 45 (1967); Braun, The Vascular Flora of Ohio 1: 344 (1967).
Leaves 3 or 4, umbels with 4–8 flowers, perianth-lobes white or yellowish with a green mid-vein and capsule more or less spherical. *SC & SE USA.* H5. Summer.

**90. CALOSCORDUM** Herbert
*V.A. Matthews*
Similar to *Nothoscordum* but umbel subtended by only one bract. Bulb with whitish papery tunic. Leaves 2–6, very narrowly linear, channelled above, shorter than flowering stem, often somewhat withered at flowering time. Stem 10–25 cm, bearing an umbel of 10–20 flowers which is subtended by a single bract. Flower-stalks unequal. Perianth bright pink, 6–8 mm, joined in the lower one-third, lobes spreading-reflexed, each with a darker central vein, perianth 5–7 mm across the lobes. Stamens borne on the perianth-tube. Ovary superior.

A genus of 1 species. Cultivation and propagation as for *Allium* (p. 233).

**1. C. neriniflorum** Herbert (*Nothoscordum neriniflorum* (Herbert) Bentham & Hooker; *Allium neriniflorum* (Herbert) Baker). Illustration: Edwards's Botanical Register 33: t. 5 (1847); Rix & Phillips, The bulb book, 145 (1981).
*N China, USSR (Pamir Mountains, SE Siberia), ?Mongolia.* H3. Late Summer.

**91. LEUCOCORYNE** Lindley
*S.G. Knees*
Herbaceous perennials. Bulbs to 2 cm, spherical to ovoid, tunic dark brown. Leaves 2–5, deciduous, basal, narrowly linear, usually channelled, 15–25 cm × 1–3 mm, appearing after the flowers. Scapes to 50 cm, bearing umbels of 2–12 flowers subtended by 2 spathes. Perianth funnel-shaped to very widely funnel-shaped, white to purple or bluish violet. Stamens 3 with dorsifixed anthers. Staminodes 3. Stigmas capitate. Ovary superior, 3-celled. Ovules many.

A genus of 5 species, most smelling strongly of garlic. All are native to Chile. Protection from extreme conditions in a cool glasshouse is recommended and cultivation requirements are similar to *Freesia* (p. 381) or *Ixia* (p. 380). Literature: Pizarro, C.M., *Flores Silvestres de Chile* (1966).

1a. Perianth purple-violet; scapes with 2–6 flowers          **1. purpurea**

b. Perianth lilac-blue fading to white in the centre; scapes with 6–9 flowers **2. ixioides\***

### 1. L. purpurea Gay (*L. ixioides* (Hooker) Lindley var. *purpurea* (Gay) Baker).
Illustration: Pizarro, Flores Silvestres de Chile, t. 9 (1966).
Scapes to 50 cm, with 2–6 flowers. Perianth-segments 1.5–2 cm, deep purple-violet throughout, broadly lanceolate with prominent mid-veins. Bracts channelled, 5–6 cm. Seeds with black testa. *Chile*. H5. Spring.

### 2. L. ixioides (Hooker) Lindley.
Illustration: Botanical Magazine, 9457 (1936); Marshall Cavendish encyclopedia of gardening **39**: 1085 (1968).
Scapes to 45 cm with 6–9 flowers. Perianth-segments 1.2–1.5 cm, lilac-blue at apex, fading to white at throat, diamond-shaped to lanceolate. Stamens and staminodes yellow. *Chile*. H5. Spring.

**\*L. odorata** Lindley. Similar to *L. ixioides*, but generally with paler, very fragrant flowers. *Chile*. H5. Spring.

## 92. IPHEION Rafinesque
*D.A. Webb*
Perennials with small bulbs and fleshy roots. Leaves basal, linear. Flowers solitary (rarely in umbels of 2), with 2 partly united bracts on the scape some distance below the flower. Perianth with a narrowly bell-shaped tube and spreading lobes. Stamens within the perianth-tube; anthers dorsifixed, opening inwards. Ovary superior. Stigma small, obscurely 3-lobed. Fruit a many-seeded capsule.

A genus of 10 or more species from temperate S America. It has, from time to time been included in many other genera, giving rise to many synonyms, but is now generally recognised as distinct.

There is only 1 species in cultivation. It should be planted in fairly dry, gritty soil in a sunny place, but even in districts where it is hardy it forms an attractive pot-plant. Literature: Traub, H.P. & Moldenke, H.N., The genus Ipheion: diagnosis, key to species and synonymy, *Plant Life* **11**: 125–30 (1955).

### 1. I. uniflorum (Graham) Rafinesque
(*Triteleia uniflora* Lindley; *Milla uniflora* Graham; *Brodiaea uniflora* (Lindley) Engler). Illustration: Botanical Magazine, n.s., 185 (1952); Perry, Flowers of the world, 173 (1972); Rix & Phillips, The bulb book, 111 (1981).
Leaves 20–25 cm × 4–7 mm, somewhat

fleshy, smelling of onion when bruised. Scape as long as leaves or slightly shorter; bracts brownish, scarious, forming a sheathing tube with 2 acute lobes. Flowers white or violet-blue, 4 cm in diameter, the lobes rather longer than the tube. Anthers and stigma just visible at the mouth of the tube. *Uruguay and warm-temperate Argentina*. H3–4. Spring.

The colour of the flowers varies from pure white to clear violet-blue, but the variants are scarcely worth naming. 'Froyle Mill', however, with flowers of a deep, dusky violet is very distinct.

## 93. BRODIAEA J.E. Smith
*S.J.M. Droop*
Perennial herbs arising from corms which are more or less spherical and have dark brown, fibrous coats. Leaves linear, crescent shaped in section, without a keel, with vein impressions weak or lacking. Flowers in an umbel subtended by scarious bracts. Perianth united into a tube below, lobes of the outer whorl narrower than those of the inner. Stamens 3, opposite the inner perianth-lobes, usually alternating with 3 staminodes; filaments joined to the perianth-tube for most of their lengths, lacking appendages; anthers basifixed. Ovary superior, stalkless, 3-celled with several ovules in each cell; stigma with 3 spreading, winged lobes. Seeds flattened, angled.

A genus of about 15 species from western N America, about 5 of which are found in cultivation. In the past, *Brodiaea* has included *Triteleia* (p. 248), *Dichelostemma* (p. 249) and *Ipheion* (above). They may be grown in open borders in suitable areas, or in pots, in a soil that is light and well-drained. Propagation is by offsets or by seed. Literature: Hoover, R.F., A revision of the genus Brodiaea, *American Midland Naturalist* **22**: 551–74 (1939); Hoover, R.F., Further observations on Brodiaea and some related genera, *Herbertia* **11**: 13–23 (1955); Niehaus, T.F., A biosystematic study of the genus Brodiaea (Amaryllidaceae), *University of California Publications, Botany* **60**: 1–66 (1971); Niehaus, T.F., The Brodiaea complex, *The Four Seasons* **6**(1): 11–21 (1980).

*Perianth.* to 2.4 cm: **1,5**; 2.4 cm or more: **2–4**.
*Perianth-tube.* Splitting in fruit: **2**; not splitting in fruit: **1,3–5**.
*Perianth-lobes.* 3 times as long as tube: **2**; less than 2½ times as long as tube: **1,3–5**. Widely spreading: **1**; ascending and recurved: **2–5**.

*Staminodes.* Shorter than anthers, flat, acute: **3**. Apex notched: **1,4**; apex 3-fid: **5**. Borne close to the anthers: **1,2,4**; borne distant from anthers: **3,5**.

1a. Perianth-tube tightly constricted above the ovary; lobes widely spreading **1. minor**
b. Perianth-tube not or slightly constricted above ovary; lobes ascending and recurved **2**
2a. Perianth-lobes about 3 times the length of the tube; tube thin, transparent and splitting in fruit **2. californica**
b. Perianth-lobes less than 2½ times as long as tube; tube thick, opaque, not splitting in fruit **3**
3a. Anthers exceeding staminodes and stigma; perianth-tube funnel-shaped; staminodes flat, acute **3. elegans**
b. Anthers and stigma exceeded by staminodes; perianth-tube ovoid to narrowly bell-shaped; margins of staminodes more or less rolled inwards, apex notched or 3-fid **4**
4a. Scape 5–25 cm; staminodes with margins conspicuously rolled inwards, apex notched **4. coronaria**
b. Scape 5–70 mm; staminodes with margins slightly rolled inwards, apex unequally 3 fid **5. terrestris**

### 1. B. minor (Bentham) Watson
(*B. grandiflora* Smith var. *minor* Bentham). Illustration: Rickett, Wild flowers of the United States **5**: t. 18 (1971).
Scape 2–30 cm. Flower-stalks 1–3 cm. Perianth 1.7–2.4 cm, the lobes widely spreading, less than 2½ times as long as tube, pink to purple; tube strongly constricted above the ovary, thick, opaque, not splitting in fruit. Staminodes with the margins rolled inwards, apex usually notched, slightly exceeding the anthers and stigma. Anthers 1–6 mm, filaments 0.5–4 mm. Ovary 3–5 mm, style 4–9 mm. *USA (California)*. H5. Summer.

### 2. B. californica Lindley (*B. grandiflora* Smith var. *elatior* Bentham). Illustration: Rickett, Wild flowers of the United States **5**: t. 16 (1971).
Scape 20–70 cm. Flower-stalks 2–13 cm. Perianth 3–4.5 cm, lobes about 3 times as long as tube, pale to deep purple or pink, ascending and recurved; tube not constricted above the ovary, thin, transparent and splitting in fruit. Staminodes 1.6–2.7 cm, erect, linear, with margins wavy or rolled inwards, reaching the same level as the anthers and stigma. Anthers

9–13 mm, filaments 7–10 mm. Ovary 7–11 mm, style 1.8–2.2 cm. *USA (northern California)*. H5. Summer.

**3. B. elegans** Hoover (*B. coronaria* (Salisbury) Engler var. *mundula* Jepson). Illustration: Rickett, Wild flowers of the United States **5**: t. 17 (1971).
Scape 10–50 cm. Flower-stalks 5–100 mm. Perianth 3–4.5 cm, lobes less than 2½ times as long as tube, deep purple, ascending and recurved; tube funnel-shaped, thick, opaque and not splitting in fruit. Staminodes 6–9 mm, erect and flat, acute, about the same level as the stigma, but shorter than the anthers. Anthers 4–10 mm, filaments 4–6 mm. Ovary 9–15 mm, style 7–15 mm. *USA (California, Oregon)*. H5. Summer.

**4. B. coronaria** (Salisbury) Engler (*B. grandiflora* Smith; *B. howellii* Eastwood). Illustration: Botanical Magazine, 2877 (1829); Rickett, Wild flowers of the United States **5**: t. 17 (1971).
Scape 5–25 cm. Flower-stalks 1–5 cm. Perianth 2.4–4 cm, lobes less than 2½ times as long as tube, pale to deep purple, ascending and recurved; tube often slightly constricted above the ovary, ovoid to narrowly bell-shaped, thick, opaque and not splitting in fruit. Staminodes close to anthers, with the margins conspicuously rolled inwards and apex notched, exceeding anthers and stigma. Anthers 4.5–7 mm, filaments 2–4.5 mm. Ovary 8–10 mm, style 7–11 mm. *USA (California, Oregon, Washington)*. H5. Summer.

**5. B. terrestris** Kellogg (*B. grandiflora* Smith var. *macropoda* Torrey; *B. coronaria* (Salisbury) Engler var. *macropoda* (Torrey) Hoover).
Scape 5–70 mm. Flower-stalks 3–15 cm. Perianth 2–2.4 cm, lobes less than 2½ times as long as tube, purplish pink, ascending and recurved; tube usually not constricted above the ovary, ovoid to narrowly bell-shaped, thick, opaque and not splitting in fruit. Staminodes distant from anthers, with the margins slightly rolled inwards, the apex unequally 3-fid with the central tooth smaller than the laterals, considerably exceeding anthers and stigma. Anthers 3–5 mm, filaments 1.5–3 mm. Ovary 4–5 mm, style 6–9 mm. *USA (California, southern Oregon)*. H5. Summer.

Subsp. **terrestris** is described above; other subspecies exist in the wild but are not generally in cultivation. Distinction between *B. terrestris* and *B. coronaria* is

difficult in wild populations, and this may be reflected in gardens.

## 94. TRITELEIA Lindley
*S.J.M. Droop*
Like *Brodiaea*, but corms flattened and with pale brown coats, leaves distinctly keeled beneath and with evident veins, staminodes 0, fertile stamens 6, anthers versatile; ovary shortly stalked, seeds more or less spherical, angled.

A genus of 17 species from western N America, mostly from northern California and southern Oregon. Seven species and 1 garden hybrid are available in cultivation. The genus was formerly included in *Brodiaea* (Subgenus *Triteleia*). In the account below measurements of the perianth are from the base of the tube to the tip of the lobe when fully extended. Cultivation as for *Brodiaea* (p. 247). Literature: Hoover, R.F., A systematic study of Triteleia, *American Midland Naturalist* **25**: 73–100 (1941); Lenz, L.W., Triteleia tubergenii, an amphidiploid of garden origin, *Aliso* **7**: 157–60 (1970); Lenz, L.W., A biosystematic study of Triteleia (Liliaceae), *Aliso* **8**: 221–58 (1975); Niehaus, T.F., The Brodiaea complex, *The Four Seasons* **6**(1): 11–21 (1980).

*Flower-stalks*. More than 1 cm: **2**.
*Perianth-tube*. Longer than lobes: **1,3,7**; equalling lobes to *c.* half as long: **2,4,6**; less than half as long as lobes: **5**. Tapered at base: **3,6,7**; rounded at base: **1**; spreading from the base: **2,4,5**.
*Stamens*. Inserted on to perianth-tube at 2 levels at least 2 mm apart: **1–3**; inserted at more or less the same level: **4–7**.
*Filaments*. Thread-like: **3,6**; narrowly expanded at base: **1,2**; broadly expanded at base: **5,7**; broad, with parallel sides and apical appendages: **4**.
*Ovary*. Yellow: **2**. Shorter than its stalk: **3,6,7**.

1a. Stamens inserted on the perianth-tube at 2 levels which are more than 2 mm apart                                2
 b. Stamens inserted on the perianth-tube at more or less the same level          4
2a. Perianth-tube rounded at the base
                                **1. grandiflora**
 b. Perianth-tube acute, tapered or spreading at the base                     3
3a. Flower-stalks usually more than 4 times as long as flowers; perianth-tube acute at base; ovary yellow
                                **2. peduncularis**
 b. Flower-stalks rarely more than 2 times as long as flowers; perianth-

tube tapered at base; ovary green, blue or white                    **3. laxa**
4a. Stamens alternately long and short; filaments broad with forked apical appendages                **4. ixioides**
 b. Stamens equal or nearly so; filaments thread-like or broad only at the base, without appendages              5
5a. Perianth bowl-shaped, broadly spreading from the base, to 1.9 cm; stalk of ovary very short; perianth-lobes *c.* 3 times longer than tube
                                **5. hyacinthina**
 b. Perianth more or less funnel-shaped, tapered at the base, more than 1.9 cm; ovary-stalk considerably longer than ovary; perianth-lobes at most 2 times longer than tube      6
6a. Perianth yellow; lobes 1.5–2 times as long as tube              **6. hendersonii**
 b. Perianth blue or pale purple; lobes usually shorter than tube
                                **7. bridgesii**

**1. T. grandiflora** Lindley (*Brodiaea grandiflora* Macbride not J.E. Smith; *B. douglasii* Piper). Illustration: Botanical Magazine, 6907 (1886).
Scape 20–70 cm. Flower-stalks 1–4 cm, usually shorter than the flowers, but elongating in fruit. Perianth 1.5–3 cm, bright blue or rarely white, tube as long as or longer than lobes, rounded at base. Stamens inserted in 2 whorls separated by more than 2 mm, filaments slightly dilated at base; anthers 2–4 mm. Ovary green or blue, usually longer than its stalk. *Western N America (British Columbia, Oregon & northern Utah)*. H5. Summer.

**2. T. peduncularis** Lindley (*Brodiaea peduncularis* (Lindley) Watson). Illustration: Rickett, Wild flowers of the United States **5**: t. 17 (1971).
Scape 10–40 cm. Flower-stalks 2–18 cm. Perianth 1.5–2.8 cm, white or pale purple, broadly funnel-shaped, the tube equalling or shorter than the lobes, broadly acute at base. Stamens inserted in 2 whorls separated by more than 2 mm, filaments slightly dilated at base, anthers 2–4 mm. Ovary yellow, as long as or longer than its stalk. *USA (California)*. H5. Summer.

**3. T. laxa** Bentham (*Brodiaea laxa* (Bentham) Watson; *B. candida* Baker; *T. candida* (Baker) Greene). Illustration: Edwards's Botanical Register **20**: t. 1685 (1834); Rickett, Wild flowers of the United States **4**: t. 8 (1970), **5**: t. 16 (1971).
Scape 10–70 cm. Flower-stalks 1–9 cm. Perianth 1.8–4.7 cm, blue or rarely white,

tube usually slightly longer than lobes, narrowly funnel-shaped, tapered at base. Stamens inserted in 2 whorls separated by more than 2 mm, filaments 3–6 mm, thread-like; anthers 2–5 mm. Ovary green, blue or white, its stalk 2–3 times longer than the ovary at flowering, relatively shorter in fruit. *USA (California & S Oregon)*. H5. Summer.

**T. × tubergenii** Lenz (*Brodiaea × tubergenii* Anon.) is a fertile hybrid between *T. peduncularis* and *T. laxa*. It occurred spontaneously in a garden, and in most characters is intermediate between its parents.

**4. T. ixioides** (Aiton) Greene (*Brodiaea ixioides* (Aiton) Watson; *B. lutea* (Lindley) Morton; *Calliprora lutea* Lindley). Illustration: Botanical Magazine, 3588 (1837); Rickett, Wild flowers of the United States **4**: t. 8 (1970), **5**: t. 16, 17 (1971). Scape 6–60 cm. Flower-stalks 1–7 cm. Perianth 1–2.2 cm, yellow, with a dark line down each lobe, tube about half as long as lobes, acute at base. Stamens inserted in more or less 1 whorl near the top of the perianth-tube, each filament broad and continued behind the anther as forked projections; anthers 1.5–2 mm. Ovary green or yellowish, longer than its stalk. *USA (California)*. H5.

A variable species divided into several subspecies, more than one of which are in cultivation; their precise identification is, however, uncertain.

**5. T. hyacinthina** (Lindley) Greene (*Brodiaea hyacinthina* (Lindley) Baker; *B. lactea* Watson; *T. lutea* Davidson & Moseley). Illustration: Journal of the Royal Horticultural Society **89**: t. 66 (1964); Rickett, Wild flowers of the United States **5**: t. 16 (1971). Scape 15–70 cm. Flower-stalks 5–50 mm. Perianth 8–18 mm, white or rarely blue or pale purple, bowl-shaped; the tube much shorter than the lobes, spreading away from the base. Stamens all inserted at the same level, filaments broadly dilated at base, anthers 1–2 mm. Ovary green, white or rarely pale blue or purple, its stalk very short at flowering, extending later. *Western N America (British Columbia to Idaho and California)*. H5. Summer.

**6. T. hendersonii** Greene (*Brodiaea hendersonii* (Greene) Watson). Illustration: Rickett, Wild flowers of the United States **5**: t. 16 (1971). Scape 12–35 cm. Flower-stalks 1.5–4 cm. Perianth 2–2.3 cm, yellow often fading to

bluish, each lobe with a dark line; lobes 1.5–2 times longer than the tube; tube narrowly funnel-shaped, tapered at base. Stamens more or less inserted at 1 level, filaments thread-like, anthers 1.5–2 mm. Ovary green or yellowish, its stalk about twice as long. *USA (SW Oregon)*. H5. Summer.

**7. T. bridgesii** (Watson) Greene (*Brodiaea bridgesii* Watson). Illustration: Rickett, Wild flowers of the United States **5**: t. 17 (1971). Scape 10–50 cm. Flower-stalks 1.5–9 cm. Perianth 2.6–4 cm, blue or pale purple, the tube usually longer than the lobes, tapered at base. Stamens inserted at more or less 1 level, filaments dilated at base, anthers 3.5–4.5 mm. Ovary green, blue or pale purple. Ovary-stalk 3–4 times as long as ovary at flowering, relatively shorter in fruit. *USA (SW Oregon, California)*. H5. Summer.

## 95. DICHELOSTEMMA Kunth
*S.J.M. Droop*

Like *Brodiaea* but leaves distinctly keeled beneath and with obvious veins; bracts coloured; fertile stamens sometimes 6; stigma with 3 short lobes.

A genus of 7 species from western N America, of which 4 are in cultivation. In the account that follows, measurements of the perianth extended from the base of the flower to the tip of the lobe when pulled straight. The genus was formerly part of *Brodiaea* (Subgenus *Dichelostemma*). Cultivation as for *Brodiaea* (p. 247). Literature: Hoover, R.F., The genus Dichelostemma, *American Midland Naturalist* **24**: 463–76 (1940); Lenz, L.W., The nature of the floral appendages in four species of Dichelostemma (Liliaceae), *Aliso* **8**: 383–9 (1976); Niehaus, T.F., The Brodiaea complex, *The Four Seasons* **6**(1): 11–21 (1980).

1a. Fertile stamens 6      **1. pulchellum**
  b. Fertile stamens 3             2
2a. Perianth-tube red, lobes green, much shorter than the tube    **2. ida-maia**
  b. Perianth entirely blue or purple, lobes as long as or longer than tube     3
3a. Flowers in umbels; staminodes entire; perianth-tube much constricted above ovary    **3. multiflorum**
  b. Flowers in short racemes; staminodes deeply bifid; perianth-tube not much constricted above ovary     **4. congestum**

**1. D. pulchellum** (Salisbury) Heller (*Brodiaea capitata* Bentham; *B. pulchella*

Greene). Illustration: Botanical Magazine, 5912 (1871); Rickett, Wild flowers of the United States **4**: t. 8 (1970), **5**: t. 16 (1971). Scape 30–60 cm. Flowers in umbels. Bracts ovate, purple; flower-stalks 1–15 mm. Perianth 1.2–1.8 cm, white to purple, the lobes equalling or exceeding the tube, not widely spreading. Stamens 6 in 2 whorls, those opposite the outer perianth-lobes smaller than the others; filaments united below to form a short tube, connective of the larger anthers extending upwards as 2 triangular appendages. Style 4–6 mm. *USA (Oregon to California, east to Utah)*. H5. Summer.

**2. D. ida-maia** (Wood) Greene (*Brodiaea ida-maia* (Wood) Greene; *B. coccinea* Gray; *Brevoortia ida-maia* Wood). Illustration: Botanical Magazine, 5857 (1870); Rickett, Wild flowers of the United States **5**: t. 18 (1971). Scape 30–90 cm. Flowers in an umbel. Perianth-tube 2–2.5 cm, bright red; lobes much shorter than the tube, greenish, recurved. Staminodes 3, each broader than long. Stamens 3, anthers 6–8 mm. Style 1.5–2 cm. *USA (Oregon, California)*. H5. Summer.

**3. D. multiflorum** (Bentham) Heller (*Brodiaea multiflora* Bentham). Illustration: Botanical Magazine, 5989 (1872); Rickett, Wild flowers of the United States **5**: t. 18 (1971). Scape 20–80 cm. Flowers in an umbel. Flower-stalks 3–15 mm. Perianth 1.6–2.1 cm, pale to dark purple, lobes about equalling the tube, widely spreading, tube much constricted above the ovary. Staminodes 3, each longer than broad, apex entire. Stamens 3, anthers 4–5 mm. Style 7–8 mm. *USA (Oregon, California)*. H5. Summer.

**4. D. congestum** (J.E. Smith) Kunth (*Brodiaea congesta* J.E. Smith). Scape 30–90 cm. Flowers in a short raceme. Perianth 1.4–2 cm, purple, lobes about equalling tube, not widely spreading; tube not much constricted above ovary. Staminodes 3, bifid. Stamens 3, anthers 4–5 mm. Style 5–6 mm. *USA (Washington to California)*. H5. Summer.

## 96. MILLA Cavanilles
*V.A. Matthews*

Perennial herbs with bulbs. Leaves linear, flat to almost terete. Flowering stem erect bearing an umbel of erect flowers, subtended by 4 chaffy spathes which do not enclose the flower buds. Perianth united to above the middle, forming a long tube. Stamens borne at the mouth of the perianth-tube, shortly projecting. Ovary superior, borne on a long stalk.

A genus of 6 species native to the southern USA and northern C America. Only 1 species is cultivated and is kept in a cool greenhouse in colder areas. In warmer regions it can be planted out-of-doors in a sunny place in the spring, and the bulbs can be lifted and stored inside over the winter. Propagation is by offsets and seed. Literature: Moore, H.E., The genus Milla and its allies, *Gentes Herbarum* **8**: 263–94 (1953).

**1. M. biflora** Cavanilles. Illustration: Edwards's Botanical Register **18**: t. 1555 (1832); Flore des Serres **4**: t. 1459 (1861); Baileya **1**: 8 (1953); Rickett, Wild flowers of the United States **3**: 27 (1969).
Leaves 2–7, narrowly linear, channelled above and rounded below, to almost terete, 10–50 cm. Flowering stem 5–45 cm, with 1–6, occasionally to 8, fragrant stalkless flowers, which appear stalked because the perianth-lobes are united into a long 10–20 cm tube. Perianth-lobes 1.5–3.5 cm, spreading flat, white or occasionally flushed with lilac, with 3, 5 or 7 green central veins, or 1 broad central stripe. Anthers longer than the filaments.
*Guatemala, Mexico and adjoining USA.* H5–G1. Late summer–autumn.

**97. CONVALLARIA** Linnaeus
*A. Leslie*
Hairless perennial herb, with creeping rhizomes. Stems erect, with several green or violet scales below and 1–4 leaves above. Leaf-blades ovate-lanceolate to elliptic, 3–23 cm × 5–110 mm, acute to acuminate. Stalks 1–24 cm, the sheathing bases simulating an elongation of the stem. Scapes solitary, sharply angled, arising from scale axils, shorter than the leaves. Inflorescence a 1-sided raceme. Flowers 5–13, nodding; flower-stalks usually curved downwards and exceeding the ovate-lanceolate bracts. Perianth spherical-bell-shaped, 5–11 × 5–11 mm, white, united for half to two-thirds of its length, lobes 6, reflexed at tips. Ovary superior, 3-celled, with 4–8 ovules in each cell. Style simple; stigma capitate. Fruit a red berry.

One very variable species distributed throughout the north temperate regions. Widely cultivated for its long lasting, highly scented flowers. Once established, it can become an aggressive weed, although it is effective as ground cover under trees and

shrubs. It grows best in semi-shade, in a humus-rich, moisture-retentive soil and whilst tolerating deep shade will not flower well in such conditions. Propagation is by division.

**1. C. majalis** Linnaeus. Illustration: Polunin, Flowers of Europe, t. 171 (1969); Keble-Martin, The concise British flora in colour, t. 84 (1965); Ross-Craig, Drawings of British Plants **29**: 16 (1972).
*North temperate regions.* H1. Spring.

There are variants in cultivation with pink flowers (var. **rosea** Reichenbach), with purple spots at the base of the filaments (forma **picta** Wilczek) and with either double white ('Flore Pleno') or double pink flowers. 'Prolificans' is a variant with a paniculate inflorescence, in which the flowers are often slightly malformed. There are also several larger-flowered or more vigorous forms as well as a number with variegated leaves, of which 'Variegata', with longitudinal creamy white stripes, is best known.

**98. SPEIRANTHA** Baker
*J. Cullen*
Stemless perennial herbs with thick rhizomes and numerous stolons. Leaves to 15 × 2.5–3 cm, erect, 6–8 in a rosette, oblanceolate, tapered slightly to the base but not stalked, apex acute. Scape with a few, small, sheathing, scale-leaves, to 15 cm. Flowers in a loose raceme, with bracts. Perianth-segments 6, free, spreading, white, 4–5 mm. Stamens 6, free. Ovary superior, 3-celled, with 3 or 4 ovules in each cell. Style 1, stigma 3-grooved. Fruit a berry.

A genus of a single species, not very frequently grown. Cultivation as for *Convallaria.*

**1. S. convallarioides** Baker. Illustration: Botanical Magazine, 4842 (1855) – as Albuca ? gardeni; Journal of the Linnaean Society, Botany **14**: t. 17 (1875).
*China (Jiangxi).* H5. Spring–summer.

J.D. Hooker (Botanical Magazine, cited above) thought, incorrectly, that the plant originated in South Africa; hence his doubtful placing of it in the genus *Albuca* (see p. 219).

**99. REINECKEA** Kunth
*P.G. Barnes*
Perennial herbs. Leaves persistent at the apex of prostrate, rhizome-like stems, linear-lanceolate, 15–35 cm, somewhat narrowed to the base. Flowers stalkless in a terminal spike on a scape 6–10 cm tall,

bracteoles 4–7 mm, triangular-ovate. Flowers pale to deep pink, perianth 8–12 mm, becoming reflexed, united at the base to form a tube 3–5 mm long. Stamens and style equal, 8–12 mm. Ovary superior. Fruit a spherical red berry, seldom formed in cultivation.

A genus of a single species which is shade-tolerant, forming mats of evergreen foliage, growing freely in acid to neutral soils but sparse-flowering in cooler areas. It is propagated by division, or seed when available.

**1. R. carnea** (Andrews) Kunth. Illustration: Botanical Magazine, 739 (1804); Illustration Horticole **9**: t. 323 (1862) – 'Variegata'.
*China, Japan.* H4. Early summer.

The cultivar 'Variegata', with leaves striped with cream, is also cultivated.

**100. ROHDEA** Roth
*J. Cullen*
Perennial herbs with rhizomes and fibrous roots. Leaves in a basal rosette, leathery, thick, variable in shape, to 45 cm. Flowers in short, dense spikes borne among the leaves and shorter than them. Perianth greenish yellow, bell-shaped, united into a tube for most of its length. Stamens 6, borne on the perianth-tube, filaments very short or absent. Ovary superior, 3-celled, with 1 ovule in each cell. Stigma peltate, style absent. Fruit a spherical, red or yellow, usually 1-seeded berry.

A genus of a single species from SE China and Japan. It has been extensively cultivated in Japan as a foliage plant, and many cultivars have been selected there, including variants with variegated or twisted and curled leaves. The plant has not proved as popular in Europe, though cultivars have been imported from time to time. It is easily grown as a house plant or in a border, but some variants require protection during the winter. Propagation is by seed.

**1. R. japonica** (Thunberg) Roth. Illustration: Botanical Magazine, 898 (1806); Kitamura et al., Coloured illustrations of herbaceous plants of Japan **3**: t. 27 (1964).
*SW China, Japan.* H3.

**101. ASPIDISTRA** Ker Gawler
*S.G. Knees*
Evergreen, stemless, perennial herbs with thick rhizomes and roots. Leaves all basal, arising singly, or in groups, at intervals along the rhizome. Leaf-blade leathery and

glossy, lanceolate to elliptic, wedge-shaped at base, acute, with sunken midrib. Leaf-stalk winged, one-third to half as long as blade, subtended by a papery sheath. Flowers bisexual, with bracts, borne individually at ground level on short stalks, rather inconspicuous and usually obscured by the foliage. Perianth united to form a spherical, bell-shaped or urceolate flower, lobes 6–8, usually dull greenish brown, often with purple spotting. Stamens 6 or 8, included in perianth-tube. Stigma peltate. Ovary 3 or 4-celled, superior. Fruit a berry.

Eight species in E Asia, from the Himalaya to Japan. As these are normally forest-floor plants they are well-adapted to cultivation as pot-plants in cool shady positions indoors, where they often thrive with minimum attention. Propagation is by division of the rhizomes, preferably in spring.

1a. Perianth 6-lobed, spherical     **3. typica**
 b. Perianth 8-lobed, bell-shaped to urceolate                          2
2a. Leaves solitary; flowers urceolate **1. elatior**
 b. Leaves often 2 or 3 per node; flowers bell-shaped               **2. lurida**

**1. A. elatior** Blume. Illustration: Davidson, Illustrated Dictionary of house plants, 39 (1983).
Leaves solitary, leaf-stalks 1–5 cm, leaf-blades ovate to lanceolate, 50–60 × 7–10 cm, with 16–20 veins, glossy dark green, sheathed at the base by a papery bract. Flowers 2.5–3 cm diameter, borne at soil level on short (5 cm) stalks arising directly from the rhizome. Perianth urceolate with 8 triangular, leathery lobes, recurved at apex, cream with purple spotting on outer face, wine-red within. Stamens 8, borne at the base of the perianth-lobes. Stigma a peltate disc, obscuring mouth of perianth. Fruit rarely produced in cultivation. *China.* H5. Spring.

'Variegata' (Illustration: Everett, Encyclopedia of horticulture, 278, 1980) has leaves with longitudinal, cream variegation.

**2. A. lurida** Ker Gawler. Illustration: Botanical Magazine, 2499 (1824); Edwards's Botanical Register **8**: 628 (1822).
Leaves lanceolate, 15–20 cm, with 5–9 main veins, 2 or 3 per node, surrounded by a papery basal sheath of 8–12 cm. Flowers almost stalkless on the rhizome, subtended by papery bracts. Perianth bell-shaped with 8, deep reddish purple, spreading lobes.

Stamens 8, attached at base of perianth-tube. Stigma peltate, not completely covering perianth throat. *China.* H5. Spring.

Plants offered under this name are often *A. elatior*.

**3. A. typica** Baillon. Illustration: Botanical Magazine, 7484 (1896).
Rhizomes stout. Leaf-blades elliptic to lanceolate with 7–9 main veins, 20–40 cm, stalks 10–25 cm. Perianth spherical, leathery, *c.* 2 cm in diameter, with 6 acute lobes, green with reddish purple spotting on outer face, purplish within. Stamens 6, attached at base of perianth-tube. Stigma peltate, 6-lobed. *China.* H5. Spring.

## 102. ASTELIA R. Brown
*J. Cullen*
Perennial herbs. Leaves evergreen, in rosettes, linear or narrowed just above the sheath, with a pronounced midrib and several other veins perceptible on the lower surface, covered with scales (at least beneath) which bear hair-like outgrowths at the base, and may fuse to form a thin skin. Flowers functionally unisexual (plants dioecious), radially symmetric, borne in a panicle which may be stalked or held within the leaf rosette; primary bracts large. Perianth 6-lobed, membranous or fleshy, persistent. Stamens 6, anthers minute in female flowers. Ovary superior, 1- or 3-celled (visible in the pistillode in male flowers), ovules few to many, axile or parietal; styles 1 or 3. Fruit a berry.

A genus of about 25 species from Australia, New Zealand, Polynesia, Réunion, Mauritius and the Falkland Islands; all the cultivated species are from New Zealand, where several of them grow epiphytically. Their leaves are covered by remarkable scales whose structure is discussed and illustrated in the paper by Moore cited below.

They are grown mainly for their striking rosettes of leaves and may be cultivated in most garden soils. Propagation is by division in spring.
Literature: Skottsberg, C., Studies in the genus Astelia Banks & Solander, *Kunglinga Svenska Vetenskapsakademiens Handlingar*, ser. 3, **14**(2): 1–106 (1934); Wheeler, J.M., Cytotaxonomy of the large asteliads of the North Island of New Zealand, *New Zealand Journal of Botany* **4**: 95–113 (1966); Moore, L.B., Australasian asteliads, *New Zealand Journal of Botany* **4**: 201–40 (1966).

1a. Upper surface of mature leaf green, scaleless (at least when mature)     2
 b. Upper surface of leaf covered by a skin of fused scales which obscures the green colour, and peels in flakes or strips as the leaf ages              3
2a. Leaf 1–2 m, narrowed just above the sheath, drooping; lateral veins beneath 3 on either side of the midrib; ovary 1-celled              **1. solandri**
 b. Leaves to 1 m, arching, stiff and leathery, lateral veins beneath about 10 on either side of the midrib; ovary 3-celled                         **4. petriei**
3a. Leaf at most 50 cm, not densely felted beneath                  **5. nivicola**
 b. Leaf 50–200 cm, densely felted beneath                              4
4a. Leaves scarcely leathery, the upper half drooping; lateral veins beneath 3 or more on either side of the midrib              **2. banksii**
 b. Leaves stiff, leathery, arched stiffly outwards; lateral veins beneath 1 or 2 on either side of the midrib              **3. nervosa**

**1. A. solandri** Cunningham
(*A. cunninghamii* J.D. Hooker; *A. solandri* subsp. *hookeriana* (Kirk) L.B. Moore). Illustration: Botanical Magazine, 5175 (1860) & 5503 (1865).
Leaves drooping, scarcely leathery, 1–2 m × 2–3.5 cm, narrowed just above the sheath; upper surface green, scaleless, lower surface whitish-scaly or green dotted with whitish scale-bases; veins usually 3 on each side of the midrib beneath. Panicle 15–40 cm, borne on a stalk 30–100 cm. Perianth yellow, pink or dark red. Ovary 1-celled, placentation parietal. Berry greenish or brownish, surrounded at the base by the membranous remains of the perianth. *New Zealand.* H4–G1. Spring.

**2. A. banksii** Cunningham.
Leaves scarcely leathery, the upper half drooping, 1–2.5 m × 3–4.5 cm, narrow and tightly folded just above the sheath; upper surface covered by a skin of fused scales which peels off in long strips; lower surface felted, white, veins 3 or more on each side of the midrib. Panicle to 50 cm, borne on a stalk *c.* 40 cm. Perianth pale greenish cream. Ovary 3-celled, placentation axile. Berry white flushed with red-purple when ripe, the base surrounded by the brownish, membranous remains of the perianth. *New Zealand.* H4–G1. Spring.

**3. A. nervosa** J.D. Hooker (*A. montana* (Cheeseman) Cockayne; *A. cockaynei* Cheeseman). Illustration: Salmon, Field guide to the alpine plants of New Zealand, 207–209 (1968); Mark & Adams, New Zealand alpine plants, t. 99 (1973).
Leaves leathery, firm, arched stiffly outwards, 50–200 × 2–6 cm, narrowed and folded just above the sheath; upper surface covered with a skin of fused scales which is smoothed or somewhat ruffled, lower surface with a layer of white or brown scales overlying a layer of hairs; veins 1 or 2 on each side of the midrib. Panicle to 40 cm, borne on a stalk of up to 40 cm. Perianth greenish brown to dark red. Ovary 3-celled, placentation axile. Berry orange to red, the perianth persisting as a fleshy cup beneath it. *New Zealand.* H4–G1. Spring.

**4. A. petriei** Cockayne. Illustration: Mark & Adams, New Zealand alpine plants, t. 98 (1973).
Leaves leathery, stiff but arching, 25–100 × 2–6.5 cm, not narrowed above the sheath; upper surface green, without scales when mature, lower surface white-scaly; veins about 10 on either side of the midrib. Panicles 3–20 cm, the females often small, stalk short. Perianth pale green becoming darker with age. Ovary 3-celled, placentation axile. Berry orange-yellow, borne on the fleshy remains of the perianth. *New Zealand.* H4–G1. Spring.

**5. A. nivicola** Cheeseman. Illustration: Mark & Adams, New Zealand alpine plants, t. 100 (1973).
Leaves 10–50 × 1–3 cm, erect or arching; upper surface with a conspicuous skin of fused scales which flakes off in older leaves, lower surface thickly felted; veins 6–many on each side of the midrib. Panicle rather small and dense, shortly stalked. Perianth greenish. Ovary 3-celled, placentation axile. Berry reddish orange. *New Zealand.* H4–G1. Spring.

**103. CLINTONIA** Rafinesque
*V.A. Matthews*
Rootstock a short rhizome. Leaves basal, sheathing at the base, entire. Flowering stem erect, unbranched, leafless. Flowers borne in an umbel or raceme, occasionally solitary, bell-shaped to star-shaped. Perianth-segments free, downy outside at least at the base. Stamens borne at the base of the perianth-segments. Ovary superior. Stigma with 2 or 3 lobes. Fruit a berry with 2 or 3 cells.
A genus of 5 species from the Himalaya,

E Asia and N America, all of which are in cultivation. They grow best in semi-shade.

1a. Flowers pinkish purple
       **4. andrewsiana**
 b. Flowers white, greenish yellow or lilac
              2
2a. Flowers 1 or 2; leaves usually hairy
 beneath      **5. uniflora**
 b. Flowers 3–30; leaves hairless except
  at the margins     3
3a. Filaments downy at the base
         **2. borealis**
 b. Filaments hairless    4
4a. Flowers in a dense umbel; fruit black, with 2 seeds in each cell
       **3. umbellulata**
 b. Flowers in a loose raceme; fruit blue or purplish black with 2–6 seeds in each cell   **1. udensis**

**1. C. udensis** Trautvetter & Meyer (*C. alpina* Baker).
Leaves 2–8, 8–35 cm, oblanceolate to obovate, hairless except for the hairy margins. Flowering stem 10–85 cm, hairy, bearing a loose raceme of 2–10 more or less erect flowers. Perianth-segments 7–15 mm, yellowish green, white or lilac. Filaments hairless. Fruit dark blue to purplish black with 2–6 seeds in each cell. *Himalaya, E Asia.* H4. Summer.

**2. C. borealis** (Aiton) Rafinesque. Illustration: Rickett, Wild flowers of the United States **1**: 33 (1966).
Leaves 2–5, 10–30 cm, elliptic to obovate, hairless except for the hairy margins. Flowering stem to 30 cm, somewhat hairy, bearing a loose umbel of 2–8 nodding flowers. Perianth-segments 1.2–2 cm, greenish yellow. Filaments downy at the base. Fruit blue. *Eastern N America.* H4. Spring–early summer.

**3. C. umbellulata** (Michaux) Morong (*C. umbellata* Torrey). Illustration: Rickett, Wild flowers of the United States **1**: 33 (1966).
Leaves 2–5, 8–30 cm, oblanceolate to obovate, hairless except for the hairy margins. Flowering stem 15–30 cm, sparsely hairy, bearing a dense umbel of 5–30 erect or scarcely nodding flowers. Perianth-segments 5–10 mm, white, usually spotted with purple and green. Filaments hairless. Fruit black with 2 seeds in each cell. *Eastern N America.* H4. Late spring–summer.

**4. C. andrewsiana** Torrey. Illustration: Rickett, Wild flowers of the United States **5**: 47 (1971).

Leaves 5 or 6, 15–25 cm, elliptic to broadly ovate, hairless except for the sparsely hairy margins. Flowering stem 25–50 cm, hairless or sparsely hairy, bearing a terminal umbel and often lateral umbels below. Perianth-segments 1–1.8 cm, pinkish purple. Filaments downy at the base. Fruit blue with 8–10 seeds in each cell. *USA (California).* H4. Summer.

**5. C. uniflora** (Schultes) Kunth. Illustration: Rickett, Wild flowers of the United States **5**: 47 (1971).
Leaves 2 or 3, 7–15 cm, oblanceolate to obovate, usually hairy beneath. Flowering stem 10–20 cm, hairy, bearing 1 or 2 erect flowers. Perianth-segments 1.8–2.5 cm, white. Filaments downy. Fruit blue with 2 or 3 seeds in each cell. *Western N America.* H4. Summer.

**104. SMILACINA** Desfontaines
*V.A. Matthews*
Perennial herbs with creeping rhizomes. Stems unbranched, leafy. Leaves alternate, stalkless or with short stalks. Flowers borne in terminal racemes or panicles. Flower-stalks jointed. Perianth-segments free. Stamens borne at base of perianth-segments. Ovary superior. Fruit a few-seeded berry.
A genus of 25 species from N America, E Asia and the Himalaya. The species grow best in semi-shade in a soil which does not dry out. Propagation is by division.

1a. Leaves hairless beneath
       **3. paniculata\***
 b. Leaves with short hairs beneath 2
2a. Flowers borne in a raceme **2. stellata**
 b. Flowers borne in a panicle
       **1. racemosa\***

**1. S. racemosa** (Linnaeus) Desfontaines (*S. amplexicaulis* Nuttall). Illustration: Peterson & McKenny, Field guide to wildflowers of north-eastern and north-central North America, 67 (1968); Hay & Synge, Dictionary of garden plants in colour f. 1380 (1969); Reader's Digest encyclopaedia of garden plants and flowers, 668 (1971); Rickett, Wild flowers of the United States **6**: 26 (1973).
Stem to 90 cm, downy. Leaves elliptic to narrowly ovate, clasping or with short 1–4 mm stalks (in lower leaves), with short hairs beneath. Flowers in many-flowered panicles to 15 cm; flower-stalks 0.5–2 mm, downy. Perianth-segments 1–3 mm, white or greenish, shorter than stamens. Berry greenish to red, often mottled with red or purple. *N America, Mexico.* H4. Summer.

**\*S. japonica** Gray. Illustration: Kitamura et al., Coloured illustrations of the herbaceous plants of Japan (Monocotyledoneae), pl. 28 f. 185 (1978). Lower leaves with stalks 1–1.5 cm, flower-stalks 2–5 mm, and usually densely hairy. *Japan, China, Korea*. H4. Late spring–summer.

**\*S. oleracea** (Baker) Hooker. Illustration: Botanical Magazine, 6313 (1877); The Garden **43**: 416 (1893). Perianth-segments 5–7 mm, white, pink or purplish red; stamens shorter than perianth-segments; berry pinkish purple with dark spots. *Himalaya, N Burma*. H4. Summer.

**2. S. stellata** (Linnaeus) Desfontaines (*S. sessilifolia* (Baker) Watson). Illustration: Braun, Vascular Flora of Ohio **1**: 366 (1967); Peterson & McKenny, Field guide to wildflowers of north-eastern and north-central North America, 67 (1968); Weber, Rocky Mountain flora, 386 (1972). Stem 20–60 cm, downy. Leaves lanceolate to oblong-lanceolate, stalkless or slightly clasping, minutely hairy beneath. Flowers in almost stalkless racemes with 6–20 flowers borne on minutely hairy stalks. Perianth-segments 3–6 mm, white or greenish, longer than stamens. Berry at first green with blue-black stripes, later turning dark red or sometimes dark blue. *N America, Mexico*. H4. Summer.

**3. S. paniculata** Martens & Galeotti. Illustration: Botanical Magazine, 8539 (1914); RHS dictionary of gardening **4**: 1966 (1951). Stem 60–100 cm, hairless. Leaves narrowly ovate, with short stalks, hairless. Flowers in a many-flowered panicle to 15 cm. Perianth-segments 3–4 mm, white or pinkish. Berry at first green, spotted with purple, later turning red. *S Mexico to Panama*. G1. Spring.

**\*S. trifolia** (Linnaeus) Desfontaines. Illustration: Rickett, Wild flowers of the United States **1**: 37 (1966); Braun, Vascular Flora of Ohio **1**: 366 (1967); Peterson & McKenny, Field guide to wildflowers of north-eastern and north-central North America, 67 (1968). Stems to 30 cm. Leaves stalkless. Racemes 4–6 cm with 4–12 flowers. *Northern N America, USSR (Siberia)*. H3. Summer.

## 105. MAIANTHEMUM Weber
*A. Leslie*

Perennial herbs, with slender creeping rhizomes, bearing solitary, long-stalked leaves. Flowering stems erect, with scales at base and 2 or 3 alternate leaves above.

Flowers white, in terminal racemes. Perianth-segments 4, free, spreading or reflexed. Flower-stalks 1 to several at a node, slender, with bracts. Stamens 4, shorter than perianth. Ovary superior, 2-celled, with 2 ovules in each cell. Style simple, short; stigma slightly 2-lobed. Fruit a 1 or 2-seeded berry.

A genus of 3 species, extending around the whole north temperate region. Only 1 of them is in general cultivation. Although invasive when grown amongst other herbaceous plants, they can provide effective ground cover under trees and shrubs, performing best in a humus rich, moisture-retentive soil. Propagation is by division.

Literature: Ingram, J., Notes on Cultivated Liliaceae: 3. Maianthemum, *Baileya* **14**: 50–9 (1966).

**1. M. bifolium** (Linnaeus) Schmidt. Illustration: Keble Martin, The concise British flora in colour, t. 84 (1965); Baileya **14**: 53 (1966); Ross-Craig, Drawings of British plants **29**: pl. 15 (1972). Stems 5–20 cm, hairy above. Leaves 3–8 × 2.5–5 cm, broadly ovate, deeply cordate at base, with a broad sinus, acute to acuminate at apex, hairy beneath; at least the lower stem-leaves distinctly stalked. Racemes 1–5 cm, with 8–20 flowers. Perianth-segments 1–3 mm. Fruit 5–6 mm, pale green and spotted at first, eventually red. *W Europe to Japan*. H1. Spring.

**\*M. canadense** Desfontaines. Illustration: Gleason, Illustrated flora of the north-eastern United States and adjacent Canada **1**: 427 (1952); Baileya **14**: 50 (1966). Stems hairless or hairy. Leaves ovate to ovate-oblong, the base shallowly cordate, with a narrow sinus, hairy or hairless beneath; stem-leaves not or very shortly stalked. Fruit pale red. *Eastern N America*. H1. Spring.

**\*M. kamtschaticum** (Chamisso) Nakai (*M. dilatatum* (Wood) Nelson & Macbride). Illustration: Baileya **14**: 55 (1966); Hitchcock et al., Vascular plants of the Pacific northwest **1**: 798 (1969). Stems 1–3.5 cm, hairless. Leaves to 20 × 10 cm, hairless; at least the lower stem-leaves distinctly stalked. Racemes 2.5–7.5 cm. *Western N America, E Asia*. H1. Spring.

## 106. STREPTOPUS Richard
*V.A. Matthews*

Herbaceous perennials with rhizomes. Stems erect, leafy. Leaves alternate, stalkless or clasping. Flowers solitary or in pairs, nodding, borne on a stalk which is fused to the stem for some distance. Flower-stalks with a downward bend at about the middle. Perianth bell-shaped, joined at the extreme base. Stamens borne at base of perianth-lobes. Ovary superior. Style usually with 3 branches. Fruit a many-seeded berry.

7–10 species from the temperate northern hemisphere. They are best planted in a shady position in soil with plenty of humus. Propagation is by division or seed.

1a. Flowers white to greenish white; leaves clasping   **1. amplexifolius***
  b. Flowers pink or pinkish purple, at least at the base; leaves stalkless but not clasping   **2. roseus***

**1. S. amplexifolius** (Linnaeus) de Candolle. Illustration: Weber, Rocky Mountain flora, 387 (1972); Grey-Wilson & Mathew, Bulbs, pl. 20 (1981).
Stem 40–100 cm, branched above, hairy or hairless. Leaves narrowly to broadly heart-shaped, clasping, hairless. Flowers in pairs on 5 cm stalks. Perianth-lobes *c.* 1 cm, greenish white, spreading widely or recurved from the middle. Anthers without horns, longer than the filaments. Style very shortly divided into 3 at the apex. Berry orange-red to blackish crimson. *C & S Europe, N Asia, Japan, N America*. H4. Summer.

**\*S. simplex** Don. Illustration: RHS Lily Year Book **31**. f. 6 (1967). Flowers white, more often solitary than in pairs; style divided almost to the middle into 3 branches. *N India, Nepal, Burma, China*. G1. Summer.

**2. S. roseus** Michaux.
Stem 25–60 cm, usually branched and often shortly hairy above. Leaves lanceolate to narrowly ovate, stalkless, margin ciliate. Flowers solitary on 1–2.5 cm stalks which may be hairy or not. Perianth-lobes 6–12 mm, pink or pinkish purple, the tips eventually recurved. Anthers 2-horned, shorter than filaments. Stigma 3-branched at the apex. Berry red. *North-eastern N America*. H3.

**\*S. streptopoides** (Ledebour) Frye & Rigg is a smaller plant 10–30 cm, with 3–4 mm perianth-lobes which are pinkish or reddish at the base and yellowish green at the recurved tips. The stigma is not divided at the apex, but has 3 faces. The berry varies from red to maroon or black. *Northern N America, USSR (E Siberia), N Japan, Korea, China*. H3.

## 107. DISPORUM Don
*V.A. Matthews*

Perennial herbs with creeping rhizomes. Stems erect, leafy, branched. Leaves alternate, stalkless or shortly stalked. Flowers terminal, solitary or in a few-flowered cluster, usually drooping. Perianth-segments free, often slightly inflated or swollen at the base. Stamens borne at the base of perianth-segments; anthers opening by slits; filaments hairless. Ovary superior, 3-celled. Style 1, with an entire or 3-branched stigma, or styles 3, free. Fruit a berry.

Approximately 10 species from N America, Himalaya, E Asia and Malaysia. They grow best in semi-shade in a moist humus-rich soil. Propagation is by seed, or division in the spring.

1a. Berry blue or black     **3. sessile***
 b. Berry orange or dark red     2
2a. Stamens equal to or longer than
    perianth     **1. hookeri***
 b. Stamens shorter than perianth     3
3a. Flowers yellowish green
                  **2. lanuginosum**
 b. Flowers white, greenish white or
    purplish     4
4a. Filaments stout; style 3-branched to
    half-way or more     **4. cantoniense**
 b. Filaments very slender; style entire or
    with 3 short branches     **1. hookeri***

**1. D. hookeri** (Torrey) Nicholson. Stem 30–100 cm. Leaves 3–14 cm, lanceolate to ovate, usually somewhat clasping, rough or shortly hairy beneath, margin with forward-pointing hairs. Flowers in clusters of 1–3, greenish white. Perianth-segments 9–20 mm, slightly inflated at the base. Stamens longer or shorter than perianth-segments; anthers hairless or hairy, shorter than filaments. Style hairless or with short hairs in the lower half. Berry red, usually with 4–6 seeds. *North-west N America*. H3. Summer.

Var. **hookeri**. Illustration: Rickett, Wild flowers of the United States 6: t. 27 (1973); Niehaus, Field guide to Pacific States wild flowers, 5 (1976). Leaves rough beneath; stamens usually shorter than perianth-segments; style usually hairless.

Var. **oreganum** (Watson) Jones (*D. oreganum* (Watson) Howell). Illustration: Clark, Wild flowers of British Columbia, 27 (1973). Leaves shortly hairy beneath; stamens usually longer than perianth-segments; style usually with short hairs in the lower half.

*****D. smithii** (Hooker) Piper. Illustration: Rickett, Wild flowers of the United States 5:

27 (1971); Niehaus, Field guide to Pacific States wildflowers, 5 (1976). Flowers in clusters of 2–6, perianth-segments 1–3 cm, stamens shorter than perianth-segments, style shortly hairy for its full length, berry orange to red with 5–9 seeds. *Western coastal N America*. H4. Spring–summer.

*****D. trachycarpum** (Watson) Bentham & Hooker. Illustration: Rickett, Wild flowers of the United States 6: 27 (1973); Spellenberg, Audubon Society field guide to North American wildflowers: western region, f. 81 (1979). Leaves hairy beneath at first, losing their hairs with age, not clasping at the base, stamens equal to or slightly longer than perianth-segments, berry orange to red with 6–12 seeds. *Western N America*. H4. Summer.

**2. D. lanuginosum** (Michaux) Nicholson. Illustration: Rickett, Wild flowers of the United States 1: 27 (1966); Peterson & McKenny, Field guide to wildflowers of northeastern and north-central North America, 371 (1968).
Stem 30–90 cm. Leaves 3–12 cm, lanceolate to narrowly ovate, hairy beneath with forward-pointing marginal hairs. Flowers in clusters of 1–3, yellowish green. Perianth-segments 1.2–2 cm, slightly inflated at the base. Stamens shorter than perianth-segments; anthers shorter than filaments. Style hairless. Berry red, usually with 2 seeds, occasionally with 4–6. *Eastern central N America*. H3. Early summer.

**3. D. sessile** (Thunberg) Don. Illustration: Kitamura et al., Coloured illustrations of the herbaceous plants of Japan (Monocotyledoneae), pl. 28, f. 187 (1978). Stem 30–60 cm. Leaves 5–15 cm, oblong to narrowly oblong, hairless. Flowers in clusters of 1–3, white, greenish towards tip. Perianth-segments 2.5–3 cm, slightly inflated at the base. Stamens more or less equal to perianth-segments; anthers shorter than filaments. Style hairless, with 3 branches. Berry blue-black. *Japan*. H4. Late spring–early summer.

'Variegatum' has white-variegated leaves.

*****D. leschenaultianum** Don. Illustration: Botanical Magazine, 6935 (1887). Flowers in clusters of 2–5, up to 1.8 cm. Stamens shorter than perianth-segments and anthers more or less equal to filaments. *S India, Sri Lanka*. G1. Early summer.

*****D. smilacina** Gray. Illustration: Kitamura et al., Coloured illustrations of the herbaceous plants of Japan (Monocotyledoneae), pl. 28, f. 188 (1978).

Flowers usually solitary, erect, 1.5–1.8 cm, not inflated at the base. Berries black, drooping. *USSR (E Siberia), Japan*. H3. Spring.

**4. D. cantoniense** (Loureiro) Merrill (*D. pullum* invalid). Illustration: Botanical Magazine, 916 (1806) – as Uvularia chinensis; Royal Horticultural Society's dictionary of gardening 2: 698 (1951). Stem up to 90 cm, hairless. Leaves 5–15 cm, lanceolate, hairless. Flowers in clusters of 3–6, white to purplish. Perianth-segments *c.* 2.5 cm. Stamens shorter than perianth-segments; anthers shorter than filaments. Style 3-branched to half-way or more. Berry red. *Himalaya, China, E Indies*. H4. Early summer.

## 108. POLYGONATUM Miller
*V.A. Matthews*

Perennial herbs with horizontal creeping rhizomes. Stems unbranched, solitary, erect at least in lower part, leafy in upper part. Leaves all borne on the stem, stalkless, alternate, opposite or whorled, often glaucous beneath. Flowers solitary or in clusters, borne in the leaf-axils. Bracts persistent or deciduous, usually small and membranous, rarely large and leafy. Perianth united into a tube, usually for more than half its length; corona absent. Stamens borne on the tube; anthers included, 2-lobed at the base. Ovary superior, 3-celled. Style slender, included. Stigma 3-lobed. Fruit a spherical berry, red, orange or blue-black when ripe.

About 50 species found in north temperate regions. They grow best in a fairly rich soil in a semi-shady or shady position. Propagation is by division. Literature: Ownbey, R.P., The Liliaceous genus Polygonatum in North America, *Annals of the Missouri Botanical Garden* 31: 373–413 (1944); Jeffrey, C., The genus Polygonatum (Liliaceae) in Eastern Asia, *Kew Bulletin* 34: 435–71 (1980).

*Upper leaves*. Opposite or whorled: 1,2,;
    alternate: 2–9. Hairless beneath
    (sometimes rough on veins): 1–3,5–9;
    hairy on veins beneath: 4,6.
*Flowers*. Pink or purplish: 1–3; white,
    greenish or yellowish: 1,3–9. Erect: 2,3.
*Stamens*. Borne about half-way up tube:
    1–4,8,9; borne towards mouth of tube:
    1,5–7.
*Filaments*. Hairless or warty: 1–4,6–9;
    hairy: 4,5,8.

1a. Flowers erect     2
 b. Flowers drooping     3

2a. Stem to 10 cm; leaves 1.5–4 cm;
    perianth-lobes 4–5 mm    **3. hookeri**
  b. Stem to 70 cm; leaves 7–15 cm;
    perianth-lobes *c.* 1.5 mm    **2. roseum**
3a. Upper leaves opposite or whorled
              **1. verticillatum***
  b. Upper leaves alternate        4
4a. Stamens borne about half-way up the
    perianth-tube           5
  b. Stamens borne towards the mouth of
    the perianth-tube        7
5a. Leaves hairy on the veins beneath
              **4. latifolium***
  b. Leaves hairless or minutely rough on
    the veins beneath        6
6a. Leaves sickle-shaped; fruit 3–4 mm in
    diameter         **8. falcatum**
  b. Leaves not sickle-shaped; fruit
    8–12 mm in diameter    **9. biflorum**
7a. Filaments downy    **5. multiflorum**
  b. Filaments hairless or warty    8
8a. Flowers 8–20 mm    **6. odoratum***
  b. Flowers 2.1–3.6 cm   **7. stenanthum**

**1. P. verticillatum** (Linnaeus) Allioni
(*P. macrophyllum* Sweet). Illustration: Ross-Craig, Drawings of British plants **29**: pl. 12 (1972); Fitter et al., Wild flowers of Britain and northern Europe, 269 (1974); Grey-Wilson & Mathew, Bulbs, pl. 20 (1981).
Stem 20–80 cm, angled, hairless or sometimes sparsely downy. Leaves opposite or whorled (lowermost alternate), 6.5–15 cm, lanceolate to narrowly ovate, minutely rough on veins beneath. Flowers drooping, solitary or in clusters of 2 or 3. Perianth 5–10 mm, white, cylindric although somewhat contracted in the middle. Stamens borne about half-way up the tube; filaments hairless. *Europe, temperate Asia.* H2. Spring–summer.

***P. stewartianum** Diels. Leaves (except lowermost) always in whorls, 5–10 cm and with a tendril at the apex. Perianth purplish pink; stamens borne near the mouth of the tube. *China.* H3. Summer.

***P. sibiricum** Delaroche. Illustration: Komarov, Flora SSSR **4**: pl. 28 (1935). Stems 40–100 cm, round in section, hairless. Upper leaves in whorls, 6.5–25 cm, each with a tendril at the apex. Flowers solitary or in clusters of up to 30. Stamens borne near the mouth of the tube. *China, Mongolia, SE USSR.* H2. Summer.

**2. P. roseum** (Ledebour) Kunth.
Illustration: Botanical Magazine, 5049 (1858).
Stem to 70 cm, round in section, hairless. Leaves alternate, opposite or in whorls of three, 7–15 cm, linear to narrowly

lanceolate, often minutely rough on veins beneath. Flowers erect, solitary or in pairs. Perianth 1–1.2 cm, pinkish, cylindric with the lobes slightly reflexed. Stamens borne about half-way up the tube; filaments hairless. *China, USSR (Soviet Central Asia, W Siberia).* H3. Summer.

**3. P. hookeri** Baker. Illustration: Hooker's Icones Plantarum **23**: pl. 2218 (1892); Bulletin of the Alpine Garden Society **36**: 313 (1968); Reader's Digest encyclopaedia of garden plants and flowers, 541 (1971). Stem to 10 cm, hairless. Leaves alternate, clustered towards stem apex, 1.5–4 cm, linear to narrowly elliptic, hairless beneath. Flowers erect, solitary. Perianth 1.1–1.2 cm, pinkish, purplish, or occasionally greenish yellow, free lobes 4–5 mm, spreading. Stamens borne about half-way up the tube; filaments hairless. *E Himalaya, China.* H3.

**4. P. hirtum** (Poiret) Pursh (*P. latifolium* (Jacquin) Desfontaines). Illustration: Reichenbach, Icones Florae Germanicae et Helveticae **7**: t. 965 (1845); Grey-Wilson & Mathew, Bulbs, pl. 20 (1981) – as *P. jacquinii.*
Stem 20–120 cm, angled, sparsely downy. Leaves alternate, 7–12.5 cm, lanceolate to ovate, sparsely downy on veins beneath. Flowers drooping, solitary or in clusters of 2–5. Perianth 1–2.5 cm, white, tipped with green, cylindric. Stamens borne about half-way up the tube; filaments hairless or sparsely downy. *EC & SE Europe, European USSR, NW Turkey.* H3. Spring–summer.

***P. humile** Maximowicz (*P. officinale* Allioni var. *humile* (Maximowicz) Baker). Illustration: Flora SSSR **4**: pl. 28 (1935). Leaves not sickle-shaped, hairy on the veins beneath. Flowers solitary or in pairs. Filaments 4–5 mm. *USSR (Siberia), China, Japan, Korea.* H3. Summer.

**5. P. multiflorum** (Linnaeus) Allioni. Illustration: Ross-Craig, Drawings of British plants **29**: pl. 13 (1972); Fitter et al., Wild flowers of Britain and northern Europe, 269 (1974).
Stem 30–90 cm, round in section, hairless. Leaves alternate, 5–15 cm, lanceolate to ovate, hairless beneath. Flowers drooping, in clusters of 2–6. Perianth 9–20 mm, white, usually tipped with green, cylindric, somewhat contracted in the middle. Stamens borne towards the mouth of the tube; filaments sparsely downy. *Europe, temperate Asia.* H2. Spring–summer.

'Variegatum' has the leaves striped with white.

A hybrid between *P. multiflorum* and *P. odoratum* which displays characters intermediate between those of the parents, is the most common *Polygonatum* in cultivation, and called **P. × hybridum** Brügger (Illustration: Reader's Digest encyclopaedia of garden plants and flowers 541, 1971; Perry, Flowers of the world, 172, 1972).

**6. P. odoratum** (Miller) Druce (*P. vulgare* Desfontaines; *P. officinale* Allioni; *P. japonicum* Morren & Decaisne; *P. odoratum* (Miller) Druce var. *thunbergii* (Morren & Decaisne) Hara). Illustration: Ross-Craig, Drawings of British plants **29**: pl. 14 (1972).
Stem 30–85 cm, angled, hairless. Leaves alternate, 3–15 cm, lanceolate to ovate, hairless beneath. Flowers drooping, fragrant, solitary or in clusters of 2–4. Perianth 8–20 mm, white, tipped with green, cylindric or widening towards the mouth. Stamens borne towards the mouth of the tube; filaments hairless, minutely warty or not. *Europe, temperate Asia.* H2. Spring–summer.

'Variegatum' has cream-striped leaves.

For the hybrid between *P. odoratum* and *P. multiflorum* which may key out under *P. odoratum*, see note under *P. multiflorum.*

***P. pubescens** (Willdenow) Pursh. Leaves finely downy on the veins beneath, flowers yellowish green, the perianth-tube contracted at the base of the spreading lobes and minutely warty inside. Filaments densely warty. *NE & NC USA, SE Canada.* H2.

**7. P. stenanthum** Nakai (*P. macranthum* (Maximowicz) Koidzumi).
Stem to 120 cm, round in section, hairless. Leaves alternate, 8.5–17.5 cm, lanceolate to ovate, hairless beneath. Flowers drooping, solitary or in clusters of 2–4. Perianth 2.1–3.6 cm, white, cylindric, the free lobes 4–7 mm. Stamens borne towards the mouth of the tube; filaments warty. *Japan, Korea.* H4.

**8. P. falcatum** Gray.
Stem 50–85 cm, hairless. Leaves alternate, 6–23 cm, narrowly lanceolate to ovate-elliptic, sickle-shaped, sometimes minutely rough on veins beneath. Flowers drooping, in clusters of 2–5. Perianth 1.1–2.2 cm, white, cylindric. Stamens borne about half-way up the tube; filaments usually hairless, occasionally hairy or warty, 5–7 mm. Fruit 3–4 mm in diameter. *Japan, Korea.* H4.

Often offered by nurserymen as 'P. pumilum'.

**9. P. biflorum** (Walter) Elliott
(*P. commutatum* (Schultes) Dietrich;
*P. giganteum* Dietrich). Illustration: Rickett,
Wild flowers of the United States **1**: 35
(1965).
Stem 40–200 cm, hairless. Leaves
alternate, 4–18 cm, narrowly lanceolate to
broadly elliptic, hairless beneath. Flowers
drooping, solitary or in clusters of up to 15.
Perianth 1.1–2.3 cm, whitish to greenish
yellow, cylindric with 3 or 4 mm lobes,
often tipped with green. Stamens borne
about half-way up the tube; filaments
hairless, sometimes minutely warty. Fruit
8–12 mm in diameter. *E USA, SC Canada.*
H3. Summer.

**109. SCOLIOPUS** Torrey
*V.A. Matthews*
Perennial, hairless herbs with a short,
slender rootstock. Leaves usually 2. Flowers
borne in an umbel at the top of the short,
underground stem, each flower thus
appearing to be solitary. Perianth-segments
6, free, in 2 dissimilar whorls. Sepals (outer
segments) broad, spreading to reflexed.
Petals (inner segments) narrow, erect.
Stamens 3, attached to the base of the
sepals. Ovary superior. Style 1, very short.
Stigmas 3, linear. Ovary 3-angled, 1-celled.
Fruit a capsule.

A genus of 2 species from W USA; both
are in cultivation. Cultivation as for *Trillium*
(below). Propagation is by seed.

**1. S. bigelovii** Torrey. Illustration: Bulletin of
the Alpine Garden Society **35**: 190 (1967);
Rickett, Wild flowers of the United States **5**:
45 (1971); Spellenberg, Audubon Society
field guide to North American wildflowers –
western region, f. 377 (1979).
Leaves 8–10 cm at flowering-time,
subsequently growing to 20 cm, broadly
elliptic, usually blotched with dark brown or
purple. Flower-stalks 10–15 cm, 3-angled,
tinged with red, curving down towards the
ground at fruiting-time. Flowers with a
strong, unpleasant smell. Sepals 1.5–2 cm,
spreading or usually reflexed, pale green
with longitudinal stripes of purple-brown.
Petals linear, erect, purple. Stamens
5–6 mm. Stigmas 5–6 mm, becoming
recurved. Capsule 1.5–1.8 cm, ellipsoid,
bearing the persistent style and stigmas at
the top. *W USA (California, SW Oregon).* H4.
Late winter–spring.
**\*S. hallii** Watson. Illustration: Abrams,
An Illustrated flora of the Pacific States **1**:
450 (1923). Flower-stalks to 8 cm, petals
8–10 mm, stamens *c.* 4 mm, stigmas *c.*
2 mm. *W USA (Oregon).* H4. Spring.

**110. MEDEOLA** Linnaeus
*V.A. Matthews*
Perennial herbs with horizontal, white
tubers to 8 cm, smelling of cucumber. Stem
simple, 20–90 cm, hairy at first, eventually
becoming hairless. Leaves in 2 whorls, the
lower whorl about half-way up the stem
with 5–11 lanceolate to obovate, shortly
acuminate leaves, the upper whorl just
below the flowers, with 3, occasionally 4 or
5 smaller, ovate leaves. Flowers 3–9,
nodding, in an umbel at the top of the
stem, flower-stalks to 2.5 cm, spreading or
recurved, becoming more or less erect at
fruiting-time. Perianth of 6, free, similar
segments, 7–8 mm, pale greenish yellow,
recurved. Stamens 6, filaments longer than
anthers. Ovary superior. Styles 3, linear,
recurved. Fruit a spherical, dark purple or
red berry with 3 cells, 4–8 mm across.

A genus of only one species from North
America. Cultivation as for *Trillium*
(below). Propagation is by seed.

**1. M. virginiana** Linnaeus. Illustration:
Rickett, Wild flowers of the United States **1**:
25 (1966); Peterson & McKenny, Field
guide to wildflowers of northeastern and
north-central North America, 103 (1968);
Niering & Olmstead, Audubon Society field
guide to North American wildflowers –
eastern region, f. 3 (1979); Dean et al.,
Wildflowers of Alabama, 17 (1983).
*Eastern N America.* H3. Early summer.

**111. PARIS** Linnaeus
*V.A. Matthews*
Perennials with creeping rhizomes. Leaves
4 or more in a whorl near the top of the
unbranched stem. Flowers solitary,
terminal. Outer perianth-segments 4–6,
green; inner perianth-segments 4–6,
yellow, very narrow. Stamens 4–10 with
short flat filaments and basifixed anthers.
Ovary superior. Styles free, usually 4. Fruit
a fleshy, berry-like capsule.

There are about 20 species in the genus,
native to temperate Eurasia. They grow
best in a shady position.

**1. P. quadrifolia** Linnaeus. Illustration:
Felsko, A book of wild flowers, pl. 150
(1956); Polunin, Flowers of Europe, f. 170
(1969); Fitter et al., Wild flowers of Britain
and northern Europe, 269 (1974); Rose,
The wild flower key, 407 (1981).
Stem 10–40 cm, hairless. Leaves usually 4,
ovate to obovate, 5–15 cm, with short
stalks. Perianth-segments usually 4,
2–3.5 cm. Fruit bluish black, spherical.
*Europe, USSR (Caucasus, Siberia).* H3.
Spring–summer.

**\*P. polyphylla** Smith. Illustration:
Hutchinson, Families of flowering plants, f.
381 (1973); Rix & Phillips, The bulb book,
105 (1981). Stem to 90 cm. Leaves 5–9,
linear to oblong-lanceolate, stalks 2–3 cm.
Perianth-segments 4–6, the inner
2–11 cm. Fruit greenish brown, splitting to
reveal red seeds. *Himalaya, Burma, China,
Thailand.* H4. Spring–summer.

**112. TRILLIUM** Linnaeus
*V.A. Matthews*
Perennial herbs with short, thick rhizomes.
Stem simple, usually erect and hairless.
Leaves 3 in a whorl at the top of the stem,
with netted veins. Flower solitary, stalked
or not. Perianth of 2 dissimilar whorls of 6
free segments. Sepals (outer whorl) 3,
usually green. Petals (inner whorl) 3,
varying in colour, rarely green. Stamens 6
with basifixed anthers. Styles 3. Ovary
superior with 3 or 6 angles or wings. Fruit
a 3-celled berry, usually spherical.

A genus of about 30 species occurring
mainly in N America, with a few species
from the W Himalaya to NE Asia. They
require a rich, moist soil and partial shade.
Propagation is by division and by seed.

Plants with aberrant flowers (e.g. with 4
sepals and 4 petals, or with stamens
changed into petals or vice versa) occur in
most species, and are deliberately cultivated
in *T. grandiflorum* and *T. ovatum*.
Literature: Samejima, J. & K., Studies on
the eastern Asiatic Trillium (Liliaceae), *Acta
Horti Gotoburgensis* **25**: 157–259 (1962);
Mitchell, R.J., The genus Trillium, *Journal of
the Scottish Rock Garden Club* **11**: 271–83
(1969), **12**: 16–20 (1970), **12**: 115–23
(1970), **13**: 13–19 (1972); Freeman, J.D.,
Revision of Trillium subgenus
Phyllantherum (Liliaceae), *Brittonia* **27**:
1–62 (1975); Christian, P., Trilliums,
*Bulletin of the Alpine Garden Society* **48**:
61–6, 139–43, 189–202, 339–42 (1980).

*Stem.* Lying along ground: 3. Velvety or
    downy: **3,8.**
*Leaves.* Mottled: **1–8,10–12**; not mottled:
    **3,5,7–10,12–27.** Stalkless or nearly so:
    **2–8,10–13,15,18–21,23–27**; with a
    stalk 5 mm or more: **1,9,14,16,17,22.**
*Flowers.* Stalkless, erect: **1–12**; stalked with
    the flower held erect or horizontally:
    **16–27**; stalked with the flower nodding:
    **13–15.**
*Petals.* Erect: **1–12**; spreading:
    **16–18,20–27**; recurved or reflexed:
    **13–15,19.** Less than 5 mm wide: **23.**
*Basic colour of petals.* White: **10,11,13,
    16–22,24,25**; pink: **11,13,14,16,18–22,**

24; brownish, purplish or deep red:
1–4,6,8,9,11,12,15,20,23,26,27; yellow
or greenish: 1,2,4,5,7,8,11,18,20,23.

1a. Flowers stalkless                                  2
 b. Flowers with a stalk                              15
2a. Leaves with stalks 5–200 mm                       3
 b. Leaves stalkless or nearly so                     4
3a. Petals clawed at the base; sepals
    reflexed; stamens 1–1.5 cm
                              **1. recurvatum**
 b. Petals tapered towards the base but
    not clawed; sepals spreading, not
    reflexed; stamens 1.3–3 cm
                              **9. petiolatum**
4a. Leaf apex acuminate                               5
 b. Leaf apex blunt to acute                          7
5a. Petals purple or brownish purple
                              **6. cuneatum**
 b. Petals yellow to greenish yellow,
    sometimes purplish at the base        6
6a. Sepals 1.5–3.8 cm, apex acute
                              **5. discolor**
 b. Sepals 2.5–6.5 cm, apex blunt
                              **7. luteum**
7a. Connectives prolonged up to 1.5 mm
    beyond anther sacs                                8
 b. Connectives prolonged 2–4 mm
    beyond anther sacs                               14
8a. Petals white or pink                              9
 b. Petals red, brownish, purple, yellow
    or greenish yellow                               10
9a. Anthers with lateral dehiscence,
    connectives green            **10. albidum**
 b. Anthers with introrse dehiscence,
    connectives purple
                            **11. chloropetalum**
10a. Stigmas recurved, at least at the tips
                                                     11
 b. Stigmas spreading to erect                       12
11a. Leaves with minute white dots on
    upper surface; filaments 3–6 mm
                              **8. viride**
 b. Leaves lacking minute white dots;
    filaments 1.5–2.5 mm
                            **4. underwoodii**
12a. Petals 2.5–5 cm, mainly pale
    yellow, never purple or red-brown
                              **5. discolor**
 b. Petals 5–10 cm, variously coloured, if
    less than 5 cm, then purple or red-
    brown                                            13
13a. Petals less than 1 cm wide; stamens
    only slightly longer than the ovary
                            **12. angustipetalum**
 b. Petals usually 1 cm or more wide;
    stamens about twice as long as the
    ovary                   **11. chloropetalum**
14a. Stem erect, hairless; stigmas erect
                              **2. sessile**
 b. Stem usually lying along the ground.

with the leaves held more or less at
ground level; stigmas spreading with
recurved tips                **3. decumbens**
15a. Petals less than 5 mm wide
                            **23. govanianum**
 b. Petals more than 5 mm wide        16
16a. Leaves stalkless or almost so        21
 b. Leaves with stalks                    17
17a. Flowers hanging, level with or below
    leaves                                           18
 b. Flowers held erect above the leaves, at
    least at first                                   19
18a. Ovary deep red; petals c. 2 cm
                              **13. cernuum**
 b. Ovary pale green; petals 2.5–5 cm
                              **14. catesbaei**
19a. Petals with bright or deep red marks
    at the base; anthers purple
                            **16. undulatum**
 b. Petals lacking red marks at the base;
    anthers yellow                                   20
20a. Flower-stalk 1.5–2.5 cm; petals
    white, 2.5–4.5 cm            **17. nivale**
 b. Flower-stalk 2.5–5 cm; petals white
    or pale pink, with pink-purple dots
    and lines, 2–2.5 cm          **22. rivale**
21a. Sepals purple to purplish green, or
    reddish green                                    22
 b. Sepals green, or if flushed with
    reddish purple than petals not purple
    or purple-brown, or if petals pale pink
    then flowers not smelling of wet dog
                                                     24
22a. Petals to 2 cm, ovate-circular to
    circular, blunt, or petals sometimes
    absent                                           23
 b. Petals 3–8 cm, broadly lanceolate to
    ovate, acute               **20. erectum**
23a. Flower-stalk 2–4 cm; sepals 1.5–2 cm
                              **26. smallii**
 b. Flower-stalk 5–7 cm; sepals 2.5–3 cm
                              **27. amabile**
24a. Anthers red to purple                25
 b. Anthers yellow or cream               26
25a. Flowers pleasantly scented, nodding;
    stigmas yellow               **15. vaseyi**
 b. Flowers smelling of wet dog, erect;
    stigmas purple             **20. erectum**
26a. Petals overlapping and erect at the
    base, spreading above
                            **18. grandiflorum**
 b. Petals spreading from the base        27
27a. Flowers smelling of wet dog
                              **20. erectum**
 b. Flowers scentless or fragrant,
    sometimes unpleasant                             28
28a. Flowers erect          **25. kamtschaticum**
 b. Flowers held in a more or less
    horizontal position                              29
29a. Sepals reflexed             **19. flexipes**
 b. Sepals spreading                      30

30a. Petals white fading to pink or red, or
    pink fading to whitish; stigmas erect
    or spreading                 **21. ovatum**
 b. Petals white fading to pale reddish
    purple; stigmas short and thick,
    recurved                   **24. tschonoskii**

**1. T. recurvatum** Beck. Illustration: Rickett,
Wild flowers of the United States 2: 33
(1967); Peterson & McKenny, Field guide to
wildflowers of northeastern and north-
central North America, 241 (1968);
Mohlenbrock, The illustrated flora of
Illinois, flowering plants – lilies to orchids,
101 (1970); Courtenay & Zimmerman,
Wildflowers and weeds, 5 (1972).
Stem 12–45 cm, hairless. Leaf-stalks
5–80 mm. Leaves lanceolate to broadly
ovate or elliptic, mottled. Flower stalkless.
Sepals 1.5–4 cm, lanceolate to ovate,
reflexed. Petals 2–5 cm, erect, dark purple
to reddish brown, or yellow, lanceolate to
ovate with a distinct claw at the base.
Stamens 1–1.5 cm with purple filaments;
anthers purple, c. 1.5 mm. Ovary purple,
6-winged. Stigmas with recurved tips.
*E USA.* H3. Spring.

**2. T. sessile** Linnaeus. Illustration: Hay &
Synge, Dictionary of garden plants in
colour, f. 889 (1969); Horticulture 50(3):
39 (1972); Mathew, The larger bulbs, 81
(1978); Rix & Phillips, The bulb book, 103,
105 (1981).
Stem 10–30 cm, hairless. Leaves stalkless,
4–12 cm, elliptic to ovate or circular, blunt
or acute, mottled. Flower stalkless, with a
spicy scent. Sepals 1.3–4 cm, lanceolate to
ovate, spreading, green, tinged with brown
at the base. Petals 1.7–4.5 cm, erect, dark
purple to greenish yellow (forma
**viridiflorum** Beyer), narrowly elliptic to
lanceolate. Stamens 9–24 mm, filaments
shorter than the purple anthers; anther-
connectives extending 2–3.5 mm beyond
the anthers. Ovary purple, with 6 angles or
wings. Stigmas erect. *NE USA.* H3.
Spring–summer.

   Many of the plants sold as *T. sessile* are
*T. cuneatum.*

**3. T. decumbens** Harbison. Illustration:
Bulletin of the American Rock Garden
Society 39: 112 (1981); Dean et al.,
Wildflowers of Alabama, 13 (1983).
Stem 5–20 cm, velvety, lying along the
ground. Leaves stalkless, resting on the
ground, 5–13 cm, ovate to almost circular,
usually mottled with silvery-green in a V-
shaped patch. Flower stalkless. Sepals
2–5 cm, lanceolate to ovate, spreading,
green with a purplish flush. Petals

3.5–8 cm, erect, dark purple or reddish brown, sometimes fading to yellow-green above, oblanceolate to lanceolate, twisted. Stamens 1–2.5 cm, filaments shorter than the anthers; anther-connectives extending more than 2 mm beyond the anthers. Ovary with 6 angles or wings. Stigmas spreading with recurved tips. Fruit blackish. *SE USA (Alabama, Georgia)*. H3. Spring.

**4. T. underwoodii** Small. Illustration: Dean et al., Wildflowers of Alabama, 15 (1983).
Stem 12–30 cm, hairless. Leaves stalkless, 5–15 cm, usually drooping, lanceolate to ovate-elliptic, acute, mottled. Flower stalkless, with a foetid smell. Sepals 2.5–5.5 cm, lanceolate to ovate, spreading. Petals 3.5–6.5 cm, dark purple, maroon, brownish purple or greenish yellow, oblanceolate to narrowly elliptic. Stamens 8–20 mm, filaments 1.5–2.5 mm, shorter than the purple anthers; anther-connectives prolonged. Ovary purple, with 6 angles or wings. Stigmas recurved onto the ovary. *SE USA (Alabama, Georgia, N Florida)*. H3. Spring.

**5. T. discolor** Hooker. Illustration: Botanical Magazine, 3097 (1831); Justice & Bell, Wild flowers of North Carolina, 18 (1968).
Stem 10–15 cm, sometimes more, hairless. Leaves stalkless, 5–14 cm, elliptic to almost circular, acute to acuminate, usually mottled with dark green. Flower stalkless, with a spicy scent. Sepals 1.5–3.8 cm, lanceolate to ovate, acute, spreading. Petals 2.5–5 cm, pale yellow with a greenish or occasionally purplish base, oblanceolate, slightly twisted. Stamens 8–20 mm, filaments shorter than the anthers, purple. Ovary purple, with 6 ridges. Stigmas spreading to nearly erect. *SE USA (Georgia, North & South Carolina)*. H3. Spring.

**6. T. cuneatum** Rafinesque (*T. hugeri* Small). Illustration: Rickett, Wild flowers of the United States 2: 33 (1967); Justice & Bell, Wild flowers of North Carolina, 17 (1968); Bulletin of the Alpine Garden Society 48: 198 (1980).
Stem 30–60 cm, hairless. Leaves stalkless, ovate to circular, acuminate, mottled. Flower stalkless, with a spicy or musky scent. Sepals green, sometimes yellow-green or purple towards the apex. Petals purple or brownish purple, *c.* 2.5 cm wide. *E USA*. H3. Spring.

Nurseries often offer plants of *T. cuneatum* under the name *T. sessile*.

**7. T. luteum** (Muhlenberg) Harbison (*T. viride* Beck var. *luteum* (Muhlenberg)

Gleason; *T. sessile* Linnaeus var. *luteum* Muhlenberg). Illustration: Rickett, Wild flowers of the United States 2: 33 (1967); Rix & Phillips, The bulb book, 103 (1981).
Stem 15–45 cm, hairless. Leaves stalkless, 5–15 cm, elliptic to broadly ovate, acuminate, more or less mottled. Flower stalkless, smelling of freesias or lemon-oil. Sepals 2.5–6.5 cm, lanceolate to narrowly oblong-elliptic, blunt, spreading. Petals 3–9 cm, erect, yellow to greenish yellow, oblanceolate to obovate or narrowly elliptic. Stamens 8–25 mm, filaments shorter than the anthers. Ovary with 6 angles or ridges. *SE USA (Tennessee, Kentucky, North Carolina, N Georgia)*. H3. Spring.

**8. T. viride** Beck. Illustration: Mohlenbrock, The illustrated flora of Illinois, flowering plants – lilies to orchids, 104 (1970); Rickett, Wild flowers of the United States 6: 29 (1973).
Stem 20–40 cm, hairless or downy above. Leaves stalkless, 8–15 cm, narrowly to broadly elliptic, acute or blunt, slightly mottled or not, upper surface with tiny white dots. Flower stalkless, smelling of rotten fruit. Sepals 3–6 cm, lanceolate, acute, spreading to slightly deflexed. Petals 3–7 cm, erect, usually greenish or yellowish, purplish at the base, or sometimes purple or yellowish throughout, oblanceolate, sometimes twisted. Stamens 1.3–2.7 cm, filaments much shorter than the anthers. Ovary with 6 angles or wings. Stigmas with recurved tips. *North-central USA (Missouri, Illinois)*. H3. Spring.

**9. T. petiolatum** Pursh. Illustration: Abrams, An illustrated flora of the Pacific States 1: 451 (1923); Rickett, Wild flowers of the United States 6: 29 (1973); Niehaus & Ripper, Field guide to Pacific States wildflowers, 257 (1973).
Stem 5–20 cm, hairless. Leaves stalked, 10–20 cm, elliptic to ovate, blunt. Flower stalkless. Sepals 2–5 cm, oblong-elliptic to oblanceolate, acute, spreading. Petals 3–5 cm, erect, incurved near tips, purple, brownish purple or greenish purple, linear to lanceolate, acute. Stamens 1.3–3 cm, filaments shorter than the anthers. Stigmas usually with recurved tips which extend between the stamens. *NW USA (Washington, Oregon, Idaho)*. H3. Spring.

**10. T. albidum** Freeman. Illustration: Rickett, Wild flowers of the United States 5: 37 (1971) – as T. chloropetalum with white petals; Brittonia 27: 50 (1975).

Stem 20–60 cm, hairless. Leaves stalkless, 7–20 cm, ovate, blunt, mottled or not. Flower stalkless, sweetly scented. Sepals 2.7–7 cm, lanceolate, spreading. Petals 4–10 cm, erect, white, sometimes pink towards the base, oblanceolate to obovate. Stamens 1.5–3 cm, filaments much shorter than the anthers, pale green; anther-connectives green. Ovary green. *W USA (Washington, Oregon, N California)*. H3. Spring.

**11. T. chloropetalum** (Torrey) Howell. Illustration: Bulletin of the Alpine Garden Society 42: 21 (1974).
Stem 25–50 cm, hairless. Leaves stalkless, 8–20 cm, ovate to diamond-shaped, mottled. Flower stalkless, with a sweet or spicy scent. Sepals 3–6.5 cm, lanceolate, spreading. Petals 5–10 cm, erect, greenish yellow to purple, narrowly oblanceolate to obovate. Stamens 1.5–3.5 cm, purple, filaments much shorter than anthers which have purple connectives. Ovary purple, more or less 6-angled. Stigmas straight, purple. Fruit reddish. *W USA (California)*. H3. Spring.

Var. **giganteum** (Hooker & Arnott) Munz has petals which vary from white to red or red-purple (Illustration: Rickett, Wild flowers of the United States 5: 37, 1971 – as *T. chloropetalum* with red petals).

**12. T. angustipetalum** (Torrey) Freeman (*T. sessile* Linnaeus var. *angustipetalum* Torrey).
Stem 20–65 cm, hairless. Leaves almost stalkless, 10–23 cm, ovate to broadly ovate, blunt, usually mottled. Flower stalkless, with a spicy or musty smell. Sepals 3–6 cm, narrowly lanceolate to narrowly oblong. Petals 4–5 cm, sometimes to 10 cm, to 1 cm, wide, erect, dark purple or red-brown, linear. Stamens 1–2.2 cm, purple, filaments much shorter than the anthers. Ovary purple or red-brown, sharply 6-angled at the top. Stigmas dark purple, erect. *W USA (California)*. H4. Spring.

**13. T. cernuum** Linnaeus. Illustration: Botanical Magazine, 954 (1806); Rickett, Wild flowers of the United States 2: 29 (1967); Courtenay & Zimmerman, Wildflowers and weeds, 5 (1972); Bulletin of the Alpine Garden Society 42: 21 (1974).
Stem 25–60 cm, hairless. Leaves shortly stalked, diamond-shaped. Flower-stalk *c.* 2.5 cm, curving downwards below the leaves. Petals *c.* 2 cm, white or pale pink, margin wavy. Anthers dark purple. Ovary

and fruit dark red. *E Canada, NE USA.* H3. Spring.

**14. T. catesbaei** Elliott (*T. nervosum* Elliott; *T. stylosum* Nuttall). Illustration: RHS Lily Year Book **25**: f. 10 (1961); Rickett, Wild flowers of the United States **2**: 31 (1967); Justice & Bell, Wild flowers of North Carolina, 19 (1968); Journal of the Scottish Rock Garden Club **13**: f. 2 (1972).
Stem 20–50 cm, hairless, red-tinged. Leaves shortly stalked, 5–7.5 cm, elliptic to ovate. Flower-stalk *c.* 5 cm, recurved, the flower borne level with or just below the leaves. Sepals recurved. Petals 2.5–5 cm, pale to deep pink, recurved. Ovary pale green. *SE USA.* H4. Spring–summer.

Much confusion surrounds the names *T. catesbaei*, *T. nervosum* and *T. stylosum*. Investigations presently being carried out may result in *T. nervosum* (*T. stylosum*) being separated from *T. catesbaei*.

**15. T. vaseyi** Harbison. Illustration: Rickett, Wild flowers of the United States **2**: 33 (1967); Justice & Bell, Wild flowers of North Carolina, 20 (1968).
Stem 30–60 cm, hairless. Leaves stalkless, 10–20 cm, broadly diamond-shaped. Flower stalked, fragrant, *c.* 10 cm in diameter, nodding. Petals maroon-red, overlapping, blunt. Stamens purple. Stigmas yellow. *SE USA (North & South Carolina, Georgia, Tennessee).* H4. Spring–summer.

**16. T. undulatum** Willdenow (*T. erythrocarpum* Michaux). Illustration: Rickett, Wild flowers of the United States **2**: 29 (1967); Justice & Bell, Wild flowers of North Carolina, 20 (1968); Bulletin of the Alpine Garden Society **42**: 21 (1974), **48**: 188 (1980); Rix & Phillips, The bulb book, 103 (1981).
Stem 10–30 cm, hairless, flushed with pink towards the base. Leaves stalked, to 15 cm or more, tapering to a sharp point. Flower borne on a short, erect stalk. Sepals *c.* 1 cm, green with red margins. Petals 2–3 cm, white or very pale pink, with a broad red band at the base which extends up the main veins, margin wavy. Filaments longer than the purple anthers. Ovary white. Fruit red. *Eastern N America.* H3. Spring–summer.

**17. T. nivale** Riddell. Illustration: Botanical Magazine, 6449 (1879); Rickett, Wild flowers of the United States **1**: 29 (1966); Courtenay & Zimmerman, Wildflowers and weeds, 5 (1972); Rix & Phillips, The bulb book, 37 (1981).
Stem 5–12 cm, hairless. Leaf-stalks 5–10 cm. Leaves 2.5–3.5 cm, broadly lanceolate, ovate or elliptic. Flower-stalk 1.5–2.5 cm, erect, reflexing with age. Sepals to 2 cm, green. Petals 2.5–4.5 cm, white. Filaments slightly shorter than the yellow anthers. Ovary green. *SE USA.* H4. Late winter–spring.

**18. T. grandiflorum** (Michaux) Salisbury. Illustration: Horticulture **35**(2): inside front cover (1957); Rickett, Wild flowers of the United States **2**: 29 (1967); Bulletin of the Alpine Garden Society **42**: 21 (1974), **48**: 178 (1980); Rix & Phillips, The bulb book, 102 (1981).
Stem 20–45 cm, hairless. Leaves almost stalkless, to 30 cm, broadly ovate to almost circular. Flower-stalk *c.* 5 cm, erect. Sepals 3–5 cm, green. Petals 4–9 cm, white, gradually fading to pink, margin wavy, oblanceolate to almost circular, overlapping and erect at the base, then spreading above. Stamens 1.5–2.5 cm with greenish filaments and yellow anthers. Styles straight, almost erect. Fruit dark red. *Eastern N America.* H3. Spring–summer.

A number of forms of *T. grandiflorum* have been recognised: forma **flore pleno** Anon. has double flowers (Illustration: RHS Lily Year Book **26**: f. 23, 1962); forma **parvum** Gates is a smaller plant than *T. grandiflorum* itself with flowers which quickly turn pink and age to purple-pink; forma **roseum** Anon. has flowers which become pink within hours of opening and age almost to red; forma **variegatum** Smith has petals which vary from having a central green line to being wholly green with a narrow white edge (Illustration: Everett, Illustrated encyclopedia of horticulture **10**: 3402, 1982).

**19. T. flexipes** Rafinesque. Illustration: Mohlenbrock, The illustrated flora of Illinois, flowering plants – lilies to orchids, 114 (1970); Rickett, Wild flowers of the United States **6**: 29 (1973).
Stem to 50 cm, hairless. Leaves stalkless, to 20 cm, acute to shortly acuminate. Flower-stalk holding the flower almost horizontally, flower usually fragrant. Sepals 2–5 cm, reflexed. Petals 2–5 cm, white or sometimes flushed with pink, spreading from the base, reflexed. Filaments shorter than the yellow or cream anthers. Ovary white or pinkish. Stigmas recurved. *Northern C USA (Kansas, Ohio, Illinois, Michigan, Missouri).* H3. Spring.

Often sold under the name *T. cernuum*.

**20. T. erectum** Linnaeus. Illustration: Rickett, Wild flowers of the United States **2**: 29 (1967); Journal of the Scottish Rock Garden Club **12**: f. 36 (1970); Bulletin of the Alpine Garden Society **42**: 21 (1974); Rix & Phillips, The bulb book, 102 (1981).
Stem 15–60 cm, hairless. Leaves stalkless, broadly ovate to almost circular. Flower-stalks to 10 cm, erect, flowers smelling of wet dog. Sepals 3–5 cm, green, flushed with red especially at the margin. Petals 3–8 cm, bright maroon-red, acute. Anthers 6–12 mm, red-purple. Ovary brown-purple. *Eastern N America.* H3. Spring–early summer.

The following variants are in cultivation: var. **album** Pursh (Illustration: Journal of the Scottish Rock Garden Club **12**: f. 37, 1970) which has white or pale pink petals and yellow anthers; forma **albiflorum** Hoffmann which has white petals tinged with green; forma **luteum** Louis-Marie with a stem to 12 cm and yellow petals lined with dark red.

**21. T. ovatum** Pursh. Illustration: Journal of the Scottish Rock Garden Club **12**: f. 5 (1970); Bulletin of the Alpine Garden Society **46**: 59 (1978); Spellenberg, Audubon Society field Guide to North American wildflowers – western region, f. 49 (1979); Rix & Phillips, The bulb book, 102 (1981).
Stem 10–50 cm, tinged with red. Leaves stalkless, 5–15 cm, diamond-shaped with 5 obvious veins, acute or shortly acuminate. Flower-stalk 2–8 cm, erect. Flowers varying in smell from spicy to unpleasant. Sepals 1.5–6 cm. Petals 2.5–6.5 cm, white, fading to pink or red, ovate, spreading from the base. Anthers cream to yellow. Stigmas erect or spreading. *Western N America.* H3. Spring.

Double-flowered variants occur occasionally as do variants with green-striped petals.

Forma **hibbersonii** Taylor & Szczawinski (*T. hibbersonii* invalid). Illustration: Bulletin of the American Rock Garden Society **31**: 46 (1973); Syesis **7**: 250 (1974); Bulletin of the Alpine Garden Society **42**: 21 (1974), **48**: 138 (1980). Smaller in all its parts, with pink petals which eventually fade to whitish.

**22. T. rivale** Watson. Illustration: Gardeners' Chronicle ser. 3, **54**: 43 (1913); Botanical Magazine, n.s., 444 (1964); Bulletin of the Alpine Garden Society **42**: 21 (1974), **48**: 138 (1980).
Stem to 25 cm, usually much less, hairless. Leaf-stalks 5–20 mm. Leaves to 3 cm, narrowly ovate to ovate, acuminate. Flower-stalk 2.5–5 cm, erect. Sepals

1.2–1.6 cm, green. Petals 2–2.5 cm, white or pale pink, splashed and dotted with pink-purple especially towards the base. Filaments white, shorter than the yellow anthers. Ovary cream. Stigmas recurved. *W USA (Oregon, California)*. H4. Spring.

**23. T. govanianum** D. Don (*Trillidium govanianum* D. Don) Kunth. Illustration: Blatter, Beautiful flowers of Kashmir **2**: pl. 61 (1928).
Stem 10–30 cm, hairless. Leaves shortly stalked, 3.5–11 cm, ovate, acute. Flower stalked. Sepals 1–2 cm, green tinged with purple, lanceolate. Petals 1–2.5 cm, less than 5 mm wide, yellow to greenish brown or purplish. Filaments reddish, anthers yellow. Stigma reddish. Fruit purple. *Himalaya (Kashmir to Bhutan)*. H3.

**24. T. tschonoskii** Maximowicz. Illustration: Acta Horti Gotoburgensis **25**: pl. 3 (1962); Journal of the Scottish Rock Garden Club **11**: 276 (1969); Rix & Phillips, The bulb book, 105 (1981).
Stem *c.* 15 cm, hairless. Leaves stalkless, 7–17 cm, diamond-shaped to circular, acuminate. Flower stalked, held horizontally. Sepals 1–3.2 cm, greenish, broadly lanceolate to narrowly ovate, acute. Petals white, fading to pale reddish purple, slightly longer than the sepals, acute. Stamens 9–16 mm. Ovary pale green, sometimes sparsely dotted with purple. Stigmas short, thick, recurved. *Japan, Korea, SW China, Nepal*. H3. Spring–summer.

**25. T. kamtschaticum** Pallas. Illustration: Acta Horti Gotoburgensis **25**: pl. 2 (1962); Journal of the Scottish Rock Garden Club **11**: 274 (1969); Bulletin of the Alpine Garden Society **42**: 21 (1974).
Stem 25–30 cm, hairless. Leaves stalkless, 7–15 cm, diamond-shaped to ovate, acuminate. Flower stalked, erect. Sepals 2–5 cm, broadly lanceolate to ovate-oblong, green. Petals 2.5–4.5 cm, white, tinged with mauve as they age, acute. Filaments much shorter than the anthers. Fruit green, stained or spotted with purple at the top. *NE Asia*. H3. Late spring–summer.

**26. T. smallii** Maximowicz. Illustration: Acta Horti Gotoburgensis **25**: pl. 4 (1962). Stem 20–40 cm, hairless. Leaves stalkless, 7–15 cm, diamond-shaped, acute. Flower-stalk 2–4 cm, flowers held horizontally at first, becoming erect with age. Sepals 1.5–2 cm, purplish green. Petals to 2 cm, thick, dark purple-brown, sometimes absent. Anthers 2.5–3.5 mm, brownish,

often converted into petals. Ovary greenish or purple. Stigmas very short. *Japan*. H3. Spring.

**27. T. amabile** Miyabe & Tatewaki. Illustration: Transactions of the Sapporo Natural History Society **15**: 137 (1938). Leaves stalkless, 12–15 cm, ovate to diamond-shaped. Flower-stalk 5–7 cm. Sepals 2.5–3 cm, broadly lanceolate, purple. Petals *c.* 2 cm, purple, broadly ovate to circular, sometimes absent. Anthers 6–7 mm, purple, often converted into petals or vice versa. Ovary pale yellowish, reddish at the top. *Japan*. H3. Spring.

The status of *T. amabile* is uncertain. Some botanists consider it to be an hybrid between *T. smallii* and *T. kamtschaticum*, or even the same as *T. smallii*.

**113. ASPARAGUS** Linnaeus
*J. Cullen*
Perennial herbs, shrubs or climbers, with rhizomes and usually with fusiform tubers. Leaves reduced to small scales at the nodes, their bases hardened and often projecting as spines. The function of the leaves is taken over by cladodes (leaf-like stems) which are either borne singly or in groups of 3–50 at the nodes in the axils of the scale-leaves; the cladodes may be flattened, leaf-like, awl-like or thread-like. Flowers bisexual or functionally unisexual, borne singly or in few- to many-flowered clusters among the cladodes, more rarely in terminal, umbel-like clusters, or in racemes borne on the older shoots; flower-stalks distinctly jointed. Perianth shortly united at the base, bell-shaped or with the 6 lobes widely spreading. Stamens 6. Ovary superior, 3-celled. Fruit usually a spherical berry containing 1–6 seeds.

A genus of 50–60 species from the Old World, absent from Australasia. Some 24 species are grown for the sake of their feathery shoots; some, wrongly called Smilax, are in demand as background material for bouquets, and 1 species is widely grown as a luxury vegetable. In using the account that follows, scale-leaves and their bases should be looked for on the main shoots.

Plants are relatively easy to grow in good garden soil, though the genus covers a wide range of hardiness; the plants grown in glasshouses should be give ample water and partial shade, though they require good ventilation. Propagation is mainly by seed, though cuttings can be used for some, and some can be divided at the roots. Literature: Jessop, J.P., The genus

Asparagus in southern Africa, *Bothalia* **9**: 31–96 (1966).

| | | |
|---|---|---|
| 1a. | Cladodes distinctly flattened | 2 |
| b. | Cladodes rounded or angled in section, not flattened | 10 |
| 2a. | Cladodes 8–20 mm wide, leaf-like, with several parallel veins, borne singly | **24. asparagoides** |
| b. | Cladodes at most 7 mm wide, not as above, with a single vein or no visible veins, usually borne in clusters of 3–50 at the nodes | 3 |
| 3a. | Plants climbing or scrambling | 4 |
| b. | Plants erect or trailing but not climbing | 8 |
| 4a. | Flowers borne singly or in clusters of 2 or 3 in the cladode-axils | 5 |
| b. | Flowers in racemes of more than 2 or 3, borne on the older shoots | 6 |
| 5a. | Plant a shrub; cladodes 2.5–5 cm | **8. lucidus** |
| b. | Plant a herb; cladodes 5–15 mm | **18. scandens** |
| 6a. | Cladodes 5–9 cm × 5–7 mm | **21. falcatus** |
| b. | Cladodes smaller | 7 |
| 7a. | Cladodes borne in a single plane on each main shoot; ridges of shoots and margins of cladodes roughened by papillae | **23. drepanophyllus** |
| b. | Cladodes borne in several planes on each main shoot; twigs and margins of cladodes smooth | **22. aethiopicus** |
| 8a. | Cladodes 1–3 at each node, the cladodes of each shoot arranged in several planes; flowers in racemes | **20. densiflorus** |
| b. | Cladodes 3–5 at each node, all the cladodes on an individual shoot arranged in a single plane; flowers solitary or in pairs or in clusters of 1–4 at the ends of the twigs | 9 |
| 9a. | Flowers 1–2 at each node, borne among the cladodes; berry spherical | **7. filicinus** |
| b. | Flowers in clusters of 1–4 at the tips of the twigs; berry 3-lobed | **17. madagascariensis** |
| 10a. | Flowers in racemes | 11 |
| b. | Flowers in clusters of 1–many, borne among the cladodes at the nodes, or terminating the twigs | 12 |
| 11a. | Plant a climber | **19. racemosus** |
| b. | Plant erect, not climbing | **20. densiflorus** |
| 12a. | Flowers in clusters of 1–4, terminating the twigs | 13 |
| b. | Flowers in clusters of 1–15, borne at the nodes | 14 |
| 13a. | Cladodes 10 mm or more | |

× 0.4–0.8 mm, 3- or 4-angled in section, the angles appearing as ridges which are papillose **11. umbellatus**

b. Cladodes to 8 × 0.1–0.2 mm, round in section, smooth **12. setaceus**

14a. Flowers 6–many in each cluster; spines on the older shoots conspicuous, projecting at about 90° to the shoot, more than 5 mm; old stems white **14. albus**

b. Flowers 1–3 in each cluster; spines smaller or absent, or, if large, then projecting obliquely downwards; old stems usually not white   15

15a. Young twigs white and deeply grooved   16

b. Young twigs not both white and deeply grooved   17

16a. Ultimate branches bearing spines **15. laricinus**

b. Ultimate branches spineless **16. retrofractus**

17a. Cladodes 2–4-angled in section, the angles appearing on the surface as low ridges   18

b. Cladodes round in section, without ridges   21

18a. Cladodes in groups of 2 or 3 at each node **10. crispus**

b. Cladodes in groups of 5 or more at each node   19

19a. Cladodes 2–10 × 0.3–0.5 mm, glaucous, hard and sharp **6. acutifolius**

b. Cladodes 10–30 × 0.1–0.2 mm, green or dark green, soft and not sharp   20

20a. Perianth 6–8 mm; flower-stalk flexuous, exceeding the cladodes among which it is borne **1. tenuifolius**

b. Perianth 2–5 mm; flower-stalk straight, not or scarcely exceeding the cladodes among which it is borne **4. trichophyllus**

21a. Main stems conspicuously grooved and ridged, the ridges sometimes papillose   22

b. Main stems smooth, neither grooved and ridged nor papillose   24

22a. Scale-leaves not hardened at the base, the base continuing the scale and lying adpressed to the stem below the point of insertion of the leaf; cladodes in groups of 3–7 at each node **9. virgatus**

b. Scale-leaves hardened at the base, forming a tough, conical projection; cladodes in groups of 4–25   23

23a. Cladodes usually 1–6 cm ×

0.5–1.2 mm, strongly 3-angled; main stem with smooth ridges **5. verticillatus**

b. Cladodes usually 5–20 × 0.2–0.5 mm, rather weakly 4-angled; main stem with papillose ridges **3. pseudoscaber**

24a. Plant erect, herbaceous; flowers 1–3 at each node **2. officinalis**

b. Plants climbing, woody; flowers 4 or more at each node **13. africanus**

**1. A. tenuifolius** Lamarck. Illustration: Bonnier, Flore complète 10: t. 596, f. 2666 (1929).

Stems to 1 m, herbaceous, erect; stem and branches smooth, the branches green. Scale-leaves with small, hardened bases which are not spine-like. Cladodes 1–3 cm × 0.1–0.2 mm, thread-like, 2–4-angled in section, soft and not sharp, green or dark green, in groups of 15–40 or more at each node. Flowers unisexual, borne in clusters of 1–3 among the cladodes; flower-stalks flexuous, exceeding the cladodes among which they are borne, jointed just below the flower. Perianth 6–8 mm. Berry 1–1.6 cm, red, containing 2–6 seeds. *Mediterranean area.* H5–G1. Summer.

**2. A. officinalis** Linnaeus. Illustration: Bonnier, Flore complète 10: t. 596, f. 2667 (1929).

Stems to 1 m, herbaceous, erect; stem and branches smooth, the branches green. Scale-leaves with small, hardened, conical bases. Cladodes usually 1–2.5 cm, rarely smaller or larger, round in section, green or pale green, stiff, borne in groups of 4–25 at each node. Flowers usually unisexual, borne in clusters of 1–3 among the cladodes; flower-stalks arching downwards, mostly longer than the cladodes among which they are borne, jointed at or about the middle or just below. Perianth 4–6 mm. Berry 6–10 mm, red, containing usually 2–4 seeds. *Europe, N Africa, SW Asia; introduced elsewhere.* H1. Summer.

Widely cultivated as a luxury vegetable; variable, with a number of cultivars available.

**3. A. pseudoscaber** (Ascherson & Graebner) Grecescu (*A. officinalis* var. *pseudoscaber* Ascherson & Graebner).

Similar to *A. officinalis* but main stem grooved and ridged, the ridges papillose; cladodes usually 5–20 × 0.2–0.5 mm, rather weakly 4-angled in section. *Yugoslavia, Romania, W USSR (Ukraine).* H1. Summer.

**4. A. trichophyllus** Bunge.

Stems erect, 2 m, branched, smooth, woody; branches numerous, green. Scale-leaves on the main shoots with conspicuous, hardened, spine-like bases which point downwards. Cladodes 1.2–2.5 cm, 2–4-angled in section, appearing weakly ridged, stiff, pointed, pale green, borne in groups of 20–30 at the nodes. Flowers mostly unisexual, in groups of 1–3 in the axils of cladodes at the nodes. Perianth 2–3 mm. Flower-stalks straight, shorter than the cladodes among which they are borne, jointed at the middle or just above. Berry 6–10 mm. *NE Asia, N China.* H2. Summer.

**5. A. verticillatus** Linnaeus.

Stem to 2.5 m, woody, branched, erect, conspicuously grooved and ridged, the ridges sometimes papillose. Scale-leaves with hardened bases which project downwards as small spines. Cladodes 1–6 cm × 0.5–1.2 mm, strongly 3-angled, hard, sharp, the angles papillose, borne in groups of 10–20 or more at each node. Flowers mostly unisexual, in groups of 1–10 at the nodes. Perianth 2.5–4 mm. Flower-stalks deflexed, jointed above the middle. Berry 5–8 mm, black, containing 1–3 seeds. *SE Europe to USSR (W Siberia).* H2. Summer.

**6. A. acutifolius** Linnaeus. Illustration: Bonnier, Flore complète 10: t. 595, f. 2664 (1929); Huxley & Taylor, Wild flowers of Greece and the Aegean, t. 378 (1977).

Stem woody, erect, to 2 m, white or grey, faintly ridged, the ridges papillose. Scale-leaves with the bases hardened to form downwardly pointing projections. Cladodes 2–10 × 0.3–0.5 mm, 2–4-angled in section, glaucous, hard and sharp, borne in groups of 10–30 or more at each node. Flowers mostly unisexual, borne in clusters of 1–4 among the cladodes. Perianth 3–4 mm. Flower-stalks about as long as the cladodes, jointed below the middle. Berry 4–10 mm, black, containing 1 or 2 seeds. *Mediterranean area.* H5. Summer.

**7. A. filicinus** D. Don. Illustration: Gardeners' Chronicle 44: f. 47, 48 (1908). Stems herbaceous, erect, to 2 m, twigs, branches and cladodes borne in horizontal planes. Bases of scale-leaves hardened but scarcely projecting. Cladodes very regularly arranged, flattened (the branches looking superficially like a fern frond), 6–20 × 1–2 mm, dark green, in clusters of 3–5 at each node. Flowers 1 or 2 (rarely more) at each node, usually unisexual, borne among the cladodes. Perianth 2–3 mm. Flower-stalks spreading, exceeding the cladodes, jointed in the upper half. Berry spherical, black. *Himalaya, SW China*. H4. Summer.

Very fern-like in appearance, and sometimes mistaken for a true fern.

**8. A. lucidus** Lindley.
Plant a woody climber to 3 m. Scale-leaves with hardened bases which project obliquely downwards as blunt spines. Cladodes 2.5–5 cm × 1–2 mm, flattened, bright green, borne in groups of 2–6 at each node, margins papillose. Flowers usually unisexual, in clusters of 1–3 among the cladodes. Perianth 2–2.5 mm. Flower-stalks straight, spreading, shorter than the cladodes, jointed in the upper half. Berry 6–8 mm, brownish. *China, Korea, Japan, Taiwan*. H3. Summer.

**9. A. virgatus** Baker. Illustration: Saunders' Refugium botanicum, t. 214 (1870); Gibson, Wild flowers of Natal (coastal region), t. 11 (1975).
Stems erect, straight, grooved and ridged, to 1 m, woody. Scale-leaves not hardened at the base, the base continuing the scale and lying adpressed to the stem below the point of leaf insertion. Cladodes round in section, to 2.5 cm, borne in groups of 3–7 at the nodes. Flowers usually bisexual, in groups of up to 6 among the cladodes. Perianth 3–4 mm. Flower-stalks 5–9 mm, jointed in the lower half. Berry red, containing a single seed. *Southern Africa*. G1.

**10. A. crispus** Lamarck. Illustration: Mason, Western Cape sandveld flowers, t. 9 (1972); Gartenpraxis, 53 (1983).
Stem herbaceous, erect or trailing, green, to 1 m. Scale-leaves with rather soft, spine-like projections at their bases. Cladodes 3–9 × 0.5–1 mm, 2–4-angled in section, borne in groups of 2 or 3 at the nodes. Flowers solitary, among the cladodes. Perianth 4–5 mm. Flower-stalks 6–11 mm, jointed in the upper half. Berry ovoid, pale, 8–15 mm, containing 3–9 seeds. *South Africa (Cape Province)*. G1.

**11. A. umbellatus** Link. Illustration: Botanical Magazine, 7733 (1900); Lagascalia 9: 83 (1979).
Stems climbing, to 5 m, grooved and ridged, the ridges papillose. Scale-leaves with hardened bases which project obliquely downwards as small, blunt spines. Cladodes 10–30 × 0.3–0.8 mm, 3 or 4-angled in section, the angles appearing as papillose ridges, borne in groups of

10–30 (rarely fewer) at the nodes. Flowers usually bisexual, borne in umbel-like clusters of 1–4, terminating the twigs. Perianth 5–7 mm. Flower-stalks 5–8 mm, jointed in the lower half. Berry 6–8 mm, yellow or orange, slightly 3-lobed, containing 1 or 2 seeds. *Canary Islands, Madeira*. G1. Summer.

Variable in its native habitat. A plant in cultivation under the name *A. myriocladus* misapplied (not *A. myriocladus* Baker, which is a synonym of *A. densiflorus*) is similar, but has thread-like, extremely fine cladodes. Its origin is unknown. For an illustration of the plant grown under this name, see Gartenpraxis, 55 (1983).

**12. A. setaceus** (Kunth) Jessop (*A. plumosus* Baker). Illustration: Gibson, Wild flowers of Natal (coastal region), t. 11 (1975).
Plant usually a woody climber or scrambler, often large, more rarely erect. Stems smooth, green. Scale-leaves with hardened bases which project obliquely downwards as small, sharp spines. Cladodes to 10 mm, thread-like, green or dark green, borne in groups of 8–20 at the nodes. Flowers usually bisexual, solitary at each node, among the cladodes, and terminating the branches. Perianth *c*. 3 mm. Flower-stalks 2–5 mm, jointed at about the middle. Berry *c*. 6 mm, red, containing 1–3 seeds. *Southern Africa*. G2. Autumn.

A commonly cultivated species, often used as background material in bouquets. It is variable, and several cultivars have been selected: the most commonly grown of these is 'Nanus' (*A. plumosus* var. *nanus* Anon.), which is of more compact, upright habit.

**13. A. africanus** Lamarck (*A. cooperi* Baker; *A. asiaticus* misapplied). Illustration: Gibson, Wild flowers of Natal (coastal region), t. 11 (1975); Clay & Hubbard, The Hawaii garden, tropical exotics, 91 (1977).
Plant a woody climber (rarely erect) to 3 m. Stems smooth. Scale-leaves with hardened bases which project obliquely downwards as small spines. Cladodes round in section, thread-like, 5–10 mm, green or dark green, borne in groups of 8–20 at the nodes. Flowers usually bisexual, in clusters of 2 or 3, borne among the cladodes. Perianth 2.5–3.5 mm. Flower-stalks straight, 5–8 mm, jointed in the lower half. Berry *c*. 6 mm, red, containing 1 seed. *Southern Africa*. G1. Spring.

**14. A. albus** Linnaeus.
Stem erect, woody, to 1 m, white, smooth or slightly ridged. Scale-leaves with

conspicuous hardened bases which project outwards at more or less 90° from the shoot as strong, sharp spines. Cladodes 5–25 × 0.5–1.5 mm, more or less 3-angled in section, ridged, borne in groups of 10–20 at the nodes. Flowers usually bisexual, in clusters of 6–20 borne among the cladodes at the nodes. Perianth 2–3 mm. Flower-stalks 3–7 mm, jointed in the lower half. Berry 4–7 mm, black, containing 1 or 2 seeds. *W Mediterranean area*. H5. Summer.

**15. A. laricinus** Burchell.
Stem erect, woody, to 2 m, grooved, white; young branches both grooved and white. Scale-leaves with bases hardened into sharp spines which project obliquely downwards; spines also borne on the ultimate branchlets. Cladodes rigid, round in section, 8–30 mm, borne in groups of 15–60 at the nodes. Flowers usually bisexual, in clusters of 3 or 4 among the cladodes at the nodes and sometimes in clusters of 7 or 8 terminating the branchlets. Perianth 2.5–3 mm. Flower-stalks to 7 mm, jointed in the lower half. Berry *c*. 6 mm, red, containing 1–3 seeds. *South Africa, Botswana*. G1. Spring–early summer.

**16. A. retrofractus** Linnaeus.
Similar to *A. laricinus* but stems scrambling or weakly climbing, spines absent or few, always absent from the ultimate branchlets, berry orange, containing 1 seed. *South Africa, South West Africa*. G1. Spring–early summer.

**17. A. madagascariensis** Baker. Illustration: Botanical Magazine, 8046 (1905).
Stems erect, woody, green, angled, to 60 cm or more. Scale-leaves with hardened bases which project as downwardly hooked spines. Cladodes arranged in 1 plane on each shoot, flattened, lanceolate, acute, 1–1.7 cm × 5–7 mm, dark green, borne in groups of 3 at the nodes. Flowers usually bisexual, borne in terminal, umbel-like clusters of usually 4. Perianth to 6 mm. Flower-stalks erect, straight, jointed at or just below the middle. Berry bright red, somewhat 3-lobed, containing 1–3 seeds. *Madagascar*. G2. Autumn.

**18. A. scandens** Thunberg. Illustration: Botanical Magazine, 7675 (1899); Gartenpraxis, 55 (1983).
Plant herbaceous, scrambling or climbing. Stems smooth or slightly ridged, green. Scale-leaves with small hardened bases which scarcely project. Cladodes flattened, 5–15 × 0.8–1.6 mm, linear-lanceolate, borne in groups of 2 or 3, when 3, 1 of

them longer than the other 2. Flowers usually bisexual, usually solitary, occasionally in groups of 2 or 3, among the cladodes. Perianth 3–4 mm. Flower-stalks 8–12 mm, jointed in the upper half. Berry red, containing 1 seed. *South Africa (Cape Province)*. G1. Summer.

**19. A. racemosus** Willdenow (*A. tetragonus* Bresler). Illustration: Botanical Magazine, 8288 (1909).

Plant a large, woody climber to 7 m or more. Main stems robust, finely ridged, papillose. Scale-leaves with hardened bases projecting as long sharp spines more or less at right angles to the shoot. Cladodes 1–2 cm × 1 mm, 3-angled in section, the angles appearing as ridges, greyish green, often strongly curved. Racemes with many flowers, borne on the older branches at nodes from which the cladodes have fallen. Perianth 2–4 mm. Berries to 6 mm, black, containing several seeds. *Himalaya, India, Sri Lanka and SW China to South Africa*. G1. Summer.

A variable species of very wide distribution.

**20. A. densiflorus** (Kunth) Jessop (*A. sprengeri* Regel; *A. meyeri* invalid; *A. myersii* invalid; *A. sarmentosus* misapplied). Illustration: Botanical Magazine, 8052 (1906); Hay et al., Dictionary of indoor plants in colour, t. 45 (1974); Gibson, Wild flowers of Natal (coastal region), t. 11 (1975); Clay & Hubbard, The Hawaii garden, tropical exotics, 93 (1977); Gartenpraxis, 53–4 (1983).

Erect or trailing shrubs to 1 m. Stem finely ridged, green or brown. Scale-leaves with small, blunt, hardened bases which project almost vertically downwards. Cladodes usually flat, linear or linear-oblong, occasionally (in some cultivars) rather rounded in section, green, 5–15 × 1–2 mm, borne in groups of 1–3 at the nodes. Racemes with several to many flowers, borne among the cladodes at the nodes. Perianth 2–5 mm. Berry red, containing 1 seed. *South Africa*. G1. Summer.

A variable species, commonly grown, in which a number of cultivars has been recognised. 'Sprengeri', with a rather trailing, loose habit, is often seen, as is 'Myersii' in which the lateral branchlets are very regularly arranged and all more or less the same size, producing shoots that look like feathery green cylinders.

**21. A. falcatus** Linnaeus. Illustration: Botanical Magazine, 8751 (1918); Gibson, Wild flowers of Natal (coastal region), t. 11 (1975).

Stems climbing, smooth, woody, grey-brown, to 12 m or more. Bases of scale-leaves hardened as large, sharp spines which project outwards and slightly downwards from the shoot. Cladodes 5–9 cm × 5–7 mm, linear-oblong, green, with a conspicuous midrib, borne in groups of 1–3 at the nodes. Flowers borne in racemes on the older shoots. Perianth 2.5–4 mm. Berry to 5 mm, containing 1 or 2 seeds. *Southern and E Africa to Sri Lanka*. G1. Summer.

**22. A. aethiopicus** Linnaeus (*A. ternifolius* J.D. Hooker). Illustration: Botanical Magazine, 7728 (1900).

Like *A. falcatus* but climbing to 7 m, cladodes 1–4 cm × 1–2 mm, borne in groups of 3–6, perianth to 3 mm, berry 5–7 mm. *C & southern Africa*. G2. Summer.

**23. A. drepanophyllus** Welwitsch. Stems woody, climbing, to 10 m, faintly ridged, the ridges papillose. Scale-leaves with hardened bases which project as small, sharp spines. Cladodes all borne in a single plane on each shoot, in groups of 3–5 at the nodes, flat, the central cladode of each group 2–7 cm, the laterals somewhat smaller, all with papillose margins. Flowers in racemes borne on the older shoots. Perianth to 4 mm. Berry red, somewhat 3-lobed, 1–1.2 cm, containing 1 seed. *Tropical W & C Africa*. G2. Winter.

**24. A. asparagoides** (Linnaeus) Druce (*Medeola asparagoides* Linnaeus; *A. medeoloides* (Linnaeus filius) Thunberg). Illustration: Botanical Magazine, 5384 (1863); Mason, Western Cape sandveld flowers, t. 9 (1972); Gartenpraxis, 55 (1983).

Stem twining, to 1.5 m, smooth or slightly ridged. Scale-leaves with hardened but scarcely projecting bases. Cladodes borne singly, leaf-like, with several parallel veins, ovate or ovate-lanceolate, 1.5–3.5 cm × 8–20 mm. Flowers solitary or in pairs in the axils of the cladodes; flower-stalks 5–8 mm, jointed in the upper half. Perianth-lobes 5–7 mm. Berry 6–8 mm, reddish, containing 1–4 seeds. *South Africa (Cape Province); naturalised in the Mediterranean area*. H5–G1. Spring.

**114. DANAË** Medikus
*P.F. Yeo*

Evergreen shrubs with short rhizomes; stems 60–120 cm, branched. Leaves papery, those of main stems mostly about 1.5–2.5 cm, ovate-lanceolate, readily shed, those of axillary branches about 2 mm, persistent, 5 or 6 on each branch,

subtending leaf-like stems (cladodes). Cladodes mostly 3–7 cm, ovate-lanceolate, asymmetric. Flowers bisexual, nearly spherical, in racemes of 5–8 at tips of branches. Perianth 2–3.5 mm, cream-coloured; segments joined for most of their length; mouth much narrower than the diameter of the flower. Stamens united into a tube enclosed within the perianth; anthers 6. Ovary superior. Style slender, with 3 stigmatic lobes, reaching the mouth of the staminal tube. Fruit an orange-red berry with 1 or rarely 2 large seeds.

One species, easily distinguished from *Ruscus* in having the racemes separate from the cladodes. It forms a dense shrub with rich green glossy foliage, but is not valuable for its berries which are small and inconspicuous. It can be grown in sun or shade, and is easily increased by division; raising it from seed is very slow. The shoots, preformed underground, emerge in spring and flower in the same season; once fully expanded their growth is ended.

**1. D. racemosa** (Linnaeus) Moench (*D. laurus* Medikus). Figure 35 (1a–1f), p. 264. Illustration: Flora SSSR 4: 445 (1935).

*W. Asia*. H2. Early summer.

**115. SEMELE** Kunth
*P.F. Yeo*

Evergreen climbers with short rhizomes. Stems 5–7 m, 1–2 cm thick, uniform in diameter, twining in upper part, green, bearing only scale-leaves, which soon become brown and chaffy, and axillary branches. Branches to about 90 cm, bearing ovate scale-leaves 4–8 mm long, up to about 40 in number, all or most subtending a cladode (a leaf-like stem). Cladodes mostly 2.5–7 × 1–5 cm and broadly ovate or sometimes heart-shaped, but the terminal ones ovate-lanceolate. Flowers unisexual, both sexes on the same plant, borne in clusters in notches on margins of cladodes; clusters 1–several per cladode. Perianth cream, united for about one-third of its length, lobes spreading; diameter of male flowers about 9 mm, of female about 6.5 mm. Stamens united into a tube with 6 anthers at the top, present also in female flowers but smaller, their anthers never opening. Ovary with a style topped by 3 stigmatic lobes emerging from the staminal tube. Fruit an orange-red berry containing 1 large seed.

The single species is a tall climber suitable for growing up a pillar; flowers

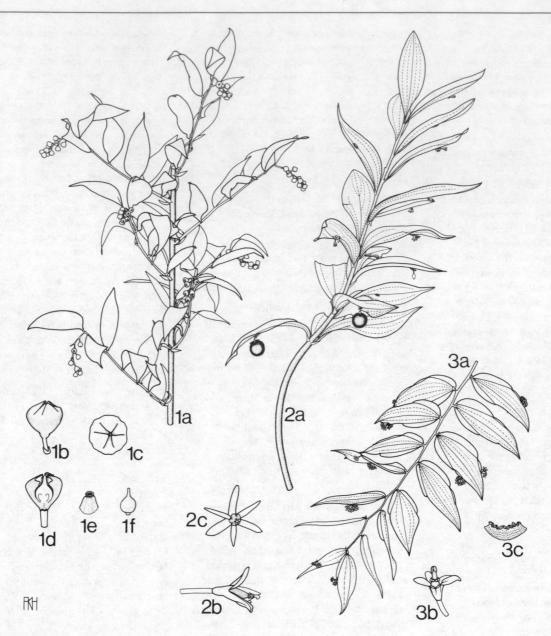

**Figure 35.** Shoots and flowers of *Danaë*, *Ruscus* and *Semele*:
1, *Danaë racemosa* (a, flowering shoot; b, flower seen from the
side; c, flower seen from above; d, flower in longitudinal section;
e, staminal tube with anthers; f, ovary). 2, *Ruscus colchicus*
(a, flowering and fruiting shoot; b, female flower, partly closed,
seen from the side; c, male flower seen from above). 3, *Semele
androgyna* (a, lateral branch in flower; b, functionally female
flower with 2 perianth-lobes removed; c, staminal tube opened
out, with sterile anthers).

moderately conspicuous but not colourful. The stems, unlike those of *Ruscus* and *Danaë*, are not completely preformed underground, and unfold over a long period. Growth begins in spring, flowering takes place 12 months later and fruiting 9–12 months after that. Control of plant size should therefore be by removal of only a proportion of the shoots of each annual generation. Propagation is by division or by seed.

**1. S. androgyna** (Linnaeus) Kunth. Figure 35 (3a–3c), p. 264. Illustration: Botanical Magazine, 1898 (1816), 3029 (1830); RHS dictionary of gardening **4**: 1935 (1951).

*Canary Islands.* G1. Early summer.

## 116. RUSCUS Linnaeus
*P.F. Yeo*

Evergreen herbs with short rhizomes and unbranched stems, or shrubs with branched stems. Leaves reduced to papery scales subtending branches or leaf-like stems (cladodes). Flowers unisexual, inconspicuous, borne in succession in clusters which are subtended by a bract and are usually solitary on the face of the cladode. Perianth-segments 6, free, the inner smaller than the outer, green, tinged or peppered with purple. Stamens united into a fleshy tube which is present in the flowers of both sexes; in the male flowers the tube is topped by stalkless anthers which are represented in the female flower by minute, papery flanges. Ovary filling the space within the staminal tube, vestigial in male flowers. Stigma more or less entire, protruding from the neck of the staminal tube. Fruit a red berry with thin flesh and 1–4 large seeds.

A genus of 6 species ranging from the Azores to the South Caspian region. *R. aculeatus* is suitable for tall ground-cover in dense shade; the dried shoots are frequently painted and used for decoration. The other species may be used for low ground-cover or as tub-plants. In species which are mainly dioecious, berries are occasionally produced on predominantly male plants. The plants are highly resistant to unfavourable conditions and neglect. They may be increased by division in autumn or spring; raising them from seed is very slow. The shoots, preformed underground, mostly emerge in spring and flower the following autumn or spring; once fully expanded, their growth is ended. Literature: Yeo, P.F., A contribution to the taxonomy of the genus Ruscus, *Notes from*

the Royal Botanic Garden, Edinburgh. **28**: 237–64 (1968).

1a. Stem with 1 or more whorls of branches     **1. aculeatus**
　b. Stem unbranched or occasionally with 1 branch     2
2a. Inflorescence-bract 3.5–13 mm wide, with 5–15 veins     **3. hypoglossum**
　b. Inflorescence-bract not more than 3.5 mm wide, with 5 or fewer veins     3
3a. Cladodes twisted at the base so that the upper side faces downwards; staminal tube top-shaped or cup-shaped, striped and lobed     **6. streptophyllum**
　b. Cladodes not twisted at the base so as to invert the blade; staminal tube barrel-shaped, club-shaped or flask-shaped, without stripes or obvious lobes     4
4a. Cladodes with bract and flowers on underside (a cladode with these on the upper side occurs occasionally); bract with edges joined to the cladode at base; staminal tube 0.5–2.5 mm or occasionally to 3.5 mm     **2. colchicus**
　b. Cladodes with bract and flowers on the upper side or varying within the plant; edges of bract not evidently joined to cladode at base; staminal tube 2.75–3.5 mm     5
5a. Stems erect; cladodes usually ovate, dark green; male staminal tube thickest at the apex, female almost cylindric     **4. hypophyllum**
　b. Stems more or less oblique; cladodes usually obovate-lanceolate or oblanceolate, pale green; plants female; staminal tube thickest near the middle     **5. × microglossum**

**1. R. aculeatus** Linnaeus. Illustration: Ross-Craig, Drawings of British plants **20**: t. 10 (1972); Graf, Exotica, edn 8, 1103 (1976). Stems to 1.25 m, with several branches arising from more than 1 node. Cladodes mostly not more than 2.5 cm, very numerous, ovate, spine-tipped, rigid, twisted edge-on to stem. Plants essentially unisexual (except certain cultivated plants which produce self-fertilising, bisexual flowers giving rise to a rich crop of berries). Bract very small, on upper surface of cladode. Flowers on very short stalks. Perianth-segments spreading, outer about 2.5 mm. Staminal tube about 2 mm, violet. Stigma spherical. *Mediterranean and Black Sea areas, extending to the Azores and N to Britain and Belgium.* H2. Autumn, spring; berries in winter.

Var. **angustifolius** Boissier is low-growing and slender and has cladodes 4–6 times as long as wide; cultivated plants are female.

**2. R. colchicus** Yeo. Figure 35 (2a–2c), p. 264. Illustration: Notes from The Royal Botanic Garden, Edinburgh **28**: 243, 244 (1968).
Stems to 60 cm, unbranched, usually oblique or arching, with up to about 30 cladodes, the upper turned to one side. Cladodes 4–13.5 × 2.5–5.3 cm, broadly to narrowly ovate or elliptic to narrowly elliptic. Plants unisexual. Bract 4.5–8 × 2–3.5 cm, normally on underside of cladode, its edges joined to cladode at the base, green, with 3–6 veins. Flower-stalks 2–8 mm. Perianth-segments spreading or slightly recurved, the outer 2.5–4 mm. Staminal tube dark violet, sometimes with slightly paler stripes, in male flowers 0.5–2.5 or occasionally to 3.5 mm, top-shaped, cylindric or ovoid, in female flowers 1.25–2 mm, ovoid. Stigma spherical. *Black Sea coast of S Caucasus and NE Turkey.* H2. Autumn, spring.

**3. R. hypoglossum** Linnaeus. Illustration: Kerner & Oliver, Natural History of plants **1**: 333 (1894); Hegi, Illustrierte Flora von Mitteleuropa **2**: 266 (1909); Notes from the Royal Botanic Garden, Edinburgh **28**: 245 (1968).
Stems to 40 cm, unbranched, arching, with up to 22 cladodes mostly turned to one side. Cladodes 3–10 × 1–3.3 cm, obovate, obovate-lanceolate or broadly ovate, tapered at the base, thick, not dark in colour. Plants unisexual. Bract 1–3 cm × 3.5–13 mm, on upper side of cladode, its edges joined to cladode at base, green, with 5–15 veins. Flower-stalks 5–8 mm. Perianth-segments spreading, the outer 3–4 mm. Staminal tube violet, with faint paler bands, in male flowers 3–3.5 mm, nearly cylindric, in female about 2.5 mm, thickest at about the middle, not distinctly lobed. Stigma spherical. *Italy & Czechoslovakia to N Turkey.* H2. Autumn, spring.

**4. R. hypophyllum** Linnaeus. Illustration: Coste, Flore de la France **3**: 356 (1905); Notes from the Royal Botanic Garden, Edinburgh **28**: 247 (1968).
Stems to 70 cm, or, in the shade, to 1 m, erect, occasionally with one branch and with up to about 16 cladodes. Cladodes mostly 5–9 × 1.2–5.5 cm, usually ovate with a long acute apex, but sometimes other shapes, dark green. Plants bisexual. Bract usually 4.5–9 × 1–2 mm, on upper

or lower surface of cladode (even on the same shoot) or sometimes in a notch at the edge, or one on each surface; its edges not joined to the cladode at the base, papery or green, with 1–3 or rarely 4 veins. Flower-stalks 5–10 mm. Perianth-segments spreading, the outer 4–4.5 mm. Staminal tube dull violet, sometimes with faint pale bands, in male flowers 3–4 mm, thickest at the tip, in female flowers 2.75–3 mm, nearly cylindric. Stigma spherical. *S Spain, N Africa, Sicily, SE France (Isle of Hyeres)*. H4.

Flowers of both sexes may be produced on the same shoot, but at times shoots may bear many flowers of one sex exclusively.

**5. R. × microglossum** Bertoloni. Illustration: Bonnier, Flore complète **10**: t. 597 (1929); Graf, Exotica, edn 8, 1103 (1976); Notes from the Royal Botanic Garden, Edinburgh **28**: 246 (1968). Stems to about 50 cm, unbranched, oblique or arching, with up to about 19 cladodes slightly turned to one side. Cladodes 4.5–10 × 1.4–3 cm, obovate-lanceolate, oblanceolate or, rarely ovate-lanceolate, with a long acute apex, rather light green. Plants usually female. Bract 5.5–11.5 × 1.5–2.5 mm, on upper or lower surface of cladode but those of 1 shoot never all on the lower surface, its edges not or scarcely joined to cladode at base, green, with 3 or 4 veins. Flower-stalks 3–6 mm. Perianth-segments spreading, the outer about 4 mm. Staminal tube (female) 2.75–3.5 mm, rich, dark violet, broadest about the middle. Stigma spherical. *Garden origin*. H2. Flowering all year.

Evidently a hybrid between *R. hypoglossum* and *R. hypophyllum* and widely cultivated, usually under the name of one of the parents; sometimes naturalised. Most cultivated material appears to belong to a single clone.

**6. R. streptophyllum** Yeo. Illustration: Botanical Magazine, 2049 (1819); Notes from The Royal Botanic Garden, Edinburgh **28**: 248 (1968); Bean, Trees and shrubs hardy in the British Isles, edn 8, 4: 242 (1980). Stems to about 60 cm, unbranched, oblique and arching, with up to *c*. 18 cladodes in 2 rows. Cladodes 5–18 × 2.2–8 cm, broadly elliptic to ovate to narrowly elliptic or lanceolate, twisted at the base so that the upper surface faces downwards. Plants bisexual. Bract 4–11 × 1.5–2.5 mm, on upper surface of cladode, its edges not joined to the cladode at the base, papery, with 1 vein and sometimes 2 additional,

faint veins. Flower-stalks 6–23 mm. Outer perianth-segments about 3.5 mm, reflexed, inner divergent or spreading. Staminal tube 2.75–3.25 mm, whitish with 12 violet stripes, in male flowers top-shaped, in female cup-shaped in profile, in either sex with a 6-lobed rim and slightly hollowed apex, the anthers in male flowers occupying only the centre of the apex. Stigma mushroom-shaped. *Madeira*. G1.

**117. OPHIOPOGON** Ker Gawler
*E. Campbell*
Herbaceous perennials with short rhizomes and sometimes stolons as well. Roots fine or thickened, sometimes swollen and tuberous. Leaves all basal, in tufts, linear, leathery. Flowers in a terminal raceme or spike on a 3-angled or flattened scape. Perianth bell-shaped, of 6 free, equal, drooping, pale purple to white segments. Stamens 6, the filaments very short, the anthers lanceolate, acute, more or less united about the style. Ovary partly inferior, 3-celled. Style simple, straight or slightly curved, stigma small. Seeds fleshy, berry-like, exposed early, spherical or oblong, usually blue and often persistent.

A genus of about 20 species from India, China and Japan south to West Malaysia, Borneo and the Philippines. Only 3 are in general cultivation, and can be used as ground-cover. They grow well in a light, sandy soil, but may require protection from cold during the winter. *O. japonicus* is sometimes grown as a submerged aquatic; it does not flower under these conditions. Propagation is by division.

1a. Tufted, leaves 1–1.5 cm broad with 9–13 veins; flower-stalks 1–1.5 cm; flowers 7–8 mm    **1. jaburan**
  b. Stoloniferous, leaves less than 7 mm broad, with up to 9 veins; flower-stalks to 1 cm; flowers 4–7 mm    **2**
2a. Leaves to 4 mm wide; scape to 12 cm; flower-stalks 2–6 mm    **2. japonicus**
  b. Leaves more than 4 mm wide; scape more than 12 cm; flower-stalks 5–10 mm    **3. planiscapus**

**1. O. jaburan** (Kunth) Loddiges (*Flueggea jaburan* Kunth; *Slateria jaburan* (Kunth) Siebold; *Convallaria japonica* Linnaeus var. *major* Thunberg; *Mondo jaburan* (Kunth) Bailey).
Plant without stolons. Roots cord-like and without tubers. Leaves 30–80 × 1–1.5 cm, linear, stiff, with 9–13 veins and with rough margins. Scape 30–50 cm, flattened. Flower-stalks 1–1.5 cm. Flowers 3–8 at

each node, in a loose raceme. Perianth 7–8 mm, white to pale purple. Anthers 4–5 mm. Seeds blue. *Japan*. H5–G1. Summer.

**2. O. japonicus** (Linnaeus) Ker Gawler (*Convallaria japonica* Linnaeus; *C. japonica* var. *minor* Thunberg; *Mondo japonicus* (Linnaeus) Farwell). Illustration: Botanical Magazine, 1063 (1808).
Roots slightly thickened and tuberous. Plant with stolons. Leaves mostly 10–20 cm × 2–4 mm, mostly curved, with 3–5 veins. Scape 7–12 cm, flattened. Flower-stalks 2–6 mm. Flowers 1–3 at each node in a loose raceme. Perianth 4–5 mm, pale purple to nearly white. Anthers 2.5 mm. Seeds deep blue. *Japan, China, Korea*. G1. Summer.

**3. O. planiscapus** Nakai (*O. japonicus* var. *wallichianus* Maximowicz).
Roots thickened, tuberous. Plant often stoloniferous. Leaves erect, 30–50 cm × 4–6 mm, with several veins. Scape 15–50 cm, weakly 3-angled near the base. Flower-stalks 5–10 mm. Flowers 1–3 at each node, in a loose raceme. Perianth 6–7 mm, pale purple or white. Anthers 2.5–3 mm. Seeds dull blue-black. *Japan*. H5–G1. Spring–summer.

'Nigrescens' is a dark, blackish green variant which is grown for its foliage, and is probably the same as the plant cultivated in America as 'Arabica'.

**118. LIRIOPE** Loureiro
*E. Campbell*
Perennial herbs with rhizomes and sometimes stolons. Leaves basal, linear, tufted. Flowers borne on a terete or flattened scape in a terminal spike or raceme; bracts small or large. Perianth of 6 free segments which are equal, ascending and persistent, forming a bell-shaped flower, purple, lilac or white. Stamens 6, filaments bent upwards in the middle, anthers oblong, obtuse. Ovary superior, 3-celled with 2 ovules in each cell. Style columnar, bent upwards in the middle, stigma small. Fruit a capsule. Seeds 1–few, spherical, rather fleshy, usually dark purple.

A genus of about 4 species from Japan, China and Korea, of which only 2 are well-known in cultivation. The genus has long been confused with *Ophiopogon* (above), and most species have names in this genus. The classification of *Liriope* is confusing, in spite of its small size.

1a. Leaves 4–7 mm broad; spike bearing

5–9 whorls of flowers, rather open; flowers pale purple to almost white

**1. spicata**

b. Leaves *c.* 2 cm broad; spike bearing more than 10 whorls of flowers, very dense; flowers purple to bluish

**2. muscari**

**1. L. spicata** (Thunberg) Loureiro (*Convallaria spicata* Thunberg; *Ophiopogon spicatus* (Thunberg) Loddiges).
Root-system shallow, roots slender, borne on a jointed stock just beneath the soil surface; some small tubers present. Leaves flexuous and grass-like, to 42 cm × 4–7 mm. Scapes slender, not exceeding the leaves. Bracts short and scarious. Flowers 20–36, in a loose spike, borne in 5–9 whorls each with 2–4 flowers. Perianth 2–5 mm across, pale purple to white. *Japan, China, Korea.* G1. Summer.

**2. L. muscari** (Decaisne) Bailey (*L. graminifolia* (Linnaeus) Baker var. *densiflora* Baker; *L. spicata* var. *densiflora* (Baker) Wright). Illustration: Botanical magazine, 5348 (1862).
Root-system rather deep, roots slender, matted, arising from an upright stock which also bears a few thick tubers. Leaves stiff and grass-like, to 60 × 2 cm. Scape thick, flattened, usually exceeding the leaves. Bracts conspicuous, more or less leaf-like. Flowers 40–70, in a dense spike, borne in more than 10 whorls of 4–7 each. Perianth purple to bluish, 5–8 mm across. *China.*

Much confusion surrounds the name *L. graminifolia*; it has certainly been applied to both of the species included here at one time or another, and it seems likely that the species (if it exists at all in the wild) is not to be found in cultivation.

**119. PELIOSANTHES** Andrews
*J. Cullen*
Herbaceous perennials with long or short, erect or creeping rhizomes. Leaves borne at intervals along the rhizome and also frequently in a terminal tuft, conspicuously stalked, linear to oblong-ovate or obovate. Inflorescence a raceme with several to many flowers, arising from the rhizome, subtended by scale-leaves at the base; flowers 1–6 to each bract. Perianth united into a tube below, green, bluish, violet or purple, lobes 6, erect or spreading. Stamens 6, borne at the mouth of the perianth-tube, their filaments united and arching over and more or less covering the ovary, the anthers borne at the apex of the tube,

slightly on the underside. Ovary usually at least partly inferior, 3-celled, with 2–4 basal ovules in each cell. Style columnar, often somewhat 3-lobed, exposed at the mouth of the staminal tube. Fruit a capsule which ruptures early, exposing the blue seeds.

A genus of a single species (see the paper by Jessop, cited below), formerly split into 2 genera (*Lourya* Baillon and *Peliosanthes* in the strict sense), with up to 12 species recognised within them. The 1 species is occasionally grown in warm glasshouses in a mixture of peat, sand and loam. Propagation is by division of the rhizomes. Literature: Jessop, J.P., A revision of Peliosanthes (Liliaceae), *Blumea* 23: 141–59 (1976).

**1. P. teta** Andrews.
Leaves 10–50 × 1–12 cm, stalks 4–50 cm. Inflorescence to 35 (rarely to 75) cm. Perianth-lobes to 8 mm. *Eastern Himalaya to SE Asia, Taiwan and China.* G2. Spring.

Jessop recognises 2 subspecies, both of which have been grown, and both of which have essentially the same distribution as the species:

Subsp. **teta.** Illustration: Andrews' Botanical Repository **10**: t. 605 (1810); Botanical Magazine, 1302 (1810). Flowers 2–6 in the axil of each bract, perianth usually green, rarely blue.

Subsp. **humilis** (Andrews) Jessop (*P. humilis* Andrews; *P. albida* Baker; *P. violacea* Baker; *Lourya campanulata* Baillon). Illustration: Andrews' Botanical Repository **10**: t. 634 (1810); Botanical Magazine, 1532 (1813), 7110 (1890), 7482 (1896), 8276 (1909). Flowers solitary in each bract-axil, often white, bluish, violet or purple, more rarely green.

**120. ALETRIS** Linnaeus
*J. Cullen*
Perennial herbs with rhizomes and fibrous roots. Leaves mostly in a basal rosette. Flowering stem erect, bearing a terminal, spike-like raceme. Flowers almost stalkless, held ascending-erect. Perianth united into a tube for most of its length, the outside of the tube warty and 6-ridged. Stamens 6. Ovary superior, 3-celled, each cell with several ovules. Fruit a beaked capsule, surrounded by the persistent but dried and shrunken perianth.

A genus of about 10 species from eastern N America and E Asia. Only 2 species, both from N America are grown. They are easily cultivated in a sunny position, and can be propagated by seed or division.

1a. Perianth tubular, to 1.2 cm, white

**1. farinosa**

b. Perianth bell-shaped, to 6 mm, yellow

**2. aurea**

**1. A. farinosa** Linnaeus. Illustration: Botanical Magazine, 1418 (1811); Rickett, Wild flowers of the United States **2**(1): t. 12 (1967).
Leaves to 20 × 1.5 cm. Stem to 1 m. Perianth to 1.2 cm, tubular, white, very warty. Anthers orange-red, exposed at the mouth of the perianth-tube. Beak of capsule exposed in fruit. *E USA.* H3. Summer.

**2. A. aurea** Walter. Illustration: Rickett, Wild flowers of the United States **2**(1): t. 12 (1967).
Leaves to 8 cm, yellowish green. Stem to 70 cm. Perianth yellow, bell-shaped, to 6 mm. Beak of capsule not exposed in fruit. *SE USA.* H3. Summer.

**121. LUZURIAGA** Ruiz & Pavon
*P.F. Yeo*
Slightly woody trailing plants with rooting stems, sometimes climbing trees, producing simple or branched lateral leafy shoots. Lowest leaves of lateral shoots often scale-leaves. Photosynthetic leaves jointed to the stems, very shortly stalked; blades with parallel veins and a few cross-veins, lower surface dark green but facing upwards, either because the shoot is pendent or because the leaf-stalk is twisted, upper surface (facing downwards) strongly glaucous except for green stripes following the veins. Flowers usually solitary, axillary, their stalks about as long as the perianth. Perianth-segments 6, free, white. Stamens 6; anthers basifixed or dorsifixed and versatile. Ovary superior, 3-celled; style simple. Floral parts becoming spotted with orange-brown when dry. Ovules few in each cell. Fruit a berry.

Four species in Peru, Chile, the Falkland Islands and New Zealand. Slightly frost-tolerant plants with beautiful white flowers. They need cool, moist conditions.

**1. L. radicans** Ruiz & Pavon. Illustration: Botanical Magazine, 6465 (1879).
Aerial stems sparsely branched. Leaves 1–3.5 cm × 3–12 mm, linear-oblong to elliptic-ovate or broadly elliptic. Flowers few, solitary or rarely 2 or 3 together, nodding, fragrant. Perianth-segments *c.* 1.2–1.5 cm × 3–8 mm, narrowly elliptic, widely spreading. Stamens with thick filaments; anthers much longer than the filaments, basifixed, erect. Berry orange or red. *Chile.* H4. Summer.

**\*L. polyphylla** (W.J. Hooker) Macbride (*L. erecta* Kunth). Illustration: Botanical Magazine, 9192 (1860). Has numerous closely set branches and numerous closely set uniformly elliptic leaves, most of them subtending flowers; filaments longer than anthers; anthers dorsifixed, reflexed. *Chile.* H5. Summer.

## 122. EUSTREPHUS R. Brown
*P.F. Yeo*

Hairless, evergreen climbers to several metres, twining when young, growing from a small tuberous rootstock. Foliage borne on branched lateral shoots arising from axils of scale-leaves on main stem. Leaves alternate, set in 2 rows, the lower ones on each branch being scale-leaves. Foliage-leaves jointed to the branch, very shortly stalked, with blades 5–10 cm, ovate to narrowly oblong, with upper surface facing upwards. Flowers bisexual, solitary or several together in axils of upper leaves, nodding, on slender stalks *c.* 3 cm long, which are jointed about the middle. Perianth-segments 6, *c.* 6 mm, ovate or lanceolate, spreading, white, pink or pale purple, inner densely fringed with long hairs. Stamens 6, slightly shorter than perianth-segments; filaments very short, united into a tube; anthers erect in centre of flower, yellow. Ovary superior, 3-celled, with several ovules in each cell. Style 1; stigma not or slightly lobed. Fruit to 1.3 cm in diameter, berry-like but eventually dehiscent, orange. Seeds black.

One species in Australasia. It requires protection from frost, and in summer, shade for part of the day and absence of very high temperatures. In hot exposed conditions it forms a bush. It is propagated by seed. Literature: Schlittler, J., Die Gattung Eustrephus R. Br. ex Sims und Geitonoplesium (R. Br.) A. Cunn. Morphologischanatomische Studie mit besonderer Berucksichtigung der systematischen, nomenclatorischen und arealgeographischen Verhaltnissen, *Berichte der Schweizerischen botanische Gesellschaft* **61**: 175–239 (1951).

**1. E. latifolius** R. Brown. Illustration: Botanical Magazine, 1245 (1809); Baglin & Mullins, Australian wildflowers in colour, 105 (1969); Wrigley & Fagg, Australian native plants, 321 (1979), fruit; Morley & Toelken, Flowering plants in Australia, 339 (1983).

*E Australia, New Guinea, New Caledonia.* G1. Spring, summer.

## 123. GEITONOPLESIUM R. Brown
*P.F. Yeo*

Slender-stemmed, hairless, evergreen plants climbing to a height of several metres by twining. Foliage borne on branched lateral shoots that arise singly or in small groups from clusters of scale-leaves on the main stem. Leaves alternate, set in 2 rows, shortly stalked, jointed to the stem, lower surface darker than upper but facing upwards, either because the shoot is pendent or because the leaf-stalk is twisted. Flowers bisexual, in sparse cymes at tips of branches. Perianth-segments 6, entire. Stamens 6; filaments much shorter than the anthers, flattened and enlarged towards base; anthers attached just above their base, opening by 2 pores. Ovary 3-celled, superior, with a few ovules in each cell. Style simple. Fruit berry-like, sometimes eventually dehiscent. Seeds black.

One very variable species, or perhaps 2, in SE Asia and Australia. A graceful, easily grown glasshouse climber requiring a minimum winter night temperature of 12°C.
Literature: Schlittler, J., Die Gattung Eustrephus R. Br. ex Sims und Geitonoplesium (R. Br.) A. Cunn. Morphologischanatomische Studie mit besonderer Berucksichtigung der systematischen, nomenclatorischen und arealgeographischen Verhaltnissen, *Berichte der Schweizerischen botanischen Gesellschaft* **61**: 175–239 (1951).

**1. G. cymosum** (R. Brown) W.J. Hooker. Illustration: Botanical Magazine, 3131 (1832); Morley & Toelken, Flowering plants in Australia, 338 (1983).
Leaves commonly 2–8 cm × 4–8 mm and narrowly lanceolate, but very variable. Flowers nodding, bell-shaped; perianth-segments *c.* 6 × 2 mm, green with whitish edges, sometimes tinged with purple or red, concealing the stamens. Fruit to 1.2 cm in diameter, black or dark blue. *SE Asia & Philippines to warm-temperate E Australia & Fiji.* G1.

## 124. PHILESIA Jussieu
*P.F. Yeo*

Shrub, 15–30 cm or, in nature, to 120 cm, with underground stolons. Main stem bearing scale-leaves and branches. Branches angled, with scale-leaves at base and shortly stalked photosynthetic leaves above. Leaf-blades pinnately veined, hard in texture, dark green above, glaucous beneath, recurved at the edges. Flowers

solitary or few at the ends of the leafy branches, nodding, subtended by several pale green scale-leaves overlapping each other and the base of the flower. Perianth of 3 outer sepal-like segments and 3 inner petal-like segments, the inner slightly united at the base and each with a basal nectarial pouch. Stamens with filaments slightly united to perianth at base and united to each other for one-quarter to two-thirds of their length; anthers basifixed, opening by slits. Ovary 1-celled with 3 parietal placentas; style club-shaped and slightly lobed at apex. Ovules many. Fruit a berry. Perianth becoming spotted with orange-brown when dry.

A single species from Chile. It requires peaty soil and, in dry districts, shade. It is propagated by division or layering, or from seed.

**1. P. magellanica** Gmelin (*P. buxifolia* Willdenow). Illustration: Botanical Magazine, 4738 (1853); Bean, Trees and shrubs hardy in the British Isles, edn 4, **3**: 148 (1976).
Leaf-blades 1.5–3.5 cm × 3–8 mm. Flowers narrowly bell-shaped; outer perianth-segments 1.5–2.2 cm, narrowly ovate, obtuse, greenish or pinkish, inner 4.5–6.5 cm, purplish red. Stamens nearly as long as perianth. Style slightly longer than stamens. *Chile.* H5. Summer.

## 125. × PHILAGERIA Masters
*P.F. Yeo*

Similar to *Lapageria* in habit but much less vigorous. Leaf-blades 3-veined. Flowers arranged as in *Philesia*, similar to those of *Lapageria* but outer perianth-segments not more than two-thirds the length of inner; stamens united to each other and to the perianth for *c.* 4 mm.

An artificial hybrid between *Lapageria* and *Philesia*. It requires peaty soil and cool moist conditions and it resents disturbance. Propagation is by layering the new basal shoots; the layers have to be treated very carefully when being transplanted.

**1. × P. veitchii** Masters. Illustration: Botanical Magazine, n.s., 92 (1950). Leaf-stalks 5–7 mm; leaf-blades 4–5.5 × 1–1.5 cm, lanceolate. Outer perianth-segments *c.* 3.8 × 1.6 cm, dark red with a bluish bloom, inner 6 × 3 cm, deep red with a slight bluish bloom. Stamens *c.* 4.3 cm. Style *c.* 4.8 cm, deep pink. *Known only in cultivation.* H5. Summer.

An artificial hybrid between *Lapageria rosea* and *Philesia magellanica*.

## 126. LAPAGERIA Ruiz & Pavon
*P.F. Yeo*

Woody climber, spreading strongly by underground stolons. Stems thick, twining, branched, to *c.* 10 m. Leaves alternate, with stalks to 1 cm; blades leathery, with parallel main veins and net-veins between them. Flowers 1–3 on short scaly shoots in axils of upper leaves, pendent. Perianth-segments 6, entire, each with a pouched nectary at the base. Stamens 6, free from one another or very slightly joined at the base; filaments subulate; anthers basifixed, opening by slits. Ovary superior, 1-celled with 3 parietal placentas; style club-shaped and slightly lobed at apex. Ovules many. Fruit a berry. Seeds pale yellow or brownish.

A handsome climber for a cool greenhouse or walls and trellises out-of-doors. It needs cool, moist conditions and is better planted out than kept in pots. Propagation is by seed, layering or pegging a detached and coiled shoot onto the surface of a sandy and peaty soil containing a little loam, and providing warm humid conditions. The berry is edible.

**1. L. rosea** Ruiz & Pavon. Illustration: Botanical Magazine, 4447 (1849); 4892 (1856) – 'Albiflora'; Gartenflora, t. 1445 (1897) – 'Illsemannii'; Parey's Blumengärtnerei **1**: 323 (1958).
Leaves 6–12 × 2–5 cm, ovate or ovate-lanceolate. Flowers bell shaped, purplish red with small pale spots, varying to nearly white. Perianth-segments 6.5–9.5 cm, inner much wider at apex than outer. Stamens four-fifths as long as perianth. Style slightly longer than stamens. *Chile.* H5. Summer–winter.

## 127. SMILAX Linnaeus
*J.C.M. Alexander*

Woody or herbaceous, evergreen or deciduous, perennial climbers or scramblers with rhizomes or tubers. Stems terete or angled, usually spiny at least below, branched. Leaves alternate, simple, sometimes shallowly lobed, papery to leathery, sometimes spiny on the margins, with 3–9 prominent veins interconnected by net-veins; lower leaves reduced to scales; leaf-stalks often very short, bearing a pair of tendrils (modified stipules) near the base, and sometimes a pair of stipule-like auricles. Flowers white to pale green, yellow or brown, lateral, in clusters, racemes, umbels, racemes of umbels, or solitary. Male and female flowers on separate plants. Perianth-segments free; male flowers with 6 free stamens borne at the base of the perianth-segments; female flowers with up to 6 staminodes and a superior ovary on which 1–3 stigmas are borne directly (styles absent). Fruit a spherical to ovoid, red, blue or black berry with 1–6 seeds.

About 200 species, mostly tropical but with some representatives in temperate Eurasia and N America. Several species yield sarsaparilla, formerly used medicinally but now only used for flavouring drinks and sweets. The Smilax of florists is *Asparagus asparagoides* (p. 260). The flowers are not beautiful, and most species are not reliably hardy, but can be effective if allowed to clamber through large trees. They are not particular as to soil-type and can be propagated by division, from seed or from half-ripe cuttings.
Literature: Killip, E.P. & Morton, C.V., A revision of the Mexican and Central American species of Smilax, *Carnegie Institute of Washington Publication* **461**, Botany of the Maya area: No. 12 (1936); Koyama, T., Materials towards a monograph of the genus Smilax, *Quarterly Journal of the Taiwan Museum* **13**: 1–61 (1960); Morton, C.V., Mexican Smilax, *Brittonia* **14**: 299–309 (1962); Duncan, W.H., The woody vines of the SE States, *Sida* **3**: 1–76 (1967).

1a. Herbaceous plants, lacking spines; flowers foetid     **1. herbacea**
  b. Woody plants, usually spiny at least at base; flowers not foetid     2
2a. Leaves glaucous beneath     3
  b. Leaves not glaucous beneath     4
3a. Leaf-bases truncate to rounded; auricles shorter than leaf-stalks     **2. glauca**
  b. Leaf-bases usually cordate; auricles as long as leaf-stalks     **3. discotis**
4a. Leaves strongly mottled with white or pale green     **7. argyrea**
  b. Leaves not strongly mottled     5
5a. Flowers in spikes or racemes, of umbels or clusters, or solitary     6
  b. Flowers in a single umbel     8
6a. Mature leaves tapered to an acute or obtuse angle at base     **4. smallii**
  b. Mature leaves rounded, truncate, cordate, sagittate or hastate at base     7
7a. Flowers in clusters on axillary or terminal branches; leaves almost always spiny on margin     **5. aspera**
  b. Flowers in racemes of umbels or solitary; leaves not spiny on margin, sometimes spiny on veins beneath     **6. regelii**

8a. Mature leaves tapered to an acute or obtuse angle at base     9
  b. Mature leaves rounded, truncate or cordate at base     11
9a. Deciduous, rarely partially evergreen; berries red     **8. walteri**
  b. Evergreen, berries black     10
10a. Leaves papery     **4. smallii**
  b. Leaves thick and leathery     **9. laurifolia**
11a. Leaf-tips rounded or shallowly notched     **10. china**
  b. Leaf-tips acute to obtuse     12
12a. Branches not spiny     13
  b. Branches spiny     14
13a. Leaf-margins smooth; berries red or white     **8. walteri**
  b. Leaf-margins rough or finely toothed; berries bluish black     **11. hispida**
14a. Stems terete with slightly raised lines; berries red     **12. excelsa**
  b. Stems 4-angled; berries blue to black     **13. rotundifolia**

**1. S. herbacea** Linnaeus. Illustration: Botanical Magazine, 1920 (1817); Rickett, Wild flowers of the United States **1**: 45 (1966); Justice & Bell, Wild flowers of North Carolina, 16 (1968).
Herbaceous. Stem to 3 m, lacking spines, much branched, dying down each winter. Leaves 5–12 × 3–12 cm, triangular to ovate or lanceolate, rounded to shortly pointed at the tip, rounded to truncate or slightly cordate at the base, leathery, with 7–9 veins. Leaf stalk to 8 cm. Flowers few to numerous, greenish, foetid; umbel-stalk 10–30 cm. Perianth-segments 3.5–6 mm. Berries almost spherical, *c.* 1 cm across, blue-black. *Eastern N America.* H2. Summer.

**2. S. glauca** Walter. Illustration: Sida **3**: 38 (1967).
Deciduous or semi-evergreen. Stems spiny, especially below, often glaucous; spines stiff and slender; young branches terete, green or brown. Leaves 4–13 × 3.5–10 cm, ovate to lanceolate, acute to obtuse at the tip, rounded to truncate or slightly cordate at the base, glaucous and papillose beneath, sometimes also above, with *c.* 7 veins, often variegated, auricles shorter than the leaf-stalks. Flowers 5–11 per umbel, yellow to brown; umbel-stalk 1–1.5 cm, arching, flattened, usually longer than the leaf-stalk. Berries spherical, 5–8 mm across, black, glaucous. *SE USA.* H2. Spring–summer.

Descriptions and illustrations in many European books and journals, e.g. Botanical Magazine, 1846 (1816), and RHS Dictionary of Gardening, differ in several respects from those in N American works. As *S. glauca* is native to N America, the American concept of the species has been followed here.

**3. S. discotis** Warburg.
Deciduous. Stems angled, with hooked spines to 4 mm. Leaves 4–10 × 2–5 cm, ovate, acute to obtuse at the tip, usually cordate at the base, glaucous beneath, with 3–5 veins. Leaf-stalk 2–4 cm; auricles large, equalling the leaf-stalks. Flowers greenish yellow; umbel-stalk to 4 cm. Berries blue-black. *China.*

**4. S. smallii** Morong (*S. lanceolata* misapplied). Illustration: Sida **3**: 39 (1967); Radford et al., Manual of the vascular flora of the Carolinas, edn 2, 288 (1968).
Evergreen. Stems to 3 m, terete and hairless, spiny below, spines 5–6 mm, upper stems spineless. Leaves to 15 × 7 cm, usually much smaller, lanceolate to ovate, abruptly acuminate at the tip, gradually tapered at the base, papery, hairless, with 5 veins, shiny dark green above, duller and paler beneath, not spiny. Leaf-stalks to 1.6 cm. Flowers few to numerous, in umbels, or spikes or racemes of umbels, green; umbel-stalks 1–10 mm, shorter than or equal to leaf-stalks. Perianth-segments: male 5–6 mm, female 3–4 mm. Styles 3. Berries spherical, 5–7 mm across, black. *SE USA, E Mexico & C America.* H5–G1. Summer.

**5. S. aspera** Linnaeus (*S. maculata* Roxburgh). Illustration: Schneider, Illustriertes Handbuch der Laubhölzkunde **2**: 863 (1912); Polunin & Smythies, Flowers of SW Europe, 63 (1973); Ceballos et al., Plantas silvestres de la peninsula Iberica, 387 (1980).
Creeping, scrambling or climbing evergreen. Stems to 15 m, angled, sparsely to densely spiny, rarely spineless. Leaves very variable in shape and size, 4–11 × 2–10 cm, narrowly to broadly lanceolate, triangular, ovate, oblong or kidney-shaped, usually abruptly narrowed at the tip, cordate, hastate, sagittate or rarely truncate at the base, usually leathery, spiny on margins and on main veins (spines sometimes very few and not on every leaf), with 5–9 veins, shiny on both surfaces. Leaf-stalk to 2 cm, usually spiny. Flowers fragrant, 5–30 in clusters on axillary and terminal branches 2–15 cm long. Perianth-segments pale green, 2–4 mm. Berries *c.* 6 mm across, black or red. *Canary Islands, S Europe to Ethiopia and India.* H5–G1. Summer–autumn.

**6. S. regelii** Killip & Morton (*S. grandifolia* Regel; *S. utilis* Hemsley; *S. saluberrima* Gilg; *S. ornata* Hooker; *S. officinalis* Hanbury & Fluckinger). Illustration: Botanical Magazine, 7054 (1889).
Stem to 15 m, sharply 4-angled and spiny below, 4-angled or winged above, spines *c.* 1 cm. Lower leaves very variable, to 30 × 21 cm, ovate to oblong, rounded or acuminate at apex, cordate or hastate at base; upper leaves much smaller, lanceolate to oblong, gradually tapered at base; all papery, often spiny on the veins beneath, with 5–7 veins. Leaf-stalks to 7 cm. Male flowers solitary or in racemes of umbels to 6.5 cm long. Female flowers solitary, stalks to 10 cm though usually shorter. Berries to 1.3 cm across, black. *Northern C America.* G1.

**7. S. argyrea** Linden & Rodigas. Illustration: Illustration Horticole **39**: pl. 152 (1892). Stems wiry, with short, stout spines. Leaves to 25 cm, lanceolate, acute or acuminate, 3-veined, dark green with white or pale green blotches. Leaf-stalks short. Flowers and fruit unknown. *Peru & Bolivia.* G1.

**8. S. walteri** Pursh. Illustration: Rickett, Wild flowers of the United States **2**: 41 (1967); Sida **3**: 39 (1967).
Deciduous, rarely partially evergreen. Stems slender, slightly angled, spiny near the base, yellowish or brownish; branches squarish in section, not spiny. Leaves 5–12 × 1.5–6.5 cm, ovate to lanceolate, obtuse or abruptly acute at the tip, broadly tapered, rounded or cordate at the base, with 5–7 veins. Flowers 6–15 in single umbels, yellow, green or brown, drying brownish orange; umbel-stalks 5–15 cm, flattened, usually shorter than leaf-stalks. Berries 8–12 mm across, red, rarely white. *E USA from New Jersey to Texas.* H5–G1. Spring–summer.

**9. S. laurifolia** Linnaeus. Illustration: Rickett, Wild flowers of the United States **2**: 41 (1967); Sida **3**: 38 (1967); Justice & Bell, Wild flowers of North Carolina, 16 (1968); Radford et al., Manual of the vascular flora of the Carolinas, edn 2, 288 (1968).
Evergreen. Stem terete, green, spiny below; branches angled, usually spineless, rarely very spiny. Leaves 5–20 × 1–7.5 cm, narrowly ovate to oblong-lanceolate, abruptly tapered at apex and base, margin often inrolled, thick and leathery, 3-veined, midrib much more prominent than lateral veins beneath. Flowers greenish, in single umbels; umbel-stalks shorter than or equal

to leaf-stalks. Berries ovoid, 6–8 mm across, black. *SE USA.* H5–G1. Summer.

**10. S. china** de Candolle. Illustration: Makino, New illustrated flora of Japan, 858 (1963).
Deciduous. Stems to 5 m, climbing or scrambling, slightly spiny or spineless. Leaves to 8 cm, broadly ovate to circular, sometimes wider than long, rounded or notched at the tip, rounded, truncate or cordate at the base, leathery or papery, with 5–7 veins. Leaf-stalks 8–25 mm. Flowers yellowish green in single umbels; umbel stalks *c.* 2.5 cm. Berries *c.* 9 mm, spherical, red. *China, Korea & Japan.* H3. Spring.

**11. S. hispida** Muhlenberg (*S. pseudo-china* misapplied; *S. tamnoides* misapplied). Illustration: Radford et al., Manual of the vascular flora of the Carolinas, edn 2, 286 (1968).
Deciduous. Stems to 15 m, terete or slightly angled, densely spiny below; branches angled, lacking spines, hairless. Leaves 5–15 × 1.5–2 cm, lanceolate to ovate or almost circular, abruptly pointed at the tip, rounded to cordate at the base, with 5–9 veins, shiny, drying grey, margins rough or finely toothed. Leaf-stalks 6–18 mm. Flowers to 25 in single umbels; umbel-stalks 1.5–10 cm, 1½ times longer than leaf-stalks or more. Berries *c.* 6 mm, bluish black. *S & C USA.* H3. Spring–summer.

**12. S. excelsa** Linnaeus. Illustration: Botanical Magazine, 9067 (1924); RHS Dictionary of Gardening, edn 2, 1966 (1956).
Deciduous scrambler or climber. Stems to 20 m, terete with slightly raised lines, spines straight, to 7 mm; branches spiny. Leaves very variable, 5–13 cm, slightly less in width, broadly ovate to circular, acuminate at the tip, rounded, cordate or truncate at the base, membranous or slightly leathery, margins rough or with fine teeth, with 5–7 veins. Leaf-stalks to 1.2 cm. Flowers green, 4–12, in single umbels; umbel-stalks 7–20 mm. Perianth-segments: male 5–7 mm, female 4–5 mm. Berries 8–10 mm across, red. *From E Bulgaria through Turkey to USSR (S shores of the Caspian Sea).* H4. Summer.

**13. S. rotundifolia** Linnaeus. Illustration; Sida **3**: 38 (1967); Radford et al., Manual of the vascular flora of the Carolinas, edn 2, 286 (1968).
Deciduous or partly evergreen. Stems to 10 m, 4-angled, with spines to 8 mm on the angles; branches spiny. Leaves

5–15 cm, broadly ovate to circular, rounded to cordate at the base, thick and leathery, sometimes spiny on the margins, 5-veined, dark, shiny green. Leaf-stalks 6–12 mm. Flowers greenish yellow, in single umbels; umbel-stalks 5–15 mm, equal to slightly longer than leaf-stalks, flattened. Berries 5–8 mm across, blue to black. *Eastern N America*. H2. Spring–summer.

# VII. AGAVACEAE

Plants of very variable form, trees with trunks, shrubs with thin, woody stems, herbs, or stemless rosette-plants, often monocarpic. Leaves in rosettes, either borne directly on the rootstocks or at the ends of the stems or branches, more rarely distributed along the stems, usually thickened or fleshy, mostly hard, leathery and persisting for several years, bases thickened and sheathing, margins often spiny or horny, apices often spine-tipped (the apex formed by the upcurving of the margins which coalesce and become a solid, hard or soft spine or rarely a cylindric projection – see figure 36, p. 272), usually very fibrous, venation often indistinct. Flowers in terminal panicles or racemes, usually bisexual, not always produced each year. Perianth usually united into a long or short tube at the base, the lobes 6, more or less equal. Stamens 6, usually borne at the top of the perianth-tube, more rarely attached to the perianth. Ovary superior or inferior, usually 3-celled; ovules usually many per cell, more rarely 1 per cell or the ovary 1-celled with 1–3 ovules. Fruit a capsule or berry, or indehiscent, dry and spongy.

A very troublesome family from the point of view of identification. Its separation from the Liliaceae (p. 117) and the Amaryllidaceae (p. 291) is based, at least in part, on cytological, chemical and anatomical characters, and this makes a clear diagnosis of the family difficult to prepare. *Aloe*, which is here retained, as is traditional, in the Liliaceae, is the genus most likely to be confused with the Agavaceae (when it is not in flower). Aloes, however, usually have fleshy leaves with soft, non-spinous tips, and can usually be recognised by this character in the vegetative state (almost all the Agavaceae with fleshy leaves have spiny tips to them).

The family is very variable, including species that seem to have little in common with each other, except that they are either woody, or have thick, fibrous leaves, or both. However, the family is recognised in most recent studies of flowering plant families, and so is included here; its content and circumscription here follows the system of Hutchinson (Families of Flowering Plants, 662–5, 1959) in including *Phormium*, rather than that of the Syllabus der Pflanzenfamilien, edn 12, in which this genus is treated as a member of the Liliaceae. The order of the genera adopted here follows suggestions by Wunderlich (Oesterreichische Botanische Zeitschrift **97**: 438–502, 1950), based on her studies of the anatomy and embryology of selected species.

The genera in the family are relatively easily defined on floral characters, though, so defined, they are vegetatively very variable. Unfortunately, flowers are usually not available in cultivation – many of the species flower only at long intervals (Agaves are known colloquially as Century Plants), or take many years to attain the flowering condition and then die; others, such as the shrubby species of *Dracaena*, are often grown as house or office plants and generally die or are discarded before they attain sufficient size for flowering to take place. Thus, a key to the genera based on flowering material is not very helpful to the gardener. The formal, dichotomous key given here makes use of both floral and vegetative characters and shows the 'received' features used in separating the genera. The informal key is based largely on vegetative characters (though some floral ones are included as well for the sake of completeness), and will be a help in identifying groups of species rather than genera as such (see figure 36, p. 272). It will not provide a definite answer in all cases, and reference must be made to the rest of the text, and particularly to the illustrations cited. In this key, the genera and species are referred to by their entry numbers, first the number of the genus and then, in parentheses, the numbers of the species within it which show the character, e.g. 1(11–13,15,16,18) means that species 11, 12, 13, 15, 16 & 18 of genus 1 (*Yucca*) show the character in question; 10(all) means that all the species of genus 10 (*Beaucarnea*) show the character.

*Habit.* Plant (as grown) with a distinct, woody trunk which makes up half or more of the total height: 1(1,3–10,17), 4(1,2), 9(all), 10(all), 11(1–3), 12(1–3), 14(1–6); plant (as grown) with short, stout, sometimes prostrate woody trunks which do not form half of the total height: 1(2,10,14,15), 4(2–6), 5(1,7,10, 12,17,24,27), 11(4,6,7), 13(1), 14(all); plant (as grown) shrubby, with slender, woody stems which are leafy in the upper part: 12(4–12), 14(7); plant (as grown) consisting essentially of a large or small rosette of persistent leaves borne directly on the soil surface, without a distinct stem or trunk: 1(11–13,15,16, 18), 2(all), 3(all), 4(7–9), 5(2–6,8, 9,11,13–16,18–23,25,26), 8(all), 11(5,7), 13(2–11); plant herbaceous with non-woody aerial stems which are sometimes leafy, or with rosettes of deciduous leaves: 6(all), 7(all).

*Leaves.* Thick, increasing in thickness towards the mid-line especially near the base, leathery, stiff or flexible, persistent or rarely dying off in winter: 1(1–17), 2(all), 3(all), 4(all), 5(all), 8(all), 12(1), 13(all), 14(all); thick but essentially parallel-surfaced, hard and leathery, not fleshy, persistent: 1(18), 12(2–9); thin, parallel-sided, leathery and persistent: 9(all), 10(all), 15(all); thin, persistent, evergreen but not parallel-sided; 12(10–15); herbaceous, dying off annually: 6(all), 7(all). Equitant: 15(all).

*Leaf-margins.* With broadly based, hard or firm spines or prickles: 4(4–9), 5(3–5,7,10,11,13–27), 11(1,2,4–7); with distinct, usually regular, saw-like teeth: 1(8,9,16–18); entire or minutely toothed to the naked eye, often rough to the touch: 3(all), 4(1–3), 5(1,6,12), 6(all), 7(all), 8(all), 9(2), 10(all), 11(3), 12(all), 13(all), 15(all); splitting off as straight or curled threads (excluding very old leaves which are disintegrating): 1(1–6,11–15), 2(all), 5(2,8,9), 9(1).

*Leaf apex.* With a distinct, hard or soft, solid, spine-like tip: 1(all), 2(all), 3(all), 4(all), 5(all), 13(all); with a solid, cylindric tip: 8(all); fraying into a number of fibres: 9(2), 11(1,2,4–6); tapering to a point which is neither solid nor fibrous, the apex often drying and turning brown rapidly: 6(all), 7(all), 9(1), 10(all), 11(3, 7), 12(all), 14(all), 15(all).

*Flowers.* All bisexual: 1(all), 2(all), 3(all), 4(all), 5(all), 6(all), 7(all), 8(all), 12(all), 13(all), 14(all), 15(all); most unisexual: 9(all), 10(all), 11(all).

*Perianth.* Radially symmetric: 1(all), 2(all), 3(all), 4(all), 5(all), 6(all), 8(all), 9(all), 10(all), 11(all), 12(all) 13(all), 14(all); bilaterally symmetric: 7(all), 15(all). More than 1 cm: 1(all), 2(all), 3(all), 4(all), 5(all), 6(all), 7(all), 8(all), 12(all),

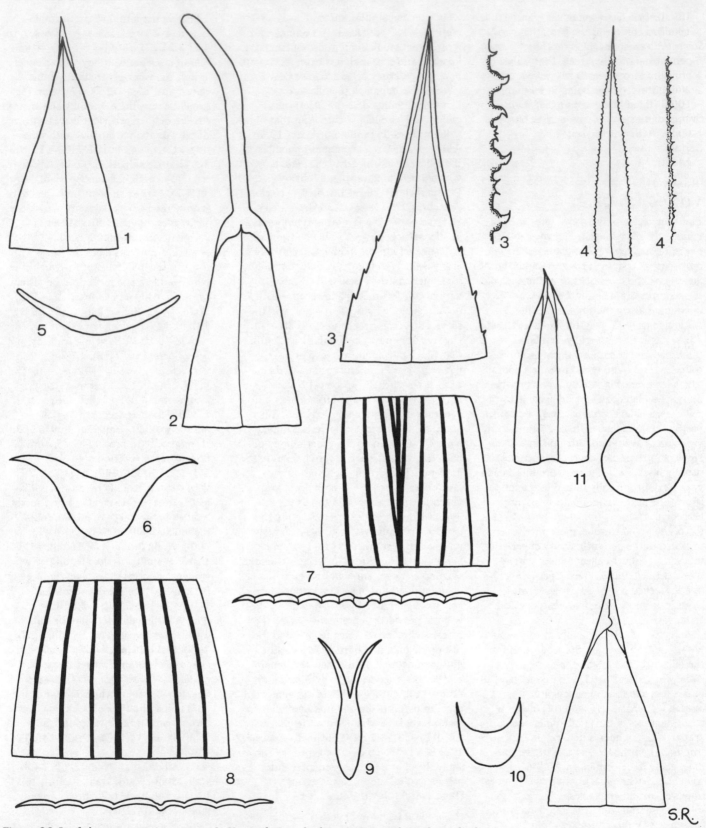

**Figure 36.** Leaf characters in Agavaceae. 1, *Yucca gloriosa*, leaf-tip. 2, *Doryanthes palmeri*, leaf-tip. 3, *Agave horrida*, leaf-tip and margin. 4, *Agave striata*, leaf-tip and margin. 5, *Beaucarnea recurvata*, section of leaf near the base. 6, *Agave attenuata*, section of leaf near the base. 7, *Cordyline indivisa*, section of leaf and details of venation. 8, *Dracaena fragrans*, section of leaf and details of vennation. 9, *Phormium tenax*, section of leaf near the base. 10, *Sansevieria trifasciata*, section of leaf and leaf-tip. 11, *S. roxburghiana*, section of leaf and leaf-tip.

13(all), 15(all); less than 1 cm: 9(all), 10(all), 11(all), 12(1, 14), 14(all).

*Stamens.* Conspicuously projecting from the perianth: 5(all), 6(all), 15(all); not projecting from the perianth: 1(all), 2(all), 3(all), 4(all), 7(all), 8(all), 9(all), 10(all), 11(all), 12(all), 13(all), 14(all).

*Ovary.* Inferior: 2(all), 3(all), 4(all), 5(all), 6(all), 7(all), 8(all); superior: 1(All), 9(all), 10(all), 11(all), 12(all), 13(all), 14(all), 15(all).

1a. Ovary inferior — 2
 b. Ovary superior — 7
2a. Perianth bilaterally symmetric; plant of herbaceous habit, with leafy stems — **7. Polianthes**
 b. Perianth radially symmetric; plants with woody stems or trunks or consisting of stemless rosettes — 3
3a. Anthers clearly projecting from the perianth — 4
 b. Anthers included in the perianth — 5
4a. Leaves dying off annually, without spines; perianth-lobes spreading or reflexed — **6. Manfreda**
 b. Leaves persisting for several years, usually with spiny margins or at least ending in a spine; perianth-lobes erect — **5. Agave**
5a. Filaments and style swollen below the middle; apex of leaf narrowed to a persistent spine — **4. Furcraea**
 b. Filaments and style not swollen below; apex of leaf tapering but not spiny, if solid and cylindric, soon drying and falling — 6
6a. Leaves at most 75 cm, tapering to an acute but not solid apex; perianth lobes erect, united into a distinct tube at the base — **3. Beschorneria**
 b. Leaves more than 1 m, apex solid, cylindric, soon drying and falling, perianth-lobes spreading, united into a very short tube at the base — **8. Doryanthes**
7a. Flowers all bisexual — 8
 b. Flowers mostly unisexual — 13
8a. Leaves folded towards the base, equitant; flowers bilaterally symmetric — **15. Phormium**
 b. Leaves neither folded towards the base nor equitant; flowers radially symmetric — 9
9a. Perianth-lobes 2.5 cm or more; filaments thickened towards their apices — **1. Yucca**
 b. Perianth-lobes always less than 2.2 cm; filaments not thickened towards their apices — 10
10a. Plant herbaceous; leaves all basal,

margins splitting off as threads — **2. Hesperaloe**
 b. Plants without the above combination of characters — 11
11a. Ovules numerous in each cell of the ovary; venation pinnate with the lateral veins parallel, branching from the pronounced midrib at an acute angle — **14. Cordyline**
 b. Ovules 1 in each cell of the ovary, or ovary 1-celled; veins all strictly parallel, not as above, or veins invisible — 12
12a. Plant tree-like, with fleshy leaves, or shrubby with thin, woody stems and thin leaves, or herbaceous with thin leaves, rhizomes absent (our species) — **12. Dracaena**
 b. Plant not tree like, usually stemless, leaves always fleshy; creeping or subterranean rhizomes present — **13. Sansevieria**
13a. Ovary 3-celled, with 2 ovules in each cell — **9. Nolina**
 b. Ovary 1-celled with 2 or 3 ovules — 14
14a. Trunk conspicuously inflated at the extreme base — **10. Beaucarnea**
 b. Trunk not conspicuously inflated at the base — **11. Dasylirion**

## 1. YUCCA Linnaeus
*J.Cullen*

Plants stemless or with short stems or tall, woody trunks. Leaves borne in rosettes at the ends of the stems or branches, persistent, stiff or flexible, fleshy, margins entire, finely toothed or splitting off as threads, bases expanded and somewhat sheathing. Inflorescence a much branched, erect or rarely hanging panicle, stalk long and conspicuous or short and hidden by the leaves. Flowers bisexual, subtended by bracts. Perianth bell-shaped, hemispherical or spherical, opening more widely at night, radially symmetric, of 6 lobes which are united at their bases for a short distance. Stamens 6, the filaments free from the perianth, or, more rarely, united to it for some distance, usually curved out towards their apices, which are swollen and broader than the anthers. Ovary superior, 3-celled, each cell with many ovules, usually tapering into the 3- or 6-lobed stigma (more rarely the style thread-like). Fruit a capsule or indehiscent and dry or fleshy. Seeds winged or wingless.

A genus of about 30 species from arid regions of the USA, Mexico, Guatemala and the West Indies. Its classification (particularly the recognition of sections) is based largely on characters of the fruit and

seeds. In general, neither of these is available in cultivation (the complex adaptations to pollination by moth larvae render fertilisation away from the native habitats unlikely) and so the present account is based mainly on features of the stems, leaves, inflorescences and flowers and is therefore artificial to some extent. The existence of many hybrids (both wild and man-made) makes identification even more difficult, and the present account (particularly the keys) must be used with caution.

The species grow well in sandy or gravelly soils, generally in sites where frost is not too severe and rainfall is limited, or, if heavy, the soil is very well drained., They are readily propagated by seed (if it is available) or by stem or root (rhizome) cuttings.

Literature: Trelease, W., The Yucceae, *Report of the Missouri Botanical Garden* 13: 27–133 (1902); Molon, G., *Le Yucche* (1914); MacKelvey, S.D., *Yuccas of the south western United States*, part 1 (1938), part 2 (1947); Webber, J.M., Yuccas of the southwest, *USDA Monograph* 17 (1953); Matuda, E. & Lujan, I.P., *Las Plantas Mexicanas del genero Yucca* (1980).

*Plant.* Stemless: 11–13,15,16,18; with a distinct stem or trunk: 1–10,11,14,15,17.

*Leaves.* At most 2 cm wide: 1,13,14,15,17,18; more than 2 cm wide: 2–12,16. Margins entire or with a few obscure teeth: 7,10; margins finely but distinctly toothed: 1,8,9,16–18; margin splitting off as straight or curled threads: 2–6,11–15.

*Inflorescence.* Erect, borne essentially between the leaves (though overtopping them), stalk hidden: 1,2,6–10; erect, borne on a distinct stalk which is clearly visible above the leaves: 3,5,11–18; hanging, but with a short but obvious stalk: 4.

*Filaments.* Not arching out towards their apices: 18.

*Style.* Thread-like, the ovary not tapering into it: 18.

*Fruit.* Indehiscent, dry or fleshy: 1–10; a dehiscent capsule: 11–18.

1a. Leaf-margins without conspicuous threads (a few threads may be formed when the leaf is very old) — 2
 b. Leaf-margin with conspicuous threads throughout most of its life — 9
2a. Plant stemless — 3
 b. Plant with a distinct stem or trunk — 5
3a. Style thread-like; filaments attached to

the perianth at their bases, not arching outwards towards their apices **18. whipplei**

b. Style stout, the ovary tapering into it; filaments not or scarcely attached to the perianth at their bases, arching outwards towards their apices **4**

4a. Panicle hairless; leaves with a yellowish or brown, finely toothed margin **16. rupicola**

b. Panicle usually downy; leaves with entire, green margins **12. flaccida**

5a. Leaf-margin conspicuously, though finely toothed **6**

b. Leaf-margin entire or rarely with a few, scattered, inconspicuous teeth **8**

6a. Leaves 1.2–1.7 cm wide, margins finely and evenly toothed **17. rostrata**

b. Leaves 2.5–7 cm wide, toothing small but coarse and uneven **7**

7a. Ovary stalked; flowers white, usually tinged with purple **8. aloifolia**

b. Ovary stalkless; flowers white or cream **9. elephantipes**

8a. Leaves channelled, blue-green, stiff and rigid, erect-spreading, surfaces rough **7. treculeana**

b. Leaves flat or pleated towards the apex, green, somewhat flexible, spreading or recurved, smooth on the surfaces **10. gloriosa***

9a. Plants without distinct stems, or with very short stems which are hidden by the leaves **10**

b. Plants with distinct, often trunk-like stems **14**

10a. Perianth-lobes almost entirely free **2. baccata**

b. Perianth-lobes united at the base into a distinct, though short tube **11**

11a. Style narrow at the base, then swollen above, ultimately narrowing, green; leaves 5–13 mm wide **15. glauca**

b. Style oblong or tapering evenly to the apex, white or cream; leaves more than 1.5 cm wide **12**

12a. Leaves distinctly narrowed towards the base, with numerous, very curled threads on the margins **13**

b. Leaves tapering from base to apex, with straight or slightly curled threads on the margins **12. flaccida**

13a. Perianth-lobes 3–5 cm; leaves stiff and ascending **13. smalliana**

b. Perianth-lobes more than 5 cm; leaves usually flexible and recurved **11. filamentosa**

14a. Leaves at most 2.5 cm wide **15**

b. Leaves more than 2.5 cm wide **18**

15a. Style narrow at the base, then swollen above, ultimately narrowing, green **15. glauca**

b. Style oblong, not as above, white or cream **16**

16a. Inflorescence erect, borne just above the leaves, its stalk hidden by them; perianth-lobes thick, incurved **1. brevifolia**

b. Inflorescence erect, borne on a stalk which overtops the leaves, or hanging; perianth-lobes thin, spreading **17**

17a. Inflorescence hanging; leaves narrowed towards the base, 2–2.5 cm wide **4. filifera***

b. Inflorescence erect or spreading; leaves tapering evenly from the base, rarely exceeding 1.8 cm wide **14. elata**

18a. Ovary (including style) 4.5 cm or more **19**

b. Ovary (including style) less than 4.5 cm **20**

19a. Leaf with thick, generally curled threads on the margins; perianth-lobes united only at their extreme bases **2. baccata**

b. Leaf with few, thin, straight threads; perianth-lobes distinctly united into a short tube at their bases **3. arizonica**

20a. Leaves rigid, yellowish green, rough on both surfaces; style clearly distinguished from the ovary **6. torreyi**

b. Leaves flexible, blue-green, smooth on both surfaces; style not clearly distinguishable from the ovary **5. schottii**

**1. Y. brevifolia** Engelmann (*Y. arborescens* (Torrey) Trelease; *Clistoyucca arborescens* (Torrey) Trelease).

Tree to 15 m, with rough bark, usually branched. Leaves in dense, elongate rosettes, spreading, 15–60 cm × 7–15 mm, rigid, lanceolate, somewhat narrowed towards the base, margins finely toothed, very sharply pointed. Inflorescence a panicle to 35 cm, erect, held just above the leaves, the stalk hidden; axis often roughly hairy. Flowers green or greenish yellow, perianth-lobes thick, incurved, 2.5–5 cm. Style oblong. Fruit indehiscent, dry and spongy, ovoid, 5–10 cm. Seeds unwinged. *USA (California to Utah)*. H5–G1. Autumn.

**2. Y. baccata** Torrey. Illustration: Illustration Horticole **20**: t. 115 (1853); Rickett, Wild flowers of the United States **4**(1): t. 12 (1970); Matuda & Lujan, Las

plantas mexicanas del genero Yucca, 66–7 (1980).

Stem distinct, though usually shorter than the leaves, thick. Leaves narrowly lanceolate, 30–70 × 3.5–5 cm, concave, rough on both surfaces, blue-green, margin splitting off as coarse, curled threads. Panicle shortly stalked, held between the leaves, to 60 cm. Flowers bell-shaped. Perianth-lobes *c.* 7.5 cm, white or cream, often tinged with purple, united only at the extreme base. Ovary (including style) more than 4.5 cm. Fruit fleshy, indehiscent, spindle-shaped, 20–25 cm. Seeds unwinged. *USA (Colorado and Utah to New Mexico), N Mexico*. H5–G1. Autumn.

**3. Y. arizonica** MacKelvey. Illustration: Matuda & Lujan, Las plantas mexicanas del genero Yucca, 74 (1980).

Stem woody, simple, to 3 m. Leaves narrowly lanceolate, 60–80 × 3–4 cm, thick, straight, smooth, pale green or yellowish green, margins splitting off as fine, straight threads. Panicle 50–100 cm, long-stalked, flowers held well above the leaves. Flowers bell-shaped, perianth-lobes 5.5–12 cm, cream, often tinged with purple, forming a distinct, though short tube at the base. Ovary (including style) more than 4.5 cm. Fruit indehiscent, fleshy, 15–20 cm, with a long or short beak. Seeds unwinged. *USA (Arizona), Mexico*. H5–G1. Autumn.

**4. Y. filifera** Chabaud (*Y. australis* (Engelmann) Trelease). Illustration: Botanical Magazine, 7197 (1891), excluding habit drawing; Matuda & Lujan, Las plantas mexicanas del genero Yucca, 100 (1980).

Stem woody, trunk-like, to 10 m or more, ultimately branched, bark rough. Leaves to 55 × 2.5 cm, lanceolate, somewhat narrowed towards the base, rigid, margins splitting off as numerous fine, curled threads. Panicle hanging from a curved stalk, oblong, very dense. Flowers bell-shaped, perianth-lobes 3.8–5.2 cm, cream. Fruit indehiscent, oblong. Seeds wingless. *Mexico*. H5–G1. Autumn.

***Y. valida** Brandegee. Illustration: Matuda & Lujan, Las plantas mexicanas del genero Yucca, 97–8 (1980). Panicle conical, more or less erect or ascending, held well above the leaves. *Mexico*. H5–G1. Autumn.

**5. Y. schottii** Engelmann. Illustration: Rickett, Wild flowers of the United States **4**(1): t. 12 (1970); Matuda & Lujan, Las plantas mexicanas del genero Yucca, 84 (1980).

Stem woody, to 5 m, rarely branched. Leaves 50–80 × 3.5–5 cm, lanceolate, flexible, smooth, blue-green, the margins splitting off as rather sparse threads. Panicle erect, somewhat exceeding the leaves, the stalk mostly hidden, rather open, axes densely woolly. Flowers almost spherical, often wider than long, perianth-lobes 2–4 cm, white, the outer 3 with brownish apices. Ovary (including style) less than 4.5 cm, the ovary itself tapering gradually into the indistinct style. Fruit indehiscent, oblong. Seeds wingless. *SW USA, Mexico*. H5–G1. Autumn.

**6. Y. torreyi** Shafer. Illustration: Rickett, Wild flowers of the United States 4(1): t. 12 (1970); Matuda & Lujan, Las plantas mexicanas del genero Yucca, 77 (1980).
Stem woody, ultimately to 4 m, usually simple. Leaves 30–100 × 3–5 cm, narrowly lanceolate, rigid, yellowish green, rough on both surfaces, the margins splitting off as tough threads which are curled at first. Panicle dense, oblong, held just above the leaves on a hidden stalk. Flowers spherical or bell-shaped, cream often tinged with purple, perianth-lobes 3.5–7.5 cm. Ovary (including style) less than 4.5 cm, the style more or less distinct from the ovary. Fruit indehiscent, seeds unwinged. *USA (Texas, New Mexico), N Mexico*. H5–G1. Autumn.

**7. Y. treculeana** Carrière (*Y. canaliculata* Hooker). Illustration: Botanical Magazine, 5201 (1860); Rickett, Wild flowers of the United States 2(1): t. 15 (1967); Matuda & Lujan, Las plantas mexicanas del genero Yucca, 76 (1980).
Stem woody, ultimately to 5 m, branched. Leaves 50–100 × 2.5–5 cm, very narrowly lanceolate, rigid, erect-spreading, channelled, rough on both surfaces, margins entire. Panicle ellipsoid, dense, borne just above the leaves, its short common stalk hidden. Flowers spherical to hemispherical, white or faintly tinged with purple, perianth-lobes 3–4.5 cm. Fruit indehiscent. Seeds unwinged. *USA (Texas), W Mexico*. H5–G1. Autumn.

**8. Y. aloifolia** Linnaeus. Illustration: Polunin & Everard, Trees and bushes of Europe, 181 (1976); Matuda & Lujan, Las plantas mexicanas del genero Yucca, 108 (1980).
Stem woody, simple or branched, to 8 m or more. Leaves 25–40 × 2.5–6 cm, rigid and flat, narrowly lanceolate, narrowed slightly towards the base, margin toothed, the teeth small and irregular. Panicle held just above the leaves, erect. Flowers bell-shaped, white

often tinged with purple, perianth-lobes to 5.5 cm. Ovary shortly stalked. Fruit indehiscent, with purple pulp. Seeds unwinged. *SE USA, Mexico, West Indies*. H5–G1. Autumn.

A very variable species; variegated and coloured-leaved cultivars are grown; in var. **draconis** (Linnaeus) Engelmann the leaves are drooping.

**9. Y. elephantipes** Regel (*Y. guatemalensis* Baker). Illustration: Botanical Magazine, 7997 (1905); Chickering, Flowers of Guatemala, 123 (1973); Matuda & Lujan, Las plantas mexicanas del genero Yucca, 105–6 (1980).
Like *Y. aloifolia* but trunk to 10 m, leaves 50–100 × 5–7 cm, flowers white or cream, ovary stalkless, fruit with greenish pulp. *Mexico, Guatemala*. G1. Autumn.

**10. Y. gloriosa** Linnaeus (*Y. acuminata* Sweet). Figure 36(1), p. 272. Illustration: Botanical Magazine, 1260 (1810); Saunders, Refugium Botanicum 5: t. 316–20 (1872).
Stem woody, to 5 m, ultimately branched. Leaves to 50 × 5 cm, glaucous when young, green when older, flat or pleated near the apex, smooth, ascending, spreading or recurved, lanceolate, somewhat narrowed towards the base, flexible, margins entire or with a few inconspicuous teeth. Panicle large, erect, borne directly above the leaves on a hidden stalk. Flowers bell-shaped, greenish white, cream or reddish, perianth-lobes 5–7 cm. Fruit indehiscent, not fleshy. Seeds unwinged. *SE USA (North Carolina to Florida)*. H5–G1. Autumn.

Probably the most widely grown of the species, and also the best-known of a complex of species which are difficult to distinguish. The following may be found in European gardens:

*****Y. recurvifolia** Salisbury (*Y. recurva* Haworth). Illustration: Saunders, Refugium Botanicum 5: t. 321 (1872). Leaves less stiff, usually recurved in their upper half. *SE USA (Georgia to Missouri)*. H5–G1. Autumn.

*****Y. desmetiana** Baker. Leaves rigid, without an obvious spine at the apex; inflorescence, flowers and fruit apparently unknown. *?Mexico*. H5–G1.

*****Y. flexilis** Carrière. Inflorescence usually with a distinct stalk, held well above the leaves. *Origin unknown*. H5–G1. Autumn.

**11. Y. filamentosa** Linnaeus (*Y. angustifolia* misapplied). Illustration: Botanical Magazine, 900 (1806).

Plant stemless or almost so, the stem, if present, usually short and hidden by the leaves. Leaves *c*. 50 × up to 2.5 cm, oblanceolate, very clearly narrowed towards the base, rather abruptly tapered to the apex, flexible and recurved, green or slightly glaucous, margin splitting off as stout, curled threads. Panicle clearly stalked, held well above the leaves. Flowers bell-shaped, white tinged with green, or cream. Perianth-lobes 5–8 cm. Fruit a dehiscent capsule. Seeds winged. *E USA (New Jersey to Florida)*. H5–G1. Autumn.

A variable species; several cultivars, including variegated forms, are available.

**12. Y. flaccida** Haworth (*Y. puberula* Haworth). Illustration: Edwards's Botanical Register 22: t. 1895 (1836); Saunders, Refugium Botanicum 5: t. 323 (1872).
Like *Y. filamentosa* but leaves tapering from base to apex, margin splitting off as straight or slightly curled threads, panicle axes usually downy. *S USA (North Carolina to Alabama)*. H5–G1. Autumn.

**13. Y. smalliana** Fernald (*Y. filamentosa* var. *smalliana* (Fernald) Ahles). Illustration: Rickett, Wild flowers of the United States 2(1): t. 15 (1967).
Like *Y. filamentosa* but leaves stiff and ascending, perianth-lobes 3–5 cm. *SE USA (South Carolina to Florida)*. H5–G1. Autumn.

**14. Y. elata** Engelmann (*Y. radiosa* (Engelmann) Trelease). Illustration: Botanical Magazine, 7650 (1899); Rickett, Wild flowers of the United States 4(1): t. 12 (1970); Matuda & Lujan, Las plantas mexicanas del genero Yucca, 132 (1980).
Plants with an obvious woody stem to 3 m. Leaves very numerous, 40–70 cm × 8–15 mm, narrowly linear, tapering slightly from base to apex, flat, pale green, flexible, the margins splitting off as fine white threads. Panicle ellipsoid, borne on a distinct stalk, held well above the leaves. Flowers bell-shaped, cream-white, sometimes tinged with green or pink, perianth-lobes 3.5–5 cm. Fruit a dehiscent capsule. Seeds winged. *USA (Arizona), Mexico*. H5–G1.

**15. Y. glauca** Fraser (*Y. angustifolia* Pursh). Illustration: Saunders, Refugium Botanicum 5: t. 315 (1872); Botanical Magazine, 2236 (1821); Rickett, Wild flowers of the United States 4(1): t. 13 (1970).
Plant stemless or with stem to 30 cm. Leaves 50–70 cm × 5–13 mm, linear or tapering slightly from the base, spreading, flexible, pale green, margins splitting off as fine threads. Panicle borne above the leaves

on a long or short stalk. Flowers bell-shaped or almost spherical, greenish white, glossy, perianth-lobes thick, 4.5–6 cm. Style green, narrowed at the base, then broadening, ultimately narrowing again. Fruit a dehiscent capsule. Seeds with broad wings. *SE USA*. H5–G1. Autumn.

It is possible that *Y. stricta* Sims (Illustration: Botanical Magazine, 2222, 1821) is the same species; if so, this name would have to be used.

**16. Y. rupicola** Scheele. Illustration: Botanical Magazine, 7172 (1891). Plant stemless. Leaves 30–60 × 2–4 cm, narrowing somewhat towards the base, flexible, dark green, often wavy or twisted, margins finely toothed. Panicle held well above the leaves on a distinct stalk, hairless. Flowers bell-shaped, white or greenish white, perianth-lobes 4–7 cm. Fruit a dehiscent capsule, beaked at the apex. Seeds usually with narrow wings. *USA (Texas)*. H5–G1. Autumn.

**17. Y. rostrata** Trelease. Illustration: Matuda & Lujan, Las plantas mexicanas del genero Yucca, 126 (1980). Stems woody, ultimately to 4.5 m, not or little branched. Leaves 40–60 × 1.2–1.7 cm, linear or slightly broadened above the base, flat, thin and flexible, margins yellow or brown, finely toothed. Panicle borne well above the leaves on a distinct stalk. Flowers spherical to bell-shaped, white, perianth-lobes 4–5 cm. Fruit a dehiscent capsule with a long beak. Seeds more or less unwinged. *USA (Texas), N Mexico*. H5–G1. Autumn.

**18. Y. whipplei** Torrey (*Hesperoyucca whipplei* (Torrey) Baker). Illustration: Botanical Magazine, 7662 (1899); Rickett, Wild flowers of the United States 4(1): t. 12 (1970). Plant stemless. Leaves 30–100 cm × *c.* 15 mm, linear, rigid, glaucous, margins yellowish, finely toothed. Panicle borne well above the leaves on a distinct stalk, oblong. Flowers cream-white, sometimes tinged purple, very fragrant. Perianth hemispherical or bell-shaped, lobes 5–7 cm. Filaments clearly attached to the perianth at the base, their apices not turned outwards. Style thread-like, much narrower than the apex of the ovary, short, stigma 3-lobed. Capsule dehiscent, cylindric, with a short point at the apex. *USA (California) and adjacent Mexico*. H5–G1.

A variable species, divided into several subspecies by American botanists; it is not known which of these is in cultivation.

**2. HESPERALOE** Engelmann
*J. Cullen*
Stemless perennial herbs with swollen, bulb-like but fibrous bases. Leaves evergreen, forming clumps, linear but tapering, the margins splitting off as coarse or fine threads. Inflorescence usually a panicle, more rarely unbranched. Flowers bisexual, radially symmetric, cylindric bell-shaped; perianth-lobes 6, greenish or pink. Stamens 6, anthers included within the perianth. Ovary superior, 3-celled, ovules numerous in each cell. Fruit a capsule. Seeds numerous, thin and flat, black.

A genus of 3 species from Mexico and the adjacent USA. Cultivation as for *Yucca* (p. 273).
Literature: Trelease, W., The Yucceae, *Report from the Missouri Botanical Garden* 13: 29–38 (1902).

1a. Flowers dark pink to orange-pink; marginal threads of the leaves very fine                         **1. parviflora**
  b. Flowers green tinged with purple; marginal thread of the leaves coarse                                  **2. funifera**

**1. H. parviflora** (Torrey) Coulter (*H. yuccifolia* Engelmann; *Yucca parviflora* Torrey). Illustration: Botanical Magazine, 7723 (1900); Rickett, Wild flowers of the United States 3(1): t. 6 (1969).
Leaves to 1.3 m × 3 cm near the base, with very fine marginal threads. Panicle to 1.2 m. Flowers pendent, dark pink to orange-pink, borne on red-pink stalks, perianth to 3 cm. *USA (Texas)*. H5–G1. Summer.

Plants with shorter, more bell-shaped flowers have been called var. **engelmanii** (Krauskopf) Trelease (*H. engelmanii* Krauskopf).

**2. H. funifera** (Lemaire) Trelease (*Yucca funifera* Lemaire).
Leaves to 2 m × 5-6 cm near the base, marginal threads coarse. Panicle to nearly 3 m. Flowers green tinged with purple, borne on similarly coloured stalks, perianth to 2.5 cm. *N Mexico*. H5–G1. Summer.

**3. BESCHORNERIA** Kunth
*C.J. Couper*
Succulent perennial herbs with short, tuberous rootstocks. Leaves mostly in a basal rosette, persistent, ascending, lanceolate, glaucous, leathery and fleshy but flaccid, margins rough, minutely toothed; stem-leaves few, decreasing in size upwards, ultimately bract-like. Raceme or panicle erect or obliquely arched; bracts ovate, conspicuous, coloured and scarious. Flowers bisexual, clustered, stalked, pendent, their stalks jointed near the apex. Perianth tubular, lobes 6, free almost to the base, radially symmetric, more or less equal, oblanceolate to lanceolate, tube very short and funnel-shaped. Stamens 6, borne at the bases of the perianth-lobes, filaments thread-like, anthers dorsifixed. Ovary inferior, club-shaped, 3-celled with many ovules in each cell, style thickened at the base, stigma capitate. Capsule more or less spherical, 3-celled. Seeds flattened, brownish black.

A genus of about 12 species from Mexico. Cultivation as for *Agave* (p. 278).

1a. Leaves to 3 cm wide; flowers in an erect raceme              **1. tubiflora**
  b. Leaves more than 4 cm wide; flowers in panicles                              2
2a. Leaves more than 60 cm; flowering stem 2.1–2.4 m      **5. dekosteriana**
  b. Leaves to 60 cm; flowering stem at most 2 m                                 3
3a. Flowering stem drooping, less than 1.2 m                        **2. yuccoides**
  b. Flowering stem erect, more than 1.2 m                                        4
4a. Leaves keeled, stiff and fleshy; flowers downy                    **3. pubescens**
  b. Leaves not keeled, thin and firm; flowers hairless              **4. bracteata**

**1. B. tubiflora** Kunth. Illustration: Botanical Magazine, 4642 (1852); Gartenflora 24: t. 851 (1875).
Stemless. Leaves 12 or more, 30–60 × 1.5–2.5 cm, linear, long-acuminate, rough on both surfaces, more or less recurved, thickened and triangular in section at the base, minutely striped. Raceme erect, axis 60–120 cm, brownish green. Bracts ovate, violet-red. Flowers 2–4 in a cluster, *c.* 3.5 cm, drooping, stalks unequal and shorter than the bracts. Perianth-lobes *c.* 2.5 cm × 4 mm, reddish green, spathulate. Ovary *c.* 1.2 cm, 6-angled, purplish green. *Mexico*. G1.

**2. B. yuccoides** Hooker. Illustration: Botanical Magazine, 5203 (1860); Gardeners' Chronicle 46: 8, 138, 139, 309, 313 (1909); Journal of the Royal Horticultural Society 38: 196 (1912); Hay & Synge, Dictionary of garden plants, t. 1460 (1969).
Stemless. Leaves *c.* 20, 30–60 × *c.* 5 cm, lanceolate, narrowed to the base, rough on the lower surface, the apex sharp. Panicle with drooping, red branches on a stalk

90–120 cm; bracts lanceolate, acuminate, deep red-pink. Flowers 5–7.5 cm, dark green tinged with yellow, hairless; perianth-lobes linear-oblong. Ovary cylindric. *Mexico*. G1.

**3. B. pubescens** Berger. Illustration: Gardeners' Chronicle **40**: t. 138 (1906).
Stemless. Leaves numerous, *c.* 60 × 5 cm, linear-lanceolate, thickly keeled, stiff and fleshy along the midrib, hardened at the apex, rough only near the tip. Panicle red, on a stalk 1.2–1.5 m; bracts triangular-lanceolate, bright red, hairless. Flowers 3 or 4 in a cluster, green, eventually becoming yellow, downy throughout, stalks 2–4 cm, hanging, hairless. Perianth-lobes linear, little spreading, hairless inside. Ovary *c.* 2.5 cm, conical-cylindric. *Mexico*. G1.

**4. B. bracteata** Jacobi. Illustration: The Garden **13**: 15 (1878); Botanical Magazine, 6641 (1882).
Stemless. Leaves 20–30, 30–60 × *c.* 5 cm, thin but firm, rough beneath, margins scarious or dry. Panicle borne on a terete, reddish brown stalk which is 1.2–2 m; bracts ovate, pale red, scarious. Flowers 3.5–5 cm, deep red, perianth-lobes *c.* 2.5 cm, oblanceolate, obtuse, incurved, greenish, becoming yellowish red outside, hairless. *Mexico*. G1.

**5. B. dekosteriana** Koch. Illustration: Botanical Magazine, 6768 (1884).
Stemless. Leaves 20 or more, 60–75 × 5–6 cm, oblanceolate, long-acuminate, base very thick and firm, smooth on the upper surface, rough on the margins and lower surface, dull green above, glaucous beneath. Panicle lax, borne on a yellow-brown stalk which is 2.1–2.4 m; bracts ovate, white tinged with bright red. Flowers in clusters of 2 or 3, 3.5–6 cm, stalks to 2.5 cm. Perianth-lobes *c.* 3.5 cm, green, overlapping. Ovary green tinged with red, protruding a little into the perianth-tube. *Mexico*. G1.

### 4. FURCRAEA Ventenat
*C.J. Couper*

Monocarpic succulents. Leaves in a rosette which is basal in trunkless species, terminal in those with a trunk, persistent, lanceolate, rigid, narrowed towards the base, margins finely toothed or spiny and with a spine at the apex. Panicle tall, loose, flowers often replaced by bulbils. Bracts small, membranous; flower-stalks jointed. Flowers bisexual, radially symmetric. Perianth bell-shaped, of 6 lobes united into a short tube at the base, the lobes spreading horizontally. Stamens 6, attached at the apex of the perianth-tube, filaments erect, swollen basally; anthers oblong, versatile. Ovary inferior, oblong, 3-celled with many ovules in each cell. Style swollen basally with 3 prominent angles, stigma capitate. Fruit a capsule with many seeds.

A genus of about 20 species from tropical and subtropical America, separated from *Agave* because of its short-tubed flowers, short, thickened filaments and stoutly-based style. The genus is sometimes referred to incorrectly as *Fourcroya* Sprengel. Cultivation as for *Agave* (p. 278).
Literature: Michotte, F., *Agaves et Fourcroyas, Culture et Exploitation*, cdn 3 (1931).

1a. Trunk present, at least 60 cm in height   2
  b. Trunk absent, or if present, less than 40 cm in height   5
2a. Leaf-margins with hooked, dark brown spines along most of their length   **4. selloa**
  b. Leaf-margins entire or minutely toothed, without spines, or rarely with a few spines near the base   3
3a. Leaves 1.2–2.5 m, bright green   **3. foetida**
  b. Leaves to 1.2 m, blue-green or glaucous   4
4a. Trunk less than 2 m; leaves rigidly spreading, glaucous   **2. bedinghausii**
  b. Trunk more than 2.5 m; leaves flaccidly spreading, blue-green   **1. roezlii**
5a. Leaves to 45 × 3.5–5 cm, dull or dark olive-green   **9. undulata**
  b. Leaves more than 50 × 5 cm, bright green   6
6a. Leaves distinctly keeled; marginal spines 2–3 mm   **6. hexapetala**
  b. Leaves not keeled; marginal spines more than 3 mm   7
7a. Trunk short but distinct, 15–20 cm; marginal spines of the leaves reddish brown   **5. tuberosa**
  b. Trunk absent or indistinct; marginal spines brown   8
8a. Leaves 60–130 × 6–7 cm; panicle-branches minutely downy   **8. pubescens**
  b. Leaves 1.5–2.3 m × 9–12.5 cm; panicle-branches hairless   **7. elegans**

**1. F. roezlii** André (*Agave argyrophylla* Anon.; *Yucca argyrophylla* Nicholson; *Roezlia regia* Anon). Illustration: Revue Horticole for 1887, 353; Jacobsen, Handbook of succulent plants, t. 610 (1960).
Trunk to 4 m × 30–35 cm. Leaves 1–1.2 m × 7.5–12 cm, gradually tapering, flaccidly leathery, fleshy, spreading, acute, concave or pleated, blue-green; margins thinner, upcurved, minutely toothed. Flowering axis 3–5 m, stout, downy; bracts large, downy. Panicle large, pyramidal, freely branched, the branches spreading and drooping at their tips. Flowers mostly in 3s on short stalks. Perianth-lobes 5–6 cm, white, softly downy inside. Capsule ellipsoid. Elongate bulbils freely produced. *Mexico*. G1.

**2. F. bedinghausii** Koch (*Roezlia bulbifera* Siebert & Voss; *Yucca bulbifera* invalid; *Beschorneria multiflora* invalid). Illustration: La Belgique Horticole **13**: 327 (1863); Botanical Magazine, 7170 (1891); The Garden **40**: 143 (1891) & **52**: 197 (1897).
Trunk 90–180 × 15–22 cm. Leaves 50 or more, 60–120 × 5–10 cm, narrowed gradually to the apex and narrowed to *c.* 2.5 cm wide near the base, rigidly outcurving, stiff, flat, striped, glaucous on both surfaces, the upper smooth, the lower rough, margins minutely toothed. Flowering axis to 6 m, the panicle occupying most of it, the branches elongate and drooping. Flower-stalks 6–8 mm. Perianth-lobes *c.* 2.5 cm, white tinged with green outside. Ovary *c.* 2.5 cm, downy. Capsule broadly oblong. Ovoid bulbils freely produced. *C Mexico*. G1.

**3. F. foetida** (Linnaeus) Haworth (*Agave foetida* Linnaeus; *F. gigantea* Ventenat). Illustration: Redouté, Liliacées 8: t. 476 (1816); Botanical Magazine, 2250 (1821); Gartenflora **1**: t. 3 (1852); Revue Horticole, ser. IV, **4**: 206, 207 (1857).
Trunk 90–120 cm. Leaves 40–50, 1.2–2.5 m × 10–20 cm, narrowed to 6–7.5 cm broad above the base, rigid, flat or channelled, the lower surface rough, all bright, glossy green, narrowed gradually to a sharp terminal spine, marginal spines absent or few and very small (to 3 mm). Flowering axis 6–12 m, the panicle occupying about half of its length, branches compound. Flower-stalks 6–12 mm. Perianth-lobes *c.* 2.5 cm. Ovary *c.* 1.8 cm, hairless. Flowers strongly scented. Bulbils freely produced. *C & northern S America, naturalised and introduced elsewhere*. G1.

**4. F. selloa** Koch (*F. flavoviridis* Hooker). Illustration: Botanical Magazine, 6148 (1875); Jacobsen, Handbook of succulent plants, t. 611–13 (1960).

Trunk to 1.5 m × 10–12.5 cm. Leaves 30–40, 90–120 × 7.5–10 cm, narrowed to c. 3.5 cm wide at the base, spreading, very rigid, concave or pleated, bright or glossy dark green, the undersurface rough; terminal spine blunt, marginal spines 5–6 mm, hooked, brown, c. 3–4 cm apart. Flowering axis to 5 m; bracts 7.5–10 cm, hairless; branches of the panicle mostly simple. Flower-stalks c. 1.2 cm. Perianth-lobes 2.5–4 cm. Ovary c. 2.5 cm. Bulbils freely produced. *Mexico, Guatemala & Colombia*. G1.

'Marginata', with pale leaf margins, is widely grown.

**5. F. tuberosa** (Miller) Aiton. Illustration: Revue Horticole for 1877, 234, and for 1880, 394–5; Report from the Missouri Botanical Garden **18**: t. 1–4 (1907). Trunk 15–20 × 5–7.5 cm. Leaves c. 30, 60–175 × 5–25 cm, firm, hairless, almost flat, smooth on the upper surface, rough on the lower, terminal spine blunt, margins upcurved towards the apex with horny, hooked, reddish brown spines which are 5–9 mm long and 2–3 cm apart, sometimes absent towards the base and apex. Flowering axis to 9 m, branches compound. Perianth-lobes 2.5–3 cm. Flowers sweetly scented. Bulbils elongate, freely produced. *West Indies*. G1–2.

**6. F. hexapetala** (Jacquin) Urban (*F. cubensis* (Jacquin) Ventenat; *Agave hexapetala* Jacquin; *A. cubensis* Jacquin). Illustration: Nicholson, Illustrated dictionary of gardening **2**: 37 (1884–8); Annales du Jardin Botanique de Buitenzorg, Supplement **3**: t. 40 (1907). Trunk very short, 7.5–10 cm thick. Leaves c. 30, 60–125 × 5–10 cm, narrowed to c. 3.5 cm wide above the base, spreading, rough and rather distinctly keeled on the lower surface, the upper surface bright, shining green; terminal spine blunt, brown, marginal spines 2–3 mm, triangular, brown, 1.6–2.5 cm apart. Flowering axis to 6 m. Panicle loose, diamond-shaped, the lower branches slightly compound. Flower-stalks very short. Perianth-lobes c. 2.5 cm, white, greenish outside. Ovary c. 1.8 cm. *Cuba, Haiti*. G1.

**7. F. elegans** Todaro. Illustration: Todaro, Hortus Botanicus Panormitanus **1**: t. 4 (1876–8); Botanical Magazine, 8461 (1912). Trunk absent or very short. Leaves 40–50, 1.5–2.3 m × 9–12.5 cm, narrowed to c. 7.5 cm wide above the base, bright green, smooth above, rough beneath, terminal spine short, brown, marginal spines large, brown. Flowering axis 6–8 m, the panicle occupying most of its length, the branches very compound. Perianth-lobes c. 3.5 cm, greenish white. Ovary c. 2.5 cm. *Mexico*. G1.

**8. F. pubescens** Todaro. Illustration: Botanical Magazine, 7250 (1892); Annales du Jardin Botanique de Buitenzorg, Supplement **3**: t. 43 (1907). Trunk absent. Leaves c. 30, 60–130 × 6–7 cm, rigid, very much thickened at the base on both surfaces, grooved towards the apex, green; terminal spine sharp, marginal spines c. 5 mm, triangular-hooked, 1.6–3 cm apart. Flowering axis to 7 m, panicle c. 5.5 m, branches ascending, curved, very compound, minutely downy. Perianth-lobes 2.5–3.5 cm. Ovary densely downy. Capsule large. Bulbils freely produced. *?Mexico*. G1.

**9. F. undulata** Jacobi (*F. pubescens* Baker not Todaro). Illustration: Botanical Magazine, 6160 (1875). Trunk absent. Leaves 20–30, c. 45 × 3.5–5 cm, narrowed to 1.8–2.5 cm above the base, wavy, recurved, keel and midrib beneath rough, dull or dark green above; terminal spine blunt, marginal spines triangular-hooked, brown. Flowering axis c. 3.5 m. Panicle 2–2.5 m, branches short, mostly simple, drooping, minutely downy. Perianth-lobes c. 3 cm, downy. Ovary hairless. *Mexico*. G1.

## 5. AGAVE Linnaeus

*C.J. Couper & J. Cullen*

Monocarpic or perennial leaf-succulents without fleshy underground stems. Stems absent or short, sometimes forming trunks. Leaves in a rosette, persisting for many years, small to very large, often very fleshy, usually hard and leathery, usually with spines or teeth on the margins (more rarely the margins unrolling as fine threads) and with a conspicuous terminal spine. Inflorescence a large spike or panicle with many bisexual, radially symmetric flowers. Perianth tubular, of 6 lobes which are united for most of their length. Stamens 6, anthers projecting from the perianth. Ovary inferior, 3-celled, with many ovules in each cell. Fruit a capsule. Seeds flattened, black.

A difficult genus of over 100 species, mostly from C America and the West Indies, but extending both north and south from this centre; a few species are naturalised in other areas (e.g. the Mediterranean). Their large, fleshy leaves and tall inflorescences render them unsuitable for the making of good herbarium specimens, and their flowering pattern – monocarpic or flowering irregularly over long periods – means that much of their study must take place in the field.

Very many species are recorded as having been in cultivation, but their identification is dubious, and there is much uncertainty. The key to the species included here is based entirely on vegetative characters (flowering material being only irregularly available) and must be used with some caution. The leaves of most species bear spines – hardened, broadly based usually distant and dark-coloured prickles which are sometimes borne on fleshy pads – along their margins; a few species have teeth rather than spines: these are small, narrowly based and evenly distributed; a few others have margins that are entire or finely roughened or saw-like; and, finally, a few have margins which split off as threads.

Agaves will grow well in relatively poor, sandy soils, as long as they are well drained. They will also grow well in pots, provided that these are large enough. Propagation is by seed or by suckers which have already rooted while still attached to the parent plant.

Literature: Jacobi, G.Z., *Von Versuch zu einer systematischen Ordnung der Agaven* (1864); Trelease, W., Species in Agave, *American Philosophical Society Proceedings* **49**: 232–237 (1910); Trelease, W., Agave in the West Indies, *Memoirs of the National Academy of Sciences* **11**: 1–55 (1913); Berger, A., *Die Agaven* (1915); Michotte, F., *Agaves et Fourcroyas, Culture et Exploitation*, edn 3 (1931); Gentry, H.S., The Agave family in Sonora, *USDA Agricultural Handbook No. 399* (1972); Gentry, H.S., The Agaves of Baja California, *Occasional Papers of the California Academy of Sciences*, *130* (1978); Gentry, H.S., *Agaves of continental North America* (1982).

1a. Leaves with margins unrolling as hair-like threads                                2
  b. Leaves with smooth, toothed, spiny or horny margins which do not unroll as threads                                4
2a. Mature rosettes to 25 cm in diameter                              **8. parviflora**
  b. Mature rosettes more than 25 cm in diameter                                3
3a. Leaves narrowed towards the base                              **2. filifera***
  b. Leaves not narrowed towards the base                              **9. toumeyana**

4a. Leaves without spines or teeth on the margins, though the margins sometimes rough    **5**

b. Leaves with obvious spines or teeth on the margins    **8**

5a. Leaves conspicuously narrowed towards the base    **6**

b. Leaves not narrowed towards the base    **7**

6a. Leaves stiff, 90–130 × 9–12 cm, terminal spine hard    **27. sisalana**

b. Leaves soft, to 75 × 12–16 cm, terminal spine soft    **1. attenuata**

7a. Leaves oblong, thick, acutely 3-angled and keeled towards the apex, dark green with white lines on the surface and margins    **6. victoriae-reginae**

b. Leaves linear, shortly tapering, flat, without white lines on the surface and margins    **12. striata***

8a. Leaves soft and flexible, with teeth on the margins but without spines    **10. albicans**

b. Leaves stiff and leathery, with spines on the margins    **9**

9a. Leaves conspicuously narrowed towards the base    **10**

b. Leaves not narrowed towards the base    **21**

10a. Rosette very large, more than 2 m in diameter when mature    **11**

b. Rosette less than 2 m in diameter when mature    **14**

11a. Terminal spine not decurrent on the leaf margins    **14. americana**

b. Terminal spine decurrent on the leaf margins    **12**

12a. Leaves very dark blackish green; plant not producing suckers    **19. atrovirens**

b. Leaves green, glaucous or whitish; plant producing suckers    **13**

13a. Leaves arching over, whitish green, 2–2.5 m    **15. franzosinii**

b. Leaves ascending or erect, green or glaucous, 1–2 m    **26. salmiana**

14a. Rosettes more than 65 cm in diameter when mature    **15**

b. Rosettes less than 65 cm in diameter when mature    **17**

15a. Mature plant producing suckers    **16**

b. Mature plant not producing suckers    **21. marmorata**

16a. Leaves 12–18 cm broad, usually reddish in part, and with paler cross bands    **17. colorata**

b. Leaves at most 10 cm broad, not coloured as above    **18. palmeri**

17a. Leaves 7 cm or more wide    **18**

b. Leaves less than 7 cm wide    **20**

18a. Rosette to 25 cm in diameter; terminal spine of the leaf 3–4.5 cm    **20. potatorum**

b. Rosette to 65 cm in diameter; terminal spine 1–3 cm    **19**

19a Leaves 20–30 × 10–12 cm, glaucous or bloomed    **22. parrasana**

b. Leaves 35–65 × 7–10 cm, green    **11. polyacantha**

20a. Leaves 2–3 cm broad, glaucous or whitish; rosette producing suckers    **25. macroacantha**

b. Leaves 4–7 cm broad, green; rosette not producing suckers    **3. horrida**

21a. Rosette not producing suckers    **7. xylonacantha**

b. Rosette producing suckers    **22**

22a. Rosette less than 65 cm in diameter    **13. utahensis**

b. Rosette more than 65 cm in diameter    **23**

23a. Leaves more than 6 cm broad    **24**

b. Leaves less than 6 cm broad    **26**

24a. Leaves 25–40 × 8–12 cm    **23. parryi**

b. Leaves 60–180 × 8–16 cm    **25**

25a. Leaf margins straight, spines 3–6 mm; trunk present    **24. fourcroydes**

b. Leaf margins sinuous, spines 8–15 mm; trunk absent    **16. scabra**

26a. Leaves 2.5–4 cm wide, margins straight, teeth generally directed towards the leaf base    **4. lechuguilla**

b. Leaves 4–5 cm wide, margins sinuous, teeth directed outwards or upcurved    **5. lophantha**

Subgenus **Littaea** (Tagliabue) Baker. Flowers in a simple raceme or spike, usually in pairs (more rarely in 4's or more) in the axils of each bract; axis of inflorescence narrower than the apex of the crown of the rosette.

**1. A. attenuata** Salm-Dyck. Figure 36 (6), p. 272. Illustration: Botanical Magazine, 5333 (1862) – as A. glaucescens; Gardeners' Chronicle **17**: t. 63–4 (1896); Lamb, Cacti and other succulents **2**: t. 267 (1959); Jacobsen, Handbook of succulent plants, t. 84 (1960).

Trunk to 1 m, prostrate, branching above ground, each branch ending in a rosette. Rosettes with 10–20 leaves. Leaves 50–70 × 12–16 cm, narrowed towards the base, ovate-elliptic, spreading, soft, glaucous, margins entire, pale, terminal spine needle-like but soft. Spike to 3.5 m, recurved. Flowers in 2s, 3.5–5 cm, greenish yellow. Spike with numerous plantlets after flowering. *Mexico.* H5–G1.

**2. A. filifera** Salm-Dyck. Illustration: Gardeners' Chronicle **7**: t. 49 (1877) & **4**: t. 135–7 (1881); Jacobsen, Handbook of succulent plants, t. 96 (1960).

Rosettes trunkless, *c.* 65 cm in diameter with 60–100 leaves, producing suckers freely. Leaves 15–30 × 2–4 cm, lanceolate, long-tapering to the apex, narrowed towards the base, stiff, spreading and somewhat upcurved, shiny dark green with 2 or 3 white lines on the surface, margins pale, horny, splitting off as 5 or 6 long, thin threads; terminal spine 1–2 cm, brown, later grey, openly grooved. Spike to 2.5 m, flowers very dense. Flowers 3–3.5 cm, greenish at first, later reddish. *Origin uncertain, probably Mexico.* H5–G1.

*****A. schidigera** Lemaire. Illustration: Illustration Horticole **9**: t. 330 (1862); Botanical Magazine, 5641 (1867). Rosette 90–100 cm in diameter, leaves 30–50 × 1.2–1.6 cm. *Mexico.* H5–G1.

**3. A. horrida** Jacobi. Figure 36(3), p. 272. Illustration: Gardeners' Chronicle **7**: t. 99 (1877); Botanical Magazine, 6511 (1880). Rosette trunkless, to 60 cm in diameter, with 50–100 leaves, not producing suckers. Leaves 20–35 × 4–7 cm, narrowed towards the base, oblanceolate-oblong, stiff and hard, glossy or yellowish green, spreading, margins grey, horny, sinuous, marginal spines 1–2 cm distant, curved in all directions and with accessory spines; terminal spine 2.5–4 cm, flattened above, red-brown becoming grey. Spike 3–4 m, dense. Flowers in 2s, 3.5–4 cm, pale yellowish green. *Mexico.* H5–G1.

Plants of this species are strictly monocarpic, and die after flowering; propagation is by seed only.

**4. A. lechuguilla** Torrey. Illustration: Report from Missouri botanical garden **7**: t. 31 (1896).

Rosettes *c.* 1 m in diameter, with 40–60 leaves, producing suckers. Leaves 25–50 × 2.5–4 cm, triangular-oblong, curved or ascending, green or bluish, but not glaucous, pale-banded beneath; margins straight, narrow, horny, with slender spines, terminal spine 1.5–4 cm, slender, recurved, flattened, grey-brown. Spike 1–4 m. Flowers 3–4.5 cm. *USA (Texas), Mexico.* H5–G1.

**5. A. lophantha** Schiede. Illustration: Lamb, Cacti and other succulents **2**: t. 270 (1959); Jacobsen, Handbook of succulent plants, t. 103 (1960).

Rosette *c.* 1 m in diameter, with 30–40 leaves, producing suckers. Leaves

30–70 × 4–5 cm, lanceolate, spreading, long-tapering, stiff, hard, glossy green, the under surface with darker lines, the upper surface sometimes with a pale, longitudinal stripe; margin horny, marginal spines stout, triangular, straight or upcurved; terminal spine 1–2.5 cm, sharp, openly grooved. Spike to 4.5 cm, dense towards the apex. Flowers in 2s, 3.5–4.5 cm, pale greenish yellow. *Mexico*. H5–G1.

A very variable species, in which a number of varieties has been described. The most commonly grown is var. **univittata** (Haworth) Anon. (*A. univittata* Haworth), in which each leaf has a pale, longitudinal stripe along the upper surface.

**6. A. victoriae-reginae** Moore (*A. ferdinandi-regis* Berger). Illustration: Gardeners' Chronicle **18**: t. 148–9 (1882); Jacobsen, Handbook of succulent plants, t. 115–17 (1960); Rowley, The illustrated encyclopaedia of cacti and succulents, 169 (1978).
Rosette trunkless, to 60 cm in diameter, with up to 200 leaves, not usually producing suckers. Leaves 10–20 × 4–6 cm, triangular-oblong, smooth, leathery, acutely 3-angled, keeled towards the apex, then narrowed into a blunt tip, margins and keel without spines, smooth with white, horny bands; dull, dark green marked with oblique, white lines; terminal spine 1–3 cm, brown, needle-like, often with additional, very short spines on either side. Spike to 5 m, loose below, dense above. Flowers usually in 3s, 3–5 cm, pale green. *Mexico*. H5–G1.

**7. A. xylonacantha** Salm-Dyck. Illustration: Botanical Magazine, 5660 (1867).
Rosettes simple, borne on a short trunk, with *c.* 20 leaves, not forming suckers. Leaves 30–90 × 5–12 cm, tapering from the base, oblong, hard, rough, dull, glaucous or grey-green (rarely yellow-green), dark-lined beneath, the margins broad, white, horny and with irregular spines which are 8–15 mm and 3–5 cm distant, borne on broad cushions; terminal spine 2.5–5 cm, straight, needle-like, grooved at the base, pale brown or grey. Spike to 6 m, sinuous. Flowers 3–8 together, 4–5 cm, greenish. *Mexico*. H5–G1.

**8. A. parviflora** Torrey. Illustration: Report from Missouri botanical garden **7**: t. 30 (1896); Lamb, Cacti and other succulents **2**: t. 271 (1959); Jacobsen, Handbook of succulent plants, t. 108–9 (1960); USDA Agricultural Handbook No. 399: t. 10 (1972).

Rosette trunkless, 15–25 cm in diameter, producing suckers very rarely. Leaves 6–10 cm × 8–10 mm, narrowed towards the base, oblong-linear, stiff and hard, green, the margin splitting off as a number of white threads, minutely toothed towards the base; terminal spine weak, 5–8 mm, brown or greyish white. Spike 1–1.8 m, slender, loose, axis reddish. Flowers in 2s, 3s or 4s, 1.3–1.5 cm, pale yellow. *USA (Arizona), Mexico*. H5–G1.

**9. A. toumeyana** Trelease. Illustration: Lamb, Cacti and other succulents **2**: t. 272 (1959).
Rosette trunkless, *c.* 30 cm in diameter, not producing suckers. Leaves 20–30 × 1.5–2 cm, linear, tapering slightly, dark green with some white markings on the upper surface, the margins splitting off, at least in the upper part, into thin, white threads; terminal spine 1–2 cm. Spike dense. Flowers to 2.5 cm, greenish or pale yellow. *USA (Arizona)*. H5–G1.

**10. A. albicans** Jacobi. Illustration: Gardeners' Chronicle **8**: t. 138 (1877); Botanical Magazine, 7207 (1891).
Trunk very short. Rosette with 20–30 leaves, not producing suckers. Leaves 30–60 × 7–13 cm, oblanceolate, narrowed towards the base, shortly tapering to the apex, soft, fleshy, fragile, pale green, glaucous or whitish, margins with weak teeth which are at first colourless, later brown; terminal spine 1–2 cm, brown, soft, bristle-like, narrowly grooved. Spike *c.* 1 m, rather few-flowered above the middle. Flowers 4–6 cm, in pairs, reddish green, shortly stalked. *Mexico*. H5–G1.

Gentry (1982) treats this as *A. celsii* Hooker var. *albicans* (Jacobi) Gentry.

**11. A. polyacantha** Jacobi. Illustration: Botanical Magazine, 5006 (1857) – as A. densiflora; Jacobsen, Handbook of succulent plants, t. 96 (1960).
Rosette trunkless, to 65 cm in diameter, with *c.* 30 leaves, at first simple, later forming mats, not producing suckers. Leaves 35–65 × 7–10 cm, narrowed towards the base, tapering above, oblanceolate, leathery, stiff and rigid, green (rarely glaucous ?), margins sinuous with spines 2–3 mm, black-brown, triangular, confluent in the upper part into a brown, horny line; terminal spine 1–2.5 cm, grooved, black-brown. Spike to 2.5 m, dense. Flowers 5–7 cm, in pairs, light green or reddish. *Mexico*. H5–G1.

**12. A. striata** Zuccarini. Figure 36(4), p. 272. Illustration: Botanical Magazine,

4950 (1856); Gardeners' Chronicle **8**: t. 109 (1877); Gartenflora **29**: 24 (1880); Jacobsen, Handbook of succulent plants, t. 112 (1960).
Trunk short, but branched when old. Rosettes *c.* 1 m or more in diameter, with 150–200 leaves. Leaves 25–60 cm × 5–10 mm, stiffly spreading, linear, tapering abruptly at the apex, 3-angled at the base, glaucous or grey-green with darker stripes, margins entire or roughened and saw-like; terminal spine 1–5 cm, needle-like, brown. Spike 3–3.6 m, flowers closely packed, *c.* 3–4 cm, greenish yellow. *N Mexico*. H5–G1.

**\*A. stricta** Salm-Dyck (*A. striata* var. *stricta* (Salm-Dyck) Baker). Illustration: Gartenflora **31**: 56 (1882); Cactus Journal **1**: 139 (1898); Rowley, The illustrated encyclopaedia of succulents, 169 (1978).
Very similar to *A. striata* but leaves 25–50 cm × 8–10 mm, terminal spine 1–2 cm, 3 or 4-angled, spike to 2.5 m, dense. *S Mexico*. H5–G1.

Differs from *A. striata* in details of its flowers as well as in the characters mentioned above.

**13. A. utahensis** Engelmann. Illustration: Report from Missouri botanical garden **7**: t. 32 (1896).
Rosette trunkless, to 40 cm in diameter, usually producing suckers. Leaves 12–30 × 1.5–3 cm, tapering from the base, stiff, erect-spreading, glaucous, margins straight or sinuous, without a horny band towards the base, teeth 2–4 mm, 1–2.5 cm apart, hooked, triangular; terminal spine 2–4 cm, grey, grooved beneath, tip brown. Spike 1.5–4.5 m, sometimes slightly branched. Flowers in 2s or 4s, 2.5–3 cm, yellow. *SW USA (California, Arizona, Nevada, Utah)*. H5–G1.

Subgenus **Agave**. Inflorescence a panicle, the flowers borne in clusters at the ends of the branches; axis of inflorescence as broad as the crown of the rosette.

**14. A. americana** Linnaeus. Illustration: Revue Horticole, t. 77 (1904); Lamb, Cacti and other succulents **2**: t. 266 (1959); Jacobsen, Handbook of succulent plants, t. 80–1 (1960).
Rosettes trunkless, 2–3 m in diameter with 20–60 leaves, producing suckers. Leaves 1–2 m × 15–30 cm, narrowed towards the base, oblanceolate-spathulate, rigidly spreading or reflexed towards the apex, smooth, leathery, usually glaucous, margins sinuous-toothed, marginal spines *c.* 8 mm, curved, blackish brown, later grey;

terminal spine 3–5 cm, not decurrent on the leaf margins, thick, obliquely flattened, rough at the base. Panicle 5–8 m, slender, with 25–30 branches. Flowers 7–10 cm, pale yellow. *Mexico; widely naturalised in the Mediterranean area and elsewhere.* H5–G1.

A very variable species; some variants have variegated leaves.

**15. A. franzosinii** Baker. Illustration: Gardeners' Chronicle 12: f. 31 (1892); Botanical Magazine, 8317 (1910); Jacobsen, Handbook of succulent plants, t. 97 (1960).
Rosette trunkless, to 4.5 m in diameter, with 40–50 leaves, producing suckers. Leaves 2–2.5 m × 30–40 cm, narrowed towards the base, oblanceolate, arching outwards, whitish grey variegated with green, very rough, margins sinuous-toothed, marginal spines *c.* 6 mm, triangular, hooked, black-brown, those at the base not confluent, the median 1.5–2 cm apart, the upper 5–6 cm apart; terminal spine 3–6 cm, needle-like, narrow, grooved, decurrent for *c.* 20 cm. Panicle to 11.5 m, elongate or widely cylindric-ovate. Flowers *c.* 8 cm, yellow, on stalks 1–2 cm. *Mexico.* H5–G1.

**16. A. scabra** Salm-Dyck. Illustration: Report from Missouri botanical garden 22: t. 75–9 (1911).
Rosette trunkless, with *c.* 30 leaves, producing suckers. Leaves 60–110 × 12–16 cm, ovate, tapering, thin and flexible, grey, smooth, margins sinuous-toothed, spines 8–16 mm, 1.5–2 cm distant, stout, straight or hooked; terminal spine 3.5–6 cm, widely grooved in the centre, decurrent. Panicle to 4 m. Flowers 6–8 cm, yellowish, shortly stalked. *Mexico.* H5–G1.

**17. A. colorata** Gentry. Illustration: USDA Agricultural Handbook No. 399: t. 44 (1972).
Trunk short. Rosette with few leaves, producing suckers while young. Leaves 25–60 × 12–18 cm, thick, firm, broadly lanceolate, narrowed towards the base, glaucous, often red-tinted and cross-banded; margins conspicuously sinuous-toothed, marginal spines 5–10 mm, slightly flexed or straight, brown to greyish; terminal spine 3–5 cm, needle-like, grooved to the lower half. Panicle 2–3 m, narrow, with dense lateral branches in the upper half. Flowers 5–7 cm, yellowish green or coppery red. *Mexico (Sonora).* H5–G1.

**18. A. palmeri** Engelmann. Illustration: Report from Missouri botanical garden 7: t. 48–52 (1896); Cactus Journal 2: (1899); USDA Agricultural Handbook No. 399: t. 35 (1972).
Rosette trunkless, 1–1.2 m in diameter, with *c.* 30 leaves, ultimately producing suckers. Leaves to 80 × 7–10 cm, narrowed towards the base, oblanceolate, rigid, pale green, glaucous or reddish, margins sinuous-toothed; marginal spines of various sizes, usually alternating larger and smaller, purple or grey; terminal spine 3–6 cm, sharp, needle-like, grooved, long-decurrent as horny strips on the margins, brown or grey. Panicle 3–5 m, elongate. Flowers 4.5–5.5 cm, greenish yellow or waxy white, reddish in bud. *USA (New Mexico, Arizona), Mexico.* H5–G1.

**19. A. atrovirens** Salm-Dyck (*A. latissima* Jacobi; *A. coccinea* misapplied). Illustration: The Garden **76**: 251, 575 (1912); Jacobsen, Handbook of succulent plants, t. 83 (1960).
Rosette trunkless, with 20–30 leaves, not producing suckers. Leaves 1.5–2 m × 25–40 cm, narrowed towards the base, oblanceolate, tapering from the upper one-third above, erect-spreading, very fleshy, dull dark green or blackish green, margins sinuous-toothed, upper spines distant, the lower confluent to a horny line; terminal spine *c.* 5 cm, narrowly grooved above, decurrent on the margins for up to 65 cm. Panicle 6–10 m, pyramidal, bearing 20–30 lateral branches. Flowers *c.* 10 cm, reddish to greenish yellow, borne on stalks to 1 cm. *Mexico.* H5–G1.

Material grown under this name may prove to be *A. salmiana* (p. 282).

**20. A. potatorum** Zuccarini. Illustration: Jacobsen, Handbook of succulent plants, t. 111 (1960).
Rosette trunkless, 20–25 cm in diameter, with up to 80 leaves, not producing suckers. Leaves 25–40 × 9–18 cm, narrowed towards the base, obovate-spathulate, glaucous or green, margins usually sinuous-toothed, spines 5–10 mm, 1.2–3 cm distant, triangular, sharply pointed, at first yellow, brown or blackish, later grey, sometimes of 2 sizes alternating; terminal spine 3–4.5 cm, broadly furrowed above, little decurrent on the margins, brown. Panicle *c.* 3.5 m, pyramidal. Flowers 5.5–8 cm, yellow-green, on stalks 1–1.5 cm. *Mexico.* H5–G1.

Var. **verschaffeltii** (Lemaire) Berger is frequently cultivated; it has greyish green leaves and red- or yellow-brown spines.

**21. A. marmorata** Roezl. Illustration: Botanical Magazine, 8442 (1912).
Rosette more or less trunkless, to 2 m in diameter, with 30–50 leaves, not producing suckers. Leaves 1–1.3 m × 20–30 cm, narrowed towards the base, oblanceolate-spathulate, spreading, very rough, light grey, bluish or almost white, margins rough, sinuous between the spines; spines 6–12 mm, 4–5 cm distant, with smaller spines between the larger ones, all rusty brown; terminal spine 1.5–3 cm, needle-like, recurved, narrowly grooved, grey-brown. Panicle to 3.7 m, elongate. Flowers 4–5 cm, golden yellow. *Mexico.* H5–G1.

**22. A. parrasana** Berger (*A. wislizenii* Engelmann). Illustration: Report from Missouri botanical garden 22: t. 80–1 (1911); Jacobsen, Handbook of succulent plants, t. 107 (1960); Lamb, Cacti and other succulents 3: t. 435 (1971).
Rosette to 60 cm in diameter, not forming suckers. Leaves 20–30 × 10–12 cm, stiff, thick, obovate, narrowed towards the base, shortly tapered to the apex, smooth, dull glaucous green or bloomed, margins sinuous-toothed, the upper half with stout curved spines which are 5–10 mm; the upper 4–8 spines are confluent at their bases, all yellowish brown, later red, ultimately grey; terminal spine 2–3 cm, brown, decurrent to the upper, confluent, marginal spines. Panicle to 3.5 m. Flowers 5–6 cm, yellow. *Mexico.* H5–G1.

**23. A. parryi** Engelmann. Illustration: Gardeners' Chronicle 12: t. 39 (1879); Report from Missouri botanical garden 22: t. 91–3 (1911).
Rosette trunkless, with 20–30 leaves, producing suckers. Leaves 25–40 × 8–12 cm, broadly oblong, acute or at least tapering, grey or glaucous, margins sinuous, marginal spines 3–7 mm, 1–2.5 cm apart, narrowly triangular, straight or recurved, brown later grey, bases of the uppermost spines confluent; terminal spine more or less linear, 1.5–3 cm, openly grooved, flat to the tip on the upper side, brown later grey, long-decurrent. Panicle 3–5 m, branches curved upwards. Flowers 6–7.5 cm, stalked, red in bud, cream-yellow when open. *SW USA, Mexico.* H5–G1.

**24. A. fourcroydes** Lemaire. Illustration: Memoirs of the National Academy of Sciences 11: t. 110–12 (1913); Botanical Magazine, 8746 (1918); Parey's Blumengärtnerei, 355 (1958).

Trunk to 1 m. Rosette 2–2.5 m in diameter, with about 100 leaves, producing suckers. Leaves 1.2–1.8 m × 8–12 cm, linear or oblanceolate, rigid, glaucous or grey, smooth, margins straight, upcurved, marginal spines 3–6 mm, regularly spaced at 2–3 cm intervals, slender, dark brown; terminal spine 2–3 cm, narrowly conical, openly and shortly grooved above the base, dark brown. Panicle 5–6 m, elongate, pyramidal. Flowers 6–7 cm, greenish yellow. *Mexico.* H5–G1.

**25. A macroacantha** Zuccarini. Illustration: Botanical Magazine, 5940 (1871) – as *A. besseriana*; Gardeners' Chronicle **8**: t. 27 (1877); Report from Missouri botanical garden **18**: t. 18–26 (1907).
Trunk very short or absent. Rosette with 30–50 leaves, producing suckers. Leaves 25–35 × 2.5–3 cm, narrowed towards the base, tapering towards the apex, oblanceolate, erect-spreading, very rigid and stiff, glaucous to whitish grey, margins finely rough and cartilaginous between the spines; spines few, 3–4 mm, hooked and curved forwards, all similar; terminal spine 3–3.5 cm, straight, 3-angled, decurrent. Panicle to 3 m, slender, with 10–15 branches. Flowers 5–5.5 cm, numerous, greenish flushed with red, shortly stalked. *Mexico.* H5–G1.

**26. A. salmiana** Salm-Dyck (*A. atrovirens* var. *salmiana* (Salm-Dyck) Trelease). Illustration: Gardeners' Chronicle **29**: t. 51 (1886); The Garden **29**: 495 (1886); Revue Horticole for 1872, t. 40–1; Parey's Blumengärtnerei, 357 (1958).
Rosette large, loosely leaved, forming suckers. Leaves 1–2 m × 20–35 cm, oblancolate, narrowed towards the base, tapered above, firm and smooth, rigid, grey-green; margins sinuous, marginal spines 1–2 cm (including the fleshy cushions on which they stand), confluent at the base of the leaf, the median spines 2–3 cm distant, the upper 7–8 cm distant; terminal spine 5–10 cm, black-brown, long-decurrent. Panicle 9–19 m. Flowers 8–11 cm, greenish yellow. *Mexico.* H5–G1.

Var. **ferox** (Koch) Gentry (*A. ferox* Koch). Illustration: Gardeners' Chronicle **20**: t. 94 (1896) & **43**: t. 173 (1908); Jacobsen, Handbook of succulent plants, t. 95 (1960). Leaves shining green. *Mexico.* H5–G1.

**27. A. sisalana** Perrine. Illustration: Die Tropenpflanze **2**: 212 (1898) & **4**: 7 (1900); Memoirs of the National Academy of Sciences **11**: t. 113–15 (1913).

Trunk to 1 m. Rosette dense, stiff and spreading in all directions, producing suckers. Leaves 90–130 × 9–12 cm, narrowed towards the base, narrowly oblanceolate, grey-green or bright green, margins without spines (when mature) but with a cartilaginous edge; terminal spine hard, to 2.5 cm, conical, obliquely flattened on the upper surface, shallowly grooved, black-brown. Panicle 6–7 m. Flowers 5–6.5 cm, green. *Mexico, widely cultivated for fibres elsewhere.* H5–G1.

**6. MANFREDA** Salisbury
*C.J. Couper*
Perennial herbs with succulent roots, fleshy underground stems or bulbous rootstocks. Stems absent or very short. Leaves in a basal rosette, dying off annually, small, pliable, without spines, margins often wavy, green, often with brown, red or purple blotches. Spike loose with several bisexual flowers. Perianth radially symmetric, lobes 6, united below and forming a usually narrow and deep cylindric tube, the lobes equal or almost so, spreading or reflexed, green, yellow or reddish purple. Stamens 6, filaments borne on the perianth-tube, long-projecting, anthers versatile. Ovary inferior, 3-celled with numerous ovules in 2 rows in each cell; style long-projecting, stigma club-shaped, 3-lobed. Fruit an erect, apically dehiscent capsule. Seeds numerous, black.

A small genus related to *Agave* and requiring similar cultivation conditions (p. 278). Propagation is by seed.

1a. Leaves oblong-spathulate, reddish-striped; filaments borne at the base of the perianth-tube     **1. virginica**
  b. Leaves narrowly linear or linear, mottled with brown; filaments borne near the top of the perianth-tube     **2**
2a. Leaves to 2 cm broad; stamens about as long as the lobes of the perianth; capsules about as broad as long     **3. maculosa**
  b. Leaves more than 2 cm broad; stamens much longer than the lobes of the perianth; capsules much longer than broad     **2. variegata**

**1. M. virginica** (Linnaeus) Salisbury (*Agave virginica* Linnaeus; *Polianthes virginica* (Linnaeus) Shinners). Illustration: Botanical Magazine, 1157 (1808); Report from Missouri botanical garden **7**: t. 26–7 (1896); Britton & Brown, Illustrated flora of the northern USA and Canada, edn 3, 445 (1952).
Rosette with 6–15 leaves. Leaves to

60 × 2.5–5 cm, oblong-spathulate, gradually tapered to the base, sharply tapered at the apex, somewhat grooved, thick, flaccid, dark green striped with red; margins wavy. Scape 80–180 cm, slender, spike 30–50 cm with *c.* 30 flowers. Flowers 2.5–5 cm, the lower on stalks 6–8 mm, greenish or brownish yellow, scented, perianth-lobes linear-oblong. Filaments borne at the base of the perianth-tube. Anthers *c.* 1.2 cm. Capsules 1.5–2 cm, longer than broad. *USA (Maryland to Missouri, Florida and Texas).* H5–G1. Summer.

**2. M. variegata** (Jacobi) Rose (*Agave variegata* Jacobi; *Polianthes variegata* (Jacobi) Shinners). Illustration: Saunders, Refugium Botanicum **5**: t. 326 (1872).
Rosette with few leaves. Leaves 20–45 × 2–4 cm, lanceolate, narrowed slightly to the base and gradually upwards to a blunt point, deeply grooved, fleshy, glaucous and with brown markings, margins curved upwards. Scape 90–130 cm. Flowers *c.* 4 cm, almost stalkless, brownish green, fragrant, lobes narrow, about as long as the tube or longer. Stamens to 5 cm, filaments borne at the top of the perianth-tube. Anthers *c.* 8 mm. Capsules much longer than broad, 1.5–2.2 cm. *USA (S Texas), Mexico.* H5–G1. Spring.

**3. M. maculosa** (Hooker) Rose (*Agave maculosa* Hooker; *Polianthes maculosa* (Hooker) Shinners). Illustration: Botanical Magazine, 5122 (1859); Gardeners' Chronicle for 1872, 1194; Bailey, Standard cyclopaedia of horticulture, t. 2319 (1916).
Rosette with few leaves. Leaves 15–30 × 1–2 cm, linear-lanceolate, recurved, concave, thick, fleshy, glaucous with brown or greenish markings, margins transparent with small, unequal teeth. Scape 90–120 cm. Spike 20–30 cm, loosely flowered. Flowers 10–18, 4–5 cm, more or less stalkless, fragrant, perianth greenish white but turning pink with age, lobes oblong, shorter than the tube. Stamens slightly projecting, filaments borne at the top of the perianth-tube, anthers 9–16 mm. Capsules about as broad as long, 2–2.5 cm. *USA (S Texas), N Mexico.* H5–G1. Spring–summer.

**7. POLIANTHES** Linnaeus
*C.J. Couper*
Small perennial herbs with short, tuberous rootstocks. Leaves succulent, deciduous, without spines, mostly basal, those on the stem small and decreasing in size upwards.

Flowers in loose racemes or spikes, bisexual, bilaterally symmetric, white or orange-red. Perianth-lobes 6, united below into a cylindric or funnel-shaped tube which is bent near the base and expanded near the middle, much longer than the lobes. Stamens 6, filaments inserted on the perianth-tube. Ovary inferior, 3-celled, with numerous ovules and a 3-lobed stigma. Fruit an ovoid capsule with the perianth persistent at its apex. Seeds flat.

A genus of 2 species from Mexico. They are easily grown in well-drained soils, though, if grown out-of-doors in areas prone to frost, the tubers should be lifted in the autumn and stored in a cool (but not cold), dry atmosphere. Propagation is mainly by suckers from the tubers.

1a. Flowers waxy white; perianth-tube *c.* 3 times longer than the lobes; each flower pair subtended by 1 bract
**1. tuberosa**

b. Flowers pale red or orange; perianth-tube more than 3 times longer than the lobes; each flower pair subtended by 1 bract and 2 bracteoles
**2. geminiflora**

**1. P. tuberosa** Linnaeus. Illustration: Saunders, Refugium Botanicum 1: t. 63 (1815); Botanical Magazine, 1817 (1816); Hay & Synge, Dictionary of garden plants, t. 630 (1969).
Tuber conical. Basal leaves 6–9, 30–45 × 1–2 cm, linear-lanceolate, smooth, entire, bright green spotted with brown on the back, reddish near the base; stem-leaves long-acuminate with clasping bases. Spike 50–100 cm, rigid, bracts leaf-like. Each flower pair subtended by 1 bract only. Flowers 3.5–6 cm, the tube 2.5–3.5 cm, the lobes 1–2 cm, all waxy white, strongly fragrant. *Mexico.* H5–G1. Summer.

Normally seen in cultivation as a double flowered variant.

**2. P. geminiflora** (Llave & Lexarza) Rose (*Bravoa geminiflora* Llave & Lexarza). Illustration: Botanical Magazine, 4741 (1853); Contributions from the US National herbarium 8: t. 4 (1903).
Tuber shallow. Basal leaves 30–50 × 1–1.3 cm, linear, subulate, green; stem-leaves lanceolate, erect and adpressed to the stem. Spike 8–40 cm, each flower pair subtended by a bract and 2 bracteoles. Flowers *c.* 2 cm, strongly bent downwards near the base, pale red or orange, the perianth-tube more than 3 times longer than the lobes, fragrant. *Mexico.* H5–G1. Summer.

**8. DORYANTHES** Correa
*C.J. Couper*
Large, monocarpic or perennial leaf-succulents, without fleshy underground stems. Trunk absent. Leaves in a dense basal rosette, persistent, sword-shaped, the apex solid and cylindric but drying and soon falling. Inflorescence to 6 m, a terminal, spherical panicle of short spikes, the axis bearing much reduced, basally sheathing leaves. Flowers bisexual, bright red, often replaced by bulbils. Perianth radially symmetric, formed from 6 lobes which are united at their extreme bases into a short tube, the lobes sickle-shaped. Stamens 6, inserted at the bases of the perianth-lobes, anthers erect, basifixed, not projecting from the perianth. Ovary inferior, 3-celled, with many ovules in vertical rows. Style elongate, stigma capitate. Capsule with numerous, flattened, brown, kidney-shaped seeds.

A small genus from Australia. Cultivation is in general as for *Agave* (p. 278), but the plants prefer a richer soil. Propagation is by seed, by bulbils, or by suckers which are sparingly produced after flowering.

1a. Leaves 1.8–2.4 m, ribbed, the cylindric point at the apex 10–15 cm
**1. palmeri**

b. Leaves 1.2–1.8 m, smooth, the cylindric point at the apex 5–7.5 cm
**2. excelsa**

**1. D. palmeri** Hill. Figure 36(2), p. 272. Illustration: Flore des Serres, ser. 2, 10: t. 2097–8 (1874); Botanical Magazine, 6665 (1883); Revue Horticole for 1891: 548; Gartenflora 61: 324 (1912).
Leaves 1.8–2.4 m × 10–15 cm, strongly ribbed, narrowed downwards to a winged, stalk-like base which is *c.* 2.5 cm broad; the cylindric apex of the leaf 10–15 cm. Flowering axis to 4 m, panicle 1.8–3 m, bearing numerous, short, ascending leaves. Bracts *c.* 5 cm, ovate, bright red. Perianth-lobes lanceolate, not longer than the ovary, bright orange-red outside, white inside. Stamens a little shorter than the perianth-lobes, anthers *c.* 1.2 cm, linear-oblong. *Australia (Queensland).* G1.

**2. D. excelsa** Correa. Illustration: Botanical Magazine, 1685 (1814); Gartenflora 42: t. 421 (1864); Revue Horticole for 1865: 465–6 and for 1891: 548; Flore des Serres, ser. 2, 8: t. 1912–13 (1869).
Leaves 1.2–1.8 m × 7.5–10 cm, smooth, narrowed gradually to a winged base which is *c.* 2.5 cm broad, the cylindric point at the apex 5–7.5 cm. Flowering axis

to 6 m, bearing numerous ascending leaves. Bracts *c.* 7.5 cm, bright red. Perianth-lobes 6–7.5 cm, linear, longer than the ovary. Stamens almost as long as perianth-lobes, anthers linear, *c.* 2.5 cm. Ovary 3-angled in cross-section. *Australia (Queensland, New South Wales).* G1.

**9. NOLINA** Michaux
*C.J. Couper*
Dioecious trees with woody trunks. Leaves many, persistent, borne in a rosette at the end of the trunk, narrowly linear, stiff, usually erect, more rarely ascending or drooping, often with rough margins. Panicle compound, at the apex of an almost leafless scape, its main branches subtended by tapering, triangular bracts. Flower-stalks jointed near their bases. Flowers mostly unisexual (occasionally a few bisexual), radially symmetric. Perianth of 6 segments which are whitish, persistent, papery, 1-nerved and small. Stamens 6, the filaments short and slender but dilated at their bases. Ovary superior, deeply 3-lobed, 3-celled, ovules 2 in each cell. Fruit a thin-walled, often inflated capsule with 1 seed per cell, often only 1 cell developing. Seeds spherical to oblong, wrinkled.

A genus of about 25 species from the southern USA and Mexico. Cultivation as for *Yucca* (p. 273).
Literature. Trelease, W.T., The Desert Group Nolineae, *Proceedings of the American Philosophical Society* 50: 404–43 (1911).

1a. Leaves 90–100 cm, tip entire, margins splitting off as threads
**1. bigelovii**

b. Leaves more than 2 m, tip frayed, margins rough
**2. longifolia**

**1. N. bigelovii** (Torrey) Watson (*Dasylirion bigelovii* Torrey; *Beaucarnea bigelovii* (Torrey) Baker). Illustration: Contributions from the US National Herbarium 16: t. 109 (1916); Jacobsen, Handbook of succulent plants, 697 (1960); Wiggins, Flora of Baja California, 833 (1980).
Trunk to 3 m, its base club-shaped, to 70 cm in diameter, rough-barked, much narrowed towards the apex, with few, short branches and leaves in rosettes at their ends. Leaves 90–100 × 2–2.5 cm, bases triangular, the rest linear, slightly tapering, not keeled on the backs; margins splitting off as brown fibres, the tip entire, not frayed and brush-like. Perianth-segments of female flowers 2.5–3 mm, the outer 3 strongly reflexed. Fruit-stalks half as long as the capsules. *USA (Arizona) to Mexico (Baja California, Sonora).* G1. Spring–summer.

**2. N. longifolia** (Schultes) Hemsley (*Yucca longifolia* Schultes; *Dasylirion longifolium* (Schultes) Zuccarini; *Beaucarnea longifolia* (Schultes) Baker). Illustration: Gardeners' Chronicle **16**: 67 (1894); Revue Horticole for 1911: 206; Jacobsen, Handbook of succulent plants, 689 (1960).
Trunk more than 2 m, rough-barked, swollen at the base with a thick, corky ring, narrowing towards the apex, with a few short branches and a dense crown of leaves. Leaves more than 2 m × 2.5 cm, long-tapering, thin, firm, margins minutely rough, the apex frayed and brush-like. Flowering axis to 2 m. Perianth-segments of female flowers *c.* 1.5 mm, white. Fruit-stalks as long as the inflated capsules. *Mexico*. G1. Spring–summer.

## 10. BEAUCARNEA Lamarck
*C.J. Couper*
Dioecious trees. Trunk elongate, irregularly scarred by the remains of leaf-bases, its base strongly inflated, more deeply scarred than the part above. Leaves in rosettes, persistent, broadly linear, stiff, acuminate, grooved, the bases dilated and stem-clasping, completely hairless, margins smooth or slightly rough. Flowers unisexual (rarely a few bisexual), radially symmetric, numerous, very small, in a wide, terminal panicle. Each flower shortly stalked, white, slightly fragrant, subtended by a bracteole, the male flowers falling quickly. Perianth-segments 6. Ovary superior, 1-celled, with 2–3 ovules. Fruit a 3-winged capsule.

A genus of about 6 species from the southern USA and Mexico, formerly included in *Nolina*. Cultivation as for *Yucca* (p. 273).

1a. Leaves 90–180 cm, green, margins
   smooth                              **1. recurvata**
 b. Leaves less than 90 cm, glaucous or
   grey, margins rough                          2
2a. Leaves 60–90 cm × 6–12 mm,
   margins yellowish, slightly rough;
   capsule *c.* 8 × 6 mm               **2. stricta**
 b. Leaves 45–50 cm × 3–5 mm, at least
   the margins glaucous, rough; capsule
   *c.* 9 × 9 mm                       **3. gracilis**

**1. B. recurvata** Lemaire (*Nolina recurvata* (Lemaire) Hemsley; *B. tuberculata* Roezl; *N. tuberculata* invalid). Figure 36(5), p. 272. Illustration: Illustration Horticole **8**: 58 (1961); Flore des Serres, ser. 2, **18**: 26 (1869); Gardeners' Chronicle **46**: 4 (1909); Jacobsen, Handbook of succulent plants, 698 (1960).
Plant to 9 m. Trunk 4–6 m, 50 cm or more

in diameter at the base, with few branches above. Leaves 90–180 × *c.* 2 cm, linear, tapering slightly, curved, thin, flat, though grooved, green, margins smooth. Flowering axis bearing flowering branches almost to the base. Perianth *c.* 1.5 mm. *Mexico*. G1.

**2. B. stricta** Lemaire (*Nolina stricta* (Lemaire) Ciferri & Giacomini; *B. glauca* Roezl).
Plant to 10 m. Leaves 60–90 cm × 6–12 mm, stiffly spreading, keeled, pale or glaucous grey, margins yellow, slightly rough, the surface with papillose grooves. Flowering axis bearing flowering branches almost to the base. Perianth *c.* 1.5 mm. Fruit 8 × 6 mm, seeds 3–4.5 mm. *Mexico*. G1.

**3. B. gracilis** Lemaire (*Nolina gracilis* (Lemaire) Ciferri & Giacomini; *B. oedipus* Rose).
Like *B. stricta* but leaves 45–50 cm × 3–6 mm, glaucous, very rough on the margins. Fruit *c.* 9 × 9 mm, seeds *c.* 3 mm. *Mexico*. G1.

## 11. DASYLIRION Zuccarini
*C.J. Couper*
Dioecious shrubs or trees. Trunk short, stout, or absent. Leaves in a dense, terminal rosette, persistent, narrowly linear, hard and striped, broad and concave at the base, margins with hooked spines, apex entire or fibrous. Flowering axis tall, bearing dry, lanceolate, bract-like leaves. Panicle narrow, dense; bracts membranous, persistent, flower-stalks short, jointed near the apex. Flowers unisexual, bell-shaped, radially symmetric. Perianth-segments 6, oblong-ovate, whitish, all equal, finely toothed. Stamens 6, projecting. Ovary superior (rudimentary in male flowers), 1-celled with 2 or 3 ovules, style short, erect, stigmas 3. Fruit dry, indehiscent, 3-angled, the angles often winged, with 1 obtusely triangular seed.

A genus of about 15 species from southern USA to Mexico, easily grown in well-drained soil. Propagation is by seed. Literature: Trelease, W.T., The desert group Nolineae, *Proceedings of the American Philosophical Society* **50**: 404–43 (1911).

1a. Trunk 80 cm or more                        2
 b. Trunk less than 80 cm, or absent
   altogether                                  4
2a. Leaves to 6 mm broad, apex entire
                              **3. longissimum**
 b. Leaves more than 8 mm broad, apex
   fibrous                                     3
3a. Leaves more than 2.5 cm broad,
   glaucous                       **1. wheeleri**

 b. Leaves less than 1.8 cm broad, pale
   green, scarcely glaucous
                              **2. acrotrichum**
4a. Leaf-apex entire        **7. glaucophyllum**
 b. Leaf-apex fibrous                          5
5a. Leaves less than 1.5 cm broad, bright
   green                        **6. graminifolium**
 b. Leaves more than 1.5 cm broad,
   glaucous                                    6
6a. Leaves rough; marginal spines
   1–3 mm, deep yellow
                              **5. serratifolium**
 b. Leaves smooth; marginal spines
   3–4 mm, orange or reddish
                              **4. leiophyllum**

**1. D. wheeleri** Watson. Illustration: Plant World **10**: 255 (1907); De Wildeman, Icones selectae Horti Thenensis **6**: t. 225 (1908); USDA Bulletin No. 728, t. 5 (1918).
Trunk at least 1 m. Leaves *c.* 100 × 2.5–4 cm, glaucous, margins finely toothed and with hooked, reddish brown spines which are 1.5–3 mm long and 1–2 cm apart. Flowering axis 2.5–5 m, panicle 1.5–2.5 m, the branches flexuous and hanging. Bracts cup-shaped, irregularly toothed, 2–3 mm. Perianth *c.* 2 mm. Fruit 7–9 × 4–8 mm, narrowly reversed heart-shaped, deeply notched at the apex. Seeds 3–4 mm, 3-angled. *USA (Arizona to Texas), N Mexico*. G1. Summer.

**2. D. acrotrichum** Zuccarini (*D. gracile* Planchon). Illustration: Botanical Magazine, 5030 (1858); Flore des Serres, ser. 2, **4**: t. 1448 (1861); Gartenflora **30**: 24 (1881).
Trunk 90–150 cm, robust. Leaves 60–100 cm × 9–18 mm, the outer recurved, all pale green, margins distinctly and finely toothed between the hooked, pale yellow, brown-tipped spines which are 1–2 mm long; apex divided into 20–30 fibres. Flowering axis 2.5–4 m. Panicle 1.2–1.5 m, dense and cylindric, bracts *c.* 2 mm, ovate, entire. Perianth *c.* 2 mm. Fruit *c.* 8 × 6–8 mm, obovoid, shallowly notched, wings broadening upwards, free from the style. *Mexico*. G1. Summer.

**3. D. longissimum** Lemaire (*D. quadrangulatum* Watson). Illustration: Parey's Blumengärtnerei, 307 (1958).
Trunk 90–200 cm. Leaves very numerous, 1.2–1.8 m × *c.* 6 mm, spreading in all directions, 4-angled (the upper and lower surfaces raised to low keels), margin smooth, apex entire. Flowering axis 1.5–5 m. Fruit 7–9 mm, scarcely notched. *Mexico*. G1. Summer.

**4. D. leiophyllum** Engelmann.
Trunk less than 80 cm. Leaves *c.*
$100 \times 2.5–3$ cm, smooth, glossy green or
somewhat glaucous, margin with spines
3–4 mm long and 1–1.5 cm apart,
recurved, becoming orange or reddish;
apex fibrous. Panicle to 3 m. Perianth
greenish, to 2 mm. Fruits $6–9 \times 2–6$ mm,
broadly ellipsoid, openly and deeply
notched at the apex, style thick. Seeds *c.*
3 mm. *USA (Texas, New Mexico), Mexico.*
G1. Summer.

**5. D. serratifolium** (Schultes) Zuccarini
(*Yucca serratifolia* Schultes; *D. laxiflorum*
Baker). Illustration: Report from the
Missouri botanical garden 14: t. 12 (1903).
Plant trunkless or almost so. Leaves
$60–100 \times 1.5–3$ cm, glaucous, rough,
margins finely toothed and with hooked,
deep yellow spines 1–3 mm long and
2–3 cm apart; apex fibrous. Panicle loose.
Perianth *c.* 2 mm. Fruit $6–8 \times 6–8$ mm,
more or less spherical, broadly winged,
apex notched, style as long as the notch.
*SE Mexico.* G1. Summer.

**6. D. graminifolium** (Zuccarini) Zuccarini
(*Yucca graminifolia* Zuccarini).
Trunk short, less than 80 cm. Leaves
$90–120 \times 1.2–1.4$ cm, linear, long-
acuminate, bright green, glossy, margin
finely toothed between the spines which are
1–2 mm long, horny and yellowish; apex
of 6–8 spreading fibres. Flowering axis
2.4–2.7 m. Panicle narrow. Perianth *c.*
2 mm. Fruit ellipsoid, 6–9 mm. *Mexico.* G1.
Summer.

**7. D. glaucophyllum** Hooker. Illustration:
Botanical Magazine, 5041 (1858); Revue
Horticole for 1911: 87.
Trunk at most 30 cm. Leaves
$60–120 \times 1.2–1.8$ cm, linear, acuminate,
intensely glaucous, marginal spines
1–2 mm, hooked, deep yellow, the margin
finely toothed between them; apex entire.
Panicle to 1.2 m, dense. Perianth *c.* 2 mm,
greenish white, lobes red at their tips. Fruit
6–8 mm broad, style half as long as the
notch at the apex. *E Mexico.* G1. Summer.

**12. DRACAENA** Linnaeus
*J.J. Bos & J. Cullen*
Plants herbaceous or woody, sometimes
forming trees. Roots usually orange. Leaves
usually spirally arranged, occasionally in
false whorls, stalked or stalkless, green or
frequently variegated with white, silver or
yellow, sometimes with red margins;
venation always parallel. Inflorescence a
panicle of varying form, with elongate

branches or the branches short and
condensed, the whole sometimes head-like;
major bracts sometimes large, flowers 2 or
more to each floral bract. Flowers bisexual,
usually small, stalkless or with short stalks,
usually fragrant and opening at night.
Perianth radially symmetric, with 6 lobes
which are united into a long or short tube
at the base, rarely all almost free, the lobes
spreading or recurving, white, greenish or
with red midribs to each lobe. Stamens 6,
inserted at the top of the perianth-tube,
filaments slender or thickened. Ovary
3-celled, each cell with 1 ovule. Fruit a
spherical or lobed, red, orange or yellow
berry containing 1–3 seeds.

A genus of about 60 species, mostly from
Africa. They are mainly grown for their
patterned or variegated leaves rather than
for their flowers, and tend to be retained
only while in their juvenile stages (many
which are popular as house plants are
capable of becoming much too large for
most sites); for several species apparently
nothing is recorded of their flowers and
fruits. Hence, the identification of cultivated
material has to be based, as far as possible,
on the leaves; this is not entirely
satisfactory. Some plants are grown merely
under cultivar names (e.g. 'Song of India')
and are, perhaps of hybrid origin.

Plants grow well in glasshouses or in
heated domestic rooms or offices; they
require rather large pots if they are to reach
a reasonable size. A compost of peat, sand
and loam is suitable, and the plants should
be placed in a position where they receive
good light. They are readily propagated by
cutting short pieces of stem and laying
them on compost in a plant house with
bottom heat.
Literature: Bos, J.J., Dracaena in West
Africa, *Belmontia* 17(80): 1–126 (1985).

1a. Leaves sword-shaped, with clasping,
     flaring bases; trees or large shrubs   2
  b. Leaves more or less elliptic, with long
     or short stalks; small shrubs          9
2a. Freely branching shrub; most leaves
     less than 40 cm                        3
  b. Single-stemmed trees or small trees,
     generally not branching before
     flowering; most leaves more than
     40 cm                                  4
3a. Leaves with distinct red margins
                              **9. marginata**
  b. Leaves without red margins
                                **7. reflexa**
4a. Leaves scarcely narrowed towards the
     flaring base, glaucous; leaf-scars
     usually with patches of dark red resin
                                    **1. draco**

  b. Leaves gradually narrowed towards
     the flaring base, green or variously
     variegated; resin absent              5
5a. Leaves to 3 cm wide with distinctly
     wavy margins towards the base
                            **3. umbraculifera**
  b. Leaves wider, margins not wavy        6
6a. Leaves with a distinct colourless (or
     rarely red) margin      **6. aletriformis**
  b. Leaves without distinct colourless or
     red margin                            7
7a. Flowers on distinct, long-persisting
     stalks in groups of 3–5 towards the
     ends of the branches; leaves not
     variegated                  **2. arborea**
  b. Flowers almost stalkless in well-
     spaced, many-flowered clusters; leaves
     variegated or not                     8
8a. Leaves dark green, most less than
     5 cm wide              **5. deremensis**
  b. Leaves light or bright green, most
     5 cm or more wide        **4. fragrans**
9a. Leaves variegated                     10
  b. Leaves not variegated                15
10a. Variegation of leaves consisting of
     irregular transverse bands of grey-
     white                    **11. goldieana**
  b. Variegation of leaves not as above
                                          11
11a. Variegation consisting of longitudinal
     bands of white or pale yellow
                              **8. sanderiana**
  b. Variegation consisting of separate or
     merging dots or rings                12
12a. Leaf-stalk almost as long as to longer
     than the blade; variegation consisting
     of few to many light green,
     transverse, oval blotches
                             **15. phrynioides**
  b. Leaf-stalk much shorter than the
     blade; variegation not as above      13
13a. Inflorescence unbranched
                               **12. surculosa**
  b. Inflorescence branched                14
14a. Inflorescence branches well
     developed, loosely or densely covered
     with clusters of flowers; leaf
     variegation of rounded, yellow dots
                                **10. elliptica**
  b. Inflorescence branches short, bearing
     distant, many-flowered clusters;
     variegation of yellow or whitish dots
                            **13. × masseffiana**
15a. Leaf-stalk not more than one-sixth of
     the blade length                     16
  b. Leaf-stalk more than one-sixth of the
     blade length                         17
16a. Inflorescence branched   **10. elliptica**
  b. Inflorescence not branched
                               **12. surculosa**

17a. Leaf-stalk narrow, terete
                           **14. aubryana**
  b. Leaf-stalk strap-shaped, gradually
     merging into the leaf sheath
                          **8. sanderiana**

**1. D. draco** (Linnaeus) Linnaeus.
Illustration: Botanical Magazine, 4571
(1881); Graf, Exotica, edn 8, 1073, 1078,
(1973); Bramwell & Bramwell, Wild flowers
of the Canary islands, t. 322 (1974);
Gartenpraxis for 1976: 560.
Tree to 18 m or more with a thick trunk,
unbranched till the first flowering, later
with radiating branches which repeat the
same pattern. Leaves stalkless, arching or
reflexed, in dense rosettes at the ends of the
branches, 30–60 × 2.5–4 cm, sword-
shaped, base flaring and clasping, apex
acute, glaucous, with no obvious midrib;
leaf-scars with patches of dark red resin.
Inflorescence a large, erect, terminal,
much-branched panicle with flowers in
groups of 4 or 5. Perianth *c.* 8 mm, lobed
almost to the base, greenish white outside,
white inside. Berries orange-yellow. *Atlantic
islands.* H5–G1.
  A striking tree, particularly when old.

**2. D. arborea** (Willdenow) Link.
Illustration: Gartenflora **46**: t. 1438
(1897); Belmontia **17**(80): 24 (1985).
Tree to 20 m with a distinct trunk,
ultimately branching above. Leaves
stalkless, clustered at the ends of the
branches, spreading or recurved,
50–150 × 4–10 cm, sword-shaped, with a
prominent midrib on the lower surface.
Panicles hanging, shortly branched, up to
2 m, flowers in 3s, conspicuously stalked.
Perianth *c.* 2 cm, creamy white, with a
distinct tube at the base. Berry orange or
red. *Tropical W Africa south to Angola.*
G1–2.

**3. D. umbraculifera** Jacquin. Illustration:
Loddiges' Botanical Cabinet 3: t. 289
(1818).
Plant tree-like, 1–10 m, trunk erect and
unbranched. Leaves stalkless, borne in a
dense, umbrella-like rosette at the top of the
trunk, spreading and recurving,
60–100 × up to 3 cm, linear-lanceolate,
gradually narrowed to the flaring and
clasping base, with a distinct midrib on
both surfaces, margins distinctly wavy
towards the base. Raceme terminal, head-
like, borne immediately above the sheaths
of the leaves, flowers in clusters at the ends
of the branches. Perianth 5–6 cm, white
tinged with red, tube longer than the lobes.
*?Mauritius.* G2.

Introduced into Europe from a plant
cultivated in Mauritius; its wild origin is
uncertain.

**4. D. fragrans** (Linnaeus) Ker-Gawler.
Figure 36(8), p. 272. Illustration: Botanical
Magazine, 1081 (1808); Graf, Exotica, edn
8, 1072 (1973); Hay et al., Dictionary of
indoor plants in colour, t. 196 (1974); Clay
& Hubbard, The Hawaii garden, tropical
exotics, 107 (1977).
Plant shrubby, erect, to 15 m but generally
much smaller in cultivation, stem
ultimately branched. Leaves borne in the
upper part of the stem, stalkless, spreading
and recurving, 50–150 × 5–10 cm,
lanceolate, gradually narrowed to the
flaring base, bright, pale green, the midrib
conspicuous only beneath. Panicle
terminal, erect, arching or hanging, its axis
conspicuously zig-zag, the short branches
bearing spherical clusters of flowers at their
ends and often a further cluster near the
base. Flowers very fragrant. Perianth
1.5–2.5 cm, white. Berry orange-red.
*Tropical Africa, from Gambia to Ethiopia,
south to Angola.* G2.
  Variegated cultivars (e.g. 'Lindenii',
'Victoria' and 'Massangeana') are most
frequently grown.

**5. D. deremensis** Engler. Illustration: Perry,
Wild flowers of the world, 17 (1972); Graf,
Exotica, edn 8, 1072, 1076–7 (1973); Hay
et al., Dictionary of indoor plants in colour,
t. 195 (1974).
Plant shrubby, erect, to 5 m, but often
much smaller in cultivation. Leaves
spreading, to 45 × 5 cm, tapering towards
the flaring base, lanceolate, dark green,
usually variegated in cultivated plants.
Inflorescence a large, erect, branched
panicle to 50 cm, the flowers in clusters of
20–30 together. Perianth *c.* 2 cm, tube *c.*
half the total length, dark red outside,
white within, strongly scented. *E tropical
Africa.* G2.
  'Warneckii', which has leaves variegated
with 2 longitudinal white stripes, and
'Bausei' with a central white band, are the
most commonly grown variants; other
cultivars are available.

**6. D. aletriformis** (Haworth) Bos (*D.
hookeriana* koch; *D. rumphii* (Hooker) Regel;
*D. latifolia* Regel). Illustration: Kaier, Indoor
plants in colour, 52 (1961); Graf, Exotica,
edn 8, 1073, 1079 (1973).
Plant shrubby, erect, to 2 m or more, often
much smaller in cultivation. Leaves
spreading and recurved, 60–75 × 5–10 cm,
lanceolate, tapering towards the flaring
base, midrib inconspicuous on the upper

surface, more prominent beneath; margins
narrow and white-translucent or rarely
reddish. Inflorescence a large, terminal
panicle with elongate branches, the flowers
borne in clusters of 2 or 3. Perianth
2.5–3 cm, greenish white, the tube up to
half the total length. Filaments expanded.
Berry orange. *SE Africa.* G1–2.
  Variable in the breadth of the leaves and
the colour of their edges. A few cultivars
are available (e.g. 'Latifolia', 'Variegata').
Plants in cultivation as *D. concinna*
probably belong to this species.

**7. D. reflexa** Lamarck. Illustration:
Botanical Magazine, 6327 (1877); Graf,
Exotica, edn 8, 1078, 1104 (1973) – as
Pleomele reflexa; Clay & Hubbard, The
Hawaii garden, tropical exotics, 111
(1977).
Plant shrubby, erect, to 5 m in the wild,
usually much less in cultivation, often
branched. Leaves very variable, borne
towards the top of the stem, spreading and
recurved, to 20 × 5 cm, narrowly lanceolate
to elliptic, tapering towards the flaring base,
midrib inconspicuous above, convex
beneath. Inflorescence a terminal panicle, its
branches racemose. Perianth 1.2–2 cm,
greenish outside, white inside, the tube to
one-quarter of the total length. Berry
orange-red. *Madagascar, Mauritius.* G2.

**8. D. sanderiana** Masters. Illustration:
Illustration Horticole **40**: t. 175 (1893) –
as D. thalioides var. foliis variėgatis; Graf,
Exotica, edn 8, 1075, 1077 (1973); Clay &
Hubbard, The Hawaii garden, tropical
exotics, 113 (1977).
Plant shrubby, erect, to 1.5 m. Leaves
spreading and recurved, 18–25 × 1–4 cm,
lanceolate, tapering into a strap-shaped
stalk, glossy green broadly variegated with
longitudinal, silver-white stripes, rarely not
variegated, midrib inconspicuous above.
*Cameroun.* G2.
  An obscure species; the variant with
variegated leaves was only once collected in
the wild, and it has not flowered in
cultivation.

**9. D. marginata** Lamarck. Illustration: Graf,
Exotica, edn 8, 1077 (1973); Hay et al.,
Dictionary of indoor plants in colour, t. 198
(1974).
Plant shrubby, erect and branched, to 2 m.
Leaves spreading and recurved, 30–40 ×
1–2 cm, linear-lanceolate, tapering
towards the flared and clasping base, red-
margined, midrib inconspicuous above,
convex beneath. Panicle terminal. Flowers
white. Berries yellow. *Île de Réunion.* G2.

A doubtful species, apparently only once collected in the wild. A variant with white-variegated as well as red-margined leaves ('Tricolor') is sometimes grown.

**10. D. elliptica** Thunberg. Illustration: Botanical Magazine, 4787 (1854).
Plant shrubby, to 1 m, erect but with spreading branches. Leaves spreading and recurved, 7–20 × 4–6 cm, ovate or ovate-lanceolate, acute, abruptly rounded into short stalks, green marked with irregular yellow spots. Inflorescence a terminal panicle with elongate branches, the flowers borne in groups of 2 or 3. Perianth 1.2–2.5 cm, yellowish green, the tube forming up to half the total length. Berries red. *India, Indonesia (Java, Sumatra).* G2.

The variant with leaves spotted with yellow ('Maculata') is most commonly grown.

**11. D. goldieana** Lindley. Illustration: Botanical Magazine, 6630 (1882); Graf, Exotica, edn 8, 1075, 1079 (1973); Clay & Hubbard, The Hawaii garden, tropical exotics, 109 (1977).
Shrub to 2 m, stem erect. Leaves ovate, ovate-lanceolate or ovate-elliptic, 15–30 × 5–13 cm, spreading, tapering into a distinct stalk which is up to one quarter of the length of the blade, dark green variegated with silvery blotches forming transverse bars. Inflorescence a congested, head-like, terminal panicle borne between the sheaths of the upper leaves so that it appears to be without a common stalk (more rarely an elongate panicle). Perianth *c.* 2.5 cm, the tube longer than the spreading lobes, white. *Nigeria to Gabon.* G2.

**12. D. surculosa** Lindley (*D. godseffiana* Masters). Illustration: Edwards's Botanical Register **14**: t. 1169 (1828); Botanical Magazine, 7584 (1898); Graf, Exotica, edn 8, 1074 (1973); Hay et al., Dictionary of indoor plants in colour, t. 197 (1974).
Shrub *c.* 1 m, the branches erect or arching. Leaves spreading, borne in false whorls of 2–4, usually elliptic, acute, 5–20 × 2–6 cm, abruptly tapered to short but distinct stalks, dark green spotted with small, round, yellow or white dots. Inflorescence a terminal raceme, the flowers borne in groups of 1–5, or all flowers in a terminal cluster. Perianth 1.5–2.5 cm, white, the tube forming about half of the total length. Berries orange-scarlet. *Tropical Africa.* G2.

There are 2 varieties in cultivation: var. **surculosa** with distinct, pure white spots is usually found under the name *D. godseffiana* (when very densely spotted, as 'Punctulata' or 'Florida Beauty') and var. **maculata** J.D. Hooker, with very pale green spots or rings.

**13. D. × maseffiana** Bos. Illustration: Gardeners' Chronicle **121**: f. 111 (1947); Notes from the Royal Botanic Garden, Edinburgh **40**: 536 (1983).
Shrub to 2 m or more, branched. Leaves elliptic, tapering towards the base to short stalks which are stem-clasping, deep green covered with small, white or yellow, rounded spots. Inflorescence a panicle with short side branches, bearing distant, many-flowered clusters. Flowers white, the lobes with green midribs. *Garden origin.* G2.

A hybrid between *D. surculosa* and *D. fragrans*, raised in Puerto Rico but currently available in Europe. All the cultivated material belongs to 'Pennock'.

**14. D. aubryana** Morren (*D. humilis* Baker; *D. thalioides* Regel). Illustration: Graf, Exotica, edn 8, 1076 – as D. kindtiana, 1078 & 1104 – as Pleomele thalioides (1973); Clay & Hubbard, The Hawaii garden, tropical exotics, 115 (1977).
Unbranched shrub. Stems erect, short, covered by the overlapping leaf-sheaths. Leaves 15–60 or more × 5–12 cm, linear-lanceolate, arching, abruptly and sometimes asymmetrically rounded into a distinct, terete stalk which is almost as long as the blade. Inflorescence terminal, erect, a 15–50 cm, spike-like raceme or sparsely branched panicle with the flowers in groups of 1–3. Perianth 1–5 cm, the tube shorter than the lobes, white with red lines outside, white within. *Tropical W Africa, south to Angola.* G2.

**15. D. phrynioides** Hooker. Illustration: Botanical Magazine, 5352 (1862); Graf, Exotica, edn 8, 1076 (1973).
Unbranched shrub. Stems short, covered by overlapping leaf sheaths. Leaves 10–75 × 3–9 cm, ovate, acuminate, abruptly narrowed into an erect stalk which is almost as long as the blade, holding the leaves well above the inflorescence. Inflorescence terminal, often nodding, borne at the top of the stem, between the sheaths of the upper leaves. Bracts ovate, overlapping, often tinged with red, sometimes abruptly tapered into long 'tails'. Perianth white, *c.* 2 cm, the tube distinctly longer than the lobes. Berries scarlet, distinctly horned. *Tropical W Africa.* G2.

This is a rather difficult species to grow well.

**13. SANSEVIERIA** Thunberg
*C.J. Couper*
Herbaceous, perennial leaf-succulents with creeping rhizomes, usually without obvious aerial stems. Leaves persistent, fleshy or leathery, flat, channelled, half-cylindric, cylindric or laterally compressed, rigid or flexible, margins entire and smooth or cartilaginous. Raceme or panicle basal, the axis with scale-like leaves or sheaths at the base. Flowers bisexual, solitary or in clusters, fragrant, borne on stalks which are jointed near the middle or towards the apex. Perianth of 6 lobes united below into a distinct tube, lobes narrow, radially symmetric. Stamens 6, projecting from the perianth. Ovary superior, 3-celled with 1 ovule in each cell. Style slender, thread-like. Fruit a berry containing 1–3 bony seeds.

A genus of over 50 species from S & E Africa, Arabia and India. They are easily grown, several of them being popular house plants which can stand a surprising amount of neglect. Propagation is by division of the rhizome or by leaf cuttings. Literature: Brown, N.E., Sansevieria, a monograph of all the known species, *Kew Bulletin* for 1915; Morgenstern, K.D., *Sansevierias* (1979).

1a. Plants with distinct, leafy stems which are 30 cm or more

    **1. arborescens**

  b. Plants without stems or stems inconspicuous, much less than 30 cm

    2

2a. Leaves terete in section, without a distinct groove running from near the apex almost to the sheath    3

  b. Leaves flat, or if terete, then with a distinct groove running from near the apex almost to the sheath    4

3a. Stem short, but branching above ground; leaves 8–12 to a rosette

    **2. gracilis**

  b. Stems absent; leaves 3 or 4 to a rosette    **4. cylindrica**

4a. Leaves terete but distinctly grooved and with flattened sides

    **3. ehrenbergii**

  b. Leaves fleshy but flat, not as above    5

5a. Leaves with a distinct, red or white, horny margin    6

  b. Leaves without a margin distinctly different from the substance of the leaf    8

6a. Rhizome very extensive (to 1 m or more), whitish or pale green above ground; leaves elliptic, oblong or broadly lanceolate, glaucous and

marked with inconspicuous bars of pale green on both surfaces
**11. grandis**

  b. Rhizomes not extensive, brownish when above ground; leaves lanceolate, dull, dark green marked with paler bands or blotches which disappear with age, on both surfaces    7

7a. Leaves to 45 cm, apex with a withered, whitish point which is *c.* 1.6 cm    **10. hyacinthoides**

  b. Leaves 45 cm or more, apex with a green, needle-like point which is 3–6 mm    **9. metallica**

8a. Leaves not tapering towards the base    **5. roxburghiana***

  b. Leaves tapering towards the base    9

9a. Stem 2.5–12.5 cm; perianth-tube pale pinkish white, lobes tinged with mauve within, dull mauve or purplish outside    **7. parva**

  b. Stem absent or completely hidden by leaf bases; flowers dull pink outside, whitish inside, or pale greenish both outside and in    10

10a. Leaves 6–20 to a rosette; flowers dull pink or purplish outside    **6. dooneri**

  b. Leaves 1–5, rarely to 6, to a rosette; flowers greenish white outside    **8. trifasciata**

**1. S. arborescens** Gerome & Labroy. Illustration: Kew Bulletin for 1915: 201; Morgenstern, Sansevierias, t. 6 (1979). Stem 90–120 cm, erect, terete, leafy throughout. Leaves 20–45 × 2–4.5 cm, narrowly lanceolate or linear-lanceolate, spreading or recurved, smooth, flat or concave, narrowing slightly to the sheathing base, grass-green with no markings; margins recurved or wavy, hardened, whitish or reddish; tip 8–25 mm, stout, needle-like, green and hardened below, sharp and pale brown at the apex. Flowers unknown. *Tropical E Africa.* G1–2.

**2. S. gracilis** N.E. Brown. Illustration: Kew Bulletin for 1915: 204; Lamb, Cacti and other succulents **4**: t. 698 (1975); Morgenstern, Sansevierias, t. 34–5 (1979).
Stems 2.5–7.5 cm, branching above ground. Leaves 8–12 to a rosette, 22–75 cm, ascending or spreading, firmly flexible, smooth, channelled for 5–12 mm at the base, the rest cylindric, terete, gradually tapering, deep green with narrow, darker bands on the younger leaves; tip 2–6 mm, very acute, white or brownish. Raceme 6–7.5 cm, loose, spike-

like. Flowers in pairs, white, the tube 2–2.5 cm, the lobes 1–1.2 cm, linear, obtuse. *Tropical E Africa.* G1–2.

**3. S. ehrenbergii** Baker. Illustration: Hooker's Icones Plantarum **23**: t. 2269 (1893); Morgenstern, Sansevierias, t. 26–9 (1979).
Stem to 25 cm. Leaves 5–9 to a rosette, 75–180 × 2.5–4 cm, terete, laterally compressed with flattened sides, rounded on the back, with a triangular channel down the face and 5–9 shallow grooves down the sides and back, tapering upwards, dark green with blackish green grooves, but without transverse markings; margins of channel acute, reddish brown with white membranous edges, the tip 6–12 mm, hard and spine-like. Panicle to 2 m. Flowers 4–7 in a cluster, the tube 5–6 mm, the lobes *c.* 7 mm, linear, obtuse. *Tropical E Africa.* G1–2.

**4. S. cylindrica** Bojer. Illustration: Botanical Magazine, 5093 (1859); Revue Horticole for 1861: t. 109 & for 1901: t. 70; Morgenstern, Sansevierias, t. 22 (1979). Stems and aerial branches absent. Leaves 3 or 4 in a rosette, 75–150 cm, cylindric or compressed-cylindric, 2-ranked, stiffly erect, faintly rough, 2–3 cm thick at the base, gradually tapering, green or whitish green, with transverse bands of darker green, the banding becoming fainter with age, and with numerous longitudinal, dark green or blackish green lines which are impressed in older leaves; tip 4–6 mm, hardened, acute, whitish. Raceme spike-like, 35–75 cm on a stalk 60–90 cm. Flowers 5 or 6 in a cluster, white or pinkish, tube 1.5–2.5 cm, lobes 1.5–2 cm, linear, obtuse, margins rolled outwards. *Angola; South Africa?* G1–2.

**5. S. roxburghiana** Schultes & Schultes. Figure 36(11), p. 272. Illustration: Morgenstern, Sansevierias, t. 81 (1979). Stemless. Leaves 6–24 in a rosette, the inner 20–60 × 1.2–2.5 cm, linear, ascending and slightly recurved, stiff, concave, channelled down the face, rounded on the back, green and transversely marked with regular, dark green bars on both surfaces, and with 6–11 longitudinal, dark green lines on the back; margins green becoming whitish; tip 6–50 mm, stout, subulate, soft, green. Raceme to 75 cm, spike-like. Flowers *c.* 4 in a cluster, the tube 6–7 mm, the lobes 8–9 mm, linear, obtuse. *India.* G1–2.

***S. zeylanica** (Linnaeus) Willdenow (*S. hyacinthoides* misapplied). Illustration:

Redouté, Liliacées **5**: t. 290 (1810); Edwards's Botanical Register **2**: t. 160 (1816); Morgenstern, Sansevierias, t. 137 (1979). Very similar to *S. roxburghiana*, but leaves longer, thicker and more rigid, lighter green and with more numerous longitudinal, dark green lines on the back. *Sri Lanka.* G2.

**6. S. dooneri** N.E. Brown. Illustration: Kew Bulletin for 1915: 232; Morgenstern, Sansevierias, t. 25 (1979). Stems absent or up to 5 cm, but hidden by the leaf bases. Leaves 6–20 in a rosette, 10–42 × 1.6–3 cm, strap-shaped or narrowly lanceolate, gradually narrowed to a somewhat stalk-like base, leathery but flexible, recurved-spreading, smooth, flat on the face, dark green with paler green, transverse bands on both surfaces, without longitudinal lines on the back; margins green; tip 5–60 mm, stout, soft, subulate, green. Raceme 30–40 cm, loose. Flowers 2 or 3 in a cluster, dull pink or pale purplish outside, whitish within, the tube *c.* 1.1 cm, the lobes 1.1–1.2 cm, linear. *Tropical E Africa.* G1–2.

**7. S. parva** N.E. Brown. Illustration: Kew Bulletin for 1915: 232; Lamb, Cacti & other succulents **4**: t. 700 (1975); Morgenstern, Sansevierias, t. 64–5 (1979). Stems 2.5–12 cm. Leaves 6–14 in a rosette, 20–45 cm × 8–28 mm, linear, linear-lanceolate or lanceolate, ascending or recurved, spreading, smooth, leathery but not flexible, channelled on the face, rounded on the back, narrowed into a stalk *c.* 5 cm long; surfaces marked with distinct transverse, alternating bands of darker and paler green, the markings becoming paler with age; margins green, the tip 3.5–7.5 cm, soft, subulate, green. Raceme *c.* 30 cm, loose and open. Flowers in pairs or borne individually, the tube 1–1.1 cm, the lobes 8–9 mm, linear, obtuse, dull mauve or purplish outside, white tinged with mauve inside. *Tropical E Africa.* G1–2.

**8. S. trifasciata** Prain (*S. guineensis* Gerome & Labroy not (Linnaeus) Willdenow). Figure 36(10), p. 272. Illustration: Parey's Blumengärtnerei, 312 (1958); Lamb, Cacti & other succulents **4**: t. 517 (1971); Hay et al., Dictionary of indoor plants in colour, t. 441 (1974); Morgenstern, Sansevierias, t. 99–103 (1979). Stemless. Leaves 30–120 × 2.5–7 cm, up to 5 or rarely to 6 in a rosette, linear-lanceolate or narrowly lanceolate, smooth, erect, stiff or firmly leathery, narrowing gradually to a concave-channelled stalk, more rarely (in some cultivars) smaller, stalkless and overlapping in a distinct, flat rosette; variously banded with pale and dark green, whitish or yellow, sometimes almost silver on both surfaces, margins of the same colour as the adjacent parts of the leaf; tip 3–36 mm, subulate, green. Raceme 30–75 cm. Flowers in clusters of 1–3, pale green or greenish white, tube 6–12 mm, the lobes 1.4–1.8 cm, linear, obtuse. *W tropical Africa.* G1–2.

Very variable. Numerous cultivars, based on the type of variegation and the form of the rosette, are available.

**9. S. metallica** Gerome & Labroy. Illustration: Kew Bulletin for 1915: 246; Morgenstern, Sansevierias, t. 60 (1979). Stemless. Leaves 1–4 in a rosette, 45–150 × 5–12.5 cm, elongate-lanceolate or broadly strap-shaped, erect or spreading, smooth, tapering into a deeply channelled stalk, all dull dark green, obscurely marked above, more distinctly beneath with transverse bands or blotches of paler green which often disappear with age; margins whitish or pale reddish brown, tip 3–6 mm, soft, subulate, green. Raceme 45–120 cm, spike-like. Flowers 2–4 in a cluster, white, tube 1.2–1.6 cm, lobes 1.6–2.2 cm, linear-oblong, margins rolled outwards. *Tropical Africa.* G2.

**10. S. hyacinthoides** (Linnaeus) Druce (*S. thyrsiflora* (Petagna) Thunberg; *S. guineensis* (Linnaeus) Willdenow). Illustration: Botanical Magazine, 1179 (1809); Redouté, Liliacées 6: t. 330 (1811); Revue Horticole for 1915: t. 174; Morgenstern, Sansevierias, t. 43 (1979). Stemless but with a creeping rhizome which is brown when above ground. Leaves 2–4 to a rosette, 15–45 × 2.5–6 cm, lanceolate, erect, smooth, tapering into a concave, channelled stalk, surfaces dull green with numerous transverse, pale green bands which disappear with age; margins brownish red or whitish, tip c. 1.8 cm, usually whitish and withered. Raceme 45–75 cm, spike-like. Flowers 2–6 to a cluster, tube c. 1.8 cm, greenish white, lobes c. 1.8 cm, linear, margins rolled outwards, whitish. *S Africa.* G1.

**11. S. grandis** Hooker. Illustration: Botanical Magazine, 7877 (1903); Hay & Synge, Dictionary of garden plants, t. 641 (1959); Morgenstern, Sansevierias, t. 38–41 (1979). Rhizome long and creeping, to 1 m, whitish or pale green when above ground. Stem

absent. Leaves 4 or 5 in a rosette, 30–60 × 8–15 cm, elliptic, oblong or broadly lanceolate, narrowed towards the base, stiffly leathery, flat with wavy, recurved margins, glaucous with regular, transverse bands of lighter green on both surfaces, though these more evident beneath; margins reddish brown, hard, with membranous white edges when young; tip 4–6 mm, acute, soft, white. Raceme c. 60 cm, compact, spike-like. Flowers white, tube c. 1.8 cm, lobes c. 1.8 cm, linear, obtuse, margins rolled outwards. *Origin unknown, perhaps South Africa.* G1–2.

Perhaps no more than a variant of *S. hyacinthoides.*

**14. CORDYLINE** R. Brown
*J. Cullen*
Plants woody, tufted or tree-like, branching usually falsely dichotomous. Leaves crowded in rosettes at the ends of the branches, persistent, usually lanceolate or sword-shaped, stalked, the lateral veins diverging from the conspicuous midrib at an acute angle. Flowers in dense, terminal panicles, which often appear lateral due to the growth of a new vegetative shoot from a nearby leaf axil. Flowers bisexual, bracteate (often with 2 bracts to each flower-stalk), stalked, radially symmetric. Perianth of 6 lobes united into a short tube at the base, the lobes usually reflexed. Stamens 6, anthers versatile, filaments flattened, attached to the top of the perianth-tube. Ovary superior, 3-celled, with several ovules in each cell. Fruit a small berry which dries with age; seeds several, curved, black.

A genus of about 20 species from SE Asia, Australasia, Polynesia and Hawaii, cultivated for their striking overall appearance. The genus is difficult to distinguish as a whole from *Dracaena* (p. 285), and the number of ovules in each cell of the ovary (several in *Cordyline*, one in *Dracaena*) is the only certain character, though the leaf venation (all veins parallel and entering the leaf from the base in *Dracaena*, parallel but branching from the midrib at a very acute angle in *Cordyline*) is usually characteristic; see figure 36(7 & 8), p. 272. Most species have names under *Dracaena* and these names are often found in catalogues.

The plants are easily cultivated in a good, well-drained soil. A few are grown as pot plants in glasshouses and rooms, and require similar soil and a position with good light. Propagation as for *Dracaena* (p. 285).

1a. Leaves 10–20 cm, 2–3 times longer than broad **7. haageana**
  b. Leaves 30–200 cm, more than 3 times longer than broad 2
2a. Mature leaves 1 m or more, 4–15 cm broad 3
  b. Mature leaves to 1 m, 1–10 cm broad 4
3a. Leaves 4–8 cm broad, not glaucous beneath **5. banksii**
  b. Leaves 10–15 cm broad, glaucous beneath **6. indivisa**
4a. Leaves less than 3 cm broad 5
  b. Leaves more than 3 cm broad 6
5a. Perianth-lobes all equal, white or pinkish, 4–5 mm; flowers very widely spaced **3. pumilio**
  b. Perianth-lobes unequal, the outer 3 lobes shorter than the inner 3, all pale purple; flowers very dense **4. stricta**
6a. Plant a tree to 20 m, usually with an obvious trunk; flowers scented
    **1. australis***
  b. Plant a shrub to 4 m, without an obvious trunk; flowers not scented
    **2. fruticosa***

**1. C. australis** (Forster) Endlicher (*Dracaena australis* Forster). Illustration: Botanical Magazine, 5636 (1867).
Tree to 20 m, the trunk ultimately massive, to 1.5 m in diameter, usually little branched below, richly branched above, bark rough and fissured. Leaves 30–100 × 3–6 cm, lanceolate, light green on both surfaces, midrib and veins indistinct. Panicle to 1.5 m × 60 cm, much branched, the branches more or less at right angles and well spaced, the axes almost hidden by the flowers. Flowers almost stalkless, sweetly scented; perianth 5–6 mm, white, lobes equal, free almost to the base, slightly keeled, reflexed. Berry c. 4 mm in diameter, spherical, whitish. *New Zealand.* H5–G1.

***C. baueri** Hooker. Trunk slender, to 3 m, leaves 15–60 × 5–6.5 cm. *New Zealand (Norfolk Island).* H5–G1.

**2. C. fruticosa** (Linnaeus) Chevalier (*C. terminalis* (Linnaeus) Kunth; *Dracaena glauca* Anon.). Illustration: Graf, Exotica, edn 8, 1070–1 (1973).
Shrub to 4 m, stems erect, slender and clustered. Leaves 30–60 × 5–10 cm, conspicuously stalked, the stalk grooved, the leaf green or variably tinged with red or purple. Panicles to 60 cm, richly branched, broad, with many flowers, the axis clearly visible between them. Flowers almost stalkless, c. 6 mm, white, yellowish or reddish, the 3 outer lobes somewhat shorter than the 3 inner, all reflexed. Fruit

*c.* 8 mm in diameter, spherical, red. *Tropical SE Asia, Australia, Hawaii.* H5–G1.

Very variable in leaf coloration; numerous cultivars have been selected on the basis of this character.

**\*C. rubra** Kunth. Shrub to 5 m, leaves 30–40 × 4–5 cm; flowers *c.* 7 mm, pale purple. *Origin unknown.* H5–G1.

**3. C. pumilio** J.D. Hooker.
Shrub to 2 m, often flowering when smaller. Stem to 1.5 cm wide, arching. Leaves 30–200 × 1–2 cm, green, midrib conspicuous at least beneath, narrowed to a channelled stalk. Panicle to 60 cm, very open and with slender branches. Flowers almost stalkless, widely spaced, perianth 4–5 mm, all the lobes similar in size, recurved, white or sometimes pink outside. Fruit 4–5 mm in diameter, spherical, bluish or flecked with blue spots. *New Zealand.* H5–G1.

**4. C. stricta** (Sims) Endlicher (*Dracaena stricta* Sims; *Dracaena congesta* Anon.).
Illustration: Morley & Everard, Wild flowers of the world, t. 142 (1970).
Shrub to 4 m. Leaves 30–60 × 1.5–2.5 cm, linear-lanceolate, green. Panicle pyramidal, to 60 cm, richly branched. Flowers almost stalkless, perianth *c.* 7 mm, pale purple, the lobes reflexed, the outer 3 conspicuously shorter than the inner 3. *Australia.* H5–G1.

**5. C. banksii** Hooker (*Dracaena beuckelarii* Koch).
Shrub to 4 m, stems several, 10–15 cm in diameter. Leaves 1–2 m × 4–8 cm, linear-lanceolate, green, stalked, midrib conspicuous above, as are several lateral veins. Panicle to 2 m, richly branched and with many flowers, axes visible between the flowers which are very fragrant. Perianth *c.* 1 cm, all lobes equal, spreading, white. Fruit 4–5 mm in diameter, spherical, white or bluish. *New Zealand.* H5–G1.

**6. C. indivisa** (Forster) Steudel (*C. hookeri* Kirk; *Dracaena indivisa* Forster). Figure 36(7), p. 272. Illustration: Illustration Horticole 7: t. 264 (1860).
Shrub to 8 m with massive, sparingly branched stems. Leaves 1–2 m × 10–15 cm, narrowly lanceolate, green above, glaucous beneath, the midrib broad and conspicuous, often red, sometimes the stronger lateral veins also red. Panicle to 1.6 m, compact, slightly branched, the axes hidden by the crowded flowers which are borne on stalks 2–3 mm long. Perianth 7–8 mm, lobes all equal, strongly recurved, whitish inside, pale

purple outside. Fruit *c.* 6 mm in diameter, spherical, bluish. *New Zealand.* H5–G1.

**7. C. haageana** Koch. Illustration: Gartenflora **20**: t. 675 (1871) & **27**: 115 (1878).
Small shrub to 1 m or more. Leaves 10–20 × *c.* 6 cm, broadly lanceolate, abruptly contracted into a long stalk, green. Panicle *c.* 30 cm, with few branches and with numerous flowers which are almost stalkless. Perianth *c.* 7 mm, lobes all equal, reflexed, white inside, purplish outside. Fruit *c.* 3 mm in diameter, spherical. *Australia (Queensland).* G1–2.

**15. PHORMIUM** Forster & Forster
*C.J. Couper*
Rhizomatous perennial plants, ultimately becoming woody at the base. Leaves persistent, linear, folded towards the base and equitant, keeled, marked with many fine, close, longitudinal stripes. Flowering axis long, bearing alternate, deciduous bracts which are scarious, the upper bracts subtending and entirely enclosing the short, alternately branched flowering branches. Flower-stalks jointed near the apex. Flowers bisexual, bilaterally symmetric. Perianth-lobes 6, more or less equal, erect, united into a tube at the base. Stamens 6, projecting from the perianth. Ovary superior, 3-celled, elongate, ovules many in each cell. Fruit a long, many-seeded capsule, the seeds flattened and almost winged, black and shiny.

A genus of 2 species from New Zealand. They are easily grown in rich soils. Propagation is by seed or by division.

1a. Leaves to 3 m, stiff; flowers predominantly dull red, tips of the inner perianth-lobes not or only slightly recurved **1. tenax**
  b. Leaves usually less than 2 m, flexible; flowers greenish with tones of orange or yellow, tips of inner perianth-lobes usually markedly recurved **2. cookianum**

**1. P. tenax** Forster & Forster. Figure 36(9), p. 272. Illustration: Botanical Magazine, 3199 (1832); Illustration Horticole 13: t. 481 (1866); Hay & Synge, Dictionary of garden plants, t. 1732 (1959); The Garden **102**: 111 (1977).
Base of plant usually brightly coloured. Leaves 1–3 m × 5–12 cm, stiff and often erect, at least in the lower part. Flowering axis usually erect, to 5 m. Flowers 2.5–5 cm, predominantly dull red, the tips of the inner perianth-lobes not or only

slightly recurved. Capsule to 10 cm, though often much less, erect, 3-sided, abruptly contracted at the apex, not twisted, remaining firm and dark in age. Seeds 9–10 × 4–5 mm, elliptic, plate-like, sometimes twisted. *New Zealand.* H5–G1. Summer.

**2. P. cookianum** Le Jolis (*P. colensoi* Hooker; *P. hookeri* Hooker). Illustration: Revue Horticole, ser. 3, **2**: 5 (1848); Illustration Horticole **19**: t. 93 (1872); Botanical Magazine, 6973 (1888).
Base of plant usually pale. Leaves mostly less than 2 m, flexible, drooping. Flowering axis to 2 m, often inclined or arching. Flowers 2.5–4 cm, usually predominantly greenish with tones of orange or yellow, tips of inner perianth-lobes markedly recurved, usually 1 more so than the other 2. Capsule often more than 10 cm, occasionally to 20 cm, hanging, cylindric, gradually narrowed to the apex, twisted and becoming pale, fibrous and spirally curved with age. Seeds 8–10 mm, elliptic, plate-like. *New Zealand.* H5–G1. Summer.

## VIII. HAEMODORACEAE
Perennial herbs with fibrous roots, rhizomes or tubers. Leaves mostly basal, linear or sword-shaped. Flowers in cymes, panicles or racemes, bisexual, radially or bilaterally symmetric; perianth-lobes 6, at least partly fused into a tube towards the base. Stamens 3 or 6. Ovary superior or inferior, 3-celled, each cell with 1 or many ovules. Fruit a capsule opening between the septa.

There are about 16 genera, native in sub-tropical or tropical habitats in the southern hemisphere, except for 2 genera in North America. Only 3 genera are known in cultivation in Europe.

1a. Perianth-tube split equally **2. Conostylis**
  b. Perianth-tube split more deeply on 1 side **2**
2a. Ovules solitary in each cell; flowers clothed with black hairs **3. Macropidia**
  b. Ovules numerous; flowers clothed in yellow or red, but not black hairs **1. Anigozanthos**

**1. ANIGOZANTHOS** Labillardière
*E.C. Nelson*
Perennial herbs. Leaves sword-shaped, arising from a rhizome, to 40 cm long, hairless. Flowers in a 1-sided raceme, densely clothed with hairs. Stamens 6.

Style 1. Ovules numerous in each cell of ovary.

A genus of about 12 species restricted to Western Australia, inhabiting the siliceous sandplain. Only 1 species is widely cultivated but new cultivars are being produced in Australia which may eventually reach Europe. *Anigozanthos* should be grown in full sun, in very well drained compost; propagation is by seed, division, or tissue culture.
Literature: Geerinck, G., Revision du genre Anigozanthos Labill. (Haemodoraceae d'Australie), *Bulletin Jardin Botanique National de Belgique* **40**: 261–76 (1970); *Growing Native Plants* **2**: 30–1 (1972).

1a. Inflorescence a spike-like raceme
  **2. manglesii**
  b. Inflorescence a panicle   **1. flavidus***

**1. A. flavidus** de Candolle. Illustration: Botanical Magazine, 1151 (1809); Erickson, Flowers and plants of Western Australia, 66 (1973); Gardener, Wildflowers of Western Australia, 16 (1973).
Robust perennial to 2 m tall. Leaves strap-like. Flower-spike to 2 m; flowers often yellow-green, sometimes pink, orange, red or green. *W Australia*. H5–G1. Summer.

This is the hardiest species but it is frost-sensitive. Other species may be cultivated in southern Europe.

***A. pulcherrima** Hooker. Illustration: Botanical Magazine, 4180 (1845); Growing Native Plants **2**: 31 (1972); Gardener, Wildflowers of Western Australia, 16 (1973). Like *A. flavidus* but with bright yellow flowers and only 1 m tall. *W Australia*. Summer.

***A. rufa** Labillardière. Illustration: Growing Native Plants **2**: 30 (1972). Like *A. pulcherrima* but with maroon flowers. *W Australia*. H5–G1. Summer.

**2. A. manglesii** D. Don. Illustration: Gardener, Wildflowers of Western Australia, 15 (1973).
Perennial with short rhizome. Leaves sword-like. Flower-spike with rich red velvet on stems and flowers, to 1 m tall. Perianth bright green with red hairs. *W Australia*. H5–G1.

A spectacular plant but short-lived and susceptible to 'ink disease'. It is best raised from seed and treated as a biennial.

**2. CONOSTYLIS** R. Brown
*E.C. Nelson*
Perennial herbs. Leaves in tufts, linear or strap-shaped. Flowers in cymes or panicles,

not one-sided. Perianth clothed in hairs, lobes united into a tube at the base, splitting equally. Stamens 6. Style 1.

There are about 20 species in this genus, restricted to Western Australia. Only 1 species is in cultivation here and it is uncommon. These plants need full sun and well-drained sandy soil and generally require greenhouse protection in Europe.

**1. C. candicans** Endlicher.
Leaves arranged in a fan or tuft, strap-shaped, grey-green, to 30 cm long. Flowering stems longer than leaves. Flowers white-yellow, about 20 in each cyme. *W Australia*. H5–G1.

**3. MACROPIDIA** Harvey
*E.C. Nelson*
Perennial herb with sword-shaped leaves to 50 cm tall. Flowers in a branched panicle. Perianth green and covered with dense black hairs; lobes only united into a tube for *c*. half their length, reflexed at apex and curled back. Stamens 6. Ovule 1 in each carpel.

A genus containing only 1 species, sometimes placed within *Anigozanthos* (as Section *Macropidia* (Harvey) Geerinck). *Macropidia* is confined to Western Australia (north of Perth) where it inhabits the woodlands and sandplain heath. It is frost-sensitive, and requires full sun and well-drained sandy soil, or a cool, frost-free glasshouse in well-drained compost. Propagation is by seed or tissue culture.

**1. M. fuliginosa** (Hooker) Druce (*Anigozanthos fuliginosus* Hooker). Illustration: Botanical magazine, 4291 (1846); Erickson, Flowers and plants of Western Australia, 96–7 (1973); Gardener, Wildflowers of Western Australia, 14 (1973).
*W Australia*. H5–G1.

---

# IX. AMARYLLIDACEAE

Perennial herbs, usually with bulbs, more rarely with rhizomes. Leaf-bases sometimes forming a false-stem or neck above the bulb. Leaves often in 2 ranks, usually flat or with the margins bent outwards when young, usually basal, more rarely borne on the stem. Inflorescence an umbel or reduced to a solitary terminal flower, subtended by usually 2 (rarely 1 or more than 2) spathes (bracts) which enclose the whole inflorescence in bud and usually persist beneath it in flower; individual flowers often with bracts which are smaller

than the spathes. Perianth radially or bilaterally symmetric, either of 6, free segments, or joined into a tube at the base and with 6 lobes which are all alike or, more rarely, those of the outer whorl different from those of the inner. Stamens 6, usually borne on the top of the perianth-tube, often deflexed and curved upwards at their tips; anthers usually dorsifixed and versatile, sometimes basifixed, opening by slits or more rarely by pores. A corona often present (see figure 37, p. 292), joining the bases of the stamens or as outgrowths of the perianth; it is funnel-shaped, cup-shaped, cylindric or disc-like (more rarely a low ridge); scales (perianth-scales) or tufts of hairs sometimes present at the bases of the filaments. Ovary completely inferior, 3-celled with 2–many ovules in each cell, placentation axile or basal. Fruit a capsule or berry with few to many seeds.

A family of 65–70 genera and about 850 species, found in most parts of the world. About half the genera are cultivated. The limits of the family are controversial, and parts of the *Agavaceae* (p. 271–290) and *Liliaceae* (subfamilies *Allioideae*, p. 230–250 and *Alstroemerioideae*, p. 208–211) are sometimes included within it. The delimitation of genera within the family is also difficult, frequently based on the presence or absence of a corona (and, if present, its nature). Different authors define the corona in different ways; in this account the small appendages (perianth-scales) or tufts of hairs found at the attachment of stamens to perianth in some genera (*Hippeastrum*, *Zephyranthes*) are not regarded as forming a corona.
Literature: Baker, J.G., *Handbook of the Amaryllideae* (1888). The American journal *Herbertia* (later incorporated into *Plant Life*), 1933 and continuing, is largely devoted to the family.

1a. Plant with a flowering stem which is leafy in the lower part   **1. Ixiolirion**
  b. Plant with a leafless scape, the leaves all basal to it   2
2a. Corona present, joining the bases of the filaments so that they are apparently borne on a tube, cup or ridge   3
  b. Corona absent, or, if present, the filaments free from it; perianth-scales or tufts of hairs may be present at the bases of the filaments   14
3a. Scape hollow   **34. Phaedranassa**
  b. Scape solid   4
4a. Anthers basifixed   **24. Chlidanthus**

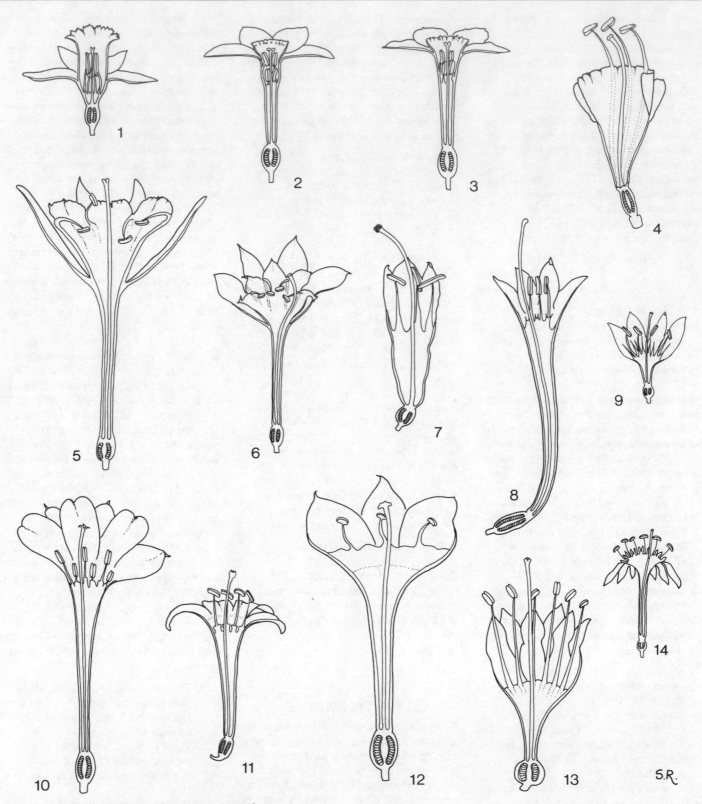

**Figure 37.** Sections of flowers of various genera of Amaryllidaceae to show the coronas. 1, *Narcissus pseudonarcissus*. 2, *N. poeticus*. 3, *N. tazetta*. 4, *N. hedraeanthus*. 5, *Hymenocallis narcissiflora*. 6, *Pamianthe peruviana*. 7, *Phaedranassa carmioli*. 8, *Stenomesson variegatum*. 9, *Vagaria parviflora*. 10, *Chlidanthus fragrans*. 11, *Caliphruria subedentata*. 12, *Eucharis grandiflora*. 13, *Urceolina peruviana*. 14, *Pancratium maritimum*. (Not to scale.)

b. Anthers dorsifixed, often versatile  5
5a. Ovules borne in 1 or 2 vertical rows in each cell of the ovary  6
b. Ovules borne side by side in each cell of the ovary  11
6a. Leaves with a broad blade and distinct stalk  7
b. Leaves linear or strap-shaped, not divided into blade and stalk  8
7a. Perianth-lobes red, orange, yellow or green, without any white
**30. Urceolina**
b. Perianth-lobes white, either completely or for the most part
**31. Eucharis**
8a. Perianth yellow, pink or reddish, if white then with green stripes  9
b. Perianth completely white, without green stripes  10
9a. Flowers 1 or 2, more or less erect, yellow or pale yellow; corona c. 8.5 cm  **27. Paramongaia**
b. Flowers in few-flowered umbels, usually pendent, not usually yellow; corona 1 cm or less
**29. Stenomesson**
10a. Stamens directed horizontally towards the centre of the flower; stigma weakly 3-lobed  **28. Pamianthe**
b. Stamens spreading; stigma capitate
**26. Pancratium**
11a. Ovules 2–8 per cell, basal
**33. Hymenocallis**
b. Ovules 2 or 3 per cell, axile  12
12a. Leaves with a conspicuous median white band  **25. Vagaria**
b. Leaves without a conspicuous median white band  13
13a. Spathes much larger than the bracts of the individual flowers; perianth-tube tapered, straight
**32. Caliphruria**
b. Spathes only slightly larger than the floral bracts; perianth-tube cylindric and curved below, expanded above
**31. Eucharis**
14a. Anthers basifixed, opening by pores  15
b. Anthers dorsifixed, usually versatile, opening by slits  17
15a. Flowers erect; filaments almost as long as anthers  **22. Lapiedra**
b. Flowers pendent on arching stalks; filaments much shorter than anthers  16
16a. Segments of the inner whorl of the perianth different from those of the outer whorl  **21. Galanthus**
b. All perianth-segments similar
**23. Leucojum**

17a. Spathes 1 or 2, united below to form a tube around the flower-stalks  18
b. Spathes 1–several (usually 2), not united below to form a tube  23
18a. Corona distinct, trumpet-, funnel- or cup-shaped, more rarely a disc or low ridge, arising at the junction of the perianth-tube and lobes
**13. Narcissus**
b. Corona absent; perianth-scales or tufts of hairs sometimes present at the bases of the filaments  19
19a. Perianth-lobes at most 1 cm
**14. Tapeinanthus**
b. Perianth-lobes 2 cm or more  20
20a. Perianth strongly bilaterally symmetric, mostly crimson, the bases of the 3 lower segments rolled to form a slightly open tube around the bases of the filaments and style
**4. Sprekelia**
b. Perianth radially symmetric or almost so, not as above  21
21a. Stigma simple, capitate or minutely 3-toothed; perianth usually yellow inside and out, rarely white
**5. Sternbergia**
b. Stigma distinctly 3-fid; perianth usually not yellow, occasionally yellow inside  22
22a. Stamens of 2 different lengths, deflexed, curved upwards at their tips
**3. Habranthus**
b. Stamens all of the same length, more or less straight and spreading
**2. Zephyranthes**
23a. Scape hollow  24
b. Scape solid  27
24a. Perianth cylindric, very narrowly funnel-shaped or bell-shaped, the tube longer than the lobes  25
b. Perianth funnel-shaped, often widely so, lobes longer than the tube  25
25a. Leaves and scape borne on a long, aerial neck which is blackish; flowers lilac-blue  **8. Worsleya**
b. Leaves and scape borne more or less at soil level, bulb without a neck or the neck very short; flowers not lilac-blue  26
26a. Perianth-lobes each with a cushion-like swelling at the base  **11. Vallota**
b. Perianth-lobes without such swellings
**9. Hippeastrum**
27a. Ovules 1–3 in each cell of the ovary  28
b. Ovules 4 or more in each cell of the ovary  31
28a. Spathes 2; perianth-lobes spreading, tube curved  29

b. Spathes several; perianth-lobes erect, tube straight  30
29a. Uppermost stamen erect, pressed against the uppermost perianth-lobe, the other 5 deflexed, curved upwards at their tips  **7. Griffinia**
b. All stamens deflexed then curved upwards at their tips, or all widely spreading  **6. Lycoris**
30a. Leaves very obviously in 2 ranks, usually thick and fleshy, without distinct midribs; rhizome absent
**18. Haemanthus**
b. Leaves not obviously in 2 ranks, thin and with distinct midribs; rhizome present below the bulb  **19. Scadoxus**
31a. Perianth-tube long, cylindric or somewhat widened towards the apex, the lobes spreading at a wide angle into a cup- or bell-shape  **15. Crinum**
b. Perianth-tube very short or absent, the lobes not spreading from it at a wide angle  32
32a. Scape obviously compressed and 2-edged; fruit a berry  **20. Clivia**
b. Scape not obviously compressed and 2-edged; fruit a capsule  33
33a. Umbel spherical with 20–60 flowers on stalks which are at least 10 cm
**16. Brunsvigia**
b. Umbel not as above, flowers usually fewer than 20, stalks always less than 10 cm  34
34a. Perianth-tube at most 5 mm, the lobes curving away from it, separate and distinct, margins often wavy
**17. Nerine**
b. Perianth-tube at least 1 cm, the lobes forming a funnel shape, overlapping or almost so, margins not wavy
**10. Amaryllis**

**1. IXIOLIRION** (Fischer) Herbert
*J. Cullen*
Perennial herbs with bulbs. Leaves mostly in a basal rosette, usually winter-persistent, also present on the stems. Flowers in a terminal umbel with or without several racemosely arranged flowers below it. Perianth radially symmetric, either of 6 free segments or united into a tube below and 6-lobed. Stamens 6, attached to the perianth-tube (if present) or to the bases of the perianth-segments or -lobes. Ovary 3-celled; ovules numerous in each cell, placentation axile. Fruit a capsule. Seeds numerous, black.

A genus of about 4 species from SW & C Asia; sometimes split into 2 genera, *Ixiolirion* in the strict sense, and *Kolpakowskia* Regel. Only 2 species are generally grown. They require a light, rich, well-drained soil and a sunny position. Literature: Traub, H.P., The Ixiolirion Tribe, *Herbertia* 9: 53–9 (1942).

  1a. Perianth-segments completely free, the filaments attached at their extreme bases    **1. tataricum**
    b. Perianth with a distinct tube below, the filaments attached to it
                **2. kolpakovskyanum**

**1. I. tataricum** (Pallas) Herbert (*I. pallasii* Fischer & Meyer; *I. montanum* (Labillardière) Herbert). Illustration: Edwards's Botanical Register 30: t. 66 (1844); Gartenflora 22: t. 775 (1873), 26: t. 910 (1877).
Bulb ovoid, to 2.5 cm in diameter. Leaves mostly winter-persistent, mostly in a basal rosette. Stem to 40 cm, bearing 2 or 3 leaves. Flowers in an umbel with 0–4 racemosely arranged flowers below it. Perianth openly funnel-shaped, of 6 blue or violet-blue segments, each segment 2–5 cm with usually 3 darker lines down the middle. Filaments attached to the segments only at the extreme base. *SW & C Asia, India (Kashmir).* H4. Spring–summer.

Very variable; several units have been recognised from time to time as species, subspecies or varieties, under the names *ledebourii* Fischer & Meyer and *montanum* (Labillardière) Herbert, but it is uncertain which of these remain in cultivation.

**2. I. kolpakovskyanum** Regel (*Kolpakowskya ixiolirioides* Regel). Similar to *I. tataricum* but flowers usually restricted to an umbel, perianth 2–2.5 cm, united into a distinct tube below, whitish violet, the filaments attached to the perianth-tube. *C Asia.* H3. Spring–summer.

**2. ZEPHYRANTHES** Herbert
*E. Campbell*
Perennial herbs with bulbs. Leaves usually present with the flowers, narrow, linear. Flowers solitary, borne on a hollow scape, subtended by a tubular spathe which is usually bifid towards the apex. Perianth usually united into a tube below, 6-lobed, the 3 inner lobes slightly narrower than the 3 outer, the tube often minutely scaly in the throat, more rarely of 6 free segments, radially symmetric. Stamens 6, erect, attached near the throat of the perianth-tube, anthers versatile. Ovary 3-celled with many ovules. Style thread-like, erect, stigma 3-fid. Fruit a capsule splitting between the septa. Seeds black.

A genus of about 60 species from warm regions of the western hemisphere, of which 6 are common in cultivation as border or pot plants.
Literature: Sealy, J.R., Zephyranthes, Pyrolirion, Habranthus and Hippeastrum, *Journal of the Royal Horticultural Society* 62: 195–209 (1937); Herklots, G.A.C., Wind Flowers part 1, Zephyranthes, *The Plantsman* 2: 8–19 (1980).

*Bulb.* More than 3 cm thick: **4**; less than 3 cm thick: **1–3,5,6**.
*Scape.* Up to 15 cm: **1,3,4,6**; usually more than 15 cm: **1,2,5**.
*Flowering time.* Spring: **1,5,6**; summer: **1,2,4,6**; autumn: **2,3**.
*Flowers.* Stalked: **2–6**; stalkless: **1**.
*Perianth-lobes,* White, sometimes tinged with green or pink: **1,2,4,5**; pink: **3,6**. Up to 4 cm: **1–3**; more than 4 cm: **4–6**.

  1a. Flowers stalkless    **1. verecunda**
    b. Flowers stalked    2
  2a. Style shorter than stamens
                **2. candida**
    b. Style longer than stamens    3
  3a. Flower-stalk longer than spathe    4
    b. Flower-stalk shorter than spathe    5
  4a. Spathes 1.3–2 cm; flowers pink; perianth-lobes 3–4 cm    **3. rosea**
    b. Spathes *c.* 2.7 cm; flowers white tinged with green at the base; perianth-lobes 4–5 cm    **4. tubispatha**
  5a. Spathes 2.5–3 cm; flowers pure white; perianth-lobes *c.* 7.5 × 1.2 cm    **5. atamasco**
    b. Spathes 4–5 cm; flowers purplish pink; perianth-lobes 6.4–7.5 × 1.3–2 cm    **6. grandiflora**

**1. Z. verecunda** Herbert (*Z. sessilis* Herbert). Illustration: Botanical Magazine, 2583 (1825).
Bulb ovoid, 2–2.5 cm across. Leaves 15–23 cm × *c.* 6 mm, recurved or almost erect. Scapes 2–3, 7.5–23 cm. Spathe *c.* 2.5 cm, cylindric in the lower half, bifid above. Flower stalkless. Perianth white tinged with pink, 3.8–5 cm, tube 2–3 cm, lobes 2.5–3.7 cm × 6–8 mm, spreading. *N Mexico.* H5–G1. Spring–summer.

**2. Z. candida** (Lindley) Herbert (*Argyropsis candida* (Lindley) Roemer). Illustration: Botanical Magazine, 2607 (1825); Loddiges' Botanical Cabinet, t. 1419 (1828); Rix & Phillips, The bulb book, 182 (1981).
Bulb ovoid, *c.* 2.5 cm across. Leaves to 38 cm × 3–5 mm, erect. Scape 15–23 cm. Spathe *c.* 2.5 cm, cylindric in the lower half, not bifid. Flower-stalk *c.* 1.5 cm. Perianth pure white or slightly tinged with red, 3.8–5 cm, tube absent, segments *c.* 1.3 cm broad, spreading. Style much shorter than stamens. *Argentina, Uruguay.* H5–G1. Summer–autumn.

**3. Z. rosea** (Sprenger) Lindley. Illustration: Edwards's Botanical Register 10: t. 821 (1824); Botanical Magazine, 2537 (1825).
Bulb spherical, *c.* 2 cm across. Leaves 10–20 cm × *c.* 5 mm, spreading. Scapes 1–3, 10–15 cm. Spathe 1.3–2 cm, cylindric, bifid at the tip only. Flower-stalk more than 2 cm. Perianth pink, 3–3.5 cm, tube *c.* 4 mm, lobes 6–8 mm broad. Style longer than stamens. *West Indies, Guatemala.* G1. Autumn.

**4. Z. tubispatha** (L'Héritier) Herbert (*Z. insularum* Hume). Illustration: Botanical Magazine, n.s., 232 (1954–55).
Bulb spherical, 3–3.5 cm across. Leaves 12–33 cm × 4–7 mm, erect. Scape 12–15 cm. Spathe *c.* 2.7 cm, cylindric in the lower half, bifid above. Flower-stalk *c.* 3.5 cm. Perianth white tinged with green, 4–5 cm, tube *c.* 4 mm, lobes *c.* 1.3 cm broad. *West Indies, Venezuela, Colombia.* G1–H5. Summer.

**5. Z. atamasco** (Linnaeus) Herbert. Illustration: Loddiges' Botanical Cabinet, t. 1899 (1832).
Bulb ovoid, to 2.5 cm across. Leaves *c.* 30 cm × 6 mm. Scape 15–30 cm. Spathe 2.5–3 cm, cylindric in the lower half, bifid above. Flower-stalk less than 2.5 cm. Perianth pure white, *c.* 7.5 cm, tube *c.* 1.2 cm, lobes *c.* 1.2 cm broad. *USA (Virginia, Florida, Alabama, Mississippi).* H5–G1. Spring.

**6. Z. grandiflora** Lindley (*Z. carinata* Herbert; *Z. tsouii* Hu). Illustration: Botanical Magazine, 2594 (1825); Edwards's Botanical Register 11: t. 902 (1825).
Bulb ovoid, *c.* 2.5 cm across. Leaves 15–30 cm × 5–8 mm, spreading. Scape 13–25 cm. Spathe 4–5 cm, cylindric in the lower half, slit on one side in the upper half. Flower-stalk *c.* 2.5 cm. Perianth pale purplish pink with a white throat, 6.3–7.5 cm, tube 3.1–3.3 cm, lobes 1.3–1.8 cm broad. Style longer than stamens. *West Indies, Cuba, Mexico, Guatemala.* H5–G1. Spring–summer.

**3. HABRANTHUS** Herbert
*J. Cullen*
Perennial herbs with bulbs. Leaves linear. Flowers solitary, terminal (in a reduced, 1-flowered umbel), subtended by a spathe (a pair of fused bracts) which is tubular in its

lower half and usually somewhat bifid above. Perianth slightly bilaterally symmetric, funnel-shaped, with a short or very short tube below. Corona absent. Stamens 6, deflexed, then curving upwards towards the apex, of 2 different lengths. Style deflexed downwards then curving upwards, exceeding the stamens, stigma 3-fid. Fruit a top-shaped capsule containing numerous seeds.

A genus of 10–20 species from temperate S America (but see under No. 3), very similar to *Zephyranthes* (p. 294), differing only in a slightly bilaterally symmetric perianth, the unequal, deflexed stamens and deflexed style. Cultivation as for *Zephyranthes*.

Literature: Sealy, J.R., Zephyranthes, Pyrolirion, Habranthus and Hippeastrum, *Journal of the Royal Horticultural Society* **62**: 195–209 (1937).

1a. Perianth 2.5–3.5 cm, bright yellow inside, coppery or purplish red outside **3. tubispathus***
   b. Perianth 6–10 cm, red, pinkish mauve or whitish flushed with pink inside and out    2
2a. Perianth 7.5–10 cm, red or pink, the tube deeper mauve-pink **1. brachyandrus**
   b. Perianth *c.* 6 cm, white flushed with pink, the tube green    **2. robustus**

**1. H. brachyandrus** (Baker) Sealy (*Hippeastrum brachyandrum* Baker). Illustration: Botanical Magazine, 7344 (1894); The Garden **49**: t. 1048 (1896); Horticulture **53**: 82 (1975).
Bulb ovoid, to 2.5 cm in diameter, dark brown, with a long neck. Leaves linear, to 25 cm. Scape to 30 cm, spathe to 5 cm, flower-stalks slightly longer than the spathe. Perianth ascending-erect, 7.5–10 cm, bright red or pinkish mauve inside and out, the tube deeper red or pinkish mauve. Stamens to 2.5 cm. *S Brazil, Paraguay.* H5. Summer.

**2. H. robustus** Herbert (*Zephyranthes robusta* (Herbert) Baker). Illustration: Botanical Magazine, 9126 (1927).
Similar to *H. brachyandrus* but leaves developing after flowering, scape at most 15 cm, perianth to 6 cm, white flushed with pink, the tube green, stamens *c.* 3 cm. *S Brazil, ?Argentina.* H5. Summer.

**3. H. tubispathus** (L'Héritier) Traub (*Amaryllis tubispatha* L'Héritier; *H. andersonii* Herbert). Illustration: Edwards's Botanical Register **16**: t. 1345 (1840); Addisonia **22**: t. 727 (1945); Plant Life **26**: 100 (1970).

Bulb ovoid, to 2.5 cm in diameter, with a short neck. Leaves 13–15 cm, narrowly linear. Scape to 15 cm. Spathe 2.5–3 cm, exceeded by the flower-stalk. Perianth erect, bright yellow inside, coppery or purplish red outside, 2.5–3.5 cm; tube very short. Stamens 1.2–1.7 cm. *S Brazil, Uruguay, E Argentina & S Chile.* H5. Summer.

The nomenclature of this species has been elucidated by Ravenna, Plant Life **26**: 99–103 (1970). The variant with the flowers coppery outside has been distinguished as *H. andersonii* var. *cupreus* Anon., but is no more than a selection.

*H. texanus** (Herbert) Steudel (*H. andersonii* var. *texanus* Herbert; *Zephyranthes texana* (Herbert) Herbert). Illustration: Botanical Magazine, 3596 (1837). Very similar, but smaller, perianth to 2.5 cm. *Southern USA (Texas).* H5. Summer.
Scarcely distinguishable from *H. tubispathus* but notable on account of its distribution. Perhaps an early introduction of *H. tubispathus* into N America.

**4. SPREKELIA** Heister
*J. Cullen*
Perennial herbs with bulbs. Leaves linear, present at flowering. Scape about as long as the leaves, hollow. Flower solitary, terminal (a reduced, 1-flowered umbel), subtended by a spathe which is tubular in its lower half, bifid above. Flower-stalk shorter than the spathe, the flower horizontal or slightly drooping. Perianth of 6 free segments, strongly bilaterally symmetric, the uppermost segment broadest, the 2 upper lateral segments sickle-shaped, arching downwards, the 3 lower segments with their basal parts rolled to form a tube (slightly open above) surrounding the lower parts of the stamens and style. Stamens deflexed, then curved upwards. Style deflexed, then curved upwards, exceeding the stamens. Stigma 3-fid. Ovary 6-angled, 3-celled with numerous ovules in each cell. Capsule 3-angled, containing many seeds.

A genus of a single species from Mexico and Guatemala. Cultivation as for *Hippeastrum* (p. 297), though it tolerates generally cooler conditions than most of the tropical species of that genus.

**1. S. formosissima** (Linnaeus) Herbert (*Amaryllis formosissima* Linnaeus). Illustration: Botanical Magazine, 47 (1788); Journal of the Royal Horticultural Society **87**: f. 80 (1962); Horticulture **48**(10): 37 (1970); Moggi & Guignolini, Flori da balcone e da giardino, f. 329 (1982).

Bulb to 5 cm in diameter, brown. Leaves to 50 × 2 cm. Spathe to 5 cm. Perianth 7–11 cm, bright crimson, the base of the segments sometimes striped or blotched with yellow. *Mexico, Guatemala.* G1. Spring.

**5. STERNBERGIA** Waldstein & Kitaibel
*B. Mathew & J.C.M. Alexander*
Perennial herbs with bulbs; tunics membranous, brown or black. Leaves basal, linear or strap-shaped to narrowly lanceolate, flat or shallowly channelled above, sometimes keeled beneath, often twisted. Flower-stems 1–several, sometimes below ground at flowering (Nos. 5 & 6), elongating and often arching in fruit. Flowers solitary, usually yellow (white in No. 4) appearing in spring or autumn. Spathe membranous, tubular below, split above. Perianth radially symmetric; tube cylindric, narrow; lobes 6, in 2 similar whorls, oblanceolate to obovate (very narrow in No. 6). Corona absent. Stamens 6, in 2 unequal whorls, borne at top of perianth-tube, shorter than perianth-lobes. Style 1, bearing an entire or minutely 3-toothed capitate stigma. Fruit a cylindric to spherical capsule containing numerous large, dark seeds often with fleshy appendages.

A genus of 7 or 8 species centred on Turkey, extending west to Spain, east to Kashmir and south to Israel. They are most likely to flourish if grown in a bulb frame where they can be given a dry dormant period in the summer to encourage the formation of flowers. *S. lutea* and *S. sicula* will also perform well in the open ground in a sheltered, sunny well-drained site. The spring-flowering species, *S. candida* and *S. fischeriana* are not hardy, and should be planted quite deeply to discourage the bulbs from dividing too often which produces a succession of small, non-flowering plants. Most species can also be grown in pots in an alpine house.

Literature: Mathew, B., A review of the genus Sternbergia, *Plantsman* **5**: 1–16 (1983).

1a. Flowers produced in spring    2
   b. Flowers produced in autumn    3
2a. Flowers yellow; spathe 2.5–4.5 cm; perianth-lobes 2–4 cm  **3. fischeriana**
   b. Flowers white; spathe 5–5.5 cm; perianth-lobes 4.3–5 cm    **4. candida**

3a. Leaves appearing at or before
flowering-time; perianth-tube 2 cm or
less                                    4

b. Leaves absent at flowering-time;
perianth-tube usually 2–6.5 cm        5

4a. Leaves 7–12 mm wide, usually bright
shiny green; perianth-lobes usually
3–5.5 × 1–2 cm                 **1. lutea**

b. Leaves 2–5 mm wide, deep green
with greyish stripe; perianth-lobes
usually 2–4 cm × 3–12 mm   **2. sicula**

5a. Leaves 8–16 mm wide; perianth-lobes
3.5–7.5 × 1 cm or more     **5. clusiana**

b. Leaves 1–4 mm wide; perianth-lobes
usually less than 3 cm × 5 mm
**6. colchiciflora**

**1. S. lutea** (Linnaeus) Sprengel. Illustration:
Botanical Magazine, 290 (1795); Hay &
Synge, Dictionary of garden plants in
colour, 111 (1969); Rix & Phillips, The
bulb book, 176 & 184 (1981); Plantsman
**5**: 1 (1983).
Bulb 2–4 cm across. Leaves 4–6,
appearing with or just before the flowers,
7–12 mm wide, narrowly lanceolate,
slightly channelled above, keeled beneath,
usually bright shiny green. Flower-stem
2.5–20 cm, longer in fruit; spathe 3–6 cm.
Flower deep yellow; tube 5–20 mm; lobes
3–5.5 × 1–2 cm, oblanceolate to obovate.
Seeds lacking an appendage. *From Spain to
Iran & USSR (Soviet Central Asia) – (seldom
truly wild)*. H4–5. Autumn.

**2. S. sicula** Gussone (*S. lutea* subsp. *sicula*
(Tineo) Webb; *S. lutea* var. *graeca*
Reichenbach; *S. lutea* var. *angustifolia*
Anon.). Illustration: Polunin & Huxley,
Flowers of the Mediterranean, f. 254
(1965); Rix & Phillips, The bulb book, 185
(1981); Plantsman **5**: 7 (1983).
Similar to *S. lutea*. Bulb 1–2.5 cm across.
Leaves 2–5 mm wide, linear, dark green
with central grey stripe. Flower-stem
3–70 mm; spathe 3–6 cm. Perianth-tube
4–10 mm; lobes 2–3.4 cm × 4–12 mm.
*Italy, Sicily, Greece, Crete, Aegean Islands &
W Turkey*. H4–5. Autumn.

**3. S. fischeriana** (Herbert) Ruprecht.
Illustration: Botanical Magazine, 7441
(1895); Journal of the Royal Horticultural
Society **98**: 442 (1973); Rix & Phillips, The
bulb book, 29 (1981); Plantsman **5**: 9
(1983).
Bulb 2.5–3.5 cm across. Leaves 4–7,
appearing before the flowers, 6–12 mm
wide, strap-shaped, flat, not or scarcely
keeled, usually dark greyish green. Flower-
stem 3–15 cm, longer in fruit; spathe
2.5–4.5 cm. Flower pale yellow; tube

3–5 mm; lobes 2–3.5 cm × 5–8 mm,
oblanceolate. Seeds with fleshy appendage.
*USSR (Azerbaijan) & S Turkey to Kashmir
(not recorded from Afghanistan)*. H4–5.
Spring.

**4. S. candida** Mathew & Baytop.
Illustration: The Garden **104**: 302 (1979);
Rix & Phillips, The bulb book, 29 (1981);
Plantsman **5**: 9 (1983).
Similar to *S. fischeriana*. Flower-stem
12–20 cm; spathe 5–5.5 cm. Flower white,
fragrant; lobes 4.3–5 cm × 9–18 mm.
*SW Turkey*. H4–5. Winter–spring.

**5. S. clusiana** (Ker Gawler) Sprengel
(*S. macrantha* Baker). Illustration: Botanical
Magazine, 7459 (1896); Mathew, Dwarf
bulbs, pl. 78 (1973); Rix & Phillips, The
bulb book, 176 (1981); Plantsman **5**: 12
(1983).
Bulb 2.5–4 cm across. Leaves 5–12,
appearing long after the flowers, 8–16 mm
wide, strap-shaped, flat, not keeled, often
twisted, greyish green. Flower-stem very
short and below ground at flowering,
raising ovary above ground in fruit. Spathe
5–10 cm. Flower bright to greenish yellow;
tube 3–6.5 cm; lobes
3.7–7.5 × 1.1–3.3 cm, obovate to
oblanceolate. Seeds with large fleshy
appendage. *Turkey & Iran south to Israel &
Jordan*. H4–5. Autumn.

**6. S. colchiciflora** Waldstein & Kitaibel.
Illustration: Polunin, Flowers of Europe, pl.
172 (1969); Mathew, Dwarf bulbs, pl. 79
(1973); Rix & Phillips, The bulb book, 29 &
184 (1981); Plantsman **5**: 12 & 15 (1983).
Bulb 5–15 mm across. Leaves 3–6,
appearing long after the flowers, 1–4 mm
wide, narrowly linear, not or scarcely
keeled, twisted, sometimes strongly, dark
green, sometimes glaucous. Flower-stem
very short and below ground at flowering,
raising ovary above ground in fruit. Spathe
2.2–4 cm. Flowers yellow; tube 3–6.5 cm;
lobes 2.3–3.3 cm × 2–5 mm, narrowly
oblanceolate. Seeds with large fleshy
appendage. *SE Spain & Italy eastwards to
Israel & Iran*. H4–5. Autumn.

**6. LYCORIS** Herbert
*E. Campbell*
Herbaceous perennials with ovoid bulbs.
Leaves basal, linear or strap-shaped,
usually appearing after the inflorescences.
Flowers in a terminal umbel borne on a
solid scape, the umbel subtended by 2 free
spathes. Perianth bilaterally or radially
symmetric, with a usually very short tube
at the base, lobes equal, spreading or

strongly recurved; perianth-scales minute.
Stamens 6, erect or ascending, borne on
the throat of the perianth-tube, deflexed,
anthers versatile. Ovary 3-celled, with 2 or
3 ovules per cell. Style thread-like, stigma
minute. Capsule almost spherical to ovoid.
Seeds black-brown.

A genus of 10–12 species from China
and Japan, of which 5 are commonly found
in cultivation as pot plants. In northern
Europe they require the protection of a
greenhouse. Cultivation as for *Hippeastrum*
(p. 297).
Literature: Traub, H.P. & Moldenke, H.N.,
*Amaryllidaceae Tribe Amarylleae*, 165–82
(1949).

1a. Leaves 1.8–2.5 cm wide; flowers
fragrant, rose pink, sometimes with
purple or blue markings       2

b. Leaves to 1.8 cm wide; flowers not
fragrant, yellow or red        3

2a. Spathes ovate; flowers rose pink,
without purple-blue markings;
perianth-tube absent     **1. sprengeri**

b. Spathes lanceolate; flowers pale rose
pink with purple or blue markings;
perianth-tube present   **2. squamigera**

3a. Flowers golden yellow; leaves
1.2–1.8 cm wide; bulb 5–6 cm in
diameter                  **3. aurea**

b. Flowers red; leaves 6–12 mm wide;
bulb to 4 cm in diameter       4

4a. Leaves 6–8 mm wide; perianth rose
red to deep red, lobes very reflexed
and with wavy margins    **4. radiata**

b. Leaves 1–1.2 cm wide; perianth
bright red, lobes slightly reflexed and
without wavy margins   **5. sanguinea**

**1. L. sprengeri** Baker.
Leaves 30–60 × *c.* 2 cm, linear. Scape *c.*
30 cm. Spathes short, ovate, acute. Flower-
stalks 2–4 cm. Flowers many, erect,
fragrant. Perianth 7–8 cm, rose-pink,
funnel-shaped, without a tube, the lobes
recurved at the tips, margins not wavy.
Stamens not projecting. *Japan*. H4.
Summer.

**2. L. squamigera** Maximowicz.
Illustration: Botanical Magazine, 7547
(1897).
Bulb ovoid, 4–5 cm in diameter. Leaves *c.*
30 × 1.8–2.5 cm, strap-shaped, produced in
spring. Scape 50–70 cm. Spathes
lanceolate, 2–4 cm. Flower-stalks 1–3 cm.
Flowers 6–8, almost erect, fragrant.
Perianth 9–10 cm, funnel-shaped, pale rose
pink with blue or purple veining; tube *c.*
2.5 cm, lobes recurved at their tips,
margins slightly wavy. *Japan*. H4. Summer.

The name *L. incarnata* Sprenger, which seems to be invalid, probably applies to *L. squamigera* or a variety of it.

**3. L. aurea** (L'Héritier) Herbert. Illustration: Botanical Magazine, 409 (1798); Edwards's Botanical Register, t. 611 (1822). Bulb 5–6 cm in diameter, ovoid. Leaves to 60 × 1.2–1.8 cm, strap-shaped, fleshy, glaucous, produced after the flowers. Scape to 60 cm. Spathes 3–5 cm, lanceolate. Flower-stalks 8–15 mm. Flowers 5 or 6, erect. Perianth 9.5–10 cm, golden yellow, funnel-shaped; tube 1.5–2 cm, lobes recurved at their tips, the margins usually very wavy. Stamens slightly projecting. *China, Japan.* H4. Spring–summer.

**4. L. radiata** (L'Héritier) Herbert. Illustration: Edwards's Botanical Register, t. 596 (1822). Bulb 2.5–3.5 cm in diameter, broadly ellipsoid. Leaves 30–60 cm × 6–8 mm, strap-shaped, dark green. Scape 30–50 cm. Spathes 2–4 cm, linear-lanceolate. Flower-stalks 6–15 mm. Flowers 4–6, drooping. Perianth 4–5 cm, rose-red to deep red; tube 6–8 mm, lobes very reflexed, margins very wavy. Stamens strongly projecting. *Japan.* H5. Summer–autumn.

*L. albiflora* Koidzumi is very similar, and is probably merely a variety.

**5. L. sanguinea** Maximowicz. Bulb 1.5–2 cm in diameter. Leaves 1–1.2 cm wide, linear, dark green. Scape 30–50 cm. Spathes 2–4 cm, lanceolate. Flower-stalks *c.* 5 cm. Flowers 4–6, erect. Perianth 5–6 cm, bright red, funnel-shaped, tube 1.2–1.5 cm, lobes slightly recurved, margins not wavy. Stamens not projecting. *China, Japan.* H4. Summer–autumn.

**7. GRIFFINIA** Ker Gawler
*J. Cullen*
Perennial herbs with bulbs. Leaves distinctly stalked (in our species) with broadly oblong-lanceolate blades. Scape solid. Umbel with several flowers; spathes 2, free. Perianth horizontal, openly funnel-shaped, distinctly bilaterally symmetric, with short tube at the base. Stamens 6, spreading. Ovary 3-celled, with 2 ovules in each cell. Style somewhat deflexed, stigma capitate. Fruit a capsule containing few seeds.

A genus of 6 species from Brazil, of which only 1 is in general cultivation. Conditions as for the tropical species of *Hippeastrum*.
Literature: Traub, H.P. & Moldenke, H.N.,

*Amaryllidaceae Tribe Amarylleae*, 153–7 (1949).

**1. G. hyacinthina** (Ker Gawler) Ker Gawler (*Amaryllis hyacinthina* Ker Gawler; *Lycoris hyacinthina* (Ker Gawler) Herbert). Illustration: Edwards's Botanical Register **2**: t. 163 (1816), **6**: t. 444 (1820); Everett, Illustrated encyclopedia of gardening **5**: opposite p. 822 (1964). Bulb 5–6 cm in diameter. Leaf-blade 15–22 × 5–7.6 cm, stalk almost as long as the blade. Umbel with 8–10 flowers, flower-stalks very short. Perianth purplish, violet-blue or white, 5–6.5 cm. *Brazil.* G2. Summer.

**8. WORSLEYA** Traub
*J. Cullen*
Bulb large, dark brown with a very long, aerial, trunk-like neck. Leaves several, evergreen, distinctly borne in 2 ranks, strap-shaped, curved downwards in a sickle shape, green with translucent, cartilaginous margins. Scape hollow, 2-edged, shorter than the leaves. Umbel with 4–14 flowers; spathes 4, 2 of them shorter and narrower than the others, free, longer than the flower-stalks. Perianth horizontal, funnel-shaped, somewhat bilaterally symmetric, with a short tube at the base. Corona absent. Stamens 6, deflexed and then curved upwards towards the apex, shorter than the perianth. Ovary inferior, 3-celled, with many ovules in each cell. Style deflexed, then curved upwards, exceeding the stamens. Stigma capitate and obscurely 3-lobed. Capsule large, wider than long.

A genus of a single species from Brazil, notable for the extremely long neck of its bulb. Cultivation as for *Hippeastrum*, but the plant is frequently short-lived.
Literature: Traub, H.P. & Moldenke, H.N., *Amaryllidaceae Tribe Amarylleae*, 22–4 (1949); Martinelli, G., Nota sobre Worsleya rayneri (J.D. Hooker) Traube & Moldenke, espécie ameaçada de extinçal, *Rodriguesia* **36**: 65–71 (1984).

**1. W. rayneri** (J.D. Hooker) Traub & Moldenke (*Amaryllis rayneri* J.D. Hooker; *A. procera* Duchartre not Salisbury; *Hippeastrum procerum* (Duchartre) Lemaire; *Worsleya procera* (Duchartre) Traub). Illustration: Botanical Magazine, 5883 (1871); The Garden **45**: 350 (1894); Everard & Morley, Wild flowers of the world, t. 190 (1970). Bulb 13–15 cm in diameter, neck to 1 m. Leaves 60–100 × 4.5–7.5 cm. Perianth

lilac-blue, 14–16 cm, tube to 2 cm. *S Brazil (Organ mountains).* G1–2. Winter–spring.

**9. HIPPEASTRUM** Herbert
*J. Cullen*
Perennial herbs with bulbs. Leaves few, usually not stalked. Scape slender or conspicuously stout, hollow. Flowers 2–10 in an umbel subtended by 2 large, equal, free spathes which enclose the whole umbel in bud and persist beneath it in flower. Flowers erect, horizontal or drooping. Perianth bilaterally symmetric, with a long or short tube. Corona absent; sometimes small scales or tufts of hairs are found at the junction of the filaments and perianth. Stamens 6, attached to the perianth-tube, deflexed and then curving upwards towards the apex. Ovary 3-celled, with few to many ovules in each cell. Style deflexed downwards then curving upwards; stigma capitate or 3-fid. Fruit a spherical or 3-lobed capsule.

In the interpretation followed here, a genus of about 80 species from C & S America. In using the name *Hippeastrum* for these plants and *Amaryllis* (p. 300) for a South African species, common European practice is followed. However, H.P. Traub (see references below) and members of The American Plant Life Society, regard all the species here included in *Hippeastrum* as belonging to the genera *Amaryllis* or *Rhodophiala*, and treat the South African plant as a species of *Brunsvigia* (p. 311). This whole problem turns on the interpretation of what Linnaeus intended when he described *Amaryllis belladonna*. No generally acceptable solution has yet been agreed, though Goldblatt (Taxon **33**: 511–16, 1984) has reviewed the arguments and has proposed the name *Amaryllis* for conservation in the sense in which it is used here.

The small-flowered species (Nos. 1–5) are (with their wild allies) sometimes regarded as forming the genus *Rhodophiala* Presl, but a broad concept of *Hippeastrum* seems appropriate here. The large-flowered species (Nos. 6–15) and their allies form a very striking group of plants, much cultivated by enthusiasts, and many hybrids have been raised, named and distributed. Lists of these names appear from time to time in the journal Plant Life, which should be consulted for further information.

Hippeastrums are relatively easily grown in good soil. All can be grown in pots, but many benefit from being planted in a border. Potted plants can be forced to produce flowers earlier than the seasons indicated here. Enthusiasts use many specialised techniques for growing show plants; the journal Plant Life, and Traub's Amaryllis Manual, cited below, should be consulted for details.

Literature: Traub, H.P. & Moldenke, H.N., *Amaryllidaceae Tribe Amarylleae* (1949); Traub, H.P., The genera Rhodophiala and Phycella, *Plant Life* 12: 67–76 (1956); Traub, H.P., *The Amaryllis Manual* (1958).

*Leaves*. Linear, less than 2 cm broad (often much less): **1–5**; oblong, more than 2 cm broad, somewhat narrowed to the base but not stalked: **6–14**; oblanceolate, more than 2 cm broad, stalked: **15**.

*Perianth*. To 6 cm long: **1–5**; more than 6 cm long: **6–15**. Tube 10–13 cm: **6**; tube 1–6 cm: **7–15**; tube less than 1 cm: **1–5**. Lobes yellowish green spotted with red: **1,14**; lobes conspicuously striped red and green: **12**; lobes mauve with a superimposed darker mauve network: **15**.

*Perianth-scales*. Present and incurved, closing the throat of the perianth-tube: **12–14**; present but not incurved: **1–5**; present and very obscure, or absent: **6–11,15**.

*Stigma*. 3-fid: **1–4,7,8,12,13**; capitate: **5,6,9–11,14,15**.

1a. Perianth up to 6 cm long; leaves linear, at most 2 cm broad, usually less than 1.3 cm broad                 **2**
 b. Perianth 7–25 cm long; leaves oblong or oblanceolate, mostly more than 2.5 cm broad                          **6**
2a. Stigma capitate            **5. pratense**
 b. Stigma 3-fid                           **3**
3a. Umbel 1–2-flowered; spathes 2.5–3.8 cm                 **4. roseum**
 b. Umbel 3–10-flowered; spathes 3.8–7.5 cm                           **4**
4a. Stamens almost as long as perianth; perianth yellow or yellow tinged or spotted with red     **1. bagnoldii**
 b. Stamens at most two-thirds of the length of the perianth; perianth yellow or red                             **5**
5a. Perianth-lobes oblong, acute; spathes 2.5–5 cm             **2. advenum**
 b. Perianth-lobes oblanceolate, clawed, obtuse; spathes 5–7.5 cm
                                 **3. bifidum**
6a. Perianth-tube 10–13 cm
                         **6. solandriflorum**
 b. Perianth-tube at most 2.5 cm      **7**
7a. Stigma 3-fid                          **8**
 b. Stigma capitate                      **11**
8a. Scales closing the throat of the perianth-tube                        **9**

 b. Scales obscure or absent, not closing the throat of the perianth-tube     **10**
9a. Each perianth-lobe mostly green, with many red stripes and a red margin
                        **12. psittacinum***
 b. Each perianth-lobe mainly red and without stripes, though with a greenish midrib         **13. aulicum**
10a. Perianth 10–15 cm, tube *c*. 2.5 cm; scales very obscure or absent
                               **7. vittatum**
 b. Perianth 7.5–10 cm, tube *c*. 2 cm; small bristles or scales present at the bases of the filaments      **8. rutilum**
11a. Perianth-scales conspicuous, closing the throat of the perianth-tube; perianth yellowish, copiously spotted with red                  **14. pardinum**
 b. Perianth-scales absent or inconspicuous, not closing the throat of the perianth-tube; perianth not as above                          **12**
12a. Leaves narrowed to an almost terete stalk; perianth reddish mauve with a superimposed network of reddish lines; ovules 1–3 per cell
                           **15. reticulatum**
 b. Leaves not narrowed to a stalk; perianth not as above; ovules numerous per cell                      **13**
13a. Perianth only very slightly bilaterally symmetric, the lobes broadly overlapping, each with a white mark at the base            **9. leopoldii**
 b. Perianth distinctly bilaterally symmetric, lobes not as above       **14**
14a. Perianth-tube 2.2–3 cm; base of each perianth-lobe yellowish green or white, this area not forming a star shape when the whole flower is viewed                **11. puniceum**
 b. Perianth-tube to 1.3 cm; each perianth-lobe with a narrowly triangular white mark at the base, forming a star shape when the whole flower is viewed           **10. reginae***

**1. H. bagnoldii** (Herbert) Baker (*Habranthus bagnoldii* Herbert; *Amaryllis bagnoldii* (Herbert) Dietrich; *Rhodophiala bagnoldii* (Herbert) Traub). Illustration: Edwards's Botanical Register 17: t. 1396 (1831). Bulb to 5 cm in diameter, almost black. Leaves linear, to 30 cm × 6 mm, somewhat glaucous, obtuse. Scape slender, to 30 cm. Umbel with 4–8 flowers; spathes 4–5 cm, flower-stalks 2–7 cm. Perianth ascending, openly funnel-shaped, yellow or yellow tinged or spotted with red, 3–5.5 cm, the tube *c*. 5 mm. Perianth-scales small, not closing the throat of the tube. Stamens

almost as long as the perianth. Stigma 3-fid, style about as long as the perianth. *Chile*. H5–G1. Winter.

**2. H. advenum** (Ker Gawler) Herbert (*Amaryllis advena* Ker Gawler; *Rhodophiala advena* (Ker Gawler) Traub). Illustration: Botanical Magazine, 1125 (1808); Edwards's Botanical Register 10: t. 849 (1824); Saunders, Refugium Botanicum 5: t. 332 (1872).
Bulb ovoid, with a short neck, to 4 cm in diameter, brown. Leaves linear, to 30 cm × 5 mm, somewhat glaucous, blunt. Scape slender, to 50 cm. Umbel with 2–6 flowers; spathes 2.5–5 cm, flower-stalks 2–7 cm. Perianth horizontal or almost so, openly funnel-shaped, pink, red or yellow, 3.5–5 cm, the tube to 5 mm, the lobes oblong, acute. Perianth-scales small, not closing the throat of the perianth-tube, margins scalloped. Stamens at most two-thirds of the length of the perianth. Style almost as long as the perianth, stigma 3-fid. *Chile*. H5–G1. Winter.

**3. H. bifidum** (Herbert) Baker (*Amaryllis bifida* (Herbert) Sprengel; *A. kermesina* Lindley; *Rhodophiala bifida* (Herbert) Traub). Illustration: Botanical Magazine, 2597 (1825), 2639 (1826); Edwards's Botanical Register 14: t. 1148 (1828), t. 1638 (1834).
Bulb ovoid, usually with a conspicuous neck, to 4 cm in diameter, dark brown. Leaves linear, to 45 × 1.3 cm, slightly glaucous, blunt. Scape 10–40 cm. Umbel with 2–7 flowers; spathes 5–7.5 cm, flower-stalks 2.5–5 cm. Perianth ascending to erect, rather narrowly funnel-shaped, bright to deep red, 4–5 cm, tube *c*. 3 mm, lobes oblanceolate, clawed. Stamens at most two-thirds of the length of the perianth. Style about as long as perianth, stigma 3-fid. *Argentina, Uruguay*. H5–G1. Winter.

Traub recognises 3 varieties (under his *Amaryllis bifida*): the above (var. **bifida**); var. **pulchra** (Herbert) Traub & Moldenke which is a small plant with rather smaller flowers; and var. **spathacea** (Herbert) Traub & Moldenke (*A. spathacea* (Herbert) Schultes) in which the bulb is darker, the spathes incompletely separated, and the stamens and style are shorter.

**4. H. roseum** (Sweet) Baker (*Habranthus roseus* Sweet; *Amaryllis rosea* (Sweet) Traub & Uphof not Lamarck; *A. barlowii* Traub & Moldenke; *Rhodophiala rosea* (Sweet) Traub). Illustration: Sweet, British flower garden, ser. 2, 2: t. 107 (1833).

Bulb ovoid, with a short neck, to 2.5 cm in diameter, dark brown. Leaves linear, to 30 cm × 5 mm, glaucous. Scape to 20 cm. Umbel with 1–2 flowers; spathes 2.5–3.5 cm, flower-stalks 2–3 cm. Perianth funnel-shaped, almost horizontal, bright red, *c.* 5 cm, the tube very short, greenish. Stamens shorter than the perianth and style, stigma 3-fid. *Chile (Chiloe)*. H5–G1. Winter.

**5. H. pratense** (Poeppig) Baker (*Amaryllis pratensis* Poeppig; *Rhodophiala pratensis* (Poeppig) Traub). Illustration: Edwards's Botanical Register **28**: t. 35 (1842); Botanical Magazine, 3961 (1842); The Garden for 1878, 51 (1878).
Bulb ovoid, with a short neck, 3–4 cm in diameter, dark brown. Leaves linear, bright green, 30–50 cm × 6–13 mm, obtuse. Scape to 60 cm. Umbel with 2–8 flowers; spathes 3–5 cm, flower-stalks 2.5–4 cm. Perianth openly funnel-shaped, horizontal, bright red or purplish, the lobes often marked with yellow lines towards the base, 5–7 cm, the tube *c.* 6 mm. Perianth-scales small. Stamens shorter than the perianth. Style about as long as the perianth, stigma capitate. *Chile*. H5–G1. Spring.

**6. H. solandriflorum** (Lindley) Herbert (*Amaryllis solandriflora* Lindley; ?*A. elegans* Sprengel). Illustration: Lindley, Collectanea Botanica, t. 11 (1821); Botanical Magazine, 2573 (1825), 3542 (1837) – as *H. ambiguum*.
Bulb ovoid, with a short neck, 7–10 cm in diameter. Leaves oblong, blunt, to 50 × 2–2.5 cm. Scape 40–60 cm. Umbel with 2–6 flowers; spathes 5–8 cm, flower-stalks to 6 cm. Perianth horizontal, funnel-shaped, 18–25 cm, tube 10–13 cm, greenish white or white, sometimes striped or tinged with red. Perianth-scales small. Stamens shorter than the perianth. Style as long as the perianth, stigma capitate though slightly 3-lobed. *Northern S America, south to Peru*. G2. Winter (sometimes also summer).

Variable in the precise colouring of the flowers; several varieties have been recognised to cover this variation. The species is easily recognised by its large, long-tubed flowers.

**7. H. vittatum** (L'Héritier) Herbert (*Amaryllis vittata* L'Héritier). Illustration: Lindley, Collectanea Botanica, t. 12 (1821); Botanical Magazine, 129 (1795).
Bulb more or less spherical, 5–8 cm in diameter. Leaves oblong, obtuse, 45–60 × 2.5–3.5 cm, bright green. Scape 60–100 cm. Umbel with 2–6 flowers; spathes to 7.5 cm, flower-stalks to 7.5 cm. Perianth drooping, funnel-shaped, white striped with red, 9–15 cm, tube to 2.5 cm. Stamens shorter than the perianth. Style about as long as the perianth, stigma 3-fid. *Peru, Brazil*. G2. Early spring.

This species was crossed in the 1790s with *H. reginae* to produce a range of popular hybrids known as *H.* × *johnsonii*.

**8. H. rutilum** (Ker Gawler) Herbert (*Amaryllis rutila* Ker Gawler; *A. striata* in the sense of Traub & Moldenke and Traub). Illustration: Edwards's Botanical Register **1**: t. 38 (1815), **3**: t. 226 (1817) & **7**: t. 534 (1821); Botanical Magazine, 2475 (1824).
Bulb more or less spherical with a short neck, 5–8 cm in diameter. Leaves oblong, blunt, to 30 cm × at least 2.5 cm, bright green. Scape to 30 cm. Umbel with 2–4 flowers; spathes 3–6.5 cm, flower-stalks 3–6 cm. Perianth horizontal or drooping, bright red, scarlet, orange or reddish orange, each lobe with a greenish white base or green stripe along the midrib; total length 7.5–11 cm, tube greenish, *c.* 2 cm. Stamens shorter than the perianth. Style about as long as the perianth, stigma 3-fid. *S & C Brazil*. G2. Winter.

Variable in flower colour; var. **rutilum** (described above) has the perianth-lobes bright crimson, each with an elongate green base; var. **fulgidum** Baker has the perianth scarlet to red-orange, each lobe with a greenish stripe towards the base; and var. **crocatum** (Ker Gawler) Anon. has the perianth-lobes orange-yellow.

**9. H. leopoldii** Dombrain (*Amaryllis leopoldii* (Dombrain) Moore). Illustration: Floral Magazine **9**: t. 475 (1870); Gardeners' Chronicle for 1870, 733.
Bulb with a short neck, more or less spherical, to 7.5 cm in diameter. Leaves oblong, 45–60 × 2.5–4 cm, green. Scape stout. Umbel with 2 flowers, flower-stalks 3.5–3.7 cm. Perianth almost radially symmetric (the lowermost lobe only slightly narrower than the others), openly funnel-shaped, horizontal, bright red; the 5 larger lobes are broad and overlapping, each with a narrowly triangular white stripe at the base, giving the appearance of a 5-pointed star when the flower is viewed from the front; total length of perianth *c.* 11 cm, tube *c.* 1 cm. Perianth-scales small, edged with whitish bristles. Stamens shorter than the perianth. Style about as long as the perianth, stigma capitate. *Bolivia*. G2. ?Winter.

This species was introduced into cultivation in 1869, and was thought to have come from Peru. It was soon lost, however, and was not refound until the 1970s, when it was rediscovered in Bolivia (see Plant Life **30**: 81–3, 1974).

**10. H. reginae** (Linnaeus) Herbert (*Amaryllis reginae* Linnaeus; *A. alberti* Lemaire). Illustration: Botanical Magazine, 453 (1799).
Bulb with a neck, more or less spherical, to 7.5 cm in diameter. Leaves to 60 × 2.5–4 cm, oblong, blunt, somewhat narrowed towards the base, green. Scape to 50 cm. Umbel with 2–4 flowers; spathes 5–7.5 cm, flower-stalks 5–7.5 cm. Perianth horizontal or drooping, funnel-shaped, bright red with a 5-pointed, greenish white star at the base, 8–12 cm, tube to 1.3 cm. Stamens slightly shorter than the perianth. Style about as long as the perianth, stigma capitate. *C & northern S America*. G2. Winter–spring.

Plants in cultivation under the name *Amaryllis miniata* Ruiz & Pavon probably belong to this species. The name *Amaryllis alberti* applies to a variant with double flowers.

**\*H. machupijchense** Vargas. Illustration: Botanical Magazine, n.s., 821 (1981). Similar, but the flowers more open and the perianth red, each lobe with a green base which extends along the lobe as green stripes. *Peru*. G2.

**11. H. puniceum** (Lamarck) Kunze (*Amaryllis belladonna* in the sense of Traub; *A. equestris* Herbert; *Hippeastrum equestre* (Aiton) Herbert). Illustration: Botanical Magazine, 305 (1795); Edwards's Botanical Register **3**: t. 234 (1817); Gartenflora **44**: t. 1418 (1895).
Bulb more or less spherical, with a short neck, to 5 cm in diameter, brownish. Leaves oblong, slightly tapered to the base, abruptly tapered to a blunt apex, 25–50 × 2.5–4.5 cm, bright green. Scape to 60 cm. Umbel with 2–4 flowers; spathes to 6 cm, flower-stalks 5–8 cm. Perianth funnel-shaped, horizontal, 10–13 cm, the tube *c.* 2.5 cm; tube green, lobes red or orange-red (rarely white), the base of each greenish white, the whitish parts not forming a star shape when the flower is viewed from the front. Stamens shorter than the perianth. Style somewhat exceeding the stamens, stigma capitate. *C & northern S America, Caribbean*. G2. Winter–spring.

Variable; a number of varieties has been described, based on flower colour and size. This is the species which Traub and his followers interpret as being the true *Amaryllis belladonna* of Linnaeus.

**12. H. psittacinum** (Ker Gawler) Herbert (*Amaryllis psittacina* Ker Gawler). Illustration: Edwards's Botanical Register 3: t. 199 (1817).
Bulb more or less spherical, with a short neck, to 10 cm in diameter. Leaves oblong, tapered abruptly to a blunt apex, 30–50 × 2.5–3.8 cm. Umbel with 2–4 flowers; spathes exceeding the flower-stalks. Perianth openly funnel-shaped, horizontal, 10–12 cm; tube greenish, very short, the lobes mainly green shading to yellowish green above, the green area with numerous red stripes which merge into the red margins. Perianth-scales rather prominent, incurved and closing the throat of the perianth. Stamens shorter than perianth. Style about as long as perianth, stigma 3-fid. *S Brazil*. G2. Spring.

**13. H. aulicum** (Ker Gawler) Herbert (*Amaryllis aulica* Ker Gawler; *A. robusta* Otto & Dietrich; *H. aulicum* var. *robustum* (Otto & Dietrich) Voss). Illustration: Botanical Magazine, 1038 (1827), 3311 (1834); Edwards's Botanical Register 6: t. 444 (1820); Hay et al., Dictionary of indoor plants in colour, 280 (1974).
Similar to *H. psittacina* but leaves broader, spathes and flower-stalks longer, flowers 13–15 cm, each perianth-lobe red with slightly darker red veining, the base of each with a large, variably shaped, green blotch. *C Brazil, Paraguay*. G2. Winter.

**14. H. pardinum** (J.D. Hooker) Lemaire (*Amaryllis pardina* J.D. Hooker). Illustration: Botanical Magazine, 5645 (1867); Pacific Horticulture **38**(3): 36 (1977).
Bulb more or less spherical, with a short neck, to 8 cm in diameter. Leaves oblong, somewhat narrowed towards the base, abruptly tapered to the blunt apex, bright green, to 60 × 5 cm. Umbel with 2 flowers; spathes *c*. 5 cm, flower-stalks 3.5–5 cm. Perianth openly funnel-shaped, horizontal or somewhat ascending, to 10 cm; tube to 1.3 cm, lobes oblanceolate, greenish or yellowish green flushed and spotted with red. Perianth-scales rather conspicuous, incurved, closing the throat of the perianth. Stamens a little shorter than the perianth. Style as long as the perianth, stigma capitate. *Bolivia*. G2. Winter–spring.
Formerly thought to originate from Peru (see Plant Life **30**: 81–3, 1974).

**15. H. reticulatum** (L'Héritier) Herbert (*Amaryllis reticulata* L'Héritier). Illustration: Botanical Magazine, 657 (1803); Andrews

Botanist's Repository, t. 179 (1811); Gartenflora **38**: t. 1297 (1889).
Bulb more or less spherical, with a short neck. Leaves thin, oblanceolate, tapered at the base into an almost terete stalk, to 30 × 2.5 cm. Umbel with 3–5 flowers; spathes to 5 cm, flower-stalks 2.5–4.5 cm. Perianth funnel-shaped, drooping, red-mauve with a network of darker veins, 10–12.5 cm, tube 1.3–2.5 cm. Stamens shorter than the perianth. Style about as long as the perianth, stigma capitate, slightly 3-lobed. Ovules 1–3 in each cell. *S Brazil*. G2. Winter.

**10. AMARYLLIS** Linnaeus
*J. Cullen*
Perennial herb with bulbs. Leaves appearing after the flowers, oblong, somewhat narrowed to the base. Scape stout, solid. Flowers 5–many in an umbel subtended by 2 large, equal, free spathes which enclose the whole umbel in bud. Flowers ascending to erect. Perianth somewhat bilaterally symmetric, with a short tube and 6 oblanceolate lobes which are somewhat clawed, acute, the 3 outer lobes each with a small, hairy, inwardly pointing appendage just below the apex. Stamens 6, deflexed and then curving upwards towards the apex. Ovary 3-celled, with few ovules in each cell. Style deflexed and then curving upwards, stigma capitate. Fruit a capsule, splitting irregularly. Seeds few.
A genus of a single species from South Africa. For details of the controversy over its nomenclature, see under *Hippeastrum* (p. 297). The 1 species has been hybridised with *Crinum moorei* (p. 310) to produce a range of hybrids bearing the name × **Amarcrinum** Coutts ( × *Crinodonna* Anon.), with species of *Brunsvigia* (p. 311), producing hybrids named × **Amarygia** Ciferri & Giacomini ( × *Brunsdonna* Anon.) and with species of *Nerine* (p. 311). Cultivation as for *Hippeastrum* (p. 297); the bulbs should not be disturbed or transplanted.

**1. A. belladonna** Linnaeus (*A. rosea* Lamarck; *Coburgia belladonna* (Linnaeus) Herbert; *Callicore rosea* (Lamarck) Hannibal; *Brunsvigia rosea* (Lamarck) Hannibal). Illustration: The Garden **43**: t. 490 (1893); Flowering plants of Africa **30**: t. 1200 (1954–5); Horticulture **51**(8): 23 (1973); Eliovson, Wild flowers of southern Africa, t. 3 (1980).
Scape to 60 cm or more. Umbel with usually 6 or more flowers; spathes to 8 cm.

Perianth 6–10 cm, purplish pink or pink, often whitish towards the base or rarely all white; tube *c*. 1.2 cm. *South Africa*. H5. Late summer–autumn.
Variable in flower colour; numerous varieties (cultivars) have been selected and named.

**11. VALLOTA** Herbert
*P.F. Yeo*
Perennial herbs with bulbs; remaining evergreen in cultivation. Leaves rather numerous, strap-shaped, obtuse or acute, somewhat sheathing at the base. Inflorescence-stalk as long as or longer than the leaves. Umbels with 2 spathes and several smaller bracts. Flowers several, bisexual, radially symmetric, widely funnel-shaped; tube gradually enlarged in its apical half, lobes broad, somewhat longer than the tube, with cushion-like swellings at the throat. Stamens attached at the throat, shorter than the perianth-lobes; filaments not enlarged at the base; anthers dorsifixed. Style slightly longer than perianth, 3-lobed. Ovary 3-celled, oblong, with numerous ovules. Seeds black, winged at the base.
One species from South Africa with showy red or occasionally pink or white flowers; sometimes included in *Cyrtanthus*, in which case it must be called *C. purpureus*. It is usually grown in pots and should be disturbed as little as possible. Repotting should be done in spring or early summer before the flower-stems are sent up and great care is necessary not to damage the roots. It is relatively inactive in winter, when it should be kept at 4–8°C, but not dried out. Propagation is by offsets or seed; seedlings take 3 years to reach maturity.

**1. V. speciosa** (Linnaeus filius) Durand & Schinz (*V. purpurea* (Aiton) Herbert; *Cyrtanthus purpureus* (Aiton) Herbert). Illustration: Botanical Magazine, 1430 (1811); Parey's Blumengärtnerei, **1**: 332 (1958); Huxley & Gilbert, Success with house plants, 390 (1979).
Leaves 20–45 × 2–6 cm, bright green. Stem 30–60 cm, hollow. Flower-stalks 2–5 cm. Flowers 7–10 cm long and about as wide, horizontal to erect, scarlet with whitish marks in throat or a white centre, or entirely white ('Alba'); tube 3–5 cm; lobes *c*. 3.5–4.5 × 2–3 cm. *South Africa*. G1. Summer–autumn.

**12. CYRTANTHUS** Linnaeus
*E. Campbell*
Herbaceous perennials with bulbs which sometimes have a neck made up of the

sheathing bases of the old leaves. Leaves linear or strap-shaped, produced with or after the flowers, channelled or flat, sometimes flexuous. Flowers in a terminal umbel borne on a hollow scape, umbel subtended by 2 or 4 spathes. Perianth with a long tube which is straight or curved; lobes more or less equal, the outer 3 each with an incurved point or tuft of hairs at the apex. Stamens 6, borne on the perianth-tube, the anthers versatile. Ovary 3-celled with many ovules. Style thread-like, distinctly or indistinctly 3-lobed at the apex. Capsule oblong, 3-celled. Seeds black.

A genus of about 40 species from southern Africa, of which 6 are commonly found in cultivation as pot plants. They are easily grown; annual re-potting is beneficial. Propagation is by seed or offsets. Literature: Dyer, R.A., A Review of the genus Cyrtanthus, *Herbertia* **6**: 65–103 (1939).

*Bulb*. More than 7 cm in diameter: **1,2**; less than 6 cm in diameter: **3–6**.
*Leaves*. More than 3 cm broad: **1**: up to 2 cm broad: **2–6**.
*Scape*. Green: **2–6**; purple: **1,5**.
*Spathes*. Four: **1**; two: **2–6**. Spotted: **4**.
*Perianth*. Red: **1,2,5,6**; yellow: **1,3**; white or cream: **3,4**. More than 2 cm wide at the throat: **2**: Fragrant: **4**.
*Style*. Projecting: **2–6**; not projecting: **1,5**.

1a. Leaves 3–6 cm broad; spathes 4
                           **1. obliquus**
  b. Leaves up to 2.5 cm broad; spathes 2
                                 2
2a. Flowers white or yellow         3
  b. Flowers red or orange-red     4
3a. Leaves 1–2 mm broad; spathes 2–2.5 cm, linear lanceolate, without spots when young; flowers 2–5, pale yellow             **3. ochroleucus**
  b. Leaves 5–10 mm broad; spathes 2.5–3.8 cm, lanceolate, spotted with red-brown when young; flowers 4–10, white or cream      **4. mackenii**
4a. Leaves 35 cm or more; scape green; perianth 5 cm or more         5
  b. Leaves 15–30 cm; scape tinged with purple; perianth to 5 cm
                     **5. macowanii\***
5a. Spathes to 8 cm; flowers 1–2 (rarely 3); throat of perianth 2.5–3 cm wide
                     **2. sanguineus**
  b. Spathes 4–5 cm; flowers 4–10; throat of perianth 8–10 mm wide
                   **6. angustifolius**

**1. C. obliquus** (Linnaeus) Aiton.
Illustration: Botanical Magazine, 1133

(1808); Loddiges' Botanical Cabinet, t. 947 (1824).
Bulb almost spherical, *c.* 10 cm in diameter. Leaves 20–60 × 3–6 cm, strap-shaped, glaucous, twisted towards the apex, obtuse, channelled. Scape 20–60 cm, green tinged with purple. Flower-stalks 5–30 mm, recurved. Spathes 4, *c.* 3 × 1 cm, ovate-oblong. Flowers 6–12, hanging. Perianth *c.* 7 cm, yellow and red, sometimes tinged with green, tube straight or curved at the base, *c.* 4.5 cm, to 2 cm wide at the throat; lobes *c.* 2.5 × 1.5 cm, not spreading. Style indistinctly 3-fid. *South Africa (Cape Province)*. H5–G1. Spring–summer.

**2. C. sanguineus** (Lindley) Hooker.
Illustration: Botanical Magazine, 5218 (1860); The Garden, t. 344 (1890).
Bulb ovoid, to 8 cm in diameter. Leaves *c.* 40 × 1–2 cm, linear-lanceolate, green or glaucous, channelled. Scape *c.* 30 cm, green. Flower-stalks to 2 cm, erect. Spathes 2, to 8 cm, lanceolate. Flowers 1 or 2 (rarely 3), almost erect. Perianth 7.5–10.5 cm, bright red, tube straight or curved, 4.5–6.5 cm, 2.5–3 cm wide at the throat; lobes 3–4 × 1.3–1.8 cm, spreading or recurved. Style projecting, 3-lobed. *South Africa (Cape Province, Natal)*. H5–G1. Summer.

Nordal (Norwegian Journal of Botany **26**: 183–192, 1979) splits this variable species into several subspecies; which of these are in cultivation is not known.

**3. C. ochroleucus** (Herbert) Burchell (*C. lutescens* Herbert).
Bulb spherical, to 2–2.5 cm in diameter. Leaves 15–30 cm × 1–2 mm, produced after flowering has started, linear to thread-like, slightly channelled. Scape 15–30 cm, green. Flower-stalks 5–25 mm, erect. Spathes 2, 2–2.5 cm, linear-lanceolate. Flowers 2–5, more or less erect. Perianth 4–5 cm, pale yellow, tube slightly curved, 3–4 cm, 5–7 mm wide at the throat; lobes to 1 cm, spreading. Style projecting, shortly 3-fid. *South Africa (Cape Province)*. H5–G1. Winter.

**4. C. mackenii** Hooker. Illustration: Botanical Magazine, 5374 (1863).
Bulb ovoid, 2.5–4 cm in diameter. Leaves 2–6, 20–30 cm × 5–10 mm, linear. Scape usually exceeding 30 cm, slightly glaucous. Flower-stalks 1–3 cm, erect. Spathes 2, 2.5–3.8 cm, lanceolate, spotted with red-brown when young. Flowers 4–10, more or less erect, fragrant. Perianth *c.* 5 cm, pure white or cream, tube slightly curved,

2–3 cm, 5 mm wide at the throat; lobes 6–7 mm, recurved. Style projecting, 3-fid. *South Africa (Natal)*. H5–G1. Summer.

**5. C. macowanii** Baker. Illustration: La Belgique Horticole **197**: t. 15 (1885).
Bulb ovoid, to 2.5 cm in diameter. Leaves 1–3, 15–30 cm × 3–7 mm, linear, dark green. Scape *c.* 30 cm, green tinged with purple. Flower-stalks to 2 cm. Spathes 2, 2.5–3.7 cm, lanceolate. Flowers 6–8, hanging. Perianth 3.5–4 cm, bright orange-red, tube curved, 2.8–3.5 cm, 4–5 mm wide at the throat; lobes 5–7 mm, spreading. Style slightly projecting, 3-toothed. *South Africa (Cape Province)*. H5–G1. Summer.

**\*C. obrienii** Baker. Leaves bright green, flowers bright, pale orange-red, style not projecting. *South Africa (Cape Province, Natal)*. H5–G1. Summer.

**6. C. angustifolius** (Linnaeus) Linnaeus. Illustration: Botanical Magazine, 271 (1794); Loddiges' Botanical Cabinet, t. 368 (1819).
Bulb ovoid, to 4.5 cm diameter. Leaves 2 or 3, *c.* 45 cm × 7–20 mm, linear to strap-shaped, slightly channelled. Scape to 45 cm, green, stout. Flower-stalks 1–5 cm, more or less erect. Spathes 2, 4–5 cm, lanceolate, acute. Flowers 4–10, hanging. Perianth 6–8 cm, red, tube curved, 5–7 cm, 8–10 mm wide at the throat; lobes to 1 cm, spreading. Style projecting, 3-lobed. *South Africa (Cape Province)*. H5–G1. Spring–summer.

## 13. NARCISSUS Linnaeus
*J. Cullen*
Perennial herbs with bulbs. Leaves 1–several from each bulb, basal, erect, spreading or prostrate, almost cylindric in section to flat and broad. Scape present. Inflorescence an umbel of 2–20 flowers or flowers solitary, subtended by a usually scarious bract (spathe). Perianth with a tube below, 6-lobed. Corona almost always present, usually conspicuous, free from the stamens. Stamens 6, usually in 2 whorls, more rarely in 1. Ovary 3-celled with many ovules. Capsule ellipsoid to almost spherical, many-seeded. Seeds sometimes with an appendage.

This genus exhibits, perhaps more acutely than any other, the taxonomic difficulties which arise from long-established cultivation, hybridisation and selection, and subsequent escape and naturalization. For many populations, especially in S France and NW Italy, it is impossible to say with any confidence whether they are native or naturalized; and

nearly all the early binomials are based on cultivated plants of which the origin was only guessed at. This not only gives rise to nomenclatural problems but it means also that the geographical background for specific delimitation is confused and defective.
D.A. Webb, Flora Europaea 5: 78, 1980

The above quotation neatly indicates the problems with *Narcissus* in terms of its classification and identification. It is a genus of perhaps 50 species (some considered as subspecies or varieties by some authors), most of which have been in cultivation, some of them for a very long time. Many thousands of hybrids have been raised, and many variants have been given cultivar names (see the International Register, cited below). They are grown for ornament, for naturalisation on banks and slopes, and as an important cut-flower crop. Most originate in Europe (especially France, Italy, Spain and Portugal) though a few are natives of N Africa and the eastern Mediterranean area.

The account given here largely follows that of Webb (Flora Europaea 5: 78–82, 1980), except that some groups treated by him as subspecies are retained here at specific level: this has been done purely for convenience in use, and users are referred to the Flora Europaea account for a biologically more appropriate arrangement.

Narcissi are extremely popular garden plants, and are easily grown in borders, on grassy banks where they can naturalise and spread, or in pots. Pot-grown plants can be forced for early flowering. A sunny, open or partially shaded situation is best, and the plants will spread by means of bulbs and seed. Bulbs can be lifted and dried during the winter; this should be done if the plants become overcrowded. Lifting should not be done until the leaves have yellowed and are dying off, and the lifted bulbs should be stored in a cool, dry place. Propagation is by bulbs and seed.
Literature: Burbidge, F.W., *The Narcissus: its history and culture* (1875); Pugsley, H.W., Narcissus poeticus and its allies, *Journal of Botany* 53: supplement 2 (1915); Pugsley, H.W., A monograph of Narcissus subgenus Ajax, *Journal of the Royal Horticultural Society* 53: 17–93 (1933); Bowles, E.A., *A Handbook of Narcissus* (1934); Fernandes, A., Sur la phylogenie des espèces du genre Narcissus L., *Boletim da Sociedade Broteriana* ser. 2, 25: 113–90 (1951); Meyer, F. G., Narcissus species and wild hybrids, *American Horticultural Magazine* 45: 47–76 (1966); Fernandes, A.,

Key to the identification of native and naturalized taxa of the genus Narcissus L., *Daffodil and Tulip Yearbook for 1968*. 33: 37–66 (1967); Royal Horticultutal Society, *Classified list and International Register of Daffodil names* (1954, 1955, 1958, 1961, 1965, 1969).

1a. Plant flowering in autumn                         2
  b. Plant flowering in winter, spring or early summer                                   5
2a. Corona rudimentary, formed by a low, white ridge which is sometimes 6-lobed; filaments longer than anthers
                                    **1. broussonetii**
  b. Corona well-developed though sometimes small, not as above; filaments shorter than anthers          3
3a. Perianth and corona green
                                    **29. viridiflorus**
  b. Perianth and corona not green        4
4a. Leaves 2–4.5 mm wide, present at flowering; corona greenish yellow
                                    **15. elegans\***
  b. Leaves to 1 mm wide, not present at flowering; corona orange
                                    **2. serotinus**
5a. Perianth-tube obconical, i.e. with a narrow base, widening evenly to the insertion of the lobes, funnel-shaped (narrowly so in some hybrids); stamens in a single whorl, the filaments as long as or longer than the anthers                             6
  b. Perianth-tube cylindric, parallel-sided for most of its length or widening a little towards the insertion of the lobes; stamens in 2 whorls, filaments of lower shorter than the anthers   19
6a. Flowers bilaterally symmetric, the stamens deflected downwards, the filaments then curving upwards their apices; corona funnel-shaped            7
  b. Flowers radially symmetric, the stamens not as above; corona flattish, bell-shaped or cylindric              11
7a. Flower white, or white faintly flushed with green            **34. cantabricus**
  b. Flower orange-yellow to pale yellow or cream, or if whitish then flushed with yellow                                  8
8a. Flower-stalk at least 9 mm (rarely shorter); stamens usually not projecting from the corona           9
  b. Flower-stalk absent or at most 5 mm; stamens usually projecting from the corona                                10
9a. Leaves mostly erect; corona not contracted towards its margin
                                    **31. bulbocodium**
  b. Leaves lying on the ground or at least

arching downwards; corona somewhat narrowed towards its margin                          **32. obesus**
10a. Perianth-lobes 3–4.5 mm broad at the base; flower-stalk absent; scape 5–8 cm, shorter than the leaves
                                    **35. hedraeanthus**
  b. Perianth-lobes 1-2.5 mm broad at the base; flower-stalk 1–5 mm; scape 15–20 cm, longer than the leaves
                                    **33. romieuxii**
11a. Perianth-tube narrowly funnel-shaped, 3–4 times as long as wide  12
  b. Perianth-tube broadly funnel-shaped, up to 1½ times as long as wide       13
12a. Perianth and corona of the same colour; flower-stalks 2–3 cm
                                    **38. × odorus**
  b. Perianth-lobes paler yellow than the corona; flower-stalks 1.2–1.5 cm
                                    **39. × incomparabilis**
13a. Perianth and corona white or cream
                                    **37. pseudonarcissus**
  b. At least the corona yellow or orange
                                                        14
14a. Perianth-lobes reflexed at an angle of almost 180° to the corona, hiding the ovary             **43. cyclamineus**
  b. Perianth-lobes spreading or erect-spreading, not hiding the ovary    15
15a. Perianth-lobes to 1.5 cm; corona constricted near the middle
                                    **42. asturiensis**
  b. Perianth-lobes more than 1.5 cm; corona not constricted near the middle                               16
16a. Spathe 6–10 cm; flower-stalks 4–9 cm          **36. longispathus**
  b. Spathe 2–6 cm; flower-stalks to 3.5 cm                                17
17a. Perianth-tube 1.5–2.5 cm
                                    **37. pseudonarcissus**
  b. Perianth-tube 8–15 mm           18
18a. Corona 1.6–2.5 cm        **41. minor**
  b. Corona 3.5–4 cm          **40. bicolor**
19a. Leaves cylindric or semi-cylindric, usually not channelled down the inner face, 1–3 mm wide; scape terete
                                                        20
  b. Leaves mostly flat and channelled down the inner face, 3–25 mm wide, scape usually compressed       29
20a. Leaves smooth and rounded on the outer surface, usually dark green   21
  b. Leaves with 2 keels on the outer surface or 4-angled, usually glaucous
                                                        26
21a. Perianth-tube straight, 2–3 cm      22
  b. Perianth-tube straight or curved, up to 2 cm                                23

22a. Leaves narrowly linear, rather flattened; corona *c.* 5 cm high, deeper yellow than the perianth-lobes **20. × tenuior**

b. Leaves almost terete; corona 2–4 mm high, of the same colour as the perianth-lobes **19. jonquilla**

23a. Longer flower-stalks longer than the spathe and usually longer than the flower **24**

b. Longer flower-stalks included in the spathe, shorter than the flower **25**

24a. Perianth-tube 1.5–2.2 cm; corona 4–5 mm high, about half the length of the perianth-lobes **23. fernandesii**

b. Perianth-tube 1–1.5 cm; corona 5–6 mm high, more than half the length of the perianth-lobes **22. willkommii**

25a. Leaves usually shorter than the scape; perianth-tube straight; corona more than 9 mm in diameter **21. assoanus**

b. Leaves usually longer than scape; perianth-tube curved; corona less than 9 mm in diameter **24. gaditanus**

26a. Flower solitary (rarely 2 together); spathe tubular for three-quarters of its length; flower-stalks 0–3 mm **27**

b. Flowers 1–5 in each umbel; spathe tubular for up to half of its length; flower-stalks 6 mm or more **28**

27a. Flower yellow; corona usually 6-lobed, rarely sinuous or almost entire **25. rupicola**

b. Flower white or cream; corona with sinuous margins **26. watieri***

28a. Leaves 4-angled, usually lying on the ground and twisted, the angles minutely rough; flower 1.2–1.7 cm in diameter **28. scaberulus**

b. Leaves not 4-angled, erect, smooth; flower 1.7–2.5 cm in diameter **27. calcicola**

29a. Perianth-lobes reflexed, hiding the ovary; corona cup-shaped, longer than broad **30. triandrus**

b. Perianth-lobes not or little reflexed, not hiding the ovary **30**

30a. Umbel with 4–20 flowers (rarely fewer); corona without a red or scarious margin **31**

b. Umbel with 1–3 flowers; corona with a red or scarious margin **42**

31a. Corona yellow or orange **32**

b. Corona white **37**

32a. Flowers to 4 cm in diameter **33**

b. Flowers 4–5 cm in diameter **35**

33a. Perianth-lobes white **3. tazetta**

b. Perianth-lobes yellow or pale yellow **34**

34a. Umbel with 1 or 2 flowers; perianth-lobes pale yellow **6. corcyrensis**

b. Umbel with 3–6 flowers; perianth-lobes bright yellow **8. × intermedius**

35a. Perianth-lobes white **7. cypri**

b. Perianth-lobes yellow or pale yellow **36**

36a. Perianth-lobes cream or pale yellow; corona yellow **4. italicus**

b. Perianth-lobes bright or golden yellow; corona deep yellow or orange **5. aureus**

37a. Flower 1.2–2 cm in diameter **38**

b. Flower 2–4 cm in diameter **40**

38a. Corona about half the length of the perianth-lobes, lobed at the margins **14. dubius**

b. Corona at most one-third of the length of the perianth-lobes, margin entire or almost so **39**

39a. Perianth-lobes ovate-oblong, more or less obtuse, about half the length of the perianth-tube; bulb 5–7 cm in diameter **12. pachybolbus**

b. Perianth-lobes ovate-lanceolate, acute, less than half the length of the perianth-tube; bulb smaller **13. canariensis**

40a. Flower 2–2.5 cm in diameter **11. panizzianus**

b. Flower 2.5–4 cm in diameter **41**

41a. Leaves glaucous; scape very compressed, 2-edged; corona usually finely scalloped **9. papyraceus**

b. Leaves green; scape less flattened, not 2-edged; corona entire **10. polyanthus**

42a. Flowers in umbels of 2–3; pollen sterile **18. × medioluteus**

b. Flower solitary; pollen fertile **43**

43a. Stamens of the lower whorl not projecting from the perianth-tube; perianth-lobes overlapping **16. poeticus**

b. Stamens all projecting from the corona; perianth-lobes not overlapping **17. radiiflorus**

Section **Aurelia** (Gay) Baker. Leaves flat. Flowers in umbels. Perianth-tube cylindric at the base, widening somewhat above; lobes spreading. Corona rudimentary. Filaments longer than the anthers, anthers dorsifixed, projecting from the perianth-tube. Flowering in autumn.

**1. N. broussonetii** Lagasca. Illustration: Botanical Magazine, 7016 (1888); Flore de l'Afrique du Nord **6**: 74 (1959).
Bulb 3–4 cm, brown. Leaves glaucous, broadly linear, to 45 × 1.3 cm, flat or scarcely channelled on the inner face, striate on the outer. Flowers 3–12 in an umbel, spathe 3–3.5 cm, whitish; flower-stalks 5–25 mm. Perianth-tube cylindric at base, widening somewhat above, straight, 1.8–2.8 cm; lobes 1.2–1.6 cm, white, elliptic, somewhat acute. Corona rudimentary, reduced to a low white ridge which is sometimes 6-lobed. Stamens and style projecting from the perianth-tube. *Morocco.* H5–G1. Autumn.

Section **Serotini** Parlatore. Leaves thread-like, absent at flowering. Flowers usually solitary. Perianth-tube cylindric. Corona short. Filaments straight, shorter than the anthers to slightly longer; anthers dorsifixed, not projecting. Flowering in autumn.

**2. N. serotinus** Linnaeus. Illustration: Mathews, Lilies of the field, 2 (1978); Huxley & Taylor, Flowers of Greece and the Aegean, t. 387 (1977).
Bulb 1.5–2 cm, dark. Leaves appearing in spring, cylindric and thread-like, 10–20 cm × *c.* 1 mm, glaucous. Scape 10–25 cm. Flowers solitary or occasionally in an umbel of 2 or 3, erect, fragrant; spathe 1.5–3.5 cm, tubular for about half its length. Flower-stalks 7–30 mm. Perianth-tube 1.2–2 cm, cylindric, slightly expanded above; lobes 1–1.6 cm × 3–7 mm, oblong-lanceolate, acute, spreading, white. Corona 1–1.5 mm high, 3–4 mm in diameter, 6-lobed, orange. *Mediterranean area.* H4. Autumn.

Section **Tazettae** de Candolle (Section *Hermione* (Salisbury) Sprengel). Leaves flat or channelled. Flowers in umbels. Perianth-tube cylindric, corona short. Filaments much shorter than anthers; anthers dorsifixed, included or those of the upper whorl projecting. Flowering mainly in spring.

**3. N. tazetta** Linnaeus. Figure 37(3), p. 292. Illustration: Botanical Magazine, 948, 1026 (1807); Polunin & Huxley, Flowers of the Mediterranean, t. 257 (1965); Huxley & Taylor, Flowers of Greece and the Aegean, t. 386 (1977).
Bulb to 5 cm, dark. Leaves flat, green, striate, 20–50 cm × 5–25 mm, obtuse. Scape robust, compressed, to 45 cm. Flowers to 4 cm in diameter, in umbels of 3–15 (rarely fewer), fragrant; flower-stalks unequal, the longest to 7.5 cm. Perianth-tube 1.2–2 cm; lobes 8–22 mm, spreading, usually ovate and somewhat overlapping at their bases, white. Corona yellow, 3–6 mm high, 6–11 mm in diameter. Stamens not projecting from the corona.
*W Mediterranean area.* H3. Winter–spring.

Variable in size; very small variants have been called *N. patulus* Loiseleur (*N. tazetta* subsp. *patulus* (Loiseleur) Baker).

**4. N. italicus** Ker Gawler (*N. tazetta* subsp. *italicus* (Ker Gawler) Baker; *N. ochroleucus* Loiseleur). Illustration: Botanical Magazine, 1188 (1809).
Similar to *N. tazetta*, but flowers to 5 cm in diameter, the perianth-lobes cream to pale yellow, not or scarcely overlapping at their bases; corona yellow. *N & E Mediterranean area*. H3. Spring.

Variable in the precise coloration of the perianth; some species (or subspecies or varieties) have been described on the basis of this variation.

**5. N. aureus** Loiseleur (*N. tazetta* subsp. *aureus* (Loiseleur) Baker; *N. bertolonii* Parlatore; *N. cupularis* (Salisbury) Schultes). Illustration: Grey-Wilson & Mathew, Bulbs, pl. 24 (1981).
Similar to *N. tazetta* but the perianth-lobes bright or golden yellow and the corona deep yellow or orange. *SE France, NW Italy, Sardinia; naturalised elsewhere*. H4. Winter–spring.

Also variable in size, colour of flower, and in sometimes having glaucous leaves.

**6. N. corcyrensis** (Herbert) Nyman (*N. tazetta* subsp. *corycyrensis* (Herbert) Baker).
Similar to *N. tazetta* but flowers sometimes reduced to 1 or 2 per umbel, perianth-lobes pale yellow, narrow, not overlapping at the base, sometimes somewhat reflexed. *S France, Italy, Balkan peninsula*. H4. Spring.

A rather doubtful species of uncertain distribution; it has been suggested that it is a natural hybrid of *N. serotinus* and *N. italicus*.

**7. N. cypri** Sweet. Illustration: Sweet, British Flower Garden, ser. 2, **1**: t. 92 (1831).
Similar to *N. tazetta* but larger, scape compressed only above, flowers *c.* 5 cm in diameter, perianth-lobes white, ovate-oblong, overlapping at their bases, corona bright yellow. *Cyprus, ?Syria; introduced in Italy*. H5. Spring.

**8. N. × intermedius** Loiseleur. Illustration: Grey-Wilson & Mathew, Bulbs, pl. 26 (1981).
Leaves 30–45 cm × 5–8 mm, almost cylindric but channelled down the inner face. Scape to 40 cm, terete. Flowers in umbels of 3–6, fragrant. Perianth-tube 1.4–2 cm, lobes 1–1.8 cm, ovate, overlapping at their bases, bright yellow. Corona 3–4 mm, slightly deeper yellow than the perianth-lobes, its margin slightly lobed. *Spain, S France, Italy*. H4. Spring.

The hybrid between *N. tazetta* and *N. jonquilla*.

**9. N. papyraceus** Ker Gawler. Illustration: Botanical Magazine, 947 (1806); Polunin & Huxley, Flowers of the Mediterranean, t. 258 (1965); Ceballos et al., Plantas silvestres de la peninsula Iberica, 391 (1980).
Bulbs to 7 cm, dark. Leaves flat, slightly glaucous, 10–45 cm × 4–12 mm, obtuse. Scape 14–50 cm, compressed. Flowers 3–20 in an umbel, fragrant; flower-stalks of varying lengths, the longest to 5 cm. Flowers 2.5–4 cm in diameter. Perianth-tube 1.2–2.5 cm; lobes 1–1.8 cm, ovate or narrowly ovate, overlapping at their bases, spreading, white. Corona pure white, *c.* 3 mm, margin usually slightly and minutely scalloped. *SW Europe & Mediterranean area*. H4. Winter–spring.

**10. N. polyanthus** Loiseleur (*N. papyraceus* Ker Gawler subsp. *polyanthus* (Loiseleur) Ascherson & Graebner).
Similar to *N. papyraceus* but leaves green, scape terete, corona entire. *S France; naturalised in Spain & Italy*. H4. Winter–spring.

**11. N. panizzianus** Parlatore (*N. papyraceus* Ker Gawler subsp. *panizzianus* (Parlatore) Arcangeli; *N. barlae* Parlatore).
Similar to *N. papyraceus* and *N. polyanthus*; scape strongly compressed and 2-edged; flowers 2–2.5 cm in diameter. *Portugal, Spain, SE France, N Italy*. H4. Winter–spring.

**12. N. pachybolbus** Durieu. Illustration: Botanical Magazine, 6825 (1885).
Bulb large, to 7 cm, dark. Leaves flat, 30–50 cm × 6–12 mm, green. Scape to 40 cm, compressed, 2-edged and ridged. Flowers 1.2–2 cm in diameter, in an umbel of 6–20; flower-stalks somewhat unequal, the longest to 5 cm. Perianth-tube 1–1.5 cm; lobes ovate-oblong, obtuse, 6–8 mm, white. Corona *c.* 3 mm, white. *Morocco, Algeria*. H5. Winter.

**13. N. canariensis** Burbidge. Illustration: Burbidge, The Narcissus, t. 48 (1875).
Similar to *N. pachybolbus* but bulb smaller, perianth-lobes ovate-lanceolate, acute, less than 6 mm. *Canary Islands*. G1. Winter–spring.

**14. N. dubius** Gouan. Illustration: Grey-Wilson & Mathew, Bulbs, pl. 24 (1981).
Bulb 2–3 cm, dark. Leaves 12–25 cm × 3–5 mm, flat, glaucous. Scape to 20 cm, strongly compressed. Flowers in umbels of 2–6; flower-stalks unequal, the longest to 4 cm. Perianth-tube 1–1.4 cm, cylindric but widening slightly above; lobes 6–8 mm, broadly elliptic, overlapping at the base, white. Corona 3–4 mm, white, the margin more or less lobed. *S France, NE Spain*. H4. Spring.

Thought to be derived by allopolyploidy from a hybrid between *N. papyraceus* and *N. assoanus*.

**15. N. elegans** (Haworth) Spach. Illustration: Grey-Wilson & Mathew, Bulbs pl. 24 (1981).
Bulb 2–3.5 cm, very dark. Leaves 8–25 cm × 2–4.5 mm, channelled, glaucous, appearing before or with the flowers. Scape to 35 cm. Flowers in umbels of 2–10, fragrant, 2–3 cm in diameter; flower-stalks unequal, the longest to 4.5 cm. Perianth-tube 1.1–2 cm, greenish; lobes 1.2–2 cm × 3–6 mm, narrowly oblong to elliptic, acute, not overlapping at the base, somewhat reflexed, white. Corona 1–2 mm high, 3–4 mm in diameter, yellowish green. Anthers not projecting from corona. *W Mediterranean area, Italy, Sicily*. H4. Autumn.

Variable in flower colour; a number of varieties has been recognised, based on this character.

\*N. × **rogendorfii** Battandier is the wild hybrid between *N. elegans* and *N. tazetta*. It is similar to *N. elegans* but has a larger corona, broader perianth-lobes and an umbel with up to 12 flowers. *Algeria*. H5. Autumn.

Section **Narcissus**. Leaves flat. Flower solitary. Perianth-tube cylindric; corona small. Filaments much shorter than anthers; anthers dorsifixed, partly projecting or all included. Flowering in spring–summer.

**16. N. poeticus** Linnaeus. Figure 37(2), p. 292. Illustration: de Wit, Plants of the world **2**: t. 136 (1963–5); Grey-Wilson & Mathew, Bulbs, pl. 24 (1981).
Bulb to 4 cm, dark. Leaves 20–40 cm × 6–10 or rarely to 12 mm, flat, more or less glaucous. Scape to 50 cm, compressed. Flower-stalk 1–4.5 cm. Perianth-tube 2–3 cm; lobes mostly 2–2.5 cm, obovate to almost circular, overlapping at their bases without distinct claws, white or pale cream. Corona disc-like or cup-shaped, to 1.4 cm in diameter, yellow with red or scarious, minutely

scalloped margin. Stamens in 2 whorls, the lower whorl included inside the perianth-tube, the upper whorl not projecting from the corona. *S Europe, from France to SW USSR.* H2. Spring–summer.

A very variable species in terms of flower size and the relative positions of the 2 whorls of stamens. Fernandes (1967) recognises 5 varieties:

Var. **poeticus**. Corona disc-like. *S France, Italy.*

Var. **verbanensis** Herbert (*N. verbanensis* (Herbert) Pugsley). Plant rather small. Corona cup-shaped, 2 mm high, 8–9 mm in diameter; flower 3.5–5 cm in diameter, perianth-lobes strongly mucronate. *Italy.*

Var. **hellenicus** (Pugsley) Fernandes (*N. hellenicus* Pugsley). Corona cup-shaped, c. 3 mm high, 1.2–1.4 cm in diameter; flower 3.5–5 cm in diameter, perianth-lobes obtuse though mucronate. *Greece.*

Var. **recurvus** (Haworth) Fernandes (*N. recurvus* Haworth). Flower 5–7 cm in diameter; corona cup-shaped, greenish yellow below the red margin. *Switzerland.*

Var. **majalis** (Curtis) Fernandes (*N. majalis* Curtis). Flower 5–7 cm in diameter; corona with a red margin and a whitish zone just below it. *France.*

As well as these (wild) varieties, many selections (including double-flowered variants) have been given cultivar names.

**17. N. radiiflorus** Salisbury (*N. angustifolius* Haworth; *N. poeticus* subsp. *radiiflorus* (Salisbury) Baker). Illustration: Botanical Magazine, 193 (1792); Grey-Wilson & Mathew, Bulbs, pl. 24 (1981).
Similar to *N. poeticus* but leaves 5–8 mm wide, perianth-lobes 2.2–3 cm, narrowly obovate, tapered to a conspicuous claw, not or scarcely overlapping at their bases; corona 2–2.5 mm high, 8–10 mm in diameter. *SC Europe & the western part of the Balkan peninsula.* H3. Spring–summer.

Not always clearly separable from *N. poeticus*, and, like it, very variable. Fernandes (1967) recognises the following varieties:

Var. **radiiflorus**. Corona cup-shaped, 2–2.5 mm high, 8–10 mm in diameter, bright yellow edged with red. *Switzerland, Austria, Yugoslavia.*

Var. **stellaris** (Haworth) Fernandes (*N. stellaris* Haworth). Corona cup-shaped, like that of var. *radiiflorus* but with a narrow, white zone within the red margin. *S France to Romania (? and farther east).*

Var. **poetarum** Burbidge & Baker. Corona flat and disc-like, wholly red. *Origin uncertain.*

Var. **exertus** (Haworth) Fernandes (*N. majalis* var. *exertus* Haworth; *N. exertus* (Haworth) Pugsley). Corona flat and disc-like, bright yellow with a scarlet margin. *Switzerland and adjacent France.*

**18. N. × medioluteus** Miller (*N. biflorus* Curtis). Illustration: Grey-Wilson & Mathew, Bulbs, pl. 26 (1981).
Bulb to 6 cm. Leaves 45–70 cm × 7–10 mm, flat, glaucous. Scape to 60 cm, compressed. Flowers usually 2 (more rarely 3 or 4) per umbel, fragrant; flower-stalks 2.5–3.5 cm. Perianth-tube 2–2.5 cm, widened towards the apex; lobes 1.8–2.2 cm, white, almost circular. Corona 3–5 mm high, 9–12 mm in diameter, bright yellow, with a red or scarious margin. *S France, widely naturalised elsewhere in Europe.* H3. Spring.

The natural hybrid between *N. poeticus* and *N. tazetta.*

Section **Jonquillae** de Candolle (Section *Apodanthi* Fernandes). Leaves narrow, usually not flat. Flowers solitary or in umbels. Perianth-tube cylindric or slightly widened above; lobes broad and overlapping. Corona small to large, its diameter greater than its height. Filaments much shorter than the anthers; anthers dorsifixed, included in the corona or partly projecting. Flowering mostly in spring.

**19. N. jonquilla** Linnaeus. Illustration: Botanical Magazine, 15 (1787); Genders, Bulbs, 273 (1973); Ceballos et al., Plantas silvestres de la peninsula Iberica, 392 (1980). Bulb 1–2.5 cm, dark. Leaves erect, dark green, almost terete, channelled on the inner surface, slightly striate on the outer, 1–4 mm wide. Flowers in umbels of 2–5, very fragrant, yellow; flower-stalks unequal, the longest to 5 cm. Flowers 2–3.5 cm in diameter. Perianth-tube straight, 2–3 cm; lobes ovate to elliptic or obovate, mucronate, 1–1.5 cm. Corona 2–4 mm high, 9–15 mm in diameter, of the same colour as the perianth-lobes, margin finely scalloped. *Spain, Portugal, widely naturalised elsewhere in Europe.* H4. Winter–spring.

Cultivated on a commercial scale in S France for its perfume.

**20. N. × tenuior** Curtis. Illustration: Botanical Magazine, 379 (1797).
Like *N. jonquilla* but leaves narrowly linear and flat, corona flatter, to 5 mm high, of a deeper yellow than the perianth-lobes. *Garden origin.* H4. Spring.

The hybrid between *N. jonquilla* and *N. poeticus.*

**21. N. assoanus** Dufour (*N. juncifolius* of authors; *N. requienii* Roemer). Illustration: Rix & Phillips, The bulb book, 108 (1981). Bulb 1–3 cm, dark. Leaves 1–2 mm wide, usually shorter than the scape, green. Scape to 25 cm. Flowers solitary or in pairs, yellow, fragrant, 1.6–2.2 cm in diameter; flower-stalks unequal, the longer to 2 cm, all included within the spathe. Perianth-tube 1.2–2 cm, straight; lobes spreading, obovate, 7–10 mm; corona 4–6 mm high, 1–1.7 cm in diameter. *S France, S & E Spain.* H5. Spring.

Fernandes (1967) recognised 2 varieties under *N. requienii*, var. *requienii* with yellow flowers and var. *pallens* (Willkomm) Fernandes (*N. pallens* Willkomm) with pale, sulphur-yellow flowers.

**22. N. willkommii** (Sampaio) Fernandes (*N. jonquilloides* Willkomm). Similar to *N. assoanus* but leaves 2–3 mm wide, flowers in umbels of 2 or 3, 1.4–1.9 cm in diameter, the longest flower-stalk exceeding the spathe, perianth-tube 1.1–1.5 cm, lobes 6–9 mm, corona 5–6 mm high, 7–11 mm in diameter. *S Portugal, SW Spain.* H5. Spring.

Hybrids with *N. gaditanus* occur in S Portugal.

**23. N. fernandesii** Pedro. Similar to *N. assoanus* and *N. willkommii*; longer flower-stalk longer than the spathe, perianth-tube 1.5–2.2 cm, lobes 9–11 mm, corona 4–5 mm. *S Portugal.* H5. Spring.

A somewhat obscure species; in many respects intermediate between *N. willkommii* and *N. gaditanus.*

**24. N. gaditanus** Boissier & Reuter (*N. minutiflorus* Willkomm). Illustration: Taylor, Wild flowers of Spain and Portugal, 19 (1972); Rix & Phillips, The bulb book, 108 (1981). Similar to *N. assoanus* but leaves usually exceeding the scape, the perianth-tube slightly to strongly curved, 8–17 mm, and the corona 4–8 mm in diameter. *S Portugal, S Spain.* H4. Winter–spring.

**25. N. rupicola** Dufour. Illustration: Botanical Magazine, 6473C (1880), n.s., 577 (1970); Ceballos et al., Plantas silvestres de la peninsula Iberica, 394 (1980); Rix & Phillips, The bulb book, 108 (1981).
Bulb 1–2.5 cm, dark. Leaves more or less glaucous, channelled on the inner face and with 2 keels on the outer, to 15 cm × 1.5–3 mm, erect or prostrate, margins smooth. Scape exceeding the

leaves, 1-flowered. Spathe conspicuous, tubular for three-quarters of its length; flower-stalk usually absent, rarely to 3 mm. Flowers yellow, not or faintly fragrant. Perianth-tube 1.3–2 cm, lobes obovate, 7–13 × 5–10 mm. Corona 2–5 mm high, 7–10 mm in diameter, usually 6-lobed, more rarely sinuous or almost entire. *N Portugal, C Spain, Morocco*. H4. Spring–summer.

Fernandes (1967) recognises a number of subspecies, of which the following are cultivated:

Subsp. **rupicola**. Whole plant glaucous; perianth-tube 1.3–2 cm. *C Spain, N Portugal*.

Subsp. **marvieri** (Jahandiez & Maire) Maire & Weiller (*N. marvieri* Jahandiez & Maire). Plant scarcely glaucous; perianth-tube 2–2.7 cm. *Morocco*.

**26. N. watieri** Maire (*N. rupicola* Dufour subsp. *watieri* (Maire) Maire & Weiller). Illustration: Botanical Magazine, 9443 (1936); Rix & Phillips, The bulb book, 108 (1981).
Similar to *N. rupicola* but flower-stalk to 1 mm, flower white or cream-white, the corona with sinuous margins. *Morocco*. H5. Spring.

**\*N. atlanticus** Stern. Illustration: Daffodil & Tulip Yearbook **16**: f. 13 (1950). Similar to both *N. rupicola* and *N. watieri* but with larger, cream-white flowers which are strongly fragrant. *Morocco*. H5. Spring.

**27. N. calcicola** Mendonça. Illustration: Boletim do Sociedade Broteriana, ser. 2, **7**: 318 (1930); Botanical Magazine, n.s., 180 (1952); Grey-Wilson & Mathew, Bulbs, pl. 25 (1981).
Similar to *N. rupicola* but spathe tubular for up to half its length, subtending usually 2–5 flowers which are 1.7–2.5 cm in diameter; perianth-tube 1.3–1.8 cm, lobes 7–14 × 5–8 mm, corona 4–8 mm high, 6–9 mm in diameter, margins obscurely scalloped. *Portugal*. H4. Spring.

**28. N. scaberulus** Henriques.
Similar to *N. rupicola* but leaves 4-angled, often prostrate or arching over and twisted, and with rough angles; spathe tubular for up to one-quarter of its length, subtending 1–3 flowers. Flower-stalks 8–15 mm. Flowers 1.2–1.7 cm in diameter, perianth-tube 1.2–1.7 cm, lobes 4.5–7 × 3–5 mm. Corona 2–5 mm high, 5–7 mm in diameter, margin entire or finely scalloped. *N Portugal*. H4. Spring.

**29. N. viridiflorus** Schousboe. Illustration: Botanical Magazine, 1687 (1815); Polunin

& Smythies, Flowers of SW Europe, t. 60 (1973); Grey-Wilson & Mathew, Bulbs, pl. 24 (1981).
Bulb 2–3 cm, dark or pale. Leaves appearing soon after the flowers, dark green, cylindric, hollow, to 30 cm × 4 mm. Flowers in umbels of 2–5, dull green with an unpleasant smell; flower-stalks very unequal, the longest to 7 cm. Perianth-tube 1–1.5 cm, narrow, widening somewhat towards the apex; lobes 1–1.6 cm, linear-oblong, acute, widely spreading or somewhat reflexed. Corona *c*. 1 mm high, 3–5 mm in diameter, 6-lobed. *Spain (Gibraltar area), Morocco*. H5. Autumn.

Hybrids with *N. serotinus* occur in the Gibraltar area.

Section **Ganymedes** Schultes & Schultes. Leaves flat to almost cylindric, narrow. Flowers solitary or in umbels. Perianth-tube cylindric; lobes narrow, reflexed. Corona more or less bell-shaped to almost cylindric, its diameter equalling or exceeding its height. Filaments very unequal, those of the lower whorl as long as the anthers, those of the upper whorl longer than the anthers; anthers dorsifixed. Flowering in spring.

**30. N. triandrus** Linnaeus. Illustration: Botanical Magazíne, 48 (1793); Taylor, Wild flowers of Spain and Portugal, 31 (1972); Polunin & Smythies, Flowers of SW Europe, t. 60 (1973); Rix & Phillips, The bulb book, 108 (1981).
Bulb 1–2 cm, dark. Leaves 15–30 cm × 1.5–3 mm, dark green, blunt. Scape about as long as to longer than the leaves, to 45 cm. Flowers solitary or up to 6 in an umbel, drooping, white to bright yellow. Perianth-tube 1–2 cm, lobes 1–2.5 cm (rarely to 3 cm), lanceolate to linear-oblong, strongly reflexed. Corona 5–25 mm high, 7–25 mm in diameter, margin sinuous. The upper 3 stamens and the style project from the corona. *Spain, Portugal, NW France*. H3. Spring.

A very variable species. Numerous attempts have been made to divide it up into species, subspecies or varieties, but none of these seems satisfactory. In general, plants with solitary, usually bright yellow flowers are referred to N. **concolor** Link (*N. pallidulus* Graells; *N. triandrus* subsp. *pallidulus* (Graells) Webb – Illustration: Ceballos et al., Plantas silvestres de la peninsula Iberica, 394, 1980); plants with 2 or 3 flowers which are pale yellow are referred to N. **triandrus** var. **cernuus** (Salisbury) Baker; and plants with pure white flowers to N. **triandrus** var. **triandrus**;

plants with the perianth-lobes a deeper yellow than the corona are referred to N. **pulchellus** Salisbury, which may be a hybrid between *N. concolor* and *N. triandrus* var. *cernuus*. This scheme has simplicity to recommend it, but it is not adequate, and the group requires further study.

Section **Bulbocodii** de Candolle (genus *Corbularia* Haworth). Bulb small. Leaves narrow. Flowers solitary, bilaterally symmetric. Perianth-lobes narrow and spreading. Corona very conspicuous, funnel-shaped. Stamens with long filaments which are deflexed downwards and then curved upwards near their apices; anthers dorsifixed.

A complex of 5 very similar species, differing in size, flower colour and flowering time. There is little agreement in the literature as to what species should be recognised, what infraspecific units they should contain, and what their precise geographical distribution is. The account given here broadly follows Fernandes (1967) in including 5 species; these are often very difficult to distinguish. For 3 of them, an outline of the infraspecific units recognised by Fernandes is given; some of these variants are very similar and the keys can only be regarded as provisional. The whole group requires much further study.

**31. N. bulbocodium** Linnaeus (*Corbularia bulbocodium* (Linnaeus) Haworth). Illustration: The Garden **101**: 351 (1976); Ceballos et al., Plantas silvestres de la peninsula Iberica, 393 (1980) – var. nivalis; Rix & Phillips, The bulb book, 107 (1981); Il Giardino Fiorito **48**: 501 (1982).
Bulb 1.5–2 cm, whitish or pale brown. Leaves 2–several from each bulb, 1.5–4 mm broad, dark green, usually erect. Scape usually exceeding the leaves. Flowers horizontal or almost so, pale to deep yellow or orange-yellow, often tinged with green. Flower-stalk at least 5 mm, usually more than 9 mm, at most 3 cm. Perianth-tube 4–25 mm; lobes 6–15 × 1.5–5 mm, linear to narrowly triangular, acute. Corona 7–25 × 9–35 mm, not at all narrowed at the margin. Stamens and style usually included in the corona, occasionally slightly projecting from it. *SW & W France, Spain, Portugal, N Africa*. H4. Winter–spring.

The most widely distributed of the species of the section. The following key provides a synopsis of those variants recognised by Fernandes which are thought to be in cultivation:

1a. Flower deep or golden yellow (subsp. **bulbocodium**) 2
  b. Flower pale yellow 5
2a. Flower-stalk to 2 cm; perianth to 3.5 cm 3
  b. Flower-stalk more than 2 cm; perianth 3.5 cm or more 4
3a. Leaves to 1.5 mm broad, the outer surface slightly striate; scape smooth (*W France, Spain, Portugal, Morocco*) var. **bulbocodium**
  b. Leaves more than 1.5 mm broad, the outer surface conspicuously striate; scape slightly ridged (*Spain, Morocco*) var. **nivalis** (Graells) Baker
4a. Leaves to 2.5 mm broad, erect, about as long as the scape; flower 3–3.5 cm, the corona *c.* 2 cm in diameter (*W France and Spain*) var. **conspicuous** (Haworth) Baker
  b. Leaves to 4 mm broad, recurved, longer than the scape; flower 3.5–5 cm, the corona *c.* 3 cm in diameter (*W Portugal*) var. **serotinus** (Haworth) Fernandes
5a. Flowers pale lemon yellow, 3.5–5 cm (subsp. **bulbocodium** – *Spain*) var. **citrinus** Baker
  b. Flowers pale yellow, to 3.5 cm 6
6a. Flowering in spring; stamens usually projecting from the corona (subsp. **bulbocodium** – *Spain*) var. **graellsii** (Webb) Baker
  b. Flowering in winter; stamens not projecting (subsp. **praecox** – *Morocco*) var. **praecox** Gattefosse & Weiller

**32. N. obesus** Salisbury (*N. bulbocodium* subsp. *obesus* (Salisbury) Maire). Illustration: Botanical Magazine, n.s., 650 (1973); Grey-Wilson & Mathew, Bulbs, pl. 25 (1981).
Very similar to *N. bulbocodium* but leaves very narrow, prostrate or arching and deflexed; perianth usually without any green tinge, the corona slightly narrowed at its margin. *Portugal, Spain*. H4. Spring.

**33. N. romieuxii** Braun-Blanquet & Maire (*N. bulbocodium* subsp. *romieuxii* (Braun-Blanquet & Maire) Emberger & Maire). Illustration: Rix & Phillips, The bulb book, 106, 107 (1981).
Bulb about 1 cm, dark brown. Leaves 2–several per bulb, as long as to somewhat shorter than the scape, striate on the outer side, dark green, more than 1 mm broad. Scape to 20 cm. Spathe whitish or dark-coloured. Flower-stalk 1–5 mm. Flowers pale yellow to almost white but with a yellow tinge or flush. Perianth-lobes 1–2.5 mm broad, acute, spreading. *Morocco*. H5. Winter–spring.

Very similar to both *N. bulbocodium* and *N. hedraeanthus*, but known only from Morocco. The following variants recognised by Fernandes may be grown:

1a. Flowers pale yellow (subsp. **romieuxii**) 2
  b. Flowers whitish, flushed or tinged with yellow (subsp. **albidus** (Emberger & Maire) Fernandes) 3
2a. Spathe whitish, scarious; flower-stalk to 3 mm, rarely more var. **romieuxii**
  b. Spathe dark-coloured; flower-stalk more than 3 mm var. **rifanus** (Emberger & Maire) Fernandes
3a. Perianth-lobes to 2.5 mm wide at the base, longer than the corona var. **albidus**
  b. Perianth-lobes to 1.5 mm wide at the base, as long as, or shorter than the corona var. **zaianicus** (Maire et al.) Fernandes

The very early-flowering variant known as var. *mesatlanticus* Emberger & Maire, is probably the same as var. *romieuxii* above.

**34. N. cantabricus** de Candolle (*N. clusii* Dunal). Illustration: Botanical Magazine, 5831 (1870) – subsp. monophyllus; Ceballos et al., Plantas silvestres de la peninsula Iberica, 392 (1980) – subsp. cantabricus; Rix & Phillips, The bulb book, 107 (1981) – subsp. monophyllus & cantabricus.
Bulb 1–2 cm, dark brown. Leaves 1 or more to each bulb, erect, 1–1.5 mm broad, dark green, as long as or longer than the scape. Scape erect, 10–25 cm. Spathe 2–3 cm, pale or dark. Flower-stalk 3–9 mm. Flowers horizontal, pure white or white flushed with green. Perianth-tube 1.2–2 cm, lobes 8–12 × 2–4.5 mm, linear-lanceolate. Corona 1.2–1.8 × 2–3.5 cm, margin expanded and frilled. Stamens and style not projecting from the corona. *S Spain, Morocco, Algeria*. H5. Winter.
Fernandes recognises 3 subspecies (as below), these further divided into varieties which are not included here, as it is uncertain which of them are in cultivation.

1a. Leaves mostly 1 to each bulb (*Spain, Morocco, Algeria*) subsp. **monophyllus** (Durieu) Fernandes
  b. Leaves 2 or more to each bulb 2
2a. Leaves flexuous, 0.5–1.5 mm broad; stamens more or less equal (*Spain, Morocco*) subsp. **cantabricus**
  b. Leaves stiff, 1–2.5 mm broad; stamens very unequal (*Morocco*) subsp. **tananicus** (Maire) Fernandes

**35. N. hebraeanthus** (Webb & Heldreich) Colmeiro (*Corbularia hedraeanthus* Webb & Heldreich). Figure 37(4), p. 292. Illustration: Botanical Magazine, n.s., 248 (1955); Rix & Phillips, The bulb book, 106 (1981); Grey-Wilson & Mathew, Bulbs, pl. 25 (1981).
Bulb *c.* 1 cm, dark brown. Leaves 1 per bulb, thread-like or terete, *c.* 1 mm broad, striate on the outer surface, dark green, usually exceeding the scape. Scape to 8 cm, emerging horizontally, later curved or erect. Spathe dark, 1 cm or more. Flower-stalk absent. Flowers horizontal, pale sulphur-yellow. Perianth-tube to 1.25 cm, lobes 3–4.5 mm broad, oblong, obtuse but with a fine, acute point, not spreading widely. Corona 8–13 mm. Stamens and style conspicuously projecting from the corona, the style exceeding the stamens. *Spain (Jaén, Ciudad Reál)*. H4. Winter–spring.

Section **Pseudonarcissi** de Candolle (Section *Ajax* (Haworth) Dumortier). Leaves flat, usually wide. Flowers usually solitary. Perianth-tube broadly funnel-shaped, the lobes spreading. Corona large, more or less cylindric, its height at least twice its diameter. Anthers usually basifixed, not projecting from the corona. Flowering in spring.

**36. N. longispathus** Pugsley. Illustration: Botanical Magazine, n.s., 246 (1955); Grey-Wilson & Mathew, Bulbs, pl. 26 (1981).
Bulb to 3 cm. Leaves at least 40 cm, often much longer, 1–1.5 cm wide. Scape exceeding the leaves, to 1.75 m. Spathe 6–10 cm, green at flowering. Flowers solitary or occasionally in pairs, ascending; stalks 4–9 cm. Perianth-tube 1–1.5 cm; lobes 2.5–3.2 cm, yellow, spreading, not twisted. Corona 2.5–3 cm, slightly deeper yellow than the lobes, its margin expanded and scalloped. *SE Spain*. H5. Spring.

**37. N. pseudonarcissus** Linnaeus. Figure 37(1), p. 292.
Bulb 2–5 cm, dark or rather pale. Leaves 8–50 cm × 5–16 mm, usually glaucous, erect or somewhat spreading, obtuse. Scape usually exceeding the leaves, to 90 cm. Spathe 2–6 cm, scarious. Perianth-tube 1.5–2.5 cm; lobes 1.5–5.5 cm, white to deep yellow, spreading, sometimes twisted. Corona 1.5–4.5 cm or more, white to deep yellow, the margin sometimes conspicuously expanded and then sometimes reflexed, minutely to deeply toothed or lobed. Stamens and style

included in the corona, the anthers close and forming a ring beneath the stigma. *W Europe, long cultivated and naturalised in many places.* H2. Spring.

The most widespread of the species, and the most difficult in terms of its classification. It has been divided up into some 20 species (Pugsley, 1933), many of them further divided into varieties. But the existence of numerous intermediates (both wild and cultivated) makes this treatment unusable for many specimens. The most reasonable account seems to be that in Flora Europaea, and that is followed here with minor modifications.

1a. Corona white to pale yellow    2
 b. Corona deep golden yellow    4
2a. Flower-stalk 1–2.5 cm; corona white to very pale yellow
                                        subsp. **moschatus**
 b. Flower-stalk 3–10 mm; corona pale yellow    3
3a. Corona 3–4 cm, distinctly expanded and usually lobed at the margin
                                        subsp. **pallidiflorus**
 b. Corona 2–3.5 cm, scarcely expanded at the margin and usually only slightly lobed    subsp. **pseudonarcissus**
4a. Perianth-lobes to 2 cm    5
 b. Perianth-lobes 2 cm or more    6
5a. Perianth-lobes twisted; corona expanded at the margin    subsp. **major**
 b. Perianth-lobes not twisted; corona not expanded at the margin
                                        subsp. **nevadensis**
6a. Perianth-lobes white or yellow, paler than the corona    7
 b. Perianth-lobes deep golden yellow, of the same colour as the corona    8
7a. Flower-stalk 8–25 mm; corona conspicuously expanded at margin; flowers horizontal or ascending
                                        subsp. **nobilis**
 b. Flower-stalk 3–12 mm; corona scarcely expanded at margin; flowers horizontal or drooping
                                        subsp. **pseudonarcissus**
8a. Leaves to 15 cm    subsp. **portensis**
 b. Leaves at least 20 cm    9
9a. Leaves 20–50 cm long, 5–15 mm broad    subsp. **major**
 b. Leaves 20–30 cm long, 6–10 mm broad    subsp. **obvallaris**

Subsp. **pseudonarcissus**. Illustration: Botanical Magazine, n.s., 216 (1953); Grey-Wilson & Mathew, Bulbs, pl. 26 (1981). Leaves 12–35 cm × 6–12 mm. Flowers horizontal or drooping, stalks 3–12 mm. Perianth-lobes 2–3.5 cm, white to yellow, more or less twisted. Corona 2–3.5 cm, deeper yellow than the lobes, its margin scarcely expanded, not or little lobed. *W Europe except for Portugal and S Spain.* H2. Spring.

Subsp. **pallidiflorus** (Pugsley) Fernandes (*N. pallidiflorus* Pugsley; *N. macrolobus* (Jordan) Pugsley; *N. pallidus* misapplied). Illustration: Polunin & Smythies, Flowers of SW Europe, t. 60 (1973); Grey-Wilson & Mathew, Bulbs, pl. 26 (1981). Leaves 15–30 cm × 5–12 mm. Flowers horizontal or drooping, stalks 3–10 mm. Perianth-lobes 3–4 cm, cream to pale yellow, more or less twisted; corona 3–4 cm, deeper yellow than the lobes but still pale, margin expanded and usually reflexed. *France, Spain (Pyrenees and northern Spanish mountains).* H3. Spring.

Plants from the Pyrenees with large flowers are sometimes recognised as *N. macrolobus.*

Subsp. **moschatus** (Linnaeus) Baker (*N. moschatus* Linnaeus; *N. tortuosus* Haworth; *N. albescens* Pugsley; *N. alpestris* Pugsley). Illustration: Botanical Magazine, 1300 (1810); Grey-Wilson & Mathew, Bulbs, pl. 26 (1981). Leaves 10–40 cm × 5–12 mm. Flowers horizontal or drooping. Perianth-lobes 2–3.5 cm, white or cream, twisted; corona 2.5–4 cm, usually the same colour as the lobes, rarely pale yellow, its margins scarcely expanded. *France, Spain (Pyrenees and northern Spanish mountains).* H4. Spring.

Plants with the flowers completely white have been recognised as *N. alpestris* and plants with flowers at the larger end of the range have been treated as *N. tortuosus* (flower-stalks 1–1.8 cm) and *N. albescens* (flower-stalks 2–2.5 cm), but all these units intergrade.

Subsp. **nobilis** (Haworth) Fernandes (*N. nobilis* Haworth). Illustration: Taylor, Wild flowers of Spain and Portugal, 99 (1972); Ceballos et al., Plantas silvestres de la peninsula Iberica, 390 (1980); Grey-Wilson & Mathew, Bulbs, pl. 26 (1981). Leaves 15–50 cm × 8–15 mm. Flowers horizontal or ascending, stalks 8–25 mm. Perianth-lobes 3–3.5 cm, pale yellow, usually twisted; corona 3–4.5 cm, deep golden yellow, margin expanded. *N Portugal, N Spain.* H3. Spring.

Plants with very large flowers which have the margin of the corona very conspicuously expanded have been recognised as *N. leonensis* Pugsley (*N. nobilis* var. *leonensis* (Pugsley) Fernandes). The name *N. gayii* Pugsley has been applied to similar plants with rather smaller, pale flowers; the origin of this variant is unknown.

Subsp. **major** (Curtis) Baker (*N.major* Curtis; *N. hispanicus* Gouan; *N. confusus* Pugsley). Leaves 20–50 cm × 5–15 mm. Flowers horizontal to almost erect, stalks 8–30 mm. Perianth-lobes 1.8–4 cm, yellow to deep yellow, twisted; corona 2–4 cm, the same colour as the lobes, its margin more or less expanded. *S France, Spain, Portugal.* H4. Spring.

Plants with long flower-stalks which are conspicuously curved at the top have been recognised as *N. hispanicus*; and plants with shorter, more or less straight flower-stalks as *N. confusus.*

Subsp. **portensis** (Pugsley) Fernandes (*N. portensis* Pugsley). Leaves 8–12 cm × 5–7 mm, much shorter than the scape. Flowers horizontal or slightly drooping, stalks 5–15 mm. Perianth-lobes 2–3 cm, not twisted, deep yellow; corona 2.5–3.5 cm, the same colour as the perianth-lobes, rather flaring, but not conspicuously expanded at the margin. *N Portugal, N & C Spain.* H3. Spring.

Subsp. **obvallaris** (Salisbury) Fernandes (*N. obvallaris* Salisbury). Leaves 20–30 cm × 8–10 mm. Flowers horizontal, stalks 1–1.5 cm. Perianth-lobes 2.5–3 cm, deep golden yellow; corona 3–3.5 cm, the same colour as the lobes, its margins expanded and sometimes reflexed. *Origin uncertain, naturalised in Britain (Wales).* H2. Spring.

Subsp. **nevadensis** (Pugsley) Fernandes (*N. nevadensis* Pugsley). Illustration: Taylor, Wild flowers of Spain and Portugal, 34 (1972); Grey-Wilson & Mathew, Bulbs, pl. 26 (1981). Leaves 12–30 cm × 5–10 mm. Flowers horizontal to almost erect, often in umbels of 2–4; stalks 2–3.5 cm. Perianth-lobes 1.5–2 cm, pale yellow, not twisted; corona 1.5–2.5 cm, bright yellow. *S Spain (Sierra Nevada).* H4. Spring.

**38. N. × odorus** Linnaeus. Illustration: Botanical Magazine, 934 (1806); Grey-Wilson & Mathew, Bulbs, pl. 26 (1981). Bulb to 3 cm. Leaves 35–50 cm × 6–8 mm, keeled, bright green. Scape to 40 cm, more or less terete. Flowers in an umbel of 1–4, usually fragrant. Perianth-tube to 2 cm, narrowly funnel-shaped; lobes to 2.5 × 1.3 cm, bright yellow. Corona 1.3–1.8 cm, 1.7–2 cm in diameter, the margin lobed or almost entire. *Garden origin; naturalised in S Europe.* H3. Spring.

The garden hybrid between *N. jonquilla* and *N. pseudonarcissus.*

**39. N. × incomparabilis** Miller. Illustration: Grey-Wilson & Mathew, Bulbs, pl. 26 (1981).

Bulb to 3 cm. Leaves 17–35 cm ×
8–12 mm, more or less flat, somewhat
glaucous. Scape to 40 cm, compressed.
Flower solitary, not or only slightly
fragrant. Perianth-tube 2–2.5 cm,
narrowly funnel-shaped; lobes
2.5–3 × 1.2–1.6 cm, spreading, pale
yellow. Corona 1.3–2.2 cm, 1.7–2 cm in
diameter, deep orange-yellow, its margin
wavy and deeply lobed. *S & C France;
naturalised elsewhere in Europe.* H3. Spring.

The natural hybrid between *N. poeticus*
and *N. pseudonarcissus*.

**40. N. bicolor** Linnaeus (*N. abscissus*
(Haworth) Roemer & Schultes). Illustration:
Botanical Magazine, 1187 (1809); Polunin
& Smythies, Flowers of SW Europe, t. 60
(1973).
Bulb to 3.5 cm, dark. Leaves
30–35 × 1–2 cm. Scape to 40 cm. Flower
solitary, horizontal, stalk 1.5–3.5 cm.
Perianth-tube 6–12 mm, lobes 3.5–4 cm,
cream or pale yellow, spreading, not
twisted. Corona 3.5–4 cm, 2.5–3 cm in
diameter, deep yellow, its margin scarcely
expanded, entire or finely scalloped. *France
& Spain (Pyrenees).* H3. Spring.

*N. abscissus* is the name that has been
applied to wild plants with narrow leaves,
narrower perianth-lobes and a corona
1.5–2 cm in diameter; these intergrade
with *N. bicolor*.

**41. N. minor** Linnaeus (*N. provincialis*
Pugsley; *N. pumilus* Salisbury; *N. nanus*
Spach; *N. parviflorus* (Jordan) Pugsley).
Illustration: Botanical Magazine, n.s., 410
(1962); Rix & Phillips, The bulb book, 107
(1981).
Bulb 2–3 cm, brown. Leaves
8–25 cm × 3–14 mm, glaucous. Scape to
25 cm. Flower solitary, horizontal or
drooping, stalk 3–20 mm. Perianth-tube
9–15 mm, lobes 1.6–2.2 cm, pale or deep
yellow, erect-spreading, sometimes twisted.
Corona 1.6–2.5 cm, deep yellow, its
margin toothed or lobed. *France, N Spain.*
H3. Spring.

Variable in size and flower colour; this
variation, which is not well correlated, has
been used for the recognition of several
species treated here as synonymous with *N.
minor*.

**42. N. asturiensis** (Jordan) Pugsley (*N.
minimus* misapplied). Illustration: Botanical
Magazine, 9495 (1937); Polunin &
Smythies, Flowers of SW Europe, t. 60
(1973); Gartenpraxis, 21 (1976); Rix &
Phillips, The bulb book, 107 (1981).
Bulb 1.5–2 cm, brown. Leaves

5–15 cm × 2–8 mm, glaucous. Scape
usually to 10 cm, rarely more. Flower
solitary, drooping, stalk 3–10 mm.
Perianth-tube 5–9 mm, lobes 7–14 mm,
erect-spreading, not twisted, yellow. Corona
8–16 mm, constricted at about the middle,
inflated above and below the constriction,
yellow, margin spreading or reflexed,
variously lobed. *N Portugal, N & C Spain.*
H4. Spring.

**43. N. cyclamineus** de Candolle.
Illustration: Botanical Magazine, 6950
(1887); Gartenpraxis, 21 (1976); Rix &
Phillips, The bulb book, 109 (1981).
Bulb to 2 cm, brownish. Leaves
15–30 cm × 4–6 mm, bright green. Scape
15–30 cm. Flower solitary, drooping or
pointing downwards almost vertically,
spathe *c.* 2 cm, green at flowering, later
scarious; flower-stalk 1.5–2.5 cm.
Perianth-tube 2–3 cm, lobes to 2 cm,
lanceolate, reflexed backwards at almost
180° to the corona, hiding the ovary,
perianth-tube and part of the flower-stalk,
deep yellow. Corona to 2 cm, cylindric but
widening slightly above, its margin toothed
or scalloped. *NW Portugal, NW Spain.* H4.
Early spring.

A parent of many popular, early-
flowering hybrids.

**14. TAPEINANTHUS** Herbert
*J. Cullen*
Very like *Narcissus*. Leaves absent at
flowering. Flowers with perianth-tube
extremely short, corona rudimentary or
absent.

A genus of a single species from Spain
and N Africa, very closely related to, and
perhaps not distinct from, *Narcissus*.
Cultivation as for *Narcissus* (p. 301).

**1. T. humilis** (Cavanilles) Herbert (*Narcissus
humilis* (Cavanilles) Traub). Illustration:
Stocken, Andalusian flowers and
countryside, 93 (1969); Polunin &
Smythies, Flowers of south-west Europe, t.
61 (1973).
Bulb to 1.5 cm. Leaves 1 or rarely 2 from
each bulb, to 20 cm × 1 mm, appearing in
spring and dying off before flowering. Scape
7–20 cm. Flower solitary (rarely 2
together), erect, yellow. Perianth-lobes to
1 cm × 2–3 mm. *SW Spain, N Africa.* H5.
Autumn.

The validity of the generic name
*Tapeinanthus* is uncertain, and this plant is
sometimes found in the literature under the
generic names *Braxireon* Rafinesque and
*Carregnoa* Boissier.

**15. CRINUM** Linnaeus
*E. Campbell*
Herbaceous perennials with bulbs which
usually have a distinct neck made up of the
remains of sheathing leaf bases. Leaves
linear or strap-shaped, channelled or flat,
usually dying off in winter. Flowers in a
terminal umbel subtended by 2 spathes,
borne on a solid, compressed scape which
arises laterally from the bulb. Perianth
radially or bilaterally symmetric, with a
long tube which may be straight or curved,
the 6 lobes almost equal, spreading widely
into a funnel- or bell-shape, white or white
with various shadings of red or pink.
Stamens 6, inserted in the throat of the
perianth-tube, filaments arching or
deflexed, anthers versatile. Ovary 3-celled,
ovules numerous. Style ascending or
deflexed. Fruit almost spherical, irregularly
dehiscent. Seeds almost spherical.

A genus of about 120 species distributed
throughout the tropics and warm
temperate regions of the world. Some of the
species will grow out-of-doors in a sunny
position; these may require protection or
lifting and drying during winter. All require
a rich but well-drained soil. Propagation is
by seeds or offsets.
Literature: Uphof, J.C., A Review of the
Species of Crinum, *Herbertia* 9: 63–69
(1942); Nordal, I., Revision of the E African
Taxa of the Genus Crinum, *Norwegian
Journal of Botany* 24: 177–94 (1977);
Verdoorn, I.C., The Genus Crinum in
southern Africa, *Bothalia* 11: 27–50
(1973–5).

*Bulb.* Neck less than 15 cm: 1,2,6,9,12–14;
neck usually more than 15 cm: 3–11.
*Leaves.* Up to 2.5 cm broad: 1,5,12; more
than 2.5 cm broad: 2–14.
*Scape.* Green: 1,3,4,6–9,11,12; green
tinged with red or purple: 2,5,10,12–14.
*Flowering time.* Spring: 5,8,10,11,13;
summer: 1–3,6–14; autumn: 4,8,11,13;
winter: 13.
*Flowers.* 12 or fewer: 1–7,12–14; 15 or
more: 2,4,5,8–11. Fragrant: 2–6,8–14;
not fragrant: 1,7.
*Perianth-tube.* Curved: 1–7, 12–14;
straight: 8–14. Green: 2,3,5–9,12,14;
green tinged with red or purple:
1,2,4,5,12,13; red or purple: 10,11.
*Perianth-lobes.* 1.5 cm or more wide: 1–7;
less than 1.5 cm wide: 8–14.

1a. Perianth-tube distinctly curved; lobes
1.5 cm or more wide; stamens and
style deflexed                               2
  b. Perianth-tube straight or very slightly

curved; lobes less than 1.5 cm wide;
stamens and style not deflexed    8
2a. Leaves folded to form a channel    3
   b. Leaves flat    6
3a. Flowers usually 4–7 per umbel;
perianth-lobes to 5 cm
     **1. campanulatum**
   b. Flowers usually 6–20 per umbel;
perianth-lobes 5 cm or more    4
4a. Bulb with neck less than 15 cm; leaf
margins without teeth    **2. latifolium**
   b. Bulb with neck c. 30 cm; leaf margins
toothed    5
5a. Leaf margins narrowly cartilaginous,
not wavy; flowering in summer
     **3. bulbispermum**
   b. Leaf margin rough, wavy; flowering
in autumn    **4. macowanii**
6a. Leaves erect, not curved or bent over;
perianth-tube green tinged with
purple; flowering in spring
     **5. zeylanicum**
   b. Leaves recurved or bent over;
perianth-tube green; flowering in
summer    7
7a. Bulb without a neck or neck
less than 30 cm, flowers fragrant
     **6. jagus**
   b. Bulb with neck 30–120 cm; flowers
not fragrant    **7. moorei**
8a. Umbels with 12 or more flowers;
perianth-lobes linear, spreading or
reflexed    9
   b. Umbels with up to 12 flowers;
perianth-lobes lanceolate, spreading
or ascending    12
9a. Perianth-tube green; lobes less than
7.5 cm    10
   b. Perianth-tube purple-red or green
tinged with purple; lobes more than
7.5 cm    11
10a. Scape to 70 cm; flower-stalks very
short; style as long as the filaments
     **8. asiaticum**
   b. Scape more than 70 cm; flower-stalks
2.5–4 cm; style much shorter than
filaments    **9. pedunculatum**
11a. Scape green tinged with purple;
flower-stalks 2.5 cm or more; leaves
60–110 cm    **10. augustum**
   b. Scape green; flowers stalkless; leaves
90–120 cm    **11. amabile**
12a. Leaves linear; perianth-tube
7.5–10 cm    **12. amoenum**
   b. Leaves strap-shaped; perianth-tube
12.5–15 cm    13
13a. Leaves 2.5–4 cm wide, wavy; scape
usually less than 30 cm
     **13. purpurascens**
   b. Leaves 5–6.5 cm wide, not wavy;
scape 45–60 cm    **14. erubescens**

**1. C. campanulatum** Herbert (*C. aquaticum*
Burchell). Illustration: Flowering plants of
Africa **137**: t. 1455 (1965).
Bulb 4 cm in diameter with a neck 5 cm or
more. Leaves 90–120 × 1.4–2.5 cm, linear,
channelled, margins narrowly cartilaginous
and toothed. Scape 30 cm or more, green.
Flower-stalks 1–4.5 cm. Flowers 4–7, bell-
shaped, not fragrant. Perianth-tube curved,
to 5 cm, reddish green. Lobes
4–5 × 1.5–2 cm, white flushed with pink,
the tips slightly recurved. Stamens deflexed.
Style deflexed, usually longer than the
stamens. *South Africa*. H5–G1. Summer.

**2. C. latifolium** Linnaeus (*C. speciosum*
Herbert; *C. insigne* Sweet). Illustration:
Botanical Magazine, 2217 (1821);
Edwards's Botanical Register **15**: t. 1297
(1824).
Bulb to 15 cm in diameter, with a short
neck. Leaves to 85 × 7.5–12 cm, strap-
shaped, channelled, with a distinct midrib,
margins rough. Scape 30–60 cm, green
tinged with purple. Flowers stalkless,
10–20 in the umbel, bell-shaped, fragrant.
Perianth-tube 7.5–12 cm, curved, green
tinged with red. Lobes to 12 × 2.5 cm,
white faintly tinged with red in their
centres, the tips reflexed. Stamens deflexed.
Style deflexed. *Tropical Asia*. G2. Summer.

**3. C. bulbispermum** (Burman) Milne-
Redhead & Schweickerdt. Illustration:
Botanical Magazine, 661 (1803); Flowering
plants of Africa **29**: t. 1150 (1939).
Bulb 7.5–13 cm in diameter, neck to
30 cm. Leaves 60–90 × 7.5–11 cm, strap-
shaped, channelled, reflexed, margins
narrowly cartilaginous, toothed. Scape
50–90 cm, green. Flower-stalks 4.5–9 cm.
Flowers 6–12 in the umbel, funnel-shaped,
fragrant. Perianth-tube 7.5–10 cm, curved,
green. Lobes 7.5–10 × c. 2.5 cm, white
with a dark red streak, or entirely suffused
with red. Stamens deflexed. Style deflexed.
*South Africa*. H5–G1. Summer.

**4. C. macowanii** Baker (*C. pedicellatum*
Pax; *C. johnstonii* Baker). Illustration:
Botanical Magazine, 6381 (1878), 7812
(1902).
Bulb 6–25 cm in diameter with neck to
30 cm. Leaves 10–130 × 2–16 cm, linear
or strap-shaped, bright green, channelled,
margins wavy, rough and toothed. Scape
10–60 cm, green. Flower-stalks 1–4 cm.
Flowers 4–20, bell-shaped, fragrant.
Perianth-tube 7–15 cm, curved, green
tinged with brownish red. Lobes
7–13 × 2–4 cm, white with a pink or red
streak, tips reflexed. Stamens deflexed. Style

deflexed, longer than the stamens. *South
Africa*. H5–G1. Autumn.

**5. C. zeylanicum** Linnaeus (*C. ornatum*
(Aiton) Bury; *C. yuccaeflorum* Salisbury; *C.
scabrum* Herbert; *C. kirkii* Baker).
Illustration: Botanical Magazine, 1171
(1809), 2180 (1820), 6512 (1880).
Bulb 5–20 cm in diameter, neck to 30 cm.
Leaves 75–120 × 2–12 cm, strap-shaped,
green, with a distinct midrib, erect or
spreading, margins often wavy, sometimes
rough or toothed. Scape 25–70 cm, green
tinged with purple. Flowers 3–24, most
often 4–8, stalkless or almost so, bilaterally
symmetric, bell-shaped, fragrant. Perianth-
tube 9–16 cm, curved, green or tinged
with purple. Lobes 7–13 × 2–4 cm, white
with a broad purple or pale purple streak,
the upper reflexed more than the lower.
Stamens and style deflexed. *Tropical Asia,
tropical E Africa*. G2. Spring.

**6. C. jagus** (Thompson) Dandy (*C.
giganteum* Andrews; *C. podophyllum* Baker;
*C. laurentii* Durand & De Wildeman; *C.
rattrayi* Anon). Illustration: Botanical
Magazine, 5206 (1860), 6483 (1880).
Bulb 2.5–20 cm in diameter, without a
neck or neck to 7 cm. Leaves
20–90 × 4–12 cm, strap-shaped, bright or
dark green, almost membranous, not
channelled, with a distinct midrib, margins
often wavy, sometimes rough. Scape
25–75 cm, green. Flowers stalkless or
almost so, 2–12, bell-shaped, fragrant.
Perianth-tube 10–20 cm, curved, green.
Lobes 7–10 × 2.4 cm, pure white or with a
pale green midrib, widely spreading and
overlapping, tips reflexed. Stamens deflexed.
Style deflexed, usually longer than the
stamens. *Tropical Africa*. G2. Summer.

**7. C. moorei** Hooker. Illustration: Botanical
Magazine, 6113 (1874); Flowering plants
of Africa **34**: t. 1351 (1961).
Bulb 15–19 cm in diameter, neck
30–120 cm. Leaves 65–150 × 6–10 cm,
strap-shaped, spreading, not channelled,
with a distinct midrib, margins often wavy.
Scape 45–70 cm, green. Flower-stalks
1–8 cm. Flowers 5–10, funnel-shaped, not
fragrant. Perianth-tube 8–10 cm, curved,
green or tinged with red. Lobes
8–10 × 2–4 cm, white usually suffused
with pink, widely spreading. Stamens
deflexed. Style much longer than the
stamens, deflexed. *South Africa (Natal, Cape
Province)*. H5-G1. Summer.

**8. C. asiaticum** Linnaeus. Illustration:
Botanical Magazine, 1073 (1807).
Bulb 10–12.5 cm in diameter with a neck

15–23 cm. Leaves 90–120 × 7.5–13 cm, strap-shaped, spreading, not channelled, margins sometimes rough. Scape 45–60 cm, green. Flower-stalks very short. Flowers *c*. 20, salver-shaped, fragrant. Perianth-tube straight, 7.5–10 cm, green. Lobes 6–7.5 cm × 6–8 mm, linear, white, spreading. Stamens not deflexed. Style ascending. *Tropical SE Asia*. G2. Spring–autumn.

**9. C. pedunculatum** Brown. Illustration: Edwards's Botanical Register **1**: t. 52 (1815).
Bulb to 10 cm in diameter, with a short neck. Leaves *c*. 120 × 10–12.5 cm, strap-shaped, glaucous, thick, recurved, channelled or not. Scape 60–90 cm, green. Flower-stalks 2.5–4 cm. Flowers 20–30, salver-shaped, fragrant. Perianth-tube straight, to 10 cm, green. Lobes 6 cm × 5–7 mm, linear, white, spreading. Stamens not deflexed. Style ascending, shorter than the stamens. *Australia (Queensland)*. H5–G1. Summer.

**10. C. augustum** Roxburgh (*C. amabile* var. *augustum* (Roxburgh) Herbert). Illustration: Botanical Magazine, 2397 (1823); Edwards's Botanical Register **8**: t. 679 (1823).
Bulb *c*. 12.5 cm in diameter, neck to 30 cm. Leaves 60–110 × 7.5–10 cm, strap-shaped, bright green, spreading, broadly channelled, with a distinct midrib. Scape to 75 cm, green tinged with purple. Flower-stalks 2.5–3 cm. Flowers 20–30, salver-shaped, fragrant. Perianth-tube 7.5–10 cm, straight, purple-red. Lobes 7.5–10 × *c*. 1.2 cm, linear, white with a pink midrib, reddish purple with white margins on the back, spreading. Stamens ascending. Style ascending, shorter than the stamens. *Mauritius, Seychelles Islands*. G2. Spring–summer.

**11. C. amabile** Don. Illustration: Botanical Magazine, 1605 (1813).
Bulb 15–30 cm in diameter, neck *c*. 30 cm. Leaves 90–120 × 7.5–10 cm, strap-shaped, thick, spreading, not channelled, margins sometimes rough. Scape 60–90 cm, green. Flowers stalkless, 20–30, salver-shaped, fragrant. Perianth-tube 7.5–10 cm, straight, purple-red. Lobes 7.5–10 × *c*. 1.2 cm, linear, white suffused with purple, midrib dark red on the back. Stamens ascending. Style ascending. *Indonesia (Sumatra)*. G2. Spring–autumn.

**12. C. amoenum** Roxburgh.
Bulb 5–10 cm in diameter, with a very short neck. Leaves 45–60 × 2.5–5 cm, linear, bright green, spreading, channelled or not, margins wavy. Scape 30–60 cm, green or tinged with purple. Flowers 6–12, stalkless, salver-shaped, fragrant. Perianth-tube 7.5–10 cm, straight or slightly curved, green or tinged with purple. Lobes 5–7.5 × *c*. 1 cm, lanceolate, white tinted on the back with pinkish red, spreading or ascending. Stamens ascending. Style ascending. *E. Himalaya, Burma*. G2. Summer.

**13. C. purpurascens** Herbert. Illustration: Botanical Magazine, 6525 (1880).
Bulb *c*. 5 cm in diameter, with a very short neck, producing stolons. Leaves 45–90 × 2.5–4 cm, strap-shaped, dark green, almost erect or recurved, channelled, margins wavy and rough. Scape usually less than 30 cm, green tinged with purple. Flowers 6–10, stalkless, salver-shaped, faintly fragrant. Perianth-tube 12.5–15 cm, straight or slightly curved, green tinged with purple. Lobes 6–7.5 cm × 8–10 mm, white tinged with purple, spreading or ascending. Stamens ascending. Style ascending, longer than the stamens. *Tropical W Africa*. G2. Can flower throughout the year, but mainly in autumn–winter.

**14. C. erubescens** Linnaeus. Illustration: Botanical Magazine, 1232 (1809); Loddiges' Botanical Cabinet, t. 31 (1817).
Bulb small, neck short. Leaves 60–90 × 5–6.5 cm, strap-shaped, spreading and recurved, channelled or flat, margins toothed and rough. Scape 45–60 cm, green tinged with purple. Flowers 6–12, almost stalkless, salver-shaped, fragrant. Perianth-tube 12.5–15 cm, straight or slightly curved, green. Lobes 6–7.5 cm × 6–10 mm, lanceolate, white sometimes tinged with pinkish red on the back, spreading or ascending. Stamens ascending. Style ascending. *Tropical S America*. G2. Summer.

**16. BRUNSVIGIA** Heister
*J. Cullen*
Perennials with bulbs, the bulb large, subterranean or exposed. Leaves large, usually withering before flowering. Scape erect, somewhat compressed and angled. Umbel spherical, with 20–60 flowers borne on very long stalks; spathes 2, free. Perianth tubular-funnel-shaped, strongly bilaterally symmetric, with a short tube below, some or all of the lobes recurved, the whole bent just above the tube. Corona absent. Stamens 6, projecting from the perianth. Ovary 3-celled, with several ovules in each cell. Fruit a large capsule.
A genus of an uncertain number of species from South Africa. The 2 species included here are remarkable objects, with their large, many-flowered umbels. They are perhaps more striking than beautiful, and are relatively infrequently grown. They require a well-drained soil and a sunny position, and should have a period free from watering between the death of the umbel and the appearance of new leaves. Literature: Dyer, R.A., Review of the genus Brunsvigia, *Plant Life* **6**: 63–83 (1950).

1a. Leaves downy above; bulb subterranean, the leaves lying flat on the ground; perianth 5.5–7 cm
**1. orientalis**
b. Leaves not downy above; bulb mostly above ground, the leaves ascending to erect; perianth 7–8.5 cm
**2. josephinae**

**1. B. orientalis** (Linnaeus) Ecklon (*Amaryllis orientalis* Linnaeus; *B. gigantea* Schultes). Illustration: Flowering plants of Africa **36**: t. 1440 (1964); Mason, Western Cape sandveld flowers, t. 17 (1972).
Bulb ovoid to spherical, mostly below ground, 10–15 cm in diameter. Leaves oblong-ovate, narrowed to the base, 30–45 × 7–12 cm or more, lying flat on the ground, downy above. Scape 30–50 cm. Umbel with 20–40 (rarely more) flowers; flower-stalks 10–20 cm. Perianth bright red or pink, 5.5–7 cm, tube *c*. 5 mm, the lobes linear-lanceolate. Ovary 3-angled. Capsule strongly 3-angled. *South Africa (Cape Province)*. G1. Autumn.

**2. B. josephinae** (Redouté) Ker Gawler. Illustration: Edwards's Botanical Register **3**: t. 192–3 (1817); Botanical Magazine, 2578 (1825); Flowering plants of Africa **31**: t. 1223 (1956); Eliovson, Wild flowers of southern Africa, frontispiece (1980).
Bulb mostly above ground, ovoid, to 30 cm in diameter, covered with matted, membranous tunics. Leaves ascending to erect, oblong, to 90 × 5–20 cm, hairless. Scape to 45 cm or more. Umbel with 20–30 (rarely to 60 in some variants) flowers; flower-stalks 15–20 cm, elongating in fruit. Perianth 7–8.5 cm, deep red with yellow spots, the tube 1.25–1.5 cm, lobes linear. Ovary and capsule scarcely angled. *South Africa (Cape Province)*. G1. Autumn.

**17. NERINE** Herbert
*J. Cullen*
Perennial herbs with bulbs, the bulb sometimes with an elongate neck. Leaves present or absent at flowering. Scape

usually exceeding the leaves. Umbel with 4–20 flowers, subtended by 2 persistent spathes. Perianth usually bilaterally symmetric, more rarely radially symmetric, with a very short tube and 6 arching and spreading lobes which often have crisped margins. Stamens 6, attached to the perianth-tube at the base, usually deflexed, more rarely erect, sometimes with appendages at the bases of the filaments. Ovary inferior, 3-celled and -lobed, with usually 4 ovules in each cell. Fruit a membranous capsule.

A genus of about 30 species from South Africa, grown for the sake of their late, usually pink flowers; most of the cultivated species with normally pink flowers have developed white-flowered variants. As well as the species, several hybrids, some of uncertain parentage, are grown. Nerines are somewhat tender in most of Europe, but can be grown either outside, or in pots in a greenhouse, in rather poor, well-drained soil in a sunny position; plants in shade do not produce flowers regularly. Propagation is by seed or bulbs.
Literature: Traub, H.P., Review of the genus Nerine Herb., *Plant Life* **23**: supplement (1967).

*Leaves.* Thread-like: 8; 1.5–5 mm wide: 3,4,9–11; 6–9 mm wide: 5–7; more than 1 cm wide: 1–3,5,6. Glaucous: 7.
*Scape and flower-stalks.* Both hairless: 1–3,5–7; scape hairless, flower-stalks hairy: 9,10; both hairy: 4,8,11.
*Perianth.* Radially symmetric: 1.
*Perianth-lobes.* More than 4 cm: 2. Usually with crisped margins: 1–6,8,10,11; margins not crisped: 7,9.
*Stamens and style.* Deflexed: 2–11; not deflexed: 1.
*Filaments.* Without appendages at the base: 1–8; with appendages: 9–11.

1a. Filaments with appendages at the base     2
  b. Filaments without appendages at the base     4
2a. Perianth pure white, lobes with smooth margins    **9. pancratioides**
  b. Perianth distinctly pink, if sometimes pale, lobes with crisped margins    3
3a. Scape hairy; filament-appendages small, lanceolate, entire
               **11. masonorum**
  b. Scape hairless; filament-appendages to 1 cm, oblong, toothed
              **10. appendiculata**
4a. Stamens and style erect, not markedly deflexed downwards    **1. sarniensis**

  b. Stamens and style markedly deflexed downwards     5
5a. Leaves thread-like, terete    **8. filifolia**
  b. Leaves flat, 1.5–30 mm wide     6
6a. Leaves 1.5–3 cm wide     7
  b. Leaves 3–13 mm wide     8
7a. Perianth-lobes 5–7.5 cm, deep pink with a deeper midrib    **2. bowdenii**
  b. Perianth-lobes 2.5–3 cm, pale pink throughout    **3. flexuosa**
8a. Leaves glaucous; perianth white, margins of the lobes not crisped
               **7. pudica**
  b. Leaves bright green; perianth pink or magenta, margins of the lobes crisped
               9
9a. Leaves 3–8 mm wide; scape and flower-stalks hairy    **4. angustifolia**
  b. Leaves 6 mm or more wide; scape and flower-stalks hairless     10
10a. Perianth-lobes 1.7–2 cm, pale pink
              **5. undulata**
  b. Perianth-lobes 2.5–3 cm, bright or rose-pink    **6. humilis**

**1. N. sarniensis** (Linnaeus) Herbert.
Illustration: Botanical Magazine, 294 (1801), 1089 & 1090 (1808) & 2124 (1820) – varieties; Flowering plants of South Africa **9**: t. 355 (1929); Eliovson, Wild flowers of southern Africa, t. 94 (1980).
Bulb more or less spherical, 2–3 cm in diameter. Leaves oblong, obtuse, bright green, 1–2 cm wide. Scape 25–50 cm, erect. Umbel with up to 20 flowers; flower-stalks unequal. Perianth radially symmetric, lobes to 3.5 cm, variously coloured, margins crisped. Filaments without appendages at their bases. Stamens and style erect, not deflexed. Ovary hairless. *South Africa (Cape Province).* H5. Autumn.

A variable species, particularly in the colour and size of the flowers. Traub recognises 8 varieties, some of them unknown in the wild. Var. **sarniensis** has bright crimson flowers; var. **plantii** Baker has dull or rosy crimson flowers and the perianth-lobes are somewhat clawed; var. **venusta** Baker has a scarlet perianth; var. **corusca** (Ker Gawler) Baker (*N. corusca* Ker Gawler) Herbert) has large leaves and large, scarlet flowers; var. **rosea** Baker has rose-pink flowers; var. **alba** Traub has white flowers; var. **curvifolia** (Jacquin) Traub (*N. curvifolia* (Jacquin) Herbert) has rather thick, curved leaves and somewhat larger, bright scarlet flowers – forma **fothergillii** (Andrews) Traub (*N. fothergillii* Andrews) is a rather more robust variant;

var. **moorei** (Leichtlin) Traub has a shorter scape and bright, scarlet flowers and may be a hybrid between var. *curvifolia* and *N. flexuosa*.

*N. sarniensis* has been naturalised on the island of Guernsey in the English channel for many years.

**2. N. bowdenii** Watson (*N. veitchii* Anon.).
Illustration: Botanical Magazine, 8117 (1907); Flowering plants of South Africa **22**: t. 841 (1942); Eliovson, Wild flowers of southern Africa, t. 100 (1980); Rix & Phillips, The bulb book, 170 (1981).
Bulb somewhat elongate, with a conspicuous neck. Leaves 8–30 × 1.5–3 cm, tapering, obtuse. Scape to 45 cm. Umbel with 4–12 flowers; flower-stalks slightly 3-angled in section, to 5 cm. Perianth bilaterally symmetric, pink to deep pink or magenta, the lobes 5–7.5 cm, each with a darker pink stripe along the midrib, margins crisped in the upper half. Stamens and style deflexed. Filaments without appendages at their bases. Ovary hairless. *South Africa (Cape Province, Natal, Orange Free State).* H5. Autumn.

Somewhat variable in flower size and colour; the most widely grown of the species.

**3. N. flexuosa** (Jacquin) Herbert.
Illustration: Flowering plants of South Africa **15**: t. 561 (1935).
Very similar to *N. bowdenii*, but perianth pale pink or white, lobes 2.5–3 cm. *South Africa (Cape Province, Natal, Orange Free State).* H5. Autumn.

**4. N. angustifolia** (Baker) Watson.
Illustration: Flowering plants of South Africa **17**: t. 658 (1937); Botanical Magazine, n.s., 244 (1954).
Bulb ovoid with a conspicuous neck. Leaves to 30 cm × 3–8 mm, slightly tapered, obtuse, bright green. Scape 40–70 cm, sparsely covered with short, spreading hairs. Umbel with up to 12 flowers; flower-stalks 1–7 cm, with a dense covering of spreading, gland-tipped hairs. Perianth magenta, bilaterally symmetric, lobes 3–5 cm, somewhat crisped, each with a darker midrib. Stamens and style deflexed. Filaments without appendages at their bases. Ovary covered with gland-tipped hairs. *South Africa.* H5. Autumn.

**5. N. undulata** (Linnaeus) Herbert.
Illustration: Botanical Magazine, 369 (1797).
Bulb ovoid. Leaves shorter than the scape, 6–13 mm wide. Scape to 50 cm. Umbel

with up to 12 flowers; flower-stalks 2–4 cm. Perianth bilaterally symmetric, pale pink, lobes 1.5–2 cm, with conspicuously crisped margins. Filaments without appendages. Ovary hairless. *South Africa*. H5. Autumn.

**6. N. humilis** (Jacquin) Herbert. Illustration: Botanical Magazine, 726 (1804); Flowering plants of South Africa **15**: t. 564 (1935).
Very similar to *N. undulata* but umbel with 10–20 bright or rose-pink flowers, perianth-lobes 2.5–3 cm, less conspicuously crisped. *South Africa (Cape Province)*. H5. Autumn.

**7. N. pudica** J.D. Hooker. Illustration: Botanical Magazine, 5901 (1871).
Bulb spherical or elongate. Leaves 20–25 cm × 6–9 mm, glaucous. Scape to 50 cm. Umbel with 4–6 flowers; flower-stalks 2.5–4 cm. Flowers white, each segment with a pink keel; perianth bilaterally symmetric, lobes 2.5–3.5 cm, margins not crisped. Stamens and style deflexed. Filaments without appendages. Ovary hairless. *Origin uncertain, presumably South Africa*. H5. Autumn.

Described from a plant of uncertain origin which was grown at Kew; the species has not been found in the wild.

**8. N. filifolia** Baker. Illustration: Botanical Magazine, 6547 (1881); Flowering plants of South Africa **15**: t. 568 (1935); Eliovson, Wild flowers of southern Africa, t. 98 (1980).
Bulb spherical with a short neck. Leaves bright green, thread-like, 15–20 cm. Scape to 30 cm, covered with short, spreading, glandular hairs. Umbel with 8–12 flowers; flower stalks 2.5–4 cm, glandular-hairy. Perianth bilaterally symmetric, pink, lobes 2.5–3 cm, margins crisped. Stamens and style deflexed. Ovary minutely hairy. *South Africa (Orange Free State)*. H5. Autumn.

**9. N. pancratioides** Baker.
Bulb ovoid with a short neck. Leaves linear, to 40 cm × *c.* 2 mm, bright green. Scape to 60 cm. Umbel with 10–20 flowers; flower-stalks 2.5–4 cm, densely covered with short, spreading hairs. Perianth bilaterally symmetric, white, lobes 2–2.5 cm, their margins not crisped. Stamens and style deflexed. Filaments alternating with oblong, bifid appendages. Ovary hairy. *South Africa (Natal)*. H5. Autumn.

**10. N. appendiculata** Baker.
Bulb ovoid. Leaves to 45 cm × 2.5–5 mm, channelled, linear. Scape to 80 cm or more.

Umbel with 10–20 flowers; flower-stalks 3.5–5 cm, densely covered with short, spreading, gland-tipped hairs. Perianth bilaterally symmetric, pale to deep pink or magenta-pink, lobes 2.5–3 cm with crisped margins. Stamens and style deflexed. Each filament with 2 white, toothed appendages which are up to 1 cm long, at the base. Ovary hairy. *South Africa (Natal)* H5. Autumn.

**11. N. masonorum** Bolus. Illustration: Flowering plants of South Africa **15**: t. 570 (1935); Eliovson, Wild flowers of southern Africa, t. 99 (1980).
Bulb with a short neck. Leaves to 20 cm × 1.5 mm, somewhat thread-like. Scape to 30 cm, with a dense covering of short, spreading hairs. Umbel with 4–15 flowers; flower-stalks 1–3 cm, densely hairy. Perianth bilaterally symmetric, pink, lobes to 1.5 cm, each with a darker pink stripe along the midrib. Stamens and style deflexed. Filaments each with 2 lanceolate, entire appendages at the base. Ovary hairless. *South Africa (Cape Province)*. H5. Autumn.

The name was originally, but incorrectly spelled 'masoniorum'.

## 18. HAEMANTHUS Linnaeus
*J.C.M. Alexander*

Deciduous or evergreen perennial herbs. Bulbs ovoid to pear-shaped, often flattened, with fleshy, cream-coloured, pinkish or green scales which are either unequal and 2-ranked with rounded or irregular margins, or almost equal with horizontally truncate, even margins. Leaves usually 2, 4 or 6, basal thick, without a mid-rib, 2-ranked, erect, arched or prostrate, spathulate or strap-shaped to lanceolate, broadly elliptic, oblong or almost circular, hairy or hairless, usually ciliate, green, occasionally with darker green or purple bars or spots. Scape to *c.* 40 cm, appearing at about the same time as the current season's leaves or when leaves are absent, erect or curved, fleshy, often flattened, sometimes ridged or spotted, solid. Umbel with 4–150 flowers, inversely conical to hemispherical or spherical. Spathes 4–13, erect, spreading or bent downwards, stiff and fleshy or membranous. Flowers white to red; perianth-tube cylindric or narrowly bell-shaped; perianth-lobes 6, longer than tube, erect to slightly spreading, oblong to narrowly lanceolate. Corona absent. Stamens 6, borne at top of perianth-tube, usually projecting from flower. Ovary 3-celled, with 1 or 2 ovules per cell. Style

slender, about equalling the stamens, minutely 3-lobed at apex. Fruit an ovoid to spherical, pulpy, white, yellow, pink or red berry; seeds fleshy, ovoid, red or green.

A genus now regarded as having 21, often rather variable species from southern Africa, grown for their striking dense shaving-brush-like heads of white, pink or red flowers. Most species do best with a minimum temperature of *c.* 10 °C during the growing season, but can also be grown under cooler conditions or as house-plants if given plenty of light. They should be planted in a mixture of peat and sandy loam with the bulbs completely covered (except Nos 3 & 4), and should only be repotted at infrequent intervals as they resent disturbance. Dilute liquid fertiliser can be applied when the young leaves are actively growing. Water should be given much more sparingly when the plants are resting, and the deciduous species (Nos 1, 2, 5–7) can be allowed to dry out completely. Propagation is easily effected from offsets.

Literature: Friis, I. & Nordal, I., Studies on the genus Haemanthus IV, *Norwegian Journal of Botany* **23**: 63–77 (1976); Snijman, D., A revision of the genus Haemanthus, *Journal of South African Botany* Supplementary Volume **12**: 1–139 (1984).

1a. Leaves present at flowering-time or immediately after, bulb ovoid to spherical, often green; scales almost equal, with horizontally truncate, even margins      2
  b. Leaves not present at flowering-time; bulb ovoid to pear-shaped, never green; scales 2-ranked with rounded or irregular margins      5
2a. Evergreen; bulbs green if exposed to light; spathes white, firm and erect   3
  b. Deciduous; bulbs not green if exposed to light; spathes pink, membranous, soon bent downwards      4
3a. Leaves floppy, 2.5–11.5 cm; scape 5–35 cm; spathes slightly separate      **3. albiflos**
  b. Leaves firm, 7–25 cm; scape less than 6 cm; spathes overlapping at least at base      **4. deformis**
4a. Stamens projecting from flower 5–15 mm; perianth-tube 2.5–3.5 mm      **2. carneus**
  b. Stamens not projecting from flower; perianth-tube 5–12 mm      **1. humilis**
5a. Spathes stiff and fleshy, erect and closely overlapping      6

b. Spathes firm but thin, erect to slightly spreading, overlapping at base only or slightly above                     7

6a. Leaves often barred with reddish purple near base; scape usually spotted or streaked; spathes shorter than to slightly longer than flowers      **5. coccineus**

b. Leaves unmmarked or slightly spotted at base; scape usually unmarked; spathes distinctly longer than flowers      **7. pubescens**

7a. Scape unmarked, furrowed, often flattened; leaves rough, unmarked      **6. sanguineus**

b. Scape usually spotted or streaked, not furrowed, at most slightly flattened; leaves smooth, usually barred with reddish purple near base      **5. coccineus**

*Plant.* Deciduous: **1, 2, 5–7**; evergreen: **3, 4.**

*Bulb-scales.* Almost equal, margins horizontal: **1–4**; unequal, 2-ranked, margins irregular or rounded: **5–7**. Becoming green in light: **3,4.**

*Leaves.* 4 or more: **3,4.** Almost square or circular: **4–6**; broader than long: **6**. White-spotted: **3**; red-barred: **5**. Rough above: **6**; with spreading, white hairs above: **7.** Glaucous: **5,7.** Margins red: **5,7**; margins cartilaginous: **6.**

*Scape.* Appearing when leaves absent: **5–7**; appearing with or immediately before leaves: **1–4**. Appearing beside new leaves (other leaves absent): **1,2**; appearing between new leaves (other leaves present): **3**; appearing between old leaves (new leaves appear later): **4**. Very hairy: **1**. Furrowed: **6.** 6 cm or less: **4**; 4–40 cm: **1–3,5–7.**

*Umbel.* Dense: **1,3–7**; loose: **2,5,6.** With 4–60 flowers: **7**; with 25–50 flowers: **1,3,4**; with 25–100 flowers: **5**; with 25–150 flowers: **2,6.**

*Spathes.* 4 or 5: **7**; 4–9: **2–5**; 7–10: **1**; 5–11: **6**. Membranous: **1,2,6**; stiff or fleshy: **3–5,7.** Spreading or curved: **1,2**; erect and overlapping: **4,5,7**; erect but separated: **3,6,7.** White: **3,4**; pink, red or purple: **1,2,5–7.** 1.3–2 cm: **2**; 1.8–4 cm: **1,3**; 2–7 cm: **5,6**; 3–8 cm: **4,7.**

*Flowers.* White: **1,3,4**; pink to red: **1,2,5–7**; red and white: **7.**

*Perianth-tube.* 3.5 mm or less: **2, 5–7**; 7 mm or more: **1,3,4,6.**

*Perianth-lobes.* Less than 1 cm: **2**; more than 2 cm: **4–7.**

*Stamens.* Not projecting from flower: **2.**

**1. H. humilis** Jacquin subsp. **hirsutus** (Baker) Snijman (*H. hirsutus* Baker; *H. candidus* Bull; *H. allisonii* Baker; *H. nelsonii* Baker; *H. carneus* misapplied). Illustration: Botanical Magazine, 9293 (1933); Pearse, Mountain splendour, 55 (1978); Eliovson, Wild flowers of southern Africa, 51 (1980); Journal of South African Botany Supplementary Volume **12**: 33 (1984).

Deciduous. Bulb to 8 cm across, ovoid to spherical; scales more or less equal, horizontally truncate. Leaves 2, rarely 3, soft, erect or arched, 15–30 × 5.5–13 cm, lanceolate to narrowly or broadly elliptic or oblong, light green, usually hairy beneath, sometimes above also, ciliate. Scape 15–30 cm, erect or slightly curved, pale green to reddish purple, usually densely hairy, appearing with the new leaves. Umbel dense, hemispherical, with 25–120 flowers; spathes 7–10, membranous, spreading or curved downwards, pink, 2–3 cm × 3–15 mm, narrowly to broadly triangular or lanceolate, acute. Flowers pale pink or white; perianth-tube 5–12 mm; lobes 1–1.7 cm, slightly spreading. Stamens projecting from flower 5–15 mm. Berries greenish white to cream or orange. *Southern Africa.* G1. Late summer.

*H. humilis* subsp. *humilis* with smaller umbels on slenderer almost hairless scapes is probably not in cultivation.

**2. H. carneus** Ker Gawler. Illustration: Botanical Register **6**: t. 509 (1821); Botanical Magazine, 3373 (1834).

Similar to *H. humilis* subsp. *hirsutus*. Leaves prostrate, rarely arched. Scape 6–27 cm, appearing with or immediately before the new leaves, erect. Umbel loose, spherical; spathes 5–7, narrowly lanceolate, 13–20 × 2–5 mm. Flowers pink; perianth-tube 2.5–3.5 mm, slightly broader above; perianth-lobes 8.5–10 mm; stamens less than half perianth-lobe length. *Eastern South Africa.* G1. Late summer.

Often regarded as part of *H. humilis.*

**3. H. albiflos** Jacquin (*H. albomaculatus* Baker; *H. pubescens* misapplied). Illustration: Botanical Magazine, 1239 (1809); Batten & Bokelmann, Wild flowers of the eastern Cape Province, pl. 17, 1 (1966); The Garden **103**: 439 (1978); Everett (ed.), Encyclopedia of horticulture, 1580 (1981).

Evergreen. Bulb to 8 cm across, sometimes flattened; scales more or less equal, horizontally truncate, turning green if exposed to light. Leaves 2, 4 or 6, erect,

arched or prostrate, 9–40 × 2.5–11.5 cm, oblong to elliptic, sometimes white-spotted, ciliate, otherwise usually hairless. Scape 5–35 cm, arising between current season's leaves, erect or slightly curved, green, hairless or very slightly hairy. Umbel dense, ovoid or inversely conical, with 25–50 flowers; spathes 4–8, stiff, erect but separated, white with green veins, 1.8–4 cm × 6–30 mm, oblong to broadly ovate, usually ciliate. Flowers white; perianth-tube 4–7 mm; lobes 1–1.8 cm, slightly separated. Stamens projecting by up to 9 mm. Berries white, orange or red. *South Africa.* G1. Early summer.

Hairy-leaved forms known as var. *pubescens* Baker are no longer considered distinct.

**4. H. deformis** Hooker (*H. baurii* Baker). Illustration: Botanical Magazine, 5903 (1871); Gibson, Wild flowers of Natal (Coastal Region), pl. 13, 2 (1975); Journal of South African Botany Supplementary Volume **12**: 65 (1984).

Similar to *H. albiflos.* Leaves 2 or 4, strongly arched, 8–26 × 7–25 cm, very broadly oblong to almost square, unspotted. Scape to 6 cm, arising between previous season's leaves, flattened. Spathes 6 or 7, overlapping, 3–6 × 1.2–4.5 cm. Perianth-tube 7.5–9 mm; lobes 2.1–2.4 cm. Berries orange to red. *Eastern South Africa.* G1. Summer.

Sometimes considered an impoverished form of *H. albiflos*, though cultivation experiments show this not to be the case.

**5. H. coccineus** Linnaeus (*H. tigrinus* Jacquin; *H. coarctus* Jacquin; not *H. coccineus* Forsskal). Illustration: Botanical Magazine, 1075 (1808); Mason, Western Cape sandveld flowers, 58 (1972); The Garden **100**: 582 (1975); Journal of South African Botany Supplementary Volume **12**: 99 & 101 (1984).

Deciduous. Bulb 3–15 cm across, ovoid to pear-shaped, flattened; scales in 2 overlapping rows, thick, with irregular or rounded margins. Leaves 2, rarely 3, firm, erect, arched or prostrate, 15–45 × 8.5–15 cm, strap-shaped to elliptic or almost circular, sometimes slightly glaucous, unmarked or barred with reddish purple, often hairy on backs, margins sometimes red, often ciliate. Scape 6–37 cm, appearing when leaves absent, erect, usually spotted or streaked, usually hairless. Umbel dense or spreading, inversely bell-shaped to conical, with 25–100 flowers; spathes 6–9, stiff and fleshy, rarely thin, erect or slightly spreading, overlapping, pink to scarlet, 2–6 × 1.5–4.8 cm, variously shaped. Flowers pink to scarlet; perianth-tube 2–5 mm, whitish; lobes 1.2–2.6 cm, narrow and spreading. Stamens projecting by up to 1 cm. Berries white to deep pink. *South & South West Africa*. G1. Late summer.

**H. × clarkei** Anon., the artificial hybrid between *H. coccineus* and *H. albiflos*, but more closely resembling the former, is sometimes grown.

**6. H. sanguineus** Jacquin (*H. rotundifolius* Ker Gawler; *H. incarnatus* Herbert). Illustration: Batten & Bokelmann, Wild flowers of the eastern Cape Province, pl. 18, 2 (1966); Jackson, Wild flowers of Table mountain, 107 (1977); Journal of South African Botany Supplementary Volume **12**: 69 (1984).
Similar to *H. coccineus*. Leaves 2, leathery, prostrate, 9–40 × 3.5–28 cm, often broader than long, dark green and usually rough above, paler, smooth, shiny and sometimes spotted below; margins cartilaginous. Scape to 27 cm, usually furrowed and compressed above, pink to dark red, unmarked and hairless. Spathes 5–11, erect but separated, thin, 20–70 × 7–50 mm, laterals keeled. Perianth-tube 2–9 mm; lobes 1–2.4 cm. *South Africa*. G1. Summer.

**7. H. pubescens** Linnaeus (*H. albiflos* Jacquin var. *pubescens* misapplied). Illustration: Mason, Western Cape sandveld flowers, 59 (1972), Journal of South African Botany Supplementary Volume **12**: 115, 119 & 121 (1984).
Deciduous. Bulb 4.5–8 cm across, ovoid to pear-shaped, flattened; scales in 2 overlapping rows, fleshy, with irregular or rounded margins, sometimes very loosely packed. Leaves 2, rarely 3, arched to prostrate, 10–20 × 1.5–4.5 cm, strap-shaped to slightly spathulate, dark green with white spreading hairs above, paler, less hairy and sometimes spotted below, sometimes glaucous and hairless all over; margin ciliate, sometimes red. Scape 4–28 cm, appearing when leaves absent, pink or red, usually unmarked and hairless. Umbel dense, inversely bell-shaped to conical, with 4–60 flowers. Spathes 4 or 5, erect, stiff and fleshy, red, rarely pink, 3–8 × 1–3 cm, abruptly pointed, sometimes ciliate. Flowers scarlet with white tube and tips, rarely pink. Perianth-tube 2–5 mm, with swellings below lobes; lobes 1.3–2.6 cm, narrow and erect. Stamens projecting by up to 1.2 cm. Berries white to

pale pink. *South & South West Africa*. G1. Summer.

There has been some confusion between non-flowering plants of *H. pubescens* and hairy-leaved forms of *H. albiflos*, previously separated as var. *pubescens* but no longer considered distinct.

**19. SCADOXUS** Rafinesque
*J.C.M. Alexander*
Perennial herbs. Bulb spherical (very narrow in No. 1), borne on a compact rhizome formed from old bulb-bases. Leaves 4–12, basal, spirally arranged (loosely 2-ranked in No. 1), erect or spreading. Leaf-stalks spotted, sheathing, forming a thick false-stem or neck above the bulb (except in No. 1), bearing a few scale-leaves. Leaf-blades elliptic to ovate, obtuse, acute or acuminate, tapered, rounded, truncate or slightly cordate at base, membranous, hairless, with a distinct midrib. Scape lateral in scale-leaf axil (apparently terminal in No. 1), arising with or before the leaves, terete below, angled above, solid, spotted. Spathes 4–many, erect and conspicuous at flowering or withering early. Flowering-head conical, hemispherical or spherical, bearing 10–200 erect (rarely pendulous) flowers. Perianth-tube cylindric; perianth-lobes 6, erect or spreading, linear to lanceolate; tips glandular-hairy. Corona absent. Stamens 6, borne at top of perianth-tube, projecting (except in No. 4); anthers yellow or red. Ovary spherical, 3-celled, with 1 or 2 ovules per cell; style longer than stamens; stigma entire or minutely 3-pointed. Berry fleshy, orange to red, with 1–3 pale seeds.

A genus of 9 species from tropical and southern Africa with 1 reaching Yemen and possibly Oman. Previously considered part of *Haemanthus*, the species are distinguished by their membranous leaves with a distinct midrib, whose bases usually form a thick false-stem. Having a more tropical distribution than *Haemanthus* they require warmer and moister conditions; techniques for culture and propagation are similar. Literature: Friis, I. & Nordal, I., Studies on the genus Haemanthus IV, *Norwegian Journal of Botany* **23**: 63–77 (1976).

1a. Leaf-stalks not sheathing, not forming a false-stem; leaves loosely 2-ranked; scape apparently terminal
        **1. cinnabarinus**
  b. Leaf-stalks sheathing, forming a false stem; leaves spirally arranged; scape lateral     2

2a. Spathes erect and conspicuous when flowers open    **2. puniceus**
  b. Spathes withered and drooping when flowers open    3
3a. Perianth-tube longer than 5 mm; perianth-lobes narrower than 5 mm; mature filaments longer than perianth-lobes    **3. multiflorus**
  b. Perianth-tube 5 mm or less; perianth-lobes broader than 5 mm; mature filaments shorter than perianth-lobes
        **4. pole-evansii**

**1. S. cinnabarinus** (Decaisne) Friis & Nordal (*Haemanthus lindenii* Brown). Illustration: Gardeners' Chronicle **8**: 437 (1890); de Wit, Plants of the world II: pl. 130 (1965); Journal of the Royal Horticultural Society **92**: 206 (1967); Flore d'Afrique Central, Amaryllidaceae, 15 (1973).
Bulb 1–2 cm across, narrow and underdeveloped. Leaves 4–8, loosely 2-ranked; leaf-stalks 5–20 cm, not forming a false-stem; leaf-blades 5–20 × 4.5–14 cm, lanceolate to broadly ovate or elliptic, acute or obtuse, attenuate, truncate or almost cordate at base. Scape 15–45 cm, appearing with the leaves and apparently terminal. Spathes numerous, 2–4 cm × 7–11 mm, drooping and soon withering. Flowering-head conical to spherical, with 15–100 orange to pale red or scarlet flowers. Perianth-tube 3–15 mm; perianth-lobes 1–3 cm × 2.5–6 mm, with 5 (rarely 4–7) veins. Berries 1–2 cm, orange. *Sierra Leone to Angola & Uganda*. G2. Spring.

**2. S. puniceus** (Linnaeus) Friis & Nordal (*Haemanthus magnificus* (Herbert) Herbert; *H. insignis* Hooker; *H. magnificus* var. *insignis* (Hooker) Baker; *H. magnificus* var. *superbus* Baker; *H. natalensis* Poppe; *H. rouperi* Anon.). Illustration: Botanical Magazine, 4745 (1853) & 5378 (1863); Flowering plants of Africa **43**: pl. 1681 (1971); The Garden **103**: 439 (1978).
Bulb c. 7 cm across. Leaves 7–12; false-stem to 50 cm; leaf-blades to 30 × 12 cm, oblong to elliptic, almost circular when young, often wavy-edged, acuminate, obtuse. Scape to c. 40 cm or more, appearing with the leaves. Spathes various; the outer 6 or 7 lanceolate, spathulate or rhombic, 3–10 × 1–5 cm, erect when flowers open, drooping later, reddish brown to green; the inner narrower and paler. Flowering-head conical, with 30–100 yellowish green to pink or scarlet flowers. Perianth-tube 3–12 mm; perianth-lobes linear, erect or almost so, 1.3–2.7 cm × 0.7–2 mm, with 3 (rarely 1–5) veins. Berries 5–10 mm, yellow. *E & southern Africa*. G2. Spring–summer.

'Konig Albert' with taller scapes, spreading perianth-lobes and erect green spathes is a selection from the cross between *S. multiflorus* subsp. *katherinae* and *S. puniceus*. These hybrids were collectively known as *Haemanthus* × *hybridus* Wittmack or *H.* × *andromeda* Laplace.

**3. S. multiflorus** (Martyn) Rafinesque. Bulb to 6.5 cm across. Leaves *c.* 5; false-stem to 60 cm; leaf-blades 11–32 × 4–20 cm, lanceolate to ovate, obtuse, acute or acuminate, tapered to almost cordate at base. Scape to 75 cm, appearing with or before the leaves. Spathes to 6 × 1.5 cm, linear to lanceolate, colourless or reddish, drooping when flowers open. Flowering-head hemispherical to spherical, with 10–200 scarlet flowers fading to pink. Perianth-tube 5–26 mm; perianth-lobes 1.2–3.2 cm × 0.5–5 mm, linear, spreading, with 3 (rarely 1–5) veins. Berries 5–10 mm, orange. *Tropical & southern Africa, Yemen, ?Oman.* G2. Spring–summer.

A widespread and very variable species which has been divided into 3 not very distinct subspecies.

Subsp. **multiflorus** (*Haemanthus sacculus* Phillips; *H. kalbreyeri* Baker; *H. tenuiflorus* Herbert; *H. lynesii* Stapf; *H. coccineus* Forsskal not Linnaeus). Illustration: Botanical Magazine, 3870 (1841) & 8975 (1923); Journal of the Royal Horticultural Society **94**: 22 (1969); Flora of Tropical East Africa, Amaryllidaceae, 6 (1982). Very small to robust. Perianth-tube usually shorter than 1.5 cm; perianth-lobes usually narrower than 2.5 mm. *Distribution as for species.*

Subsp. **katherinae** (Baker) Friis & Nordal (*Haemanthus katherinae* Baker). Illustration: Botanical Magazine, 6778 (1884); Graf, Tropica, 63 (1978); Hay et al., Dictionary of indoor plants, pl. 262 (1983). Plants robust, to 1.2 m. Leaves wavy. Perianth-tube longer than 1.6 cm; perianth-lobes 2.2–4 mm broad. *Eastern S Africa.*

Subsp. **longitubus** (Wright) Friis & Nordal (*Haemanthus longitubus* Wright; *H. mannii* Baker). Illustration: Botanical Magazine, 6364 (1878). Plants to 65 cm. Perianth-tube longer than 1.5 cm; perianth-lobes 1.4–3.5 mm broad. *W Africa.* G2.

**4. S. pole-evansii** Obermeyer. Illustration: Flowering plants of Africa **37**: pl. 1452 (1965); Journal of the Royal Horticultural Society **87**: 310 (1962); Botanical Magazine, n.s., 572 (1970).

False-stem to 85 cm; leaf-blades 24–52 × 3–17 cm. Scape to 1.2 m, bearing 30–70 pink or scarlet flowers. Perianth-tube 4–5 mm; perianth-lobes elliptic, spreading, 2–2.5 cm × 5–10 mm. Stamens 1.2–1.3 cm, shorter than perianth-lobes. Berries 1–2 cm, scarlet. *E Zimbabwe.* G2. Summer.

**20. CLIVIA** Lindley
*A. Leslie*
Perennials with short rhizomes and thick, fleshy roots. Leaves usually clustered at the end of the rhizome (at the apex of a distinct stem in one species), more or less in 2 ranks, strap-shaped, evergreen. Scapes solid, compressed, 2-edged, bearing umbels of numerous erect or pendulous flowers. Bracts several, the outer spathe-like. Flowers straight or curved, the outer perianth-lobes usually much narrower than the inner; perianth-tube very short. Stamens attached at the throat of tube, curved at base; anthers opening by slits. Stigma shortly 3-lobed. Fruit a red berry. Seeds few, spherical-angular.

A genus of 4 species from South Africa, one of which (*C. miniata*) is popular as a house or conservatory plant, valued for its bold evergreen foliage and umbels of striking orange flowers. They are sometimes listed as species of *Imantophyllum*.

Clivias will succeed in a good, well-drained soil and, once established, an annual top-dressing is all that is required. During their resting period only a minimum of heat and water is necessary, but both should be increased at the onset of new growth. Flowering is reputed to be enhanced if the plants are underpotted. Propagation is effected by division after flowering or by seed.

The apparent distinctions between species Nos. **1–3** are not always clear-cut in cultivated specimens. Natural or artificial hybridization may be involved. Species **2** and **3** are reputed not always to be distinguished readily in the wild. Literature: Wittmack, L., Clivia (Imantophyllum) cyrtanthiflora van Houtte (Clivia nobilis and Clivia miniata), *Gartenflora* **53**: 225–8 (1904).

1a. Plants with a distinct stem
                                        **1. caulescens**
  b. Plants usually lacking a distinct stem
                                                      2

2a. Flowers erect to spreading, broadly
      funnel-shaped            **4. miniata**
  b. Flowers spreading to pendulous,
      narrowly funnel-shaped                3

3a. Leaves gradually tapered at apex;
      umbels usually with 12–20 flowers
                                        **2. gardenii**
  b. Leaves abruptly contracted at apex;
      umbels often with 40–60 flowers
                                        **3. nobilis**

**1. C. caulescens** Dyer. Illustration: The Flowering Plants of South Africa **23**: t. 891 (1943).
Stems to 45 cm. Leaves 30–90 × 3–5 cm, margin entire, gradually narrowed at apex. Scapes *c.* 30 cm, with umbels of 15–20 pendulous flowers. Perianth narrowly funnel-shaped, straight or slightly curved; lobes 3.5–4 cm, red or pinkish red with yellowish margins and green tips. Stamens equalling or exceeding the perianth, anthers 2–2.5 mm. *South Africa.* G1. Spring.

**2. C. gardenii** J.D. Hooker. Illustration: Botanical Magazine, 4895 (1856). Plant without stems. Leaves 45–75 × 2.5–4 cm, margin entire or slightly toothed, gradually tapered at apex. Scapes 30–60 cm, with umbels of usually 10–20 pendulous flowers. Perianth narrowly funnel-shaped, often strongly curved; lobes 3.5–5 cm, dull orange or brick red, with yellowish margins above and spreading green tips. Stamens equalling or exceeding the perianth; anthers 2.5–3 mm. *South Africa.* G1. Winter–spring.

**3. C. nobilis** Lindley (*Imantophyllum aitonii* J.D. Hooker). Illustration: Edwards's Botanical Register, t. 1182 (1828); Botanical Magazine, 2856 (1828); Loddiges' Botanical Cabinet, t. 1906 (1833).
Plant without stems. Leaves 30–60 × 2.5–5 cm, margins toothed, abruptly contracted at apex, sometimes slightly notched. Scapes 30–50 cm, with umbels of often 40–60 pendulous flowers. Perianth narrowly funnel-shaped, straight or slightly curved; lobes 2–4 cm, red or yellowish red, with yellow margins and erect or slightly incurved green tips. Stamens equalling or exceeding perianth; anthers *c.* 2 mm. *South Africa.* G1. Spring–summer.

**4. C. miniata** (Lindley) Bosse. Illustration: Botanical Magazine, 4783 (1854); The Flowering Plants of South Africa **1**: t. 13 (1921), **11**: t. 411 (1931) – var. flava. Plant without stems. Leaves 40–60 × 3.5–6.5 cm, margin slightly toothed, gradually tapered at apex. Scapes 30–45 cm, with umbels of 12–20 erect to spreading flowers. Perianth broadly funnel-shaped, straight; lobes 5–7 cm, bright red or orange (rarely yellow – var. **flava** Phillips), with a yellow or white and yellow base, tips spreading. Stamens shorter than or equalling perianth; anthers 3.5–4.5 mm. *South Africa.* G1. Spring–summer.

There are several cultivars, selected on the basis of flower colour or improved flower size; plants with variegated foliage ('Striata') are also known.

**C. × cyrtanthiflora** (Van Houtte) Voss (*C. miniata × C. nobilis*) has arisen in cultivation. It usually has pendulous flowers, ranging in colour from yellow to red, which are intermediate in size and shape between those of the parents.

## 21. GALANTHUS Linnaeus
*C.D. Brickell*

Perennial herbs with bulbs. Leaves basal, 2, linear, strap-shaped or oblanceolate, enclosed in a tubular, membranous sheath at the base. Spathe of 2 fused bracts. Flowers pendent. Perianth segments free, the outer 3 acute to more or less obtuse, spathulate or oblanceolate to narrowly obovate, shortly clawed, erect-spreading; the inner half to two-thirds as long as the outer segments, usually notched, oblong, spathulate or oblanceolate, tapered to the base, erect, with a green patch around the notch and sometimes also at the base. Stamens inserted at the base of the perianth, shorter than the inner segments. Filaments much shorter than the anthers. Anthers basifixed, opening by terminal pores. Style slender, exceeding the anthers; stigma capitate. Capsule ellipsoid, or almost spherical, opening by 3 flaps. Seeds light brown, each with an appendage.

A genus of some 12 species from Europe, Turkey, Iran and the Caucasus, which grow well in woodland conditions in alkaline or acid soils given a top-dressing of leaf-mould annually. Many hybrids have occurred in cultivation and numerous clones have been selected and named. Literature: Beck, G., Die Schneeglöckchen, eine monographische Skizze der Gattung Galanthus, *Wiener Illustrierte Garten Zeitung* **19**: 45–58 (1894); Gottlieb-Tannenhain, P. von, Studien über die Formen der Gattung Galanthus, *Abhandlungen der Zoologische-botanische Gesellschaft in Wien* **2**(4): 1–95 (1904); Traub, H.P. & Moldenke, H.N., The tribe Galantheae, *Herbertia* **14**: 85–114 (1947); Stern, F.C., *Snowdrops and snowflakes* (1956); Schwarz, O., Tentative keys to the wild species of Galanthus L.,

*Bulletin of the Alpine Garden Society* **31**: 131–6 (1963); Artiushenko, Z.T., Kriticheskiy obzor roda Galanthus L., *Botanicheskii Zhurnal* **51**: 1437–51 (1966); Artiushenko, Z.T., Taxonomy of the genus Galanthus L., *RHS Daffodil & Tulip Year Book* **32**: 62–82 (1966); Artiushenko, Z.T., Galanthus L. (Amaryllidaceae) in Greece, *Annales Musei Goulandris* **2**: 9–21 (1974); Webb, D.A., The European species of Galanthus L., *Botanical Journal of the Linnaean Society* **76**: 307–13 (1978); Kamari, G., A biosystematic study of the genus Galanthus L. in Greece, Part II, *Botanika Chronika* **1**: 60–98 (1981), Part I, *Botanische Jahrbücher* **103**: 107–35 (1982).

1a. Leaves with margins folded back in bud, at least at the base as the leaves first develop    2
  b. Leaves flat or rolled in bud    3
2a. Flowers produced from October to February; leaves linear, 3–5 mm broad when mature, upper surface dull green with a distinct, glaucous central stripe    **1. reginae-olgae**
  b. Flowers produced from January to April; leaves narrowly oblanceolate to strap-shaped, 5–21 mm broad when mature, upper surface glaucous    **2. plicatus**
3a. Leaves green, usually glossy, never glaucous    4
  b. Leaves glaucous, at least on the lower surface, never entirely green    6
4a. Leaves flat in bud, linear, 3–10 mm broad when mature    **7. rizehensis***
  b. Leaves rolled in bud, narrowly oblanceolate or strap-shaped, 1–3 cm broad when mature    5
5a. Outer surface of inner perianth-segments with green patches at apex and base; leaves narrowly oblanceolate    **8. fosteri**
  b. Outer surface of inner perianth-segments with a green patch at the apex only; leaves strap-shaped to narrowly oblanceolate    **9. ikariae***
6a. Leaves rolled in bud, oblanceolate, 5–32 mm broad when mature    7
  b. Leaves flat in bud, linear to very narrowly oblanceolate, 3–10 mm broad when mature    8
7a. Inner perianth-segments with separate green patches at base and apex (these sometimes joined at maturity); leaves more or less upright at maturity    **3. elwesii**
  b. Inner perianth-segments with a green patch at the apex only; leaves recurved at maturity    **4. caucasicus**

8a. Outer surface of inner perianth-segments with green patches at apex and base; leaves frequently twisted    **5. gracilis**
  b. Outer surface of inner perianth-segments with a green patch at the apex only; leaves not twisted    9
9a. Upper surface of leaves dull green with a distinct, central, glaucous stripe, lower surface glaucous    **1. reginae-olgae**
  b. Both surfaces of the leaves glaucous    **6. nivalis**

**1. G. reginae-olgae** Orphanides (*G. olgae* Boissier; *G. nivalis* Linnaeus subsp. *reginae-olgae* (Orphanides) Gottlieb-Tannenhain; *G. corcyrensis* (Beck) Stern). Illustration: Stern, Snowdrops and snowflakes, pl. 4 & f. 6, 7 (1956); Rix & Phillips, The bulb book, 183 (1981).

Bulb ovoid (rarely almost spherical), 1.5–3 × 1.5–2 cm. Leaves linear, flat but with margins slightly folded back at the base as the leaves emerge, absent or to 11 cm × 3–4 mm at flowering, to 30 × 3.5 cm and recurved at maturity, obtuse at apex, flat, upper surface dull green with a central, glaucous stripe, lower surface glaucous. Scape 5–22 cm. Outer perianth-segments convex, elliptic to narrowly obovate, 1.5–2.5 cm × 5–11 mm, inner segments flat, not flared at the apex, narrowly obovate, 9–12 × 5–7 mm, each with a green patch at the apex. Filaments 0.5–1 mm, anthers 3–4 mm. Capsule spherical to broadly ellipsoid, to 1.4 × 1.2–1.4 cm. *W Turkey, Balkans, Sicily.* H3. Autumn–early spring.

Two poorly defined subspecies are recognised, differing only in flowering period. Subsp. **reginae-olgae** from W Turkey, Greece, Sicily and Albania flowers October–November; subsp. **vernalis** Kamari, from Yugoslavia, N Greece and Sicily is spring-flowering. In some seasons the flowering periods overlap. In subsp. *reginae-olgae* the leaves may appear with the flowers or after them.

**2. G. plicatus** Bieberstein.
Bulb almost spherical to ovoid, 2–3 × 1.7–2 cm. Leaves narrowly oblanceolate to strap-shaped, margins folded back in bud, sometimes only slightly so at maturity, 5–21 × 1–1.5 cm at flowering, to 30 × 2 cm and upright at maturity, glaucous, often with a paler central band on the upper surface, apex obtuse, hooded. Scape 9–22 cm. Outer perianth-segments convex, elliptic to obovate, 1.7–2.8 cm × 6–12 mm, inner

segments flat, not flared at the apex, oblong-spathulate, 8–10 × 4–7 mm, with separate green patches at base and/or apex, these sometimes joined in the centre. Filaments 0.5–1 mm, anthers 6–7 mm. Capsule ellipsoid to almost spherical, to 1.6 × 1.2 cm. *Turkey, E Romania, USSR (Crimea).* H2. Early spring.

Two poorly defined subspecies are recognised:

Subsp. **plicatus**. Illustration: Stern, Snowdrops and snowflakes, f. 10 (1956); Rix & Phillips, The bulb book, 12 (1981); Grey-Wilson & Mathew, Bulbs, pl. 22 (1981). Inner perianth-segments with apical green patch only. *E Romania, USSR (Crimea).*

Subsp. **byzantinus** (Baker) Webb (*G. byzantinus* Baker). Illustration: Stern, Snowdrops and snowflakes, f. 11 (1956); Rix & Phillips, The bulb book, 15 (1981); Mathew & Baytop, Bulbous plants of Turkey, f. 1 (1984). Inner perianth-segments with both apical and basal green patches. *Turkey.*

**3. G. elwesii** J.D. Hooker (*G. graecus* Boissier; *G. maximus* Velenovsky; *G. elwesii* var. *whittallii* Moon). Illustration: Botanical Magazine, 6166 (1875); The Garden **57**: 45 (1900); Rix & Phillips, The bulb book, 12, 13 (1981).
Bulb almost spherical to ovoid, 2–3.5 × 0.5–2.5 cm. Leaves narrowly oblanceolate, rolled in bud, 7.5–32 cm × 6–13 mm at flowering, to 36 × 3.2 cm and upright at maturity, sometimes twisted, glaucous, apex obtuse, hooded. Scape 12–28 cm. Outer perianth-segments convex, elliptic to broadly obovate, 2–2.7 cm x 9–19 mm, inner perianth segment with separate green patches at base and apex, these sometimes joined in the centre. Filaments 1 mm, anthers 5–6 mm. Capsule broadly ellipsoid to spherical, 1–2 × 1–2 cm. *N Greece, Bulgaria, W Turkey.* H2. Early spring.

**4. G. caucasicus** (Baker) Grossheim (*G. nivalis* Linnaeus subsp. *caucasicus* Baker). Illustration: Stern, Snowdrops and snowflakes, f. 14 (1956); Rix & Phillips, The bulb book, 16 (1981).
Bulb spherical or almost so, 2–3 × 1.5–3 cm. Leaves strap-shaped or narrowly oblanceolate, rolled in bud, 4–11 × 1–1.8 cm at flowering, to 20 × 2.2 cm and recurved at maturity, glaucous, apex obtuse to acute, flat or slightly hooded. Scape to 15 cm. Outer perianth-segments convex, obovate to broadly obovate, 1.7–2 cm × 9–12 mm, inner segments flat,

not flared at the apex, narrowly oblong-ovate to obovate, 9–12 × 5–6 mm, each with a green patch at the apex. Filaments 1 mm, anthers 5–6 mm. Capsule spherical or almost so, 1–2 × 1–2 cm. *USSR (Caucasus).* H2. Late winter–early spring.

Early flowering variants, blooming in November and December, have been recognised as var. **hiemalis** Stern. The closely related **G. alpinus** Sosnovsky from the Caucasus, with shorter, broader leaves and smaller, more spherical flowers is sometimes cultivated. Another related Caucasian species, **G. bortkewitschianus** Koss, is also occasionally grown; its glaucous leaves are strongly hooded at the apex, and are narrower than those of *G. alpinus.*

**G. × grandiflorus** Baker is the name given to hybrids between *G. plicatus* and *G. caucasicus*; they are intermediate between the parents, with glaucous leaves whose margins are narrowly bent back in bud, and inner perianth-segments with apical green marks.

**5. G. gracilis** Celakovsky (*G. bulgaricus* Velenovsky; *G. graecus* misapplied; *G. elwesii* J.D. Hooker subsp. *minor* Webb; ? *G. elwesii* var. *stenophyllus* Kamari). Illustration: Stern, Snowdrops and snowflakes, f. 9 (1956) – as G. graecus; Rix & Phillips, The bulb book, 13f (1981).
Bulb ovoid to almost spherical, 1.5–2.8 cm × 8–20 mm. Leaves linear, occasionally very narrowly oblanceolate, flat in bud, 5–15 cm × 3–7 mm at flowering, to 27 × 3.9 cm at maturity, frequently distinctly twisted (occasionally only slightly so), glaucous, apex obtuse, flat. Scape 8–14 cm. Outer perianth-segments convex, narrowly obovate-spathulate, 1.5–2.6 cm × 7–10 mm, inner segments flat, usually noticeably flared at apex, narrowly oblong-spathulate, 7–11 × 3.5 mm, with separate green patches at apex and base which sometimes join in the centre. Filaments 1–2 mm, anthers 4–5 mm. Capsule ovoid to almost spherical, 1.1–1.6 × 1–1.4 cm. *Bulgaria, Greece, Turkey.* H2. Early spring.

**6. G. nivalis** Linnaeus (*G. angustifolius* Koss).
Bulb ovoid to almost spherical, 1.5–2.7 × 1–2.4 cm. Leaves linear, flat in bud, 9–21 cm × 6–9 mm at flowering, to 30 × 1 cm at maturity, glaucous, apex obtuse, flat. Scape 11–18 cm. Outer perianth-segments convex, elliptic to oblanceolate or narrowly obovate, 1.8–2.8 cm × 6–10 mm, inner segments

flat, not flared at the apex, narrowly obovate, 7–11 × 5 mm, each with a green patch at the apex. Filaments 1–2 mm, anthers 5 mm. Capsule spherical or almost so, 1.4 × 1.2–1.4 cm.

Two subspecies are commonly grown:

Subsp. **nivalis**. Illustration: Polunin, Flowers of Europe, f. 173 (1959). Leaves 9–16 cm at flowering and not recurved; flowering mainly from February to April. *Most of Europe (wild and naturalised).* H2. Early spring.

Subsp. **cilicicus** (Baker) Gottlieb-Tannenhain (*G. cilicicus* Baker). Illustration: Gardeners' Chronicle **23**: 79 (1898). Leaves 16–18 cm at flowering, recurved. Flowering mainly from November to March. *SW Turkey.* H4. Winter–spring.

This species is extremely variable. Variants with large flowers from Italy have been described as subsp. **imperati** (Bertoloni) Baker; a number of variants with double flowers are commonly naturalised. Variants with separate spathes, sometimes elongated, have been described as *G. scharlockii* Caspary. Variants with yellow ovaries and yellow markings on the inner segments have been called var. *lutescens* Harpur-Crewe and var. *flavescens* Allen.

**7. G. rizehensis** Stern. Illustration: Stern, Snowdrops and snowflakes, f. 8 (1956); Rix & Phillips, The bulb book, 16 (1981); Mathew & Baytop, Bulbous plants of Turkey, f. 2 (1984).
Bulb ovoid to almost spherical, 1.5–2 × 1.1–1.5 cm. Leaves linear, flat in bud, 9–16 cm × 3–10 mm at flowering, to 20 × 1 cm and recurved at maturity, deep green, apex obtuse, flat. Scape 10–20 cm. Outer perianth-segments slightly convex, oblong-elliptic, 1.5–2 cm × 6–8 mm, inner segments flat, not flared at the apex, oblong-spathulate, 8–9 × 4–5 mm, each with a green patch at apex. Filaments 0.5 mm, anthers 5 mm. Capsule almost spherical, 7–8 × 5–6 mm. *N Turkey, N Iran.* H3. Early spring.

*****G. lagodechianus** Kemularia-Natadze (*G. cabardensis* Voss; *G. ketzkhovellii* Kemularia-Natadze; *G. kemulariae* Kuthatheladze). Leaves bright, glossy green. *USSR (Caucasus).* H3. Early spring.

*G. transcaucasicus* Fomin, from the Caucasus and Iran, is very similar to *G. rizehensis* and is almost certainly the same species.

**8. G. fosteri** Baker (*G. latifolius* Ruprecht forma *fosteri* (Baker) Beck). Illustration: Wiener Illustrierte Garten Zeitung **19**: t. 17

f. 17 (1894); Addisonia 16: pl. 513 (1931); Rix & Phillips, The bulb book, 16 (1981). Bulb almost spherical to ovoid, 2–2.5 × 1.5–2.2 cm. Leaves narrowly oblanceolate to strap-shaped, rolled in bud, 6–15 cm × 5–20 mm at flowering, to 25 × 2.2 cm and recurved at maturity, bright, deep green, apex obtuse, not or slightly hooded. Scape 8–23 cm. Outer perianth-segments slightly convex, elliptic to narrowly obovate, 1.9–2.6 cm × 9–12 mm, inner segments flat, slightly flared at the apex, oblong-spathulate, 8–12 × 3–6 mm, with separate green patches at base and apex. Filaments 1 mm, anthers 4–5 mm. Capsule almost spherical to obovoid, 1–1.3 × 1 cm. *Turkey, Lebanon.* H5. Early spring.

Bulb-frame treatment or an open, well-drained, sunny site is required for successful cultivation.

**9. G. ikariae** Baker (*G. latifolius* Ruprecht; *G. ikariae* subsp. *latifolius* Stern; *G. platyphyllus* Traub & Moldenke; *G. woronowii* Losina-Losinskaya; *G. ikariae* subsp. *snogerupii* Kamari). Illustration: Botanical Magazine 9474 (1937); Stern, Snowdrops and snowflakes, f. 12, 13 (1956); Rix & Phillips, The bulb book, 14, 15 (1981).
Bulb ovoid, 1.8–3.5 × 1.6–2.8 cm. Leaves strap-shaped to narrowly oblanceolate, rolled in bud, 5–16 cm × 5–20 mm at flowering, 25–30 cm × 5–30 mm at maturity, often recurved, bright or deep green, apex obtuse, hooded. Scape 11–25 cm. Outer perianth-segments convex, oblong-elliptic, 1.7 2.6 cm × 6 11 mm, inner segments flat, not flared at apex, narrowly obovate, 9 × 5 mm, each with a green patch at apex. Filaments 0.5–1.5 mm, anthers 5–6 mm. Capsule spherical or almost so, 1.5 × 1.4–1.5 cm. *Aegean Islands, N Turkey, N Iran, USSR (Caucasus).* H3. Spring.

Various attempts have been made to separate the Caucasian populations (as *G. latifolius* Ruprecht, *G. ikariae* subsp. *latifolius* Stern or *G. platyphyllus* Traub & Moldenke) from those found on the Aegean islands of Ikaria, Andros, Naxos, Tinos and Skyros. The differences claimed in leaf shape and colour, the size of the perianth-segments and the presence of the apical green patches on the inner segments do not appear tenable on present evidence.
**\*G. krasnovii** Khokhrjakov. Very similar, but inner perianth-segments said to be un-notched. *USSR (Caucasus).*

**\*G. allenii** Baker. Illustration: Stern, Snowdrops and snowflakes, pl. 5 (1956). Leaves duller green and slightly glaucous; flowers with an almond fragrance. *USSR (Caucasus).*

## 22. LAPIEDRA Lagasca
*J. Cullen*
Perennial herb with a bulb. Leaves few, basal. Flowers erect in an umbel of 4–9; spathe of 2 bracts. Perianth-segments all similar, free, spreading. Filaments as long as anthers; anthers basifixed, opening by pores. Capsule flattened-spherical, 3-lobed, surrounded by the persistent, withered perianth. Seeds few.

A genus of 2 species from the west Mediterranean area. Cultivation as for *Galanthus* (p. 317).

**1. L. martinezii** Lagasca. Illustration: Polunin & Smythies, Flowers of south-west Europe, 389 (1973).
Bulb to 5 cm in diameter. Leaves to 25 × 1 cm, appearing after the flowers, each with a pale band on the upper surface. Perianth-segments 8–12 mm, white, each with a green stripe on the outside. *S Spain, N Africa.* H5. Late summer.

## 23. LEUCOJUM Linnaeus
*D.A. Webb*
Perennial herbs with bulbs. Flowers bell-shaped, nodding, white or pink, solitary or in umbels of up to 5 (rarely 7), subtended by a spathe of 1 or 2 bracts. Perianth-tube and corona absent. Perianth-segments free, all more or less alike. Anthers blunt at the tip, not pointed as in *Galanthus*, opening by pores. Capsule erect, more or less spherical; seeds numerous.

About 10 species in S, W & C Europe, NW Africa and SW Asia. *L. vernum* is easily grown in sun or semi-shade in a reasonably water-retentive soil. *L. aestivum* needs more moisture than do other species, and should be grown in semi-shade in a fairly heavy soil. The other species, though hardy in the warmest parts of Europe, are, on account of their small size and delicate habit, best cultivated in a cool greenhouse in pans of light, sandy soil.
Literature: Stern, F.C., *Snowdrops and Snowflakes* (1956).

*Flowering season.* Late winter to early summer: **1–4**; late summer or autumn: **5,6**.
*Leaves.* 7–25 mm broad: **1,2**; not more than 3 mm broad: **3–6**.
*Spathe.* 1 bract: **1,2,5**; 2 bracts: **3,4,6**.

1a. Leaves at least 7 mm broad; scape stout, usually hollow    **2**
 b. Leaves not more than 3 mm broad; scape very slender, solid    **3**
2a. Flowers in umbels of 2–7 (very rarely solitary); seeds black, without an appendage    **1. aestivum**
 b. Flowers solitary or in pairs; seeds white, with an appendage    **2. vernum**
3a. Flowering in late summer or autumn; leaves appearing during or after the flowering period    **4**
 b. Flowering in late winter or spring; leaves well developed before the flowers open    **5**
4a. Flowers usually solitary; flower-stalk not more than 5 mm; spathes 2    **6. roseum**
 b. Flowers often in umbels of 2 or 3; longer flower-stalks 1.5–2.5 cm; spathe 1    **5. autumnale**
5a. Perianth-segments 1.3–2.3 cm    **3. trichophyllum**
 b. Perianth-segments 8–12 mm    **4. nicaeense\***

**1. L. aestivum** Linnaeus. Illustration: Loddiges' Botanical Cabinet, 1478 (1828); Ross-Craig, Drawings of British plants 29: t. 8 (1972); Rix & Phillips, The bulb book, 77 (1981).
Bulb to 4 cm. Leaves strap-shaped, c. 40- × 1.5 cm. Scape about as long as leaves, hollow, 2-winged. Spathe 1, 3–5 cm, green. Umbel with 3–5 (rarely 7) flowers; perianth-segments 1.5–2 cm, white tipped with green. Seeds black, without an appendage. *W, C & S Europe, eastwards to Turkey and the Caucasian region.* H2. Late spring–early summer.

The above description applies to the typical plant, subsp. *aestivum*. Subsp. **pulchellum** (Salisbury) Briquet, which is perhaps commoner in cultivation, is less robust and smaller in most of its parts, with only 2–4 flowers in the umbel and perianth-segments 1–1.4 cm. It is native to the W Mediterranean region, and flowers 3–4 weeks earlier than subsp. *aestivum*. A robust, large-flowered clone of subsp. *aestivum* is cultivated under the name of 'Gravetye'.

**2. L. vernum** Linnaeus. Illustration: Botanical Magazine, 1993 (1817); Javorka & Czapody, Iconographia florae hungaricae, f. 736 (1934); Rix & Phillips, The bulb book, 12 (1981).
Bulb c. 2.5 cm. Leaves to 25 cm × 8–25 mm, appearing with or just after the flowers. Scape 15–30 cm, slightly

hollow, narrowly 2-winged, bearing usually a single flower (less often 2) and a spathe of 1 bract about equalling the flower-stalk. Perianth 1.5–2.5 cm, white, with a green or yellow spot near the tip of each segment. Seeds 7 mm, whitish, with an appendage. *Europe, from Belgium and Poland to the Pyrenees and Yugoslavia.* H2. Spring.

The plants with yellow spots on the perianth are sometimes distinguished as var. *carpathicum* Sweet. Var. *vagneri* Stapf is a robust variant with normally 2 flowers on each scape.

**3. L. trichophyllum** Schousboe (*L. grandiflorum* de Candolle). Illustration: Botanical Magazine, 9585 (1940); Maire, Flore de l'Afrique du Nord **6**: 14 (1959); Rix & Phillips, The bulb book, 37 (1981). Bulb *c.* 1.5 cm. Leaves to 30 cm × *c.* 1.5 mm. Scape slightly shorter, very slender, bearing 2–4 flowers. Flower-stalks to 5 cm, usually longer than the 2 spathes. Perianth 1.3–2.5 cm, white, variably tinged with pink, especially at the base, but sometimes pure white. Seeds 1.5 mm, black, without an appendage. *Spain, Portugal & Morocco.* H4–5. Spring.

Wild plants vary greatly in the size of their flowers, but the variation is continuous. Those in cultivation usually represent the large-flowered variant formerly distinguished as var. *grandiflorum* (de Candolle) Willkomm.

**4. L. nicaeense** Ardoino (*L. hiemale* de Candolle in part). Illustration: Botanical Magazine, 6711 (1883); Bicknell, Flowering plants of the Riviera, t. 70 (1885). Bulb *c.* 2 cm. Leaves to 30 cm × 2 mm. Scape somewhat shorter, slender, bearing 1 flower (rarely 2 or 3). Flower-stalk *c.* 1.5 cm, equalling or shorter than the 2 spathes. Perianth 9–13 mm, white. Seeds 2–3 mm, black, with an appendage. *Coastal region of SE France, from Nice eastwards (almost extinct).* H5. Spring.

*L. longifolium** (Roemer) Grenier. Differs chiefly in its smaller bulb, longer scape with 1–4 flowers, flower-stalks usually exceeding the spathe, and seeds without an appendage. *Corsica.* H5.

**5. L. autumnale** Linnaeus. Illustration: Botanical Magazine, 960 (1806); Gartenflora **36**: t. 1261 (1886). Bulb *c.* 10 mm. Leaves to 15 cm, very slender, appearing during or just after the flowering period. Scape to 15 (rarely 25) cm, slender, bearing 1–4 flowers. Flower-

stalks to 2.5 cm, usually exceeding the spathe. Perianth 9–15 mm, white, or tinged with pink, especially at the base. Seeds 1 mm, black, without an appendage. *W Mediterranean region.* H4–5. Autumn.

**6. L. roseum** Martin. Illustration: Stern, Snowdrops and snowflakes, t. 13 (1956). Like *L. autumnale*, but with spathes 2, shorter flower-stalks and smaller flowers, which are usually solitary and always pink. *Corsica, Sardinia.* H5. Autumn.

**24. CHLIDANTHUS** Herbert
*S.G. Knees*
Perennial herbs with bulbs and thick, fleshy roots, mature bulbs 3–4 cm in diameter. Leaves 5 or 6, linear, obtuse, 15–40 cm × 6–9 mm, appearing after flowers, surrounded by a papery sheath that is often up to one-third as long as the leaves. Scape solid, 12–30 cm, bracts 2, 7–9 cm, papery, soon withering, subtending an umbel of 3–5 stalked flowers, with stalks 9–12 mm. Flowers fragrant, bisexual, almost radially symmetric, perianth 4–7 cm with 6 yellow lobes, a tube at base, and corona small. Anthers basifixed. Ovary 3-celled, style 3-branched, projecting from the perianth. Fruit a capsule containing several black, flattened seeds.

A genus of 1 (possibly 3) species, native to southern Peru, tender in much of northern Europe; hardy in southern Europe. Bulbs should be potted in spring in a well-drained compost and then plunged in a frame. The pots should be lifted in autumn and kept dry in frost-free conditions until the following spring. Plants in permanent sites will benefit from a top dressing in spring. Propagation is by seeds or offsets.
Literature: Ravenna, P.F., Contributions to South American Amaryllidaceae VI, *Plant Life* **30**: 71–3 (1974).

**1. C. fragrans** Herbert. Figure 37(10), p. 292. Illustration: Edwards's Botanical Register, t. 640 (1822); Flore des Serres, t. 326 (1848).
*Southern Peru.* H4. Spring.

**25. VAGARIA** Herbert
*S.G. Knees*
Perennial herbs with white fleshy roots and narrow, flask-shaped bulbs that are clustered together and covered to the neck with a papery brown tunic, often sheathing the scape and leaf bases for up to 15 cm. Leaves 4–7, linear, 10–20 cm × 8–12 mm. extending after flowering to 40–60 cm,

obtuse and with a conspicuous median white band. Scape 15–25 cm, somewhat flattened and slightly 4-angled, solid. Bracts 2–4, papery, 3–5 cm long. Flowers bisexual, 6–9 per umbel, stalks 5–20 mm, apparently extending after flowering. Perianth-tube *c.* 1 cm, lobes 6, each narrowly elliptic to lanceolate with hooded apex, 2.5–3.5 cm, white with a conspicuous green midrib, somewhat keeled beneath. Corona arising at throat of perianth, consisting of 12 linear, acute teeth, occurring in pairs at the bases of the filaments. Anthers 6, versatile. Ovary 3-celled, ovules 2 or 3 per cell, placentation axile; stigma capitate. Fruit a capsule containing several 3-sided, glossy black seeds.

A genus of 4 species, 3 in N America and 1 in the Middle East, this being the only 1 cultivated. The tunicated bulbs and leaves suggest that the bulbs are often deeply buried in the wild, and it may be desirable to plant the bulbs well below soil level in cultivation, to help overwintering. They should be kept dry and frost-free in the winter and may either be repotted or given a suitable top-dressing in spring. Propagation is by offsets or seed (though this rarely ripens in northen Europe). Literature: Herklots, G.A.C., Eurycles and Vagaria, *The Plantsman* **3**: 220–9 (1982).

**1. V. parviflora** Herbert. Figure 37(9), p. 292. Illustration: Botanical Magazine, 9406 (1935); The Plantsman **3**: 226, 227 (1982).
*Syria and Israel.* H4. Summer.

**26. PANCRATIUM** Linnaeus
*D.A. Webb*
Perennial herbs with large bulbs and several broadly linear to strap-shaped basal leaves which are more or less in 2 ranks. Scape bearing an umbel of 3–15 flowers (rarely a solitary flower), subtended by usually 2 scarious spathes. Flowers large, white, fragrant. Perianth-tube present; perianth-lobes linear-lanceolate, spreading or almost erect; corona conspicuous, united to the lower part of the filaments, which thus appear to be inserted on its margin. Anthers dorsifixed. Stigma capitate. Ovary 3-celled with ovules in 1 or 2 vertical rows in each cell. Fruit a many-seeded capsule; seeds black, angular, dry.

A genus of about 16 species, extending from the Canary Islands through the Mediterranean region to tropical Asia, and southwards through W Africa to Namibia. It differs from *Hymenocallis* in its numerous seeds with a thin, dry, black testa. Many tropical species formerly included under *Pancratium* are now assigned to *Hymenocallis* (p. 323).

*P. zeylanicum* requires hot-house conditions. The other species can be cultivated out-of-doors in favoured regions of W & S Europe, but they require a very sunny position in well-drained soil. In the case of *P. maritimum* the bulbs should be planted at a depth of about 30 cm and surrounded by sand.

1a. Free part of filament much longer than the anther | 2
b. Free part of filament equalling or shorter than the anther | 3
2a. Flowers in umbels; spathes 2 | **1. illyricum**
b. Flower solitary; spathe 1 | **4. zeylanicum**
3a. Flower-stalks 2–3 cm; perianth-tube 1.5–2.5 cm; corona one-third as long as perianth | **2. canariense**
b. Flower-stalks 5–10 mm; perianth-tube to 7.5 cm; corona two-thirds as long as perianth | **3. maritimum**

**1. P. illyricum** Linnaeus. Illustration: Botanical Magazine, 718 (1804); Coste, Flore de la France 3: 382 (1906). Bulb-scales purplish black. Leaves *c.* 50 × 1.5–3 cm, glaucous, about equalling the scape. Flower-stalks 1–1.5 cm. Perianth-tube *c.* 2 cm, fairly stout. Perianth-lobes 5 cm, linear-oblong to narrowly elliptic. Corona much shorter than perianth, with pairs of long, narrow teeth alternating with the stamens. Free part of filament much longer than anther. *Islands of W Mediterranean (Corsica, Sardinia, Capraia).* H4–5. Late spring–early summer.

**2. P. canariense** Ker Gawler. Illustration: Edwards's Botanical Register 2: t. 174 (1816); Bramwell, Wild flowers of the Canary Islands, f. 111 (1974). Differs from *P. illyricum* chiefly in its somewhat broader leaves, much longer flower-stalks (up to 3 cm), shorter teeth to the corona, and free part of filament scarcely as long as the anther. *Canary Islands.* H5. Early autumn.

**3. P. maritimum** Linnaeus. Figure 37(14), p. 292. Illustration: Bonnier, Flore Complète 11: t. 604 (1930); Flore de l'Afrique du Nord 6: 36 (1959); Goulandris et al., Wild flowers of Greece, t. 68 (1968). Like *P. illyricum*, but with pale bulb-scales, a very long neck to the bulb, and somewhat longer and narrower leaves. Flower-stalks

5–10 mm; perianth-tube to 7.5 cm, very slender. Corona two-thirds as long as perianth, with short, triangular teeth between the stamens. Free part of filament scarcely longer than the anther. *Mediterranean region and SW Europe.* H5. Summer.

**4. P. zeylanicum** Linnaeus. Illustration: Botanical Magazine, 2538 (1825). Differs from all the above in having a solitary flower with a single spathe, and leaves of a fresh green, not glaucous. Flower-stalk very short; perianth-tube long and slender. Corona and filaments much as in *P. illyricum*. *Tropical Asia.* G2. Summer.

## 27. PARAMONGAIA Velarde
*J.C.M. Alexander*

Deciduous perennial herbs. Bulb to 6.5 cm across; leaf-bases not forming a false-stem or neck above the bulb. Leaves 6–8, all basal, bright green, sometimes glaucous, 2-ranked, to 75 × 5 cm, narrowly strap-shaped, stalkless, reaching full size after flowering. Scape erect, solid, to *c.* 75 cm, bearing 1 (rarely 2) long-lasting, yellow, fragrant flowers, to 18 cm across. Perianth-tube *c.* 10 cm; lobes to 9.5 × 4 cm, not thickened at base, spreading; corona *c.* 8.5 × 2 cm. Stamens 6, filaments borne well below corona margin, anthers versatile, not projecting. Style free from perianth-tube, not winged. Ovary 3-celled with ovules in 1 or 2 vertical rows in each cell. Fruit a capsule *c.* 4 cm long, with 50–75 seeds in rows (superposed).

A single species native to only 2 areas in coastal and montane Peru. This species is not easy to grow. The bulbs should be planted deeply in a mixture of loam, sand and peat with added leaf-mould and charcoal. In the wild they are as much as 30 cm below the surface. In the dormant season (summer) the bulbs should be baked at 21–4 °C, preferably in the pots in which they are being grown. When the leaves appear in about October, water should be freely given and similar temperatures maintained with a night minimum of about 18 °C. Plants do best in full sunshine and mature bulbs will produce one or two offsets per year.

**1. P. weberbaueri** Velarde. Illustration: American Horticulturist 54(6): 30 (1975). *Peru.* G2. Autumn.

## 28. PAMIANTHE Stapf
*P.F. Yeo*

Perennial herbs with bulbs. Leaves strap-shaped, evergreen or deciduous, sheathing

one another at the base to form a false-stem. Flowers bisexual, radially symmetric, in umbels on an angled stalk; bracts conspicuous. Perianth-tube slender, straight, scarcely widened towards the throat; perianth-lobes obovate to oblanceolate, white. Stamens borne on a large white corona, attached in deep notches and directed horizontally towards the centre of the flower; anthers dorsifixed. Style weakly 3-lobed. Ovary 3-celled, with many ovules in 1 or 2 vertical rows in each cell. Seeds flattened, winged at the base.

Three species from tropical South America, of which *P. peruviana* is grown for its spectacular and exquisitely scented flowers; it is sometimes evergreen but can apparently be induced to become dormant in winter by drying it off. It may be grown in pots and requires a loam-based compost. Propagation is by offsets or, preferably, by seed which takes 12–15 months to mature on the plant but then germinates rapidly if sown at once; seedlings grow rapidly in humid conditions.

**1. P. peruviana** Stapf. Figure 37(6), p. 292. Illustration: Botanical Magazine, 9315 (1933); Hay et al., Dictionary of indoor plants in colour, 368 (1974); Everett, Encyclopedia 7: 2478 (1981). Plant 40–50 cm high. Pseudo-stem to 30 cm, composed of scale-leaves and leaf-bases; leaf blades to 50 cm, 2–4 cm wide, spreading and drooping. Inflorescence stalk with its visible part *c.* 10 cm. Flowers 2–4, *c.* 13 cm in diameter, shortly stalked, spreading. Perianth-tube 12–13 cm × 7–10 mm, green, solid at the base for *c.* 1 cm, divided into 3 by lengthwise partitions for the next 6–7 cm. Perianth-lobes to 12 × 3 cm, oblanceolate, divergent. Corona to 8 cm, bell-shaped, with pointed lobes; notches from which the stamens arise much deeper than those between the stamens. Style emerging on lower side of flower. *Peru.* G1. Spring–early summer.

## 29. STENOMESSON Herbert
*S.G. Knees*

Perennial herbs with bulbs, the brown tunic often exceeding the neck for up to 25 cm. Leaves 3–8, deciduous, linear to lanceolate, 20–35 × 2–3 cm at flowering time, often extending to 60 or 70 cm after flowering. Scape solid, sometimes 4-angled, 30–60 cm, bearing 2 broadly ovate spathes and 1–4 leaf-like bracts. Flowers shortly stalked, usually pendulous, 1–6 per umbel. Perianth tubular, with 6 ovate, hooded lobes, variously coloured and usually with a small, toothed corona arising just above the base of filaments. Anthers versatile, including in perianth, style projecting, stigma capitate. Ovary 3-angled, ovules many in 1 or 2 vertical rows in each cell. Fruit a capsule containing flattened black seeds.

A genus of about 20 species mostly native to the high Andes. Although several species were grown in the past there is little evidence of these now being widely cultivated, even in specialist collections, and most of what is grown is probably *S. variegatum*. This has been shown by Sealy (reference below) to be extremely variable and includes plants formerly grown as *S. incarnatum* (Humboldt, Bonpland & Kunth) Baker, *S. luteoviride* Baker and *S. fulvum* Herbert. An open compost and a winter minimum of 5 °C are 2 major requirements for success with *Stenomesson*. Bulbs should never be allowed to dry out completely and liberal quantities of water are required during active growth. Propagation is normally by offsets. Literature: Sealy, J.R., *Botanical Magazine*, n.s., 503 (1967); Ravenna, P.F., Contributions to the South American Amaryllidaceae IV and V, *Plant Life* 27: 73–84 (1971).

**1. S. variegatum** (Ruiz & Pavon) Macbride. Figure 37(8), p. 292. Illustration: Edwards's Botanical Register, t. 1497 (1832); Botanical Magazine, n.s., 503 (1967); Marshall Cavendish encyclopedia of gardening 7: 2142 (1967); Rix, Growing bulbs, f. 2.0 (1983).
Scape usually 4-angled. Flowers white, yellow, pink or reddish with broad green banding along the midrib near tips of the lobes. Perianth slender, tubular, 10–13 cm, with spreading lobes. *Bolivia, Ecuador, Peru.* H5. Spring–summer.

**30. URCEOLINA** Reichenbach
*S.G. Knees*
Perennial herbs with bulbs 4–7 × 1.5–5 cm and fibrous roots. Leaves 1 or 2, 35–50 × 1–15 cm, stalked for about one-quarter of their length, appearing with or after the flowers. Leaf-blades ovate to oblong, flattened, acute. Leaf-stalks *c.* 5 mm wide, almost winged. Scape solid, 12–25 cm, bearing umbels of 3–6 flowers subtended by 2 spathes. Perianth urceolate, with projecting style and stamens. Corona small. Anthers versatile. Ovary 3-celled with numerous ovules in vertical rows. Fruit a capsule.

A genus of 2 species, both native to

S America, flowering more freely in pots than in the open ground. In northern Europe they require glasshouse protection and need to be kept dry during the dormant season. Propagation is by offsets which are freely produced.

1a. Leaves linear, 1–5 cm wide, appearing after the flowers; perianth orange or red **1. peruviana**
  b. Leaves ovate to oblong, 8–15 cm wide, appearing with the flowers; perianth yellow **2. urceolata**

**1. U. peruviana** (Presl) Macbride (*U. miniata* (Herbert) Bentham). Figure 37(13), p. 292. Illustration: Edwards's Botanical Register, t. 68 (1839); Botanical Magazine, n.s., 399 (1962); Mathew, The larger bulbs, 140 (1978).
Leaves flattened, strap-like, 40 × 1–5 cm, usually appearing after the flowers. Fused part of perianth 3–3.5 cm long, narrow at base for 1–1.3 cm, urceolate above, lobes spreading, 9–11 mm; bright red or orange, stamens projecting for *c.* 1.5 cm beyond the perianth. *Peru & Bolivia.* H5. Spring–summer.

**2. U. urceolata** (Ruiz & Pavon) M.L. Green (*U. pendula* (Herbert) Herbert). Illustration: Botanical Magazine, 5464 (1864); Everard & Morley, Wild flowers of the world, t. 190 (1970); Mathew, The larger bulbs, 140 (1978).
Leaf-blades ovate to oblong, 30–50 × 9–15 cm, usually appearing with the flowers. Perianth yellow, fused part 7.5–10 cm, narrow at the base for 2–3 cm, broadly urceolate above, lobes recurved, acute, green-tipped. *Peru.* H5. Spring–summer.

**31. EUCHARIS** Planchon & Linden
*P.F. Yeo*
Perennial herbs with bulbs. Leaves evergreen with a stalk and a distinct ovate or elliptic blade with numerous arching veins; stalks not sheathing one another. Flowers bisexual, radially symmetric, in an umbel on a stalk about as long as the leaves; bracts all similar in length. Perianth-tube long, slender, curved, somewhat widened above; perianth-lobes shortly oblong to broadly ovate (not linear), white. Stamens borne on a corona which usually bears a lobe on either side of each stamen, but may form a narrow, unlobed rim. Anthers dorsifixed. Style 3-lobed. Ovary 3-celled; ovules numerous, in vertical rows, more rarely 2 per cell. Seeds rounded, black or blue.

About 10 species from tropical S America, providing showy white flowers accompanied by glossy dark green leaves. The flowers of *E. amazonica* are suitable for cutting. Several hybrids have arisen in cultivation. The plants require shade, high humidity and high temperatures (15–20 °C, rising to 25–30 °C in summer) and are often planted under benches and near heating pipes in hot-houses, but may also be grown in pots or tubs. They are propagated by careful removal of offsets or by seeds sown as soon as ripe.

Meerow considers that *Eucharis* and *Caliphruria* should be united, but for the purposes of this work it seems better to separate them.
Literature: Meerow, A., Two new species of pancratioid Amaryllidaceae from Peru and Ecuador, *Brittonia* 36: 19–25 (1984); Meerow, A.W. & Dehgan, B., Re-establishment and lectotypification of Eucharis amazonica Linden ex Planchon (Amaryllidaceae), *Taxon* 33: 416–22 (1984).

1a. Corona in the form of a narrow rim at throat of flower, without lobes or teeth **1. sanderi**
  b. Corona 6 mm or more, lobed or toothed **2**
2a. Flowering bulb with more than 1 leaf: perianth-lobes not recurved, the outer similar to the inner, broadly ovate **2. amazonica***
  b. Flowering bulb with only 1 leaf: perianth-lobes slightly recurved, the outer about twice as long as wide, much wider than inner **3. candida**

**1. E. sanderi** Baker. Illustration: Botanical Magazine, 6676 (1883), 6831B (1885). Plant to *c.* 45 cm. Leaves 2 on each flowering bulb. Leaf-stalks to 15 cm. Leaf-blades to 25 cm, with *c.* 6–10 veins on either side of the midrib. Inflorescence-stalk to *c.* 45 cm, with 2–6 flowers. Flower-stalks *c.* 5 mm. Flowers 4–6 cm in diameter, scentless, drooping by the curvature of the tube, which is *c.* 6 cm long and enlarged beneath the limb. Perianth-lobes broadly ovate, spreading or merely divergent, overlapping, inner and outer similar. Throat of flower cream-coloured within and striped with pale yellow. Corona a mere rim, not lobed, cream-coloured. Filaments 6–8 mm. Style much longer than the stamens. Ovules many in each cell of the ovary. *Colombia.* G2.

**2. E. amazonica** Planchon (*E. grandiflora* misapplied). Figure 37(12), p. 292.

Illustration: Botanical Magazine, 4971 (1857); Flore des Serres 12: 69 (1857); Hay et al., Dictionary of indoor plants in colour, 220 (1974); Everett, Encyclopedia of horticulture 4: 1280 (1981).
Plant to c. 60 cm. Leaves several on each flowering bulb. Leaf-stalks 12–25 cm. Leaf-blades to c. 30 × 14 cm, with c. 20 veins on each side of midrib, dark glossy green. Inflorescence-stalk to c. 60 cm, with about 8 flowers. Flower-stalks c. 1.5–2 cm at first, lengthening to 3 cm. Flowers 10–12 cm in diameter, fragrant, drooping by the curvature of the tube, which is 6–8 cm long and gradually enlarged just beneath the limb; perianth-lobes c. 4 × 3 cm, broadly ovate, spreading, overlapping, inner and outer similar. Corona c. 12 mm × 2–2.5 cm, with an obtuse lobe on either side of each filament and a narrow sinus between adjoining lobes; white with a pale greenish yellow stripe in line with each filament. Filaments c. 6 mm, narrowly triangular. Style slightly exceeding the stamens. Ovules 9–12 in each cell of the ovary. *Peru, Ecuador*. G2. Summer, winter and occasionally at other times.

The closely related *E. grandiflora* Planchon & Linden, from Colombia, is believed to have been lost from cultivation. It has long been confused with *E. amazonica*.

*E. mastersii* Baker. Illustration: Botanical Magazine, 6831A (1885). Similar to *E. amazonica* but corona half as long, with 12 equal broadly triangular teeth. *Colombia*. G2.

3. E. candida Planchon & Linden. Illustration: Flore des Serres 8: 107 (1852–3).
Similar to *E. amazonica* but with only one leaf per bulb at flowering time, with smaller scentless flowers, c. 7 cm in diameter, which have recurved perianth-lobes, the outer differing from the inner in being oblong, about twice as long as wide, with a narrower, less distinctly lobed, almost completely yellow corona cut nearly to the base, and with only 2 ovules in each cell of the ovary. *Colombia*. G2.

32. CALIPHRURIA Herbert
*P.F. Yeo*
Perennial herbs with bulbs. Leaves 2 or more to each bulb, evergreen with a stalk and a distinct elliptic to obovate blade; stalks not sheathing one another to form a pseudo-stem. Flowers bisexual, radially symmetric, in umbels on a stalk as long as or longer than the leaves, with 2 spathes concealing the floral bracts. Perianth-tube gradually widened from the base, straight, greenish. Perianth-lobes somewhat shorter than tube, approximately ovate, white. Stamens attached to throat of flower; filaments abruptly widened towards base, the enlargement on either side sometimes produced into a tooth; anthers dorsifixed. Style 3-lobed. Ovary 3-celled, with 2 or 3 ovules in each cell.

Three or perhaps more species from Colombia, sometimes included in *Eucharis*, but differing from it in the smaller flowers with a small corona. They are greenhouse subjects which, if grown in pots or tubs, may be placed out-of-doors in summer. Propagation is by offsets.

1. C. subedentata Baker (*Eucharis subedentata* (Baker) Bentham & Hooker). Figure 37(11), p. 292. Illustration: Botanical Magazine, 6289 (1877); Nicholson, Dictionary of gardening 1: 241 (1884).
Plant to c. 45 cm. Leaf-stalk c. 30 cm. Leaf-blade c. 20 cm, more than half as wide as long, with about 10 veins on either side of the pale midrib. Flower-stalks 1–1.5 cm. Flowers to c. 4 cm in diameter, horizontal or nodding, funnel-shaped; tube 3.5 cm; perianth-lobes c. 2 cm, slightly recurved at the tips. Stamens about half as long as perianth-lobes; filaments with the enlargement entire, toothed on one side or toothed on both, all conditions occurring in the same flower. Style not exceeding perianth-lobes. *Colombia*. G1.

*C. hartwegiana* Herbert (*Eucharis hartwegiana* (Herbert) Nicholson). Illustration: Botanical Magazine, 6259 (1876). Smaller in leaves and flowers than *C. subedentata* by about one-third. Leaf-blades thicker and firmer with inconspicuous veins. Leaf-stalks and inflorescence-stalks grey-green. Filaments with the basal enlargement produced on each side into a tooth about as long as the narrow part of the filament. Style exceeding perianth-lobes. *Colombia*. G1.

33. HYMENOCALLIS Salisbury
*J.C.M. Alexander*
Perennial herbs with spherical to ovoid bulbs. Leaves evergreen or deciduous, basal, usually 2-ranked, erect to arching, strap-shaped to sword-shaped or narrowly ovate, stalked or gradually narrowed to a stalkless base; bases sometimes forming a false-stem or neck above the bulb. Scape erect, 2-edged, solid, often bluish, bearing 1–several stalkless or shortly stalked, often fragrant flowers, whose bases are enclosed in chaffy bracts. Perianth-tube narrow; perianth-lobes 6, linear to narrowly lanceolate, ascending, arched or spreading (adpressed to corona in No. 13). Corona funnel-shaped, remaining erect (subgenera *Hymenocallis* & *Ismene*) or becoming sharply bent (subgenus *Elisena*); margin erect or spreading, often complexly lobed or toothed. Lower parts of filaments fused to corona; free parts erect or hanging outwards (subgenera *Hymenocallis* & *Elisena*) or inturned (subgenus *Ismene*). Style long, equalling or projecting from the corona. Ovary 3-celled with 2 (rarely 3) or 4–8 basal ovules side by side in each cell. Fruit a fleshy capsule; seeds green, fleshy, often only 1 developing per cell or even per fruit.

A genus of 35–40 species, native to the tropics and subtropics of the New World, with a few species naturalised in E and W Africa and Sri Lanka. Many of the species were originally considered under *Pancratium*, a name now restricted to Old World species with many ovules in vertical rows, developing into black angular seeds. The genera *Ismene* Salisbury and *Elisena* Herbert are here included in *Hymenocallis* as subgenera. The differences between them are small and hybridisation is quite common.

All species perform well in light, fertile, well-drained soil, with plenty of moisture during the growing season. In general, the evergreen species require cool or warm greenhouse conditions and may need shading from strong sun. Many of the deciduous species can be grown out-of-doors in sheltered, sunny sites. In areas with frost it may be wise to lift the bulbs in autumn and keep them dry and frost-free before planting them out when the soil has warmed up. They are easily propagated from seed or offsets.

*Leaves.* Clearly stalked: 7,8; stalkless, but narrower and parallel-sided towards the base: 10–15; stalkless, not narrow and parallel-sided towards the base: 1–6,11,15. Strap-shaped to sword-shaped, c. 5 times as long as broad or more: 1–6,10–15; elliptic to ovate, c. 3 times as long as broad or less: 7–9.
*Scape.* 30 cm or less: 10; 30–60 cm: 1–5,7–9,11–15; 60 cm or more: 5,6,12,14.
*Flowers.* Yellow: 1. Stalked: 1–6,8; stalkless: 1,7,9–15. 4 or fewer per umbel: 1–3,6,10; 5–10 per umbel: 1–5,7–9,12,14; 10 or more per umbel: 7,8,13,14.

*Corona*. Longer than perianth-tube: 3,5,6; shorter than perianth-tube: 1–4,7–15. Bending forward and becoming flattened: 5,6.

*Stamens*. Free part less than half corona length: 1–4; more than half corona length: 5–15. Sharply inturned, not projecting from corona: 1–3; projecting from corona: 3–15; incurved over corona: 4; parallel and more or less horizontal (upper 3 sometimes bending downwards across corona): 5,6; erect, spreading or hanging outwards: 7–15.

*Style*. Not or scarcely projecting from corona: 1–3; projecting 4 cm or more from corona: 4–15.

1a. Free parts of stamens less than half corona length, all sharply inturned and not projecting from corona, or all incurved and converging above corona (subgenus *Ismene*) 2
  b. Free parts of stamens longer than, equal to or slightly shorter than corona, not all inturned or converging above corona 5
2a. Flowers yellow **1. amancaes**
  b. Flowers white or greenish white 3
3a. Perianth-tube 3.8–5 cm **3. pedunculata**
  b. Perianth-tube 7.6–10 cm 4
4a. Flowers shortly stalked; filaments *c.* 2.5 cm, converging above corona **4. × macrostephana**
  b. Flowers stalkless; filaments *c.* 1.3 cm, sharply inturned, not projecting from corona **2. narcissiflora**
5a. Corona longer than perianth-tube, deeply cup-shaped, horizontal or downturned, bending downwards and becoming elliptic to oblong in cross-section before flower withers; stamens parallel or the upper 3 becoming deflexed (subgenus *Elisena*) 6
  b. Corona shorter than perianth-tube, funnel-shaped, erect or ascending, remaining circular in cross-section; stamens erect, arched or hanging outwards (subgenus *Hymenocallis*) 7
6a. Perianth-tube *c.* 1 cm; filaments 7.5 cm, remaining parallel **5. longipetala**
  b. Perianth-tube 4–5 cm; filaments 2.5–4 cm, the upper 3 becoming bent downwards **6. × festalis**
7a. Leaves clearly divided into blade and stalk 8
  b. Leaves stalkless, though often much narrower near base 10
8a. Perianth-tube 12.5–20 cm; ovules 4 or 5 per cell **7. tubiflora**

  b. Perianth-tube 9 cm or less; ovules 2 per cell 9
9a. Perianth-tube 6.5–9 cm; leaf-blades at least 3 times longer than broad, 26–66 × 8–15.5 cm **8. speciosa**
  b. Perianth-tube 4.5–5 cm; leaf-blades *c.* 2 times longer than broad, 10.5–30 × 4–15 cm **9. ovata**
10a. Corona 1.5 cm or less; leaves 3–5 **10. harrisiana**
  b. Corona 2 cm or more; leaves usually 7 or more 11
11a. Free parts of stamens 2–3 cm; corona narrow and green at base; perianth-lobes spreading at right angles to tube **15. rotata**
  b. Not this combination of characters 12
12a. Ovules 2 per cell; narrow part of mature leaf at least 2.5 cm wide 13
  b. Ovules 4 or more per cell; narrow part of mature leaf 2.1 cm wide or less 14
13a. Perianth-tube 4–6.5 cm, shorter than lobes **11. caribaea**
  b. Perianth-tube 10–15 cm, longer than lobes **12. latifolia**
14a. Broad part of mature leaf less than 3.8 cm wide; perianth-lobes adpressed to corona below, arched above; corona margin spreading **13. littoralis**
  b. Leaf usually more than 5 cm wide; perianth-lobes ascending but not adpressed to corona; corona margin erect **14. pedalis**

**1. H. amancaes** (Ruiz & Pavon) Nicholson (*Ismene amancaes* (Ruiz & Pavon) Herbert). Illustration: Botanical Magazine, 1224 (1809).
Bulb spherical, 3.5–5 cm in diameter; neck 15–25 cm. Leaves 4 or 5, strap-shaped, bright green, erect to arching, 25–60 × 2.5–5 cm. Scape 11–33 cm, bearing 2–6 stalkless or shortly stalked, bright to deep yellow flowers. Perianth-tube 5–7.5 cm, slightly curved, greenish yellow. Perianth-lobes linear to narrowly lanceolate, 6–7.5 cm, greenish at tips. Corona funnel-shaped, 5–6 × 6–8.5 cm, irregularly 12-toothed. Free parts of filaments 1–2 cm, sharply inturned. Style equal to or slightly projecting from corona. Ovules 2 per cell. *Peru.* H5–G1. Summer.

**2. H. narcissiflora** (Jacquin) MacBride (*H. calathina* (Ker Gawler) Nicholson; *Ismene calathina* (Ker Gawler) Herbert; not *I. narcissiflora* Roemer). Figure 37(5), p. 292. Illustration: Botanical Magazine, 2685 (1826); RHS Dictionary of gardening, edn 2, 1029 (1956).

Bulb spherical; neck 20–30 cm. Leaves 6–8, loosely 2-ranked, strap-shaped to sword-shaped, dark green, acute to obtuse, narrowed at base, 20–75 × 3.5–6 cm. Flowering stem 30–60 cm, bearing 1–5, fragrant, stalked flowers. Perianth-tube 7.5–10 cm, green. Perianth-lobes 7–9 × *c.* 1.3 cm, lanceolate, white, Corona 6–7.5 cm, bell-shaped to funnel-shaped with ragged, spreading lobes, white. Free parts of filaments 1–1.5 cm, sharply inturned. Style projecting, *Peru & Bolivia.* H5–G1. Summer.

**3. H. pedunculata** (Herbert) MacBride (*Ismene macleana* Herbert; *H. macleana* (Herbert) Nicholson; *Ismene virescens* Lindley). Illustration: Botanical Magazine, 3675 (1838).
Similar to *H. narcissiflora*. Neck *c.* 15 cm. Leaves 23–45 × 2–5.5 cm. Perianth-tube 3–5 cm; perianth-lobes 4–5.5 cm; corona 4–5 cm, green with white, spreading lobes. *Peru.* H5. Summer.

**4. H. × macrostephana** Baker. Illustration: Botanical Magazine, 6436 (1879); The Garden 89: 209 (1925).
Bulb ovoid, *c.* 5 cm in diameter; neck long. Leaves 8 or 9 in a loose rosette, evergreen, oblanceolate, obtuse, bright green, 50–90 × 6–8 cm, *c.* 2.5 cm wide at base. Scape 30–45 × 2.5 cm, bearing 6–10 shortly stalked, fragrant flowers. Perianth-tube 6.5–8.5 cm, green below, white above. Perianth-lobes 9–11 cm, linear-lanceolate, white to pale greenish yellow, outer lobes with thick green tips. Corona funnel-shaped, 5.5–6 × 5–7.5 cm, obtusely 12-toothed, white. Free parts of filaments *c.* 2.5 cm; style projecting *c.* 4.5 cm from corona. *?Garden origin.* G2. Spring.
  This is probably the result of a cross between *H. narcissiflora* and *H. speciosa*.

**5. H. longipetala** (Lindley) MacBride (*Elisena longipetala* Lindley). Illustration: Botanical Magazine, 3873 (1841).
Bulb 3.5–5 cm; neck to 50 cm. Leaves 6–8, 2-ranked, stiff, strap-shaped, acute, 45–55 × *c.* 4 cm. Scape 60–90 cm, bearing 5–10, white flowers on stalks *c.* 6 mm long. Perianth-tube *c.* 1 cm × 1 cm; perianth-lobes 7.5–10 cm, linear, green at base, wavy. Corona *c.* 3 cm, becoming flattened and bent forward, marginal teeth bent inwards. Free parts of filaments 6–7 cm, parallel, upturned at tips. Style projecting 1–2 cm beyond stamens. *Ecuador & Peru.* H5–G1. Summer.

**6. H. × festalis** (Worsley) Schmarse (*Ismene festalis* Worsley). Illustration: Gartenflora 82: 337 (1933).

Neck *c.* 4.5 cm. Leaves *c.* 9, loosely 2-ranked, strap-shaped, 60–90 × 5–7 cm. Scape *c.* 120 cm, bearing *c.* 4, white, stalked, fragrant flowers. Perianth-tube 4–5 cm; perianth-lobes 10–12 cm, the outer narrower and more deeply channelled. Corona *c.* 5 cm, becoming flattened and bent forward; marginal teeth bent. Free parts of stamens 2.5–4 cm, parallel or converging at first, the upper 3 become bent downwards. Style projecting *c.* 7.5 cm from corona. *Garden origin.*

The hybrid between *H. narcissiflora* and *H. longipetala.*

**7. H. tubiflora** Salisbury (*H. undulata* (Humboldt et al.) Herbert; *H. moritziana* Kunth; *H. guianensis* (Ker Gawler) Herbert). Illustration: Edwards's Botanical Register, 265 (1818); Gardener's Chronicle **27**: 89 (1900).
Bulb ovoid, 7.5–10 cm in diameter; neck short. Leaves 7–12, stalked, evergreen; blades elliptic to ovate, abruptly pointed, tapered to rounded at base, 20–38 cm × 7.5–15 cm; stalks 15–30 cm. Scape to 60 cm, bearing 5–20, stalkless, white flowers. Perianth-tube 12–20 cm, greenish; perianth-lobes 8.5–14 cm × 4–6 mm. Corona 1.8–2.5 cm, funnel-shaped, margin erect or slightly spreading. Free parts of stamens 4–6.5 cm, ascending to erect. Style projecting beyond stamens, arching. *NE South America.* G2. Spring & autumn.

**8. H. speciosa** (Salisbury) Salisbury. Illustration: Botanical Magazine, 1453 (1812).
Bulb spherical, 7.5–10 cm in diameter. Leaves 12–20, stalked, evergreen; blades broadly elliptic to oblong, acuminate, tapering into the stalk, 25–65 × 8–16 cm; stalks 9–17 cm. Scape 30–45 cm, bearing 7–12, stalked, fragrant, white to greenish flowers; stalks to 1 cm. Perianth-tube 6.5–9 cm, greenish; perianth-lobes 9.5–15 cm × 5–7 mm, white, recurved. Corona funnel-shaped, 2.5–5 cm, toothed. Free parts of stamens 3–5 cm, erect to hanging, filaments greenish white. Style bending downwards, almost as long as perianth-lobes. Ovules 2. *West Indies.* G1. Autumn–winter.

**9. H. ovata** (Miller) Sweet (*H. amoena* misapplied). Illustration: Botanical Magazine, 1467 (1812).
Similar to *H. speciosa.* Leaves 5–8. Flowers 5 or 6, stalkless; perianth-tube 4.5–6 cm; perianth-lobes 5.5–10 cm. Filaments green. *?West Indies.*

**10. H. harrisiana** Herbert. Illustration: Botanical Magazine, 6562 (1881).
Bulb spherical to ovoid, *c.* 4 cm in diameter. Leaves 3–5, deciduous, oblanceolate, much narrowed below though not with a true stalk, 20–34 cm overall, obtuse to acute; broad part 19–30 × 3–5.1 cm, oblanceolate to oblong; narrow part *c.* 7 cm. Scape 20–25 cm, bearing 1–3, stalkless, white flowers. Perianth-tube 8–13 cm, greenish below; perianth-lobes 6–7.5 cm, linear. Corona funnel-shaped, 1.3–1.8 cm, margins spreading, 6-toothed. Free parts of stamens 2.5–3.5 cm, erect to hanging. Ovules 2. *Mexico.* G1. Spring–summer.

**11. H. caribaea** (Linnaeus) Herbert (*Pancratium caribaeum* Linnaeus, in part; *H. amoena* (Salisbury) Herbert). Illustration: Botanical Magazine, 826 (1805).
Bulb spherical, 7.5–10 cm in diameter. Leaves many, evergreen, shining, loosely 2-ranked, strap-shaped to sword-shaped, acute or obtuse, slightly narrowed below, stalkless, 30–90 × 5–10 cm. Scape 30–60 cm, bearing 8–10, stalkless, fragrant, white flowers. Perianth-tube 4–6.5 cm, greenish; perianth-lobes 8–12 cm, linear. Corona funnel-shaped, 2–3 cm, margin erect. 12-toothed. Free parts of stamens 3–5 cm, erect to hanging. Ovules 2 per cell. *West Indies.* G2. Summer–autumn.

**12. H. latifolia** (Miller) Roemer (*Pancratium caribaeum* Linnaeus, in part; *H. caribaea* misapplied; *H. caymanensis* Herbert; *H. keyensis* Small).
Very similar to *H. caribaea.* Scape 60–80 cm; perianth-tube 10–15 cm, always longer than the lobes. *USA (Florida) & West Indies.* G2. Summer.

**13. H. littoralis** (Jacquin) Salisbury (*H. senegambica* Kunth & Bouché, in part; *H. americana* (Miller) Roemer). Illustration: Botanical Magazine, 825 (1805) & 2621 (1826); Grey, Hardy Bulbs **2**: 47 (1938); Everett, Encyclopedia of horticulture, 1758 (1981).
Bulb spherical, 7.5–10 cm in diameter. Leaves numerous, evergreen, 2-ranked, stalkless, 40–120 cm; upper part strap-shaped, 2–4 cm wide, acute, tapering to a narrower basal part 1–2 cm wide. Scape 45–60 cm, bearing 6–11, white, stalkless, fragrant flowers. Perianth-tube 14–17 (rarely 10–20) cm, straight; perianth-lobes 7.5–8 (rarely to 12.5) cm, adpressed to corona below, arched above. Corona 2.5–3 (rarely 2–3.5) cm, funnel-shaped below,

widely flared above. Free parts of stamens 4–6 cm, erect to hanging. Ovules 4 or 5 (rarely 8) per cell. *Mexico, Guatemala & Northern S America; naturalised in W Africa & Sri Lanka.* G1. Spring.

**14. H. pedalis** Herbert (*H. senegambica* Kunth & Bouché in part). Illustration: Edwards's Botanical Register, t. 1641 (1834).
Similar to *H. littoralis.* Leaves 2.5–7 cm wide in upper part, tapering to 1.5–2 cm in lower part. Perianth-tube slightly curved; perianth-lobes wavy-edged. Corona funnel-shaped, margins erect. *Colombia & Brazil; naturalised in W and possibly E Africa.* G2. Spring.

**15. H. rotata** (Ker Gawler) Herbert (*H. mexicana* misapplied; *H. lacera* Salisbury, in part). Illustration: Botanical Magazine, 827 (1805) & 1082 (1808).
Bulb ovoid, 3.5–5 cm in diameter, bearing stolons; neck *c.* 4 cm. Leaves 7 or 8, deciduous, linear to strap-shaped or sword-shaped, obtuse, 30–68 × 1.1–3.6 cm, tapered at base, or sometimes from near the middle to a narrower basal portion 7–17 mm wide. Scape 45–60 cm, bearing 2–4 (rarely to 8), stalkless, white, fragrant flowers. Perianth-tube 6–9.5 cm; perianth-lobes 6.2–9.5 cm, spreading at right angles to tube. Corona 3–5 cm; narrow and green below, funnel-shaped and white above, margin spreading. Free parts of stamens 2–3 cm, erect. Ovary with 2 (rarely 4–6) ovules per cell. *Jamaica & southern USA.* H5–G1. Spring.

## 34. PHAEDRANASSA Herbert
*S.G. Knees*

Perennial herbs with bulbs. Leaves 1–4, stalked, appearing with, or a little after, the flowers. Leaf-blades lanceolate to elliptic, 15–30 × 3.5–7 cm, stalks 5–18 cm, somewhat channelled. Scape hollow, 2–70 cm, widest at the base, decreasing to less than half its diameter at the apex. Flowers pendulous, in umbellate clusters of 4–11. Perianth tubular, richly coloured, with combinations of green, yellow, crimson and scarlet. Corona a ring of small transparent teeth at the base of the filaments. Stamens projecting from the perianth, arising about 1 cm from its base, filaments unequal in length, anthers versatile. Ovary 3-celled, broadly triangular, style projecting from the perianth, stigma capitate. Fruit a capsule, deeply grooved and containing several black, flattened seeds.

A genus of about 6 species, all native to S America. None is widely cultivated and their morphological similarity has caused difficulty.

A good loamy soil and a position in a warm sunny site are both essential. If pot-grown, plants require a resting period once the leaves have died down, and should be kept dry until the following spring. Flowering periods may be rather erratic. Propagation is normally by offsets, although, occasionally, viable seed may be produced.

Literature: Ravenna, P.F., Contributions to South American Amaryllidaceae III, *Plant Life* **25**: 55–62 (1969).

**1. P. carmioli** Baker. Illustration: Botanical Magazine, n.s., 71 (1949); Graf, Exotica, edn 3, 102 (1963).

Bulb 5 cm in diameter, with thin brown tunic. Leaves usually 2, 30–50 cm, elliptic with midrib almost keeled beneath. Scape 50 cm, flowers 7–9. Perianth tubular, 3.5–4.5 × 1 cm, green at base, tube scarlet, lobes green with yellow markings on margins, somewhat reflexed at the apex. Filaments white, anthers greenish yellow. Ovary 3-angled, 6–8 × 4–5 mm. Capsule 1.5 cm × 5 mm. Seeds 1–1.2 cm × 4–6 mm. *Possibly native of Peru, but most collections originate from Costa Rica where it is widely grown*. H4. Spring–summer.

**\*P. lehmannii** Regel. Plants very similar to *P. carmioli* but differing in their shorter perianth-tube (3–3.5 cm) that is almost entirely red. The stamens project for 1.5–2 cm.

**\*P. dubia** (Humboldt, Bonpland & Kunth) Macbride. This species has a longer perianth-tube (4.5–5 cm) and stamens project for only 7 mm.

*P. chloracra* (Herbert) Herbert, *P. obtusa* (Herbert) Lindley, *P. ventricosa* Baker and possibly also *P. schizantha* Baker are probably the same as *P. dubia*. *P. viridiflora* Baker may just be another colour form.

# X. TECOPHILAEACEAE

Herbs with corms or tubers. Leaves basal, linear or ovate, with parallel veins. Flowers bisexual, radially symmetric, solitary or in a raceme or panicle. Perianth-segments 3, united at base into a short tube, or not united. Fertile stamens 6, or 3 with 3 infertile staminodes. Ovary half-inferior, 3-celled, ovules numerous. Fruit a capsule.

A family of 6 genera and about 20 species native in California, S & C America, and South Africa. Only 1 genus is in general cultivation.

## 1. TECOPHILAEA Colla
*E.C. Nelson*

Corms with fibrous tunics. Leaves 2 or 3, linear, enclosed in a sheath when young, to 10 cm long, green, hairless. Perianth blue to purple with darker veins and white throat, 4 cm long.

A crocus-like plant with elegant blue flowers. The soil should be moist at flowering time, but dry for the rest of the year. Propagation is by offsets. Two species have been described, but only 1 is in cultivation.

**1. T. cyanocrocus** Leybold. Illustration: Botanical Magazine, 8987 (1923). Hay, Reader's Digest encyclopaedia of garden plants and flowers, 696 (1973). *Chile (Andes)*. H5.

Two cultivars are generally listed: 'Leichtlinii' with prominent white throat and 'Violacea' with deep purple flowers.

# XI. HYPOXIDACEAE

Perennial herbs with rhizomes or corms, and fleshy as well as fibrous roots. Leaves often pleated, usually hairy, stalked or not, usually all basal. Flowers solitary or in the axils of bracts in racemes, occasionally contracted into a dense head, rarely borne at ground level. Perianth of 6 lobes usually united into a (sometimes short) tube below, the lobes spreading, usually at least the outer 3 hairy on their backs. Stamens 6, borne on the perianth-tube. Ovary inferior, 3-celled, with numerous, axile ovules in each cell (the ovary is occasionally subterranean, borne between the leaf-sheaths, and the perianth is raised into the air by a beak on top of the ovary). Fruit a capsule opening by a lid, a berry, or, when the ovary is subterranean, capsule-like but thin-walled and breaking up irregularly. Seeds usually black.

A family of 3–8 genera (different authors vary in the number of genera recognised) from most of the warmer parts of the world. Its classification is confused and identification can be difficult. Fortunately, very few species are cultivated, and they are arranged here in 3 broadly circumscribed genera; sufficient synonymy is given, however, so that names in other genera are easily available.

1a. Flowers in a head; leaves distinctly and conspicuously stalked
          **1. Curculigo**
  b. Flowers solitary or in racemes; leaves not or obscurely stalked    2

2a. Flowers yellow or orange, rarely white when each perianth-lobe has a dark purple blotch at the base
          **2. Hypoxis**
  b. Flowers red, pink or white, perianth-lobes without dark blotches at the base    **3. Rhodohypoxis**

## 1. CURCULIGO Gaertner
*J. Cullen*

Rhizomatous herbaceous perennials. Leaves large, conspicuously stalked, the blades conspicuously pleated, hairy or not, borne in a tuft. Flowers in a dense head-like raceme, each flower subtended by a large bract, the head borne at soil-level or on an elongate, aerial stalk which is shorter than the leaves. Perianth of 6 lobes united into a tube at the base (sometimes short). Stamens 6, borne on the perianth-tube. Ovary inferior, 3-celled, with many ovules in each cell. Fruit a berry containing several seeds.

A genus of 35 species from tropical Asia, Africa and America. Some of them are placed in the genus *Molineria* Colla by some authors (see Geerinck, Bulletin du Jardin Botanique National de Belgique **69**: 70–2, 1969, and Hilliard & Burtt, Notes from the Royal Botanic Garden, Edinburgh **36**: 72–6, 1978). Only 2 species are widely grown; they are easily cultivated, requiring warm, humid conditions and a peaty soil. Propagation is by offsets or division of the rhizome.

1a. Head of flowers borne at soil-level; bracts green, hairless    **2. latifolia**
  b. Head of flowers borne on an elongate, erect, aerial stalk; bracts brown, hairy
          **1. capitulata**

**1. C. capitulata** (Loureiro) Kuntze (*C. recurvata* Dryander; *Molineria recurvata* (Dryander) Herbert). Illustration: Edwards's Botanical Register 9: t. 770 (1823); Parey's Blumengärtnerei 1: 363 (1958); Iconographia Cormophytorum Sinicorum 5: 547 (1976).

Rhizome oblique. Leaves large, erect then arching, stalk 30–60 cm, hairy, blade oblanceolate, acute, 60–90 × 5–15 cm, hairless or with some hairs on the veins. Scape 7–25 cm, hairy, bearing a deflexed head which is 2.5–7 cm long. Bracts brownish, hairy, 1.2–3 cm. Perianth with a very short tube at the base, yellow, lobes 6–8 mm. Berry spherical, 6–8 mm, seeds black. *Himalaya & SE Asia to Australia*. G2.

Plants with variegated leaves have been grown as 'Variegata' and 'Striata'.

**2. C. latifolia** Dryander. Illustration: Edwards's Botanical Register **9**: t. 754 (1823); Botanical Magazine, 2034 (1818); Keng, Orders and families of Malayan seed plants, f. 177 (1969).

Rhizome creeping. Leaves erect, then arching, stalk 7–30 cm, blade lanceolate or oblong-lanceolate, acuminate, 30–60 × 5–10 cm, hairless or somewhat hairy beneath. Scape absent, the flower-head borne at soil-level. Bracts green, hairless, 1.2–3.6 cm. Perianth with a distinct tube at the base, yellow, lobes 8–12 mm, densely hairy outside. Berry oblong, to 1.2 cm, seeds black and shining. *Burma, Malaysia, Indonesia & Borneo.* G2.

## 2. HYPOXIS Linnaeus
*J. Cullen*

Plants with corms and thick, fleshy roots. Leaves mostly basal, sometimes borne on the stem, ridged or not, often hairy, their bases sheathing and overlapping. Scape present, bearing a solitary flower or a 2–12-flowered raceme. Perianth-lobes 6, the outer 3 slightly broader than the inner and greenish and usually hairy outside. Stamens 6. Ovary inferior, 3-celled, with several ovules in each cell. Style present or very short, stigma 3-lobed. Capsule containing several seeds, opening by means of a lid formed from the ovary wall. Seeds black or brown, spherical or elongate.

A genus of perhaps 150 species from N America, Africa, tropical Asia and Australia. Only a small number of species is now grown, though many more were in cultivation in the late nineteenth century. The classification of the genus, and the identification of the species are difficult, and much further study is needed: species No. 1 included here (and the species mentioned under it) are often placed in the genus *Spiloxene* Salisbury, which differs from *Hypoxis* in the strict sense in being totally hairless, possessing linear anthers, and in having the 3 lobes of the stigma more distinct.

The plants are easily grown in good soils, and can be propagated by seeds or offsets.

*Plant.* Completely hairless: **1**; hairy in at least some part (leaves, scape, backs of perianth-lobes, ovary): **2–6**.
*Leaves.* Hairless: **1,2**; hairy: **3–6**. More than 3 cm broad: **2,3**; to 1 cm broad: **1,4–6**.
*Flower.* To 3 cm in diameter: **1,2,5,6**; more than 3 cm in diameter: **1,3,4**.
*Perianth-lobes.* Each with a dark spot at the base: **1**. Outer lobes hairless or very sparsely hairy outside: **1,5**; outer lobes densely hairy outside: **2–4,6**.

1a. Whole plant completely hairless
  **1. capensis***
  b. At least some part of the plant hairy
  **2**
2a. Leaves 3–5 cm broad **3**
  b. Leaves to 1 cm broad **4**
3a. Leaves completely hairless; flowers 2.5–3 cm in diameter; style very short, the stigma borne almost directly on the ovary **2. latifolia**
  b. Leaves hairy, at least along the veins; flowers 3–3.6 cm in diameter; style evident **3. hemerocallidea**
4a. Flowers 3–3.6 cm in diameter; leaves rigidly leathery, deeply channelled, 30–45 cm **4. longifolia**
  b. Flowers to 3 cm in diameter; leaves not as above, to 30 cm **5**
5a. Leaves thread-like, *c.* 1 mm wide; outer perianth-lobes hairless or very sparsely hairy **5. hygrometrica**
  b. Leaves flat, 2–10 mm wide; outer perianth-lobes very densely hairy **6. hirsuta***

**1. H. capensis** (Linnaeus) Druce (*H. stellata* Linnaeus filius; *Spiloxene capensis* (Linnaeus) Garside). Illustration: Botanical Magazine, 662 (1803), 1223 (1809); Flowering plants of South Africa **39**: t. 1557A (1969).

Whole plant hairless. Corm 1–2 cm. Leaves mostly basal, 10–30 cm × 3–9 mm, the outer at least with pronounced midribs and V-shaped in section; stem leaves few, sheathing. Stem 5–25 cm, terminated by a solitary flower. Perianth-lobes widely spreading, 9–30 mm, yellow or white, each with an iridescent, purple spot at the base. Style short, stigma as long as the stamens. Seeds spherical. *South Africa (Cape Province).* G1. Spring–summer.

**\*H. neocanaliculata** Geerinck (*Spiloxene canaliculata* Garside, not *H. canaliculata* Baker). Illustration: Flowering plants of South Africa **39**: t. 1557B (1969). Outer leaves without midribs, U-shaped in section; perianth-lobes orange, each with a dark, non-iridescent spot at the base; seeds elongate and curved. *South Africa (Cape Province).* G1. Spring–summer.

**2. H. latifolia** J.D. Hooker. Illustration: Botanical Magazine, 4817 (1854).
Corm spherical, 6–8 cm in diameter. Leaves mostly borne on the stem, their bases sheathing and conspicuously overlapping, leathery, ribbed, hairless, 30–60 × 3–5 cm. Racemes borne in the leaf-axils, with 10–12 flowers. Flowers 2.5–3 cm in diameter. Perianth-lobes spreading, yellow, the outer 3 densely hairy outside, to 1.2 cm. Style very short, the stigmas borne almost directly on the ovary. *South Africa (Natal).* G1. Spring–summer.

*H. colchicifolia* Baker is probably the same species; it is reputed to have a shorter, fewer-flowered, more corymbose raceme. It originates from the same general area.

**3. H. hemerocallidea** Fischer & Meyer (*H. elata* J.D. Hooker). Illustration: Botanical Magazine, 5690 (1868).
Corm spherical, 8–10 cm in diameter. Leaves mostly borne at the base of the scape, slightly leathery, ribbed, hairy along the veins and margins, 45–60 × 3–5 cm. Racemes borne in the leaf-axils, loose, with 6–12 flowers; axis hairy. Flowers 3–3.6 cm in diameter. Perianth-lobes spreading, yellow, the outer 3 very densely hairy outside, to 1.8 cm. Style evident. *South Africa.* G1. Spring–summer.

**4. H. longifolia** Baker. Illustration: Botanical Magazine, 6035 (1873).
Corm oblong, 4–5 cm in diameter. Leaves mostly borne at the base of the scape, leathery, deeply channelled, 30–45 cm × 3–4 mm, shortly hairy beneath. Raceme corymbose with 2–4 flowers. Flowers 3–3.6 cm in diameter. Perianth-lobes spreading, yellow, 1.5–1.8 cm, the outer 3 densely hairy outside. Stigma evident. *South Africa.* G1. Spring–summer.

**5. H. hygrometrica** Labillardière.
Corm oblong, 6–10 mm. Leaves basal, thread-like, 6–15 cm × *c.* 1 mm, with sparse hairs on the outer side. Flowers solitary or 2–3 in a corymbose raceme; scape hairy towards the base. Perianth-lobes spreading, yellow, 1–1.5 cm, the outer 3 hairless or very sparsely hairy outside. Style elongate. *E Australia (Queensland to Tasmania).* G1. Spring–summer.

**6. H. hirsuta** (Linnaeus) Coville. Illustration: Rickett, Wild flowers of the United States **1**(1): t. 10 (1966).
Corm oblong, small. Leaves mostly basal, flat, to 30 cm × 2–10 mm, ribbed, hairy. Scapes borne between the leaves, densely hairy, bearing corymbose racemes of 3–7 flowers. Perianth-lobes 8–11 mm, yellow, the outer 3 densely hairy outside. Style evident. *Eastern N America.* H3. Spring–summer.

**\*H. micrantha** Pollard. Leaves 2–6 mm broad, raceme with 1–2 flowers. *SE USA.* G1. Spring–summer.

## 3. RHODOHYPOXIS Nel
*J. Cullen*

Perennial herbs with fleshy and fibrous roots. True stem absent. Leaves all basal, usually hairy. Scape present or absent (when absent the ovary of the flower is within the leaf-sheaths at or below ground-level). Flowers solitary or few. Perianth of 6 lobes united into a tube below, the lobes spreading, the 3 inner conspicuously clawed and each with a pronounced, inflexed curve at the junction of claw and blade, which closes the mouth of the tube. Stamens 6, borne on the perianth-tube at 2 levels. Ovary 3-celled, with numerous ovules in each cell; when borne at ground-level, the ovary has a beak which raises the perianth into the air. Fruit a capsule opening by a lid, or, if borne at ground-level, breaking up irregularly to release the seeds. Seeds black.

A genus of 6 species, all from southern Africa. Only 1 species is frequently grown, and numerous cultivars have been developed from it. It is usually easily grown in a light soil, with abundant moisture during the growing season (though care must be taken to prevent overwatering, as the long hairs on the leaves retain moisture). Mathew (reference below) gives a key to and descriptions of, other species which may appear in specialist collections. Propagation is by seed.
Literature: Hilliard, O.M. & Burtt, B.L., Notes on some plants from southern Africa: VII, *Notes from the Royal Botanic Garden Edinburgh* **36**: 43–76 (1978); Mathew, B., Rhodohypoxis, *The Plantsman* **6**: 49–59 (1984).

**1. R. baurii** (Baker) Nel (*Hypoxis baurii* Baker).
Leaves mostly 4–10, 2.5–11 cm × 1.2–12 mm, linear-lanceolate to lanceolate, hairy, especially on the upper surface. Scape to 15 cm, hairy. Flowers solitary or in pairs, white, pink or red; perianth-tube to 2.5 mm, outer lobes 1–2 cm, somewhat hairy towards the base. *South Africa.*

Hilliard & Burtt (reference above) recognise 3 varieties, of which 2 probably occur in cultivation:

Var. **baurii**. Illustration: Botanical Magazine, 9412, 1 (1935); Eliovson, Wild flowers of southern Africa, t. 104 (1980); Rix & Phillips, The bulb book, 151 (1981). Perianth-lobes red or deep, bright pink. *South Africa (Cape, Natal, Transkei) Lesotho.* H5–G1. Late spring.

Var. **platypetala** (Baker) Nel (*Hypoxis platypetala* Baker; *H. baurii* var. *platypetala* (Baker) Baker; *R. baurii* forma *platypetala* (Baker) Milne-Redhead; *R. platypetala* (Baker) Gray). Illustration: Botanical Magazine, 9412, 2 (1935); Gibson, Wild flowers of Natal (inland region), t. 29 (1978); Eliovson, Wild flowers of southern Africa, t. 105 (1980); Rix & Phillips, The bulb book, 151 (1981). Perianth white or rarely pale pink. *South Africa (Natal, Cape, Transkei).* H5–G1. Late spring.

***R. rubella** (Baker) Nel. Leaves thread-like, 1–4.5 cm × 1–1.5 mm, sparsely hairy; flowers solitary, the ovary subterranean and hidden by the leaf-sheaths, the perianth raised on a long ovary beak; fruits thin-walled, breaking up irregularly to release the seeds. *South Africa (Natal, Cape).* G1?.

Reputedly in cultivation, but more interesting than attractive.

---

## XII. VELLOZIACEAE

Herbs or shrubs or sometimes tree-like. Leaves usually evergreen, often borne in 3 ranks. Flowers solitary or in 2s or 3s, terminal, radially symmetric, without obvious bracteoles. Perianth of 6 segments, free or united into a tube below. Stamens 6 or many in 6 bundles, free from the perianth. Ovary inferior, 3-celled with many ovules in each cell. Fruit a capsule containing many seeds. Seeds compressed, black.

A family of 2 genera, *Vellozia* and *Barbacenia* (other segregate genera are recognised by some authors); unfortunately, there are two ways of circumscribing these genera, neither of which has won complete acceptance. If the genera are defined using the character of whether or not the perianth is united into a tube below, then those species with a perianth-tube are placed in *Barbacenia* and those without in *Vellozia* (including the species below, which is then correctly named *Vellozia elegans*). On this basis, *Vellozia* occurs in S America and in Africa and *Barbacenia* is restricted to S America. On the other hand, the genera can be separated using the number of stamens as the defining character. This means that those species with 6 stamens are placed in *Barbacenia* (including the species described below, which is then correctly named *Barbacenia elegans*) – so defined, the genus occurs in S America and Africa; while

those species with many stamens are placed in the genus *Vellozia*, which, so defined, is restricted to S America. Further work on this problem is necessary; meanwhile, the 1 cultivated species can be referred to correctly as either *B. elegans* or *V. elegans*. In this account, the genus is defined as having 6 stamens and the name *Barbacenia* is used.

## 1. BARBACENIA Vandelli
*J. Cullen*

Perennial herbs or shrubs (ours). Leaves more or less linear, usually thinly leathery and evergreen. Perianth of 6 more or less equal, free segments. Stamens 6, anthers opening by longitudinal slits, filaments usually very short. Fruit a capsule.

A genus of about 90 species (see above) from Arabia, Africa and S America. Several species have been in cultivation from time to time, but only 1 seems to persist. It is easily grown in a freely draining compost in a warm place; in a glasshouse it can form dense mats and act as a ground cover. Propagation is by seed or by division. Literature: Creves, S., A revision of the Old World species of Vellozia, *Journal of Botany* **59**: 273–84 (1921).

**1. B. elegans** (J.D. Hooker) Pax (*Vellozia elegans* J.D. Hooker; *V. talboti* invalid; *Talbotia elegans* invalid). Illustration: Botanical Magazine, 5803 (1869); Gibson, Wild flowers of Natal (inland region), t. 17 (1975).
Plant slightly woody, forming dense mats to 30 cm high when well grown. Leaves in the upper part of the stem in 3 ranks, 9–20 cm × 6–20 mm, linear-oblong, tapered towards the apex where the margins are slightly toothed, widely V-shaped in section, keeled beneath. Buds pale purple, flowers white. Perianth-segments 1–1.8 cm, acute, somewhat spreading, enlarging and turning green after flowering. Ovary 3-winged. *South Africa (Natal).* H5–G1. Summer.

---

## XIII. TACCACEAE

Perennial herbs with rhizomes. Leaves basal, usually with long stalks which are ribbed and have sheathing bases. Flowers borne in cymose umbels surrounded by usually 4 involucral bracts in 2 whorls, the outer ones persisting after flowering. Floral bracts thread-like, equal in number to the flowers and falling after flowering. Flowers bisexual, radially symmetric, with 6 perianth-lobes in 2 whorls, joined into a short, broad tube. Stamens 6, attached to the perianth-tube, with short, flattened filaments. Ovary inferior, 1-celled; ovules numerous, parietal. Fruit berry-like, with 6 ribs, containing 10–many seeds.

A family of a single genus.
Literature: Drenth, E., A revision of the family Taccaceae, *Blumea* **20**: 376–406 (1972).

## 1. TACCA Forster & Forster
*V.A. Matthews*
The only genus of the family; for description, see above. It is a pantropical genus with 10 species.

Taccas must be grown in a warm greenhouse and require a winter resting period during which water should be almost withheld. Propagation is by division of the rhizomes.

1a. Leaves lobed; involucral bracts 4–12  
            **1. leontopetaloides**
  b. Leaves unlobed; involucral bracts 4  2
2a. Fruit *c.* 1 cm, pale yellow; floral bracts to 8 cm      **3. plantaginea**
  b. Fruit 2–5 cm, green, orange-red, purple or blackish; floral bracts to 25 cm            3
3a. Perianth-lobes usually reflexed, falling after flowering, floral bracts white or yellowish green, darker at the base        **2. integrifolia**
  b. Perianth-lobes spreading, persisting until fruiting; floral bracts pale green to violet-green    **4. chantrieri**

**1. T. leontopetaloides** (Linnaeus) Kuntze (*T. pinnatifida* Forster & Forster; *T. artocarpifolia* Seeman). Illustration: Botanical Magazine, 6124 (1874) & 7299, 7300 (1893); Hooker's Icones Plantarum **26**: t. 2515, 2516 (1897) – as T. viridis. Leaves 1–3, the blades to 70 × 120 cm, broadly obovate to ovate, 3-lobed with each lobe itself lobed or dissected; stalks 17–150 cm. Scape 20–170 cm. Umbel with 20–40 flowers. Involucral bracts 4–12, of different sizes. Floral bracts to 25 cm, dark purple to blackish brown. Flowers 6–17 mm, on drooping stalks to 6 cm, pale yellow, yellow-green or dark purplish green. Perianth-lobes persisting until fruiting. Fruit more or less spherical, 1.5–2.5 cm, green, becoming orange when ripe. *Old World tropics.* G2. Flowering at any time of the year.

**2. T. integrifolia** Ker Gawler (*T. aspera* Roxburgh; *T. laevis* Roxburgh; *T. cristata* Jack; *Ataccia cristata* (Jack) Kunth).

Illustration: Botanical Magazine, 1488 (1812); Baillon, Histoire des plantes **13**: 167 (1894); Everard & Morley, Flowers of the world, t. 125 (1970).
Leaves 2–13, the blades 7–65 × 3–24 cm, very variable in shape, usually oblong or lanceolate but sometimes oblong-obovate or linear-lanceolate; stalks 4–40 cm. Umbel with up to 30 flowers. Involucral bracts 4. Floral bracts to 25 cm, white or yellowish green with a darker base. Flowers 1.4–2.7 cm, on stalks 5–40 cm, green, greenish violet or brownish purple, darkening with age. Perianth-lobes usually reflexed, falling after flowering. Fruit 2.5–5 cm, circular or triangular in cross-section, green to black tinged with purple. *NE India, Malaysia, Indonesia (Sumatra, Java), Borneo & Thailand.* G2. Mainly summer.

**3. T. plantaginea** (Hance) Drenth. Illustration: Das Pflanzenreich **92**: 12 (1928) – as Schizocapsa plantaginea. Leaves 3–8, the blades 8–36 × 2–9 cm, oblong or lanceolate; stalks 5–30 cm. Scape 7–25 cm. Umbel with 6–20 flowers. Involucral bracts 4. Floral bracts to 8 cm. Flowers 9–17 mm, on stalks to 3 cm, white to green tinged with brownish purple. Perianth-lobes spreading, persisting until fruiting. Fruit *c.* 1 cm, triangular in cross-section, pale yellow. *S China, Thailand, N Vietnam & Laos.* G2.

**4. T. chantrieri** André. Illustration: Graf, Exotica, edn 8, 1468 (1976); Clay & Hubbard, The Hawaii garden, tropical exotics, 140 (1977).
Leaves 3–12, the blades 17–55 × 4–22 cm, ovate, elliptic or lanceolate; stalks 11–45 cm. Scape 6–60 cm. Umbel with up to 25 flowers. Involucral bracts 4. Floral bracts to 20 cm, pale green or violet-green. Flowers 1–2.5 cm on stalks 1–4 cm, greenish white when young, aging to red, purple or blackish. Perianth-lobes reflexed, persisting until fruiting. Fruit 2–4 cm, triangular or circular in cross-section, green, orange-red or purple. *NE India & SE Asia.* G2.

---

# XIV. DIOSCOREACEAE

Herbaceous or slightly woody climbers with slender, twining stems arising from tubers or rhizomes. Aerial tubers sometimes present. Leaves usually alternate, sometimes opposite, entire, usually heart-shaped, sometimes palmately lobed, net-veined with 3–11 primary veins extending

upwards from the junction with the leaf-stalk. Flowers small, greenish, usually unisexual (the plants then dioecious), in axillary spikes, racemes or panicles. Perianth of 6 lobes in 2 whorls, usually united at the base. Male flowers with 6 stamens in 2 whorls, sometimes 1 whorl reduced to staminodes or absent; anthers opening by slits. Female flowers with inferior ovaries of 3 united carpels; placentation usually axile, ovules usually 2 in each cell; styles 3 or 1 with 3 stigmas. Fruit a capsule, often 3-winged, or a berry. Seeds usually winged.

A family of 6 genera and 630 species, mainly from tropical and warm temperate regions. In these areas several species of *Dioscorea* are grown for their edible tubers (yams) but in Europe the genus is grown only for ornament.

The following terms are used for the direction of twining as seen when looking down on the plant from above: 'clockwise' indicates movement like the hands of a clock, as in the Hop, *Humulus lupulus*; 'anticlockwise' indicates movement in the opposite direction as in the Runner Bean, *Phaseolus coccineus*.
Literature: Knuth, R., Dioscoreaceae, *Das Pflanzenreich*. **87**: 1–387 (1924); Knuth, R., Dioscoreaceae in Engler, A. & Prantl, K., *Die Natürlichen Pflanzenfamilien*. edn 2, **15a**: 438–62 (1930).

1a. Fruit a berry      **1. Tamus**
  b. Fruit a capsule    **2. Dioscorea**

## 1. TAMUS Linnaeus
*C.J. King*
Perennial, dioecious herbs with large, cylindric to ovoid tubers. Leaves alternate, heart-shaped. Flowers in axillary racemes. Perianth bell-shaped. Stamens 6, rudimentary in the female flower. Style with 3 bilobed stigmas. Fruit a berry with few, unwinged, spherical seeds.

A genus of 5 species from the Atlantic islands, Europe, N Africa and SW Asia, 1 of which is occasionally cultivated for the attractive but poisonous red berries. It thrives well in moist, well-drained soils in half-shade, and may become a pestilent weed. Propagation is by seed or by division of the rootstock.

**1. T. communis** Linnaeus. Illustration: Bonnier, Flore complète **10**: t. 597 (1929); Hegi, Illustriertes flora von Mitteleuropa **2**: t. 67 (1939); Ross-Craig, Drawings of British plants **29**: t. 9 (1972); Phillips, Wild flowers of Britain, 83(f) & 173 (f) (1977).

Root-tuber ovoid, blackish, to 20 cm or more. Stem twining clockwise, to 4 m. Leaves 8–15 × 4–11 cm, deeply cordate at base, long-acuminate at apex, dark shining green, with 3–9 main veins diverging from the base. Flowers 3–6 mm in diameter, greenish yellow. Berry ovoid-ellipsoid, 1–1.2 cm, red, rarely yellowish, with 1–6, pale yellow, wrinkled seeds. *W & S Europe, N Africa, SW Asia.* H4. Summer.

## 2. DIOSCOREA Linnaeus
*C.J. King*

Dioecious climbers with twining stems arising from tuberous roots, sometimes bearing aerial tubers in the leaf axils. Perianth of male flowers bell-shaped. Perianth of female flowers deeply 6-lobed. Fruit a 3-angled or 3-winged capsule; seeds flat, usually winged.

About 600 species, mainly in tropical and subtropical regions. Several of the hardier species are grown in European gardens for their leafy, twining habit, and some of the tender ones, often with handsome coloured leaves, are cultivated as house plants. The hardier species can be grown in any garden soil with good drainage, but they require a sunny position. The tenderer species should be kept moderately dry during the winter resting period but need plenty of water during growth. Propagation is chiefly by division of the rootstock during the dormant season, or by aerial tubers where these are produced, but seed may also be used. Variegated-leaved kinds are grown from cuttings.

*Root-tubers.* Partially exposed: **3,9.** Large: **3,7,9.**
*Aerial tubers.* Large: **1.**
*Stems.* Twining clockwise: **1,4,9;** twining anticlockwise: **2,7,8.** Angled: **3,5,7;** terete: **1,4,8.**
*Leaves.* Bicoloured or multicoloured above: **5,6.** Always opposite: **7.** Narrow: **8.**

1a. Leaves uniformly green above        2
  b. Leaves bicoloured or multicoloured above        8
2a. Root-tubers small or absent; aerial tubers large        **1. bulbifera**
  b. Root-tubers small or large; aerial tubers small or absent        3
3a. Root-tubers partially exposed        4
  b. Root-tubers not exposed        5
4a. Leaves more than 5 cm broad        **3. macrostachya**
  b. Leaves less than 5 cm broad        **9. elephantipes**
5a. Leaves opposite        **7. batatas**

  b. Leaves mostly alternate (occasionally some opposite in **2** and **4**)        6
6a. Leaves mostly linear or linear-lanceolate, sometimes hastate        **8. hastifolia**
  b. Leaves ovate-lanceolate to broadly ovate        7
7a. Stem twining clockwise        **4. balcanica**
  b. Stem twining anticlockwise        **2. cotinifolia**
8a. Leaves heart-shaped to ovate-lanceolate, 7-veined; male flowers stalkless        **5. amarantoides**
  b. Leaves heart-shaped to circular, 9–11-veined; male flowers stalked        **6. discolor**

**1. D. bulbifera** Linnaeus. Illustration: Parey's Blumengärtnerei **1:** 366 (1958); Graf, Exotica, edn 8, 710 (1976); Graf, Tropica, 393 (1978).
Root-tubers small or absent. Stem terete, twining clockwise to 6 m, bearing aerial, axillary, almost spherical or angled tubers to 10 cm across. Leaves alternate or opposite, ovate or heart-shaped, cuspidate, to 25 × 18 cm; leaf-stalks 6–14 cm. Male flowers in spikes 3–10 cm; female flowers in spikes 10–25 cm. Capsule to 2.5 × 1.5 cm. *Tropical Africa & Asia.* G2. Late summer.

**2. D. cotinifolia** Kunth (*D. malifolia* Baker). Illustration: Batten & Bokelmann, Wild flowers of the eastern Cape Province, t. 21, f. 3 (1966).
Root-tubers to 9 × 5 cm. Stem twining anticlockwise. Leaves alternate, sometimes opposite, broadly ovate, truncate or slightly cordate at base, to 8 × 5 cm; leaf-stalks 1.8–3 cm. Male flowers in racemes 5–7.5 cm; female flowers in few-flowered racemes which are 7.5–15 cm. Capsule to 3 cm. *South Africa.* G2. Summer.

**3. D. macrostachya** Bentham (*Testudinaria macrostachya* (Bentham) Rowley). Illustration: Graf, Tropica, 392, 393 (1978) – as D. macrostachys.
Root-tubers 20 cm or more across, partially exposed, the surface deeply grooved into irregular segments. Stem slightly angled, probably twining anticlockwise. Leaves alternate, ovate, acuminate at apex, cordate at base, to 20 × 18 cm, with long stalks. Male flowers in clusters of 2 or 3 along racemes 15–30 cm; female flowers solitary along racemes. Capsule to 2.5 cm. *C America, from SC Mexico to Panama.* G2.

**4. D. balcanica** Kosanin. Illustration: Osterreichische Botanische Zeitschrift **64:** t. 3 (1914); Jelitto & Schacht, Die Freiland-Schmuckstauden **1:** 164 (1963).

Root-tubers to 2 cm in diameter. Stem terete, twining clockwise to 1.5 m or more. Leaves alternate or almost opposite, ovate or heart-shaped, shortly acuminate, to 7 × 6 cm, 9-veined, with long stalks. Fruiting spike to 7 cm, pendent. Capsule 2–2.5 cm across. *SW Yugoslavia & N Albania.* H4. Early summer.

**5. D. amarantoides** Presl.
Stem slightly angled. Leaves alternate, heart-shaped to ovate-lanceolate, 7-veined, to 10 × 5 cm. Male flowers stalkless, in dense spikes arranged in a panicle up to 40 cm. Capsule 2 × 1.3 cm. *Peru.* G2.

The typical plant is probably not in general cultivation, but a number of non-flowering cultivars, often placed under the name *D. multicolor*, appear to belong here. They have somewhat larger leaves (up to 13 × 8 cm) which are reddish purple beneath, but on the upper side are boldly patterned in various colours, usually with a metallic sheen (Illustrations: Illustration Horticole **18:** 52, 1871; Graf, Exotica, edn 8, 710, 1976). Those most frequently grown are 'Argyraea', 'Chrysophylla', 'Eldorado', 'Melanoleuca' and 'Metallica'.

**6. D. discolor** Kunth. Illustration: Parey's Blumengärtnerei **1:** 367 (1958); Graf, Tropica, 392, 393 (1978).
Root-tubers to 5–7 cm across. Leaves alternate, heart-shaped, cuspidate, 13–15 cm, dark green above with areas of lighter green, whitish along the midrib, reddish purple beneath. Male flowers in racemes; female flowers in spikes. *Tropical S America.* G2. Late summer.

*D. discolor* is very like *D. dodecaneura* Vellozo, native from Guyana to Paraguay, and may be only a form of that species.

**7. D. batatas** Decaisne (*D. opposita* misapplied). Illustration: Flore des Serres **10:** 7 (1854–5).
Root-tubers to 90 cm. Stem slightly angled, twining anticlockwise to 3 m and bearing small, aerial, axillary tubers. Leaves opposite, ovate, cordate at the base, 4–8 cm. Flowers in spikes, cinnamon-scented. *Temperate E Asia.* H2. Late summer.

**8. D. hastifolia** Nees (*Testudinaria hastifolia* invalid). Illustration: Erickson et al. (eds), Flowers and plants of Western Australia, 37 (1973).
Root-tubers to 12 × 3 cm. Stem terete, twining anticlockwise to 2 m. Leaves alternate, mostly linear or linear-lanceolate but sometimes hastate, to 8 cm. Male flowers in racemes 5–8 cm; female

inflorescences very short with usually only 2 or 3 flowers. Capsule *c.* 2 cm across. *W Australia.* G2. Autumn.

**9. D. elephantipes** (L'Héritier) Engler (*Testudinaria elephantipes* (L'Héritier) Lindley). Illustration: Botanical Magazine, 1347 (1811); Graf, Tropica, 393 (1978). Root-tubers to 90 cm in diameter, partially exposed, the surface deeply grooved into irregular segments (the well-known 'elephant's foot'). Stem twining clockwise to 6 m. Leaves alternate, heart-shaped to kidney-shaped, 2–5 cm broad. Male flowers in racemes 5–7.5 cm; female flowers in racemes. Capsule *c.* 1.5 × 1.3 cm. *South Africa.* G1. Late summer.

# XV. PONTEDERIACEAE

Annual or perennial aquatic herbs with rhizomes or stolons. Leaves alternate, linear or with a distinct stalk and blade. Flowers bisexual, solitary or paired or in spikes or racemes which are subtended by a spathe, radially or bilaterally symmetric. Perianth-segments usually 6. Stamens 3 or 6, or occasionally only 1. Ovary superior, of 3 united carpels; ovules 3–many, placentation axile or parietal. Style 1. Fruit a capsule or nutlet.

There are 9 genera, all native in tropical or subtropical aquatic habitats. None of the species is native in Europe, although 4 are now naturalised; 4 genera are represented in cultivation.

1a. Stamens 3; flowers stalkless
  **2. Heteranthera**
 b. Stamens 6; flowers stalked or not  2
2a. One stamen longer than the others
  **3. Monochoria**
 b. All stamens equal  3
3a. Plants floating; leaf-stalks inflated at the base  **1. Eichhornia**
 b. Plants usually rooted in the substrate; leaf-stalks not inflated  4
4a. Perianth distinctly 2-lipped, hairy outside  **4. Pontederia**
 b. Perianth not 2-lipped, hairless outside
  **1. Eichhornia**

## 1. EICHHORNIA Kunth
*E.C. Nelson*

Annual or perennial herbs. Stems floating or creeping; stolons often present. Leaves submerged, floating or emergent; stalks sometimes inflated, blades expanded. Flowers in a spike or panicle. Perianth blue, tubular, with 6 lobes. Stamens 6, inserted on the perianth at 2 levels; anthers more or less equal. Styles 1 or 3. Fruit a capsule with numerous seeds.

A genus of 7 species, now widely dispersed through the tropics, though the majority were originally native in tropical America. One species is commonly grown in tropical aquaria and is now a serious weed in aquatic habitats in tropical regions. The plants should be grown in shallow water over a rich, loamy soil. Propagation is by division.

1a. Base of leaf-stalk inflated, bulbous in appearance  **1. crassipes**
 b. Base of leaf-stalk not inflated
  **2. azurea**

**1. E. crassipes** (Martius) Solms-Laubach. Illustration: Gartenflora **37**: t. 1271 (1888); Cook, Water plants of the world, 484 (1974). Plants usually floating; roots conspicuously clothed in fine, blue hairs. Leaves with markedly swollen stalks, the swelling containing spongy, air-filled cells; blade ovate, to 8 cm. Spike to 15 cm with 8–12 flowers which are 5–7 cm in diameter. Perianth blue, lobes 6. *Brazil, now widely naturalised in the tropics elsewhere.* G1. Summer.

Very easy to grow in warm water. Several cultivars are available, some of which have yellow or lilac-red flowers.

**2. E. azurea** (Swartz) Kunth (*Pontederia azurea* Swartz). Illustration: Botanical Magazine, 6487 (1880). Like *E. crassipes*, but the leaf-stalks not swollen. Perianth-lobes joined at the base and apex of the tube, the tube in between often split. *Subtropical and tropical America.* G2. Summer.

## 2. HETERANTHERA Ruiz & Pavon
*E.C. Nelson*

Annual or perennial herbs with submerged, floating or creeping stems. Leaves linear or with expanded blades and distinct stalks. Inflorescence a spike with many flowers. Perianth-lobes 6. Stamens 3, 1 usually much larger than the others. Fruit a capsule with numerous seeds.

A genus of about 10 species from tropical and subtropical America and Africa. They generally grow in shallow water or on mud. There are 3 species which are occasionally grown in aquaria, and 1 is naturalised in Europe. Propagation is by cuttings.

1a. Flowers yellow; all leaves submerged, linear  **1. dubia**
 b. Flowers blue; some leaves floating and not linear  2

2a. Floating leaves kidney-shaped
  **2. reniformis**
 b. Floating leaves elliptic
  **3. zosterifolia**

**1. H. dubia** (Jacquin) MacMillan. Illustration: Britton & Brown, Illustrated flora of the United States and Canada, edn 2, **1**: 464 (1913). Submerged perennial. Leaves stalkless, linear, to 15 cm. Flowers yellow, lying flat on the water surface. All stamens equal. *N & S America.* G1. Summer.

An easily cultivated plant for tropical aquaria. It does not attract algal growth, but is a poor oxygenator.

**2. H. reniformis** Ruiz & Pavon. Submerged, creeping perennial. Floating leaves kidney-shaped, broader than long, to 3 × 3.5 cm; stalk to 10 cm. Flowers 2–5 in a spike, blue. Posterior stamen with a large, green anther, the other 2 stamens with smaller, yellow anthers. *N & S America.* H5–G1. Summer.

May be grown out-of-doors in the mildest areas, or in indoor aquaria in shallow water. It can spread rapidly, and can become a weed.

**3. H. zosterifolia** Martius. Illustration: Cook, Water plants of the world, 486 (1974). Submerged, creeping perennial. Submerged leaves stalkless, linear, to 5 cm × 7 mm; floating leaves with stalks to 8 cm, blades elliptic, to 4 cm. Flowering spikes usually with 2 flowers. Perianth blue. All stamens alike, anthers pale blue. *Brazil, Bolivia.* G2. Summer.

This is a good oxygenator for tropical aquaria. In good light and warm water it will grow lushly and provide shade.

## 3. MONOCHORIA Presl
*E.C. Nelson*

Annual or perennial herbs, stems erect or creeping. Leaves stalked, emergent. Flowers blue, in elongate racemes. Perianth-lobes 6. Stamens 6, inserted on the perianth-tube. Anthers unequal, 5 small and yellow, the sixth larger and blue. Fruit a capsule with numerous seeds.

A genus of about 5 species from tropical habitats in Africa, Asia and Australia. Only 1 species is cultivated in Europe. Propagation is by division or by cuttings.

**1. M. vaginalis** Presl. Illustration: Cook, Water plants of the world, 487 (1974). Stems creeping, rooted in the substrate. Leaves on long stalks, blades arrowhead-shaped. Raceme erect, to 60 cm. *Africa, India.* G2. Summer.

**4. PONTEDERIA** Linnaeus
*E.C. Nelson*

Perennial herbs with submerged, floating or creeping stems. Leaves submerged floating or emergent, linear or stalked. Flowers in a spike, blue. Perianth tubular, hairy outside, 2-lipped, each lip with 3 lobes. Stamens 6, anthers equal. Fruit a nutlet with 1 seed.

A genus of about 5 species from America, generally found growing in shallow water or on mud. Only 1 species is widely cultivated in Europe, where it is also naturalised; it is suitable for pond margins in very shallow water.
Literature: Lowden, R.M., Revision of the genus Pontederia L., *Rhodora* 75: 426–87 (1973).

**1. P. cordata** Linnaeus. Illustration: Botanical Magazine, 1156 (1808), 2932 (1829) & 3753 (1839); Cook, Water plants of the world, 488 (1974).
A robust perennial to 1 m. Leaves erect; stalk to 25 cm, blades triangular to ovate, with a cordate base, to 20 cm broad. Flowers blue, densely arranged in a spike. Perianth to 8 mm in diameter. *N & S America*. H5. Summer.

One variant is reported to be taller, with a longer spike and lanceolate leaves, but is not as hardy as the species; it probably belongs to var. **lancifolia** (Muhlenberg) Torrey.

---

# XVI. IRIDACEAE

Herbs (rarely shrubby) with corms or rhizomes. Leaves basal and borne on the stem, often erect and equitant, usually in 2 ranks. Flowers solitary or borne in spikes or panicles, each subtended by a bract and sometimes a bracteole, occasionally more than 1 flower to a bract. Large bracts (spathes) may subtend the whole or part of the inflorescence (figure 38, p. 333). Perianth radially or bilaterally symmetric, with a tube at the base or the segments completely free. Stamens usually 3. Ovary inferior, usually 3-celled, occasionally 1-celled; ovules numerous, placentation axile. Style usually 3-branched above, the branches sometimes further divided or broad and petaloid. Fruit a capsule splitting between the septa.

A family of about 70 genera and 1500 species from most parts of the world; the cultivated species are mainly from the Mediterranean area and South Africa. The genera are often difficult to distinguish, and care must be taken using the key below.

1a. Perianth united into a tube at the base, which may be very short (some genera with extremely short perianth-tubes are keyed out under 1b as well as here)                                                         2
  b. Perianth-segments completely free (but including some genera with extremely short, inconspicuous perianth-tubes)                                     38
2a. Leaves neither arranged in 2 ranks nor equitant                                                                   3
  b. Leaves arranged in 2 ranks, usually equitant                                                                        7
3a. Leaves each with a white stripe on the upper surface              **16. Crocus**
  b. Leaves without such a stripe        4
4a. Styles petaloid                            **1. Iris**
  b. Styles not petaloid                        5
5a. Filaments united, at least at the base                                         **18. Galaxia**
  b. Filaments free                                 6
6a. Perianth-tube shorter than the lobes                                           **19. Romulea**
  b. Perianth-tube longer than the lobes                                           **17. Syringodea**
7a. Perianth bilaterally symmetric (sometimes due only to the curved tube)                                        8
  b. Perianth radially symmetric (tube always straight)                       19
8a. Style branches 3, each bifid near the tip                                              9
  b. Style unbranched or branches 3, each entire                                      12
9a. Inflorescence a spike, its axis bent, or at least curved, at about 90° at the level of the lowest flower, the flowers pointing upwards from the more or less horizontal axis       **34. Freesia**
  b. Inflorescence a spike, not as above, or a panicle                              10
10a. Perianth-tube conspicuously widening upwards; stamens regularly arranged                       **37. Watsonia**
  b. Perianth-tube cylindric, not or scarcely widening upwards; stamens usually asymmetrically arranged     11
11a. Basal leaves exceeding the stem-leaves; stem-leaves with undulate-crisped margins; corms with hard, woody tunics            **35. Lapeirousia**
  b. Basal leaves not exceeding the stem-leaves; stem-leaves not as above; corms with fibrous tunics            **36. Anomatheca**
12a. Bilateral symmetry of the perianth mainly due to the curved tube, the lobes more or less radially symmetric                                        13

  b. Bilateral symmetry of the flower due not only to the curved tube, but due also to the lobes, which are unequal in size and bilaterally symmetric     17
13a. Leaves conspicuously pleated; leaves and stems densely hairy   **38. Babiana**
  b. Leaves usually not pleated; leaves and stems usually not hairy (occasionally leaves only with a few fine hairs)   14
14a. Stamens symmetrically arranged
                                      **30. Hesperantha**
  b. Stamens asymmetrically arranged, arching to the upper side of the flower
                                                          15
15a. Perianth-tube slender throughout or with a very slender basal part, abruptly expanded above
                                        **39. Crocosmia**
  b. Perianth-tube gradually widening from the base                              16
16a. Style exceeding the stamens
                                          **41. Tritonia**
  b. Style about as long as the stamens
                                          **44. Gladiolus**
17a. The 3 lower lobes of the perianth very small and inconspicuous; bracts almost as long as the perianth-tubes
                                        **46. Anomalesia**
  b. The 3 lower lobes of the perianth shorter than the others but still conspicuous; bracts much shorter than the perianth-tubes            18
18a. The 3 lower segments of the perianth not united to each other beyond the mouth of the perianth-tube; perianth red, orange or red and yellow
                                        **45. Chasmanthe**
  b. The 3 lower lobes of the perianth united to each other beyond the mouth of the perianth-tube; perianth lavender to deep purple, with yellow stripes                **43. Synnotia**
19a. Perianth-lobes distinctly in 2 whorls, the lobes of the outer whorl differing from those of the inner in shape and size                                         20
  b. Perianth-lobes all similar, apparently in 1 whorl, or if obviously in 2 whorls then the lobes all similar in size and appearance                                22
20a. Inner perianth-lobes differing from the outer in size, shape, posture and often in colour; style-branches petaloid   21
  b. Inner perianth-lobes differing from the outer only in width; style-branches not petaloid      **40. Melasphaerula**
21a. Rootstock consisting of finger-like tubers; inner perianth-lobes greenish yellow, blade of outer lobes dark purple-brown; ovary 1-celled
                                        **4. Hermodactylus**

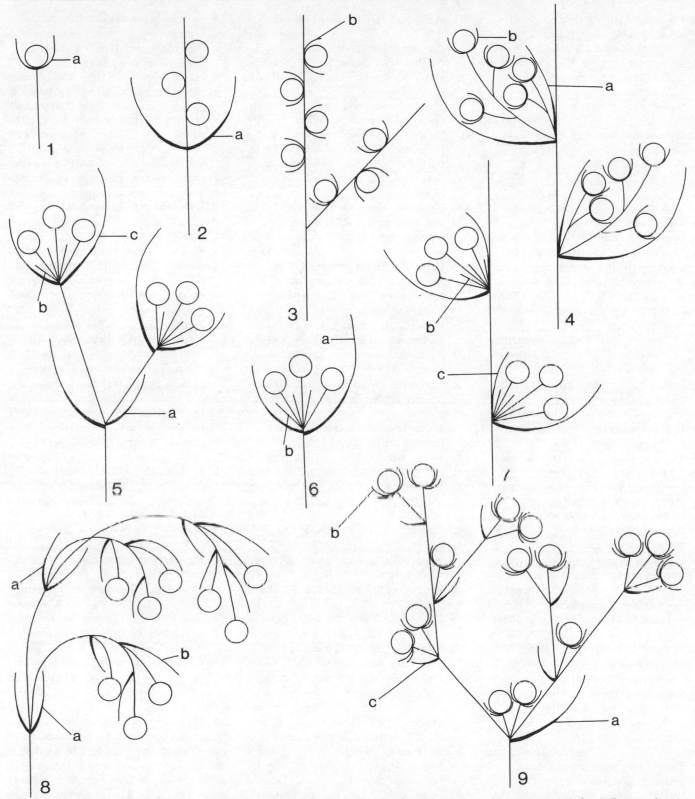

**Figure 38.** Inflorescences of Iridaceae. 1, Solitary, terminal flower with 2 spathes (a), e.g. most *Iris* species. 2, Spike with 2 spathes (a), flowers without bracts, e.g. some *Iris* species. 3, Spike without spathes, each flower with 2 bracts (b), e.g. *Lapeirousia*. 4, Panicle, each group of flowers with 2 spathes (a), – each flower with 2 bracts (b), e.g. *Orthrosanthus*. 5, Panicle with primary spathes (a), each group of flowers with secondary spathes (c), each flower with 1 bract (b), e.g. some *Sisyrinchium* species. 6, Terminal inflorescence with 2 spathes (a), each flower with 1 bract (b), e.g. some *Sisyrinchium* species. 7, Panicle without primary spathes, each group of flowers like those in No. 5 but stalkless with secondary spathes (c), each flower with 1 bract (b), e.g. some *Sisyrinchium* species. 8, Arching panicle, each group of flowers with 2 spathes (a), each flower with 1 bract (b) at the base of the stalk, e.g. *Dierama*. 9, Complex panicle with 1 primary spathe (a), each branch with 1 secondary spathe (c), each flower with 3 bracts (b), e.g. *Nivenia*.

b. Rootstock a bulb or rhizome; flowers never with the above colour combination; ovary 3-celled **1. Iris**

22a. Style-branches fringed **14. Ferraria**

b. Style-branches not fringed, or style unbranched 23

23a. Style-branches 3, each bifid at the tip **35. Lapeirousia**

b. Style-branches 3, each entire, or style unbranched 24

24a. Filaments united 25

b. Filaments free 28

25a. Style not branched, stigma entire or 3-toothed **26. Solenomelus**

b. Style 3-branched, stigmas 3 26

26a. Flowers stalkless **33. Ixia**

b. Flowers with stalks (which may be concealed within bracts) 27

27a. Style-branches club-shaped, with rounded tips **20. Gelasine**

b. Style-branches narrowing from the base, tips acute or capitate **25. Sisyrinchium**

28a. Plant woody **28. Nivenia**

b. Plant not woody 29

29a. Leaves conspicuously pleated; stems and leaves densely hairy **38. Babiana**

b. Leaves not pleated; stems and leaves not hairy 30

30a. Stems arching over; bracts scarious, conspicuous; flowers pendent **32. Dierama**

b. Stems not arching over; bracts inconspicuous, green or scarious; flowers not pendent 31

31a. Plants with rhizomes 32

b. Plants with corms 34

32a. Flowers scarlet, crimson or pink **29. Schizostylis**

b. Flowers blue or purplish 33

33a. Perianth-lobes twisting spirally after flowering **27. Aristea**

b. Perianth-lobes not twisting spirally after flowering **24. Orthrosanthus**

34a. Corm-tunics hard and woody 35

b. Corm-tunics papery or fibrous 36

35a. Style longer than perianth-tube, branches usually short, recurved, exceeding the anthers **31. Geissorhiza**

b. Style shorter than or equal to perianth-tube, branches long, ascending, often not reaching the anther-tip **30. Hesperantha**

36a. Capsule thin-walled, the outlines of the seeds visible through it **33. Ixia**

b. Capsule with firm, membranous walls, the outlines of the seeds not visible 37

37a. Perianth-tube funnel-shaped; bracts entire or raggedly toothed near the tip **42. Sparaxis**

b. Perianth-tube cylindric; bracts notched near the tip **41. Tritonia**

38a. Fertile stamens 2 **21. Diplarrhena**

b. Fertile stamens 3 39

39a. Style-branches petaloid 40

b. Style-branches not petaloid 45

40a. Flower with an apparent perianth-tube which is, in fact, a beak on top of the ovary **3. Gynandriris**

b. Flower without a beak on top of the ovary 41

41a. Flowering stem with many long branches; claw of outer perianth-segments marked with transverse bands **2. Pardanthopsis**

b. Flowering stem unbranched or with a few short branches; claw of outer perianth-segments not marked with transverse bands 42

42a. Filaments united, at least at the base; rootstock a corm **5. Moraea**

b. Filaments usually free; rootstock a rhizome or bulb 43

43a. Leaves usually all basal; perianth-tube present, though sometimes extremely short **1. Iris**

b. Leaves basal and also borne on the stem; perianth-tube absent 44

44a. Leaves pleated **7. Cypella**

b. Leaves not pleated **6. Dietes**

45a. Filaments free 46

b. Filaments united at least at the base 54

46a. Perianth-segments distinctly in 2 whorls, the segments of the outer whorl differing from those of the inner in shape or size 47

b. Perianth-segments all similar, in 1 or 2 whorls 51

47a. Each perianth-segment with a band of hairs down the centre **13. Alophia**

b. Perianth-segments hairless, or if hairy, then not as above 48

48a. Flowers white to greenish yellow, to 1.8 cm in diameter **40. Melasphaerula**

b. Flowers bright yellow, blue or violet-blue, if white or pale yellow, then 2–10 cm in diameter 49

49a. Leaves pleated **7. Cypella**

b. Leaves not pleated 50

50a. Flowers 2–2.5 cm in diameter; outer perianth-segments yellow with spots at the base **8. Trimezia**

b. Flowers 3–10 cm in diameter; outer perianth-segments white, blue or violet-blue, if yellow then with bars at the base **9. Neomarica**

51a. Perianth-segments twisting spirally after flowering 52

b. Perianth-segments not twisting spirally after flowering 53

52a. Flowers blue **27. Aristea**

b. Flowers yellow or orange-red with red or purple spots **23. Belamcanda**

53a. Flowers stalkless; leaves not stiff, often curled **31. Geissorhiza**

b. Flowers shortly stalked; leaves stiff and leathery **24. Orthrosanthus**

54a. Style-branches fringed **14. Ferraria**

b. Style-branches not fringed 55

55a. Style-branches divided, bifid or notched 56

b. Style-branches entire 59

56a. Flowers 3–15 cm in diameter; stigmas on radii alternating with the anthers **10. Tigridia**

b. Flowers smaller; stigmas on the same radii as the anthers 57

57a. Perianth red, orange or yellow; segments coiling spirally after flowering **11. Rigidella**

b. Perianth purple or bluish; segments not coiling spirally after flowering 58

58a. Each perianth-segment with a band of hairs down the centre; anther-connective broad, fiddle-shaped **13. Alophia**

b. Perianth-segments hairless; anther-connective narrow, linear **12. Herbertia**

59a. Rootstock a corm 60

b. Rootstock a rhizome or just a cluster of fibrous roots 61

60a. Flowers blue; corm-tunics membranous **20. Gelasine**

b. Flowers not blue; corm-tunics netted **15. Homeria**

61a. Segments of the outer whorl of the perianth smaller than those of the inner whorl **22. Libertia**

b. Perianth-segments all of the same size **25. Sisyrinchium**

**1. IRIS** Linnaeus

*V.A. Matthews & B. Mathew*

Rhizomatous or bulbous perennial herbs. Leaves usually basal and in 2 ranks, flat, channelled, 4-angled or nearly cylindric. Flowers 1–several, borne within 2 spathes (figure 38 (1 & 2), p. 333). Perianth radially symmetric, united at the base into a tube (hypanthial tube). Falls (outer lobes) narrowed towards the base into a claw (haft), the blade sometimes bearded. Standards (inner lobes) usually erect or arching, sometimes horizontal or deflexed, narrowed towards the base into a claw,

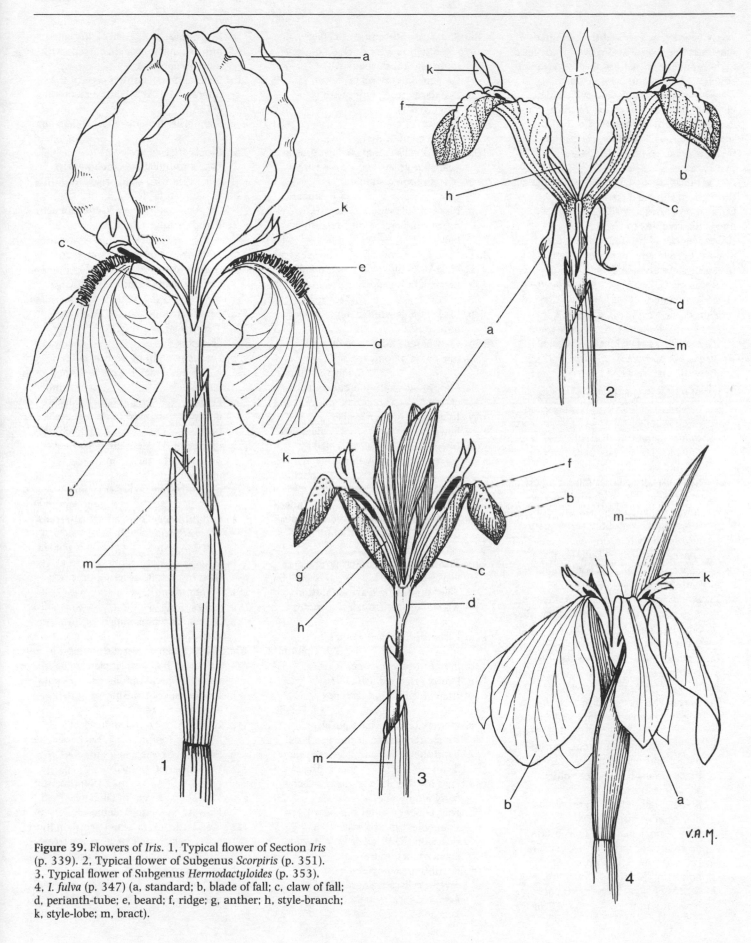

**Figure 39.** Flowers of *Iris*. 1, Typical flower of Section *Iris* (p. 339). 2, Typical flower of Subgenus *Scorpiris* (p. 351). 3, Typical flower of Subgenus *Hermodactyloides* (p. 353). 4, *I. fulva* (p. 347) (a, standard; b, blade of fall; c, claw of fall; d, perianth-tube; e, beard; f, ridge; g, anther; h, style-branch; k, style-lobe; m, bract).

rarely bearded or very reduced. Stamens 3, filaments free, borne at the base of the falls. Style-branches 3, each 2-lobed beyond the stigma, coloured and petaloid, each covering a stamen. Capsule cylindric to ellipsoid, more or less round to triangular in cross-section, often with 3 or 6 ribs; seeds numerous, sometimes bearing a fleshy appendage (aril). See figure 39, p. 335.

A genus of about 250 species distributed throughout the northern hemisphere. Cultivation varies according to the species and is too complex to discuss here. Reference should be made to Mathew, B., The Iris (1981) where the requirements of the species are discussed.

Literature: Dykes, W.R., *The Genus Iris* (1915), reprinted 1974; Lenz, L.W., A revision of the Pacific coast irises, *Aliso* 4: 1–72 (1958); Cohen, V.A., *A guide to the Pacific coast irises* (1967); Grey-Wilson, C., *The genus Iris, subsection Sibiricae* (1971); Mathew, B., *The Iris* (1981).

1a. Plants with a bulb; leaves 4-angled or almost cylindric in cross-section, or channelled                                    2
  b. Plants usually with a rhizome, never bulbous; leaves flat, sometimes very narrow                                           19
2a. Leaves 4-angled or almost cylindric in cross-section, if channelled then standards erect and perianth-tube more than 5 cm; bulb-tunics netted and strongly fibrous                                  3
  b. Leaves channelled; bulb-tunics papery to tough and leathery, never netted or fibrous                                     8
3a. Standards very reduced, bristle-like, 3–5 mm                          **92. danfordiae**
  b. Standards not bristle-like, more than 2 cm                                          4
4a. Blade of falls pale yellow, spotted with green             **93. winogradowii**
  b. Blade of falls not pale yellow           5
5a. Leaves almost cylindric with 8 ribs; falls whitish with a dark violet apex
                                          **89. bakeriana**
  b. Leaves 4-angled; falls not whitish with a dark violet apex                    6
6a. Blade of falls pale to deep blue, sometimes violet-blue                      7
  b. Blade of falls whitish, greyish lilac, lilac-blue, purplish brown, violet-blue or purple, if blue then bracts green, not papery      **88. reticulata***
7a. Leaves more or less absent at flowering time           **90. histrioides**
  b. Leaves present, often to 30 cm at flowering time              **91. histrio**

8a. Standards horizontal or deflexed, if erect then flowers smelling of cloves; dormant bulbs usually with thickened fleshy roots              9
  b. Standards erect; bulbs with thin fibrous roots                             15
9a. Stem to 15 cm                            10
  b. Stem 15 cm or more                      12
10a. Blade of falls cream, yellow, greenish yellow or green, sometimes with pinkish brown staining
                                          **83. caucasica***
  b. Blade of falls whitish, greyish, brownish, lilac, purple, reddish or blue                                        11
11a. Falls 2.5–5 cm             **84. persica***
  b. Falls 5–8 cm               **81. planifolia***
12a. Blade of falls yellow or greenish
                                          **85. bucharica***
  b. Blade of falls whitish, blue, or lilac to purple                                     13
13a. Claw of falls unwinged or with a narrow wing on each side
                                          **86. albomarginata***
  b. Claw of falls with a broad wing on each side                                      14
14a. Internodes of stem visible at flowering time        **87. magnifica***
  b. Internodes of stem concealed by leaves at flowering time
                                          **82. aucheri***
15a. Flowers yellow; perianth-tube 3.5–5 cm                    **79. juncea**
  b. Flowers blue to violet, reddish violet or bronze, if yellow or white then perianth-tube to 3 cm          16
16a. Plant leafless in winter, producing leaves in the spring      **76. latifolia**
  b. Plant producing leaves in autumn which persist through the winter
                                          17
17a. Perianth-tube 1–10 mm
                                          **77. xiphium***
  b. Perianth-tube 1–3 cm                    18
18a. Flowers rich reddish violet; standards blunt at the apex
                                          **78. filifolia**
  b. Flowers blue to violet-purple; standards pointed at the apex, or if blunt then falls with a sparse yellow beard              **80. tingitana***
19a. Falls bearing an obvious beard of fairly long hairs                       20
  b. Falls lacking an obvious beard, but sometimes with cockscomb-like ridges on the blade or very short downy hairs on the claw           50
20a. Beard-hairs composed of several cells; stems branched or not; seed lacking a fleshy appendage                              21

  b. Beard-hairs composed of one cell only; stems unbranched; seeds with a fleshy appendage                  37
21a. Stem almost absent, if present then unbranched, 5–30 cm, sometimes to 60 cm                                   22
  b. Stem branched, usually 15–125 cm
                                          28
22a. Bracts sharply keeled               23
  b. Bracts rounded or sometimes the outer one with a slight keel          24
23a. Perianth-tube 1.5–2.5 cm
                                          **5. reichenbachii**
  b. Perianth-tube 3–4.5 cm
                                          **6. suaveolens**
24a. Stem absent or up to 1 cm; plant leafless in winter                       25
  b. Stem at least 3 cm; leaves present throughout the winter              26
25a. Leaves strongly curved, less than 1 cm wide; flowers 3.5–4.5 cm in diameter                        **2. attica**
  b. Leaves straight or slightly curved, 1–1.5 cm wide; flowers 5–6 cm in diameter                      **1. pumila**
26a. Flowers 8–9 cm in diameter; leaves abruptly narrowed to an incurved tip                            **10. albicans**
  b. Flowers 4–8 cm in diameter; leaves not abruptly narrowed to an incurved tip                          27
27a. Perianth-tube 5–7.5 cm
                                          **4. pseudopumila**
  b. Perianth-tube 2–5 cm   **3. lutescens***
28a. All bracts completely papery at flowering time          **11. pallida***
  b. Inner bracts green or purplish at least in part at flowering time        29
29a. Standards dirty yellow to clear yellow                              30
  b. Standards white, bluish, or lavender to purple                        32
30a. Bracts green or purplish at least in the lower half, or completely greenish-membranous  **8. variegata***
  b. Bracts papery-translucent at the apex and margin                     31
31a. Bracts sharply keeled; lobes of style-branches toothed    **5. reichenbachii**
  b. Bracts not keeled; lobes of style-branches not toothed
                                          **12. schachtii***
32a. Flowers 9–15 cm in diameter       33
  b. Flowers 5–8 cm in diameter         34
33a. Standards with a small beard on the lower part; bracts to 11 cm, narrow
                                          **13. kashmiriana**
  b. Standards beardless; bracts to 5.5 cm                 **9. germanica***
34a. Standards whitish with violet veins
                                          **8. variegata***

b. Standards lavender to violet or purple    35

35a. Bracts sharply keeled    **5. reichenbachii**

b. Bracts not keeled    36

36a. Bracts transparent towards margin and apex    **12. schachtii***

b. Bracts not transparent towards margin and apex    **7. aphylla**

37a. Flowers lilac, purple or blue, mottled and blotched darker    38

b. Flowers of various colours, not mottled or blotched although sometimes veined or finely spotted    39

38a. Stem to 3 cm, rarely to 7 cm; perianth-tube 5–7.5 cm    **25. kamaonensis**

b. Stem 5–15 cm; perianth-tube 2–3 cm    **26. hookeriana***

39a. Both falls and standards bearded    40

b. Only the falls bearded    42

40a. Falls with a dark signal patch; beard of standards greenish or almost absent    **22. korolkowii***

b. Falls lacking a dark signal patch; beard of standards yellow, lilac, purple or blue    41

41a. Falls lilac-blue; beard yellow    **23. hoogiana**

b. Falls deep purplish-blue, brownish or purplish, sometimes shading to blue in the centre; beard yellow, lilac or blue; if falls lilac-blue then beard purple    **24. stolonifera***

42a. Stem with 1–3 flowers; flowers yellow, 3–5 cm in diameter    **14. bloudowii***

b. Stem bearing 1 flower; flowers of various colours but rarely yellow, at least 5 cm in diameter    43

43a. Ground colour of falls deep violet to blackish purple or purple-brown    **20. atropurpurea***

b. Ground colour of falls white, cream, yellow, greenish, greyish, lilac-pink or bluish    44

44a. Beard of falls whitish, yellow, yellowish brown or reddish brown, sometimes with purple-tipped hairs, or purple hairs at edge of beard    45

b. Beard of falls purple or purplish brown    48

45a. Plant to 30 cm at flowering time    46

b. Plant more than 30 cm at flowering time, if less then leaves 1–1.5 cm wide or flowers yellow, spotted with brown    47

46a. Beard of falls with yellowish brown or reddish brown hairs, becoming purple at edge of beard    **18. samariae**

b. Beard of falls with yellow hairs, sometimes whitish or tipped with purple    **17. sari***

47a. Flowers 13–20 cm in diameter, the overall appearance greyish, brown or purple    **15. gatesii***

b. Flowers 6–9 cm in diameter    **16. lortetii***

48a. Flowering stem to 25 cm, sometimes to 30 cm    49

b. Flowering stem 30 cm or more    **15. gatesii***

49a. Falls and standards circular, more or less equal in size    **19. iberica**

b. Falls and standards differing in size, the falls usually smaller and narrower    **21. acutiloba***

50a. Blade of falls with a central ridge or crest (occasionally with 3 ridges) which may be dissected or not    51

b. Blade of falls lacking a central ridge or crest    60

51a. Stem more or less absent, or if present then unbranched    52

b. Stem branched    54

52a. Leaves 2–5 mm wide    **75. decora**

b. Leaves 5 mm or more wide    53

53a. Leaves 1–3 cm wide; falls with 3 ridges; perianth-tube 4–6 cm    **67. cristata**

b. Leaves usually less than 1 cm wide; falls with 1 ridge; perianth-tube to 2 cm    **68. lacustris***

54a. Flowers 3–5 cm in diameter    55

b. Flowers 5.5–10 cm in diameter    59

55a. Stem 10–35 cm; ridge on falls undissected, or if dissected then leaves only 2–5 mm wide; lobes of style-branches not fringed, or if fringed then bracts fused at the base    56

b. Stem 45–100 cm; ridge on falls dissected or frilly; lobes of style-branches fringed    58

56a. Bracts fused at the base; ridge on falls mainly white, yellow at the apex    **71. gracilipes**

b. Bracts not fused at the base; ridge on falls yellow or orange-yellow, sometimes white or purple at the apex    57

57a. Leaves 2–5 mm wide; upper half of standards bent downwards    **75. decora**

b. Leaves 1–1.5 cm wide; standards wholly erect    **69. tenuis**

58a. Falls with a fringed margin    **72. japonica**

b. Falls with a scalloped-wavy margin    **73. confusa**

59a. Flowering stem 25–35 cm, slightly branched; perianth-tube *c.* 2.5 cm    **70. tectorum**

b. Flowering stem 30–200 cm, much-branched; perianth-tube *c.* 1.2 cm    **74. wattii***

60a. Seeds scarlet    **62. foetidissima**

b. Seeds not scarlet    61

61a. Style-branches bearing golden yellow glands on the upper surface towards the margin    **66. unguicularis***

b. Style-branches lacking golden yellow glands    62

62a. Standards much reduced, very narrowly lanceolate    **30. setosa**

b. Standards not reduced    63

63a. Flowers pale yellow or greenish yellow, the blade of the falls with an orange signal patch; bracts to 18 cm; rhizome usually vertical, covered with spiny, needle-like fibres    **65. grant-duffii**

b. Flowers of various colours, if pale yellow then lacking an orange signal patch on the falls; bracts 3–12 cm; rhizome lacking spiny fibres    64

64a. Stem hollow, if solid then leaves glossy on 1 side, grey-green on the other    65

b. Stem solid    72

65a. Flowers yellow    66

b. Flowers violet, blue-violet or reddish purple, occasionally white    67

66a. Flowering stem 35–40 cm; leaves shorter than flowering stem; flowers 5–6 cm in diameter with erect standards    **33. forrestii**

b. Flowering stem 60–75 cm; leaves more or less equal in height to flowering stem; flowers 6–8 cm in diameter, the standards held at an oblique angle    **34. wilsonii**

67a. Stem solid    **38. clarkei**

b. Stem hollow    68

68a. Bracts brown and papery at flowering time; leaves green on both sides    **32. sibirica**

b. Bracts green or reddish at flowering time although sometimes brownish at the tip; leaves grey-green, or with 1 side glossy green    69

69a. Leaves equal in length to the flowering stem or more or less so    70

b. Leaves shorter than the flowering stem    **35. delavayi**

70a. Leaves glossy green on one side, grey-green on the other    **36. bulleyana**

b. Leaves grey-green on both sides   71
71a. Standards erect          **31. sanguinea**
  b. Standards held obliquely
                      **37. chrysographes***
72a. Falls *c.* 1 cm      **27. minutoaurea**
  b. Falls 3 cm or more                    73
73a. Falls basically white, lilac to purple,
    pinkish, orange-red or red          74
  b. Falls basically cream or yellow   108
74a. Leaves 1–5 mm wide                75
  b. Leaves 6 mm or more wide          85
75a. Blade of falls with a central patch or
    stripe of orange, yellow or whitish
                                          76
  b. Blade of falls lacking such a patch or
    stripe                                80
76a. Flowering stem 30–90 cm; perianth-
    tube 2–20 mm                          77
  b. Flowering stem 4–50 cm, if taller
    then perianth-tube 3.5–6.5 cm    78
77a. Flowers 8–15 cm in diameter; claw
    of falls yellow, the yellow spreading
    upwards to the base of the blade
                        **50. ensata**
  b. Flowers 5.5–7 cm in diameter; claw
    of falls greenish white with violet
    veins              **54. prismatica**
78a. Perianth-tube 2–8 cm                79
  b. Perianth-tube 6–15 mm
                      **45. missouriensis***
79a. Blade of falls with a cream or
    yellowish central band, or band
    absent; apex of rhizome with
    remains of shiny brown leaf bases
                        **63. tenuifolia***
  b. Blade of falls with an orange stripe;
    old leaf bases not present at apex of
    rhizome                **28. verna**
80a. Capsule rounded, not ribbed; bracts
    with pink margins    **29. ruthenica**
  b. Capsule ribbed; bracts lacking pink
    margins                               81
81a. Capsule with 3 ribs                 82
  b. Capsule with 6 ribs                 83
82a. Perianth-tube 5–30 mm
                      **41. innominata***
  b. Perianth-tube 2–3 mm
                      **54. prismatica**
83a. Ribs of capsule evenly spaced       84
  b. Ribs of capsule arranged in 3 pairs
                      **57. sintenisii***
84a. Capsule with a beak; perianth-tube
    2–3 mm with a slender beak at the
    base which is part of the ovary
                        **64. lactea**
  b. Capsule lacking a beak; perianth-
    tube *c.* 1 cm, lacking a beak at the
    base              **45. missouriensis***
85a. Flowering stem to 6 cm; falls *c.*
    3.5 cm                **28. verna**

b. Flowering stem 10 cm or more; falls
    4 cm or more, if less then blade
    lacking a central orange stripe      86
86a. Falls red, coppery-red or orange-red
                        **52. fulva**
  b. Falls basically white or shades of
    pink or lilac to purple              87
87a. Blade of falls with a central patch or
    stripe of orange or yellow           88
  b. Blade of falls lacking such a patch or
    stripe, or patch, if present, whitish
                                         101
88a. Flowering stem branched             89
  b. Flowering stem unbranched           97
89a. Capsule with 3 ribs                 90
  b. Capsule with 6 ribs                 94
90a. Stigma triangular   **39. douglasiana**
  b. Stigma 2-lobed                      91
91a. Flowers 6–8 cm in diameter         92
  b. Flowers 8–15 cm in diameter        93
92a. Blade of falls 3–4 cm wide, with a
    hairy yellow central patch
                        **48. virginica**
  b. Blade of falls 1.5–3 cm wide, central
    patch yellowish green, not hairy
                        **49. versicolor**
93a. Leaves 4–12 mm wide with a
    prominent midrib        **50. ensata**
  b. Leaves 1.5–4 cm wide, lacking a
    prominent midrib      **51. laevigata**
94a. Flowers white except for a yellow
    patch on the blade of the falls
                        **59. orientalis**
  b. Flowers blue to lilac or purple     95
95a. Leaves *c.* 2.5 cm wide
                        **53. brevicaulis***
  b. Leaves 3–12 mm wide                 96
96a. Capsule with 6 evenly spaced ribs
                      **45. missouriensis***
  b. Capsule with 6 ribs arranged in 3
    pairs                  **55. spuria**
97a. Capsule with 3 ribs                 98
  b. Capsule with 6 ribs                 99
98a. Leaves 4–12 mm wide with a
    prominent midrib        **50. ensata**
  b. Leaves 1.5–4 cm wide, lacking a
    prominent midrib      **51. laevigata**
99a. Capsule with 6 evenly spaced ribs
                                         100
  b. Capsule with 6 ribs arranged in 3
    pairs                  **55. spuria**
100a. Flowers 2 or 3 in each set of bracts;
    leaves usually overtopping the
    flowering stem which is usually
    branched          **45. missouriensis***
  b. Flowers 3–8 in each set of bracts;
    leaves equal in height to flowering
    stem or shorter; stem usually
    unbranched          **46. longipetala**
101a. Flowering stem branched           102
  b. Flowering stem unbranched          104

102a. Leaves 6–12 mm wide; capsule with
    6 ribs arranged in 3 pairs
                        **55. spuria**
  b. Leaves 2–7 mm wide; capsule with
    3 or 6 evenly spaced ribs           103
103a. Perianth-tube 2–3 mm; capsule with
    3 ribs              **54. prismatica**
  b. Perianth-tube *c.* 10 mm; capsule
    with 6 ribs        **45. missouriensis***
104a. Capsule with 3 ribs
                      **43. macrosiphon***
  b. Capsule with 6 ribs                105
105a. Ribs of capsule evenly spaced     106
  b. Ribs of capsule in pairs at each of
    the 3 corners                       107
106a. Flowers 2 or 3 in each set of bracts;
    leaves usually overtopping the
    flowering stem which is usually
    branched          **45. missouriensis***
  b. Flowers 3–8 in each set of bracts;
    leaves equal in height to flowering
    stem or shorter ; stem usually
    unbranched          **46. longipetala**
107a. Flowering stem 20–40 cm,
    compressed or 2-winged
                        **56. graminea**
  b. Flowering stem 30–90 cm, not
    compressed or 2-winged    **55. spuria**
108a. Flowers 12–18 cm in diameter   109
  b. Flowers 10 cm in diameter or less
                                         110
109a. Flowers lemon yellow; lobes of style-
    branches broadly triangular, 4–
    5 mm, strongly recurved
                        **60.  × monnieri**
  b. Flowers golden yellow; lobes of style-
    branches narrowly triangular,
    *c.* 1 cm, not strongly recurved
                        **61. crocea**
110a. Leaves 1–3 cm wide                111
  b. Leaves up to 1 cm wide             112
111a. Capsule with 3 ribs; flowers
    7–10 cm in diameter
                      **47. pseudacorus**
  b. Capsule with 6 ribs; flowers to 7 cm
    in diameter          **55. spuria**
112a. Flowers 3–4 cm in diameter; capsule
    round in cross-section
                        **29. ruthenica**
  b. Flowers 4 cm or more in diameter;
    capsule ribbed, or if round then
    perianth-tube 5–8 cm               113
113a. Perianth-tube to 1 cm             114
  b. Perianth-tube 1.5–12 cm           117
114a. Stigma triangular; capsule with 3
    ribs                               115
  b. Stigma 2-lobed; capsule with 6 ribs
                                        116
115a. Stem to 30 cm, unbranched;
    perianth-tube 5–10 mm
                        **40. bracteata***

b. Stem 30–80 cm, usually branched; perianth-tube 2–3 mm     **54. prismatica**

116a. Falls dingy yellow to golden yellow, usually with darker veining; flowers to 7 cm in diameter     **55. spuria**

b. Falls cream to pale yellow, lacking darker veining; flowers 7–10 cm in diameter     **58. kerneriana**

117a. Leaves persisting through the winter; capsule with 3 ribs     118

b. Leaves not persisting through the winter; capsule never with 3 ribs     **63. tenuifolia***

118a. Leaves with pink, red or purplish coloration at the base     119

b. Leaves lacking pink, red or purplish coloration at the base     120

119a. Leaves 2–4 mm wide; perianth-tube 1.5–3 cm     **41. innominata***

b. Leaves 7–9 mm wide; perianth-tube 3–6 cm     **44. purdyi***

120a. Perianth-tube widened at the apex into a bowl-like shape; style-branches longer than the toothed lobes     **43. macrosiphon***

b. Perianth tube not widened into a bowl-like shape; style-branches more or less equal in length to the narrow entire lobes     **42. chrysophylla***

Subgenus **Iris**. Falls bearing a beard of fairly long hairs.

Section **Iris** (Section *Pogoniris* (Spach) Baker). Rhizomes stout, giving rise to fans of sword-shaped, usually fairly broad leaves. Stems simple or branched, usually with 2 or more flowers. Falls and standards well-developed, the falls bearing a prominent beard with each hair made up of several cells. Seeds lacking a fleshy appendage.

**1. I. pumila** Linnaeus. Illustration: Dykes, The genus Iris, pl. 32 (1913); Everard & Morley, Flowers of the world, pl. 42 (1970); Schauer & Caspari, Pflanzenführer, 119 (1978); Grey-Wilson & Mathew, Bulbs, pl. 27 (1981).
Stem more or less absent; plant 10–15 cm at flowering time. Leaves grey-green, 10–15 × 1–1.5 cm, almost straight to somewhat curved, absent in winter. Bracts 5–10 cm, the outer usually green and often slightly keeled, the inner rather membranous, not keeled. Flowers solitary, scented, 5–6 cm in diameter, yellow, blue, purple or a mixture of these colours. Perianth-tube 4–9 cm. Falls 3.5–6 cm; beard yellow or bluish. Standards 4–8 cm. *Eastern C & SE Europe to the Urals.* H2. Spring.

*I. pumila* has been used frequently in the production of the often-grown dwarf bearded cultivars.

**2. I. attica** Boissier & Heldreich (*I. pumila* Linnaeus subsp. *attica* (Boissier & Heldreich) Hayek). Illustration: Huxley & Taylor, Flowers of Greece and the Aegean, f. 406–8 (1977); Mathew, The iris, f. 3 (1981).
Similar to *I. pumila* but smaller, only 5–10 cm at flowering time. Leaves 4–7 cm × 4–7 mm, strongly curved. Flowers 3.5–4.5 cm in diameter, whitish, yellow, blue, purple or bicoloured. *Greece, S Yugoslavia & NW Turkey.* H2. Spring.

Regarded by some botanists as a subspecies of *I. pumila*.

**3. I. lutescens** Lamarck (*I. chamaeiris* Bertoloni). Illustration: Botanical Magazine, 2861 (1828); Polunin, Flowers of Europe, f. 1692 (1969); Grey-Wilson & Mathew, Bulbs, pl. 27 (1981); Mathew, The iris, f. 5 (1981).
Plant 5–35 cm, stem unbranched. Leaves to 30 cm × 5–25 mm, straight or slightly curved, present throughout the winter. Bracts 3–5.5 cm, green or sometimes brownish and papery at the apex. Flowers 1 or 2, 6–7 cm in diameter, yellow, violet or a mixture of the two, occasionally white. Perianth-tube 2–3.5 cm. Falls 5–7.5 cm; beard yellow, at least the part on the blade of the falls. Standards 5.5–7.5 cm. *NE Spain, S France & Italy.* H2. Spring.

***I. scariosa** Link. Bracts dry and more or less transparent. Flowers 4–5 cm in diameter, reddish-violet, blue-violet, yellow or almost white. Falls to 4.5 cm. Standards to 3.5 cm. *USSR (Urals eastward to the Tien Shan).* H2. Spring.

***I. subbiflora** Brotero (*I. lutescens* Lamarck subsp. *subbiflora* (Brotero) Webb & Chater). Illustration: Botanical Magazine, 1130 (1808); Dykes, The genus Iris, pl. 33 (1913); Polunin & Smythies, Flowers of south-west Europe, pl. 62 (1973). Flowers 7–8 cm in diameter, deep violet with a perianth-tube 3.5–5 cm. Falls to 3.5 cm; beard white or violet, at least the part on the blade of the falls. *SW Spain, Portugal.* H2. Summer.

**4. I. pseudopumila** Tineo.
Stem *c.* 3 cm; plant 15–25 cm at flowering time. Leaves grey-green, to 20 × 1–1.5 cm, slightly curved, present throughout the winter. Bracts to 12 cm. Flowers solitary, 6–8 cm in diameter, white, yellow, purple or bicoloured. Perianth-tube 5–7.5 cm. Beard of white hairs with yellow tips.

*SE Italy, Sicily, Malta & W Yugoslavia.* H2. Spring.

**5. I. reichenbachii** Heuffel. Illustration: Dykes, The genus Iris, pl. 34 (1913); Botanical Magazine, 8812 (1919); Grey-Wilson & Mathew, Bulbs, pl. 27 (1981); Mathew, The Iris, f. 3 (1981).
Plant 10–30 cm, stems unbranched or sometimes with a lateral branch. Leaves 8–35 × 1–1.5 cm. Bracts 3–5.5 cm, green, membranous on margin and at apex, sharply keeled. Flowers 1 or 2, 5–6.5 cm in diameter, greenish yellow, brownish purple or violet, often with darker veins. Perianth-tube 1.5–2.5 cm. Falls 4–6 cm; beard of yellow or bluish-purple hairs with white tips. Standards 4–6 cm. *SE Europe.* H2. Spring–early summer.

**6. I. suaveolens** Boissier & Reuter (*I. mellita* Janka; *I. rubromarginata* Baker).
Plant 8–15 cm, stems unbranched. Leaves sometimes purple-edged, to 22 cm × 4–10 mm, curved. Bracts 3.5–7 cm, sharply keeled. Flowers 1 or 2, 4.5–5.5 cm in diameter, yellow, purple, brownish purple or bicoloured. Perianth-tube 3–4.5 cm. Falls 3–5.5 cm; beard yellow or bluish. Standards 3.5–6 cm. *SE Europe, NW Turkey.* H2. Spring early summer.

**7. I. aphylla** Linnaeus (*I. benacensis* Stapf; *I. nudicaulis* Lamarck). Illustration: Botanical Magazine, 5806 (1869); Dykes, A handbook of garden Irises, pl. 21 (1924); Grey-Wilson & Mathew, Bulbs, pl. 27 (1981).
Plant 15–30 cm, stems branched usually from below the middle or from the base. Leaves 15–40 cm × 5–20 mm, curved. Bracts 3–6 cm, green, sometimes tinged with purple towards the tip, somewhat inflated. Flowers 1–5, 6–7 cm in diameter, dark purple to violet-blue. Perianth-tube 1.5–2.2 cm. Falls 4–6.5 cm; beard of white or bluish hairs, sometimes with yellow tips. Standards 4–6.5 cm, the claw often veined with brown. *C & E Europe, western USSR.* H2. Spring, occasionally again in autumn.

**8. I. variegata** Linnaeus (*I. reginae* Horvat; *I. rudskyi* Horvat). Illustration: Botanical Magazine, 16 (1787); Grey-Wilson & Mathew, Bulbs, pl. 27 (1981); Mathew, The iris, 29 (1981); Mathew, P.J. Redouté: Lilies and related flowers, 189 (1981).
Plant 20–45 cm, stem branched in the upper half. Leaves deep green, 12–30 × 1–3 cm, slightly curved, ribbed. Bracts 3–5.5 cm, green, often purple-tinged, inflated. Flowers 3–6, 5–7 cm in diameter. Perianth-tube 1.8–2.5 cm. Falls

4.5–6 cm, ground colour whitish or pale yellow, heavily veined with reddish brown or violet, sometimes the veining running together into a complete brown-purple stain; beard yellow. Standards yellow, or sometimes whitish with violet veins. *C & SE Europe*. H1. Late spring—summer.

There is a group of cultivars with white standards and coloured falls which are known collectively as **I. amoena** de Candolle. It is likely that they arose by selection or hybridisation from *I. variegata*.

*****I. imbricata** Lindley. Illustration: Edwards's Botanical Register **31**: t. 35 (1845); Botanical Magazine, 7701 (1900); Journal of the Royal Horticultural Society **90**: f. 28 (1965); Mathew, The iris, f. 2 (1981). Leaves grey-green, usually straight. Flowers 2 or 3, 7–9 cm in diameter, dull or pale yellow. Falls 5–6.5 cm, usually with brownish veins on the claw. *Iran, USSR (Transcaucasia)*. H2. Spring–summer.

*****I. junonia** which is 50–65 cm tall with leaves 3–5 cm wide and flowers which vary from white to cream, yellow, pale blue or purple, may key out here (see under 9. *I. germanica*).

### 9. I. germanica Linnaeus.
Illustration: Botanical Magazine, 670 (1805); Polunin, Flowers of Europe, f. 1693 (1969); Grey-Wilson & Mathew, Bulbs, pl. 27 (1981); Mathew, P.J. Redouté: Lilies and related flowers, 165 (1981).

Plant 50–120 cm, stems with 1 or 2 branches in the upper half. Leaves somewhat grey-green, 30–50 × 2–4.5 cm, almost straight, present throughout the winter. Bracts 3.5–5.5 cm, often tinged with purple in the lower half, membranous above. Flowers 3–5, scented, 9–10 cm in diameter, bluish violet. Perianth-tube 1.7–2.5 cm. Falls 5.5–9 cm; beard yellow. Standards 5.5–9 cm, often rather paler than falls. *Origin unknown, but probably native to the E Mediterranean area; widely naturalised*. H1. Late spring.

There are several named variants of *I. germanica* which are offered for sale, differing mainly in the colour of the flowers.

The following 5 similar plants have been described as species but may in fact be cultivars with a restricted distribution. It is doubtful whether they are truly wild, being usually recorded from cemeteries or near houses.

*****I. biliottii** M. Foster. Stems 60–80 cm with 2 or 3 branches. Falls reddish purple with a beard of white hairs tipped with yellow. Standards bluish purple. *N Turkey*. H2. Early summer.

*****I. cypriana** Baker & M. Foster. Outer bract brown and papery. Flowers *c*. 15 cm in diameter, lilac-blue, with falls which broaden towards the apex. Perianth tube *c*. 2.5 cm. *Cyprus*. H2. Late spring.

*****I. junonia** Schott. Stems 50–65 cm with *c*. 4 branches. Flowers smaller than those of *I. germanica*, varying from white, cream or yellow to pale blue or purple. Claw of falls often whitish with brownish purple veins. *S Turkey*. H2. Early summer.

*****I. mesopotamica** Dykes. Similar to *I. cypriana* but outer bract brown and papery in the upper third only. Perianth-tube 1.3–1.8 cm. Falls lavender-blue, the white claw with purple-brown veins; beard white and orange. *S Turkey, Syria & Israel*. H2. Late spring–early summer.

*****I. trojana** Stapf. Illustration: Dykes, The genus Iris, pl. 37 (1913). Bracts rather narrow, flushed with purple at the base. Flowers *c*. 10 cm in diameter, with reddish purple falls and paler, bluer standards. Beard of white yellow-tipped hairs. The veining on the claw of the falls is finer than in the rest of this group. *W Turkey*. H2. Early summer.

*****I. × sambucina** Linnaeus (including *I. × lurida* Aiton, *I. × squalens* Linnaeus, *I. × neglecta* Hornemann). Illustration: Botanical Magazine, 669 (1803), 986 (1807); Mathew, P.J. Redouté: Lilies and related flowers, 175, 185 (1981). Stem *c*. 45 cm. Bracts green, flushed with purple. Falls reddish or brownish purple, the yellowish claw strongly veined, beard yellow or orange. Standards dull purple, often with a yellowish tinge. *N Italy, NW Yugoslavia; naturalised elsewhere*. H2. Spring.

All the variable hybrids between *I. pallida* and *I. variegata* are included under the name *I. × sambucina*.

*****I. × kochii** Stapf grows to *c*. 45 cm and has red-purple or blue-purple flowers. The falls have a yellow beard and brown-veined claws. *N Italy*.

Thought to be a hybrid between *I. germanica* and *I. pallida* subsp. *cengialtii*.

### 10. I. albicans Lange.
Illustration: Grey-Wilson & Mathew, Bulbs, pl. 27 (1981); Mathew, P.J. Redouté: Lilies and related flowers, 161 (1981).

Plant 30–60 cm, stems unbranched or with 1 stalkless lateral flower-head. Leaves grey-green, 1.5–2.5 cm wide, abruptly narrowed to an incurved tip, present throughout the winter. Bracts green or tinged with purple in lower part, papery-transparent towards the tip. Flowers 1–3,

sweetly scented, 8–9 cm in diameter, white or blue. Falls with the claw flushed with greenish yellow; beard of white, yellow-tipped hairs. *Saudi Arabia, Yemen; naturalised elsewhere*. H2. Late spring.

The blue-flowered variant is known under the cultivar name 'Madonna'.

### 11. I. pallida Lamarck
(*I. dalmatica* Anon.; *I. pallida* var. *dalmatica* Anon.). Illustration: Botanical Magazine, 685 (1803); Grey-Wilson & Mathew, Bulbs, pl. 27 (1981); Mathew, P.J. Redouté: Lilies and related flowers, 171 (1981).

Plant 15–120 cm, stems branched above the middle. Leaves grey-green, 20–60 × 1–4 cm, some usually persisting throughout the winter. Bracts 2–3.5 cm, silvery-papery. Perianth-tube 8–11 mm. Flowers 3–6, scented, 9–11 cm in diameter, lilac-blue. Falls 5–7.5 cm; beard yellow. Standards 5–7.5 cm. *W Yugoslavia*. H2. Late spring–early summer.

Subsp. **cengialtii** (Ambrosi) M. Foster (*I. cengialtii* Ambrosi). Illustration: Mathew, The iris, f. 1 (1981). Plant only to 45 cm. Leaves greener than subsp. *pallida*, usually withering in the winter. Bracts brownish-papery. Flowers usually 2, deep bluish purple. Beard of falls of white hairs tipped with yellow or orange. *NE Italy*. H2. Late spring-early summer.

*I. pallida* 'Variegata' (Illustration: Hay & Synge, Dictionary of garden plants in colour, f. 1197, 1969) comes in two forms: 'Argentea' has the leaves striped with white and in 'Aurea' the leaves are yellow-striped.

*****I. florentina** Linnaeus. Flowers white, slightly flushed with blue. *Origin unknown*. H2. Late spring.

This plant has been much confused with *I. albicans* from which it differs in the flowers being neither pure white nor blue, and the bracts being brown-papery. Lateral flowers, if present, are borne on short stalks. It is probably an albino cultivar close to *I. germanica*, and perhaps best regarded as *I. germanica* 'Florentina'.

### 12. I. schachtii Markgraf.
Illustration: Mathew, The iris, f. 6, 7 (1981).

Plant 10–30 cm, stem with 1–3 branches. Leaves grey-green, to 1.5 cm wide. Bracts green, sometimes tinged with purple, transparent towards the margins and apex, slightly inflated. Flowers 5–6 cm in diameter, yellowish or purple. Perianth-tube 1.5–2 cm. Falls 5–6.5 cm. Standards 3.5–6.5 cm. *C Turkey*. H3. Spring.

*****I. albertii** Regel. Illustration: Botanical Magazine, 7020 (1888); Dykes, The genus

Iris, pl. 38 (1913). Plant 30–70 cm, with leaves 2–3 cm wide. Flowers 6–8 cm in diameter, lavender to violet. Claw of falls veined with brownish red. *USSR (Kazakstan)*. H2. Late spring.

**13. I. kashmiriana** Baker. Illustration: Botanical Magazine, 6869 (1886), 9378 (1934); Mathew, The iris, 29 (1981). Plant 75–125 cm, stems with 1 or branches. Leaves pale grey-green, to 60 × 2–3 cm, ribbed, straight. Bracts to 11 cm, green. Flowers 2 or 3 per branch, scented. 10–12 cm in diameter, white or pale lilac-blue. Perianth-tube 2.2–2.5 cm. Falls 6.5–9.5 cm; beard of dense white, yellow-tipped hairs. Standards 7–9 cm, with a similar but smaller beard on the lower part. *India (Kashmir)*. H4. Early summer.

Section **Psammiris** (Spach) J. Taylor. Rhizomatous. Stems unbranched with 1–3 flowers. Flowers 3–5 cm in diameter, yellow, beard of falls made up of hairs composed of one cell only. Seeds with a fleshy appendage.

**14. I. bloudowii** Bunge. Rhizome non-stoloniferous, with numerous brown fibres around the growing point. Plant 15–35 cm. Leaves to 30 cm × 7–13 mm, straight or slightly curved. Bracts very inflated. Flowers 2 or 3 per bract, *c.* 5 cm in diameter, yellow with a brownish or purple coloration on the lower part of the falls and standards. Beard yellow. *C & E USSR; ? China*. H2. Spring.

*I. humilis Georgi (*I. arenaria* Waldstein & Kitaibel; *I. flavissima* Pallas). Illustration: Edwards's Botanical Register, t. 549 (1821); Grey-Wilson & Mathew, Bulbs, pl. 28 (1981). Plant 5–25 cm. Leaves 5–17 cm × 2–7 mm. Flowers 1–3, 3–4 cm in diameter. Falls yellow with purple veins and an orange beard, the blade spreading horizontally. Standards yellow or purple, veined with purple. *E Europe, USSR*. H2. Spring.

Section **Oncocylus** (Siemssen) Baker. Rhizomatous. Stems unbranched, 1-flowered. Flowers 5 cm or more in diameter, rarely yellow, beard of falls made up of hairs composed of 1 cell only. Seeds with a fleshy appendage almost as large as the seed itself and often whitish.

**15. I. gatesii** M. Foster. Illustration: Botanical Magazine, 7867 (1902); Bulletin of the Alpine Garden Society **39**: 287 (1971); Mathew, The iris, f. 18 (1981). Rhizome compact, stout. Plant 45–60 cm. Leaves 5–7, grey-green, 5–9 mm wide, straight. Bracts 7.5–11.5 cm, pale green or pinkish brown. Flowers 13–20 cm in diameter, appearing greyish, brown or purplish depending on the colour and amount of veining and spotting. Perianth-tube 3–6 cm. Falls 8–10 cm, often with a small brown or blackish signal patch; beard yellow, brownish or purplish. Standards 8–10 cm. All perianth-lobes whitish or pale yellow, with fine veins and tiny spots of reddish brown or blackish. Style-branches 5–7 cm, lobes reflexed. *N Iraq, SE Turkey*. H4. Summer.

*I. bismarckiana Damman & Sprenger (*I. nazarena* (Herbert) Dinsmore). Illustration: Botanical Magazine, 7986 (1904). Rhizome stoloniferous. Leaves 2–2.5 cm wide. Flowers 10–12 cm in diameter. Falls cream or yellowish, densely spotted and veined with reddish brown or purple, and with a large blackish purple signal patch; beard dark purple. Standards white, speckled and veined with purple or blue. *NE Israel, S Syria*. H4. Spring.

*I. kirkwoodii Chaudhary. Illustration: Botaniska Notiser **125**: 498 (1972), **128**: 387 (1975). Plant to 75 cm. Leaves 1–1.5 cm wide, curved and often drooping. Flowers 8–12 cm in diameter. Falls white, pale bluish or pale greenish, densely covered with deep purple veins and dots; beard with long brownish or purple hairs; signal patch dark purple. Standards similar but less heavily marked. *N Syria, S Turkey*. H4. Late spring.

*I. westii Dinsmore (*I. susiana* Linnaeus forma *westii* (Dinsmore) Sealy). Illustration: Botanical Magazine, n.s., 550 (1969). Flowers 12–15 cm in diameter. Falls yellowish, heavily speckled and veined with deep brown or purplish; beard of long, sparse, purple hairs; signal patch deep velvety brown. Standards pale lilac, speckled and veined with darker lilac-blue. *Lebanon*. H4. Late spring.

*I. sofarana M. Foster (*I. susiana* Linnaeus forma *sofarana* (M. Foster) Sealy). Illustration: Botanical Magazine, n.s., 550 (1969). Flowers 10–13 cm in diameter. Falls creamy-white, densely speckled and veined with brownish purple or violet; beard of sparse, dark purple hairs; signal patch blackish purple. Standards white, speckled and veined with violet or reddish purple. *Lebanon*. H4. Late spring.

*I. susiana Linnaeus. Illustration: Botanical Magazine, 91 (1790). Flowers 10–12 cm in diameter. Falls greyish, heavily veined with deep purple; beard of sparse, deep purple hairs; signal patch velvety black. Standards greyish, heavily veined with deep purple. *Origin unknown, but probably Lebanon*. H4. Late spring.

*I. haynei (Baker) Mallet. Illustration: Gardeners' Chronicle **35**: 266 (1904); Journal of the Royal Horticultural Society **29**: proceedings f. 112 (1905). Plant to 40 cm. Flowers 10–12 cm in diameter, fragrant, with dark veins and spots on a pale ground giving an overall grey-lilac colour, with the standards more purplish. Perianth-tube *c.* 2.5 cm. Signal patch blackish brown. *N Israel*. H4. Spring.

**16. I. lortetii** Boissier. Illustration: Botanical Magazine, 7251 (1892); Dykes, The genus Iris, pl. 27 (1913); Everard & Morley, Flowers of the world, pl. 51 (1970); Mathew, The iris, f. 12, 13 (1981). Plant 30–50 cm. Leaves grey-green, 1–1.5 cm wide, straight, abruptly narrowed at the apex. Flowers 8–9 cm in diameter. Perianth-tube to 5 cm. Falls to 7.5 cm, whitish or pale lilac with pink or maroon speckling and veining, the margin sometimes minutely toothed; beard reddish or yellowish, rather sparse; signal patch deep maroon. Standards similar in colour to falls, but veined rather than speckled. *S Lebanon*. H4. Late spring.

*I. auranitica Dinsmore. Illustration: Journal of the Royal Horticultural Society **87**; f. 117 (1962); Warburton & Hamblen (eds), The world of irises, 18 (1978). Flowers yellow, spotted with brown giving a bronze appearance. Falls with a maroon signal patch and a dense beard of yellow hairs which are tipped with purple. *Syria*. H4. Late spring.

**17. I. sari** Baker (*I. lupina* M. Foster). Illustration: Botanical Magazine, 7904 (1903); Dykes, The genus Iris, pl. 26 (1913); Bulletin of the Alpine Garden Society **45**: 113 (1977); Mathew, The iris, f. 16, 17 (1981). Rhizome compact, stout. Plant 10–30 cm. Leaves 5–7, to 30 cm × 3–9 mm, curved or almost straight. Bracts 5–9.5 cm, green. Flowers 7–10 cm in diameter. Perianth-tube 2–3 cm. Falls 3.5–5.5 cm, cream, yellowish or greenish, veined with brownish purple or crimson, often minutely toothed and wavy on the margin; beard yellow; signal patch velvety brown or dark red. Standards 4–8.5 cm, similar in colour to falls or more bluish purple. Style-branches 3.5–5.5 cm, lobes scalloped. *C & S Turkey*. H4. Late spring.

*I. barnumae Baker & M. Foster. Illustration: Botanical Magazine, 7050 (1889); Mathew, The iris, f. 6 (1981). Leaves to 5 mm wide. Flowers purple, rather indistinctly veined; beard yellow, the hairs sometimes purple-tipped; signal patch small and only slightly darker than the falls. *E Turkey, NE Iraq & adjacent Iran*. H4. Late spring.

Forma **urmiensis** (Hoog) Mathew & Wendelbo (*I. urmiensis* Hoog). Illustration: Botanical Magazine, 7784 (1901); Rechinger (ed), Flora Iranica: Iridaceae, pl. 14 (1975). Flowers yellow. *SE Turkey, NW Iran*.

Subsp. **demavendica** (Bornmüller) Mathew & Wendelbo (*I. demavendica* (Bornmüller) Dykes). Illustration: Botanical Magazine, n.s., 448 (1964); Rechinger (ed), Flora Iranica: Iridaceae, pl. 14 (1975). Leaves 4–7 mm wide. Flowers bluish violet; beard white or cream. *Iran (Elburz Mountains)*.

\*I. **meda** Stapf. Illustration: Botanical Magazine, 7040 (1889); Journal of the Royal Horticultural Society 88: f. 58 (1963); Mathew, The iris, f. 4 (1981). Leaves 1.5–4 mm wide. Flowers 5–7 cm in diameter, cream or whitish, suffused and veined with golden brown, especially towards the irregularly wavy edge of the segments. Beard yellow; signal patch purplish or brown. Style-branches 2–3 cm. *W Iran*. H4. Spring.

\*I. **sprengeri** Siehe (*I. elizabethae* Siehe). Illustration: Gardeners' Chronicle ser. 3, **36**: 50 (1904). Rhizome creeping, stoloniferous. Plant 5–15 cm. Leaves 3–5 mm wide. Perianth-tube 1–1.5 cm. Falls 5.5–6 cm, yellow, occasionally whitish, veined and spotted with purplish red; beard yellow; signal patch dark purple. Standards white with blackish and purplish red veins. Style-branches 2.5–3.5 cm. *C Turkey*. H4. Late spring.

\*I. **paradoxa** which has standards 8–10 cm and much smaller falls which are almost entirely covered with a dense beard, may key out here (see under 20. *I. atropurpurea*).

18. I. **samariae** Dinsmore. Illustration: Pacific Horticulture 40(3): front cover (1979).
Rhizome stout with many bristle-like fibres at the growing point. Plant 25–30 cm. Leaves to 35 × 1 cm, almost straight, clothing the stem throughout its length. Flowers 9–10 cm in diameter. Falls 4–5 cm, creamy white spotted and veined with purple; beard of sparse reddish brown or yellowish brown hairs which become shorter and purple towards the beard edge; signal patch dark brown. Standards usually pinkish purple. Style-branches *c.* 5 cm. *NW Jordan*. H4. Late spring.

19. I. **iberica** Hoffmann.
Rhizome compact, not stoloniferous. Plant 15–20 cm. Leaves grey-green, strongly curved, 2–6 mm wide. Bracts 3–7 cm, greenish. Perianth-tube 2–3.5 cm. Falls 5.5–6.5 cm, whitish, heavily spotted and veined with brown, blade deflexed at an angle of *c.* 60° from the horizontal; beard purplish brown, signal patch brown or blackish. Standards 4.5–8.5 cm, white, cream or pale bluish, less veined than the falls, sometimes with sparse hairs at the base. Style-branches 3.5–7 cm, lobes scalloped. *SW Asia*. H4. Late spring.

Subsp. **elegantissima** (Sosnowsky) Fedorov & Takhtadjan (*I. elegantissima* Sosnowsky). Illustration: Botanical Magazine, 5847 (1870); Bulletin of the Alpine Garden Society **35**: 362 (1967); Rechinger (ed), Flora Iranica: Iridaceae, pl. 12 (1975); Mathew, The iris, f. 14, 15 (1981). Plant usually 20–30 cm. Falls cream or pale yellow, heavily spotted and veined with brownish purple, blade sharply deflexed to an almost vertical position. Standards completely white or slightly veined at the base with brown. *NE Turkey, NW Iran, USSR (Armenia)*. H4. Spring.

Subsp. **lycotis** (Woronow) Takhtadjan (*I. lycotis* Woronow). Illustration: Botanical Magazine, n.s., 580 (1970); Rechinger (ed), Flora Iranica: Iridaceae, pl. 12 (1975); Mathew, The iris, f. 8 (1981). Falls and standards more or less equally heavily spotted and veined on a whitish ground, the falls with a concave blade standing out at an angle. *NE Iraq, SE Turkey, NW & W Iran, USSR (Armenia)*. H4. Spring.

20. I. **atropurpurea** Baker. Illustration: Gartenflora, t. 1361 (1891); Mathew, The iris, f. 7 (1981).
Rhizomes often stoloniferous. Plant 15–25 cm. Leaves somewhat grey-green, *c.* 15 cm, curved. Flowers *c.* 8 cm in diameter. Perianth-tube *c.* 3 cm. Falls to 5 cm, blackish purple, the blade veined with greenish yellow, the claw veined with reddish purple; beard yellow; signal patch blackish. Standards to 7.5 cm, reddish purple with faint blackish veins. *Israel*. H4. Spring.

\*I. **atrofusca** Baker. Illustration: Botanical Magazine, 7379 (1894); Warburton & Hamblen (eds), The world of irises, 193 (1978). Leaves pale green, straight. Perianth-tube 5–6.5 cm. Falls dark purple-brown, veined and dotted with red-black. Standards usually paler and more of a deep red colour. *Israel*. H4. Spring.

\*I. **nigricans** Dinsmore. Flowers 8–10 cm in diameter. Perianth-tube *c.* 6.5 cm. Falls

dark brownish purple to deep purple, with darker veins and dots; beard dark, purple-tipped on a whitish ground. Standards whitish, very heavily veined with dark purple. *Jordan*. H4. Spring.

\*I. **paradoxa** Steven. Illustration: Botanical Magazine, 7081 (1889); Journal of the Royal Horticultural Society 90: f. 21 (1965); Everard & Morley, Flowers of the world, pl. 51 (1970); Mathew, The iris, f. 5 (1981). Rhizome not stoloniferous. Flowers variable in colour; names have been given to the various forms which are listed below. Falls 2.5–4 cm, almost horizontal, oblong, very small in comparison with the large rounded standards, almost entirely covered with a dense beard. Standards 7–10 cm, notched at the apex. Style-branches 2–3 cm, lobes minutely scalloped. *E Turkey, N Iran, USSR (Transcaucasia)*. H4. Late spring.

Forma **paradoxa**. Falls blackish; standards deep violet.

Forma **atrata** Grossheim. Falls and standards blackish violet.

Forma **choschab** (Hoog) Mathew & Wendelbo. Falls blackish violet; standards white or pale lilac, veined with purple.

Forma **mirabilis** Gawrilenko. Falls golden yellow covered with a darker yellow to orange beard; standards pale yellow or pale blue.

\*I. **barnumae** which has yellow, bluish violet or purple flowers with less obvious veining, may key out here (see under 17, *I. sari*).

21. I. **acutiloba** Meyer.
Rhizome slender, much-branched. Plant 8–25 cm. Leaves grey-green, curved, 2–6 mm wide. Bracts 5–8 cm, green or slightly pinkish. Flowers 5–7 cm in diameter, whitish, strongly veined with brown or grey, the segments rather pointed. Perianth-tube 1.5–2 cm. Falls 4–7 cm, with a sparse purplish or brownish beard, and 2 dark red-brown or blackish spots, 1 in the centre and 1 at the apex. Standards 4–8 cm. Style-branches 2–4 cm. *S USSR, Iran*. H3. Spring–early summer.

Subsp. **lineolata** (Trautvetter) Mathew & Wendelbo (*I. ewbankiana* M. Foster; *I. helena* (K. Koch) K. Koch). Illustration: Botanical Magazine, 9333 (1933); Journal of the Royal Horticultural Society 88: f. 56 (1963), **93**: f. 204 (1968); Hay & Synge, Dictionary of garden plants in colour, f. 791 (1969). Falls with only 1 dark spot. *USSR (Transcaucasia – S of R. Kura), NW & NE Iran, adjacent Turkmenistan*.

**\*I. mariae** Barbey (*I. helenae* Barbey). Flowers 8–10 cm in diameter, lilac or pinkish veined with reddish brown. Claw of falls, beard and signal patch deep purple. *Israel.* H4. Late spring.

**\*I. paradoxa** which has standards 8–10 cm and much smaller falls which are almost entirely covered with a dense beard, may key out here (see under 20. *I. atropurpurea*).

Section **Regelia** Lynch. Rhizomatous. Stems unbranched, usually bearing 2 flowers (1 in *I. afghanica*). Falls and standards both bearded, each hair of the beard composed of 1 cell only. Seeds with a fleshy appendage.

**22. I. korolkowii** Regel. Illustration: Gartenflora, t. 766 (1873); Botanical Magazine, 7025 (1888); Iris year book 1968: f. 6 (1968); Mathew, The iris, f. 10 (1981).
Rhizome thick, slightly stoloniferous, with fibrous remains of old leaves. Plant 40–60 cm. Leaves tinged with purple at the base, 5–10 mm wide. Bracts 9–10 cm. Flowers 2 or 3, 6–8 cm in diameter. Perianth-tube 2.5–3 cm. Falls 7–11 cm, creamy white veined with dark maroon, occasionally purplish with green or purple veins, blade deflexed, pointed; beard dark; signal patch blackish, brown or deep green. Standards 7–11 cm, similar in colour to falls, pointed, beard very sparse. Style-branches *c.* 4.5 cm with scalloped lobes. *USSR (Soviet Central Asia – Tien Shan & Pamir Alai), NE Afghanistan.* H4. Early summer.

*I. korolkowii* has been crossed with *I. stolonifera.*

**\*I. afghanica** Wendelbo. Illustration: Notes from the Royal Botanic Garden Edinburgh 31: 339 (1972); Botanical Magazine, n.s., 668 (1974); Mathew, The iris, f. 9 (1981); Heywood & Chant, Popular encyclopedia of plants, 181 (1982). Plant 15–35 cm. Leaves 2–5 mm wide. Flowers 1, sometimes 2, 8–9 cm in diameter. Falls cream or white, heavily veined with purplish brown; beard dark; signal patch purple. Standards 6–7.5 cm, pale yellow with a greenish beard. *NE Afghanistan.* H4. Spring.

**23. I. hoogiana** Dykes. Illustration: Botanical Magazine, 8844 (1920); Dykes, A handbook of garden irises, pl. 19 (1924); Mathew, The iris, f. 20 (1981).
Rhizomes thick, stoloniferous. Plant 40–60 cm. Leaves to 50 × 1–1.5 cm, straight or slightly curved, tinged with

purple. Flowers 2 or 3, scented, 7–10 cm in diameter, lilac-blue. Perianth-tube *c.* 2.5 cm. Falls and standards to 7.5 cm; beard yellow. *USSR (Soviet Central Asia – Pamir Alai).* H3. Late spring.

Several cultivars have been selected and named. *I. hoogiana* has been hybridised with *I. stolonifera.*

**24. I. stolonifera** Maximowicz (*I. leichtlinii* Regel; *I. vaga* M. Foster). Illustration: Botanical Magazine, 7861 (1902); Dykes, The genus iris, pl. 29 (1913).
Rhizomes stoloniferous. Plant 30–60 cm. Leaves bluish green, to 60 cm × 5–15 mm, prominently nerved. Flowers 2 or 3, 7–8 cm in diameter, a mixture of brown and purple, the perianth-lobes usually brownish at the wavy margins, shading to blue towards the centre; beard yellowish or blue. Falls to 7.5 cm. Standards to 6.5 cm. *USSR (Soviet Central Asia – Pamir Alai).* H3. Late spring.

A species which is very variable in stature and flower colour; some of the forms are commercially available and have been named.

**\*I. heweri** Grey-Wilson & Mathew. Illustration: Journal of the Royal Horticultural Society **99:** f. 87 (1974); Rechinger (ed), Flora Iranica: Iridaceae, pl. 3, 15 (1975). Plant 10–15 cm, occasionally to 30 cm. Leaves green, 2–5 mm wide, strongly curved. Flowers *c.* 5 cm in diameter, deep purplish blue; beard lilac. Falls and standards 3.5–4.5 cm. *NE Afghanistan.* H4. Late spring.

**\*I. darwasica** Regel. Leaves grey-green, to 45 cm × 4–8 mm, straight. Flowers 5–6 cm in diameter, lilac with purple veins; beard purple. *USSR (Southern Soviet Central Asia).* H3. Spring.

Section **Pseudoregelia** Dykes. Rhizomatous. Stems unbranched with 1 or 2 flowers. Flowers lilac or purple, with darker blotches on the falls, beard of the falls made up of hairs composed of 1 cell only. Seeds with a fleshy appendage.

**25. I. kamaonensis** D. Don. Illustration: Botanical Magazine, 6957 (1887); Dykes, The genus iris, pl. 30 (1913); Everard & Morley, Flowers of the world, pl. 107 (1970); Mathew, The iris, f. 11 (1981).
Rhizome thick and knobbly, producing rather fleshy roots. Plant to 7 cm. Leaves to 45 cm × 2–10 mm, overtopping the flowers. Flowers 1 or 2, stemless, scented, 4–5 cm in diameter. Perianth-tube 5–7.5 cm. Falls and standards 3.5–4 cm,

lilac-purple, mottled and blotched darker, standards slightly paler than falls. Beard of white hairs with yellow tips. Style-branches *c.* 2.5 cm. *Himalaya from Kashmir to W China.* H3. Late spring.

The name is frequently misspelled *kumaonensis.*

**26. I. hookeriana** M. Foster. Illustration: Botanical Magazine, 7276 (1893); Journal of the Royal Horticultural Society **93:** f. 128 (1968).
Plant to 15 cm. Leaves to 30 × 1–2.5 cm. Flowers usually 2, scented, on stems 5–12 cm. Perianth-tube 2–3 cm. Falls and standards *c.* 5 cm, lilac, purple or bluish, sometimes white. *W Himalaya.* H3. Late spring.

**\*I. tigridia** Ledebour. Leaves to 10 cm × 1–4 mm. Flowers 1 or 2, on stems 10–15 cm, lilac to deep blue with purple mottling. Beard white, or hairs yellow-tipped, the area of the blade surrounding the beard being white, veined with purple. *SE USSR, Mongolia, NW China.* H3.

Subgenus **Limniris** (Tausch) Spach. Falls lacking a beard, but often with cockscomb-like ridges on the blade, or very short, downy hairs on the claw.

Section **Limniris**. Rhizomes often compact, clump-forming. Flower colour various. Falls with no crest-like ridge in the centre, occasionally with short, downy hairs on the claw.

Series **Chinensis** (Diels) Lawrence. Rhizomes thin, wiry, often far-creeping. Leaves with obvious ribs. Flowers yellow, flattish in appearance due to spreading falls and standards. Capsules triangular in cross-section.

**27. I. minutoaurea** Makino (*I. minuta* Franchet & Savatier). Illustration: Botanical Magazine, 8293 (1910); Mathew, The iris, f. 14 (1981).
Rhizomes thin, wiry. Plant 8–10 cm. Leaves 12–15 cm × 2–3 mm, elongating after flowering to *c.* 40 cm. Flowers solitary, 2–2.5 cm in diameter. Perianth-tube 2–2.5 cm. Falls *c.* 1 cm, yellow, often sparsely spotted with purple, spreading. Standards shorter than falls, paler yellow, spreading, the claw brownish. *Origin unknown: possibly Korea or China.* H2. Spring.

Series **Vernae** (Diels) Lawrence. Rhizomatous. Flowers white or lilac-blue. Capsule triangular in cross-section. Seeds spherical, with a fleshy appendage which dries out quickly after the seed is shed from the capsule.

**28. I. verna** Linnaeus. Illustration: Sweet, British flower garden ser. 1, 1: 68 (1824); Rickett, Wild flowers of the United States 1: pl. 11 (1966); Hay & Synge, Dictionary of garden plants in colour, f. 95 (1969); Mathew, The iris, 82 (1981).
Plant 4–6 cm. Leaves green or grey-green, to 15 cm × 3–15 mm. Flowers 1 or 2, 3–5 cm in diameter. Perianth-tube 2–5 cm. Falls *c.* 3.5 cm, bright lilac-blue, sometimes white, the blade with a central orange stripe, claw orange or yellow. Standards lilac-blue, erect. *SE USA.* H3. Spring.

Series **Ruthenicae** (Diels) Lawrence. Rhizomatous. Flowers basically whitish. Capsule round in cross-section, the segments quickly opening and curling back. Seeds pear-shaped with a fleshy appendage.

**29. I. ruthenica** Ker Gawler. Illustration: Botanical Magazine, 1123 (1808), 1393 (1811); Dykes, The genus Iris, pl. 13 (1913); Grey-Wilson & Mathew, Bulbs, 161 (1981). Rhizome shortly creeping. Plant 3–15 cm. Leaves green, to 30 cm × 2–5 mm, erect. Bracts 3–5 cm, membranous, greenish with pink margins. Flowers usually 1, occasionally 2, fragrant, 3–4 cm in diameter. Perianth-tube 8–10 mm. Falls 4.5–5 cm, whitish with bluish lavender or violet margins and veins, blade held horizontally. Standards 4–4.5 cm, almost erect, bluish lavender or violet. *E Europe, through C Asia to China and Korea.* H2. Late spring.

Series **Tripetalae** (Diels) Lawrence. Rhizomatous. Flowers basically blue to purple, standards narrowly lanceolate.

**30. I. setosa** Link. Illustration: Edwards's Botanical Register, t. 10 (1847); Dykes, The genus Iris, pl. 23 (1913); Clark, Wild flowers of the Pacific northwest, 86 (1976); Mathew, The iris, 85, f. 15 (1981). Rhizomes fairly stout, often clad in the fibrous remains of old leaf bases. Plant 15–90 cm, stem usually branched. Leaves 20–50 × 1–2.5 cm, the bases often red-tinged. Bracts green, often with purple margins. Flowers 2 or 3, sometimes solitary, 6–9 cm in diameter. Perianth-tube 5–10 mm. Falls with a blue-purple to blue blade which narrows abruptly into a narrow claw with blue or purple veins on a whitish or pale yellow ground. Standards blue-purple to blue, narrowly lanceolate, sharply pointed, sometimes with 2 lobes at the base of the blade. Style-branches with toothed lobes. *Northeast N America, E USSR, China, N Korea, Japan, Aleutian, Sakhalin & Kurile Islands to Alaska.* H1. Summer.

An extremely variable species in which several variants have been described. These are summarised below and are listed alphabetically, as the naming is somewhat chaotic and needs to be sorted out.

Forma **alpina** Komarov. A Siberian form with a very short stem.

Var. **arctica** (Eastwood) Dykes (*I. arctica* Eastwood). A dwarf Alaskan variety, the purple flowers variegated with white.

Subsp. **canadensis** (M. Foster) Hultén (*I. hookeri* G. Don). A dwarf form from eastern N America, the stem unbranched, almost leafless and bearing 1, or rarely 2, lavender-blue flowers.

Subsp. **hondoensis** Honda. Stems to 75 cm bearing large purple flowers. *Japan.*

Subsp. **interior** (Anderson) Hultén. An Alaskan plant with narrow leaves and papery, violet bracts which are shorter than normal.

Var. **nasuensis** Hara. Stems to 1 m bearing very large flowers and wide leaves. *Japan.*

Forma **platyrhyncha** Hultén. An Alaskan form whose solitary flowers have standards which are larger and wider than usual.

Forma **serotina** Komarov. A Siberian form with solitary, stemless flowers.

Series **Sibiricae** (Diels) Lawrence. Rhizomes stout. Stems hollow (except in *I. clarkei*). Leaves not persisting through the winter. Falls with 2 flanges at the base of the claw. Stigma triangular. Capsule almost round to triangular in cross-section. Seeds D-shaped to almost cubic.

**31. I. sanguinea** Donn (*I. orientalis* Thunberg). Illustration: Dykes, The genus Iris, pl. 1 (1913); Grey-Wilson, The genus Iris, subsection Sibiricae, 5 (1971). Plant 30–75 cm, stems hollow, usually unbranched. Leaves slightly grey-green, 5–12 mm wide, more or less equalling stems, or slightly longer. Bracts green or reddish, sometimes brown and papery at the tip. Flowers 6–8 cm in diameter, borne on stalks which are almost equal in length. Blade of falls white with a network of reddish purple veins which merge into a solid band of colour near the margin, claw yellowish or orange, finely veined with purple. Standards smaller, erect. *SE USSR, Korea, Japan.* H2. Early summer.

'Alba' has white flowers, usually with some purple veins and 'Violacea' has deep violet flowers with larger than normal standards. *I. sanguinea* has been hybridised with *I. sibirica*.

**32. I. sibirica** Linnaeus. Illustration: Perry, Flowers of the world, 144 (1972); Grey-Wilson & Mathew, Bulbs, pl. 28 (1981); Mathew, The iris, 90 (1981); Mathew, P.J. Redouté: Lilies and related flowers, 179 (1981).
Plant 50–120 cm, stem branched, hollow. Leaves green, 25–80 cm × up to 9 mm, shorter than the flowering stems. Bracts 3–5 cm, brown and papery. Flowers 6–7 cm in diameter, borne on unequal stalks. Falls 3–6 cm, blue or blue-violet, the obovate or oblong blade with a white central area strongly veined with violet, blade narrowing abruptly to a paler claw with dark veining. Standards 2.5–5.5 cm. *C & E Europe, NE Turkey, USSR (Caucasus, European Russia to Lake Baikal).* H1. Late spring–summer.

Many cultivars are available, the flowers varying in colour from white to shades of blue or deep violet-blue.

**33. I. forrestii** Dykes. Illustration: Dykes, The genus Iris, pl. 3 (1913); Grey-Wilson, The genus Iris, subsection Sibiricae, 11 (1971); Mathew, The iris, 90 (1981). Plant 35–40 cm, stems unbranched, hollow. Leaves glossy on one side, grey-green on the other, much shorter than the flowering stem. Bracts green. Flowers 5–6 cm in diameter, scented. Falls yellow with brownish purple lines on the claw. Standards yellow, erect. *W China (Yunnan, Sichuan), N Burma.* H2. Summer.

**34. I. wilsonii** Wright. Illustration: Botanical Magazine, 8340 (1910); Dykes, The genus Iris, pl. 2 (1913); Grey-Wilson, The genus Iris, subsection Sibiricae, 13 (1971). Plant 60–75 cm, stems unbranched, hollow. Leaves grey-green, more or less equalling the flowering stems and often rather drooping. Bracts green. Flowers 6–8 cm in diameter. Falls pale yellow, veined and dotted with purplish brown especially in the centre of the blade and on the claw. Standards pale yellow, held at an oblique angle, margins rather wavy. *W China.* H2. Summer.

**35. I. delavayi** Micheli. Illustration: Botanical Magazine, 7661 (1899); Hay & Synge, Dictionary of garden plants in colour, f. 1187 (1969); Grey-Wilson, The genus Iris, subsection Sibiricae, 15 (1971).

Plant 90–150 cm, stems branched, hollow. Leaves grey-green, shorter than the flowering stems. Bracts green, papery at the tips. Flowers 7–9 cm in diameter. Falls pale or dark purple, the circular blade notched at the apex, with a large white patch in the centre. Standards similar in ground colour, held at an oblique angle. *W China (Yunnan, Sichuan)*. H2. Summer.

*I. delavayi* and *I. wilsonii* have been crossed to produce plants with purple flowers whose falls have a yellow ground, veined with bluish purple.

**36. I. bulleyana** Dykes. Illustration: Dykes, The genus Iris, pl. 6 (1913); Grey-Wilson, The genus Iris, subsection Sibiricae, 17 (1971).
Plant 35–45 cm, stems unbranched, hollow. Leaves glossy green on one side, grey-green on the other, more or less equal in length to the flowering stems. Bracts green, papery at the tips. Flowers 6–8 cm in diameter. Falls spreading to drooping, violet at the tips, the centre white or yellowish, dotted and streaked with deep violet, the claw greenish yellow. Standards violet, held at an oblique angle. *W China, N Burma*. H2. Summer.

Formerly considered to be a hybrid between *I. forrestii* and *I. chrysographes*, but recent Chinese work suggests it is probably a good species.

**37. I. chrysographes** Dykes. Illustration: Dykes, The genus Iris, pl. 4 (1913); Grey-Wilson, The genus Iris, subsection Sibiricae, 21 (1971).
Plant 35–45 cm, stems unbranched, hollow. Leaves grey-green, 1–1.5 cm wide, more or less equal in length to the flowering stems. Bracts green. Flowers 6–7 cm in diameter, scented. Falls reddish violet, the blade drooping vertically, with or without gold streaks in the centre. Standards reddish violet, held at an oblique angle. *W China (Yunnan, Sichuan)*. H2. Summer.

'Rubella' has very dark flowers.
*I. chrysographes* hybridises with both *I. forrestii* and *I. sibirica*.
***I. dykesii** Stapf. Illustration: Grey-Wilson, The genus Iris, subsection Sibiricae, 19 (1971). Similar, but the leaves sheath the stem for most of its length. Flowers larger, deep violet-purple, the centre of the blade of the falls with white and yellow veins. *Origin unknown*. H2. Summer.

A plant whose origin and status are shrouded in mystery. Probably it is Chinese. Grey-Wilson suggests that it may be a hybrid, but the parentage is so far unknown.

**38. I. clarkei** Baker. Illustration: Botanical Magazine, 8323 (1910); Dykes, The genus Iris, pl. 5 (1913); Grey-Wilson, The genus Iris, subsection Sibiricae, 23 (1971).
Plant to 60 cm, stems branched, solid. Leaves glossy green on one side, grey-green on the other, 1.3–2 cm wide, shorter than the flowering stems. Bracts green. Flowers 7–7.5 cm in diameter. Falls blue-violet to deep blue or reddish purple, the centre of the blade with a large, white area, veined with violet, claw somewhat yellowish. Standards spreading outwards. *E Himalaya*. H2. Early summer.

Series **Californicae** (Diels) Lawrence. Rhizomes tough, roots wiry. Stems usually unbranched, generally bearing 1 or 2 flowers. Falls usually with a horizontally-spreading blade. Stigma triangular (except in *I. purdyi*). Capsule triangular to rounded in cross-section, with 3 ribs.

**39. I. douglasiana** Herbert. Illustration: Botanical Magazine, 6083 (1874); Dykes, The genus Iris, pl. 8 (1913); Cohen, A guide to the Pacific coast irises, 32 (1967); Hay & Synge, Dictionary of garden plants in colour, f. 93 (1969).
Plant 15–70 cm, stems branched. Leaves to 1 m × 2 cm, stained with red at the base, persisting through the winter. Bracts 6–12 cm, held together or divergent. Flowers 2 or 3, 7–10 cm in diameter. Perianth-tube 1.5–2.8 cm. Falls 5–8.5 cm, lavender to purple, the blade with a yellowish central area and darker veins; sometimes the falls are nearly white with darker veins. Standards 4.5–7 cm. Style-branches 1.5–3.5 cm, the lobes with toothed margins. *W USA (S Oregon, California)*. H2. Summer.

**40. I. bracteata** Watson. Illustration: Botanical Magazine, 8640 (1915); Journal of the Royal Horticultural Society 91: f. 92 (1966); Cohen, A guide to the Pacific coast irises, 34 (1967); Rickett, Wild flowers of the United States 5: pl. 20 (1971).
Plant 20–30 cm, stems unbranched. Leaves glossy green, often red at the base, to 1 cm wide, persisting through the winter; flowering stems bearing short bract-like leaves to the top, usually tinged with pink or purple-red. Bracts 5–9 cm, held closely around the flower-stalk and ovary. Flowers 2, 6–7.5 cm in diameter. Perianth-tube 5–10 mm. Falls 4.5–8 cm, cream to yellow, with brown or reddish purple veins, the blade with a central deeper yellow area. Standards 4.5–7.5 cm, erect, slightly less veined than the falls.

Style-branches 2–3 cm. *W USA (S Oregon, N California)*. H2. Early summer.
***I. hartwegii** Baker. Illustration: Dykes, The genus Iris, pl. 10 (1913); Aliso 4(1): 32 (1958); Cohen, A guide to the Pacific coast irises, 10 (1967); Rickett, Wild flowers of the United States 4: pl. 16 (1970). Leaves lacking red at the base, not persisting through the winter. Bracts divergent and separated from one another. Flowers pale yellow or lavender. *W USA (California)*. H3. Summer.
***I. tenax** Lindley. Illustration: Journal of the Royal Horticultural Society 91: f. 95 (1966); Rickett, Wild flowers of the United States 5: pl. 19 (1971); Clark, Wild flowers of the Pacific northwest, 58 (1976); Mathew, The iris, 99 (1981). Leaves 3–5 mm wide, not persisting through the winter. Bracts divergent. Flowers 7–9 cm in diameter, white, cream, yellow or lavender to deep purple-blue, the falls, if purple, with a central patch of white or yellow. *NW USA (SW Washington, W Oregon)*. H2. Early summer.

**41. I. innominata** Henderson. Illustration: Journal of the Royal Horticultural Society 91: f. 93, 94 (1966); Hay & Synge, Dictionary of garden plants in colour, f. 1190 (1969); Rickett, Wild flowers of the United States 5: pl. 21 (1971); Mathew, The iris, 97, f. 16 (1981).
Plant 15–25 cm, stems unbranched. Leaves dark green, purplish at the base, 2–4 mm wide, persisting through the winter. Bracts 3.5–6 cm, held closely around the perianth-tube, margins papery. Flowers 1 or 2, 6.5–7.5 cm in diameter, cream, yellow, orange, lilac pink or pale bluish purple to dark purple. Perianth-tube 1.5–3 cm. Falls 4.5–6.5 cm, often veined, margin frilly. Standards slightly shorter, margin frilly. Style-branches 2–2.5 cm, lobes reflexed, toothed. *W USA (SW Oregon, NW California)*. H2. Summer.

***I. hartwegii** which has leaves which do not persist through the winter, divergent bracts and perianth-tube 5–10 mm, may key out here (see under 40. *I. bracteata*).
***I. tenax** which has leaves which do not persist through the winter, flowers 7–9 cm in diameter with a perianth-tube 6–10 mm, may key out here (see under 40. *I. bracteata*).

**42. I. chrysophylla** Howell. Illustration: Cohen, A guide to the Pacific coast irises, 26 (1967); Rickett, Wild flowers of the United States 5: pl. 21 (1971).
Plant almost stemless or to 20 cm. Leaves often slightly grey-green, 3–5 mm wide,

longer than the flowering stems, persisting through the winter. Bracts 5–8.5 cm, held close together. Flowers usually 2, 6–7 cm in diameter, cream or pale yellow with deeper yellow or lilac veins. Perianth-tube 4.5–12 cm. Falls 4.5–6.5 cm, rather narrow, pointed. Standards 3–5.5 cm. Style-branches 1.5–2.5 cm, lobes more or less entire. *W USA (Oregon, N California)*. H3. Late spring–summer.

**\*I. tenuissima** Dykes. Illustration: Aliso 4(1): 69 (1958); Journal of the Royal Horticultural Society **91**: f. 90 (1966); Rickett, Wild flowers of the United States **5**: pl. 20 (1971). Plant 15–30 cm. Perianth-tube abruptly widened to form a conspicuous throat in the upper one-quarter to half. Flowers cream, with brown or purple veins on the falls. *W USA (N California)*. H4. Early summer.

**43. I. macrosiphon** Torrey. Illustration: Dykes, The genus Iris, pl. 12 (1913); Journal of the Royal Horticultural Society **91**: f. 88 (1966); Cohen, A guide to the Pacific coast irises, 18, f. 1 (1967); Rickett, Wild flowers of the United States **5**: pl. 20 (1971).
Plant almost stemless or to 25 cm, stem unbranched. Leaves grey-green, with no pink colour at the base, *c.* 6 mm wide, longer than the flowering stems, persisting through the winter. Bracts 4–4.5 cm, green, held closely together. Flowers 2, often fragrant, 5–6 cm in diameter, white, cream, yellow, lavender or purple, veined especially on the falls. Perianth-tube 3.5–8.5 cm, widened at the apex into a bowl-like shape. Falls 4–7 cm, the centre of the blade often whitish. Standards slightly shorter than falls. Style-branches 2–3.5 cm, with reflexed toothed lobes. *W USA (California)*. H4. Summer.

**\*I. munzii** R.C. Foster. Illustration: Aliso 4(1): 55 (1958); Cohen, A guide to the Pacific coast irises, 16 (1967). Plant to 75 cm. Leaves 1.5–2 cm wide. Bracts 6.5–11 cm, divergent, often separated from one another. Flowers 2–4, 6–7.5 cm in diameter, pale blue, or lavender to deep reddish purple, often with darker veins. Perianth-tube 7–10 mm. Falls and standards with wavy or frilly margins. *W USA (California)*. H5. Summer.

**\*I. hartwegii** which has leaves which do not persist through the winter, divergent bracts, flowers 6–8 cm in diameter and perianth-tube only 5–10 mm, may key out here (see under 40. *I. bracteata*).

**44. I. purdyi** Eastwood. Illustration: Dykes, The genus Iris, pl. 11 (1913); Aliso 4(1):

59 (1958); Cohen, A guide to the Pacific coast irises, 30 (1967); Rickett, Wild flowers of the United States **5**: pl. 20 (1971).
Plant 20–35 cm. Leaves dark glossy green, tinged with pink or red at the base, to 8 mm wide; flowering stems bearing short bract-like leaves to the top. Bracts 5–7.5 cm, often flushed with red. Flowers 2, *c.* 8 cm in diameter, white or cream, the falls often tinged with lavender, and veined and spotted with purplish pink or brown. Perianth-tube 3.5 cm. Falls 5.5–8.5 cm, spreading. Standards 5–7 cm, spreading. Style-branches 2–3 cm. Stigma truncate to 2-lobed. *W USA (coastal N California)*. H3. Summer.

**\*I. fernaldii** R.C. Foster. Illustration: Aliso 4(1): 28, 29 (1958); Journal of the Royal Horticultural Society **91**: f. 89 (1966); Cohen, A guide to the Pacific coast irises, 24, f. 2 (1967). Leaves grey-green, tinged with purple at the base. Flowers pale yellow, the falls with a deeper yellow central line, and sometimes with a pale purple tinge or veining. Stigma triangular. *W USA (C California)*. H3. Summer.

Series **Longipetalae** (Diels) Lawrence. Rhizomes thick, clad in remains of old leaves. Stems persisting for 1 or more years after flowering. Flowers whitish to blue or lavender. Stigma 2-lobed. Capsule tapering at top and bottom, 6-ribbed.

**45. I. missouriensis** Nuttall (*I. tolmieana* Herbert). Illustration: Botanical Magazine, 6579 (1881); Rickett, Wild flowers of the United States **4**: pl. 16 (1970); Spellenberg, Audubon Society field guide to North American wildflowers – western region, f. 606 (1979).
Plant to 50 cm, stem usually branched. Leaves pale green, 3–7 mm wide, usually longer than the flowering stems. Bracts 4–7 cm, green at the base, papery above. Flowers 2 or 3, pale blue or lilac to deep blue or lavender. Perianth-tube *c.* 1 cm. Falls *c.* 6 cm, strongly veined, blade often with a yellow signal patch. Standards shorter than falls. Style-branches to 2.5 cm, with toothed lobes. *Western N America*. H3. Late spring–summer.

**\*I. pontica** which differs in growing only to 10 cm and having solitary violet flowers and completely membranous bracts, may key out here (see under 57. *I. sintenisii*).

**\*I. tenax** which has leaves coloured pink at the base, and an unbranched stem carrying flowers which vary from pale lavender to deep blue-purple, through

white to yellow, may key out here (see under 40. *I. bracteata*).

**46. I. longipetala** Herbert. Illustration: Journal of the Royal Horticultural Society **88**: f. 111 (1963); Rickett, Wild flowers of the United States **5**: pl. 20 (1971); Mathew, The iris, f. 17 (1981).
Leaves dark green, equal to or shorter than flowering stems. Bracts 7–12 cm. Flowers 3–8, whitish veined with lilac-purple. Falls 6–10 cm, often with a slightly yellowish signal patch. Standards notched. *W USA (coastal C California)*. H3. Late spring–summer.

Series **Laevigatae** (Diels) Lawrence. Rhizomes stout. Flowers 6–15 cm in diameter. Stigma 2-lobed. Capsule triangular to nearly round in cross-section, 3-ribbed, thin-walled, breaking up irregularly rather than splitting into 3 cells. Seeds shiny.

**47. I. pseudacorus** Linnaeus (*I. pseudacorus* var. *acoriformis* (Boreau) Lynch). Illustration: Polunin, Flowers of Europe, f. 1690 (1969); Fitter et al., Wild flowers of Britain and northern Europe, 273 (1974); Grey-Wilson & Mathew, Bulbs, pl. 28 (1981); Mathew, P.J. Redouté: Lilies and related flowers, 173 (1981).
Rhizome stout, clad with fibrous remains of old leaf bases. Plant 73–160 cm, stems branched. Leaves 50–90 × 1–3 cm, grey-green, with a prominent midrib. Bracts 4–10 cm, green with a papery margin. Flowers 1–3 per branch, 7–10 cm in diameter, yellow. Perianth-tube 1–1.5 cm. Falls 5–7.5 cm, blade broadly ovate to almost circular, with a deeper yellow central area and usually with some violet or brown veining. Standards 2–3 cm, narrowly oblong. Style-branches 3.5–4.5 cm. *Europe to W Siberia, Caucasus, Turkey, Iran, N Africa*. H1. Summer.

'Alba' has cream flowers. 'Variegata' has the leaves striped with yellow. Var. **bastardii** (Boreau) Lynch has the blade of the falls lacking the darker yellow central patch.

**48. I. virginica** Linnaeus. Illustration: Botanical Magazine, 703 (1804); Rickett, Wild flowers of the United States **1**: pl. 11 (1966).
Plant 30–100 cm, stems often arching, branched. Leaves 1–3 cm wide, rather soft, drooping at the tips, as long as the flowering stems or longer. Flowers 1–4, 6–8 cm in diameter, blue, violet, lilac or pinkish lavender. Falls *c.* 7.5 cm, spreading,

blade with a central yellow, hairy patch surrounded by a white, purple-veined area. Standards *c*. 5 cm, erect. *Southeastern USA*. H2. Summer.

Very similar to the next species: some botanists consider it to be the same.

**49. I. versicolor** Linnaeus. Illustration: Botanical Magazine, 21 (1787); Rickett, Wild flowers of the United States 1: pl. 11 (1966); Niering & Olmstead, Audubon Society field guide to North America wildflowers – eastern region, f. 620 (1979); Mathew, The iris, f. 19 (1981).
Plants 20–80 cm, stems branched. Leaves 35–60 × 1–2 cm, often pinkish at the base, as long as the flowering stems or shorter. Bracts 3–8 cm, green. Flowers 6–8 cm in diameter, violet, blue-purple, lavender or dull greyish purple. Falls 4–6 cm, widely spreading, blade often with a central greenish yellow blotch, surrounded by a white, purple-veined area which continues down the claw. Standards 2.5–4 cm. *Eastern N America*. H2. Summer.
'Kermesina' has reddish purple flowers.

**50. I. ensata** Thunberg (*I. kaempferi* Lemaire). Illustration: Botanical Magazine, 6132 (1874); Dykes, The genus Iris, pl. 19 (1913).
Plant 60–90 cm, stems usually unbranched. Leaves 20–60 cm × 4–12 mm, shorter than the flowering stems, midrib prominent. Flowers 8–15 cm in diameter, purple or reddish purple. Perianth-tube 1–2 cm. Falls *c*. 7.5 cm, claw yellow, the yellow spreading upwards to the base of the ovate or elliptic blade. Standards *c*. 5 cm, erect. *E USSR, China, Japan*. H2. Summer.

Many named cultivars have been developed which may have larger flowers, double flowers, a more extensive colour range, and often flowers with spreading standards producing a flattish appearance (Illustration: Perry, Flowers of the world, 145, 1982.).

**51. I. laevigata** Fischer. Illustration: Perry, Flowers of the world, 145 (1972); Kitamura et al., Coloured illustrations of herbaceous plants of Japan (Monocotyledoneae), pl. 21 f. 143 (1978); Mathew, The iris, f. 18 (1981).
Plant to 40 cm, stems often with 1 branch. Leaves 1.5–4 cm wide. Flowers 8–10 cm in diameter, white or bluish purple. Perianth-tube to 2 cm. Falls to 6.5 cm, claw yellow or whitish, this colour spreading upwards to the obovate or ovate-elliptic blade. Standards much smaller than falls, erect. *E Asia*. H2. Summer–autumn.

There are many named cultivars available, some of which extend the colour range. 'Variegata' has striped leaves.

Series **Hexagonae** (Diels) Lawrence. Rhizomatous. Flowers 5.5–12 cm in diameter. Capsules 6-ribbed. Seeds very large with a corky coat.

**52. I. fulva** Ker Gawler. Illustration: Dykes, The genus Iris, pl. 21 (1913); Journal of the Royal Horticultural Society 88: f. 113 (1963); Rickett, Wild flowers of the United States 2: pl. 22 (1967); Everard & Morley, Wild flowers of the world, pl. 166 (1970).
Plant 45–80 cm, stems straight or very slightly zig-zag. Leaves 1.5–2.5 cm wide, drooping at the tips. Flowers 5.5–6.5 cm in diameter, red, orange-red or coppery-red. Perianth-tube to 2.5 cm. Falls to 6.5 cm, drooping. Standards *c*. 5 cm, drooping, notched at the apex. Style-branches held at an oblique angle so they stand above the rest of the flower. *S USA (Mississippi valley)*. H3. Summer.
**I. × fulvala** Dykes with purple-red flowers is occasionally grown. It is a hybrid between *I. fulva* and *I. brevicaulis*.

**53. I. brevicaulis** Rafinesque (*I. foliosa* Mackenzie & Bush). Illustration: Dykes, The genus Iris, pl. 20 (1913); Rickett, Wild flowers of the United States 1: pl. 11 (1966); Hay & Synge, Dictionary of garden plants in colour, f. 1189 (1969).
Plant 30–50 cm, stems zig-zag. Stem leaves *c*. 2.5 cm wide, overtopping the flowers. Flowers terminal and borne in the leaf axils, 6–10 cm in diameter, bright blue-violet. Falls reflexed, blade broadly ovate, with a central yellow band usually surrounded by white, claw usually with greenish white veins. Standards much smaller than falls, widely spreading. *S USA (Mississippi valley)*. H3. Summer.
**\*I. hexagona** Walter. Illustration: Botanical Magazine, 6787 (1884); Rickett, Wild flowers of the United States 3: pl. 11 (1969). Stems not zig-zag. Flowers 10–12 cm in diameter, bluish purple or lavender. Blade of falls with a yellow signal patch. Standards erect. *SE USA*. H3. Summer.

Series **Prismaticae** (Diels) Lawrence. Rhizomes thin, far-creeping. Capsule triangular in cross-section with 1 rib at each corner. Seeds with smooth coats.

**54. I. prismatica** Ker Gawler. Illustration: Botanical Magazine, 1504 (1813); Dykes, The genus Iris, pl. 7 (1913); Rickett, Wild

flowers of the United States 1: pl. 11 (1966).
Rhizomes far-creeping. Plant 30–80 cm, stems wiry, rarely straight, usually with 1 branch. Leaves 50–70 cm × 2–7 mm, bluish green. Bracts papery-brown. Flowers 2 or 3 at top of stem, 1 on the branch, 5.5–7 cm in diameter, borne on stalks to 4 cm. Perianth-tube 2–3 mm. Falls spreading widely, pale violet or violet-blue, blade whitish in the centre with violet veins, claw greenish white with violet veins. Standards violet, almost erect. Style-branches 2–3 cm, lobes recurved. *USA (E coast and S Appalachian mountains)*. H2. Summer.

A variant with creamy white flowers is occasionally grown.

Series **Spuriae** (Diels) Lawrence. Rhizomes woody, roots wiry. Stems unbranched or closely branched. Outside of perianth-tube with 3 drops of nectar at the base of the perianth-lobes. Stigma 2-lobed. Capsule triangular in cross-section with 2 ribs at each corner, and an apical beak. Seeds hard, with a loose, shiny, papery coat.

**55. I. spuria** Linnaeus.
A variable species of which the following subspecies are cultivated.
Subsp. **spuria**. Illustration: Grey-Wilson & Mathew, Bulbs, pl. 28 (1981); Mathew, P.J. Redouté: Lilies and related flowers, 183 (1981). Plant 50–80 cm. Bracts almost transparent at the tip. Flowers 6–8 cm in diameter, lilac or violet-blue veined with violet. Falls 4.5–6 cm, blade with a central yellow stripe, claw longer than the blade. *C Europe*. H2. Summer.
Subsp. **carthalinae** (Fomin) Mathew. Plant to 90 cm. Flowers white or sky blue, overtopping the leaves. *USSR (Georgia)*. H2. Summer.
Subsp. **halophila** (Pallas) Mathew & Wendelbo (*I. halophila* Pallas). Plant 40–90 cm. Flowers 6–7 cm in diameter, dingy pale yellow to bright golden yellow, usually with darker veins. Falls 4–6 cm, the claw longer than the blade. *S Romania, USSR (Ukraine, Moldavia, Caucasus, W Siberia)*. H2. Summer.
Subsp. **maritima** (Lamarck) Fournier. Plant 30–50 cm. Bracts green. Falls 3–4.5 cm, cream with purple veins, the blade dark purple, claw with a greenish stripe, claw longer than the blade. *SW Europe*. H2. Summer.
Subsp. **musulmanica** (Fomin) Takhtadjan (*I. klattii* Kemularia-Nathadze; *I. musulmanica* Fomin; *I. violacea* Klatt).

Illustration: Rechinger (ed), Flora Iranica: Iridaceae, pl. 12 (1975); Mathew, The iris, f. 20 (1981). Plant 40–90 cm. Flowers pale violet to deep lavender-violet with darker veins. Falls 5.5–8 cm, the blade with a central yellow stripe and sometimes a yellow suffusion towards the base, claw equal to or slightly longer than the blade. *E Turkey, N & NW Iran, USSR (S Caucasus)*. H2. Summer.

Subsp. **notha** (Bieberstein) Ascherson & Graebner. Plant 70–90 cm. Flowers violet-blue to bright blue. Claw of falls with a broad yellow stripe, claw equal to or longer than the blade. *USSR (S Caucasus)*. H2. Summer.

**56. I. graminea** Linnaeus. Illustration: Berrisford, Irises, pl. 29 (1961); Polunin, Flowers of Europe, f. 1686 (1969); Grey-Wilson & Mathew, Bulbs, pl. 28 (1981); Mathew, P.J. Redouté: Lilies and related flowers, 167 (1981).
Plant 20–40 cm, stem strongly flattened or 2 winged. Leaves 35–100 cm × 5–15 mm, the uppermost stem leaf overtopping the flowers. Bracts unequal, the lower larger and rather leaf-like. Flowers 1 or 2, fragrant, 7–8 cm in diameter. Falls 3–5 cm, violet, the centre of the blade with a white area veined with violet, claw conspicuously winged, often tinged with green or brown. Standards 2.5–4 cm, purple. Style-branches purple, often tinged with green or brown in the lower part. *NE Spain, through C Europe to W USSR, N & W Caucasus*. H2. Summer.

Var. **pseudocyperus** (Schur) Beck is a larger plant with unscented flowers. *Czechoslovakia, Romania*.

**57. I. sintenisii** Janka. Illustration: Cave, The iris, 81 (1959); Goulandris & Goulimis, Wild flowers of Greece, 185 (1969); Grey-Wilson & Mathew, Bulbs, pl. 28 (1981); Mathew, The iris, 111, f. 21 (1981).
Rhizome slender, clad in the brown fibrous remains of leaf bases. Plant 10–30 cm, stem unbranched, cylindric to slightly flattened, but not winged. Leaves 20–50 cm × 2–5 mm, usually longer than the flowering stems. Bracts 4–7 cm, papery, keeled. Flowers usually 1, sometimes 2, 5–6 cm in diameter with rather narrow perianth-lobes. Falls 3.8–4.5 cm, veined violet on a white ground giving a metallic violet-blue appearance. Standards 3–4 cm. Styles 2.5–3 cm, lobes recurved. *SE Europe, N Turkey, SW USSR*. H2. Summer.

*\*I. pontica** Zapalowicz. Plant *c*. 10 cm. Flowers solitary, violet, often veined darker.

Falls 4.5–5.5 cm, blade almost circular, usually with a whitish or pale greenish yellow, violet-veined area in the centre, claw separated by a definite constriction. Standards 3.5–5 cm. *Romania, USSR (W Ukraine, Caucasus, ?C Asia)*. H2. Late spring.

**58. I. kerneriana** Baker. Illustration: Cave, The iris, 97 (1959); Thomas, Perennial garden plants, pl. III (1976); Mathew, The iris, 111 (1981).
Rhizome slender, clad with the fibrous remains of the old leaf bases. Plant 20–30 cm, sometimes to 55 cm, stem unbranched. Leaves usually to 5 mm wide, sometimes to 1 cm, generally slightly shorter than the flowering stems. Bracts rather inflated, becoming chaffy as the fruit develops. Flowers 2–4, 7–10 cm in diameter, deep cream to pale yellow. Perianth-tube 7–10 mm. Falls 5–6.5 cm, blade with a deep yellow central blotch, strongly recurved so that the tip sometimes touches the stem. Standards 4.5–5.5 cm, erect, wavy-margined, notched at the tip. Style-branches following the curve of the falls. *N & C Turkey*. H2. Summer.

**59. I. orientalis** Miller (*I. ochroleuca* Linnaeus; *I. spuria* Linnaeus subsp. *ochroleuca* (Linnaeus) Dykes). Illustration: Botanical Magazine, 61 (1793); Goulandris & Goulimis, Wild flowers of Greece, 183 (1968); Hay & Synge, Dictionary of garden plants in colour, f. 1196 (1969); Perry, Flowers of the world, 145 (1972).
Plant 40–90 cm, stem usually with 1 branch, sometimes more. Leaves 1–2 cm wide. Bracts papery. Flowers 2 or 3, 8–10 cm in diameter, white. Falls with a circular blade 2.5–3.5 cm wide, with a large yellow signal patch. Standards to 8.5 cm, erect. *NE Greece, W Turkey*. H2. Summer.

**60. I. × monnieri** de Candolle.
Plant to 120 cm, stems stout. Leaves *c*. 60 cm. Flowers lemon-yellow, fragrant. Falls with a circular blade 3.5 cm wide and notched at the apex. Style-branches with strongly recurved, broadly triangular, 4–5 mm lobes. *Turkey*. H2. Summer.

Thought to be a hybrid between *I. orientalis* and *I. xanthospuria* Mathew & Baytop.

**61. I. crocea** R.C. Foster (*I. aurea* Lindley). Illustration: Edwards's Botanical Register **33**: t. 59 (1847); Dykes, The genus Iris, pl. 16 (1913).
Plant to 1.5 m, stems usually branched. Leaves to 75 × 1.5–2 cm. Flowers 2 or 3

per branch, 12–18 cm in diameter, deep golden yellow. Falls 7.5–8 cm, blade with a wavy margin and longer than the claw. Style-branches with narrowly triangular lobes *c*. 1 cm, not strongly recurved. *Origin unknown, possibly Kashmir*. H2. Summer.

Series **Foetidissimae** (Diels) Mathew. Rhizomatous. Leaves persisting through the winter. Seeds bright red, remaining attached to the capsule for a long time after it has split.

**62. I. foetidissima** Linnaeus. Illustration: Hay & Synge, Dictionary of garden plants in colour, f. 1188 (1969) – fruit only; Fitter et al., The wild flowers of Britain and northern Europe, 273 (1974); Grey-Wilson & Mathew, Bulbs, pl. 28 (1981); Mathew, P.J., Redouté: Lilies and related flowers, 163 (1981).
Plant 30–90 cm, stem branched, rather flattened. Leaves 30–70 × 1–2.5 cm, dark green. Bracts 5–10 cm, green. Flowers 1–5 per branch, 5–7 cm in diameter. Perianth-tube *c*. 1 cm. Falls 3–5 cm, blade lilac-grey veined with purple, fading to white in the centre, claw brownish, winged, sometimes as wide as the blade. Standards 2–4 cm, brownish, flushed with lilac. Style-branches brownish or yellowish. Seeds scarlet. *S & W Europe, N Africa, Atlantic islands*. H2. Summer.

Yellow-flowered variants are known and occur occasionally in cultivation. There is also a form with variegated leaves, and a very rare variant with white seeds.

Series **Tenuifoliae** (Diels) Lawrence. Rhizome small, almost vertical, apex with persistent shiny, brown remains of old leaf bases. Stigma 2-lobed. Capsule more or less rounded to triangular in cross-section, if 6-ribbed then with a rib at each corner and on each face. Seeds cubic to pear-shaped, wrinkled.

**63. I. tenuifolia** Pallas.
Plant 10–30 cm, stems with brown fibrous leaf remains around the base. Leaves to 40 cm × 1–3 mm. Bracts green, 7–10 cm. Flowers 1 or 2, scented, 4–6 cm in diameter. Perianth-tube 5–8 cm. Falls 4–6 cm, the blade with violet or lilac veins on a cream ground and with a cream or yellowish central stripe. Standards slightly shorter than falls, bluish violet or lilac. Style-branches with narrow, sharply pointed lobes. *SE USSR through C Asia to Mongolia & W China*. H2. Late spring.

*\*I. songarica** Schrenk. Illustration: Mathew, The iris, f. 22 (1981). Flowers

5–7 cm in diameter, up to 6 in each set of bracts, greyish lavender-blue, spotted and veined darker, the standards 6.5 cm or more. *C & E Asia*. H2.

Series **Ensatae** (Diels) Lawrence. Rhizomatous, clump-forming. Ovary with a slender apical beak beneath a very short perianth-tube, giving the perianth-lobes the appearance of being almost free from one another. Capsule 6-ribbed, beaked. Seeds spherical, shiny.

**64. I. lactea** Pallas (*I. biglumis* Vahl). Illustration: Edwards's Botanical Register **26**: t. 1 (1840); Mathew, The iris, 126 (1981).
Rhizomes tough, with wiry roots, clad with reddish or purple-brown fibrous remains of old leaf bases. Plant 6–40 cm. Leaves grey-green, 3–5 mm wide, equalling or longer than flowering stems, strongly ribbed. Flowers 2 or 3, scented, 4–6 cm in diameter, blue, blue-violet or purple. Perianth-tube 2–3 mm, with a slender beak at the base which is part of the ovary. Falls rather narrow, claw whitish or yellowish with dark veins. *S USSR, China, Mongolia, Korea, Himalaya from Afghanistan to Xizang*. H2. Early summer.

Series **Syriacae** (Diels) Lawrence. Rhizome almost vertical, the rather swollen leaf bases giving a bulb like appearance.

**65. I. grant-duffii** Baker. Illustration: Botanical Magazine, 7604 (1898); Cave, The iris, 32 (1959).
Rhizome clad in remains of leaf bases, which become spiny. Plant 25–30 cm, stems unbranched. Leaves grey-green, to 35 cm × 5–10 mm, prominently veined, margin white. Bracts to 18 cm, greenish, margin papery. Flowers solitary, slightly scented, *c*. 6 cm in diameter. Perianth-tube *c*. 7 mm. Falls 6–7 cm, sulphur-yellow or greenish yellow, blade with an orange signal patch and sometimes with a few black streaks, claw often with purplish veins. Standards *c*. 7 cm, similar in colour to the falls, or paler. Style-branches *c*. 4.5 cm, pale yellow. *Israel, Syria, ?Lebanon*. H4. Late spring.

Series **Unguiculares** (Diels) Lawrence. Rhizomes much-branched, forming large patches which eventually die in the middle to leave a circle of growth. Leaves persisting through the winter. Flowers stemless, but with a long perianth-tube. Style-branches joined at the base into a slender tube, and bearing golden yellow glands on the upper surface, towards the margin.

**66. I. unguicularis** Poiret (*I. stylosa* Desfontaines). Illustration: Hay & Synge, Dictionary of garden plants in colour, f. 1203 (1969); Perry, Flowers of the world, 146 (1972); Grey-Wilson & Mathew, Bulbs, pl. 28 (1981).
Stem absent or very short. Leaves 45–60 × to 1 cm, borne in erect, compact tufts. Bracts more or less green, 6–13 cm. Flowers solitary, fragrant, almost stemless, lavender. Perianth-tube 6–20 cm. Falls 7–8 cm, the blade with a yellow central band, usually surrounded by a white area veined with lavender, claw with dark veins. Standards 7–8 cm. Style-branches with golden yellow glands on the upper surface, towards the margin. *Algeria, Tunisia, S & W Turkey, W Syria; ?Israel*. H2. Winter–spring.

Several cultivars are available with white, deep violet, pinkish or variegated flowers. The following is the most distinct and often regarded as a separate species.

'Cretensis' (*I. cretensis* Janka; *I. cretica* Baker). Illustration: Journal of the Royal Horticultural Society **101**: 336 (1976); Huxley & Taylor, Flowers of Greece and the Aegean, f. 409, 410 (1977); Gartenpraxis 1980: 113 (1980); Mathew, The iris, f. 23 (1981). Leaves 1–3 mm wide. Flowers with narrow segments less than 5.5 cm. Blade of falls violet or deep lavender at the apex, the rest of the blade and the claw white, veined with violet; centre of blade with an orange stripe. *S Greece, Crete*.

**\*I. lazica** Albov (*I. unguicularis* Poiret var. *lazica* (Albov) Dykes). Illustration: Journal of the Royal Horticultural Society **90**: f. 12 (1965). Leaves 15–30 cm × 8–15 mm, borne in distinct spreading fans. Falls lavender-blue, claw white-veined and spotted with darker lavender, the spots continuing onto the blade which has a pale yellow central stripe. *NE Turkey, USSR (SW Caucasia)*. H2. Autumn and spring.

Section **Lophiris** (Tausch) Tausch. Rhizomes often creeping with widely spreading stolons. Flowers white, lilac or blue. Falls with a ridge which is dissected (entire in *I. tenuis* and *I. speculatrix*), and sometimes has an extra ridge on each side.

*I. tenuis* and *I. speculatrix* do not fit well into this section, and probably do not belong here, but at the moment botanists are unsure of their position. In this account they are left, in the absence of further research, in Section *Lophiris*.

**67. I. cristata** Solander. Illustration: Journal of the Royal Horticultural Society **88**: f. 115 (1963); Rickett, Wild flowers of the United States **2**: pl. 22 (1967); Green Scene **3**(5): back cover (1975); Mathew, The iris, 72, f. 12 (1981).
Rhizomes much branched, small. Leaves to 15 × 1–3 cm at flowering time, often lengthening later. Flowers almost stemless, 3–4 cm in diameter, usually lilac-blue, sometimes white, lilac, purple or blue. Perianth-tube 4–6 cm. Blade of falls with a central white patch, along which run 3 crisped ridges which continue down the claw; on or around the ridges is a variable amount of yellow or orange-brown. Standards narrower than falls. *Eastern N America*. H2. Early summer.

**68. I. lacustris** Nuttall. Illustration: Rickett, Wild flowers of the United States **1**: pl. 11 (1966); Hay & Synge, Dictionary of garden plants in colour, f. 94 (1969).
Very similar to *I. cristata* but smaller and with a stem to 5 cm. Leaves to 1 cm wide. Flowers sky-blue, occasionally white. Perianth-tube to 2 cm. Blade of falls with a central white patch and a yellow central minutely toothed ridge. *N America (Great Lakes region)*. H1. Late spring–summer.

**\*I. speculatrix** Hance. Illustration: Botanical Magazine, 6306 (1877); Walden & Hu, Wild flowers of Hong Kong, pl. 35 (1977). Flowering stem 20–25 cm. Leaves to 1 m × 5–10 mm, with distinct cross-veining. Perianth-tube *c*. 1 cm, below which is a solid beak from the top of the ovary. Falls *c*. 2.5 cm, lilac, the blade with a central yellow ridge surrounded by a white area, speckled with purple, which in turn is surrounded by a dark purple area; claw violet-veined. Standards lilac, spreading obliquely. *SE China, Hong Kong*. H5. Spring–summer.

**69. I. tenuis** Watson. Illustration: Abrams, An illustrated flora of the Pacific states **1**: 463 (1923); Aliso **4**: 313 (1959); Rickett, Wild flowers of the United States **5**: pl. 19 (1971).
Rhizome far-creeping. Plant 30–35 cm, stem branched. Leaves *c*. 30 × 1–1.5 cm. Flowers 1 per branch, 3–4 cm in diameter. Falls to 3 cm, whitish, veined with blue-purple and blotched with yellow, central ridge yellow, undissected. Standards erect, notched at the tip. Style-lobes entire. *W USA (Oregon)*. H2. Late spring.

**70. I. tectorum** Maximowicz. Illustration: Botanical Magazine, 6118 (1874); Dykes, The genus Iris, pl. 24 (1913); Journal of

the Royal Horticultural Society 88: f. 116 (1963); Hay & Synge, Dictionary of garden plants in colour, f. 1202 (1969).
Rhizome fat, greenish. Plant 25–35 cm, stems slightly branched. Leaves to 30 × 2.5–5 cm, ribbed. Flowers 2 or 3 per branch, 8–10 cm in diameter. Perianth-tube c. 2.5 cm. Falls to 5 cm, lilac with some darker veining and blotching, only slightly deflexed, blade wavy-margined with a white dissected ridge bearing a few dark spots. Standards spreading obliquely, margin wavy. Style-lobes irregularly toothed. *C & SW China, ?Burma; naturalised in Japan.* H2. Early summer.

'Alba' has flowers which are white except for a few yellow veins around the ridge and on the claw of the falls.

**71. I. gracilipes** A. Gray. Illustration: Botanical Magazine, 7926 (1903); Cave, The iris, 45 (1959); Kitamura et al., Coloured illustrations of herbaceous plants of Japan (Monocotyledoneae), pl. 20 f. 139 (1978); Mathew, The iris, 73 (1981).
Rhizome with short stolons. Plant 10–15 cm, stems branched. Leaves 5–10 mm wide. Bracts fused at the base, the fused part sheathing the ovary and the base of the perianth-tube. Flowers 1 per branch, 3–4 cm in diameter. Perianth-tube 1–1.5 cm. Falls c. 2.5 cm, lilac to lilac-blue, the centre of the blade with a white area veined with violet, and a white ridge which is yellow at the apex. *China, Japan.* H2. Early summer.

'Alba' has white flowers.

**72. I. japonica** Thunberg (*I. chinensis* Curtis). Illustration: Botanical Magazine, 373 (1797); Kitamura et al., Coloured illustrations of herbaceous plants of Japan (Monocotyledoneae), pl. 20 f. 138 (1978); Mathew, The iris, 70 (1981); Mathew, P.J. Redouté: Lilies and related flowers, 159 (1981).
Rhizome stoloniferous. Plant 45–80 cm, stems branched. Leaves green and glossy, to 45 cm. Flowers 3 or 4 per branch, 4–5 cm in diameter, appearing flattish because of the spreading perianth-lobes, white to pale bluish lavender. Perianth-tube to 2 cm. Falls with a fringed margin, the blade with an orange ridge surrounded by purple blotches. Standards to 3 cm, margin fringed. Style-branches with fringed lobes. *C China, Japan.* H3. Spring.

'Variegata' (Illustration: Mathew, The iris, f. 13, 1981), has leaves striped with cream.

*I. japonica* has been crossed with *I. confusa* and *I. wattii.*

**73. I. confusa** Sealy. Illustration: Mathew, The iris, 70 (1981).
Rhizomes with short stolons. Plant to 1 m, stems branched. Flowers 4–5 cm in diameter, flattish like those of *I. japonica,* white, the blade of the falls with a yellow ridge surrounded by yellow and sometimes purple spots. Falls and standards scalloped-wavy on the margin. Style-branches with fringed lobes. *W China (Yunnan).* H4. Late winter–spring.

**74. I. wattii** Baker. Illustration: Dykes, A handbook of garden irises, pl. 13 (1924); Everard & Morley, Wild flowers of the world, pl. 107 (1970).
Plant 1–2 m, stems branched. Leaves 60–90 × 3.5–7.5 cm, ribbed. Flowers 2 or 3 per branch, 5.5–8 cm in diameter. Falls to 5 cm, the blade deflexed to a vertical position, lilac-blue with a central whitish area spotted with darker lilac and deep yellow, and a central prominent whitish ridge with deep yellow spots, margin of blade crisped. Standards to 5 cm, spreading, lavender-blue, margins wavy. Style-branches usually with fringed lobes. *India (Manipur), W China (Yunnan).* H4. Spring–summer.

There is some doubt as to whether the plants in cultivation as described above are true *I. wattii;* they may be of hybrid origin (see Iris year book 62–6, 1979). Further study of both wild and cultivated plants is necessary to clarify the confusion.

**\*I. milesii** M. Foster. Illustration: Botanical Magazine, 6889 (1886).
Rhizomes fat, greenish. Plant 30–75 cm. Falls to 3 cm, the blade spreading or reflexed, pinkish lilac blotched with purple, central ridge toothed. Standards pinkish lilac. *Himalaya.* H3. Summer.

**Subgenus Nepalensis** (Dykes) Lawrence. Rhizome tiny, consisting of a small growing point to which the plant dies back completely in winter; roots swollen at least in part. Flowers very short-lived, lasting less than a day. Blade of falls with a linear ridge.

**75. I. decora** Wallich (*I. nepalensis* D. Don). Illustration: Sweet, British flower garden ser. 2, **1:** t. 11 (1829); Dykes, The genus Iris, pl. 39 (1913); Journal of the Royal Horticultural Society **94:** f. 100 (1969); Mathew, The iris, f. 24 (1981).
Rhizome covered in bristly fibres. Plant 10–30 cm, stem branched or not. Leaves 2–5 mm wide, longer than flowering stems, strongly ribbed. Flowers usually 2, slightly scented, 4–5 cm in diameter, pale bluish

lavender to deep reddish purple. Perianth-tube 3.5–5 cm. Falls to 3.5 cm, spreading, the blade with an orange-yellow central ridge which becomes white or purple at the apex, claw whitish with purple veins. Standards spreading, the upper half bending downwards. Style-branches with broad lobes which have toothed or crisped margins. *Himalaya from Kashmir to China.* H3. Summer.

**Subgenus Xiphium** (Miller) Spach. Bulbs with papery or leathery tunics; roots not swollen and fleshy. Leaves channelled. Stem unbranched with 1–3 flowers. Falls beardless (except in *I. boissieri* which has a sparsely hairy band on the blade). Standards erect, well-developed (but narrow in *I. serotina*).

**76. I. latifolia** (Miller) Voss (*I. anglica* Steudel; *I. xiphioides* Ehrhart). Illustration: Readers' Digest encyclopaedia of garden plants and flowers, 370 (1971); Perry, Flowers of the world, 143 (1972); Grey-Wilson & Mathew, Bulbs, pl. 29 (1981); Rix & Phillips, The bulb book, 153 (1981).
Plant to 60 cm. Leaves 25–60 cm × 5–8 mm, greyish white on upper surface, not persisting through the winter. Flowers usually 2 in each set of bracts, violet-blue. Perianth-tube c. 5 mm. Falls 6–7.5 cm, blade ovate-oblong, not sharply delimited from the claw and with a central yellow patch, claw widely winged. Standards 4–6 cm, oblanceolate. *NW Spain & Pyrenees.* H3. Summer.

Many cultivars are available with white, blue, violet or purple flowers.

**77. I. xiphium** Linnaeus (*I. hispanica* Steudel). Illustration: Polunin & Huxley, Flowers of the Mediterranean, f. 263 (1965); Hay & Synge, Dictionary of garden plants in colour, f. 785 (1969); Grey-Wilson & Mathew, Bulbs, pl. 29 (1981); Mathew, The iris, 139 (1981).
Plant 40–60 cm. Leaves 20–70 cm × 3–5 mm, greyish green, persisting through the winter. Bracts 6–10 cm, green. Flowers 1 or 2, usually blue or violet, occasionally white, yellow, bronze or bicoloured. Perianth-tube 1–3 mm. Falls 4.5–6.5 cm, blade almost circular with a central orange or yellow blotch, claw longer than blade, unwinged. Standards 4.5–6.5 cm, oblanceolate. *Spain, Portugal, SW France, S Italy, Corsica, Morocco, Algeria & Tunisia.* H3. Spring–early summer.

A variable plant which has been introduced to gardens from all parts of its range. Some of the variants are available under the following names:

'Battandieri' (*I. battandieri* M. Foster). A N African form with white flowers which have an orange ridge on the blades of the falls.

'Lusitanica' (*I. lusitanica* Ker Gawler). Illustration: Botanical Magazine, 679 (1803). Flowers yellow or bronze, found in Portugal.

'Praecox' (var. *praecox* Dykes). An early-flowering variety from the Gibraltar area, with large flowers.

'Taitii' (*I. taitii* M. Foster). Flowers pale blue, appearing in early summer.

*I. xiphium* is widely grown for the cut-flower trade.

**\*I. serotina** Willkomm. Illustration: Botanical Magazine, n.s., 733 (1977); Grey-Wilson & Mathew, Bulbs, pl. 29 (1981); Mathew, The iris, f. 25 (1981). Leaves dying away before flowering time. Flowers 2 or 3, violet-blue. Perianth-tube 5–10 mm. Falls 3.5–4.5 cm, blade with a yellow central line. Standards 7–14 mm, very narrow, almost bristle-like. *SE Spain, ?Morocco.* H3. Late summer.

**78. I. filifolia** Boissier. Illustration: Botanical Magazine, 5929 (1871); Dykes, The genus Iris, pl. 44 (1913); Journal of the Royal Horticultural Society 89: f. 8 (1964); Grey-Wilson & Mathew, Bulbs, pl. 29 (1981).

Plant 25–45 cm. Leaves 0.5–3 mm wide, appearing in autumn and persisting through the winter, the emerging young shoot being enclosed in a purple- and white-blotched sheath. Flowers 1 or 2, reddish-violet. Perianth-tube 1–3 cm. Blade of falls almost circular with a central orange patch. *S Spain, NW Africa.* H3. Summer.

**79. I. juncea** Poiret. Illustration: Botanical Magazine, 5890 (1871); Dykes, A handbook of garden irises, pl. 12 (1924); Grey-Wilson & Mathew, Bulbs, 164 (1981). Bulb covered with tough leaf bases which split into fibrous points at the apex. Plant 30–40 cm. Leaves 0.5–3 mm wide, appearing in autumn and persisting through the winter. Flowers 2, fragrant, bright yellow. Perianth-tube 3.5–5 cm. *N Africa, S Spain, Sicily.* H3. Summer.

The following varieties have been in cultivation and may still be grown: var. **mermieri** Lynch with sulphur-yellow flowers, var. **numidica** Anon. with lemon-yellow flowers and var. **pallida** Lynch with large, soft yellow flowers.

**80. I. tingitana** Boissier & Reuter. Illustration: Botanical Magazine, 6775 (1884); Hay & Synge, Dictionary of garden plants in colour, f. 784 (1969). Plant stout, to 50 cm. Leaves to 45 cm, arching, silvery-green, appearing in autumn and persisting through the winter. Flowers 1–3 in each set of bracts, pale to deep blue. Perianth-tube 1–2.5 cm. Falls to 7.5 cm, blade obovate to almost circular with a central orange-yellow ridge. Standards to 10 cm. *NW Africa.* H3. Late winter–spring.

Var. **fontanesii** (Godron) Maire (*I. fontanesii* Godron) flowers later and is a more slender plant with flowers of deep violet-blue.

There are many garden hybrids between *I. xiphium* and *I. tingitana*, possibly also with influence from *I. latifolia*. These are sometimes grouped under the name I. × **hollandica** Anon. and known as Dutch Iris. The flowers vary in colour from white, yellow and bronze to pale blue or purple (Illustration: Hay & Synge, Dictionary of garden plants in colour, f. 788, 789, 790, 1969; Rix & Phillips, The bulb book, 152, 153, 1981).

**\*I. boissieri** Henriques. Illustration: Botanical Magazine, 7097 (1890); Grey-Wilson & Mathew, Bulbs, pl. 29 (1981). Plant 30–40 cm. Flowers solitary, deep purple, with a yellow, sparsely bearded stripe in the centre of the blade of the falls. *N Portugal, NW Spain.* H3. Summer.

Subgenus **Scorpiris** Spach (Section *Juno* (Roemer & Schultes) Bentham). Bulbs with papery tunics; roots fleshy or thickened. Leaves channelled, held in 1 plane, often grey-green with a white margin. Falls beardless but with a ridge in the centre of the blade, which is often wavy or dissected. Standards much reduced, usually horizontal or deflexed (except in *I. cycloglossa*).

**81. I. planifolia** (Miller) Fiori & Paoletti (*I. alata* Poiret). Illustration: Hay & Synge, Dictionary of garden plants in colour, f. 779 (1969); Grey-Wilson & Mathew, Bulbs, pl. 29 (1981); Mathew, P.J. Redouté: Lilies and related flowers, 181 (1981); Rix & Phillips, The bulb book, 33 (1981). Bulb large with fleshy, but not swollen roots. Plant 10–15 cm. Leaves shiny green, 10–30 × 1–3 cm, concealing the stem. Bracts *c.* 10 cm, almost hidden by the leaves. Flowers 1–3, 6–7 cm in diameter. Perianth-tube 8–18 cm. Falls 5–8 cm, bluish violet, the blade with a yellow central ridge surrounded by violet veining,

claw winged. Standards 2–2.3 cm, horizontal, oblanceolate, usually toothed. Stigma 2-lobed. *S Spain & Portugal, Sardinia, Sicily, Crete, N Africa.* H3. Winter.

There are variants which have white flowers, or flowers of dark blue with the blade of the falls edged with white.

**\*I. palaestina** (Baker) Boissier. Similar but smaller with greenish, whitish or bluish flowers. Standards 1–1.8 cm. Stigma not 2-lobed. *Coastal Israel & Lebanon, ?S Syria.* H4. Winter.

**82. I. aucheri** (Baker) Sealy (*I. fumosa* Boissier; *I. sindjarensis* Boissier). Illustration: Hay & Synge, Dictionary of garden plants in colour, f. 774 (1969); Mathew, Dwarf bulbs, 148 (1973); Mathew, The iris, f. 26 (1981); Rix & Phillips, The bulb book, 59 (1981).

Bulbs with fleshy, not very swollen roots. Plants 15–40 cm. Leaves green, to 25 × 2.5–4.5 cm, concealing the stem at flowering time. Bracts 7–9 cm, long-pointed. Flowers 3–6, usually blue, rarely almost white. Perianth-tube 5–6.5 cm. Falls 4–6 cm, blade with a central yellow ridge and a wavy margin, claw winged. Standards 2–3.5 cm, horizontal or deflexed, obovate. *SE Turkey, N Iraq, N Syria, Jordan, NW Iran.* H4. Late winter–spring.

Usually offered in the trade under the name *I. sindiarensis*, and under this name it has been crossed with *I. persica* to produce the hybrid *I.* × **sindpers** Hoog (Illustration: Bulletin of the Alpine Garden Society 18: 148, 1950; Botanical Magazine, n.s., 419, 1963; Rix & Phillips, The bulb book, 53, 1981). The plants are only *c.* 25 cm and the flowers are blue or blue-mauve with a prominent orange or deep yellow ridge on the blade of the falls which may have a darker apex. Another hybrid is *I.* 'Sindpur' in which the other parent is *I. galatica* (*I. purpurea*); this resembles *I. aucheri* in general habit and size and shape of flower but is a translucent mauve colour.

**\*I. graeberiana** Sealy. Illustration: Botanical Magazine, n.s., 126 (1950). Leaves glossy green above, greyish below, margins whitish. Flowers blue, sometimes tinged with violet. Blade of falls with a white central ridge surrounded by a white area which is strongly veined. *USSR (Soviet Central Asia).* H4. Spring.

**\*I. willmottiana** M. Foster. Illustration: Rix & Phillips, The bulb book, 59 (1981). Flowers lavender or light purple, blotched with white, the blade of the falls marked with deeper purple, and with a white central ridge. Standards *c.* 1.5 cm, deflexed, diamond-shaped to 3-lobed. *USSR (Soviet Central Asia – Pamir Alai)*. H3. Spring.

*I. willmottiana* is probably no longer in cultivation in Britain although it may still be grown in other parts of Europe. Plants grown under the name *I. willmottiana* 'Alba' are probably a form of *I. bucharica*.

\***I. pseudocaucasica** which differs in having leaves grey-green below, yellowish green or blue flowers and bracts 4–5 cm, may key out here (see under 83. *I. caucasica*).

**83. I. caucasica** Hoffmann. Illustration: Rechinger (ed), Flora Iranica: Iridaceae, pl. 17 (1975); Mathew, The iris, f. 30 (1981); Rix & Phillips, The bulb book, 60, 61 (1981).
Bulb with only slightly thickened roots. Plant 10–15 cm. Leaves 5–7, grey-green, 10–12 × to 2 cm, concealing the stem at flowering time, margin white. Bracts 6–7.5 cm. Flowers 1–4, greenish or yellow-green. Perianth-tube 3–4.5 cm. Falls 3–4 cm, the blade with a central, yellow, scalloped ridge, claw winged. Standards 1.3–2.5 cm, horizontal, oblanceolate. *C & NE Turkey, NE Iraq, NW Iran, USSR (Caucasus)*. H3. Late winter–spring.

\***I. baldschuanica** Fedtschenko. Illustration: Mathew, Dwarf bulbs, pl. 52 (1973). Roots of bulb very swollen. Perianth-tube 8–10 cm. Flowers cream, the blade of the falls with a central yellow ridge, the style-branches tinged with pinkish brown. Claw of falls with downturned margins. Standards horizontal to slightly deflexed, lanceolate. *USSR (S Tadjikistan), NW Afghanistan*. H4. Spring.

\***I. doabensis** Mathew. Illustration: Botanical Magazine, n.s., 620 (1972); Rechinger (ed), Flora Iranica: Iridaceae, pl. 23 (1975); Mathew, The iris, f. 29 (1981). Roots of bulb very swollen. Leaves shiny green. Flowers deep yellow. Perianth-tube 7–8 cm. Falls with a deeper yellow wavy ridge on the blade, often surrounded by purple veins, claw with downturned margins. Standards *c*. 8 cm, slightly reflexed. *NE Afghanistan*. H4. Spring.

\***I. fosteriana** Aitchison & Baker. Illustration: Journal of the Royal Horticultural Society **90**: f. 211 (1965); Mathew, Dwarf bulbs, pl. 53 (1973); Rechinger (ed), Flora Iranica: Iridaceae, pl. 19 (1975); Mathew, The iris, f. 26 (1981). Leaves green with white margins. Falls creamy yellow with a white band round the margin of the blade which has a yellow central ridge surrounded by a darker yellow

area veined with brown, claw unwinged. Standards purple, sharply reflexed. *USSR (S Turkmenistan), NE Iran, NW Afghanistan*. H4. Spring.

\***I. linifolia** (Regel) Fedtschenko. Illustration: Journal of the Royal Horticultural Society **93**: f. 5 (1968). Bulb with very swollen roots. Plant 5–10 cm. Leaves 4–7 mm wide, revealing the stem at flowering time. Flowers pale yellow, the blade of the falls deeper with a central whitish or yellowish toothed ridge, claw unwinged. Standards *c*. 1 cm, horizontal to deflexed, more or less 3-lobed. *USSR (Soviet Central Asia – Tien Shan, Pamir Alai)*. H4. Spring.

\***I. pseudocaucasica** Grossheim (*I. caucasica* misapplied). Illustration: Botanical Magazine, n.s., 405 (1962); Rix & Phillips, The bulb book, 33 (1981); Mathew, The iris, f. 32 (1981). Leaves 3–4. Bracts 4–5 cm. Flowers yellowish green or blue, the blade of the falls with a yellow wavy crest, the claw broadly winged, narrowing abruptly to the blade. *SE Turkey, NE Iraq, N & NW Iran, USSR (S Transcaucasia)*. H4. Early summer.

\***I. palaestina** which has shiny green leaves, bracts *c*. 10 cm and greenish, whitish or bluish flowers with a perianth-tube 8–18 cm, may key out here (see under 81. *I. planifolia*).

\***I. drepanophylla** which differs in having tiny bristle-like standards only *c*. 3 mm, may key out here (see under 85. *I. bucharica*).

**84. I. persica** Linnaeus. Illustration: Botanical Magazine, 1 (1787), 8059 (1906); Mathew, Dwarf bulbs, 148 (1973); Mathew, The iris, f. 29 (1981); Rix & Phillips, The bulb book, 58 (1981). Bulb with rather thin roots. Plant to 10 cm. Leaves 3–4, green above, greyish beneath, margin whitish, to 10 cm × 5–15 mm. Outer bract 8–9 cm, rigid, green, inner bract shorter, whitish-papery. Flowers 1–4, translucent, silvery-grey, yellowish, brownish or grey-green. Perianth-tube 6–8 cm. Falls 3.5–4.5 cm, usually with a purplish or brownish blade, with a prominent central yellow ridge often brown- or purple-spotted, claw with upturned wings. Standards 1.5–2.5 cm, horizontal to deflexed, oblanceolate or more or less 3-lobed. Style-branches 3.5–4.5 cm. *S & SE Turkey, N Syria, NE Iraq*. H4. Winter–spring.

Several varieties have been described based on variation in flower colour, but may be no longer in cultivation. Recent

work on the species does not recognise these varieties.

\***I. galatica** Siehe (*I. purpurea* (Anon.) Siehe). Flowers 1 or 2 with a perianth-tube 4–6 cm, wholly reddish purple, or greenish yellow with a purple blade to the falls, or greenish purple with darker falls. Falls 2.5–5 cm, blade with a prominent yellow or orange central ridge, claw winged. Standards 1–2 cm, horizontal to slightly deflexed, spathulate to 3-lobed. Style-branches 2.5–3.5 cm. *Central N Turkey*. H4. Spring.

\***I. nicolai** Vvedensky. Illustration: Botanical Magazine, n.s., 483 (1965); Journal of the Royal Horticultural Society **91**: f. 10 (1966); Mathew, The iris, f. 30 (1981). Bulb with swollen, radish-like roots. Plant 12–15 cm. Leaves scarcely developed at flowering time, the tips incurved, enclosing the stem in a tubular sheath, to 25 × 5–6 cm at fruiting time. Perianth-tube 8–11 cm. Falls whitish or lilac, the blade velvety deep violet at the apex, and with a central orange ridge with 2 deep violet veins on each side, claw margins downturned. Standards pale blue-lilac to whitish, more or less horizontal, obovate with inrolled margins. Style-branches pale blue-lilac to whitish, tinged with dull purple. *USSR (Soviet Central Asia – Pamir Alai), NE Afghanistan*. H4. Spring.

\***I. rosenbachiana** Regel. Illustration: Botanical Magazine, 7135 (1890); Mathew, Dwarf bulbs, 148 (1973). Similar to *I. nicolai* but leaves further developed at flowering time, the tips curving outwards, basic colour of flowers reddish purple. *USSR (Soviet Central Asia – Pamir Alai)*. H4. Spring.

*I. rosenbachiana* and *I. nicolai* are really only colour forms of one species: in Flora Iranica they have been united under the earlier name of *I. rosenbachiana*.

\***I. stenophylla** Baker. Illustration: Botanical Magazine, 7734 (1900), 7793 (1901); Rix & Phillips, The bulb book, 58 (1981). Bulb with fleshy roots. Leaves very short at flowering time, lengthening eventually to 10–20 cm × 5–10 mm. Flowers usually solitary, violet-blue or lilac-blue. Falls usually with a darker blade, with a central yellow ridge surrounded by a violet-spotted whitish area. *S Turkey*. H4. Early spring.

Subsp. **allisonii** Mathew (Illustration: Mathew, The iris, 166, 1981) has more leaves which are 1.5–1.8 cm wide when fully grown, with wavy margins. The flowers are a bluer colour and along the claw of the falls is a line of hairs.

**\*I. microglossa** which differs in having leaves with ciliate margins and pale lavender-blue to nearly white flowers may key out here (see under 87. *I. magnifica*).

**\*I. pseudocaucasica** which differs in having yellowish green or blue flowers with the perianth-tube 3–4 cm may key out here (see under 83. *I. caucasica*).

**85. I. bucharica** M. Foster (*I. orchioides* misapplied). Illustration: Synge, Collins guide to bulbs, pl. XI (1961); Hay & Synge, Dictionary of garden plants in colour, f. 776 (1969); Rechinger (ed), Flora Iranica: Iridaceae, pl. 20, 21 (1975); Rix & Phillips, The bulb book, 60, 62 (1981).
Bulb with rather thin roots. Plant 20–40 cm. Leaves bright glossy green, curved, to 20 × 2.5–3.5 cm, with whitish margins. Bracts 7.5–9 cm, green. Flowers 2–6, golden yellow to almost white. Perianth-tube 4.5–5 cm. Falls *c*. 4 cm, blade yellow with blotches of green, brownish or dull violet on either side of the central yellow, scalloped ridge. Standards 1.5–2 cm, spreading to deflexed. Style-branches *c*. 3.5 cm. *USSR (Tadjikistan), NE Afghanistan*. H3. Spring.

**\*I. aitchisonii** (Baker) Boissier. Illustration: Rechinger (ed), Flora Iranica: Iridaceae, pl. 17 (1975). Leaves grey-green, to 40 cm × 4–8 mm. Bracts 5–6 cm. Flowers 1–3, yellow, violet or bicoloured. Perianth-tube 2.5–3.5 cm. Claw of falls broadly winged. *E Afghanistan, Pakistan*. H3. Early spring–early summer.

**\*I. drepanophylla** Aitchison & Baker. Illustration: Iris year book f. 9 (1968); Journal of the Royal Horticultural Society **93**: f. 206 (1968), **99**: f. 85 (1974); Rix & Phillips, The bulb book, 62 (1981). Roots of bulb very swollen. Bracts 5–6.5 cm, membranous. Flowers 2–8, yellowish or greenish. Perianth-tube 3.5–4 cm. Falls with a yellow ridge on the blade, claw with downturned margins. Standards *c*. 3 mm, very narrow, almost bristle-like, slightly deflexed. *USSR (Turkmeniya), NE Iran, N & NW Afghanistan*. H4. Spring.

**86. I. albomarginata** R.C. Foster (*I. caerulea* Fedtschenko). Illustration: Mathew, The iris, f. 27 (1981).
Plant 20–30 cm. Leaves with white margin; internodes of stem visible. Flowers 2–5, blue. Perianth-tube *c*. 4 cm. Falls to 3.5 cm, blade with a white central ridge surrounded by a yellowish area, claw not broadly winged. Standards 2–2.5 cm, oblanceolate, deflexed. *USSR (Soviet Central Asia – Tien Shan, Alayskiy Khrebet)*. H4. Spring.

**\*I. warleyensis** M. Foster. Illustration: Botanical Magazine, 7956 (1904); Journal of the Royal Horticultural Society 91: f. 159 (1966); Mathew, Dwarf bulbs, pl. 54 (1973). Flowers purplish blue to deep violet, the blade of the falls darker, bordered with white, the central ridge whitish to yellow, toothed or crisped, surrounded by a yellow area. Standards 1–2 cm, narrowly linear to almost 3-lobed. *USSR (Soviet Central Asia – Pamir Alai)*. H3. Spring.

*I. warleyensis* has been crossed with *I. aucheri* (*I. sindjarensis*) to produce *I.* 'Warlsind' in which the falls are yellow, blotched with purple-brown on the blade which has a yellow ridge, and the standards and styles are whitish (Illustration: Gartenpraxis 9: 438, 1978).

**87. I. magnifica** Vvedensky. Illustration: Gartenpraxis 9: 437 (1978); Rix & Phillips, The bulb book, 60, 61 (1981); Greenhouse & Garden 9(5): 26 (1984).
Plant 30–60 cm. Leaves shiny green, 3–5 cm wide; internodes of stem visible. Flowers 3–7, to 7.5 cm in diameter, pale lilac or sometimes white. Perianth-tube 4.5–5 cm, blade with a white undissected ridge surrounded by an area of yellow, claw with broad wings. Standards 2–3 cm, obovate, horizontal to deflexed. *USSR (Soviet Central Asia)*. H3. Spring.

**\*I. cycloglossa** Wendelbo. Illustration: Rechinger (ed), Flora Iranica: Iridaceae, pl. 17 (1975); Botanical Magazine, n.s., 708 (1976); Rix & Phillips, The bulb book, 7 (1981). Flowers 1–3, 8–10 cm in diameter, clove-scented, blue-violet. Perianth-tube 3.5–4.5 cm. Falls 6–7 cm, white towards the centre of the blade and with a yellow blotch but no ridge. Standards 4–4.5 cm, at first erect, becoming horizontal as the flower ages. *NW Afghanistan*. H3. Spring.

**\*I. microglossa** Wendelbo. Illustration: Journal of the Royal Horticultural Society **91**: f. 8 (1966); Mathew, Dwarf bulbs, pl. 55 (1973); Rechinger (ed), Flora Iranica: Iridaceae, pl. 18 (1975); Rix & Phillips, The bulb book, 62 (1981). Leaves grey-green, 1.5–2.5 cm wide, margins white, ciliate. Flowers 1–4, 4.5–5.5 cm in diameter, pale lavender-blue to almost white. Perianth-tube 3–4.5 cm. Falls 4–4.5 cm, central ridge of blade white or pale yellow. Standards 1.5–2 cm, oblanceolate. *NE Afghanistan*. H4. Spring.

**\*I. aitchisonii** which is only 15–30 cm tall, with leaves 4–8 mm wide, and deep yellow, violet or sometimes bicoloured flowers with a perianth-tube 2.5–3.5 cm, may key out here (see under 85. *I. bucharica*).

Subgenus **Hermodactyloides** Spach (Section *Reticulata* Dykes). Bulbs with fibrous, netted tunics, many bulblets often produced. Leaves usually 4-angled in cross-section (nearly cylindric in *I. bakeriana*, channelled in *I. kolpakowskiana*). Stems unbranched, flowers solitary, of typical Iris type (except *I. danfordiae* which has very reduced standards). Capsules more or less stalkless (except in *I. pamphylica*).

**88. I. reticulata** Bieberstein. Illustration: Synge, Collins guide to bulbs, pl. 17 (1961); Hay & Synge, Dictionary of garden plants in colour, f. 780 (1969); Mathew, The iris, f. 34 (1981); Rix & Phillips, The bulb book, 52 (1981).
Plant 7–15 cm. Leaves 4-angled, very variable in their development at flowering time, hardly developed to overtopping the flowers. Bracts green. Flowers solitary, pale blue to deep violet-blue or reddish purple. Perianth-tube 4–7 cm. Falls 4–5 cm, blade usually with a central yellow ridge. Standards 3.5–5 cm. Style-branches 3–4 cm, lobes sometimes toothed. Stigma 2-lobed. *N & S Turkey, NE Iraq, N & W Iran, USSR (Caucasus)*. H2. Early spring.

Many named cultivars of *I. reticulata* are available, differing mainly in the colour of their flowers (Illustration: Synge, Collins guide to bulbs, pl. 17, 1961; Hay & Synge, Dictionary of garden plants in colour, f. 781, 782, 783, 1969; Rix & Phillips, The bulb book, 54, 1981). Hybrids have been produced between *I. reticulata* and *I. bakeriana*; they can be identified by their leaves which have *c*. 6 ribs.

**\*I. kolpakowskiana** Regel. Illustration: Botanical Magazine, 6489 (1880); Mathew, Dwarf bulbs, 148 (1973); Rix & Phillips, The bulb book, 55 (1981). Leaves channelled like those of Section *Scorpiris*, short at flowering time, lengthening later to 25 cm. Flowers pale lilac to pale purple, blade of falls dark reddish purple with an orange-yellow ridge. Stigma entire. *USSR (Soviet Central Asia – Tien Shan)*. H3. Late winter.

**\*I. pamphylica** Hedge. Illustration: Journal of the Royal Horticultural Society **96**: f. 150 (1971); Botanical Magazine, n.s., 648 (1973); Rix & Phillips, The bulb book, 57 (1981); Mathew, The iris, f. 37 (1981).
Plant 10–20 cm. Leaves 4-angled, 17–25 cm at flowering time, lengthening later to up to 55 cm. Falls 3.5–4 cm, blade brownish purple with a central slightly raised yellow area spotted with purple, claw greenish with purple veins. Standards *c*. 4 cm, pale blue, the lower part green spotted with purple. Capsules on a

stalk, pendulous. *S Turkey*. H3. Late winter–spring.

*I. vartanii M. Foster. Illustration: Botanical Magazine, 6942 (1887). Flowers usually greyish lilac to white, smelling of almonds. Blade of falls with darker veins and a pale yellow central ridge. Lobes of style-branches very long and narrow. *Israel, ?Syria*. H3. Autumn–winter.

**89. I. bakeriana** M. Foster. Illustration: Synge, Collins guide to bulbs, pl. 17 (1961); Hay & Synge, Dictionary of garden plants in colour, f. 775 (1969); Gartenpraxis 9: 437 (1978); Rix & Phillips, The bulb book, 56 (1981).
Similar to *I. reticulata* but the leaves are almost cylindric with 8 ribs. Flowers whitish, the falls with a dark violet apex, and violet spotting and veining around the cream-coloured ridge on the blade, and on the claw. Standards and style-branches bluish lilac. *SE Turkey, N Iraq, W Iran*. H2. Late winter–spring.

Very like *I. reticulata*; in the wild intermediate plants have been found. It is probably best regarded as a variety of *I. reticulata* as in Flora Iranica (1975) where it appears as var. *bakeriana* (M. Foster) Mathew & Wendelbo.

**90. I. histrioides** (G.F. Wilson) Arnott. Illustration: Dykes, The genus Iris, pl. 46 f. 1 (1913); Botanical Magazine, 9341 (1934); Mathew, The iris, f. 35 (1981).
Leaves 4-angled, more or less absent at flowering time, lengthening later to up to 50 cm. Bracts papery. Flowers 6–7 cm in diameter, blue. Blade of falls with a yellow ridge surrounded by a blue-spotted pale area, blade obviously distinct from claw. *Central N Turkey*. H2. Early spring.

Var. **sophenensis** (M. Foster) Dykes has deep violet-blue flowers with relatively little spotting, and narrower segments.

Several cultivars of *I. histrioides* are available, differing in the shade of blue or violet-blue. There is confusion surrounding 'Major' which varies in colour according to the nursery selling it.

*I. histrioides* has been hybridised with either *I. danfordiae* or *I. winogradowii* to produce *I. 'Katharine Hodgkin'* which has flowers of a curious grey-yellow-lilac mixture (Illustration: Rix & Phillips, The bulb book, 53, 1981). 'Frank Elder' is a similar hybrid, in which *I. winogradowii* is a definite parent.

**91. I. histrio** Reichenbach. Illustration: Synge, Collins guide to bulbs, pl. 17 (1961); Journal of the Royal Horticultural Society **91**: f. 187 (1966); Mathew, The iris, f. 36 (1981); Rix & Phillips, The bulb book, 52 (1981).
Leaves 4-angled, to 30 cm at flowering time, lengthing later to up to 60 cm. Bracts papery. Flowers 6–8 cm in diameter, pale blue. Blade of falls with a yellow ridge surrounded by a whitish blue-spotted area, blade not obviously distinct from claw. *S Turkey, Syria, Lebanon*. H2. Early spring.

Var. **aintabensis** G.P. Baker. Illustration: Synge, Collins guide to bulbs, pl. 17 (1961); Mathew, Dwarf bulbs, pl. 51 (1975); Rix & Phillips, The bulb book, 52 (1981). A small variant whose leaves at flowering time are less advanced than those of *I. histrio* itself. Flower smaller, paler blue.

**92. I. danfordiae** (Baker) Boissier. Illustration: Synge, Collins guide to bulbs, pl. 17 (1961); Readers' Digest encyclopaedia of garden plants and flowers, 360 (1971); Mathew, The iris, f. 38 (1981); Rix & Phillips, The bulb book, 54 (1981).
Plant 6–7 cm. Leaves 4-angled, varying at flowering time from 1 cm to overtopping flowers. Flowers bright yellow. Perianth-tube 3–6 cm. Falls 2.5–4 cm, blade with a deep yellow or orange central ridge surrounded by sparse green spots. Standards 3–5 mm, very reduced, bristle-like. Style-branches 2–3.5 cm with irregularly toothed lobes. *Turkey*. H2. Early spring.

**93. I. winogradowii** Fomin. Illustration: Synge, Collins guide to bulbs, pl. 17 (1961); Readers' Digest encyclopaedia of garden plants and flowers, 370 (1971); Mathew, Dwarf bulbs, 148 (1973); Rix & Phillips, The bulb book, 53 (1981).
Plant to 20 cm. Leaves 4-angled, elongating to 40 cm. Flowers pale yellow. Perianth-tube *c*. 1.8 cm. Falls *c*. 5 cm, blade with an orange central ridge, surrounded by green spots which extend down the claw. Standards to 5 cm, oblanceolate. *USSR (Abkhazia)*. H2. Early spring.

The following names appear in current commerical catalogues:
**I. fimbriata.** Plants offered under this name may be *I. japonica* or *I. tectorum*.
**I. intermedia.** A name of no botanical validity which is sometimes used for bearded irises of hybrid origin which grow 35–65 cm tall.

### 2. PARDANTHOPSIS Lenz
*V.A. Matthews & B. Mathew*
Related to and resembling *Iris*. Stems much-branched with many small, short-lived flowers. Perianth-segments free, each fall with transverse bands on the claw. Perianth-segments twisting spirally after flowering.

The genus contains only 1 species which, until recently, was included in *Iris*. Easily grown in a sunny position with plenty of moisture during the summer. Propagation is by seed, and seedlings should be raised to replace the parent plants which are not long-lived.

**1. P. dichotoma** (Pallas) Lenz (*Iris dichotoma* Pallas). Illustration: Edwards's Botanical Register, t. 246 (1817); Sweet, British flower garden 1: t. 96 (1825); Botanical Magazine, 6428 (1879).
Rhizome small with a mass of thick roots. Stems 40–100 cm, much-branched. Leaves 6–8, borne in a fan, to 30 × 2.5 cm. Spathes papery, 5- or 6-flowered. Flowers 3.5–4.5 cm in diameter, opening in the afternoon. Falls usually creamy white with purplish brown spots and bands in the centre and lower part, or deep reddish purple with whitish, purple-spotted area in the centre of the blade and with banding on the claw. Standards smaller than falls. Style-branches with 2 narrow lobes. *USSR (Siberia), N China, Mongolia*. H2. Summer.

*P. dichotoma* has been crossed with *Belamcanda chinensis* to produce × **Pardancanda norrisii** Lenz. These plants are variable, but generally resemble *P. dichotoma* in flower shape, with reduced style-branches. The flowers are in shades of salmon and apricot.

### 3. GYNANDRIRIS Parlatore
*D.A. Webb & V.A. Matthews*
Rootstock a corm, surrounded by fibrous tunics. Leaves 1 or 2, linear, channelled. Flowers in terminal and axillary cymes, short-lived, like those of *Iris* or *Moraea* in appearance with petaloid style-branches, and free perianth-segments differentiated into falls (outer) and standards (inner). Stamens partly united, closely adherent to the underside of the style-branches. Ovary extended upwards as a slender, sterile beak, on which the perianth is directly inserted; perianth-tube absent. Capsule with many seeds.

A genus of 9 species, from southern Africa and the Mediterranean area. The Mediterranean species need summer baking if they are to flower and do best in a bulb-frame in areas where the summer is not completely warm and dry. *G. setifolia* should be started into growth in the autumn and dried off the following spring. They grow best in a well-drained soil and can be propagated by seed or corm division. Literature: Goldblatt, P., Systematics of Gynandriris (Iridaceae), a Mediterranean–southern African disjunct, *Botaniska Notiser* 133: 239–60 (1980).

**1. G. sisyrinchium** (Linnaeus) Parlatore (*Iris sisyrinchium* Linnaeus). Illustration: Coste, Flore de la France 3: 366 (1906); Polunin & Huxley, Flowers of the Mediterranean, f. 269 (1965); Taylor, Wild flowers of Spain and Portugal, 23 (1972); Reisigl & Danetsch, Flore Mediterranéenne, 111 (1979).
Stem 10–40 cm. leaves usually 2, with a long, sheathing base; blade flexuous, usually lying on the ground. Stem bearing 1–4 compact cymes, each with 1–6 flowers. Bracts silvery or transparent. Flowers 3–4 cm in diameter, violet-blue or lavender. Falls 2–4 cm, the blade with a white and/or orange signal patch. Standards 2–3 cm, erect. Anthers 4.5–10 mm. Beak of ovary 2.5 cm. *From Portugal through the Mediterranean region and SW Asia to Pakistan.* H4. Spring.

\*G. monophylla Klatt. Illustration: Botaniska Notiser 133: 255 (1980). Stem only to 5 cm and usually only 1 leaf present. Flowers 2–2.5 cm in diameter, pale slate-blue. Falls 1–2 cm, the blade with an orange signal patch surrounded by a white area. Anthers 2–4 mm. *E Mediterranean area.* H4. Spring.

\*G. setifolia (Linnaeus filius) R.C. Foster. Illustration: Botaniska Notiser 133: 249 (1980). Stem 5–20 cm. Leaves 1 or 2. Flowers pale lavender-blue, the blade of the falls with a yellow signal patch. Standards reflexed. Anthers 2.5–4 mm. *South Africa (SW Cape).* H4. Winter.

## 4. HERMODACTYLUS Miller
*D.A. Webb*
Rootstock of 2–4 palmately branched tubers. Leaves in 2 ranks, narrowly linear, 4-angled, longer than the stem. Flowering stem to 40 cm. Flowers radially symmetric, terminal, solitary, partly enclosed by a pale green bract, yellowish green except for the blade of the falls which is very dark purple-brown to almost black. Falls 4–5 cm; standards 2–2.5 cm. Perianth-tube short, funnel-shaped. Ovary with a short, slender sterile beak at the apex.
A genus of 1 species, differing from *Iris* in

its tuberous rootstock and 1-celled ovary with parietal placentation. It can be grown in any sunny, well-drained site, but in regions where the spring weather is likely to damage the flowers it is better under glass.

**1. H. tuberosa** (Linnaeus) Miller (*Iris tuberosus* Linnaeus). Illustration: Polunin, Flowers of Europe. f. 1683 (1969); Huxley & Taylor. Flowers of Greece and the Aegean, f. 411 (1977); Grey-Wilson & Mathew, Bulbs, pl. 30 (1981); Rix & Phillips, The bulb book, 57 (1981).
*E & C Mediterranean region.* H4. Spring.

## 5. MORAEA Miller
*V.A. Matthews*
Rootstock a corm with membranous, fibrous or netted tunics. Leaves 1 to several, usually linear. Stem with reduced bract-like sheathing leaves above. Flowers 1 to several, borne within bracts (spathes) which may be green or dry and papery. Flowers often short-lived. Perianth radially symmetric, segments free. Falls (outer segments) larger than standards (inner segments), reflexed or spreading, each bearing a conspicuous nectar guide. Standards reflexed to erect, entire or 3-lobed. Stamens opposite outer perianth-segments, adpressed to style-branches. Filaments completely joined, to joined at the base only. Style-branches petaloid with 2 lobes (crests). Fruit a capsule containing angled seeds.
A genus of about 100 species found in Africa south of the Sahara. They grow best in a well-drained soil and should be kept dry during cold weather. The species from tropical Africa are not hardy and require protection from frost; those from the SW Cape grow during the winter and flower in spring, hence they also need winter protection. Species from the E Cape are summer-flowering and better for outdoor cultivation in Europe. Propagation is by seed.
Literature: Goldblatt, P., Contributions to the knowledge of Moraea (Iridaceae) in the summer rainfall region of South Africa, *Annals of the Missouri Botanical Garden* 60: 204–59 (1973); Goldblatt, P., The genus Moraea in the winter rainfall region of Southern Africa, *Annals of the Missouri Botanical Garden* 63: 657–786 (1976); Goldblatt, P., Systematics of Moraea (Iridaceae) in tropical Africa, *Annals of the Missouri Botanical Garden* 64: 243–95 (1977).

*Main colour of flowers.* White: 2–4,6,7; yellow: 1,2,4,8; lilac to purple, or brown: 1–4,7; red or orange: 2,5.
*Nectar guide.* Yellow, brownish or bluish, sometimes spotted: 2–4,8; green or dark blue: 5. Edged with a ring of a different colour: 2,6,7.
*Stem.* Hairless: 1–4,6–8; downy to long-hairy: 2,5,7.
*Standards.* 3-lobed: 3–7; not 3-lobed: 1–3,5,8. Reflexed: 1–3; erect or spreading: 2–8.

1a. Leaves absent or 1    2
  b. Leaves 2 or more    7
2a. Main colour of flowers orange to red    **5. neopavonia**
  b. Main colour of flowers white, yellow, lilac, purple or brown    3
3a. Standards 3-lobed    4
  b. Standards not 3-lobed    6
4a. Stem finely downy to shaggily hairy; anthers 6–15 mm    **7. villosa\***
  b. Stem hairless; anthers 3–5 mm    5
5a. Standards 1.5–2 cm    **6. aristata**
  b. Standards to 1.2 cm    **4. tricuspidata\***
6a. Main colour of flowers white or lilac, if yellow then stem with a long lower internode; falls 1.5–4 cm; standards 1.4–3.5 cm; anthers 4–8 mm    **2. fugax\***
  b. Main colour of flowers yellow; falls 3.5–7.5 cm; standards 3–6 cm; anthers 7–15 mm    **8. spathulata\***
7a. Anthers *c*. 1 cm    **1. ramosissima\***
  b. Anthers 2.5–8 mm    8
8a. Filaments joined only at the base, or united for up to half their length    **2. fugax\***
  b. Filaments joined for more than half their length or completely united    **3. bipartita**

**1. M. ramosissima** (Linnaeus filius) Druce (*M. ramosa* (Thunberg) Ker Gawler). Illustration: Botanical Magazine, 771 (1804); Courtenay-Latimer et al., The flowering plants of the Tsitsikama Forest and Coastal National Park, pl. 17 (1967); Annals of the Missouri Botanical Garden 63: 690 (1976).
Leaves many, curved, to 30 × 1.5 cm. Flowering stems 50–120 cm, much-branched, hairless. Flowers yellow, perianth-segments reflexed, falls *c*. 3 cm with a brownish, spotted nectar guide, standards somewhat shorter. Filaments 1.2–1.5 cm, free or joined near the base; anthers *c*. 1 cm. Style-branches 2–2.5 cm. *South Africa (SW Cape).* H4. Spring.

\*M. polystachya (Thunberg) Ker Gawler. Illustration: Marloth, Flora of South Africa

4: pl. 39A (1915); Flowering plants of Africa 35: t. 1385 (1962); Batten & Bokelmann, Wild flowers of the eastern Cape Province, pl. 34 f. 1 (1966). Flowers lilac, falls to 5.5 cm with a yellow or orange nectar guide. Filaments to 1 cm, joined to the apex. Style branches *c.* 1 cm, lobes to 2 cm. *Southern Africa.* H5. Autumn.

**2. M. fugax** (de la Roche) Jacquin (*M. edulis* (Linnaeus filius) Ker Gawler; *M. longifolia* (Schneevoogt) Sweet; *M. odora* Salisbury). Illustration: Botanical Magazine, 613 (1803), 1238 (1809); Rice & Compton, Wild flowers of the Cape of Good Hope, pl. 188 (1950); Mason, Western Cape sandveld flowers, 77 (1972); Annals of the Missouri Botanical Garden 63: 726 (1976).
Leaves usually 1, sometimes 2, trailing, usually much longer than stem. Flowering stems 12–40 cm, branched, with a conspicuous long, lower internode. Flower white, violet or yellow, strongly scented. Falls 2.3–4 cm, the yellow ones with a darker yellow nectar guide, the white or violet ones with the yellow nectar guide surrounded by a bluish colour; claw downy. Standards 2–3.5 cm, erect to slightly reflexed. Filaments 6–10 mm, joined in the lower half; anthers 4–8 mm. Style-branches 1.5–2 cm, lobes 6–16 mm. *South Africa (SW Cape).* H4. Spring.

**\*M. gawleri** Sprengel (*M. sulphurea* Baker; *M. undulata* Ker Gawler). Illustration: Botanical Magazine, 7658 (1899); Annals of the Missouri Botanical Garden 63: 690 (1976). Leaves 2, erect or spreading, somewhat coiled, margin flat or wavy. Flowers cream, yellow or brick red. Falls 1.2–2.8 cm with a dark yellow nectar guide spotted and lined with brown. Standards reflexed. Anthers 2–3 mm. *South Africa (SW Cape).* H4. Spring.

**\*M. vegeta** Linnaeus (*M. tristis* (Linnaeus filius) Ker Gawler). Illustration: Jacquin, Icones Plantarum Rariorum, t. 225 (1795); Botanical Magazine, 577 (1802); Annals of the Missouri Botanical Garden 63: 694 (1976). Leaves several. Stem minutely downy. Flowers dull yellow to brownish, flushed with blue or purple. Falls 2–2.5 cm with a yellow, spotted nectar guide. Standards reflexed. Filaments 5–6 mm, joined in the lower one-third; anthers 3–4 mm. Style-branches 7–8 mm. *South Africa (SW Cape).* H4. Spring.

**\*M. papilionacea** (Linnaeus filius) Ker Gawler. Illustration: Botanical Magazine, 750 (1804); Marloth, Flora of South Africa

4: pl. 39F (1915); Rice & Compton, Wild flowers of the Cape of Good Hope, pl. 187 (1950); Annals of the Missouri Botanical Garden 63: 698 (1976). Leaves 2–4, usually downy on lower surface, margin ciliate. Flowering stem 10–15 cm, downy to long-hairy. Flowers yellow or brick red, scented. Falls *c.* 2.5 cm, the yellow nectar guides outlined with green or red. Standards reflexed. Filaments 3–6 mm, united at the base. Style-branches to 8 mm. *South Africa (SW Cape).* H4. Spring.

**\*M. ciliata** (Linnaeus filius) Ker Gawler. Illustration: Annals of the Missouri Botanical Garden 63: 712 (1976); le Roux & Schelpe, Namaqualand & Clanwilliam – South African wild flower guide No. 1: 47 (1981). Leaves 2–6, slightly downy, margin ciliate, often wavy. Flowers white, yellow, lilac or pale brownish, scented. Falls 2–3.5 cm, with a yellow nectar guide. Standards erect or spreading. Filaments united at the base. Style-branches 7–15 mm. *South Africa (SW Cape).* H4. Spring.

**\*M. tricolor** Andrews (*M. ciliata* (Linnaeus filius) Ker Gawler var. *barbigera* (Salisbury) Baker). Illustration: Andrews, Botanist's repository, t. 83 (1800); Botanical Magazine, 1012 (1807). Leaves usually 3, hairless, margin ciliate, flat or slightly wavy. Flowers red. Falls 2–2.5 cm, with a yellow, bearded nectar guide. Standards erect to spreading. Filaments 3–4 mm, joined at the base; anthers 2.5–3 mm. Style-branches to 5 mm. *South Africa (SW Cape).* H4. Spring.

**\*M. thomsonii** Baker. Illlustration: Botanical Magazine, 7976 (1904); Moriarty, Wild flowers of Malawi, pl. 5 (1975); Annals of the Missouri Botanical Garden 64: 263 (1977); Cribb & Leedal, The mountain flowers of Tanzania, 180 (1982). Leaf 1, trailing, dry or absent at flowering time. Flowers pale lilac. Falls 1.5–2 cm with yellow to orange nectar guides, which are often brown-spotted. Standards erect, becoming spreading. Filaments 4–5 mm, joined in lower two-thirds. Style-branches *c.* 7 mm, lobes minute or to 5 mm. *E Africa from South Africa to Ethiopia.* G1. Summer.

**\*M. stricta** Baker. Illustration: Letty, Wild flowers of the Transvaal, pl. 36 f. 1 (1962); Pearse, Mountain splendour, 75 f. 3 (1978). Leaf 1, usually dry or absent at flowering time. Flowers pale lilac to mauve. Falls *c.* 2 cm, spotted with yellow on the blade. Standards erect. Filaments *c.* 3.5 mm, joined at the base. Style-branches 7–8 mm, lobes *c.* 3 mm. *South Africa (E Cape).* H4. Spring–summer.

**3. M. bipartita** L. Bolus (*M. polyanthos* misapplied). Illustration: Annals of the Missouri Botanical Garden 63: 748 (1976) – as M. polyanthos.
Leaves usually 3, 2–6 mm wide. Flowering stem to 30 cm, branched. Flowers lilac, with reflexed perianth-segments. Falls 2–3 cm with a yellow nectar guide. Standards 1.5–2.5 cm, reflexed. Filaments 4–6 mm, joined in the lower two-thirds to three-quarters; anthers *c.* 5 mm. Style-branches *c.* 5 mm, lobes 5–8 mm. *South Africa (SW Cape).* H4. Late spring–summer.

**\*M. polyanthos** Linnaeus filius (*Homeria lilacina* L. Bolus). Very similar but with filaments completely united and anthers *c.* 6 mm. Style-branches very narrow, without lobes. *South Africa.*

**\*M. carsonii** Baker. Illustration: Annals of the Missouri Botanical Garden 64: 253 (1977); Tredgold & Biegel, Rhodesian wild flowers, pl. 7 (1979). Leaves 2, sometimes 3. Flowers mauve, with spreading perianth-segments. Falls 2–3 cm with yellow, purple-spotted nectar guides. Filaments 7–8 mm, joined in the lower two-thirds. Style-branches *c.* 8 mm, lobes 7–10 mm. *Zimbabwe, Zambia, Malawi & S Zaire.* G1. Spring–summer.

**\*M. fergusoniae** L. Bolus (*M. fimbriata* Klatt). Illustration: Annals of the Missouri Botanical Garden 63: 698 (1976). Leaves 3–6, margin ciliate. Flowers white to lilac. Standards erect, lanceolate or 3-lobed. Filaments completely united or free in the upper quarter. *South Africa (SW Cape).* H4. Spring.

**4. M. tricuspidata** (Linnaeus filius) Lewis (*M. tricuspis* (Thunberg) Ker Gawler; *Iris tricuspis* Thunberg). Illustration: Botanical Magazine, 696 (1803); Annals of the Missouri Botanical Garden 63: 768 (1976). Leaf 1, often trailing, 4–7 mm wide. Flowering stem 25–60 cm, simple or branched. Flowers white. Falls 2.5–3 cm with a yellow, brown or bluish, speckled nectar guide, claw downy. Standards *c.* 1.2 cm, erect, 3-lobed. Filaments 6–7 mm, joined in the lower half; anthers *c.* 5 mm. Style-branches 6–7 mm, lobes 5–7 mm. *South Africa (SW Cape).* H4. Spring–early summer.

**\*M. bellendenii** (Sweet) N.E. Brown (*M. pavonia* (Linnaeus filius) Ker Gawler var. *lutea* (Ker Gawler) Baker). Illustration: Botanical Magazine, 772 (1804); Mason, Western Cape sandveld flowers, 77 (1972); Annals of the Missouri Botanical Garden 63: 673, 768 (1976). Leaf to 1 cm wide. Flowering stem 50–100 cm, branched.

Flowers yellow, the falls with a speckled nectar guide. Filaments 3–5 mm, joined except for the upper 1 mm. *South Africa (SW Cape)*. H4. Spring.

*M. unguiculata Ker Gawler. Illustration: Botanical Magazine, 593 (1802), 1047 (1807); Annals of the Missouri Botanical Garden **63**: 760 (1976). Flowers whitish, sometimes yellow, brown or violet. Falls 1–2.2 cm with yellow, speckled nectar guides. Filaments 4–6 mm, joined almost to the apex; anthers 3–4 mm. Style-lobes 2.5–4 mm. *South Africa (SW Cape)*. H4. Spring–summer.

**5. M. neopavonia** R.C. Foster (*M. pavonia* (Linnaeus filius) Ker Gawler). Illustration: Botanical Magazine, 1247 (1809); Journal of the Royal Horticultural Society **75**: f. 154 (1950) – as M. pavonia var. magnifica; Annals of the Missouri Botanical Garden **63**: 770 (1976).
Leaf 1, basal, sparsely downy on undersurface, exceeding inflorescence. Flowering stems 30–60 cm, simple or with 1 branch, downy. Flowers orange to red. Falls 2.2–4 cm, spreading, nectar guide dark blue or green, often speckled. Standards 1.5–2.5 cm, lanceolate to 3-lobed, erect to spreading. Filaments *c.* 4 mm, joined except for the upper 1 mm; anthers *c.* 1 cm, longer than style-lobes. Style branches *c.* 5 mm, lobes 1–2 mm. *South Africa (SW Cape)*. H4. Spring–summer.

*M. neopavonia* has been hybridised with *M. villosa*.

**6. M. aristata** (de la Roche) Ascherson & Graebner (*M. glaucopsis* (de Candolle) Draplez). Illustration: Botanical Magazine, 168 (1791); Redouté, Les Liliacées, t. 42 (1803); Sweet, The British flower garden ser. 2: t. 249 (1834); Annals of the Missouri Botanical Garden **63**: 776 (1976). Leaf 1, basal, hairless, exceeding inflorescence. Flowering stem 25–35 cm, simple or sometimes with 1 branch, hairless. Flowers white. Falls 3–3.5 cm, spreading, nectar guide composed of concentric crescents of green, blue-violet or black, claw bearded. Standards 1.5–2 cm, 3-lobed, erect. Filaments 3–4 mm, joined for most of their length; anthers *c.* 5 mm. Style-branches *c.* 7 mm, lobes *c.* 7 mm. *South Africa (SW Cape)*. H4. Late spring–summer.

**7. M. villosa** (Ker Gawler) Ker Gawler (*M. pavonia* (Linnaeus filius) Ker Gawler var. *villosa* (Ker Gawler) Baker). Illustration: Botanical Magazine, 571 (1802), n.s., 112

(1950); Journal of the Royal Horticultural Society **75**: f. 154 (1950); Morley, Wild flowers of the world, pl. 87 (1970); Annals of the Missouri Botanical Garden **63**: 776 (1976).
Leaf 1, basal, downy on the undersurface, exceeding inflorescence. Flowering stem 20–40 cm, simple or branched, hairy. Flowers usually purple. Falls 3–4 cm, spreading, nectar guide yellow, surrounded by a dark contrasting colour, often blue or green, claw bearded. Standards 1.6–3 cm, spreading, 3-lobed. Filaments *c.* 5 mm, joined for most of their length; anthers 6–10 mm. Style-branches 5–7 mm, lobes 5–8 mm. *South Africa (SW Cape)*. H4. Spring.

*M. gigandra L. Bolus. Illustration: Gardeners' Chronicle **119**: f. 95, 99 (1946); Botanical Magazine, n.s., 188 (1952); Annals of the Missouri Botanical Garden **63**: 770 (1976); Veld & Flora **68**: 88 (1982). Flowers creamy-white or purple; creamy-white falls with an orange nectar guide spotted with green and surrounded by a band of blue; purple falls with a black basal blotch surrounded by a band of blue; claw indistinct, hairless. Standards creamy-yellow, greenish at the base with a purplish blue band. Anthers 1.3–1.5 cm, much longer than style-lobes. *South Africa (SW Cape)*. H4. Spring.

**8. M. spathulata** (Linnaeus filius) Klatt (*M. spathacea* (Thunberg) Ker Gawler). Illustration: Batten & Bokelman, Wild flowers of the eastern Cape Province, pl. 28 f. 1 (1966); Annals of the Missouri Botanical Garden **64**: 267 (1977); Rix & Phillips, The bulb book, 152 (1981). Leaf 1, basal, hairless, exceeding inflorescence. Flowering stem to 1 m, usually simple. Flowers pale to deep yellow, scented. Falls 3.5–7.5 cm, reflexed, nectar guides orange yellow. Standards 3–6 cm, erect. Filaments 8–15 mm, joined in the lower two-thirds; anthers 7–15 mm, purple. Style-branches 1.2–1.8 cm, lobes to 1 cm. *Southern Africa*. H4. Spring–summer.

*M. huttonii (Baker) Obermeyer. Illustration: Botanical Magazine, 6174 (1875); Batten & Bokelman, Wild flowers of the eastern Cape Province, pl. 34 f. 5 (1966); Flowering Plants of Africa, 1581 (1970). Nectar guide of falls edged with dark brown lines. Style-lobes 1–1.3 cm with a brown or purple basal blotch. *South Africa (E Cape)*, *Lesotho*. H4. Summer.

*M. moggii N.E. Brown. Illustration: Letty, Wild flowers of the Transvaal, pl. 36 f. 2 (1962); Botanical Magazine, n.s., 469

(1965). Nectar guide of falls with purple or brownish red lines of spots. Filaments joined in the lower half; anthers cream. Style-lobes 1–2 cm. *South Africa (E Cape)*. H4. Summer.

## 6. DIETES Klatt
*V.A. Matthews*

Rhizome thick, creeping. Leaves several, in 2 ranks, leathery, linear to sword-shaped. Stem erect, hairless, branched above, bearing leaves at the lower nodes and bracts at the upper nodes. Flowers short-lived, in groups, each group subtended by a spathe-bract. Perianth radially symmetric, segments free, the outer larger than the inner, with an ascending claw and spreading blade and with a nectar guide at the base of the blade. Filaments usually free. Style short, divided into 3 large, flattened, petaloid branches which divide at the tips into paired style crests; stigma-lobes borne on the lower surface. Fruit a 3-celled capsule with many seeds.

A genus of 6 species from E, C and SE tropical Africa and coastal Southern Africa, with 1 outlying species on Lord Howe Island. Cultivation as for *Moraea* (p. 355). Literature: Goldblatt, P., Systematics, physiology and evolution of Dietes (Iridaceae), *Annals of the Missouri Botanical Garden* **68**: 132–53 (1981).

1a. Flowers yellow; leaves with a distinct, usually double central vein  **2. bicolor**
  b. Flowers white; leaves lacking a distinct central vein  2
2a. Leaves 3–5 cm at widest point; claws of inner perianth-segments unmarked  **1. robinsoniana**
  b. Leaves up to 2 cm at widest point; claws of inner perianth-segments marked with dark brown towards the base or dotted with orange, if unmarked then outer perianth-segments 3.5 cm or less  3
3a. Outer perianth-segments 2.5–3.5 cm; stamens 8–15 mm; style-branches 7–9 mm; flowers lasting 1 day  **3. iridioides**
  b. Outer perianth-segments 4.5–6 cm; stamens 1.7–2.3 cm; style-branches 1.2–2 cm; flowers lasting 3 days  **4. grandiflora**

**1. D. robinsoniana** (Mueller) Klatt (*Moraea robinsoniana* (Mueller) Bentham & Mueller). Illustration: Botanical Magazine, 7212 (1892); Annals of the Missouri Botanical Garden **68**: 142 (1981).
Stem 1–1.5 m. Leaves 3–5 cm at widest

point. Flowers white, Outer perianth-segments *c.* 4.5 cm, claw *c.* 1.5 cm, blade *c.* 3 cm wide, with an orange nectar guide. Filaments to *c.* 1 cm; anthers 5–6 mm. Style *c.* 4 mm, branches *c.* 1 cm. *Australia (Lord Howe Island).* H5. Spring–summer.

**2. D. bicolor** (Steudel) Klatt (*Moraea bilcolor* Steudel). Illustration: Flore des Serres ser. 1, **7**: t. 744 (1852); Batten & Bokelmann, Wild flowers of the eastern Cape Province, pl. 34 f. 6 (1966); Flowering Plants of Africa **39**: t. 1525 (1968); Annals of the Missouri Botanical Garden **68**: 144 (1981). Stem 80–120 cm. Leaves 6–12 mm at widest point. Flowers yellow. Outer perianth-segments *c.* 3.5 cm, claw *c.* 1.2 cm, bearded and spotted with orange, blade to 2.3 cm wide, usually with a dark brown nectar guide. Filaments *c.* 6 mm; anthers 4–8 mm. Style *c.* 2 mm, branches 8–10 mm. *South Africa (E Cape).* H4. Spring–summer.

**3. D. iridioides** (Linnaeus) Klatt (*Moraea iridioides* Linnaeus; *M. catenulata* Lindley; *Dietes vegeta* misapplied). Illustration: Edwards's Botanical Register, t. 1074 (1827); Batten & Bokelmann, Wild flowers of the eastern Cape Province, pl. 31 f. 2 (1966); Flowering Plants of Africa **39**: t. 1524 (1968); Annals of the Missouri Botanical Garden **68**: 146 (1981). Stem 15–60 cm. Leaves 6–20 mm at widest point, spreading. Flowers white. Outer perianth-segments 2.5–3.5 cm, claw *c.* 1.5 cm, hairy in centre and papillate, blade 1.2–1.6 cm wide, with a yellow nectar guide, claw often with orange dots. Filaments 5–9 mm; anthers 3–6 mm. Style 2–3 mm, branches 7–9 mm, blue or white, flushed with blue-violet. *Southern South Africa, north through eastern Africa to Kenya.* H5. Spring–summer.

'Johnsonii' has longer, erect leaves and larger flowers.

**4. D. grandiflora** N.E. Brown. Illustration: Batten & Bokelmann, Wild flowers of the eastern Cape Province, pl. 34 f. 3 (1966); Jeppe, Natal wild flowers, pl. 18f (1975); Annals of the Missouri Botanical Garden **68**: 151 (1981). Stem 1–1.5 m. Leaves 1–2 cm at widest point. Flowers white. Outer perianth-segments 4.5–6 cm, claw 2–2.5 cm with a yellow beard, blade 2.5–3.5 cm wide, with a yellow nectar guide; claws of inner perianth-segments marked with dark brown towards the base. Filaments 1–1.3 cm, anthers 7–10 mm. Style *c.* 5 mm, branches 1.2–2 cm, flushed with

mauve. *South Africa (E Cape through Transkei to Natal).* H5. Spring–summer.

**7. CYPELLA** Herbert
*V.A. Matthews*
Rootstock a long bulb with pleated tunics. Leaves pleated. Flowers short-lived, with the 3 outer perianth-segments large, spreading, and the 3 inner much smaller, usually erect and often hairy on the claw; segments free. Filaments free. Style longer than stamens, the branches sometimes petaloid. Fruit a capsule containing angular seeds.

A genus of about 15 species occurring from Mexico to Argentina. They are not generally hardy and so the bulbs should be lifted and stored in sand in a frost-free place during the winter. Propagation is by seed or offsets.

1a. Flowers yellow      **1. herbertii***
  b. Flowers blue or violet-blue

                        **2. plumbea***

**1. C. herbertii** (Lindley) Herbert. Illustration: Botanical Magazine, 2599 (1825); Sweet, British Flower Garden ser. 2, **4**: t. 33 (1830); Flore des Serres ser. 1, **5**: t. 537 (1849); Rix & Phillips, The bulb book, 173 (1981). Stem 30–50 cm. Leaves 15–30 × *c.* 2 cm, linear. Inflorescence usually much-branched. Flowers 4–6 cm across. Outer perianth-segments dull orange-yellow with a violet central stripe, claw concave; inner perianth-segments much smaller, yellow spotted with purple, rolled towards the tip. Stigma purple. *Brazil, Uruguay, Argentina; ?Paraguay.* H5–G1. Summer.

**\*C. peruviana** Baker. Illustration: Botanical Magazine, 6213 (1876). Differs in the inflorescence usually being unbranched, except in vigorous specimens. Flowers bright yellow, the lower part of each segment forming a cup banded with purple or brownish spots; inner segments very hairy on the central part. Stigma yellow, petaloid. *Peru, Bolivia.* G1. Winter.

**2. C. coelestis** (Lehmann) Diels (*C. plumbea* Lindley). Illustration: Botanical Magazine, 3710 (1839); Garden **60**: 10 (1901). Stem 40–75 cm. Leaves basal, to 45 × 2.5 cm. Inflorescence unbranched. Flowers 5–8 cm across, dull blue with brownish shading and yellow-brown spots in the cup formed by the bases of the segments; inner segments with a yellow patch, and with hairs on the claw. Style bluish lilac. *Brazil, Uruguay, Argentina; ?Paraguay.* G1. Late summer.

**\*C. herrerae** R.C. Foster. Differs in having narrower basal leaves (7–8 mm) and short stem leaves bearing bulbils in the axils. Flowers deep blue to violet-blue, the central cup whitish with red-brown spots. *Peru.* G1. Winter.

**8. TRIMEZIA** Herbert
*V.A. Matthews*
Rhizome bulb-like, covered with the fibrous remains of old leaf bases. Leaves linear to linear-lanceolate. Perianth radially symmetric, segments free, in 2 dissimilar whorls, shortly clawed at the base. Stamens borne opposite the style-branches; filaments free. Fruit a dry capsule on a long stalk. Seeds black, angled.

A genus of 5 species from central and tropical South America and the West Indies. The plants grow best in a well-drained compost and need winter protection. Propagation is by seed.

**1. T. martinicensis** (Jacquin) Herbert. Illustration: Botanical Magazine, 416 (1797); Fournet, Flore illustrée des phanerogames de Guadeloupe et de Martinique, 246 (1978). Stem to 20 cm, rarely more in cultivation. Leaves to 30 cm, linear, in a basal fan. Flowers short-lived, produced in succession from overlapping bracts, 2–2.5 cm across. Outer perianth-segments yellow with brownish or purplish spots at the base, to 2.5 cm, erect or incurved; inner much smaller, incurved, yellow. *S. America, West Indies; naturalised elsewhere in the Tropics.* H5. Mainly spring, spasmodically throughout the year.

**9. NEOMARICA** Sprague
*V.A. Matthews*
Rhizomes short, creeping. Leaves borne in basal fan, sword-shaped, with many veins. Flowering stems erect with a large leaf-like bract. Flowers short-lived, borne in clusters, the stalks flat and winged. Perianth radially symmetric, segments free, in 2 dissimilar whorls, clawed at the base, outer spreading, inner smaller, usually erect, abruptly reflexed at the tip. Filaments free. Style-branches 2-fid or 3-fid, opposite the stamens. Fruit a capsule.

A genus of about 15 species from tropical America and West Africa. The illegitimate name *Marica* (Ker Gawler) Herbert was used in the past for this genus and still occasionally occurs in the literature. Ravenna has recently indicated (in Prance, G.T. & Elias, T.S. (eds), Extinction is forever, 257 1977) that *Neomarica* and *Trimezia* overlap to such an extent that they should not be maintained as separate genera; however as his final work has not yet been published, these genera are retained in this Flora. The species are tender and require protection from frost. They thrive in any well-drained soil and should be kept fairly dry throughout the winter. Propagation is by seed or division of the rhizomes.

1a. Outer perianth-segments yellow
    **4. brachypus***
  b. Outer perianth-segments white, blue or blue-lilac   **2**
2a. Leaves ribbed; outer perianth-segments white; style-branches 3-fid   **3**
  b. Leaves not obviously ribbed; outer perianth-segments blue or blue-lilac; style-branches 2-fid   **3. caerulea**
3a. Plant to 90 cm; flowers 6–10 cm across   **1. northiana**
  b. Plants to 60 cm; flowers to 5 cm across   **2. gracilis**

**1. N. northiana** (Schneevoogt) Sprague. Illustration: Botanical Magazine, 654 (1803); Everett, Encyclopedia of horticulture **7**: 2302 (1981). Stems to 90 cm. Leaves ribbed, to 60 × 5 cm. Flowers 6–10 cm across, fragrant. Outer perianth-segments white, inner segments violet towards the apex, both yellowish marked with red at the base. Style-branches 3-fid, the 2 lateral teeth erect, the central 1 often reflexed. *Brazil*. G1. Spring–summer.

This species may be viviparous, producing young plants on the old flowering stems.

**2. N. gracilis** (Hooker) Sprague. Illustration: Botanical Magazine, 3713 (1839). Similar to *N. northiana* but shorter (to 60 cm) and with flowers less than 6 cm across. The style-branches have 3 erect teeth. *S Mexico to N Brazil*. G2. Summer.

**3. N. caerulea** (Ker Gawler) Sprague. Illustration: Edwards's Botanical Register, t. 713 (1823); Hooker, Exotic Flora, f. 222 (1827); Botanical Magazine, 5612 (1866). Stems to 60 cm. Leaves not obviously ribbed, 90–180 × *c.* 3.5 cm. Flowers to 10 cm across. Outer perianth-segments blue or blue-lilac, inner deeper blue, both barred with yellow and brown at the base. Style-branches 2-fid, each with a horn-like appendage at the base. *Brazil*. G2. Summer.

**4. N. brachypus** (J.G. Baker) Sprague. Illustration: Botanical Magazine, 6380 (1878).

Leaves 40–50 × 2.5–4 cm. Flowers 7–8 cm across. Perianth-segments yellow, barred with reddish brown at the base. Style-branches 3-fid. *West Indies*. G1. Summer.

***N. longifolia** (Link & Otto) Sprague. Illustration: Everett, Encyclopedia of horticulture **7**: 2302 (1981). Leaves *c.* 30 cm; flowers to 5 cm across. *Brazil*. G1. Summer.

### 10. TIGRIDIA Jussieu
*J. Cullen*
Perennial herbs with bulbs. Leaves mostly basal, equitant, in a fan. Spathes 2, enclosing 2–several flowers. Perianth of 6 free segments, the outer larger than the inner, with their bases overlapping to form a cup, the blades spreading or reflexed. Inner segments spreading, mostly ovate, the margins often upcurved near the middle so that the segment appears fiddle-shaped, with a nectary on the inner surface. Stamens 3, filaments united into a tube around the style. Stigmas 3, each bifid, on radii alternating with the anthers. Fruit a capsule.

A genus of 23 species from Mexico and Guatemala. In cool areas the bulbs should be planted in rich, well-drained soil in spring. In autumn, after flowering, the plants should be lifted and dried off, when the bulbs can be stored in a cool, frost-free place. In warmer areas the bulbs may be left in the ground. Propagation is by bulbs or by seed.
Literature: Molseed, E., The Genus Tigridia (Iridaceae) of Mexico and Central America, *University of California Publications, in Botany* **54** (1970); Guglielmi, L., Tigridia, l'occloxchitl degli Aztechi, *Il Giardino Fiorito* **49**: 380–1 (1983).

1a. Flower 10–15 cm in diameter; staminal tube 4–7 cm   **1. pavonia**
  b. Flower at most 6 cm in diameter; staminal tube 3–5 mm   **2. violacea**

**1. T. pavonia** (Linnaeus filius) de Candolle (*T. pringlei* Watson). Illustration: Botanical Magazine, 532 (1801), n.s., 544 (1969). Bulb 3–5 × 1.5–4 cm. Stem to 1.5 m. Basal leaves 20–50 × 1.5–6 cm, lanceolate. Stem-leaves 1–several. Flowers 10–15 cm in diameter, pink, red, orange, yellow or white, often spotted or blotched with darker colours. Outer perianth-segments 7–10 cm, the blade spreading or reflexed. Inner segments much shorter, the blade spreading or reflexed. Staminal tube 4–7 cm, projecting conspicuously beyond the cup of the flower; anthers erect and incurving. Capsule 3–7 cm. *Mexico*. H5–G1. Summer.

This species originated in Mexico and has been cultivated there for a long time and is now naturalised in much of C & S America. It is very variable in flower colour and size, and many variants have been named and described in the past (some as 'varieties', others as species, e.g. *T. conchiflora* Sweet, *T. grandiflora* Salisbury, etc.).

**2. T. violacea** Schlechtendahl, Otto & Dietrich. Illustration: Botanical Magazine, 7536 (1894).
Bulb 3–4 × 1.5–2 cm. Stems to 30 cm. Basal leaves short at flowering time, later to 45 cm, linear. Stem-leaf 1. Flowers 3–5 cm in diameter, purple to violet, the cup paler and spotted, its margins yellow. Outer perianth-segments with oblong, obtuse but shortly pointed blades, spreading. Inner segments ovate, markedly pouched in the middle. Staminal tube 3–5 mm, not projecting from the cup; anthers spreading. Capsule 1.3–2 cm. *Mexico*. H5–G1. Summer.

### 11. RIGIDELLA Lindley
*J. Cullen*
Perennial herbs with bulbs. Leaves pleated, mostly basal. Flowers in cymes, mostly enclosed by 2 spathes. Flowers pendent or erect, fruits stiffly erect. Perianth of 6 free segments, the 3 outer large, overlapping and forming a cup at the base, the blades reflexed, all coiling spirally as the flower dies; the 3 inner segments smaller or much smaller, adpressed to the staminal column, each with a gland (nectary?) on the inner surface. Stamens 3, filaments united into a tube which projects conspicuously beyond the cup formed by the outer perianth-segments. Style enclosed within the staminal tube; stigmas 3, on the same radii as the anthers, each deeply bifid. Fruit a papery capsule.

A genus of 4 species from Mexico and Guatemala. They may be grown in pots, or, in favourable areas, in an open border. They require a rich, well-drained soil and much sunshine. Staking is generally necessary. The plants can be treated as half-hardy, the bulbs being lifted and dried off after growth ceases. The flowers each last only one day, the outer perianth-segments coiling spirally at the end of this, but are produced in succession over a fairly long period.
Literature: Cruden, R.W., The systematics of Rigidella, *Brittonia* **23**: 217–25 (1971).

1a. Flowers erect; inner perianth-segments 2.2–3 cm   **3. orthantha**
  b. Flowers pendent though fruits erect; inner perianth-segments 9–12 mm   **2**

2a. Flowers with dark spots or lines on the rim of the cup formed by the outer perianth-segments; basal leaves developed only after flowering

**2. flammea**

b. Flowers without dark spots or lines; basal leaves present at flowering

**1. immaculata**

**1. R. immaculata** Herbert. Illustration: Edwards's Botanical Register **27**: t. 68 (1841); Flore des Serres, ser. 1, **5**: t. 502 (1840), ser. 2, **11**: t. 2215 (1875).
Bulb 4–5.5 × 2–3 cm. Stem to 1 m but often less. Basal leaves present at flowering, 60–90 × 2.5–5.5 cm. Stem-leaves usually 2. Flowers pendent. Outer perianth-segments 4.5–5.5 cm, the cup 1.3–1.5 cm, all scarlet. Inner segments 9–12 mm, flat, yellow with red tips. Style-branches 5–6 mm, without small projections between them. Capsule to 4 cm. *Mexico, Guatemala.* H5–G1. Spring–early Summer.

**2. R. flammea** Lindley. Illustration: Edwards's Botanical Register **26**: t. 16 (1840).
Bulb to 5 × 2.5 cm. Stems to 1 m. Basal leaves developing after flowering, ultimately to 1 m or more. Stem-leaves usually few and reduced. Flowers pendent. Outer perianth-segments 5–6.5 cm, the cup 1.8–2 cm, scarlet and marked with dark spots or lines around the rim of the cup. Inner perianth-segments 9–11 mm, yellow with red tips, margins inrolled over the gland. Style-branches 3–4 mm, with short projections between each. Capsule 1.5–3.5 cm. *Mexico.* H5–G1. Spring–early summer.

**3. R. orthantha** Lemaire. Illustration: Flore des Serres, ser. 1, **1**: t. 107 (1845); Botanical Magazine, n.s., 667 (1974).
Bulb to 10 × 5 cm. Stems to 1 m. Basal leaves to 100 × 8.5 cm. Stem-leaves 1–2. Flowers erect. Outer perianth-segments 5–6 cm, scarlet, cup 1–1.2 cm. Inner segments 2.2–3 cm, flat, scarlet. Style-branches 5–7.5 cm, without projections between them. Capsule to 4 cm. *Mexico, Guatemala.* H5–G1. Spring–early summer.

This species has been hybridised with *Tigridia pavonia* (p. 359); the hybrid, which is apparently without a name, is sometimes grown.

**12. HERBERTIA** Sweet
*J. Cullen*
Perennial herbs with corms. Stems erect, simple or with few branches, the stem or the individual branches ending in many 1-flowered inflorescences. Leaves mostly basal, linear-lanceolate, tapered towards both ends, pleated. Spathes 2, completely hiding the flower-stalk and ovary, the inner longer than the outer. Perianth of 6 free segments, the 3 outer much larger than the 3 inner, all spreading or reflexed. Stamens 3, filaments united into a tube, the anthers spreading, connective narrow, linear. Stigmas 3, on the same radii as the stamens, each with a notched apex beyond which the anther projects. Fruit a many-seeded capsule.

A genus of perhaps 7 species, mostly from temperate S America. Its classification is confused, and information about the species in the wild is scattered. It has been much confused with *Alophia*, and, at one time, bore the name *Trifurcia* Herbert (see Goldblatt, P., Brittonia **27**: 373–85, 1976; Annals of the Missouri Botanical Garden **64**: 378–9, 1978). The flowers are very short-lived, lasting individually for only one day, but they are produced successively over a period of some weeks. The plants are uncommon in cultivation, and, in most areas, require cool greenhouse treatment, though the corms may be planted out in a sunny border in spring and lifted and dried off when the leaves begin to wither. In general, whether in pots or in a border, a well-drained soil or compost is necessary. Propagation is by seed.

1a. Stem 30–50 cm; leaves 18–20 cm × 5–8 mm    **1. amatorum**
  b. Stems 6–15 cm; leaves 4–10 cm × 3–5 mm    2
2a. Outer perianth-segments *c.* 2.8 × 1.2 cm, inner perianth-segments *c.* 8 × 2 mm    **3. pulchella**
  b. Outer perianth-segments *c.* 1.5 cm × 6–7 mm, inner perianth-segments *c.* 5.5 × 2 mm    **2. lahue**

**1. H. amatorum** Wright (*Trifurcia amatorum* (Wright) Goldblatt). Illustration: Botanical Magazine, 8175 (1908).
Corm spherical, brown. Stems 30–50 cm, little branched. Leaves linear-lanceolate, 18–20 cm × 5–8 mm. Outer perianth-segments reflexed, obovate, *c.* 3.5 × 2–2.5 cm, purple with a white median stripe and a hairy, 2-lobed, yellow nectary at the base. Inner segments purple, spreading, *c.* 1.5 cm × 5 mm, the claw hairy. Anthers yellow, stigmas pale purple. *Southern S America (Uruguay only ?).* H5–G1. Spring.

**2. H. lahue** (Molina) Goldblatt (*Alophia lahue* (Molina) Espinosa; *Trifurcia lahue* (Molina) Goldblatt). Illustration: Correa, Flora Patagonica **2**: 170 (1969).
Corm ovoid to spherical, brown. Stem 6–15 cm, simple or with 1 branch. Leaves linear-lanceolate, often sickle-shaped, 4–8 cm × 3–4 mm. Outer perianth-segments spreading, oblanceolate, violet marked with dark blue towards the base, *c.* 1.5 cm × 6–7 mm. Inner perianth-segments *c.* 5.5 × 2 mm, oblanceolate, violet-purple, glandular towards the base within. Anthers yellow, stigmas pale purple. *S Chile, S Argentina.* H5–G1. Spring.

The description and distribution given above are of the southern American plant, which is subsp. *lahue*. Subsp. **caerulea** (Herbert) Goldblatt (*Alophia caerulea* Herbert) occurs in the southern USA (Texas and Louisiana) and is taller, with longer leaves; it has been in cultivation.

**3. H. pulchella** Sweet (*Trifurcia pulchella* (Sweet) Goldblatt). Illustration: Sweet, British Flower Garden **2**: t. 222 (1840); Botanical Magazine, 3862 (1841).
Corm ovoid, brown, Stems erect, 7.5–15 cm, not branched or with 1 branch. Leaves linear-lanceolate, *c.* 10 cm × 5 mm. Outer perianth-segments spreading or reflexed, blue or pinkish violet, sometimes with a median white stripe, dark violet or spotted with purplish blue at the base, *c.* 2.8 × 1.2 cm. Inner perianth-segments spreading, pale purple. Anthers yellow, stigmas reddish. *Southern S America as far north as S Brazil.* H5–G1. Spring.

**13. ALOPHIA** Herbert
*J. Cullen*
Perennial herbs with corms. Stems branched and bearing several leaves. Leaves linear-lanceolate, pleated. Inflorescences terminal, with few flowers, subtended by 2 spathes, the inner longer than the outer. Perianth of 6 free segments, the outer 3 a little longer than the inner 3, all ovate, each with a band of hairs down the centre. Stamens 3, filaments usually partly united, anthers spreading, each with a broad, fiddle-shaped connective. Stigmas 3, on the same radii as the stamens, each deeply divided into 2 ascending, thread-like branches. Fruit a many-seeded capsule.

A genus of a few species extending from the southern part of the USA through C America to S America. Only 1 species is cultivated, requiring conditions like those for *Herbertia*.

**1. A. drummondii** (Graham) Foster (*A. drummondiana* Herbert; *Gelasine punctata* Herbert). Illustration: Botanical Magazine, 3779 (1840).
Corm ovoid, Stems 15–45 cm. Leaves

linear-lanceolate, tapered towards both ends, to 30 × 1.5 cm. Perianth-segments purple or purple-blue, spotted with brown near the centre, the outer 3, 2–2.5 cm, the inner 3, c. 1.5 cm with margins inrolled over the band of hairs. *S USA (Arkansas, Louisiana, Texas, Oklahoma), Mexico.* H5–G1. Early Summer.

## 14. FERRARIA Burman
*D.A. Webb*
Rootstock a corm. Leaves in 2 ranks. Flowers in small cymes with numerous bracts. Perianth radially symmetric, tube short, lobes all similar. Stamens united for most of their length to form a tube around the style. Style-branches enlarged, fringed, somewhat petaloid, bifid. Capsule linear; seeds numerous.

A genus of about 10 species, from southern and tropical Africa.
Literature: de Vos, M.P., The African genus Ferraria, *Journal of South African Botany* **45**: 295–376 (1979).

**1. F. crispa** Burman (*F. undulata* Linnaeus). Illustration: Botanical Magazine, 144 (1791); Rice & Compton, Wild flowers of the Cape of Good Hope, t. 191 (1950); Flowering plants of Africa 34: t. 1316 (1959).
Stem 20–50 cm. Lower leaves 15–30 cm, including the long sheathing base, the upper ones shorter; blade more or less erect, sword-shaped to sickle-shaped. Cymes numerous, mostly 2-flowered, in the axils of leaf-like bracts. Bracts c. 4 × 3 cm, boat-shaped, leaf-like. Perianth-lobes c. 2 cm, consisting of an oblong, erect claw and a slightly longer, spreading or slightly deflexed, strongly wavy blade, the whole dark brown with pale lines and spots, or pale yellow with brown lines and spots. Style-branches fringed with numerous long, hair-like processes. Flowers very short-lived, somewhat foetid. *South Africa (Cape Province).* H5–G1. Spring.

Hardy in favoured spots in S Europe, if the bulbs are planted deep, but usually grown under glass.

## 15. HOMERIA Ventenat
*V.A. Matthews*
Corm covered with netted tunics. Leaves solitary or several, linear. Stem erect, hairless, usually branched. Inflorescence with several flowers, each flower subtended by 2 leafy bracts. Perianth radially symmetric, segments free, almost equal or the inner ones slightly smaller, usually with a basal claw, the blade with a basal

nectary which may extend down the claw. Stamens opposite the outer perianth-segments, filaments united. Style-branches 3, entire, flattened, with or without crests. Fruit a cylindric capsule, with a flat top or beaked.

A genus of 31 species from southern Africa, the cultivated ones coming from Cape Province in South Africa. Cultivation is as for *Ixia* (p. 380).
Literature: Goldblatt, P., Systematics and biology of Homeria (Iridaceae), *Annals of the Missouri Botanical Garden* **68**: 413–503 (1981).

1a. Leaves 2 or more    **1. ochroleuca**\*
  b. Leaves solitary    **2**
2a. Plant 15–35 cm; filaments 5–6 mm, downy at least at the base   **3. collina**\*
  b. Plant 30–75 cm; filaments 7–10 mm, hairless, or if sparsely downy in lower half then outer perianth-segments 3.5–4 cm    **3**
3a. Each perianth-segment with a diffuse nectary in the middle of the claw; anthers 4.5–8 mm; flowers with an unpleasant smell   **1. ochroleuca**\*
  b. Outer perianth-segments each with a distinct nectary at the base; anthers 8–11 mm; flowers not smelling unpleasant    **2. flaccida**

**1. H. ochroleuca** Salisbury (*H. collina* (Thunberg) Salisbury var. *ochroleuca* (Salisbury) Baker). Illustration: Botanical Magazine, 1103, 1108 (1808); Annals of the Missouri Botanical Garden 68: 454 (1981); Everett, Encyclopedia of horticulture 5: 1698 (1981).
Stem 35–75 cm, once to several times branched. Leaves usually 1, occasionally 2 or 3, 6–15 mm wide. Flowers yellow, sometimes orange in the centre, strongly and rather unpleasantly scented. Outer perianth-segments 3–4 cm, inner slightly smaller, nectaries diffuse. Filaments 7–10 mm, smooth and hairless; anthers 4.5–8 mm, divergent. Style divided just above the level of the bases of the anthers, branches 5–6.5 mm. *South Africa (SW Cape).* H4. Summer.

\***H. miniata** (Andrews) Sweet (*H. lineata* Sweet). Illustration: Andrews, Botanists Repository, t. 404 (1804); Sweet, British Flower Garden 2: pl. 152, 178 (1826); Annals of the Missouri Botanical Garden 68: 486 (1981). Differs in having 2 or 3 leaves and flowers which may be reddish, pink, yellow or white. Outer perianth-segments 1.3–2.2 cm, with triangular nectaries dotted with green, the claw downy above the midline. Filaments

downy at the base; anthers c. 2 mm, not diverging. *South Africa (SW Cape).* H4. Summer.

**2. H. flaccida** Sweet (*H. collina* (Thunberg) Salisbury var. *aurantiaca* (Zuccagni) Baker; *H. breyniana* (Linnaeus) Lewis var. *aurantiaca* (Zuccagni) Lewis). Illustration: Botanical Magazine, 1612 (1814); Annals of the Missouri Botanical Garden 68: 465 (1981).
Stem 30–60 cm with several branches. Leaf 1, 9–13 mm wide. Flowers yellow to salmon-orange with deep yellow nectaries. Outer perianth-segments 3.5–4 cm, claws smooth or with tiny papillae in the lower part. Filaments 7–8 mm, smooth or with sparse papillae, downy below; anthers 8–11 mm, divergent towards the top. Style divided c. 3 mm above the level of the bases of the anthers, branches to 7 mm. *South Africa (SW Cape).* H4. Summer.

**3. H. collina** (Thunberg) Salisbury (*H. breyniana* (Linnaeus) Lewis). Illustration: Annals of the Missouri Botanical Garden 68: 472 (1981).
Stem 18–35 cm, unbranched or with few branches. Leaf 1, usually trailing, 4–10 mm wide. Flowers pale yellow or salmon-pink, fragrant, nectaries usually deep yellow edged with green. Outer perianth-segments 3–3.5 cm, claw often with papillae near the base. Filaments 5–6 mm, sparsely downy below; anthers 5–6 mm, divergent towards the top. Style divided slightly above the level of the bases of the anthers, branches 5–6 mm. *South Africa (SW Cape).* H4. Summer.

\***H. elegans** (Jacquin) Sweet. Illustration: Annals of the Missouri Botanical Garden 68: 463 (1981). Differs in having yellow flowers with a large green blotch in the central or upper one-third of each outer perianth-segment, which is usually shaded orange towards the top. Anthers 8–10 mm, divergent. *South Africa (SW Cape).* H4. Summer.

## 16. CROCUS Linnaeus
*B. Mathew*
Corm enclosed by several fibrous, papery or leathery tunics. Sheathing leaves up to 5, enclosing the aerial shoot. Leaves all basal, linear, not in 2 ranks, present at flowering time or absent and appearing later; upper surface green or grey-green with a central whitish stripe. Flowers 1 to several, produced in autumn, winter or spring, each carried on a short subterranean flower-stalk which is sometimes subtended by a membranous spathe (prophyll – the 'basal

spathe' of various authors). Bracts 1 or 2, membranous, subtending the ovary and sheathing the perianth-tube (the 'floral spathe' of various authors). Perianth radially symmetric with a long narrow tube; lobes 6, usually equal or inner whorl sometimes smaller. Style with 3 or more branches. Ovary subterranean. Capsule cylindric or ellipsoid, maturing above ground level by elongation of the flower-stalk; seeds numerous. See figure 40, p. 363.

A genus of 80 species from Europe and W & C Asia, easily distinguished from all other genera by the funnel-shaped flowers with a very long perianth-tube. Most species can be cultivated in sunny situations in a wide range of soil types providing the drainage is good, allowing a dryish period during dormancy in summer. A few species tolerate or even prefer semi-shady conditions. Propagation is by seed or by natural vegetative corm division. Literature: Maw, G., *A monograph of the genus Crocus* (1886); Bowles, E.A., *A Handbook of Crocus and Colchicum for Gardeners* (1924), revised edition (1952); Mathew, B., *The Crocus, a revision of the genus Crocus (Iridaceae)* (1982).

The following key is constructed to identify plants which are representative of the species as they are generally known in cultivation; mutations such as albinos do not necessarily key out satisfactorily.

1a. Flowers produced from January to April    2
  b. Flowers produced from September to December    22
2a. Flowers wholly pale creamy yellow to deep yellow or orange, sometimes striped or stained externally    3
  b. Flowers white, blue, lilac, purple or violet, sometimes with a yellow throat, often striped externally    8
3a. Style with 6 or more branches    **15. olivieri**
  b. Style with 3 branches, these sometimes expanded and lobed at the apex    4
4a. Leaves usually numbering 10–20 from mature flowering-sized corms    **16. korolkowii**
  b. Leaves 8 or fewer per corm    5
5a. Leaves 2.5–4 mm wide, dark green; anthers markedly arrow-shaped    **14. flavus**
  b. Leaves very slender, usually 0.5–2 mm wide, usually grey-green; anthers not markedly arrow-shaped    6

6a. Perianth-lobes strongly striped, suffused or speckled with purplish brown on the exterior; corm-tunic netted-fibrous    **10. angustifolius**
  b. Perianth-lobes either unmarked on the exterior or if striped or suffused with purplish brown then corm-tunic not fibrous but papery or leathery with horizontal rings    7
7a. Corm-tunic fibrous, the fibres strongly netted; perianth-lobes obtuse to rounded    **11. ancyrensis**
  b. Corm-tunic membranous or papery, with rings at the base; perianth-lobes subacute to obtuse    **12. chrysanthus**
8a. Anthers whitish    **18. laevigatus**
  b. Anthers yellow or rarely blackish    9
9a. Style divided into many slender orange to scarlet branches    **17. fleischeri**
  b. Style divided into 3 branches, sometimes each branch expanded and frilled or fringed at the apex    10
10a. Throat of perianth pale to deep yellow or orange    11
  b. Throat of perianth with no yellowish coloration    16
11a. Corm-tunic papery, leathery or eggshell-like, with horizontal rings at the base    **13. biflorus**
  b. Corm-tunic fibrous or papery but with no rings at the base    12
12a. Leaves markedly greyish green above    **6. versicolor**
  b. Leaves deep green above    13
13a. Perianth-lobes creamy, buff, yellowish gold or silvery on the outside, usually with dark longitudinal stripes or veins    14
  b. Perianth-lobes not creamy, buff, yellowish gold or silvery outside and usually with no conspicuous longitudinal stripes or veins    **9. sieberi**
14a. Flowers mid to deep purple inside, with a deep yellow or orange, hairless throat; corm-tunic with parallel fibres    **7. imperati**
  b. Flowers mid to pale lilac or violet inside with a rather pale yellow, hairy throat; corm-tunic with netted fibres    15
15a. Bract solitary, clearly visible, sheathing the perianth-tube; leaves usually 2–6 mm wide at flowering time    **3. etruscus**
  b. Bracts 1 or 2, more or less equal and clearly visible, sheathing the perianth-tube; leaves usually 1–2 mm wide at flowering time    **8. dalmaticus**
16a. Flowers white, sometimes flushed or

strongly stained or striped with violet-blue    17
  b. Flowers pale lilac to deep purple or reddish purple, sometimes with conspicuous narrow stripes outside    18
17a. Leaves 3–8, 0.5–3.5 mm wide, grey-green; bracts 2    **13. biflorus**
  b. Leaves 2–4, usually 4–8 mm wide, green; bract solitary    **1. vernus**
18a. Flowers irregularly striped and veined on a paler ground    **1. vernus**
  b. Flowers uniformly striped and veined outside, sometimes only at the base, or flowers unstriped    19
19a. Flowers unstriped and more or less uniform in colour on the outside, or with only fine inconspicuous veining    20
  b. Flowers with conspicuous stripes externally    21
20a. Flowers lilac-blue, purplish or reddish purple, usually with a white or very pale tube; leaves 2–3 mm wide    **2. tommasinianus**
  b. Flowers lilac, purple or violet, usually with the tube the same colour as or darker than the segments; leaves usually 4–8 mm wide    **1. vernus**
21a. Corm-tunics with parallel fibres only; perianth-lobes usually 2–2.7 cm; tube usually 5–11 cm; style-branches yellow to orange    **5. minimus**
  b. Corm-tunics netted-fibrous at apex; perianth-lobes usually 2–3.5 cm; tube usually 3.5–7 cm; style-branches orange to red    **4. corsicus**
22a. Inner and outer perianth-lobes very unequal in shape and size; style-branches lilac    **35. banaticus**
  b. Inner and outer perianth-lobes equal, or nearly so    23
23a. Anthers blackish maroon    **13. biflorus**
  b. Anthers yellow or creamy white    24
24a. Anthers creamy white    25
  b. Anthers yellow    30
25a. Leaves absent at flowering time    26
  b. Leaves present at flowering time, but sometimes only the tips visible    27
26a. Throat either marked with 2 yellow blotches at the base of each lobe (occasionally merging into a V-shape), or throat white    **29. kotschyanus**
  b. Throat with a continuous deep yellow zone    **28. pulchellus**
27a. Style divided into 3 yellow branches; flowers creamy white with no conspicuous stripes on the outside    **30. ochroleucus**

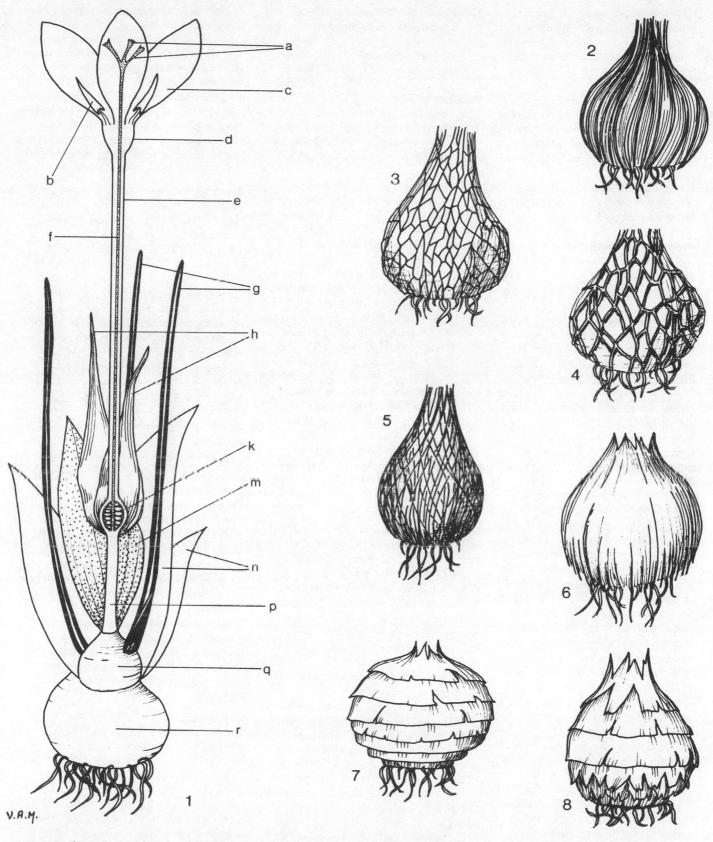

**Figure 40.** Characters used in the identification of *Crocus* species. 1, Diagram showing the parts of a *Crocus* plant (a, style-branch; b, anther; c, perianth-lobe; d, throat; e, perianth-tube; f, style; g, leaf; h, bract; k, ovary; m, spathe; n, sheathing leaf; p, scape; q, new corm; r, old corm). 2–8, Types of corm-tunics. 2, Parallel fibres. 3, Finely netted fibres. 4, Coarsely netted fibres. 5, Interwoven fibres. 6, Papery with parallel fibres. 7, Non-fibrous, with rings at the base. 8, Non-fibrous splitting into teeth at the base.

b. Style divided into many slender, yellow, orange or reddish branches **28**

28a. Corm-tunics tough and splitting into triangular teeth at the base; flowers lilac or white usually striped or suffused with violet outside **18. laevigatus**

b. Corm-tunics papery, splitting into soft fibres at the base; flowers white or lilac-blue, only occasionally dark-veined **29**

29a. Flowers usually lilac-blue, remaining open in dull weather and at night; filaments strongly hairy **20. tournefortii**

b. Flowers usually white, closing at night and in dull weather; filaments minutely hairy **19. boryi**

30a. Style clearly divided into more than 3 branches **31**

b. Style divided into only 3 branches, these sometimes expanded at apex **36**

31a. Leaves present at flowering time but sometimes with only the tips visible; throat usually pale to deep yellow **32**

b. Leaves absent at flowering time and not visible for some time after the flowers have withered; throat either with no trace of yellow colour, or sometimes pale yellow **33**

32a. Style divided into few (usually 3) orange to bright red branches, each shortly subdivided or lobed at the apex; throat deep yellow; flowers white or faintly lilac **22. niveus**

b. Style divided into 6 or more slender orange branches; throat white or pale yellow; flowers pale to deep lilac-blue or purple **26. serotinus**

33a. Flowers deep purple or rarely lilac-purple without conspicuous darker veining; corms stoloniferous with parallel-fibrous tunics **25. nudiflorus**

b. Flowers white, lilac-blue, purple-blue, violet-blue or deep purple, usually with conspicuous darker veining, especially towards the base of the lobes **34**

34a. Perianth-lobes with dark longitudinal veins and conspicuous but finer cross-veining, usually speckled on the outside; corm-tunics non-fibrous, with rings at the base **27. speciosus**

b. Perianth-lobes with conspicuous longitudinal veins only, usually towards the base of the segments; corm-tunic coarsely netted-fibrous **35**

35a. Flowers white to deep lilac-blue with a pale yellow throat **34. cancellatus**

b. Flowers deep purple or sometimes lilac with no yellow in the throat **24. medius**

36a. Throat yellow **37**

b. Throat not yellow **40**

37a. Bract silvery white; leaves grey-green, 0.5–1 mm wide **32. hadriaticus**

b. Bract greenish; leaves deep green, 1–3 mm wide **38**

38a. Flowers mid to deep purple inside, prominently dark-veined and/or buff outside **23. longiflorus**

b. Flowers white to pale or pinkish lilac, not conspicuously dark-veined or buff outside **39**

39a. Style-branches undivided at the apex; perianth-lobes 6–15 mm wide; corm-tunics papery, splitting lengthways **33. caspius**

b. Style-branches subdivided or lobed at the apex; perianth-lobes 1.5–3.5 cm wide; corm-tunics finely netted-fibrous **22. niveus**

40a. Flowers white, sometimes stained violet or brownish at the base **32. hadriaticus**

b. Flowers pale to deep lilac-purple **41**

41a. Style-branches very long, deep red **31. sativus***

b. Style-branches short, white to orange **21. goulimyi**

**1. C. vernus** (Linnaeus) Hill.
Corm-tunic with fine parallel or slightly netted fibres. Leaves 2–4, usually 4–8 mm wide, often much shorter than the flower, green. Flowers white, purple or striped; throat white or purplish, hairy or hairless. Spathe present. Bract solitary. Anthers yellow; style divided into 3 deep yellow to orange-red, rarely whitish, branches, each much-expanded and fringed at the apex.

Subsp. **vernus** (*C. napolitanus* Mordant & Loiseleur; *C. purpureus* Weston). Illustration: Mathew, Dwarf bulbs, 88 (1973) – as C. scepusiensis; Rix & Phillips, The bulb book, 26a–e (1981); Mathew, The crocus, pl. 1a (1982). Flowers often pale to deep purple, or striped; perianth-lobes usually 3–5.5 cm × 9–20 mm. Style usually equalling or exceeding the stamens. *C, S & E Europe from Italy to W Russia.* H2. Spring.

The plants occurring at the eastern end of the range of distribution often have dark V-shaped markings at the tips of the perianth-lobes; they have been named

*C. heuffelianus* Herbert and *C. scepusiensis* (Rehmann & Woloszczak) Borbas.

Subsp. **albiflorus** (Schultes) Ascherson & Graebner (*C. albiflorus* Schultes; *C. caeruleus* Weston). Illustration: Grey-Wilson & Mathew, Bulbs, 34 (1981) – subsp. albiflorus; Mathew, The crocus, pl. 1b (1982). Flowers often white; perianth-lobes usually 1.5–3 cm × 4–10 mm. Style usually much shorter than the stamens. *W, C & S Europe from Czechoslovakia to the Pyrenees.* H2. Spring.

**2. C. tommasinianus** Herbert. Illustration: Synge, Collins' guide to bulbs, pl. 7 & 10 (1971); Hay & Beckett, Reader's Digest encyclopaedia of garden plants and flowers, 190 (1971); Mathew, Dwarf bulbs, 73 (1973); Rix & Phillips, The bulb book, 19 (1981).
Corm-tunic with fine parallel or slightly netted fibres. Leaves 3 or 4, 2–3 mm wide, equalling or exceeding the flowers, green. Flowers pale lilac to deep reddish purple, often silvery or creamy on the exterior, sometimes with darker tips to the lobes, rarely whitish-tipped; throat white, sparsely hairy. Spathe present. Bract solitary. Perianth-lobes 2.5–4.5 cm × 8–20 mm. Anthers yellow; style divided into 3 orange branches, each expanded and fringed at the apex. *S Yugoslavia to S Hungary.* H2. Spring.

**3. C. etruscus** Parlatore. Illustration: Synge, Collins' guide to bulbs, pl. 7 (1971); Grey-Wilson & Mathew, Bulbs, pl. 34 (1981); Rix & Phillips, The bulb book, 24c (1981); Mathew, The crocus, pl. 3 (1982). Corm-tunic with distinctly netted fibres. Leaves 3 or 4, 2–6 mm wide, usually about equal to the flower, green. Flower with pale purple-blue interior, creamy or silvery and veined violet on the outside; throat pale yellow, hairy. Spathe present. Bract solitary. Perianth-lobes 3–4 cm × 9–13 mm. Anthers yellow; style divided into 3 orange branches, each expanded at the apex. *NW Italy.* H2–3. Spring.

**4. C. corsicus** Vanucci. Illustration: Synge, Collins' guide to bulbs, pl. 7 (1971); Grey-Wilson & Mathew, Bulbs, pl. 32 (1981); Rix & Phillips, The bulb book, 29i (1981); Mathew, The crocus, pl. 12 (1982). Corm-tunic with fine fibres, netted in the upper part. Leaves 3 or 4, 1–1.5 mm wide, slightly shorter than or about equalling the flower, green. Flowers pale to deep purple on the interior, often yellowish with 1–3 violet stripes and finer veining on the exterior; throat white or pale purple, hairless. Spathe present. Bract solitary. Perianth-lobes usually 2–3.5 cm × 7–13 mm. Anthers yellow; style divided into 3 orange or red branches, each expanded or slightly subdivided at the apex. *C & N Corsica*. H3. Spring.

**5. C. minimus** de Candolle. Illustration: Synge, Collins' guide to bulbs, pl. 7 (1971); Grey-Wilson & Mathew, Bulbs, pl. 33 (1981); Rix & Phillips, The bulb book, 21i (1981); Mathew, The crocus, pl. 11 (1982).
Corm-tunic fibrous, with parallel fibres. Leaves 3–5, 0.5–1 mm wide, usually exceeding the flowers, green. Flowers mid to deep purple on the interior, usually yellowish with conspicuous violet veins, or stained violet, on the outside; throat white, hairless. Spathe present. bracts 2, very unequal, or rarely solitary. Perianth-lobes usually 2–2.7 cm × 4–8 mm. Anthers yellow; style divided into 3 yellow or orange branches, each obscurely lobed or expanded and frilled at the apex. *Sardinia & S Corsica*. H3. Spring.

**6. C. versicolor** Ker Gawler. Illustration: Synge, Collins' guide to bulbs, pl. 7 (1971); Rix & Phillips, The bulb book, 24a (1981); Mathew, The crocus, pl. 9 (1982).
Corm tunics membranous, the outer ones becoming wholly fibrous with parallel fibres. Leaves 3–5, 1.5–3 mm wide, usually reaching the base of the flowers, markedly grey-green. Flowers white to mid-purple, usually conspicuously dark-striped on the exterior, sometimes on a yellowish ground; throat usually pale yellow, hairless. Spathe present. Bract solitary or accompanied by a narrow bracteole. Perianth-lobes 2.5–3.5 cm × 7–10 mm. Anthers yellow; style divided into 3 yellow or orange branches, each expanded and sometimes slightly lobed at the apex. *Mountains of SE France, NW Italy*. H3. Spring.

The most frequently cultivated variant of this species is a white-flowered, conspicuously striped cultivar known as 'Picturatus'.

**7. C. imperati** Tenore. Illustration: Hay & Beckett, Reader's Digest encyclopaedia of garden plants and flowers, 189 (1971); Grey-Wilson & Mathew, Bulbs, pl. 33 (1981); Rix & Phillips, The bulb book, 26d (1981).
Corm-tunic membranous, or the outer becoming fibrous, with parallel fibres. Leaves 3–6, 2–3 mm wide, equalling or exceeding the flowers, green. Flowers mid

to deep purple on the interior, and yellowish with 1–5 conspicuous violet stripes on the outside; throat yellow to orange, hairless. Spathe present. Bracts 1 or 2. Perianth-lobes 2.3–4 cm × 7–18 mm. Anthers yellow; style divided into 3 deep orange or reddish branches, each expanded and shortly lobed.

Subsp. **imperati**. Bracts 2. Filaments 6–9 mm; anthers 1.2–2.1 cm. Style-branches deep orange or red. *W Italy (Naples southwards)*. H3. Spring.

Subsp. **suaveolens** (Bertoloni) Mathew. Bract solitary. Filaments 3–5 mm; anthers 8–12 mm. Style-branches yellow or orange. *W Italy (N of Naples, to Rome)*. H3. Spring.

Some of the commercial stocks of *C. imperati* are referable to subsp. *suaveolens*.

**8. C. dalmaticus** Visiani. Illustration: Synge, Collins' guide to bulbs, pl. 7 (1971); Grey-Wilson & Mathew, Bulbs, pl. 33 (1981); Rix & Phillips, The bulb book, 26c (1981); Mathew, The crocus, pl. 36 (1982).
Corm-tunic fibrous, the fibres markedly netted. Leaves 3–5, usually 1–2 mm wide, shorter than or equalling the flower, dark green. Flowers pale violet on the interior, usually yellowish or silvery with fine purple lines on the outside; throat yellow, sparsely hairy. Spathe absent. Bracts 2. Perianth-lobes 1.5–3.5 × 1–1.5 cm. Anthers yellow; style divided into 3 orange branches, each slightly expanded and fringed at the apex. *SW Yugoslavia, NW Albania*. H2. Spring.

The commercially available clone of *C. dalmaticus* has wider leaves than wild plants and may be of hybrid origin; stocks of it often show virus symptoms with the flowers spotted and streaked on the exterior.

**9. C. sieberi** Gay. Illustration: Journal of Royal Horticultural Society **74**: f. 6 (1949) – subsp. atticus and f. 7b – forma tricolor; Grey-Wilson & Mathew, Bulbs, pl. 33 (1981); Rix & Phillips, The bulb book, 18g (1981) – subsp. atticus and 24d – subsp. sieberi; Mathew, The crocus, pl. 37a (1982).
Corm-tunic fibrous, the fibres markedly netted. Leaves 2–8, 1–6 mm wide, shorter than to equalling the flower, dark green. Flowers pale lilac to deep lilac-purple, or white with purple zones, or stripes, on the outside (subsp. **sieberi** from Crete); throat yellow to orange, hairless or hairy. Spathe absent. Bracts 2. Perianth-lobes

2–4 cm × 7–15 mm. Anthers yellow; style divided into 3 yellow to orange-red branches, each much-expanded and frilled or lobed at the apex. *S Albania, S Bulgaria, S Yugoslavia, Greece, Crete*. H2. Spring.

The most commonly cultivated variant of *C. sieberi* is subsp. **atticus** (Boissier & Orphanides) Mathew which has lilac flowers and a rather coarsely netted corm-tunic; the white-flowered subsp. **sieberi** (*C. sieberi* var. *heterochromus* Halacsy; *C. sieberi* var. *versicolor* Boissier & Heldreich) from Crete is rarely seen in cultivation but a hybrid between it and subsp. *atticus*, called 'Hubert Edelsten' is often offered in catalogues; forma **tricolor** Burtt (a form of subsp. **sublimis** (Herbert) Mathew) has striking lilac-blue flowers with a white zone surrounding the yellow throat.

**10. C. angustifolius** Weston (*C. susianus* Ker Gawler). Illustration: Hay & Beckett, Reader's Digest encyclopaedia of garden plants and flowers, 191 (1971); Grey-Wilson & Mathew, Bulbs, pl. 34 (1981); Rix & Phillips, The bulb book, 24f (1981); Mathew, The crocus, pl. 43 (1982).
Corm-tunic fibrous, the fibres strongly netted. Leaves 3–6, 0.5–1.5 mm wide, usually equalling or exceeding the flower at flowering time, grey-green. Flowers yellow, heavily suffused or striped outside with purplish brown; throat yellow, hairless or minutely hairy. Spathe absent. Bracts 2. Perianth-lobes 1.7–3.4 cm × 6–13 mm. Anthers yellow; style divided into 3 yellow to deep orange-red branches. *SW Russia (Crimea, S Ukraine, Armenia)*. H2. Spring.

A hybrid called **C. × stellaris** Haworth, somewhat similar in appearance to *C. angustifolius*, is sometimes cultivated. It has yellow flowers with dark stripes on the exterior; the probable parentage is *C. angustifolius × C. flavus*.

**11. C. ancyrensis** (Herbert) Maw. Illustration: Hay & Beckett, Reader's Digest encyclopaedia of garden plants and flowers, 187 (1971); Synge, Collins' guide to bulbs, pl. 7 (1971); Rix & Phillips, The bulb book, 25j (1981); Mathew, The crocus, pl. 44 (1982).
Corm-tunic fibrous, the fibres strongly netted. Leaves 3–6, usually 0.5–1 mm wide, shorter than to just exceeding the flower, greyish green. Flowers bright yellow or orange, sometimes with a purplish tube; throat yellow, hairless. Spathe absent. Bracts 2. Perianth-lobes 1.5–3 cm × 9–13 mm. Anthers yellow; style divided

into 3 orange or orange-red branches. *C & N Turkey.* H2. Spring.

This is usually seen in cultivation as the selection known as 'Golden Bunch' which produces several flowers per corm.

**12. C. chrysanthus** (Herbert) Herbert. Illustration: Hay & Beckett, Reader's Digest encyclopaedia of garden plants and flowers, 187, 188 (1971); Synge, Collins' guide to bulbs, pl. 7 (1971); Grey-Wilson & Mathew, Bulbs, pl. 34 (1981); Rix & Phillips, The bulb book, 20b & c (1981). Corm-tunic papery, leathery or eggshell-like, splitting into rings at the base. Leaves 3–7, 0.5–2.5 mm wide, shorter than to exceeding the flower, usually greyish-green. Flowers yellow to orange-yellow, sometimes striped or suffused with bronze or purple on the outside, occasionally creamy white; throat yellow, hairless. Spathe absent. Bracts 2. Perianth-lobes usually 1.5–3.5 cm × 5–11 mm. Anthers yellow, sometimes with blackish basal lobes; style divided into 3 yellow to orange branches. *Albania, Bulgaria, Greece, S Yugoslavia, E Roumania, W, C & S Turkey.* H2. Spring.

Variants of *C. chrysanthus* have been selected to give a larger flower size and a greater range of colour than in the wild forms. Some of the plants listed as "*C. chrysanthus* cultivars" are however forms of the related *C. biflorus* or hybrids between the two species.

**13. C. biflorus** Miller. Corm-tunic papery, leathery or eggshell-like, splitting into rings at the base. Leaves usually 3–8, 0.5–3.5 mm wide, usually grey-green. Flowers with a white, lilac or blue ground colour, often conspicuously striped and veined with purple on the outside; throat yellow or white, hairless or finely hairy. Spathe absent. Bracts 2. Perianth-lobes 1.8–3 cm × 4–13 mm. Anthers yellow, sometimes with blackish basal lobes, or wholly blackish maroon; style divided into 3 yellow to orange branches. *S Europe & W Asia, from Sicily east to Crimea, Turkey, Caucasus and Iran.* H2. Spring or autumn (subsp. *melantherus* only).

A very variable and widespread species in which 14 subspecies are recognised but only a few of these are in general cultivation.

Subsp. **biflorus** (*C. pusillus* Tenore; *C. argenteus* Sabine; *C. praecox* Haworth; *C. biflorus* var. *parkinsonii* Sabine). Illustration: Rix & Phillips, The bulb book, 27h (1981); Mathew, The crocus, pl. 50a (1982). Flowers appearing in spring, white or lilac, usually conspicuously striped with purple or brownish on the outside; throat yellow. *Italy, Sicily, Rhodes, NW Turkey.*

Subsp. **weldenii** (Hoppe & Furnrohr) Mathew (*C. weldenii* Hoppe & Furnrohr). Illustration: Hay & Beckett, Reader's Digest encyclopaedia of garden plants and flowers, 188 (1971). Flowers appearing in spring, plain white, sometimes flushed with pale blue at the base or rarely all over the exterior of the lobes; throat not yellow. *W Yugoslavia, N Albania.*

Subsp. **alexandri** (Velenovsky) Mathew (*C. alexandri* Velenovsky). Illustration: Rix & Phillips, The bulb book, 24b (1981). Flowers appearing in spring, white, heavily stained with violet on the outside; throat not yellow. *SW Bulgaria, S Yugoslavia.*

The cultivar 'Lady Killer' is very similar to this subspecies.

Subsp. **melantherus** (Boissier & Orphanides) Mathew ("*C. crewei*" of many authors but not Hooker). Illustration: Rix & Phillips, The bulb book, 174e (1981); Annales Musei Goulandris **6**: 73 (1983). Flowers autumnal, white with conspicuous purple or greyish stripes or speckling on the outside; throat yellow. Anthers usually blackish maroon. *S Greece.*

Subsp. **pulchricolor** (Herbert) Mathew ("*C. aerius*" of many authors but not Herbert). Illustration: Hay & Beckett, Reader's Digest encyclopaedia of garden plants and flowers, 188 (1971); Mathew, The crocus, pl. 50m (1982). Flowers appearing in spring, usually rich blue-violet, often darker towards the base of the lobes, normally unstriped on the outside; throat deep yellow. *NW Turkey.*

The cultivars 'Blue Bird' and 'Blue Pearl' almost certainly belong to this subspecies.

In addition to selected forms of *C. biflorus* some hybrids have been raised between it and *C. chrysanthus*, for example 'Advance', which has a curious yellow and blue flower coloration.

**14. C. flavus** Weston (*C. aureus* Sibthorp & Smith; *C. lagenaeflorus* Salisbury; *C. luteus* Lamarck; *C. maesiacus* Ker Gawler; *C. lacteus* Sabine). Illustration: Bowles, Handbook of Crocus and Colchicum, edn 2, 98 (1952); Hay & Beckett, Reader's Digest encyclopaedia of garden plants and flowers, 187 (1971); Grey-Wilson & Mathew, Bulbs, pl. 33 (1981); Mathew, The crocus, pl. 61a (1982).

Corm-tunic papery, splitting at the base into parallel fibres. Leaves 4–8, 2.5–4 mm wide, shorter than to equalling the flower, green. Flowers yellow to orange-yellow; throat yellow, hairless or hairy. Spathe absent. Bracts 2. Perianth-lobes 2–3.5 cm × 6–12 mm. Anthers yellow; style divided into about 3 short, pale yellow to orange branches, usually shorter than the anthers. *S Yugoslavia, C & N Greece, Bulgaria, Roumania, NW Turkey.* H2. Spring.

The true species is not often cultivated but the most commonly seen yellow crocus in gardens (the large 'Dutch Yellow' or 'Golden Yellow') is very similar in appearance; this is a hybrid between *C. flavus* and *C angustifolius* of considerable antiquity. In the nineteenth century many variants and hybrids of *C. flavus* were known in gardens but these have now almost disappeared from cultivation.

**15. C. olivieri** Gay. Corm-tunic membranous, splitting at the base into coarse triangular teeth. Leaves 1–4, usually 2–5 mm wide, slightly shorter than to just exceeding the flowers at flowering time, green. Flowers orange-yellow, sometimes stained or striped with purplish bronze on the outside; throat yellow, hairless or hairy. Spathe absent. Bracts 2. Perianth-lobes 1.5–3.5 cm × 4–12 mm. Anthers yellow; style divided into 6–15 slender yellow or orange branches.

Subsp. **olivieri** (*C. suterianus* Herbert). Illustration: Synge, Collins' guide to bulbs, pl. 7 (1971); Rix & Phillips, The bulb book, 20g (1981); Mathew, The crocus, pl. 63a (1982). Flower usually unmarked on the outside; style divided into about 6 branches. *S Yugoslavia, SE Roumania, S Bulgaria, Albania, Greece, Turkey.* H2. Spring.

Subsp. **balansae** (Baker) Mathew (*C. balansae* Baker). Illustration: Mathew, The crocus, pl. 63b (1982). Flower usually striped or suffused with purplish brown or mahogany on the outside; style divided into 12–15 branches. *W Turkey, Aegean islands.* H2. Spring.

**16. C. korolkowii** Maw. Illustration: Synge, Collins' guide to bulbs, pl. 7 (1971); Mathew, Dwarf bulbs, 73 (1973); Rix & Phillips, The bulb book, 18b (1981); Mathew, The crocus, pls. 60 (1982). Corm-tunic papery, splitting into many fine parallel fibres. Leaves usually 10–20, 1–2.5 mm wide, usually shorter than the flower at flowering time, green. Flowers yellow, usually conspicuously speckled or stained outside with brown, purple or black, sometimes on a buff ground; throat yellow, hairless. Spathe absent. Bracts 2.

Perianth-lobes 2–3.5 cm × 6–12 mm. Anthers yellow; style divided into 3 orange branches. *N & E Afghanistan, N Pakistan, USSR (Soviet Central Asia).* H2. Spring.

**17. C. fleischeri** Gay. Illustration: Synge, Collins' guide to bulbs, pl. 7 (1971); Rix & Phillips, The bulb book, 20f (1981); Mathew, The crocus, pl. 73 (1982). Corm-tunic fibrous, the fibres interwoven. Leaves 5–8, 0.5–1 mm wide, usually longer than the flower at flowering time, green or grey-green. Flowers white with a purplish or brownish base, sometimes with a median central stripe on the outside of each outer lobe; throat yellow, hairless. Spathe absent. Bracts 2. Perianth-lobes 1.7–3.1 cm × 4–6 mm. Anthers yellow; style divided into numerous slender orange to scarlet branches. *S & W Turkey, Aegean islands.* H3. Spring.

**18. C. laevigatus** Bory & Chaubard (*C. fontenayi* Reuter). Illustration: Synge, Collins' guide to bulbs, pl. 6 (1971); Rix & Phillips, The bulb book, 174g & h (1981); Mathew, The crocus, pl. 76 (1982). Corm-tunic smooth and tough, splitting at the base into long, narrowly triangular teeth. Leaves usually 3 or 4, 1–2.5 mm wide, usually about equalling the flower at flowering time, green. Flowers with a white or lilac ground colour, on the outside sometimes overlaid with silvery cream, buff or yellow and usually marked with 1–3 deep violet or purple stripes, rarely wholly suffused with dark purple outside; throat yellow, hairless. Spathe absent. Bracts 2. Perianth-lobes 1.3–3 cm × 4–18 mm. Anthers creamy white; style divided into many yellow to deep orange branches. *C & S Greece, Crete.* H3. Autumn or winter.

An extremely variable species, usually cultivated as the clone 'Fontenayi' which has conspicuously striped lilac flowers.

**19. C. boryi** Gay (*C. cretensis* Kornicke). Illustration: Grey-Wilson & Mathew, Bulbs, pl. 31 (1981); Rix & Phillips, The bulb book, 179m (1981); Mathew, The crocus, pl. 77 (1982). Corm-tunic papery, splitting at the base into many parallel fibres. Leaves 3–7, 1–3.5 mm wide, shorter than to much exceeding the flower at flowering time, dark green. Flowers creamy white, occasionally faintly veined purple on the outside; throat deep yellow, hairless or finely hairy. Spathe absent. Bracts 2. Perianth-lobes 2–5 cm × 7–23 mm. Anthers creamy white; style divided into many slender

orange or reddish branches. *W & S Greece, SE Crete.* H3. Autumn.

**20. C. tournefortii** Gay. Illustration: Synge, Collins' guide to bulbs, pl. 6 (1971); Rix & Phillips, The bulb book, 179 l (1981); Mathew, The crocus, pl. 78 (1982). Corm-tunic papery, splitting at the base into many parallel fibres. Leaves 5–10, 1–2.5 mm wide, shorter than or equalling the flower, dark green. Flowers remaining open in dull weather and at night, usually lilac-blue, sometimes veined darker towards the base, occasionally white (in Crete); throat usually deep yellow, hairless or hairy. Spathe absent. Bracts 2. Perianth-lobes 1.5–3.6 cm × 4–13 mm. Anthers creamy white; style divided into many slender orange or reddish branches, usually widely spreading and very conspicuous. *S Greece (Cyclades and Aegean islands), NE Crete.* H3. Autumn.

**21. C. goulimyi** Turrill. Illustration: Botanical Magazine, n.s., 354 (1960); Grey-Wilson & Mathew, Bulbs, pl. 3 (1981); Rix & Phillips, The bulb book, 182h (1981); Mathew, The crocus, pl. 19 (1982). Corm-tunic smooth and tough, splitting at the base into narrow triangular teeth. Leaves 4–6, 1–2.5 mm wide, often nearly as long as the flower at flowering time, green. Flowers pale to deep lilac-purple, the inner lobes often paler than the outer; throat white, hairy. Spathe present. Bract solitary. Perianth-lobes 1.6–3.8 × 1.1–1.8 cm. Anthers yellow; style divided into 3 white to orange branches, each slightly expanded or lobed at the apex. *S Greece (S Peloponnese).* H3. Autumn.

**22. C. niveus** Bowles. Illustration: Botanical Magazine, n.s., 146 (1951); Synge, Collins' guide to bulbs, pl. 6 (1971); Grey-Wilson & Mathew, Bulbs, pl. 31 (1981); Mathew, The crocus, pl. 18 (1982). Corm-tunic fibrous, the fibres finely netted. Leaves 5–8, 1–2 mm wide, usually nearly as long as the flowers but sometimes with only the tips visible, dark green. Flowers usually white but sometimes pale lilac or the inner lobes white and the outer lilac; throat deep yellow, sparsely hairy. Spathe present. Bracts 2. Perianth-lobes 3–6 × 1.5–3.5 cm; anthers deep yellow; style divided into 3 orange to bright red branches, each subdivided into many shorter branches or expanded and slightly lobed at the apex. *S Greece (S Peloponnese).* H3. Autumn.

**23. C. longiflorus** Rafinesque (*C. odorus* Bivona). Illustration: Synge, Collins' guide to bulbs, pl. 6 (1971); Rix & Phillips, The bulb book, 174f (1981); Grey-Wilson & Mathew, Bulbs, pl. 31 (1981); Mathew, The crocus, pl. 16 (1982). Corm-tunic fibrous, the fibres netted, but sometimes rather obscurely so. Leaves 4–6, 1–3 mm wide, usually shorter than the flower at flowering time but sometimes only the tips visible, green. Flowers lilac to purple, usually dark-veined outside, sometimes on a yellowish external ground colour; throat yellow, hairless or sparsely hairy. Spathe present. Bract solitary. Perianth-lobes 2.2–4.3 cm × 7–16 mm. Anthers yellow; style divided into 3 orange-red or rarely yellow branches, each expanded and frilled or lobed at the apex. *SW Italy, Sicily, Malta.* H3. Autumn.

**24. C. medius** Balbis. Illustration: Synge, Collins' guide to bulbs, pl. 6 (1971); Hay & Beckett, Reader's Digest encyclopaedia of garden plants and flowers, 191 (1971); Rix & Phillips, The bulb book, 183r (1981); Mathew, The crocus, pl. 17 (1982). Corm-tunic fibrous, the fibres strongly netted. Leaves 2 or 3, 2.5–4 mm wide, absent at flowering time and not appearing until long after, dark green. Flowers lilac to deep purple, veined darker towards the base; throat white, often veined with purple, hairless. Spathe present. Bract solitary. Perianth-lobes 2.5–5 × 1.2–1.7 cm; anthers yellow; style divided into many slender deep orange to scarlet branches. *NW Italy, SE France.* H2. Autumn.

**25. C. nudiflorus** Smith. Illustration: Synge, Collins' guide to bulbs, pl. 6 (1971); Grey-Wilson & Mathew, Bulbs, pl. 32 (1981); Rix & Phillips, The bulb book, 175i (1981); Mathew, The crocus, pl. 14 (1982). Corm usually producing stolons; tunics papery, splitting into parallel fibres and often becoming wholly fibrous. Leaves 3 or 4, 2–4 mm wide, absent at flowering time and not appearing until long after, dark green. Flowers deep purple or more rarely paler lilac-purple, not conspicuously veined; throat not yellow, usually hairless. Spathe present. Bract solitary. Perianth-lobes 3–6 cm × 9–20 mm. Anthers yellow; style divided into many slender orange branches. *SW France, N & E Spain.* H2. Autumn.

**26. C. serotinus** Salisbury. Corm-tunic papery with parallel fibres or wholly fibrous with the fibres finely to coarsely netted. Leaves 3–7, 0.5–2 mm wide, usually shorter than the flower but sometimes equalling it or occasionally absent, but if so then developing immediately the flowers wither; green in colour. Flowers pale to deep lilac or occasionally purple, sometimes veined darker; throat white or pale yellow, hairy or hairless. Spathe present. Bract usually solitary but occasionally with a small bracteole. Perianth-lobes 2.5–3.8 cm × 8–13 mm. Anthers yellow; style divided into many slender orange branches.

Subsp. **serotinus**. Illustration: Grey-Wilson & Mathew, Bulbs, pl. 32 (1981); Mathew, The crocus, pl. 15a (1982). Corm-tunic coarsely netted. Leaves 3 or 4. Perianth-lobes 2.5–3.8 cm × 8–11 mm. *C & S Portugal*. H3. Autumn. This subspecies is rarely seen in cultivation.

Subsp. **clusii** (Gay) Mathew (*C. clusii* Gay). Corm-tunic finely netted. Leaves 4–7. Perianth-lobes usually 2.5–3.8 cm × 8–13 mm. *C & N Portugal, NW & SW Spain*. H3. Autumn.

Subsp. **salzmannii** (Gay) Mathew (*C. salzmannii* Gay; *C. asturicus* Herbert; *C. granatensis* Maw). Illustration: Synge, Collins' guide to bulbs, pl. 6 (1971) – subsp. salzmannii; Rix & Phillips, The bulb book, 175k (1981) – subsp. salzmannii; Mathew, The crocus, pl. 15c (1982). Corm-tunic papery, splitting into parallel fibres. Leaves 5–7. Perianth-lobes usually 3.5–5.5 × 1–1.7 cm. *N Africa, S, C & N Spain, Gibraltar*. H3. Autumn.

This is the most commonly cultivated variant of *C. serotinus*. In N Africa there occur very vigorous forms with leaves well-developed at flowering time; the cultivar known as *C. salzmannii* 'Erectophyllus' is very similar to these. The deep-coloured cultivar *C. asturicus* 'Atropurpureus' is probably referable to subsp. *salzmannii*, although it may be of hybrid origin.

**27. C. speciosus** Bieberstein. Illustration: Synge, Collins' guide to bulbs, pl. 6 & pl. 10 (1971); Hay & Beckett, Reader's Digest encyclopaedia of garden plants and flowers, 190 (1971); Rix & Phillips, The bulb book, 175j (1981); Mathew, The crocus, pl. 74a (1982).
Corm-tunic leathery or eggshell-like, splitting into rings at the base. Leaves usually 4, 3–5 mm wide, absent at flowering time and not appearing until long after, deep green. Flowers lilac-blue to purple- or violet-blue, conspicuously veined

darker and often speckled on the outside; throat whitish or sometimes faintly yellow, hairless. Spathe absent. Bracts 2. Perianth-lobes usually 3.7–6 × 1–2.2 cm. Anthers yellow; style divided into many slender yellow to deep orange branches. *USSR (Crimea, Caucasus), N Iran, N & C Turkey*. H2. Autumn.

There are several commonly cultivated clones, mainly differing in the depth of flower colour, and an albino, 'Albus'. The species occasionally hybridises with *C. pulchellus* to produce intermediates.

**28. C. pulchellus** Herbert. Illustration: Synge, Collins' guide to bulbs, pl. 6 (1971); Rix & Phillips, The bulb book, 182a (1981); Grey-Wilson & Mathew, Bulbs, pl. 32 (1981); Mathew, The crocus, pl. 75 (1982).
Corm-tunic papery or leathery, splitting into rings at the base. Leaves usually 4, 4–5 mm wide, absent at flowering time and not appearing until long after, dark green. Flowers clear pale to mid bluish-lilac with darker longitudinal veins; throat deep yellow, hairless or slightly hairy. Spathe absent. Bracts 2. Perianth-lobes 1.8–4 cm × 8–20 mm. Anthers creamy white; style divided into many yellow or orange branches. *S Yugoslavia, S Bulgaria, N Greece, NW Turkey*. H3. Autumn.

**29. C. kotschyanus** Koch (*C. zonatus* Gay). Illustration: Hay & Beckett, Reader's Digest encyclopaedia of garden plants and flowers, 189 (1971); Synge, Collins' guide to bulbs, pl. 6 (1971); Rix & Phillips, The bulb book, 176c (1981); Mathew, The crocus, pl. 23a (1982).
Corm often flattened and rather misshapen; tunic thinly papery with the fibres parallel near the base and weakly netted at the apex. Leaves 4–6, 1.5–4 mm wide, absent at flowering time and not appearing until long after, green. Flowers pale to mid bluish lilac with conspicuous darker veins; throat whitish, usually with 2 yellow blotches near the base of each lobe, these sometimes merging to form a larger V-shaped zone, or sometimes with no yellow in the throat. Spathe present. Bracts 2. Perianth-lobes 3–4.5 cm × 5–18 mm. Anthers creamy white; style divided into 3 creamy yellow to yellow branches, sometimes each slightly subdivided at the apex. *C & S Turkey, NW Syria, C & N Lebanon*. H3. Autumn.

The frequently cultivated var. **leucopharynx** Burtt lacks any yellow markings in the throat, which has a large clearly defined white zone; this clone is often erroneously referred to as

*C. karduchorum* Maw which is a related but distinct species, very rarely cultivated.

**30. C. ochroleucus** Boissier & Gaillardot. Illustration: Synge, Collins' guide to bulbs, pl. 6 (1971); Hay & Beckett, Reader's Digest encyclopaedia of garden plants and flowers, 190 (1971); Rix & Phillips, The bulb book, 174c (1981); Mathew, The crocus, pl. 26 (1982).
Corm-tunic thinly papery with the fibres parallel near the base and weakly netted at the apex. Leaves 3–6, 1–1.5 mm wide, usually visible but rather short at flowering time, green. Flowers creamy white; throat pale to deep yellow, hairy. Spathe present. Bracts 2. Perianth-lobes 2.3–3.7 cm × 7–10 mm. Anthers creamy white; style divided into 3 deep yellow branches, each shortly lobed or frilled at the apex. *SW Syria, Lebanon, N Israel*. H3. Autumn.

**31. C. sativus** Linnaeus (*C. orsinii* Parlatore; *C. sativus* var. *cashmirianus* Royle). Illustration: Synge, Collins' guide to bulbs, pl. 6 (1971); Rix & Phillips, The bulb book, 183n (1981); Mathew, The crocus, pl. 29a (1982).
Corm-tunic fibrous, the fibres finely netted. Leaves 7–12, 0.5–1.5 mm wide, usually equalling or exceeding the flower at flowering time, grey-green. Flowers lilac-purple, conspicuously veined darker and stained darker towards the base; throat lilac-purple, hairy. Spathe present. Bracts 2. Perianth-lobes 3.5–5 × 1–1.5 cm. Anthers yellow; style divided into 3 long red branches. *Unknown as a wild plant but occasionally persisting in semi-wild situations as a relic of cultivation*. H3. Autumn.

*C. sativus* is a sterile triploid of unknown origin. It was formerly widely cultivated in S Europe and Asia as the source of the spice Saffron, prepared from its dried style-branches. It is still grown commercially to some extent in a few Mediterranean countries and in other parts of Europe as an ornamental.

**\*C. cartwrightianus** Herbert. Illustration: Mathew, The crocus, pl. 29 (1982). Differs mainly in having smaller flowers which may be white or lilac-purple. *S Greece*. H3. Autumn.

**32. C. hadriaticus** Herbert. Illustration: Synge, Collins' guide to bulbs, pl. 6 (1971); Rix & Phillips, The bulb book, 182i (1981); Mathew, The crocus, pl. 33 (1982).
Corm-tunic fibrous, the fibres finely netted. Leaves 5–9, 0.5–1 mm wide, usually shorter than the flowers at flowering time

but occasionally equalling them or sometimes very short with only the tips visible, grey-green. Flowers white, often stained brownish or violet at the base; throat deep yellow or rarely white, hairy. Spathe present. Bracts 2. Perianth-lobes 2–4.5 cm × 7–20 mm. Anthers yellow; style divided into 3 orange or red branches. *W & S Greece*. H3. Autumn.

Plants cultivated under the name *C. cartwrightianus* 'Albus' are usually referable to *C. hadriaticus*. The absence of yellow in the throat of the former is usually sufficient to distinguish between the two species. The name *C. hadriaticus* var. *chrysobelonicus* Herbert also appears in literature and catalogues but this is fairly typical of the species and does not merit a separate name.

**33. C. caspius** Fischer & Meyer. Illustration: Mathew, Dwarf bulbs, 73 (1973); Rix & Phillips, The bulb book, 179o (1981); Mathew, The crocus, pl. 57 (1982). Corm-tunics papery, splitting longitudinally but not markedly fibrous. Leaves usually 4–6, 1–2 mm wide, equalling or exceeding the flowers, dark green. Flowers white to pinkish lilac; throat deep yellow, hairy. Spathe absent. Bracts 2. Perianth-lobes 2–4 cm × 6–15 mm. Anthers yellow; style divided into 3 orange branches, each expanded at the apex. *USSR, N Iran (Caspian region)*. H3. Autumn.

**34. C. cancellatus** Herbert. Corm-tunic fibrous, the fibres conspicuously netted. Leaves 4–7, 1–2 mm wide, normally absent at flowering time but rarely with the tips visible, grey-green. Flowers white or pale to deep lilac-blue, usually veined darker; throat pale yellow, hairy or hairless. Spathe absent. Bracts 2. Perianth-lobes 3–5.5 cm × 7–18 mm. Anthers yellow; style divided into many slender orange branches.

Two of the 5 subspecies are cultivated to a small extent:

Subsp. **cancellatus** (*C. cilicicus* Maw; *C. cancellatus* var. *cilicicus* (Maw) Maw). Illustration: Synge, Collins' guide to bulbs, pl. 6 (1971); Grey-Wilson & Mathew, Bulbs, pl. 32 (1981); Rix & Phillips, The bulb book, 182j (1981); Mathew, The crocus, pl. 39a (1982). Flowers pale to mid lilac-blue, usually veined with violet, rather slender. Perianth-tube 3–8 cm; lobes 3–4.2 cm × 6–11 mm, oblanceolate or narrowly elliptic, usually acute or almost so. *S Turkey, W Syria, Lebanon, N Israel*. H3. Autumn.

Subsp. **mazziaricus** (Herbert) Mathew. Illustration: Mathew, The crocus, pl. 39c (1982). Flowers creamy white to deep lilac-blue, often veined darker, usually rather goblet-shaped. Perianth-tube 5–15 cm; lobes usually 3–5.5 × 1–1.8 cm, obovate or oblanceolate, obtuse or almost acute. *S Yugoslavia, Greece, S & W Turkey*. H3. Autumn.

**35. C. banaticus** Gay (*C. byzantinus* Herbert; *C. iridiflorus* Reichenbach). Illustration: Synge, Collins' guide to bulbs, pl. 6 (1971); Mathew, Dwarf bulbs, 83 (1973); Rix & Phillips, The bulb book, 182 l (1981); Mathew, The crocus, pl. 79 (1982). Corm-tunic finely fibrous, the fibres parallel at the base, netted at the apex. Leaves 1–3, 5–7 mm wide, absent at flowering time and not appearing until long after, green. Flowers lilac to purple, the inner lobes often paler than the outer and markedly smaller. Spathe present. Bract solitary. Outer perianth-lobes 3.7–5 × 1.3–2.5 cm, the inner 2.3–3 × 1.2–1.3 cm; anthers yellow; style divided into many slender, widely spreading, lilac branches. *C, W & NW Roumania, NE Yugoslavia, SW Russia*. H2. Autumn.

**17. SYRINGODEA** Hooker
*J.C.M. Alexander*
Corm compressed or spherical with a pointed base; tunics hard and woody, the outer often cracked or split. Leaves 1–several, not 2-ranked, thread-like, terete or shallowly grooved on inner surface. Inflorescence-stalk short and subterranean. Bracts membranous, the inner smaller and bifid. Flowers solitary, rarely 2, funnel-shaped. Perianth-tube long and slender, straight or slightly curved; lobes spreading, entire or notched. Filaments free; anthers erect. Style slender, 3-branched; branches entire or irregularly divided. Capsule spherical.

A genus of about 7 South African herbs, similar to *Crocus* and *Romulea*. Corms should be planted in May, about 5 cm deep, in a well-drained, rich, sandy loam and given plenty of water when in full growth. Plants should be given less water once growth ceases and protected in winter with a sheet of glass. The blooms last longer if the plants are grown in an alpine house. Literature: de Vos, M.P., Die suid-Afrikaanse genus Syringodea, *Journal of South African Botany* 40(3): 201–54 (1974).

**1. S. pulchella** Hooker. Illustration: Botanical Magazine, 6072 (1873); Gardeners' Chronicle 81: 79 (1927); Journal of South African Botany 40(3): 230 (1974). Corm *c.* 1.2 cm in diameter, spherical to ovoid. Leaves 4–6, stiff, arched, 7.5–12 cm, hairless, bases enclosed in membranous sheaths. Flowers purple, *c.* 2.5 cm in diameter; perianth-tube 3–5 cm, cylindric; lobes *c.* 1.2 cm, narrowly triangular to obovate, deeply notched. Spathe-bracts 1.2–2 cm, lanceolate. Stigmas linear, undivided, barely exceeding the anthers. *South Africa*. H5–G1. Autumn.

**18. GALAXIA** Thunberg
*V.A. Matthews*
Corms with fibrous tunics which are often vertically ribbed. Stem underground or slightly above ground level, elongating in fruit. Leaves borne at the top of the stem, not 2-ranked. Perianth with a long tube which raises the flower above the leaves; lobes more or less equal, spreading. Filaments partly to completely joined, inserted at the throat of the perianth-tube. Stigmas entire or fringed. Ovary more or less stalkless, ovoid. Fruit a capsule containing many small, angled seeds.

A genus of 12 species from SW South Africa. The plants are best grown in a sandy soil, with the corms kept dry throughout the winter. Propagation is by seed or offsets.
Literature: Goldblatt, P., Biology and systematics of Galaxia (Iridaceae), *Journal of South African Botany* 45: 385–423 (1979).

1a. Flowers completely yellow; stigmas fringed **1. ovata***
  b. Flowers purple or pinkish usually with a yellow throat; stigmas entire **2. versicolor**

**1. G. ovata** Thunberg. Illustration: Botanical Magazine, 1208 (1809); Journal of South African Botany 45: 413 (1979). Leaves ovate, apex obtuse with a tiny point, margin thickened, minutely ciliate. Flowers yellow, perianth-tube 5–20 mm; perianth-lobes 1.6–2 cm. Filaments joined, 4–5 mm; anthers erect, 2–3 mm. Style extending well above anthers; stigmas fringed. *South Africa (SW Cape)*. H4. Late spring–summer.

**\*G. fugacissima** (Linnaeus filius) Druce (*G. graminea* Thunberg). Illustration: Botanical Magazine, 1292 (1810); Marloth, The Flora of South Africa 4: pl. 45 f. F (1915); Journal of South African Botany 45: 416 (1979). Differs in having linear leaves, filaments sometimes free in the upper third and diverging anthers. The flowers are scented. *SW South Africa*. H4. Summer.

**2. G. versicolor** Klatt (*G. ovata* Thunberg var. *purpurea* Ker Gawler). Illustration: Botanical Magazine, 1516 (1812); Marloth, The Flora of South Africa **4**: pl. 45 f. E (1915); Journal of South African Botany **45**: 400 (1979).
Leaves lanceolate, often with a wavy margin which is minutely ciliate. Flowers purple or pinkish, usually with a yellow throat; perianth-tube 1–2.5 cm; perianth-lobes 1.5–2 cm. Filaments free at the tips, 5–6 mm; anthers curved, 1.5–3 mm. Style extending well above anthers; stigmas entire, often with a wavy margin. *South Africa (SW Cape)*. H4. Summer.

## 19. ROMULEA Maratti
*E.M. Rix*
Perennial herbs with corms. Corms with hard, brown tunics, usually asymmetric at the base. Basal leaves usually 2, rarely 1 or up to 6, usually 4-grooved. Stem-leaves 1–6; all leaves linear, usually hairless and 4-grooved, not in 2 ranks. Flowering stems erect or recurved (especially in fruit), rarely absent. Flowers upright, surrounded by an outer bract and an inner bracteole. Perianth united into a short tube below, lobes all similar. Stamens 3, exceeding or shorter than the style, sometimes aborted. Style 1 with 3 bifid stigmas, the branches thread-like. Capsule borne on an elongate stem. Seeds spherical.

A genus with about 10 species in the Mediterranean area (extending northwards to SW England) and about 70 species in South Africa, mostly in Cape Province. The genus is similar to *Crocus* (p. 361), but is distinguished mainly by the asymmetric base of the corm and the long-stemmed flowers (absent in *R. macowanii* var. *alticola* – see p. 372).

The European and Asiatic species are mostly hardy and succeed well in very sandy soil, kept dry in summer. The African species generally need protection (at least in northern Europe), though the alpine *R. macowanii* and its varieties should be frost-hardy. Propagation is by seed or by offsets.
Literature: Beguinot, A., Revisione monografica del genere Romulea Maratti, *Malphighia* **21**: 49–122, 364–478 (1907), **22**: 377–469 (1908), **23**: 55–117, 185–239, 257–96 (1909); De Vos, M., The genus Romulea in South Africa, *Journal of South African Botany*, supplementary volume **9** (1972).

*Leaves in section*. Terete with 4 grooves: 1–6,8–11,13,14; terete with 8 sharp ridges: 7; 4-winged: 12.
*Bracteole*. Largely membranous, with green or brown keel: 1–5,8,13,14; largely green, with membranous margins: 6,7,9–12.
*Flowers*. Mostly more than 2 cm: 1–4,7,9–14; mostly less than 2 cm: 4–6,8,9.
*Perianth-lobes*. Lilac or purple: 1,2–8,13; white: 1,8,9,13; red, pinkish or magenta: 1,6,9–11,13; yellow: 1,12–14.
*Throat of perianth*. Yellow: 1,6–14; white or purple: 2–6,8; with dark blotches: 11.
*Anthers*. Overtopping or equalling stigmas: 4–6,8–14; not reaching level of stigmas: 1–3,7,9.

1a. Perianth more than 2 cm  2
 b. Perianth less than 2 cm  16
2a. Stigmas overtopping anthers  3
 b. Stigmas equalling or shorter than anthers  8
3a. Perianth-tube more than one-third of the total length of the perianth  **2. tempskyana**
 b. Perianth-tube less than one-third of the total length of the perianth  4
4a. Bracteole mostly green, with membranous margins  5
 b. Bracteole mostly membranous, with green keel  6
5a. Leaves long and slender, 4-grooved  **6. ramiflora**
 b. Leaves stiff, erect and thick, 8-ridged  **7. nivalis**
6a. Perianth deep violet, lobes obtuse; anthers reaching two-thirds or more of the length of the perianth  **3. requienii**
 b. Perianth violet, magenta, pink, yellow or white; anthers reaching at most two-thirds of the length of the perianth  7
7a. Basal leaves 2  **1. bulbocodium**
 b. Basal leaves several  **9. rosea**
8a. Perianth yellow  9
 b. Perianth purple to pink or white  11
9a. Leaves winged, 2–5 mm wide, usually ciliate  **12. hirta**
 b. Leaves terete, 0.6–3 mm wide, hairless  10
10a. Stem longer than the flower  **13. flava**
 b. Stem shorter than the flower, ovary usually subterranean  **14. macowanii**
11a. Flowering stem dichotomously branched  **10. dichotoma**
 b. Flowering stem simple or umbel-like  12

12a. Perianth scarlet or red, with a black blotch on each lobe  **11. sabulosa**
 b. Perianth-lobes without black blotches  13
13a. Perianth with yellowish throat  14
 b. Perianth with white or purplish throat  15
14a. Corm rounded at the base  **9. rosea**
 b. Corm flattened at the base, with a crescent-shaped ridge  **13. flava**
15a. Anthers reaching less than half the length of the perianth  **4. ligustica**
 b. Anthers reaching more than half the length of the perianth  **5. linaresii**
16a. Corm rounded at the base  **9. rosea**
 b. Corm obliquely narrowed at the base  17
17a. Bracteole largely green  **6. ramiflora**
 b. Bracteole largely membranous  18
18a. Throat of perianth purple  **5. linaresii**
 b. Throat of perianth white or yellow  **8. columnae**

**1. R. bulbocodium** (Linnaeus) Sebastiani & Mauri (*R. grandiflora* Todaro; *R. clusiana* (Lange) Nyman). Illustration: Rix & Phillips, The bulb book, 31 (1981).
Leaves mostly recurved, 0.8–2 mm broad. Stem with 1–6 flowers. Bracts green with a narrow (*c*. 1 mm), membranous margin. Bracteole with only the central one-third green or purplish, the rest membranous. Perianth 2–3.5 (rarely to 5.5) cm, purplish, pinkish white or yellow. Tube 3.5–8 mm, the throat yellow, orange or white. Filaments hairy below. Top of anthers not exceeding the white styles. *Mediterranean area, Portugal, NW Spain & S Bulgaria*. H3. Spring.

Several varieties occur; the 3 most common in cultivation are: var. **crocea** (Boissier & Heldreich) Baker, with yellow flowers, var. **leichtliniana** (Heldreich) Beguinot with white flowers which have a yellow throat, and var. **subpalustris** Baker with purplish flowers and a white throat.

**2. R. tempskyana** Freyn. Illustration: Rix & Phillips, The bulb book, 31 (1981).
Leaves mostly recurved, *c*. 1 mm broad. Stem 1-flowered. Bract green with narrow (1.5 mm), membranous margin. Bracteole membranous with a narrow, green, central strip. Perianth 2–3.7 cm, deep purple; tube narrow, 8.5–17 mm, lobes *c*. 5 mm wide, lanceolate, reflexed when the flower is open wide; throat purple. Filaments hairless; top of anthers reaching about halfway up the perianth-lobes, not overtopping the purple style. *Greece (Rhodes), SW Turkey, Cyprus, Israel*. H3. Spring.

**3. R. requienii** Parlatore. Illustration: Botanical Magazine, 9555 (1939).
Leaves recurved, *c*. 1 mm broad. Stem with 1–3 (rarely to 6) flowers. Bract green with narrow, membranous margins. Bracteole membranous with a narrow, green, central strip. Perianth 2–2.5 cm, deep purple; tube 5–8 mm; lobes 5–6 mm wide, obovate-oblong, rounded at apex, not reflexed; throat purple or whitish. Top of anthers reaching two-thirds or more up the perianth-lobes, not overtopping the style. *Corsica, Sardinia & Italy*. H4. Spring.

**4. R. ligustica** Parlatore. Illustration: Jordan & Fourreau, Icones ad Floram Europae **2**: t. 342 (1903).
Leaves erect or recurved, 1–1.5 mm broad. Stem with 1 flower. Bract green with narrow membranous margin. Bracteole almost entirely membranous. Perianth 1.9–3.5 cm, lilac or violet; tube 5–7.5 mm, lobes obovate-elliptic, rounded to almost acute at the apex, not reflexed; throat not yellow. Filaments hairy at the base; top of anthers less than halfway up the perianth-lobes, overtopping the style. *Corsica, Sardinia & Italy*. H4. Spring.

**5. R. linaresii** Parlatore. Illustration: Jordan & Fourreau, Icones ad Floram Europae **2**: t. 337 bis (1903); Mathew & Baytop, The bulbous plants of Turkey, t. 58 (1984).
Leaves mostly recurved, 1–1.5 mm broad. Stem with 1–2 flowers. Bract green with narrow (1 mm wide), membranous margins. Bracteole with soft greenish centre, otherwise membranous. Perianth 1.3–2.6 cm, violet-purple, tube 4–7 mm, lobes not reflexed; throat deep purple. Filaments hairy at the base, top of the anthers reaching more than halfway up the perianth, overtopping the style. *Sicily, Greece, W Turkey*. H4. Spring.

Two subspecies occur: subsp. **linaresii** from coastal sands in Sicily, has the perianth 2–2.6 cm with blunt lobes, whereas subsp. **graeca** Beguinot from further east has the perianth 1.3–2 cm with acute lobes.

**6. R. ramiflora** Tenore. Illustration: Jordan & Fourreau, Icones ad Floram Europae **2**: t. 335 bis (1903).
Leaves erect or recurved, 0.75–1.5 mm. Stem with 1–4 flowers, elongating to 30 cm in fruit; flower-stalks elongating to 10 cm in fruit. Bract green, strongly veined. Bracteole green with a membranous margin. Perianth 1–2.2 cm, lilac or pink, tube 2.5–7 mm, lobes oblanceolate, acute: throat white or yellow.

Filaments hairy or not; top of anthers reaching about two-thirds of the way up the perianth, at about the same level as the top of the style. *Mediterranean area*. H3. Spring.

Two subspecies occur: subsp. **ramiflora** which has a perianth 1–1.8 cm, sometimes shorter than the bract and the style reaching to slightly below the tops of the anthers; and subsp. **gaditana** (Kunze) Marais, confined to the west and south of the Iberian peninsula, which has the perianth 2–3 cm, lilac or pink, green outside, and the style slightly exceeding the tops of the anthers.

**7. R. nivalis** (Boissier & Kotschy) Klatt. Illustration: Rix & Phillips, The bulb book, 31 (1981).
Leaves erect, stiff, *c*. 2 mm wide, with 8 equal, narrow ridges. Stem with 1–3 flowers, not elongating in fruit. Bract green with a narrow (*c*. 1 mm) membranous margin. Bracteole green with wide membranous margin. Perianth 2–2.5 cm, lilac; tube *c*. 3 mm, lobes lanceolate, blunt; throat yellow. Top of anthers about halfway up the perianth, not overtopping the style. *Lebanon*. H4. Spring.

**8. R. columnae** Sebastiani & Mauri. Illustration: Ross-Craig, Drawings of British plants **29**: t. 3 (1972).
Leaves recurved, 0.6–1 mm wide. Stem with 1–3 flowers, flower-stalks short. Bract green with narrow membranous margins. Bracteole almost entirely membranous. Perianth 1–1.5 cm, whitish to pale lilac with darker veins, tube 2.5–5.5 mm, lobes lanceolate or oblanceolate, acute; throat yellow to white. Filaments often hairless; top of anthers reaching about halfway up the perianth, overtopping the style. *W Europe, from SW England to the Mediterranean area*. H3. Spring.

**9. R rosea** (Linnaeus) Ecklon. Illustration: Journal of South African Botany, supplementary volume **9**: f. 87 (1972).
Corm rounded at the base, the tunic splitting into bent teeth. Basal leaves several, erect or recurved, 1–2.5 mm wide. Stem short, with several flowers, not elongating in fruit; flower-stalks to 15 cm. Bract green with a very narrow membranous margin. Bracteole with wide membranous margins, with brown streaks. Perianth 1.5–4.5 cm, magenta, pinkish or white, tube 2–5 mm, lobes 3–10 mm wide; throat yellowish, sometimes with a bluish zone. Filaments hairy below; top of anthers reaching about halfway up the perianth,

shorter than or exceeding the style. *South Africa, naturalised in S France and Guernsey and also in places in the southern hemisphere*. H4. Spring.

A very variable species; var. **rosea** has flowers 2.2–4.5 cm and anthers shorter than the style; var. **australis** (Ewart) De Vos has flowers 1.5–2.2 cm, usually pale pink, and anthers exceeding the style.

**10. R. dichotoma** (Thunberg) Baker. Illustration: Journal of South African Botany, supplementary volume **9**: f. 57, 61 (1972).
Basal leaf 1 (rarely 2), erect or sometimes bent, 1–2 mm wide. Stem leaves 1–2. Stem 4–35 cm, erect, rigid, dichotomously branched near the top, with several flowers. Bract green with a very narrow membranous margin. Bracteole green with wide membranous margins, reddish towards the top. Perianth 2–4 cm, pink with deeper veins in the yellow throat; tube 3–5 mm, lobes 6–10 mm wide. Filaments minutely hairy below; top of anthers reaching less than halfway up the perianth, equalling or just overtopping the stigmas. *South Africa (Cape Province)*. H4. Spring.

**11. R. sabulosa** Beguinot. Illustration: Journal of South African Botany, supplementary volume **9**: f. 79, 93 (1972).
Corm rounded at the base, the tunic splitting into bent teeth. Basal leaves almost erect to recurved, *c*. 1 mm wide. Stem with 1–4 flowers, short, hidden by the leaf bases. Flower-stalks 4–14 cm. Bract stiff, green, with narrow, brown, membranous margin. Bracteole 2-keeled, green with membranous margin which is white below, brown above. Perianth 3–5 cm (rarely more), lobes 1.2–2 cm wide, acute, pointed or obtuse, shining scarlet or red, with a black, pale-edged blotch in the throat; tube 2–4 mm. Filaments hairless or slightly hairy below; top of anthers well below the middle of the perianth, overtopping the stigmas. *South Africa (Cape Province)*. H4. Spring.

**12. R. hirta** Schlechter. Illustration: Flowering plants of Africa **29**: t. 1137 (1952).
Corm rounded at base, the tunic splitting into bent teeth. Basal leaves 3–6, 4-winged, wings ciliate or not, the whole leaf 2–5 mm wide. Stem with 1–4 flowers, short, hidden by the leaf bases. Flower-stalks 4–20 cm. Bract green with very narrow membranous margin. Bracteole with wider membranous margins and tip. Perianth 1.8–3.5 cm, lobes 4–8 mm wide,

more or less obtuse to acute, pale yellow, often with a brownish transverse band at about the middle; tube 4–5 mm. Filaments minutely hairy below; top of anthers below the middle of the perianth, overtopping the stigmas (sometimes only slightly). *South Africa (Cape Province)*. H4. Spring.

**13. R. flava** (Lamarck) De Vos (*R. bulbocodioides* misapplied). Illustration: Botanical Magazine, 1392 (1811) – as Trichonema caulescens.
Basal leaf 1, minutely ciliate or hairless, 0.8–3 mm wide. Stem with 1–4 or more flowers, hidden by the leaf base or elongate and evident. Flower-stalks 1.5–7 cm. Bract green with very narrow membranous margin. Bracteole membranous, sometimes with a brownish central strip. Perianth 2–4 cm, lobes 3–12 mm wide, acute to obtuse, usually yellow, sometimes white, rarely blue or pinkish; throat golden yellow; tube 3–7 mm. Filaments hairy below; top of anthers below the middle of the perianth, about level with the stigmas. *South Africa (Cape Province)*. H4. Winter–spring.

**14. R. macowanii** Baker (*R. longituba* Bolus). Illustration: Botanical Magazine, n.s., 515 (1967).
Basal leaves 3–6, more or less erect or recurved, 0.6–1 mm wide. Stem with 1–3 flowers, hidden by the leaf bases. Flower-stalks 1–8 cm, often below ground. Bract green, membranous below. Bracteole more membranous, with wide, white margin, tip green. Perianth 2.2–10.5 cm, lobes 5–15 mm wide, acute to obtuse, golden yellow; tube 2–6.5 cm, throat often orange-yellow. Filaments hairy below; tops of anthers about level with the middle of the perianth, level with the stigmas. *South Africa, Lesotho*. H4. Summer, rarely spring.
 Three varieties occur: var. **alticola** (Burtt) De Vos, from over 2000 m in the Drakensberg in South Africa and Lesotho, has the perianth-tube 3.5–6.5 cm. Var. **macowanii** and var. **oreophila** De Vos have perianth-tubes to 3.3 cm; var. **macowanii** from the Karroo has larger flowers and wider perianth-lobes than var. **oreophila**, which is an alpine variant from 2400–2700 m in the northeast Cape and Lesotho.

**20. GELASINE** Herbert
*J.C.M. Alexander*
Corm *c.* 8 mm in diameter, almost spherical, with membranous tunics. Basal leaves 45–60 × 2.5 cm, 2-ranked, pleated, narrowly lanceolate, acuminate. Stem to

60 cm, bearing 3 or 4 reduced sheathing leaves and a terminal cluster of 2–4 funnel-shaped flowers. Spathe-bracts green, unequal. Flower-stalks short. Perianth radially symmetric, blue above, white with black spots at base; tube very short; lobes *c.* 1.8 cm, similar, obovate, each with a small point. Filament tube shorter than the triangular, blue, ascending anthers. Style short, divided into 3 short linear stigmas with rounded tips.
 A genus of 1 species from eastern S America. *Gelasine* does well in light, well-drained soil in full sun, but will need protection from frost. If grown in a cool greenhouse, the foliage will continue to grow through the winter. It is easily propagated by cormlets or from seed which should be sown as soon as it is ripe.

**1. G. azurea** Herbert. Illustration: Botanical Magazine, 3779 (1840); Journal of Horticulture 26: 488 (1874).
*S Brasil & Uruguay*. H5–G1. Spring.

**21. DIPLARRHENA** Labillardière
*V.A. Matthews*
Rhizome short. Leaves mostly basal, linear, in 2 ranks. Stem erect, hairless, simple or occasionally branched. Flowers short-lived, 2 or more within the terminal bracts. Perianth free, of 6 segments, the outer larger than the inner. Fertile stamens 2; filaments free. Style short, divided into 3 flattened branches. Fruit a 3-celled capsule containing flattened seeds.
 A genus of 2 species from Australia and Tasmania, both of which are in cultivation. The plants should be cultivated as for *Libertia* but it should be remembered that they are slightly less hardy. Propagation is by seed, or division of the rhizome.

1a. Flowers 3–4 cm in diameter, usually 2 or 3 in each set of bracts
                                         **1. moraea**
  b. Flowers 6–15 cm in diameter, 5 or 6 in each set of bracts        **2. latifolia**

**1. D. moraea** Labillardière. Illustration: King & Burns, Wild flowers of Tasmania, 103 (1969); Rotherham et al., Flowers and plants of New South Wales and southern Queensland, f. 289 (1975).
Stems 45–65 cm. Leaves 30–45 cm × 8–12 mm, shorter than the stems. Bracts 5–7 cm, each set enclosing 2 or 3 (occasionally more) flowers. Flowers 3–4 cm in diameter, fragrant. Outer perianth-segments reflexed-spreading, almost circular, one larger than the other two, *c.* 3 cm; inner segments erect, shorter,

narrower, white flushed or veined with lilac or yellow towards the tips. *SE Australia, Tasmania*. H4. Summer.

**2. D. latifolia** Bentham. Illustration: Curtis, The endemic Flora of Tasmania 6: pl. 120 (1978).
More robust than *D. moraea*, with leaves to 2.5 cm wide. Flowers 6–15 cm in diameter, 5 or 6 in each set of bracts. *Tasmania*. H4. Summer.

**22. LIBERTIA** Sprengel
*J.C.M. Alexander & D.A. Webb*
Perennials with short, creeping rhizomes and fibrous roots. Leaves numerous, mostly basal, equitant, linear, overlapping, flat and rigid (except no. 1). Flowering stems erect, bearing a few reduced leaves. Flowers radially symmetric, in loose clusters or panicles borne terminally and in the axils of the stem-leaves. Perianth-segments entirely free, spreading, the inner segments usually longer than the outer. Stamens 3; filaments slightly fused at base. Style with 3 entire, linear, spreading branches; fruit a many-seeded, 3-celled capsule.
 A genus of up to 20 species from Australasia and the temperate or montane regions of S America, which does well in light peaty soil in a warm sunny border. Most of the species are hardy, though *L. caerulescens* (no. 5) requires protection from frost. Plants can be propagated from seed or by careful division in spring.

1a. Flowers blue          **5. caerulescens**
  b. Flowers white                              2
2a. Flower-clusters dense; flower-stalks shorter than bracts        **4. formosa**
  b. Flower-clusters loose; flower-stalks longer than bracts                        3
3a. Leaves soft, clear green; capsule spherical or broader than long
                                         **1. pulchella**
  b. Leaves stiff, brownish green; capsule longer than broad                        4
4a. Leaves mostly 20–30 cm × 4–6 mm; capsule 5–7 mm       **2. ixioides**
  b. Leaves mostly 30–50 cm × 7–12 mm; capsule 10–15 mm       **3. grandiflora**

**1. L. pulchella** Sprengel (*L. micrantha* Cunningham). Illustration: Galbraith, Wild flowers of SE Australia, pl. 51 (1977).
Basal leaves linear, soft, uniformly green, 5–15 cm × 2–4 mm. Stem 8–30 cm with a single reduced leaf in the lower half. Flowers 2–8 in a stalkless terminal cluster (rarely 1 or 2 lateral clusters also present). Spathe-bracts lanceolate, green, 6–12 mm. Flower-stalks 18–30 mm, slender, downy.

Perianth-segments 4–6 mm, oblong to ovate, white, the outer almost as large as the inner. Stamens *c.* as long as perianth. Capsule almost spherical, 3–4 mm in diameter. *SE Australia, Tasmania, New Zealand and Papua New Guinea.* H4. Spring.

**2. L. ixioides** (Forster) Sprengel (not *L. ixioides* Gay or *L. ixioides* Klatt). Illustration: Moore & Irwin, Oxford Book of New Zealand plants, 187 (1978). Basal leaves numerous, tufted, linear, rigid and leathery with cartilagenous margins, green, midrib pale, 20–30 cm or more × 4–6 mm. Stem 30–60 cm, branched in upper half. Inflorescence a broad panicle of many, stalked umbels, each with 2–10 flowers. Bracts many, minute, green, lanceolate. Segments 6–9 mm, circular to oblong, white; the outer slightly shorter, oblong, white tinged with brown or green. Stamens *c.* 4 mm. Capsule conical to oblong, 5–7 mm. *New Zealand & Chatham Islands.* H4. Summer.

**3. L. grandiflora** (R. Brown) Sweet. Illustration: Gardeners' Chronicle **43**: 2 (1908); Kirk, British garden flora, 500 (1927); Flora **125**: 22 (1930). Similar to *L. ixioides.* Leaves 30–75 cm × 7–12 mm. Umbels with 3–6 flowers, stalked. Inner perianth-segments 12–15 mm; the outer *c.* 6 mm, with a greenish brown keel. Capsule 10–15 mm, yellow. *New Zealand.* H4. Summer.

**4. L. formosa** Graham (*L. chilensis* Klotsch; *L. ixioides* Klatt). Illustration: Botanical Magazine, 3294 (1834); Garden **70**: 175 (1906) & **75**: 545 (1911). Basal leaves linear, rigid, dark green, 15–45 cm × 6–12 mm. Stem 50–120 cm, with 1 or 2 reduced sheathing leaves below the inflorescence. Flowers in many dense stalkless terminal umbels, on a simple or branched axis. Outer bract of each umbel large, obovate, membranous; inner bracts smaller, oblong. Flower-stalks shorter than outer bracts, 6–12 mm. Inner perianth-segments white or pale yellow, obovate to wedge-shaped, 12–18 mm; outer perianth-segments greenish brown, oblong, 6–9 mm. Stamens 6–10 mm. Capsule spherical, *c.* 6 mm in diameter. *Chile.* H3. Spring.

**5. L. caerulescens** Kunth. Illustration: Gartenflora **22**: t. 759 (1873). Basal leaves linear, rigid, green, to 30 cm or more. Stem 30–60 cm, bearing 2–4 reduced leaves. Inflorescence 10–15 cm, of numerous many-flowered stalkless umbels

on a single or sparsely branched axis. Outer bracts firm, ovate to lanceolate; inner bracts ovate, membranous. Flower-stalks very short. Inner perianth-segments pale blue, oblong, *c.* 6 mm; outer perianth-segments much smaller, greenish brown. Stamens as long as perianth. Capsule very small, spherical. *Chile.* H5–G1. Spring.

**23. BELAMCANDA** Adanson
*V.A. Matthews & B. Mathew*
Rhizomatous perennial with branched stems and leaves borne in fans. Perianth-segments more or less alike, not differentiated into outer and inner, free, twisting spirally after flowering. Stamens free, not held against style. Styles 3, slender, not expanded and petaloid. Capsule splitting into 3 segments which reflex exposing many shiny black seeds attached to the central axis.

A genus of 1, possibly 2, species, best grown in any good soil, in sun or semi-shade with plenty of moisture during the growing period. It is not long-lived. Propagation is by seed or division.

**1. B. chinensis** (Linnaeus) de Candolle (*Pardanthus chinensis* (Linnaeus) Ker Gawler). Illustration: Botanical Magazine, 171 (1791); Bruggeman, Tropical plants, pl. 82 (1957); Everett, Encyclopedia of horticulture **2**: 395 (1981); Kitamura et al., Coloured illustrations of herbaceous plants of Japan (Monocotyledoneae), pl. 20 f. 136 (1981). Stems 60–100 cm. Leaves to 25 × 1–2 cm. Flowers 3–12, 4–5 cm in diameter, yellowish or orange-red spotted with red or dark purple. *E USSR (Ussuri region), China, Japan, Taiwan, N India.* H2. Summer.

**24. ORTHROSANTHUS** Sweet
*J.C.M. Alexander*
Rhizome very short. Stem erect, bearing *c.* 12 stiff, leathery, acute, linear, equitant, 2-ranked leaves near the base and 1 or 2 reduced leaves higher up. Flowers in stalked or stalkless clusters forming a loose panicle, one per cluster open at a time, each lasting for only a few hours. Clusters each subtended by a pair of lanceolate to oblong spathe-bracts (see figure 38(4), p. 333). Flowers each with a similar but smaller pair of bracteoles; flower-stalks very short. Perianth blue, radially symmetric; tube short; lobes oblong, spreading, similar. Filaments free, attached to top of perianth-tube; anthers linear. Style very short; branches entire, linear to oblong. Young capsule partly enclosed by spathe-bracts.

A genus of about 7 species, 2 from tropical America, the rest from Australia.

1a. Leaf-margin rough to touch; flowers purplish blue, on stalks 2.5–4 cm
**1. chimboracensis**
b. Leaf-margin smooth; flowers pale blue on very short, barely exposed stalks
**2. multiflorus**

**1. O. chimboracensis** (Humboldt, Bonpland & Kunth) Baker (*O. occisapungum* Ruiz). Illustration: Botanical Magazine, 8731 (1917); RHS Dictionary of gardening, edn 2, 1450 (1956 and reprints). Rhizome short, stout, woody. Basal leaves to 40 cm × 6–10 mm, ribbed; margins finely toothed. Stem 30–60 cm, terete, sparsely branched. Flowers to 4 cm wide, purplish blue, in clusters of 3 or 4, forming a loose panicle. Outer spathe-bract *c.* 2.5 cm, lanceolate, acuminate, with a membranous margin and a sharp ciliate keel; the inner *c.* 1 cm. Flower-stalks 2.5–5 cm; bracteoles obtuse. Perianth-tube short; lobes *c.* 1.5 cm. Filaments *c.* 5 mm; anthers *c.* 5 mm. Style-branches *c.* 8 mm. Capsule 12–18 mm, 3-sided, dull brown. *Mexico to Peru.* H5–G1. Summer.

**2. O. multiflorus** Sweet (*Sisyrinchium cyaneum* Lindley). Illustration: Edwards's Botanical Register, 1090 (1821); RHS Dictionary of gardening, edn 2, 1450 (1956 and reprints). Basal leaves 30–45 cm × 3–4 mm, linear, acute, rigid. Stem slender, equal to or longer than leaves. Flowers blue, in a narrow panicle of 5–8 clusters, some clusters stalkless. Spathe-bracts 12–15 mm, green with chaffy tips, enclosing several smaller bracteoles. Flower-stalks very short. Fully open flower 2.5–4 cm across; lobes all similar, obovate, obtuse to rounded, with a darker mid-vein. Filaments free; anthers projecting beyond flower and stigmas. Capsule *c.* 8 mm. *SW Australia.* H4. Spring–summer.

**25. SISYRINCHIUM** Linnaeus
*J.C.M. Alexander*
Rhizome absent or very short; roots fleshy or fibrous. Stem erect, simple or branched, terete or flattened, often winged. Leaves mostly basal, 2-ranked, linear to strap-shaped or very narrowly lanceolate, acute, pale to dark or bluish green, often blackening on drying. Stem-leaves broader and shorter. Spathe-bracts usually in pairs, lanceolate, acute, often with a white translucent margin. In branched species a lower (primary) pair of spathe-bracts gives

rise to inflorescence-stalks, on the latter are further (secondary) pairs of spathe-bracts, from which the flower-stalks arise. In unbranched species, a single pair of spathe-bracts gives rise directly to the flower-stalks. Flower-stalks terete or flattened, often arched, lengthening in fruit. Flowers solitary or in clusters (see figure 38(5–7), p. 333). Perianth radially symmetric, very shortly fused at base, lobes oblong, spreading, often notched with a short point in the notch. Stamens arising from base of perianth; filaments united into a cylindric or flask-shaped tube for some or all of their length. Ovary 3-celled, spherical to ovoid; ovules many. Style mostly hidden within filament-tube, with 3 acute or capitate, entire, stigmatic branches.

Variously estimated to contain between 70 and 200 hardy and half-hardy species from the New World. Some authors consider that 1 or 2 species may also be native to W Europe. Many have small flowers and are of little horticultural value. The cultivated species thrive on well-drained sandy soil in full sunlight and are easily propagated from seed. The generic names *Bermudiana*, *Marica* and *Olsynium* have been applied to species now considered under *Sisyrinchium*. The classification and naming of several of the species is very confused. Many names have been consistently misused which makes use of references particularly hazardous. The situation is especially bad among the North American blue-flowered species. Parent (1980) considers that the widely used names *S. bermudiana* and *S. angustifolium* were unclear from the time of their original publication and should be abandoned. His suggestions have been followed in this account.
Literature: Parent, G.H., Le Genre Sisyrinchium L. (Iridaceae) en Europe: un Bilan Provisoire, *Lejeunia* **99**: 1–40 (1980).

1a. Flowers bright yellow, sometimes tinged with green                      2
  b. Flowers white, very pale yellow, off-white, blue, purple or violet         3
2a. Stem unbranched, leafless
                              **1. californicum**
  b. Stem usually branched and bearing leaves                **2. graminifolium\***
3a. Flowers in distant clusters on an unbranched stem           **3. striatum\***
  b. Flower clusters 1 per stem or branch
                                               4
4a. Stems round in section                5
  b. Stems flattened or winged             6
5a. Perianth-lobes less than 1.3 cm, white with pinkish purple veins,

yellow at base; filament-tube flask-shaped; stigmas not protruding
                              **4. filifolium**
  b. Perianth-lobes *c.* 1.8 cm, pinkish purple (rarely pure white); filament-tube cylindric; stigmas protruding
                              **5. douglasii**
6a. Stems unbranched                         7
  b. Stems branched                          8
7a. Stems 4 mm or more in diameter
                              **7. montanum**
  b. Stems 3 mm or less in diameter
                              **6. idahoense\***
8a. Flowers white, off-white, very pale blue or very pale yellow              9
  b. Flowers blue or bluish purple          10
9a. Stem 30–45 cm; perianth-lobes *c.* 12 mm; filament-tube inflated
                              **8. iridifolium\***
  b. Stem 7.5–20 cm; perianth-lobes 4–6 mm; filament-tube cylindric
                              **9. micranthum**
10a. Inflorescence-stalks thread-like, slender                  **10. chilense**
  b. Inflorescence-stalks winged or flattened              **11. graminoides\***

**1. S. californicum** (Ker Gawler) Dryander (*Marica californica* Ker Gawler; *S. brachypus* (Bicknell) Henry; *S. boreale* (Bicknell) Henry; *S. convolutum* misapplied). Illustration: Botanical Magazine, 983 (1807); Hay & Beckett, Reader's Digest encyclopaedia of garden plants and flowers, 667 (1971); Rickett, Wild flowers of the United States **5**(1): 73 (1971).
Stem to 60 cm, unbranched, flattened and winged, to 4 (rarely 6) mm in diameter, slightly glaucous, drying blackish. Leaves all basal, usually shorter than the stem, to 6 mm wide, linear, flat, obtuse. Spathe-bracts lanceolate, obtuse; margins translucent; the outer to 5 cm, the inner a little shorter. Flowers 3–7 in a single cluster; flower-stalks 1–4 cm, usually longer than the bracts. Perianth-lobes 1.2–1.8 cm, oblong, obtuse to acute, bright yellow with 5–7 darker veins. Stamens *c.* 7 mm; anthers orange to yellow, 3–5 mm. Mature capsule 7–12 × 6–8 mm, ellipsoid, 3-angled, blackish; seeds numerous, to 1.5 mm in diameter. *Western N America (Vancouver to California)*. H3. Summer.

The name *S. brachypus* has been given to plants from the northern part of the range with short flower-stalks (1–2 cm) hidden in the spathe-bracts and seeds less than 1 cm in diameter. If this is accepted as a separate species then the name *S. brachypus* should be applied to many plants cultivated as *S. californicum*.

**2. S. graminifolium** Lindley (*S. majale* Link, Klotsch & Otto). Illustration: Edwards's Botanical Register, 1067 (1827).
Roots fleshy; stem to 45 cm, branched, flattened and winged, minutely hairy. Basal leaves linear, grass-like, shorter than stems; stem-leaves with broad inflated sheathing-bases. Spathe-bracts oblong to lanceolate, with translucent margins, inflated; the outer, 1.8–2.5 cm. Flower-clusters 3 or 4; flowers 4–8 per cluster; flower-stalks 1.5–2.5 cm. Perianth-lobes *c.* 1.3 cm, oblong to obovate, acute, yellow, often with a brownish purple spot near the base. Filaments united below into a cylindric tube. Capsule *c.* 8 mm. *Chile*. H5–G1. Spring.

Var. **maculatum** (Hooker) Baker (*S. maculatum* Hooker). Illustration: Botanical Magazine, 3197 (1832). Spots more pronounced; brownish purple patches also present in outer parts of alternate perianth-lobes.

Subsp. **nanum** (Philippi) Ravenna (*S. nanum* Philippi). Illustration: Edwards's Botanical Register, 1914, text 1915 (1836). Smaller in stature with broader leaves.

**\*S. convolutum** Nocca (*S. graminifolium* misapplied; *S. alatum* misapplied; *S. iridifolium* misapplied). Illustration: Redouté, Liliacées **1**: pl. 47 (1803); Sanchez, Flora del valle de Mexico, f. 63 (1969). Densely tufted, roots fibrous. Basal leaves 15–20 cm × 4–8 mm. Stem broadly winged. Spathe-bracts 2.5–3 cm. Flower-stalks arched in fruit. Filaments fused only at base. *Mexico to Peru*. G2.

**\*S. tenuifolium** Humboldt & Bonpland (*S. hartwegii* Baker). Illustration: Botanical Magazine, 2117 (1819) & 2313 (1822); Sanchez, Flora del valle de Mexico, f. 63 (1969). Roots fleshy. Basal leaves linear to terete, 8–15 cm × 2–4 mm. Stem narrowly winged, 10–35 cm. Spathe-bracts 1.2–1.8 cm. Flowers 3 or 4 per cluster, on stalks protruding from the spathe-bracts and lengthening in fruit. Perianth-lobes 6–10 mm, yellow, striped with green, bending downwards when flower fully open. Capsule oblong, 6–9 mm, erect at maturity. *Mexico*. H4.

**3. S. striatum** Smith (*S. graminifolium*, *S. iridifolium*, *S. lutescens* and *S. pachyrhizum* misapplied). Illustration: Botanical Magazine, 701 (1803); Hay & Synge, Dictionary of garden plants, 175 (1969); Hay & Beckett, Reader's Digest encyclopaedia of garden plants and flowers, 667 (1971).

Root fibres slender. Stem to 75 cm, erect, unbranched, flattened and narrowly winged, greyish green. Basal leaves 8–10, linear to narrowly lanceolate, 20–35 × 1.2–1.8 cm, greyish green; stem-leaves 1 or 2, narrowly lanceolate with sheathing bases, to 30 × 1.5 cm. Flowers pale greenish yellow, in several stalkless clusters spaced along stem; each cluster of 10–20 flowers enclosed in a pair of ovate, slightly chaffy spathe-bracts. Flower-stalks equalling the bracts in length. Perianth-lobes 1.5–1.8 cm, oblanceolate, strongly veined with purplish brown. Filaments united only at base. Capsule spherical, c. 6 mm in diameter. *Chile & Argentina; naturalised in Isles of Scilly (SW Britain).* H5–G1. Summer.

A variegated form is available with yellow-striped leaves.

**\*S. arenarium** Poeppig (*S. cuspidatum* Poeppig; *S. striatum* misapplied). Illustration: Correa (ed), Flora Patagonica 2: 174 (1969). Roots fleshy. Stem scarcely flattened, hairy between flower-clusters. Basal leaves many, 10–40 cm × 3–6 mm. Flowers yellow with greyish brown veins; perianth-lobes 11–13 × c. 6 mm. Ovary very yellow-hairy. *Chile & Argentina.* H5–G1.

**4. S. filifolium** Gaudichaud-Beaupré (*S. junceum* Presl subsp. *filifolium* (Gaudichaud Beaupré) Ravenna). Illustration: Botanical Magazine, 6829 (1885); Vallentin & Cotton, Plants of the Falkland Islands, pl. 55 (1921); Muñoz Pizarro, Sinopsis de la Flora Chilena, t. 223 (1959).
Roots slender, slightly fleshy. Stem 15–35 cm, unbranched, terete. Basal leaves merging with stem-leaves, 3–6, linear, acute, 10–15 cm, 3-channelled; stem-leaves smaller. Spathe-bracts 5–12 cm. Flowers 2–8 in 1 or 2 terminal clusters, overtopped by final stem-leaf; clusters often hanging; flower-stalks c. 2.5 cm. Flowers c. 2.5 cm across, white with reddish purple veins, yellow at base. Perianth-lobes obovate, c. 1.2 cm. Filaments united at base. Capsule spherical to ovoid. *S Chile & Falkland Islands.* H2. Spring.

Subsp. **junceum** (Presl) Parent (*S. junceum* Presl; *S. juncifolium* Herbert). Illustration: Muñoz Pizarro, Sinopsis de la Flora Chilena, t. 224 (1959); Correa (ed), Flora Patagonica 2: 181 (1969). Basal leaves usually 2, terete, rarely flattened. Spathe-bracts 2–2.5 cm. Flowers 13–29 mm wide; perianth-lobes c. 1 cm.

Filament-tube narrowly flask-shaped. *S Peru, Bolivia, Chile & Argentina; naturalised in Canada (Ontario).*

**5. S. douglasii** Dietrich (*S. grandiflorum* Douglas; *S. inalatum* Nelson; *S. inflatum* (Suksdorf) St John; *S. angustifolium* misapplied). Illustration: Botanical Magazine, 3509 (1836); Hay & Beckett, Reader's Digest encyclopaedia of garden plants and flowers, 667 (1971); Rickett, Wild flowers of the United States 5(1): 69 (1971) & 6(1): 62 & 63 (1973).
Stem 10–30 cm, unbranched, terete. Basal leaves to 10 cm, linear with sheathing bases; stem-leaves few. Spathe-bracts very unequal; the outer to 12.5 cm; the inner shorter than the slender flower-stalks. Flowers usually 2 or 3 per cluster, 4–5 cm wide, purple, pink or white, sometimes with a basal green blotch, often slightly hanging. Perianth-lobes c. 1.8 cm. Filament-tube somewhat inflated. Style projecting, as long as perianth. *N America: British Columbia south to N California & Utah.* H2. Spring–summer.

Plants from the eastern part of the range with pink flowers and strongly inflated filament-tubes have been called *S. inflatum*.

**6. S. idahoense** Bicknell (*S. macounii* Bicknell; *S. anceps* misapplied; *S. bellum* misapplied; *S. montanum* misapplied). Illustration: Haskin, Wild flowers of the Pacific coast, 56 (1934); Rickett, Wild flowers of the United States 5(1): 69 (1971).
Leaves all basal, 7–30 cm × c. 3 mm, linear to narrowly lanceolate, acute. Stem 20–50 cm, narrow, unbranched, bluish green, winged and often twisted. Spathe-bracts usually very unequal; the outer 3–6 cm, narrowly lanceolate, acute; the inner usually much smaller. Flower-stalks equal to or shorter than inner spathe-bract. Flowers 1–6 per cluster, dark purplish blue, yellow at base. Perianth-lobes c. 1.3 cm. *Western N America; naturalised in eastern Sweden.* H2. Summer.

Considered by some authors to be part of *S. graminoides*. Smaller plants (to 20 cm) with fewer, larger flowers and a large stem-bract have been called *S. sarmentosum* Greene (*S. birameum* Piper); if these are considered part of *S. idahoense*, then the name *S. sarmentosum* being earlier would apply to the whole species.

**\*S. mucronatum** Michaux (*S. bermudiana* Linnaeus var. *mucronatum* (Michaux) Gray; *S. gramineum* Lamarck; *S. junceum* misapplied). Illustration: Botanical Magazine, 464 (1799); Britton & Brown,

Illustrated flora of the northern states & Canada, edn 2, 544 (1913). Leaves usually 2 mm wide or less. Stem stiffly erect, unbranched, almost thread-like. Flower-stalks longer than inner spathe-bract, spreading or arching. Spathe-bracts often reddish purple. Flowers violet. *Eastern N America.* H2. Spring–summer.

**7. S. montanum** Greene (*S. angustifolium* misapplied). Illustration: Bulletin of the Torrey Botanical Club 23: pl. 265 (1896); Rickett, Wild flowers of the United States 6(1): 62 & 63 (1973).
Stem to 60 cm, erect, unbranched, stiff, flattened and winged, pale green or bluish green. Leaves shorter than stem, to 4 mm wide, pale green, remaining green when dry. Spathe-bracts sometimes fused at base for c. 6 mm; the outer bract c. twice as long as the inner and very finely tapered. Flower-stalks shorter than inner spathe-bract, erect, making an angle with the stem of less than 30°. Flowers c. 1.3 cm across, deep purplish blue. Filaments almost completely fused. *N America from the Rocky Mountains east to the Atlantic Coast & Greenland; widely naturalised in Europe.* H2. Spring–summer.

Var. **crebrum** Fernald (*S. angustifolium* misapplied) is commonly grown and has dark green foliage often tinged with purple. *Eastern N America from Ontario to Virginia; Greenland.*

**8. S. iridifolium** Kunth (*S. chilense* misapplied; *S. iridioides* misapplied). Illustration: Edwards's Botanical Register 8: t. 646 (1822); Loddiges' Botanical Cabinet 20: t. 1979 (1833); Healy & Edgar, Flora of New Zealand 3: f. 22 (1980).
Stem to 45 cm, branched, flattened and winged with bristly edges. Basal leaves shorter than stem, to 1 cm wide, linear to narrowly lanceolate, with bristly edges. Stem-leaves 2–4, similar but smaller. Spathe-bracts 2.5–3.8 cm, with bristly edges, similar in size and shape, or the inner a little smaller. Flowers in terminal clusters of 4–6, white to cream or pale yellow with purple veins merging into a dark purple centre. Flower-stalks c. equalling bract-length, becoming longer in fruit. Filaments more than half fused into a hairy, urn-shaped tube. Perianth-lobes c. 1.3 cm, oblong with a short, acute point, not overlapping when flower fully open. *Tropical S America (? Chile).* H5–G1. Summer.

**\*S. laxum** Sims. Illustration: Botanical Magazine, 2312 (1822). Similar, though not bristly; perianth-lobes c. 7 mm, ovate, overlapping when flower fully open. *Temperate S America*. H4. Summer.

**9. S. micranthum** Cavanilles. Illustration: Botanical Magazine, 2116 (1819); Burbidge & Gray, Flora of the Australian Capital Territory, 109 (1970).
Stem 7.5–20 cm, flattened and very narrowly winged, branched below, bearing several stem-leaves. Basal leaves less than half stem-length, linear, acute. Flowers in small terminal clusters, occasionally some in lower axils, 2–6 per cluster. Spathe-bracts similar, lanceolate, 1.2–2 cm. Flower-stalks equal to or slightly shorter than spathe-bracts. Perianth-lobes off-white to pale yellow, reddish purple at base, 4–6 mm. Filaments fused into a very short, cylindric tube. *C American to NW Argentina; naturalised in Australia*. H3. Summer.

**10. S. chilense** Hooker (*S. geniculatum* misapplied). Illustration: Botanical Magazine, 2786 (1827).
Roots fibres slender. Stem 15–45 cm, branched, flattened and narrowly winged, 2–3 mm wide; nodes prominent. Basal leaves linear to sword-shaped, 8–30 cm × 2–4 mm, veins prominent. Flowers in 4–6 clusters (rarely as many as 20), each cluster with 3–6 flowers. Inflorescence-stalks 10–12.5 cm, slender, thread-like, solitary in axils or in terminal groups of 4–6. Spathe-bracts similar, 1.8–2.5 cm, rigid, linear to lanceolate, acute. Flower-stalks very slender, 2.5–3.5 cm. Perianth-lobes *c.* 1.3 cm, purplish blue, yellow at base, oblong, notched at the top with a narrow tooth in the notch. Filaments almost totally united into a sparsely hairy cylindric tube. *Mexico to Uruguay & Chile; naturalised in Mauritius and possible S France*. H4. Summer.

This species may not be in cultivation but is included here because its name has been widely misapplied to plants that are cultivated. Many other illustrations are available but they nearly all represent plants which do not fit Hooker's original description.

**11. S. graminoides** Bicknell (*S. angustifolium* Miller, ambiguous; *S. bermudiana* Linnaeus in part, ambiguous; *S. anceps* Cavanilles; *S. gramineum* Curtis, not Lamarck; *S. arenicola* Bicknell; *S. montanum* misapplied; *S. chilense* misapplied; *S. birameum* misapplied). Illustration: Botanical Magazine, 464 (1799); Rickett, Wild flowers of the United States 2(1): 79 (1967); Fitter et al., Wild flowers of Britain and northern Europe, 273 (1974).

Stem to 50 cm, almost always branched, flattened and broadly winged, flexuous, not stiff. Leaves equal to or shorter than the stem, to *c.* 6 mm wide, dark bluish green, turning brown when dry. Inflorescence stalks stout, flattened and winged. Spathe-bracts more or less equal, sometimes the outer longer. Flower-stalks longer than bracts, spreading, making an angle with the stem of at least 60°. Flowers 1.2–1.8 cm across, pale to dark purplish blue, yellow at base, greyish outside. Filaments fused into a slender cylindric tube. *Eastern N America (area imprecise through misidentification and misuse of names); ?Greenland; Ireland, SW Britain, Spain & France*. H2. Spring–summer.

Many authors consider the European population to be naturalised rather than native. Hybrids are recorded with *S. montanum* (no. 7).

Similar plants with slender, once-branched stems and dense tufts of old leaf-bases have been named *S. arenicola* Bicknell, a name sometimes found in horticultural literature and catalogues.

**\*S. bellum** Watson (*S. angustifolium* Miller var. *bellum* (Watson) Baker). Illustration: Rickett, Wild flowers of the United States 5(1): 68 & 69 (1971). Stems narrowly winged; leaves *c.* half stem-length, *c.* 4 mm wide. Spathe-bracts very unequal. Flowers violet-blue; perianth-lobes *c.* 1.2 cm. *California & New Mexico*. H3. Plants grown under this name are usually *S. idahoense* (no. 6).

**\*S. atlanticum** Bicknell. Leaves narrow, much shorter than stems, not darkening on drying. Stems slender, often branched more than once, nodes prominent. *Eastern N America*.

**\*S. iridioides** Curtis (*S. bermudiana* Linnaeus in part, ambiguous). Illustration: Botanical Magazine, 94 (1789). Flowers up to 2 times as large as those of *S. graminoides*, bright purplish blue. Spathe-bracts very broad, slightly unequal. *Bermuda*.

### 26. SOLENOMELUS Miers
*J.C.M. Alexander*
Herbs with short rhizomes. Leaves 2-ranked, mostly basal or clustered near stem base, a few higher up. Flowers many, on short stalks, in clusters enclosed by a pair of spathe-bracts. Perianth yellow or blue, radially symmetric, with a slender cylindric tube, lobes spreading, similar. Filaments attached at throat, completely fused into a tube; anthers oblong. Style simple; stigma entire or with 3 very short teeth.

A genus of 2 species from temperate S America, closely related to *Sisyrinchium* but differing in its almost undivided or entire style. Cultivation as for *Sisyrinchium* (p. 373).

1a. Flowers yellow; perianth-tube *c.* 2.5 cm; leaves linear
　　　　　　　　　　**1. pedunculatus**
　b. Flowers blue; perianth-tube *c.* 5 mm; leaves almost terete
　　　　　　　　　　**2. sisyrinchium**

**1. S. pedunculatus** (Hooker) Hochreutiner (*Sisyrinchium pedunculatum* Hooker; *Solenomelus chilensis* Miers; not *Sisyrinchium chilense* Hooker). Illustration: Botanical Magazine, 2965 (1830); Transactions of the Linnaean Society 19: t. 8 (1845). Basal leaves 15–30 cm, linear, grass-like, gradually tapered. Stem 30–45 cm, bearing a few reduced leaves and 1–3 long-stalked clusters of flowers in their axils. Spathe-bracts *c.* 2.5 cm, concave, ovate to oblong with white membranous margins. Flower-stalks very short. Perianth yellow; tube *c.* 2.5 cm; lobes *c.* 2.5 cm, oblanceolate to obovate, obtuse to rounded. Anthers much shorter than staminal tube. *Chile*. H3. Summer.

**2. S. sisyrinchium** (Grisebach) Diels (*Lechlera sisyrinchium* Grisebach; *S. lechleri* Baker). Illustration: Correa, Flora Patagonica 8(2): 185 (1969).
Densely tufted. Basal leaves many, 10–30 cm × 1–1.5 mm, terete, acute, firm. Stem leaves fewer and smaller. Stem 10–30 cm, bearing 1 or 2 short-lived flowers. Spathe-bracts 2.5–3 cm, the outer lanceolate, rigid, green with white margins, the inner smaller. Flower-stalks to 1 mm. Perianth blue; tube *c.* 5 mm, very narrow; lobes 1.3–1.4 cm × 4 mm, oblanceolate. Anthers equal to or a little shorter than staminal tube. Stigma entire forming a short tube at style apex. *Chile*. H3. Summer.

### 27. ARISTEA Aiton
*J.C.M. Alexander*
Herbs with creeping rhizomes bearing fibrous roots. Leaves mostly basal, 2-ranked, linear. Flowers usually blue, radially symmetric, in spikes, racemes or corymbs, on a long, erect, inflorescence-stalk which may bear a few reduced leaves. Spathe-bracts small. Perianth free almost to base; lobes similar, twisting spirally after flowering. Filaments free, inserted at top of the short perianth-tube. Style thread-like, with 3 small obovate stigmatic lobes. Fruit

ovoid to cylindric, obtuse or acutely 3-angled.

A genus of about 50 species, 40 of which are in southern Africa, the rest further north in tropical Africa. Plants thrive in cold-house beds. Most of the cultivated species will tolerate a few degrees of frost but very rarely survive transplanting at anything beyond the seedling stage; it is useless to establish plants in a cool-house and then transplant them to an open bed. For out-of-door culture, seedlings should be pricked out into gritty, well-drained soil in perishable pots and planted directly into sunny sheltered positions. Seeds germinate freely.

1a. Leaves 3 mm wide or less, usually less than 15 cm long; flowers 1–4 per cluster     **1. cyanea**
  b. Leaves 6 mm wide or more, usually more than 15 cm long; flowers many per cluster     **2**
2a. Leaves 45 cm or less; stem to 45 cm, flattened     **2. ecklonii**
  b. Leaves 60 cm or more; stem to 120 cm, not flattened     **3. confusa\***

**1. A. cyanea** Aiton (*Ixia africana* Linnaeus; *A. africana* invalid; not *A. cyanea* de Wildeman or *A. africana* Hoffmannsegg). Illustration: Botanical Magazine, 458 (1799); Marloth, Flora of South Africa 4: pl. 44 (1915); Jackson, Wild flowers of Table Mountain, 102 (1977).
Leaves 7.5–15 cm × 2–3 mm, linear, acute, somewhat stiff. Inflorescence-stalk 5–15 cm, bearing 1 or 2 reduced leaves and 1–4 clusters of short-stalked, blue flowers with yellow throats. Outer spathe-bract ovate, *c.* 8 mm, fringed; inner lanceolate. Perianth-lobes *c.* 1.2 cm, obovate, shallowly notched. Capsule oblong, acutely angled, *c.* 8 mm. *South Africa.* H5–G1. Spring and autumn.

**2. A. ecklonii** Baker. Illustration: Wood & Evans, Natal Plants 1: pl. 68 (1899); Macmillan, Tropical gardening & planting, edn 3, 190 (1925) & edn 5, 183 (1952); Eliovson, Wild flowers of southern Africa, frontispiece (1980).
Leaves 15–45 cm × 6–12 mm, linear, acute, dark green, red at base, not rigid. Stem flattened, 30–60 cm, bearing several reduced leaves. Flowers in a loose panicle of many 3- or 4-flowered clusters; flower-stalks 6–12 mm. Outer spathe-bracts lanceolate, green, to 6 mm, with membranous margin. Perianth blue, 8–12 mm; outer lobes oblong, inner lobes

ovate. Capsule cylindric, 12–18 mm. *Southern Africa.* H5–G1. Summer–autumn.

**3. A. confusa** Goldblatt (*A. capitata* misapplied and confused). Illustration: Botanical Magazine, 605 (1802); Marloth, Flora of South Africa 4: pl. 44 (1915); Hamer, Wild flowers of the Cape, 36 (1926).
Plant robust, to 1.5 m; rhizome thick. Basal leaves 50–100 × 1.5 cm, rigid. Stem-leaves 3–5, to 45 cm. Flowers in a narrow, much branched panicle, fragrant; spathe-bracts ovate, brown and chaffy. Perianth deep blue; lobes *c.* 1.5 cm, ovate to wedge-shaped. Capsule 1.8–2.5 cm, oblong to cylindric, acutely angled. *SW South Africa.* H5–G1. Summer.

**\*A. major** Andrews (*A. thyrsiflora* misapplied and confused; *A. capitata* misapplied). Illustration: Rice & Compton, Wild flowers of the Cape of Good Hope, 192 (1950); Journal of the Royal Horticultural Society **93**: 471 (1968). Basal leaves to 3 cm wide, with brown margins. Spathe-bracts linear-lanceolate, to 1.2 cm, entire with white margin. *SW South Africa.* H5–G1. Spring–summer.

**28. NIVENIA** Ventenat
*J.C.M. Alexander*
Evergreen, perennial shrubs. Stems branching, rarely unbranched, woody below. Leaves linear, 2-ranked, acute, overlapping at base. Flowering-stem flat and 2-edged, ending in a spike, corymb or umbel subtended by a single spathe. Each flower or pair of flowers subtended by 3 bracts (see figure 38(9), p. 333). Flowers radially symmetric, stalkless, falling entire on fading. Perianth with a distinct tube and 6 similar lobes. Flowers of 2 types (dimorphic): either with long stamens and short styles or with long styles and short stamens. In some species both types occur, but on separate plants. Filaments free, borne at top of perianth-tube. Style-branches small, entire. Capsule spherical to cylindric.

A genus of 8 species from the South African Cape, unusual for the Iridaceae in their shrubby habit and dimorphic flowers; not to be confused with *Nivenia* Brown in the Proteaceae (correctly called *Paranomus* Salisbury). Plants perform well in full sun in light sandy soil, requiring plenty of moisture when actively growing. They can withstand some frost but not if conditions are damp. Easily propagated from seed. Literature: Weimark, H., A revision of the genus Nivenia Vent., *Svensk Botanisk Tidskrift* **34**: 355–72 (1940).

**1. N. corymbosa** (Ker Gawler) Baker. Illustration: Botanical Magazine, 895 (1805); Grey, Hardy bulbs, 324 (1937); Everett, Encyclopedia of horticulture 7: 2331 (1981).
Stem to 1.2 m, much branched; branches flattened. Bases of stem and branches densely leafy. Leaves linear, rigid, 5–20 cm × 2–5 mm, ascending at *c.* 45° to stem or branch. Flowering-stems 5–15 cm, flattened; flowers blue in dense corymbose panicles. Spathes 5–6 mm, oblong, brown. Perianth-tube 6–8.5 mm, emerging from bracts; lobes 6–10 × 3–6 mm, each with a small white basal spot. Stamens either concealed or protruding; style either concealed or 12.5–15 mm and protruding. *South Africa.* H5. Autumn.

**\*N. fruticosa** (Linnaeus filius) Baker. Illustration: Svensk Botanisk Tidskrift **34**: 360 (1940). Dwarf undershrub, rarely exceeding 20 cm. Upper parts of branches leafy. Leaves 3–9 cm × 1.5–2.5 mm. Flowering-stem to 2.5 cm; flowers blue in a single cluster; tube 1–1.3 cm; lobes 1–1.3 cm. Stamens projecting *c.* 6 mm from flower; style concealed in tube. *South Africa.* H5. Autumn.

**29. SCHIZOSTYLIS** Backhouse & Harvey
*J.C.M. Alexander & D.A. Webb*
Rootstock short, rhizome-like (rarely producing a corm); roots fibrous. Basal leaves 2–4, 2-ranked, 30–45 cm × 6–12 mm, linear, keeled; midrib distinct. Stem 30–90 cm, slender, terete, erect, bearing *c.* 3 reduced leaves. Flowers 4–14 in a 2-sided spike, radially symmetric, each with an unequal pair of green lanceolate bracts. Perianth-tube 2.5–3 cm, narrowly conical, yellowish or reddish green; perianth-lobes *c.* 2.5 cm, similar, ovate to oblong, acute, overlapping, scarlet to crimson. Stamens 3; filaments slender, free, attached to top of perianth-tube; anthers bright yellow. Style equalling perianth-tube, divided into 3, entire, slender, red branches *c.* 1.8 cm long.

A genus from South Africa now generally thought to contain only 1 species; commonly grown for cut flowers. It is easily cultivated in wet peaty or sandy loam in full sun. An annual mulch of peat or compost helps to retain moisture and encourages growth. Propagation is mostly by division.

**1. S. coccinea** Backhouse & Harvey (*S. pauciflora* Klatt). Illustration: Botanical Magazine, 5422 (1864); Synge, Collins' guide to bulbs, pl. 32 (1961); Hay &

Beckett, Reader's Digest encyclopaedia of garden plants and flowers, 652 (1971). *Eastern South Africa*. H4. Summer–autumn.

The most commonly seen cultivars are 'Viscountess Byng', a vigorous plant with slender, pale pink flowers and 'Mrs Hegarty' with rose pink flowers (Illustration: Synge, Collins' guide to bulbs, pl. 32, 1961; Hay & Synge, Dictionary of garden plants, 110, 1969). There is some confusion between these cultivars and plants given the name *S. pauciflora*.

## 30. HESPERANTHA Ker Gawler
*J.C.M. Alexander*

Corm spherical to ovoid, asymmetric; tunics woody, all at same level, or the older ones displaced upwards with bases irregularly split or evenly divided into overlapping segments. Basal leaves 2–4, in 2 ranks, linear or lanceolate, acute, hairless, often curved. Stem simple or branched, round in section, erect, bearing 1–3 leaves with inflated sheathing bases. Flowers few, in loose 1 or 2-sided spikes, each with a pair of leaf-like, chaffy or transparent bracts. Perianth-tube narrow, sometimes strongly curved; lobes 6, in 2 whorls, usually similar in size, shape and colour. Stamens symmetrically arranged, borne at top of perianth-tube; filaments free. Style shorter than or equal to perianth-tube; branches ascending, entire, long but not usually exceeding anthers.

A genus of about 55 species from Africa south of the Sahara, most strongly represented in South Africa. Most of the species have previously been included in *Ixia* or *Geissorhiza*. The flowers open in late afternoon and are deliciously scented. Cultivation conditions are similar to those for *Geissorhiza* (p. 379).
Literature: Foster, R.C., Studies in the Iridaceae V, *Contributions from the Gray Herbarium* 166: 3–27 (1948); Goldblatt, P., Corm Morphology in Hesperantha, *Annals of the Missouri Botanical Garden* 69: 370–8 (1982).

1a. Perianth-lobes more than 2 cm      2
  b. Perianth-lobes 2 cm or less        3
2a. Perianth-tube *c*. 8 mm, strongly curved; flowers white, pointing downwards                   **1. angusta**
  b. Perianth-tube 3–5 mm, not strongly curved; flowers yellow or yellow and purple, not pointing downwards
                                          **6. inflexa**
3a. Perianth-tube curved at base or above
                                          4
  b. Perianth-tube straight              5

4a. Basal leaves 4; stem 14–17 cm; perianth-tube 6–7 mm
                                **4. graminifolia**
  b. Basal leaves 5–6; stem to 45 cm; perianth-tube to 1.2 cm    **5. radiata**
5a. Leaves lanceolate, 5–8 mm wide
                                     **3. falcata**
  b. Leaves linear, 4–5 mm wide
                                      **2. buhrii**

*Basal Leaves*. Linear: **1,2,4–6**; lanceolate: **3**.
*Stem*. Branched: **1–3,6**; unbranched: **3–5**.
*Flower-spike*. 1-sided: **4,5**; 2-sided: **1–3,6**.
*Flower colour*. Mostly white: **1–5**; mostly yellow: **4,6**; red, brown or purple on back of inner perianth-lobes: **2,3,5**; purple patch at base and apex of perianth-lobe: **6**.
*Outer spathe-bract*. Less than 1 cm: **3,4**; 1–2 cm: **2,5**; more than 2.5 cm: **1,6**.
*Perianth-tube*. Curved: **1,4,5**; straight: **3,6**. Less than 1 cm: **1,3,4,6**; 1 cm or more: **2,5**.
*Perianth-lobes*. Less than 1 cm: **4,5**; 1–2 cm: **2,3**; more than 2 cm: **1,6**.

**1. H. angusta** (Jacquin) Ker Gawler. Illustration: Batten & Bokelmann, Wild flowers of the eastern Cape Province, pl. 27, 5 (1966); Roux & Schelpe, Namaqualand & Clanwilliam, 46 (1981).
Basal leaves 2 or 3, to 32 cm × 3 mm, linear, acute, hairless, sheathing at base, slightly thickened at midrib and edges. Stem to 40 cm, branched; stem-leaves 1 or 2, to 22 cm. Main stem with 2–6 flowers in a loose 2-sided spike; side branches with 1 or 2 flowers. Flowers white. Outer bracts to 2.8 cm, membranous, notched at tip; inner to 2.5 cm, chaffy, 2-ridged, bifid. Perianth-tube *c*. 8 mm, strongly curved in upper part (causing flower to point downwards); lobes *c*. 2–4 cm × 7 mm, reflexed, ovate to oblong, acute or minutely notched. Stamens 4–7 mm. Style equal to perianth-tube; branches to 1.4 cm. *South Africa (Cape Province)*. H5–G1. Spring.

Foster (1948) maintains that almost all the wild specimens collected as *H. angusta* should be called *H. bachmannii* Baker. This may also apply to plants grown as *H. angusta*.

**2. H. buhrii** L. Bolus.
Basal leaves 3, linear, acute, hairless, 14–21 cm × 4–5 mm. Stem 22–27 cm, branched; stem- and branch-leaves 9.5–11 cm, with inflated sheaths. Main stem with 6 or 7 flowers in a loose, 2-sided spike; branches each with 4–7 flowers. Outer bracts to 1.8 cm, leaf-like, membranous at tip, keeled; inner to 7 mm,

membranous, 2-keeled. Perianth-tube 1–1.4 cm; lobes 1.5–2 cm × 6–9 mm, obovate, obtuse to acute, white; outer lobes flushed with pink or purple on backs. Stamens *c*. 1.3 cm. Ovary 3–5 mm. Style 7–8 mm; branches to 7 mm. *South Africa*. H5–G1. Spring–summer.

**3. H. falcata** Ker Gawler. Illustration: Botanical Magazine, 566 (1802); Marloth, Flora of South Africa 4: pl. 42 (1915); Mason, Western Cape sandveld flowers, 68 (1972).
Basal leaves 2–4, lanceolate, hairless, to 8 cm. Stem simple or once branched, to 30 cm, with 1 or 2 small sheathing leaves. Flowers 2–10 in a loose, 2-sided spike. Outer bracts to 8 mm, oblong, leaf-like with chaffy edges. Perianth-tube straight, *c*. 8 mm; lobes to 1.2 cm, oblong, white, outer lobes red or purple on backs. Anthers *c*. 6 mm. Style-branches *c*. 4 mm. *SW South Africa*. H5–G1. Spring.

**4. H. graminifolia** Sweet. Illustration: Botanical Magazine, 1254 (1810); Hamer, Wild flowers of the Cape, 78 (1926).
Basal leaves 4, the lowest rudimentary, the others 7–14 cm × 2.5 mm, linear, curved, acute, hairless; midrib thick. Stem 14–17 cm, bearing 1 small sheathing leaf 2–3.5 cm. Flowers 3–9 in a loose, 1-sided spike. Outer bracts to 8 mm, oblong to ovate, acute, leaf-like; inner to 1 cm, entire or bifid, 2-keeled. Perianth-tube 6–7 mm, curved from base; lobes 5–6 × 2 mm, obovate; the outer greenish, sometimes tinged with pink; the inner white to pale yellow. Stamens 4–5 mm. Ovary 2 mm, ovoid to oblong. Style *c*. 4 mm; branches *c*. 3 mm. *South Africa*. H5–G1. Autumn.

**5. H. radiata** Ker Gawler. Illustration: Botanical Magazine, 573 (1802); Trauseld, Wild flowers of the Natal Drakensberg, 41 (1969).
Basal leaves 5 or 6. Stem to 45 cm, unbranched; outer bracts to 1.8 cm. Perianth-tube to 1.2 cm, strongly curved; lobes to 8 mm; the inner white; the outer purplish red on backs. Style-branches *c*. 6 mm. *South Africa*. H5–G1. Summer–autumn.

**6. H. vaginata** (Sweet) Goldblatt (*Geissorhiza vaginata* Sweet; *H. inflexa* misapplied; *H. inflexa* (Delaroche) Foster var. *stanfordiae* (L. Bolus) Foster; *H. stanfordiae* L. Bolus). Illustration: Sweet, British Flower Garden, pl. 138 (1826).
Basal leaves 2, linear, curved, acute, hairless, to 15 cm × 8 mm, midrib slightly prominent. Stem to 32 cm, branched at

base and once higher up, bearing 3 leaves 18–23 cm × 9–13 mm, with inflated sheathing bases. Flowers 2 or 3 in a loose, 2-sided spike. Outer bracts to 3 cm, oblong, keeled, membranous with translucent edges; inner to 2.5 cm, 2-keeled. Perianth-tube 3–5 mm, straight, abruptly widened at throat; lobes 3–3.5 × *c*. 1.5 cm, obovate to spathulate, notched, yellow with purple patches at base and tip, or pure yellow. Stamens *c*. 2 cm. Ovary 2–3 mm. Style 3–4 mm; branches 8–9 mm. *South Africa (Cape Province)*. H5–G1. Spring.

Goldblatt (1970) has shown that the name *H. inflexa* does not apply to this species.

## 31. GEISSORHIZA Ker Gawler
*J.C.M. Alexander*

Corm ovoid, asymmetric, flat-based; tunics of hard, woody, overlapping layers, splitting longitudinally. Basal leaves lanceolate, 2-ranked, linear or thread-like, often curled, with margin or midvein often thickened. Stem simple or branched, erect, bearing a few reduced leaves. Flowers radially symmetric in a 1- or 2-sided spike, each with a pair of leaf-like bracts. Perianth-tube short, erect, widening above; lobes 6, in 2 similar whorls, spreading. Stamens borne at lobe-bases; filaments free. Style slender, longer than perianth-tube; branches short, entire, recurved.

A genus of 60–70 species from southern Africa, most of which were originally considered part of *Ixia*. It is similar to *Hesperantha* from which it differs in its longer style with shorter branches. Most species are capable of withstanding a few degrees of frost, particularly if protected with sacking, straw, or glass. Plants do best in full sun in a well-drained sandy loam, and can be propagated by offsets or seed. Literature: Foster, R., Studies in the Iridaceae II. A revision of Geissorhiza, *Contributions from the Gray Herbarium* **135**: 3–78 (1941).

1a. Veins of leaf all equally conspicuous
  2
  b. Midrib (and sometimes 2 other veins) more conspicuous than other veins  3
2a. Flowers yellow; stamens *c*. half perianth length          **1. humilis**
  b. Flowers purple; stamens almost as long as perianth          **2. radians**
3a. Basal leaves 1.5 cm wide or more; perianth-lobes yellow with red midrib, longer than 2.5 cm          **6. grandis**
  b. Basal leaves 1 cm wide or less; perianth-lobes reddish or bluish

purple (in part), shorter than 2.5 cm
  4
4a. Stem-leaves 3          **3. aspera**
  b. Stem-leaf 1          5
5a. Flower-spike 2-sided; inner perianth-lobes white          **5. ovata**
  b. Flower-spike 1-sided; all perianth-lobes reddish purple          **4. erosa**

*Basal leaves.* Less than 10 cm: **5**. Less than 3 mm wide: **1,2**; 5–12 mm wide: **3–5**; more than 15 mm wide: **6**.
*Stem-leaves.* Bristly: **3**. Solitary: **1,2,4,5**; 2 or more: **3,6**. More than 15 cm: **3**.
*Stem.* Unbranched: **1,2,4–6**; once-branched: **4,5**; much-branched: **3**.
*Flowers.* Yellow: **1,6**; red, blue or purple: **2–4**; reddish purple and white: **5**.
*Outer bract.* Less than 1.5 cm: **1–3,5**; 2–2.5 cm: **3**; *c*. 4 cm: **6**.
*Perianth-tube.* Less than 1 cm: **1–4**; more than 1 cm: **5,6**.
*Perianth-lobes.* Less than 2 cm: **1–5**; 2–3 cm: **4**; more than 4 cm: **6**.

**1. G. humilis** (Thunberg) Ker Gawler. Illustration: Botanical Magazine, 1255 (1810).
Basal leaves 2, linear, acute, hairless, to 15 cm × 1 mm, thicker at margins and midrib. Stem-leaf 1, to 5 cm, with sheathing base 1–2 cm. Stem unbranched, 10–14 cm, round in section, hairless, purplish. Flowers yellow, *c*. 3 in a loose, usually 1-sided spike. Outer bracts *c*. 1 cm, oblong to ovate, obtuse to truncate, leaf-like, white and chaffy near the apex. Inner bracts similar, slightly notched at tip. Perianth-tube to 4.5 mm; perianth-lobes *c*. 1.6 cm × 8 mm, obovate to elliptic, acute. Stamens *c*. 1 cm; anthers about equal to filaments. Ovary to 4 mm, conical. Style *c*. 1.3 cm. *South Africa (Cape Province)*. H5. Summer.

**2. G. radians** (Thunberg) Goldblatt (*Ixia rochensis* invalid; *I. radians* Thunberg; *Geissorhiza rochensis* invalid). Illustration: Botanical Magazine, 598 (1802); Graf, Tropica, 525 (1978); Eliovson, Wild flowers of southern Africa, edn 6, 50 (1980).
Basal leaves 2, longer than inflorescence, *c*. 1.5 mm wide, thread-like, sometimes 4-angled, hairless. Stem-leaf 1, to 8.5 cm, sheath very inflated. Stem to *c*. 15 cm, unbranched, round in section, hairless. Flowers 1 or 2, purple, centres red, ringed with white. Outer bracts to 1.1 cm, ovate-lanceolate, leaf-like, with a membranous, 3-lobed apex; inner bracts a little smaller. Perianth-tube to 7 mm; perianth-lobes to 1.8 × 1.2 cm, obovate, obtuse. Filaments *c*.

8 mm; anthers 4–5 mm. Ovary to 3 mm. Style top level with anthers; the stigmas recurved, exceeding them. *South Africa (Cape Province)*. H5. Spring.

**3. G. aspera** Goldblatt (*Ixia secunda* misapplied; not *I. secunda* Delaroche; *G. secunda* misapplied). Illustration: Botanical Magazine, 597 (1802); Marloth, Flora of South Africa 4: pl. 43 (1915); Rice & Compton, Wild flowers of the Cape of Good Hope, pl. 197 (1950).
Basal leaves 2, to 28 cm × 6 mm, linear, acute, tapering towards the base, hairless; margins, midrib and sometimes 2 other veins prominent. Stem-leaves 3, 17–25 cm, sparsely bristly, the lower 2 with sheathing bases. Stem to 40 cm or more, much branched, slender, bristly. Flowers bluish purple, 3–6 in a loose, 1-sided spike. Outer bracts to 1.2 cm, oblong to ovate, leaf-like, brownish above; inner bracts to 8 mm. Perianth-tube 1–2 mm; perianth-lobes to 1.2 cm × 4 mm, elliptic to obovate. Stamens *c*. 7 mm. Ovary 2–3 mm, ovoid to conical. Style 8 mm; stigmas 1.5 mm. *South Africa*. H5. Spring.

**4. G. erosa** (Salisbury) Foster (*Ixia erosa* Salisbury). Illustration: Eliovson, Wild flowers of southern Africa, 50 (1980).
Basal leaves 2, linear, narrowed at base, to 23 cm × 7 mm; margins, midrib and sometimes 2 other veins prominent and hairy. Stem-leaf 1, to 14 cm; sheath inflated, hairy on veins. Stem to *c*. 35 cm, unbranched. Flowers 3–7 in a loose, 1-sided spike, reddish purple. Outer bracts to 2.3 cm, entire or 3-lobed, leafy below, brown and membranous above; inner bracts slightly shorter. Perianth-tube 2–4 mm; perianth-lobes to 2.4 × 1 cm, obovate, rounded. Stamens *c*. 1.5 cm. Ovary to 4 mm, conical. Style *c*. 1.4 cm; stigmas *c*. 3 mm. *South Africa*. H5. Spring.

Var. **kermesina** (Klatt) Foster (*G. hirta* (Thunberg) Ker Gawler) is more commonly grown and is usually called *G. hirta*. It is smaller, sometimes once-branched, with 2–4 reddish or bluish flowers per spike, and perianth-lobes to 15 × 8 mm. Illustration: Eliovson, Wild flowers of southern Africa, edn 6, 50 (1980).

**5. G. ovata** (Burman) Ascherson & Graebner (*Ixia ovata* Burman; *G. excisa* (Linnaeus) Ker Gawler). Illustration: Rice & Compton, Wild flowers of the Cape of Good Hope, pl. 194 (1950); Jackson, Wild flowers of Table Mountain, 94 (1977).
Basal leaves 2 or 3, to 7 × 1 cm, ovate to narrowly elliptic, acute, black-dotted,

leathery with prominent margins and midrib. Stem-leaf 1, with sheathing base; free portion to 1.8 cm. Stem to 25 cm, simple, rarely once-branched. Flowers 1–7 in a loose, 2-sided spike, reddish purple and white. Outer bracts to 1.4 cm, oblong to ovate, acute or truncate, leaf-like, with reddish purple edges and tip; inner bracts slightly shorter. Perianth-tube to 2 cm, usually shorter; perianth-lobes to 18 × 6 mm, the outer reddish purple, the inner white. Stamens *c.* 6 mm. Ovary to 7 mm, ovoid to conical. Style to 2.6 cm; stigmas 2.5 mm. *South Africa.* H5. Spring.

**6. G. grandis** Hooker. Illustration: Botanical Magazine, 5877 (1870).
Basal leaves 2 or 3, to 20 × 2 cm, linear to narrowly elliptic; midrib prominent. Stem-leaves usually smaller. Stem to 30 cm, unbranched. Flowers *c.* 6 in a loose, 1-sided spike, yellow. Outer bracts *c.* 4 cm, ovate to elliptic, leaf-like; inner bracts to 2.5 cm. Perianth-tube to 2.5 cm, slender and curved; perianth-lobes to 4.5 × 1 cm, elliptic to obovate, obtuse, with red midribs. Stamens *c.* 1.5 cm. Ovary *c.* 8 mm, ovoid. Style *c.* 4 cm; stigmas *c.* 9 mm. *South Africa.* H5. Spring.

**32. DIERAMA** Koch
*D.A. Webb*
Rootstock a corm. Leaves mostly basal, linear, stiff, 2-ranked. Stems exceeding the leaves, arching over towards the tip, each ending in a loose, 1-sided panicle with numerous, very slender primary branches, each of which terminates in a pendulous, crowded, spike-like, raceme of 3–6 radially symmetric flowers (figure 38(8), p. 333); bracts scarious. Perianth bell-shaped, its lobes *c.* twice as long as the tube. Filaments free. Stigma 3-lobed, lobes spathulate. Capsule spherical, with 6 seeds.

A genus of about 12 species (different authors give 2–25), confined to southern and E tropical Africa; only 2 are in cultivation. In older books and catalogues they may be found under *Sparaxis*. Although rather slow to become established, they flourish without much attention in any well-drained, sunny site. *D. pulcherrima* is particularly effective if planted at the edge of a pool where its arching stems overhang the water.

1a. Leaves 8–10 mm wide; perianth-lobes 2.5–3 cm, usually deep purple **1. pulcherrima**
  b. Leaves 6–8 mm wide; perianth-lobes *c.* 2 cm, usually pink **2. pendula**

**1. D. pulcherrima** (Hooker) Baker.
Illustration: Botanical Magazine, 5555 (1866); Perry, Flowers of the world, 150 (1972).
Densely tufted, eventually forming a large clump. Leaves to 100 cm × 8–10 mm. stems *c.* 1.5 m. Bracts linear, leaf-like; bracteoles spathe-like, scarious, lanceolate, whitish except for a brown base. Perianth usually deep purple, the lobes 2.5–3 cm. *South Africa.* H4. Late summer.

**2. D. pendula** (Linnaeus) Baker.
Illustration: Journal of the Royal Horticultural Society **39**: f. 9 (1913).
Differs from *D. pulcherrima* in its more slender habit and in being smaller in all its parts (stem to 100 cm; leaves *c.* 60 cm × 7–8 mm; perianth-lobes *c.* 2 cm) and in having pink, not purple flowers. Colour-variants are, however, known in both species. *S & E tropical Africa, northwards to Ethiopia.* H4. Summer.

Much confusion has been caused by Edwards's Botanical Register, t. 1360 (1830), which depicts another species (*D. pansa*) with lilac flowers, but labels it *Sparaxis pendula*.

**33. IXIA** Linnaeus
*J. Cullen*
Perennial herbs with corms with papery or fibrous tunics. Stems simple or little branched. Leaves few, mostly basal, 2-ranked, linear, lanceolate or thread-like, usually with a false midrib. Inflorescence a spike or a panicle of spikes with few, usually erect branches; spikes with few to many flowers borne in 2 ranks or spirally arranged. Bracts short but exceeding the ovaries, usually 3-toothed at their apices. Perianth of 6 equal or almost equal lobes, joined below into a short or long tube. Stamens attached at the throat of the tube or within it, filaments free or united into a short tube. Style thread-like, exceeding the stamens, stigma 3-fid. Fruit a thin-walled capsule opening by 3 flaps and containing many angular seeds.

A genus of about 45 species, all from the Cape Province of South Africa, where many are now rare. They are mostly grown in pots, though can be grown out-of-doors in sunny, frost-free areas. The corms should be planted in autumn in a well-drained compost; little water should be given during the winter, and the plants are best kept in a sunny, well-ventilated place. Watering can be increased in the spring, and should be continued after flowering until the leaves show signs of dying down.

Most of the Ixias grown today are hybrids of uncertain parentage (though *I. maculata* has made a significant contribution). The species are known to hybridise freely, and many different colour variants of many species were introduced to Europe (mainly through Holland) at the end of the eighteenth and the beginning of the nineteenth centuries. Much hybridisation took place then, and a wide range of flower colour is now available. Many variants were selected and named (see Gardeners' Chronicle, 1873). Literature: Lewis, G.J., South African Iridaceae: the genus Ixia L., *Journal of South African Botany* **28**: 45–195 (1962).

*Leaves.* To 2 mm wide: **1**.
*Perianth.* Predominant colour orange-yellow or yellow: **7**; cream, sometimes tinged with red: **8**; green: **2**; white, pale purple or bluish: **1,3**; pink, red or purple: **4,5**.
*Perianth-tube.* More than 4 cm: **8**; 1 cm or more: **2,7**; 3–10 mm: **1–4,6**; less than 3 mm: **5**.
*Perianth-lobes.* Much shorter than tube: **8**; about as long as tube: **7**; 2–4 times as long as tube: **1–4**; 6–8 times as long as tube; **5**.

1a. Perianth-tube 4–7 cm, much longer than the lobes **8. paniculata**
  b. Perianth-tube to 1.8 cm, equalling lobes or shorter than them **2**
2a. Leaves thread-like, to 2 mm wide; stamens borne near the base of the perianth-tube **1. capillaris**
  b. Leaves not thread-like, 2 mm or more wide; stamens borne in the throat of the perianth-tube **3**
3a. Perianth-lobes green with a purple or blackish base **3. viridiflora**
  b. Perianth-lobes variously coloured, not as above **4**
4a. Filaments completely free from each other **5**
  b. Filaments united into a tube, at least at their base **7**
5a. Perianth-tube 2–3 mm; leaves with the false midrib and margins somewhat thickened **5. campanulata**
  b. Perianth-tube 4 mm or more; leaves with false midrib and margins not thickened **6**
6a. Perianth-lobes to 2 times longer than the tube; perianth white, mauve or bluish; flowers faintly scented **2. polystachya**
  b. Perianth-lobes 4 or more times longer than the tube; perianth pink, red or purple, rarely white; flowers not scented **4. patens**

7a. Flowers orange or yellow, perianth-lobes 1½–2 times longer than the tube
**6. maculata**

b. Flowers variously coloured, rarely orange or yellow, the lobes of the perianth up to 1½ times longer than the tube
**7. monadelpha**

**1. I. capillaris** Linnaeus filius. Illustration: Botanical Magazine, 570 (1803).
Corm 1–1.5 cm in diameter. Stems to 50 cm. Leaves usually 3, erect, thread-like, 6–18 cm × 0.5–2 mm, often spirally twisted. Spike with 1–4 white or pale blue flowers which are sometimes green at the throat. Perianth-tube 5–7 mm, slender below, swollen and cup-like above, lobes 1–1.5 cm × 4–6 mm. Stamens arising near the base of the perianth-tube, projecting, filaments free, 4–7 mm, anthers 3–5 mm. *South Africa (Cape Province)*. H5–G1. Spring–summer.

**2. I. polystachya** Linnaeus. Illustration: Flowering plants of South Africa **25**: t. 968 (1945–6).
Corm 8–25 mm in diameter. Stem 30–100 cm. Leaves 5–8, 15–50 cm × 2–8 mm (rarely broader). Spike or panicle loose or dense, with few to many flowers which are white, mauve or bluish with green, mauve or purple centres, sometimes tinged with yellow, faintly scented. Perianth-tube slender, 5–14 mm, lobes 1–2.8 cm × 5–9 mm. Stamens arising at the throat of the perianth-tube, filaments free, 3–4 mm, anthers 4–7 mm. *South Africa (Cape Province)*. H5–G1. Summer.

**3. I. viridiflora** Lamarck (*I. maculata* Linnaeus var. *viridis* Jacquin). Illustration: Botanical Magazine, 549 (1801); Eliovson, Wild flowers of southern Africa, 51 (1955).
Corm 1–3 cm in diameter. Stem 50–100 cm. Leaves 5–7, 40–55 cm × 2–4 mm, with 4–6 conspicuous veins. Spike with 12–many flowers which are green with dark purple to almost blackish centres. Perianth-tube slender, 6–9 mm, lobes 1.6–2.5 cm × 7–11 mm. Stamens arising at the throat of the perianth-tube, filaments free, purple, 3–4 mm, anthers 9–13 mm, purple or yellow. *South Africa (Cape Province)*. H5–G1. Spring–summer.

**4. I. patens** Aiton. Illustration: Botanical Magazine, 522 (1801); Gartenflora **11**: t. 356 (1862).
Corms 1.5–2 cm in diameter. Stem 20–50 cm. Leaves 5–7, 10–35 cm × 6–12 mm, firm, with several prominent veins and cartilaginous margins. Spike

loose with 5–15 flowers which are various shades of pink, red or purple, rarely white, occasionally greenish in the centre. Perianth-tube slender, 4–6 mm; lobes with prominent veins, widely spreading, 1.6–2.4 cm × 5–12 mm. Stamens arising at the throat of the perianth-tube, filaments free, 4–6 mm, anthers 6–8 mm. *South Africa (Cape Province)*. H5–G1. Spring.

This species is perhaps threatened with extinction in its native habitat.

**5. I. campanulata** Houttuyn (*I. speciosa* Andrews; *I. crateroides* Ker Gawler). Illustration: Botanical Magazine, 594 (1803).
Corm 1–1.8 cm in diameter. Stems 10–15 cm. Leaves 5–10 cm × 2–5 mm, usually with 3–6 veins, the false midrib and margins somewhat thickened. Spike with usually 3–6 flowers which are usually red, each lobe with a whitish band outside, occasionally entirely white or deep red. Perianth-tube slender, 2–3 mm, lobes 1.2–2.5 cm × 8–12 mm. Stamens arising at the throat of the perianth-tube, filaments free, 3–5 mm, anthers 6–8 mm. *South Africa (Cape Province)*. H5–G1. Summer.

**6. I. maculata** Linnaeus (*I. conica* Salisbury). Illustration: Botanical Magazine, 539 (1801); Flowering plants of South Africa **9**: t. 329 (1929); Journal of South African Botany **28**: 139 (1962).
Corm 1–2 cm in diameter. Stem 15–50 cm. Leaves 10–35 cm × 2–7 mm, often spirally twisted, with several fine veins. Spike short, the 4–many flowers crowded, orange or yellow-orange with dark brown, purple or blackish centres. Perianth-tube slender, 5–8 mm; lobes 1.5–3 cm × 8–12 mm. Stamens arising at the throat of the perianth-tube; filaments 3–5 mm, united into a tube at the base; anthers 7–9 mm. *South Africa (Cape Province)*. H5–G1. Spring–summer.

**7. I. monadelpha** Delaroche (*I. columnaris* Salisbury). Illustration: Botanical Magazine, 607 (1803); Flowering plants of South Africa **8**: t. 317 (1928).
Corms 1.3–2 cm in diameter. Stems 15–40 cm. Leaves 4–7, 8–30 cm × 3–10 mm, often spirally twisted, with several veins. Spike with 4–12 densely packed flowers which are blue, violet, purple, pink or white, usually with a green or brown centre which is often outlined in another colour. Perianth-tube slender, 1–1.8 cm, lobes 1.3–2 cm × 6–11 mm. Stamens arising at the throat of the perianth-tube; filaments

3–6 mm, united for at least half of their length into a tube; anthers 7–8 mm, erect and almost touching each other. *South Africa (Cape Province)*. H5–G1. Spring–summer.

**8. I. paniculata** Delaroche (*I. longiflora* Bergius). Illustration: Botanical Magazine, 256 (1794).
Corm 1–1.5 cm in diameter. Stem 30–100 cm. Leaves 15–60 cm × 3–12 mm. Spike loose or dense, with 5–18 flowers which are usually pale cream tinged or flushed with red, or occasionally pale yellowish. Perianth-tube slender, slightly and gradually widening above, 4–7 cm; lobes 1.5–2.5 cm × 3–8 mm. Stamens arising in the perianth-tube, filaments free, 5–6 mm, anthers 6–8 mm, within the perianth-tube or partly or wholly projecting from it. *South Africa (Cape Province)*. H5–G1. Spring.

## 34. FREESIA Klatt

*J. Cullen*

Perennial herbs with corms with fibrous coats. Leaves in 2 ranks, mostly basal, with a few prominent veins and usually a false midrib. Stems usually unbranched or occasionally with 1 or 2 inflorescence-bearing branches. Inflorescence a spike, its axis bent sharply at an angle of about 90° at the level of the lowest flower, the flowers borne on the upper side of the axis. Bracts 2, membranous or green, just exceeding the ovary. Perianth consisting of a very narrow tube at the base, this gradually widening to a broader tube, ending in 6 lobes; the tube is curved. Stamens 3, free, not or scarcely projecting from the perianth-tube. Stigmas 3, each deeply divided into 2 lobes. Fruit a few-seeded capsule.

A genus of 19 species, all from the Cape Province of South Africa. Most of the Freesias grown today, either as garden ornamentals or as an important cut-flower crop, are hybrids (**F. × hybrida** Bailey). Many of them are of uncertain parentage, due to the misapplication of names when the plants were introduced to Europe. All of the species included here may have been involved in hybridisation, but *F. lactea* (flowers size and scent), *F. corymbosa* (flower colour and scent) and *F. armstrongii* (flower colour) have probably contributed most. There are surprisingly few good illustrations of the species.

Cultivation is generally as for most South African Iridaceae. The plants can be grown in pots in a cool greenhouse, or planted out in spring in a sunny border, the corms being lifted and dried off in autumn. For cultivation as a cut-flower crop, see the books by Mackenzie and Smith cited below. Literature: Brown, N.E., Freesia Klatt and its History, *Journal of South African Botany* **1**: 1–31 (1935); Mackenzie, W.F., *Freesias* (1957); Smith, D., *Freesias* (1979); Kragtwijk, G., Freesia-sortiment, *Tuinbouwgids* **20**: 380–1 (1965).

1a. Bracts entirely or mostly brownish, membranous, often with the tips dark brown or almost black                    2
  b. Bracts green, sometimes with the margins minutely membranous, tips never dark brown or blackish           4
2a. Perianth-lobes entirely rose-pink
                                   **1. armstrongii**
  b. Perianth-lobes yellow or greenish, some of them blotched with other colours                                          3
3a. Perianth pale to intense yellow, lower lobes each with an orange blotch
                                     **2. corymbosa**
  b. Perianth brownish or greenish yellow, the backs of the lobes suffused with violet, the lower lobes each with a brownish blotch          **3. refracta**
4a. Perianth 2.5–3.2 cm; bracts 4–5 mm
                                    **4. sparrmannii**
  b. Perianth 3.4–6.5 cm; bracts 6–12 mm                                      5
5a. Flowers 5–6.5 cm, entirely white except for the yellowish base of the tube, without blotches on the lobes
                                        **6. lactea**
  b. Flowers 3.6–4.6 cm, whitish or yellow, each of the 3 lower lobes, or at least the lowest, with a yellow or orange blotch          **5. xanthospila**

**1. F. armstrongii** Watson. Stems 20–40 cm. Leaves 5–13 cm × 6–16 mm, linear-lanceolate or linear-oblong. Bracts 4–6 mm, mostly membranous, the upper part purplish brown. Flowers not fragrant. Perianth 2.8–3.4 cm, the lobes 1–1.6 cm, the broad part of the tube 1.8–2.2 cm, the narrow basal part 6–8 mm; all rose-pink except for the basal part of the tube, which is yellow. *South Africa (Cape Province)*. H5–G1. Spring.

**2. F. corymbosa** (Burman) N.E. Brown (*F. odorata* (Loddiges) Klatt; *F. refracta* var. *odorata* (Loddiges) Voss). Stems 10–45 cm. Leaves 5–40 cm × 3–9 mm, at first flat on the ground, later erect, linear but gradually

tapering. Bracts 4–6 mm, mostly membranous, tips dark brown to blackish. Flowers fragrant. Perianth 2.4–3.8 cm, the lobes 8–12 mm, the broad part of the tube 1.6–2.6 cm, the narrow basal part 6–14 mm; all yellow (entirely orange-yellow in var. **aurea** (Gumbleton) N.E. Brown) sometimes slightly flushed with purple, the throat and at least the lowermost lobe orange. *South Africa (Cape Province)*. H5–G1. Spring.

**3. F. refracta** Klatt. Illustration: Edwards's Botanical Register **2**: t. 135 (1816). Very like *F. corymbosa*, but flowers brownish or greenish yellow suffused with violet on the backs of the lobes, the lower lobes each with a brown or orange-brown blotch. *South Africa (Cape Province)*. H5–G1. Spring.

Though the name *F. refracta* is widely used in the literature, the species to which it applies is one of the least attractive and least widely grown. The widespread use of the name is due to confusion with *F. corymbosa*, *F. sparrmannii* and *F. lactea*.

**4. F. sparrmannii** (Thunberg) N.E. Brown (*F. refracta* var. *alba* misapplied). Illustration: Flowering Plants of South Africa **1**: t. 11 (1920). Stems 10–20 cm. Leaves ascending, 5–15 cm × 4–8 mm, linear but gradually tapering. Bracts 4–5 mm, green throughout. Perianth 2.4–3 cm, lobes *c.* 1 cm, broad part of tube *c.* 1.8 cm, narrow basal part to 1 cm; all white or creamy white (pale yellow in var. **flava** N.E. Brown). *South Africa (Cape Province)*. H5–G1. Spring.

**5. F. xanthospila** Klatt. Stems 15–25 cm. Leaves erect or ascending, 7.5–20 cm × 6–11 mm, linear or linear-lanceolate, gradually tapering. Bracts 7–8 mm, green throughout. Flower scented. Perianth 3.6–4.6 cm, the lobes 1.4–1.6 cm, the broad part of the tube 2–2.3 cm, the narrow basal part 8–10 mm; white with blotches of yellow on the lower lobes and sometimes on the base of the tube or (in var. **leichtlinii** (Klatt) N.E. Brown) the tube yellowish and the blotches orange. *South Africa (Cape Province)*. H5–G1. Spring.

**6. F. lactea** Klatt (*F. alba* Foster; *F. refracta* var. *alba* (Foster) Anon.). Plant 20–30 cm. Leaves ascending, 7–20 cm × 4–8 mm, linear, gradually tapering. Bracts 8–12 mm, green throughout. Flowers strongly and sweetly scented. Perianth 5–6.5 cm, lobes

1.2–1.4 cm, broad part of the tube 3–3.6 cm, narrow basal part 1.2–1.6 cm; all white or creamy white, except for the lower part of the tube, which is yellowish. *South Africa (Cape Province)*. H5–G1. Spring.

This species, which has contributed its scent and large flower size to many hybrids, is usually found in the literature under the name *F. refracta* var. *alba*. This name is sometimes attributed to Baker or to other authors; it has also been used for *F. sparrmannii*.

## 35. LAPEIROUSIA Pourret
*J. Cullen*

Perennial herbs with corms covered by hard, woody coats. Stems erect, branched or not, usually bearing leaves. Basal leaves 1 or 2, usually longer than the stem-leaves; stem-leaves 2-ranked, usually with undulate-crisped margins, often bearing a keel or 2 keels on the lower surface, with branches in their axils. Flowers in a corymbose panicle, or in spikes at the ends of the stem and the branches; bracts 2, the outer larger than the inner (figure 38 (3), p. 333). Perianth with a long, narrow tube and 6 symmetrically or asymmetrically disposed lobes. Stamens 3, attached to the throat of the perianth-tube, placed to one side of the flower or symmetric. Stigmas 3, each deeply divided. Fruit an oblong-elliptic, membranous capsule with many angled or rounded seeds.

A genus of about 30 species from Africa, most of them from South Africa. Its name is frequently misspelled 'Lapeyrousia'. Several species traditionally included in it are treated here under *Anomatheca* (p. 383). Cultivation as for *Ixia* (p. 380). Literature: Lawrence, G.H.M., Notes on Lapeirousia in Cultivation, *Baileya* **3**: 131–6 (1955); Goldblatt, P., A Revision of the Genera Lapeirousia Pourret and Anomatheca Ker in the Winter Rainfall Region of South Africa, *Contributions from the Bolus Herbarium* **4** (1972).

1a. Inflorescence a corymbose panicle of spikes; flowers mostly blue, perianth-tube to 1 cm               **1. corymbosa**
  b. Inflorescence a simple spike (ending each branch); flowers red, violet, lavender, cream or white, perianth-tube 2 cm or more           **2. pyramidalis**\*

**1. L. corymbosa** (Linnaeus) Ker. Illustration: Botanical Magazine, 595 (1801); Marloth, Flora of South Africa **4**: t. 23 (1915). Corms *c.* 1 cm in diameter, dark brown to black, the coats woody, finally split longitudinally. Stems 5–45 cm, somewhat flattened and 2-winged. Basal leaves 8–10 cm, linear or linear-lanceolate, usually sickle-shaped. Inflorescence a corymbose panicle with many flowers, bracts 3–10 mm. Perianth more or less symmetric, the tube 4–10 mm, the lobes spreading, about as long as the tube, blue, each with an inverted V-shaped white mark near the base, the marks on the 6 lobes forming a star shape. *South Africa (Cape Province)*. H5–G1. Spring.

Several subspecies exist in South Africa, differing in size and perianth colour, but only subsp. *corymbosa* (described above) appears to have been cultivated.

**2. L. pyramidalis** (Lamarck) Goldblatt (*L. fissifolia* (Jacquin) Ker; *L. bracteata* (Thunberg) Ker). Illustration: Botanical Magazine, 1246 (1810); Flowering plants of South Africa 6: t. 239 (1926). Corm bell-shaped, *c.* 1 cm in diameter. Stem at most 12 cm, terete, usually without branches. Basal leaf usually 1, linear to lanceolate; stem-leaves several, shorter and broader than the basal leaf, margins entire or undulate-crisped. Inflorescence a simple, few-flowered spike; bracts green and leaf-like. Perianth irregular, the tube 2–4 cm, the lobes unequal, the uppermost the largest, *c.* 8 mm, forming a hood over the anthers and stigma, all white, white flushed with purple or with reddish pink, the lower 3 lobes sometimes marked with a contrasting colour at the base. Anthers to the upper side of the flower, borne under the uppermost perianth-lobe. *South Africa (Cape Province)*. H5–G1. Spring.

**\*L. jacquinii** N.E. Brown (*L. anceps* misapplied). Illustration: Sweet, British flower garden 1: t. 143 (1826); Le Roux & Schelpe, Namaqualand and Clanwilliam, South African wild flower guide 1: 56 (1981). Plant small but often branched; stems 3-angled; stem leaves 2-keeled beneath; the angles of the stem and the keels of the leaves are undulate-crisped to irregularly toothed; perianth dark violet. *South Africa (Cape Province)*. H5–G1. Spring.

**\*L. fabricii** (Delaroche) Ker (*L. anceps* misapplied; *Peyrousia aculeata* Sweet; *Lapeirousia aculeata* Sweet; *L. compressa* Pourret). Illustration: Sweet, British flower garden 1: t. 39 (1826); Le Roux & Schelpe, Namaqualand and Clanwilliam, South African wild flower guide 1: 56 (1981). Plant small, usually branched; stems

2-angled and winged; stem leaves 1-keeled beneath; angles of stem and keels of leaves undulate-crisped to irregularly toothed; perianth cream or cream flushed with pink, sometimes with red spots on the lower lobes. *South Africa (Cape Province)*. H5–G1. Spring.

The name *L. anceps* has been incorrectly used for both the above species. *L. anceps* (Linnaeus filius) Ker is a quite separate species which is not now cultivated; it has stems to 30 cm, with many spreading, elongate branches and the perianth is white to pink.

## 36. ANOMATHECA Ker
*J. Cullen*
Herbaceous perennials with conical corms which have fibrous coats. Stems erect, branched or not. Basal leaves several, soft, with conspicuous false midribs, shorter than the 2-ranked stem-leaves. Inflorescence a simple spike with few flowers (in ours). Perianth with a tube and 6 lobes of which the uppermost is somewhat larger than the others, rendering the flower slightly bilaterally symmetric. Stamens 3, asymmetrically disposed. Stigmas 3, each forked and recurved. Fruit a rough-surfaced capsule containing many rounded seeds.

A genus of 4 species from southern Africa, formerly included in *Lapeirousia* (p. 382), but sufficiently distinct in corm, leaf and habit characters. Cultivation as for Ixia (p. 380).
Literature: Lawrence, G.H.M., Notes on Lapeirousia in Cultivation, *Baileya* 3: 131–6 (1955); Goldblatt, P., A Revision of the Genera Lapeirousia Pourret and Anomatheca Ker in the Winter Rainfall Region of South Africa, *Contributions from the Bolus Herbarium* 4 (1972).

1a. Stems flattened and winged; flowers green **4. viridis**
  b. Stems terete; flowers not green **2**
2a. Leaves to 10 cm, much shorter than the inflorescence, obtuse but with a small point **1. verrucosa**
  b. Leaves 10–60 cm, about as long as the inflorescence, acute **3**
3a. Perianth-lobes less than half as long as the tube, held at right angles to it **2. laxa**
  b. Perianth-lobes more than half as long as the tube, curving outwards and forming a cup-shape **3. grandiflora**

**1. A. verrucosa** (Vogel) Goldblatt (*Lapeirousia juncea* (Linnaeus filius) Ker; *Anomatheca juncea* (Linnaeus filius) Ker).

Illustration: Botanical Magazine, 606 (1802).
Corms conical. Stems terete, 8–20 cm. Basal leaves to 10 cm, much shorter than the inflorescence, linear-oblong, obtuse but with a small point and with a conspicuous false midrib. Inflorescence of several spikes. Perianth slightly irregular, pink, fragrant, tube 1–1.5 cm, the lobes *c.* 9 × 4 mm, the lower marked with red-purple spots. *South Africa (Cape Province)*. H5–G1. Spring.

**2. A. laxa** (Thunberg) Goldblatt (*Lapeirousia laxa* (Thunberg) N.E. Brown; *A. cruenta* Lindley; *Lapeirousia cruenta* (Lindley) Baker). Illustration: Edwards's Botanical Register 16: t. 1369 (1830); Eliovson, Wild flowers of southern Africa, ii (1955); Hay & Synge, Dictionary of garden plants in colour, 100 (1969).
Corms conical, *c.* 1 cm in diameter. Stem terete, erect, 10–30 cm, not or little branched. Leaves mostly basal, equalling or exceeding the inflorescence, acute, with conspicuous false midribs. Inflorescence a simple spike with up to 6 flowers; bracts 6–8 mm. Perianth usually red, occasionally white or bluish, tube 2–3 cm, the lobes to 1.3 cm, slightly irregularly disposed, spreading at right angles to the tube, each of the 3 lower lobes with a dark red or purple spot near the base. Stamens borne towards the upper side of the flower, the 3 anthers touching. *South Africa, Moçambique*. H5–G1. Mostly spring.

**3. A. grandiflora** Baker (*Lapeirousia grandiflora* (Baker) Baker). Illustration: Botanical Magazine, 6924 (1887); Lindley & Baker, Flowers of the veld, f. 18 (1972); Moriarty, Wild flowers of Malawi, t. 7 (1975).
Similar to *A. laxa* but stems and leaves to 60 cm, perianth brownish or purplish red, the 3 lower lobes each with a dark spot at the base, the tube 3–3.5 cm, the lobes 2.5–4 cm, curving outwards, forming a cup-shape. *Southern Africa north to Malawi*. H5–G1. Summer.

**4. A. viridis** (Aiton) Goldblatt (*Lapeirousia viridis* (Aiton) Bolus). Illustration: Mason, Western Cape sandveld flowers, t. 20 (1972).
Corm 1–2 cm in diameter. Stem to 35 cm, flattened and winged. Leaves much shorter than the inflorescence, margins smooth or undulate-crisped. Inflorescence a simple spike with up to 6 (rarely more) flowers; bracts *c.* 1 cm. Perianth-tube 2–3 cm, the lobes 1–2 cm, sharply reflexed from the

apex of the tube, green. *South Africa (Cape Province)*. H5–G1. Spring.

## 37. WATSONIA Miller
*E. Campbell*

Herbaceous perennials with more or less spherical corms. Leaves basal and borne on the stem, 2-ranked, usually rigid, sword-shaped to linear, each usually with a prominent main vein. Flowers in 2 ranks in a simple or branched spike. Perianth forming a curved tube at the base, the upper part gradually or abruptly widened, the lobes more or less equal or the 3 inner slightly wider than the 3 outer, all oblong or lanceolate. Stamens 3, arising below the throat of the perianth-tube. Ovary 3-celled, each cell with many ovules; style thread-like with 3 bifid stigmas. Fruit an oblong capsule.

A genus of about 60 species from South Africa; about 14 of them are in general cultivation today, though many more have been grown, and may still persist here and there. They are best grown in a well-drained soil in a sunny position, and require a good deal of water while growing. The corms of the spring-flowering species may be lifted and dried off for the winter. Propagation is by seeds or offsets.

Literature: Roux, J.P., Studies in the genus Watsonia Miller, *Journal of South African Botany* **46**: 365–78 (1980).

*Leaves*. Up to 90 cm: **1–9,11**; 90 cm or more: **10**.
*Stem*. Up to 90 cm: **2,5,7–9**; 90 cm or more: **1,3–6,10–12**.
*Spike*. Simple: **2,7,9**; branched: **1–6,8–12**.
*Perianth*. White: **7**; purplish red or lilac: **6,11**; pink, red or orange-red: **1–5,7–10,12**. Up to 6 cm: **2–4,6–8**; 6 cm or more: **1,4,5,8–12**.
*Perianth-tube*. Up to 4 cm: **2–4,6–8**; 4 cm or more: **1,5,8–12**.

1a. Plant usually bearing bulbils in the axils of the leaves and bracts after flowering     **1. bulbillifera**
 b. Plant without bulbils     2
2a. Perianth to 4.5 cm, its tube to 2 cm     3
 b. Perianth 4.5 cm or more, its tube 2 cm or more     4
3a. Perianth-tube 1.4–2 cm; bracts 2–3 cm     **2. galpinii**
 b. Perianth-tube to 1.3 cm; bracts 1–2 cm     **3. marginata**
4a. Perianth-tube almost as long as the lobes     **4. pyramidata**
 b. Perianth-tube longer than the lobes     5

5a. Perianth-lobes up to 2.5 cm     6
 b. Perianth-lobes more than 2.5 cm     10
6a. Perianth-tube 5 cm or more     **5. beatricis**
 b. Perianth-tube less than 5 cm     7
7a. Stem 1–1.5 m     **6. wilmaniae**
 b. Stem at most 90 cm     8
8a. Leaves 60–90 cm; spike compact     **7. densiflora**
 b. Leaves 15–60 cm; spike loose     9
9a. Leaves to 1.2 cm wide; spike branched     **8. fulgens**
 b. Leaves 1.8–2.5 cm broad; spike simple     **9. meriana***
10a. Leaves more than 75 cm; perianth-tube to 4.3 cm     **10. tabularis**
 b. Leaves to 75 cm; perianth-tube 5 cm or more     11
11a. Basal leaves 4, stem leaves usually 5; perianth lilac-purple     **11. wordsworthiana**
 b. Basal leaves usually 8, stem leaf 1; perianth orange-red     **12. fourcadei**

**1. W. bulbillifera** Mathews & Bolus. Illustration: Flowering plants of South Africa **19**: t. 726 (1939).
Basal leaves 5 or 6, to 60 × 6 cm. Bulbils present in the axils of the stem-leaves and bracts after flowering. Stem 1.2–1.5 m. Flowers few, in loose, branched spikes. Bracts to 2.2 cm, shorter than the internodes. Perianth brick red, pale outside, longitudinally striped with white inside; tube to 4.3 cm, lobes to 2.4 × 1.3 cm, the outer 3 oblong, the inner 3 obovate. *South Africa (Cape Province)*. H5–G1. Spring–summer.

**2. W. galpinii** Bolus. Illustration: Flowering plants of South Africa **2**: t. 45 (1922).
Leaves 4–8, all basal, to 35 × 1.5 cm. Stem to 75 cm, occasionally branched. Flowers many in a fairly dense spike. Bracts 2–3 cm, longer than the internodes. Perianth bright red, tube to 2 cm, lobes to 1.5 × 1 cm, oblong to oblong-ovate. *South Africa (Cape Province)*. H5–G1. Summer–autumn.

**3. W. marginata** (Ecklon) Ker Gawler. Illustration: Botanical Magazine, 608 (1802); Flowering plants of South Africa **19**: t. 748 (1939).
Basal leaves 3–4, 45–80 cm × 4–5 mm, stem-leaves 3. Stem 1.5–1.8 m. Flowers many, in a dense, branched spike. Bracts 1–2 cm, longer than the internodes. Perianth rose-pink, tube *c.* 1.3 cm, lobes 2–3 cm, ovate-oblong. *South Africa (Cape Province)*. H5–G1. Spring–summer.

**4. W. pyramidata** (Andrews) Stapf (*W. rosea* Ker Gawler). Illustration: Botanical Magazine, 1072 (1807); Flowering plants of South Africa **25**: t. 974 (1945).
Basal leaves 5, to 73 × 2–2.5 cm, stem-leaves 6. Stem 1.2–1.6 m. Flowers many in a loose, branched spike. Bracts *c.* 1 cm, shorter than the internodes. Perianth rose-pink, tube 3–3.5 cm, lobes 3–3.5 cm, elliptic. *South Africa (Cape Province)*. H5–G1. Summer.

**5. W. beatricis** Mathews & Bolus. Illustration: Botanical Magazine, 9139 (1928); Flowering plants of South Africa **19**: t. 744 (1939); Journal of the Royal Horticultural Society **75**: f. 229 (1950); Gledhill, Eastern Cape sandveld flowers, t. 17 (1977).
Basal leaves 3, to 75 × 3 cm, stem-leaves about 8. Stem to 90 cm, simple. Flowers many in a dense, branched spike. Bracts 1.5–3.5 cm, equal to or longer than the internodes. Perianth deep orange-red, tube *c.* 5 cm, lobes to 2.5 × 1–1.6 cm, oblong to ovate-oblong. *South Africa (Cape Province)*. H5–G1. Summer–autumn.

**6. W. wilmaniae** Mathews & Bolus. Illustration: Flowering plants of South Africa **19**: t. 743 (1939).
Basal leaves 6, to 50 × 4 cm. Stem to 1.5 m. Flowers many in a fairly dense, 4-branched spike. Bracts to 2 cm, slightly longer than the internodes. Perianth purplish pink with darker stripes, tube to 4 cm, lobes to 2 × 1 cm, obovate. *South Africa (Cape Province)*. H5–G1. Summer–autumn.

**7. W. densiflora** Baker. Illustration: Botanical Magazine, 6400 (1878).
Basal leaves 60–90 × 1.2–1.8 cm, stem-leaves present. Stem 60–90 cm. Flowers many in a dense, simple spike. Bracts 1.2–2.4 cm, longer than the internodes. Perianth rose-pink or sometimes whitish, tube *c.* 2.5 cm, lobes *c.* 1.8 cm, oblong. *South Africa (Transvaal, Natal, Orange Free State)*. H5–G1. Summer–autumn.

**8. W. fulgens** (Andrews) Persoon (*W. angusta* Ker Gawler). Illustration: Botanical Magazine, 600 (1802).
Basal leaves several, 30–60 × 1.7–2.5 cm, stem-leaves present. Stem 60–90 cm, usually branched. Flowers many, in a loose, branched spike. Bracts to 2.5 cm, shorter than or sometimes slightly longer than the internodes. Perianth scarlet, tube 3.7–5 cm, sharply bent, lobes 1.8–2 cm, oblanceolate. *South Africa (Cape Province)*. H5–G1. Summer.

**9. W. meriana** (Linnaeus) Miller (*W. ardernei* Sander). Illustration: Gledhill, Eastern Cape sandveld flowers, t. 17 (1977); Gibson, Wild flowers of Natal (inland region), t. 39 (1978).

Basal leaves 3–4, 30–60 cm × 5–12 mm, stem-leaves present. Stem 30–90 cm, simple or branched. Flowers few to several in a loose spike. Bracts *c.* 2.5 cm. Perianth bright rose-red, rarely scarlet or whitish, tube to 5 cm, lobes to 2.5 cm, oblong to oblong-ovate. *South Africa (Cape Province).* H5–G1. Summer.

*\*W. coccinea* Herbert. Leaves less than 30 cm, flowers 4–6, in a simple spike, perianth bright crimson. *South Africa (Cape Province).* H5–G1. Summer.

*\*W. humilis* Miller. Illustration: Botanical Magazine, 631 (1803). Leaves to 30 cm × 2.5–5 mm, flowers 4–6 in a simple spike, perianth bright pink. *South Africa (Cape Province, Natal).* H5–G1. Summer.

These 2 species may well be merely varieties of *W. meriana*.

**10. W. tabularis** Mathews & Bolus. Illustration: Flowering plants of South Africa 5: 238 (1926).

Basal leaves 4–5, to 100 × 5 cm, stem-leaves 2–3. Stem 1.5–1.8 m. Flowers many on a 2–3-branched spike. Bracts 1–1.8 cm, shorter than the internodes. Perianth deep orange-red on the tube and the outsides of the outer lobes, paler salmon- or rose-pink within and on the outsides of the inner lobes; tube to 4.5 cm, lobes to 3.5 × 2.5 cm, oblong. *South Africa (Cape Province).* H5–G1. Summer.

**11. W. wordsworthiana** Mathews & Bolus. Illustration: Flowering plants of South Africa 19: t. 742 (1939).

Basal leaves 4, to 60 × 2.5–3.5 cm, stem-leaves 5, to 55 cm. Stem to 1.6 m. Flowers many in a much-branched spike. Bracts 1.8–2.3 cm, shorter than the internodes. Perianth lilac-purple, tube *c.* 5 cm, lobes to 3 × 2 cm, obovate-oblong. *South Africa (Cape Province).* H5–G1. Summer.

**12. W. fourcadei** Mathews & Bolus. Illustration: Flowering plants of South Africa 5: 235 (1926).

Basal leaves 8, to 60 × 4 cm, stem-leaf 1, to 15 cm. Stem 1–1.5 m. Flowers many in a much-branched spike. Bracts 1–1.5 cm, shorter than the internodes. Perianth orange-red, tube to 5.5 cm, lobes 2.5–3 cm, oblong-linear to obovate. *South Africa (Cape Province).* H5–G1. Summer.

**38. BABIANA** Ker Gawler
*E. Campbell*

Perennials with ovoid bulbs which nearly always extend upwards as a long neck. Leaves basal, 2-ranked, with sheathing stalks below ground and sometimes extending above, blades lanceolate, pleated, usually hairy. Flowers in a 2-ranked spike borne on an erect or ascending scape. Perianth bilaterally or radially symmetric, with a short or long tube. Stamens 3, free, arising near the throat of the perianth-tube. Ovary 3-celled with few to several ovules; style simple, stigmas short. Fruit a 3-celled oblong to spherical capsule.

A genus of about 60 species from South and Southwest Africa, of which about 5 are in cultivation, usually as pot plants. They are easily grown in well-drained soils in a sunny position.

Literature: Lewis, G.J., The genus Babiana, *Journal of South African Botany*, supplementary volume 3: 1–149 (1960).

1a. Perianth-tube more than 6 cm
        **1. tubulosa**
  b. Perianth-tube less than 5.5 cm   **2**
2a. Flowers radially symmetric with equal or almost equal perianth-lobes; ovary hairy or rough   **3**
  b. Perianth bilaterally symmetric, the 3 upper perianth-lobes longer than the others, or, if all lobes are more or less equal, then ovary hairless   **4**
3a. Perianth-lobes scarlet below, blue above, broad; stigmas flattened, 5 mm or more   **2. rubro-cyanea**
  b. Perianth-lobes various but not as above; stigmas not flattened, to 3 mm
        **3. stricta**
4a. Flowers 3.3–5 cm; ovary hairless
        **4. sambucina**
  b. Flowers 5–8.5 cm; ovary hairy
        **5. plicata**

**1. B. tubulosa** (Burman) Ker Gawler (*B. tubata* (Jacquin) Sweet; *B. tubiflora* Linnaeus filius). Illustration: Andrews' Botanical Repository 1: 5 (1797).

Leaves 6–8, blades 4–15 (rarely to 20) cm × 4–22 mm. Scape ascending, with 1–3 branches, below ground 7–12 cm, the aerial part short. Flowers 6–12 in a dense spike. Perianth cream with red markings on the lower 3 lobes, the tube flushed with purple outside and the outer lobes nearly always with a reddish median line outside; tube 6.5–9 cm; upper 3 lobes 3–3.5 cm × 8–14 mm, lower lobes 1.8–2.3 cm × 5–9 mm. *South Africa.* H5–G1. Spring.

**2. B. rubro-cyanea** (Jacquin) Ker Gawler (*B. stricta* (Aiton) Ker Gawler var. *rubro-cyanea* (Ker Gawler) Baker; *B. rubrocaerulea* Reichenbach). Illustration: Botanical Magazine, 410 (1798).

Leaves 5–7, blades 6–15 cm × 6–20 mm. Scape ascending, 5–20 cm. Flowers 5–10 in a lax to fairly dense spike. Perianth-tube 1.5–2 cm; lobes red in the lower part, blue in the upper, 2–2.4 × 1–1.4 cm. Ovary downy or silky; stigmas 5 mm or more, flattened. *South Africa (Cape Province).* H5–G1. Spring.

**3. B. stricta** (Aiton) Ker Gawler. Illustration: Botanical Magazine, 621 (1802).

Leaves 6–8, blade 4–12 cm × 6–12 mm. Scape erect, 10–20 cm or rarely more. Flowers 4–8 in loose spike, sometimes slightly fragrant. Perianth pale cream or dark purple, blue or mauve-blue, sometimes with dark, blood-red centre, rarely entirely sulphur yellow. Perianth-tube 1–1.8 cm; lobes 1.5–2.5 cm × 6–14 mm. Ovary densely hairy or silky; stigmas 2.5–3 mm, not flattened. *South Africa.* H5–G1. Spring.

A variable species divided into 5 varieties by Lewis (see above), differing in flower colour, perianth-tube length and characters of the stem and leaves; it is not known which of these are in cultivation.

**4. B. sambucina** (Jacquin) Ker Gawler (*B. spathacea* Ker Gawler; *B. stellata* Schlechter). Illustration: Botanical Magazine, 638 (1803), 1019 (1807); Hay et al., Dictionary of indoor plants, t. 51 (1974).

Leaves 5–6, 4–14 cm × 3–8 mm. Scape rarely extending more than 2 cm above ground, sometimes with 1 or 2 very short branches arising below ground. Flowers 2–6 in a compact spike, fragrant. Perianth pale blue-mauve, violet or purple, with a white mark near the middle on the lower lateral lobes; tube 3–5 cm; lobes all similar, 2–3.5 cm × 5–12 mm. Ovary hairless. *South Africa.* H5–G1. Summer.

A variable species, divided by Lewis into 3 varieties.

**5. B. plicata** Ker Gawler (*B. disticha* Ker Gawler). Illustration: Botanical Magazine, 576 (1802).

Leaves 6–8, 8–12 × 1–2 cm. Scape erect, 7–20 cm. Flowers 4–10 in a dense spike, fragrant. Perianth varying from violet to pale blue or rarely white, the lower half of the lateral lobes usually with a small or large yellow area marked with 2 purple dots at the base; tube 1.8–2.5 cm; the 3 upper lobes slightly larger than the others, 1.5–2.5 cm × 7–10 mm, the outer 3 lobes pointed. Ovary densely hairy or silky. *South Africa*. H5–G1. Summer.

## 39. CROCOSMIA Planchon
*J. Cullen*

Perennial herbs with small, flattened corms and creeping stolons which readily produce new corms. Leaves mostly basal, 2-ranked, conspicuously ribbed, sometimes pleated. Inflorescence usually a panicle of spikes, more rarely a simple spike, overtopping the leaves. Bracts small, scarcely exceeding the ovaries. Perianth with a curved tube, slender throughout or slender at the base and then abruptly expanded, the lobes spreading, all similar or the uppermost slightly longer than the others. Stamens 3, free, borne on the perianth-tube, arching to the upperside of the flower, the anthers either just projecting from the mouth of the perianth-tube, or with very long filaments, the anthers borne at the level of the tips of the perianth-lobes or beyond this. Style somewhat longer than the stamens, 3-fid at the apex. Fruit a capsule opening by 3 splits and containing many seeds.

A genus of about 7 species from southern Africa. Its limits are somewhat uncertain, and, as recognised at present, it contains species which have formerly been placed in *Antholyza*, *Curtonus*, *Montbretia* and *Tritonia*. They are relatively easily grown in a well-drained soil in a sunny position, and can be propagated by corms, stolons or seed. Lifting of the corms is not generally advisable. The one commonly grown hybrid (see below) can be invasive in some places.
Literature: Kostelijk, P., Crocosmia in gardens, *The Plantsman* **5**: 246–53 (1984); de Vos, M.P., The African genus Crocosmia Planchon, *Journal of South African Botany* **50**: 463–502 (1984).

1a. Leaves conspicuously pleated, 3–7 cm wide                                                       2
  b. Leaves ribbed but not pleated, up to 3 cm wide                                                3
2a. Ribs of the leaves bearing fine hairs; axis of the spike conspicuously zig-zag; perianth-tube 3–4 cm
                                                                                                **1. paniculata**
  b. Ribs of the leaves hairless; axis of the spike not zig-zag; perianth-tube less than 3 cm            **2. masonorum**
3a. Flowers borne on 2 sides of the spike axis; perianth-tube slender, parallel-sided, the lobes 2–2.6 cm × 5–7 mm, spreading from its apex; stamens

projecting beyond the perianth-tube                                                              **4. aurea***
  b. Spikes 1-sided; perianth-tube slender at the base then abruptly expanded, the lobes 8–13 × 5–6 mm, not very widely spreading; only the anthers projecting from the mouth of the perianth-tube                                          **3. pottsii**

**1. C. paniculata** (Klatt) Goldblatt (*Antholyza paniculata* Klatt; *Curtonus paniculatus* (Klatt) N.E. Brown). Illustration: Eliovson, Wild flowers of southern Africa, xix (1955); Rix & Phillips, The bulb book, 169 (1981).
Plant to 1 m. Leaves to 90 × 6 cm, tapered to the base and apex, conspicuously pleated, with fine hairs on all the veins. Flowering stem erect. Inflorescence a panicle of spikes, the branches ascending to erect, each spike with its axis conspicuously zig-zag, especially towards the base. Flowers borne on 2 sides of the spike axis, facing outwards. Perianth 4.5–6 cm, orange, the tube curved, slender at the base then abruptly swollen, the lobes spreading, 1–1.5 cm. Stamens slightly projecting beyond the perianth-lobes. *South Africa*. H5–G1. Summer.

This species has been hybridised with *C. masonorum* and some of these hybrids are currently available.

**2. C. masonorum** (Bolus) N.E. Brown (*Tritonia masonorum* Bolus). Illustration: Synge, Collins' guide to bulbs, t. iv (1961); Horticulture **41**: 20 (1963); The Garden **101**: 463 (1976).
Plant 1–1.25 m. Leaves to 1 m × 7 cm, tapered to the base and apex, conspicuously pleated, hairless. Flowering stem erect, inflorescence a simple spike or panicle of spikes each of which arches to the horizontal; spike axis straight. Flowers borne in 2 rows on the upper side of the spike, so that they face upwards. Perianth 4–5 cm, reddish orange, the tube slender at the base then abruptly expanded, the lobes widely spreading. Stamens projecting beyond the perianth-lobes. *South Africa (Transkei)*. H5–G1. Summer–autumn.

**3. C. pottsii** (Baker) N.E. Brown (*Tritonia pottsii* Baker; *Montbretia pottsii* (Baker) Baker). Illustration: Botanical Magazine, 6722 (1833).
Plant to 90 cm. Leaves to 90 cm × 8–16 mm, long-tapered to the apex, somewhat tapered to the base, ribbed but not pleated, hairless. Inflorescence a panicle of spikes (rarely a simple spike), the branches erect, not zig-zag. Flowers numerous, borne on 1 side of the spike,

facing outwards or occasionally upwards. Perianth 2.5–3.5 cm, orange to scarlet, the tube slender at the base then abruptly expanded, the lobes not widely spreading, 8–13 × 5–6 mm. Anthers borne at the mouth of the perianth-tube or just beyond it. *South Africa*. H5–G1. Summer.

**4. C. aurea** (Hooker) Planchon (*Tritonia aurea* Hooker). Illustration: Botanical Magazine, 4335 (1847); Eliovson, Wild flowers of southern Africa, xix (1955). Plant to 1 m. Leaves to 90 × 1–2.2 cm, tapered to the apex, somewhat tapered to the base, ribbed but not pleated, hairless. Inflorescence a single, erect spike or a panicle of spikes with erect or ascending branches. Flowers borne on 2 sides of the straight spike axis, facing outwards. Perianth 3–5 cm, orange or yellowish, the tube slender and parallel-sided throughout, the lobes 2–2.6 cm × 5–7 mm, very widely spreading. Stamens projecting beyond the perianth-lobes. *South Africa*. H5–G1. Spring–early summer.

A variable species; a variant with a dark spot at the base of each perianth-lobe is sometimes grown as 'Maculata'.

***C. × crocosmiiflora** (Burbidge & Dean) N.E. Brown (*Montbretia crocosmiiflora* Lemoine). Illustration: Flowering plants of South Africa **4**: t. 152 (1924). This is the hybrid between *C. aurea* and *C. pottsii* and is more widely grown than either of its parents. It resembles *C. aurea* in general, but the flowers have a tube which is slightly widened above, the lobes do not spread as widely, and the stamens are about the same length as the perianth-lobes or shorter. The hybrid is very variable, and many cultivars have been selected, named and illustrated. They are often invasive, and require control.

## 40. MELASPHAERULA Ker Gawler
*J.C.M. Alexander*

Corm to 1.5 cm wide with hard black tunic. Basal leaves 6 or 7 in 2 ranks, 10–30 × *c.* 1.5 cm, linear to narrowly triangular, sheathing at base; midrib pale. Stem 30–75 cm, branched, slender, wiry, twisted and drooping. Flowers to 1.8 cm wide, stalkless in loose terminal and lateral spikes, 3–7 per spike. Bracts with chaffy margins. Perianth radially symmetric, with a very short tube; lobes lanceolate, acuminate, white to greenish yellow, sometimes streaked with purple; the inner 3 broader. Stamens arising at lobe-bases; filaments free. Style 3-branched, arched under inner perianth-lobes. Ovary sharply 3-angled.

A genus of a single species. Originally considered part of *Gladiolus*. Cultivation as for *Ixia* (p. 380).

**1. M. ramosa** (Linnaeus) N.E. Brown (*M. graminea* (Linnaeus filius) Ker Gawler; *M. parviflora* Sweet). Illustration: Botanical Magazine, 615 (1803); Rice & Compton, Wild flowers of the Cape of Good Hope, pl. 188 (1951).
*South Africa.* H5–G1. Spring.

## 41. TRITONIA Ker Gawler
*J. Cullen*

Perennial herbs with ovoid to spherical corms with papery or fibrous tunics. Leaves mostly basal, 2-ranked, linear, with false midribs. Inflorescence a simple spike. Bracts exceeding the ovaries. Perianth with a long or short tube gradually widening from the base or cylindric, the lobes spreading or widely spreading, all more or less equal. Stamens 3, attached to the perianth-tube, filaments free, either curving variously so that the anthers are distant, or parallel, with the anthers close or touching and all lying on the lower side of the flower. Style exceeding the stamens, stigma 3-fid. Capsule ellipsoid, membranous, opening by 3 flaps and containing many seeds.

A South African genus of about 30 species. Cultivation as for *Gladiolus* (p. 388).
Literature: de Vos, M.P., The African Genus Tritonia: part 1, *Journal of South African Botany* 48: 105–63 (1982).

1a. Perianth orange or pinkish orange, the lobes considerably longer than the tube; stamens variously curved, the anthers distant **1. crocata**
 b. Perianth pink, the lobes about as long as the tube; stamens with parallel filaments curved to the lower side of the flower, the anthers close or touching **2. rubrolucens**

**1. T. crocata** (Linnaeus) Ker Gawler (*T. hyalina* (Linnaeus filius) Baker; *T. fenestrata* (Jacquin) Ker Gawler). Illustration: Botanical Magazine, 184 (1793), 704 (1804); Eliovson, Wild flowers of southern Africa, 83 (1980); Journal of South African Botany 48: 133 (1982).
Plant 25–50 cm. Corm ovoid or spherical, 1.5–3 cm. Leaves 5–30 cm × 4–15 mm, spreading, tapered to the apex and slightly so to the base. Spike with 6–10 flowers borne in 1 or 2 rows. Perianth usually bright orange, sometimes pinkish orange, the lower segments sometimes with a yellowish stripe at the base, tube 8–15 mm,

lobes 2–2.8 cm, very rounded at the apex, the claws with translucent margins. Stamens curved and variously spreading, the anthers distant. *South Africa (Cape Province).* H5–G1. Spring–early summer.

Variable in flower colour; there is, however, no reason for separating *T. hyalina* and *T. crocata*.

**2. T. rubrolucens** Foster (*T. rosea* Klatt). Illustration: Botanical Magazine, 7280 (1893).
Plant to 60 cm, though often less. Corm ovoid, *c.* 2.5 cm. Leaves to 55 cm × 8–15 mm, spreading, tapered to the apex, slightly so to the base. Spike with *c.* 8 flowers, somewhat 1-sided. Perianth pink, tube to *c.* 1.2 cm, lobes about the same length, obtusely pointed, the 3 lower with a yellow stripe at the base, claws without translucent margins. Stamens just projecting from the perianth-tube, the anthers close or touching, on the lower side of the flower. *South Africa (Cape Province, Natal).* H5–G1. Summer–autumn.

## 42. SPARAXIS Ker
*J. Cullen*

Perennial herbs with corms with papery or fibrous tunics, the stems sometimes bearing small corms after flowering. Stems erect, rigid. Leaves mostly basal, in a fan, with several veins and a conspicuous false midrib. Spike with few flowers; bracts large, entire or raggedly toothed near the apex. Perianth radially symmetric; tube short; lobes equal. Stamens 3, symmetrically or asymmetrically arranged; filaments free, anthers sometimes coiled spirally. Style exceeding the stamens, the stigma 3-fid, the tips of the branches sometimes expanded. Capsule spherical, firm, containing several large, smooth and often shiny seeds.

A genus of 6 species, all from the Cape Province of South Africa. Following Goldblatt (reference below), *Streptanthera* Sweet is included in it. Cultivation as for *Ixia* (p. 380).
Literature: Goldblatt, P., The Genus Sparaxis, *Journal of South African Botany* 35: 219–52 (1969); The Species of Sparaxis and their Geography, *Veld & Flora* 65: 7–9 (1978).

1a. Stamens symmetrically arranged, style straight; bracts entire, more than 1 cm 2
 b. Stamens asymmetric, lying to the upper side of the flower, the style curved to the lower; bracts raggedly toothed, the solid part less than 1 cm 3

2a. Anthers coiled spirally, purple **1. elegans**
 b. Anthers straight, yellow **2. tricolor**
3a. Stems producing many obvious, small corms after flowering, usually branched and bearing 1 leaf **4. bulbifera**
 b. Stems not producing small corms, or, if so, these small and not obvious, usually unbranched and without leaves **3. grandiflora**

**1. S. elegans** (Sweet) Goldblatt (*Streptanthera elegans* Sweet; *Streptanthera cuprea* Sweet). Illustration: Veld & Flora **65**: (1): cover (1978).
Corm 1–1.7 cm in diameter. Stem 10–30 cm, bearing small corms at the lowest nodes. Leaves tapered to the apex, 8–25 cm × 5–14 mm. Spike with 1–5 flowers; bracts to 2.2 cm, entire. Perianth-tube yellow, narrow at the base then widening, 6–8 mm; lobes ovate, obtuse, 1.8–2.2 × 1.4–1.7 cm, reddish orange fading to pink, rarely white, usually with a violet band at the base. Stamens symmetric, filaments yellow, 5–6 mm, anthers dark purple, coiled around the style, shorter than the filaments. Tips of the style-branches expanded. *South Africa (Cape Province).* H5–G1. Spring.

**2. S. tricolor** (Schneevogt) Ker (*Ixia tricolor* Schneevogt). Illustration: Botanical Magazine, 381 (1797), 1482 (1812); Eliovson, Wild flowers of southern Africa, 83 (1955).
Corm 1–2 cm in diameter. Stem 10–40 cm. Leaves tapered to the apex, 5–30 × 1–2 cm. Spike loose with 2–5 flowers; bracts 2.5–3 cm, entire or slightly toothed at the tip. Perianth-tube *c.* 8 mm, yellow, narrow at the base, expanded above; lobes lanceolate, 2.5–3.5 × *c.* 1 cm, orange, red or purple, marked with black to dark red at the base. Stamens symmetric, filaments 6–7 mm, anthers 8–9 mm, yellow or white, straight. *South Africa (Cape Province).* H5–G1. Spring.

A striking species which has hybridised with others to produce a range of hybrids with variously coloured flowers, generally showing a 3-coloured effect.

**3. S. bulbifera** (Linnaeus) Ker (*Ixia bulbifera* Linnaeus). Illustration: Flowering plants of South Africa 9: t. 338 (1929); Veld & Flora 65(1): cover (1978).
Corms 9–16 mm in diameter. Stems 5–50 cm, usually with 1 branch and 1 leaf, bearing numerous, obvious corms in the leaf-axil after flowering. Leaves to

40 cm × 4–10 mm. Spike loose with 1–6 flowers; bracts deeply and raggedly toothed, the solid part less than 1 cm. Perianth-tube 1.4–1.6 cm, yellow, green outside at the base; lobes 2.5–2.8 cm, cream or white inside, the outside similar or with purple streaks, rarely entirely purple. Stamens asymmetric, filaments white, 7–8 mm, anthers white, curved, 7–8 mm. *South Africa (Cape Province)*. H5–G1. Spring.

**4. S. grandiflora** (Delaroche) Ker (*Ixia grandiflora* Delaroche).
Corms 6–15 mm. Stem erect, 8–45 cm, bearing a few small corms after flowering. Leaves 3–30 cm × 4–13 mm, tapered above. Spike loose, with 1–6 flowers; bracts deeply and raggedly toothed, the solid part less than 1 cm. Perianth-tube 1–1.4 cm, yellow, purple or blackish; lobes variously coloured, 2.4–3 × 1.2–1.6 cm. Stamens asymmetric, filaments white or yellow, 7–9 mm, anthers white or yellow, a little longer than the filaments. *South Africa (Cape Province)*. H5–G1. Spring.

A variable species; Goldblatt recognises 4 subspecies, 2 of which have been grown:

Subsp. **grandiflora**. Illustration: Botanical Magazine, 541 (1804); Flowering plants of South Africa **2**: t. 60 (1922). Perianth deep reddish purple, lobes often with a white base and whitish margins.

Subsp. **fimbriata** (Lamarck) Goldblatt (*Ixia fimbriata* Lamarck; *I. liliago* de Candolle; *S. grandiflora* var. *liliago* (de Candolle) Ker). Illustration: Botanical Magazine, 779 (1804); Edwards's Botanical Register **3**: t. 258 (1818); Veld & Flora **65**(1): cover (1978). Perianth-lobes narrower, the whole perianth creamy white or pale yellow, each lobe with a purple stripe outside and deep purple blotches at the base inside.

**43. SYNNOTIA** Sweet
*J. Cullen*
Perennial herbs with corms with fibrous coats. Stems erect, usually with 1–3 branches. Leaves 7–10, mostly basal, 2-ranked, each with several veins but without a conspicuous false midrib. Inflorescence a spike with few flowers; bracts membranous or brownish, irregularly toothed or torn at the apex, concealing the ovaries. Perianth bilaterally symmetric, the tube cylindric and narrow for most of its length, broadening above and curved where it broadens; lobes 6, the uppermost the largest, erect and hood-like over the stamens and stigmas; the 2 upper lateral lobes spreading; the 3 lower lobes

are united to each other beyond the mouth of the tube, forming a lower lip, and these lobes are themselves deflexed. Stamens 3, free. Stigmas 3. Fruit a capsule with many spherical or angled seeds.

A genus of five species from the Cape Province of South Africa. The genus name was originally spelled '*Synnetia*' (Sweet, British flower garden **2**: t. 150, 1826), but Sweet himself corrected it to *Synnotia* at the end of the text to t. 154 of the same volume. Cultivation as for *Ixia* (p. 380). Literature: Lewis, G.J., A Revision of the Genus Synnotia, *Annals of the South African Museum* **40**: 137–51 (1956).

**1. S. variegata** Sweet. Illustration: Sweet, British flower garden **2**: t. 150 (1826); Rice & Compton, Wild flowers of the Cape of Good Hope, t. 194 (1950); le Roux & Schelpe, Namaqualand and Clanwilliam, South African wild flower guide **1**: 55 (1981).
Corm ovoid or spherical, 1–2.2 cm in diameter. Stem 8–40 cm. Leaves 7–8, 2–15 cm × 5–20 mm, oblong, acute. Spike with 2–7 flowers; bracts 2–2.5 cm. Perianth lavender to deep purple with yellow stripes on the lower lip and in the tube (rarely these areas whitish), uppermost lobe 2–3 cm, lateral lobes 1.5–2 cm. *South Africa (Cape Province)*. H5–G1. Spring.

**44. GLADIOLUS** Linnaeus
*P.F. Yeo*
Herbaceous, usually with a corm, often producing cormlets at base or on stolons. Lowest leaves reduced to sheaths, main leaves 2-ranked, usually equitant, linear to narrowly lanceolate, upper leaves much shorter, stem-clasping. Flowering stem terminating the leafy shoot or arising separately, simple or sparingly branched; flowers bisexual, stalkless, each with a bract and bracteole. Perianth in cultivated species bilaterally symmetric, with a tube at the base. Stamens arising on the tube, arching to the upper side of perianth. Style arching like the stamens and as long, 3-branched near apex.

About 180 species (including *Acidanthera* Hochstetter) in Africa, Madagascar and the Mediterranean Region, extending to N Europe. Many species have been cultivated but it is unlikely that more than 30 are to be found in Europe today. The Mediterranean and W Asian species are relatively hardy, being dormant in winter and flowering in early summer; they need warm conditions, and in N Europe some

need to be grown in a bulb frame. The tropical African and South African summer rainfall species are mostly not hardy in most of Europe, and must be planted out in late spring for summer and early autumn flowering. The South African winter rainfall species are suited to the Mediterranean and N European Atlantic climate, but elsewhere must be grown in frost-free glasshouses.

There is a vast number of cultivars within *G.* × *hortulanus*, the florist's and border gladiolus. Being derived mainly from summer rainfall species, these are treated in the same way. There are also many hybrids (mostly not described here) less remotely derived from the South African species; their cultural treatment depends on the their parentage.
Literature: Lewis, G.J., et al., Gladiolus, a revision of the South African species, *Journal of South African Botany*, Supplementary Volume **10** (1972); Hamilton, A.P., A history of the garden gladiolus, *The Garden* **101**: 424–8 (1976).

Length of flower includes the tube; recurved lobes must be straightened for measurement. The flower markings described in the informal key are not repeated in the descriptions of the species.

*Dormancy and flowering.* Dormant winter, flowering spring, early summer: **1–5**; dormant (usually kept in storage) winter, flowering spring, summer, autumn: **6–11**; leafy in winter, flowering spring to summer: **12–17**; leafy in winter, flowering autumn (without leaves): **12**.
*Leaf width.* Usually not more than 1 cm: **1,3,4,13–17**; usually more than 1 cm: **2,5–13**.
*Upper middle perianth-lobe.* Narrow and arched: **16**.
*Perianth-lobes.* Wavy: **6,12,17**.
*Flower scent production.* At night: **14,17**; during day: **16**.
*Anther length in relation to filament length.* Longer: **1,2**; equal or slightly shorter: **3–5,11,14,17**; much shorter: **6–10,12,14–16**.
*Principal colour of flowers.* White: **6,11–16**; pale pink to violet: **6,10,12,13,15**; purplish pink to purplish red: **1–6,13**; red: **6–9**; yellow: **6,9,10,14,17**; brownish: **9,16**; greenish: **9,12,16**.
*Flower markings.* Three lower perianth-lobes, or 2 of them, with a fusiform white stripe bordered by a darker shade of the principal colour: **1–5,13**.
*Flower length, including tube.* Not more than 5.5 cm: **1–5,10,16**.

1a. Corm more than 4 cm diameter; widest leaves at least 2 cm wide; flowers 12–20, crowded
      **6. × hortulanus**
  b. Corm and leaves narrower; flowers fewer; if these characters do not apply the flowers are not crowded    **2**
2a. Anthers longer than filaments    **3**
  b. Anthers equalling or shorter than filaments    **4**
3a. Flowers closely set in 1 row, dark dull violet; perianth-tube curved
      **1. atroviolaceus**
  b. Flowers loosely set in 2 rows, purplish pink to purplish red; perianth-tube nearly straight    **2. italicus**
4a. Flowers bright pinkish red to reddish purple; the 3 lower perianth-lobes with a dark-bordered cream-coloured central stripe; plant dormant in winter    **5**
  b. Flowers differently coloured or, if as described above, plant growing in winter    **7**
5a. Flowers in 1 row    **3. imbricatus\***
  b. Flowers in 2 rows    **6**
6a. Plant usually more than 50 cm; inflorescence often branched; flowers 10 or more    **5. communis**
  b. Plant usually less than 50 cm; inflorescence unbranched; flowers 10 or fewer    **4. illyricus**
7a. Upper middle perianth-lobe narrow, nearly parallel-sided, arched; perianth-tube 1–1.5 cm
      **16. orchidiflorus**
  b. Upper middle lobe broader, not nearly parallel-sided, not distinctly arched; perianth-tube more than 1.5 cm    **8**
8a. Main leaves 4-winged    **14. tristis**
  b. Main leaves flat or merely ribbed    **9**
9a. Main leaves to 3 mm wide (perhaps to 6 mm in cultivation)    **17. liliaceus**
  b. Main leaves mostly more than 5 mm wide    **10**
10a. Leaves 5–12, some of them 3–4 cm wide; flowers obliquely trumpet-shaped with strongly overlapping, abruptly pointed lobes, of which the lower have recurved tips
      **9. natalensis**
  b. Leaves not more than 9, not more than 2.8 cm wide; flowers differently shaped    **11**
11a. Perianth-tube more than 4 cm    **12**
  b. Perianth-tube not more than 4 cm  **13**
12a. Perianth-tube more than 7 cm, limb white with red marks in throat
      **11. callianthus**
  b. Perianth-tube not more than 7 cm, limb white or pink with pale-centred

purplish or red marks on lower lobes near the middle    **12. undulatus\***
13a. Perianth yellow or white to pink or violet    **14**
  b. Perianth mainly brilliant red or intense purplish red    **17**
14a. Perianth-lobes abruptly spreading from the top of the tube
      **15. carneus**
  b. Perianth-lobes gradually spreading  **15**
15a. Flowers 4–5.5 cm, limb bell-like with abruptly tapered lobes    **10. papilio**
  b. Flowers about 6.5–10 cm, trumpet-like with gradually tapered lobes    **16**
16a. Main leaves not emerging until flowering has begun, and then separately from the flowering stems
      **12. undulatus\***
  b. Main leaves emerging before flowering; flowering stems emerging from leafy shoots    **13. × colvillei**
17a. Leaves 5–9; flowers pure bright red
      **17**
  b. Leaves usually 3 or 4; flowers purplish red    **13. × colvillei**
18a. Perianth-lobes all about equally divergent, the lower 3 with an elongate central pale mark
      **7. cardinalis**
  b. Upper middle perianth-lobes projecting forward as a hood, 3 or all of the others spreading widely to form a semicircular lip, the lower 3 with white basal area extending across most of their width    **8. saundersii**

**1. G. atroviolaceus** Boissier. Illustration: Rix & Phillips, The bulb book, 150 (1981).
Plant 30–70 cm, growing in spring. Leaves 3, the lower 30–40 cm × 4–8 mm. Flowers 4–10, 4–4.5 cm, closely set in 1 row, dark dull violet; tube distinctly curved. Anthers longer than filaments. Seeds not winged. *Greece, Turkey, Iraq, Iran.* H2. Early summer.

**2. G. italicus** Miller (*G. segetum* Ker Gawler). Illustration: Botanical Magazine, 719 (1804); Grey-Wilson & Mathew, Bulbs, pl. 30 (1981).
Plant 50–110 cm, growing in spring. Leaves 3–5, the lower 18–50 cm × 5–17 mm. Flowers 6–16, 4–5 cm long, loosely spaced in 2 rows, purplish pink to purplish red; tube nearly straight; lobes scarcely overlapping. Anthers longer than filaments. Seeds not winged. *S Europe to Afghanistan, extending to NW Africa and Canary Isles.* H2. Early summer.

**3. G. imbricatus** Linnaeus. Illustration: Mathew, The larger bulbs, 65 (1978); Grey-Wilson & Mathew, Bulbs, pl. 30 (1981); Rix & Phillips, The bulb book, 151 (1981).
Plant 30–80 cm, growing in spring. Leaves 2 or 3, the lower 15–35 × 1–1.5 cm. Flowers 4–12, *c.* 3 cm, closely set in 1 row, pinkish red to reddish purple; tube strongly curved near throat; lobes usually overlapping. Anthers not longer than filaments. Seeds winged. *C & E Europe (N to the Baltic states of the USSR, naturalised in Finland north almost to the arctic circle), Turkey.* H2. Early summer.

**\*G. palustris** Gaudin, also with reddish purple flowers in 1 row, differs in its smaller size (to 50 cm) and loose spike of 2–6 flowers. *C, S & SE Europe.* H2. Early summer.

**4. G. illyricus** Koch. Illustration: Ross-Craig, Drawings of British plants **29:** pl. 5 (1972); Grey-Wilson & Mathew, Bulbs, pl. 30 (1981).
Plant 25–50 cm, growing in spring. Leaves 4 or 5, the lower 10–40 cm × 4–10 mm. Flowers 3–10, 3.5–5.5 cm, loosely spaced in 2 rows, reddish purple; tube slightly curved; lobes 6–16 mm wide, sometimes overlapping. Anthers not longer than filaments. Seeds winged. *S & W Europe.* H2. Early summer.

**5. G. communis** Linnaeus.
Plant 50–100 cm, growing in spring. Leaves 3–5, the lower 30–70 cm × 5–25 mm. Stem often with 1–3 branches. Flowers 10–20, 4–5.5 cm, in 2 rows, pink to deep purplish red; tube slightly curved; lobes more or less overlapping. Anthers not longer than the filaments. Seeds winged. Early summer.

Subsp. **communis**. Illustration: Botanical Magazine, 86 (1789); Bonnier, Flore Complète **11:** pl. 601 (1931); Huxley & Taylor, Flowers of Greece and the Aegean, f. 413 (1977). Plant not more than 80 cm; lower leaves 30–50 cm × 1–15 mm. Flowers rather loosely spaced, usually pink; lobes 10–20 mm wide. *C & S Europe, NW Africa.* H2.

Subsp. **byzantinus** (Miller) Hamilton (*G. byzantinus* Miller). Illustration: Botanical Magazine, 874 (1805); Grey-Wilson & Mathew, Bulbs, pl. 30 (1981). Lower leaves 30–70 × 1–2.5 cm. Flowers rather closely set, usually deep purplish red; lobes 1.5–2.5 cm wide. *S Spain, Sicily, NW Africa.* H3.

The most commonly cultivated European gladiolus; the cultivated variant is slightly different from the wild plant, is nearly sterile and has 90 chromosomes instead of 120.

**6. G.  × hortulanus** Bailey. Illustration: Synge, Collins' guide to bulbs, pl. 15 (1961).
Corm usually more than 4 cm in diameter. Plant growing in summer. Largest leaves usually 2 cm or more wide. Flowers 12–20, closely set, mostly 6–15 cm wide, variously coloured, in the smaller cultivars more often with definite blotches on the lobes, in the larger more often uniform or bicoloured; lobes broad and strongly overlapping. *Garden origin.* H5. Summer.

This name covers the commonly grown garden and florist's hybrids; although selection has been mainly for large size, smaller cultivars have also been raised (Butterfly, Peacock and Primulinus gladioli); these make it difficult to separate *G.  × hortulanus* from the species by means of a key. Nos 7–11, 14 and 15, among others, are parents of *G.  × hortulanus*. The hybrid which served as the foundation for the development of the larger-flowered *G.  × hortulanus* cultivars was *G.  × gandavensis* Van Houtte (*G. natalensis  × G. oppositiflorus* Herbert).

**7. G. cardinalis** Curtis. Illustration: Botanical Magazine, 135 (1790); Flore des Serres 19: 109 (1873); Bolus et al., A second book of South African flowers, 150 (1936); Lewis, Gladiolus, 24 (1972).
Corm to 3 cm in diameter. Plant 60–115 cm, growing in summer; stem often arched. Leaves 5–9, the lower 40–90 × 1.5–2.8 cm. Flowers *c.* 6, *c.* 8 cm long, closely spaced, indistinctly 2-rowed, bright red, the lower 3 lobes with a central whitish mark; tube 3–4 cm; lobes obtuse, upper slightly hooded, others widely spreading. Anthers much shorter than filaments. *South Africa (SW Cape Province).* H5. Summer.

When crossed with *G. carneus* this species gave rise to **G.  × insignis** Paxton, from which was developed, probably by crossing with other species and hybrids, a group of cultivars known as Nanus Hybrids (not *G. nanus* Andrews, which is *Babiana nana* (Andrews) Sprengel).

**8. G. saundersii** Hooker. Illustration: Botanical Magazine, 5873 (1870); Flowering plants of South Africa 5: t. 182 (1925); Lewis, Gladiolus, 24 (1972).
Corm 2.5–4 cm in diameter. Plant 40–90 cm, growing in summer. Leaves 7 or 8, the lower 25–60 cm × 6–26 mm. Flowers 3–8 or rarely up to 12, 6.5–11 cm long, red, the basal halves of lower 3 lobes whitish and speckled with red towards throat, midrib sometimes white to the tip; tube 3–3.8 cm, strongly curved, widened gradually; lobes with recurved tips, the upper hooded, the 3 lower and, usually, the 2 upper laterals, widely spreading and overlapping to form a semicircular lip. Anthers much shorter than filaments. *South Africa.* H5. Summer.

**9. G. natalensis** (Ecklon) W.J. Hooker (*G. psittacinus* J.D. Hooker; *G. quartinianus* Richard; *G. primulinus* Baker). Illustration: Botanical Magazine, 3032 (1830), 5884 (1871), 6202 (1875), 6739 (1884), 8080 (1906); The Garden 101: 427 (1976).
Corms to 5 cm in diameter, sometimes producing stolons. Plant to 1.5 m, growing in spring and summer. Leaves 5–12, some to 60 × 1–3 cm, others to 30 × 1–4 cm. Flowers usually 8–12 cm, up to 25, rather loosely spaced, indistinctly 2-rowed, yellow, yellow streaked with orange-red, orange-red or red with central yellow areas on lobes, often with transitional red streaks or peppering, or coloured in similar combinations of green and brown; tube about half the length of the flower, curved, enlarging gradually; lobes acute, overlapping, diverging gradually, the 3 upper or the upper central forming a hood, the 3 lower with recurved tips, the middle one about twice as wide as the laterals. Anthers much shorter than filaments. *South Africa (E Cape Province) to Ethiopia and Arabia.* H5. Spring–autumn.

A late-flowering, red and yellow form with long stolons, 'Hookeri' (Illustration: Flowering plants of South Africa 3: pl. 116, 1923; Lewis, Gladiolus, 33, 1972), is widely cultivated in warm-temperate countries.

**10. G. papilio** J.D. Hooker (*G. purpureo-auratus* J.D. Hooker). Illustration: Botanical Magazine, 5565 (1866), 5944 (1872); Flore des Serres 19: 107 (1873); Lewis, Gladiolus, 48 (1972).
Corm 1–3 cm diameter, usually with stolons. Plant 50–90 cm, growing in summer. Leaves 7 or 8, the lower 20–100 cm × 5–20 mm or sometimes to 2.5 cm. Flowers 3–10, 4–5.5 cm, loosely arranged in 2 more or less distinct rows, bright or dull yellow, sometimes tinged with purple outside, or mainly pale pinkish violet; throat sometimes and lower lobes always blotched with brown or purplish red; tube 1.5–2.3 cm, bent near the throat; lobes more or less obtuse, overlapping, forming a usually drooping, bell-shaped limb. Anthers much shorter than filaments. *South Africa (Transkei to Transvaal), Lesotho, Swaziland.* H3. Summer.

**11. G. callianthus** Marais (*Acidanthera bicolor* Hochstetter; *A. murielae* invalid; *A. tubergenii* invalid). Illustration: The Garden 47: 342 (1895); Synge, Collins' guide to bulbs, pl. 5 f. 1 (1961).
Corms to 2.5 cm in diameter. Plant to 1.1 m, growing in spring and summer. Leaves slightly shorter than the stem, to 2.5 cm wide. Flowers up to 10, to 14 cm long, fragrant, loosely set, opening singly, white, all lobes except the upper with a purplish red rhomboidal mark near the base; tube to 10 cm, slender, curved; limb nodding, with nearly equal, divergent lobes. Anthers rather large. *E Africa (Ethiopia to Malawi).* H5. Autumn.

**12. G. undulatus** Linnaeus (*G. cuspidatus* Jacquin). Illustration: Botanical Magazine, 582, 591 (1802); Bolus et al., A second book of South African flowers, 154 (1936).
Corms to 30 cm in diameter. Plant 30–95 cm, growing in winter. Leaves about 5, 25–75 cm × 5–20 mm. Flowers 4–9, to *c.* 13 cm long, loosely set in 2 rows, greenish white, white, cream or pink, lower lobes with elongated or spade-shaped, pale-centred, red or purple marks; tube 5–7 cm, slender; lobes lanceolate, gradually divergent, with more or less wavy and recurved tips. Anthers much shorter than filaments. *South Africa (SW Cape Province).* H5. Spring–summer.

***G. carmineus** Wright. Illustration: Botanical Magazine, 8068 (1906); Bolus et al., A second book of South African flowers, 149 (1936). Growing in winter. Main leaves appearing on shoots separate from the flowering shoots and shortly after them. Flowers erect, 6.5–10 cm; tube only 3–4.5 cm, widened upwards; lobes ovate to obovate, not wavy and recurved. *South Africa (SW Cape Province).* H5. Autumn.

**13. G.  × colvillei** Sweet. Illustration: Sweet, British flower garden, t. 155 (1826); Rix & Phillips, The bulb book, 151 (1981).
Corm *c.* 2 cm in diameter. Plant *c.* 50 cm, growing in winter. Leaves 3 or 4, *c.* 25 × 1.2 cm. Flowers *c.* 8, *c.* 7 cm long, closely set in 2 rows, white to purplish red or pale pinkish violet, lower lobes with elongate pale marks which may be dark-edged; perianth-lobes gradually divergent. Anthers shorter than filaments. *Garden origin.* H3. Early–late summer.

A cross between *G. cardinalis* and

*G. tristis*, first raised in England in 1823; a number of cultivars has been named.

**14. G. tristis** Linnaeus. Illustration: Botanical Magazine, 272 (1794), 1098 (1808); Flowering plants of South Africa **5**: pl. 175 (1925); Bolus et al., A first book of South African flowers, 138 (1928); Lewis, Gladiolus, 161 (1972); The Garden **101**: 426 (1976).

Corm to 3 cm diameter. Plant 40–150 cm, more usually 50–70 cm, growing in winter. Leaves 2–4, the lower about as long as the stem, 1.5–5 mm wide, twisted, 4-winged. Flowers 1–8 (vars. **tristis** and **concolor** (Salisbury) Baker) or 7–10 or at most 20 (var. **aestivalis** (Ingram) Lewis), 6–9.5 cm long, loosely set in 1 row, fragrant at night, white to buff or dull yellow (var. **concolor**) and flushed or peppered with violet to purplish brown outside, and sometimes along midribs of upper lobes inside; midribs often greenish (vars. **tristis** and **aestivalis**); tube 4–6 cm, curved above the middle, widened gradually above; limb gradually flared, lobes nearly equal, slightly twisted, gradually tapered to acute, recurved tips. Anthers shorter than filaments. *South Africa (S Cape Province)*. H5. Spring–summer.

**15. G. carneus** Delaroche (*G. blandus* Aiton). Illustration: Botanical Magazine, 625, 645, 648 (1803), 1665 (1814), 8923 (1938) – a small, slender form; Bolus et al., A second book of South African flowers, 153 (1936); Lewis, Gladiolus, 96, 99 (1972).

Corm to 3 cm in diameter. Plant variable, 20–100 cm tall, growing in winter. Leaves up to 5, usually to 30 cm × 5–10 mm but sometimes to 60 × 2 cm. Flowers usually 2–8, to 8 cm long, loosely spaced in 2 rows, white, cream, pale pink or pale pinkish violet, the three lower lobes usually with a reddish or yellowish patch, the throat sometimes blotched with red; tube 2–4 cm; lobes abruptly spreading. Anthers much shorter than filaments. *South Africa (SW Cape Province)*. H5. Spring and early summer.

**16. G. orchidiflorus** Andrews. Illustration: Botanical Magazine, 688 (1803); Sweet, British flower garden, 156 (1826); Flowering plants of South Africa 4: pl. 165 (1924); Lewis, Gladiolus, 144 (1972); Rice & Compton, Wild flowers of the Cape of Good Hope, pl. 203, f. 2 (1951); Mason, Western Cape sandveld flowers, 83, f. 4 (1972).

Corm to 2.5 cm in diameter. Plant to 40 cm, growing in winter. Leaves 3–8, the largest about as tall as stem and to 5 mm wide. Stem often branched, flexuous above. Flowers 5–15, 3–5 cm, loosely or closely spaced in 1 row or 2, fragrant, greyish white to greyish green or brownish cream, usually with a purplish sheen, the 3 upper with a brownish purple band, sometimes divided by a greenish stripe, the 3 lower with a basal yellow or yellow-green area bordered with brownish purple; tube 1–1.5 cm long; upper lobe narrow and arched, the others wide and spreading. Stamens lying close under perianth-lobe; filaments several times longer than anthers. *South Africa (W Cape Province to Orange Free State)*. H5. Spring.

**17. G. liliaceus** Houttuyn (*G. grandis* Thunberg). Illustration: Botanical Magazine, 1042, lower figure (1807); Marloth, Flora of South Africa 4: 157 (1915); Batten & Bokelmann, Wild flowers of the eastern Cape Province, pl. 29, f. 7 (1966).

Corm to 2.5 cm diameter. Plant to 60 cm or rarely to 90 cm, growing in winter. Leaves usually 3, the lowest about as tall as stem, 3 mm wide. Flowers 1–3 or rarely 5, 7–11.5 cm long, loosely set in 1 row, fragrant at night, dull yellow, densely streaked or mottled with pinkish brown or purple except on the basal halves of the lower lobes, sometimes red in throat, upper lobes sometimes translucent near base, the colour changing at night to dark purplish violet or violet; tube curved, 3.8–6 cm, enlarged upwards; lobes nearly equal, progressively spreading, tips recurved, wavy and tapered to an acute, obtuse or notched apex. Anthers slightly shorter than filaments. *South Africa (S Cape Province)*. H5. Early spring–summer.

**45. CHASMANTHE** N.E. Brown
*J. Cullen*

Perennial herbs with large, flattened-spherical corms which have fibrous 'necks'. Leaves mostly basal, in 2 ranks, blade-like, with a false midrib or with 2 or 3 equal primary veins and no false midrib. Inflorescence usually a simple, erect spike, occasionally with 1 erect branch from the base, the flowers facing outwards. Bracts acute, longer than the ovaries. Perianth bilaterally symmetric, conspicuously curved, tubular below, the lower part of the tube very slender and sometimes twisted, broadening into the upper part; lobes 6, the upper much longer than the others and directed forwards, the 2 upper lateral lobes parallel to the uppermost, the 3 lower lobes deflexed. Stamens 3, attached to the perianth-tube where it widens, the anthers borne beneath the upper perianth-lobe. Style slightly longer than the stamens, stigma 3-fid. Capsule spherical, exceeding the bracts, opening by 3 splits and containing many ovoid, brightly coloured seeds.

There are 9 species from South and South West Africa. Cultivation as for *Gladiolus* (p. 388).

1a. Perianth-tube with the narrow base conspicuously twisted, broadening abruptly, the base of the broader part formed by 3 pouches    **2**
  b. Perianth-tube with the narrow base not twisted, broadening gradually, the base of the broader part without pouches    **3**
2a. Spike with up to 7 flowers borne on 1 side    **1. aethiopica**
  b. Spike with more than 7 flowers borne in 2 rows    **2. vittigera**
3a. Perianth bright red, the tube 1.8–2.5 cm    **3. caffra**
  b. Perianth orange or scarlet and yellow, the tube 3.8–4 cm    **4. floribunda**

**1. C. aethiopica** (Linnaeus) N.E. Brown (*Antholyza aethiopica* Linnaeus).

Plant 20–70 cm. Leaves 20–55 × 1–2 cm, linear, tapering to the apex, narrowing slightly to the base, sometimes 1 side narrowing rather abruptly, with a false midrib. Spike usually simple, occasionally with 1 branch; flowers up to 7, borne on 1 side. Perianth scarlet and yellow, the tube *c*. 2.8 cm, its lower part narrow and conspicuously twisted, broadening abruptly into an upper part whose base is formed by 3 pouches; upper lobe *c*. 2.5 cm × 7 mm. *South Africa (Cape Province)*. H5–G1. Spring–early Summer.

**2. C. vittigera** (Salisbury) N.E. Brown (*Antholyza vittigera* Salisbury; *A. aethiopica* var. *vittigera* (Salisbury) Baker). Illustration: Botanical Magazine, 1172 (1809).

Very similar to *C. aethiopica* but leaves broader, spike with more than 7 flowers borne in 2 rows on the axis; perianth with dark stripes towards the base. *South Africa (Cape Province)*. H5–G1. Spring–early summer.

**3. C. caffra** (Baker) N.E. Brown (*Antholyza caffra* Baker). Illustration: Botanical Magazine, 9470 (1937).

Plant to 50 cm. Leaves long-tapered to the apex, somewhat narrowed to the base, to 35 cm × 6 mm, without a false midrib but with 2 or 3 equally prominent primary veins. Spike usually simple with 12–20 flowers borne in 2 rows. Perianth bright red; tube 1.8–2.5 cm, broadening gradually from a basal part which is not twisted; upper lobe 2.5–3.2 cm. *South Africa (Cape Province, Natal)*. H5–G1. Spring–early Summer.

**4. C. floribunda** (Salisbury) N.E. Brown (*Antholyza floribunda* Salisbury; *A. praealta* Redouté). Illustration: Botanical Magazine, 561 (1801); Eliovson, Wild flowers of southern Africa, xviii (1955).
Plant to 1 m or more. Leaves 20–65 × 2.5–5.5 cm, tapering to the apex and often abruptly and 1-sidedly to the base, with a conspicuous false midrib. Spike simple with 20–30 or more orange-red or scarlet and yellow flowers borne in 2 rows. Perianth-tube 3.8–4 cm, its basal part not

twisted, broadening gradually above; upper lobe to 3.3 cm × 8–9 mm. *South Africa*. H5–G1. Summer.

**46. ANOMALESIA** N.E. Brown
*J. Cullen*
Similar to *Chasmanthe* but corms small, producing stolons which end in new corms, leaves with 2–3 equal primary veins (no false midrib), bracts almost as long as the perianth-tubes, perianth-tube with a slender base which is not twisted, broadening gradually above, the 6 lobes very unequal, the uppermost the longest, the 2 upper lateral lobes spreading upwards, above the upper lobe, the 3 lower lobes very small and inconspicuous. Stigmas 3. Fruit a capsule. Seeds flattened, winged.
A genus of 2 species from South Africa. Cultivation as for *Gladiolus* (p. 388).

**1. A. cunonia** (Linnaeus) N.E. Brown (*Antholyza cunonia* Linnaeus). Illustration: Botanical Magazine, 343 (1796).
Plant to 90 cm, though often much less. Leaves linear, long-tapering to the apex, to 30 × 1 cm. Spike with up to 7 flowers borne in 2 rows. Perianth scarlet to dark red, the lowermost lobe greenish, turned inwards, hidden by the others. *South Africa*. H5–G1. Spring.

**\*A. splendens** (Sweet) N.E. Brown (*Anisanthus splendens* Sweet; *Antholyza splendens* (Sweet) Steudel). Illustration: Sweet, British flower garden, ser. 2, 1: t. 84 (1833). Leaves broader, perianth-tube greenish, the lobes mostly bright red, the lowermost deflexed downwards and outwards, bright red at the tip. *South Africa*. H5–G1. Spring.

# Addenda to the LILIACEAE

**NECTAROSCORDUM** Lindley
Differs from *Allium* (p. 233) in that the outer perianth-segments have 3–7 veins, the flower-stalks are swollen below the flowers and the ovary contains many ovules.
A genus of 2–3 species from S Europe and W Asia, only 1 of which is cultivated; it is closely related to *Allium* and included in it by some botanists. It will grow in sun or shade and has no soil preferences. Propagation is by seed.

**1. N. siculum** (Ucria) Lindley (*Allium siculum* Ucria). Illustration: Edwards's Botanical Register 22: t. 1913 (1836); Synge, Collins guide to bulbs, pl. 1, f. 2 (1971); Mathew, The larger bulbs, 30–f. 4 (1978); Rix & Phillips, The bulb book, 152 e (1981).
Stem to 120 cm. Leaves 3 or 4, basal, with a sharp keel beneath. Flowers up to 30, in a loose terminal umbel subtended by a deciduous spathe; flower-stalks unequal, drooping at first, becoming erect as the fruit develops. Perianth-segments 1.4–2.5 cm, whitish or cream, tinged with green and often pale pink outside. *W. Mediterranean region*. H3. Late spring–summer.
The above description refers to subsp. **siculum**.

Subsp. **bulgaricum** (Janka) Stearn (*N. bulgaricum* Janka; *N. dioscoridis* (Sibthorp & Smith) Zahariadi; *Allium bulgaricum* (Janka) Prodan). Illustration: Botanical Magazine, n.s., 257 (1955); Rix & Phillips, The bulb book, 152 d (1981); Mathew & Baytop, The bulbous plants of Turkey, f. 104 (1984). The greenish perianth-segments are strongly flushed with purple and edged with white. *SE Europe, NW Turkey, USSR (Crimea)*.

**BESSERA** Schultes
Differs from *Nothoscordum* (p. 246) in having drooping flowers with the stamens projecting, and a staminal tube which is toothed at the top. The perianth-lobes are 3–4 cm, each scarlet with a green mid-vein outside, and creamy-white inside except for the margins and midrib which are scarlet.
A Mexican genus of 1, or possibly 2 species. Cultivation as for *Brodiaea* (p. 247).

**1. B. elegans** Schultes. Illustration: Edwards's Botanical Register 25: t. 38 (1839); Botanical Magazine, n.s., 270 (1956). *SW & SC Mexico*. H5–G1. Summer.

**BLOOMERIA** Kellogg
Differs from *Nothoscordum* (p. 246) in having the flower-stalks jointed at the top, and the filaments broadened at the base

into a cup-like structure with 2 short teeth at the top.
A genus of 3 species from SW USA and N Mexico. Cultivation as for *Brodiaea* (p. 247). Literature: Ingram, J., A monograph of the genera Bloomeria and Muilla (Liliaceae), *Madroño* 12: 19–27 (1953).

**1. B. crocea** (Torrey) Coville (*B. aurea* Kellogg). Illustration: Parsons, The wild flowers of California, 155 (1897); Abrams, Illustrated flora of the Pacific States 1: 398 (1923); Rix, Growing bulbs, 71 (1983). Perianth-segments 8–12 mm, spreading, each with a darker central stripe. *USA (S California), Mexico (Baja California)*. H5. Spring.

**MUILLA** Watson
Differs from *Nothoscordum* (p. 246) in having almost terete leaves, the umbel subtended by more than one spathe and the ovary with many ovules.
The genus contains 5 species native to SW USA and N Mexico. Cultivation as for *Brodiaea* (p. 247).
Literature: see *Bloomeria* (above).

**1. M. maritima** (Torrey) Watson. Illustration: Abrams, Illustrated flora of the Pacific States 1: 397 (1923). Perianth-segments 3–6 mm, greenish-white with a brownish mid-vein. *USA (S California), Mexico (Baja California)*. H5. Summer.

# GLOSSARY

*abscission-zone.* A predetermined layer at which leaves or fern-fronds break off.

*achene.* A small, dry, indehiscent, 1-seeded fruit, in which the fruit-wall is of membranous consistency and free from the seed.

*aciculus.* see *Nomocharis* p. 207.

*acuminate.* With a long, slender point.

*adpressed.* Closely applied to a leaf or stem and lying parallel to its surface, but not adherent to it.

*adventitious.* (1) Of roots: arising from a stem or leaf, not from the primary root derived from the radicle of the seedling. (2) Of buds: arising somewhere other than in the axil of a leaf.

*aggregate fruit.* A collection of small fruits, each derived from a single free carpel, closely associated on a common receptacle, but not united. *Ranunculus* and *Rubus* provide familiar examples.

*alternate.* Arising singly, 1 at each node, not opposite or whorled (figure 41(2), p. 394).

*anastomosing.* Describes veins of leaves or fern-fronds which rejoin after branching from each other or from the main vein or midrib.

*anatropous.* Describes an ovule which turns through 180° in the course of development, so that the micropyle is near the base of the funicle (figure 44(2), p. 397).

*annual.* A plant which completes its life-cycle from seed to seed in less than 1 year.

*annulus.* The row of thickened cells which runs round the margin of a fern-sporangium (figure 1(7), p. 2).

*anther.* The uppermost part of a stamen, containing the pollen (figure 43(3 & 4), p. 396).

*antheridium.* The male sexual organ of the gametophyte of lower plants, in which are developed the antherozoids.

*antherozoid.* A male gamete capable of motion in a liquid by means of flagella. Of the plants dealt with in this work only pteridophytes and a few gymnosperms produce antherozoids.

*apical.* Describes the attachment of an ovule to the apex of a 1-celled ovary (figure 44(8), p. 397).

*apiculate.* With a small point.

*apomictic.* Reproducing by asexual means, though often by the agency of seeds, which are produced without the usual sexual nuclear fusion.

*arachnoid.* Describes hairs which are soft, long and entangled, suggestive of cobwebs.

*archegonium.* The female sexual organ of the gametophyte of lower plants; it is normally flask-shaped and contains a single female gamete, the *ovum*.

*aril.* An outgrowth from the region of the hilum, which partly or wholly envelops the seed; it is usually fleshy.

*ascending.* Prostrate for a short distance at the base, but then curving upwards so that the remainder is more or less erect; sometimes used less precisely to mean pointing obliquely upwards.

*attenuate.* Drawn out to a fine point.

*auricle.* A lobe, normally 1 of a pair, at the base of the blade of a leaf, bract, sepal or petal.

*awn.* A slender but stiff bristle on a sepal or fruit.

*axil.* The upper angle between a leaf-base or leaf-stalk and the stem that bears it (figure 41(1), p. 394).

*axile.* A form of placentation in which the cavity of the ovary is divided by septa into 2 or more cells, the placentas being situated on the central axis (figure 44(10), p. 397).

*axillary.* Situated in or arising from an axil (figure 41(1), p. 394).

*back-cross.* A cross between a hybrid and a plant similar to one of its parents.

*basal.* (1) Of leaves: arising from the stem at or very close to its base. (2) Of placentation: describes the attachment of an ovule to the base of a 1-celled ovary (figure 44(6 & 7), p. 397).

*basifixed.* Attached to its stalk or supporting organ by its base, not by its back (figure 43(3), p. 396).

*berry.* A fleshy fruit containing 1 or more seeds embedded in pulp, as in the genera *Berberis*, *Ribes* and *Phoenix*. Many fruits (such as those of *Ilex*) which look like berries and are usually so called in popular speech, are, in fact, drupes.

*biennial.* A plant which completes its life-cycle from seed to seed in a period of more than 1 year but less than 2.

*bifid.* Forked; divided into 2 lobes or points at the tip.

*bilaterally symmetric.* Capable of division into similar halves along 1 plane and 1 only (figure 43(9), p. 396).

*bipinnate.* Of a leaf: with the blade divided pinnately into separate leaflets which are themselves pinnately divided (figure 41(18), p. 394).

*bract.* A leaf-like or chaffy organ bearing flower in its axil or forming part of an inflorescence, differing from a foliage-leaf in size, shape, consistency or colour (figure 42(2), p. 395).

*bracteole.* A small, bract-like organ which occurs on the flower-stalk, above the bract, in some plants.

*bulb.* A seasonally dormant underground bud, usually fairly large, consisting of a number of fleshy leaves or leaf-bases.

*bulbil.* A small bulb, especially one borne in a leaf-axil or in an inflorescence.

*bulblet.* A small bulb developing from a larger one.

*calyx.* The sepals; the outer whorl of a perianth (figure 43(1), p. 396).

*campylotropous.* Describes an ovule which becomes curved during development and lies with its long axis at right angles to the funicle (figure 44(4), p. 397).

*capitate.* Compact and approximately spherical, head-like.

*capitulum.* An inflorescence consisting of small flowers (florets), usually numerous, closely grouped together so as to form a 'head', and often provided with an involucre.

*capsule.* A dry, dehiscent fruit derived from 2 or more united carpels and usually containing numerous seeds.

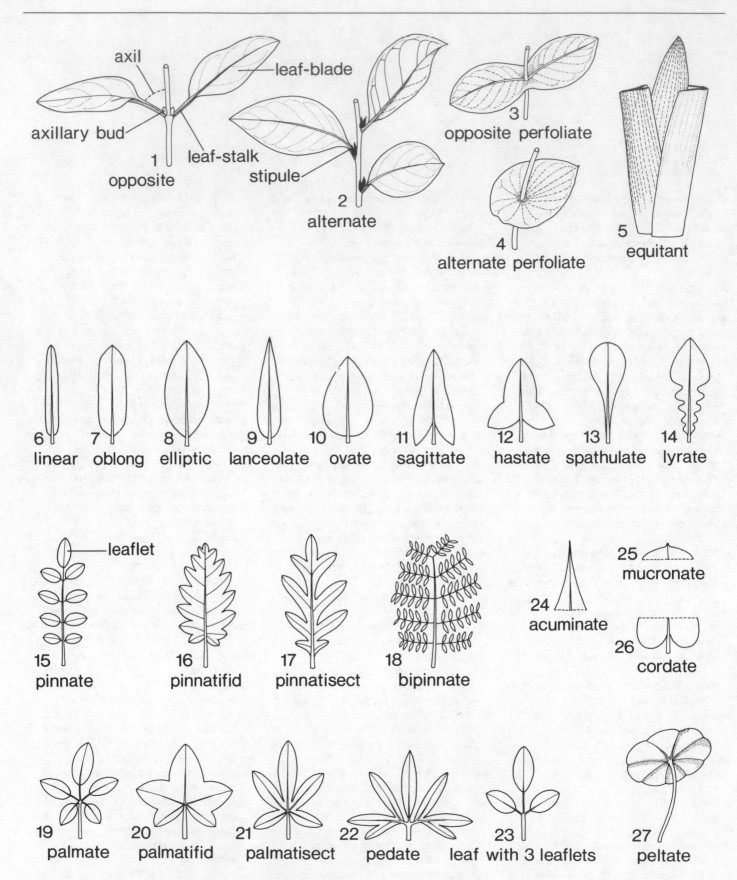

**Figure 41.** Leaves. 1–5, Leaf insertion types. 6–14, leaf-blade outlines. 15–23, Leaf dissection types. 24–26, Leaf apex and base shapes. 27, Attachment of leaf-stalk to leaf-blade.

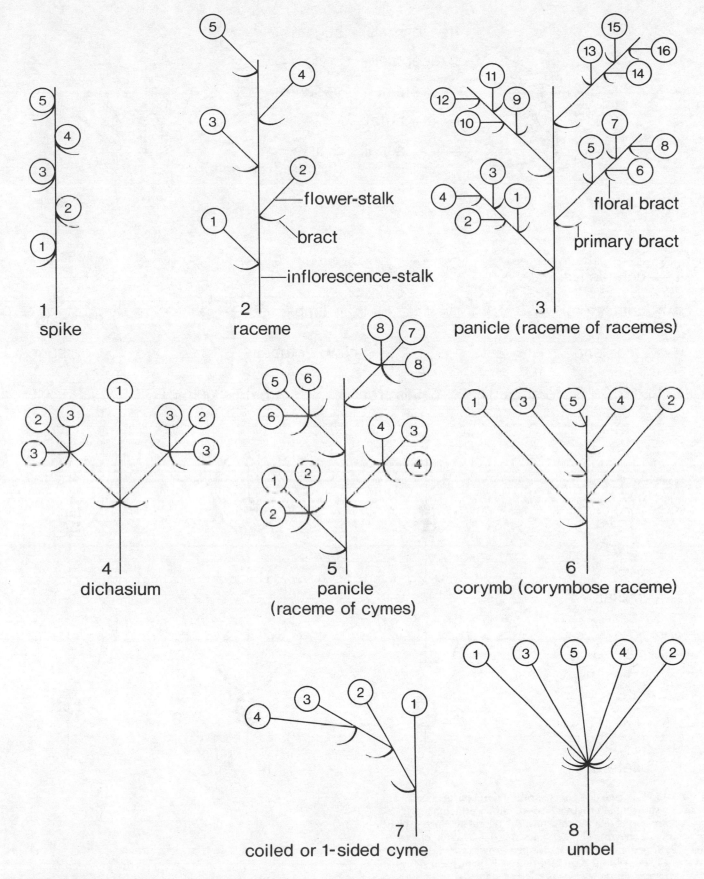

**Figure 42.** Inflorescences.

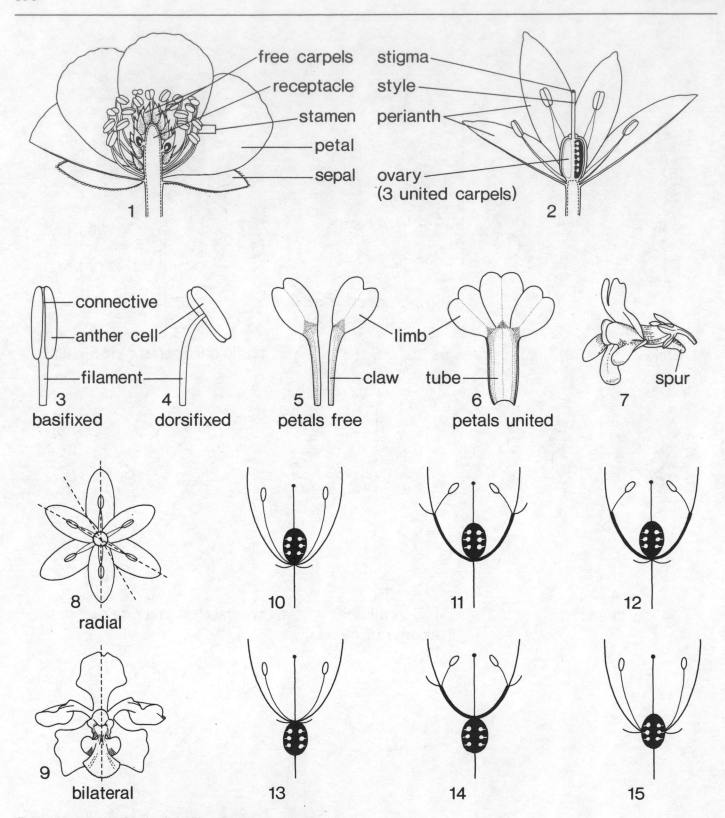

**Figure 43.** 1,2, Two flowers illustrating floral parts. 3,4, Two stamens showing alternative types of anther attachment. 5–7, Some terms relating to petals. 8,9, Floral symmetry, planes of symmetry shown by broken lines. 10–15, Position of ovary. 10–12, Superior ovaries. 13,14, Inferior ovaries. 15, Half-inferior ovary. 11, Perigynous zone bearing sepals, petals and stamens. 12, Perigynous zone bearing petals and stamens. 14, Epigynous zone bearing sepals, petals and stamens.

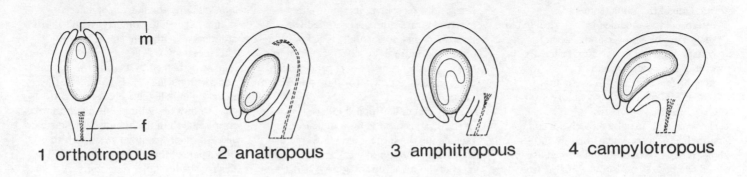

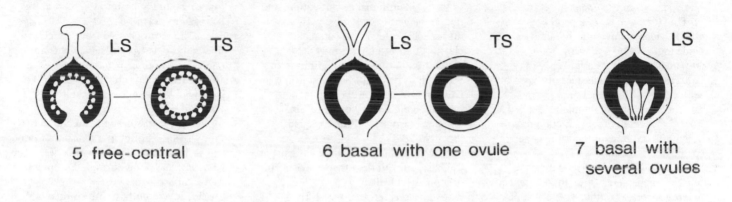

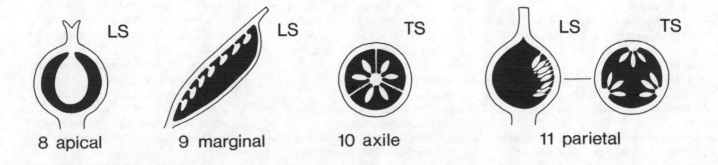

**Figure 44.** Ovules and placentation. 1–4, Ovule forms (f, funicle; m, micropyle). 5–11, Placentation types (LS, longitudinal section; TS, transverse section).

*carpel.* One of the units (sometimes interpreted as modified leaves) situated in the centre of a flower and together constituting the gynaecium or female part of the flower. If more than 1, they may be free or united. They contain ovules and bear a stigma (figure 43(1 & 2), p. 396).

*caruncle.* A soft, usually oil-rich appendage attached to a seed near the hilum.

*chromosome.* One of the small, thread-like or rod-like bodies consisting of nucleic acid and containing the genes, which appear in a cell nucleus shortly before cell division.

*ciliate.* Fringed on the margin with usually fine hairs.

*circinate.* Coiled at the tip, so as to resemble a crozier.

*cladode.* A branch which takes on the functions of a leaf (the leaves being usually vestigial). It may be flattened, as in *Ruscus* (p. 265), or needle-like, as in many species of *Asparagus* (p. 260).

*claw.* The narrow base of a petal or sepal, which widens above into the limb or blade (figure 43(5), p. 396).

*cleistogamous.* Describes a flower with reduced corolla, which does not open but sets seed by self-pollination.

*clone.* The sum-total of the plants derived from the vegetative reproduction of an individual, all having the same genetic constitution.

*compound.* (1) Of a leaf: divided into separate leaflets. (2) Of an inflorescence: bearing secondary inflorescences in place of single flowers. (3) Of a fruit: derived from more than 1 flower.

*compressed.* Flattened from side to side.

*comus.* A tuft of infertile flowers or bracts sometimes found at the top of a dense inflorescence.

*cone.* A compact, cylindrical or shortly conical inflorescence or fruiting inflorescence with closely overlapping bracts or scales.

*connective.* The tissue which separates the 2 lobes of an anther, and to which the filament is attached (figure 43(3), p. 396).

*cordate.* Describes the base of a leaf-blade which has a rounded lobe on either side of the central sinus (figure 41(26), p. 394).

*corm.* An underground, thickened stem-base, often surrounded by papery leaf-bases, and superficially resembling a bulb.

*corolla.* The petals; the inner whorl of the perianth (figure 43(1), p. 396).

*corona.* (1) A tubular or ring-like structure attached to the inside of the perianth (or perigynous or epigynous zone), either external to the stamens or united with their filaments; it is usually lobed or dissected (figure 37, p. 292). (2) A ring-like structure or circlet of appendages on the outside of a tube formed by united filaments.

*corymb.* A broad, flat-topped inflorescence. In the strict sense the term indicates a raceme in which the lowest flowers have stalks long enough to bring them to the level of the upper ones (figure 42(6), p. 395), but the term *corymbose* is often used to indicate a flat-topped cymose inflorescence.

*cotyledon.* One of the leaves preformed in the seed.

*crisped.* (1) Of hairs: strongly curved, so that the tip lies near the point of attachment. (2) Of leaves, leaflets or petals: finely and complexly wavy.

*cultivar.* A variant of horticultural interest or value, maintained in cultivation, and not conveniently equatable with an infraspecific category in botanical classification. A cultivar may arise in cultivation or may be brought in from the wild. Its distinguishing name can be Latin in form, e.g. 'Alba', but is more usually in a modern language, e.g. 'Madame Lemoine', 'Frühlingsgold', 'Beauty of Bath'.

*cupule.* A group of bracts, united at least at the base, surrounding the base of a fruit or a group of fruits.

*cyme.* An inflorescence in which the terminal flower opens first, other flowers being borne on branches arising below it (figure 42(5 & 7), p. 395).

*decumbent.* More or less horizontal for most of its length, but erect or semi-erect near the tip.

*decurrent.* Continued down the stem below the point of attachment as a ridge or ridges.

*dehiscent.* Splitting, when ripe, along 1 or more predetermined lines of weakness.

*dichasium.* A form of cyme in which each node bears 2 equal lateral branches (figure 42(4), p. 395).

*dichotomous.* Dividing into 2 equal branches; regularly forked.

*dioecious.* With male and female flowers or cones on separate plants.

*diploid.* Possessing in its normal vegetative cells 2 similar sets of chromosomes.

*disc.* A variously contoured, ring-shaped or circular area (sometimes lobed) within a flower, from which nectar is secreted.

*dissected.* Deeply divided into lobes or segments.

*distylic.* Having the flowers of different plants either with long styles and shorter stamens or with long stamens and shorter styles.

*dorsifixed.* Attached to its stalk or supporting organ by its back, usually near the middle (figure 43(4), p. 396).

*double.* Of flowers: with petals much more numerous than in the normal wild state.

*drupe.* An indehiscent fruit in which the outer part of the wall is soft and usually fleshy, but the inner part stony. A drupe may be 1-seeded as in *Prunus* or *Juglans*, or may contain several seeds, as in *Ilex*. In the latter case each seed is enclosed in a separate stony endocarp and constitutes a pyrene.

*drupelet.* A miniature drupe forming part of an aggregate fruit.

*ellipsoid.* As elliptic but applied to a solid body.

*elliptic.* About twice as long as broad, tapering equally both to the tip and the base (figure 41(8), p. 394).

*embryo.* The part of a seed from which the new plant develops; it is distinct from the endosperm and seed-coat.

*endocarp.* The inner, often stony layer of a fruit-wall in those fruits in which the wall is distinctly 3-layered.

*endosperm.* A food-storage tissue found in many seeds, but not in all, distinct from the embryo and serving to nourish it and the young seedling during germination and establishment.

*entire.* With a smooth, uninterrupted margin; not lobed or toothed.

*epicalyx.* A group of bracts attached to the flower-stalk immediately below the calyx and sometimes partly united with it.

*epigeal.* The mode of germination in which the cotyledons appear above ground and carry on photosynthesis during the early stages of establishment.

*epigynous.* Describes a flower, or preferably the petals, sepals and stamens (or perianth and stamens) of a flower in which the ovary is inferior (figure 43(13 & 14), p. 396).

*epigynous zone.* A rim or cup of tissue on which the sepals, petals and stamens are borne in some flowers with inferior ovaries (figure 43(14), p. 396).

*epiphyte.* A plant which grows on another plant but does not derive any nutriment from it.

*equitant.* Used of leaves folded so that they are V-shaped in section at the base, the bases overlapping regularly, as in many Iridaceae (figure 41(5), p. 394).

*exocarp.* The outer, skin-like layer of a fruit-wall in those fruits in which the wall is distinctly 3-layered.

*fall.* See Iridaceae (figure 39, p. 335).

*farina.* The flour-like wax present on the stem and leaves of many species of *Primula* and of a few other plants.

*fascicle.* A bunch of leaves often enclosed at the base by a sheath.

*fastigiate.* With all branches more or less erect, giving the plant a narrow tower-like outline.

*filament.* The stalk of a stamen, bearing the anther at its tip (figure 43(3 & 4) p. 396).

*filius.* Used with authority names to distinguish between parent and offspring when both have given names to species, e.g. Linnaeus (C. Linnaeus, 1707–1778), Linnaeus filius (C. Linnaeus, 1741–1783, son of the former).

*flagellum.* A whip-like structure whose beating motion propels antherozoids. (Also used in zoology).

*floret.* A small flower, aggregated with others into a compact inflorescence.

*follicle.* A dry dehiscent fruit derived from a single free carpel, and with a single line of dehiscence.

*free.* Not united to any other organ except by its basal attachment.

*free-central.* A form of placentation in which the ovules are attached to the central axis of a 1-celled ovary (figure 44(5), p. 397).

*frond.* The leaf of a fern or cycad.

*fruit.* The structure into which the gynaecium is transformed during the ripening of the seeds; a *compound fruit* is derived from the gynaecia of more than one flower. The term 'fruit' is often extended to include structures which are derived in part from the receptacle (*Fragaria*), epigynous zone (*Malus*) or inflorescence-stalk (*Ficus*) as well as from the gynaecium.

*funicle.* The stalk of an ovule (figure 44(1), p. 397).

*fusiform.* Spindle-shaped; cylindric, but tapered gradually at both ends.

*gamete.* A single sex-cell which fuses with one of the opposite sex during sexual reproduction.

*gametophyte.* The sexual generation in plants which have an alternation of generations (see sporophyte). In pteridophytes it is free-living but much smaller than the sporophyte; in seed plants it is totally contained within the ovule and pollen-grain.

*gland-dotted.* With minute patches of secretory tissue usually appearing as pits

on the surface, as translucent dots when held up to the light, or both.

*glandular.* (1) Of a hair: bearing at the tip a usually spherical knob of secretory tissue. (2) Of a tooth: similarly knobbed or swollen at the tip.

*glaucous.* Green strongly tinged with bluish grey; with a greyish waxy bloom.

*graft-hybrid.* A plant which, as a consequence of grafting, contains a mixture of tissues from 2 different species. Normally the tissues of 1 species are enclosed in a 'skin' of tissue from the other species.

*gynaecium.* The female organs (carpels) of a single flower, considered collectively, whether they are free or united.

*half-inferior.* Of an ovary: with its lower part inferior and its upper part superior (figure 43(15), p. 396).

*haploid.* Possessing in its normal vegetative cells only a single set of chromosomes.

*hastate.* With 2 acute, divergent lobes at the base, as in a mediaeval halberd (figure 41(12), p. 394).

*haustorium.* The organ with which a parasitic plant penetrates its host and draws nutriment from it.

*herb.* A plant in which the stems do not become woody, or, if somewhat woody at the base, do not persist from year to year.

*herbaceous.* Of a plant: possessing the qualities of a herb as defined above.

*heterostylic.* Having flowers in which the length of the style relative to that of the stamens varies from one plant to another.

*hilum.* The scar-like mark on a seed indicating the point at which it was attached to the funicle.

*hybrid.* A plant produced by the crossing of parents belonging to 2 different named groups (e.g. genera, species, subspecies etc.). An $F_1$ hybrid is the primary product of such a cross. An $F_2$ hybrid is a plant arising from a cross between 2 $F_1$ hybrids (or from the self-pollination of an $F_1$ hybrid).

*hydathode.* A water-secreting gland immersed in the tissue of a leaf near its margin.

*hypocotyl.* That part of the stem of a seedling which lies between the top of the radicle and the attachment of the cotyledon(s).

*hypogeal.* The mode of germination in which the cotyledons remain in the seed-coat and play no part in photosynthesis.

*hypogynous.* Describes a flower, or, preferably the petals, sepals and stamens

(or perianth and stamens) of a flower in which the ovary is superior and the petals, sepals and stamens (or perianth and stamens) arise as individual whorls on the receptacle (figure 43(10), p. 396).

*incised.* With deep, narrow spaces between the teeth or lobes.

*included.* Not projecting beyond the organs which enclose it.

*indefinite.* More than 12 and possibly variable in number.

*indehiscent.* Without preformed lines of splitting; opening, if at all, irregularly by decay.

*indusium.* A membranous flap or disc, covering, at least during development, the sori of many ferns (figure 1(6), p. 2).

*inferior.* Of an ovary: borne beneath the sepals, petals and stamens (or perianth and stamens) so that these appear to arise from its top (figure 43(13 & 14), p. 396).

*inflorescence.* A number of flowers which are sufficiently closely grouped together to form a structural unit (figure 42, p. 395).

*infraspecific.* Denotes any category below species level, such as subspecies, variety and form. To be distinguished from *subspecific*, which means relating to subspecies only.

*integument.* The covering of an ovule, later developing into the seed-coat. Some ovules have a single integument, others 2.

*internode.* The part of a stem between 2 successive nodes.

*involucre.* A compact cluster of bracts around the stalk at or near the base of some flowers or inflorescences or around the base of a capitulum; sometimes reduced to a hair or ring of hairs.

*keel.* A narrow ridge, suggestive of the keel of a boat, developed along the midrib (or rarely other veins) of a leaf, petal or sepal.

*laciniate.* With the margin deeply and irregularly divided into narrow and unequal teeth.

*lanceolate.* 3–4 times as long as wide and tapering more gradually towards the tip (figure 41(9), p. 394).

*layer.* To propagate by pegging down on the ground a branch from near the base of a shrub or tree, so as to induce the formation of adventitious roots.

*leaflet.* One of the leaf-like components of a compound leaf (figure 41(15), p. 394).

*lenticel.* A small, slightly raised interruption of the surface of the bark (or of the outer corky layer of a fruit) through which air can penetrate to the inner tissues.

*ligule*. A small membranous flap (more rarely a line or hairs) at the base of a leaf-blade.

*limb*. A broadened part, furthest from the base, of a petal, corolla or similar organ, which has a relatively narrow basal part – the claw or tube (figure 43(5 & 6), p. 396).

*linear*. Parallel-sided and many times longer than broad (figure 41(6), p. 394).

*lip*. A major division of the apical part of a bilaterally symmetric calyx or corolla in which the petals or sepals are united; there is normally an upper and a lower lip, but either may be missing.

*lyrate*. Pinnatifid or pinnatisect, with a large terminal and small lateral lobes.

*median*. Describes the position of fern sori which are about halfway between the mid-vein and the margin.

*medifixed*. Of a hair: lying parallel to the surface on which it is borne and attached to it by a stalk (usually short) at its mid-point.

*megasporangium*. A spore-producing organ containing only megaspores.

*megaspore*. The larger type of spore produced by a plant which has spores of 2 sizes; it gives rise to a female gametophyte.

*mericarp*. A carpel, usually 1-seeded, released by the break-up at maturity of a fruit formed from 2 or more joined carpels.

*mesocarp*. The central, often fleshy layer of a fruit-wall in those fruits in which the wall is distinctly 3-layered.

*micropyle*. A pore in the integument(s) of an ovule and the coat of a seed (figure 44(1), p. 397)

*microsporangium*. A spore-producing organ producing only microspores; a pollen-sac.

*microspore*. The smaller type of spore produced by a plant which has spores of 2 sizes; it gives rise to a male gametophyte.

*monocarpic*. Flowering and fruiting once and then dying.

*monoecious*. With separate male and female flowers or cones on the same plant; in flowering plants male flowers may contain non-functional carpels (and vice versa).

*mucronate*. Provided with a short narrow point at the apex (figure 41(25), p. 394).

*mycorrhiza*. A symbiotic association between the roots of a green plant and a fungus.

*nectary*. A nectar-secreting gland.

*neuter*. Without either functional male or female parts.

*node*. The point at which 1 or more leaves or flower parts are attached to an axis.

*nut*. A 1-seeded indehiscent fruit with a woody or bony wall.

*nutlet*. A small nut, usually a component of an aggregate fruit.

*obconical*. Shaped like a cone, but attached at the narrow end.

*oblanceolate*. As lanceolate, but attached at the more gradually tapered end.

*oblong*. With more or less parallel sides and about 2–5 times as long as broad (figure 41(7), p. 394).

*obovate*. As ovate, but attached at the narrower end.

*obovoid*. As ovoid, but attached at the narrower end.

*opposite*. Describes 2 leaves, branches or flowers attached on opposite sides of the axis at the same node.

*orthotropous*. Describes an ovule which stands erect and straight (figure 44(1), p. 397).

*ovary*. The lower part of a carpel, containing the ovule(s) (i.e. excluding style and stigma); the lower, ovule-containing part of a gynaecium in which the carpels are united (figure 43(2), p. 396).

*ovate*. With approximately the outline of a hen's egg (though not necessarily blunt-tipped) and attached at the broader end (figure 41(10), p. 394).

*ovoid*. As ovate, but applied to a solid body.

*ovule*. The small body from which a seed develops after pollination (figure 44, p. 397).

*palmate*. Describes a compound leaf composed of more than 3 leaflets, all arising from the same point, as in the leaf of *Aesculus* (figure 41(19), p. 394).

*palmatifid*. Lobed in a palmate manner, with the incisions pointing to the place of attachment, but not reaching much more than halfway to it (figure 41(20), p. 394).

*palmatisect*. Deeply lobed in a palmate manner, with the incisions almost reaching the base (figure 41(21), p. 394).

*panicle*. A compound raceme, or any freely branched inflorescence of similar appearance (figure 42(3 & 5), p. 395).

*papillose*. Covered with small blunt protuberances (papillae).

*paraphysis*. A hair-like structure found among the sporangia in the sori of some ferns.

*parietal*. A form of placentation in which the placentas are borne on the inner surface of the walls of a 1-celled ovary (figure 44(11), p. 397).

*pectinate*. With leaves or leaflets in 2 opposite, regular, eye-lash-like rows.

*pedate*. With a terminal lobe or leaflet, and on either side of it an axis curving outwards and backwards, bearing lobes or leaflets on the outer side of the curve (figure 41(22), p. 394).

*peltate*. Describes a leaf or other flat structure with a stalk attached other than at the margin (figure 41(27), p. 394).

*perennial*. Persisting for more than 2 years.

*perfoliate*. Describes a pair of stalkless opposite leaves of which the bases are united, or a single leaf in which the auricles are united, so that the stem appears to pass through the leaf or leaves (figure 41(3 & 4), p. 394).

*perianth*. The calyx and corolla considered collectively, used especially when there is no clear differentiation between calyx and corolla; used also to denote a calyx or corolla when the other is absent (figure 43(2), p. 396).

*perigynous*. Describing a flower, or, preferably the petals, sepals and stamens (or perianth and stamens) of a flower in which the ovary is superior and the petals, sepals and stamens (or perianth and stamens) are borne on the margins of a rim or cup which itself is borne on the receptacle below the ovary (it often appears as though the sepals, petals and stamens, or perianth and stamens, are united at their bases) (figure 43(11 & 12), p. 396).

*perigynous zone*. The rim or cup of tissue on which the sepals, petals and stamens (or perianth and stamens) are borne in a perigynous flower (figure 43(11 & 12), p. 396).

*petal*. A member of the inner perianth-whorl (corolla) used mainly when this is clearly differentiated from the calyx. The petals usually function in display and often provide an alighting place for pollinators (figure 43(1), p. 396).

*petaloid*. Like a petal in texture and colour.

*phyllode*. A leaf-stalk taking on the function and, to a variable extent, the form of a leaf-blade.

*phyllopodium*. See Filicopsida (p. 6 & figure 1(11), p. 2).

*pinna*. The primary division of a pinnate frond or leaf; it may be simple or itself divided (figure 1(2), p. 2).

*pinnate*. Describes a compound leaf or frond in which distinct leaflets or pinnae are arranged on either side of the axis or rachis (figure 41(15), p. 394). If these leaflets are themselves of a similar compound structure the leaf or frond is termed bipinnate (similarly, tripinnate, etc.).

*pinnatifid*. Lobed in a pinnate manner, with the incision reaching not much more than halfway to the axis or rachis (figure 41(16), p. 394).

*pinnatisect*. Deeply lobed in a pinnate manner, with the incisions almost reaching the axis or rachis (figure 41(17), p. 394).

*pinnule*. One of the divisions of a pinna in a bipinnate fern-frond; one of the ultimate divisions of a more-than-twice pinnate frond (figure 1(3), p. 2).

*pistillode*. A sterile ovary in a male flower.

*placenta*. A part of the ovary, often in the form of a cushion or ridge, to which ovules are attached.

*placentation*. The manner of arrangement of the placentas.

*pollen-sac*. One of the cavities in an anther in which the pollen is formed; in flowering plants each anther normally contains 4 pollen sacs, 2 on either side of the connective and separated by a partition which shrivels at maturity.

*polyploid*. Possessing in its normal vegetative cells more than 2 sets of chromosomes.

*protandrous*. With anthers beginning to shed their pollen before the stigmas in the same flower are receptive.

*prothallus*. See Pteridophyta (p. 1).

*protogynous*. With stigmas becoming receptive before the anthers in the same flower shed their pollen.

*pulvinus*. A swollen region at the base of a leaflet, leaf-blade or leaf-stalk.

*pyrene*. A small nut-like body enclosing a seed, 1 or more of which, surrounded by fleshy tissue, make up the fruit of, for example, *Ilex*.

*raceme*. An inflorescence consisting of stalked flowers arranged on a single axis, the lower opening first (figure 42(2), p. 395).

*rachis*. Of a compound fern frond or pinna: the central axis.

*radially symmetric*. Capable of division into 2 similar halves along 2 or more planes of symmetry (figure 43(8), p. 396).

*radicle*. The root preformed in the seed and normally the first visible root of a seedling.

*raphe*. A perceptible ridge or stripe, at one end of which is the hilum, on some seeds.

*receptacle*. The tip of an axis to which the floral parts or perigynous zone (when present) are attached (figure 43(1), p. 396).

*reflexed*. Bent sharply backwards from the base.

*rhizome*. A horizontal stem, situated underground or on the surface, serving the purpose of food storage or vegetative reproduction or both; roots or stems arise from some or all of its nodes.

*rootstock*. The compact mass of tissue from which arise the new shoots of a herbaceous perennial. It usually consists mainly of stem tissue, but is more compact than is generally understood by rhizome.

*runner*. A slender above-ground stolon with very long internodes.

*sagittate*. With a backwardly directed basal lobe on either side, like an arrow-head (figure 41(11), p. 394).

*saprophytic*. Dependent for its nutrition on soluble organic compounds in the soil. Saprophytic plants do not photosynthesise and lack chlorophyll; some plants, however, are *partially saprophytic* and combine the two modes of nutrition.

*scale-leaf*. A reduced leaf, usually not photosynthetic.

*scape*. A leafless flower-stalk or inflorescence-stalk arising usually from ground level.

*scarious*. Dry and papery, often translucent.

*schizocarp*. A fruit which, at maturity, splits into its constituent mericarps.

*scion*. A branch cut from one plant to be grafted on the rooted stock of another.

*seed*. A reproductive body adapted for dispersal, developed from an ovule and consisting of a protective covering (the seed-coat), an embryo and, usually, a food reserve.

*semi-parasite*. A plant which obtains only part of its nourishment by parasitism.

*sepal*. A member of the outer perianth whorl (calyx) when 2 whorls are clearly differentiated as calyx and corolla, or when comparison with related plants shows that the corolla is absent. The sepals most often function in protection and support of other floral parts (figure 43(1), p. 396).

*septum*. An internal partition.

*sheath*. The part of a leaf or leaf-stalk which surrounds the stem, being either tubular or with free but overlapping edges.

*shrub*. A woody plant with several stems or branches from near the base, and of smaller stature than a tree.

*simple*. Not divided into separate parts.

*sinus*. The gap or indentation between 2 lobes or teeth.

*sorus*. A group of sporangia with more or

less clearly defined boundaries, often more or less circular, but in some cases linear (figure 1(5), p. 2).

*spathe*. A large bract at the base of a flower or inflorescence and wholly or partly enclosing it. The term is used in some families to denote collectively 2 or 3 such bracts.

*spathulate*. With a narrow basal part, which towards the apex is gradually expanded into a broad blunt blade.

*spike*. An inflorescence or subdivision of an inflorescence consisting of stalkless flowers arranged on a single axis (figure 42(1), p. 395).

*sporangiophore*. See Sphenopsida (p. 5).

*sporangium*. A hollow structure in which spores are developed.

*spore*. A small, asexual reproductive body, usually consisting of a single haploid cell. In pteridophytes it germinates to form a prothallus (gametophyte).

*sporocarp*. A closed structure with a woody or membranous wall containing a number of megasporangia and microsporangia.

*sporophyll*. A more or less modified leaf, bearing sporangia (term used in this book only for Cycadaceae).

*sporophyte*. The asexual generation in plants which have an alternation of generations (see gametophyte). In the plants covered by this book it is the predominant generation.

*spur*. An appendage or prolongation, more or less conical or cylindric, often at the base of an organ. The spur of a corolla or single petal or sepal is usually hollow and often contains nectar (figure 43(7), p. 396)

*stamen*. The male organ, producing pollen, generally consisting of an anther borne on a filament (figure 43(1), p. 396).

*staminode*. An infertile stamen, often reduced or rudimentary or with a changed function.

*standard*. See Iridaceae (figure 39, p. 335).

*stellate*. Star-like, particularly of branched hairs.

*stigma*. The part of a style to which pollen adheres, normally differing in texture from the rest of the style (figure 43(2), p. 396).

*stipe*. The stalk of a fern-frond.

*stipule*. An appendage, usually 1 of a pair, found beside the base of the leaf-stalk in many flowering plants, sometimes falling early, leaving scars. In some cases the 2 stipules are united; in others they are partly united to the leaf-stalk.

*stock*. A rooted plant, often with the upper

part removed, on to which a scion may be grafted.

*stolon.* A far-creeping, more or less slender, above-ground or underground rhizome giving rise to a new plant at its tip and sometimes at intermediate nodes.

*stoma.* A microscopic ventilating pore in the surface of a leaf or other herbaceous part of a plant.

*style.* The usually slender upper part of a carpel or gynaecium, bearing the stigma (figure 43(2), p. 396).

*subtend(ed).* Used of any structure (e.g. a flower) which occurs in the axil of another organ (e.g. a bract); in this case the bract subtends the flower.

*subulate.* Narrowly cylindric, and somewhat tapered to the tip.

*sucker.* An erect shoot originating from a bud on a root or a rhizome, sometimes at some distance from the stem of the parent plant.

*superior.* Of an ovary: borne at the morphological apex of the flower so that the petals, sepals and stamens (or perianth and stamens) arise on the receptacle below the ovary (figure 43(10–12), p. 396).

*suture.* A line marking an apparent junction of neighbouring organs.

*synangium.* See Psilopsida (p. 3).

*tendril.* A thread-like structure which by its coiling growth can attach a shoot to something else for support.

*terete.* Approximately circular in cross-section; not necessarily perfectly cylindric, but without grooves or ridges.

*tessellated.* With a chequered pattern of light and dark squares.

*tetraploid.* Possessing in its normal vegetative cells 4 similar sets of chromosomes.

*throat.* The part of a calyx or corolla transitional between tube and limb or lobes.

*triploid.* Possessing in its normal vegetative cells 3 similar sets of chromosomes. Most triploid plants are highly sterile.

*tristylic.* Having the flowers of different plants with long, short, or intermediate-length styles; the stamens of each flower are of 2 lengths which are not the same as the style-length of that flower.

*truncate.* As though with the tip or base cut off at right angles.

*tuber.* A swollen underground stem or root used for food-storage.

*tubercle.* A small, blunt, wart-like protuberance.

*tunic.* The dead covering of a bulb or corm.

*turion.* A specialised perennating bud in some aquatic plants, consisting of a short shoot covered in closely packed leaves, which persists through the winter at the bottom of the water.

*umbel.* An inflorescence in which the flower-stalks arise together from the top of an inflorescence-stalk. This is a *simple*

*umbel* (figure 42(8), p. 395); in a *compound umbel* the several stalks arising from the top of the inflorescence-stalk terminate not in flowers but in secondary umbels.

*undivided.* Without major divisions or incisions, though not necessarily entire.

*urceolate.* Shaped like a pitcher or urn, hollow and contracted at or just below the mouth.

*vascular bundle.* A strand of conducting tissue, usually surrounded by softer tissue.

*vein.* A vascular strand, usually in leaves or floral parts and visible externally.

*venation.* The pattern formed by the veins.

*versatile.* Of an anther: flexibly attached to the filament by its approximate mid-point so that a rocking motion is possible.

*vessel.* A microscopic water-conducting tube formed by a sequence of cells not separated by end-walls.

*viviparous.* Bearing young plants, bulbils or leafy buds which can take root; they can occur anywhere on the plant and may be interspersed with, or wholly replace, the flowers in an inflorescence.

*whorl.* A group of more than 2 leaves or floral organs inserted at the same node.

*wing.* A thin, flat extension of a fruit, seed, sepal or other organ.

*xerophytic.* Drought tolerant. Can also describe the environment in which drought-tolerant plants live.

# INDEX

Synonyms, and names mentioned only in observations, are printed in *italic* type.